Electrical and Electronics Reference Manual

for the Electrical and Computer PE Exam

John A. Camara, PE

The Power to Pass®
www.ppi2pass.com

Professional Publications, Inc. • Belmont, California

Benefit by Registering This Book with PPI

- Get book updates and corrections.
- Hear the latest exam news.
- Obtain exclusive exam tips and strategies.
- Receive special discounts.

Register your book at **www.ppi2pass.com/register**.

Report Errors and View Corrections for This Book

PPI is grateful to every reader who notifies us of a possible error. Your feedback allows us to improve the quality and accuracy of our products. You can report errata and view corrections at **www.ppi2pass.com/errata**.

ELECTRICAL AND ELECTRONICS REFERENCE MANUAL FOR THE ELECTRICAL AND COMPUTER PE EXAM

Current printing of this edition: 1

Printing History

edition number	printing number	update
1	1	New book.

Printed in the United States of America.

PPI
1250 Fifth Avenue, Belmont, CA 94002
(650) 593-9119
www.ppi2pass.com

ISBN: 978-1-59126-166-7

Library of Congress Control Number: 2010910401

PHYSICAL, FUNDAMENTAL, AND MATHEMATICAL CONSTANTS

quantity	symbol	customary U.S.	SI
Charge			
electron	e		-1.60218×10^{-19} C
proton	p		$+1.60218 \times 10^{-19}$ C
Density			
air [STP] [32°F, (0°C)]		0.0805 lbm/ft^3	1.29 kg/m^3
air [70°F, (20°C), 1 atm]		0.0749 lbm/ft^3	1.20 kg/m^3
earth [mean]		345 lbm/ft^3	5520 kg/m^3
mercury		849 lbm/ft^3	1.360×10^4 kg/m^3
seawater		64.0 lbm/ft^3	1025 kg/m^3
water [mean]		62.4 lbm/ft^3	1000 kg/m^3
Distance [mean]			
earth radius		2.09×10^7 ft	6.371×10^6 m
earth-moon separation		1.26×10^9 ft	3.844×10^8 m
earth-sun separation		4.89×10^{11} ft	1.49×10^{11} m
moon radius		5.71×10^6 ft	1.737×10^6 m
sun radius		2.28×10^9 ft	6.96×10^8 m
first Bohr radius	a_0	1.736×10^{-10} ft	5.292×10^{-11} m
Gravitational Acceleration			
earth [mean]	g	32.174 ft/sec^2	9.8067 m/s^2
moon [mean]		5.321 ft/sec^2	1.622 m/s^2
Mass			
atomic mass unit	u	3.66×10^{-27} lbm	1.6605×10^{-27} kg
earth		4.11×10^{23} slugs	5.97×10^{24} kg
earth [customary U.S.]		1.32×10^{25} lbm	n.a.
electron [rest]	m_e	2.008×10^{-30} lbm	9.1094×10^{-31} kg
moon		1.623×10^{23} lbm	7.347×10^{22} kg
neutron [rest]	m_n	3.693×10^{-27} lbm	1.6749×10^{-27} kg
proton [rest]	m_p	3.688×10^{-27} lbm	1.6726×10^{-27} kg
sun		4.387×10^{30} lbm	1.989×10^{30} kg
Pressure, atmospheric		14.696 lbf/in^2	1.01323×10^5 Pa
Temperature, standard		32°F (492°R)	0°C (273K)
Velocity			
earth escape		3.67×10^4 ft/sec	1.12×10^4 m/s
light [vacuum]	c	9.84×10^8 ft/sec	2.9979×10^8 m/s
sound [air, STP]	a	1090 ft/sec	331 m/s
[air, 70°F (20°C)]	a	1130 ft/sec	344 m/s
Volume, molal ideal gas [STP]		359 ft^3/lbmol	22.41 m^3/kmol
Fundamental Constants			
Avogadro's number	N_A		6.0221×10^{23} mol^{-1}
Bohr magneton	μ_B		9.2740×10^{-24} J/T
Boltzmann constant	k	5.65×10^{-24} ft-lbf/°R	1.38065×10^{-23} J/K
Faraday constant	F		96 485 C/mol
gravitational constant	g_c	32.174 lbm-ft/lbf-sec^2	
gravitational constant	G	3.44×10^{-8} $\text{ft}^4/\text{lbf-sec}^4$	6.674×10^{-11} $\text{N·m}^2/\text{kg}^2$
nuclear magneton	μ_N		5.050×10^{-27} J/T
permeability of a vacuum	μ_0		1.2566×10^{-6} N/A^2 (H/m)
permittivity of a vacuum	ϵ_0		8.8542×10^{-12} $\text{C}^2/\text{N·m}^2$ (F/m)
Planck's constant	h		6.6260×10^{-34} J·s
Rydberg constant	R_∞		1.0974×10^7 m^{-1}
specific gas constant, air	R	53.3 ft-lbf/lbm-°R	287.03 J/kg·K
Stefan-Boltzmann constant		1.71×10^{-9} $\text{Btu/ft}^2\text{-hr-°R}^4$	5.670×10^{-8} $\text{W/m}^2\text{·K}^4$
triple point, water		32.02°F, 0.0888 psia	0.01109°C, 0.61173 kPa
universal gas constant	R^*	1545 ft-lbf/lbmol-°R	8314 J/kmol·K
	R^*	1.986 Btu/lbmol-°R	0.08206 atm·L/mol·K

Mathematical Constants		
Archimedes' number (pi)	π	3.14159 26536
base of natural logs	e	2.71828 18285
Euler constant	γ	0.57721 56649

Topics

Topic I: Mathematics

Topic II: Basic Theory

Topic III: Field Theory

Topic IV: Circuit Theory

Topic V: Power Generation

Topic VI: Protection and Safety

Topic VII: Power System Management

Topic VIII: Measurement and Instrumentation

Topic IX: Electronics

Topic X: Digital Systems

Topic XI: Communications

Topic XII: Control Systems

Topic XIII: Electrical Materials

Topic XIV: Codes and Standards

Topic XV: Professional

Where do I find practice problems to test what I've learned in this Reference Manual?

The *Electrical and Electronics Reference Manual* provides a knowledge base that will prepare you for the Electrical PE Electrical and Electronics exam. But there's no better way to exercise your skills than to practice solving problems. To test your knowledge, you need *Electrical and Electronics Practice Problems for the Electrical and Computer PE Exam*. This essential study aid will challenge you with 440 practice problems, each with a complete, step-by-step solution.

Electrical and Electronics Practice Problems for the Electrical and Computer PE Exam may be purchased from PPI at **www.ppi2pass.com** or from your favorite bookstore.

Table of Contents

Appendices Table of Contents vii

Preface ix

Acknowledgments xi

Codes and References xiii

Introduction xv

Topic I: Mathematics

Systems of Units 1-1
Energy, Work, and Power 2-1
Engineering Drawing Practice 3-1
Algebra 4-1
Linear Algebra 5-1
Vectors 6-1
Trigonometry 7-1
Analytic Geometry 8-1
Differential Calculus 9-1
Integral Calculus 10-1
Differential Equations 11-1
Probability and Statistical Analysis of Data 12-1
Computer Mathematics 13-1
Numerical Analysis 14-1
Advanced Engineering Mathematics 15-1

Topic II: Basic Theory

Electromagnetic Theory 16-1
Electronic Theory 17-1
Communication Theory 18-1
Acoustic and Transducer Theory 19-1

Topic III: Field Theory

Electrostatics 20-1
Electrostatic Fields 21-1
Magnetostatics 22-1
Magnetostatic Fields 23-1
Electrodynamics 24-1
Maxwell's Equations 25-1

Topic IV: Circuit Theory

DC Circuit Fundamentals 26-1
AC Circuit Fundamentals 27-1
Transformers 28-1
Linear Circuit Analysis 29-1
Transient Analysis 30-1
Time Response 31-1
Frequency Response 32-1

Topic V: Power Generation

Batteries, Fuel Cells, and Power Supplies 33-1

Topic VI: Protection and Safety

Protection and Safety 34-1

Topic VII: Power System Management

Power System Management 35-1

Topic VIII: Measurement and Instrumentation

Measurement and Instrumentaion 36-1

Topic IX: Electronics

Electronics Fundamentals 37-1
Junction Transistors 38-1
Field Effect Transistors 39-1
Electrical and Electronic Devices 40-1

Topic X: Digital Systems

Digital Logic 41-1
Logic Network Design 42-1
Synchronous Sequential Networks 43-1
Digital Systems 44-1

Topic XI: Communications

High-Frequency Transmission 45-1
Antenna Theory 46-1
Communication Links 47-1
Signal Formats 48-1
Signal Multiplexing 49-1
Communications Systems and Channels 50-1

Topic XII: Control Systems

Analysis of Control Systems 51-1

Topic XIII: Electrical Materials

Electrical Materials 52-1

Topic XIV: Codes and Standards

National Electrical Code 53-1
National Electrical Safety Code 54-1

Topic XV: Professional

Engineering Economic Analysis 55-1
Engineering Law 56-1
Engineering Ethics 57-1
Electrical Engineering Frontiers 58-1
Engineering Licensing in the United States 59-1

Appendices A-1

Index I-1

Appendices Table of Contents

1.A Conversion Factors A-1
1.B Common SI Unit Conversion Factors A-3
8.A Mensuration of Two-Dimensional Areas A-7
8.B Mensuration of Three-Dimensional Volumes . . . A-9
10.A Abbreviated Table of Indefinite Integrals A-10
11.A Laplace Transforms A-11
12.A Areas Under the Standard Normal Curve A-12
15.A Gamma Function Values A-13
15.B Bessel Functions J_0 and J_1 . . . . A-14
15.C Properties of Fourier Series for Periodic Signals A-15
15.D Properties of Fourier Transform for Aperiodic Signals A-16
15.E Fourier Transform Pairs A-17
15.F Properties of Laplace Transform A-18
15.G Laplace Transforms A-19
15.H Properties of Discrete-Time Fourier Series for Periodic Signals A-20
15.I Properties of Discrete-Time Fourier Transform for Aperiodic Signals A-21
15.J Properties of z-Transforms A-22
15.K z-Transforms A-23
17.A Room Temperature Properties of Silicon, Germanium, and Gallium Arsenide (at 300K) A-24
18.A Electromagnetic Spectrum A-25
25.A Comparison of Electric and Magnetic Equations A-26
27.A Impedance of Series-Connected Circuit Elements A-28
27.B Impedance of Parallel-Connected Circuit Elements A-29
29.A Two-Port Parameter Conversions A-31
30.A Resonant Circuit Formulas A-32
37.A Periodic Table of Elements A-33
37.B Semiconductor Symbols and Abbreviations . . A-34
40.A Standard Zener Diodes A-38
45.A Smith Chart A-40
48.A Error Function Values A-41
48.B Error Function Values A-42
52.A Periodic Table: Materials' Properties Summary A-43
52.B Table of Relative Atomic Weights A-44
53.A Comparison of Wire (and Sheet-Metal Gage) Diameters (Thickness) in Inches A-45
53.B Conductor Ampacities A-47
53.C Conductor Properties A-48
53.D Conductor AC Properties A-49
55.A Standard Cash Flow Factors and Formulas . . . A-50
55.B Cash Flow Equivelant Factors A-51

Preface

The *Electrical and Electronics Reference Manual for the Electrical and Computer PE Exam* is the most comprehensive resource available for the National Council of Examiners for Engineering and Surveying (NCEES) Principles and Practice of Engineering (PE) Electrical and Computer–Electrical and Electronics Exam. This reference includes myriad topics, encompassing the fundamentals of electrical engineering, including those topics appearing on the NCEES Electrical and Electronics exam.

The scope of this book is beyond that of the Electrical and Electronics exam, as it is intended to be. When the exam is complete and you are a licensed electrical engineer, you will need a resource for electrical engineering information, either in your work or simply to satisfy the curiosity that makes us thinking humans. In both the exam and your career, I hope this book serves you well.

WHY I WROTE THIS BOOK

The *Electrical and Electronics Reference Manual* descended from the eighth edition of the *Electrical Engineering Reference Manual for the Power, Electrical and Electronics, and Computer PE Exams.* The first edition of the *Electrical Engineering Reference Manual* was published in 1978. Over its many years in print, it has evolved with the industry and with changes to the Electrical and Computer PE exam. In April 2009, NCEES made a fundamental shift in the focus of this exam, converting the single morning session and three specialized afternoon sessions into three independent full-day specialized exams—for Power, for Electrical and Electronics, and for Computer Engineering. I have written the *Electrical and Electronics Reference Manual* (as well as two other specialized reference manuals) to expand and focus the electrical and electronics coverage of the *Electrical Engineering Reference Manual* into a text supplying what you need to succeed on the Electrical and Electronics exam.

My goal for the *Electrical and Electronics Reference Manual* is to provide future electrical and electronics engineers with the depth of information required to gain a solid understanding of theory and of practical applications. My hope is that the *Electrical and Electronics Reference Manual* allows you to master the fundamentals necessary for the exam, and to become one of those rare individuals who understands how advanced technology works.

WHAT'S IN THIS BOOK

In selecting and preparing the contents, I followed the NCEES Electrical and Electronics exam specifications. I have endeavored throughout the text to include the exact terminology used by these specifications.

Along with revising sections from the best-selling aforementioned reference, I added new chapters on power system management and protection and safety. I also expanded the chapters on AC circuit fundamentals; batteries, fuel cells, and power supplies; measurement and instrumentation; electrical and electronic devices; and signal formats.

While I wrote the book with the NCEES specifications in mind, the contents of the *Electrical and Electronics Reference Manual* are not limited to the specifications of the Electrical and Electronics exam. This reference also contains the fundamental blocks you need to build a complete understanding of electronics and electrical engineering. With an understanding of the basics, you will comprehend the majority of the exam questions, as well as real-world electrical engineering problems. In each chapter, I explore mathematical, theoretical, and practical applications of the chapter's topic. You can focus your study on any or all of these types of applications.

The sources I consulted while developing this book's content include: (1) information on the Electrical and Electronics exam published by NCEES; (2) PE Electrical and Computer: Electrical and Electronics Sample Questions and Solutions, published by NCEES; (3) electrical engineering curricula at leading colleges and universities; (4) current literature in the field of electrical engineering; (5) electrical engineering websites; and (6) survey comments from those who have recently taken the PE exam and/or who purchased the *Electrical Engineering Reference Manual.*

A number of features enhance the usefulness of this text. The introduction gives details on how to use the book effectively. Lists of codes and references recommended for additional study are provided. Appendices provide information you need in your exam preparation, consolidate data you may need to quickly reference during the exam, and exhibit mathematical, basic theoretical, and practical data. I have constructed a comprehensive Index to aid in the search for specific information. The publisher has also provided an excellent resource in support of your success: you may access online updates, report errors, and participate in discussions with other prospective electrical engineers through PPI's website, **www.ppi2pass.com**.

FOR WHOM THIS BOOK IS WRITTEN

The *Electrical and Electronics Reference Manual* is written for you. It is written to be a reference at all times in your career, whether you are an undergraduate or an experienced engineer. If you are an exam candidate, the *Electrical and Electronics Reference Manual* is an efficient resource for exploring exam topics systematically and exhaustively. If you are a practicing electrical and electronics engineer, it functions as a comprehensive reference, discussing all aspects of this field in a realistic manner, and incorporating the most common formulas and data. Finally, for the engineering student, it presents a thorough review of the fundamentals of electrical engineering.

LOOKING FORWARD TO THE FUTURE

Future editions of this book will be very much shaped by what you and others want to see in it. I am braced for the influx of comments and suggestions from those readers who (1) want more topics, (2) want more detail in existing topics, and (3) think that continuing to include some topics is just plain lame. Computer hardware and programming change quickly, as does the grand unified (atomic) theory of everything. I expect several chapters to generate some kind of debate.

Should you find an error in this book, know that it is mine, and that I regret it. Beyond that, I hope two things happen. First, please let me know about the error by using the error reporting form on the PPI website, found at **www.ppi2pass.com/errata**. Second, I hope you learn something from the error—I know I will! I appreciate suggestions for improvement, additional questions, and recommendations for expansion so that new editions or similar texts will better meet the needs of future examinees.

Read carefully, prepare well, and you will triumph over the exam!

John A. Camara, PE

Acknowledgments

It is with enduring gratitude that I thank Michael R. Lindeburg, PE, who, with the sixth edition of the *Electrical Engineering Reference Manual*, allowed me to realize my dream of authoring a significant engineering text. In taking the reins from the title's previous author, the late Raymond Yarbrough, I was entrusted with the responsibility of shaping that text as the electrical and computer PE exam and the engineering profession evolved. It is my hope that the *Electrical and Electronics Reference Manual for the Electrical and Computer PE Exam*—which is based on the *Electrical Engineering Reference Manual*—will match the comprehensive and cohesive quality of the best of PPI's family of engineering books, which I have admired for three decades.

The *Electrical and Electronics Reference Manual* would not have been realized without the professionalism and meticulous attention to detail shown by Megan Synnestvedt, copy editor. Without her spreadsheets, I would have lost track of chapter numbers right from the beginning! In addition, my thanks go out to the wonderfully responsive team at PPI, including Meaghan Banks, editorial project manager; Sarah Hubbard, director of new product development, whose guidance from the beginning has been most appreciated; Cathy Schrott, director of production, who keeps it all moving forward; copy editors Tyler Hayes, Jenny Lindeburg King, Scott Marley, Magnolia Molcan, Sesa Pabalan, and Trina Tom; acquisitions assistant Katie Throckmorton for obtaining permissions; Tom Bergstrom, technical illustrator; Alex Foster and Kate Hayes, the vitally important typesetters; and Amy Schwertman, who designed the cover.

Gregg Wagener, PE, reviewed each of the chapters of this book, providing valuable insight, correcting numerous errors and, in general, teaching me a great deal along the way.

Michael Lindeburg provided Chapters 1–14 on mathematics, as well as the chapters on economic analysis, law, ethics, and engineering licensure. He also provided sections of the illumination chapter. These contributions saved me considerable time, and I am grateful to use material of this quality. John Goularte, Jr., taught me firsthand the applications of biomedical engineering, and the real-world practicality of this topic is attributable to him.

The knowledge expressed in this book represents years of training, instruction, and self-study in electrical, electronic, mechanical, nuclear, marine, space, and a variety of other types of engineering—all of which I find fascinating. I would not understand any of it were it not for the teachers, instructors, mentors, family, and friends who've spent countless hours with me and from whom I've learned so much. A few of them include: Mrs. Mary Avila, Capt. C. E. Ellis, Mr. Claude Estes, Capt. Karl Hasslinger, Mr. Jerome Herbeck, Mr. Jack Hunnicutt, Mr. Ralph Loya, Mr. Harry Lynch, Mr. Harold Mackey, Mr. Michael O'Neal, Mr. Mark Richwine, Mr. Charles Taylor, Mrs. Abigail Thyarks, and Mr. Jim Triguerio. There are others who have touched my life in special ways: Jim, Marla, Lora, Todd, Tom, and Rick. There are many others, though here they will remain nameless, to whom I am indebted. Thanks.

A very special thanks to my son, Jac Camara, whose annotated periodic table became the basis for the Electrical Materials chapter. Additionally, our discussions about quantum mechanics and a variety of other topics helped stir the intellectual curiosity I thought was fading.

Thanks also to my daughter, Cassiopeia, who has been a constant source of inspiration, challenge, and pride.

Finally, thanks to my wife, Becky, and my mother, Arlene, both of whom have made me a better person. In addition to providing valuable content guidance and assistance with the Index, Becky was, more importantly, patient and supportive through the entire process, even when I wasn't.

John A. Camara, PE

Codes and References

Electrical and Electronics Reference Manual is based on the following codes and references. The Electrical and Electronics PE exam is consistent with the last edition of the code issued before the year of the exam. For more information about the variety of codes related to electrical engineering, refer to the following organizations and their websites.

American National Standards Institute (www.ansi.org)
Electronic Industries Alliance (www.eia.org)
Federal Communications Commission (www.fcc.gov)
Institute of Electrical and Electronic Engineers (www.ieee.org)
Instrument Society of America (www.isa.org)
International Organization for Standardization (www.iso.org)
National Electrical Manufacturer's Association (www.nema.org)
National Fire Protection Agency (www.nfpa.org)

The PPI website (**www.ppi2pass.com/eefaq**) provides the dates of the codes, standards, and regulations that the current exams are based on, according to NCEES. Use this information to decide which editions of these books should be part of your exam preparation.

The minimum recommended library of materials to bring to the Electrical and Electronics exam consists of this book, the applicable code book, a standard handbook of electrical engineering, and one or two textbooks that cover fundamental circuit theory (both electrical and electronic). The following list includes codes necessary for taking the exam, as well as textbooks and handbooks that you may be familiar with and that would be useful resources while taking and preparing for the exam. The books listed with "reference" brackets are those that should prove useful to you after the exam.

CODES

Federal Communications Commission (FCC). *Code of Federal Regulations*, "FCC Rules": Title 47-Telecommunications, Ch. 73. "Radio Broadcast Rules," (47CFR73). [Communications]

Institute of Electrical and Electronics Engineers (IEEE). 2008 *National Electrical Safety Code*, (NESC). [Power]

National Fire Protection Association (NFPA). 2008 *National Electrical Code*, (NFPA 70 or NEC). [Power]

REFERENCES

An Introduction to Digital and Analog Integrated Circuits and Applications. Sanjit K. Mitra. Harper & Row, Publishers. [Digital Circuit Fundamentals textbook]

Applied Electromagnetics. Martin A. Plonus. McGraw-Hill. [Electromagnetic Theory textbook]

CRC Materials Science and Engineering Handbook. James F. Shackelford, William Alexander, eds. CRC Press, Inc. [General engineering handbook]

CRC Standard Mathematical Tables and Formulae. William H. Beyer, ed. CRC Press, Inc. [General engineering reference]

Electronics Engineers' Handbook. Donald Christiansen, ed. McGraw-Hill. [Electrical and Electronics handbook]

Introduction to Computer Engineering: Hardware and Software Design. Taylor L. Booth. John Wiley & Sons. [Computer Design Basics textbook]

Linear Circuits. M. E. Van Valkenburg and B. K. Kinariwala. Prentice Hall, Inc. [AC/DC Fundamentals textbook]

McGraw-Hill Dictionary of Scientific and Technical Terms. Sybil P. Parker, ed. McGraw-Hill. [General engineering reference]

McGraw-Hill Internetworking Handbook. Ed Taylor. McGraw-Hill. [Computer handbook]

Microelectronic Circuit Design. Richard C. Jaeger, Travis Blalock. McGraw-Hill. [Electronic Fundamentals textbook]

Microelectronics. Jacob Millman, Arvin Millman. McGraw-Hill. [Electronic Fundamentals textbook]

National Electrical Code 2011 Handbook. National Fire Protection Association. [Power handbook]

Process/Industrial Instruments and Controls Handbook. Gregory K. McMillan, Douglas Considine, eds. McGraw-Hill. [Power and Electrical and Electronics handbook]

Standard Handbook for Electrical Engineers. Donald G. Fink and H. Wayne Beaty. McGraw-Hill. [Power and Electrical and Electronics handbook]

The Biomedical Engineering Handbook. Joseph D. Bronzino, ed. CRC Press. [Electrical and Electronics handbook]

The Communications Handbook. Jerry D. Gibson, ed. CRC Press, Inc. [Electrical and Electronics handbook]

The Computer Science and Engineering Handbook. Allen B. Tucker, Jr., ed. CRC Press, Inc. [Computer reference]

The Internet for Scientists and Engineers, Brian J. Thomas. SPIE Press. [Computer reference]

Wireless and Cellular Communications. William C.Y. Lee. McGraw-Hill. [Electrical and Electronics handbook]

Introduction

Part 1: How You Can Use This Book xv
Quickstart xv
For the PE Candidate xvi
If You Are a Practicing Engineer or an Engineering Student xvii
If You Are an Instructor xviii

Part 2: Everything You Ever Wanted to Know About the PE Exam xix
What is the Format of the PE Exam? xix
What Subjects Are on the PE Exam? xix
What is the Typical Problem Format? xx
What Does "Most Nearly" Really Mean? xx
How Much Mathematics Is Needed for the Exam? . xx
How About Engineering Economics? xxi
What About Professionalism and Ethics? xxi
Is the Exam Tricky? xxi
What Makes the Problems Difficult? xxi
Does the PE Exam Use SI Units? xxi
Why Does NCEES Reuse Some Problems? xxi
Does NCEES Use the Exam to Pre-Test Future Problems? xxi
Are the Practice Problems Representative of the Exam? xxi
What Reference Material is Permitted in the Exam? xxii
How Many Books Should You Bring? xxii
May Tabs Be Placed on Pages? xxii
What About Electronic Devices? xxii
How is the Exam Graded and Scored? xxiii
How You Should Guess xxiii
What is the Passing Rate? xxiii
How is the Cut Score Established? xxiii
Cheating and Exam Subversion xxiii

Part 3: How to Prepare for and Pass the Electrical and Electronics PE Exam xxiv
What Should You Study? xxiv
Do You Need a Classroom Review Course? xxiv
How Long Should You Study? xxiv
What the Well-Heeled Electrical Engineer Should Begin Accumulating xxiv
Additional Review Material xxv
What You Won't Need xxv
Should You Look for Old Exams? xxv
What Should You Memorize? xxv
Do You Need a Review Schedule? xxv
A Simple Planning Suggestion xxvi
How You Can Make Your Review Realistic xxvi
What to Do a Few Days Before the Exam xxvi
Prepare Your Car xxvi
Prepare Your Exam Kits xxvii
Prepare for the Worst xxviii
What to Do the Day Before the Exam xxviii
What to Do the Day of the Exam xxviii
What to Do During the Exam xxviii
What About Eating and Drinking in the Exam Room? xxix
How to Solve Multiple-Choice Problems xxix
Solve Problems Carefully xxix
Should You Talk to Other Examinees After the Exam? xxix
After the Exam xxx
And Then There's the Wait xxx
And When You Pass xxxi

PART 1: HOW YOU CAN USE THIS BOOK

QUICKSTART

If you're the type of reader who desires information distilled into concise bits and you're ready to begin your study, then this is the book for you; read this section and get started!

I say "this is the book for you" for the following reasons.

- The practice followed in writing this book was to take complex topics and parse them into condensed, readable, and understandable chunks.
- The PE exam consists of problems that require (on average) six minutes to solve. The exam problems focus on small portions of complex electrical engineering topics. Your study should follow the same pattern, and this book is written around that pattern.
- Once the exam is over, this book will remain a useful reminder of the basics. And once you understand the basics, much more complex problems can be solved by breaking them into their constituent parts.

Because the chapters in this book are independent, you can focus your study on your weakest areas first. Review the Table of Contents to familiarize yourself with the scope of the book, then use Table 1 in this Introduction as a guide for focusing your study. The Index is a great starting place to search for specific information and to plan your review of the topics covered by the NCEES specifications. The mathematics, theory, and specialty topics, while not directly listed in the NCEES requirements, supply fundamental background on topics that may be necessary for the occasional problem on the exam. The Appendices are succinct references for common electrical information. I have included the Professional chapters to refine your professional engineering knowledge and round out the knowledge base you have available for your career.

Commence study immediately. Six months is not an unreasonable timeframe to prepare, though your learning curve will vary depending upon your background. Start with your weakest areas and work toward your strongest. Review every week.

Work problems three to five times a week. The key to success on the exam is to work as many problems as possible. Use all the PPI resources available to you: this manual's companion, *Electrical and Electronics Practice Problems for the Electrical and Computer PE Exam*, Exam Cafe, and *Electrical Engineering Sample Examinations for the Power, Electrical and Electronics, and Computer PE Exams*. Browse the PPI website; other products are always under development in the continuing expansion of exam materials, and you may find the ideal resource designed to help you pass the exam.

This book focuses on the basics that are the building blocks of engineering knowledge. It is the basics that are necessary to solve complex problems, which can always be divided into simpler, individual concepts. A thorough review of the basics, terminology, and problems in this text and the companion *Electrical and Electronics Practice Problems* book will enable the studious reader to solve the variety of problems that might appear on the PE exam or on the job. Continue to study until the exam date. Enjoy the learning process and revel in the knowledge gained.

If you are the type of reader that desires a more thorough review and plan, and a deeper understanding of the PE exam before commencing your studies, proceed to the rest of the Introduction.

FOR THE PE CANDIDATE

Make sure you meet the requirements before taking the PE exam—this is the first step toward licensure. PPI offers several texts and online resources to prepare you for this crucial exam. The *FE Review Manual* is written for engineers who need rapid preparation for the FE Exam, but who understand the fundamentals of engineering. The clear writing style and easy-to-understand explanations in PPI's *Core Engineering Concepts for Students and Professionals* make it the perfect cross-disciplinary engineering handbook for expanding your knowledge in other engineering fields.

While you are preparing for the Electrical and Electronics exam, the following suggestions may help.

- Concentrate primarily on the fundamentals of electrical engineering. Take time to review the principles of any problem you find difficult.
- Use the Index extensively. Every significant term, law, theorem, and concept has been indexed. If you don't recognize a term, look for it in the Index and read the related text for a more comprehensive overview. The Index is the quickest way to find material in an exam situation, so you should familiarize yourself with it before exam day. General terms or concepts, which are covered in a variety of subject areas, may not be indexed, though the specifics of the concepts will be listed.
- Some subjects appear in more than one chapter. Use the Index liberally and learn all there is to know about a particular subject.
- Know which subjects in this book are covered on the PE exam, and which are provided as support material. Chapters 1 through 25 and 55-59 are supportive and do not cover exam subjects. They do, however, provide background for the other chapters, enhance understanding of fundamentals, and are useful following the exam for quick reference. (The exam subjects known at the time of this printing are provided in Table 1. These are subject to change; the most up-to-date information can be found on PPI's website at **www.ppi2pass.com/eefaqs**.)
- Become intensely familiar with this book. This means knowing the order of the chapters, the approximate location of important figures and tables, what appendices are available, and so on.
- After reviewing a chapter of this book, solve each of the related problems in the companion volume, *Electrical and Electronics Practice Problems for the Electrical and Computer PE Exam.*
- Practice in problem solving is the key to success on the exam. Review examples and solve as many problems as possible in the companion volume. It's a good idea to begin with problems that fall in your weakest areas and work toward your stronger areas.
- Use the answers to the practice problems to check your work. If your answer doesn't match the given solution, study the solution to determine why. The majority of this book uses SI units, except where customary U.S. (English) units are the standard. Some equations in the U.S. system use the term g/g_c. For calculations at standard gravity, 1.00 is the numerical value of this fraction. Therefore, it is necessary to incorporate this quantity only in problems with a non-standard gravity, or when being meticulous with units.
- Use Table 3, Schedule for Self-Study, as your guide for focusing your exam preparation. Table 3 is found at the end of this Introduction.
- Practice by completing as many problems as possible. Remember that you'll have an average of only six minutes per problem on the exam, so detailed design is not possible. Rather, it is more likely that you will need to solve a portion of a design problem, complete a piece of an analysis puzzle, or demonstrate an understanding of general requirements or standard practice.

Table 1 Detailed Analysis of Tested Subjects

ELECTRICAL AND ELECTRONICS EXAMINATION

General Electrical Engineering (50%)

Circuit analysis (25%): passive components; DC circuits; sinusoidal analysis; transient analysis; power and energy calculations; battery characteristics and ratings; power supply

Measurement and instrumentation (10%): transducer system characteristics; data evaluation; operational amplifiers

Safety and design limits (4%): interface applications; failure limits and circuit protection; safety grounding; electromagnetic interference and exposure; reliability; electric shock and burns

Signal processing (11%): sampling theory; transforms and applications; analog-to-digital and digital-to-analog conversions

Digital Systems (11%)

Digital logic (5%): Boolean algebra; combinational and sequential logic

Digital components (6%): digital devices, memory devices; programmable logical devices; microcontrollers/embedded systems

Electric and Magnetic Field Theory and Applications (9%)

Electromagnetic fields (4%): theory; EMI/EMC

Transmission lines and guided waves (2.5%): transmission lines; waveguides

Antennas (2.5%): gain, patterns, and polarization; impedance

Electronics (12.5%)

Electronic circuit theory (6.5%): small-signal and large-signal models; active networks and filters; nonlinear circuits; sinusoidal steady-state analysis; transient analysis; power and energy calculations

Electronic components and circuits (6%): solid-state power devices and power electronics applications; battery characteristics and ratings; power supplies; oscillators and phase-locked loop characteristics; amplifiers; modulators and demodulators; diodes; circuit protection and safety; transistors and applications

Control System Fundamentals (7.5%)

Block diagrams; Characteristic equations; Frequency response; Time response; Control system design and implementation; Stability

Communications (10%)

Modulation (4%): analog modulation; digital modulation; spread spectrum modulation characteristics

Noise and interference (2%): signal-to-noise ratio; quantization noise; noise figure and temperature; interference; coding, error detection, and correction

Telecommunications (4%): wireline communications; wireless communications; optical communications; multiplexing; traffic and switching

IF YOU ARE A PRACTICING ENGINEER OR AN ENGINEERING STUDENT

If you are a practicing engineer or an engineering major and have obtained this book as a general-reference handbook, the following suggestions and information may be of assistance.

- Both the Index and Table of Contents are extensive and meant as the starting point for any information desired. Every significant term, law, theorem, and concept has been indexed in every conceivable way—backwards and forwards.
- Chapters are grouped into interrelated topics, each offering a broad overview of a given area of electrical

engineering. If you want a quick introduction to a given topic, refer to that portion of the book.

- The mathematics chapters can be used as quick reviews when working any engineering problems.
- The Advanced Engineering Mathematics chapter provides, in simple but accurate terms, the information necessary to understand the importance and use of various advanced mathematical concepts. This should allow you to more fully comprehend and appreciate professional journals or engineering-focused presentations that use such concepts without resorting to a complete review of undergraduate math texts.
- The Advanced Engineering Mathematics chapter is also meant as the starting point for delving more deeply into advanced mathematical concepts by providing the names and defining the variables of often-used mathematical formulas. This is done to help you through those situations where you know you need more mathematical information, but don't know where to begin, what questions to ask, or what mathematical concept is involved.
- The Appendices are meant as quick references for common electrical information.

This book is intended as a broad overview of the fundamentals of electrical and computer engineering. Here you should find the topics and terms you deal with daily, as well as many that you rarely encounter. All are discussed in enough depth that this book may be your first and last source of information. At the very least, you will find the term you are looking for, its relative place in the spectrum of electrical and computer engineering, and enough information to perform a search through other sources for more details.

IF YOU ARE AN INSTRUCTOR

If you are teaching a review course for the PE exam without the benefit of recent, firsthand exam experience, you can use the material in this book as a guide to prepare your lectures. Your students should use the book as a textbook. You should emphasize the subjects in each chapter and avoid subjects that were omitted. Subjects that do not appear in this book have rarely, if ever, appeared on the PE exam.

In an attempt to over-prepare students, I suggest assigning homework with practice problems that are more difficult and more varied than actual exam problems. It is often more efficient to cover several procedural steps in one practice problem than to ask simple "one liners" or definition problems.

"Capacity assignment" should be the goal in a successful review course. To do all the homework for some chapters requires approximately 15 to 20 hours of preparation per week. One student may be able to put in 20 hours, while another may only be able to put in 10 hours; assigning the full 20 hours of homework ensures that each student is working to his or her capacity. After the PE exam, both students will honestly say that they could not have prepared any more than they did in your course.

Instead of individually grading students' homework, permit students to make use of existing solutions to check their own work and learn approaches to the problems in their homework set. This will allow you to address special needs or questions written on the assignments, rather than simply marking the work as right or wrong. Solved practice problems are available in the accompanying *Electrical and Electronics Practice Problems* text.

Lecture coverage of some exam subjects is necessarily brief; other subjects are not covered at all. These omissions are intentional; they are not the result of scheduling. Why? First, time is not on your side in a review course. Second, some subjects rarely appear on the exam. Third, there are some subjects that are not well-received by the students, because they are not subjects that are pertinent to all students' objectives. Unless you have six months in which to teach your PE review, your students' time can be better spent covering topics that apply to the majority of examinees. Table 1 lists the approximate percentage of problems on the exam that will be devoted to a given topic. Using this percentage and the depth of coverage in this book, you can decide which topics to focus on in your review course.

In organizing the review course format outlined in Table 2, I have tried to order the subjects in a logical, progressive manner, keeping my eye on "playing the high-probability subjects." I cover early in the course those subjects that everyone needs to learn. I leave the more specific subjects to the end.

Use the formulas, illustrations, and tables of data in the text to minimize chalkboard time. Avoid proofs and emphasize problem solving. The more problems solved, the higher the chances of exam success.

The practice problems in the companion volume, *Electrical and Electronics Practice Problems for the Electrical and Computer PE Exam*, span the spectrum from quick-answer to in-depth probing of knowledge. The more in-depth problems can be used as homework or timed problems.

One possible review course format is outlined in Table 2. The sessions are weekly, with the course lasting three and a half months. Most sessions could be covered in three hours of lecture. All the skipped chapters and any related practice problems are presented as floating assignments to be made up in the students' "free time."

I strongly believe in exposing students to a realistic sample exam, but since the review course usually ends only a few days before the real PE exam, I hesitate to make students sit for several hours in the late evening to take an in-class exam. Instead, I recommend distributing and assigning a take-home sample exam at the first meeting of the review course.

Caution your students not to peek at the sample exam prior to taking it. Looking at the sample exam or otherwise using it to direct their review will produce

Table 2 Typical PE Review Course Format

session	subject covered	chapter(s)
1	Introduction: PE Examination; Mathematics; Professional	Introduction, 1-15, 55-58, 59
2	Basic Theory and Field Theory	17, 20, 21
3	Linear Circuit Theory	29, 30, 31
4	Electronics Fundamentals	37
5	Junction Transistors	38
6	Field Effect Transistors	39
7	Electrical and Electronic Devices	40
8	Frequency Response	32
9	Digital Logic	41
10	Logic Network Design	42
11	Synchronous Sequential Networks	43
12	Digital Systems	36, 44
13	Controls	57
14	Communications	46, 47, 50

unwarranted specialization in subjects contained in the sample exam. For the sample exam to be realistic practice, it needs to mimic as much as possible the challenges of the actual exam.

There are many ways to organize a PE review course, depending on your available time, budget, intended audience, facilities, and enthusiasm. However, all good course formats have the same result: students struggle with the workload during the course, and then they breeze through the exam afterward.

PART 2: EVERYTHING YOU EVER WANTED TO KNOW ABOUT THE PE EXAM

WHAT IS THE FORMAT OF THE PE EXAM?

The NCEES PE exam in electrical and computer engineering consists of two four-hour sessions separated by a one-hour lunch period. The exam is an open-book exam. All problems are multiple choice. Upon registering for the exam, examinees choose to work one of three exam modules: (1) Computer Exam, (2) Electrical and Electronics Exam, or (3) Power Exam. Each exam consists of 80 problems (40 in the morning and 40 in the afternoon) that test knowledge in the areas specified. Examinees must work all 80 problems in the exam of their choice.

For more information on NCEES and the exams necessary to become a licensed electrical engineer, refer to Chap. 59, Engineering Licensing in the United States.

WHAT SUBJECTS ARE ON THE PE EXAM?

NCEES publishes PE exam specifications—a list of the knowledge areas covered and their approximate distribution on the exam. Despite the published exam structure, the exact number of problems that will appear in each subject area cannot be predicted reliably. There is no guarantee that any single subject will occur in any quantity.

Any problem may be categorized in a variety of areas and subjects. For example, a problem regarding MOSFETs may be considered an Electronics problem in the area of digital components; a Computer problem in the area of memory devices; or depending on the wording, a Power problem in the area of load sharing.

Table 1 describes the exam subjects in detail. Most examinees find the list formidable. The percentage breakdowns in Table 1 are according to NCEES, but these percentages are approximate. NCEES adds,

> The examination is developed with questions that require a variety of approaches and methodologies, including design, analysis, and application. Some questions may require knowledge of engineering economics. The knowledge areas specified as

examples of kinds of knowledge are not exclusive or exhaustive categories.

As you can see, there is no distinction between subjects in the morning and afternoon sessions.

WHAT IS THE TYPICAL PROBLEM FORMAT?

Almost all of the problems stand alone—that is, they are completely independent. NCEES allows that some sets of problems may start with a statement of a "situation" that will apply to (typically) two to five following problems. Use of such grouped problems is expected to be increasingly rare, however. Since the problems are multiple-choice in design, all required data should appear in the situation statement. Values calculated for one problem will not generally be required to correctly solve another problem. Problems are independent or start with new given data. A mistake on one of the problems shouldn't cause you to get a subsequent problem wrong. However, considerable time may be required to repeat previous calculations with a new set of given data. Some problem statements may contain superfluous information (data not relevant to solving the problem).

Each problem will have four answer choices, labeled "A," "B," "C," and "D." One of the answer choices is correct (or, "most nearly correct," as described in the following section). The remaining answer choices are incorrect, and may be logical distractors (e.g., plausible answers, or the results of common miscalculations).

WHAT DOES "MOST NEARLY" REALLY MEAN?

One of the more disquieting aspects of these problems is that the available answer choices are seldom exact. Answer choices generally have only two or three significant digits. Exam problems ask, "Which answer choice is most nearly the correct value?" or they instruct you to complete the sentence, "The value is approximately…" A lot of self-confidence is required to move on to the next problem when you don't find an exact match for the answer you calculated, or if you have had to split the difference because no available answer choice is close.

NCEES has described it like this:

> Many of the questions on NCEES exams require calculations to arrive at a numerical answer. Depending on the method of calculation used, it is very possible that examinees working correctly will arrive at a range of answers. The phrase "most nearly" is used to accommodate answers that have been derived correctly but that may be slightly different from the correct answer choice given on the exam. You should use good engineering judgment when selecting your choice of answer. For example, if the question asks you to calculate an electrical current or determine the load on a beam, you should literally select the answer option that is most nearly what you calculated, regardless of whether it is more or less than your calculated value. However, if the question asks you to select a fuse or circuit breaker to protect against a calculated current or to size a beam to carry a load, you should select an answer option that will safely carry the current or load. Typically, this requires selecting a value that is closest to but larger than the current or load.

The difference is significant. Suppose you are asked to calculate "most nearly" the current drawn by a given electrical load. Suppose, also, that the result of the calculation is 11 A. If the provided answer choices were (A) 5 A, (B) 10 A, (C) 15 A, and (D) 20 A, you would go with answer choice (B), because this is most nearly what you calculated. If, however, you were asked to select the breaker that provides the required overcurrent protection for this load, you would select answer choice (C).

HOW MUCH MATHEMATICS IS NEEDED FOR THE EXAM?

Generally, only simple algebra, trigonometry, and geometry are needed on the PE exam. You will need to use the trigonometric, logarithm, square root, and similar buttons on your calculator. There is no need to use any other method for these functions. For functions I recommend you program into your calculator, see the section "What about Electronic Devices?"

There are no pure mathematics problems (algebra, geometry, trigonometry, etc.) on the exam. However, you will need to apply your knowledge of these subjects to the exam problems. Except for simple quadratic equations, you will probably not need to find the roots of higher-order equations. Occasionally, it will be convenient to use the equation-solving capability of an advanced calculator. However, other solution methods will always exist.

There is little or no use of calculus on the exam. Rarely, you may need to take a simple derivative to find a maximum or minimum of some function. Even rarer is the need to integrate to find an average or moment of inertia.

There is essentially no need to solve differential equations. On the exam, there are simple linear circuits and components that can be solved or explained with differential equations. Often these applications, and others that use such mathematics, can be solved by other means.

Basic statistical analysis of observed data may be necessary. Statistical calculations are generally limited to finding means, medians, standard deviations, variances, percentiles, and confidence limits. The only population distribution you need to be familiar with is the normal curve. Probability, reliability, hypothesis testing, and statistical quality control are not explicit exam subjects.

The Electrical and Electronics exam is concerned with numerical answers, not with proofs or derivations. You will not be asked to prove or derive formulas.

Occasionally, a calculation may require an iterative solution method. Generally, there is no need to complete more than two iterations. You will not need to program your calculator to obtain an "exact" answer. Nor will you generally need to use complex numerical methods.

HOW ABOUT ENGINEERING ECONOMICS?

For most of the early years of engineering licensing, problems on engineering economics appeared frequently on the exams. This is no longer the case. However, in its outline of exam subjects, NCEES notes: "Some problems may require knowledge of engineering economics." What this means is that engineering economics can constitute anything from nothing to several problems on the exam. While the degree of required engineering economics knowledge may have decreased somewhat, the basic economic concepts (e.g., time value of money, present worth, non-annual compounding, comparison of alternatives, etc.) are still valid test subjects.

WHAT ABOUT PROFESSIONALISM AND ETHICS?

For many years, NCEES has discussed adding professionalism and ethics problems to the PE exam. These subjects are not currently part of the specifications, and there has yet to be a problem on these subjects in the exam.

IS THE EXAM TRICKY?

Other than providing superfluous data, the PE exam is not a "tricky exam." The exam does not overtly try to get you to fail. Examinees manage to fail on a regular basis with perfectly straightforward problems. The exam problems are difficult in their own right; NCEES does not need to provide misleading or conflicting statements. However, you will find that commonly made mistakes are represented in the available answer choices. Thus, the alternative answers (known as *distractors*) will be logical.

Problems are generally practical, dealing with common and plausible situations that you might experience in your job. Circuit design is restricted to connecting electrical components to create a desired outcome. You will not be asked to design a new IC chip or build an equivalent vacuum tube circuit.

WHAT MAKES THE PROBLEMS DIFFICULT?

Problems can be difficult for any number of reasons, the most common being a lack of knowledge in a given area.

Beyond that fundamental cause, some problems are difficult because the pertinent theory is not obvious. Also, data needed may be hard to find. For example, building requirements are in the *National Electrical Code*; without the code, and knowledge of where given information is located within the code, certain problems will be nearly impossible to solve.

Some problems are difficult because they defy imagination. For example, problems pertaining to the theoretical functioning of certain electronic components are more easily solved if the physical layout of the component can be visualized. Some problems are difficult because the computational time is lengthy. Some problems are difficult because the terminology is obscure. This can occur in almost any subject.

DOES THE PE EXAM USE SI UNITS?

The PE exam primarily uses the SI system. A few topics utilize the customary U.S. system where they correspond to commonly accepted industry standards. The example that comes to mind is problems involving the *National Electrical Code*, where, until recently, most units were given in customary U.S. units with SI provided in parenthesis. Even in this topic, SI is now foremost and customary U.S units are now in parentheses.

WHY DOES NCEES REUSE SOME PROBLEMS?

NCEES reuses some of the more reliable problems from each exam. The percentage of repeat problems isn't high—no more than 25% of the exam. NCEES repeats problems in order to equate the performance of one group of examinees with the performance of an earlier group. The repeated problems are known as *equaters*.

Occasionally, a new problem appears on the exam that very few of the examinees do well on. Usually, the reason for this is that the subject is too obscure or too difficult. Also, there have been cases where a low percentage of the examinees get the answer correct because the problem was inadvertently stated in a poor or confusing manner. These, as well as questions that everyone gets correct, are considered defective and are generally removed from the exam scoring process.

NCEES tracks the usage and "success" of each of the exam problems. Such "rogue" problems are not repeated without modification. This is one of the reasons historical analysis of problem types shouldn't be used as the basis of your review.

DOES NCEES USE THE EXAM TO PRE-TEST FUTURE PROBLEMS?

NCEES does not use the PE exam to "pre-test" or qualify future problems. (It does use this procedure on the FE exam, however.) All of the problems you work will contribute toward your final score (unless a problem is deemed defective).

ARE THE PRACTICE PROBLEMS REPRESENTATIVE OF THE EXAM?

The practice problems in the companion book, *Electrical and Electronics Problems for the Electrical and Computer PE Exam*, cover exam subjects. However, they are generally more comprehensive and complex than actual exam problems. For example, some practice problems are marked "Time limit: one hour." Compared to the six-minute problems on the PE exam, such one-hour problems are considerably more time-consuming and difficult to work.

WHAT REFERENCE MATERIAL IS PERMITTED IN THE EXAM?

The PE exam is an open-book exam. Most states do not have any limits on the numbers and types of books you can use. Personal notes in a three-ring binder and other semi-permanent covers can usually be used. You cannot share books with other examinees during the exam.

Some states use a "shake test" to eliminate loose papers from binders. Make sure that nothing escapes from your binders when they are inverted and shaken.

A few states do not permit collections of solved problems such as *Schaum's Outline Series*, sample exams, and solutions manuals. Some states maintain a formal list of banned books.

Strictly speaking, loose paper and scratch pads are not permitted in the exam. Certain types of preprinted graphs and logarithmically scaled graph papers may be used, but are almost never needed. If you choose to use these papers, they should be three-hole punched and brought in a three-ring binder. An exception to this restriction may be made for laminated and oversize charts, graphs, and tables that are commonly needed for particular types of problems.

HOW MANY BOOKS SHOULD YOU BRING?

You won't use many books in the exam. The trouble is, you can't know in advance which ones you will need, which is why many examinees show up with boxes and boxes of books. Without a doubt, there are things that you will need that are not in the *Electrical and Electronics Reference Manual*. But there are not so many of these omissions that you need to bring your company's entire library. The exam is very fast-paced. You will not have time to use books with which you are not thoroughly familiar. The exam doesn't require you to know obscure solution methods or to use difficult-to-find data.

It really is unnecessary to bring a large quantity of books with you. Essential books are identified in the Codes and References section in this text, and you should be able to decide which additional support you need for the areas that give you the most trouble. This book and five to ten other references of your choice should be sufficient for most of the problems you encounter.

MAY TABS BE PLACED ON PAGES?

It is common to tab pages in your books in an effort to reduce the time required to locate useful sections. Inasmuch as some states consider Post-it notes to be "loose paper," your tabs should be of the more permanent variety. Although you can purchase tabs with gummed attachment points, it is also possible simply to use transparent tape to attach the Post-its you have already placed in your books.

WHAT ABOUT ELECTRONIC DEVICES?

The exam requires use of a scientific calculator. However, it may not be obvious that you should bring a spare calculator with you to the exam. It is always unfortunate when an examinee is not able to finish because his or her calculator was dropped or stolen or stopped working for some unknown reason.

NCEES has banned communicating and text-editing calculators from the exam site. Only select types of calculators are permitted. Check the current list of permissible devices at the PPI website (**www.ppi2pass.com/calculators**). Nomographs and specialty slide rules are permitted.

The exam has not been optimized for any particular brand or type of calculator. In fact, for most calculations, a \$15 scientific calculator will produce results as satisfactory as those from a \$200 calculator. There are definite benefits to having built-in statistical functions, unit-conversion, and equation-storing capabilities. However, these benefits are not so great as to give anyone an unfair advantage.

It is essential that a calculator used for the PE exam have the following functions.

- trigonometric and inverse trigonometric functions
- π
- $\sqrt{x}$ and x^2
- both common and natural logarithms
- y^x and e^x

For maximum speed, your calculator should also have or be programmed for the following functions.

- converting between polar (phasor) and rectangular vectors
- performing simultaneous linear equations
- performing rms calculations
- simplifying Smith chart calculations
- calculating determinants of matrices
- linear regression
- calculating factors for economic analysis problems

You may not share calculators with other examinees. Be sure to take your calculator with you whenever you leave the exam room for any length of time.

Laptop computers are generally not permitted in the exam. Their use has been considered, and some states may actually permit them. However, considering the nature of the exam problems, it is very unlikely that laptops would provide any advantage.

You may not use a walkie-talkie, cell phone, or other communications device during the exam.

HOW IS THE EXAM GRADED AND SCORED?

The maximum number of points you can earn on the Electrical and Electronics PE exam is 80. The minimum number of points for passing (known as "the cut score" by NCEES) varies from one exam administration to another. The cut score is determined through a rational procedure, without the benefit of knowing examinees' performance on the exam. That is, the exam is not graded on a curve. The cut score is selected based on what you are expected to know, not based on passing a certain percentage of engineers.

Each of the problems is worth one point. Grading is straightforward, since a computer grades your score sheet. Either you get the problem right or you don't. There is no deduction for incorrect answers, so guessing is encouraged. However, if you mark two or more answers for the same problem, no credit is given for the problem.

You will receive the results of your exam from your state board (not NCEES) by mail. Allow at least four months for notification. Your score may or may not be revealed to you, depending on your state's procedure. Even when the score is reported to you, it may have been scaled or normalized to a portion of 100%.

HOW YOU SHOULD GUESS

NCEES produces defensible licensing exams. As a result, there is no pattern to the placement of correct responses. Therefore, it most likely will not help you to guess all "A," "B," "C," or "D."

The proper way to guess is as an engineer. You should use your knowledge of the subject to eliminate illogical answer choices. Illogical answer choices are those that violate good engineering principles, that are outside normal operating ranges, or that require extraordinary assumptions. Of course, this requires you to have some basic understanding of the subject in the first place. Otherwise, it's back to random guessing. That's the reason that the minimum passing score is higher than 25%.

You won't get any points using the "test-taking skills" that helped you in college—the skills that helped with tests prepared by subject experts, rather than testing experts. You won't be able to eliminate any [verb] answer choices from "Which [noun] ..." problems. You won't find problems with options of the "more than 50" and "less than 50" variety. You won't find one answer choice among the four that has a different number of significant digits, or has a verb in a different tense, or has some singular/plural discrepancy with the stem. The distractors will always match the stem, and they will be logical.

WHAT IS THE PASSING RATE?

NCEES cites the passing rate for first-time examinees as being 63%. The rate for first-time examinees differs from repeat examinees. The rate from one exam to the next differs. Because of the variability, and because of periodic exam format changes, historical passing rates are essentially irrelevant. The most important considerations determining your probability of success are preparedness, problem solving ability, experience, and knowledge of the fundamentals.

HOW IS THE CUT SCORE ESTABLISHED?

The raw cut score may be established by NCEES before the exam is administered, or after. Final adjustments may be made following the exam date.

NCEES uses a process known as the *modified Angoff procedure* to establish the cut score. This procedure starts with a small group (the cut score panel) of professional engineers selected by NCEES. Each individual in the group reviews each problem and estimates its difficulty by giving their opinion of the number of minimally qualified engineers out of a hundred examinees who should know the correct answer to the problem. (This is equivalent to predicting the percentage of minimally qualified engineers who will answer correctly.)

Next, the panel assembles, and the estimates for each problem are openly compared and discussed. Eventually, a consensus value is obtained for each. When the panel has established a consensus value for every problem, the values are summed and divided by 100 to establish the cut score.

Various minor adjustments can be made after the exam is administered to account for examinee population (as characterized by the average performance on equater problems) and any flawed problems.

CHEATING AND EXAM SUBVERSION

There aren't very many ways to cheat on an open-book test. The proctors are well trained in spotting the few ways that do exist. You should not talk to other examinees in the room, nor should you pass notes back and forth. The number of people who are released to use the restroom may be limited to prevent discussions.

NCEES regularly reuses "good" problems that have appeared on previous exams. Therefore, exam security is a serious issue with NCEES, which goes to great lengths to make sure nobody copies the problems. You may not keep your exam booklet, enter the text of problems into your calculator, or copy problems into your own material.

The proctors are concerned about exam subversion, which generally means activity that might invalidate the exam or the exam process. The most common form of exam subversion involves trying to copy exam problems for future use.

PART 3: HOW TO PREPARE FOR AND PASS THE ELECTRICAL AND ELECTRONICS PE EXAM

WHAT SHOULD YOU STUDY?

Strictly speaking, you don't have to study every subject on the exam in order to pass. However, the more subjects you study, the more you'll improve your chances of passing. You should use Table 3, the Schedule for Self Study, to help you plan which subjects you are going to study, and when. It is best to decide early in the preparation process. The strategy you use will depend on your background. There are four common strategies; while you must choose the strategy that works best for you, taking a broad approach to exam preparation, the first strategy listed, tends to be the most successful.

Many successful examinees can attest that taking the broad approach to studying was their key to passing the exam. This strategy is to review the fundamentals in a broad range of undergraduate subjects (which means studying all or most of the chapters in this book). The exam includes enough fundamentals problems to make this strategy worthwhile. Overall, it's the best approach.

Practicing engineers who have little time for preparation often concentrate on the subject areas in which they hope to find the most problems. By studying the list of exam subjects, some have been able to focus on those subjects that will give them the highest probability of finding enough problems that they can answer. This strategy works as long as the exam cooperates and supplies enough of the types of problems they need. Too often, though, examinees who pick and choose subjects to review can't find enough workable problems to ensure they pass the exam.

Engineers who have been away from the classroom for a long time tend to concentrate on the subjects in which they have extensive experience, in the hope that the exam will feature lots of problems in those subjects. This method is seldom successful.

Some engineers plan on modeling their solutions from similar problems they have found in textbooks, collections of solutions, and old exams. These engineers often spend a lot of time compiling and indexing the example and sample problem types in all of their books, rather than in reviewing the material. This is not a legitimate preparation method, and it is almost never successful.

DO YOU NEED A CLASSROOM REVIEW COURSE?

Approximately 60% of first-time PE examinees take a review course of some form. Classroom, audio, video, correspondence, and internet courses are available for some or all of the exam topics. Courses provide several significant advantages over self-directed study, some of which may apply to you.

- A course structures and paces your review. It ensures that you keep going forward without getting bogged down in one subject.
- A course focuses you on a limited amount of material. Without a course, you might not know which subjects to study.
- A course provides you with the problems you need to solve. You won't have to spend time looking for them.
- A course spoon-feeds you the material. You may not need to read the whole book!
- The course instructor can answer your problems when you are stuck.

You probably already know if any of these advantages apply to you. A review course will be less valuable if you are inherently thorough, self-motivated, and highly disciplined.

HOW LONG SHOULD YOU STUDY?

We've all heard stories of the person who didn't crack a book until the week before the exam and still passed it with flying colors. Yes, these people really exist. However, I'm not one of them, and you probably aren't either. In fact, after having taught engineers in my own classes, I'm convinced that these people are as rare as the ones who have taken the exam five times and still can't pass it.

A thorough review takes approximately 300 hours. Most of this time is spent solving problems. Some of it may be spent in class; some is spent at home. Some examinees spread this time over a year. Others cram it all into two months. Most classroom review courses last for three or four months. The best time to start studying will depend on how much time you can afford to spend on exam preparation each week.

WHAT THE WELL-HEELED ELECTRICAL ENGINEER SHOULD BEGIN ACCUMULATING

While the *Electrical and Electronics Reference Manual* offers a comprehensive review of Electrical and Electronics exam subjects, there are many references and resources that you should begin to assemble for review and for use in the exam.

It is unlikely that you could pass the PE exam without accumulating other books and resources. There certainly isn't much margin for error if you show up with only one book. There are many problems that require knowledge, data, and experience that are presented and described only in books dedicated to a single subject. You would have to be truly lucky to go in "bare," find the right mix of problems, and pass.

Few examinees are able to accumulate all of the references needed to support the exam's entire body of knowledge. The accumulation process is too expensive and time-consuming, and the sources are too diverse. Like purchasing an insurance policy, what you end up with will be more a function of your budget than of your needs. In some cases, one book will satisfy several needs.

The books and other items listed in the Codes and References section of this book are regularly cited by examinees as having been particularly useful to them. This listing only includes the major "named" books that have become standard references in the industry. These books are in addition to any textbooks or resources that you might choose to bring.

ADDITIONAL REVIEW MATERIAL

In addition to this book and its accompanying *Electrical and Electronics Problems for the Electrical and Computer PE Exam*, PPI can provide you with many targeted references and study aids, some of which are listed here.

- *Electrical Engineering Sample Examinations for the Power, Electrical and Electronics, and Computer PE Exams*
- *NCEES Electrical and Electronics Sample Questions and Solutions*
- *Engineering Unit Conversions*
- *Electrical PE Exam Cafe* (available as a 30- or 60-day subscription)
- *Electrical PE Electrical and Electronics Passing Zone*

For a complete listing of available sample problems, other books, calculator software, videos, and other exam related products, please go to PPI's website at **www.ppi2pass.com/eeproducts**.

WHAT YOU WON'T NEED

Generally, people bring too many things to the exam. A general rule is that you shouldn't bring books that you have not looked at during your review. If you didn't need a book to understand the examples in this text, you likely won't need it during the exam.

There are other things that you won't need.

- Books on basic and introductory subjects: You won't need books that cover trigonometry, geometry, or calculus.
- Books on non-exam subjects: Such subjects as materials science, statics, dynamics, mechanics of materials, drafting, history, the English language, geography, and philosophy are not part of the exam.
- Books on mathematical analysis, or extensive mathematics tabulations: Mathematics is a tool for success on the PE exam, it is not the exam's focus.
- Obscure books and materials: Books that are in a different language, doctoral theses, and papers presented at technical societies won't be needed during the exam.
- Old textbooks or obsolete, rare, and ancient books: NCEES exam committees are aware of which textbooks are in use. Material that is available only in out-of-print publications and old editions won't be used.
- Handbooks in other disciplines: You probably won't need a civil, mechanical, or industrial engineering handbook.
- Crafts- and trades-oriented books: The exam does not expect you to have detailed knowledge of trades or of manufacturing operations.
- Manufacturer's literature and catalogs: No part of the exam requires you to be familiar with products that are proprietary to any manufacturer.
- U.S. government publications: With the exceptions of the publications mentioned in this book, no government publications are necessary for the PE exam.
- Your state's laws: The PE exam is a national exam. Nothing unique to your state will appear on it.

SHOULD YOU LOOK FOR OLD EXAMS?

The traditional approach to preparing for standardized tests includes working sample tests. However, NCEES does not release old tests or problems after they are used. Therefore, there are no official problems or tests available from legitimate sources. NCEES has published a booklet of sample problems and solutions to illustrate the format of the exam. These problems have been assembled from various previous exams, and the publication is not a true "old exam."

WHAT SHOULD YOU MEMORIZE?

You get lucky here, because it isn't necessary to actually memorize anything. The exam is open-book, so you can look up any procedure, formula, or piece of information that you need. You can speed up your problem-solving response time significantly if you don't have to look up the conversion from W to ft-lbf/sec, or the definition of the sine of an angle, but you don't have to memorize these kinds of things. As you work practice problems in the companion book, you are likely to absorb the information that you come across more than a few times.

DO YOU NEED A REVIEW SCHEDULE?

It is important that you develop and adhere to a review outline and schedule. Once you have decided which subjects you are going to study, you can allocate the available time to those subjects in a manner that makes sense to you. If you are not taking a review course (where the order of preparation is determined by the lectures), you should make an outline of subjects for self-study and use it to schedule your preparation. A fill-in-the-dates schedule, listing every chapter and the exam subject(s) covered in each, is provided in Table 3 at the end of this Introduction. An interactive, adjustable, and customizable study schedule is also available on the PPI website as part of Passing Zone, **www.ppi2pass.com/ppi/PPIPZ**.

A SIMPLE PLANNING SUGGESTION

Designate some location (a drawer, a corner, a cardboard box, or even a paper shopping bag left on the floor) as your "exam catch-all." Use your catch-all during the months before the exam when you have revelations about things you should bring with you. For example, you might realize that the plastic ruler marked off in tenths of an inch that is normally kept in the kitchen junk drawer can help you. Or, you might decide that a certain book is particularly valuable. Or that it would be nice to have dental floss after lunch. Or that large rubber bands are useful for holding books open.

It isn't actually necessary to put these treasured items in the catch-all during your preparation. You can, of course, if it's convenient. But if these items will have other functions during the time before the exam, at least write yourself a note and put the note into the catch-all. When you go to pack your exam kit a few days before the exam, you can transfer some items immediately, and the notes will be your reminders for the other items that are back in the kitchen drawer.

HOW YOU CAN MAKE YOUR REVIEW REALISTIC

In the exam, you must be able to quickly recall solution procedures, formulas, and important data. You must remain sharp for eight hours or more. When you played a sport back in school, your coach tried to put you in game-related situations. Preparing for the PE exam isn't much different from preparing for a big game. Some part of your preparation should be realistic and representative of the exam environment.

There are several things you can do to make your review more representative. For example, if you gather most of your review resources (i.e., books) in advance and try to use them exclusively during your review, you will become more familiar with them. This is also an excellent means of finding the inadequacies of your references. If you find the inadequacies during your exam preparation, you can add to or change your references. This is of course not possible if you wait until exam day to evaluate your resources.

Learning to use your time wisely is one of the most important lessons available during your review. You will undoubtedly encounter problems that end up taking much longer than you expected. In some instances, you will cause your own delays by spending too much time looking through books for things you need (or just by looking for the books themselves!). Other times, the problems will entail too much work. Learn to recognize these situations so that you can make an intelligent decision about skipping such problems on the exam.

WHAT TO DO A FEW DAYS BEFORE THE EXAM

There are a few things you should do a week or so before the exam. If necessary, you should arrange for childcare and transportation. Since the exam does not always start or end at the designated time, make sure that your child-care and transportation arrangements are flexible.

Check PPI's website for last-minute updates and errata to any PPI books you might have and are planning to bring to the exam.

If you haven't already done so, read the "Advice from Examinees" section of PPI's website, **ww.ppi2pass.com/eeadvice**.

If you haven't been following along on the Engineering Exam Forum on PPI's website, **www.ppi2pass.com/forums**, use the search function to locate discussions on this bulletin board.

If it's convenient, visit the exam location in order to find the building, parking areas, exam room, and restrooms. If it's not convenient, you may find driving directions and/or site maps on the web.

Take the battery cover off your calculator and check to make sure you are bringing the correct size replacement batteries. Some calculators require a different kind of battery for their "permanent" memories. Put the cover back on and secure it with a piece of masking tape. Write your name on the tape to identify your calculator.

If your spare calculator is not the same as your primary calculator, spend a few minutes familiarizing yourself with how it works. In particular, you should verify that your spare calculator is functional and has adequate battery charge.

PREPARE YOUR CAR

[ ] Gather snow chains, shovel, and tarp to lie on while installing chains.

[ ] Check tire pressures.

[ ] Check your spare tire.

[ ] Check for tire installation tools.

[ ] Verify that you have the vehicle manual.

[ ] Check fluid levels (oil, gas, water, brake fluid, transmission fluid, window-washing solution).

[ ] Fill up with gas.

[ ] Check battery and charge if necessary.

[ ] Know something about your fuse system (where they are, how to replace them, etc.).

[ ] Assemble all required maps.

[ ] Fix anything that might slow you down (missing wiper blades, etc.).

[ ] Check your taillights.

[ ] Affix the recently arrived DMV license sticker.

[ ] Fix anything that might get you pulled over on the way to the exam (burned-out taillight or headlight, broken lenses, bald tires, missing license plate, noisy muffler).

[ ] Treat the inside windows with anti-fog solution.

[ ] Put a roll of paper towels in the back seat.

[ ] Gather exact change for any bridge tolls or toll roads.

[ ] Put $20 in your glove box.

[ ] Check for current registration and proof of insurance.

[ ] Locate a spare key.

[ ] Find your AAA or other roadside-assistance cards and phone numbers.

[ ] Plan out alternate routes.

PREPARE YOUR EXAM KITS

Second in importance to your scholastic preparation is the preparation of your two exam kits. The first kit consists of a bag, box (plastic milk crates hold up better than cardboard in the rain), or wheeled travel suitcase containing items to be brought with you into the exam room.

[ ] letter admitting you to the exam

[ ] photographic identification (e.g., driver's license)

[ ] this book

[ ] other textbooks and reference books

[ ] regular dictionary

[ ] scientific/engineering dictionary

[ ] review course notes in a three-ring binder

[ ] cardboard boxes or plastic milk crates to use as a bookcase

[ ] primary calculator

[ ] spare calculator

[ ] instruction booklets for your calculators

[ ] extra calculator batteries

[ ] straightedge and rulers

[ ] compass

[ ] protractor

[ ] scissors

[ ] stapler

[ ] transparent tape

[ ] magnifying glass

[ ] small (jeweler's) screwdriver for fixing your glasses or for removing batteries from your calculator

[ ] unobtrusive (quiet) snacks or candies, already unwrapped

[ ] two small plastic bottles of water

[ ] travel pack of tissue (keep in your pocket)

[ ] handkerchief

[ ] headache remedy

[ ] personal medication

[ ] $3.00 in miscellaneous change

[ ] light, comfortable sweater

[ ] loose shoes or slippers

[ ] cushion for your chair

[ ] earplugs

[ ] wristwatch with alarm

[ ] several large trash bags ("raincoats" for your boxes of books)

[ ] roll of paper towels

[ ] wire coat hanger (to hang up your jacket or to get back into your car in an emergency)

[ ] extra set of car keys on a string around your neck

The second kit consists of the following items and should be left in a separate bag or box in your car in case they are needed.

[ ] copy of your application

[ ] proof of delivery

[ ] light lunch

[ ] a beverage

[ ] sunglasses

[ ] extra pair of prescription glasses

[ ] raincoat, boots, gloves, hat, and umbrella

[ ] street map of the exam area

[ ] parking permit

[ ] battery-powered desk lamp

[ ] your cell phone

[ ] piece of rope

The following items cannot be used during the exam and should be left at home.

[ ] personal pencils and erasers (NCEES distributes mechanical pencils at the exam)

[ ] fountain pens

[ ] radio or tape/CD player

[ ] battery charger

[ ] extension cords

[ ] scratch paper

[ ] note pads

PREPARE FOR THE WORST

All of the occurrences listed in this section happen to examinees on a regular basis. Granted, you cannot prepare for every eventuality. But, even though each of these occurrences taken individually is a low-probability event, taken together, they are worth considering in advance.

- Imagine getting a flat tire, getting stuck in traffic, or running out of gas on the way to the exam.
- Imagine arriving late. Can you get into the exam without having to make two trips from your car?
- Imagine having to park two blocks from the exam site. How are you going to get everything to the exam room? Can you actually carry everything that far? Could you use a furniture dolly, a supermarket basket, or perhaps a helpmate?
- Imagine rain and snow as you are carrying your cardboard boxes of books into the exam room. Would plastic trash bags be helpful?
- Imagine a delay of an hour while they find someone to unlock the building, turn on the heat, or wait for the head proctor to bring instructions.
- Imagine a Star Trek convention, square-dancing contest, construction site, or auction in the next room.
- Imagine a site next to an airport with frequent takeoffs, or next to a construction site with a pile driver, or next to the NHRA's Drag Racing Championship.
- Imagine a seat where someone nearby chews gum with an open mouth; taps his pencil or drums her fingers; or wheezes, coughs, and sneezes for eight hours.
- Imagine the distraction of someone crying or of proctors evicting yelling and screaming examinees who have been found cheating. Imagine the tragedy of another examinee's serious medical emergency.
- Imagine a site without any heat, with poor lighting, or with sunlight streaming directly into your eyes.
- Imagine a hard folding chair and a table with one short leg.
- Imagine a power outage occurring sometime during the exam.
- Imagine a proctor who (a) tells you that one of your favorite books can't be used in the exam, (b) accuses you of cheating, or (c) calls "time up" without giving you any warning.
- Imagine not being able to get your lunch out of your car or find a restaurant.
- Imagine someone stealing your calculator during lunch.
- Imagine getting sick or nervous in the exam.

Now that you have imagined these possibilities, plan how you would deal with them. Preparing yourself for the worst possible experience will reduce any surprise situations and enable you to address any real issues that may happen.

WHAT TO DO THE DAY BEFORE THE EXAM

Take the day before the exam off from work to relax. Do not cram the last night. A good night's sleep is the best way to start the exam. If you live a considerable distance from the exam site, consider getting a hotel room in which to spend the night.

Practice setting up your exam work environment. Carry your boxes to the kitchen table. Arrange your "bookcases" and supplies. Decide what stays on the floor in boxes and what gets an "honored position" on the tabletop.

Use your checklist to make sure you have everything. Make sure your exam kit is packed and ready to go. Wrap your boxes in plastic bags in case it's raining when you carry them from the car to the exam room.

Calculate your wake-up time and set the alarms on two bedroom clocks. Select and lay out your clothing items. (Dress in layers.) Select and lay out your breakfast items.

If it's going to be hot on exam day, put your (plastic) bottles of water in the freezer.

Make sure you have gas in your car and money in your wallet.

WHAT TO DO THE DAY OF THE EXAM

Turn off the quarterly and hourly alerts on your wristwatch. Bring or buy a morning newspaper.

You should arrive at least 30 minutes before the exam starts. This will allow time for finding a convenient parking place, bringing your materials to the exam room, making room and seating changes, and calming down. Be prepared, though, to find that the exam room is not open or ready at the designated time.

Once you have arranged the materials around you on your table, take out your morning newspaper and look cool. (Only nervous people work crossword puzzles.)

WHAT TO DO DURING THE EXAM

All of the procedures typically associated with timed, proctored, computer-graded assessment tests will be in effect when you take the PE exam.

The proctors will distribute the exam booklets and answer sheets if these papers are not already on your tables. You should not open the booklets until instructed to do so. You may read the information on the front and back covers, and you should write your name in the appropriate blank spaces.

Listen carefully to everything the proctors say. Do not ask your proctors any engineering questions. Even if they are knowledgeable in engineering, they will not be permitted to answer your questions.

Answers to problems are recorded on an answer sheet contained in the test booklet. The proctors will guide you through the process of putting your name and other biographical information on this sheet when the time comes, which will take approximately 15 minutes. The time it takes to complete the form portion of the answer sheet is not included as part of the four hours you will have to complete the exam. You will be given the full four hours to answer problems.

The common suggestions to "completely fill the bubbles and erase completely" apply here. NCEES provides each examinee with a mechanical pencil with HB lead. Use of ballpoint pens and felt-tip markers is prohibited.

If you finish the exam early and there are still more than 30 minutes remaining, you will be permitted to leave the room. If you finish less than 30 minutes before the end of the exam, you may be required to remain until the end. This is done to be considerate of the people who are still working.

When you leave, you must return your exam booklet. You may not keep the exam booklet for later review.

If there are any problems that you think were flawed, in error, or unsolvable, ask a proctor for a "reporting form" on which you can submit your comments. Follow your proctor's advice in preparing this document.

WHAT ABOUT EATING AND DRINKING IN THE EXAM ROOM?

The official rule is probably the same in every state: no eating or drinking in the exam. That makes sense, for a number of reasons. Some exam sites don't want (or don't permit) stains and messes. Others don't want crumbs to attract ants and rodents. Your table partners don't want spills or smells. Nobody wants the distractions. Most important, you don't want the possible repercussions: your proctors can't give you a new exam booklet when the first one is ruined with coffee.

How this rule is administered varies from site to site and from proctor to proctor. Some proctors enforce the letter of law, threatening to evict you from the exam room when they see you chewing gum. Others may permit you to have bottled water, as long as you store the bottles on the floor where any spills will not harm what's on the table. No one is going to let you crack peanuts while you work on the exam, but I can't see anyone complaining about a hard candy melting away in your mouth. You'll just have to find out when you get there.

HOW TO SOLVE MULTIPLE-CHOICE PROBLEMS

When you begin each session of the exam, observe the following suggestions:

- Use only the pencil provided.
- Do not spend an inordinate amount of time on any single problem. If you have not answered a problem in a reasonable amount of time, make a note of it and move on to another.
- Set your wristwatch alarm for five minutes before the end of each four-hour session, and use that remaining time to guess at all of the remaining problems. Odds are that you will be successful with about 25% of your guesses, and these points will more than make up for the few points that you might earn by working during the last five minutes.
- Make mental notes about any problems for which you cannot find a correct response, that appears to have two correct responses, or that you believe have some technical flaw. Errors in the exam are rare, but they do occur. Such errors are usually discovered during the scoring process and discounted from the exam, so it is not necessary to tell your proctor, but be sure to mark the one best answer before moving on.
- Make sure all of your responses on the answer sheet are dark and completely fill the bubbles.

SOLVE PROBLEMS CAREFULLY

Many points are lost to carelessness. Keep the following items in mind when you are solving problems. Hopefully, these suggestions will be automatic in the exam.

[ ] Did you recheck your mathematical equations?

[ ] Do the units cancel out in your calculations?

[ ] Did you convert between radius and diameter?

[ ] Did you recheck all data obtained from other sources, tables, and figures?

[ ] Did you convert to the requested units?

SHOULD YOU TALK TO OTHER EXAMINEES AFTER THE EXAM?

The jury is out on this question. People react quite differently to the exam experience. Some people are energized. Most are exhausted. Some people need to unwind by talking with other examinees, describing every detail of their experience, and dissecting every exam problem. Other people need lots of quiet space, and prefer to just get into a hot tub to soak and sulk. Most engineers, apparently, are in this latter category.

Since everyone who took the exam has seen it, you will not be violating your "oath of silence" if you talk about the details with other examinees. It's difficult not to ask how someone else approached a problem that had you completely stumped. However, keep in mind that it is very disquieting to think you answered a problem correctly, only to have someone tell you where you went wrong.

NCEES requires examinees to sign an agreement stating that they "will not reveal in whole or in part any exam questions, answers or solutions to anyone during or after the exam, whether orally, in writing, or in any Internet chat rooms or otherwise." Discussing the test, even to examinees who took the same test at the same time, could prompt NCEES to invalidate your score.

AFTER THE EXAM

Yes, there is something to do after the exam. Most people come home, throw their exam "kits" into the corner, and collapse. A week later, when they can bear to think about the experience again, they start integrating their exam kits back into their normal lives. The calculators go back into the desk, the books go back on the shelves, the $3.00 in change goes back into the piggy bank, and all of the miscellaneous stuff you brought with you to the exam is put back wherever it came from.

Here's what I suggest you do as soon as you get home, before you collapse.

- [ ] Thank your spouse and children for helping you during your preparation.
- [ ] Take any paperwork you received on exam day out of your pocket, purse, or wallet. Put this inside your *Electrical and Electronics Reference Manual* for safekeeping.
- [ ] Reflect on any statements regarding exam secrecy that you signed.
- [ ] Visit the PPI website and complete the after-exam survey to help PPI improve the quality of its service and products.
- [ ] If you participated in a PPI Passing Zone, log on one last time to thank the instructors. (Passing Zones remain posted for a week after the exam.)
- [ ] Call your employer and tell him/her that you need to take a mental health day off on Monday.

A few days later, when you can face the world again, do the following.

- [ ] Make notes about anything you would do differently if you had to take the exam over again.
- [ ] Consolidate all of your application paperwork, correspondence to/from your state, and any paperwork that you received on exam day.
- [ ] If you took a live review course, call the instructor (or write a note) to say "Thanks."
- [ ] Visit the Engineering Exam Forum on PPI's website and see what other people are saying about the exam you took.
- [ ] Return any books you borrowed.
- [ ] Write thank-you notes to all of the people who wrote letters of recommendation or reference for you.
- [ ] Find and read the chapter in this book that covers ethics. There were no ethics problems on your PE exam, but it doesn't make any difference. Ethical behavior is expected of a PE in any case. Spend a few minutes reflecting on how your performance (your obligations, attitude, presentation, behavior, appearance, etc.) might be about to change once you are licensed. Consider how you are going to be a role model for others around you.
- [ ] Put all of your review books, binders, and notes someplace where they will be out of sight.

AND THEN THERE'S THE WAIT ...

Waiting for the exam results is its own form of mental torture.

Yes, I know the exam is 100% multiple-choice, and grading should be almost instantaneous. Nevertheless, you are going to wait. There are many reasons for the delay.

Although the actual machine grading "only takes seconds," consider the following facts: (a) NCEES prepares multiple exams for each administration, in case one becomes unusable (i.e., is inappropriately released) before the exam date. (b) Since the actual version of the exam used is not known until after it is finally given, the cut-score determination occurs after the exam date.

I wouldn't be surprised to hear that NCEES receives dozens, if not hundreds, of claims from well-meaning examinees who were 100% certain that the exams they took were fatally flawed to some degree—that there was not a correct answer for such-and-such problem—that there were two answers for such-and-such problem—or even, perhaps, that such-and-such problem was missing from their exam booklet altogether. Each of these claims must be considered as a potential adjustment to the cut-score.

Then, the exams must actually be graded. Since grading nearly 100,000 exams (counting all the FE and PE exams) requires specialized equipment, software, and training not normally possessed by the average employee, as well as time to do the work (also not normally possessed by the average employee), grading is invariably outsourced.

Outsourced grading cannot begin until all of the states have returned their score sheets to NCEES and NCEES has sorted, separated, organized, and consolidated the score sheets into whatever "secret sauce sequence" is best.

During grading, some of the score sheets "pop out" with any number of abnormalities that demand manual scoring.

After the individual exams are scored, the results are analyzed in a variety of ways. Some of the analysis looks at passing rates by such delineators as degree, major, university, site, and state. Part of the analysis looks for similarities between physically adjacent examinees (to look for cheating). Part of the analysis looks for exam sites that have statistically abnormal group performance. And some of the analysis looks for exam problems that have a disproportionate fraction of successful or unsuccessful examinees. Anyway, you get the idea: It's not merely putting your exam sheet in an electronic reader. All of these steps have to be completed for 100% of the examinees before any results can go out.

Once NCEES has graded your test and notified your state, when you receive results depends on when the work is done by your state. Some states have to approve the results at a board meeting; others prepare the certificates before sending out notifications. Some states are more computerized than others. Some states have 50 examinees, while others have 10,000. Some states are shut down by blizzards and hurricanes, others are administratively challenged—understaffed, inadequately trained, or over budget.

There is no pattern to the public release of results. None. The exam results are not released to all states simultaneously. (The states with the fewest examinees often receive their results soonest.) They are not released by discipline. They are not released alphabetically by state or examinee name. The people who failed are not notified first (or last). Your coworker might receive his or her notification today, and you might wait another three weeks for yours.

Some states post the names of the successful examinees on their official state websites before the results go out. Others update their websites after the results go out. Some states don't list much of anything on their websites. Remember, too, that the size or thickness of the envelope you receive from your state does not mean anything. Some states send a big congratulations package and certificate. Others send a big package with a new application to repeat the exam. Some states send a postcard. Some send a one-page letter. Some states simply send you an invoice for your license fees. (Ahh, what a welcome bill!) You just have to open it to find out.

Regularly check the Engineering Exam Forum at **www.ppi2pass.com/forums** to find out which states have released their results. You will find many other anxious examinees there. And any number of amusing conspiracy theories and rumors.

While you are waiting, I hope you will become a "Forum" regular. Log on often and help other examinees by sharing your knowledge, experiences, and wisdom. And, if you hear any good jokes at work, I hope you will share them as well.

AND WHEN YOU PASS ...

[ ] Celebrate.

[ ] Notify the people who wrote letters of recommendation or reference for you.

[ ] Read "FAQs About What Happens After You Pass the Exam" on PPI's website.

[ ] Ask your employer for a raise.

[ ] Tell the folks at PPI (who have been rootin' for you all along) the good news.

[ ] Start preparing for your career as a licensed Electrical Engineer.

Table 3 *Schedule for Self-Study*

chapter number	chapter name	start date	finish date
1	Systems of Units		
2	Energy, Work, and Power		
3	Engineering Drawing Practice		
4	Algebra		
5	Linear Algebra		
6	Vectors		
7	Trigonometry		
8	Analytic Geometry		
9	Differential Calculus		
10	Integral Calculus		
11	Differential Equations		
12	Probability and Statistical Analysis of Data		

(continued)

Table 3 Schedule for Self-Study (continued)

chapter number	chapter name	start date	finish date
13	Computer Mathematics		
14	Numerical Analysis		
15	Advanced Engineering Mathematics		
16	Electromagnetic Theory		
17	Electronic Theory		
18	Communication Theory		
19	Acoustic and Transducer Theory		
20	Electrostatics		
21	Electrostatic Fields		
22	Magnetostatics		
23	Magnetostatic Fields		
24	Electrodynamics		
25	Maxwell's Equations		
26	DC Circuit Fundamentals		
27	AC Circuit Fundamentals		
28	Transformers		
29	Linear Circuit Analysis		
30	Transient Analysis		
31	Time Response		
32	Frequency Response		
33	Batteries, Fuel Cells, and Power Supplies		
34	Protection and Safety		
35	Power System Management		
36	Measurement and Instrumentation		
37	Electronics' Fundamentals		
38	Junction Transistors		
39	Field Effect Transistors		
40	Electrical and Electronic Devices		
41	Digital Logic		
42	Logic Network Design		
43	Synchronous Sequential Networks		
44	Digital Systems		

(continued)

Table 3 Schedule for Self-Study (continued)

chapter number	chapter name	start date	finish date
45	High-Frequency Transmission		
46	Antenna Theory		
47	Communication Links		
48	Signal Formats		
49	Signal Multiplexing		
50	Communications Systems and Channels		
51	Analysis of Control Systems		
52	Electrical Materials		
53	National Electrical Code		
54	National Electrical Safety Code		
55	Engineering Economic Analysis		
56	Engineering Law		
57	Engineering Ethics		
58	Electrical Engineering Frontiers		
59	Engineering Licensing in the United States		

Topic I: Mathematics

Chapter

1. Systems of Units
2. Energy, Work, and Power
3. Engineering Drawing Practice
4. Algebra
5. Linear Algebra
6. Vectors
7. Trigonometry
8. Analytic Geometry
9. Differential Calculus
10. Integral Calculus
11. Differential Equations
12. Probability and Statistical Analysis of Data
13. Computer Mathematics
14. Numerical Analysis
15. Advanced Engineering Mathematics

1 Systems of Units

1. Introduction 1-1
2. Common Units of Mass 1-1
3. Mass and Weight 1-1
4. Acceleration of Gravity 1-2
5. Consistent Systems of Units 1-2
6. The English Engineering System 1-2
7. Other Formulas Affected by Inconsistency 1-3
8. Weight and Weight Density 1-3
9. The English Gravitational System 1-3
10. The Absolute English System 1-4
11. Metric Systems of Units 1-4
12. The cgs System 1-4
13. SI Units (The mks System) 1-5
14. Rules for Using SI Units 1-6
15. Primary Dimensions 1-7
16. Dimensionless Groups 1-8
17. Lineal and Board Foot Measurements 1-8
18. Dimensional Analysis 1-8

1. INTRODUCTION

The purpose of this chapter is to eliminate some of the confusion regarding the many units available for each engineering variable. In particular, an effort has been made to clarify the use of the so-called English systems, which for years have used the *pound* unit both for force and mass—a practice that has resulted in confusion even for those familiar with it.

2. COMMON UNITS OF MASS

The choice of a mass unit is the major factor in determining which system of units will be used in solving a problem. It is obvious that one will not easily end up with a force in pounds if the rest of the problem is stated in meters and kilograms. Actually, the choice of a mass unit determines more than whether a conversion factor will be necessary to convert from one system to another (e.g., between SI and English units). An inappropriate choice of a mass unit may actually require a conversion factor *within* the system of units.

The common units of mass are the gram, pound, kilogram, and slug.[1] There is nothing mysterious about these units. All represent different quantities of matter, as Fig. 1.1 illustrates. In particular, note that the pound and slug do not represent the same quantity of matter.[2]

Figure 1.1 *Common Units of Mass*

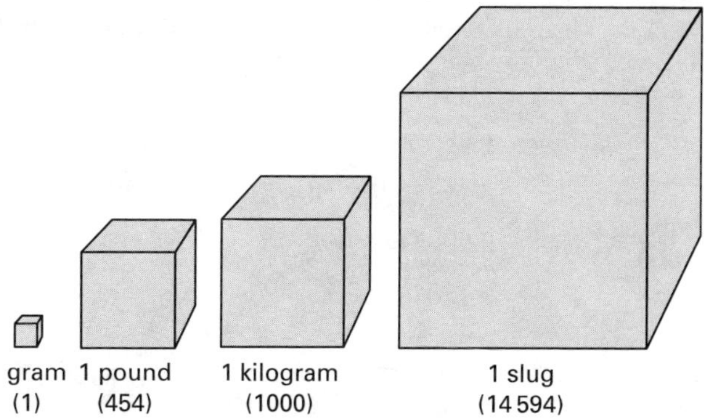

3. MASS AND WEIGHT

In SI, *kilograms* are used for mass and *newtons* for weight (force). The units are different, and there is no confusion between the variables. However, for years the term *pound* has been used for both mass and weight. This usage has obscured the distinction between the two: mass is a constant property of an object; weight varies with the gravitational field. Even the conventional use of the abbreviations *lbm* and *lbf* (to distinguish between pounds-mass and pounds-force) has not helped eliminate the confusion.

It is true that an object with a mass of one pound will have an earthly weight of one pound, but this is true only on the earth. The weight of the same object will be much less on the moon. Therefore, care must be taken when working with mass and force in the same problem.

The relationship that converts mass to weight is familiar to every engineering student.

$$W = mg \qquad 1.1$$

Equation 1.1 illustrates that an object's weight will depend on the local acceleration of gravity as well as the object's mass. The mass will be constant, but gravity will depend on location. Mass and weight are not the same.

[1]Normally, one does not distinguish between a unit and a multiple of that unit, as is done here with the gram and the kilogram. However, these two units actually are bases for different consistent systems.

[2]A slug is approximately equal to 32.1740 pounds-mass.

4. ACCELERATION OF GRAVITY

Gravitational acceleration on the earth's surface is usually taken as 32.2 ft/sec^2 or 9.81 m/s^2. These values are rounded from the more exact values of 32.1740 ft/sec^2 and 9.8066 m/s^2. However, the need for greater accuracy must be evaluated on a problem-by-problem basis. Usually, three significant digits are adequate, since gravitational acceleration is not constant anyway but is affected by location (primarily latitude and altitude) and major geographical features.

The term *standard gravity*, g_0, is derived from the acceleration at essentially any point at sea level and approximately 45° N latitude. If additional accuracy is needed, the gravitational acceleration can be calculated from Eq. 1.2. This equation neglects the effects of large land and water masses. ϕ is the latitude in degrees.

$$\begin{aligned} g_{\text{surface}} &= g'(1 + (5.305 \times 10^{-3})\sin^2\phi \\ &\quad - (5.9 \times 10^{-6})\sin^2 2\phi) \\ g' &= 32.0881 \text{ ft/sec}^2 \\ &= 9.78045 \text{ m/s}^2 \end{aligned} \qquad 1.2$$

If the effects of the earth's rotation are neglected, the gravitational acceleration at an altitude h above the earth's surface is given by Eq. 1.3. R_e is the earth's radius.

$$\begin{aligned} g_h &= g_{\text{surface}}\left(\frac{R_e}{R_e + h}\right)^2 \\ R_e &= 3960 \text{ mi} \\ &= 6.37 \times 10^6 \text{ m} \end{aligned} \qquad 1.3$$

5. CONSISTENT SYSTEMS OF UNITS

A set of units used in a calculation is said to be *consistent* if no conversion factors are needed.[3] For example, a moment is calculated as the product of a force and a lever arm length.

$$M = Fd \qquad 1.4$$

A calculation using Eq. 1.4 would be consistent if M was in newton-meters, F was in newtons, and d was in meters. The calculation would be inconsistent if M was in ft-kips, F was in kips, and d was in inches (because a conversion factor of 1/12 would be required).

The concept of a consistent calculation can be extended to a system of units. A *consistent system of units* is one in which no conversion factors are needed for any calculation. For example, Newton's second law of motion can be written without conversion factors. Newton's second law simply states that the force required to accelerate an object is proportional to the acceleration of the object. The constant of proportionality is the object's mass.

$$F = ma \qquad 1.5$$

Notice that Eq. 1.5 is $F = ma$, not $F = Wa/g$ or $F = ma/g_c$. Equation 1.5 is consistent: it requires no conversion factors. This means that in a consistent system where conversion factors are not used, once the units of m and a have been selected, the units of F are fixed. This has the effect of establishing units of work and energy, power, fluid properties, and so on.

It should be mentioned that the decision to work with a consistent set of units is desirable but unnecessary, depending on tradition and environment. Problems in fluid flow and thermodynamics are routinely solved in the United States with inconsistent units. This causes no more of a problem than working with inches and feet when calculating a moment. It is necessary only to use the proper conversion factors.

6. THE ENGLISH ENGINEERING SYSTEM

Through common and widespread use, pounds-mass (lbm) and pounds-force (lbf) have become the standard units for mass and force in the *English Engineering System*. (The English Engineering System is used in this book.)

There are subjects in the United States where the practice of using pounds for mass is firmly entrenched. For example, most thermodynamics, fluid flow, and heat transfer problems have traditionally been solved using the units of lbm/ft^3 for density, Btu/lbm for enthalpy, and Btu/lbm-°F for specific heat. Unfortunately, some equations contain both lbm-related and lbf-related variables, as does the steady flow conservation of energy equation, which combines enthalpy in Btu/lbm with pressure in lbf/ft^2.

The units of pounds-mass and pounds-force are as different as the units of gallons and feet, and they cannot be canceled. A mass conversion factor, g_c, is needed to make the equations containing lbf and lbm dimensionally consistent. This factor is known as the *gravitational constant* and has a value of 32.1740 ft-lbm/lbf-sec^2. The numerical value is the same as the standard acceleration of gravity, but g_c is not the local gravitational acceleration, g.[4] g_c is a conversion constant, just as 12.0 is the conversion factor between feet and inches.

The English Engineering System is an inconsistent system as defined according to Newton's second law.

[3]The terms *homogeneous* and *coherent* are also used to describe a consistent set of units.

[4]It is acceptable (and recommended) that g_c be rounded to the same number of significant digits as g. Therefore, a value of 32.2 for g_c would typically be used.

$F = ma$ cannot be written if lbf, lbm, and ft/sec^2 are the units used. The g_c term must be included.

$$F \text{ in lbf} = \frac{(m \text{ in lbm})\left(a \text{ in } \frac{\text{ft}}{\text{sec}^2}\right)}{g_c \text{ in } \frac{\text{ft-lbm}}{\text{lbf-sec}^2}} \qquad 1.6$$

It is important to note in Eq. 1.6 that g_c does more than "fix the units." Since g_c has a numerical value of 32.1740, it actually changes the calculation numerically. A force of 1.0 pound will not accelerate a 1.0-pound mass at the rate of 1.0 ft/sec^2.

In the English Engineering System, work and energy are typically measured in ft-lbf (mechanical systems) or in British thermal units, Btu (thermal and fluid systems). One Btu is equal to 778.17 ft-lbf.

Example 1.1

Calculate the weight in lbf of a 1.00 lbm object in a gravitational field of 27.5 ft/sec^2.

Solution

From Eq. 1.6,

$$F = \frac{ma}{g_c} = \frac{(1.00 \text{ lbm})\left(27.5 \frac{\text{ft}}{\text{sec}^2}\right)}{32.2 \frac{\text{ft-lbm}}{\text{lbf-sec}^2}}$$
$$= 0.854 \text{ lbf}$$

7. OTHER FORMULAS AFFECTED BY INCONSISTENCY

It is not a significant burden to include g_c in a calculation, but it may be difficult to remember when g_c should be used. Knowing when to include the gravitational constant can be learned through repeated exposure to the formulas in which it is needed, but it is safer to carry the units along in every calculation.

The following is a representative (but not exhaustive) listing of formulas that require the g_c term. In all cases, it is assumed that the standard English Engineering System units will be used.

- kinetic energy

$$E = \frac{m\text{v}^2}{2g_c} \quad \text{(in ft-lbf)} \qquad 1.7$$

- potential energy

$$E = \frac{mgz}{g_c} \quad \text{(in ft-lbf)} \qquad 1.8$$

- pressure at a depth

$$p = \frac{\rho g h}{g_c} \quad \text{(in lbf/ft}^2\text{)} \qquad 1.9$$

Example 1.2

A rocket that has a mass of 4000 lbm travels at 27,000 ft/sec. What is its kinetic energy in ft-lbf?

Solution

From Eq. 1.7,

$$E_k = \frac{m\text{v}^2}{2g_c} = \frac{(4000 \text{ lbm})\left(27{,}000 \frac{\text{ft}}{\text{sec}}\right)^2}{(2)\left(32.2 \frac{\text{ft-lbm}}{\text{lbf-sec}^2}\right)}$$
$$= 4.53 \times 10^{10} \text{ ft-lbf}$$

8. WEIGHT AND WEIGHT DENSITY

Weight is a force exerted on an object due to its placement in a gravitational field. If a consistent set of units is used, Eq. 1.1 can be used to calculate the weight of a mass. In the English Engineering System, however, Equation 1.10 must be used.

$$W = \frac{mg}{g_c} \qquad 1.10$$

Both sides of Eq. 1.10 can be divided by the volume of an object to derive the *weight density*, γ, of the object. Equation 1.11 illustrates that the weight density (in units of lbf/ft^3) can also be calculated by multiplying the mass density (in units of lbm/ft^3) by g/g_c. Since g and g_c usually have the same numerical values, the only effect of Eq. 1.12 is to change the units of density.

$$\frac{W}{V} = \left(\frac{m}{V}\right)\left(\frac{g}{g_c}\right) \qquad 1.11$$

$$\gamma = \frac{W}{V} = \left(\frac{m}{V}\right)\left(\frac{g}{g_c}\right) = \frac{\rho g}{g_c} \qquad 1.12$$

Weight does not occupy volume. Only mass has volume. The concept of weight density has evolved to simplify certain calculations, particularly fluid calculations. For example, pressure at a depth is calculated from Eq. 1.13. (Compare this with Eq. 1.9.)

$$p = \gamma h \qquad 1.13$$

9. THE ENGLISH GRAVITATIONAL SYSTEM

Not all English systems are inconsistent. Pounds can still be used as the unit of force as long as pounds are not used as the unit of mass. Such is the case with the consistent *English Gravitational System.*

If acceleration is given in ft/sec^2, the units of mass for a consistent system of units can be determined from Newton's second law. The combination of units in Eq. 1.14 is known as a *slug.* g_c is not needed at all since this system is consistent. It would be needed only to convert slugs to another mass unit.

$$\begin{aligned}\text{units of } m &= \frac{\text{units of } F}{\text{units of } a} \\ &= \frac{\text{lbf}}{\frac{\text{ft}}{\text{sec}^2}} = \frac{\text{lbf-sec}^2}{\text{ft}}\end{aligned} \qquad 1.14$$

Slugs and pounds-mass are not the same, as Fig. 1.1 illustrates. However, both are units for the same quantity: mass. Equation 1.15 will convert between slugs and pounds-mass.

$$\text{no. of slugs} = \frac{\text{no. of lbm}}{g_c} \qquad 1.15$$

It is important to recognize that the number of slugs is not derived by dividing the number of pounds-mass by the local gravity. g_c is used regardless of the local gravity. The conversion between feet and inches is not dependent on local gravity; neither is the conversion between slugs and pounds-mass.

Since the English Gravitational System is consistent, Eq. 1.16 can be used to calculate weight. Notice that the local gravitational acceleration is used.

$$W \text{ in lbf} = (m \text{ in slugs})\left(g \text{ in } \frac{\text{ft}}{\text{sec}^2}\right) \qquad 1.16$$

10. THE ABSOLUTE ENGLISH SYSTEM

The obscure *Absolute English System* takes the approach that mass must have units of pounds-mass (lbm) and the units of force can be derived from Newton's second law. The units for F cannot be simplified any more than they are in Eq. 1.17. This particular combination of units is known as a *poundal*.[5] A poundal is not the same as a pound.

$$\begin{aligned}\text{units of } F &= (\text{units of } m)(\text{units of } a) \\ &= (\text{lbm})\left(\frac{\text{ft}}{\text{sec}^2}\right) \\ &= \frac{\text{lbm-ft}}{\text{sec}^2}\end{aligned} \qquad 1.17$$

Poundals have not seen widespread use in the United States. The English Gravitational System (using slugs for mass) has greatly eclipsed the Absolute English System in popularity. Both are consistent systems, but there seems to be little need for poundals in modern engineering. Figure 1.2 shows the poundal in comparison to other common units of force.

[5]A poundal is equal to 0.03108 pounds-force.

Figure 1.2 *Common Force Units*

11. METRIC SYSTEMS OF UNITS

Strictly speaking, a *metric system* is any system of units that is based on meters or parts of meters. This broad definition includes *mks systems* (based on meters, kilograms, and seconds) as well as *cgs systems* (based on centimeters, grams, and seconds).

Metric systems avoid the pounds-mass versus pounds-force ambiguity in two ways. First, a unit of weight is not established at all. All quantities of matter are specified as mass. Second, force and mass units do not share a common name.

The term *metric system* is not explicit enough to define which units are to be used for any given variable. For example, within the cgs system there is variation in how certain electrical and magnetic quantities are represented (resulting in the ESU and EMU systems). Also, within the mks system, it is common practice in some industries to use kilocalories as the unit of thermal energy, while the SI unit for thermal energy is the joule. This shows a lack of uniformity even within the metricated engineering community.[6]

The "metric" parts of this book use SI, which is the most developed and codified of the so-called metric systems.[7] There will be occasional variances with local engineering custom, but it is difficult to anticipate such variances within a book that must itself be consistent.[8]

12. THE cgs SYSTEM

The *cgs system* is used widely by chemists and physicists. It is named for the three primary units used to construct its derived variables: the centimeter, the gram, and the second.

[6]In the "field test" of the metric system conducted over the past 200 years, other conventions are to use kilograms-force (kgf) instead of newtons and kgf/cm^2 for pressure (instead of pascals).

[7]SI units are an outgrowth of the *General Conference of Weights and Measures*, an international treaty organization that established the *Système International d'Unités (International System of Units)* in 1960. The United States subscribed to this treaty in 1975.

[8]Conversion to pure SI units is essentially complete in Australia, Canada, New Zealand, and South Africa. The use of nonstandard metric units is more common among European civil engineers, who have had little need to deal with the inertial properties of mass. However, even the American Society of Civil Engineers declared its support for SI units in 1985.

When Newton's second law is written in the cgs system, the following combination of units results.

$$\begin{aligned}\text{units of force} &= (m \text{ in g})\left(a \text{ in } \frac{\text{cm}}{\text{s}^2}\right) \\ &= \text{g·cm/s}^2\end{aligned} \quad 1.18$$

This combination of units for force is known as a *dyne*. Energy variables in the cgs system have units of dyne·cm or, equivalently, $\text{g·cm}^2/\text{s}^2$. This combination is known as an *erg*. There is no uniformly accepted unit of power in the cgs system, although calories per second is frequently used.

The fundamental volume unit in the cgs system is the cubic centimeter (cc). Since this is the same volume as one thousandth of a liter, units of milliliters (mL) are also used.

13. SI UNITS (THE mks SYSTEM)

SI units comprise an *mks system* (so named because it uses the meter, kilogram, and second as base units). All other units are derived from the base units, which are completely listed in Table 1.1. This system is fully consistent, and there is only one recognized unit for each physical quantity (variable).

Table 1.1 *SI Base Units*

quantity	name	symbol
length	meter	m
mass	kilogram	kg
time	second	s
electric current	ampere	A
temperature	kelvin	K
amount of substance	mole	mol
luminous intensity	candela	cd

Two types of units are used: base units and derived units. The *base units* are dependent only on accepted standards or reproducible phenomena. The *derived units* (Table 1.2 and Table 1.3) are made up of combinations of base units. The old *supplementary units* (Table 1.4) were classified as derived units in 1995.

In addition, there is a set of non-SI units that may be used. This concession is primarily due to the significance and widespread acceptance of these units. Use of the non-SI units listed in Table 1.5 will usually create an inconsistent expression requiring conversion factors.

The SI unit of force can be derived from Newton's second law. This combination of units for force is known as a *newton*.

$$\begin{aligned}\text{units of force} &= (m \text{ in kg})\left(a \text{ in } \frac{\text{m}}{\text{s}^2}\right) \\ &= \text{kg·m/s}^2\end{aligned} \quad 1.19$$

Table 1.2 *Some SI Derived Units with Special Names*

quantity	name	symbol	expressed in terms of other units
frequency	hertz	Hz	1/s
force	newton	N	kg·m/s^2
pressure, stress	pascal	Pa	N/m^2
energy, work, quantity of heat	joule	J	N·m
power, radiant flux	watt	W	J/s
quantity of electricity, electric charge	coulomb	C	
electric potential, potential difference, electromotive force	volt	V	W/A
electric capacitance	farad	F	C/V
electric resistance	ohm	Ω	V/A
electric conductance	siemens	S	A/V
magnetic flux	weber	Wb	V·s
magnetic flux density	tesla	T	Wb/m^2
inductance	henry	H	Wb/A
luminous flux	lumen	lm	
illuminance	lux	lx	lm/m^2

Energy variables in SI units have units of N·m or, equivalently, $\text{kg·m}^2/\text{s}^2$. Both of these combinations are known as a *joule*. The units of power are joules per second, equivalent to a *watt*.

Example 1.3

A 10 kg block hangs from a cable. What is the tension in the cable? (Standard gravity equals 9.81 m/s^2.)

Solution

$$\begin{aligned}F &= mg \\ &= (10\text{ kg})\left(9.81\ \frac{\text{m}}{\text{s}^2}\right) \\ &= 98.1\ \text{kg·m/s}^2 \quad (98.1\text{ N})\end{aligned}$$

Example 1.4

A 10 kg block is raised vertically 3 m. What is the change in potential energy?

Solution

$$\begin{aligned}\Delta E_p &= mg\Delta h \\ &= (10\text{ kg})\left(9.81\ \frac{\text{m}}{\text{s}^2}\right)(3\text{ m}) \\ &= 294\ \text{kg·m}^2/\text{s}^2 \quad (294\text{ J})\end{aligned}$$

Table 1.3 *Some SI Derived Units*

quantity	description	symbol
area	square meter	m^2
volume	cubic meter	m^3
speed—linear	meter per second	m/s
speed—angular	radian per second	rad/s
acceleration—linear	meter per second squared	m/s^2
acceleration—angular	radian per second squared	rad/s^2
density, mass density	kilogram per cubic meter	kg/m^3
concentration (of amount of substance)	mole per cubic meter	mol/m^3
specific volume	cubic meter per kilogram	m^3/kg
luminance	candela per square meter	cd/m^2
absolute viscosity	pascal second	Pa·s
kinematic viscosity	square meters per second	m^2/s
moment of force	newton meter	N·m
surface tension	newton per meter	N/m
heat flux density, irradiance	watt per square meter	W/m^2
heat capacity, entropy	joule per kelvin	J/K
specific heat capacity, specific entropy	joule per kilogram kelvin	J/kg·K
specific energy	joule per kilogram	J/kg
thermal conductivity	watt per meter kelvin	W/m·K
energy density	joule per cubic meter	J/m^3
electric field strength	volt per meter	V/m
electric charge density	coulomb per cubic meter	C/m^3
surface density of charge, flux density	coulomb per square meter	C/m^2
permittivity	farad per meter	F/m
current density	ampere per square meter	A/m^2
magnetic field strength	ampere per meter	A/m
permeability	henry per meter	H/m
molar energy	joule per mole	J/mol
molar entropy, molar heat capacity	joule per mole kelvin	J/mol·K
radiant intensity	watt per steradian	W/sr

Table 1.4 *SI Supplementary Units**

quantity	name	symbol
plane angle	radian	rad
solid angle	steradian	sr

*classified as derived units in 1995

Table 1.5 *Acceptable Non-SI Units*

quantity	unit name	symbol or abbreviation	relationship to SI unit
area	hectare	ha	1 ha = 10 000 m^2
energy	kilowatt-hour	kW·h	1 kW·h = 3.6 MJ
mass	metric ton[a]	t	1 t = 1000 kg
plane angle	degree (of arc)	°	1° = 0.017 453 rad
speed of rotation	revolution per minute	r/min	1 r/min = $2\pi/60$ rad/s
temperature interval	degree Celsius	°C	1°C = 1K
time	minute	min	1 min = 60 s
	hour	h	1 h = 3600 s
	day (mean solar)	d	1 d = 86 400 s
	year (calendar)	a	1 a = 31 536 000 s
velocity	kilometer per hour	km/h	1 km/h = 0.278 m/s
volume	liter[b]	L	1 L = 0.001 m^3

[a]The international name for metric ton is *tonne*. The metric ton is equal to the *megagram* (Mg).
[b]The international symbol for liter is the lowercase l, which can be easily confused with the numeral 1. Several English-speaking countries have adopted the script ℓ or uppercase L (as does this book) as a symbol for liter in order to avoid any misinterpretation.

14. RULES FOR USING SI UNITS

In addition to having standardized units, the set of SI units also has rigid syntax rules for writing the units and combinations of units. Each unit is abbreviated with a specific symbol. The following rules for writing and combining these symbols should be adhered to.

- The expressions for derived units in symbolic form are obtained by using the mathematical signs of multiplication and division. For example, units of velocity are m/s. Units of torque are N·m (not N-m or Nm).
- Scaling of most units is done in multiples of 1000.
- The symbols are always printed in roman type, regardless of the type used in the rest of the text. The only exception to this is in the use of the symbol for liter, where the use of the lower case "el" (l) may be confused with the numeral one (1). In this case, "liter" should be written out in full, or the script ℓ or L used. (L is used in this book.)
- Symbols are not pluralized: 1 kg, 45 kg (not 45 kgs).
- A period after a symbol is not used, except when the symbol occurs at the end of a sentence.
- When symbols consist of letters, there must always be a full space between the quantity and the symbols: 45 kg (not 45kg). However, for planar angle designations, no space is left: 32°C (not 32° C or 32 °C); or 42°12′45″; (not 42 ° 12 ′ 45 ″).

- All symbols are written in lowercase, except when the unit is derived from a proper name: m for meter; s for second; A for ampere, Wb for weber, N for newton, W for watt.
- Prefixes are printed without spacing between the prefix and the unit symbol (e.g., km is the symbol for kilometer). Table 1.6 lists common prefixes, their symbols, and their values.
- In text, when no number is involved, the unit should be spelled out. Example: Carpet is sold by the square meter, not by the m^2.
- Where a decimal fraction of a unit is used, a zero should always be placed before the decimal marker: 0.45 kg (not .45 kg). This practice draws attention to the decimal marker and helps avoid errors of scale.
- A practice in some countries is to use a comma as a decimal marker, while the practice in North America, the United Kingdom, and some other countries is to use a period (or dot) as the decimal marker. Furthermore, in some countries that use the decimal comma, a dot is frequently used to divide long numbers into groups of three. Because of these differing practices, spaces must be used instead of commas to separate long lines of digits into easily readable blocks of three digits with respect to the decimal marker: 32 453.246 072 5. A space (half-space preferred) is optional with a four-digit number: 1 234 or 1234.
- Some confusion may arise with the word "tonne" (1000 kg). When this word occurs in French text of Canadian origin, the meaning may be a ton of 2000 pounds.

Table 1.6 *SI Prefixes**

prefix	symbol	value
exa	E	10^{18}
peta	P	10^{15}
tera	T	10^{12}
giga	G	10^{9}
mega	M	10^{6}
kilo	k	10^{3}
hecto	h	10^{2}
deka (or "deca")	da	10^{1}
deci	d	10^{-1}
centi	c	10^{-2}
milli	m	10^{-3}
micro	μ	10^{-6}
nano	n	10^{-9}
pico	p	10^{-12}
femto	f	10^{-15}
atto	a	10^{-18}

*There is no "B" (billion) prefix. In fact, the word "billion" means 10^9 in the United States but 10^{12} in most other countries. This unfortunate ambiguity is handled by avoiding the use of the term billion.

15. PRIMARY DIMENSIONS

Regardless of the system of units chosen, each variable representing a physical quantity will have the same *primary dimensions.* For example, velocity may be expressed in miles per hour (mph) or meters per second (m/s), but both units have dimensions of length per unit time. Length and time are two of the primary dimensions, as neither can be broken down into more basic dimensions. The concept of primary dimensions is useful when converting little-used variables between different systems of units, as well as in correlating experimental results (i.e., dimensional analysis).

There are three different sets of primary dimensions in use.[9] In the $ML\theta T$ system, the primary dimensions are mass (M), length (L), time (θ), and temperature (T). Notice that all symbols are uppercase. In order to avoid confusion between time and temperature, the Greek letter theta is used for time.[10]

All other physical quantities can be derived from these primary dimensions.[11] For example, work in SI units has units of N·m. Since a newton is a kg·m/s^2, the primary dimensions of work are ML^2/θ^2. The primary dimensions for many important engineering variables are shown in Table 1.7. If it is more convenient to stay with traditional English units, it may be more desirable to work in the $FML\theta TQ$ system (sometimes called the *engineering dimensional system*). This system adds the primary dimensions of force (F) and heat (Q). Work (ft-lbf in the English system) has the primary dimensions of FL. (Compare this with the primary dimensions for work in the $ML\theta T$ system.) Thermodynamic variables are similarly simplified.

Dimensional analysis will be more conveniently carried out when one of the four-dimension systems ($ML\theta T$ or $FL\theta T$) is used. Whether the $ML\theta T$, $FL\theta T$, or $FML\theta TQ$ system is used depends on what is being derived and who will be using it, and whether or not a consistent set of variables is desired. Conversion constants such as g_c and J will almost certainly be required if the $ML\theta T$ system is used to generate variables for use in the English systems. It is also much more convenient to use the $FML\theta TQ$ system when working in the fields of thermodynamics, fluid flow, heat transfer, and so on.

[9]One of these, the $FL\theta T$ system, is not discussed here.

[10]This is the most common usage. There is a lack of consistency in the engineering world about the symbols for the primary dimensions in dimensional analysis. Some writers use t for time instead of θ. Some use H for heat instead of Q. And, in the worst mix-up of all, some have reversed the use of T and θ.

[11]A *primary dimension* is the same as a *base unit* in the SI set of units. The SI units add several other base units, as shown in Table 1.1, to deal with variables that are difficult to derive in terms of the four primary base units.

Table 1.7 *Dimensions of Common Variables*

variable	dimensional system $ML\theta T$	$FL\theta T$	$FMLT\theta Q$
mass (m)	M	$F\theta^2/L$	M
force (F)	ML/θ^2	F	F
length (L)	L	L	L
time (θ)	θ	θ	θ
temperature (T)	T	T	T
work (W)	ML^2/θ^2	FL	FL
heat (Q)	ML^2/θ^2	FL	Q
acceleration (a)	L/θ^2	L/θ^2	L/θ^2
frequency (N)	$1/\theta$	$1/\theta$	$1/\theta$
area (A)	L^2	L^2	L^2
coefficient of thermal expansion (β)	$1/T$	$1/T$	$1/T$
density (ρ)	M/L^3	$F\theta^2/L^4$	M/L^3
dimensional constant (g_c)	1.0	1.0	$ML/\theta^2 F$
specific heat at constant pressure (c_p); at constant volume (c_v)	$L^2/\theta^2 T$	$L^2/\theta^2 T$	Q/MT
heat transfer coefficient (h); overall (U)	$M/\theta^3 T$	$F/\theta LT$	$Q/\theta L^2 T$
power (P)	ML^2/θ^3	FL/θ	FL/θ
heat flow rate ($\dot{Q}$)	ML^2/θ^3	FL/θ	Q/θ
kinematic viscosity (ν)	L^2/θ	L^2/θ	L^2/θ
mass flow rate ($\dot{m}$)	M/θ	$F\theta/L$	M/θ
mechanical equivalent of heat (J)	–	–	FL/Q
pressure (p)	$M/L\theta^2$	F/L^2	F/L^2
surface tension (σ)	M/θ^2	F/L	F/L
angular velocity (ω)	$1/\theta$	$1/\theta$	$1/\theta$
volumetric flow rate ($\dot{m}/\rho = \dot{V}$)	L^3/θ	L^3/θ	L^3/θ
conductivity (k)	$ML/\theta^3 T$	$F/\theta T$	$Q/L\theta T$
thermal diffusivity (α)	L^2/θ	L^2/θ	L^2/θ
velocity (v)	L/θ	L/θ	L/θ
viscosity, absolute (μ)	$M/L\theta$	$F\theta/L^2$	$F\theta/L^2$
volume (V)	L^3	L^3	L^3

16. DIMENSIONLESS GROUPS

A *dimensionless group* is derived as a ratio of two forces or other quantities. Considerable use of dimensionless groups is made in certain subjects, notably fluid mechanics and heat transfer. For example, the Reynolds number, Mach number, and Froude number are used to distinguish between distinctly different flow regimes in pipe flow, compressible flow, and open channel flow, respectively.

Table 1.8 contains information about the most common dimensionless groups used in fluid mechanics and heat transfer.

17. LINEAL AND BOARD FOOT MEASUREMENTS

The term *lineal* is often mistaken as a typographical error for *linear*. Although "lineal" has its own specific meaning slightly different from "linear," the two are often used interchangeably by engineers.[12] The adjective *lineal* is often encountered in the building trade (e.g., 12 lineal feet of lumber), where the term is used to distinguish it from board feet measurement.

A *board foot* (abbreviated bd-ft) is not a measure of length. Rather, it is a measure of volume used with lumber. Specifically, a board foot is equal to 144 in^3 (2.36×10^{-3} m^3). The name is derived from the volume of a board 1 foot square and 1 inch thick. In that sense, it is parallel in concept to the acre-foot. Since lumber cost is directly related to lumber weight and volume, the board foot unit is used in determining the overall lumber cost.

18. DIMENSIONAL ANALYSIS

Dimensional analysis is a means of obtaining an equation that describes some phenomenon without understanding the mechanism of the phenomenon. The most serious limitation is the need to know beforehand which variables influence the phenomenon. Once these are known or assumed, dimensional analysis can be applied by a routine procedure.

The first step is to select a system of primary dimensions. (See Sec. 1.15.) Usually the $ML\theta T$ system is used, although this choice may require the use of g_c and J in the final results.

The second step is to write a functional relationship between the dependent variable and the independent variable, x_i.

$$y = f(x_1, x_2, \ldots, x_m) \qquad 1.20$$

This function can be expressed as an exponentiated series. The C_1, a_i, b_i, …, z_i in Eq. 1.21 are unknown constants.

$$y = C_1 x_1^{a_1} x_2^{b_1} x_3^{c_1} \cdots x_m^{z_1} + C_2 x_1^{a_2} x_2^{b_2} x_3^{c_2} \cdots x_m^{z_2} + \cdots \qquad 1.21$$

The key to solving Eq. 1.21 is that each term on the right-hand side must have the same dimensions as y. Simultaneous equations are used to determine some of the a_i, b_i, c_i, and z_i. Experimental data are required to determine the C_i and remaining exponents. In most analyses, it is assumed that the $C_i = 0$ for $i \geq 2$.

Since this method requires working with m different variables and n different independent dimensional quantities (such as M, L, θ, and T), an easier method is desirable. One simplification is to combine the m

[12] *Lineal* is best used when discussing a line of succession (e.g., a lineal descendant of a particular person). *Linear* is best used when discussing length (e.g., a linear dimension of a room).

Table 1.8 *Common Dimensionless Groups*

name	symbol	formula	interpretation
Biot number	Bi	$\frac{hL}{k_s}$	$\frac{\text{surface conductance}}{\text{internal conduction of solid}}$
Cauchy number	Ca	$\frac{\mathrm{v}^2}{\frac{B_s}{\rho}} = \frac{\mathrm{v}^2}{a^2}$	$\frac{\text{inertia force}}{\text{compressive force}} = \text{Mach number}^2$
Eckert number	Ec	$\frac{\mathrm{v}^2}{2c_p\Delta T}$	$\frac{\text{temperature rise due to energy conversion}}{\text{temperature difference}}$
Eötvös number	Eo	$\frac{\rho g L^2}{\sigma}$	$\frac{\text{buoyancy}}{\text{surface tension}}$
Euler number	Eu	$\frac{\Delta p}{\rho \mathrm{v}^2}$	$\frac{\text{pressure force}}{\text{inertia force}}$
Fourier number	Fo	$\frac{kt}{\rho c_p L^2} = \frac{\alpha t}{L^2}$	$\frac{\text{rate of conduction of heat}}{\text{rate of storage of energy}}$
Froude number*	Fr	$\frac{\mathrm{v}^2}{gL}$	$\frac{\text{inertia force}}{\text{gravity force}}$
Graetz number*	Gz	$\left(\frac{D}{L}\right)\left(\frac{\mathrm{v}\rho c_p D}{k}\right)$	$\frac{(\text{Re})(\text{Pr})}{L/D}$ $\frac{\text{heat transfer by convection in entrance region}}{\text{heat transfer by conduction}}$
Grashof number*	Gr	$\frac{g\beta\Delta T L^3}{\nu^2}$	$\frac{\text{buoyancy force}}{\text{viscous force}}$
Knudsen number	Kn	$\frac{\lambda}{L}$	$\frac{\text{mean free path of molecules}}{\text{characteristic length of object}}$
Lewis number*	Le	$\frac{\alpha}{D_c}$	$\frac{\text{thermal diffusivity}}{\text{molecular diffusivity}}$
Mach number	M	$\frac{\mathrm{v}}{a}$	$\frac{\text{macroscopic velocity}}{\text{speed of sound}}$
Nusselt number	Nu	$\frac{hL}{k}$	$\frac{\text{temperature gradient at wall}}{\text{overall temperature difference}}$
Péclet number	Pé	$\frac{\mathrm{v}\rho c_p D}{k}$	(Re)(Pr) $\frac{\text{heat transfer by convection}}{\text{heat transfer by conduction}}$
Prandtl number	Pr	$\frac{\mu c_p}{k} = \frac{\nu}{\alpha}$	$\frac{\text{diffusion of momentum}}{\text{diffusion of heat}}$
Reynolds number	Re	$\frac{\rho \mathrm{v} L}{\mu} = \frac{\mathrm{v}L}{\nu}$	$\frac{\text{inertia force}}{\text{viscous force}}$
Schmidt number	Sc	$\frac{\mu}{\rho D_c} = \frac{\nu}{D_c}$	$\frac{\text{diffusion of momentum}}{\text{diffusion of mass}}$
Sherwood number*	Sh	$\frac{k_D L}{D_c}$	$\frac{\text{mass diffusivity}}{\text{molecular diffusivity}}$
Stanton number	St	$\frac{h}{\mathrm{v}\rho c_p} = \frac{h}{c_p G}$	$\frac{\text{heat transfer at wall}}{\text{energy transported by stream}}$
Stokes number	Sk	$\frac{\Delta p L}{\mu \mathrm{v}}$	$\frac{\text{pressure force}}{\text{viscous force}}$
Strouhal number*	Sl	$\frac{L}{t\mathrm{v}} = \frac{L\omega}{\mathrm{v}}$	$\frac{\text{frequency of vibration}}{\text{characteristic frequency}}$
Weber number	We	$\frac{\rho \mathrm{v}^2 L}{\sigma}$	$\frac{\text{inertia force}}{\text{surface tension force}}$

*Multiple definitions exist.

variables into dimensionless groups called *pi-groups*. (See Table 1.7.)

If these dimensionless groups are represented by π_1, π_2, π_3, ..., π_k, the equation expressing the relationship between the variables is given by the *Buckingham* π-*theorem*.

$$f(\pi_1, \pi_2, \pi_3, \ldots, \pi_k) = 0 \qquad 1.22$$

$$k = m - n \qquad 1.23$$

The dimensionless pi-groups are usually found from the m variables according to an intuitive process.

Example 1.5

A solid sphere rolls down a submerged incline. Find an equation for the velocity, v.

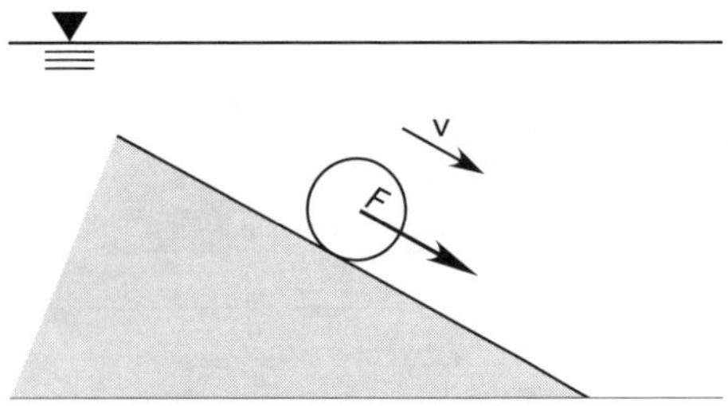

Solution

Assume that the velocity depends on the force, F, due to gravity, the diameter of the sphere, D, the density of the fluid, ρ, and the viscosity of the fluid, μ.

$$\mathrm{v} = f(F, D, \rho, \mu) = CF^aD^b\rho^c\mu^d$$

This equation can be written in terms of the primary dimensions of the variables.

$$\frac{L}{\theta} = C\left(\frac{ML}{\theta^2}\right)^a L^b \left(\frac{M}{L^3}\right)^c \left(\frac{M}{L\theta}\right)^d$$

Since L on the left-hand side has an implied exponent of one, a necessary condition is

$$1 = a + b - 3c - d \quad (L)$$

Similarly, the other necessary conditions are

$$-1 = -2a - d \quad (\theta)$$

$$0 = a + c + d \quad (M)$$

Solving simultaneously yields

$$b = -1$$

$$c = a - 1$$

$$d = 1 - 2a$$

$$\mathrm{v} = CF^aD^{-1}\rho^{a-1}\mu^{1-2a} = C\left(\frac{\mu}{D\rho}\right)\left(\frac{F\rho}{\mu^2}\right)^a$$

C and a must be determined experimentally.

2 Energy, Work, and Power

1. Energy of a Mass 2-1
2. Law of Conservation of Energy 2-1
3. Work 2-2
4. Potential Energy of a Mass 2-3
5. Kinetic Energy of a Mass 2-3
6. Spring Energy 2-3
7. Pressure Energy of a Mass 2-4
8. Internal Energy of a Mass 2-4
9. Work-Energy Principle 2-4
10. Conversion Between Energy Forms 2-5
11. Power 2-6
12. Efficiency 2-7

Nomenclature

c	specific heat	Btu/lbm-°F	J/kg·°C
C	molar specific heat	Btu/lbmol-°F	J/kmol·°C
E	energy	ft-lbf	J
F	force	lbf	N
g	gravitational acceleration, 32.2	ft/sec^2	m/s^2
g_c	gravitational constant, 32.2	lbm-ft/lbf-sec^2	–
h	height	ft	m
hp	horsepower	ft-lbf/sec	kW
I	mass moment of inertia	lbm-ft^2	kg·m^2
J	Joule's constant, 7.78	ft-lbf/Btu	–
k	spring constant	lbf/ft	N/m
m	mass	lbm	kg
MW	molecular weight	lbm/lbmol	kg/kmol
p	pressure	lbf/ft^2	Pa
P	power	ft-lbf/sec	kW
q	heat	Btu/lbm	J/kg
Q	heat	Btu	J
r	radius	ft	m
R	resistance	Ω	Ω
s	distance	ft	m
t	time	sec	s
T	temperature	°F	°C
T	torque	ft-lbf	N·m
u	specific energy	ft-lbf/lbm	J/kg
U	internal energy	Btu	J
v	velocity	ft/sec	m/s
W	work	ft-lbf	J
x	displacement	ft	m

Symbols

δ	displacement	ft	m
η	efficiency	–	–
θ	angular position	rad	rad
ρ	mass density	lbm/ft^3	kg/m^3
υ	specific volume	ft^3/lbm	m^3/kg
ϕ	angle	deg	deg
ω	angular velocity	rad/sec	rad/s

Subscripts

f	frictional
p	constant pressure
v	constant volume

1. ENERGY OF A MASS

The *energy* of a mass represents the capacity of the mass to do work. Such energy can be stored and released. There are many forms that it can take, including mechanical, thermal, electrical, and magnetic energies. Energy is a positive, scalar quantity (although the change in energy can be either positive or negative).

The total energy of a body can be calculated from its mass, m, and its *specific energy*, u (i.e., the energy per unit mass).[1]

$$E = mu \tag{2.1}$$

Typical units of mechanical energy are foot-pounds and joules. (A joule is equivalent to the units of N·m and kg·m^2/s^2.) In traditional English-unit countries, the *British thermal unit* (Btu) is used for thermal energy, whereas the kilocalorie (kcal) is still used in some applications in SI countries. *Joule's constant*, or the *Joule equivalent* (778.26 ft-lbf/Btu, usually shortened to 778, three significant digits), is used to convert between English mechanical and thermal energy units.

$$\text{energy in Btu} = \frac{\text{energy in ft-lbf}}{J} \tag{2.2}$$

Two other units of large amounts of energy are the therm and the quad. A *therm* is 10^5 Btu (1.055×10^8 J). A *quad* is equal to a quadrillion (10^{15}) Btu. This is 1.055×10^{18} J, or roughly the energy contained in 200 million barrels of oil. (Chapter 35 has more information on the units of energy.)

2. LAW OF CONSERVATION OF ENERGY

The *law of conservation of energy* says that energy cannot be created or destroyed. However, energy can

[1]The use of symbols E and U for energy is not consistent in the engineering field.

be converted into different forms. Therefore, the sum of all energy forms is constant.

$$\sum E = \text{constant} \qquad 2.3$$

3. WORK

Work, W, is the act of changing the energy of a particle, body, or system. For a mechanical system, *external work* is work done by an external force, whereas *internal work* is done by an internal force. Work is a signed, scalar quantity. Typical units are inch-pounds, foot-pounds, and joules. Mechanical work is seldom expressed in British thermal units or kilocalories. (See Chap. 35 for more information on the units of work.)

For a mechanical system, work is positive when a force acts in the direction of motion and helps a body move from one location to another. Work is negative when a force acts to oppose motion. (Friction, for example, always opposes the direction of motion and can do only negative work.) The work done on a body by more than one force can be found by superposition.

From a thermodynamic standpoint, work is positive if a particle or body does work on its surroundings. Work is negative if the surroundings do work on the object (e.g., blowing up a balloon represents negative work to the balloon). Although this may be a difficult concept to grasp, it is consistent with the conservation of energy, because the sum of negative work and the positive energy increase is zero (i.e., no net energy change in the system).[2]

The work performed by a variable force or torque is calculated from the dot products of Eq. 2.4 and Eq. 2.5.

$$W_{\text{variable force}} = \int \mathbf{F} \cdot d\mathbf{s} \quad \text{[linear systems]} \qquad 2.4$$

$$W_{\text{variable torque}} = \int \mathbf{T} \cdot d\theta \quad \text{[rotational systems]} \qquad 2.5$$

The work done by a force or torque of constant magnitude is

$$W_{\text{constant force}} = \mathbf{F} \cdot \mathbf{s} = Fs\cos\phi \quad \text{[linear systems]} \qquad 2.6$$

$$W_{\text{constant torque}} = \mathbf{T} \cdot \theta = Fr\theta\cos\phi \quad \text{[rotational systems]} \qquad 2.7$$

The nonvector forms illustrate that only the component of force or torque in the direction of motion contributes to work. (See Eq. 2.6 and Eq. 2.7.)

Common applications of the work done by a constant force are frictional work and gravitational work. (See Fig. 2.1.) The work to move an object a distance s against a frictional force of F_f is

$$W_{\text{friction}} = F_f s \qquad 2.8$$

[2]This is just a partial statement of the *first law of thermodynamics*.

Figure 2.1 *Work of a Constant Force*

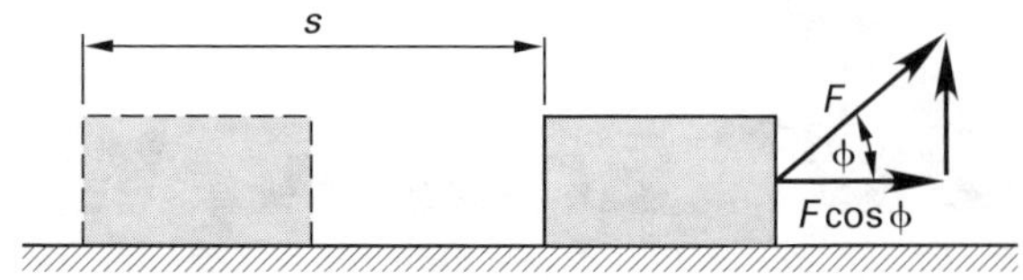

The sign of the work depends on the direction of the motion. The work done *by* gravity when a mass m changes in elevation from h_1 to h_2 is

$$W_{\text{gravity}} = mg(h_2 - h_1) \quad \text{[SI]} \qquad 2.9(a)$$

$$W_{\text{gravity}} = \left(\frac{mg}{g_c}\right)(h_2 - h_1) \quad \text{[U.S.]} \qquad 2.9(b)$$

The sign of the work depends on the direction of the motion. The work done by or on a *linear spring* whose length or deflection changes from δ_1 to δ_2 is given by Eq. 2.10.[3] It does not make any difference whether the spring is a compression spring or an extension spring.

$$W_{\text{spring}} = \tfrac{1}{2}k\left(\delta_2^2 - \delta_1^2\right) \qquad 2.10$$

Example 2.1

A lawn mower engine is started by pulling a cord wrapped around a sheave. The sheave radius is 8.0 cm. The cord is wrapped around the sheave two times. If a constant tension of 90 N is maintained in the cord during starting, what work is done on the engine?

Solution

The starting torque on the engine is

$$T = Fr = (90\ \text{N})\left(\frac{8.0\ \text{cm}}{100\ \frac{\text{cm}}{\text{m}}}\right) = 7.2\ \text{N·m}$$

The cord wraps around the sheave two times; therefore, $(2)(2\pi) = 12.6$ rad. From Eq. 2.7, the work done by a constant torque is

$$W = T\theta = (7.2\ \text{N·m})(12.6\ \text{rad}) = 90.7\ \text{J}$$

Example 2.2

A 200 lbm crate is pushed 25 ft at constant velocity across a warehouse floor. There is a frictional force of 60 lbf between the crate and floor. What work is done by the frictional force on the crate?

[3]A *linear spring* is one for which the linear relationship $F = kx$ is valid.

Solution

From Eq. 2.8,

$$\begin{aligned} W_{\text{friction}} &= -F_f s = (60 \text{ lbf})(25 \text{ ft}) \\ &= -1500 \text{ ft-lbf} \end{aligned}$$

The work is opposite to the direction of motion.

4. POTENTIAL ENERGY OF A MASS

Potential energy(gravitational energy) is a form of mechanical energy possessed by a body due to its relative position in a gravitational field. Potential energy is lost when the elevation of a body decreases. The lost potential energy usually is converted to kinetic energy or heat.

$$E_{\text{potential}} = mgh \quad \text{[SI]} \qquad 2.11(a)$$

$$E_{\text{potential}} = \frac{mgh}{g_c} \quad \text{[U.S.]} \qquad 2.11(b)$$

In the absence of friction and other nonconservative forces, the change in potential energy of a body is equal to the work required to change the elevation of the body.

$$W = \Delta E_{\text{potential}} \qquad 2.12$$

5. KINETIC ENERGY OF A MASS

Kinetic energy is a form of mechanical energy associated with a moving or rotating body. The kinetic energy of a body moving with instantaneous linear velocity, v, is

$$E_{\text{kinetic}} = \tfrac{1}{2}m\text{v}^2 \quad \text{[SI]} \qquad 2.13(a)$$

$$E_{\text{kinetic}} = \frac{m\text{v}^2}{2g_c} \quad \text{[U.S.]} \qquad 2.13(b)$$

A body can also have rotational kinetic energy.

$$E_{\text{rotational}} = \tfrac{1}{2}I\omega^2 \quad \text{[SI]} \qquad 2.14(a)$$

$$E_{\text{rotational}} = \frac{I\omega^2}{2g_c} \quad \text{[U.S.]} \qquad 2.14(b)$$

According to the *work-energy principle* (see Sec. 2.9), the kinetic energy is equal to the work necessary to initially accelerate a stationary body or to bring a moving body to rest.

$$W = \Delta E_{\text{kinetic}} \qquad 2.15$$

Example 2.3

A solid disk flywheel ($I = 200$ kg·m^2) is rotating with a speed of 900 rpm. What is its rotational kinetic energy?

Solution

The angular rotational velocity is

$$\omega = \frac{(900 \text{ rpm})\left(2\pi \ \frac{\text{rad}}{\text{rev}}\right)}{60 \ \frac{\text{s}}{\text{min}}} = 94.25 \text{ rad/s}$$

From Eq. 2.14, the rotational kinetic energy is

$$\begin{aligned} E &= \tfrac{1}{2}I\omega^2 = (\tfrac{1}{2})(200 \text{ kg·m}^2)\left(94.25 \ \frac{\text{rad}}{\text{s}}\right)^2 \\ &= 888 \times 10^3 \text{ J} \quad (888 \text{ kJ}) \end{aligned}$$

6. SPRING ENERGY

A spring is an energy storage device, since the spring has the ability to perform work. In a perfect spring, the amount of energy stored is equal to the work required to compress the spring initially. The stored spring energy does not depend on the mass of the spring. Given a spring with spring constant (stiffness) k, the *spring energy* is

$$E_{\text{spring}} = \tfrac{1}{2}k\delta^2 \qquad 2.16$$

Example 2.4

A body of mass m falls from height h onto a massless, simply supported beam. The mass adheres to the beam. If the beam has a lateral stiffness k, what will be the deflection, δ, of the beam?

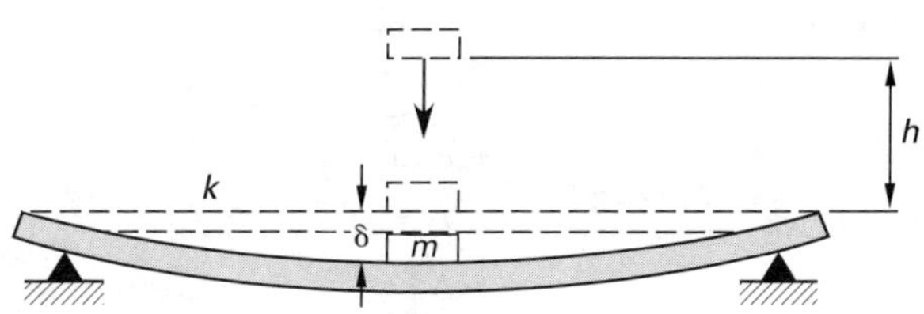

Solution

The initial energy of the system consists of only the potential energy of the body. Using consistent units, the change in potential energy is

$$E = mg(h + \delta) \quad \text{[consistent units]}$$

All of this energy is stored in the spring. Therefore,

$$\tfrac{1}{2}k\delta^2 = mg(h + \delta)$$

Solving for the deflection,

$$\delta = \frac{mg \pm \sqrt{mg(2hk + mg)}}{k}$$

7. PRESSURE ENERGY OF A MASS

Since work is done in increasing the pressure of a system (e.g., it takes work to blow up a balloon), mechanical energy can be stored in pressure form. This is known as *pressure energy, static energy, flow energy, flow work,* and *p-V work (energy)*. For a system of pressurized mass m, the pressure energy is

$$E_{\text{flow}} = \frac{mp}{\rho} = mpv \quad [v = \text{specific volume}] \tag{2.17}$$

8. INTERNAL ENERGY OF A MASS

The total internal energy, usually given the symbol U, of a body increases when the body's temperature increases.[4] In the absence of any work done on or by the body, the change in internal energy is equal to the heat flow, Q, into the body. Q is positive if the heat flow is into the body and negative otherwise.

$$U_2 - U_1 = Q \tag{2.18}$$

The property of internal energy is encountered primarily in thermodynamics problems. Typical units are British thermal units, joules, and kilocalories.

An increase in internal energy is needed to cause a rise in temperature. Different substances differ in the quantity of heat needed to produce a given temperature increase. The ratio of heat, Q, required to change the temperature of a mass, m, by an amount, ΔT, is called the *specific heat (heat capacity)* of the substance, c.

Because specific heats of solids and liquids are slightly temperature dependent, the mean specific heats are used when evaluating processes covering a large temperature range.

$$Q = mc\Delta T \tag{2.19}$$

$$c = \frac{Q}{m\Delta T} \tag{2.20}$$

The lowercase c implies that the units are Btu/lbm-°F or J/kg·°C. Typical values of specific heat are given in Table 2.1. The *molar specific heat*, designated by the symbol C, has units of Btu/lbmol-°F or J/kmol·°C.

$$C = (\text{MW}) \times c \tag{2.21}$$

For gases, the specific heat depends on the type of process during which the heat exchange occurs. Specific heats for constant-volume and constant-pressure processes are designated by c_v and c_p, respectively.

$$Q = mc_v\Delta T \quad [\text{constant-volume process}] \tag{2.22}$$

$$Q = mc_p\Delta T \quad [\text{constant-pressure process}] \tag{2.23}$$

[4]The *thermal energy*, represented by the body's enthalpy, is the sum of internal and pressure energies.

Approximate values of c_p and c_v for solids and liquids are essentially the same. However, the designation c_p is often encountered for solids and liquids. Table 2.1 gives values of c_p for selected liquids and solids.

Table 2.1 *Approximate Specific Heats of Selected Liquids and Solids*[*]

	c_p	
substance	Btu/lbm-°F	kJ/kg·°C
aluminum, pure	0.23	0.96
aluminum, 2024-T4	0.2	0.84
ammonia	1.16	4.86
asbestos	0.20	0.84
benzene	0.41	1.72
brass, red	0.093	0.39
bronze	0.082	0.34
concrete	0.21	0.88
copper, pure	0.094	0.39
Freon-12	0.24	1.00
gasoline	0.53	2.20
glass	0.18	0.75
gold, pure	0.031	0.13
ice	0.49	2.05
iron, pure	0.11	0.46
iron, cast (4% C)	0.10	0.42
lead, pure	0.031	0.13
magnesium, pure	0.24	1.00
mercury	0.033	0.14
oil, light hydrocarbon	0.5	2.09
silver, pure	0.06	0.25
steel, 1010	0.10	0.42
steel, stainless 301	0.11	0.46
tin, pure	0.055	0.23
titanium, pure	0.13	0.54
tungsten, pure	0.032	0.13
water	1.0	4.19
wood (typical)	0.6	2.50
zinc, pure	0.088	0.37

(Multiply Btu/lbm-°F by 4.1868 to obtain kJ/kg·°C.)

[*]Values in Btu/lbm-°F are the same as cal/g·°C. Values in kJ/kg·°C are the same as kJ/kg·K.

9. WORK-ENERGY PRINCIPLE

Since energy can neither be created nor destroyed, external work performed on a conservative system goes into changing the system's total energy. This is known as the *work-energy principle* (or *principle of work and energy*).

$$W = \Delta E = E_2 - E_1 \tag{2.24}$$

Generally, the term *work-energy principle* is limited to use with mechanical energy problems (i.e., conversion of work into kinetic or potential energies). When energy is limited to kinetic energy, the work-energy principle is a direct consequence of Newton's second law but is valid for only inertial reference systems.

By directly relating forces, displacements, and velocities, the work-energy principle introduces some simplifications into many mechanical problems.

- It is not necessary to calculate or know the acceleration of a body to calculate the work performed on it.
- Forces that do not contribute to work (e.g., are normal to the direction of motion) are irrelevant.
- Only scalar quantities are involved.
- It is not necessary to individually analyze the particles or component parts in a complex system.

Example 2.5

A 4000 kg elevator starts from rest, accelerates uniformly to a constant speed of 2.0 m/s, and then decelerates uniformly to a stop 20 m above its initial position. Neglecting friction and other losses, what work was done on the elevator?

Solution

By the work-energy principle, the work done on the elevator is equal to the change in the elevator's energy. Since the initial and final kinetic energies are zero, the only mechanical energy change is the potential energy change.

Taking the initial elevation of the elevator as the reference (i.e., $h_1 = 0$),

$$\begin{aligned} W &= E_{2,\text{potential}} - E_{1,\text{potential}} = mg(h_2 - h_1) \\ &= (4000\ \text{kg})\left(9.81\ \frac{\text{m}}{\text{s}^2}\right)(20\ \text{m}) \\ &= 785 \times 10^3\ \text{J} \quad (785\ \text{kJ}) \end{aligned}$$

10. CONVERSION BETWEEN ENERGY FORMS

Conversion of one form of energy into another does not violate the conservation of energy law. However, most problems involving conversion of energy are really just special cases of the work-energy principle. For example, consider a falling body that is acted upon by a gravitational force. The conversion of potential energy into kinetic energy can be interpreted as equating the work done by the constant gravitational force to the change in kinetic energy.

In general terms, *Joule's law* states that one energy form can be converted without loss into another. There are two specific formulations of Joule's law. As related to electricity, $P = I^2R = V^2/R$ is the common formulation of Joule's law. As related to thermodynamics and ideal gases, Joule's law states that "the change in internal energy of an ideal gas is a function of the temperature change, not of the volume." This latter form can also be stated more formally as "at constant temperature, the internal energy of a gas approaches a finite value that is independent of the volume as the pressure goes to zero."

Example 2.6

A 2.0 lbm projectile is launched straight up with an initial velocity of 700 ft/sec. Neglecting air friction, calculate the (a) kinetic energy immediately after launch, (b) kinetic energy at maximum height, (c) potential energy at maximum height, (d) total energy at an elevation where the velocity has dropped to 300 ft/sec, and (e) maximum height attained.

Solution

(a) From Eq. 2.13, the kinetic energy is

$$\begin{aligned} e_{\text{kinetic}} &= \frac{mv^2}{2g_c} = \frac{(2\ \text{lbm})\left(700\ \frac{\text{ft}}{\text{sec}}\right)^2}{(2)\left(32.2\ \frac{\text{ft-lbm}}{\text{lbf-sec}^2}\right)} \\ &= 15{,}217\ \text{ft-lbf} \end{aligned}$$

(b) The velocity is zero at the maximum height. Therefore, the kinetic energy is zero.

(c) At the maximum height, all of the kinetic energy has been converted into potential energy. Therefore, the potential energy is 15,217 ft-lbf.

(d) Although some of the kinetic energy has been transformed into potential energy, the total energy is still 15,217 ft-lbf.

(e) Since all of the kinetic energy has been converted into potential energy, the maximum height can be found from Eq. 2.11.

$$\begin{aligned} E_{\text{potential}} &= \frac{mgh}{g_c} \\ 15{,}217\ \text{ft-lbf} &= \frac{(2\ \text{lbm})\left(32.2\ \frac{\text{ft}}{\text{sec}^2}\right)h}{32.2\ \frac{\text{ft-lbm}}{\text{lbf-sec}^2}} \\ h &= 7609\ \text{ft} \end{aligned}$$

Example 2.7

A 4500 kg ore car rolls down an incline and passes point A traveling at 1.2 m/s. The ore car is stopped by a spring bumper that compresses 0.6 m. A constant friction force of 220 N acts on the ore car at all times. What spring constant is required?

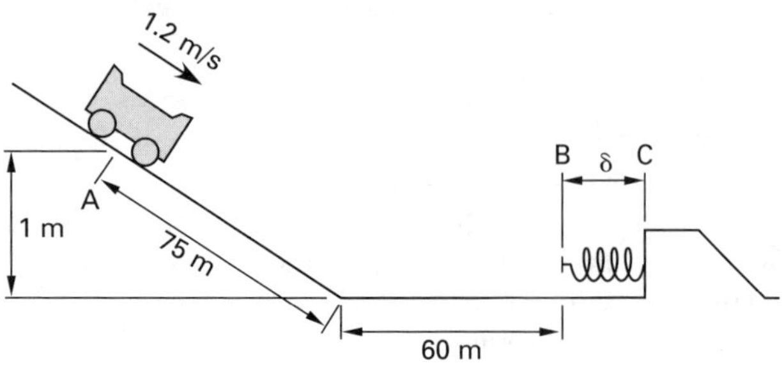

(not to scale)

Solution

The car's total energy at point A is the sum of the kinetic and potential energies.

$$\begin{aligned} E_{\text{total,A}} &= E_{\text{kinetic}} + E_{\text{potential}} \\ &= \tfrac{1}{2}m\text{v}^2 + mgh \\ &= \left(\tfrac{1}{2}\right)(4500\ \text{kg})\left(1.2\ \frac{\text{m}}{\text{s}}\right)^2 \\ &\quad + (4500\ \text{kg})\left(9.81\ \frac{\text{m}}{\text{s}^2}\right)(1\ \text{m}) \\ &= 47\,385\ \text{J} \end{aligned}$$

At point B, the potential energy has been converted into additional kinetic energy. However, except for friction, the total energy is the same as at point A. Since the frictional force does negative work, the total energy remaining at point B is

$$\begin{aligned} E_{\text{total,B}} &= E_{\text{total,A}} - W_{\text{friction}} \\ &= 47\,385\ \text{J} - (220\ \text{N})(75\ \text{m} + 60\ \text{m}) \\ &= 17\,685\ \text{J} \end{aligned}$$

At point C, the maximum compression point, the remaining energy has gone into compressing the spring a distance $\delta = 0.6$ m and performing a small amount of frictional work.

$$\begin{aligned} E_{\text{total,B}} = E_{\text{total,C}} &= W_{\text{spring}} + W_{\text{friction}} \\ &= \tfrac{1}{2}k\delta^2 + F_f\delta \end{aligned}$$

$$17\,685\ \text{J} = \tfrac{1}{2}k(0.6\ \text{m})^2 + (220\ \text{N})(0.6\ \text{m})$$

The spring constant can be determined directly.

$$k = 97\,520\ \text{N/m} \quad (97.5\ \text{kN/m})$$

11. POWER

Power is the amount of work done per unit time. It is a scalar quantity. (Although power is calculated from two vectors, the vector dot-product operation is seldom needed.)

$$P = \frac{W}{\Delta t} \qquad 2.25$$

For a body acted upon by a force or torque, the instantaneous power can be calculated from the velocity.

$$P = F\text{v} \quad \text{[linear systems]} \qquad 2.26$$

$$P = T\omega \quad \text{[rotational systems]} \qquad 2.27$$

For a fluid flowing at a rate of $\dot{m}$, the unit of time is already incorporated into the flow rate (e.g., lbm/sec). If the fluid experiences a specific energy change of Δu, the power generated or dissipated will be

$$P = \dot{m}\Delta u \qquad 2.28$$

Typical basic units of power are ft-lbf/sec and watts (J/s), although *horsepower* is widely used. (Chapter 35 contains further information on the units of power.) Table 2.2 can be used to convert units of power.

Table 2.2 *Useful Power Conversion Formulas*

$$\begin{aligned} 1\ \text{hp} &= 550\ \text{ft-lbf/sec} \\ &= 33{,}000\ \text{ft-lbf/min} \\ &= 0.7457\ \text{kW} \\ &= 0.7068\ \text{Btu/sec} \end{aligned}$$

$$\begin{aligned} 1\ \text{kW} &= 737.6\ \text{ft-lbf/sec} \\ &= 44{,}250\ \text{ft-lbf/min} \\ &= 1.341\ \text{hp} \\ &= 0.9483\ \text{Btu/sec} \end{aligned}$$

$$\begin{aligned} 1\ \text{Btu/sec} &= 778.26\ \text{ft-lbf/sec} \\ &= 46{,}680\ \text{ft-lbf/min} \\ &= 1.415\ \text{hp} \end{aligned}$$

Example 2.8

When traveling at 100 km/h, a car supplies a constant horizontal force of 50 N to the hitch of a trailer. What tractive power (in horsepower) is required for the trailer alone?

Solution

From Eq. 2.26, the power being generated is

$$\begin{aligned} P = F\text{v} &= \frac{(50\ \text{N})\left(100\ \frac{\text{km}}{\text{h}}\right)\left(1000\ \frac{\text{m}}{\text{km}}\right)}{\left(60\ \frac{\text{s}}{\text{min}}\right)\left(60\ \frac{\text{min}}{\text{h}}\right)\left(1000\ \frac{\text{W}}{\text{kW}}\right)} \\ &= 1.389\ \text{kW} \end{aligned}$$

Using a conversion from Table 2.2, the horsepower is

$$\begin{aligned} P &= \left(1.341\ \frac{\text{hp}}{\text{kW}}\right)(1.389\ \text{kW}) \\ &= 1.86\ \text{hp} \end{aligned}$$

12. EFFICIENCY

For energy-using systems (such as cars, electrical motors, elevators, etc.), the *energy-use efficiency*, η, of a system is the ratio of an ideal property to an actual property. The property used is commonly work, power, or, for thermodynamics problems, heat. When the rate of work is constant, either work or power can be used to calculate the efficiency. Otherwise, power should be used. Except in rare instances, the numerator and denominator of the ratio must have the same units.[5]

$$\eta = \frac{P_{\text{ideal}}}{P_{\text{actual}}} \quad [P_{\text{actual}} \geq P_{\text{ideal}}] \qquad 2.29$$

For energy-producing systems (such as electrical generators, prime movers, and hydroelectric plants), the *energy-production efficiency* is

$$\eta = \frac{P_{\text{actual}}}{P_{\text{ideal}}} \quad [P_{\text{ideal}} \geq P_{\text{actual}}] \qquad 2.30$$

The efficiency of an *ideal machine* is 1.0 (100%). However, all *real machines* have efficiencies of less than 1.0.

[5]The *energy-efficiency ratio* used to evaluate refrigerators, air conditioners, and heat pumps, for example, has units of Btu per watt-hour (Btu/W-hr).

3 Engineering Drawing Practice

1. Introduction 3-1
2. Normal Views of Lines and Planes 3-1
3. Intersecting and Perpendicular Lines 3-1
4. Types of Views 3-1
5. Principal (Orthographic) Views 3-2
6. Auxiliary (Orthographic) Views 3-2
7. Oblique (Orthographic) Views 3-2
8. Axonometric (Orthographic Oblique) Views 3-3
9. Perspective Views 3-3
10. Sections 3-4
11. Tolerances 3-4
12. Surface Finish 3-4
13. Electrical Schematics 3-5

1. INTRODUCTION[1]

This chapter is designed as a brief introduction to the reading of drawings and schematics to which the electrical engineer may be exposed. As many mechanical-electrical interfaces exist, Sec. 4.2 through Sec. 4.12 present the basic conventions used in such drawings. Section 3.13 is an overview of the electrical symbols used in schematics.[2] The various symbols are discussed throughout this book.

2. NORMAL VIEWS OF LINES AND PLANES

A *normal view* of a line is a perpendicular projection of the line onto a viewing plane parallel to the line. In the normal view, all points of the line are equidistant from the observer. Therefore, the true length of a line is viewed and can be measured.

Generally, however, a line will be viewed from an oblique position and will appear shorter than it actually is. The normal view can be constructed by drawing an auxiliary view from the orthographic view.[3] (See Sec. 3.6.)

Similarly, a normal view of a plane figure is a perpendicular projection of the figure onto a viewing plane parallel to the plane of the figure. All points of the plane are equidistant from the observer. Therefore, the true size and shape of any figure in the plane can be determined.

3. INTERSECTING AND PERPENDICULAR LINES

A single orthographic view is not sufficient to determine whether two lines intersect. However, if two or more views show the lines as having the same common point (i.e., crossing at the same position in space), then the lines intersect. In Fig. 3.1, the subscripts F and T refer to front and top views, respectively.

Figure 3.1 *Intersecting and Non-Intersecting Lines*

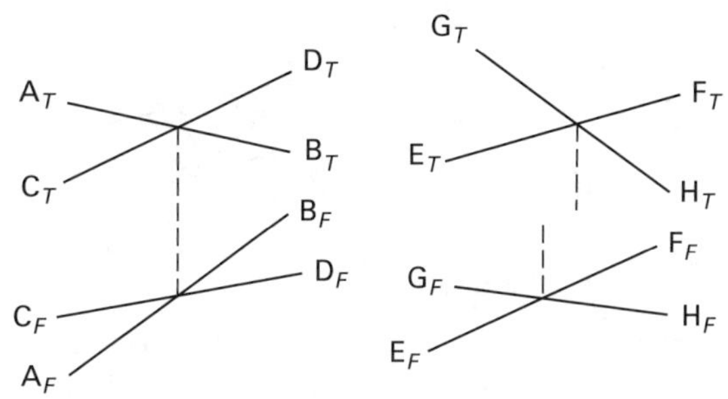

According to the *perpendicular line principle*, two perpendicular lines appear perpendicular only in a normal view of either one or both of the lines. Conversely, if two lines appear perpendicular in any view, the lines are perpendicular only if the view is a normal view of one or both of the lines.

4. TYPES OF VIEWS

Objects can be illustrated in several different ways depending on the number of views, the angle of observation, and the degree of artistic latitude taken for the purpose of simplifying the drawing process.[4] Table 3.1 categorizes the types of views.

The different types of views are easily distinguished by their *projectors* (i.e., projections of parallel lines on the object). For a cube, there are three sets of projectors corresponding to the three perpendicular axes. In an *orthographic (orthogonal) view*, the projectors are

[1]This chapter is not meant to show "how to do it" as much as it is to present the conventions and symbols of engineering drawing.

[2]A drawing is the surface portrayal of a form or figure in line. A schematic is a presentation of the element-by-element relationship of all parts of a system.

[3]The technique for constructing a normal view is covered in engineering drafting texts.

[4]The omission of perspective from a drawing is an example of a step taken to simplify the drawing process.

parallel. In a *perspective (central) view*, some or all of the projectors converge to a point. Figure 3.2 illustrates the orthographic and perspective views of a block.

Table 3.1 *Types of Views of Objects*

orthographic views
- principal views
- auxiliary views
- oblique views
  - cavalier projection
  - cabinet projection
  - clinographic projection
  - axonometric views
    - isometric
    - dimetric
    - trimetric

perspective views
- parallel perspective
- angular perspective

Figure 3.2 *Orthographic and Perspective Views of a Block*

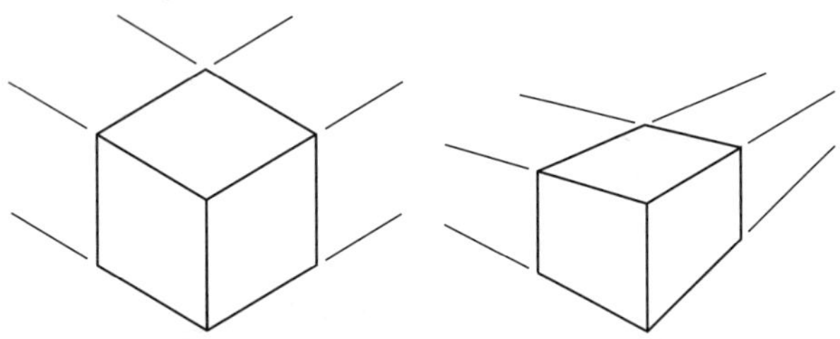

5. PRINCIPAL (ORTHOGRAPHIC) VIEWS

In a *principal view* (also known as a *planar view*), one of the sets of projectors is normal to the view. That is, one of the planes of the object is seen in a normal view. The other two sets of projectors are orthogonal and are usually oriented horizontally and vertically on the paper. Because background details of an object may not be visible in a principal view, it is necessary to have at least three principal views to completely illustrate a symmetrical object. At most, six principal views will be needed to illustrate complex objects.

The relative positions of the six views have been standardized and are shown in Fig. 3.3, which also defines the *width* (also known as *depth*), *height*, and *length* of the object. The views that are not needed to illustrate features or provide dimensions (i.e., *redundant views*) can be omitted. The usual combination selected consists of the top, front, and right side views.

It is common to refer to the front, side, and back views as *elevations* and to the top and bottom views as *plan views*. These terms are not absolute since any plane can be selected as the front.

Figure 3.3 *Positions of Standard Orthographic Views*

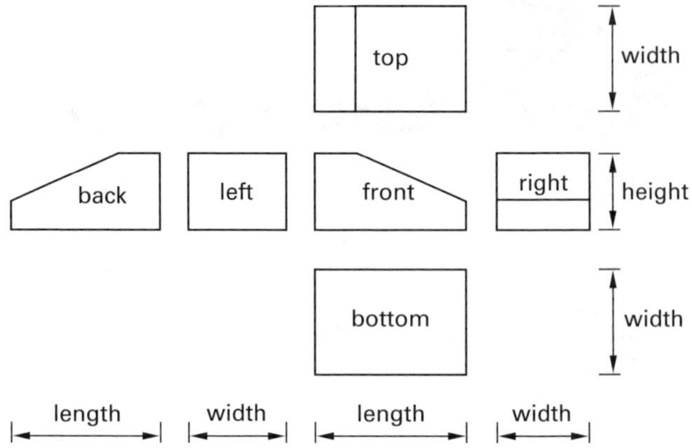

6. AUXILIARY (ORTHOGRAPHIC) VIEWS

An *auxiliary view* is needed when an object has an inclined plane or curved feature or when there are more details than can be shown in the six principal views. As with the other orthographic views, the auxiliary view is a normal (face-on) view of the inclined plane. Figure 3.4 illustrates an auxiliary view.

Figure 3.4 *Auxiliary View*

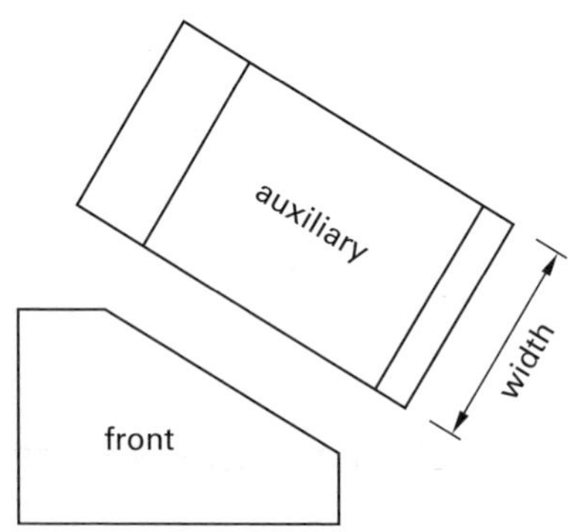

The projectors in an auxiliary view are perpendicular to only one of the directions in which a principal view is observed. Accordingly, only one of the three dimensions of width, height, and depth can be measured (scaled). In a *profile auxiliary view*, the object's width can be measured. In a *horizontal auxiliary view (auxiliary elevation)*, the object's height can be measured. In a *frontal auxiliary view*, the depth of the object can be measured.

7. OBLIQUE (ORTHOGRAPHIC) VIEWS

If the object is turned so that three principal planes are visible, it can be completely illustrated by a single *oblique view*.[5] In an oblique view, the direction from

[5]Oblique views are not unique in this capability—perspective drawings share it. Oblique and perspective drawings are known as *pictorial drawings* because they give depth to the object by illustrating it in three dimensions.

which the object is observed is not (necessarily) parallel to any of the directions from which principal and auxiliary views are observed.

In two common methods of oblique illustration, one of the view planes coincides with an orthographic view plane. Two of the drawing axes are at right angles to each other; one of these is vertical, and the other (the *oblique axis*) is oriented at 30° or 45° (originally chosen to coincide with standard drawing triangles). The ratio of scales used for the horizontal, vertical, and oblique axes can be 1:1:1 or 1:1:$^1/_2$. The latter ratio helps to overcome the visual distortion due to the absence of perspective in the oblique direction.

Cavalier (45° oblique axis and 1:1:1 scale ratio) and *cabinet* (45° oblique axis and 1:1:$^1/_2$ scale ratio) *projections* are the two common types of oblique views that incorporate one of the orthographic views. If an angle of 9.5° is used (as in illustrating crystalline lattice structures), the technique is known as *clinographic projection*. Figure 3.5 illustrates cavalier and cabinet oblique drawings.

Figure 3.5 *Cavalier and Cabinet Oblique Drawings*

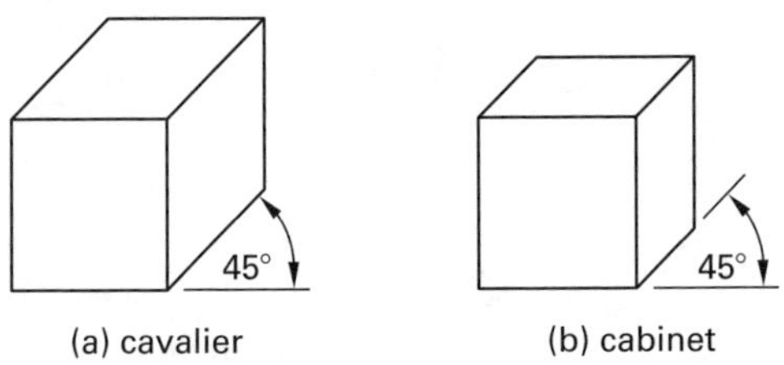

Figure 3.6 *Types of Axonometric Views*

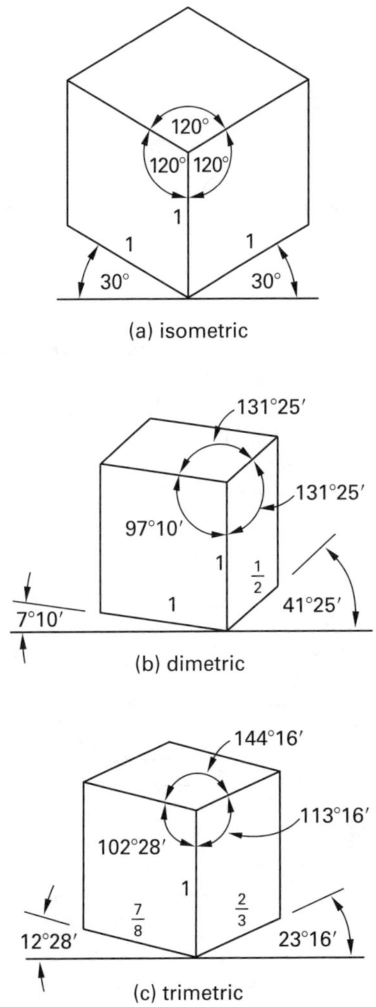

8. AXONOMETRIC (ORTHOGRAPHIC OBLIQUE) VIEWS

In axonometric views, the view plane is not parallel to any of the principal orthographic planes. Figure 3.6 illustrates types of axonometric views. Axonometric views and axonometric drawings are not the same. In a *view (projection)*, one or more of the face lengths is foreshortened. In a *drawing*, the lengths are drawn full length, resulting in a distorted illustration. Table 3.2 lists the proper ratios that should be observed.

In an *isometric view*, the three projectors intersect at equal angles (120°) with the plane. This simplifies construction with standard 30° drawing triangles. All of the faces are foreshortened an equal amount, to $\sqrt{2/3}$, or approximately 81.6% of the true length. In a *dimetric view*, two of the projectors intersect at equal angles, and only two of the faces are equally reduced in length. In a *trimetric view*, all three intersection angles are different, and all three faces are reduced different amounts.

Table 3.2 *Axonometric Foreshortening*

view	projector intersection angles	proper ratio of sides
isometric	120°, 120°, 120°	0.82:0.82:0.82
dimetric	131°25′, 131°25′, 97°10′	1:1:$^1/_2$
	103°38′, 103°38′, 152°44′	$^3/_4$:$^3/_4$:1
trimetric	102°28′, 144°16′, 113°16′	1:$^2/_3$:$^7/_8$
	138°14′, 114°46′, 107°	1:$^3/_4$:$^7/_8$

9. PERSPECTIVE VIEWS

In a *perspective view*, one or more sets of projectors converge to a fixed point known as the *center of vision*. In the *parallel perspective*, all vertical lines remain vertical in the picture; all horizontal frontal lines remain horizontal. Therefore, one face is parallel to the observer and only one set of projectors converges. In the *angular perspective*, two sets of projectors converge. In the little-used *oblique perspective*, all three sets of projectors converge. Figure 3.7 illustrates types of perspective views.

Figure 3.7 Types of Perspective Views

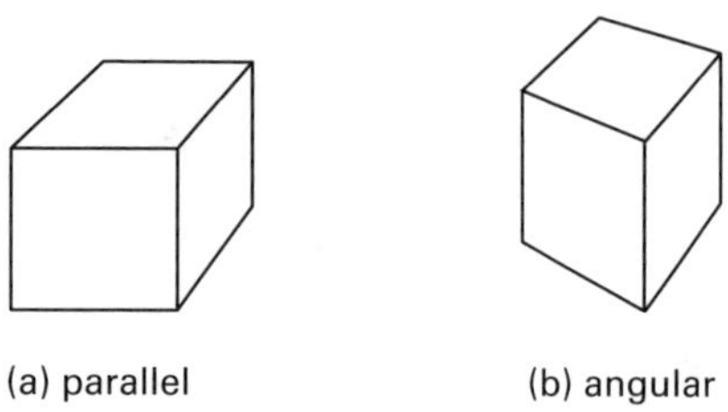

10. SECTIONS

A *section* is an imaginary cut taken through an object to reveal the shape or interior construction.[6] Figure 3.8 illustrates the standard symbol for a *sectioning cut* and the resulting sectional view. Section arrows are perpendicular to the cutting plane and indicate the viewing direction.

Figure 3.8 Sectioning Cut Symbol and Sectional View

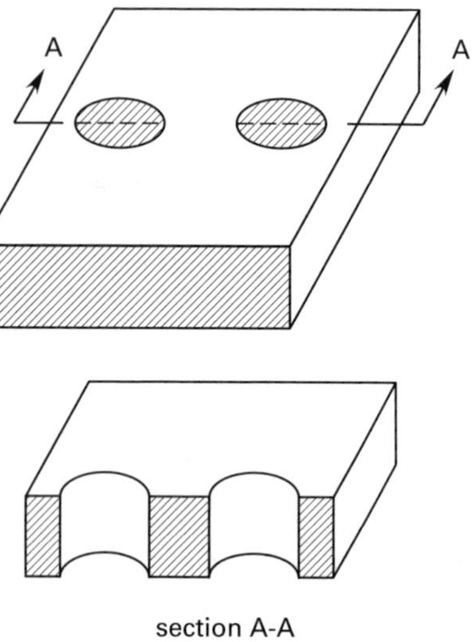

11. TOLERANCES

The *tolerance* for a dimension is the total permissible variation or difference between the acceptable limits. The tolerance for a dimension can be specified in two ways: either as a general rule in the title block (e.g., ± 0.001 in unless otherwise specified) or as specific limits that are given with each dimension (e.g., 2.575 in ± 0.005 in).

12. SURFACE FINISH

ANSI B46.1 specifies surface finish by a combination of parameters.[7] The basic symbol for designating these factors is shown in Fig. 3.9. In the symbol, A is the maximum *roughness height index*, B is the optional minimum *roughness height*, C is the peak-to-valley *waviness height*, D is the optional peak-to-valley *waviness spacing (width)* rating, E is the optional *roughness width cutoff (roughness sampling length)*, F is the *lay*, and G is the *roughness width*. Unless minimums are specified, all parameters are maximum allowable values, and all lesser values are permitted.

Since the roughness varies, the waviness height is an arithmetic average within a sampled square, and the designation A is known as the *roughness weight*, R_a.[8] Values are normally given in microns (μm) or microinches (μin) in SI or customary U.S. units, respectively. A value for the roughness width cutoff of 0.80 mm (0.03 in) is assumed when E is not specified. Other standard values in common use are 0.25 mm (0.010 in) and 0.08 mm (0.003 in). The lay symbol, F, can be = (parallel to indicated surface), ⊥ (perpendicular), C (circular), M (multidirectional), P (pitted), R (radial), or X (crosshatch).

If a small circle is placed at the A position, no machining is allowed and only cast, forged, die-cast, injection-molded, and other unfinished surfaces are acceptable.

Figure 3.9 Surface Finish Designations

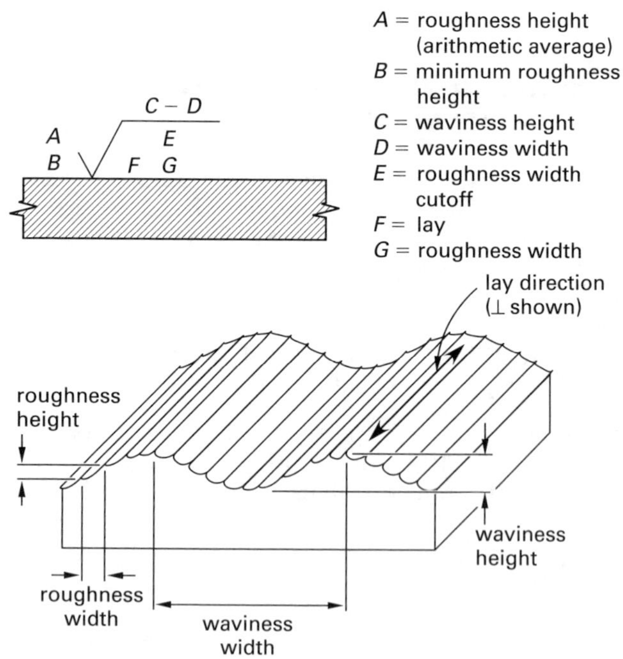

[6]The term *section* is also used to mean a *cross section*—a slice of finite but negligible thickness that is taken from an object to show the cross section or interior construction at the plane of the slice.

[7]Specification does not indicate appearance (i.e., color, luster) or performance (i.e., hardness, corrosion resistance, microstructure).

[8]The symbol R_a is the same as the AA (arithmetic average) and CLA (centerline average) terms used in other (and earlier) standards.

13. ELECTRICAL SCHEMATICS

Although their use is no longer widespread, tubes are still used in certain applications. Basic vacuum tube symbols are shown in Table 3.3.

Table 3.3 *Tube Symbols*

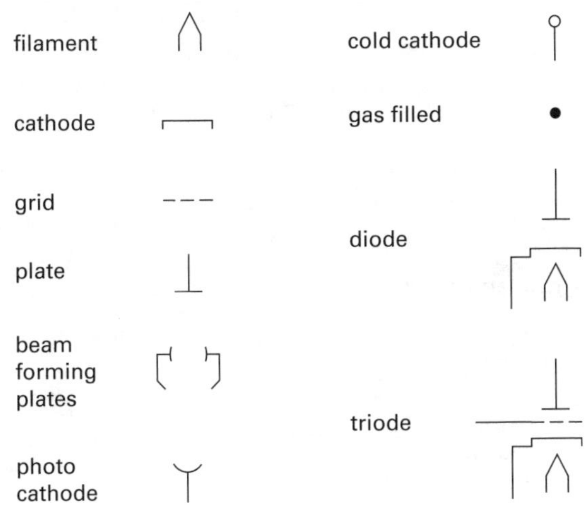

Symbols for common types of basic circuit elements (e.g., resistors, inductors, and capacitors) are shown in Table 3.4.

Table 3.4 *Basic Circuit Element Symbols*

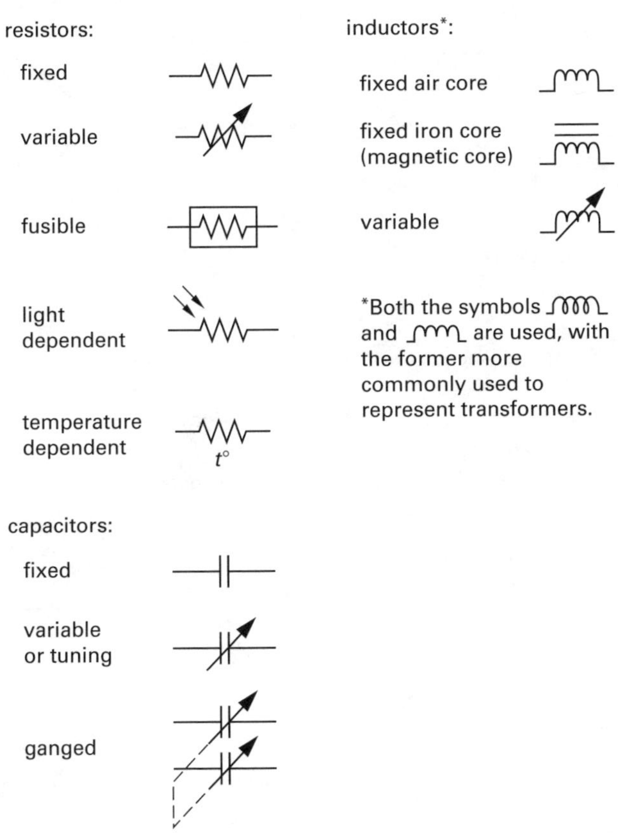

One-line electrical diagrams, often associated with power equipment or transmission line representations, have a set of standard symbols. These symbols are present on electrical and electronic schematics, especially those for ground connections, connections in general, and electrical rotating machinery. Table 3.5 shows some of these symbols.

Table 3.5 *One-Line Electrical Diagram Symbols*

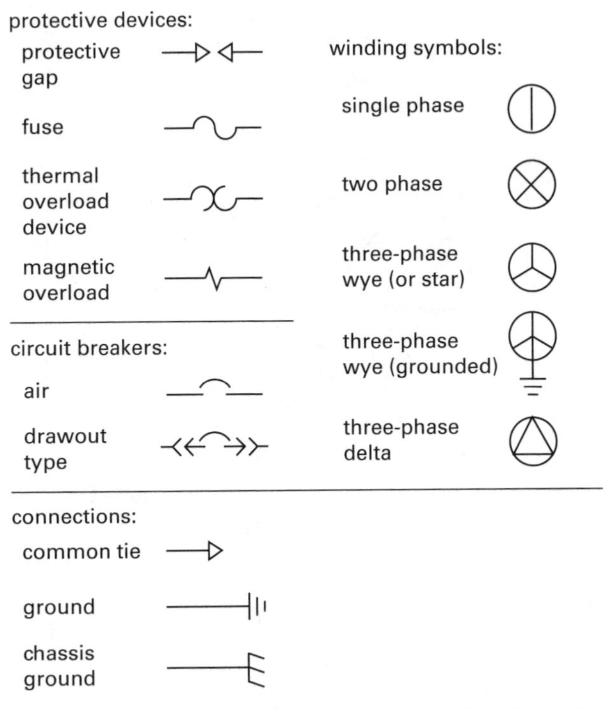

The basic symbols for semiconductor devices are displayed in Table 3.6.

Always refer to the legend of a particular drawing or schematic to ensure understanding of the symbols utilized.

Mathematics

Table 3.6 Semiconductor Symbols

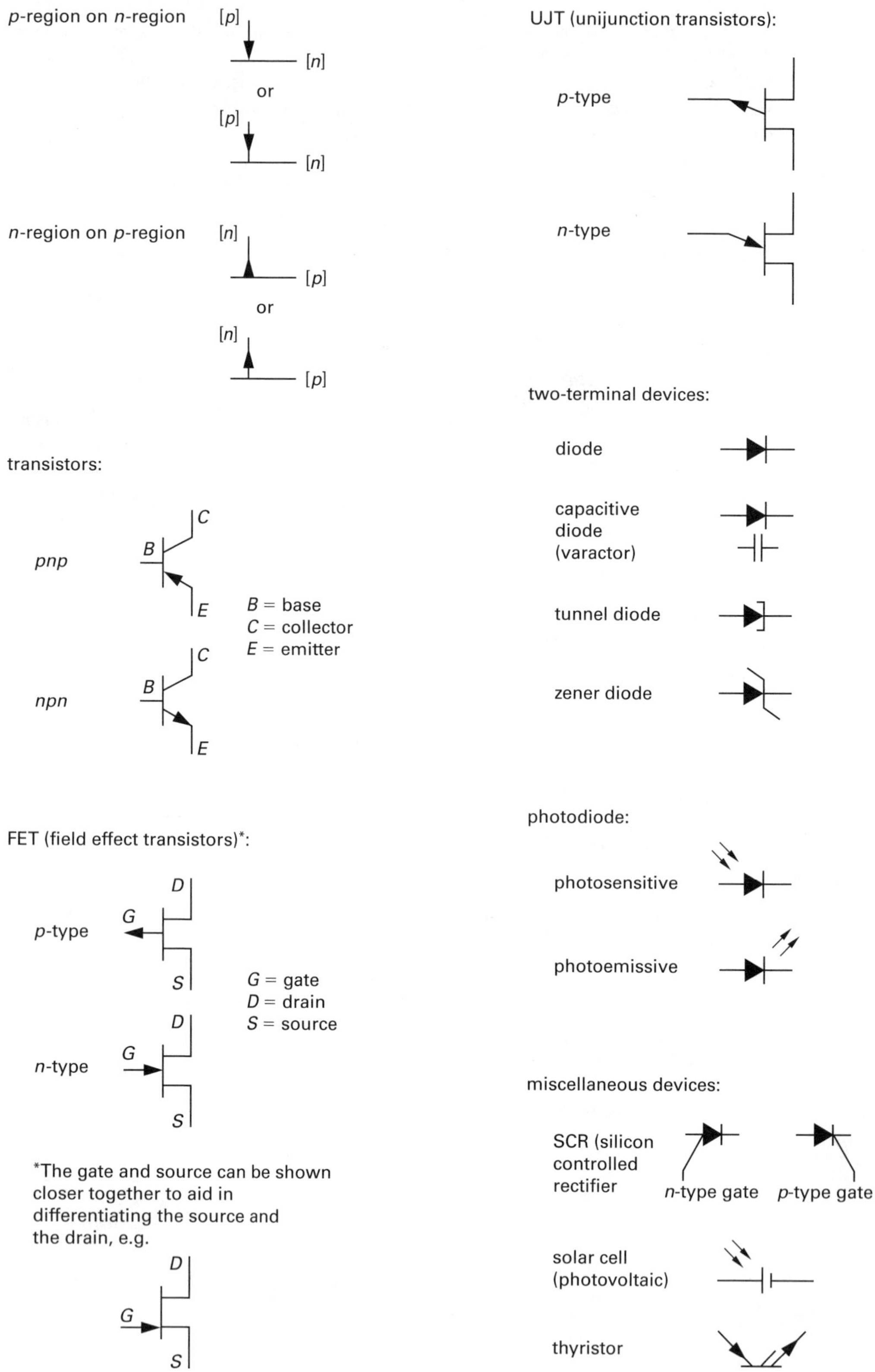

4 Algebra

1. Introduction 4-1
2. Symbols Used in This Book 4-1
3. Greek Alphabet 4-1
4. Types of Numbers 4-1
5. Significant Digits 4-1
6. Equations 4-2
7. Fundamental Algebraic Laws 4-3
8. Polynomials 4-3
9. Roots of Quadratic Equations 4-3
10. Roots of General Polynomials 4-4
11. Extraneous Roots 4-4
12. Descartes' Rule of Signs 4-4
13. Rules for Exponents and Radicals 4-5
14. Logarithms 4-5
15. Logarithm Identities 4-5
16. Partial Fractions 4-6
17. Simultaneous Linear Equations 4-7
18. Complex Numbers 4-8
19. Operations on Complex Numbers 4-8
20. Limits 4-9
21. Sequences and Progressions 4-10
22. Standard Sequences 4-11
23. Series 4-12
24. Tests for Series Convergence 4-12
25. Series of Alternating Signs 4-13

1. INTRODUCTION

Engineers working in design and analysis encounter mathematical problems on a daily basis. Although algebra and simple trigonometry are often sufficient for routine calculations, there are many instances when certain advanced subjects are needed. This chapter and the following, in addition to supporting the calculations used in other chapters, consolidate the mathematical concepts most often needed by engineers.

2. SYMBOLS USED IN THIS BOOK

Many symbols, letters, and Greek characters are used to represent variables in the formulas used throughout this book. These symbols and characters are defined in the nomenclature section of each chapter. However, some of the other symbols in this book are listed in Table 4.2.

3. GREEK ALPHABET

Table 4.1 lists the Greek alphabet.

***Table 4.1** The Greek Alphabet*

A	α	alpha	N	ν	nu
B	β	beta	Ξ	ξ	xi
Γ	γ	gamma	O	o	omicron
Δ	δ	delta	Π	π	pi
E	ϵ	epsilon	P	ρ	rho
Z	ζ	zeta	Σ	σ	sigma
H	η	eta	T	τ	tau
Θ	θ	theta	Υ	υ	upsilon
I	ι	iota	Φ	ϕ	phi
K	κ	kappa	X	χ	chi
Λ	λ	lambda	Ψ	ψ	psi
M	μ	mu	Ω	ω	omega

4. TYPES OF NUMBERS

The *numbering system* consists of three types of numbers: real, imaginary, and complex. *Real numbers*, in turn, consist of rational numbers and irrational numbers. *Rational real numbers* are numbers that can be written as the ratio of two integers (e.g., 4, $^2/_5$, and $^1/_3$).[1] *Irrational real numbers* are nonterminating, nonrepeating numbers that cannot be expressed as the ratio of two integers (e.g., π and $\sqrt{2}$). Real numbers can be positive or negative.

Imaginary numbers are square roots of negative numbers. The symbols i and j are both used to represent the square root of -1.[2] For example, $\sqrt{-5} = \sqrt{5}\sqrt{-1} = \sqrt{5}i$. *Complex numbers* consist of combinations of real and imaginary numbers (e.g., $3 - 7i$).

5. SIGNIFICANT DIGITS

The significant digits in a number include the leftmost, nonzero digits to the rightmost digit written. Final answers from computations should be rounded off to the number of decimal places justified by the data. The answer can be no more accurate than the least accurate number in the data. Of course, rounding should be done on final calculation results only. It should not be done on interim results.

[1]Notice that 0.3333333 is a nonterminating number, but as it can be expressed as a ratio of two integers (i.e., $1/3$), it is a rational number.
[2]The symbol j is used to represent the square root of -1 in electrical calculations to avoid confusion with the current variable, i.

Table 4.2 Symbols Used in This Book

symbol	name	use	example
Σ	sigma	series summation	$\sum_{i=1}^{3} x_i = x_1 + x_2 + x_3$
π	pi	3.1415927...	$p = \pi D$
e	base of natural logs	2.71828...	
Π	pi	series multiplication	$\prod_{i=1}^{3} x_i = x_1 x_2 x_3$
Δ	delta	change in quantity	$\Delta h = h_2 - h_1$
$\bar{\ }$	over bar	average value	$\bar{x}$
$\dot{\ }$	over dot	per unit time	$\dot{m}$ = mass flowing per second
!	factorial[a]		$x! = x(x-1)(x-2)\cdots(2)(1)$
$\lvert\ \rvert$	absolute value[b]		$\lvert -3 \rvert = +3$
$\sim$	similarity		$\Delta\text{ABC} \sim \Delta\text{DEF}$
$\approx$	approximately equal to		$x \approx 1.5$
$\cong$	congruency		$\text{ST} \cong \text{UV}$
$\propto$	proportional to		$x \propto y$
$\equiv$	equivalent to		$a + bi \equiv re^{i\theta}$
∞	infinity		$x \to \infty$
log	base 10 logarithm		$\log 5.74$
ln	natural logarithm		$\ln 5.74$
exp	exponential power		$\exp(x) = e^x$
rms	root-mean-square	$\sqrt{\frac{1}{n}\sum_{i=1}^{n} x_i^2}$	V_{rms}
$\angle$	phasor or angle		$\angle 53°$

[a]*Zero factorial* (0!) is frequently encountered in the form of $(n - n)!$ when calculating permutations and combinations. Zero factorial is defined as 1.
[b]The notation abs(x) is also used to indicate the absolute value.

There are two ways that significant digits can affect calculations. For the operations of multiplication and division, the final answer is rounded to the number of significant digits in the least significant multiplicand, divisor, or dividend. So, $2.0 \times 13.2 = 26$ since the first multiplicand (2.0) has two significant digits only.

For the operations of addition and subtraction, the final answer is rounded to the position of the least significant digit in the addenda, minuend, or subtrahend. So, $2.0 + 13.2 = 15.2$ because both addenda are significant to the tenth's position; but $2 + 13.4 = 15$ since the 2 is significant only in the ones' position.

Table 4.3 Examples of Significant Digits

number as written	number of significant digits	implied range
341	3	340.5–341.5
34.1	3	34.05–34.15
0.00341	3	0.003405–0.003415
341×10^7	3	340.5×10^7–341.5×10^7
3.41×10^{-2}	3	3.405×10^{-2}–3.415×10^{-2}
3410	3	3405–3415
3410*	4	3409.5–3410.5
341.0	4	340.95–341.05

*It is permitted to write "3410." to distinguish the number from its 3-significant digit form, although this is rarely done.

The multiplication rule should not be used for addition or subtraction, as this can result in strange answers. For example, it would be incorrect to round 1700 + 0.1 to 2000 simply because 0.1 has only one significant digit. Table 4.3 gives examples of significant digits.

6. EQUATIONS

An *equation* is a mathematical statement of equality, such as $5 = 3 + 2$. *Algebraic equations* are written in terms of *variables*. In the equation $y = x^2 + 3$, the value of variable y depends on the value of variable x. Therefore, y is the *dependent variable* and x is the *independent variable*. The dependency of y on x is clearer when the equation is written in *functional form*: $y = f(x)$.

A *parametric equation* uses one or more independent variables (*parameters*) to describe a function.[3] For example, the parameter θ can be used to write the parametric equations of a unit circle.

$$x = \cos\theta \qquad 4.1$$

$$y = \sin\theta \qquad 4.2$$

[3]As used in this section, there is no difference between a parameter and an independent variable. However, the term *parameter* is also used as a descriptive measurement that determines or characterizes the form, size, or content of a function. For example, the radius is a parameter of a circle, and mean and variance are parameters of a probability distribution. Once these parameters are specified, the function is completely defined.

A unit circle can also be described by a *nonparametric equation*.[4]

$$x^2 + y^2 = 1 \qquad 4.3$$

7. FUNDAMENTAL ALGEBRAIC LAWS

Algebra provides the rules that allow complex mathematical relationships to be expanded or condensed. Algebraic laws may be applied to complex numbers, variables, and real numbers. The general rules for changing the form of a mathematical relationship are given as follows.

- commutative law for addition:

$$A + B = B + A \qquad 4.4$$

- commutative law for multiplication:

$$AB = BA \qquad 4.5$$

- associative law for addition:

$$A + (B + C) = (A + B) + C \qquad 4.6$$

- associative law for multiplication:

$$A(BC) = (AB)C \qquad 4.7$$

- distributive law:

$$A(B + C) = AB + AC \qquad 4.8$$

8. POLYNOMIALS

A *polynomial* is a rational expression—usually the sum of several variable terms known as *monomials*—that does not involve division. The *degree of the polynomial* is the highest power to which a variable in the expression is raised. The following *standard polynomial forms* are useful when trying to find the roots of an equation.

$$(a + b)(a - b) = a^2 - b^2 \qquad 4.9$$

$$(a \pm b)^2 = a^2 \pm 2ab + b^2 \qquad 4.10$$

$$(a \pm b)^3 = a^3 \pm 3a^2b + 3ab^2 \pm b^3 \qquad 4.11$$

$$(a^3 \pm b^3) = (a \pm b)(a^2 \mp ab + b^2) \qquad 4.12$$

$$(a^n - b^n) = (a - b)\begin{pmatrix} a^{n-1} + a^{n-2}b + a^{n-3}b^2 \\ + \cdots + b^{n-1} \end{pmatrix}$$

$$[n \text{ is any positive integer}] \qquad 4.13$$

$$(a^n + b^n) = (a + b)\begin{pmatrix} a^{n-1} - a^{n-2}b + a^{n-3}b^2 \\ - \cdots + b^{n-1} \end{pmatrix}$$

$$[n \text{ is any positive odd integer}] \qquad 4.14$$

The *binomial theorem* defines a polynomial of the form $(a + b)^n$.

$$(a + b)^n = \underset{[i=0]}{a^n} + \underset{[i=1]}{na^{n-1}b} + \underset{[i=2]}{C_2a^{n-2}b^2} + \cdots + C_ia^{n-i}b^i + \cdots + nab^{n-1} + b^n \qquad 4.15$$

$$C_i = \frac{n!}{i!(n-1)!} \quad [i = 0, 1, 2, \ldots, n] \qquad 4.16$$

The coefficients of the expansion can be determined quickly from *Pascal's triangle*—each entry is the sum of the two entries directly above it. (See Fig. 4.1.)

The values r_1, r_2, …, r_n of the independent variable x that satisfy a polynomial equation $f(x) = 0$ are known as *roots* or *zeros* of the polynomial. A polynomial of degree n with real coefficients will have at most n real roots, although they need not all be distinctly different.

Figure 4.1 *Pascal's Triangle*

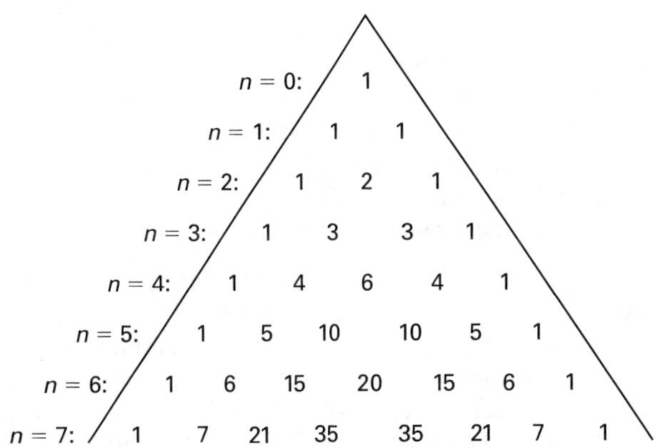

9. ROOTS OF QUADRATIC EQUATIONS

A *quadratic equation* is an equation of the general form $ax^2 + bx + c = 0$ $[a \neq 0]$. The *roots*, x_1 and x_2, of the equation are the two values of x that satisfy it.

$$x_1, x_2 = \frac{-b \pm \sqrt{b^2 - 4ac}}{2a} \qquad 4.17$$

$$x_1 + x_2 = -\frac{b}{a} \qquad 4.18$$

$$x_1x_2 = \frac{c}{a} \qquad 4.19$$

[4]Since only the coordinate variables are used, this equation is also said to be in *Cartesian equation form*.

The types of roots of the equation can be determined from the *discriminant* (i.e., the quantity under the radical in Eq. 4.17).

- If $(b^2 - 4ac) > 0$, the roots are real and unequal.
- If $(b^2 - 4ac) = 0$, the roots are real and equal. This is known as a *double root.*
- If $(b^2 - 4ac) < 0$, the roots are complex and unequal.

10. ROOTS OF GENERAL POLYNOMIALS

It is more difficult to find roots of cubic and higher-degree polynomials because few general techniques exist.

- *inspection*: Finding roots by inspection is equivalent to making reasonable guesses about the roots and substituting into the polynomial.
- *graphing*: If the value of a polynomial $f(x)$ is calculated and plotted for different values of x, an approximate value of a root can be determined as the value of x at which the plot crosses the x-axis.
- *numerical methods*: If an approximate value of a root is known, numerical methods (bisection method, Newton's method, etc.) can be used to refine the value. The more efficient techniques are too complex to be performed by hand.
- *factoring*: If at least one root (say, $x = r$) of a polynomial $f(x)$ is known, the quantity $(x - r)$ can be factored out of $f(x)$ by long division. The resulting quotient will be lower by one degree, and the remaining roots may be easier to determine. This method is particularly applicable if the polynomial is in one of the standard forms presented in Sec. 4.8.
- *special cases*: Certain polynomial forms can be simplified by substitution or solved by standard formulas if they are recognized as being special cases. (The standard solution to the quadratic equation is such a special case.) For example, $ax^4 + bx^2 + c = 0$ can be reduced to a polynomial of degree 2 if the substitution $u = x^2$ is made.

11. EXTRANEOUS ROOTS

With simple equalities, it may appear possible to derive roots by basic algebraic manipulations.[5] However, multiplying each side of an equality by a power of a variable may introduce *extraneous roots.* Such roots do not satisfy the original equation even though they are derived according to the rules of algebra. Checking a calculated root is always a good idea, but is particularly necessary if the equation has been multiplied by one of its own variables.

[5]In this sentence, *equality* means a combination of two expressions containing an equal sign. Any two expressions can be linked in this manner, even those that are not actually equal. For example, the expressions for two non-intersecting ellipses can be equated even though there is no intersection point. Finding extraneous roots is more likely when the underlying equality is false to begin with.

Example 4.1

Use algebraic operations to determine a value that satisfies the following equation. Determine if the value is a valid or extraneous root.

$$\sqrt{x-2} = \sqrt{x} + 2$$

Solution

Square both sides.

$$x - 2 = x + 4\sqrt{x} + 4$$

Subtract x from each side, and combine the constants.

$$4\sqrt{x} = -6$$

Solve for x.

$$x = \left(\frac{-6}{4}\right)^2 = \frac{9}{4}$$

Substitute $x = 9/4$ into the original equation.

$$\sqrt{\frac{9}{4} - 2} = \sqrt{\frac{9}{4}} + 2$$

$$\frac{1}{2} = \frac{7}{2}$$

Since the equality is not established, $x = 9/4$ is an extraneous root.

12. DESCARTES' RULE OF SIGNS

Descartes' rule of signs determines the maximum number of positive (and negative) real roots that a polynomial will have by counting the number of sign reversals (i.e., changes in sign from one term to the next) in the polynomial. The polynomial $f(x) = 0$ must have real coefficients and must be arranged in terms of descending powers of x.

- The number of positive roots of the polynomial equation $f(x) = 0$ will not exceed the number of sign reversals.
- The difference between the number of sign reversals and the number of positive roots is an even number.
- The number of negative roots of the polynomial equation $f(x) = 0$ will not exceed the number of sign reversals in the polynomial $f(-x)$.
- The difference between the number of sign reversals in $f(-x)$ and the number of negative roots is an even number.

Example 4.2

Determine the possible numbers of positive and negative roots that satisfy the following polynomial equation.

$$4x^5 - 5x^4 + 3x^3 - 8x^2 - 2x + 3 = 0$$

Solution

There are four sign reversals, so up to four positive roots exist. To keep the difference between the number of positive roots and the number of sign reversals an even number, the number of positive real roots is limited to zero, two, and four.

Substituting $-x$ for x in the polynomial results in

$$-4x^5 - 5x^4 - 3x^3 - 8x^2 + 2x + 3 = 0$$

There is only one sign reversal, so the number of negative roots cannot exceed one. There must be exactly one negative real root in order to keep the difference to an even number (zero in this case).

13. RULES FOR EXPONENTS AND RADICALS

In the expression $b^n = a$, b is known as the *base* and n is the *exponent* or *power*. In Eq. 4.20 through Eq. 4.33, a, b, m, and n are any real numbers with limitations listed.

$$b^0 = 1 \quad [b \neq 0] \qquad 4.20$$

$$b^1 = b \qquad 4.21$$

$$b^{-n} = \frac{1}{b^n} = \left(\frac{1}{b}\right)^n \quad [b \neq 0] \qquad 4.22$$

$$\left(\frac{a}{b}\right)^n = \frac{a^n}{b^n} \quad [b \neq 0] \qquad 4.23$$

$$(ab)^n = a^n b^n \qquad 4.24$$

$$b^{m/n} = \sqrt[n]{b^m} = (\sqrt[n]{b})^m \qquad 4.25$$

$$(b^n)^m = b^{nm} \qquad 4.26$$

$$b^m b^n = b^{m+n} \qquad 4.27$$

$$\frac{b^m}{b^n} = b^{m-n} \quad [b \neq 0] \qquad 4.28$$

$$\sqrt[n]{b} = b^{1/n} \qquad 4.29$$

$$(\sqrt[n]{b})^n = (b^{1/n})^n = b \qquad 4.30$$

$$\sqrt[n]{ab} = \sqrt[n]{a}\sqrt[n]{b} = a^{1/n}b^{1/n} = (ab)^{1/n} \qquad 4.31$$

$$\sqrt[n]{\frac{a}{b}} = \frac{\sqrt[n]{a}}{\sqrt[n]{b}} = \left(\frac{a}{b}\right)^{1/n} \quad [b \neq 0] \qquad 4.32$$

$$\sqrt[m]{\sqrt[n]{b}} = \sqrt[mn]{b} = b^{1/mn} \qquad 4.33$$

14. LOGARITHMS

Logarithms can be considered to be exponents. For example, the exponent n in the expression $b^n = a$ is the logarithm of a to the base b. Therefore, the two expressions $\log_b a = n$ and $b^n = a$ are equivalent.

The base for *common logs* is 10. Usually, "log" will be written when common logs are desired, although "$\log_{10}$" appears occasionally. The base for *natural* (*Napierian*) *logs* is 2.71828..., a number which is given the symbol e. When natural logs are desired, usually "ln" will be written, although "$\log_e$" is also used.

Most logarithms will contain an integer part (the *characteristic*) and a decimal part (the *mantissa*). The common and natural logarithms of any number less than one are negative. If the number is greater than one, its common and natural logarithms are positive. Although the logarithm may be negative, the mantissa is always positive. For negative logarithms, the characteristic is found by expressing the logarithm as the sum of a negative characteristic and a positive mantissa.

For common logarithms of numbers greater than one, the characteristics will be positive and equal to one less than the number of digits in front of the decimal. If the number is less than one, the characteristic will be negative and equal to one more than the number of zeros immediately following the decimal point.

If a negative logarithm is to be used in a calculation, it must first be converted to *operational form* by adding the characteristic and mantissa. The operational form should be used in all calculations and is the form displayed by scientific calculators.

The logarithm of a negative number is a complex number.

Example 4.3

Use logarithm tables to determine the operational form of $\log_{10} 0.05$.

Solution

Since the number is less than one and there is one leading zero, the characteristic is found by observation to be -2. From a book of logarithm tables, the mantissa of 5.0 is 0.699. Two ways of combining the mantissa and characteristic are possible.

method 1: $\bar{2}.699$

method 2: $8.699 - 10$

The operational form of this logarithm is $-2 + 0.699 = -1.301$.

15. LOGARITHM IDENTITIES

Prior to the widespread availability of calculating devices, logarithm identities were used to solve complex calculations by reducing the solution method to table look-up, addition, and subtraction. Logarithm identities are still useful in simplifying expressions containing exponentials and other logarithms. In Eq. 4.34 through Eq. 4.45, $a \neq 1$, $b \neq 1$, $x > 0$, and $y > 0$.

$$\log_b b = 1 \qquad 4.34$$

$$\log_b 1 = 0 \qquad 4.35$$

$$\log_b b^n = n \qquad 4.36$$

$$\log x^a = a \log x \qquad 4.37$$

$$\log \sqrt[n]{x} = \log x^{1/n} = \frac{\log x}{n} \qquad 4.38$$

$$b^{n \log_b x} = x^n = \text{antilog}\left(n \log_b x\right) \qquad 4.39$$

$$b^{\log_b x/n} = x^{1/n} \qquad 4.40$$

$$\log xy = \log x + \log y \qquad 4.41$$

$$\log \frac{x}{y} = \log x - \log y \qquad 4.42$$

$$\log_a x = \log_b x \log_a b \qquad 4.43$$

$$\ln x = \log_{10} x \ln 10 \approx 2.3026 \log_{10} x \qquad 4.44$$

$$\log_{10} x = \ln x \log_{10} e \approx 0.4343 \ln x \qquad 4.45$$

Example 4.4

The surviving fraction, x, of a radioactive isotope is given by $x = e^{-0.005t}$. For what value of t will the surviving percentage be 7%?

Solution

$$x = 0.07 = e^{-0.005t}$$

Take the natural log of both sides.

$$\ln 0.07 = \ln e^{-0.005t}$$

From Eq. 4.36, $\ln e^x = x$. Therefore,

$$-2.66 = -0.005t$$

$$t = 532$$

16. PARTIAL FRACTIONS

The method of *partial fractions* is used to transform a proper polynomial fraction of two polynomials into a sum of simpler expressions, a procedure known as *resolution*.[6,7] The technique can be considered to be the act of "unadding" a sum to obtain all of the addends.

[6]To be a *proper polynomial fraction*, the degree of the numerator must be less than the degree of the denominator. If the polynomial fraction is improper, the denominator can be divided into the numerator to obtain whole and fractional polynomials. The method of partial fractions can then be used to reduce the fractional polynomial.

[7]This technique is particularly useful for calculating integrals and inverse Laplace transforms in subsequent chapters.

Suppose $H(x)$ is a proper polynomial fraction of the form $P(x)/Q(x)$. The object of the resolution is to determine the partial fractions u_1/v_1, u_2/v_2, and so on, such that

$$H(x) = \frac{P(x)}{Q(x)} = \frac{u_1}{\text{v}_1} + \frac{u_2}{\text{v}_2} + \frac{u_3}{\text{v}_3} + \cdots \qquad 4.46$$

The form of the denominator polynomial $Q(x)$ will be the main factor in determining the form of the partial fractions. The task of finding the u_i and v_i is simplified by categorizing the possible forms of $Q(x)$.

case 1: $Q(x)$ factors into n different linear terms.

$$Q(x) = (x - a_1)(x - a_2)\cdots(x - a_n) \qquad 4.47$$

Then,

$$H(x) = \sum_{i=1}^{n} \frac{A_i}{x - a_i} \qquad 4.48$$

case 2: $Q(x)$ factors into n identical linear terms.

$$Q(x) = (x - a)(x - a)\cdots(x - a) \qquad 4.49$$

Then,

$$H(x) = \sum_{i=1}^{n} \frac{A_i}{(x - a)^i} \qquad 4.50$$

case 3: $Q(x)$ factors into n different quadratic terms, $x^2 + p_i x + q_i$. Then,

$$H(x) = \sum_{i=1}^{n} \frac{A_i x + B_i}{x^2 + p_i x + q_i} \qquad 4.51$$

case 4: $Q(x)$ factors into n identical quadratic terms, $x^2 + px + q$. Then,

$$H(x) = \sum_{i=1}^{n} \frac{A_i x + B_i}{(x^2 + px + q)^i} \qquad 4.52$$

Once the general forms of the partial fractions have been determined from inspection, the *method of undetermined coefficients* is used. The partial fractions are all cross-multiplied to obtain $Q(x)$ as the denominator, and the coefficients are found by equating $P(x)$ and the cross-multiplied numerator.

Example 4.5

Resolve $H(x)$ into partial fractions.

$$H(x) = \frac{x^2 + 2x + 3}{x^4 + x^3 + 2x^2}$$

Solution

Here, $Q(x) = x^4 + x^3 + 2x^2$ factors into $x^2(x^2 + x + 2)$. This is a combination of cases 2 and 3.

$$H(x) = \frac{A_1}{x} + \frac{A_2}{x^2} + \frac{A_3 + A_4 x}{x^2 + x + 2}$$

Cross-multiplying to obtain a common denominator yields

$$\frac{(A_1 + A_4)x^3 + (A_1 + A_2 + A_3)x^2 + (2A_1 + A_2)x + 2A_2}{x^4 + x^3 + 2x^2}$$

Since the original numerator is known, the following simultaneous equations result.

$$\begin{aligned} A_1 + A_4 &= 0 \\ A_1 + A_2 + A_3 &= 1 \\ 2A_1 + A_2 &= 2 \\ 2A_2 &= 3 \end{aligned}$$

The solutions are $A_1 = 0.25$; $A_2 = 1.50$; $A_3 = -0.75$; and $A_4 = -0.25$.

$$H(x) = \frac{1}{4x} + \frac{3}{2x^2} - \frac{x + 3}{4(x^2 + x + 2)}$$

17. SIMULTANEOUS LINEAR EQUATIONS

A *linear equation* with n variables is a polynomial of degree 1 describing a geometric shape in n-space. A *homogeneous linear equation* is one that has no constant term, and a *nonhomogeneous linear equation* has a constant term.

A solution to a set of simultaneous linear equations represents the intersection point of the geometric shapes in n-space. For example, if the equations are limited to two variables (e.g., $y = 4x - 5$), they describe straight lines. The solution to two simultaneous linear equations in 2-space is the point where the two lines intersect. The set of the two equations is said to be a *consistent system* when there is such an intersection.[8]

Simultaneous equations do not always have unique solutions, and some have none at all. In addition to crossing in 2-space, lines can be parallel or they can be the same line expressed in a different equation format (i.e., dependent equations). In some cases, parallelism and dependency can be determined by inspection. In most cases, however, matrix and other advanced methods must be used to determine whether a solution exists. A set of linear equations with no simultaneous solution is known as an *inconsistent system.*

[8]A homogeneous system always has at least one solution: the *trivial solution*, in which all variables have a value of zero.

Several methods exist for solving linear equations simultaneously by hand.[9]

- *graphing:* The equations are plotted and the intersection point is read from the graph. This method is possible only with two-dimensional problems.
- *substitution:* An equation is rearranged so that one variable is expressed as a combination of the other variables. The expression is then substituted into the remaining equations wherever the selected variable appears.
- *reduction:* All terms in the equations are multiplied by constants chosen to eliminate one or more variables when the equations are added or subtracted. The remaining sum can then be solved for the other variables. This method is also known as *eliminating the unknowns.*
- *Cramer's rule:* This is a procedure in linear algebra that calculates determinants of the original coefficient matrix **A** and of the n matrices resulting from the systematic replacement of column **A** by the constant matrix **B**.

Example 4.6

Solve the following set of linear equations by (a) substitution and (b) reduction.

$$2x + 3y = 12 \quad \text{[Eq. 1]}$$

$$3x + 4y = 8 \quad \text{[Eq. 2]}$$

Solution

(a) From Eq. 1, solve for variable x.

$$x = 6 - 1.5y \quad \text{[Eq. 3]}$$

Substitute $6 - 1.5y$ into Eq. 2 wherever x appears.

$$\begin{aligned} (3)(6 - 1.5y) + 4y &= 8 \\ 18 - 4.5y + 4y &= 8 \\ y &= 20 \end{aligned}$$

Substitute 20 for y in Eq. 3.

$$x = 6 - (1.5)(20) = -24$$

The solution $(-24, 20)$ should be checked to verify that it satisfies both original equations.

(b) Eliminate variable x by multiplying Eq. 1 by 3 and Eq. 2 by 2.

$$3 \times \text{Eq. 1: } 6x + 9y = 36 \quad \text{[Eq. 1']}$$

$$2 \times \text{Eq. 2: } 6x + 8y = 16 \quad \text{[Eq. 2']}$$

[9]Other methods exist, but they require a computer.

Subtract Eq. 2′ from Eq. 1′.

$$y = 20 \quad [\text{Eq. } 1' - \text{Eq. } 2']$$

Substitute $y = 20$ into Eq. 1′.

$$6x + (9)(20) = 36$$
$$x = -24$$

The solution (−24, 20) should be checked to verify that it satisfies both original equations.

18. COMPLEX NUMBERS

A *complex number*, **Z**, is a combination of real and imaginary numbers. When expressed as a sum (e.g., $a + bi$), the complex number is said to be in *rectangular* or *trigonometric form.* The complex number can be plotted on the real-imaginary coordinate system known as the *complex plane*, as illustrated in Fig. 4.2.

Figure 4.2 *A Complex Number in the Complex Plane*

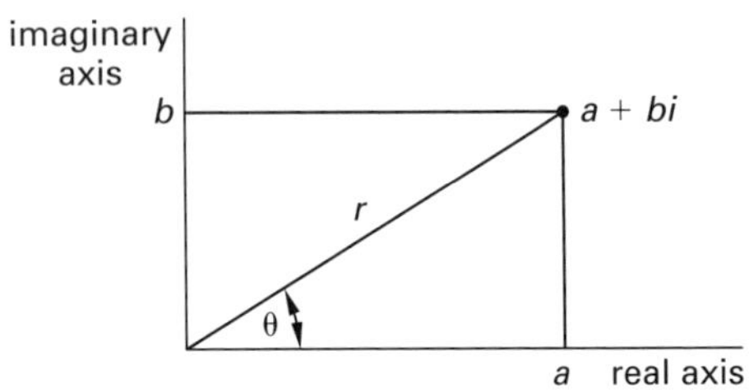

The complex number $\mathbf{Z} = a + bi$ can also be expressed in *exponential form.*[10] The quantity r is known as the *modulus* of **Z**; θ is the *argument.*

$$a + bi \equiv re^{i\theta} \qquad 4.53$$

$$r = \text{mod}\,\mathbf{Z} = \sqrt{a^2 + b^2} \qquad 4.54$$

$$\theta = \arg\mathbf{Z} = \arctan\frac{b}{a} \qquad 4.55$$

Similarly, the *phasor form* (also known as the *polar form*) is

$$\mathbf{Z} = r\angle\theta \qquad 4.56$$

The *rectangular form* can be determined from r and θ.

$$a = r\cos\theta \qquad 4.57$$

$$b = r\sin\theta \qquad 4.58$$

$$\mathbf{Z} = a + bi = r\cos\theta + ir\sin\theta$$
$$= r(\cos\theta + i\sin\theta) \qquad 4.59$$

[10]The terms *polar form*, *phasor form*, and *exponential form* are all used somewhat interchangeably.

The *cis form* is a shorthand method of writing a complex number in rectangular (trigonometric) form.

$$a + bi = r(\cos\theta + i\sin\theta) = r\,\text{cis}\,\theta \qquad 4.60$$

Euler's equation as shown in Eq. 4.61, expresses the equality of complex numbers in exponential and trigonometric form.

$$e^{i\theta} = \cos\theta + i\sin\theta \qquad 4.61$$

Related expressions are

$$e^{-i\theta} = \cos\theta - i\sin\theta \qquad 4.62$$

$$\cos\theta = \frac{e^{i\theta} + e^{-i\theta}}{2} \qquad 4.63$$

$$\sin\theta = \frac{e^{i\theta} - e^{-i\theta}}{2i} \qquad 4.64$$

Example 4.7

What is the exponential form of the complex number $\mathbf{Z} = 3 + 4i$?

Solution

$$r = \sqrt{a^2 + b^2} = \sqrt{3^2 + 4^2} = \sqrt{25} = 5$$

$$\theta = \arctan\frac{b}{a} = \arctan\frac{4}{3} = 0.927 \text{ rad}$$

$$\mathbf{Z} = re^{i\theta} = 5e^{i(0.927)}$$

19. OPERATIONS ON COMPLEX NUMBERS

Most algebraic operations (addition, multiplication, exponentiation, etc.) work with complex numbers, but notable exceptions are the inequality operators. The concept of one complex number being less than or greater than another complex number is meaningless.

When adding two complex numbers, real parts are added to real parts, and imaginary parts are added to imaginary parts.

$$(a_1 + ib_1) + (a_2 + ib_2) = (a_1 + a_2) + i(b_1 + b_2) \qquad 4.65$$

$$(a_1 + ib_1) - (a_2 + ib_2) = (a_1 - a_2) + i(b_1 - b_2) \qquad 4.66$$

Multiplication of two complex numbers in rectangular form is accomplished by the use of the algebraic distributive law, remembering that $i^2 = -1$.

Division of complex numbers in rectangular form requires use of the *complex conjugate.* The complex conjugate of the complex number $(a + bi)$ is $(a - bi)$. By multiplying the numerator and the denominator by the complex conjugate, the denominator will be converted to the real number $a^2 + b^2$. This technique is known as *rationalizing* the denominator and is illustrated in Ex. 4.8(c).

Multiplication and division are often more convenient when the complex numbers are in exponential or phasor forms, as Eq. 4.67 and Eq. 4.68 show.

$$(r_1e^{i\theta_1})(r_2e^{i\theta_2}) = r_1r_2e^{i(\theta_1+\theta_2)} \quad 4.67$$

$$\frac{r_1e^{i\theta_1}}{r_2e^{i\theta_2}} = \left(\frac{r_1}{r_2}\right)e^{i(\theta_1-\theta_2)} \quad 4.68$$

Taking powers and roots of complex numbers requires *de Moivre's theorem*, Eq. 4.69 and Eq. 4.70.

$$\mathbf{Z}^n = (re^{i\theta})^n = r^ne^{in\theta} \quad 4.69$$

$$\sqrt[n]{\mathbf{Z}} = (re^{i\theta})^{1/n} = \sqrt[n]{r}e^{i(\theta+k360°/n)} \quad [k = 0, 1, 2, \ldots, n-1] \quad 4.70$$

Example 4.8

Perform the following complex arithmetic.

(a) $(3+4i)+(2+i)$

(b) $(7+2i)(5-3i)$

(c) $\dfrac{2+3i}{4-5i}$

Solution

(a)

$$\begin{aligned}(3+4i)+(2+i) &= (3+2)+(4+1)i\\ &= 5+5i\end{aligned}$$

(b)

$$\begin{aligned}(7+2i)(5-3i) &= (7)(5)-(7)(3i)+(2i)(5)\\ &\quad -(2i)(3i)\\ &= 35-21i+10i-6i^2\\ &= 35-21i+10i-(6)(-1)\\ &= 41-11i\end{aligned}$$

(c) Multiply the numerator and denominator by the complex conjugate of the denominator.

$$\begin{aligned}\frac{2+3i}{4-5i} &= \frac{(2+3i)(4+5i)}{(4-5i)(4+5i)}\\ &= \frac{-7+22i}{(4)^2+(5)^2}\\ &= \frac{-7}{41}+i\frac{22}{41}\end{aligned}$$

20. LIMITS

A *limit* (*limiting value*) is the value a function approaches when an independent variable approaches a target value. For example, suppose the value of $y = x^2$ is desired as x approaches 5. This could be written as

$$\lim_{x\to 5} x^2 \quad 4.71$$

The power of limit theory is wasted on simple calculations such as this but is appreciated when the function is undefined at the target value. The object of limit theory is to determine the limit without having to evaluate the function at the target. The general case of a limit evaluated as x approaches the target value a is written as

$$\lim_{x\to a} f(x) \quad 4.72$$

It is not necessary for the actual value $f(a)$ to exist for the limit to be calculated. The function $f(x)$ may be undefined at point a. However, it is necessary that $f(x)$ be defined on both sides of point a for the limit to exist. If $f(x)$ is undefined on one side, or if $f(x)$ is discontinuous at $x = a$ (as in Fig. 4.3(c) and Fig. 4.3(d)), the limit does not exist.

Figure 4.3 *Existence of Limits*

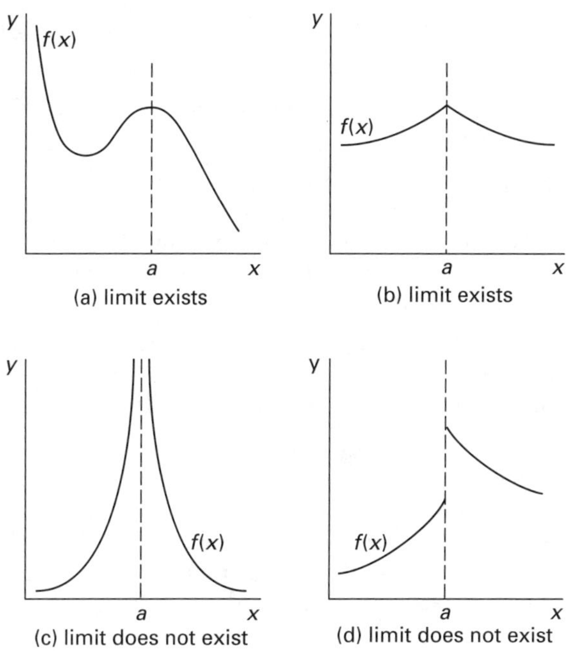

The following theorems can be used to simplify expressions when calculating limits.

$$\lim_{x\to a} x = a \quad 4.73$$

$$\lim_{x\to a}(mx+b) = ma+b \quad 4.74$$

$$\lim_{x\to a} b = b \quad 4.75$$

$$\lim_{x\to a}(kF(x)) = k\lim_{x\to a} F(x) \quad 4.76$$

$$\lim_{x\to a}\left(F_1(x)\begin{Bmatrix}+\\-\\\times\\\div\end{Bmatrix}F_2(x)\right) = \lim_{x\to a}(F_1(x))\begin{Bmatrix}+\\-\\\times\\\div\end{Bmatrix}\lim_{x\to a}(F_2(x)) \quad 4.77$$

The following identities can be used to simplify limits of trigonometric expressions.

$$\lim_{x\to 0}\sin x = 0 \quad 4.78$$

$$\lim_{x\to 0}\left(\frac{\sin x}{x}\right) = 1 \quad 4.79$$

$$\lim_{x\to 0}\cos x = 1 \quad 4.80$$

The following standard methods (tricks) can be used to determine limits.

- If the limit is taken to infinity, all terms can be divided by the largest power of x in the expression. This will leave at least one constant. Any quantity divided by a power of x vanishes as x approaches infinity.
- If the expression is a quotient of two expressions, any common factors should be eliminated from the numerator and denominator.
- *L'Hôpital's rule*, Eq. 4.81, should be used when the numerator and denominator of the expression both approach zero or both approach infinity.[11] $P^k(x)$ and $Q^k(x)$ are the kth derivatives of the functions $P(x)$ and $Q(x)$, respectively. (L'Hôpital's rule can be applied repeatedly as required.)

$$\lim_{x\to a}\left(\frac{P(x)}{Q(x)}\right) = \lim_{x\to a}\left(\frac{P^k(x)}{Q^k(x)}\right) \quad 4.81$$

Example 4.9

Evaluate the following limits.

(a) $\lim_{x\to 3}\left(\frac{x^3-27}{x^2-9}\right)$

(b) $\lim_{x\to\infty}\left(\frac{3x-2}{4x+3}\right)$

(c) $\lim_{x\to 2}\left(\frac{x^2+x-6}{x^2-3x+2}\right)$

[11] L'Hôpital's rule should not be used when only the denominator approaches zero. In that case, the limit approaches infinity regardless of the numerator.

Solution

(a) Factor the numerator and denominator. (L'Hôpital's rule can also be used.)

$$\lim_{x\to 3}\left(\frac{x^3-27}{x^2-9}\right) = \lim_{x\to 3}\left(\frac{(x-3)(x^2+3x+9)}{(x-3)(x+3)}\right)$$
$$= \lim_{x\to 3}\left(\frac{x^2+3x+9}{x+3}\right) = \frac{(3)^2+(3)(3)+9}{3+3}$$
$$= 9/2$$

(b) Divide through by the largest power of x. (L'Hôpital's rule can also be used.)

$$\lim_{x\to\infty}\left(\frac{3x-2}{4x+3}\right) = \lim_{x\to\infty}\left(\frac{3-\frac{2}{x}}{4+\frac{3}{x}}\right)$$
$$= \frac{3-\frac{2}{\infty}}{4+\frac{3}{\infty}}$$
$$= \frac{3-0}{4+0} = \frac{3}{4}$$

(c) Use L'Hôpital's rule. (Factoring can also be used.) Take the first derivative of the numerator and denominator.

$$\lim_{x\to 2}\left(\frac{x^2+x-6}{x^2-3x+2}\right) = \lim_{x\to 2}\left(\frac{2x+1}{2x-3}\right)$$
$$= \frac{(2)(2)+1}{(2)(2)-3}$$
$$= \frac{5}{1} = 5$$

21. SEQUENCES AND PROGRESSIONS

A *sequence* is an ordered *progression* of numbers, a_i, such as 1, 4, 9, 16, 25, ... The *terms* in a sequence can be all positive, all negative, or of alternating signs. a_n is known as the *general term* of the sequence.

$$\{A\} = \{a_1, a_2, a_3, \ldots, a_n\} \quad 4.82$$

A sequence is said to *diverge* (i.e., be *divergent*) if the terms approach infinity or if the terms fail to approach any finite value, and it is said to *converge* (i.e., be *convergent*) if the terms approach any finite value (including zero). That is, the sequence converges if the limit defined by Eq. 4.83 exists.

$$\lim_{n\to\infty} a_n \begin{cases} \text{converges if } L \text{ is finite} \\ \text{diverges if } L \text{ is infinite} \\ \quad \text{or does not exist} \end{cases} \qquad 4.83$$

The main task associated with a sequence is determining the next (or the general) term. If several terms of a sequence are known, the next (unknown) term must usually be found by intuitively determining the pattern of the sequence. In some cases, though, the method of *Rth-order differences* can be used to determine the next term. This method consists of subtracting each term from the following term to obtain a set of differences. If the differences are not all equal, the next order of differences can be calculated.

Example 4.10

What is the general term of the sequence A?

$$\{A\} = \left\{3, \frac{9}{2}, \frac{27}{6}, \frac{81}{24}, \ldots\right\}$$

Solution

The solution is purely intuitive. The numerator is recognized as a power series based on the number 3. The denominator is recognized as the factorial sequence. The general term is

$$a_n = \frac{3^n}{n!}$$

Example 4.11

Find the sixth term in the sequence $\{7, 16, 29, 46, 67, a_6\}$.

Solution

The sixth term is not intuitively obvious, so the method of Rth-order differences is tried. The pattern is not obvious from the first order differences, but the second order differences are all 4.

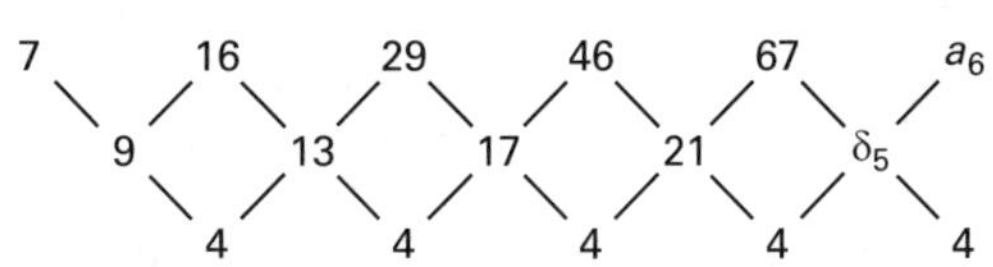

$$\delta_5 - 21 = 4$$

$$\delta_5 = 25$$

$$a_6 - 67 = \delta_5 = 25$$

$$a_6 = 92$$

Example 4.12

Does the sequence with general term e^n/n converge or diverge?

Solution

See if the limit exists.

$$\lim_{n\to\infty}\left(\frac{e^n}{n}\right) = \frac{\infty}{\infty}$$

Since ∞/∞ is inconclusive, apply L'Hôpital's rule. Take the derivative of both the numerator and the denominator with respect to n.

$$\lim_{n\to\infty}\left(\frac{e^n}{1}\right) = \frac{\infty}{1}$$

$$= \infty$$

The sequence diverges.

22. STANDARD SEQUENCES

There are four standard sequences: the geometric, arithmetic, harmonic, and p-sequence.

- *geometric sequence:* The geometric sequence converges for $-1 < r \le 1$ and diverges otherwise. a is known as the *first term*; r is known as the *common ratio.*

$$a_n = ar^{n-1} \quad \left[\begin{array}{c} a \text{ is a constant} \\ n = 1, 2, 3, \ldots, \infty \end{array}\right] \qquad 4.84$$

example: $\{1, 2, 4, 8, 16, 32\} \quad (a = 1, \; r = 2)$

- *arithmetic sequence:* The arithmetic sequence always diverges.

$$a_n = a + (n-1)d \quad \left[\begin{array}{c} a \text{ and } d \text{ are constants} \\ n = 1, 2, 3, \ldots, \infty \end{array}\right] \qquad 4.85$$

example: $\{2, 7, 12, 17, 22, 27\} \quad (a = 2, \; d = 5)$

- *harmonic sequence:* The harmonic sequence always converges.

$$a_n = \frac{1}{n} \quad [n = 1, 2, 3, \ldots, \infty] \qquad 4.86$$

example: $\{1, 1/2, 1/3, 1/4, 1/5, 1/6\}$

- *p-sequence:* The p-sequence converges if $p \ge 0$ and diverges if $p < 0$. (This is different than the p-series.)

$$a_n = \frac{1}{n^p} \quad [n = 1, 2, 3, \ldots, \infty] \qquad 4.87$$

example: $\{1, 1/4, 1/9, 1/16, 1/25, 1/36\} \quad (p = 2)$

23. SERIES

A *series* is the sum of terms in a sequence. There are two types of series. A *finite series* has a finite number of terms, and an *infinite series* has an infinite number of terms.[12] The main tasks associated with series are determining the sum of the terms and whether the series converges. A series is said to *converge* (be *convergent*) if the sum, S_n, of its term exists.[13] A finite series is always convergent.

The performance of a series based on standard sequences (defined in Sec. 4.22) is well known.

- *geometric series:*

$$S_n = \sum_{i=1}^{n} ar^{i-1} = \frac{a(1-r^n)}{1-r} \quad \text{[finite]} \tag{4.88}$$

$$S_n = \sum_{i=1}^{\infty} ar^{i-1} = \frac{a}{1-r} \quad \begin{bmatrix} \text{infinite} \\ -1 < r < 1 \end{bmatrix} \tag{4.89}$$

- *arithmetic series:* The infinite series diverges unless $a = d = 0$.

$$S_n = \sum_{i=1}^{n} (a + (i-1)d) = \frac{n(2a + (n-1)d)}{2} \quad \text{[finite]} \tag{4.90}$$

- *harmonic series:* The infinite series diverges.
- *p-series:* The infinite series diverges if $p \leq 1$. The infinite series converges if $p > 1$. (Notice that this is different than the *p*-sequence.)

24. TESTS FOR SERIES CONVERGENCE

It is obvious that all *finite series* (i.e., series having a finite number of terms) converge. That is, the sum, S_n, defined by Eq. 4.91 exists.

$$S_n = \sum_{i=1}^{n} a_i \tag{4.91}$$

Convergence of an infinite series can be determined by taking the limit of the sum. If the limit exists, the series converges; otherwise, it diverges.

$$\lim_{n\to\infty} S_n = \lim_{n\to\infty} \sum_{i=1}^{n} a_i \tag{4.92}$$

In most cases, the expression for the general term a_n will be known, but there will be no simple expression for the sum S_n. Therefore, Eq. 4.92 cannot be used to determine convergence. It is helpful, but not conclusive, to look at the limit of the general term. If L, as defined in Eq. 4.93, is nonzero, the series diverges. If L equals zero, the series may either converge or diverge. Additional testing is needed in that case.

$$\lim_{n\to\infty} a_n \begin{cases} = 0 & \text{inconclusive} \\ \neq 0 & \text{diverges} \end{cases} \tag{4.93}$$

Two tests can be used independently or after Eq. 4.93 has proven inconclusive: the ratio and comparison tests. The *ratio test* calculates the limit of the ratio of two consecutive terms.

$$\lim_{n\to\infty} \frac{a_{n+1}}{a_n} \begin{cases} < 1 & \text{converges} \\ = 1 & \text{inconclusive} \\ > 1 & \text{diverges} \end{cases} \tag{4.94}$$

The *comparison test* is an indirect method of determining convergence of an unknown series. It compares a standard series (geometric and *p*-series are commonly used) against the unknown series. If all terms in a positive standard series are smaller than the terms in the unknown series and the standard series diverges, the unknown series must also diverge. Similarly, if all terms in the standard series are larger than the terms in the unknown series and the standard series converges, then the unknown series also converges.

In mathematical terms, if A and B are both series of positive terms such that $a_n < b_n$ for all values of n, then (a) B diverges if A diverges, and (b) A converges if B converges.

Example 4.13

Does the infinite series A converge or diverge?

$$A = 3 + \frac{9}{2} + \frac{27}{6} + \frac{81}{24} + \cdots$$

Solution

The general term was found in Ex. 4.10 to be

$$a_n = \frac{3^n}{n!}$$

Since limits of factorials are not easily determined, use the ratio test.

$$\lim_{n\to\infty}\left(\frac{a_{n+1}}{a_n}\right) = \lim_{n\to\infty}\left(\frac{\frac{3^{n+1}}{(n+1)!}}{\frac{3^n}{n!}}\right) = \lim_{n\to\infty}\left(\frac{3}{n+1}\right)$$

$$= \frac{3}{\infty} = 0$$

Since the limit is less than 1, the infinite series converges.

[12]The term *infinite series* does not imply the sum is infinite.
[13]This is different from the definition of convergence for a sequence where only the last term was evaluated.

Example 4.14

Does the infinite series A converge or diverge?

$$A = 2 + \frac{3}{4} + \frac{4}{9} + \frac{5}{16} + \cdots$$

Solution

By observation, the general term is

$$a_n = \frac{1+n}{n^2}$$

The general term can be expanded by partial fractions to

$$a_n = \frac{1}{n} + \frac{1}{n^2}$$

However, $1/n$ is the harmonic series. Since the harmonic series is divergent and this series is larger than the harmonic series (by the term $1/n^2$), this series also diverges.

25. SERIES OF ALTERNATING SIGNS[14]

Some series contain both positive and negative terms. The ratio and comparison tests can both be used to determine if a series with alternating signs converges. If a series containing all positive terms converges, then the same series with some negative terms also converges. Therefore, the all-positive series should be tested for convergence. If the all-positive series converges, the original series is said to be *absolutely convergent.* (If the all-positive series diverges and the original series converges, the original series is said to be *conditionally convergent.*)

Alternatively, the ratio test can be used with the absolute value of the ratio. The same criteria apply.

$$\lim_{n\to\infty}\left|\frac{a_{n+1}}{a_n}\right| \begin{cases} <1 & \text{converges} \\ =1 & \text{inconclusive} \\ >1 & \text{diverges} \end{cases} \qquad 4.95$$

[14]This terminology is commonly used even though it is not necessary that the signs strictly alternate.

5 Linear Algebra

1. Matrices 5-1
2. Special Types of Matrices 5-1
3. Row Equivalent Matrices 5-2
4. Minors and Cofactors 5-2
5. Determinants 5-3
6. Matrix Algebra 5-4
7. Matrix Addition and Subtraction 5-4
8. Matrix Multiplication 5-4
9. Transpose 5-5
10. Singularity and Rank 5-5
11. Classical Adjoint 5-5
12. Inverse 5-6
13. Writing Simultaneous Linear Equations in Matrix Form 5-6
14. Solving Simultaneous Linear Equations 5-7
15. Eigenvalues and Eigenvectors 5-8

1. MATRICES

A *matrix* is an ordered set of *entries* (*elements*) arranged rectangularly and set off by brackets.[1] The entries can be variables or numbers. A matrix by itself has no particular value—it is merely a convenient method of representing a set of numbers.

The size of a matrix is given by the number of rows and columns, and the nomenclature $m \times n$ is used for a matrix with m rows and n columns. For a *square matrix*, the number of rows and columns will be the same, a quantity known as the *order* of the matrix.

Bold uppercase letters are used to represent matrices, while lowercase letters represent the entries. For example, a_{23} would be the entry in the second row and third column of matrix **A**.

$$\mathbf{A} = \begin{bmatrix} a_{11} & a_{12} & a_{13} \\ a_{21} & a_{22} & a_{23} \\ a_{31} & a_{32} & a_{33} \end{bmatrix}$$

A *submatrix* is the matrix that remains when selected rows or columns are removed from the original matrix.[2] For example, for matrix **A**, the submatrix remaining after the second row and second column have been removed is

$$\begin{bmatrix} a_{11} & a_{13} \\ a_{31} & a_{33} \end{bmatrix}$$

An *augmented matrix* results when the original matrix is extended by repeating one or more of its rows or columns or by adding rows and columns from another matrix. For example, for the matrix **A**, the augmented matrix created by repeating the first and second columns is

$$\left[\begin{array}{ccc|cc} a_{11} & a_{12} & a_{13} & a_{11} & a_{12} \\ a_{21} & a_{22} & a_{23} & a_{21} & a_{22} \\ a_{31} & a_{32} & a_{33} & a_{31} & a_{32} \end{array}\right]$$

2. SPECIAL TYPES OF MATRICES

Certain types of matrices are given special designations.

- *cofactor matrix:* the matrix formed when every entry is replaced by the cofactor (see Sec. 5.4) of that entry
- *column matrix:* a matrix with only one column
- *complex matrix:* a matrix with complex number entries
- *diagonal matrix:* a square matrix with all zero entries except for the a_{ij} for which $i = j$
- *echelon matrix:* a matrix in which the number of zeros preceding the first nonzero entry of a row increases row by row until only zero rows remain. A *row-reduced echelon matrix* is an echelon matrix in which the first nonzero entry in each row is a 1 and all other entries in the columns are zero.
- *identity matrix:* a diagonal (square) matrix with all nonzero entries equal to 1, usually designated as **I**, having the property that $\mathbf{AI} = \mathbf{IA} = \mathbf{A}$
- *null matrix:* the same as a zero matrix
- *row matrix:* a matrix with only one row
- *scalar matrix:*[3] a diagonal (square) matrix with all diagonal entries equal to some scalar k

[1]The term *array* is synonymous with *matrix*, although the former is more likely to be used in computer applications.
[2]By definition, a matrix is a submatrix of itself.

[3]Although the term *complex matrix* means a matrix with complex entries, the term *scalar matrix* means more than a matrix with scalar entries.

- *singular matrix:* a matrix whose determinant is zero (see Sec. 5.10)
- *skew symmetric matrix:* a square matrix whose transpose (see Sec. 5.9) is equal to the negative of itself (i.e., $\mathbf{A} = -\mathbf{A}^t$)
- *square matrix:* a matrix with the same number of rows and columns (i.e., $m = n$)
- *symmetric(al) matrix:* a square matrix whose transpose is equal to itself (i.e., $\mathbf{A}^t = \mathbf{A}$), which occurs only when $a_{ij} = a_{ji}$
- *triangular matrix:* a square matrix with zeros in all positions above or below the diagonal
- *unit matrix:* the same as the identity matrix
- *zero matrix:* a matrix with all zero entries

Figure 5.1 shows examples of special matrices.

Figure 5.1 *Examples of Special Matrices*

$$\begin{bmatrix} 9 & 0 & 0 & 0 \\ 0 & -6 & 0 & 0 \\ 0 & 0 & 1 & 0 \\ 0 & 0 & 0 & 5 \end{bmatrix}$$
(a) diagonal

$$\begin{bmatrix} 2 & 18 & 2 & 18 \\ 0 & 0 & 1 & 9 \\ 0 & 0 & 0 & 9 \\ 0 & 0 & 0 & 0 \end{bmatrix}$$
(b) echelon

$$\begin{bmatrix} 1 & 9 & 0 & 0 \\ 0 & 0 & 1 & 0 \\ 0 & 0 & 0 & 1 \\ 0 & 0 & 0 & 0 \end{bmatrix}$$
(c) row-reduced echelon

$$\begin{bmatrix} 1 & 0 & 0 & 0 \\ 0 & 1 & 0 & 0 \\ 0 & 0 & 1 & 0 \\ 0 & 0 & 0 & 1 \end{bmatrix}$$
(d) identity

$$\begin{bmatrix} 3 & 0 & 0 & 0 \\ 0 & 3 & 0 & 0 \\ 0 & 0 & 3 & 0 \\ 0 & 0 & 0 & 3 \end{bmatrix}$$
(e) scalar

$$\begin{bmatrix} 2 & 0 & 0 & 0 \\ 7 & 6 & 0 & 0 \\ 9 & 1 & 1 & 0 \\ 8 & 0 & 4 & 5 \end{bmatrix}$$
(f) triangular

3. ROW EQUIVALENT MATRICES

A matrix **B** is said to be *row equivalent* to a matrix **A** if it is obtained by a finite sequence of *elementary row operations* on **A**:

- interchanging the ith and jth rows
- multiplying the ith row by a nonzero scalar
- replacing the ith row by the sum of the original ith row and k times the jth row

However, two matrices that are row equivalent as defined do not necessarily have the same determinants. (See Sec. 5.5.)

Gauss-Jordan elimination is the process of using these elementary row operations to row-reduce a matrix to echelon or row-reduced echelon forms, as illustrated in Ex. 5.8. When a matrix has been converted to a row-reduced echelon matrix, it is said to be in *row canonical form.* The phrases *row-reduced echelon form* and *row canonical form* are synonymous.

4. MINORS AND COFACTORS

Minors and cofactors are determinants of submatrices associated with particular entries in the original square matrix. The *minor* of entry a_{ij} is the determinant of a submatrix resulting from the elimination of the single row i and the single column j. For example, the minor corresponding to entry a_{12} in a 3 × 3 matrix **A** is the determinant of the matrix created by eliminating row 1 and column 2.

$$\text{minor of } a_{12} = \begin{vmatrix} a_{21} & a_{23} \\ a_{31} & a_{33} \end{vmatrix} \qquad 5.1$$

The *cofactor* of entry a_{ij} is the minor of a_{ij} multiplied by either +1 or −1, depending on the position of the entry. (That is, the cofactor either exactly equals the minor or it differs only in sign.) The sign is determined according to the following positional matrix.[4]

$$\begin{bmatrix} +1 & -1 & +1 & \cdots \\ -1 & +1 & -1 & \cdots \\ +1 & -1 & +1 & \cdots \\ \vdots & \vdots & \vdots & \end{bmatrix}$$

For example, the cofactor of entry a_{12} in matrix **A** (described in Sec. 5.4) is

$$\text{cofactor of } a_{12} = -\begin{vmatrix} a_{21} & a_{23} \\ a_{31} & a_{33} \end{vmatrix} \qquad 5.2$$

Example 5.1

What is the cofactor corresponding to the −3 entry in the following matrix?

$$\mathbf{A} = \begin{bmatrix} 2 & 9 & 1 \\ -3 & 4 & 0 \\ 7 & 5 & 9 \end{bmatrix}$$

Solution

The minor's submatrix is created by eliminating the row and column of the −3 entry.

$$\mathbf{M} = \begin{bmatrix} 9 & 1 \\ 5 & 9 \end{bmatrix}$$

The minor is the determinant of **M**.

$$|\mathbf{M}| = (9)(9) - (5)(1) = 76$$

The sign corresponding to the −3 position is negative. Therefore, the cofactor is −76.

[4]The sign of the cofactor a_{ij} is positive if $(i + j)$ is even and is negative if $(i + j)$ is odd.

5. DETERMINANTS

A *determinant* is a scalar calculated from a square matrix. The determinant of matrix $\mathbf{A}$ can be represented as $D\{\mathbf{A}\}$, $\text{Det}(\mathbf{A})$, $\Delta\mathbf{A}$, or $|\mathbf{A}|$.[5] The following rules can be used to simplify the calculation of determinants.

- If $\mathbf{A}$ has a row or column of zeros, the determinant is zero.
- If $\mathbf{A}$ has two identical rows or columns, the determinant is zero.
- If $\mathbf{B}$ is obtained from $\mathbf{A}$ by adding a multiple of a row (column) to another row (column) in $\mathbf{A}$, then $|\mathbf{B}| = |\mathbf{A}|$.
- If $\mathbf{A}$ is triangular, the determinant is equal to the product of the diagonal entries.
- If $\mathbf{B}$ is obtained from $\mathbf{A}$ by multiplying one row or column in $\mathbf{A}$ by a scalar k, then $|\mathbf{B}| = k|\mathbf{A}|$.
- If $\mathbf{B}$ is obtained from the $n \times n$ matrix $\mathbf{A}$ by multiplying by the scalar matrix k, then $|\mathbf{kA}| = k^n|\mathbf{A}|$.
- If $\mathbf{B}$ is obtained from $\mathbf{A}$ by switching two rows or columns in $\mathbf{A}$, then $|\mathbf{B}| = -|\mathbf{A}|$.

Calculation of determinants is laborious for all but the smallest or simplest of matrices. For a 2×2 matrix, the formula used to calculate the determinant is easy to remember.

$$\mathbf{A} = \begin{bmatrix} a & b \\ c & d \end{bmatrix}$$

$$|\mathbf{A}| = \begin{vmatrix} a & b \\ c & d \end{vmatrix} = ad - bc \qquad 5.3$$

Two methods are commonly used for calculating the determinant of 3×3 matrices by hand. The first uses an augmented matrix constructed from the original matrix and the first two columns (as shown in Sec. 5.1).[6] The determinant is calculated as the sum of the products in the left-to-right downward diagonals less the sum of the products in the left-to-right upward diagonals.

$$\mathbf{A} = \begin{bmatrix} a & b & c \\ d & e & f \\ g & h & i \end{bmatrix}$$

$$\text{augmented } \mathbf{A} = \left[\begin{array}{ccc|cc} a & b & c & a & b \\ d & e & f & d & e \\ g & h & i & g & h \end{array}\right] \quad (+\ +\ +\ -\ -\ -) \qquad 5.4$$

$$|\mathbf{A}| = aei + bfg + cdh - gec - hfa - idb \qquad 5.5$$

The second method of calculating the determinant is somewhat slower than the first for a 3×3 matrix but illustrates the method that must be used to calculate determinants of 4×4 and larger matrices. This method is known as *expansion by cofactors.* One row (column) is selected as the base row (column). The selection is arbitrary, but the number of calculations required to obtain the determinant can be minimized by choosing the row (column) with the most zeros. The determinant is equal to the sum of the products of the entries in the base row (column) and their corresponding cofactors.

$$\mathbf{A} = \begin{bmatrix} a & b & c \\ d & e & f \\ g & h & i \end{bmatrix}$$

$$|\mathbf{A}| = a\begin{vmatrix} e & f \\ h & i \end{vmatrix} - d\begin{vmatrix} b & c \\ h & i \end{vmatrix} + g\begin{vmatrix} b & c \\ e & f \end{vmatrix} \qquad 5.6$$

Example 5.2

Calculate the determinant of matrix $\mathbf{A}$ (a) by cofactor expansion, and (b) by the augmented matrix method.

$$\mathbf{A} = \begin{bmatrix} 2 & 3 & -4 \\ 3 & -1 & -2 \\ 4 & -7 & -6 \end{bmatrix}$$

Solution

(a) Since there are no zero entries, it does not matter which row or column is chosen as the base. Choose the first column as the base.

$$\begin{aligned} |\mathbf{A}| &= 2\begin{vmatrix} -1 & -2 \\ -7 & -6 \end{vmatrix} - 3\begin{vmatrix} 3 & -4 \\ -7 & -6 \end{vmatrix} + 4\begin{vmatrix} 3 & -4 \\ -1 & -2 \end{vmatrix} \\ &= (2)(6 - 14) - (3)(-18 - 28) + (4)(-6 - 4) \\ &= 82 \end{aligned}$$

[5]The vertical bars should not be confused with the square brackets used to set off a matrix, nor with absolute value.

[6]It is not actually necessary to construct the augmented matrix, but doing so helps avoid errors.

(b)

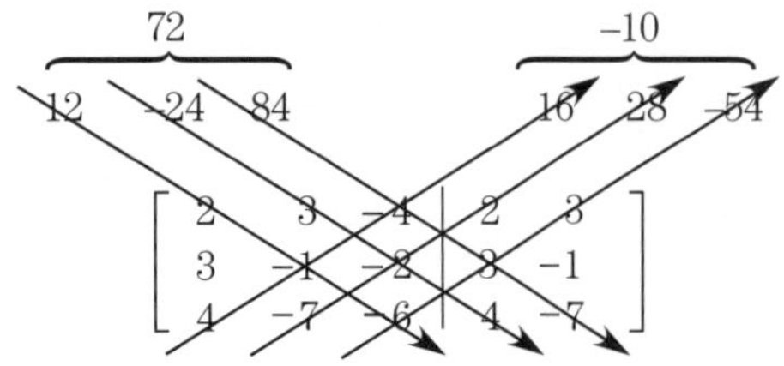

$|\mathbf{A}| = 72 - (-10) = 82$

6. MATRIX ALGEBRA[7]

Matrix algebra differs somewhat from standard algebra.

- *equality:* Two matrices, **A** and **B**, are equal only if they have the same numbers of rows and columns *and* if all corresponding entries are equal.
- *inequality:* The > and < operators are not used in matrix algebra.
- *commutative law of addition:*

$$\mathbf{A} + \mathbf{B} = \mathbf{B} + \mathbf{A} \qquad 5.7$$

- *associative law of addition:*

$$\mathbf{A} + (\mathbf{B} + \mathbf{C}) = (\mathbf{A} + \mathbf{B}) + \mathbf{C} \qquad 5.8$$

- *associative law of multiplication:*

$$(\mathbf{AB})\mathbf{C} = \mathbf{A}(\mathbf{BC}) \qquad 5.9$$

- *left distributive law:*

$$\mathbf{A}(\mathbf{B} + \mathbf{C}) = \mathbf{AB} + \mathbf{AC} \qquad 5.10$$

- *right distributive law:*

$$(\mathbf{B} + \mathbf{C})\mathbf{A} = \mathbf{BA} + \mathbf{CA} \qquad 5.11$$

- *scalar multiplication:*

$$k(\mathbf{AB}) = (k\mathbf{A})\mathbf{B} = \mathbf{A}(k\mathbf{B}) \qquad 5.12$$

It is important to recognize that, except for trivial and special cases, matrix multiplication is not commutative. That is,

$$\mathbf{AB} \neq \mathbf{BA}$$

7. MATRIX ADDITION AND SUBTRACTION

Addition and subtraction of two matrices is possible only if both matrices have the same numbers of rows and columns (i.e., order). They are accomplished by adding or subtracting the corresponding entries of the two matrices.

[7]Since matrices are used to simplify the presentation and solution of sets of linear equations, matrix algebra is also known as *linear algebra.*

8. MATRIX MULTIPLICATION

A matrix can be multiplied by a scalar, an operation known as *scalar multiplication,* in which case all entries of the matrix are multiplied by that scalar. For example, for the 2 × 2 matrix **A**,

$$k\mathbf{A} = \begin{bmatrix} ka_{11} & ka_{12} \\ ka_{21} & ka_{22} \end{bmatrix}$$

A matrix can be multiplied by another matrix, but only if the left-hand matrix has the same number of columns as the right-hand matrix has rows. *Matrix multiplication* occurs by multiplying the elements in each left-hand matrix row by the entries in each right-hand matrix column, adding the products, and placing the sum at the intersection point of the participating row and column.

Matrix division can only be accomplished by multiplying by the inverse of the denominator matrix. There is no specific division operation in matrix algebra.

Example 5.3

Determine the product matrix **C**.

$$\mathbf{C} = \begin{bmatrix} 1 & 4 & 3 \\ 5 & 2 & 6 \end{bmatrix} \begin{bmatrix} 7 & 12 \\ 11 & 8 \\ 9 & 10 \end{bmatrix}$$

Solution

The left-hand matrix has three columns, and the right-hand matrix has three rows. Therefore, the two matrices can be multiplied.

The first row of the left-hand matrix and the first column of the right-hand matrix are worked with first. The corresponding entries are multiplied, and the products are summed.

$$c_{11} = (1)(7) + (4)(11) + (3)(9) = 78$$

The intersection of the top row and left column is the entry in the upper left-hand corner of the matrix **C**.

The remaining entries are calculated similarly.

$$c_{12} = (1)(12) + (4)(8) + (3)(10) = 74$$
$$c_{21} = (5)(7) + (2)(11) + (6)(9) = 111$$
$$c_{22} = (5)(12) + (2)(8) + (6)(10) = 136$$

The product matrix is

$$\mathbf{C} = \begin{bmatrix} 78 & 74 \\ 111 & 136 \end{bmatrix}$$

Mathematics

9. TRANSPOSE

The *transpose*, $\mathbf{A}^t$, of an $m \times n$ matrix $\mathbf{A}$ is an $n \times m$ matrix constructed by taking the ith row and making it the ith column. The diagonal is unchanged. For example,

$$\mathbf{A} = \begin{bmatrix} 1 & 6 & 9 \\ 2 & 3 & 4 \\ 7 & 1 & 5 \end{bmatrix}$$

$$\mathbf{A}^t = \begin{bmatrix} 1 & 2 & 7 \\ 6 & 3 & 1 \\ 9 & 4 & 5 \end{bmatrix}$$

Transpose operations have the following characteristics.

$$(\mathbf{A}^t)^t = \mathbf{A} \qquad 5.13$$

$$(k\mathbf{A})^t = k(\mathbf{A}^t) \qquad 5.14$$

$$\mathbf{I}^t = \mathbf{I} \qquad 5.15$$

$$(\mathbf{AB})^t = \mathbf{B}^t\mathbf{A}^t \qquad 5.16$$

$$(\mathbf{A}+\mathbf{B})^t = \mathbf{A}^t + \mathbf{B}^t \qquad 5.17$$

$$|\mathbf{A}^t| = |\mathbf{A}| \qquad 5.18$$

10. SINGULARITY AND RANK

A *singular matrix* is one whose determinant is zero. Similarly, a *nonsingular* matrix is one whose determinant is nonzero.

The *rank* of a matrix is the maximum number of linearly independent row or column vectors.[8] A matrix has rank r if it has at least one nonsingular square submatrix of order r but has no nonsingular square submatrix of order more than r. While the submatrix must be square (in order to calculate the determinant), the original matrix need not be.

The rank of an $m \times n$ matrix will be, at most, the smaller of m and n. The rank of a null matrix is zero. The ranks of a matrix and its transpose are the same. If a matrix is in echelon form, the rank will be equal to the number of rows containing at least one nonzero entry. For a 3×3 matrix, the rank can either be 3 (if it is nonsingular), 2 (if any one of its 2×2 submatrices is nonsingular), 1 (if it and all 2×2 submatrices are singular), or 0 (if it is null).

The determination of rank is laborious if done by hand. Either the matrix is reduced to echelon form by using elementary row operations, or exhaustive enumeration is used to create the submatrices and many determinants are calculated. If a matrix has more rows than columns and row-reduction is used, the work required to put the matrix in echelon form can be reduced by working with the transpose of the original matrix.

[8]The *row rank* and *column rank* are the same.

Example 5.4

What is the rank of matrix **A**?

$$\mathbf{A} = \begin{bmatrix} 1 & -2 & -1 \\ -3 & 3 & 0 \\ 2 & 2 & 4 \end{bmatrix}$$

Solution

Matrix **A** is singular because $|\mathbf{A}| = 0$. However, there is at least one 2×2 nonsingular submatrix:

$$\begin{vmatrix} 1 & -2 \\ -3 & 3 \end{vmatrix} = (1)(3) - (-3)(-2) = -3$$

Therefore, the rank is 2.

Example 5.5

Determine the rank of matrix **A** by reducing it to echelon form.

$$\mathbf{A} = \begin{bmatrix} 7 & 4 & 9 & 1 \\ 0 & 2 & -5 & 3 \\ 0 & 4 & -10 & 6 \end{bmatrix}$$

Solution

By inspection, the matrix can be row-reduced by subtracting two times the second row from the third row. The matrix cannot be further reduced. Since there are two nonzero rows, the rank is 2.

$$\begin{bmatrix} 7 & 4 & 9 & 1 \\ 0 & 2 & -5 & 3 \\ 0 & 0 & 0 & 0 \end{bmatrix}$$

11. CLASSICAL ADJOINT

The *classical adjoint* is the transpose of the cofactor matrix. The resulting matrix can be designated as $\mathbf{A}_{\text{adj}}$, $\text{adj}\{\mathbf{A}\}$, or $\mathbf{A}^{\text{adj}}$.

Example 5.6

What is the classical adjoint of matrix **A**?

$$\mathbf{A} = \begin{bmatrix} 2 & 3 & -4 \\ 0 & -4 & 2 \\ 1 & -1 & 5 \end{bmatrix}$$

Solution

The matrix of cofactors is

$$\begin{bmatrix} -18 & 2 & 4 \\ -11 & 14 & 5 \\ -10 & -4 & -8 \end{bmatrix}$$

The transpose of the matrix of cofactors is

$$\mathbf{A}_{\text{adj}} = \begin{bmatrix} -18 & -11 & -10 \\ 2 & 14 & -4 \\ 4 & 5 & -8 \end{bmatrix}$$

12. INVERSE

The product of a matrix **A** and its inverse, $\mathbf{A}^{-1}$, is the identity matrix, **I**. Only square matrices have inverses, but not all square matrices are invertible. A matrix has an inverse if and only if it is nonsingular (i.e., its determinant is nonzero).

$$\mathbf{AA}^{-1} = \mathbf{A}^{-1}\mathbf{A} = \mathbf{I} \tag{5.19}$$

$$(\mathbf{AB})^{-1} = \mathbf{B}^{-1}\mathbf{A}^{-1} \tag{5.20}$$

The inverse of a 2 × 2 matrix is easily determined by formula.

$$\mathbf{A} = \begin{bmatrix} a & b \\ c & d \end{bmatrix}$$

$$\mathbf{A}^{-1} = \frac{\begin{bmatrix} d & -b \\ -c & a \end{bmatrix}}{|\mathbf{A}|} \tag{5.21}$$

For a 3 × 3 or larger matrix, the inverse is determined by dividing every entry in the classical adjoint by the determinant of the original matrix.

$$\mathbf{A}^{-1} = \frac{\mathbf{A}_{\text{adj}}}{|\mathbf{A}|} \tag{5.22}$$

Example 5.7

What is the inverse of matrix **A**?

$$\mathbf{A} = \begin{bmatrix} 4 & 5 \\ 2 & 3 \end{bmatrix}$$

Solution

The determinant is calculated as

$$|\mathbf{A}| = (4)(3) - (2)(5) = 2$$

Using Eq. 5.22, the inverse is

$$\mathbf{A}^{-1} = \frac{\begin{bmatrix} 3 & -5 \\ -2 & 4 \end{bmatrix}}{2} = \begin{bmatrix} \frac{3}{2} & -\frac{5}{2} \\ -1 & 2 \end{bmatrix}$$

Check.

$$\mathbf{AA}^{-1} = \begin{bmatrix} 4 & 5 \\ 2 & 3 \end{bmatrix} \begin{bmatrix} \frac{3}{2} & -\frac{5}{2} \\ -1 & 2 \end{bmatrix} = \begin{bmatrix} 6-5 & -10+10 \\ 3-3 & -5+6 \end{bmatrix}$$

$$= \begin{bmatrix} 1 & 0 \\ 0 & 1 \end{bmatrix} = \mathbf{I} \quad \text{[OK]}$$

13. WRITING SIMULTANEOUS LINEAR EQUATIONS IN MATRIX FORM

Matrices are used to simplify the presentation and solution of sets of simultaneous linear equations. For example, the following three methods of presenting simultaneous linear equations are equivalent:

$$a_{11}x_1 + a_{12}x_2 = b_1$$
$$a_{21}x_1 + a_{22}x_2 = b_2$$

$$\begin{bmatrix} a_{11} & a_{12} \\ a_{21} & a_{22} \end{bmatrix} \begin{bmatrix} x_1 \\ x_2 \end{bmatrix} = \begin{bmatrix} b_1 \\ b_2 \end{bmatrix}$$

$$\mathbf{AX} = \mathbf{B}$$

In the second and third representations, **A** is known as the *coefficient matrix*, **X** as the *variable matrix*, and **B** as the *constant matrix.*

Not all systems of simultaneous equations have solutions, and those that do may not have unique solutions. The existence of a solution can be determined by calculating the determinant of the coefficient matrix. These rules are summarized in Table 5.1.

- If the system of linear equations is homogeneous (i.e., **B** is a zero matrix) and $|\mathbf{A}|$ is zero, there are an infinite number of solutions.
- If the system is homogeneous and $|\mathbf{A}|$ is nonzero, only the trivial solution exists.
- If the system of linear equations is nonhomogeneous (i.e., **B** is not a zero matrix) and $|\mathbf{A}|$ is nonzero, there is a unique solution to the set of simultaneous equations.
- If $|\mathbf{A}|$ is zero, a nonhomogeneous system of simultaneous equations may still have a solution. The requirement is that the determinants of all substitutional matrices (see Sec. 5.14) are zero, in which case there will be an infinite number of solutions. Otherwise, no solution exists.

Table 5.1 Solution Existence Rules for Simultaneous Equations

	$\mathbf{B}=0$	$\mathbf{B}\neq 0$
$\vert\mathbf{A}\vert=0$	infinite number of solutions (linearly dependent equations)	either an infinite number of solutions or no solution at all
$\vert\mathbf{A}\vert\neq 0$	trivial solution only ($x_i=0$)	unique nonzero solution

14. SOLVING SIMULTANEOUS LINEAR EQUATIONS

Gauss-Jordan elimination can be used to obtain the solution to a set of simultaneous linear equations. The coefficient matrix is augmented by the constant matrix. Then, elementary row operations are used to reduce the coefficient matrix to canonical form. All of the operations performed on the coefficient matrix are performed on the constant matrix. The variable values that satisfy the simultaneous equations will be the entries in the constant matrix when the coefficient matrix is in canonical form.

Determinants are used to calculate the solution to linear simultaneous equations through a procedure known as *Cramer's rule.*

The procedure is to calculate determinants of the original coefficient matrix $\mathbf{A}$ and of the n matrices resulting from the systematic replacement of a column in $\mathbf{A}$ by the constant matrix $\mathbf{B}$. For a system of three equations in three unknowns, there are three substitutional matrices, $\mathbf{A}_1$, $\mathbf{A}_2$, and $\mathbf{A}_3$, as well as the original coefficient matrix, for a total of four matrices whose determinants must be calculated.

The values of the unknowns that simultaneously satisfy all of the linear equations are

$$x_1=\frac{|\mathbf{A}_1|}{|\mathbf{A}|} \qquad 5.23$$

$$x_2=\frac{|\mathbf{A}_2|}{|\mathbf{A}|} \qquad 5.24$$

$$x_3=\frac{|\mathbf{A}_3|}{|\mathbf{A}|} \qquad 5.25$$

Example 5.8

Use Gauss-Jordan elimination to solve the following system of simultaneous equations.

$$2x+3y-4z=1$$
$$3x-y-2z=4$$
$$4x-7y-6z=-7$$

Solution

The augmented matrix is created by appending the constant matrix to the coefficient matrix.

$$\left[\begin{array}{ccc|c} 2 & 3 & -4 & 1 \\ 3 & -1 & -2 & 4 \\ 4 & -7 & -6 & -7 \end{array}\right]$$

Elementary row operations are used to reduce the coefficient matrix to canonical form. For example, two times the first row is subtracted from the third row. This step obtains the 0 needed in the a_{31} position.

$$\left[\begin{array}{ccc|c} 2 & 3 & -4 & 1 \\ 3 & -1 & -2 & 4 \\ 0 & -13 & 2 & -9 \end{array}\right]$$

This process continues until the following form is obtained.

$$\left[\begin{array}{ccc|c} 1 & 0 & 0 & 3 \\ 0 & 1 & 0 & 1 \\ 0 & 0 & 1 & 2 \end{array}\right]$$

$x=3$, $y=1$, and $z=2$ satisfy this system of equations.

Example 5.9

Use Cramer's rule to solve the following system of simultaneous equations.

$$2x+3y-4z=1$$
$$3x-y-2z=4$$
$$4x-7y-6z=-7$$

Solution

The determinant of the coefficient matrix is

$$|\mathbf{A}|=\begin{vmatrix} 2 & 3 & -4 \\ 3 & -1 & -2 \\ 4 & -7 & -6 \end{vmatrix}=82$$

The determinants of the substitutional matrices are

$$|\mathbf{A}_1| = \begin{vmatrix} 1 & 3 & -4 \\ 4 & -1 & -2 \\ -7 & -7 & -6 \end{vmatrix} = 246$$

$$|\mathbf{A}_2| = \begin{vmatrix} 2 & 1 & -4 \\ 3 & 4 & -2 \\ 4 & -7 & -6 \end{vmatrix} = 82$$

$$|\mathbf{A}_3| = \begin{vmatrix} 2 & 3 & 1 \\ 3 & -1 & 4 \\ 4 & -7 & -7 \end{vmatrix} = 164$$

The values of x, y, and z that will satisfy the linear equations are

$$x = \frac{246}{82} = 3$$

$$y = \frac{82}{82} = 1$$

$$z = \frac{164}{82} = 2$$

15. EIGENVALUES AND EIGENVECTORS

Eigenvalues and eigenvectors (also known as *characteristic values* and *characteristic vectors*) of a square matrix $\mathbf{A}$ are the scalars k and matrices $\mathbf{X}$ such that

$$\mathbf{AX} = k\mathbf{X} \qquad 5.26$$

The scalar k is an eigenvalue of $\mathbf{A}$ if and only if the matrix $(k\mathbf{I} - \mathbf{A})$ is singular; that is, if $|k\mathbf{I} - \mathbf{A}| = 0$. This equation is called the *characteristic equation* of the matrix $\mathbf{A}$. When expanded, the determinant is called the *characteristic polynomial.* The method of using the characteristic polynomial to find eigenvalues and eigenvectors is illustrated in Ex. 5.10.

If all of the eigenvalues are unique (i.e., nonrepeating), then Eq. 5.27 is valid.

$$[k\mathbf{I} - \mathbf{A}]\mathbf{X} = 0 \qquad 5.27$$

Example 5.10

Find the eigenvalues and nonzero eigenvectors of the matrix $\mathbf{A}$.

$$\mathbf{A} = \begin{bmatrix} 2 & 4 \\ 6 & 4 \end{bmatrix}$$

Solution

$$k\mathbf{I} - \mathbf{A} = \begin{bmatrix} k & 0 \\ 0 & k \end{bmatrix} - \begin{bmatrix} 2 & 4 \\ 6 & 4 \end{bmatrix} = \begin{bmatrix} k-2 & -4 \\ -6 & k-4 \end{bmatrix}$$

The characteristic polynomial is found by setting the determinant $|k\mathbf{I} - \mathbf{A}|$ equal to zero.

$$(k-2)(k-4) - (-6)(-4) = 0$$
$$k^2 - 6k - 16 = (k-8)(k+2) = 0$$

The roots of the characteristic polynomial are $k = +8$ and $k = -2$. These are the eigenvalues of $\mathbf{A}$.

Substituting $k = 8$,

$$k\mathbf{I} - \mathbf{A} = \begin{bmatrix} 8-2 & -4 \\ -6 & 8-4 \end{bmatrix} = \begin{bmatrix} 6 & -4 \\ -6 & 4 \end{bmatrix}$$

The resulting system can be interpreted as the linear equation $6x_1 - 4x_2 = 0$. The values of x that satisfy this equation define the eigenvector. An eigenvector $\mathbf{X}$ associated with the eigenvalue $+8$ is

$$\mathbf{X} = \begin{bmatrix} x_1 \\ x_2 \end{bmatrix} = \begin{bmatrix} 4 \\ 6 \end{bmatrix}$$

All other eigenvectors for this eigenvalue are multiples of $\mathbf{X}$. Normally $\mathbf{X}$ is reduced to smallest integers.

$$\mathbf{X} = \begin{bmatrix} 2 \\ 3 \end{bmatrix}$$

Similarly, the eigenvector associated with the eigenvalue -2 is

$$\mathbf{X} = \begin{bmatrix} x_1 \\ x_2 \end{bmatrix} = \begin{bmatrix} +4 \\ -4 \end{bmatrix}$$

Reducing this to smallest integers gives

$$\mathbf{X} = \begin{bmatrix} +1 \\ -1 \end{bmatrix}$$

6 Vectors

1. Introduction 6-1
2. Vectors in n-Space 6-1
3. Unit Vectors 6-2
4. Vector Representation 6-2
5. Conversion Between Systems 6-2
6. Vector Addition 6-3
7. Multiplication by a Scalar 6-3
8. Vector Dot Product 6-3
9. Vector Cross Product 6-4
10. Mixed Triple Product 6-4
11. Vector Triple Product 6-5
12. Vector Functions 6-5

1. INTRODUCTION

A physical property or quantity can be described by a scalar, vector, or tensor. A *scalar* has only magnitude. Knowing its value is sufficient to define a scalar. Mass, enthalpy, density, and speed are examples of scalars.

Force, momentum, displacement, and velocity are examples of vectors. A *vector* is a directed straight line with a specific magnitude and is specified completely by its direction (consisting of the vector's *angular orientation* and its *sense*) and magnitude. A vector's *point of application* (*terminal point*) is not needed to define the vector.[1] Two vectors with the same direction and magnitude are said to be *equal vectors* even though their *lines of action* may be different.[2]

A vector can be designated by a boldface variable (as in this book) or as a combination of the variable and some other symbol. For example, the notations $\mathbf{V}$, $\overline{V}$, $\hat{V}$, $\vec{V}$, and $\underline{V}$ are used by different authorities to represent vectors. In this book, the magnitude of a vector can be designated by either $|\mathbf{V}|$ or V (italic but not bold).

Stress, dielectric constant, and magnetic susceptibility are examples of tensors. A *tensor* has magnitude in a specific direction but the direction is not unique. Tensors are frequently associated with *anisotropic materials* that have different properties in different directions. A tensor in three-dimensional space is defined by nine components, compared with the three that are required to define vectors. These components are written in matrix form. Stress, σ, at a point, for example, would be defined by the following tensor matrix.

$$\sigma \equiv \begin{bmatrix} \sigma_{xx} & \sigma_{xy} & \sigma_{xz} \\ \sigma_{yx} & \sigma_{yy} & \sigma_{yz} \\ \sigma_{zx} & \sigma_{zy} & \sigma_{zz} \end{bmatrix}$$

2. VECTORS IN n-SPACE

In some cases, a vector, $\mathbf{V}$, will be designated by its two endpoints in n-dimensional vector space. The usual vector space is three-dimensional force-space. Usually, one of the points will be the origin, in which case the vector is said to be "based at the origin," "origin-based," or "zero-based."[3] If one of the endpoints is the origin, specifying a terminal point P would represent a force directed from the origin to point P.

If a coordinate system is superimposed on the vector space, a vector can be specified in terms of the n coordinates of its two endpoints. The magnitude of the vector $\mathbf{V}$ is the distance in vector space between the two points, as given by Eq. 6.1. Similarly, the direction is defined by the angle the vector makes with one of the axes. Figure 6.1 illustrates a vector in two dimensions.

$$|\mathbf{V}| = \sqrt{(x_2 - x_1)^2 + (y_2 - y_1)^2} \qquad 6.1$$

$$\phi = \arctan \frac{y_2 - y_1}{x_2 - x_1} \qquad 6.2$$

Figure 6.1 *Vector in Two-Dimensional Space*

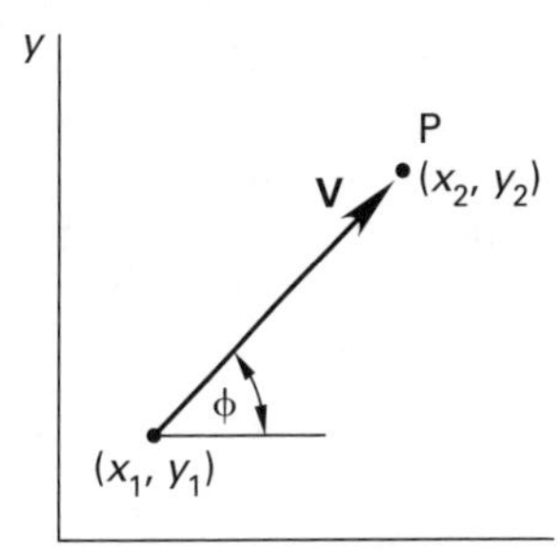

[1]A vector that is constrained to act at or through a certain point is a *bound vector (fixed vector)*. A *sliding vector (transmissible vector)* can be applied anywhere along its line of action. A *free vector* is not constrained and can be applied at any point in space.

[2]A distinction is sometimes made between equal vectors and equivalent vectors. *Equivalent vectors* produce the same effect but are not necessarily equal.

[3]Any vector directed from P_1 to P_2 can be transformed into a zero-based vector by subtracting the coordinates of point P_1 from the coordinates of terminal point P_2. The transformed vector will be equivalent to the original vector.

The *components* of a vector are the projections of the vector on the coordinate axes. (For a zero-based vector, the components and the coordinates of the endpoint are the same.) Simple trigonometric principles are used to resolve a vector into its components. A vector reconstructed from its components is known as a *resultant vector*. Figure 6.2 shows the location of direction angles.

$$V_x = |\mathbf{V}|\cos\phi_x \quad 6.3$$

$$V_y = |\mathbf{V}|\cos\phi_y \quad 6.4$$

$$V_z = |\mathbf{V}|\cos\phi_z \quad 6.5$$

$$|\mathbf{V}| = \sqrt{V_x^2 + V_y^2 + V_z^2} \quad 6.6$$

In Eq. 6.3 through Eq. 6.5, ϕ_x, ϕ_y, and ϕ_z are the *direction angles*—the angles between the vector and the x-, y-, and z-axes, respectively. The cosines of these angles are known as *direction cosines*. The sum of the squares of the direction cosines is equal to 1.

$$\cos^2\phi_x + \cos^2\phi_y + \cos^2\phi_z = 1 \quad 6.7$$

Figure 6.2 *Direction Angles of a Vector*

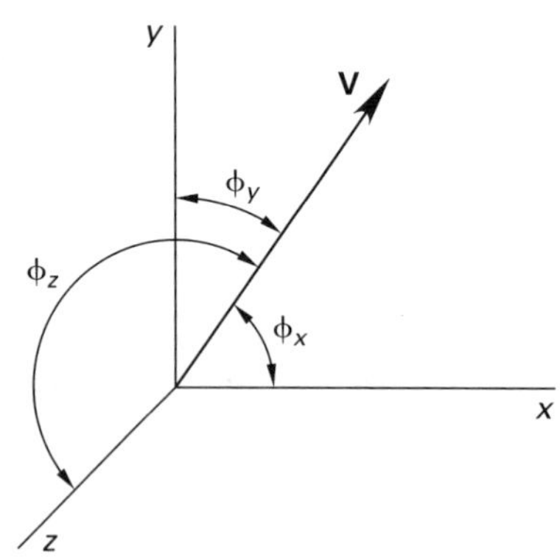

3. UNIT VECTORS

Unit vectors are vectors with unit magnitudes (i.e., magnitudes of 1). They are represented in the same notation as other vectors. (Unit vectors in this book are written in boldface type.) Although they can have any direction, the standard unit vectors (the *Cartesian unit vectors* **i**, **j**, and **k**) have the directions of the x-, y-, and z-coordinate axes and constitute the *Cartesian triad*, as illustrated in Fig. 6.3.

A vector **V** can be written in terms of unit vectors and its components.

$$\mathbf{V} = |\mathbf{V}|\mathbf{a} = V_x\mathbf{i} + V_y\mathbf{j} + V_z\mathbf{k} \quad 6.8$$

Figure 6.3 *Cartesian Unit Vectors*

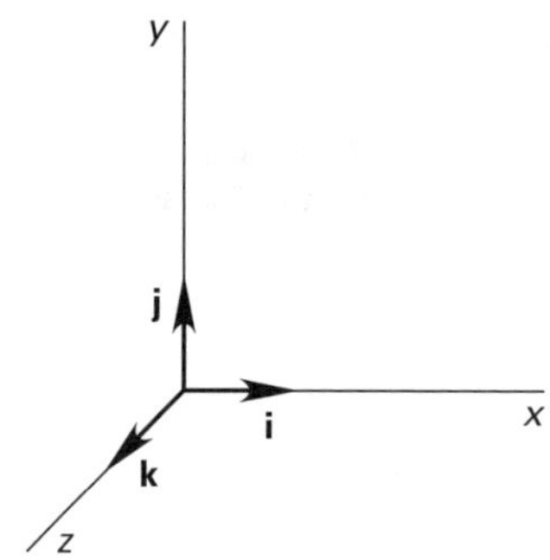

The unit vector, **a**, has the same direction as the vector **V** but has a length of 1. This unit vector is calculated by dividing the original vector, **V**, by its magnitude, |**V**|.

$$\mathbf{a} = \frac{\mathbf{V}}{|\mathbf{V}|} = \frac{V_x\mathbf{i} + V_y\mathbf{j} + V_z\mathbf{k}}{\sqrt{V_x^2 + V_y^2 + V_z^2}} \quad 6.9$$

4. VECTOR REPRESENTATION

The most common method of representing a vector is by writing it in *rectangular form*—a vector sum of its orthogonal components. In rectangular form, each of the orthogonal components has the same units as the resultant vector.

$$\mathbf{A} \equiv A_x\mathbf{i} + A_y\mathbf{j} + A_z\mathbf{k} \quad \text{[three dimensions]}$$

However, the vector is also completely defined by its magnitude and associated angle. These two quantities can be written together in *phasor form*, sometimes referred to as *polar form*.

$$\mathbf{A} \equiv |\mathbf{A}|\angle\phi = A\angle\phi$$

5. CONVERSION BETWEEN SYSTEMS

The choice of the **ijk** triad may be convenient but is arbitrary. A vector can be expressed in terms of any other set of unit vectors, **uvw**.

$$\mathbf{V} = V_x\mathbf{i} + V_y\mathbf{j} + V_z\mathbf{k} = V'_x\mathbf{u} + V'_y\mathbf{v} + V'_z\mathbf{w} \quad 6.10$$

The two representations are related.

$$V'_x = \mathbf{V}\cdot\mathbf{u} = (\mathbf{i}\cdot\mathbf{u})V_x + (\mathbf{j}\cdot\mathbf{u})V_y + (\mathbf{k}\cdot\mathbf{u})V_z \quad 6.11$$

$$V'_y = \mathbf{V}\cdot\mathbf{v} = (\mathbf{i}\cdot\mathbf{v})V_x + (\mathbf{j}\cdot\mathbf{v})V_y + (\mathbf{k}\cdot\mathbf{v})V_z \quad 6.12$$

$$V'_z = \mathbf{V}\cdot\mathbf{w} = (\mathbf{i}\cdot\mathbf{w})V_x + (\mathbf{j}\cdot\mathbf{w})V_y + (\mathbf{k}\cdot\mathbf{w})V_z \quad 6.13$$

Equation 6.11 through Eq. 6.13 can be expressed in matrix form. The dot products are known as the *coefficients of transformation*, and the matrix containing them is the *transformation matrix*.

$$\begin{pmatrix} V'_x \\ V'_y \\ V'_z \end{pmatrix} = \begin{pmatrix} \mathbf{i}\cdot\mathbf{u} & \mathbf{j}\cdot\mathbf{u} & \mathbf{k}\cdot\mathbf{u} \\ \mathbf{i}\cdot\mathbf{v} & \mathbf{j}\cdot\mathbf{v} & \mathbf{k}\cdot\mathbf{v} \\ \mathbf{i}\cdot\mathbf{w} & \mathbf{j}\cdot\mathbf{w} & \mathbf{k}\cdot\mathbf{w} \end{pmatrix} \begin{pmatrix} V_x \\ V_y \\ V_z \end{pmatrix} \quad 6.14$$

6. VECTOR ADDITION

Addition of two vectors by the *polygon method* is accomplished by placing the tail of the second vector at the head (tip) of the first. The sum (i.e., the *resultant vector*) is a vector extending from the tail of the first vector to the head of the second, as shown in Fig. 6.4. Alternatively, the two vectors can be considered as the two sides of a parallelogram, while the sum represents the diagonal. This is known as addition by the *parallelogram method*.

Figure 6.4 *Addition of Two Vectors*

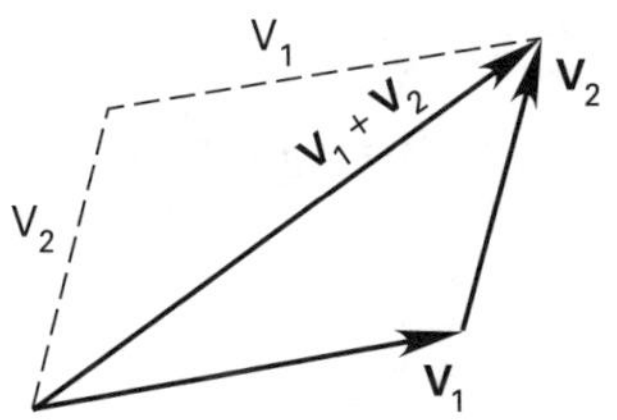

The components of the resultant vector are the sums of the components of the added vectors (that is, $V_{1x} + V_{2x}$, $V_{1y} + V_{2y}$, $V_{1z} + V_{2z}$).

Vector addition is both commutative and associative.

$$\mathbf{V}_1 + \mathbf{V}_2 = \mathbf{V}_2 + \mathbf{V}_1 \quad 6.15$$

$$\mathbf{V}_1 + (\mathbf{V}_2 + \mathbf{V}_3) = (\mathbf{V}_1 + \mathbf{V}_2) + \mathbf{V}_3 \quad 6.16$$

7. MULTIPLICATION BY A SCALAR

A vector, **V**, can be multiplied by a scalar, c. If the original vector is represented by its components, each of the components is multiplied by c.

$$c\mathbf{V} = c|\mathbf{V}|\mathbf{a} = cV_x\mathbf{i} + cV_y\mathbf{j} + cV_z\mathbf{k} \quad 6.17$$

Scalar multiplication is distributive.

$$c(\mathbf{V}_1 + \mathbf{V}_2) = c\mathbf{V}_1 + c\mathbf{V}_2 \quad 6.18$$

8. VECTOR DOT PRODUCT

The *dot product* (*scalar product*), $\mathbf{V}_1 \cdot \mathbf{V}_2$, of two vectors is a scalar that is proportional to the length of the projection of the first vector onto the second vector, as illustrated in Fig. 6.5.[4]

Figure 6.5 *Vector Dot Product*

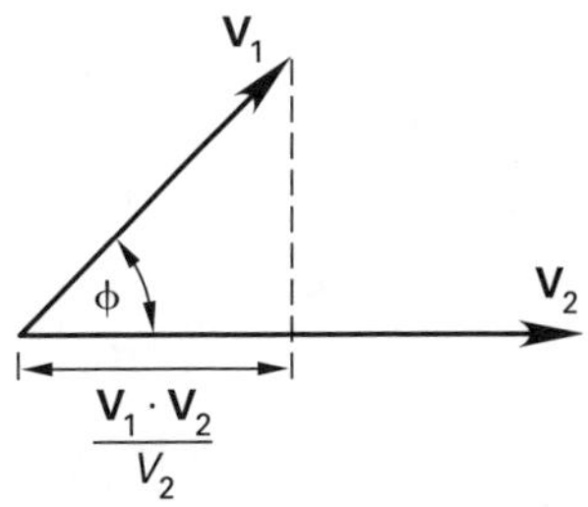

The dot product is commutative and distributive.

$$\mathbf{V}_1 \cdot \mathbf{V}_2 = \mathbf{V}_2 \cdot \mathbf{V}_1 \quad 6.19$$

$$\mathbf{V}_1 \cdot (\mathbf{V}_2 + \mathbf{V}_3) = \mathbf{V}_1 \cdot \mathbf{V}_2 + \mathbf{V}_1 \cdot \mathbf{V}_3 \quad 6.20$$

The dot product can be calculated in two ways, as Eq. 6.21 indicates. ϕ is limited to 180° and is the angle between the two vectors.

$$\begin{aligned} \mathbf{V}_1 \cdot \mathbf{V}_2 &= |\mathbf{V}_1|\,|\mathbf{V}_2|\cos\phi \\ &= V_{1x}V_{2x} + V_{1y}V_{2y} + V_{1z}V_{2z} \end{aligned} \quad 6.21$$

When Eq. 6.21 is solved for the angle between the two vectors, ϕ, it is known as the *Cauchy-Schwartz theorem.*

$$\cos\phi = \frac{V_{1x}V_{2x} + V_{1y}V_{2y} + V_{1z}V_{2z}}{|\mathbf{V}_1|\,|\mathbf{V}_2|} \quad 6.22$$

The dot product can be used to determine whether a vector is a unit vector and to show that two vectors are orthogonal (perpendicular). For any unit vector, **u**,

$$\mathbf{u} \cdot \mathbf{u} = 1 \quad 6.23$$

For two non-null orthogonal vectors,

$$\mathbf{V}_1 \cdot \mathbf{V}_2 = 0 \quad 6.24$$

Equation 6.23 and Eq. 6.24 can be extended to the Cartesian unit vectors.

$$\mathbf{i} \cdot \mathbf{i} = 1 \quad 6.25$$

$$\mathbf{j} \cdot \mathbf{j} = 1 \quad 6.26$$

$$\mathbf{k} \cdot \mathbf{k} = 1 \quad 6.27$$

$$\mathbf{i} \cdot \mathbf{j} = 0 \quad 6.28$$

$$\mathbf{i} \cdot \mathbf{k} = 0 \quad 6.29$$

$$\mathbf{j} \cdot \mathbf{k} = 0 \quad 6.30$$

[4]The dot product is also written in parentheses without a dot, that is, $(\mathbf{V}_1\mathbf{V}_2)$.

Example 6.1

What is the angle between the zero-based vectors $\mathbf{V}_1 = (-\sqrt{3}, 1)$ and $\mathbf{V}_2 = (2\sqrt{3}, 2)$?

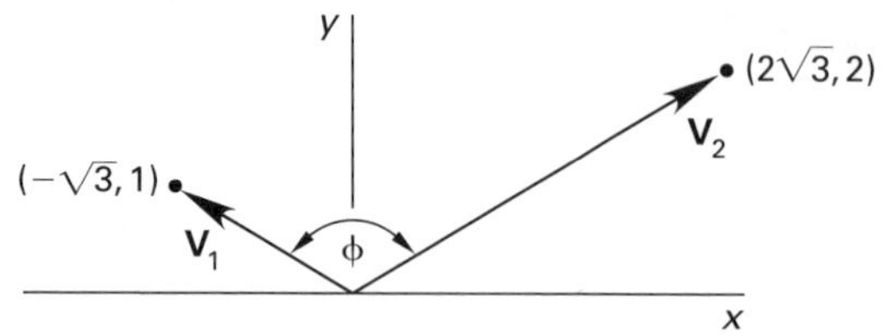

Solution

From Eq. 6.22,

$$\cos\phi = \frac{V_{1x}V_{2x} + V_{1y}V_{2y}}{|\mathbf{V}_1|\,|\mathbf{V}_2|} = \frac{V_{1x}V_{2x} + V_{1y}V_{2y}}{\sqrt{V_{1x}^2 + V_{1y}^2}\sqrt{V_{2x}^2 + V_{2y}^2}}$$

$$= \frac{(-\sqrt{3})(2\sqrt{3}) + (1)(2)}{\sqrt{(-\sqrt{3})^2 + (1)^2}\sqrt{(2\sqrt{3})^2 + (2)^2}}$$

$$= \frac{-4}{8} = -\frac{1}{2}$$

$$\phi = \arccos\left(-\frac{1}{2}\right) = 120°$$

9. VECTOR CROSS PRODUCT

The *cross product* (*vector product*), $\mathbf{V}_1 \times \mathbf{V}_2$, of two vectors is a vector that is orthogonal (perpendicular) to the plane of the two vectors.[5] The unit vector representation of the cross product can be calculated as a third-order determinant. Figure 6.6 illustrates the vector cross product.

$$\mathbf{V}_1 \times \mathbf{V}_2 = \begin{vmatrix} \mathbf{i} & V_{1x} & V_{2x} \\ \mathbf{j} & V_{1y} & V_{2y} \\ \mathbf{k} & V_{1z} & V_{2z} \end{vmatrix} \qquad 6.31$$

The direction of the cross-product vector corresponds to the direction a right-hand screw would progress if vectors $\mathbf{V}_1$ and $\mathbf{V}_2$ are placed tail-to-tail in the plane they define and $\mathbf{V}_1$ is rotated into $\mathbf{V}_2$. The direction can also be found from the *right-hand rule.*

Figure 6.6 *Vector Cross Product*

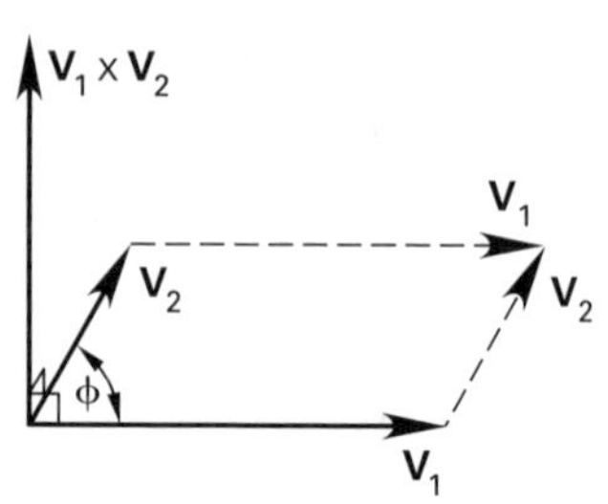

[5]The cross product is also written in square brackets without a cross, that is, $[\mathbf{V}_1\mathbf{V}_2]$.

The magnitude of the cross product can be determined from Eq. 6.32, in which ϕ is the angle between the two vectors and is limited to 180°. The magnitude corresponds to the area of a parallelogram that has $\mathbf{V}_1$ and $\mathbf{V}_2$ as two of its sides.

$$|\mathbf{V}_1 \times \mathbf{V}_2| = |\mathbf{V}_1||\mathbf{V}_2|\sin\phi \qquad 6.32$$

Vector cross multiplication is distributive but not commutative.

$$\mathbf{V}_1 \times \mathbf{V}_2 = -(\mathbf{V}_2 \times \mathbf{V}_1) \qquad 6.33$$

$$c(\mathbf{V}_1 \times \mathbf{V}_2) = (c\mathbf{V}_1) \times \mathbf{V}_2 = \mathbf{V}_1 \times (c\mathbf{V}_2) \qquad 6.34$$

$$\mathbf{V}_1 \times (\mathbf{V}_2 + \mathbf{V}_3) = \mathbf{V}_1 \times \mathbf{V}_2 + \mathbf{V}_1 \times \mathbf{V}_3 \qquad 6.35$$

If the two vectors are parallel, their cross product will be zero.

$$\mathbf{i} \times \mathbf{i} = \mathbf{j} \times \mathbf{j} = \mathbf{k} \times \mathbf{k} = 0 \qquad 6.36$$

Equation 6.31 and Eq. 6.33 can be extended to the unit vectors.

$$\mathbf{i} \times \mathbf{j} = -\mathbf{j} \times \mathbf{i} = \mathbf{k} \qquad 6.37$$

$$\mathbf{j} \times \mathbf{k} = -\mathbf{k} \times \mathbf{j} = \mathbf{i} \qquad 6.38$$

$$\mathbf{k} \times \mathbf{i} = -\mathbf{i} \times \mathbf{k} = \mathbf{j} \qquad 6.39$$

Example 6.2

Find a unit vector orthogonal to $\mathbf{V}_1 = \mathbf{i} - \mathbf{j} + 2\mathbf{k}$ and $\mathbf{V}_2 = 3\mathbf{j} - \mathbf{k}$.

Solution

The cross product is a vector orthogonal to $\mathbf{V}_1$ and $\mathbf{V}_2$.

$$\mathbf{V}_1 \times \mathbf{V}_2 = \begin{vmatrix} \mathbf{i} & 1 & 0 \\ \mathbf{j} & -1 & 3 \\ \mathbf{k} & 2 & -1 \end{vmatrix} = -5\mathbf{i} + \mathbf{j} + 3\mathbf{k}$$

Check to see whether this is a unit vector.

$$|\mathbf{V}_1 \times \mathbf{V}_2| = \sqrt{(-5)^2 + (1)^2 + (3)^2} = \sqrt{35}$$

Since its length is $\sqrt{35}$, the vector must be divided by $\sqrt{35}$ to obtain a unit vector.

$$\mathbf{a} = \frac{-5\mathbf{i} + \mathbf{j} + 3\mathbf{k}}{\sqrt{35}}$$

10. MIXED TRIPLE PRODUCT

The *mixed triple product* (*triple scalar product* or just *triple product*) of three vectors is a scalar quantity representing the volume of a parallelepiped with the three vectors making up the sides. It is calculated as a determinant. Since Eq. 6.40 can be negative, the absolute value must be used to obtain the volume in that case. Figure 6.7 shows a mixed triple product.

$$\mathbf{V}_1 \cdot (\mathbf{V}_2 \times \mathbf{V}_3) = \begin{vmatrix} V_{1x} & V_{1y} & V_{1z} \\ V_{2x} & V_{2y} & V_{2z} \\ V_{3x} & V_{3y} & V_{3z} \end{vmatrix} \qquad 6.40$$

The mixed triple product has the property of *circular permutation*, as defined by Eq. 6.41.

$$\mathbf{V}_1 \cdot (\mathbf{V}_2 \times \mathbf{V}_3) = (\mathbf{V}_1 \times \mathbf{V}_2) \cdot \mathbf{V}_3 \qquad 6.41$$

Figure 6.7 *Vector Mixed Triple Product*

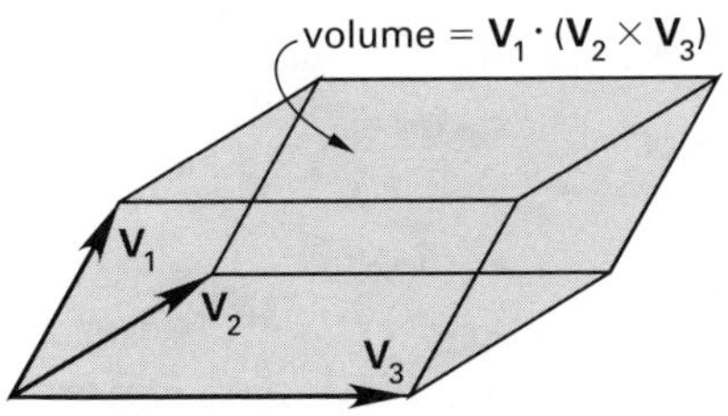

11. VECTOR TRIPLE PRODUCT

The *vector triple product* is a vector defined by Eq. 6.42. The quantities in parentheses on the right-hand side are scalars.

$$\mathbf{V}_1 \times (\mathbf{V}_2 \times \mathbf{V}_3) = (\mathbf{V}_1 \cdot \mathbf{V}_3)\mathbf{V}_2 - (\mathbf{V}_1 \cdot \mathbf{V}_2)\mathbf{V}_3 \qquad 6.42$$

12. VECTOR FUNCTIONS

A vector can be a function of another parameter. For example, a vector **V** is a function of variable t when its V_x, V_y, and V_z are functions of t.

$$\mathbf{V}(t) = (2t - 3)\mathbf{i} + (t^2 + 1)\mathbf{j} + (-7t + 5)\mathbf{k} \qquad 6.43$$

When the functions of t are differentiated (or integrated) with respect to t, the vector itself is differentiated (integrated).[6]

$$\frac{d\mathbf{V}(t)}{dt} = \frac{dV_x}{dt}\mathbf{i} + \frac{dV_y}{dt}\mathbf{j} + \frac{dV_z}{dt}\mathbf{k} \qquad 6.44$$

Similarly, the integral of the vector is

$$\int \mathbf{V}(t)\,dt = \mathbf{i}\int V_x\,dt + \mathbf{j}\int V_y\,dt + \mathbf{k}\int V_z\,dt \qquad 6.45$$

[6]This is particularly valuable when converting among position, velocity, and acceleration vectors.

7 Trigonometry

1. Degrees and Radians 7-1
2. Plane Angles 7-1
3. Triangles 7-2
4. Right Triangles 7-2
5. Circular Transcendental Functions 7-2
6. Small Angle Approximations 7-3
7. Graphs of the Functions 7-3
8. Signs of the Functions 7-3
9. Functions of Related Angles 7-3
10. Trigonometric Identities 7-3
11. Inverse Trigonometric Functions 7-4
12. Hyperbolic Transcendental Functions 7-4
13. Hyperbolic Identities 7-5
14. General Triangles 7-5
15. Spherical Trigonometry 7-5
16. Solid Angles 7-6

1. DEGREES AND RADIANS

Degrees and *radians* are two units for measuring angles. One complete circle is divided into 360 degrees (written 360°) or 2π radians (abbreviated *rad*).[1] The conversions between degrees and radians are

multiply	by	to obtain
radians	$\frac{180}{\pi}$	degrees
degrees	$\frac{\pi}{180}$	radians

The number of radians in an angle, θ, corresponds to two times the area within a circular sector with arc length θ and a radius of one, as shown in Fig. 7.1. Alternatively, the area of a sector with central angle θ radians is $\theta/2$ for a *unit circle* (i.e., a circle with a radius of one unit).

2. PLANE ANGLES

A *plane angle* (usually referred to as just an *angle*) consists of two intersecting lines and an intersection point known as the *vertex*. The angle can be referred to by a capital letter representing the vertex (e.g., B in Fig. 7.2), a letter representing the angular measure (e.g., B or β), or by three capital letters, where the middle letter is the vertex and the other two letters are two points on different lines, and either the symbol ∠ or ∢ (e.g., ∢ABC).

Figure 7.1 *Radians and Area of Unit Circle*

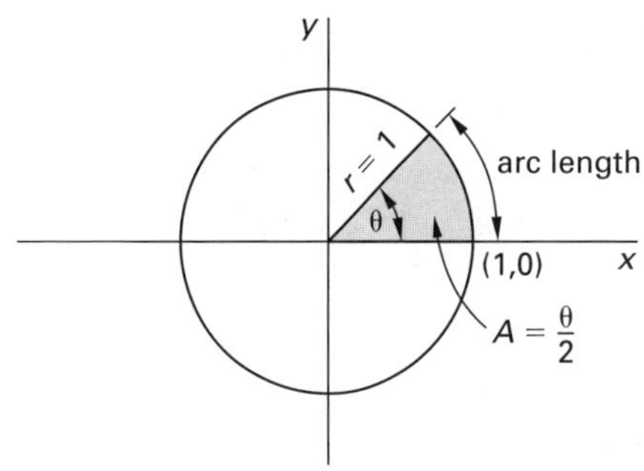

Figure 7.2 *Angle*

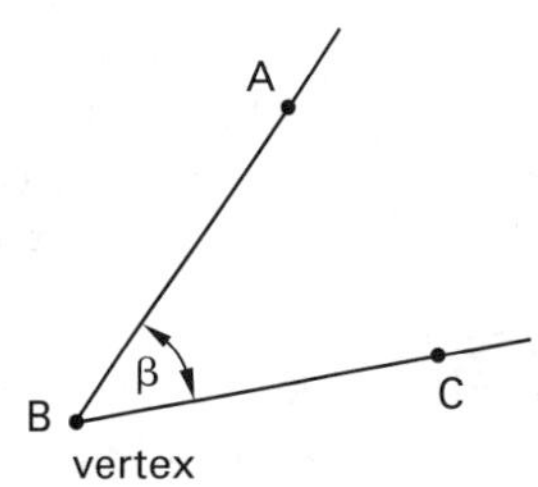

The angle between two intersecting lines generally is understood to be the smaller angle created.[2] Angles have been classified as follows.

- *acute angle:* an angle less than 90° ($\pi/2$ rad)
- *obtuse angle:* an angle more than 90° ($\pi/2$ rad) but less than 180° (π rad)
- *reflex angle:* an angle more than 180° (π rad) but less than 360° (2π rad)
- *related angle:* an angle that differs from another by some multiple of 90° ($\pi/2$ rad)
- *right angle:* an angle equal to 90° ($\pi/2$ rad)
- *straight angle:* an angle equal to 180° (π rad), that is, a straight line

Complementary angles are two angles whose sum is 90° ($\pi/2$ rad). *Supplementary angles* are two angles whose sum is 180° (π rad). *Adjacent angles* share a common

[1]The abbreviation *rad* is also used to represent *radiation absorbed dose*, a measure of radiation exposure.

[2]In books on geometry, the term *ray* is used instead of *line*.

vertex and one (the interior) side. Adjacent angles are supplementary only if their exterior sides form a straight line.

Vertical angles are the two angles with a common vertex and with sides made up by two intersecting straight lines, as shown in Fig. 7.3. Vertical angles are equal.

Angle of elevation and *angle of depression* are surveying terms referring to the angle above and below the horizontal plane of the observer, respectively.

Figure 7.3 *Vertical Angles*

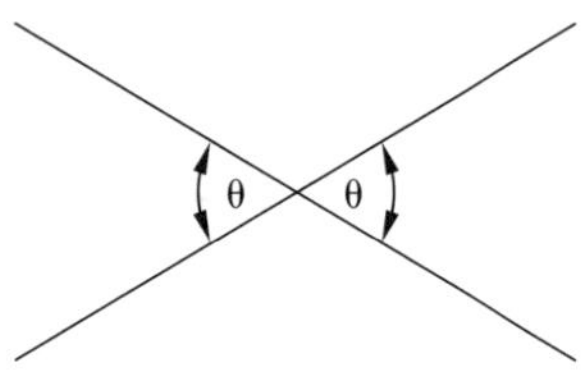

3. TRIANGLES

A *triangle* is a three-sided closed polygon with three angles whose sum is 180° (π rad). Triangles are identified by their vertices and the symbol Δ (e.g., ΔABC in Fig. 7.4). A side is designated by its two endpoints (e.g., AB in Fig. 7.4) or by a lowercase letter corresponding to the capital letter of the opposite vertex (e.g., *c*).

In *similar triangles*, the corresponding angles are equal and the corresponding sides are in proportion. (Since there are only two independent angles in a triangle, showing that two angles of one triangle are equal to two angles of the other triangle is sufficient to show similarity.) The symbol for similarity is ~. In Fig. 7.4, ΔABC ~ ΔDEF (i.e., ΔABC is similar to ΔDEF).

Figure 7.4 *Similar Triangles*

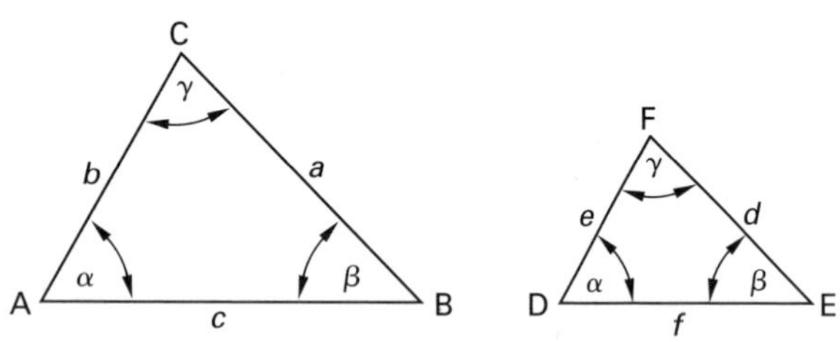

4. RIGHT TRIANGLES

A *right triangle* is a triangle in which one of the angles is 90° (π/2 rad). The remaining two angles are complementary. If one of the acute angles is chosen as the reference, the sides forming the right angle are known as the *adjacent side*, x, and the *opposite side*, y. The longest side is known as the *hypotenuse*, r. The *Pythagorean theorem* relates the lengths of these sides.

$$x^2 + y^2 = r^2 \tag{7.1}$$

In certain cases, the lengths of unknown sides of right triangles can be determined by inspection.[3] This occurs when the lengths of the sides are in the ratios of 3:4:5, 1:1:$\sqrt{2}$, 1:$\sqrt{3}$:2, and 5:12:13. Figure 7.5 illustrates a 3:4:5 triangle.

Figure 7.5 *3:4:5 Right Triangle*

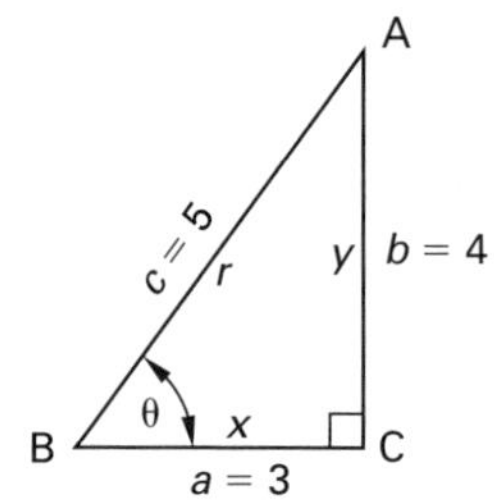

5. CIRCULAR TRANSCENDENTAL FUNCTIONS

The *circular transcendental functions* (usually referred to as the *transcendental functions*, *trigonometric functions*, or *functions of an angle*) are calculated from the sides of a right triangle. Equation 7.2 through Eq. 7.7 refer to Fig. 7.5.

$$\text{sine: } \sin\theta = \frac{y}{r} = \frac{\text{opposite}}{\text{hypotenuse}} \tag{7.2}$$

$$\text{cosine: } \cos\theta = \frac{x}{r} = \frac{\text{adjacent}}{\text{hypotenuse}} \tag{7.3}$$

$$\text{tangent: } \tan\theta = \frac{y}{x} = \frac{\text{opposite}}{\text{adjacent}} \tag{7.4}$$

$$\text{cotangent: } \cot\theta = \frac{x}{y} = \frac{\text{adjacent}}{\text{opposite}} \tag{7.5}$$

$$\text{secant: } \sec\theta = \frac{r}{x} = \frac{\text{hypotenuse}}{\text{adjacent}} \tag{7.6}$$

$$\text{cosecant: } \csc\theta = \frac{r}{y} = \frac{\text{hypotenuse}}{\text{opposite}} \tag{7.7}$$

Three of the transcendental functions are reciprocals of the others. However, while the tangent and cotangent

[3]These cases are almost always contrived examples. There is nothing intrinsic in nature to cause the formation of triangles with these proportions.

functions are reciprocals of each other, the sine and cosine functions are not.

$$\cot\theta = \frac{1}{\tan\theta} \quad 7.8$$

$$\sec\theta = \frac{1}{\cos\theta} \quad 7.9$$

$$\csc\theta = \frac{1}{\sin\theta} \quad 7.10$$

The trigonometric functions correspond to the lengths of various line segments in a right triangle with a unit hypotenuse. Figure 7.6 shows such a triangle inscribed in a unit circle.

Figure 7.6 *Trigonometric Functions in a Unit Circle*

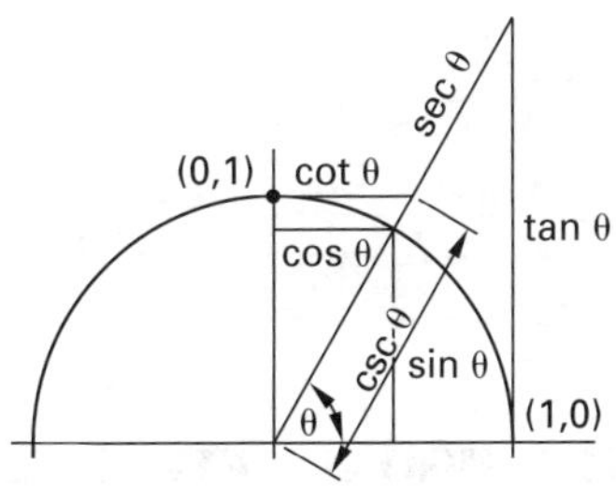

6. SMALL ANGLE APPROXIMATIONS

When an angle is very small, the hypotenuse and adjacent sides are essentially equal in length, and certain approximations can be made. (The angle θ must be expressed in radians in Eq. 7.11 and Eq. 7.12.)

$$\sin\theta \approx \tan\theta \approx \theta\big|_{\theta<10^\circ\ (0.175\ \text{rad})} \quad 7.11$$

$$\cos\theta \approx 1\big|_{\theta<5^\circ\ (0.0873\ \text{rad})} \quad 7.12$$

7. GRAPHS OF THE FUNCTIONS

Figure 7.7 illustrates the periodicity of the sine, cosine, and tangent functions.[4]

Figure 7.7 *Graphs of Sine, Cosine, and Tangent Functions*

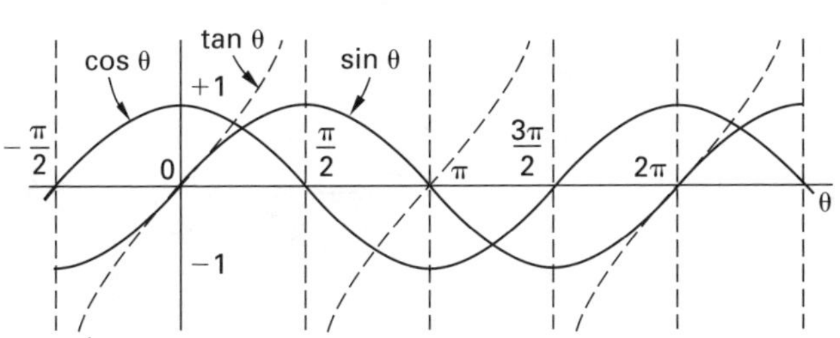

[4]The remaining functions, being reciprocals of these three functions, are also periodic.

8. SIGNS OF THE FUNCTIONS

Table 7.1 shows how the sine, cosine, and tangent functions vary in sign with different values of θ. All three functions are positive for angles $0° \le \theta \le 90°$ ($0 \le \theta \le \pi/2$ rad), but only the sine is positive for angles $90° < \theta \le 180°$ ($\pi/2$ rad $< \theta \le \pi$ rad). The concept of quadrants is used to summarize the signs of the functions: angles up to 90° ($\pi/2$ rad) are in quadrant I, between 90° and 180° ($\pi/2$ and π rad) are in quadrant II, and so on.

Table 7.1 *Signs of the Functions by Quadrant*

	quadrant			
function	I	II	III	IV
sine	+	+	−	−
cosine	+	−	−	+
tangent	+	−	+	−

9. FUNCTIONS OF RELATED ANGLES

Figure 7.7 shows that the sine, cosine, and tangent curves are symmetrical with respect to the horizontal axis. Furthermore, portions of the curves are symmetrical with respect to a vertical axis. The values of the sine and cosine functions repeat every 360° (2π rad), and the absolute values repeat every 180° (π rad). This can be written as

$$\sin(\theta + 180°) = -\sin\theta \quad 7.13$$

Similarly, the tangent function repeats every 180° (π rad), and its absolute value repeats every 90° ($\pi/2$ rad).

Table 7.2 summarizes the functions of the related angles.

Table 7.2 *Functions of Related Angles*

$f(\theta)$	$-\theta$	$90°-\theta$	$90°+\theta$	$180°-\theta$	$180°+\theta$
sin	$-\sin\theta$	$\cos\theta$	$\cos\theta$	$\sin\theta$	$-\sin\theta$
cos	$\cos\theta$	$\sin\theta$	$-\sin\theta$	$-\cos\theta$	$-\cos\theta$
tan	$-\tan\theta$	$\cot\theta$	$-\cot\theta$	$-\tan\theta$	$\tan\theta$

10. TRIGONOMETRIC IDENTITIES

There are many relationships between trigonometric functions. For example, Eq. 7.14 through Eq. 7.16 are well known.

$$\sin^2\theta + \cos^2\theta = 1 \quad 7.14$$

$$1 + \tan^2\theta = \sec^2\theta \quad 7.15$$

$$1 + \cot^2\theta = \csc^2\theta \quad 7.16$$

Other relatively common identities are listed as follows.[5]

- *double-angle formulas*

$$\sin 2\theta = 2\sin\theta\cos\theta = \frac{2\tan\theta}{1+\tan^2\theta} \quad 7.17$$

$$\cos 2\theta = \cos^2\theta - \sin^2\theta = 1 - 2\sin^2\theta$$

$$= 2\cos^2\theta - 1 = \frac{1-\tan^2\theta}{1+\tan^2\theta} \quad 7.18$$

$$\tan 2\theta = \frac{2\tan\theta}{1-\tan^2\theta} \quad 7.19$$

$$\cot 2\theta = \frac{\cot^2\theta - 1}{2\cot\theta} \quad 7.20$$

- *two-angle formulas*

$$\sin(\theta \pm \phi) = \sin\theta\cos\phi \pm \cos\theta\sin\phi \quad 7.21$$

$$\cos(\theta \pm \phi) = \cos\theta\cos\phi \mp \sin\theta\sin\phi \quad 7.22$$

$$\tan(\theta \pm \phi) = \frac{\tan\theta \pm \tan\phi}{1 \mp \tan\theta\tan\phi} \quad 7.23$$

$$\cot(\theta \pm \phi) = \frac{\cot\phi\cot\theta \mp 1}{\cot\phi \pm \cot\theta} \quad 7.24$$

- *half-angle formulas* ($\theta < 180°$)

$$\sin\frac{\theta}{2} = \sqrt{\frac{1-\cos\theta}{2}} \quad 7.25$$

$$\cos\frac{\theta}{2} = \sqrt{\frac{1+\cos\theta}{2}} \quad 7.26$$

$$\tan\frac{\theta}{2} = \sqrt{\frac{1-\cos\theta}{1+\cos\theta}} = \frac{\sin\theta}{1+\cos\theta} = \frac{1-\cos\theta}{\sin\theta} \quad 7.27$$

- *miscellaneous formulas* ($\theta < 90°$)

$$\sin\theta = 2\sin\frac{\theta}{2}\cos\frac{\theta}{2} \quad 7.28$$

$$\sin\theta = \sqrt{\frac{1-\cos 2\theta}{2}} \quad 7.29$$

$$\cos\theta = \cos^2\frac{\theta}{2} - \sin^2\frac{\theta}{2} \quad 7.30$$

$$\cos\theta = \sqrt{\frac{1+\cos 2\theta}{2}} \quad 7.31$$

$$\tan\theta = \frac{2\tan\frac{\theta}{2}}{1-\tan^2\frac{\theta}{2}}$$

$$= \frac{2\sin\frac{\theta}{2}\cos\frac{\theta}{2}}{\cos^2\frac{\theta}{2} - \sin^2\frac{\theta}{2}} \quad 7.32$$

$$\tan\theta = \sqrt{\frac{1-\cos 2\theta}{1+\cos 2\theta}}$$

$$= \frac{\sin 2\theta}{1+\cos 2\theta} = \frac{1-\cos 2\theta}{\sin 2\theta} \quad 7.33$$

$$\cot\theta = \frac{\cot^2\frac{\theta}{2} - 1}{2\cot\frac{\theta}{2}}$$

$$= \frac{\cos^2\frac{\theta}{2} - \sin^2\frac{\theta}{2}}{2\sin\frac{\theta}{2}\cos\frac{\theta}{2}} \quad 7.34$$

$$\cot\theta = \sqrt{\frac{1+\cos 2\theta}{1-\cos 2\theta}}$$

$$= \frac{1+\cos 2\theta}{\sin 2\theta} = \frac{\sin 2\theta}{1-\cos 2\theta} \quad 7.35$$

11. INVERSE TRIGONOMETRIC FUNCTIONS

Finding an angle from a known trigonometric function is a common operation known as an *inverse trigonometric operation*. The inverse function can be designated by adding "inverse," "arc-," or the superscript −1 to the name of the function. For example,

$$\text{inverse}\sin 0.5 = \arcsin 0.5 = \sin^{-1} 0.5 = 30°$$

12. HYPERBOLIC TRANSCENDENTAL FUNCTIONS

Hyperbolic transcendental functions (normally referred to as *hyperbolic functions*) are specific equations containing combinations of the terms e^θ and $e^{-\theta}$. These combinations appear regularly in certain types of problems (e.g., analysis of cables and heat transfer from fins) and are given specific names and symbols to simplify presentation.[6]

$$\text{hyperbolic sine: } \sinh\theta = \frac{e^\theta - e^{-\theta}}{2} \quad 7.36$$

$$\text{hyperbolic cosine: } \cosh\theta = \frac{e^\theta + e^{-\theta}}{2} \quad 7.37$$

$$\text{hyperbolic tangent: } \tanh\theta = \frac{e^\theta - e^{-\theta}}{e^\theta + e^{-\theta}} = \frac{\sinh\theta}{\cosh\theta} \quad 7.38$$

[5]It is an idiosyncrasy of the trade that these formulas are conventionally referred to as *formulas*, not *identities*.

[6]The hyperbolic sine and cosine functions are pronounced (by some) as "sinch" and "cosh," respectively.

hyperbolic cotangent: $\coth\theta = \frac{e^\theta + e^{-\theta}}{e^\theta - e^{-\theta}} = \frac{\cosh\theta}{\sinh\theta}$ 7.39

hyperbolic secant: $\operatorname{sech}\theta = \frac{2}{e^\theta + e^{-\theta}} = \frac{1}{\cosh\theta}$ 7.40

hyperbolic cosecant: $\operatorname{csch}\theta = \frac{2}{e^\theta - e^{-\theta}} = \frac{1}{\sinh\theta}$ 7.41

Hyperbolic functions cannot be related to a right triangle, but they are related to a rectangular (equilateral) hyperbola, as shown in Fig. 7.8. The shaded area has a value of $\theta/2$ and is sometimes given the units of *hyperbolic radians.*

$$\sinh\theta = \frac{y}{a} \quad 7.42$$

$$\cosh\theta = \frac{x}{a} \quad 7.43$$

$$\tanh\theta = \frac{y}{x} \quad 7.44$$

$$\coth\theta = \frac{x}{y} \quad 7.45$$

$$\operatorname{sech}\theta = \frac{a}{x} \quad 7.46$$

$$\operatorname{csch}\theta = \frac{a}{y} \quad 7.47$$

Figure 7.8 *Equilateral Hyperbola and Hyperbolic Functions*

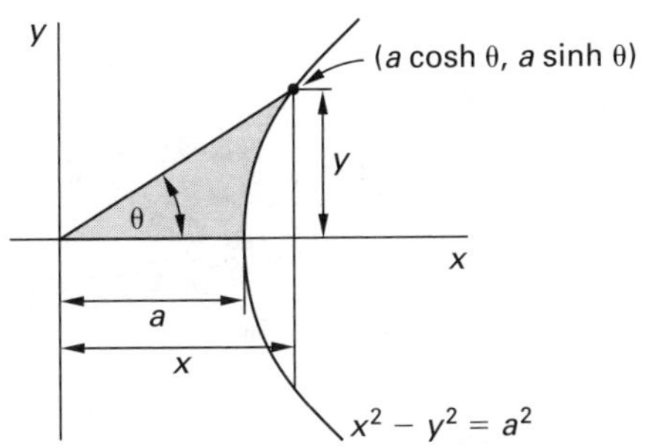

13. HYPERBOLIC IDENTITIES

The hyperbolic identities are different from the standard trigonometric identities. Some of the most important identities are presented as follows.

$$\cosh^2\theta - \sinh^2\theta = 1 \quad 7.48$$

$$1 - \tanh^2\theta = \operatorname{sech}^2\theta \quad 7.49$$

$$1 - \coth^2\theta = -\operatorname{csch}^2\theta \quad 7.50$$

$$\cosh\theta + \sinh\theta = e^\theta \quad 7.51$$

$$\cosh\theta - \sinh\theta = e^{-\theta} \quad 7.52$$

$$\sinh(\theta \pm \phi) = \sinh\theta\cosh\phi \pm \cosh\theta\sinh\phi \quad 7.53$$

$$\cosh(\theta \pm \phi) = \cosh\theta\cosh\phi \pm \sinh\theta\sinh\phi \quad 7.54$$

$$\tanh(\theta \pm \phi) = \frac{\tanh\theta \pm \tanh\phi}{1 \pm \tanh\theta\tanh\phi} \quad 7.55$$

14. GENERAL TRIANGLES

A *general triangle* (also known as an *oblique triangle*) is one that is not specifically a right triangle, as shown in Fig. 7.9. Equation 7.56 calculates the area of a general triangle.

$$\text{area} = \tfrac{1}{2}ab\sin C = \tfrac{1}{2}bc\sin A = \tfrac{1}{2}ca\sin B \quad 7.56$$

The *law of sines*, as shown in Eq. 7.57, relates the sides and the sines of the angles.

$$\frac{\sin A}{a} = \frac{\sin B}{b} = \frac{\sin C}{c} \quad 7.57$$

The *law of cosines* relates the cosine of an angle to an opposite side. (Equation 7.58 can be extended to the two remaining sides.)

$$a^2 = b^2 + c^2 - 2bc\cos A \quad 7.58$$

The *law of tangents* relates the sum and difference of two sides. (Equation 7.59 can be extended to the two remaining sides.)

$$\frac{a-b}{a+b} = \frac{\tan\left(\frac{A-B}{2}\right)}{\tan\left(\frac{A+B}{2}\right)} \quad 7.59$$

Figure 7.9 *General Triangle*

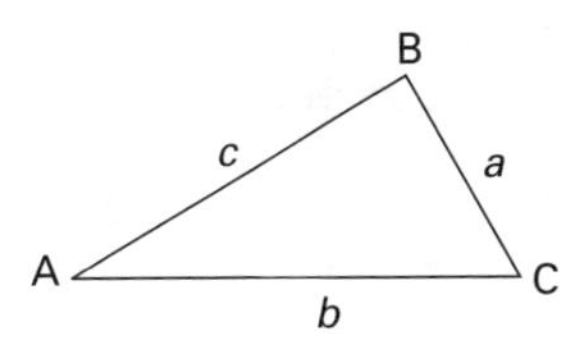

15. SPHERICAL TRIGONOMETRY

A *spherical triangle* is a triangle that has been drawn on the surface of a sphere, as shown in Fig. 7.10. The *trihedral angle* O–ABC is formed when the vertices A, B, and C are joined to the center of the sphere. The *face angles* (BOC, COA, and AOB in Fig. 7.10) are used to measure the sides (a, b, and c in Fig. 7.10). The *vertex angles* are A, B, and C. Angles are used to measure both vertex angles and sides.

Figure 7.10 Spherical Triangle ABC

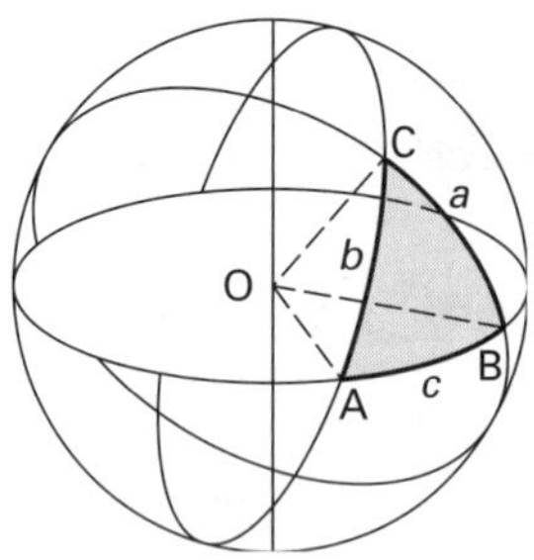

The following rules are valid for spherical triangles for which each side and angle is less than 180°.

- The sum of the three vertex angles is greater than 180° and less than 540°.

$$180° < A + B + C < 540° \qquad 7.60$$

- The sum of any two sides is greater than the third side.
- The sum of the three sides is less than 360°.

$$0° < a + b + c < 360° \qquad 7.61$$

- If the two sides are equal, the corresponding angles opposite are equal, and the converse is also true.
- If two sides are unequal, the corresponding angles opposite are unequal. The greater angle is opposite the greater side.

The *spherical excess*, ϵ, is the amount by which the sum of the vertex angles exceeds 180°. The *spherical defect*, d, is the amount by which the sum of the sides differs from 360°.

$$\epsilon = A + B + C - 180° \qquad 7.62$$

$$d = 360° - (a + b + c) \qquad 7.63$$

There are many trigonometric identities that define the relationships between angles in a spherical triangle. Some of the more common identities are presented as follows.

- *law of sines*

$$\frac{\sin A}{\sin a} = \frac{\sin B}{\sin b} = \frac{\sin C}{\sin c} \qquad 7.64$$

- *first law of cosines*

$$\cos a = \cos b \cos c + \sin b \sin c \cos A \qquad 7.65$$

- *second law of cosines*

$$\cos A = -\cos B \cos C + \sin B \sin C \cos a \qquad 7.66$$

16. SOLID ANGLES

A *solid angle*, ω, is a measure of the angle subtended at the vertex of a cone, as shown in Fig. 7.11. The solid angle has units of *steradians* (abbreviated *sr*). A steradian is the solid angle subtended at the center of a unit sphere (i.e., a sphere with a radius of one) by a unit area on its surface. Since the surface area of a sphere of radius r is r^2 times the surface area of a unit sphere, the solid angle is equal to the area cut out by the cone divided by r^2.

$$\omega = \frac{\text{surface area}}{r^2} \qquad 7.67$$

Figure 7.11 Solid Angle

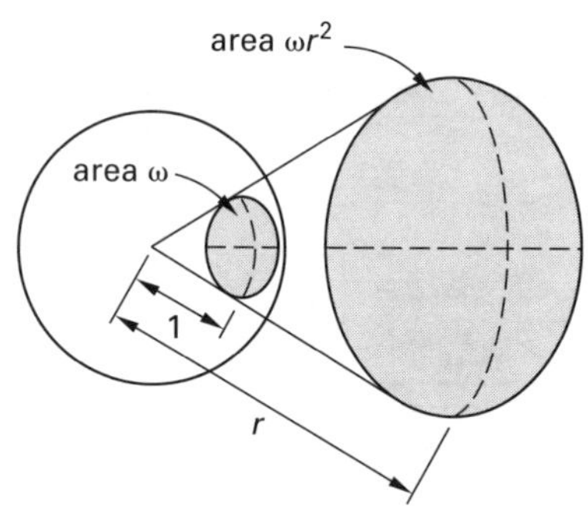

8 Analytic Geometry

1. Mensuration of Regular Shapes . . . 8-1
2. Areas with Irregular Boundaries . . . 8-1
3. Geometric Definitions . . . 8-2
4. Concave Curves . . . 8-2
5. Convex Regions . . . 8-3
6. Congruency . . . 8-3
7. Coordinate Systems . . . 8-3
8. Curves . . . 8-3
9. Symmetry of Curves . . . 8-4
10. Straight Lines . . . 8-4
11. Direction Numbers, Angles, and Cosines . . . 8-5
12. Intersection of Two Lines . . . 8-6
13. Planes . . . 8-6
14. Distances Between Geometric Figures . . . 8-7
15. Angles Between Geometric Figures . . . 8-8
16. Conic Sections . . . 8-8
17. Circle . . . 8-9
18. Parabola . . . 8-10
19. Ellipse . . . 8-10
20. Hyperbola . . . 8-11
21. Sphere . . . 8-12
22. Helix . . . 8-12

1. MENSURATION OF REGULAR SHAPES

The dimensions, perimeter, area, and other geometric properties constitute the *mensuration* (i.e., the measurements) of a geometric shape. Appendix 8.A and App. 8.B contain formulas and tables used to calculate these properties.

Example 8.1

In the study of open channel fluid flow, the hydraulic radius is defined as the ratio of flow area to wetted perimeter. What is the hydraulic radius of a 6 in inside diameter pipe filled to a depth of 2 in?

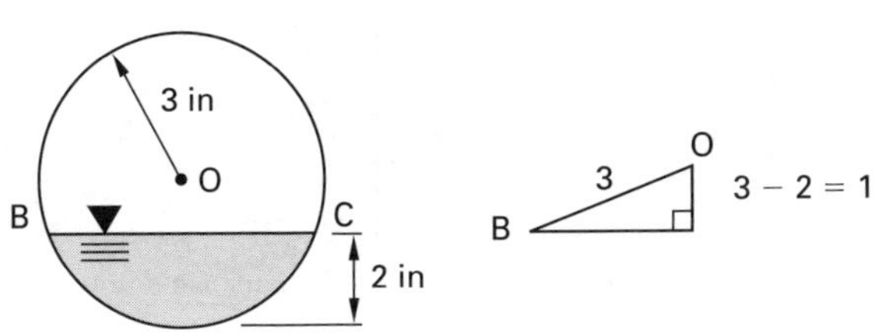

Solution

Points O, B, and C constitute a circular segment and are used to find the central angle of the circular segment.

$$\tfrac{1}{2}\angle\text{BOC} = \arccos\tfrac{1}{3} = 70.53^\circ$$

$$\phi = \angle\text{BOC} = \frac{(2)(70.53^\circ)(2\pi)}{360^\circ} = 2.462 \text{ rad}$$

From App. 8.A, the area in flow and arc length are

$$\begin{aligned} A &= \tfrac{1}{2}r^2(\phi - \sin\phi) \\ &= \left(\tfrac{1}{2}\right)(3 \text{ in})^2\big(2.462 \text{ rad} - \sin(2.462 \text{ rad})\big) \\ &= 8.251 \text{ in}^2 \\ s &= r\phi \\ &= (3 \text{ in})(2.462 \text{ rad}) = 7.386 \text{ in} \end{aligned}$$

The hydraulic radius is

$$\begin{aligned} r_h &= \frac{A}{s} \\ &= \frac{8.251 \text{ in}^2}{7.386 \text{ in}} = 1.12 \text{ in} \end{aligned}$$

2. AREAS WITH IRREGULAR BOUNDARIES

Areas of sections with irregular boundaries (such as creek banks) cannot be determined precisely, and approximation methods must be used. If the irregular side can be divided into a series of cells of equal width, either the trapezoidal rule or Simpson's rule can be used. Figure 8.1 shows an example of an irregular area.

Figure 8.1 *Irregular Areas*

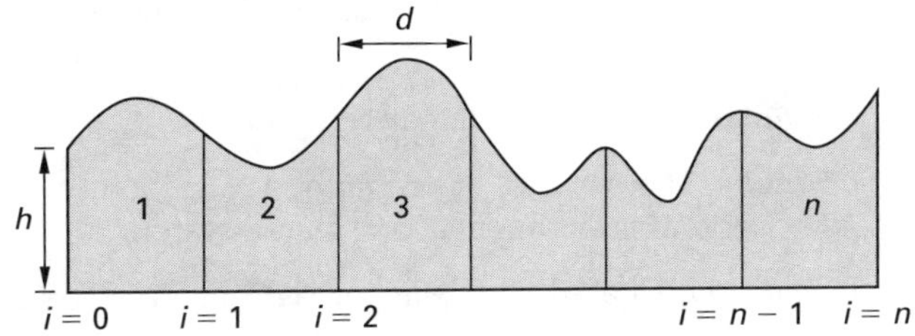

If the irregular side of each cell is fairly straight, the *trapezoidal rule* is appropriate.

$$A = \frac{d}{2}\left(h_0 + h_n + 2\sum_{i=1}^{n-1} h_i\right) \tag{8.1}$$

If the irregular side of each cell is curved (parabolic), *Simpson's rule* should be used. (n must be even to use Simpson's rule.)

$$A = \frac{d}{3}\left(h_0 + h_n + 4\sum_{\substack{i\ \text{odd}\\ i=1}}^{n-1} h_i + 2\sum_{\substack{i\ \text{even}\\ i=2}}^{n-2} h_i\right) \tag{8.2}$$

3. GEOMETRIC DEFINITIONS

The following terms are used in this book to describe the relationship or orientation of one geometric figure to another. Figure 8.2 illustrates some of the following geometric definitions.

- *abscissa:* the horizontal coordinate, typically designated as x in a rectangular coordinate system
- *asymptote:* a straight line that is approached but not intersected by a curved line
- *asymptotic:* approaching the slope of another line; attaining the slope of another line in the limit
- *center:* a point equidistant from all other points
- *collinear:* falling on the same line
- *concave:* curved inward (in the direction indicated)[1]
- *convex:* curved outward (in the direction indicated)
- *convex hull:* a closed figure whose surface is convex everywhere
- *coplanar:* falling on the same plane
- *inflection point:* a point where the second derivative changes sign or the curve changes from concave to convex; also known as a *point of contraflexure*
- *locus of points:* a set or collection of points having some common property and being so infinitely close together as to be indistinguishable from a line
- *node:* a point on a line from which other lines enter or leave
- *normal:* rotated 90°; being at right angles
- *ordinate:* the vertical coordinate, typically designated as y in a rectangular coordinate system
- *orthogonal:* rotated 90°; being at right angles
- *saddle point:* a point in three-dimensional space where all adjacent points are higher in one direction (the direction of the saddle) and lower in an orthogonal direction (the direction of the sides)
- *tangent point:* having equal slopes at a common point

[1]This is easily remembered since one must go inside to explore a cave.

Figure 8.2 Geometric Definitions

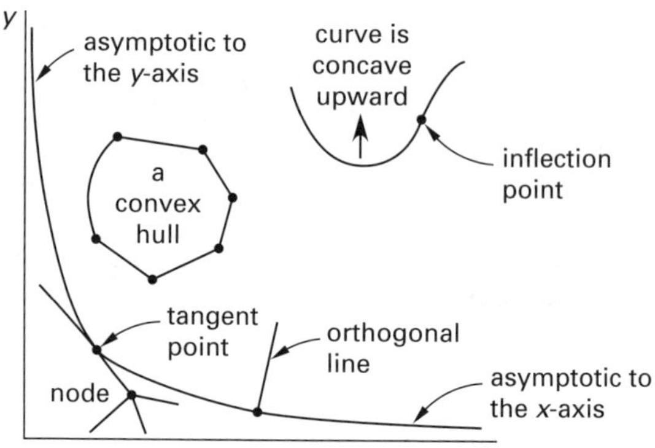

4. CONCAVE CURVES

Concavity is a term that is applied to curved lines. A *concave up curve* is one whose function's first derivative increases continuously from negative to positive values. Straight lines drawn tangent to concave up curves are all below the curve. The graph of such a function may be thought of as being able to "hold water."

The first derivative of a *concave down curve* decreases continuously from positive to negative. A graph of a concave down function may be thought of as "spilling water." (See Fig. 8.3.)

Figure 8.3 Concave Curves

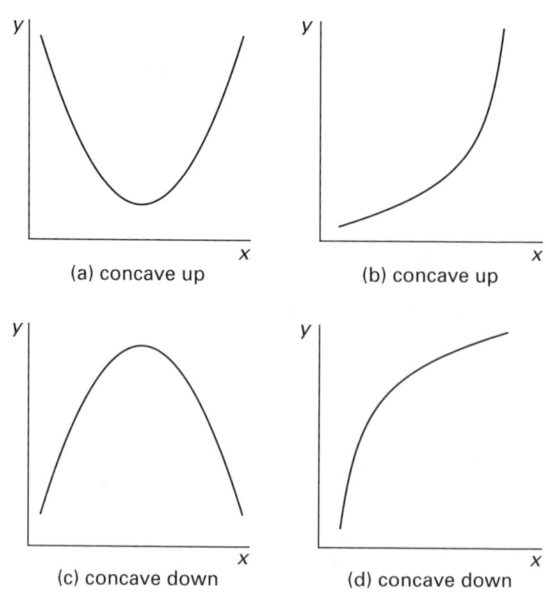

5. CONVEX REGIONS

Convexity is a term that is applied to sets and regions.[2] It plays an important role in many mathematics subjects. A set or multidimensional region is *convex* if it contains the line segment joining any two of its points; that is, if a straight line is drawn connecting any two points in a convex region, that line will lie entirely within the region. For example, the interior of a parabola is a convex region, as is a solid sphere. The *void* or *null region* (i.e., an empty set of points), single points, and straight lines are convex sets. A convex region bounded by separate, connected line segments is known as a *convex hull.* (See Fig. 8.4.)

Figure 8.4 *Convexity*

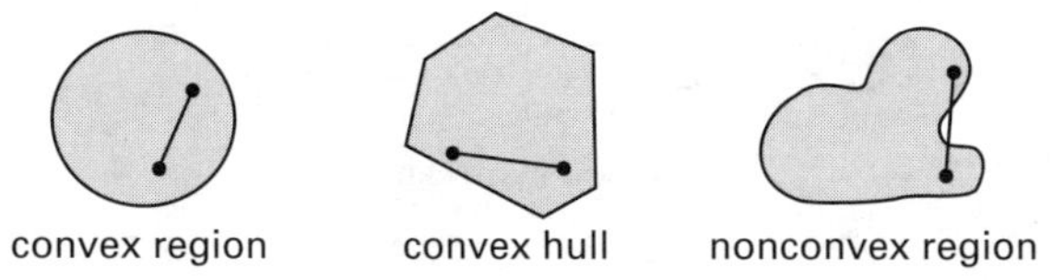

Within a convex region, a local maximum is also the global maximum. Similarly, a local minimum is also the global minimum. The intersection of two convex regions is also convex.

6. CONGRUENCY

Congruence in geometric figures is analogous to *equality* in algebraic expressions. Congruent line segments are segments that have the same length. Congruent angles have the same angular measure. Congruent triangles have the same vertex angles and side lengths.

In general, *congruency*, indicated by the symbol $\cong$, means that there is one-to-one correspondence between all points on two objects. This correspondence is defined by the *mapping function* or *isometry*, which can be a translation, rotation, or reflection. Since the identity function is a valid mapping function, every geometric shape is congruent to itself.

Two congruent objects can be in different spaces. For example, a triangular area in three-dimensional space can be mapped into a triangle in two-dimensional space.

7. COORDINATE SYSTEMS

The manner in which a geometric figure is described depends on the coordinate system that is used. The three-dimensional system (also known as the *rectangular coordinate system* and *Cartesian coordinate system*) with its x-, y-, and z-coordinates is the most commonly used in engineering. Table 8.1 summarizes the components needed to specify a point in the various coordinate systems. Figure 8.5 illustrates the use of and conversion between the coordinate systems.

[2]It is tempting to define regions that fail the convexity test as being "concave." However, it is more proper to define such regions as "nonconvex." In any case, it is important to recognize that convexity depends on the reference point: An observer within a sphere will see the spherical boundary as convex; an observer outside the sphere may see the boundary as nonconvex.

Table 8.1 *Components of Coordinate Systems*

name	dimensions	components
rectangular	2	x, y
rectangular	3	x, y, z
polar	2	r, θ
cylindrical	3	r, θ, z
spherical	3	r, θ, ϕ

Figure 8.5 *Different Coordinate Systems*

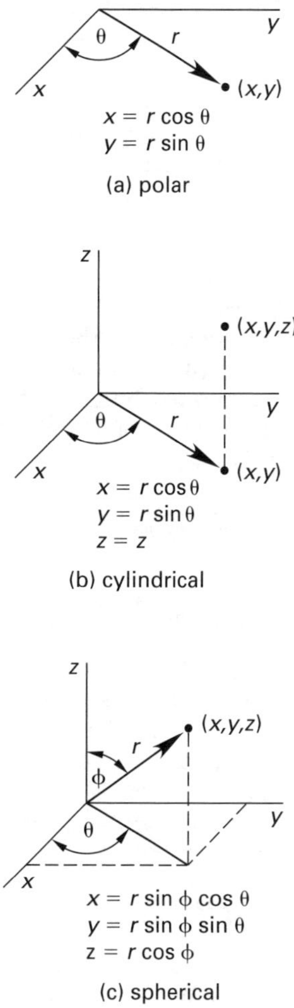

8. CURVES

A *curve* (commonly called a *line*) is a function over a finite or infinite range of the independent variable. When a curve is drawn in two- or three-dimensional space, it is known as a *graph of the curve.* It may or may not be possible to describe the curve mathematically. The *degree of a curve* is the highest exponent in the function. For example, Eq. 8.3 is a fourth-degree curve.

$$f(x) = 2x^4 + 7x^3 + 6x^2 + 3x + 9 = 0 \qquad 8.3$$

An *ordinary cycloid* ("wheel line") is a curve traced out by a point on the rim of a wheel that rolls without slipping. (See Fig. 8.6.) Cycloids that start with the tracing point down (i.e., on the x-axis) are described in parametric form by Eq. 8.4 and Eq. 8.5 and in rectangular form by Eq. 8.6. In Eq. 8.4 through Eq. 8.6, using the minus sign results in a bottom (downward) cusp at the origin; using the plus sign results in a vertex (trough) at the origin and a top (upward) cusp.

$$x = r(\theta \pm \sin\theta) \qquad 8.4$$

$$y = r(1 \pm \cos\theta) \qquad 8.5$$

$$x = r\arccos\frac{r-y}{r} \pm \sqrt{2ry - y^2} \qquad 8.6$$

Figure 8.6 *Cycloid (cusp at origin shown)*

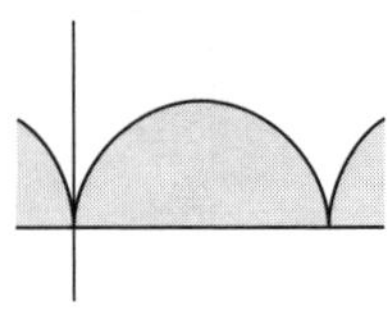

An *epicycloid* is a curve generated by a point on the rim of a wheel that rolls on the outside of a circle. A *hypocycloid* is a curve generated by a point on the rim of a wheel that rolls on the inside of a circle. The equation of a hypocycloid of four cusps is

$$x^{2/3} + y^{2/3} = r^{2/3} \quad [\text{4 cusps}] \qquad 8.7$$

9. SYMMETRY OF CURVES

Two points, P and Q, are symmetrical with respect to a line if the line is a perpendicular bisector of the line segment PQ. If the graph of a curve is unchanged when y is replaced with $-y$, the curve is symmetrical with respect to the x-axis. If the curve is unchanged when x is replaced with $-x$, the curve is symmetrical with respect to the y-axis.

Repeating waveforms can be symmetrical with respect to the y-axis. A curve $f(x)$ is said to have *even symmetry* if $f(x) = f(-x)$. (Alternatively, $f(x)$ is said to be a *symmetrical function.*) With even symmetry, the function to the left of $x = 0$ is a reflection of the function to the right of $x = 0$. (In effect, the y-axis is a mirror.) The cosine curve is an example of a curve with even symmetry.

A curve is said to have *odd symmetry* if $f(x) = -f(-x)$. (Alternatively, $f(x)$ is said to be an *asymmetrical function.*[3]) The sine curve is an example of a curve with odd symmetry.

A curve is said to have *rotational symmetry* (*half-wave symmetry*) if $f(x) = -f(x + \pi)$.[4] Curves of this type are identical except for a sign reversal on alternate half-cycles. (See Fig. 8.7.)

Table 8.2 describes the type of function resulting from the combination of two functions.

Figure 8.7 *Waveform Symmetry*

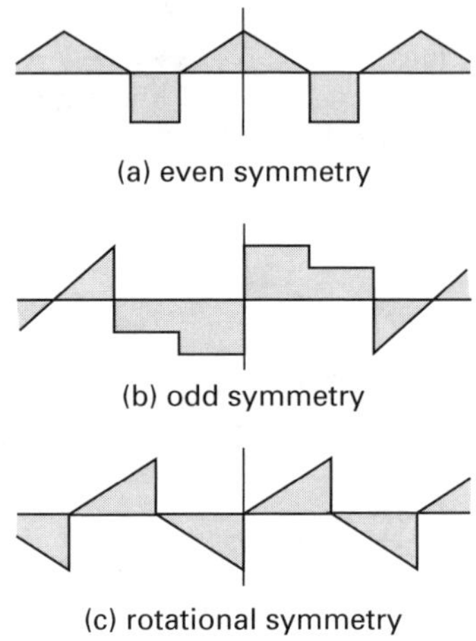

Table 8.2 *Combinations of Functions*

	operation			
	+	−	×	÷
$f_1(x)$ even, $f_2(x)$ even	even	even	even	even
$f_1(x)$ odd, $f_2(x)$ odd	odd	odd	even	even
$f_1(x)$ even, $f_2(x)$ odd	neither	neither	odd	odd

10. STRAIGHT LINES

Figure 8.8 illustrates a straight line in two-dimensional space. The *slope* of the line is m, the *y-intercept* is b, and the *x-intercept* is a. The equation of the line can be represented in several forms. The procedure for finding the equation depends on the form chosen to represent the line. In general, the procedure involves substituting one or more known points on the line into the equation in order to determine the coefficients.

Figure 8.8 *Straight Line*

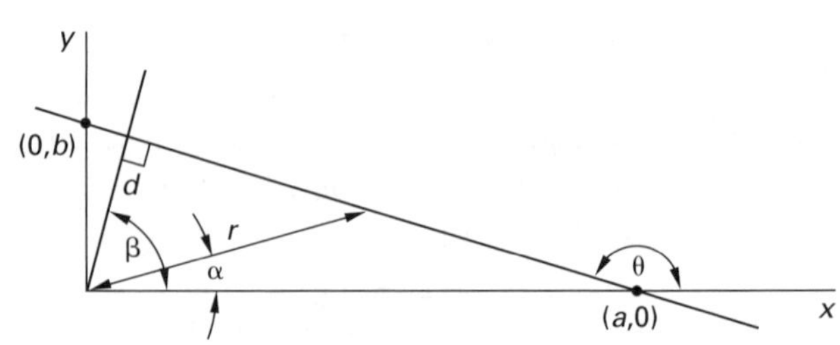

[3]Although they have the same meaning, the semantics of "even symmetry" and "asymmetrical function" are contradictory.

[4]The symbol π represents half of a full cycle of the waveform, not the value 3.141...

- *general form*

$$Ax + By + C = 0 \quad 8.8$$

$$A = -mB \quad 8.9$$

$$B = \frac{-C}{b} \quad 8.10$$

$$C = -aA = -bB \quad 8.11$$

- *slope-intercept form*

$$y = mx + b \quad 8.12$$

$$m = \frac{-A}{B} = \tan\theta = \frac{y_2 - y_1}{x_2 - x_1} \quad 8.13$$

$$b = \frac{-C}{B} \quad 8.14$$

$$a = \frac{-C}{A} \quad 8.15$$

- *point-slope form*

$$y - y_1 = m(x - x_1) \quad 8.16$$

- *intercept form*

$$\frac{x}{a} + \frac{y}{b} = 1 \quad 8.17$$

- *two-point form*

$$\frac{y - y_1}{x - x_1} = \frac{y_2 - y_1}{x_2 - x_1} \quad 8.18$$

- *normal form*

$$x\cos\beta + y\sin\beta - d = 0 \quad 8.19$$

(d and β are constants; x and y are variables.)

- *polar form*

$$r = \frac{d}{\cos(\beta - \alpha)} \quad 8.20$$

(d and β are constants; r and α are variables.)

11. DIRECTION NUMBERS, ANGLES, AND COSINES

Given a directed line from (x_1, y_1, z_1) to (x_2, y_2, z_2), the *direction numbers* are

$$L = x_2 - x_1 \quad 8.21$$

$$M = y_2 - y_1 \quad 8.22$$

$$N = z_2 - z_1 \quad 8.23$$

The distance between two points is

$$d = \sqrt{L^2 + M^2 + N^2} \quad 8.24$$

The *direction cosines* are

$$\cos\alpha = \frac{L}{d} \quad 8.25$$

$$\cos\beta = \frac{M}{d} \quad 8.26$$

$$\cos\gamma = \frac{N}{d} \quad 8.27$$

Note that

$$\cos^2\alpha + \cos^2\beta + \cos^2\gamma = 1 \quad 8.28$$

The *direction angles* are the angles between the axes and the lines. They are found from the inverse functions of the direction cosines.

$$\alpha = \arccos\frac{L}{d} \quad 8.29$$

$$\beta = \arccos\frac{M}{d} \quad 8.30$$

$$\gamma = \arccos\frac{N}{d} \quad 8.31$$

The direction cosines can be used to write the equation of the straight line in terms of the unit vectors. The line **R** would be defined as

$$\mathbf{R} = d(\mathbf{i}\cos\alpha + \mathbf{j}\cos\beta + \mathbf{k}\cos\gamma) \quad 8.32$$

Similarly, the line may be written in terms of its direction numbers.

$$\mathbf{R} = L\mathbf{i} + M\mathbf{j} + N\mathbf{k} \quad 8.33$$

Example 8.2

A line passes through the points (4, 7, 9) and (0, 1, 6). Write the equation of the line in terms of its (a) direction numbers and (b) direction cosines.

Solution

(a) The direction numbers are

$$L = 4 - 0 = 4$$
$$M = 7 - 1 = 6$$
$$N = 9 - 6 = 3$$

Using Eq. 8.33,

$$\mathbf{R} = 4\mathbf{i} + 6\mathbf{j} + 3\mathbf{k}$$

(b) The distance between the two points is

$$d = \sqrt{(4)^2 + (6)^2 + (3)^2} = 7.81$$

The line in terms of its direction cosines is

$$\mathbf{R} = \frac{4\mathbf{i} + 6\mathbf{j} + 3\mathbf{k}}{7.81} = 0.512\mathbf{i} + 0.768\mathbf{j} + 0.384\mathbf{k}$$

12. INTERSECTION OF TWO LINES

The intersection of two lines is a point. The location of the intersection point can be determined by setting the two equations equal and solving them in terms of a common variable. Alternatively, Eq. 8.34 and Eq. 8.35 can be used to calculate the coordinates of the intersection point.

$$x = \frac{B_2C_1 - B_1C_2}{A_2B_1 - A_1B_2} \quad 8.34$$

$$y = \frac{A_1C_2 - A_2C_1}{A_2B_1 - A_1B_2} \quad 8.35$$

13. PLANES

A *plane* in three-dimensional space (see Fig. 8.9) is completely determined by one of the following:

- three noncollinear points
- two nonparallel vectors $\mathbf{V}_1$ and $\mathbf{V}_2$ and their intersection point P_0
- a point P_0 and a vector, $\mathbf{N}$, normal to the plane (i.e., the *normal vector*)

Figure 8.9 *Plane in Three-Dimensional Space*

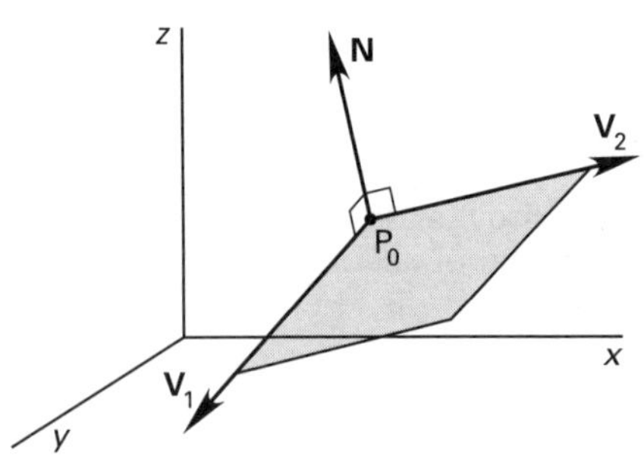

The plane can be specified mathematically in one of two ways: in rectangular form or as a parametric equation. The general form is

$$A(x - x_0) + B(y - y_0) + C(z - z_0) = 0 \quad 8.36$$

x_0, y_0, and z_0 are the coordinates of the intersection point of any two vectors in the plane. The coefficients A, B, and C are the same as the coefficients of the normal vector, $\mathbf{N}$.

$$\mathbf{N} = \mathbf{V}_1 \times \mathbf{V}_2 = A\mathbf{i} + B\mathbf{j} + C\mathbf{k} \quad 8.37$$

Equation 8.36 can be simplified as follows.

$$Ax + By + Cz + D = 0 \quad 8.38$$

$$D = -(Ax_0 + By_0 + Cz_0) \quad 8.39$$

The following procedure can be used to determine the equation of a plane from three noncollinear points, P_1, P_2, and P_3, or from a normal vector and a single point.

step 1: (If the normal vector is known, go to step 3.) Determine the equations of the vectors $\mathbf{V}_1$ and $\mathbf{V}_2$ from two pairs of the points. For example, determine $\mathbf{V}_1$ from points P_1 and P_2, and determine $\mathbf{V}_2$ from P_1 and P_3. Express the vectors in the form $A\mathbf{i} + B\mathbf{j} + C\mathbf{k}$.

$$\mathbf{V}_1 = (x_2 - x_1)\mathbf{i} + (y_2 - y_1)\mathbf{j} + (z_2 - z_1)\mathbf{k} \quad 8.40$$

$$\mathbf{V}_2 = (x_3 - x_1)\mathbf{i} + (y_3 - y_1)\mathbf{j} + (z_3 - z_1)\mathbf{k} \quad 8.41$$

step 2: Find the normal vector, N, as the cross product of the two vectors.

$$\mathbf{N} = \mathbf{V}_1 \times \mathbf{V}_2 = \begin{vmatrix} \mathbf{i} & (x_2 - x_1) & (x_3 - x_1) \\ \mathbf{j} & (y_2 - y_1) & (y_3 - y_1) \\ \mathbf{k} & (z_2 - z_1) & (z_3 - z_1) \end{vmatrix} \quad 8.42$$

step 3: Write the general equation of the plane in rectangular form (see Eq. 8.36) using the coefficients A, B, and C from the normal vector and any one of the three points as P_0.

The parametric equations of a plane also can be written as a linear combination of the components of two vectors in the plane. Referring to Fig. 8.9, the two known vectors are

$$\mathbf{V}_1 = V_{1x}\mathbf{i} + V_{1y}\mathbf{j} + V_{1z}\mathbf{k} \quad 8.43$$

$$\mathbf{V}_2 = V_{2x}\mathbf{i} + V_{2y}\mathbf{j} + V_{2z}\mathbf{k} \quad 8.44$$

If s and t are scalars, the coordinates of each point in the plane can be written as Eq. 8.45 through Eq. 8.47. These are the parametric equations of the plane.

$$x = x_0 + sV_{1x} + tV_{2x} \quad 8.45$$

$$y = y_0 + sV_{1y} + tV_{2y} \quad 8.46$$

$$z = z_0 + sV_{1z} + tV_{2z} \quad 8.47$$

Example 8.3

The following points are coplanar.

$$P_1 = (2, 1, -4)$$

$$P_2 = (4, -2, -3)$$

$$P_3 = (2, 3, -8)$$

Determine the equation of the plane in (a) general form and (b) parametric form.

Solution

(a) Use the first two points to find a vector, $\mathbf{V}_1$.

$$\begin{aligned}\mathbf{V}_1 &= (x_2 - x_1)\mathbf{i} + (y_2 - y_1)\mathbf{j} + (z_2 - z_1)\mathbf{k}\\ &= (4-2)\mathbf{i} + (-2-1)\mathbf{j} + \big(-3-(-4)\big)\mathbf{k}\\ &= 2\mathbf{i} - 3\mathbf{j} + 1\mathbf{k}\end{aligned}$$

Similarly, use the first and third points to find $\mathbf{V}_2$.

$$\begin{aligned}\mathbf{V}_2 &= (x_3 - x_1)\mathbf{i} + (y_3 - y_1)\mathbf{j} + (z_3 - z_1)\mathbf{k}\\ &= (2-2)\mathbf{i} + (3-1)\mathbf{j} + \big(-8-(-4)\big)\mathbf{k}\\ &= 0\mathbf{i} + 2\mathbf{j} - 4\mathbf{k}\end{aligned}$$

From Eq. 8.42, determine the normal vector as a determinant.

$$\mathbf{N} = \begin{vmatrix}\mathbf{i} & 2 & 0\\ \mathbf{j} & -3 & 2\\ \mathbf{k} & 1 & -4\end{vmatrix}$$

Expand the determinant across the top row.

$$\begin{aligned}\mathbf{N} &= \mathbf{i}(12-2) - 2(-4\mathbf{j} - 2\mathbf{k})\\ &= 10\mathbf{i} + 8\mathbf{j} + 4\mathbf{k}\end{aligned}$$

The rectangular form of the equation of the plane uses the same constants as in the normal vector. Use the first point and write the equation of the plane in the form of Eq. 8.36.

$$(10)(x-2) + (8)(y-1) + (4)(z+4) = 0$$

The three constant terms can be combined by using Eq. 8.39.

$$D = -\big((10)(2) + (8)(1) + (4)(-4)\big) = -12$$

The equation of the plane is

$$10x + 8y + 4z - 12 = 0$$

(b) The parametric equations based on the first point and for any values of s and t are

$$\begin{aligned}x &= 2 + 2s + 0t\\ y &= 1 - 3s + 2t\\ z &= -4 + 1s - 4t\end{aligned}$$

The scalars s and t are not unique. Two of the three coordinates can also be chosen as the parameters. Dividing the rectangular form of the plane's equation by 4 to isolate z results in an alternate set of parametric equations.

$$\begin{aligned}x &= x\\ y &= y\\ z &= 3 - 2.5x - 2y\end{aligned}$$

14. DISTANCES BETWEEN GEOMETRIC FIGURES

The smallest distance, d, between various geometric figures is given by the following equations.

- between two points in (x, y, z) format:

$$d = \sqrt{(x_2 - x_1)^2 + (y_2 - y_1)^2 + (z_2 - z_1)^2} \qquad 8.48$$

- between a point (x_0, y_0) and a line $Ax + By + C = 0$:

$$d = \frac{|Ax_0 + By_0 + C|}{\sqrt{A^2 + B^2}} \qquad 8.49$$

- between a point (x_0, y_0, z_0) and a plane $Ax + By + Cz + D = 0$:

$$d = \frac{|Ax_0 + By_0 + Cz_0 + D|}{\sqrt{A^2 + B^2 + C^2}} \qquad 8.50$$

- between two parallel lines $Ax + By + C = 0$:

$$d = \left|\frac{|C_2|}{\sqrt{A_2^2 + B_2^2}} - \frac{|C_1|}{\sqrt{A_1^2 + B_1^2}}\right| \qquad 8.51$$

Example 8.4

What is the minimum distance between the line $y = 2x + 3$ and the origin $(0, 0)$?

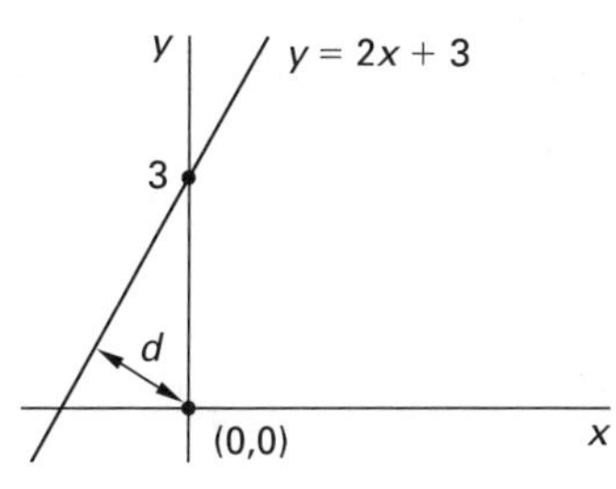

Solution

Put the equation in general form.

$$Ax + By + C = 2x - y + 3 = 0$$

Use Eq. 8.49 with $(x, y) = (0, 0)$.

$$d = \frac{|Ax + By + C|}{\sqrt{A^2 + B^2}} = \frac{(2)(0) + (-1)(0) + 3}{\sqrt{(2)^2 + (-1)^2}} = \frac{3}{\sqrt{5}}$$

15. ANGLES BETWEEN GEOMETRIC FIGURES

The angle, ϕ, between various geometric figures is given by the following equations.

- between two lines in $Ax + By + C = 0$, $y = mx + b$, or direction angle formats:

$$\phi = \arctan\left(\frac{A_1B_2 - A_2B_1}{A_1A_2 + B_1B_2}\right) \quad \text{8.52}$$

$$\phi = \arctan\left(\frac{m_2 - m_1}{1 + m_1m_2}\right) \quad \text{8.53}$$

$$\phi = |\arctan(m_1) - \arctan(m_2)| \quad \text{8.54}$$

$$\phi = \arccos\left(\frac{L_1L_2 + M_1M_2 + N_1N_2}{d_1d_2}\right) \quad \text{8.55}$$

$$\phi = \arccos\left(\begin{array}{c}\cos\alpha_1\cos\alpha_2 + \cos\beta_1\cos\beta_2 \\ + \cos\gamma_1\cos\gamma_2\end{array}\right) \quad \text{8.56}$$

If the lines are parallel, then $\phi = 0$.

$$\frac{A_1}{A_2} = \frac{B_1}{B_2} \quad \text{8.57}$$

$$m_1 = m_2 \quad \text{8.58}$$

$$\alpha_1 = \alpha_2;\ \beta_1 = \beta_2;\ \gamma_1 = \gamma_2 \quad \text{8.59}$$

If the lines are perpendicular, then $\phi = 90°$.

$$A_1A_2 = -B_1B_2 \quad \text{8.60}$$

$$m_1 = \frac{-1}{m_2} \quad \text{8.61}$$

$$\alpha_1 + \alpha_2 = \beta_1 + \beta_2 = \gamma_1 + \gamma_2 = 90° \quad \text{8.62}$$

- between two planes in $A\mathbf{i} + B\mathbf{j} + C\mathbf{k} = 0$ format, the coefficients A, B, and C are the same as the coefficients for the normal vector. (See Eq. 8.37.) ϕ is equal to the angle between the two normal vectors.

$$\cos\phi = \frac{|A_1A_2 + B_1B_2 + C_1C_2|}{\sqrt{A_1^2 + B_1^2 + C_1^2}\sqrt{A_2^2 + B_2^2 + C_2^2}} \quad \text{8.63}$$

Example 8.5

Use Eq. 8.52, Eq. 8.53, and Eq. 8.54 to find the angle between the lines.

$$y = -0.577x + 2$$

$$y = +0.577x - 5$$

Solution

Write both equations in general form.

$$-0.577x - y + 2 = 0$$

$$0.577x - y - 5 = 0$$

(a) From Eq. 8.52,

$$\phi = \arctan\left(\frac{A_1B_2 - A_2B_1}{A_1A_2 + B_1B_2}\right) = \arctan\left(\frac{(-0.577)(-1) - (0.577)(-1)}{(-0.577)(0.577) + (-1)(-1)}\right) = 60°$$

(b) Use Eq. 8.53.

$$\phi = \arctan\left(\frac{m_2 - m_1}{1 + m_1m_2}\right) = \arctan\left(\frac{0.577 - (-0.577)}{1 + (0.577)(-0.577)}\right) = 60°$$

(c) Use Eq. 8.54.

$$\phi = |\arctan(m_1) - \arctan(m_2)| = |\arctan(-0.577) - \arctan(0.577)| = |-30° - 30°| = 60°$$

16. CONIC SECTIONS

A *conic section* is any one of several curves produced by passing a plane through a cone as shown in Fig. 8.10. If α is the angle between the vertical axis and the cutting plane and β is the cone generating angle, Eq. 8.64 gives the *eccentricity*, ϵ, of the conic section. Values of the eccentricity are given in Fig. 8.10.

$$\epsilon = \frac{\cos\alpha}{\cos\beta} \quad \text{8.64}$$

All conic sections are described by second-degree polynomials (i.e., *quadratic equations*) of the following form.[5]

$$Ax^2 + Bxy + Cy^2 + Dx + Ey + F = 0 \quad \text{8.65}$$

This is the *general form*, which allows the figure axes to be at any angle relative to the coordinate axes. The *standard forms* presented in the following sections pertain to figures whose axes coincide with the coordinate axes, thereby eliminating certain terms of the general equation.

Figure 8.11 can be used to determine which conic section is described by the quadratic function. The quantity $B^2 - 4AC$ is known as the *discriminant*. Figure 8.11

[5]One or more straight lines are produced when the cutting plane passes through the cone's vertex. Straight lines can be considered to be quadratic functions without second-degree terms.

determines only the type of conic section; it does not determine whether the conic section is degenerate (e.g., a circle with a negative radius).

Figure 8.10 *Conic Sections Produced by Cutting Planes*

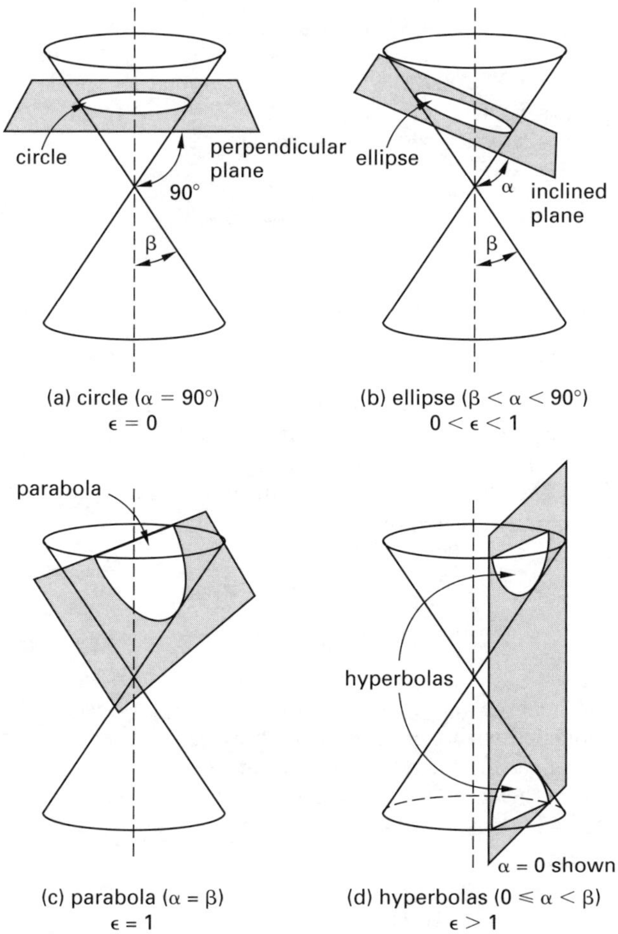

Example 8.6

What geometric figures are described by the following equations?

(a) $4y^2 - 12y + 16x + 41 = 0$

(b) $x^2 - 10xy + y^2 + x + y + 1 = 0$

(c) $x^2 + 4y^2 + 2x - 8y + 1 = 0$

(d) $x^2 + y^2 - 6x + 8y + 20 = 0$

Solution

(a) Referring to Fig. 8.11, $B = 0$ since there is no xy term, $A = 0$ since there is no x^2 term, and $AC = (0)(4) = 0$. This is a parabola.

(b) $B \neq 0$; $B^2 - 4AC = (-10)^2 - (4)(1)(1) = +96$. This is a hyperbola.

(c) $B = 0$; $A \neq C$; $AC = (1)(4) = +4$. This is an ellipse.

(d) $B = 0$; $A = C$; $A = C = 1$ ($\neq 0$). This is a circle.

Figure 8.11 *Determining Conic Sections from Quadratic Equations*

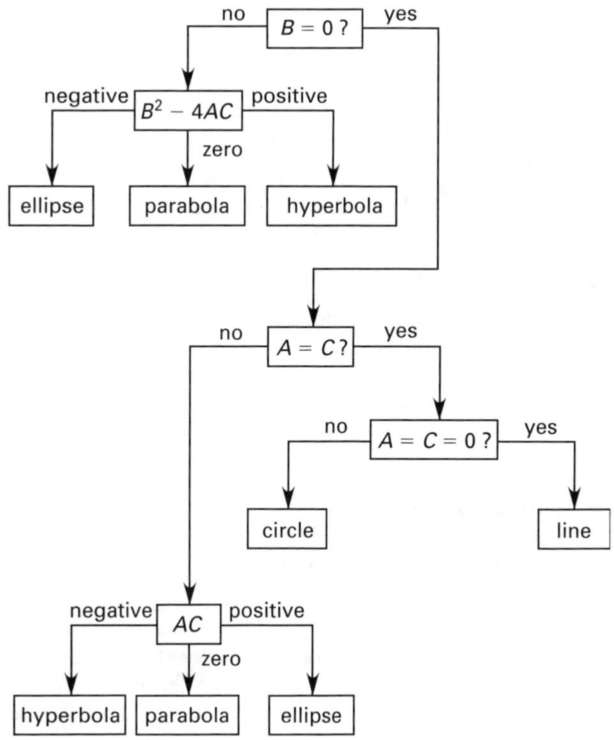

17. CIRCLE

The general form of the equation of a circle, as illustrated in Fig. 8.12, is

$$Ax^2 + Ay^2 + Dx + Ey + F = 0 \qquad 8.66$$

Figure 8.12 *Circle*

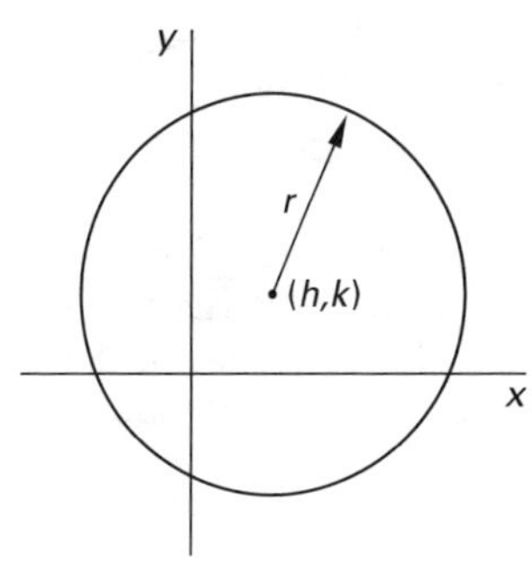

The *center-radius form* of the equation of a circle with radius r and center at (h, k) is

$$(x - h)^2 + (y - k)^2 = r^2 \qquad 8.67$$

The two forms can be converted by use of Eq. 8.68 through Eq. 8.70.

$$h = \frac{-D}{2A} \quad 8.68$$

$$k = \frac{-E}{2A} \quad 8.69$$

$$r^2 = \frac{D^2 + E^2 - 4AF}{4A^2} \quad 8.70$$

If the right-hand side of Eq. 8.70 is positive, the figure is a circle. If it is zero, the circle shrinks to a point. If the right-hand side is negative, the figure is imaginary. A *degenerate circle* is one in which the right-hand side is less than or equal to zero.

18. PARABOLA

A *parabola* is the locus of points equidistant from the *focus* (point F in Fig. 8.13) and a line called the *directrix*. A parabola is symmetric with respect to its *parabolic axis*. The line normal to the parabolic axis and passing through the focus is known as the *latus rectum*. The eccentricity of a parabola is 1.

Figure 8.13 Parabola

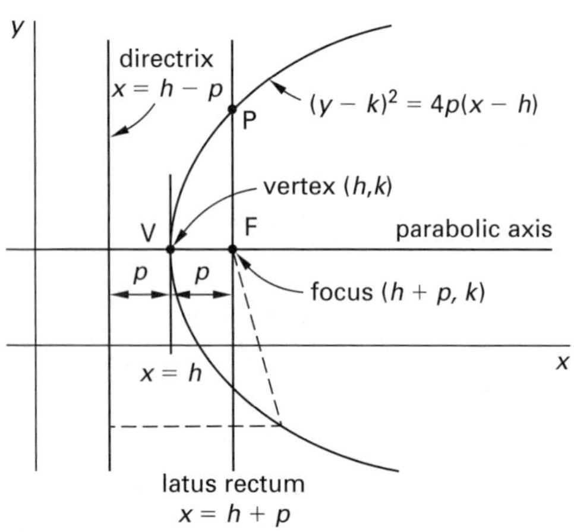

There are two common types of parabolas in the Cartesian plane—those that open right and left, and those that open up and down. Equation 8.65 is the general form of the equation of a parabola. With Eq. 8.71, the parabola points horizontally to the right if $CD > 0$ and to the left if $CD < 0$. With Eq. 8.72, the parabola points vertically up if $AE > 0$ and down if $AE < 0$.

$$Cy^2 + Dx + Ey + F = 0\Big|_{\substack{C,D\neq 0\\ \text{opens horizontally}}} \quad 8.71$$

$$Ax^2 + Dx + Ey + F = 0\Big|_{\substack{A,E\neq 0\\ \text{opens vertically}}} \quad 8.72$$

The *standard form* of the equation of a parabola with vertex at (h, k), focus at $(h + p, k)$, and directrix at $x = h - p$, and that opens to the right or left is given by Eq. 8.73. The parabola opens to the right (points to the left) if $p > 0$ and opens to the left (points to the right) if $p < 0$.

$$(y - k)^2 = 4p(x - h)\Big|_{\text{opens horizontally}} \quad 8.73$$

$$y^2 = 4px\Big|_{\substack{\text{vertex at origin}\\ h=k=0}} \quad 8.74$$

The *standard form* of the equation of a parabola with vertex at (h, k), focus at $(h, k + p)$, and directrix at $y = k - p$, and that opens up or down is given by Eq. 8.75. The parabola opens up (points down) if $p > 0$ and opens down (points up) if $p < 0$.

$$(x - h)^2 = 4p(y - k)\big|_{\text{opens vertically}} \quad 8.75$$

$$x^2 = 4py\big|_{\text{vertex at origin}} \quad 8.76$$

The general and vertex forms of the equations can be reconciled with Eq. 8.77 through Eq. 8.79. Whether the first or second forms of these equations are used depends on whether the parabola opens horizontally or vertically (i.e., whether $A = 0$ or $C = 0$), respectively.

$$h = \begin{cases} \dfrac{E^2 - 4CF}{4CD} & \text{[opens horizontally]} \\ \dfrac{-D}{2A} & \text{[opens vertically]} \end{cases} \quad 8.77$$

$$k = \begin{cases} \dfrac{-E}{2C} & \text{[opens horizontally]} \\ \dfrac{D^2 - 4AF}{4AE} & \text{[opens vertically]} \end{cases} \quad 8.78$$

$$p = \begin{cases} \dfrac{-D}{4C} & \text{[opens horizontally]} \\ \dfrac{-E}{4A} & \text{[opens vertically]} \end{cases} \quad 8.79$$

19. ELLIPSE

An *ellipse* has two foci separated along the *major axis* by a distance $2c$. The line perpendicular to the major axis passing through the center of the ellipse is the *minor axis*. The lines perpendicular to the major axis passing through the foci are the *latus recta*. The distance between the two vertices is $2a$. The ellipse is the locus of points such that the sum of the distances from the two foci is $2a$. Referring to Fig. 8.14,

$$F_1P + PF_2 = 2a \qquad 8.80$$

Figure 8.14 *Ellipse*

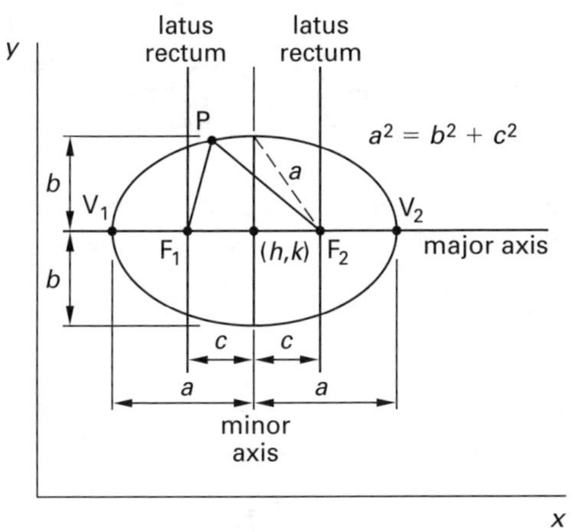

Equation 8.81 is the standard equation used for an ellipse with axes parallel to the coordinate axes, while Eq. 8.65 is the general form. F is not independent of A, C, D, and E for the ellipse.

$$Ax^2 + Cy^2 + Dx + Ey + F = 0\Big|_{\substack{AC>0\\ A\neq C}} \qquad 8.81$$

Equation 8.82 gives the standard form of the equation of an ellipse centered at (h, k). Distances a and b are known as the *semimajor distance* and *semiminor distance*, respectively.

$$\frac{(x-h)^2}{a^2} + \frac{(y-k)^2}{b^2} = 1 \qquad 8.82$$

The distance between the two foci is $2c$.

$$2c = 2\sqrt{a^2 - b^2} \qquad 8.83$$

The *aspect ratio* of the ellipse is

$$\text{aspect ratio} = \frac{a}{b} \qquad 8.84$$

The *eccentricity*, ϵ, of the ellipse is always less than 1. If the eccentricity is zero, the figure is a circle (another form of a *degenerative ellipse*).

$$\epsilon = \frac{\sqrt{a^2 - b^2}}{a} < 1 \qquad 8.85$$

The standard and center forms of the equations of an ellipse can be reconciled by using Eq. 8.86 through Eq. 8.89.

$$h = \frac{-D}{2A} \qquad 8.86$$

$$k = \frac{-E}{2C} \qquad 8.87$$

$$a = \sqrt{C} \qquad 8.88$$

$$b = \sqrt{A} \qquad 8.89$$

20. HYPERBOLA

A *hyperbola* has two foci separated along the *transverse axis* by a distance $2c$. Lines perpendicular to the transverse axis passing through the foci are the *conjugate axes*. The distance between the two vertices is $2a$, and the distance along a conjugate axis passing through each vertex between two points on the asymptotes is $2b$. The hyperbola is the locus of points such that the difference in distances from the two foci is $2a$. Referring to Fig. 8.15,

$$F_2P - PF_1 = 2a \qquad 8.90$$

Figure 8.15 *Hyperbola*

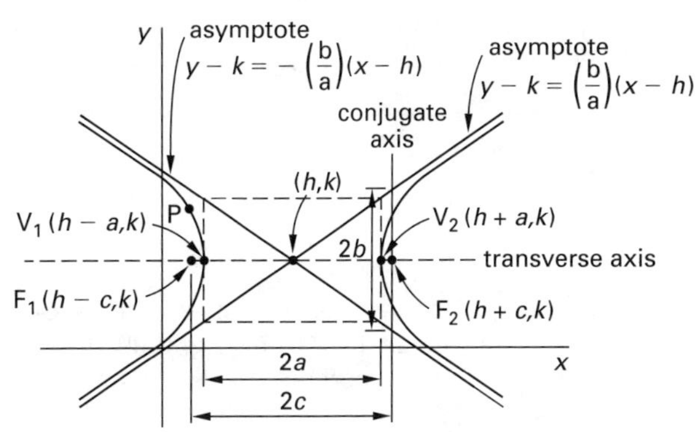

Equation 8.91 is the standard equation of a hyperbola. Coefficients A and C have opposite signs.

$$Ax^2 + Cy^2 + Dx + Ey + F = 0\big|_{AC<0} \qquad 8.91$$

Equation 8.92 gives the standard form of the equation of a hyperbola centered at (h, k) and opening to the left and right.

$$\frac{(x-h)^2}{a^2} - \frac{(y-k)^2}{b^2} = 1\Big|_{\text{opens horizontally}} \qquad 8.92$$

Equation 8.93 gives the standard form of the equation of a hyperbola that is centered at (h, k) and is opening up and down.

$$\frac{(y-k)^2}{a^2} - \frac{(y-k)^2}{b^2} = 1 \Bigg|_{\text{opens vertically}} \qquad 8.93$$

The distance between the two foci is $2c$.

$$2c = 2\sqrt{a^2 + b^2} \qquad 8.94$$

The *eccentricity*, ϵ, of the hyperbola is calculated from Eq. 8.95 and is always greater than 1.

$$\epsilon = \frac{c}{a} = \frac{\sqrt{a^2 + b^2}}{a} > 1 \qquad 8.95$$

The hyperbola is asymptotic to the lines given by Eq. 8.96 and Eq. 8.97.

$$y = \pm\frac{b}{a}(x - h) + k \Bigg|_{\text{opens horizontally}} \qquad 8.96$$

$$y = \pm\frac{a}{b}(x - h) + k \Bigg|_{\text{opens vertically}} \qquad 8.97$$

For a *rectangular (equilateral) hyperbola*, the asymptotes are perpendicular, $a = b$, $c = \sqrt{2}a$, and the eccentricity is $\epsilon = \sqrt{2}$. If the hyperbola is centered at the origin (i.e., $h = k = 0$), then the equations are $x^2 - y^2 = a^2$ (opens horizontally) and $y^2 - x^2 = a^2$ (opens vertically).

If the asymptotes are the x- and y-axes, the equation of the hyperbola is simply

$$xy = \pm\frac{a^2}{2} \qquad 8.98$$

The general and center forms of the equations of a hyperbola can be reconciled by using Eq. 8.99 through Eq. 8.103. Whether the hyperbola opens left and right or up and down depends on whether M/A or M/C is positive, respectively, where M is defined by Eq. 8.99.

$$M = \frac{D^2}{4A} + \frac{E^2}{4C} - F \qquad 8.99$$

$$h = \frac{-D}{2A} \qquad 8.100$$

$$k = \frac{-E}{2C} \qquad 8.101$$

$$a = \begin{cases} \sqrt{-C} & \text{[opens horizontally]} \\ \sqrt{-A} & \text{[opens vertically]} \end{cases} \qquad 8.102$$

$$b = \begin{cases} \sqrt{A} & \text{[opens horizontally]} \\ \sqrt{C} & \text{[opens vertically]} \end{cases} \qquad 8.103$$

21. SPHERE

Equation 8.104 is the general equation of a sphere. The coefficient A cannot be zero.

$$Ax^2 + Ay^2 + Az^2 + Bx + Cy + Dz + E = 0 \qquad 8.104$$

Equation 8.105 gives the standard form of the equation of a sphere centered at (h, k, l) with radius r.

$$(x-h)^2 + (y-k)^2 + (z-l)^2 = r^2 \qquad 8.105$$

The general and center forms of the equations of a sphere can be reconciled by using Eq. 8.106 through Eq. 8.109.

$$h = \frac{-B}{2A} \qquad 8.106$$

$$k = \frac{-C}{2A} \qquad 8.107$$

$$l = \frac{-D}{2A} \qquad 8.108$$

$$r = \sqrt{\frac{B^2 + C^2 + D^2}{4A^2} - \frac{E}{A}} \qquad 8.109$$

22. HELIX

A *helix* is a curve generated by a point moving on, around, and along a cylinder such that the distance the point moves parallel to the cylindrical axis is proportional to the angle of rotation about that axis. (See Fig. 8.16.) For a cylinder of radius r, Eq. 8.110 through Eq. 8.112 define the three-dimensional positions of points along the helix. The quantity $2\pi k$ is the *pitch* of the helix.

$$x = r\cos\theta \qquad 8.110$$

$$y = r\sin\theta \qquad 8.111$$

$$z = k\theta \qquad 8.112$$

Figure 8.16 Helix

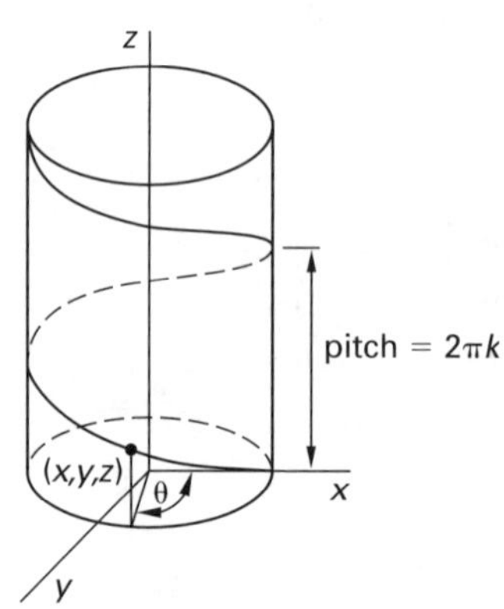

9 Differential Calculus

1. Derivative of a Function 9-1
2. Elementary Derivative Operations 9-1
3. Critical Points 9-2
4. Derivatives of Parametric Equations 9-3
5. Partial Differentiation 9-4
6. Implicit Differentiation 9-4
7. Tangent Plane Function 9-5
8. Gradient Vector 9-5
9. Directional Derivative 9-6
10. Normal Line Vector 9-6
11. Divergence of a Vector Field 9-7
12. Curl of a Vector Field 9-7
13. Taylor's Formula 9-7
14. Maclaurin Power Approximations 9-8

1. DERIVATIVE OF A FUNCTION

In most cases, it is possible to transform a continuous function, $f(x_1, x_2, x_3, \ldots)$, of one or more independent variables into a derivative function.[1] In simple cases, the *derivative* can be interpreted as the slope (tangent or rate of change) of the curve described by the original function. Since the slope of the curve depends on x, the derivative function will also depend on x. The derivative, $f'(x)$, of a function $f(x)$ is defined mathematically by Eq. 9.1. However, limit theory is seldom needed to actually calculate derivatives.

$$f'(x) = \lim_{\Delta x \to 0} \frac{\Delta f(x)}{\Delta x} \qquad 9.1$$

The derivative of a function $f(x)$, also known as the *first derivative*, is written in various ways, including

$$f'(x), \frac{df(x)}{dx}, \frac{df}{dx}, \mathbf{D}f(x), \mathbf{D}_x f(x), \dot{f}(x), sf(s)$$

A *second derivative* may exist if the derivative operation is performed on the first derivative—that is, a derivative is taken of a derivative function. This is written as

$$f''(x), \frac{d^2f(x)}{dx^2}, \frac{d^2f}{dx^2}, \mathbf{D}^2 f(x), \mathbf{D}_x^2 f(x), \ddot{f}(x), s^2 f(s)$$

A *regular* (*analytic* or *holomorphic*) *function* possesses a derivative. A point at which a function's derivative is undefined is called a *singular point*, as Fig. 9.1 illustrates.

[1]A function, $f(x)$, of one independent variable, x, is used in this section to simplify the discussion. Although the derivative is taken with respect to x, the independent variable can be anything.

Figure 9.1 *Derivatives and Singular Points*

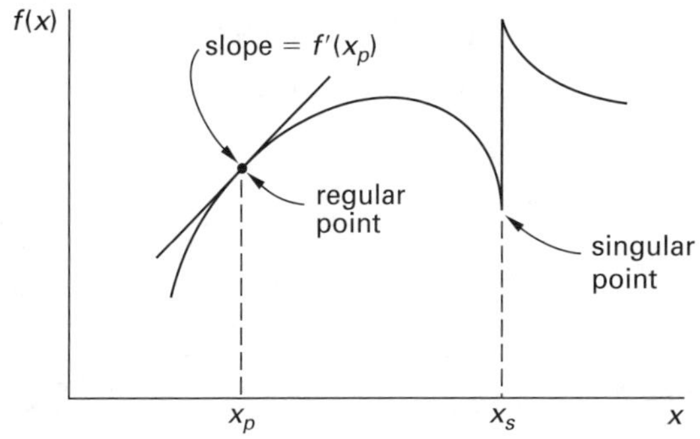

2. ELEMENTARY DERIVATIVE OPERATIONS

Equation 9.2 through Eq. 9.5 summarize the elementary derivative operations on polynomials and exponentials. Equation 9.2 and Eq. 9.3 are particularly useful. (a, n, and k represent constants. $f(x)$ and $g(x)$ are functions of x.)

$$\mathbf{D}k = 0 \qquad 9.2$$

$$\mathbf{D}x^n = nx^{n-1} \qquad 9.3$$

$$\mathbf{D}\ln x = \frac{1}{x} \qquad 9.4$$

$$\mathbf{D}e^{ax} = ae^{ax} \qquad 9.5$$

Equation 9.6 through Eq. 9.17 summarize the elementary derivative operations on transcendental (trigonometric) functions.

$$\mathbf{D}\sin x = \cos x \qquad 9.6$$

$$\mathbf{D}\cos x = -\sin x \qquad 9.7$$

$$\mathbf{D}\tan x = \sec^2 x \qquad 9.8$$

$$\mathbf{D}\cot x = -\csc^2 x \qquad 9.9$$

$$\mathbf{D}\sec x = \sec x \tan x \qquad 9.10$$

$$\mathbf{D}\csc x = -\csc x \cot x \qquad 9.11$$

$$\mathbf{D}\arcsin x = \frac{1}{\sqrt{1 - x^2}} \qquad 9.12$$

$$\mathbf{D}\arccos x = -\mathbf{D}\arcsin x \qquad 9.13$$

$$\mathbf{D}\arctan x = \frac{1}{1+x^2} \quad 9.14$$

$$\mathbf{D}\,\text{arccot}\, x = -\mathbf{D}\arctan x \quad 9.15$$

$$\mathbf{D}\,\text{arcsec}\, x = \frac{1}{x\sqrt{x^2-1}} \quad 9.16$$

$$\mathbf{D}\,\text{arccsc}\, x = -\mathbf{D}\,\text{arcsec}\, x \quad 9.17$$

Equation 9.18 through Eq. 9.23 summarize the elementary derivative operations on hyperbolic transcendental functions. Derivatives of hyperbolic functions are not completely analogous to those of the regular transcendental functions.

$$\mathbf{D}\sinh x = \cosh x \quad 9.18$$

$$\mathbf{D}\cosh x = \sinh x \quad 9.19$$

$$\mathbf{D}\tanh x = \text{sech}^2\, x \quad 9.20$$

$$\mathbf{D}\coth x = -\text{csch}^2\, x \quad 9.21$$

$$\mathbf{D}\,\text{sech}\, x = -\text{sech}\, x \tanh x \quad 9.22$$

$$\mathbf{D}\,\text{csch}\, x = -\text{csch}\, x \coth x \quad 9.23$$

Equation 9.24 through Eq. 9.29 summarize the elementary derivative operations on functions and combinations of functions.

$$\mathbf{D}kf(x) = k\mathbf{D}f(x) \quad 9.24$$

$$\mathbf{D}(f(x) \pm g(x)) = \mathbf{D}f(x) \pm \mathbf{D}g(x) \quad 9.25$$

$$\mathbf{D}(f(x)\cdot g(x)) = f(x)\mathbf{D}g(x) + g(x)\mathbf{D}f(x) \quad 9.26$$

$$\mathbf{D}\left(\frac{f(x)}{g(x)}\right) = \frac{g(x)\mathbf{D}f(x) - f(x)\mathbf{D}g(x)}{(g(x))^2} \quad 9.27$$

$$\mathbf{D}(f(x))^n = n(f(x))^{n-1}\mathbf{D}f(x) \quad 9.28$$

$$\mathbf{D}f(g(x)) = \mathbf{D}_g f(g)\mathbf{D}_x g(x) \quad 9.29$$

Example 9.1

What is the slope at $x = 3$ of the curve $f(x) = x^3 - 2x$?

Solution

The derivative function found from Eq. 9.3 determines the slope.

$$f'(x) = 3x^2 - 2$$

The slope at $x = 3$ is

$$f'(3) = (3)(3)^2 - 2 = 25$$

Example 9.2

What are the derivatives of the following functions?

(a) $f(x) = 5\sqrt[3]{x^5}$

(b) $f(x) = \sin x \cos^2 x$

(c) $f(x) = \ln(\cos e^x)$

Solution

(a) Using Eq. 9.3 and Eq. 9.24,

$$\begin{aligned} f'(x) &= 5\mathbf{D}\sqrt[3]{x^5} = 5\mathbf{D}\left(x^5\right)^{1/3} \\ &= (5)\left(\tfrac{1}{3}\right)\left(x^5\right)^{-2/3}\mathbf{D}x^5 \\ &= (5)\left(\tfrac{1}{3}\right)\left(x^5\right)^{-2/3}(5)\left(x^4\right) \\ &= \frac{25x^{2/3}}{3} \end{aligned}$$

(b) Using Eq. 9.26,

$$\begin{aligned} f'(x) &= \sin x\mathbf{D}\cos^2 x + \cos^2 x\mathbf{D}\sin x \\ &= (\sin x)(2\cos x)(\mathbf{D}\cos x) + \cos^2 x\cos x \\ &= (\sin x)(2\cos x)(-\sin x) + \cos^2 x\cos x \\ &= -2\sin^2 x\cos x + \cos^3 x \end{aligned}$$

(c) Using Eq. 9.29,

$$\begin{aligned} f'(x) &= \left(\frac{1}{\cos e^x}\right)\mathbf{D}\cos e^x \\ &= \left(\frac{1}{\cos e^x}\right)(-\sin e^x)\mathbf{D}e^x \\ &= \left(\frac{-\sin e^x}{\cos e^x}\right)e^x \\ &= -e^x\tan e^x \end{aligned}$$

3. CRITICAL POINTS

Derivatives are used to locate the local *critical points* of functions of one variable—that is, *extreme points* (also known as *maximum* and *minimum* points) as well as the *inflection points* (*points of contraflexure*). The plurals *extrema*, *maxima*, and *minima* are used without the word "points." These points are illustrated in Fig. 9.2. There is usually an inflection point between two adjacent local extrema.

Figure 9.2 *Extreme and Inflection Points*

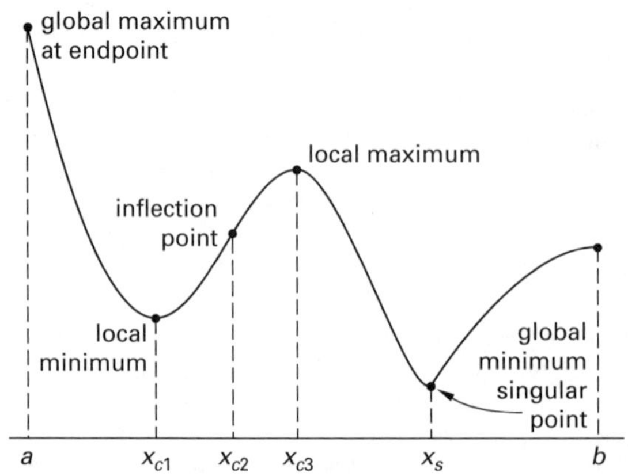

The first derivative is calculated to determine the locations of the critical points. The second derivative is calculated to determine whether a critical point is a local maximum, minimum, or inflection point, according to the following conditions. With this method, no distinction is made between local and global extrema. Therefore, the extrema should be compared with the function values at the endpoints of the interval, as illustrated in Ex. 9.3.[2] Note that $f'(x) \neq 0$ at an inflection point.

$$f'(x_c) = 0 \text{ at any extreme point, } x_c \quad 9.30$$

$$f''(x_c) < 0 \text{ at a maximum point} \quad 9.31$$

$$f''(x_c) > 0 \text{ at a minimum point} \quad 9.32$$

$$f''(x_c) = 0 \text{ at an inflection point} \quad 9.33$$

Example 9.3

Find the global extrema of the function $f(x)$ on the interval $[-2, +2]$.

$$f(x) = x^3 + x^2 - x + 1$$

Solution

The first derivative is

$$f'(x) = 3x^2 + 2x - 1$$

Since the first derivative is zero at extreme points, set $f'(x)$ equal to zero and solve for the roots of the quadratic equation.

$$3x^2 + 2x - 1 = (3x - 1)(x + 1) = 0$$

The roots are $x_1 = {}^1\!/_3$, $x_2 = -1$. These are the locations of the two extrema.

The second derivative is

$$f''(x) = 6x + 2$$

Substituting x_1 and x_2 into $f''(x)$,

$$f''(x_1) = (6)\left(\frac{1}{3}\right) + 2 = 4$$

$$f''(x_2) = (6)(-1) + 2 = -4$$

Therefore, x_1 is a local minimum point (because $f''(x_1)$ is positive), and x_2 is a local maximum point (because $f''(x_2)$ is negative). The inflection point between these two extrema is found by setting $f''(x)$ equal to zero.

$$f''(x) = 6x + 2 = 0 \text{ or } x = -\frac{1}{3}$$

[2]It is also necessary to check the values of the function at singular points (i.e., points where the derivative does not exist).

Since the question asked for the global extreme points, it is necessary to compare the values of $f(x)$ at the local extrema with the values at the endpoints.

$$f(-2) = -1$$

$$f(-1) = 2$$

$$f\left(\frac{1}{3}\right) = 22/27$$

$$f(2) = +11$$

Therefore, the actual global extrema are the endpoints.

4. DERIVATIVES OF PARAMETRIC EQUATIONS

The derivative of a function $f(x_1, x_2, \ldots, x_n)$ can be calculated from the derivatives of the parametric equations $f_1(s), f_2(s), \ldots, f_n(s)$. The derivative will be expressed in terms of the parameter, s, unless the derivatives of the parametric equations can be expressed explicitly in terms of the independent variables.

Example 9.4

A circle is expressed parametrically by the equations

$$x = 5\cos\theta$$

$$y = 5\sin\theta$$

Express the derivative dy/dx (a) as a function of the parameter θ and (b) as a function of x and y.

Solution

(a) Taking the derivative of each parametric equation with respect to θ,

$$\frac{dx}{d\theta} = -5\sin\theta$$

$$\frac{dy}{d\theta} = 5\cos\theta$$

Then,

$$\frac{dy}{dx} = \frac{\frac{dy}{d\theta}}{\frac{dx}{d\theta}} = \frac{5\cos\theta}{-5\sin\theta} = -\cot\theta$$

(b) The derivatives of the parametric equations are closely related to the original parametric equations.

$$\frac{dx}{d\theta} = -5\sin\theta = -y$$

$$\frac{dy}{d\theta} = 5\cos\theta = x$$

$$\frac{dy}{dx} = \frac{\frac{dy}{d\theta}}{\frac{dx}{d\theta}} = \frac{-x}{y}$$

5. PARTIAL DIFFERENTIATION

Derivatives can be taken with respect to only one independent variable at a time. For example, $f'(x)$ is the derivative of $f(x)$ and is taken with respect to the independent variable x. If a function, $f(x_1, x_2, x_3, \ldots)$, has more than one independent variable, a *partial derivative* can be found, but only with respect to one of the independent variables. All other variables are treated as constants. Symbols for a partial derivative of f taken with respect to variable x are $\partial f/\partial x$ and $f_x(x, y)$.

The geometric interpretation of a partial derivative $\partial f/\partial x$ is the slope of a line tangent to the surface (a sphere, ellipsoid, etc.) described by the function when all variables except x are held constant. In three-dimensional space with a function described by $z = f(x, y)$, the partial derivative $\partial f/\partial x$ (equivalent to $\partial f/\partial y$) is the slope of the line tangent to the surface in a plane of constant y. Similarly, the partial derivative $\partial f/\partial y$ (equivalent to $\partial z/\partial y$) is the slope of the line tangent to the surface in a plane of constant x.

Example 9.5

What is the partial derivative $\partial z/\partial x$ of the following function?

$$z = 3x^2 - 6y^2 + xy + 5y - 9$$

Solution

The partial derivative with respect to x is found by considering all variables other than x to be constants.

$$\frac{\partial z}{\partial x} = 6x - 0 + y + 0 - 0 = 6x + y$$

Example 9.6

A surface has the equation $x^2 + y^2 + z^2 - 9 = 0$. What is the slope of a line that lies in a plane of constant y and is tangent to the surface at $(x, y, z) = (1, 2, 2)$?[3]

Solution

Solve for the dependent variable. Then, consider variable y to be a constant.

$$z = \sqrt{9 - x^2 - y^2}$$

$$\begin{aligned}\frac{\partial z}{\partial x} &= \frac{\partial(9 - x^2 - y^2)^{1/2}}{\partial x} \\ &= \left(\tfrac{1}{2}\right)(9 - x^2 - y^2)^{-1/2}\left(\frac{\partial(9 - x^2 - y^2)}{\partial x}\right) \\ &= \left(\tfrac{1}{2}\right)(9 - x^2 - y^2)^{-1/2}(-2x) \\ &= \frac{-x}{\sqrt{9 - x^2 - y^2}}\end{aligned}$$

At the point $(1, 2, 2)$, $x = 1$ and $y = 2$.

$$\left.\frac{\partial z}{\partial x}\right|_{(1,2,2)} = \frac{-1}{\sqrt{9 - (1)^2 - (2)^2}} = -\frac{1}{2}$$

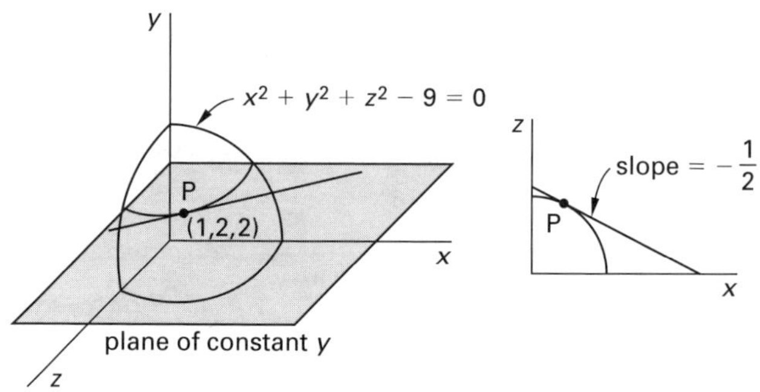

[3]Although only implied, it is required that the point actually be on the surface (i.e., it must satisfy the equation $f(x, y, z) = 0$).

6. IMPLICIT DIFFERENTIATION

When a relationship between n variables cannot be manipulated to yield an explicit function of $n - 1$ independent variables, that relationship implicitly defines the nth variable. Finding the derivative of the implicit variable with respect to any other independent variable is known as *implicit differentiation.*

An implicit derivative is the quotient of two partial derivatives. The two partial derivatives are chosen so that dividing one by the other eliminates a common differential. For example, if z cannot be explicitly extracted from $f(x, y, z) = 0$, the partial derivatives $\partial z/\partial x$ and $\partial z/\partial y$ can still be found as follows.

$$\frac{\partial z}{\partial x} = \frac{-\frac{\partial f}{\partial x}}{\frac{\partial f}{\partial z}} \qquad 9.34$$

$$\frac{\partial z}{\partial y} = \frac{-\frac{\partial f}{\partial y}}{\frac{\partial f}{\partial z}} \qquad 9.35$$

Example 9.7

Find the derivative dy/dx of

$$f(x, y) = x^2 + xy + y^3$$

Solution

Implicit differentiation is required because x cannot be extracted from $f(x,y)$.

$$\frac{\partial f}{\partial x} = 2x + y$$

$$\frac{\partial f}{\partial y} = x + 3y^2$$

$$\frac{dy}{dx} = \frac{-\frac{\partial f}{\partial x}}{\frac{\partial f}{\partial y}} = \frac{-(2x+y)}{x+3y^2}$$

Example 9.8

Solve Ex. 9.6 using implicit differentiation.

Solution

$$f(x,y,z) = x^2 + y^2 + z^2 - 9 = 0$$

$$\frac{\partial f}{\partial x} = 2x$$

$$\frac{\partial f}{\partial z} = 2z$$

$$\frac{\partial z}{\partial x} = \frac{-\frac{\partial f}{\partial x}}{\frac{\partial f}{\partial z}} = \frac{-2x}{2z} = -\frac{x}{z}$$

At the point $(1,2,2)$,

$$\frac{\partial z}{\partial x} = -\frac{1}{2}$$

7. TANGENT PLANE FUNCTION

Partial derivatives can be used to find the equation of a plane tangent to a three-dimensional surface defined by $f(x,y,z) = 0$ at some point, P_0.

$$\begin{aligned} T(x_0,y_0,z_0) = (x-x_0)\left.\frac{\partial f(x,y,z)}{\partial x}\right|_{\mathrm{P}_0} \\ + (y-y_0)\left.\frac{\partial f(x,y,z)}{\partial y}\right|_{\mathrm{P}_0} \\ + (z-z_0)\left.\frac{\partial f(x,y,z)}{\partial z}\right|_{\mathrm{P}_0} \\ = 0 \end{aligned} \qquad 9.36$$

The coefficients of x, y, and z are the same as the coefficients of **i**, **j**, and **k** of the normal vector at point P_0. (See Sec. 9.10.)

Example 9.9

What is the equation of the plane that is tangent to the surface defined by $f(x,y,z) = 4x^2 + y^2 - 16z = 0$ at the point $(2,4,2)$?

Solution

First, calculate the partial derivatives and substitute the coordinates of the point.

$$\left.\frac{\partial f(x,y,z)}{\partial x}\right|_{\mathrm{P}_0} = 8x|_{(2,4,2)} = (8)(2) = 16$$

$$\left.\frac{\partial f(x,y,z)}{\partial y}\right|_{\mathrm{P}_0} = 2y|_{(2,4,2)} = (2)(4) = 8$$

$$\left.\frac{\partial f(x,y,z)}{\partial z}\right|_{\mathrm{P}_0} = -16|_{(2,4,2)} = -16$$

$$\begin{aligned} T(2,4,2) &= (16)(x-2) + (8)(y-4) - (16)(z-2) \\ &= 2x + y - 2z - 4 \end{aligned}$$

Next, substitute into Eq. 9.36.

$$2x + y - 2z - 4 = 0$$

8. GRADIENT VECTOR

The slope of a function is the change in one variable with respect to a distance in a chosen direction. Usually, the direction is parallel to a coordinate axis. However, the maximum slope at a point on a surface may not be in a direction parallel to one of the coordinate axes.

The *gradient vector function* $\nabla f(x,y,z)$ (pronounced "del f") gives the maximum rate of change of the function $f(x,y,z)$.

$$\begin{aligned} \nabla f(x,y,z) = \frac{\partial f(x,y,z)}{\partial x}\mathbf{i} + \frac{\partial f(x,y,z)}{\partial y}\mathbf{j} \\ + \frac{\partial f(x,y,z)}{\partial z}\mathbf{k} \end{aligned} \qquad 9.37$$

Example 9.10

A two-dimensional function is defined as

$$f(x,y) = 2x^2 - y^2 + 3x - y$$

(a) What is the gradient vector for this function? (b) What is the direction of the line passing through the point $(1,-2)$ that has a maximum slope? (c) What is the maximum slope at the point $(1,-2)$?

Solution

(a) It is necessary to calculate two partial derivatives in order to use Eq. 9.37.

$$\frac{\partial f(x,y)}{\partial x} = 4x + 3$$

$$\frac{\partial f(x,y)}{\partial y} = -2y - 1$$

$$\nabla f(x,y) = (4x+3)\mathbf{i} + (-2y-1)\mathbf{j}$$

(b) Find the direction of the line passing through $(1, -2)$ with maximum slope by inserting $x = 1$ and $y = -2$ into the gradient vector function.

$$\begin{aligned}\mathbf{V} &= \Big((4)(1)+3\Big)\mathbf{i} + \Big((-2)(-2)-1\Big)\mathbf{j} \\ &= 7\mathbf{i} + 3\mathbf{j}\end{aligned}$$

(c) The magnitude of the slope is

$$|\mathbf{V}| = \sqrt{(7)^2 + (3)^2} = 7.62$$

9. DIRECTIONAL DERIVATIVE

Unlike the gradient vector (covered in Sec. 9.8), which calculates the maximum rate of change of a function, the *directional derivative*, indicated by $\nabla_u f(x,y,z)$, $D_u f(x,y,z)$, or $f'_u(x,y,z)$, gives the rate of change in the direction of a given vector, $\mathbf{u}$ or $\mathbf{U}$. The subscript u implies that the direction vector is a unit vector, but it does not need to be, as only the direction cosines are calculated from it.

$$\begin{aligned}\nabla_u f(x,y,z) = &\left(\frac{\partial f(x,y,z)}{\partial x}\right)\cos\alpha \\ &+ \left(\frac{\partial f(x,y,z)}{\partial y}\right)\cos\beta \\ &+ \left(\frac{\partial f(x,y,z)}{\partial z}\right)\cos\gamma\end{aligned} \qquad 9.38$$

$$\mathbf{U} = U_x\mathbf{i} + U_y\mathbf{j} + U_z\mathbf{k} \qquad 9.39$$

$$\cos\alpha = \frac{U_x}{|\mathbf{U}|} = \frac{U_x}{\sqrt{U_x^2 + U_y^2 + U_z^2}} \qquad 9.40$$

$$\cos\beta = \frac{U_y}{|\mathbf{U}|} \qquad 9.41$$

$$\cos\gamma = \frac{U_z}{|\mathbf{U}|} \qquad 9.42$$

Example 9.11

What is the rate of change of $f(x,y) = 3x^2 + xy - 2y^2$ at the point $(1, -2)$ in the direction $4\mathbf{i} + 3\mathbf{j}$?

Solution

The direction cosines are given by Eq. 9.40 and Eq. 9.41.

$$\cos\alpha = \frac{U_x}{|\mathbf{U}|} = \frac{4}{\sqrt{(4)^2 + (3)^2}} = \frac{4}{5}$$

$$\cos\beta = \frac{U_y}{|\mathbf{U}|} = \frac{3}{5}$$

The partial derivatives are

$$\frac{\partial f(x,y)}{\partial x} = 6x + y$$

$$\frac{\partial f(x,y)}{\partial y} = x - 4y$$

The directional derivative is given by Eq. 9.38.

$$\nabla_u f(x,y) = \left(\frac{4}{5}\right)(6x+y) + \left(\frac{3}{5}\right)(x-4y)$$

Substituting the given values of $x = 1$ and $y = -2$,

$$\begin{aligned}\nabla_u f(1,-2) &= \left(\frac{4}{5}\right)\Big((6)(1)-2\Big) + \left(\frac{3}{5}\right)\Big(1-(4)(-2)\Big) \\ &= \frac{43}{5} = 8.6\end{aligned}$$

10. NORMAL LINE VECTOR

Partial derivatives can be used to find the vector normal to a three-dimensional surface defined by $f(x,y,z) = 0$ at some point P_0. Notice that the coefficients of $\mathbf{i}$, $\mathbf{j}$, and $\mathbf{k}$ are the same as the coefficients of x, y, and z calculated for the equation of the tangent plane at point P_0. (See Sec. 9.7.)

$$\begin{aligned}\mathbf{N} = &\left.\frac{\partial f(x,y,z)}{\partial x}\right|_{P_0}\mathbf{i} + \left.\frac{\partial f(x,y,z)}{\partial y}\right|_{P_0}\mathbf{j} \\ &+ \left.\frac{\partial f(x,y,z)}{\partial z}\right|_{P_0}\mathbf{k}\end{aligned} \qquad 9.43$$

Example 9.12

What is the vector normal to the surface of $f(x,y,z) = 4x^2 + y^2 - 16z = 0$ at the point $(2, 4, 2)$?

Solution

The equation of the tangent plane at this point was calculated in Ex. 9.9 to be

$$T(2,4,2) = 2x + y - 2z - 4 = 0$$

A vector that is normal to the tangent plane through this point is

$$\mathbf{N} = 2\mathbf{i} + \mathbf{j} - 2\mathbf{k}$$

11. DIVERGENCE OF A VECTOR FIELD

The *divergence*, div **F**, of a vector field $\mathbf{F}(x, y, z)$ is a scalar function defined by Eq. 9.44 through Eq. 9.46.[4] The divergence of **F** can be interpreted as the *accumulation* of flux (i.e., a flowing substance) in a small region (i.e., at a point). One of the uses of the divergence is to determine whether flow (represented in direction and magnitude by **F**) is compressible. Flow is incompressible if div $\mathbf{F} = 0$, since the substance is not accumulating.

$$\mathbf{F} = P(x, y, z)\mathbf{i} + Q(x, y, z)\mathbf{j} + R(x, y, z)\mathbf{k} \qquad 9.44$$

$$\text{div}\,\mathbf{F} = \frac{\partial P}{\partial x} + \frac{\partial Q}{\partial y} + \frac{\partial R}{\partial z} \qquad 9.45$$

It may be easier to calculate divergence from Eq. 9.46.

$$\text{div}\,\mathbf{F} = \nabla \cdot \mathbf{F} \qquad 9.46$$

The vector del operator, ∇ is defined as

$$\nabla = \frac{\partial}{\partial x}\mathbf{i} + \frac{\partial}{\partial y}\mathbf{j} + \frac{\partial}{\partial z}\mathbf{k} \qquad 9.47$$

If there is no divergence, then the dot product calculated in Eq. 9.46 is zero.

Example 9.13

Calculate the divergence of the following vector function.

$$\mathbf{F}(x, y, z) = xz\mathbf{i} + e^x y\mathbf{j} + 7x^3 y\mathbf{k}$$

Solution

From Eq. 9.45,

$$\begin{aligned}\text{div}\,\mathbf{F} &= \frac{\partial}{\partial x}xz + \frac{\partial}{\partial y}e^x y + \frac{\partial}{\partial z}7x^3 y \\ &= z + e^x + 0 = z + e^x\end{aligned}$$

12. CURL OF A VECTOR FIELD

The *curl*, curl **F**, of a vector field $\mathbf{F}(x, y, z)$ is a vector field defined by Eq. 9.49 and Eq. 9.50. The curl **F** can be interpreted as the *vorticity* per unit area of flux (i.e., a flowing substance) in a small region (i.e., at a point). One of the uses of the curl is to determine whether flow (represented in direction and magnitude by **F**) is rotational. Flow is irrotational if curl $\mathbf{F} = 0$.

$$\mathbf{F} = P(x, y, z)\mathbf{i} + Q(x, y, z)\mathbf{j} + R(x, y, z)\mathbf{k} \qquad 9.48$$

$$\begin{aligned}\text{curl}\,\mathbf{F} = &\left(\frac{\partial R}{\partial y} - \frac{\partial Q}{\partial z}\right)\mathbf{i} + \left(\frac{\partial P}{\partial z} - \frac{\partial R}{\partial x}\right)\mathbf{j} \\ &+ \left(\frac{\partial Q}{\partial x} - \frac{\partial P}{\partial y}\right)\mathbf{k}\end{aligned} \qquad 9.49$$

It may be easier to calculate the curl from Eq. 9.50. (The vector del operator, ∇, was defined in Eq. 9.47.)

$$\begin{aligned}\text{curl}\,\mathbf{F} &= \nabla \times \mathbf{F} \\ &= \begin{vmatrix} \mathbf{i} & \mathbf{j} & \mathbf{k} \\ \frac{\partial}{\partial x} & \frac{\partial}{\partial y} & \frac{\partial}{\partial z} \\ P(x, y, z) & Q(x, y, z) & R(x, y, z) \end{vmatrix}\end{aligned} \qquad 9.50$$

If the velocity vector is **V**, then the vorticity is

$$\omega = \nabla \times \mathbf{V} = \omega_x\mathbf{i} + \omega_y\mathbf{j} + \omega_z\mathbf{k} \qquad 9.51$$

The circulation is the line integral of the velocity ,**V**, along a closed curve.

$$\Gamma = \oint V \cdot ds = \oint \omega \cdot dA \qquad 9.52$$

Example 9.14

Calculate the curl of the following vector function.

$$\mathbf{F}(x, y, z) = 3x^2\mathbf{i} + 7e^x y\mathbf{j}$$

Solution

Using Eq. 9.50,

$$\text{curl}\,\mathbf{F} = \begin{vmatrix} \mathbf{i} & \mathbf{j} & \mathbf{k} \\ \frac{\partial}{\partial x} & \frac{\partial}{\partial y} & \frac{\partial}{\partial z} \\ 3x^2 & 7e^x y & 0 \end{vmatrix}$$

Expand the determinant across the top row.

$$\begin{aligned}&\mathbf{i}\left(\frac{\partial}{\partial y}(0) - \frac{\partial}{\partial z}(7e^x y)\right) - \mathbf{j}\left(\frac{\partial}{\partial x}(0) - \frac{\partial}{\partial z}(3x^2)\right) \\ &\quad + \mathbf{k}\left(\frac{\partial}{\partial x}(7e^x y) - \frac{\partial}{\partial y}(3x^2)\right) \\ &= \mathbf{i}(0 - 0) - \mathbf{j}(0 - 0) + \mathbf{k}(7e^x y - 0) = 7e^x y\mathbf{k}\end{aligned}$$

13. TAYLOR'S FORMULA

Taylor's formula (*series*) can be used to expand a function around a point (i.e., approximate the function at one point based on the function's value at another point). The approximation consists of a series, each term composed of a derivative of the original function and a polynomial. Using Taylor's formula requires that the original function be continuous in the interval $[a,b]$ and have the required number of derivatives. To expand

[4]A bold letter, **F**, is used to indicate that the vector is a function of x, y, and z.

a function, $f(x)$, around a point, a, in order to obtain $f(b)$, Taylor's formula is

$$f(b) = f(a) + \frac{f'(a)}{1!}(b-a) + \frac{f''(a)}{2!}(b-a)^2 + \cdots + \frac{f^n(a)}{n!}(b-a)^n + R_n(b) \quad \text{9.53}$$

In Eq. 9.53, the expression f^n designates the nth derivative of the function $f(x)$. To be a useful approximation, point a must satisfy two requirements: It must be relatively close to point b, and the function and its derivatives must be known or easy to calculate. The last term, $R_n(b)$, is the uncalculated remainder after n derivatives. It is the difference between the exact and approximate values. By using enough terms, the remainder can be made arbitrarily small. That is, $R_n(b)$ approaches zero as n approaches infinity.

It can be shown that the remainder term can be calculated from Eq. 9.54, where c is some number in the interval $[a, b]$. With certain functions, the constant c can be completely determined. In most cases, however, it is possible only to calculate an upper bound on the remainder from Eq. 9.55. M_n is the maximum (positive) value of $f^{n+1}(x)$ on the interval $[a, b]$.

$$R_n(b) = \frac{f^{n+1}(c)}{(n+1)!}(b-a)^{n+1} \quad \text{9.54}$$

$$|R_n(b)| \le M_n \frac{|(b-a)^{n+1}|}{(n+1)!} \quad \text{9.55}$$

14. MACLAURIN POWER APPROXIMATIONS

If $a = 0$ in the Taylor series, Eq. 9.53 is known as the *Maclaurin series*. The Maclaurin series can be used to approximate functions at some value of x between 0 and 1. The following common approximations may be referred to as Maclaurin series, Taylor series, or power series approximations.

$$\sin x \approx x - \frac{x^3}{3!} + \frac{x^5}{5!} - \frac{x^7}{7!} + \cdots + (-1)^n \frac{x^{2n+1}}{(2n+1)!} \quad \text{9.56}$$

$$\cos x \approx 1 - \frac{x^2}{2!} + \frac{x^4}{4!} - \frac{x^6}{6!} + \cdots + (-1)^n \frac{x^{2n}}{(2n)!} \quad \text{9.57}$$

$$\sinh x \approx x + \frac{x^3}{3!} + \frac{x^5}{5!} + \frac{x^7}{7!} + \cdots + \frac{x^{2n+1}}{(2n+1)!} \quad \text{9.58}$$

$$\cosh x \approx 1 + \frac{x^2}{2!} + \frac{x^4}{4!} + \frac{x^6}{6!} + \cdots + \frac{x^{2n}}{(2n)!} \quad \text{9.59}$$

$$e^x \approx 1 + x + \frac{x^2}{2!} + \frac{x^3}{3!} + \cdots + \frac{x^n}{n!} \quad \text{9.60}$$

$$\ln(1+x) \approx x - \frac{x^2}{2} + \frac{x^3}{3} - \frac{x^4}{4} + \cdots + (-1)^{n+1}\frac{x^n}{n} \quad \text{9.61}$$

$$\frac{1}{1-x} \approx 1 + x + x^2 + x^3 + \cdots + x^n \quad \text{9.62}$$

10 Integral Calculus

1. Integration 10-1
2. Elementary Operations 10-1
3. Integration by Parts 10-2
4. Separation of Terms 10-3
5. Double and Higher-Order Integrals 10-3
6. Initial Values 10-4
7. Definite Integrals 10-4
8. Average Value 10-4
9. Area 10-4
10. Arc Length 10-5
11. Pappus' Theorems 10-5
12. Surface of Revolution 10-5
13. Volume of Revolution 10-6
14. Moments of a Function 10-6
15. Fourier Series 10-6
16. Fast Fourier Transforms 10-8
17. Integral Functions 10-8

1. INTEGRATION

Integration is the inverse operation of differentiation. For that reason, *indefinite integrals* are sometimes referred to as *antiderivatives*.[1] Although expressions can be functions of several variables, integrals can only be taken with respect to one variable at a time. The *differential term* (dx in Eq. 10.1) indicates that variable. In Eq. 10.1, the function $f'(x)$ is the *integrand*, and x is the variable of integration.

$$\int f'(x)dx = f(x) + C \qquad 10.1$$

While most of a function, $f(x)$, can be "recovered" through integration of its derivative, $f'(x)$, a constant term will be lost. This is because the derivative of a constant term vanishes (i.e., is zero), leaving nothing to recover from. A *constant of integration*, C, is added to the integral to recognize the possibility of such a constant term.

2. ELEMENTARY OPERATIONS

Equation 10.2 through Eq. 10.8 summarize the elementary integration operations on polynomials and exponentials.[2]

Equation 10.2 and Eq. 10.3 are particularly useful. (C and k represent constants. $f(x)$ and $g(x)$ are functions of x.)

$$\int k\,dx = kx + C \qquad 10.2$$

$$\int x^m\,dx = \frac{x^{m+1}}{m+1} + C \quad [m \neq -1] \qquad 10.3$$

$$\int \frac{1}{x}\,dx = \ln|x| + C \qquad 10.4$$

$$\int e^{kx}\,dx = \frac{e^{kx}}{k} + C \qquad 10.5$$

$$\int xe^{kx}\,dx = \frac{e^{kx}(kx-1)}{k^2} + C \qquad 10.6$$

$$\int k^{ax}\,dx = \frac{k^{ax}}{a\ln k} + C \qquad 10.7$$

$$\int \ln x\,dx = x\ln x - x + C \qquad 10.8$$

Equation 10.9 through Eq. 10.20 summarize the elementary integration operations on transcendental functions.

$$\int \sin x\,dx = -\cos x + C \qquad 10.9$$

$$\int \cos x\,dx = \sin x + C \qquad 10.10$$

$$\begin{aligned}\int \tan x\,dx &= \ln|\sec x| + C \\ &= -\ln|\cos x| + C\end{aligned} \qquad 10.11$$

$$\int \cot x\,dx = \ln|\sin x| + C \qquad 10.12$$

$$\begin{aligned}\int \sec x\,dx &= \ln|\sec x + \tan x| + C \\ &= \ln\left|\tan\left(\frac{x}{2}+\frac{\pi}{4}\right)\right| + C\end{aligned} \qquad 10.13$$

$$\begin{aligned}\int \csc x\,dx &= \ln|\csc x - \cot x| + C \\ &= \ln\left|\tan\frac{x}{2}\right| + C\end{aligned} \qquad 10.14$$

$$\int \frac{dx}{k^2+x^2} = \frac{1}{k}\arctan\frac{x}{k} + C \qquad 10.15$$

$$\int \frac{dx}{\sqrt{k^2-x^2}} = \arcsin\frac{x}{k} + C \quad [k^2 > x^2] \qquad 10.16$$

[1]The difference between an indefinite and definite integral (covered in Sec. 10.7) is simple: An *indefinite integral* is a function, while a *definite integral* is a number.

[2]More extensive listings, known as *tables of integrals*, are widely available. (See App. 10.A.)

Mathematics

$$\int \frac{dx}{x\sqrt{x^2-k^2}} = \frac{1}{k}\,\text{arcsec}\,\frac{x}{k} + C \quad [x^2 > k^2] \qquad 10.17$$

$$\int \sin^2 x\,dx = \tfrac{1}{2}x - \tfrac{1}{4}\sin 2x + C \qquad 10.18$$

$$\int \cos^2 x\,dx = \tfrac{1}{2}x + \tfrac{1}{4}\sin 2x + C \qquad 10.19$$

$$\int \tan^2 x\,dx = \tan x - x + C \qquad 10.20$$

Equation 10.21 through Eq. 10.26 summarize the elementary integration operations on hyperbolic transcendental functions. Integrals of hyperbolic functions are not completely analogous to those of the regular transcendental functions.

$$\int \sinh x\,dx = \cosh x + C \qquad 10.21$$

$$\int \cosh x\,dx = \sinh x + C \qquad 10.22$$

$$\int \tanh x\,dx = \ln|\cosh x| + C \qquad 10.23$$

$$\int \coth x\,dx = \ln|\sinh x| + C \qquad 10.24$$

$$\int \text{sech}\,x\,dx = \arctan(\sinh x) + C \qquad 10.25$$

$$\int \text{csch}\,x\,dx = \ln\left|\tanh\frac{x}{2}\right| + C \qquad 10.26$$

Equation 10.27 through Eq. 10.31 summarize the elementary integration operations on functions and combinations of functions.

$$\int kf(x)dx = k\int f(x)dx \qquad 10.27$$

$$\int (f(x)+g(x))dx = \int f(x)dx + \int g(x)dx \qquad 10.28$$

$$\int \frac{f'(x)}{f(x)}dx = \ln|f(x)| + C \qquad 10.29$$

$$\begin{aligned}\int f(x)dg(x) &= f(x)\int dg(x) - \int g(x)df(x) + C\\ &= f(x)g(x) - \int g(x)df(x) + C\end{aligned} \qquad 10.30$$

Example 10.1

Find the integral with respect to x of

$$3x^2 + \tfrac{1}{3}x - 7 = 0$$

Solution

This is a polynomial function, and Eq. 10.3 can be applied to each of the three terms.

$$\int \left(3x^2 + \tfrac{1}{3}x - 7\right)dx = x^3 + \tfrac{1}{6}x^2 - 7x + C$$

3. INTEGRATION BY PARTS

Equation 10.30, repeated here, is known as *integration by parts*. $f(x)$ and $g(x)$ are functions. The use of this method is illustrated by Ex. 10.2.

$$\int f(x)dg(x) = f(x)g(x) - \int g(x)df(x) + C \qquad 10.31$$

Example 10.2

Find the following integral.

$$\int x^2 e^x\,dx$$

Solution

x^2e^x is factored into two parts so that integration by parts can be used.

$$\begin{aligned} f(x) &= x^2\\ dg(x) &= e^x\,dx\\ df(x) &= 2x\,dx\end{aligned}$$

$$g(x) = \int dg(x) = \int e^x\,dx = e^x$$

From Eq. 10.31, disregarding the constant of integration (which cannot be evaluated),

$$\int f(x)dg(x) = f(x)g(x) - \int g(x)df(x)$$

$$\int x^2e^x\,dx = x^2e^x - \int e^x(2x)\,dx$$

The second term is also factored into two parts, and integration by parts is used again. This time,

$$\begin{aligned} f(x) &= x\\ dg(x) &= e^x\,dx\\ df(x) &= dx\end{aligned}$$

$$g(x) = \int dg(x) = \int e^x\,dx = e^x$$

From Eq. 10.31,

$$\begin{aligned}\int 2xe^x\,dx &= 2\int xe^x\,dx\\ &= 2\left(xe^x - \int e^x\,dx\right)\\ &= 2(xe^x - e^x)\end{aligned}$$

Then, the complete integral is

$$\begin{aligned}\int x^2e^x\,dx &= x^2e^x - 2(xe^x - e^x) + C\\ &= e^x(x^2 - 2x + 2) + C\end{aligned}$$

4. SEPARATION OF TERMS

Equation 10.28 shows that the integral of a sum of terms is equal to a sum of integrals. This technique is known as *separation of terms*. In many cases, terms are easily separated. In other cases, the technique of *partial fractions* can be used to obtain individual terms. These techniques are illustrated by Ex. 10.3 and Ex. 10.4.

Example 10.3

Find the following integral.

$$\int \frac{(2x^2+3)^2}{x}\,dx$$

Solution

$$\begin{aligned}\int \frac{(2x^2+3)^2}{x}\,dx &= \int \frac{4x^4+12x^2+9}{x}\,dx\\ &= \int\left(4x^3+12x+\frac{9}{x}\right)dx\\ &= x^4+6x^2+9\ln|x|+C\end{aligned}$$

Example 10.4

Find the following integral.

$$\int \frac{3x+2}{3x-2}\,dx$$

Solution

The integrand is larger than 1, so use long division to simplify it.

$$\begin{array}{r l} & 1 \text{ rem } 4 \quad \left(1+\dfrac{4}{3x-2}\right)\\ 3x-2 & \overline{)\,3x+2}\\ & \underline{3x-2}\\ & 4 \text{ remainder}\end{array}$$

$$\begin{aligned}\int \frac{3x+2}{3x-2}\,dx &= \int\left(1+\frac{4}{3x-2}\right)dx\\ &= \int dx + \int \frac{4}{3x-2}\,dx\\ &= x + \tfrac{4}{3}\ln|(3x-2)| + C\end{aligned}$$

5. DOUBLE AND HIGHER-ORDER INTEGRALS

A function can be successively integrated. (This is analogous to successive differentiation.) A function that is integrated twice is known as a *double integral*; if integrated three times, it is a *triple integral*; and so on. Double and triple integrals are used to calculate areas and volumes, respectively.

The successive integrations do not need to be with respect to the same variable. Variables not included in the integration are treated as constants.

There are several notations used for a multiple integral, particularly when the product of length differentials represents a differential area or volume. A double integral (i.e., two successive integrations) can be represented by one of the following notations.

$$\iint f(x,y)\,dx\,dy,\ \int_{R^2} f(x,y)\,dx\,dy,$$
$$\text{or } \iint_{R^2} f(x,y)\,dA$$

A triple integral can be represented by one of the following notations.

$$\iiint f(x,y,z)\,dx\,dy\,dz,$$
$$\int_{R^3} f(x,y,z)\,dx\,dy\,dz,$$
$$\text{or } \iiint_{R^3} f(x,y,z)\,dV$$

Example 10.5

Find the following double integral.

$$\iint (x^2+y^3x)\,dx\,dy$$

Solution

$$\int (x^2+y^3x)\,dx = \tfrac{1}{3}x^3 + \tfrac{1}{2}y^3x^2 + C_1$$

$$\int \left(\tfrac{1}{3}x^3 + \tfrac{1}{2}y^3x^2 + C_1\right)dy = \tfrac{1}{3}yx^3 + \tfrac{1}{8}y^4x^2 + C_1y + C_2$$

So,

$$\iint (x^2 + y^3 x)\,dx\,dy = \tfrac{1}{3}yx^3 + \tfrac{1}{8}y^4x^2 + C_1y + C_2$$

6. INITIAL VALUES

The constant of integration, C, can be found only if the value of the function $f(x)$ is known for some value of x_0. The value $f(x_0)$ is known as an *initial value* or *initial condition*. To completely define a function, as many initial values, $f(x_0)$, $f'(x_0)$, $f''(x_0)$, and so on, as there are integrations are needed.

Example 10.6

It is known that $f(x) = 4$ when $x = 2$ (i.e., the initial value is $f(2) = 4$). Find the original function.

$$\int (3x^3 - 7x)\,dx$$

Solution

The function is

$$f(x) = \int (3x^3 - 7x)\,dx = \tfrac{3}{4}x^4 - \tfrac{7}{2}x^2 + C$$

Substituting the initial value determines C.

$$\begin{aligned} 4 &= \left(\tfrac{3}{4}\right)(2)^4 - \left(\tfrac{7}{2}\right)(2)^2 + C \\ 4 &= 12 - 14 + C \\ C &= 6 \end{aligned}$$

The function is

$$f(x) = \tfrac{3}{4}x^4 - \tfrac{7}{2}x^2 + 6$$

7. DEFINITE INTEGRALS

A *definite integral* is restricted to a specific range of the independent variable. (Unrestricted integrals of the types shown in all preceding examples are known as *indefinite integrals*.) A definite integral restricted to the region bounded by *lower* and *upper limits* (also known as *bounds*), x_1 and x_2, is written as

$$\int_{x_1}^{x_2} f(x)\,dx$$

Equation 10.32 indicates how definite integrals are evaluated. It is known as the *fundamental theorem of calculus*.

$$\int_{x_1}^{x_2} f'(x)\,dx = f(x)\Big|_{x_1}^{x_2} = f(x_2) - f(x_1) \qquad 10.32$$

A common use of a definite integral is the calculation of work performed by a force, F, that moves an object from position x_1 to x_2.

$$W = \int_{x_1}^{x_2} F\,dx \qquad 10.33$$

Example 10.7

Evaluate the following definite integral.

$$\int_{\pi/4}^{\pi/3} \sin x\,dx$$

Solution

From Eq. 10.32,

$$\begin{aligned} \int_{\pi/4}^{\pi/3} \sin x\,dx &= -\cos x\Big|_{\pi/4}^{\pi/3} \\ &= -\cos\tfrac{\pi}{3} - \left(-\cos\tfrac{\pi}{4}\right) \\ &= -0.5 - (-0.707) = 0.207 \end{aligned}$$

8. AVERAGE VALUE

The average value of a function $f(x)$ that is integrable over the interval $[a, b]$ is

$$\text{average value} = \frac{1}{b-a}\int_a^b f(x)\,dx \qquad 10.34$$

9. AREA

Equation 10.35 calculates the area, A, bounded by $x = a$, $x = b$, $f_1(x)$ above and $f_2(x)$ below. ($f_2(x) = 0$ if the area is bounded by the x-axis.) This is illustrated in Fig. 10.1.

$$A = \int_a^b (f_1(x) - f_2(x))\,dx \qquad 10.35$$

Figure 10.1 *Area Between Two Curves*

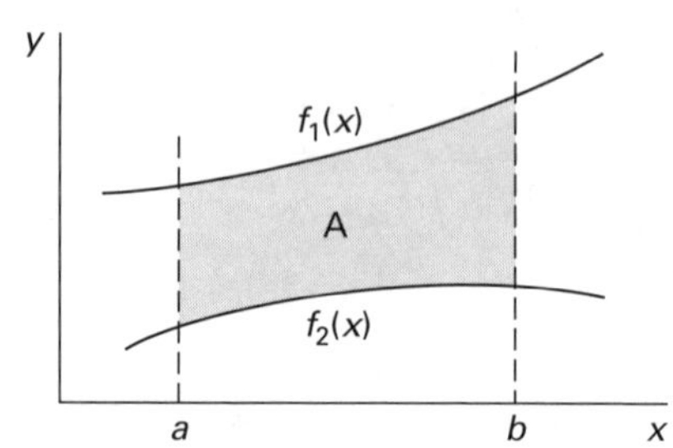

Example 10.8

Find the area between the x-axis and the parabola $y = x^2$ in the interval $[0, 4]$.

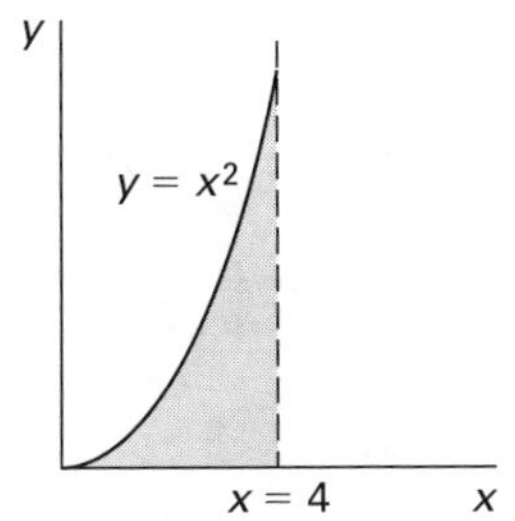

Solution

Referring to Eq. 10.35,

$$f_1(x) = x^2$$
$$f_2(x) = 0$$

$$A = \int_a^b (f_1(x) - f_2(x))\,dx = \int_0^4 x^2\,dx$$
$$= \frac{x^3}{3}\Big|_0^4 = 64/3$$

10. ARC LENGTH

Equation 10.36 gives the length of a curve defined by $f(x)$ whose derivative exists in the interval $[a, b]$.

$$\text{length} = \int_a^b \sqrt{1 + \left(f'(x)\right)^2}\,dx \qquad 10.36$$

11. PAPPUS' THEOREMS[3]

The first and second theorems of Pappus are:[4]

- *First Theorem:* Given a curve, C, that does not intersect the y-axis, the area of the *surface of revolution* generated by revolving C around the y-axis is equal to the product of the length of the curve and the circumference of the circle traced by the centroid of curve C.

$$A = \text{length} \times \text{circumference}$$
$$= \text{length} \times 2\pi \times \text{radius} \qquad 10.37$$

- *Second Theorem:* Given a plane region, R, that does not intersect the y-axis, the *volume of revolution* generated by revolving R around the y-axis is equal to the product of the area and the circumference of the circle traced by the centroid of area R.

$$V = \text{area} \times \text{circumference}$$
$$= \text{area} \times 2\pi \times \text{radius} \qquad 10.38$$

12. SURFACE OF REVOLUTION

The surface area obtained by rotating $f(x)$ about the x-axis is

$$A = 2\pi \int_{x=a}^{x=b} f(x)\sqrt{1 + \left(f'(x)\right)^2}\,dx \qquad 10.39$$

The surface area obtained by rotating $f(y)$ about the y-axis is

$$A = 2\pi \int_{y=c}^{y=d} f(y)\sqrt{1 + \left(f'(y)\right)^2}\,dy \qquad 10.40$$

Example 10.9

The curve $f(x) = {}^1\!/\!_2 x$ over the region $x = [0, 4]$ is rotated about the x-axis. What is the surface of revolution?

Solution

The surface of revolution is

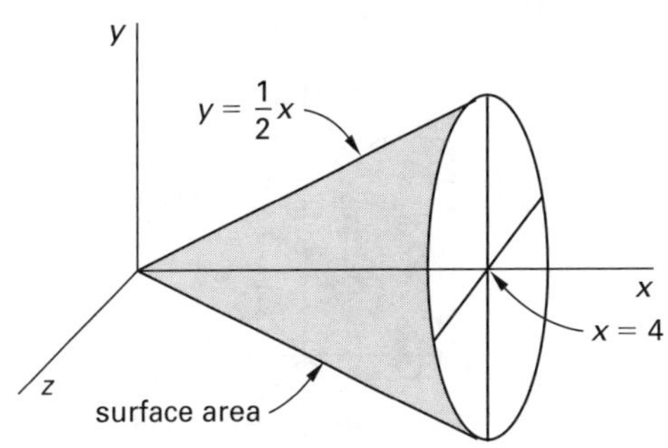

Since $f(x) = {}^1\!/\!_2 x$, $f'(x) = {}^1\!/\!_2$. From Eq. 10.39, the area is

$$A = 2\pi \int_{x=a}^{x=b} f(x)\sqrt{1 + \left(f'(x)\right)^2}\,dx$$
$$= 2\pi \int_0^4 \tfrac{1}{2}x\sqrt{1 + \left(\tfrac{1}{2}\right)^2}\,dx$$
$$= \frac{\sqrt{5}}{2}\pi \int_0^4 x\,dx$$
$$= \frac{\sqrt{5}}{2}\pi \frac{x^2}{2}\Big|_0^4$$
$$= \frac{\sqrt{5}}{2}\pi \left(\frac{(4)^2 - (0)^2}{2}\right)$$
$$= 4\sqrt{5}\pi$$

[3]This section is an introduction to surfaces and volumes of revolution but does not involve integration.
[4]Some authorities call the first theorem the second and vice versa.

13. VOLUME OF REVOLUTION

The volume obtained by rotating $f(x)$ about the x-axis is given by Eq. 10.41. $f^2(x)$ is the square of the function, not the second derivative. Equation 10.41 is known as the *method of discs*.

$$V = \pi \int_{x=a}^{x=b} f^2(x)dx \tag{10.41}$$

The volume obtained by rotating $f(x)$ about the y-axis can be found from Eq. 10.41 (i.e., using the method of discs) by rewriting the limits and equation in terms of y, or alternatively, the *method of shells* can be used, resulting in the second form of Eq. 10.42.

$$\begin{aligned} V &= \pi \int_{y=c}^{y=d} f^2(y)dy \\ &= 2\pi \int_{x=a}^{x=b} xf(x)dx \end{aligned} \tag{10.42}$$

Example 10.10

The curve $f(x) = x^2$ over the region $x = [0, 4]$ is rotated about the x-axis. What is the volume of revolution?

Solution

The volume of revolution is

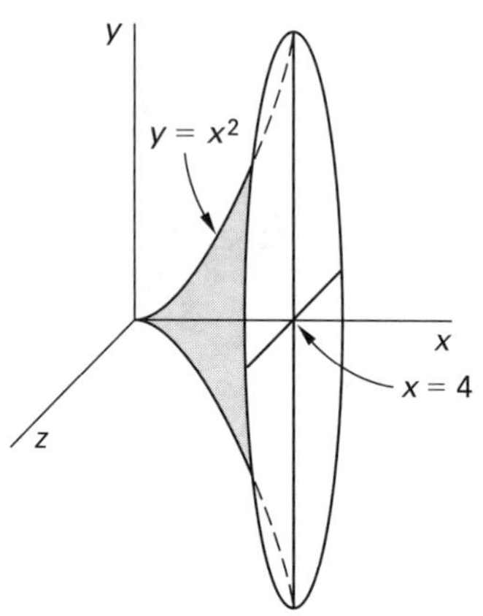

$$\begin{aligned} V &= \pi \int_a^b f^2(x)dx = \pi \int_0^4 (x^2)^2 dx \\ &= \pi \frac{x^5}{5}\bigg|_0^4 = \pi\left(\frac{1024}{5} - 0\right) = 204.8\pi \end{aligned}$$

14. MOMENTS OF A FUNCTION

The *first moment of a function* is a concept used in finding centroids and centers of gravity. Equation 10.43 and Eq. 10.44 are for one- and two-dimensional problems, respectively. It is the exponent of x (1 in this case) that gives the moment its name.

$$\text{first moment} = \int xf(x)dx \tag{10.43}$$

$$\text{first moment} = \iint xf(x, y)dx\,dy \tag{10.44}$$

The *second moment of a function* is a concept used in finding moments of inertia with respect to an axis. Equation 10.45 and Eq. 10.46 are for two- and three-dimensional problems, respectively. Second moments with respect to other axes are analogous.

$$(\text{second moment})_x = \iint y^2 f(x, y)dy\,dx \tag{10.45}$$

$$(\text{second moment})_x = \iiint (y^2 + z^2)f(x, y, z)dy\,dz\,dx \tag{10.46}$$

15. FOURIER SERIES

Any periodic waveform can be written as the sum of an infinite number of sinusoidal terms, known as *harmonic terms* (i.e., an infinite series). Such a sum of terms is known as a *Fourier series*, and the process of finding the terms is *Fourier analysis*. (Extracting the original waveform from the series is known as *Fourier inversion*.) Since most series converge rapidly, it is possible to obtain a good approximation to the original waveform with a limited number of sinusoidal terms.

Fourier's theorem is Eq. 10.47.[5] The object of a Fourier analysis is to determine the coefficients a_n and b_n. The constant a_0 can often be determined by inspection since it is the average value of the waveform.

$$\begin{aligned} f(t) = a_0 &+ a_1 \cos \omega t + a_2 \cos 2\omega t + \cdots \\ &+ b_1 \sin \omega t + b_2 \sin 2\omega t + \cdots \end{aligned} \tag{10.47}$$

ω is the *natural (fundamental) frequency* of the waveform. It depends on the actual waveform period, T.

$$\omega = \frac{2\pi}{T} \tag{10.48}$$

To simplify the analysis, the time domain can be normalized to the radian scale. The normalized scale is obtained by dividing all frequencies by ω. Then the Fourier series becomes

$$\begin{aligned} f(t) = a_0 &+ a_1 \cos t + a_2 \cos 2t + \cdots \\ &+ b_1 \sin t + b_2 \sin 2t + \cdots \end{aligned} \tag{10.49}$$

[5]The independent variable used in this section is t, since Fourier analysis is most frequently used in the time domain.

The coefficients a_n and b_n are found from the following relationships.

$$a_0 = \frac{1}{2\pi}\int_0^{2\pi} f(t)\,dt$$
$$= \frac{1}{T}\int_0^{T} f(t)\,dt \qquad 10.50$$

$$a_n = \frac{1}{\pi}\int_0^{2\pi} f(t)\cos nt\,dt$$
$$= \frac{2}{T}\int_0^{T} f(t)\cos nt\,dt \quad [n \geq 1] \qquad 10.51$$

$$b_n = \frac{1}{\pi}\int_0^{2\pi} f(t)\sin nt\,dt$$
$$= \frac{2}{T}\int_0^{T} f(t)\sin nt\,dt \quad [n \geq 1] \qquad 10.52$$

While Eq. 10.51 and Eq. 10.52 are always valid, the work of integrating and finding a_n and b_n can be greatly simplified if the waveform is recognized as being symmetrical. Table 10.1 summarizes the simplifications.

Example 10.11

Find the first four terms of a Fourier series that approximates the repetitive step function illustrated.

$$f(t) = \begin{Bmatrix} 1 & 0 < t < \pi \\ 0 & \pi < t < 2\pi \end{Bmatrix}$$

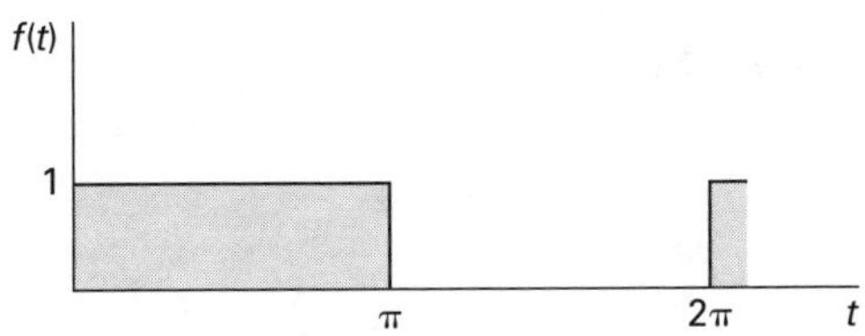

Solution

From Eq. 10.50,

$$a_0 = \frac{1}{2\pi}\int_0^{\pi} (1)\,dt + \frac{1}{2\pi}\int_{\pi}^{2\pi} (0)\,dt = \frac{1}{2}$$

This value of 1/2 corresponds to the average value of $f(t)$. It could have been found by observation.

$$a_1 = \frac{1}{\pi}\int_0^{\pi} (1)\cos t\,dt + \frac{1}{\pi}\int_{\pi}^{2\pi} (0)\cos t\,dt$$
$$= \frac{1}{\pi}\sin t\Big|_0^{\pi} + 0 = 0$$

***Table 10.1** Fourier Analysis Simplifications for Symmetrical Waveforms*

	even symmetry $f(-t) = f(t)$ 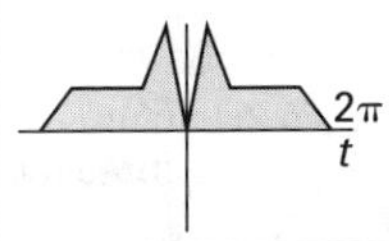	odd symmetry $f(-t) = -f(t)$ 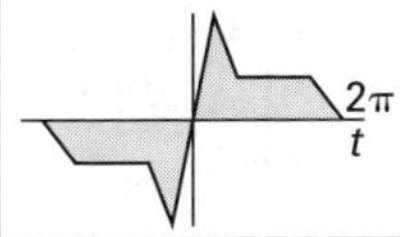
full-wave symmetry* $f(t+2\pi) = f(t)$ 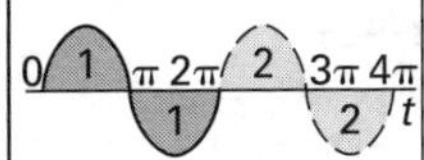 $\lvert A_2\rvert = \lvert A_1\rvert$ $\lvert A_{total}\rvert = \lvert A_1\rvert$ *any repeating wave form	$b_n = 0$ [all n] $a_n = \frac{1}{\pi}\int_0^{2\pi} f(t)\cos nt\,dt$ [all n]	$a_0 = 0$ $a_n = 0$ [all n] $b_n = \frac{1}{\pi}\int_0^{2\pi} f(t)\sin nt\,dt$ [all n]
half-wave symmetry* $f(t+\pi) = -f(t)$ 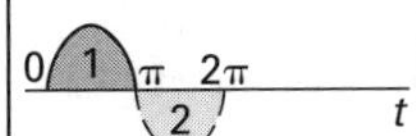 $\lvert A_2\rvert = \lvert A_1\rvert$ $\lvert A_{total}\rvert = 2\lvert A_1\rvert$ *same as rotational symmetry	$a_n = 0$ [even n] $b_n = 0$ [all n] $a_n = \frac{2}{\pi}\int_0^{\pi} f(t)\cos nt\,dt$ [odd n]	$a_0 = 0$ $a_n = 0$ [all n] $b_n = 0$ [even n] $b_n = \frac{2}{\pi}\int_0^{\pi} f(t)\sin nt\,dt$ [odd n]
quarter-wave symmetry $f(t+\pi) = -f(t)$ 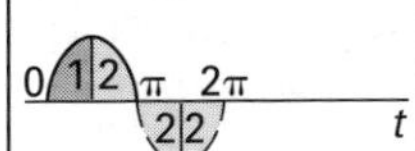 $\lvert A_2\rvert = \lvert A_1\rvert$ $\lvert A_{total}\rvert = 4\lvert A_1\rvert$	$a_0 = 0$ $a_n = 0$ [even n] $b_n = 0$ [all n] $a_n = \frac{4}{\pi}\int_0^{\frac{\pi}{2}} f(t)\cos nt\,dt$ [odd n]	$a_0 = 0$ $a_n = 0$ [all n] $b_n = 0$ [even n] $b_n = \frac{4}{\pi}\int_0^{\frac{\pi}{2}} f(t)\sin nt\,dt$ [odd n]

In general,

$$a_n = \frac{1}{\pi}\frac{\sin nt}{n}\Big|_0^{\pi} = 0$$

$$b_1 = \frac{1}{\pi}\int_0^{\pi} (1)\sin t\,dt + \frac{1}{\pi}\int_{\pi}^{2\pi} (0)\sin t\,dt$$
$$= \frac{1}{\pi} - \cos t\Big|_0^{\pi} = \frac{2}{\pi}$$

In general,

$$b_n = \frac{1}{\pi}\frac{-\cos nt}{n}\Big|_0^{\pi} = \begin{cases} 0 & \text{for } n \text{ even} \\ \frac{2}{\pi n} & \text{for } n \text{ odd} \end{cases}$$

The series is

$$f(t) = \tfrac{1}{2} + \tfrac{2}{\pi}(\sin t + \tfrac{1}{3}\sin 3t + \tfrac{1}{5}\sin 5t + \cdots)$$

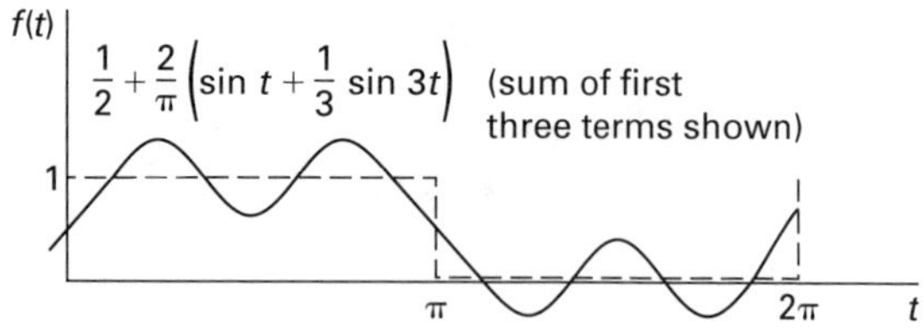

16. FAST FOURIER TRANSFORMS

Many mathematical operations are needed to implement a true Fourier transform. While the terms of a Fourier series might be slowly derived by integration, a faster method is needed to analyze real-time data. The *fast Fourier transform* (FFT) is a computer algorithm implemented in *spectrum analyzers (signal analyzers* or *FFT analyzers)* and replaces integration and multiplication operations with table look-ups and additions.[6]

Since the complexity of the transform is reduced, the transformation occurs more quickly, enabling efficient analysis of waveforms with little or no periodicity.[7]

Using a spectrum analyzer requires choosing the frequency band (e.g., 0–20 kHz) to be monitored. (This step automatically selects the sampling period. The lower the frequencies sampled, the longer the sampling period.) If they are not fixed by the analyzer, the numbers of time-dependent input variable samples (e.g., 1024) and frequency-dependent output variable values (e.g., 400) are chosen.[8] There are half as many frequency lines as data points because each line contains two pieces of information—real (amplitude) and imaginary (phase). The *resolution* of the resulting frequency analysis is

$$\text{resolution} = \frac{\text{frequency bandwidth}}{\text{no. of output variable values}} \qquad 10.53$$

17. INTEGRAL FUNCTIONS

Integrals that cannot be evaluated as finite combinations of elementary functions are called *integral functions.* These functions are evaluated by series expansion. Some of the more common functions are listed as follows.[9,10]

- *integral sine function:*

$$\mathrm{Si}(x) = \int_0^x \frac{\sin x}{x}\,dx$$
$$= x - \frac{x^3}{3\cdot 3!} + \frac{x^5}{5\cdot 5!} - \frac{x^7}{7\cdot 7!} + \cdots \qquad 10.54$$

- *integral cosine function:*

$$\mathrm{Ci}(x) = \int_{-\infty}^x \frac{\cos x}{x}\,dx = -\int_x^{\infty} \frac{\cos x}{x}\,dx$$
$$= C_E + \ln x - \frac{x^2}{2\cdot 2!} + \frac{x^4}{4\cdot 4!} - \cdots \qquad 10.55$$

- *integral exponential function:*

$$\mathrm{Ei}(x) = \int_{-\infty}^x \frac{e^x}{x}\,dx = -\int_{-x}^{\infty} \frac{e^{-x}}{x}\,dx$$
$$= C_E + \ln x + x + \frac{x^2}{2\cdot 2!} + \frac{x^3}{3\cdot 3!} + \cdots \qquad 10.56$$

- *error function:*

$$\mathrm{erf}(x) = \frac{2}{\sqrt{\pi}} \int_0^x e^{-x^2}\,dx$$
$$= \left(\frac{2}{\sqrt{\pi}}\right)\left(\frac{x}{1\cdot 0!} - \frac{x^3}{3\cdot 1!} + \frac{x^5}{5\cdot 2!} - \frac{x^7}{7\cdot 3!} + \cdots\right) \qquad 10.57$$

- *Fresnel integrals:*

$$C(x) = \int_0^x \cos(t^2)\,dt \qquad 10.58$$

$$S(x) = \int_0^x \sin(t^2)\,dt \qquad 10.59$$

[6] *Spectrum analysis,* also known as *frequency analysis, signature analysis,* and *time-series analysis,* develops a relationship (usually graphical) between some property (e.g., amplitude or phase shift) versus frequency.

[7] Hours and days of manual computations are compressed into milliseconds.

[8] Two samples per time-dependent cycle (at the maximum frequency) is the lower theoretical limit for sampling, but the practical minimum rate is approximately 2.5 samples per cycle. This will ensure that *alias components* (i.e., low-level frequency signals) do not show up in the frequency band of interest.

[9] Other integral functions include the gamma function and elliptic integral.

[10] C_E in Eq. 10.55 and Eq. 10.56 is *Euler's constant.*

$$C_E = \int_{+\infty}^0 e^{-x}\ln x\,dx$$
$$= \lim_{m\to\infty}\left(1 + \frac{1}{2} + \frac{1}{3} + \cdots + \frac{1}{m} - \ln m\right)$$
$$= 0.577215665$$

11 Differential Equations

1. Types of Differential Equations 11-1
2. Homogeneous, First-Order Linear Differential Equations with Constant Coefficients 11-2
3. First-Order Linear Differential Equations . . 11-2
4. First-Order Separable Differential Equations 11-3
5. First-Order Exact Differential Equations . . 11-3
6. Homogeneous, Second-Order Linear Differential Equations with Constant Coefficients 11-3
7. Nonhomogeneous Differential Equations . . . 11-3
8. Named Differential Equations 11-5
9. Laplace Transforms 11-5
10. Step and Impulse Functions 11-6
11. Algebra of Laplace Transforms 11-6
12. Convolution Integral 11-6
13. Using Laplace Transforms 11-7
14. Third- and Higher-Order Linear Differential Equations with Constant Coefficients . . . 11-8
15. Application: Engineering Systems 11-8
16. Application: Mixing 11-8
17. Application: Exponential Growth and Decay 11-9
18. Application: Epidemics 11-10
19. Application: Surface Temperature 11-10
20. Application: Evaporation 11-10

1. TYPES OF DIFFERENTIAL EQUATIONS

A *differential equation* is a mathematical expression combining a function (e.g., $y = f(x)$) and one or more of its derivatives. The *order* of a differential equation is the highest derivative in it. *First-order differential equations* contain only first derivatives of the function, *second-order differential equations* contain second derivatives (and may contain first derivatives as well), and so on.

A *linear differential equation* can be written as a sum of products of multipliers of the function and its derivatives. If the multipliers are scalars, the differential equation is said to have *constant coefficients*. If the function or one of its derivatives is raised to some power (other than one) or is embedded in another function (e.g., y embedded in $\sin y$ or e^y), the equation is said to be *nonlinear*.

Each term of a *homogeneous differential equation* contains either the function (y) or one of its derivatives; that is, the sum of derivative terms is equal to zero. In a *nonhomogeneous differential equation*, the sum of derivative terms is equal to a nonzero *forcing function* of the independent variable (e.g., $g(x)$). In order to solve a nonhomogeneous equation, it is often necessary to solve the homogeneous equation first. The homogeneous equation corresponding to a nonhomogeneous equation is known as a *reduced equation* or *complementary equation*.

The following examples illustrate the types of differential equations.

$y' - 7y = 0$	homogeneous, first-order linear, with constant coefficients
$y'' - 2y' + 8y = \sin 2x$	nonhomogeneous, second-order linear, with constant coefficients
$y'' - (x^2 - 1)y^2 = \sin 4x$	nonhomogeneous, second-order, nonlinear

An *auxiliary equation* (also called the *characteristic equation*) can be written for a homogeneous linear differential equation with constant coefficients, regardless of order. This auxiliary equation is simply the polynomial formed by replacing all derivatives with variables raised to the power of their respective derivatives.

The purpose of solving a differential equation is to derive an expression for the function in terms of the independent variable. The expression does not need to be explicit in the function, but there can be no derivatives in the expression. Since, in the simplest cases, solving a differential equation is equivalent to finding an indefinite integral, it is not surprising that *constants of integration* must be evaluated from knowledge of how the system behaves. Additional data are known as *initial values*, and any problem that includes them is known as an *initial value problem*.[1]

Most differential equations require lengthy solutions and are not efficiently solved by hand. However, several types are fairly simple and are presented in this chapter.

[1]The term *initial* implies that time is the independent variable. While this may explain the origin of the term, initial value problems are not limited to the time domain. A *boundary value problem* is similar, except that the data come from different points. For example, additional data in the form $y(x_0)$ and $y'(x_0)$ or $y(x_0)$ and $y'(x_1)$ that need to be simultaneously satisfied constitute an initial value problem. Data of the form $y(x_0)$ and $y(x_1)$ constitute a boundary value problem. Until solved, it is difficult to know whether a boundary value problem has zero, one, or more than one solution.

Example 11.1

Write the complementary differential equation for the following nonhomogeneous differential equation.

$$y'' + 6y' + 9y = e^{-14x}\sin 5x$$

Solution

The complementary equation is found by eliminating the forcing function, $e^{-14x}\sin 5x$.

$$y'' + 6y' + 9y = 0$$

Example 11.2

Write the auxiliary equation to the following differential equation.

$$y'' + 4y' + y = 0$$

Solution

Replacing each derivative with a polynomial term whose degree equals the original order, the auxiliary equation is

$$r^2 + 4r + 1 = 0$$

2. HOMOGENEOUS, FIRST-ORDER LINEAR DIFFERENTIAL EQUATIONS WITH CONSTANT COEFFICIENTS

A homogeneous, first-order linear differential equation with constant coefficients will have the general form of Eq. 11.1.

$$y' + ky = 0 \qquad 11.1$$

The auxiliary equation is $r + k = 0$ and it has a root of $r = -k$. Equation 11.2 is the solution.

$$y = Ae^{rx} = Ae^{-kx} \qquad 11.2$$

If the initial condition is known to be $y(0) = y_0$, the solution is

$$y = y_0 e^{-kx} \qquad 11.3$$

3. FIRST-ORDER LINEAR DIFFERENTIAL EQUATIONS

A first-order linear differential equation has the general form of Eq. 11.4. $p(x)$ and $g(x)$ can be constants or any function of x (but not of y). However, if $p(x)$ is a constant and $g(x)$ is zero, it is easier to solve the equation as shown in Sec. 11.2.

$$y' + p(x)y = g(x) \qquad 11.4$$

The *integrating factor* (which is usually a function) to this differential equation is

$$u(x) = \exp\left(\int p(x)\,dx\right) \qquad 11.5$$

The closed-form solution to Eq. 11.4 is

$$y = \frac{1}{u(x)}\left(\int u(x)g(x)\,dx + C\right) \qquad 11.6$$

For the special case where $p(x)$ and $g(x)$ are both constants, Eq. 11.4 becomes

$$y' + ay = b \qquad 11.7$$

If the initial condition is $y(0) = y_0$, then the solution to Eq. 11.7 is

$$y = \frac{b}{a}(1 - e^{-ax}) + y_0 e^{-ax} \qquad 11.8$$

Example 11.3

Find a solution to the following differential equation.

$$y' - y = 2xe^{2x} \qquad y(0) = 1$$

Solution

This is a first-order linear equation with $p(x) = -1$ and $g(x) = 2xe^{2x}$. The integrating factor is

$$u(x) = \exp\left(\int p(x)\,dx\right) = \exp\left(\int -1\,dx\right) = e^{-x}$$

The solution is given by Eq. 11.6.

$$\begin{aligned} y &= \frac{1}{u(x)}\left(\int u(x)g(x)\,dx + C\right) \\ &= \frac{1}{e^{-x}}\left(\int e^{-x}2xe^{2x}\,dx + C\right) \\ &= e^{x}\left(2\int xe^{x}\,dx + C\right) \\ &= e^{x}(2xe^{x} - 2e^{x} + C) \\ &= e^{x}(2e^{x}(x-1) + C) \end{aligned}$$

From the initial condition,

$$\begin{aligned} y(0) &= 1 \\ e^0\big((2)(e^0)(0-1) + C\big) &= 1 \\ 1\big((2)(1)(-1) + C\big) &= 1 \end{aligned}$$

Therefore, $C = 3$. The complete solution is

$$y = e^{x}\big(2e^{x}(x-1) + 3\big)$$

4. FIRST-ORDER SEPARABLE DIFFERENTIAL EQUATIONS

First-order separable differential equations can be placed in the form of Eq. 11.9. For clarity and convenience, y' is written as dy/dx.

$$m(x) + n(y)\frac{dy}{dx} = 0 \quad 11.9$$

Equation 11.9 can be placed in the form of Eq. 11.10, both sides of which are easily integrated. An initial value will establish the constant of integration.

$$m(x)\,dx = -n(y)\,dy \quad 11.10$$

5. FIRST-ORDER EXACT DIFFERENTIAL EQUATIONS

A *first-order exact differential equation* has the form

$$f_x(x,y) + f_y(x,y)y' = 0 \quad 11.11$$

Notice that $f_x(x, y)$ is the exact derivative of $f(x, y)$ with respect to x, and $f_y(x, y)$ is the exact derivative of $f(x, y)$ with respect to y. The solution is

$$f(x,y) - C = 0 \quad 11.12$$

6. HOMOGENEOUS, SECOND-ORDER LINEAR DIFFERENTIAL EQUATIONS WITH CONSTANT COEFFICIENTS

Homogeneous second-order linear differential equations with constant coefficients have the form of Eq. 11.13. They are most easily solved by finding the two roots of the auxiliary equation. (See Eq. 11.14.)

$$y'' + k_1y' + k_2y = 0 \quad 11.13$$

$$r^2 + k_1r + k_2 = 0 \quad 11.14$$

There are three cases. If the two roots of Eq. 11.14 are real and different, the solution is

$$y = A_1e^{r_1x} + A_2e^{r_2x} \quad 11.15$$

If the two roots are real and the same, the solution is

$$y = A_1e^{rx} + A_2xe^{rx} \quad 11.16$$

$$r = \frac{-k_1}{2} \quad 11.17$$

If the two roots are imaginary, they will be of the form $(\alpha + i\omega)$ and $(\alpha - i\omega)$, and the solution is

$$y = A_1e^{\alpha x}\cos\omega x + A_2e^{\alpha x}\sin\omega x \quad 11.18$$

In all three cases, A_1 and A_2 must be found from the two initial conditions.

Example 11.4

Solve the following differential equation.

$$y'' + 6y' + 9y = 0$$

$$y(0) = 0 \qquad y'(0) = 1$$

Solution

The auxiliary equation is

$$r^2 + 6r + 9 = 0$$

$$(r+3)(r+3) = 0$$

The roots to the auxiliary equation are $r_1 = r_2 = -3$. Therefore, the solution has the form of Eq. 11.16.

$$y = A_1e^{-3x} + A_2xe^{-3x}$$

The first initial condition is

$$\begin{aligned} y(0) &= 0 \\ A_1e^0 + A_2(0)e^0 &= 0 \\ A_1 + 0 &= 0 \\ A_1 &= 0 \end{aligned}$$

To use the second initial condition, the derivative of the equation is needed. Making use of the known fact that $A_1 = 0$,

$$y' = \frac{d}{dx}A_2xe^{-3x} = -3A_2xe^{-3x} + A_2e^{-3x}$$

Using the second initial condition,

$$\begin{aligned} y'(0) &= 1 \\ -3A_2(0)e^0 + A_2e^0 &= 1 \\ 0 + A_2 &= 1 \\ A_2 &= 1 \end{aligned}$$

The solution is

$$y = xe^{-3x}$$

7. NONHOMOGENEOUS DIFFERENTIAL EQUATIONS

A nonhomogeneous equation has the form of Eq. 11.19. $f(x)$ is known as the *forcing function*.

$$y'' + p(x)y' + q(x)y = f(x) \quad 11.19$$

The solution to Eq. 11.19 is the sum of two equations. The *complementary solution*, y_c, solves the complementary (i.e., homogeneous) problem. The *particular solution*, y_p, is any specific solution to the nonhomogeneous Eq. 11.19 that is known or can be found. Initial values are used to evaluate any unknown coefficients in the complementary solution *after* y_c and y_p have been combined. (The particular solution will not have any unknown coefficients.)

$$y = y_c + y_p \quad \text{11.20}$$

Two methods are available for finding a particular solution. The *method of undetermined coefficients*, as presented here, can be used only when $p(x)$ and $q(x)$ are constant coefficients and $f(x)$ takes on one of the forms in Table 11.1.

The particular solution can be read from Table 11.1 if the forcing function is of one of the forms given. Of course, the coefficients A_i and B_i are not known—these are the *undetermined coefficients*. The exponent s is the smallest nonnegative number (and will be 0, 1, or 2), which ensures that no term in the particular solution, y_p, is also a solution to the complementary equation, y_c. s must be determined prior to proceeding with the solution procedure.

Once y_p (including s) is known, it is differentiated to obtain y_p' and y_p'', and all three functions are substituted into the original nonhomogeneous equation. The resulting equation is rearranged to match the forcing function, $f(x)$, and the unknown coefficients are determined, usually by solving simultaneous equations.

If the forcing function, $f(x)$, is more complex than the forms shown in Table 11.1, or if either $p(x)$ or $q(x)$ is a function of x, the method of *variation of parameters* should be used. This complex and time-consuming method is not covered in this book.

Table 11.1 *Particular Solutions**

form of $f(x)$	form of y_p
$P_n(x) = a_0x^n + a_1x^{n-1} + \cdots + a_n$	$x^s\left(A_0x^n + A_1x^{n-1} + \cdots + A_n\right)$
$P_n(x)e^{\alpha x}$	$x^s\left(A_0x^n + A_1x^{n-1} + \cdots + A_n\right)e^{\alpha x}$
$P_n(x)e^{\alpha x}\begin{Bmatrix}\sin\omega x\\ \cos\omega x\end{Bmatrix}$	$x^s\left(\left(A_0x^n + A_1x^{n-1} + \cdots + A_n\right) \times e^{\alpha x}\cos\omega x + \left(B_0x^n + B_1x^{n-1} + \cdots + B_n\right) \times e^{\alpha x}\sin\omega x\right)$

*$P_n(x)$ is a polynomial of degree n.

Example 11.5

Solve the following nonhomogeneous differential equation.

$$y'' + 2y' + y = e^x \cos x$$

Solution

step 1: Find the solution to the complementary (homogeneous) differential equation.

$$y'' + 2y' + y = 0$$

Since this is a differential equation with constant coefficients, write the auxiliary equation.

$$r^2 + 2r + 1 = 0$$

The auxiliary equation factors in $(r + 1)^2 = 0$ with two identical roots at $r = -1$. Therefore, the solution to the homogeneous differential equation is

$$y_c(x) = C_1e^{-x} + C_2xe^{-x}$$

step 2: Use Table 11.1 to determine the form of a particular solution. Since the forcing function has the form $P_n(x)e^{\alpha x}\cos\omega x$ with $P_n(x) = 1$ (equivalent to $n = 0$), $\alpha = 1$, and $\omega = 1$, the particular solution has the form

$$y_p(x) = x^s(Ae^x \cos x + Be^x \sin x)$$

step 3: Determine the value of s. Check to see if any of the terms in $y_p(x)$ will themselves solve the homogeneous equation. Try $Ae^x\cos x$ first.

$$\frac{d}{dx}(Ae^x \cos x) = Ae^x \cos x - Ae^x \sin x$$

$$\frac{d^2}{dx^2}(Ae^x \cos x) = -2Ae^x \sin x$$

Substitute these quantities into the homogeneous equation.

$$\begin{aligned} y'' + 2y' + y &= 0 \\ -2Ae^x \sin x + 2Ae^x \cos x & \\ - 2Ae^x\sin x + Ae^x\cos x &= 0 \\ 3Ae^x \cos x - 4Ae^x \sin x &= 0 \end{aligned}$$

Disregarding the trivial (i.e., $A = 0$) solution, $Ae^x\cos x$ does not solve the homogeneous equation.

Next, try $Be^x\sin x$.

$$\frac{d}{dx}(Be^x \sin x) = Be^x \cos x + Be^x \sin x$$

$$\frac{d^2}{dx^2}(Be^x \sin x) = 2Be^x \cos x$$

Substitute these quantities into the homogeneous equation.

$$y'' + 2y' + y = 0$$

$$2Be^x \cos x + 2Be^x \cos x + 2Be^x \sin x + Be^x \sin x = 0$$

$$3Be^x \sin x + 4Be^x \cos x = 0$$

Disregarding the trivial ($B = 0$) case, $Be^x \sin x$ does not solve the homogeneous equation.

Since none of the terms in $y_p(x)$ solve the homogeneous equation, $s = 0$, and a particular solution has the form

$$y_p(x) = Ae^x \cos x + Be^x \sin x$$

step 4: Use the method of unknown coefficients to determine A and B in the particular solution. Drawing on the previous steps, substitute the quantities derived from the particular solution into the nonhomogeneous equation.

$$y'' + 2y' + y = e^x \cos x$$

$$\begin{aligned} -2Ae^x \sin x + 2Be^x \cos x & \\ + 2Ae^x \cos x - 2Ae^x \sin x & \\ + 2Be^x \cos x + 2Be^x \sin x & \\ + Ae^x \cos x + Be^x \sin x &= e^x \cos x \end{aligned}$$

Combining terms,

$$(-4A + 3B)e^x \sin x + (3A + 4B)e^x \cos x = e^x \cos x$$

Equating the coefficients of like terms on either side of the equal sign results in the following simultaneous equations.

$$-4A + 3B = 0$$

$$3A + 4B = 1$$

The solution to these equations is

$$A = \frac{3}{25}$$

$$B = \frac{4}{25}$$

A particular solution is

$$y_p(x) = \left(\tfrac{3}{25}\right)(e^x \cos x) + \left(\tfrac{4}{25}\right)(e^x \sin x)$$

step 5: Write the general solution.

$$\begin{aligned} y(x) &= y_c(x) + y_p(x) \\ &= C_1 e^{-x} + C_2 x e^{-x} + \left(\tfrac{3}{25}\right)(e^x \cos x) \\ &\quad + \left(\tfrac{4}{25}\right)(e^x \sin x) \end{aligned}$$

The values of C_1 and C_2 would be determined at this time if initial conditions were known.

8. NAMED DIFFERENTIAL EQUATIONS

Some differential equations with specific forms are named after the individuals who developed solution techniques for them.

- *Bessel equation of order ν*

$$x^2 y'' + xy' + (x^2 - \nu^2)y = 0 \qquad \text{11.21}$$

- *Cauchy equation*

$$a_0 x^n \frac{d^n y}{dx^n} + a_1 x^{n-1} \frac{d^{n-1} y}{dx^{n-1}} + \cdots + a_{n-1} x \frac{dy}{dx} + a_n y = f(x) \qquad \text{11.22}$$

- *Euler equation*

$$x^2 y'' + \alpha x y' + \beta y = 0 \qquad \text{11.23}$$

- *Gauss' hypergeometric equation*

$$x(1 - x)y'' + (c - (a + b + 1)x)y' - aby = 0 \qquad \text{11.24}$$

- *Legendre equation of order λ*

$$(1 - x^2)y'' - 2xy' + \lambda(\lambda + 1)y = 0 \quad [-1 < x < 1] \qquad \text{11.25}$$

9. LAPLACE TRANSFORMS

Traditional methods of solving nonhomogeneous differential equations by hand are usually difficult and/or time consuming. *Laplace transforms* can be used to reduce many solution procedures to simple algebra.

Every mathematical function, $f(t)$, for which Eq. 11.26 exists has a Laplace transform, written as $\mathcal{L}(f)$ or $F(s)$. The transform is written in the s-domain, regardless of the independent variable in the original function.[2] (The variable s is equivalent to a derivative operator, although it may be handled in the equations as a simple

[2]It is traditional to write the original function as a function of the independent variable t rather than x. However, Laplace transforms are not limited to functions of time.

variable.) Equation 11.26 converts a function into a Laplace transform.

$$\mathcal{L}(f(t)) = F(s) = \int_0^\infty e^{-st} f(t)\,dt \qquad 11.26$$

Equation 11.26 is not often needed because tables of transforms are readily available. (Appendix 11.A contains some of the most common transforms.)

Extracting a function from its transform is the *inverse Laplace transform* operation. Although other methods exist, this operation is almost always done by finding the transform in a set of tables.[3]

$$f(t) = \mathcal{L}^{-1}(F(s)) \qquad 11.27$$

Example 11.6

Find the Laplace transform of the following function.

$$f(t) = e^{at} \qquad [s > a]$$

Solution

Applying Eq. 11.26,

$$\mathcal{L}(e^{at}) = \int_0^\infty e^{-st} e^{at}\,dt = \int_0^\infty e^{-(s-a)t}\,dt$$

$$= -\frac{e^{-(s-a)t}}{s-a}\Bigg|_0^\infty = \frac{1}{s-a} \qquad [s > a]$$

10. STEP AND IMPULSE FUNCTIONS

Many forcing functions are sinusoidal or exponential in nature; others, however, can only be represented by a step or impulse function. A *unit step function*, u_t, is a function describing the disturbance of magnitude 1 that is not present before time t but is suddenly there after time t. A step of magnitude 5 at time $t = 3$ would be represented as $5u_3$. (The notation $5u(t-3)$ is used in some books.)

The *unit impulse function*, δ_t, is a function describing a disturbance of magnitude 1 that is applied and removed so quickly as to be instantaneous. An impulse of magnitude 5 at time 3 would be represented by $5\delta_3$. (The notation $5\delta(t-3)$ is used in some books.)

Example 11.7

What is the notation for a forcing function of magnitude 6 that is applied at $t = 2$ and that is completely removed at $t = 7$?

Solution

The notation is $f(t) = 6(u_2 - u_7)$.

[3]Other methods include integration in the complex plane, convolution, and simplification by partial fractions.

Example 11.8

Find the Laplace transform of u_0, a unit step at $t = 0$.

$$f(t) = 0 \text{ for } t < 0$$

$$f(t) = 1 \text{ for } t \geq 0$$

Solution

Since the Laplace transform is an integral that starts at $t = 0$, the value of $f(t)$ prior to $t = 0$ is irrelevant.

$$\mathcal{L}(u_0) = \int_0^\infty e^{-st}(1)\,dt = -\frac{e^{-st}}{s}\Bigg|_0^\infty$$

$$= 0 - \frac{-1}{s} = \frac{1}{s}$$

11. ALGEBRA OF LAPLACE TRANSFORMS

Equations containing Laplace transforms can be simplified by applying the following principles.

- *linearity theorem* (c is a constant.)

$$\mathcal{L}(cf(t)) = c\mathcal{L}(f(t)) = cF(s) \qquad 11.28$$

- *superposition theorem* ($f(t)$ and $g(t)$ are different functions.)

$$\begin{aligned}\mathcal{L}(f(t) \pm g(t)) &= \mathcal{L}(f(t)) \pm \mathcal{L}(g(t)) \\ &= F(s) \pm G(s)\end{aligned} \qquad 11.29$$

- *time-shifting theorem* (*delay theorem*)

$$\mathcal{L}(f(t-b)u_b) = e^{-bs}F(s) \qquad 11.30$$

- *Laplace transform of a derivative*

$$\begin{aligned}\mathcal{L}(f^n(t)) &= -f^{n-1}(0) - sf^{n-2}(0) - \cdots \\ &\quad - s^{n-1}f(0) + s^nF(s)\end{aligned} \qquad 11.31$$

- *other properties*

$$\mathcal{L}\left(\int_0^t f(u)\,du\right) = \frac{1}{s}F(s) \qquad 11.32$$

$$\mathcal{L}(tf(t)) = -\frac{dF}{ds} \qquad 11.33$$

$$\mathcal{L}\left(\frac{1}{t}f(t)\right) = \int_0^\infty F(u)\,du \qquad 11.34$$

12. CONVOLUTION INTEGRAL

A complex Laplace transform, $F(s)$, will often be recognized as the product of two other transforms, $F_1(s)$ and $F_2(s)$, whose corresponding functions $f_1(t)$ and $f_2(t)$ are known. Unfortunately, Laplace transforms cannot be computed with ordinary multiplication. That is, $f(t) \neq f_1(t)f_2(t)$ even though $F(s) = F_1(s)F_2(s)$.

However, it is possible to extract $f(t)$ from its *convolution*, $h(t)$, as calculated from either of the *convolution integrals* in Eq. 11.35. This process is demonstrated in Ex. 11.9. χ is a dummy variable.

$$\begin{aligned} f(t) &= \mathcal{L}^{-1}(F_1(s)F_2(s)) \\ &= \int_0^t f_1(t-\chi)f_2(\chi)d\chi \\ &= \int_0^t f_1(\chi)f_2(t-\chi)d\chi \end{aligned} \qquad 11.35$$

Example 11.9

Use the convolution integral to find the inverse transform of

$$F(s) = \frac{3}{s^2(s^2+9)}$$

Solution

$F(s)$ can be factored as

$$F_1(s)F_2(s) = \left(\frac{1}{s^2}\right)\left(\frac{3}{s^2+9}\right)$$

As the inverse transforms of $F_1(s)$ and $F_2(s)$ are $f_1(t) = t$ and $f_2(t) = \sin 3t$, respectively, the convolution integral from Eq. 11.35 is

$$\begin{aligned} f(t) &= \int_0^t (t-\chi)\sin 3\chi d\chi \\ &= \int_0^t (t\sin 3\chi - \chi \sin 3\chi)d\chi \\ &= t\int_0^t \sin 3\chi d\chi - \int_0^t \chi \sin 3\chi d\chi \end{aligned}$$

Expand using integration by parts.

$$\begin{aligned} f(t) &= -\tfrac{1}{3}t\cos 3\chi + \tfrac{1}{3}\chi\cos 3\chi - \tfrac{1}{9}\sin 3\chi\Big|_0^t \\ &= \frac{3t - \sin 3t}{9} \end{aligned}$$

13. USING LAPLACE TRANSFORMS

Any nonhomogeneous linear differential equation with constant coefficients can be solved with the following procedure, which reduces the solution to simple algebra. A complete table of transforms simplifies or eliminates step 5.

step 1: Put the differential equation in standard form (i.e., isolate the y'' term).

$$y'' + k_1y' + k_2y = f(t) \qquad 11.36$$

step 2: Take the Laplace transform of both sides. Use the linearity and superposition theorems. (See Eq. 11.28 and Eq. 11.29.)

$$\mathcal{L}(y'') + k_1\mathcal{L}(y') + k_2\mathcal{L}(y) = \mathcal{L}(f(t)) \qquad 11.37$$

step 3: Use Eq. 11.38 and Eq. 11.39 to expand the equation. (These are specific forms of Eq. 11.31.) Use a table to evaluate the transform of the forcing function.

$$\mathcal{L}(y'') = s^2\mathcal{L}(y) - sy(0) - y'(0) \qquad 11.38$$

$$\mathcal{L}(y') = s\mathcal{L}(y) - y(0) \qquad 11.39$$

step 4: Use algebra to solve for $\mathcal{L}(y)$.

step 5: If needed, use partial fractions to simplify the expression for $\mathcal{L}(y)$.

step 6: Take the inverse transform to find $y(t)$.

$$y(t) = \mathcal{L}^{-1}(\mathcal{L}(y)) \qquad 11.40$$

Example 11.10

Find $y(t)$ for the following differential equation.

$$y'' + 2y' + 2y = \cos t$$

$$y(0) = 1 \qquad y'(0) = 0$$

Solution

step 1: The equation is already in standard form.

step 2: $\mathcal{L}(y'') + 2\mathcal{L}(y') + 2\mathcal{L}(y) = \mathcal{L}(\cos t)$

step 3: Use Eq. 11.38 and Eq. 11.39. Use App. 11.A to find the transform of $\cos t$.

$$\begin{aligned} &s^2\mathcal{L}(y) - sy(0) - y'(0) + 2s\mathcal{L}(y) - 2y(0) + 2\mathcal{L}(y) \\ &\quad = \frac{s}{s^2+1} \end{aligned}$$

But, $y(0) = 1$ and $y'(0) = 0$.

$$s^2\mathcal{L}(y) - s + 2s\mathcal{L}(y) - 2 + 2\mathcal{L}(y) = \frac{s}{s^2+1}$$

step 4: Combine terms and solve for $\mathcal{L}(y)$.

$$\mathcal{L}(y)(s^2 + 2s + 2) - s - 2 = \frac{s}{s^2+1}$$

$$\begin{aligned} \mathcal{L}(y) &= \frac{\frac{s}{s^2+1} + s + 2}{s^2 + 2s + 2} \\ &= \frac{s^3 + 2s^2 + 2s + 2}{(s^2+1)(s^2+2s+2)} \end{aligned}$$

step 5: Expand the expression for $\mathcal{L}(y)$ by partial fractions.

$$\begin{aligned}\mathcal{L}(y) &= \frac{s^3+2s^2+2s+2}{(s^2+1)(s^2+2s+2)} \\ &= \frac{A_1 s+B_1}{s^2+1}+\frac{A_2 s+B_2}{s^2+2s+2} \\ &= \frac{\begin{array}{c} s^3(A_1+A_2)+s^2(2A_1+B_1+B_2) \\ +\,s(2A_1+2B_1+A_2)+(2B_1+B_2)\end{array}}{(s^2+1)(s^2+2s+2)}\end{aligned}$$

The following simultaneous equations result.

$$\begin{aligned} A_1 + A_2 \quad\quad\quad\quad &= 1 \\ 2A_1 \quad + B_1 + B_2 &= 2 \\ 2A_1 + A_2 + 2B_1 \quad &= 2 \\ 2B_1 + B_2 &= 2 \end{aligned}$$

These equations have the solutions $A_1 = 1/5$, $A_2 = 4/5$, $B_1 = 2/5$, and $B_2 = 6/5$.

step 6: Refer to App. 11.A and take the inverse transforms. (The numerator of the second term is rewritten from $(4s+6)$ to $((4s+4)+2)$.)

$$\begin{aligned} y &= \mathcal{L}^{-1}\Big(\mathcal{L}(y)\Big) \\ &= \mathcal{L}^{-1}\left(\frac{\left(\frac{1}{5}\right)(s+2)}{s^2+1}+\frac{\left(\frac{1}{5}\right)(4s+6)}{s^2+2s+2}\right) \\ &= \left(\tfrac{1}{5}\right)\left(\begin{array}{l} \mathcal{L}^{-1}\left(\dfrac{s}{s^2+1}\right)+2\mathcal{L}^{-1}\left(\dfrac{1}{s^2+1}\right) \\ +\,4\mathcal{L}^{-1}\left(\dfrac{s-(-1)}{\big(s-(-1)\big)^2+1}\right) \\ +\,2\mathcal{L}^{-1}\left(\dfrac{1}{\big(s-(-1)\big)^2+1}\right)\end{array}\right) \\ &= \left(\tfrac{1}{5}\right)(\cos t+2\sin t+4e^{-t}\cos t+2e^{-t}\sin t) \end{aligned}$$

14. THIRD- AND HIGHER-ORDER LINEAR DIFFERENTIAL EQUATIONS WITH CONSTANT COEFFICIENTS

The solutions of third- and higher-order linear differential equations with constant coefficients are extensions of the solutions for second-order equations of this type. Specifically, if an equation is homogeneous, the auxiliary equation is written and its roots are found. If the equation is nonhomogeneous, Laplace transforms can be used to simplify the solution.

Consider the following homogeneous differential equation with constant coefficients.

$$y^n + k_1 y^{n-1} + \cdots + k_{n-1}y' + k_n y = 0 \tag{11.41}$$

The auxiliary equation to Eq. 11.41 is

$$r^n + k_1 r^{n-1} + \cdots + k_{n-1}r + k_n = 0 \tag{11.42}$$

For each real and distinct root r, the solution contains the term

$$y = Ae^{rx} \tag{11.43}$$

For each real root r that repeats m times, the solution contains the term

$$y = (A_1 + A_2 x + A_3 x^2 + \cdots + A_m x^{m-1})e^{rx} \tag{11.44}$$

For each pair of complex roots of the form $r = \alpha \pm i\omega$ the solution contains the terms

$$y = e^{\alpha x}(A_1 \sin\omega x + A_2 \cos\omega x) \tag{11.45}$$

15. APPLICATION: ENGINEERING SYSTEMS

There is a wide variety of engineering systems (mechanical, electrical, fluid flow, heat transfer, and so on) whose behavior is described by linear differential equations with constant coefficients.

16. APPLICATION: MIXING

A typical mixing problem involves a liquid-filled tank. The liquid may initially be pure or contain some solute. Liquid (either pure or as a solution) enters the tank at a known rate. A drain may be present to remove thoroughly mixed liquid. The concentration of the solution (or, equivalently, the amount of solute in the tank) at some given time is generally unknown. (See Fig. 11.1.)

Figure 11.1 *Fluid Mixture Problem*

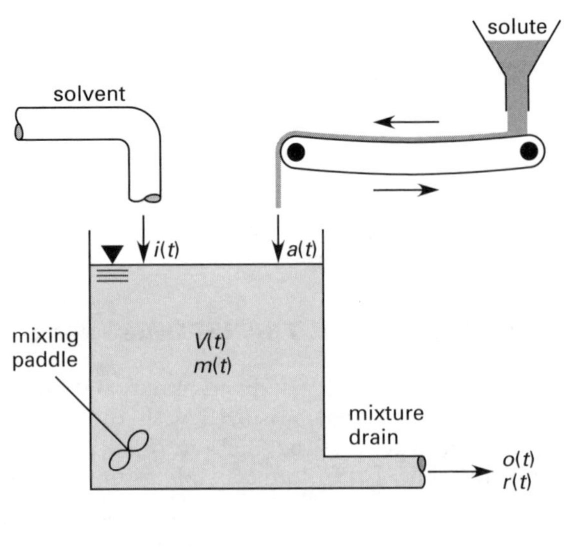

If $m(t)$ is the mass of solute in the tank at time t, the rate of solute change will be $m'(t)$. If the solute is being added at the rate of $a(t)$ and being removed at the rate of $r(t)$, the rate of change is

$$\begin{aligned} m'(t) &= \text{rate of addition} - \text{rate of removal} \\ &= a(t) - r(t) \end{aligned} \qquad 11.46$$

The rate of solute addition $a(t)$ must be known and, in fact, may be constant or zero. However, $r(t)$ depends on the concentration, $c(t)$, of the mixture and volumetric flow rates at time t. If $o(t)$ is the volumetric flow rate out of the tank, then

$$r(t) = c(t)o(t) \qquad 11.47$$

However, the concentration depends on the mass of solute in the tank at time t. Recognizing that the volume, $V(t)$, of the liquid in the tank may be changing with time,

$$c(t) = \frac{m(t)}{V(t)} \qquad 11.48$$

The differential equation describing this problem is

$$m'(t) = a(t) - \frac{m(t)o(t)}{V(t)} \qquad 11.49$$

Example 11.11

A tank contains 100 gal of pure water at the beginning of an experiment. Pure water flows into the tank at a rate of 1 gal/min. Brine containing $\frac{1}{4}$ lbm of salt per gallon enters the tank from a second source at a rate of 1 gal/min. A perfectly mixed solution drains from the tank at a rate of 2 gal/min. How much salt is in the tank 8 min after the experiment begins?

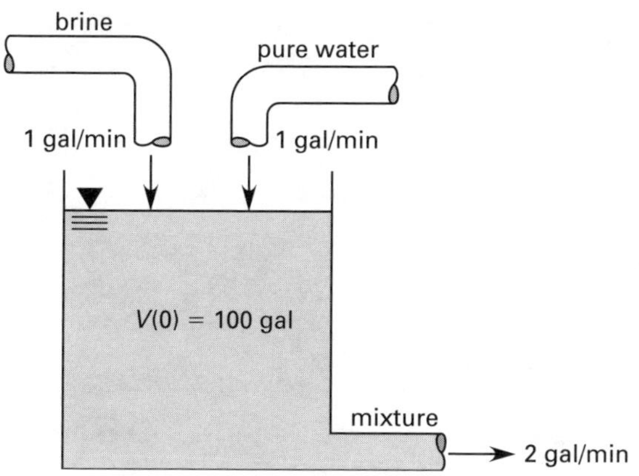

Solution

Let $m(t)$ represent the mass of salt in the tank at time t. 0.25 lbm of salt enters the tank per minute (that is, $a(t) = 0.25$ lbm/min). The salt removal rate depends on the concentration in the tank. That is,

$$\begin{aligned} r(t) &= o(t)c(t) \\ &= \left(2\ \frac{\text{gal}}{\text{min}}\right)\left(\frac{m(t)}{100\ \text{gal}}\right) = \left(0.02\ \frac{1}{\text{min}}\right)m(t) \end{aligned}$$

From Eq. 11.46, the rate of change of salt in the tank is

$$\begin{aligned} m'(t) &= a(t) - r(t) \\ &= 0.25\ \frac{\text{lbm}}{\text{min}} - \left(0.02\ \frac{1}{\text{min}}\right)m(t) \end{aligned}$$

$$m'(t) + \left(0.02\ \frac{1}{\text{min}}\right)m(t) = 0.25\ \text{lbm/min}$$

This is a first-order linear differential equation of the form of Eq. 11.7. Since the initial condition is $m(0) = 0$, the solution is

$$\begin{aligned} m(t) &= \left(\frac{0.25\ \frac{\text{lbm}}{\text{min}}}{0.02\ \frac{1}{\text{min}}}\right)\left(1 - e^{-\left(0.02\ \frac{1}{\text{min}}\right)t}\right) \\ &= (12.5\ \text{lbm})\left(1 - e^{-\left(0.01\ \frac{1}{\text{min}}\right)t}\right) \end{aligned}$$

At $t = 8$,

$$\begin{aligned} m(t) &= (12.5\ \text{lbm})\left(1 - e^{-\left(0.02\ \frac{1}{\text{min}}\right)(8\ \text{min})}\right) \\ &= (12.5\ \text{lbm})(1 - 0.852) \\ &= 1.85\ \text{lbm} \end{aligned}$$

17. APPLICATION: EXPONENTIAL GROWTH AND DECAY

Equation 11.50 describes the behavior of a substance whose quantity, $m(t)$, changes at a rate proportional to the quantity present. The constant of proportionality, k, will be negative for decay (e.g., radioactive decay) and positive for growth (e.g., compound interest).

$$m'(t) = km(t) \qquad 11.50$$

$$m'(t) - km(t) = 0 \qquad 11.51$$

If the initial quantity of substance is $m(0) = m_0$, then Eq. 11.51 has the solution

$$m(t) = m_0 e^{kt} \qquad 11.52$$

If $m(t)$ is known for some time t, the constant of proportionality is

$$k = \frac{1}{t}\ln\left(\frac{m(t)}{m_0}\right) \qquad 11.53$$

For the case of a decay, the *half-life*, $t_{1/2}$, is the time at which only half of the substance remains. The relationship between k and $t_{1/2}$ is

$$kt_{1/2} = \ln\left(\tfrac{1}{2}\right) = -0.693 \qquad 11.54$$

18. APPLICATION: EPIDEMICS

During an epidemic in a population of n people, the density of sick (contaminated, contagious, affected, etc.) individuals is $\rho_s(t) = s(t)/n$, where $s(t)$ is the number of sick individuals at a given time, t. Similarly, the density of well (uncontaminated, unaffected, susceptible, etc.) individuals is $\rho_w(t) = w(t)/n$, where $w(t)$ is the number of well individuals. Assuming there is no quarantine, the population size is constant, individuals move about freely, and sickness does not limit the activities of individuals, the rate of contagion, $\rho_s'(t)$, will be $k\rho_s(t)\rho_w(t)$, where k is a proportionality constant.

$$\rho_s'(t) = k\rho_s(t)\rho_w(t) = k\rho_s(t)(1-\rho_s(t)) \qquad 11.55$$

This is a separable differential equation that has the solution

$$\rho_s(t) = \frac{\rho_s(0)}{\rho_s(0) + (1-\rho_s(0))e^{-kt}} \qquad 11.56$$

19. APPLICATION: SURFACE TEMPERATURE

Newton's law of cooling states that the surface temperature, T, of a cooling object changes at a rate proportional to the difference between the surface and ambient temperatures. The constant k is a positive number.

$$T'(t) = -k(T(t) - T_{\text{ambient}}) \quad [k>0] \qquad 11.57$$

$$T'(t) + kT(t) - kT_{\text{ambient}} = 0 \quad [k>0] \qquad 11.58$$

This first-order linear differential equation with constant coefficients has the following solution (from Eq. 11.8).

$$T(t) = T_{\text{ambient}} + (T(0) - T_{\text{ambient}})e^{-kt} \qquad 11.59$$

If the temperature is known at some time t, the constant k can be found from Eq. 11.60.

$$k = \frac{-1}{t}\ln\left(\frac{T(t) - T_{\text{ambient}}}{T(0) - T_{\text{ambient}}}\right) \qquad 11.60$$

20. APPLICATION: EVAPORATION

The mass of liquid evaporated from a liquid surface is proportional to the exposed surface area. Since quantity, mass, and remaining volume are all proportional, the differential equation is

$$\frac{dV}{dt} = -kA \qquad 11.61$$

For a spherical drop of radius r, Eq. 11.61 reduces to

$$\frac{dr}{dt} = -k \qquad 11.62$$

$$r(t) = r(0) - kt \qquad 11.63$$

For a cube with sides of length s, Eq. 11.61 reduces to

$$\frac{ds}{dt} = -2k \qquad 11.64$$

$$s(t) = s(0) - 2kt \qquad 11.65$$

12 Probability and Statistical Analysis of Data

1. Set Theory 12-1
2. Combinations of Elements 12-2
3. Permutations 12-2
4. Probability Theory 12-2
5. Joint Probability 12-3
6. Complementary Probabilities 12-4
7. Conditional Probability 12-4
8. Probability Density Functions 12-4
9. Binomial Distribution 12-4
10. Hypergeometric Distribution 12-5
11. Multiple Hypergeometric Distribution 12-5
12. Poisson Distribution 12-5
13. Continuous Distribution Functions 12-5
14. Exponential Distribution 12-6
15. Normal Distribution 12-6
16. Application: Reliability 12-7
17. Analysis of Experimental Data 12-8
18. Measures of Central Tendency 12-10
19. Measures of Dispersion 12-10
20. Central Limit Theorem 12-11
21. Confidence Level 12-12
22. Application: Confidence Limits 12-12
23. Application: Basic Hypothesis Testing 12-12
24. Application: Statistical Process Control 12-13
25. Measures of Experimental Adequacy 12-13
26. Linear Regression 12-13

1. SET THEORY

A *set* (usually designated by a capital letter) is a population or collection of individual items known as *elements* or *members*. The *null set*, Ø, is empty (i.e., contains no members). If A and B are two sets, A is a *subset* of B if every member in A is also in B. A is a *proper subset* of B if B consists of more than the elements in A. These relationships are denoted as follows.

$$A \subseteq B \quad \text{[subset]}$$

$$A \subset B \quad \text{[proper subset]}$$

The *universal set*, U, is one from which other sets draw their members. If A is a subset of U, then A' (also designated as A^{-1}, $\tilde{A}$, $-A$, and $\overline{A}$) is the *complement* of A and consists of all elements in U that are not in A. This is illustrated in a *Venn diagram* in Fig. 12.1(a).

The *union of two sets*, denoted by $A \cup B$ and shown in Fig. 12.1(b), is the set of all elements that are either in A or B or both. The *intersection of two sets*, denoted by $A \cap B$ and shown in Fig. 12.1(c), is the set of all elements that belong to both A and B. If $A \cap B = \emptyset$, A and B are said to be *disjoint sets*.

Figure 12.1 *Venn Diagrams*

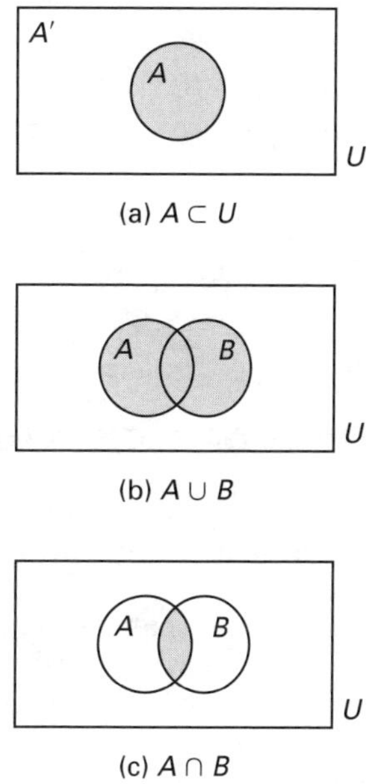

If A, B, and C are subsets of the universal set, the following laws apply.

- *identity laws*

$$A \cup \emptyset = A \qquad 12.1$$

$$A \cup U = U \qquad 12.2$$

$$A \cap \emptyset = \emptyset \qquad 12.3$$

$$A \cap U = A \qquad 12.4$$

- *idempotent laws*

$$A \cup A = A \qquad 12.5$$

$$A \cap A = A \qquad 12.6$$

- *complement laws*

$$A \cup A' = U \qquad 12.7$$
$$(A')' = A \qquad 12.8$$
$$A \cap A' = \emptyset \qquad 12.9$$
$$U' = \emptyset \qquad 12.10$$

- *commutative laws*

$$A \cup B = B \cup A \qquad 12.11$$
$$A \cap B = B \cap A \qquad 12.12$$

- *associative laws*

$$(A \cup B) \cup C = A \cup (B \cup C) \qquad 12.13$$
$$(A \cap B) \cap C = A \cap (B \cap C) \qquad 12.14$$

- *distributive laws*

$$A \cup (B \cap C) = (A \cup B) \cap (A \cup C) \qquad 12.15$$
$$A \cap (B \cup C) = (A \cap B) \cup (A \cap C) \qquad 12.16$$

- *de Morgan's laws*

$$(A \cup B)' = A' \cap B' \qquad 12.17$$
$$(A \cap B)' = A' \cup B' \qquad 12.18$$

2. COMBINATIONS OF ELEMENTS

There are a finite number of ways in which n elements can be combined into distinctly different groups of r items. For example, suppose a farmer has a hen, a rooster, a duck, and a cage that holds only two birds. The possible *combinations* of three birds taken two at a time are (hen, rooster), (hen, duck), and (rooster, duck). The birds in the cage will not remain stationary, and the combination (rooster, hen) is not distinctly different from (hen, rooster). That is, the groups are not *order conscious.*

The number of combinations of n items taken r at a time is written $C(n, r)$, C_r^n, ${}_nC_r$, or $\binom{n}{r}$ (pronounced "n choose r") and given by Eq. 12.19. It is sometimes referred to as the *binomial coefficient.*

$$\binom{n}{r} = C(n,r) = \frac{n!}{(n-r)!r!} \quad [\text{for } r \le n] \qquad 12.19$$

Example 12.1

Six people are on a sinking yacht. There are four life jackets. How many combinations of survivors are there?

Solution

The groups are not order-conscious. From Eq. 12.19,

$$C(6,4) = \frac{n!}{(n-r)!r!} = \frac{6!}{(6-4)!4!} = \frac{6\cdot5\cdot4\cdot3\cdot2\cdot1}{(2\cdot1)(4\cdot3\cdot2\cdot1)} = 15$$

3. PERMUTATIONS

An order-conscious subset of r items taken from a set of n items is the *permutation* $P(n, r)$, also written P_r^n and ${}_nP_r$. The permutation is order conscious because the arrangement of two items (say a_i and b_i) as a_ib_i is different from the arrangement b_ia_i. The number of permutations is

$$P(n,r) = \frac{n!}{(n-r)!} \quad [\text{for } r \le n] \qquad 12.20$$

If groups of the entire set of n items are being enumerated, the number of permutations of n items taken n at a time is

$$P(n,n) = \frac{n!}{(n-n)!} = \frac{n!}{0!} = n! \qquad 12.21$$

A *ring permutation* is a special case of n items taken n at a time. There is no identifiable beginning or end, and the number of permutations is divided by n.

$$P_{\text{ring}}(n,n) = \frac{P(n,n)}{n} = (n-1)! \qquad 12.22$$

Example 12.2

A pianist knows four pieces but will have enough stage time to play only three of them. Pieces played in a different order constitute a different program. How many different programs can be arranged?

Solution

The groups are order conscious. From Eq. 12.20,

$$P(4,3) = \frac{n!}{(n-r)!} = \frac{4!}{(4-3)!} = \frac{4\cdot3\cdot2\cdot1}{1} = 24$$

Example 12.3

Seven diplomats from different countries enter a circular room. The only furnishings are seven chairs arranged around a circular table. How many ways are there of arranging the diplomats?

Solution

All seven diplomats must be seated, so the groups are permutations of seven objects taken seven at a time. Since there is no head chair, the groups are ring permutations. From Eq. 12.22,

$$P_{\text{ring}}(7,7) = (7-1)! = 6\cdot5\cdot4\cdot3\cdot2\cdot1 = 720$$

4. PROBABILITY THEORY

The act of conducting an experiment (trial) or taking a measurement is known as *sampling. Probability theory* determines the relative likelihood that a particular event will occur. An *event*, e, is one of the possible outcomes of the *trial*. Taken together, all of the possible events constitute a finite *sample space*, $E = [e_1, e_2, \ldots, e_n]$. The trial is drawn from the *population* or *universe*. Populations can be finite or infinite in size.

Events can be numerical or nonnumerical, discrete or continuous, and dependent or independent. An example of a nonnumerical event is getting tails on a coin toss. The number from a roll of a die is a discrete numerical event. The measured diameter of a bolt produced from an automatic screw machine is a numerical event. Since the diameter can (within reasonable limits) take on any value, its measured value is a continuous numerical event.

An event is *independent* if its outcome is unaffected by previous outcomes (i.e., previous runs of the experiment) and *dependent* otherwise. Whether or not an event is independent depends on the population size and how the sampling is conducted. Sampling (a trial) from an infinite population is implicitly independent. When the population is finite, *sampling with replacement* produces independent events, while *sampling without replacement* changes the population and produces dependent events.

The terms *success* and *failure* are loosely used in probability theory to designate obtaining and not obtaining, respectively, the tested-for condition. "Failure" is not the same as a *null event* (i.e., one that has a zero probability of occurrence).

The *probability* of event e_1 occurring is designated as $p\{e_1\}$ and is calculated as the ratio of the total number of ways the event can occur to the total number of outcomes in the sample space.

Example 12.4

There are 380 students in a rural school—200 girls and 180 boys. One student is chosen at random and is checked for gender and height. (a) Define and categorize the population. (b) Define and categorize the sample space. (c) Define the trials. (d) Define and categorize the events. (e) In determining the probability that the student chosen is a boy, define success and failure. (f) What is the probability that the student is a boy?

Solution

(a) The population consists of 380 students and is finite.

(b) In determining the gender of the student, the sample space consists of the two outcomes $E = [\text{girl, boy}]$. This sample space is nonnumerical and discrete. In determining the height, the sample space consists of a range of values and is numerical and continuous.

(c) The trial is the actual sampling (i.e., the determination of gender and height).

(d) The events are the outcomes of the trials (i.e., the gender and height of the student). These events are independent if each student returns to the population prior to the random selection of the next student; otherwise, the events are dependent.

(e) The event is a success if the student is a boy and is a failure otherwise.

(f) From the definition of probability,

$$p\{\text{boy}\} = \frac{\text{no. of boys}}{\text{no. of students}} = \frac{180}{380} = \frac{9}{19} = 0.47$$

5. JOINT PROBABILITY

Joint probability rules specify the probability of a combination of events. If n mutually exclusive events from the set E have probabilities $p\{e_i\}$, the probability of any one of these events occurring in a given trial is the sum of the individual probabilities. The events in Eq. 12.23 come from a single sample space and are linked by the word *or*.

$$p\{e_1 \text{ or } e_2 \text{ or } \cdots \text{ or } e_k\} = p\{e_1\} + p\{e_2\} + \cdots + p\{e_k\} \qquad 12.23$$

When given two independent sets of events, E and G, Eq. 12.24 will give the probability that either event e_i or g_i, or both, will occur. The events in Eq. 12.24 come from two different sample spaces and are linked by the word *or*.

$$p\{e_i \text{ or } g_i\} = p\{e_i\} + p\{g_i\} - p\{e_i\}p\{g_i\} \qquad 12.24$$

When given two independent sets of events, E and G, Eq. 12.25 will give the probability that events e_i and g_i will both occur. The events in Eq. 12.25 come from two different sample spaces and are linked by the word *and*.

$$p\{e_i \text{ and } g_i\} = p\{e_i\}p\{g_i\} \qquad 12.25$$

Example 12.5

A bowl contains five white balls, two red balls, and three green balls. What is the probability of getting either a white ball or a red ball in one draw from the bowl?

Solution

Since the two possible events are mutually exclusive and come from the same sample space, Eq. 12.23 can be used.

$$p\{\text{white or red}\} = p\{\text{white}\} + p\{\text{red}\} = \frac{5}{10} + \frac{2}{10} = \frac{7}{10}$$

Example 12.6

One bowl contains five white balls, two red balls, and three green balls. Another bowl contains three yellow balls and seven black balls. What is the probability of getting a red ball from the first bowl and a yellow ball from the second bowl in one draw from each bowl?

Solution

Equation 12.25 can be used because the two trials are independent.

$$p\{\text{red and yellow}\} = p\{\text{red}\}p\{\text{yellow}\} = \left(\frac{2}{10}\right)\left(\frac{3}{10}\right) = \frac{6}{100}$$

6. COMPLEMENTARY PROBABILITIES

The probability of an event occurring is equal to one minus the probability of the event not occurring. This is known as *complementary probability.*

$$p\{e_i\} = 1 - p\{\text{not } e_i\} \qquad 12.26$$

Equation 12.26 can be used to simplify some probability calculations. Specifically, calculation of the probability of numerical events being "greater than" or "less than" or quantities being "at least" a certain number can often be simplified by calculating the probability of complementary event.

Example 12.7

A fair coin is tossed five times.[1] What is the probability of getting at least one tail?

Solution

The probability of getting at least one tail in five tosses could be calculated as

$$p\{\text{at least 1 tail}\} = p\{\text{1 tail}\} + p\{\text{2 tails}\} + p\{\text{3 tails}\} + p\{\text{4 tails}\} + p\{\text{5 tails}\}$$

However, it is easier to calculate the complementary probability of getting no tails (i.e., getting all heads).

$$p\{\text{at least 1 tail}\} = 1 - p\{\text{0 tails}\} = 1 - (0.5)^5 = 0.96875$$

7. CONDITIONAL PROBABILITY

Given two dependent sets of events, E and G, the probability that event e_k will occur given the fact that the dependent event g has already occurred is written as $p\{e_k|g\}$ and given by *Bayes' theorem*, Eq. 12.27.

$$p\{e_k|g\} = \frac{p\{e_k \text{ and } g\}}{p\{g\}} = \frac{p\{g|e_k\}p\{e_k\}}{\sum_{i=1}^{n} p\{g|e_i\}p\{e_i\}} \qquad 12.27$$

[1]It makes no difference whether one coin is tossed five times or five coins are each tossed once.

8. PROBABILITY DENSITY FUNCTIONS

A *density function* is a nonnegative function whose integral taken over the entire range of the independent variable is unity. A *probability density function* (PDF) is a mathematical formula that gives the probability of a discrete numerical event. A *numerical event* is an occurrence that can be described (usually) by an integer. For example, 27 cars passing through a bridge toll booth in an hour is a discrete numerical event. Figure 12.2 shows a graph of a typical probability density function.

A probability density function, $f(x)$, gives the probability that discrete event x will occur. That is, $p\{x\} = f(x)$. Important discrete probability density functions are the binomial, hypergeometric, and Poisson distributions.

Figure 12.2 *Probability Density Function*

9. BINOMIAL DISTRIBUTION

The *binomial probability density function* is used when all outcomes can be categorized as either successes or failures. The probability of success in a single trial is $\hat{p}$, and the probability of failure is the complement, $\hat{q} = 1 - \hat{p}$. The population is assumed to be infinite in size so that sampling does not change the values of $\hat{p}$ and $\hat{q}$. (The binomial distribution can also be used with finite populations when sampling with replacement.)

Equation 12.28 gives the probability of x successes in n independent *successive trials*. The quantity $\binom{n}{x}$ is the binomial coefficient, identical to the number of combinations of n items taken x at a time.

$$p\{x\} = f(x) = \binom{n}{x}\hat{p}^x\hat{q}^{n-x} \qquad 12.28$$

$$\binom{n}{x} = \frac{n!}{(n-x)!x!} \qquad 12.29$$

Equation 12.28 is a true (discrete) distribution, taking on values for each integer value up to n. The mean, μ, and variance, σ^2 (see Sec. 12.22 and Sec. 12.23), of this distribution are

$$\mu = n\hat{p} \qquad 12.30$$

$$\sigma^2 = n\hat{p}\hat{q} \qquad 12.31$$

Example 12.8

Five percent of a large batch of high-strength steel bolts purchased for bridge construction are defective. (a) If seven bolts are randomly sampled, what is the probability that exactly three will be defective? (b) What is the probability that two or more bolts will be defective?

Solution

(a) The bolts are either defective or not, so the binomial distribution can be applied.

$$\hat{p} = 0.05 \quad [\text{success} = \text{defective}]$$

$$\hat{q} = 1 - 0.05 = 0.95 \quad [\text{failure} = \text{not defective}]$$

From Eq. 12.28,

$$\begin{aligned} p\{3\} = f(3) &= \binom{n}{x}\hat{p}^x\hat{q}^{n-x} = \binom{7}{3}\hat{p}^3\hat{q}^{7-3} \\ &= \left(\frac{7\cdot 6\cdot 5\cdot 4\cdot 3\cdot 2\cdot 1}{4\cdot 3\cdot 2\cdot 1\cdot 3\cdot 2\cdot 1}\right)(0.05)^3(0.95)^4 \\ &= 0.00356 \end{aligned}$$

(b) The probability that two or more bolts will be defective could be calculated as

$$p\{x \geq 2\} = p\{2\} + p\{3\} + p\{4\} + p\{5\} + p\{6\} + p\{7\}$$

This method would require six probability calculations. It is easier to use the complement of the desired probability.

$$p\{x \geq 2\} = 1 - p\{x \leq 1\} = 1 - (p\{0\} + p\{1\})$$

$$p\{0\} = \binom{n}{x}\hat{p}^x\hat{q}^{n-x} = \binom{7}{0}(0.05)^0(0.95)^7 = (0.95)^7$$

$$\begin{aligned} p\{1\} &= \binom{n}{x}\hat{p}^x\hat{q}^{n-x} = \binom{7}{1}(0.05)^1(0.95)^6 \\ &= (7)(0.05)(0.95)^6 \end{aligned}$$

$$\begin{aligned} p\{x \geq 2\} &= 1 - \left((0.95)^7 + (7)(0.05)(0.95)^6\right) \\ &= 1 - (0.6983 + 0.2573) \\ &= 0.0444 \end{aligned}$$

10. HYPERGEOMETRIC DISTRIBUTION

Probabilities associated with sampling from a finite population without replacement are calculated from the *hypergeometric distribution.* If a population of finite size M contains K items with a given characteristic (e.g., red color, defective construction), then the probability of finding x items with that characteristic in a sample of n items is

$$p\{x\} = f(x) = \frac{\binom{K}{x}\binom{M-K}{n-x}}{\binom{M}{n}} \quad [\text{for } x \leq n] \qquad 12.32$$

11. MULTIPLE HYPERGEOMETRIC DISTRIBUTION

Sampling without replacement from finite populations containing several different types of items is handled by the *multiple hypergeometric distribution.* If a population of finite size M contains K_i items of type i (such that $\Sigma K_i = M$), the probability of finding x_1 items of type 1, x_2 items of type 2, and so on, in a sample size n (such that $\Sigma x_i = n$) is

$$p\{x_1, x_2, x_3, \ldots\} = \frac{\binom{K_1}{x_1}\binom{K_2}{x_2}\binom{K_3}{x_3}\cdots}{\binom{M}{n}} \qquad 12.33$$

12. POISSON DISTRIBUTION

Certain events occur relatively infrequently but at a relatively regular rate. The probability of such an event occurring is given by the *Poisson distribution.* Suppose an event occurs, on the average, λ times per period. The probability that the event will occur x times per period is

$$p\{x\} = f(x) = \frac{e^{-\lambda}\lambda^x}{x!} \quad [\lambda > 0] \qquad 12.34$$

λ is both the mean and the variance of the Poisson distribution.

Example 12.9

The number of customers arriving at a hamburger stand in the next period is a Poisson distribution having a mean of eight. What is the probability that exactly six customers will arrive in the next period?

Solution

$\lambda = 8$ and $x = 6$. From Eq. 12.34,

$$p\{6\} = \frac{e^{-\lambda}\lambda^x}{x!} = \frac{e^{-8}(8)^6}{6!} = 0.122$$

13. CONTINUOUS DISTRIBUTION FUNCTIONS

Most numerical events are *continuously distributed* and are not constrained to discrete or integer values. For example, the resistance of a 10% 1 Ω resistor may be any value between 0.9 Ω and 1.1 Ω. The probability of an exact numerical event is zero for continuously distributed variables. That is, there is no chance that a numerical event will be *exactly* x.[2] It is possible to determine only the probability that a numerical event will be less than x, greater than x, or between the values of x_1 and x_2, but not exactly equal to x.

While an expression, $f(x)$, for a probability density function can be written, it is used to derive the *continuous distribution function* (CDF), $F(x_0)$, which gives the probability of numerical event x_0 or less occurring, as illustrated in Fig. 12.3.

$$p\{X < x_0\} = F(x_0) = \int_0^{x_0} f(x)dx \qquad 12.35$$

$$f(x) = \frac{dF(x)}{dx} \qquad 12.36$$

Figure 12.3 Continuous Distribution Function

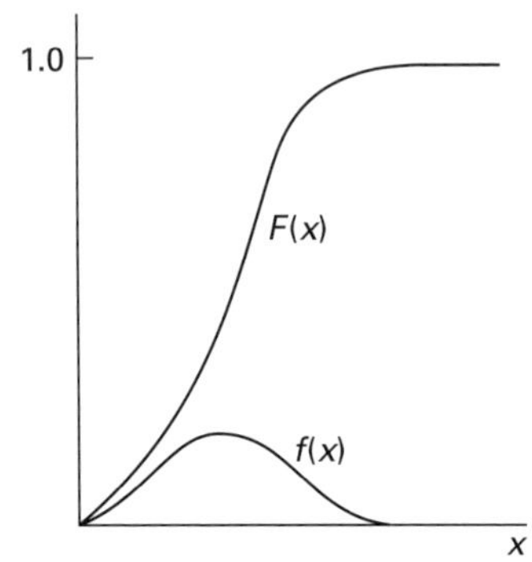

14. EXPONENTIAL DISTRIBUTION

The continuous *exponential distribution* is given by its probability density and continuous distribution functions.

$$f(x) = \lambda e^{-\lambda x} \qquad 12.37$$

$$p\{X < x\} = F(x) = 1 - e^{-\lambda x} \qquad 12.38$$

The mean and variance of the exponential distribution are

$$\mu = \frac{1}{\lambda} \qquad 12.39$$

$$\sigma^2 = \frac{1}{\lambda^2} \qquad 12.40$$

15. NORMAL DISTRIBUTION

The *normal distribution* (*Gaussian distribution*) is a symmetrical distribution commonly referred to as the *bell-shaped curve*, which represents the distribution of outcomes of many experiments, processes, and phenomena. (See Fig. 12.4.) The probability density and continuous distribution functions for the normal distribution with mean μ and variance σ^2 are

$$f(x) = \frac{e^{-\frac{1}{2}\left(\frac{x-\mu}{\sigma}\right)^2}}{\sigma\sqrt{2\pi}} \quad [-\infty < x < +\infty] \qquad 12.41$$

$$\begin{aligned} p\{\mu < X < x_0\} &= F(x_0) \\ &= \frac{1}{\sigma\sqrt{2\pi}}\int_0^{x_0} e^{-\frac{1}{2}\left(\frac{x-\mu}{\sigma}\right)^2} dx \end{aligned} \qquad 12.42$$

Figure 12.4 Normal Distribution

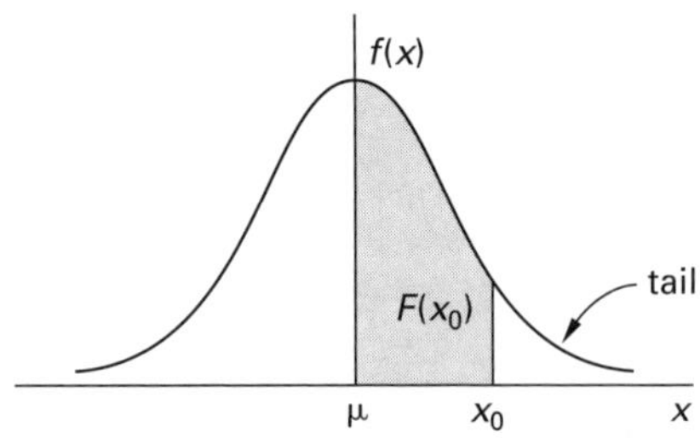

Since $f(x)$ is difficult to integrate, Eq. 12.41 is seldom used directly, and a *standard normal table* is used instead. (See App. 12.A.) The standard normal table is based on a normal distribution with a mean of zero and a standard deviation of 1. Since the range of values from an experiment or phenomenon will not generally correspond to the standard normal table, a value, x_0, must be converted to a *standard normal value*, z. In Eq. 12.43, μ and σ are the mean and standard deviation, respectively, of the distribution from which x_0 comes. For all practical purposes, all normal distributions are completely bounded by $\mu \pm 3\sigma$.

$$z = \frac{x_0 - \mu}{\sigma} \qquad 12.43$$

Numbers in the standard normal table, as given by App. 12.A, are the probabilities of the normalized x being between zero and z and represent the areas under the curve up to point z. When x is less than μ, z will be negative. However, the curve is symmetrical, so the table value corresponding to positive z can be used. The probability of x being greater than z is the complement of the table value. The curve area past point z is known as the *tail of the curve.*

The *error function*, $\text{erf}(x)$, and its complement, $\text{erfc}(x)$, are defined by Eq. 12.44 and Eq. 12.45. The error function is used to determine the probable error of a measurement.

$$\text{erf}(x_0) = \frac{2}{\sqrt{\pi}}\int_0^{x_0} e^{-x^2} dx \qquad 12.44$$

$$\text{erfc}(x_0) = 1 - \text{erf}(x_0) \qquad 12.45$$

[2]It is important to understand the rationale behind this statement. Since the variable can take on any value and has an infinite number of significant digits, we can infinitely continue to increase the precision of the value. For example, the probability is zero that a resistance will be exactly 1 Ω because the resistance is really 1.03 or 1.0260008 or 1.02600080005, and so on.

Example 12.10

The mass, m, of a particular hand-laid fiberglass (Fibreglas™) part is normally distributed with a mean of 66 kg and a standard deviation of 5 kg. (a) What percent of the parts will have a mass less than 72 kg? (b) What percent of the parts will have a mass in excess of 72 kg? (c) What percent of the parts will have a mass between 61 kg and 72 kg?

Solution

(a) The 72 kg value must be normalized, so use Eq. 12.43. The standard normal variable is

$$z = \frac{x-\mu}{\sigma} = \frac{72\,\text{kg} - 66\,\text{kg}}{5\,\text{kg}} = 1.2$$

Reading from App. 12.A, the area under the normal curve is 0.3849. This represents the probability of the mass, m, being between 66 kg and 72 kg (i.e., z being between 0 and 1.2). However, the probability of the mass being less than 66 kg is also needed. Since the curve is symmetrical, this probability is 0.5. Therefore,

$$p\{m < 72\,\text{kg}\} = p\{z < 1.2\} = 0.5 + 0.3849 = 0.8849$$

(b) The probability of the mass exceeding 72 kg is the area under the tail past point z.

$$p\{m > 72\,\text{kg}\} = p\{z > 1.2\} = 0.5 - 0.3849 = 0.1151$$

(c) The standard normal variable corresponding to $m = 61$ kg is

$$z = \frac{x-\mu}{\sigma} = \frac{61\,\text{kg} - 66\,\text{kg}}{5\,\text{kg}} = -1$$

Since the two masses are on opposite sides of the mean, the probability will have to be determined in two parts.

$$\begin{aligned} p\{61 < m < 72\} &= p\{61 < m < 66\} + p\{66 < m < 72\} \\ &= p\{-1 < z < 0\} + p\{0 < z < 1.2\} \\ &= 0.3413 + 0.3849 = 0.7262 \end{aligned}$$

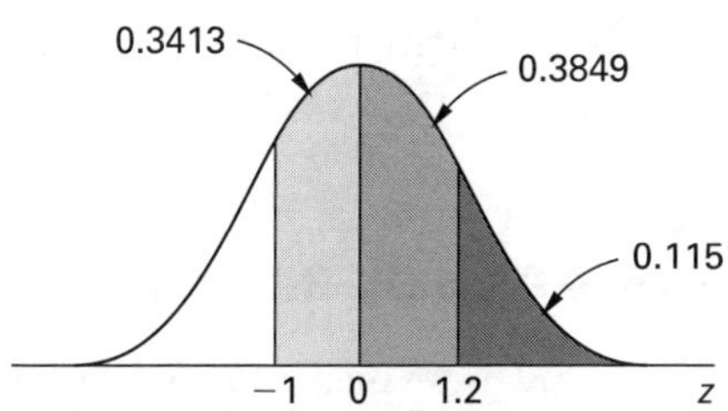

16. APPLICATION: RELIABILITY

Introduction

Reliability, $R\{t\}$, is the probability that an item will continue to operate satisfactorily up to time t. The *bathtub distribution*, Fig. 12.5, is often used to model the probability of failure of an item (or, the number of failures from a large population of items) as a function of time. Items initially fail at a high rate, a phenomenon known as *infant mortality*. For the majority of the operating time, known as the *steady-state operation*, the failure rate is constant (i.e., is due to random causes). After a long period of time, the items begin to deteriorate and the failure rate increases. (No mathematical distribution describes all three of these phases simultaneously.)

Figure 12.5 *Bathtub Reliability Curve*

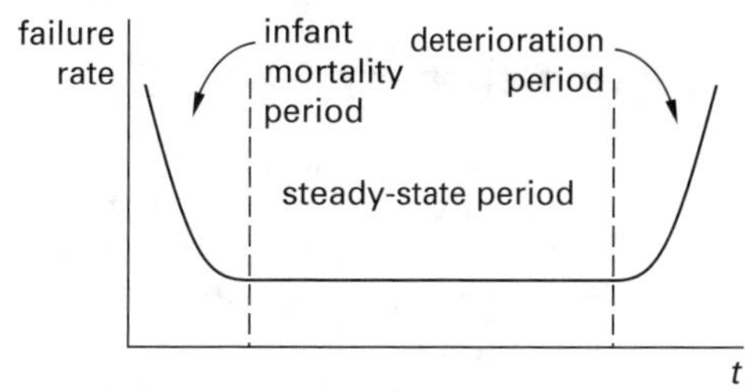

The *hazard function*, $z\{t\}$, represents the *conditional probability of failure*—the probability of failure in the next time interval, given that no failure has occurred thus far.[3]

$$z\{t\} = \frac{f(t)}{R(t)} = \frac{\frac{dF(t)}{dt}}{1 - F(t)} \qquad 12.46$$

Exponential Reliability

Steady-state reliability is often described by the *negative exponential distribution*. This assumption is appropriate whenever an item fails only by random causes and does not experience deterioration during its life. The parameter λ is related to the *mean time to failure* (MTTF) of the item.[4]

$$R\{t\} = e^{-\lambda t} = e^{-t/\text{MTTF}} \qquad 12.47$$

$$\lambda = \frac{1}{\text{MTTF}} \qquad 12.48$$

Equation 12.47 and the exponential continuous distribution function, Eq. 12.38, are complementary.

$$R\{t\} = 1 - F(t) = 1 - \left(1 - e^{-\lambda t}\right) = e^{-\lambda t} \qquad 12.49$$

The hazard function for the negative exponential distribution is

$$z\{t\} = \lambda \qquad 12.50$$

Therefore, the hazard function for exponential reliability is constant and does not depend on t (i.e., on the age of the item). In other words, the expected future life of an item is independent of the previous history (length of operation). This lack of memory is consistent with the

[3]The symbol $z\{t\}$ is traditionally used for the hazard function and is not related to the standard normal variable.
[4]The term "mean time *between* failures" is improper. However, the term *mean time before failure* (MTBF) is acceptable.

assumption that only random causes contribute to failure during steady-state operations. And since random causes are unlikely discrete events, their probability of occurrence can be represented by a Poisson distribution with mean λ. That is, the probability of having x failures in any given period is

$$p\{x\} = \frac{e^{-\lambda}\lambda^x}{x!} \qquad 12.51$$

Serial System Reliability

In the analysis of system reliability, the binary variable X_i is defined as one if item i operates satisfactorily and zero if otherwise. Similarly, the binary variable Φ is one only if the entire system operates satisfactorily. Therefore, Φ will depend on a *performance function* containing the X_i.

A *serial system* is one for which all items must operate correctly for the system to operate. Each item has its own reliability, R_i. For a serial system of n items, the performance function is

$$\Phi = X_1X_2X_3 \cdots X_n = \min(X_i) \qquad 12.52$$

The probability of a serial system operating correctly is

$$p\{\Phi = 1\} = R_{\text{serial system}} = R_1R_2R_3 \cdots R_n \qquad 12.53$$

Parallel System Reliability

A *parallel system* with n items will fail only if all n items fail. Such a system is said to be *redundant* to the nth degree. Using redundancy, a highly reliable system can be produced from components with relatively low individual reliabilities.

The performance function of a redundant system is

$$\begin{aligned}\Phi &= 1 - (1 - X_1)(1 - X_2)(1 - X_3)\cdots(1 - X_n) \\ &= \max(X_i)\end{aligned} \qquad 12.54$$

The reliability of the parallel system is

$$\begin{aligned}R &= p\{\Phi = 1\} \\ &= 1 - (1 - R_1)(1 - R_2)(1 - R_3)\cdots(1 - R_n)\end{aligned} \qquad 12.55$$

Example 12.11

The reliability of an item is exponentially distributed with mean time to failure (MTTF) of 1000 hr. What is the probability that the item will not have failed after 1200 hr of operation?

Solution

The probability of not having failed before time t is the reliability. From Eq. 12.48 and Eq. 12.49,

$$\lambda = \frac{1}{\text{MTTF}} = \frac{1}{1000 \text{ hr}} = 0.001 \text{ hr}^{-1}$$

$$R\{1200\} = e^{-\lambda t} = e^{(-0.001 \text{ hr}^{-1})(1200 \text{ hr})} = 0.3$$

Example 12.12

What are the reliabilities of the following systems?

(a)

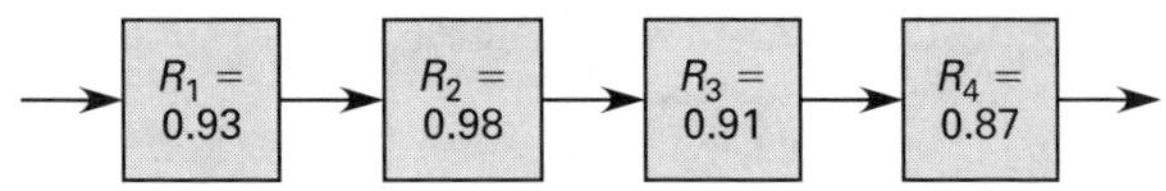

(b)

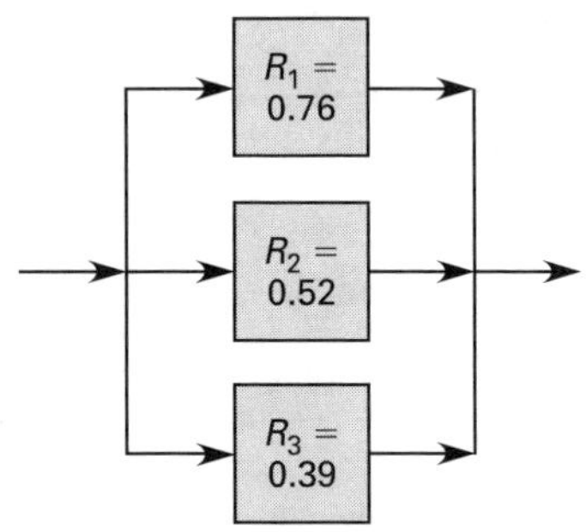

Solution

(a) This is a serial system. From Eq. 12.53,

$$\begin{aligned}R &= R_1R_2R_3R_4 = (0.93)(0.98)(0.91)(0.87) \\ &= 0.72\end{aligned}$$

(b) This is a parallel system. From Eq. 12.55,

$$\begin{aligned}R &= 1 - (1 - R_1)(1 - R_2)(1 - R_3) \\ &= 1 - (1 - 0.76)(1 - 0.52)(1 - 0.39) \\ &= 0.93\end{aligned}$$

17. ANALYSIS OF EXPERIMENTAL DATA

Experiments can take on many forms. An experiment might consist of measuring the mass of one cubic foot of concrete or measuring the speed of a car on a roadway. Generally, such experiments are performed more than once to increase the precision and accuracy of the results.

Both systematic and random variations in the process being measured will cause the observations to vary, and the experiment would not be expected to yield the same result each time it was performed. Eventually, a collection of experimental outcomes (observations) will be available for analysis.

The *frequency distribution* is a systematic method for ordering the observations from small to large, according to some convenient numerical characteristic. The *step interval* should be chosen so that the data are presented in a meaningful manner. If there are too many intervals, many of them will have zero frequencies; if there are too few intervals, the frequency distribution will have little value. Generally, 10 to 15 intervals are used.

Once the frequency distribution is complete, it can be represented graphically as a *histogram*. The procedure in drawing a histogram is to mark off the interval limits (also known as *class limits*) on a number line and then draw contiguous bars with lengths that are proportional to the frequencies in the intervals and that are centered on the midpoints of their respective intervals. The continuous nature of the data can be depicted by a *frequency polygon*. The number or percentage of observations that occur up to and including some value can be shown in a *cumulative frequency table.*

Example 12.13

The number of cars that travel through an intersection between 12 noon and 1 p.m. is measured for 30 consecutive working days. The results of the 30 observations are

79, 66, 72, 70, 68, 66, 68, 76, 73, 71, 74, 70, 71, 69, 67, 74, 70, 68, 69, 64, 75, 70, 68, 69, 64, 69, 62, 63, 63, 61

(a) What are the frequency and cumulative distributions? (Use a distribution interval of two cars per hour.) (b) Draw the histogram. (Use a cell size of two cars per hour.) (c) Draw the frequency polygon. (d) Graph the cumulative frequency distribution.

Solution

(a)

cars per hour	frequency	cumulative frequency	cumulative percent
60–61	1	1	3
62–63	3	4	13
64–65	2	6	20
66–67	3	9	30
68–69	8	17	57
70–71	6	23	77
72–73	2	25	83
74–75	3	28	93
76–77	1	29	97
78–79	1	30	100

(b)

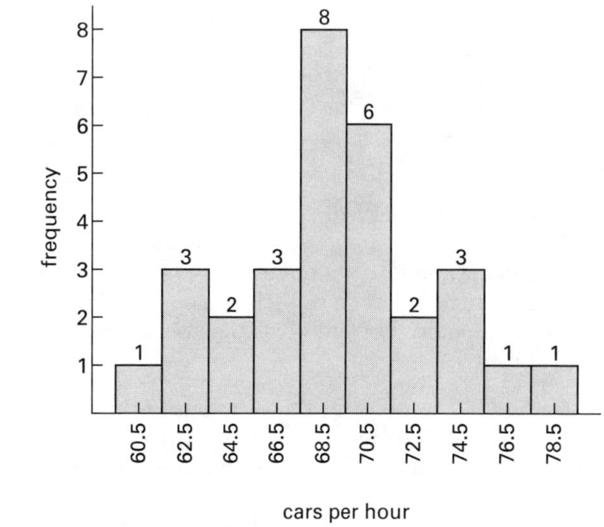

(c)

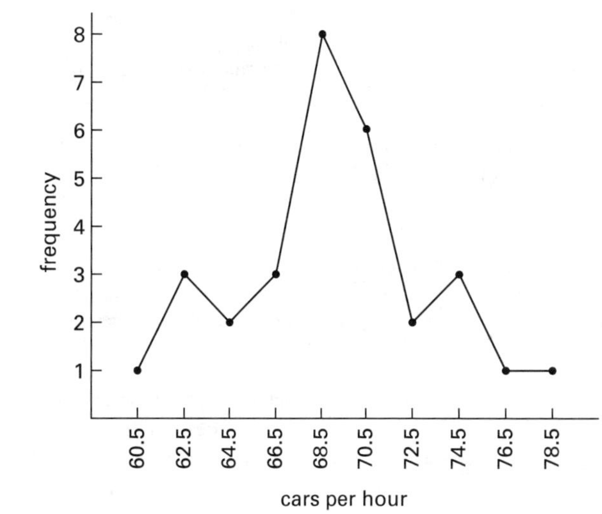

(d)

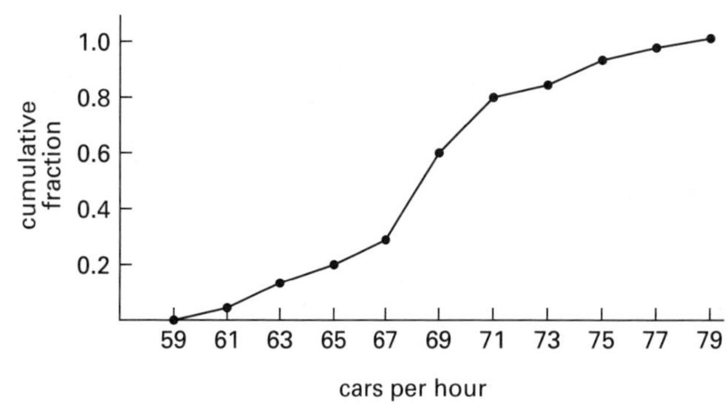

18. MEASURES OF CENTRAL TENDENCY

It is often unnecessary to present the experimental data in their entirety, either in tabular or graphical form. In such cases, the data and distribution can be represented by various parameters. One type of parameter is a measure of *central tendency*. Mode, median, and mean are measures of central tendency.

The *mode* is the observed value that occurs most frequently. The mode may vary greatly between series of observations. Therefore, its main use is as a quick measure of the central value since little or no computation is required to find it. Beyond this, the usefulness of the mode is limited.

The *median* is the point in the distribution that partitions the total set of observations into two parts containing equal numbers of observations. It is not influenced by the extremity of scores on either side of the distribution. The median is found by counting up (from either end of the frequency distribution) until half of the observations have been accounted for.

For even numbers of observations, the median is estimated as some value (i.e., the average) between the two center observations.

Similar in concept to the median are *percentiles* (*percentile ranks*), *quartiles*, and *deciles*. The median could also have been called the *50th percentile* observation. Similarly, the 80th percentile would be the observed value (e.g., the number of cars per hour) for which the cumulative frequency was 80%. The quartile and decile points on the distribution divide the observations or distribution into segments of 25% and 10%, respectively.

The *arithmetic mean* is the arithmetic average of the observations. The sample mean, $\overline{x}$, can be used as an unbiased estimator of the population mean, μ. The *mean* may be found without ordering the data (as was necessary to find the mode and median). The mean can be found from the following formula.

$$\overline{x} = \left(\frac{1}{n}\right)(x_1 + x_2 + \cdots + x_n) = \frac{\sum x_i}{n} \qquad 12.56$$

The *geometric mean* is used occasionally when it is necessary to average ratios. The geometric mean is calculated as

$$\text{geometric mean} = \sqrt[n]{x_1 x_2 x_3 \cdots x_n} \quad [x_i > 0] \qquad 12.57$$

The *harmonic mean* is defined as

$$\text{harmonic mean} = \frac{n}{\frac{1}{x_1} + \frac{1}{x_2} + \cdots + \frac{1}{x_n}} \qquad 12.58$$

The *root-mean-squared* (rms) *value* of a series of observations is defined as

$$x_{\text{rms}} = \sqrt{\frac{\sum x_i^2}{n}} \qquad 12.59$$

Example 12.14

Find the mode, median, and arithmetic mean of the distribution represented by the data given in Ex. 12.13.

Solution

First, resequence the observations in increasing order.

61, 62, 63, 63, 64, 64, 66, 66, 67, 68, 68, 68, 68, 69, 69, 69, 69, 70, 70, 70, 70, 71, 71, 72, 73, 74, 74, 75, 76, 79

The mode is the interval 68–69, since this interval has the highest frequency. If 68.5 is taken as the interval center, then 68.5 would be the mode.

The 15th and 16th observations are both 69, so the median is

$$\frac{69 + 69}{2} = 69$$

The mean can be found from the raw data or from the grouped data using the interval center as the assumed observation value. Using the raw data,

$$\overline{x} = \frac{\sum x}{n} = \frac{2069}{30} = 68.97$$

19. MEASURES OF DISPERSION

The simplest statistical parameter that describes the variability in observed data is the *range*. The range is found by subtracting the smallest value from the largest. Since the range is influenced by extreme (low probability) observations, its use as a measure of variability is limited.

The *standard deviation* is a better estimate of variability because it considers every observation. That is, in Eq. 12.60, N is the total population size, not the sample size, n.

$$\sigma = \sqrt{\frac{\sum (x_i - \mu)^2}{N}} = \sqrt{\frac{\sum x_i^2}{N} - \mu^2} \qquad 12.60$$

The standard deviation of a sample (particularly a small sample) is a biased (i.e., is not a good) estimator of the population standard deviation. An *unbiased estimator* of the population standard deviation is the *sample standard deviation, s*.[5]

$$s = \sqrt{\frac{\sum (x_i - \overline{x})^2}{n-1}} = \sqrt{\frac{\sum x_i^2 - \frac{\left(\sum x_i\right)^2}{n}}{n-1}} \qquad 12.61$$

If the sample standard deviation, s, is known, the standard deviation of the sample, σ_{sample}, can be calculated.

$$\sigma_{\text{sample}} = s\sqrt{\frac{n-1}{n}} \qquad 12.62$$

The *variance* is the square of the standard deviation. Since there are two standard deviations, there are two variances. The *variance of the sample* is σ^2, and the *sample variance* is s^2.

The *relative dispersion* is defined as a measure of dispersion divided by a measure of central tendency. The *coefficient of variation* is a relative dispersion calculated from the sample standard deviation and the mean.

$$\text{coefficient of variation} = \frac{s}{\overline{x}} \qquad 12.63$$

Skewness is a measure of a frequency distribution's lack of symmetry.

$$\begin{aligned} \text{skewness} &= \frac{\overline{x} - \text{mode}}{s} \\ &\approx \frac{3(\overline{x} - \text{median})}{s} \end{aligned} \qquad 12.64$$

Example 12.15

For the data given in Ex. 12.13, calculate (a) the sample range, (b) the standard deviation of the sample, (c) an unbiased estimator of the population standard deviation, (d) the variance of the sample, and (e) the sample variance.

Solution

$$\sum x_i = 2069$$

$$\left(\sum x_i\right)^2 = (2069)^2 = 4{,}280{,}761$$

$$\sum x_i^2 = 143{,}225$$

$$n = 30$$

$$\overline{x} = \frac{2069}{30} = 68.967$$

(a)
$$R = x_{\text{max}} - x_{\text{min}} = 79 - 61 = 18$$

(b)
$$\begin{aligned} \sigma &= \sqrt{\frac{\sum x_i^2}{n} - (\overline{x})^2} = \sqrt{\frac{143{,}225}{30} - \left(\frac{2069}{30}\right)^2} \\ &= \sqrt{17.766} = 4.215 \end{aligned}$$

(c)
$$\begin{aligned} s &= \sqrt{\frac{\sum x_i^2 - \frac{\left(\sum x_i\right)^2}{n}}{n-1}} \\ &= \sqrt{\frac{143{,}225 - \frac{4{,}280{,}761}{30}}{29}} \\ &= \sqrt{18.378} = 4.287 \end{aligned}$$

(d)
$$\sigma^2 = 17.77$$

(e)
$$s^2 = 18.38$$

20. CENTRAL LIMIT THEOREM

Measuring a sample of n items from a population with mean μ and standard deviation σ is the general concept of an experiment. The sample mean, $\overline{x}$, is one of the parameters that can be derived from the experiment. This experiment can be repeated k times, yielding a set of averages $(\overline{x}_1, \overline{x}_2, \ldots, \overline{x}_k)$. The k numbers in the set themselves represent samples from distributions of averages. The average of averages, $\overline{\overline{x}}$, and sample standard deviation of averages, $s_{\overline{x}}$ (known as the *standard error of the mean*), can be calculated.

The *central limit theorem* characterizes the distribution of the sample averages. The theorem can be stated in several ways, but the essential elements are the following points.

1. The averages, $\overline{x}_i$, are normally distributed variables, even if the original data from which they are calculated are not normally distributed.

2. The grand average, $\overline{\overline{x}}$ (i.e., the average of the averages), approaches and is an unbiased estimator of μ.

$$\mu \approx \overline{\overline{x}} \qquad 12.65$$

[5]There is a subtle yet significant difference between *standard deviation of the sample*, σ (obtained from Eq. 12.60 for a finite sample drawn from a larger population) and the *sample standard deviation*, s (obtained from Eq. 12.61). While σ can be calculated, it has no significance or use as an estimator. It is true that the difference between σ and s approaches zero when the sample size, n, is large, but this convergence does nothing to legitimize the use of σ as an estimator of the true standard deviation. (Some people say "large" is 30, others say 50 or 100.)

The standard deviation of the original distribution, σ, is much larger than the standard error of the mean.

$$\sigma \approx \sqrt{n} s_{\overline{x}} \qquad 12.66$$

21. CONFIDENCE LEVEL

The results of experiments are seldom correct 100% of the time. Recognizing this, researchers accept a certain probability of being wrong. In order to minimize this probability, the experiment is repeated several times. The number of repetitions depends on the desired level of confidence in the results.

If the results have a 5% probability of being wrong, the *confidence level*, C, is 95% that the results are correct, in which case the results are said to be *significant*. If the results have only a 1% probability of being wrong, the confidence level is 99%, and the results are said to be *highly significant*. Other confidence levels (90%, 99.5%, etc.) are used as appropriate.

22. APPLICATION: CONFIDENCE LIMITS

As a consequence of the central limit theorem, sample means of n items taken from a normal distribution with mean μ and standard deviation σ will be normally distributed with mean μ and variance σ^2/n. The probability that any given average, $\overline{x}$, exceeds some value, L, is

$$p\{\overline{x} > L\} = p\left\{ z > \left| \frac{L-\mu}{\frac{\sigma}{\sqrt{n}}} \right| \right\} \qquad 12.67$$

L is the *confidence limit* for the confidence level $1 - p\{\overline{x} > L\}$ (expressed as a percent). Values of z are read directly from the standard normal table. As an example, $z = 1.645$ for a 95% confidence level since only 5% of the curve is above that z in the upper tail. Similar values are given in Table 12.1. This is known as a *one-tail confidence limit* because all of the probability is given to one side of the variation.

Table 12.1 *Values of z for Various Confidence Levels*

confidence level, C	one-tail limit, z	two-tail limit, z
90%	1.28	1.645
95%	1.645	1.96
97.5%	1.96	2.17
99%	2.33	2.575
99.5%	2.575	2.81
99.75%	2.81	3.00

With *two-tail confidence limits*, the probability is split between the two sides of variation. There will be upper and lower confidence limits, UCL and LCL, respectively.

$$p\{\text{LCL} < \overline{x} < \text{UCL}\} = p\left\{ \frac{\text{LCL}-\mu}{\frac{\sigma}{\sqrt{n}}} < z < \frac{\text{UCL}-\mu}{\frac{\sigma}{\sqrt{n}}} \right\} \qquad 12.68$$

23. APPLICATION: BASIC HYPOTHESIS TESTING

A *hypothesis test* is a procedure that answers the question, "Did these data come from [a particular type of] distribution?" There are many types of tests, depending on the distribution and parameter being evaluated. The simplest hypothesis test determines whether an average value obtained from n repetitions of an experiment could have come from a population with known mean, μ, and standard deviation, σ. A practical application of this question is whether a manufacturing process has changed from what it used to be or should be. Of course, the answer (i.e., "yes" or "no") cannot be given with absolute certainty—there will be a confidence level associated with the answer.

The following procedure is used to determine whether the average of n measurements can be assumed (with a given confidence level) to have come from a known population.

step 1: Assume random sampling from a normal population.

step 2: Choose the desired confidence level, C.

step 3: Decide on a one-tail or two-tail test. If the hypothesis being tested is that the average has or has not *increased* or *decreased*, choose a one-tail test. If the hypothesis being tested is that the average has or has not *changed*, choose a two-tail test.

step 4: Use Table 12.1 or the standard normal table to determine the z-value corresponding to the confidence level and number of tails.

step 5: Calculate the actual standard normal variable, z'.

$$z' = \frac{\overline{x}-\mu}{\frac{\sigma}{\sqrt{n}}} \qquad 12.69$$

step 6: If $z' \geq z$, the average can be assumed (with confidence level C) to have come from a different distribution.

Example 12.16

When it is operating properly, a cement plant has a daily production rate that is normally distributed with a mean of 880 tons/day and a standard deviation of 21 tons/day. During an analysis period, the output is measured on 50 consecutive days, and the mean output is found to be 871 tons/day. With a 95% confidence level, determine whether the plant is operating properly.

Solution

step 1: Given.

step 2: $C = 0.95$ is given.

step 3: Since a specific direction in the variation is not given (i.e., the example does not ask whether the average has decreased), use a two-tail hypothesis test.

step 4: From Table 12.1, $z = 1.96$.

step 5: From Eq. 12.69,

$$z' = \left|\frac{\overline{x} - \mu}{\frac{\sigma}{\sqrt{n}}}\right| = \left|\frac{871 - 880}{\frac{21}{\sqrt{50}}}\right| = 3.03$$

Since $3.03 > 1.96$, the distributions are not the same. There is at least a 95% probability that the plant is not operating correctly.

24. APPLICATION: STATISTICAL PROCESS CONTROL

All manufacturing processes contain variation due to random and nonrandom causes. Random variation cannot be eliminated. *Statistical process control* (SPC) is the act of monitoring and adjusting the performance of a process to detect and eliminate nonrandom variation.

Statistical process control is based on taking regular (hourly, daily, etc.) samples of n items and calculating the mean, $\overline{x}$, and range, R, of the sample. To simplify the calculations, the range is used as a measure of the dispersion. These two parameters are graphed on their respective x-bar and *R-control charts*, as shown in Fig. 12.6.[6] Confidence limits are drawn at $\pm 3\sigma/\sqrt{n}$. From a statistical standpoint, the control chart tests a hypothesis each time a point is plotted. When a point falls outside these limits, there is a 99.75% probability that the process is out of control. Until a point exceeds the control limits, no action is taken.[7]

25. MEASURES OF EXPERIMENTAL ADEQUACY

An experiment is said to be *accurate* if it is unaffected by experimental error. In this case, *error* is not synonymous with *mistake*, but rather includes all variations not within the experimenter's control.

[6]Other charts (e.g., the *sigma chart*, *p-chart*, and *c-chart*) are less common but are used as required.

[7]Other indications that a correction may be required are seven measurements on one side of the average and seven consecutively increasing measurements. Rules such as these detect shifts and trends.

Figure 12.6 *Typical Statistical Process Control Charts*

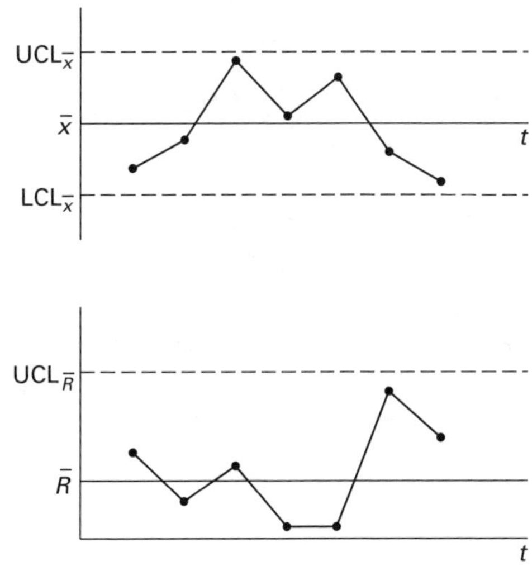

For example, suppose a gun is aimed at a point on a target and five shots are fired. The mean distance from the point of impact to the sight in point is a measure of the alignment accuracy between the barrel and sights. The difference between the actual value and the experimental value is known as *bias.*

Precision is not synonymous with accuracy. Precision is concerned with the repeatability of the experimental results. If an experiment is repeated with identical results, the experiment is said to be precise.

The average distance of each impact from the centroid of the impact group is a measure of the precision of the experiment. It is possible to have a highly precise experiment with a large bias.

Most of the techniques applied to experiments in order to improve the accuracy (i.e., reduce bias) of the experimental results (e.g., repeating the experiment, refining the experimental methods, or reducing variability) actually increase the precision.

Sometimes the word *reliability* is used with regard to the precision of an experiment. In this case, a "reliable estimate" is used in the same sense as a "precise estimate."

Stability and *insensitivity* are synonymous terms. A stable experiment will be insensitive to minor changes in the experimental parameters. For example, suppose the centroid of a bullet group is 2.1 in from the target point at 65°F and 2.3 in away at 80°F. The sensitivity of the experiment to temperature change would be

$$\text{sensitivity} = \frac{\Delta x}{\Delta T} = \frac{2.3\text{ in} - 2.1\text{ in}}{80^\circ\text{F} - 65^\circ\text{F}} = 0.0133\text{ in}/^\circ\text{F}$$

26. LINEAR REGRESSION

If it is necessary to draw a straight line ($y = mx + b$) through n data points $(x_1, y_1), (x_2, y_2), \ldots, (x_n, y_n)$, the following method based on the *method of least squares* can be used.

step 1: Calculate the following nine quantities.

$$\sum x_i \quad \sum x_i^2 \quad \left(\sum x_i\right)^2 \quad \overline{x} = \frac{\sum x_i}{n} \quad \sum x_i y_i$$

$$\sum y_i \quad \sum y_i^2 \quad \left(\sum y_i\right)^2 \quad \overline{y} = \frac{\sum y_i}{n}$$

step 2: Calculate the slope, m, of the line.

$$m = \frac{n\sum(x_i y_i) - \left(\sum x_i\right)\left(\sum y_i\right)}{n\sum x_i^2 - \left(\sum x_i\right)^2} \qquad \text{12.70}$$

step 3: Calculate the y-intercept, b.

$$b = \overline{y} - m\overline{x} \qquad \text{12.71}$$

step 4: To determine the goodness of fit, calculate the correlation coefficient, r.

$$r = \frac{n\sum(x_i y_i) - \left(\sum x_i\right)\left(\sum y_i\right)}{\sqrt{\left(n\sum x_i^2 - \left(\sum x_i\right)^2\right)\left(n\sum y_i^2 - \left(\sum y_i\right)^2\right)}} \qquad \text{12.72}$$

If m is positive, r will be positive; if m is negative, r will be negative. As a general rule, if the absolute value of r exceeds 0.85, the fit is good; otherwise, the fit is poor. r equals 1.0 if the fit is a perfect straight line.

A low value of r does not eliminate the possibility of a nonlinear relationship existing between x and y. It is possible that the data describe a parabolic, logarithmic, or other nonlinear relationship. (Usually this will be apparent if the data are graphed.) It may be necessary to convert one or both variables to new variables by taking squares, square roots, cubes, or logarithms, to name a few of the possibilities, in order to obtain a linear relationship. The apparent shape of the line through the data will give a clue to the type of variable transformation that is required. The curves in Fig. 12.7 may be used as guides to some of the simpler variable transformations.

Figure 12.8 illustrates several common problems encountered in trying to fit and evaluate curves from experimental data. Figure 12.8(a) shows a graph of clustered data with several extreme points. There will be moderate correlation due to the weighting of the extreme points, although there is little actual correlation at low values of the variables. The extreme data should be excluded, or the range should be extended by obtaining more data.

Figure 12.8(b) shows that good correlation exists in general, but extreme points are missed and the overall correlation is moderate. If the results within the small

Figure 12.7 Nonlinear Data Curves

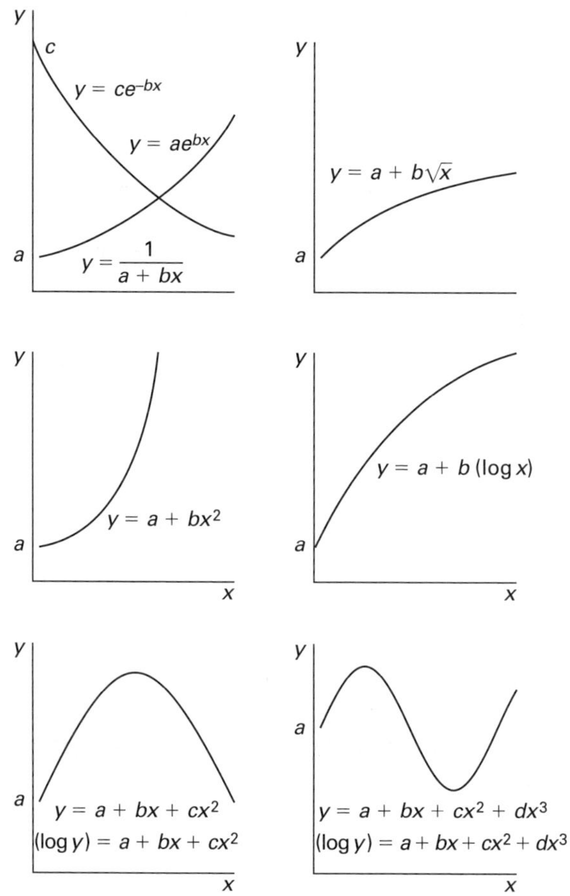

Figure 12.8 Common Regression Difficulties

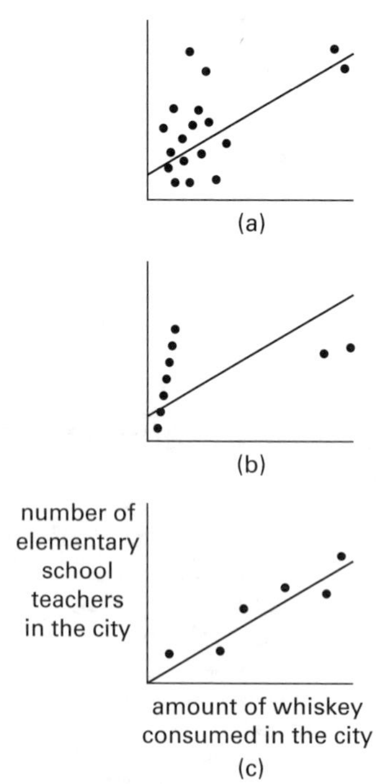

linear range can be used, the extreme points should be excluded. Otherwise, additional data points are needed, and curvilinear relationships should be investigated.

Figure 12.8(c) illustrates the problem of drawing conclusions of cause and effect. There may be a predictable relationship between variables, but that does not imply a cause and effect relationship. In the case shown, both variables are functions of a third variable, the city population. But there is no direct relationship between the plotted variables.

Example 12.17

An experiment is performed in which the dependent variable y is measured against the independent variable x. The results are as follows.

x	y
1.2	0.602
4.7	5.107
8.3	6.984
20.9	10.031

(a) What is the least squares straight line equation that represents this data? (b) What is the correlation coefficient?

Solution

(a)

$$\begin{aligned}\sum x_i &= 35.1\\ \sum y_i &= 22.72\\ \sum x_i^2 &= 529.23\\ \sum y_i^2 &= 175.84\\ \left(\sum x_i\right)^2 &= 1232.01\\ \left(\sum y_i\right)^2 &= 516.38\\ \overline{x} &= 8.775\\ \overline{y} &= 5.681\\ \sum x_i y_i &= 292.34\\ n &= 4\end{aligned}$$

From Eq. 12.70, the slope is

$$m = \frac{(4)(292.34) - (35.1)(22.72)}{(4)(529.23) - (35.1)^2} = 0.42$$

From Eq. 12.71, the y-intercept is

$$b = 5.681 - (0.42)(8.775) = 2.0$$

The equation of the line is

$$y = 0.42x + 2.0$$

(b) From Eq. 12.72, the correlation coefficient is

$$r = \frac{(4)(292.34) - (35.1)(22.72)}{\sqrt{\begin{array}{l}\big((4)(529.23) - 1232.01\big)\\ \quad\times\big((4)(175.84) - 516.38\big)\end{array}}} = 0.914$$

Example 12.18

Repeat Ex. 12.17 assuming the relationship between the variables is nonlinear.

Solution

The first step is to graph the data. Because the graph has the appearance of the fourth case, it can be assumed that the relationship between the variables has the form of $y = a + b \log x$. Therefore, the variable change $z = \log x$ is made, resulting in the following set of data.

z	y
0.0792	0.602
0.672	5.107
0.919	6.984
1.32	10.031

If the regression analysis is performed on this set of data, the resulting equation and correlation coefficient are

$$y = -0.036 + 7.65z$$

$$r = 0.999$$

This is a very good fit. The relationship between the variable x and y is approximately

$$y = -0.036 + 7.65 \log x$$

13 Computer Mathematics

1. Positional Numbering Systems 13-1
2. Converting Base-b Numbers to Base-10 13-1
3. Converting Base-10 Numbers to Base-b 13-1
4. Binary Number System 13-1
5. Octal Number System 13-2
6. Hexadecimal Number System 13-2
7. Conversions Among Binary, Octal, and Hexadecimal Numbers 13-3
8. Complement of a Number 13-3
9. Application of Complements to Computer Arithmetic 13-4
10. Computer Representation of Negative Numbers 13-4
11. Boolean Algebra Fundamental Postulates 13-5
12. Boolean Algebra Laws 13-5
13. Boolean Algebra Theorems 13-5

1. POSITIONAL NUMBERING SYSTEMS

A *base-b number*, N_b, is made up of individual *digits*. In a *positional numbering system*, the position of a digit in the number determines that digit's contribution to the total value of the number. Specifically, the position of the digit determines the power to which the *base* (also known as the *radix*), b, is raised. For *decimal numbers*, the radix is 10, hence the description *base-10 numbers*.

$$(a_n a_{n-1} \cdots a_2 a_1 a_0)_b = a_n b^n + a_{n-1} b^{n-1} + \cdots + a_2 b^2 + a_1 b + a_0 \qquad 13.1$$

The leftmost digit, a_n, contributes the greatest to the number's magnitude and is known as the *most significant digit* (MSD). The rightmost digit, a_0, contributes the least and is known as the *least significant digit* (LSD).

2. CONVERTING BASE-*b* NUMBERS TO BASE-10

Equation 13.1 converts base-b numbers to base-10 numbers. The calculation of the right-hand side of Eq. 13.1 is performed in the base-10 arithmetic and is known as the *expansion method.*[1]

[1] Equation 13.1 works with any base number. The *double-dabble (double and add)method* is a specialized method of converting from base-2 to base-10 numbers.

Converting base-b numbers (i.e., decimals) to base-10 is similar to converting whole numbers and is accomplished by Eq. 13.2.

$$(0.a_1 a_2 \cdots a_m)_b = a_1 b^{-1} + a_2 b^{-2} + \cdots + a_m b^{-m} \qquad 13.2$$

3. CONVERTING BASE-10 NUMBERS TO BASE-*b*

The *remainder method* is used to convert base-10 numbers to base-b numbers. This method consists of successive divisions by the base, b, until the quotient is zero. The base-b number is found by taking the remainders in the reverse order from which they were found. This method is illustrated in Ex. 13.1 and Ex. 13.3.

Converting a base-10 fraction to base-b requires multiplication of the base-10 fraction and subsequent fractional parts by the base. The base-b fraction is formed from the integer parts of the products taken in the same order in which they were determined. This is illustrated in Ex. 13.2(d).

4. BINARY NUMBER SYSTEM

There are only two *binary digits (bits)* in the *binary number system:* zero and one.[2] Thus, all binary numbers consist of strings of bits (i.e., zeros and ones). The leftmost bit is known as the *most significant bit* (MSB), and the rightmost bit is the *least significant bit* (LSB).

As with digits from other numbering systems, bits can be added, subtracted, multiplied, and divided, although only digits 0 and 1 are allowed in the results. The rules of bit addition are

$$0 + 0 = 0$$

$$0 + 1 = 1$$

$$1 + 0 = 1$$

$$1 + 1 = 0 \text{ carry } 1$$

Example 13.1

(a) Convert $(1011)_2$ to base-10. (b) Convert $(75)_{10}$ to base-2.

[2] Alternatively, the binary states may be called *true* and *false, on* and *off, high* and *low*, or *positive* and *negative*.

Solution

(a) Using Eq. 13.1 with $b = 2$,

$$(1)(2)^3 + (0)(2)^2 + (1)(2)^1 + 1 = 11$$

(b) Use the remainder method. (See Sec. 13.3.)

$$75 \div 2 = 37 \text{ remainder } 1$$
$$37 \div 2 = 18 \text{ remainder } 1$$
$$18 \div 2 = 9 \text{ remainder } 0$$
$$9 \div 2 = 4 \text{ remainder } 1$$
$$4 \div 2 = 2 \text{ remainder } 0$$
$$2 \div 2 = 1 \text{ remainder } 0$$
$$1 \div 2 = 0 \text{ remainder } 1$$

The binary representation of $(75)_{10}$ is $(1001011)_2$.

5. OCTAL NUMBER SYSTEM

The *octal (base-8) system* is one of the alternatives to working with long binary numbers. Only the digits 0 through 7 are used. The rules for addition in the octal system are the same as for the decimal system except that the digits 8 and 9 do not exist. For example,

$$7 + 1 = 6 + 2 = 5 + 3 = (10)_8$$
$$7 + 2 = 6 + 3 = 5 + 4 = (11)_8$$
$$7 + 3 = 6 + 4 = 5 + 5 = (12)_8$$

Example 13.2

Perform the following operations.

(a) $(2)_8 + (5)_8$

(b) $(7)_8 + (6)_8$

(c) Convert $(75)_{10}$ to base-8.

(d) Convert $(0.14)_{10}$ to base-8.

(e) Convert $(13)_8$ to base-10.

(f) Convert $(27.52)_8$ to base-10.

Solution

(a) The sum of 2 and 5 in base-10 is 7, which is less than 8 and, therefore, is a valid number in the octal system. The answer is $(7)_8$.

(b) The sum of 7 and 6 in base-10 is 13, which is greater than 8 (and, therefore, needs to be converted). Using the remainder method (see Sec. 13.3),

$$13 \div 8 = 1 \text{ remainder } 5$$
$$1 \div 8 = 0 \text{ remainder } 1$$

The answer is $(15)_8$.

(c) Use the remainder method (see Sec. 13.3).

$$75 \div 8 = 9 \text{ remainder } 3$$
$$9 \div 8 = 1 \text{ remainder } 1$$
$$1 \div 8 = 0 \text{ remainder } 1$$

The answer is $(113)_8$.

(d) Refer to Sec. 13.3.

$$0.14 \times 8 = 1.12$$
$$0.12 \times 8 = 0.96$$
$$0.96 \times 8 = 7.68$$
$$0.68 \times 8 = 5.44$$
$$0.44 \times 8 = \text{etc.}$$

The answer, $(0.1075\ldots)_8$, is constructed from the integer parts of the products.

(e) Use Eq. 13.1.

$$(1)(8) + 3 = (11)_{10}$$

(f) Use Eq. 13.1 and Eq. 13.2.

$$\begin{aligned}(2)(8)^1 + (7)(8)^0 & \\ + (5)(8)^{-1} + (2)(8)^{-2} &= 16 + 7 + \frac{5}{8} + \frac{2}{64} \\ &= (23.656)_{10}\end{aligned}$$

6. HEXADECIMAL NUMBER SYSTEM

The *hexadecimal (base-16) system* is a shorthand method of representing the value of four binary digits at a time.[3] Since 16 distinctly different characters are needed, the capital letters A through F are used to represent the decimal numbers 10 through 15. The progression of hexadecimal numbers is illustrated in Table 13.1.

Example 13.3

(a) Convert $(4D3)_{16}$ to base-10. (b) Convert $(1475)_{10}$ to base-16. (c) Convert $(0.8)_{10}$ to base-16.

Solution

(a) The hexadecimal number D is 13 in base-10. Using Eq. 13.1,

$$(4)(16)^2 + (13)(16)^1 + 3 = (1235)_{10}$$

[3]The term *hex number* is often heard.

Table 13.1 Binary, Octal, Decimal, and Hexadecimal Equivalents

binary	octal	decimal	hexadecimal
0	0	0	0
1	1	1	1
10	2	2	2
11	3	3	3
100	4	4	4
101	5	5	5
110	6	6	6
111	7	7	7
1000	10	8	8
1001	11	9	9
1010	12	10	A
1011	13	11	B
1100	14	12	C
1101	15	13	D
1110	16	14	E
1111	17	15	F
10000	20	16	10

(b) Use the remainder method. (See Sec. 13.3.)

$$1475 \div 16 = 92 \text{ remainder } 3$$
$$92 \div 16 = 5 \text{ remainder } 12$$
$$5 \div 16 = 0 \text{ remainder } 5$$

Since $(12)_{10}$ is $(\text{C})_{16}$, (or hex C), the answer is $(5\text{C}3)_{16}$.

(c) Refer to Sec. 13.3.

$$0.8 \times 16 = 12.8$$
$$0.8 \times 16 = 12.8$$
$$0.8 \times 16 = \text{etc.}$$

Since $(12)_{10} = (\text{C})_{16}$, the answer is $(0.\text{CCCCC}\ldots)_{16}$.

7. CONVERSIONS AMONG BINARY, OCTAL, AND HEXADECIMAL NUMBERS

The octal system is closely related to the binary system since $(2)^3 = 8$. Conversion from a binary to an octal number is accomplished directly by starting at the LSB (right-hand bit) and grouping the bits in threes. Each group of three bits corresponds to an octal digit. Similarly, each digit in an octal number generates three bits in the equivalent binary number.

Conversion from a binary to a hexadecimal number starts by grouping the bits (starting at the LSB) into fours. Each group of four bits corresponds to a hexadecimal digit. Similarly, each digit in a hexadecimal number generates four bits in the equivalent binary number.

Conversion between octal and hexadecimal numbers is easiest when the number is first converted to a binary number.

Example 13.4

(a) Convert $(5431)_8$ to base-2. (b) Convert $(1001011)_2$ to base-8. (c) Convert $(1011111101111001)_2$ to base-16.

Solution

(a) Convert each octal digit to binary digits.

$$(5)_8 = (101)_2$$
$$(4)_8 = (100)_2$$
$$(3)_8 = (011)_2$$
$$(1)_8 = (001)_2$$

The answer is $(101100011001)_2$.

(b) Group the bits into threes starting at the LSB.

1 001 011

Convert these groups into their octal equivalents.

$$(1)_2 = (1)_8$$
$$(001)_2 = (1)_8$$
$$(011)_2 = (3)_8$$

The answer is $(113)_8$.

(c) Group the bits into fours starting at the LSB.

1011 1111 0111 1001

Convert these groups into their hexadecimal equivalents.

$$(1011)_2 = (\text{B})_{16}$$
$$(1111)_2 = (\text{F})_{16}$$
$$(0111)_2 = (7)_{16}$$
$$(1001)_2 = (9)_{16}$$

The answer is $(\text{BF}79)_{16}$.

8. COMPLEMENT OF A NUMBER

The *complement*, N^*, of a number, N, depends on the machine (computer, calculator, etc.) being used. Assuming that the machine has a maximum number, n, of digits per integer number stored, the b's and $(b-1)$'s complements are

$$N_b^* = b^n - N \qquad 13.3$$
$$N_{b-1}^* = N_b^* - 1 \qquad 13.4$$

For a machine that works in base-10 arithmetic, the *tens* and *nines complements* are

$$N_{10}^* = 10^n - N \qquad 13.5$$
$$N_9^* = N_{10}^* - 1 \qquad 13.6$$

For a machine that works in base-2 arithmetic, the *twos* and *ones complements* are

$$N_2^* = 2^n - N \qquad 13.7$$

$$N_1^* = N_2^* - 1 \qquad 13.8$$

9. APPLICATION OF COMPLEMENTS TO COMPUTER ARITHMETIC

Equation 13.9 and Eq. 13.10 are the practical applications of complements to computer arithmetic.

$$(N^*)^* = N \qquad 13.9$$

$$M - N = M + N^* \qquad 13.10$$

The binary ones complement is easily found by switching all of the ones and zeros to zeros and ones, respectively. It can be combined with a technique known as *end-around carry* to perform subtraction. End-around carry is the addition of the *overflow bit* to the sum of N and its ones complement.

Example 13.5

(a) Simulate the operation of a base-10 machine with a capacity of four digits per number and calculate the difference $(18)_{10} - (6)_{10}$ with tens complements.

(b) Simulate the operation of a base-2 machine with a capacity of five digits per number and calculate the difference $(01101)_2 - (01010)_2$ with twos complements.

(c) Solve part (b) with a ones complement and end-around carry.

Solution

(a) The tens complement of 6 is

$$(6)_{10}^* = (10)^4 - 6 = 10{,}000 - 6 = 9994$$

Using Eq. 13.10,

$$18 - 6 = 18 + (6)_{10}^* = 18 + 9994 = 10{,}012$$

However, the machine has a maximum capacity of four digits. Therefore, the leading 1 is dropped, leaving 0012 as the answer.

(b) The twos complement of $(01010)_2$ is

$$\begin{aligned} N_2^* &= (2)^5 - N = (32)_{10} - N \\ &= (100000)_2 - (01010)_2 \\ &= (10110)_2 \end{aligned}$$

From Eq. 13.10,

$$\begin{aligned} (01101)_2 - (01010)_2 &= (01101)_2 + (10110)_2 \\ &= (100011)_2 \end{aligned}$$

Since the machine has a capacity of only five bits, the leftmost bit is dropped, leaving $(00011)_2$ as the difference.

(c) The ones complement is found by reversing all the digits.

$$(01010)_1^* = (10101)_2$$

Adding the ones complement,

$$(01101)_2 + (10101)_2 = (100010)_2$$

The leading bit is the overflow bit, which is removed and added to give the difference.

$$(00010)_2 + (1)_2 = (00011)_2$$

10. COMPUTER REPRESENTATION OF NEGATIVE NUMBERS

On paper, a minus sign indicates a negative number. This representation is not possible in a machine. Hence, one of the n digits, usually the MSB, is reserved for sign representation. (This reduces the machine's capacity to represent numbers to $n-1$ bits per number.) It is arbitrary whether the sign bit is 1 or 0 for negative numbers as long as the MSB is different for positive and negative numbers.

The ones complement is ideal for forming a negative number since it automatically reverses the MSB. For example, $(00011)_2$ is a five-bit representation of decimal 3. The ones complement is $(11100)_2$, which is recognized as a negative number because the MSB is 1.

Example 13.6

Simulate the operation of a six-digit binary machine that uses ones complements for negative numbers.

(a) What is the machine representation of $(-27)_{10}$? (b) What is the decimal equivalent of the twos complement of $(-27)_{10}$? (c) What is the decimal equivalent of the ones complement of $(-27)_{10}$?

Solution

(a) $(27)_{10} = (011011)_2$. The negative of this number is the same as the ones complement: $(100100)_2$.

(b) The twos complement is one more than the ones complement. (See Eq. 13.7 and Eq. 13.8.) Therefore, the twos complement is

$$(100100)_2 + 1 = (100101)_2$$

This represents $(-26)_{10}$.

(c) From Eq. 13.9, the complement of a complement of a number is the original number. Therefore, the decimal equivalent is -27.

11. BOOLEAN ALGEBRA FUNDAMENTAL POSTULATES

Boolean algebra is a system of mathematics that deals with the algebraic treatment of logic. When the variables of the system are limited to only one of two values, the system is called two-valued. *Switching algebra* is the term often used to describe two-valued Boolean algebra. The basic postulates of switching algebra follow.

- Postulate 1: A Boolean variable x has two possible exclusive values, 0 and 1. That is,

$$\text{if } x = 0 \text{ then } x \neq 1, \text{ and}$$
$$\text{if } x = 1 \text{ then } x \neq 0$$

- Postulate 2: The NOT operation, "$\overline{x}$" or x', is defined as

$$\overline{0} = 1 \quad \overline{1} = 0$$

- Postulate 3: The logical operations AND ($\wedge$ or $\cdot$), and OR ($\vee$ or $+$) are defined as

$$0 \cdot 0 = 0 \quad 0 + 0 = 0$$
$$0 \cdot 1 = 0 \quad 0 + 1 = 1$$
$$1 \cdot 0 = 0 \quad 1 + 0 = 1$$
$$1 \cdot 1 = 1 \quad 1 + 1 = 1$$

Sometimes the symbol for the AND operation is dropped completely and $x_1 \cdot x_2$ is represented as $x_1 x_2$.

12. BOOLEAN ALGEBRA LAWS

The process of reducing logical expressions to simpler forms is aided by the use of the following relationships, or laws, of switching algebra.[4] In the following, x can represent a single variable or a general logical function.

- Special properties of 0 and 1

$$0 + x = x \quad 0 \cdot x = 0$$
$$1 + x = 1 \quad 1 \cdot x = x$$

- Idempotence laws

$$x + x = x \quad x \cdot x = x$$

- Complementation laws

$$x + \overline{x} = 1 \quad x \cdot \overline{x} = 0$$

- Involution

$$\overline{(\overline{x})} = x$$

- Commutative laws

$$x + y = y + x \quad x \cdot y = y \cdot x$$

- Associative laws

$$x + (y + z) = (x + y) + z$$
$$x \cdot (y \cdot z) = (x \cdot y) \cdot z$$

- Distributive laws

$$x \cdot (y + z) = (x \cdot y) + (x \cdot z)$$
$$x + (y \cdot z) = (x + y) \cdot (x + z)$$

- Absorption laws

$$x + (x \cdot y) = x \qquad x \cdot (x + y) = x$$
$$x + (\overline{x} \cdot y) = x + y \quad x \cdot (\overline{x} + y) = x \cdot y$$

13. BOOLEAN ALGEBRA THEOREMS

The simplification of logical expressions is further enhanced by a set of relationships that exist as theorems, although one takes the name of a law. These theorems follow.

- Simplification theorems

$$xy + x\overline{y} = x \quad (x + y)(x + \overline{y}) = x$$

The AND symbol is often assumed between parentheses.

$$x + xy = x \qquad x(x + y) = x$$
$$(x + \overline{y})y = xy \quad x\overline{y} + y = x + y$$

- de Morgan's laws

$$\overline{(x + y + z + \cdots)} = \overline{x}\,\overline{y}\,\overline{z}\cdots$$
$$\overline{(xyz\cdots)} = \overline{x} + \overline{y} + \overline{z}\cdots$$
$$\overline{(f(x_1, x_2, x_3, \cdots, x_n, 0, 1, +, \cdot))}$$
$$= f(\overline{x_1}, \overline{x_2}, \overline{x_3}, \cdots, \overline{x_n}, 1, 0, \cdot, +)$$

- Duality

$$(x + y + z + \cdots)^{\text{D}} = xyz\cdots$$
$$(xyz\cdots)^{\text{D}} = x + y + z + \cdots$$
$$\big(f(x_1, x_2, x_3, \cdots, x_n, 0, 1, +, \cdot)\big)^{\text{D}}$$
$$= f(x_1, x_2, x_3, \cdots, x_n, 1, 0, \cdot, +)$$

- Consensus theorem

$$xy + yz + \overline{x}z = xy + \overline{x}z$$
$$(x + y)(y + z)(\overline{x} + z) = (x + y)(\overline{x} + z)$$
$$(x + y)(\overline{x} + z) = xz + \overline{x}y$$

[4]The reduction of such expressions is desired to minimize circuit complexity and costs when realizing a given expression with electronic logic circuits.

14 Numerical Analysis

1. Numerical Methods 14-1
2. Finding Roots: Bisection Method 14-1
3. Finding Roots: Newton's Method 14-2
4. Nonlinear Interpolation: Lagrangian Interpolating Polynomial 14-2
5. Nonlinear Interpolation: Newton's Interpolating Polynomial 14-3

1. NUMERICAL METHODS

Although the roots of second-degree polynomials are easily found by a variety of methods (by factoring, completing the square, or using the quadratic equation), easy methods of solving cubic and higher-order equations exist only for specialized cases. However, cubic and higher-order equations occur frequently in engineering, and they are difficult to factor. Trial and error solutions, including graphing, are usually satisfactory for finding only the general region in which the root occurs.

Numerical analysis is a general subject that covers, among other things, iterative methods for evaluating roots to equations. The most efficient numerical methods are too complex to present and, in any case, work by hand. However, some of the simpler methods are presented here. Except in critical problems that must be solved in real time, a few extra calculator or computer iterations will make no difference.[1]

2. FINDING ROOTS: BISECTION METHOD

The *bisection method* is an iterative method that "brackets" (also known as "straddles") an interval containing the *root* or *zero* of a particular equation.[2] The size of the interval is halved after each iteration. As the method's name suggests, the best estimate of the root after any iteration is the midpoint of the interval. The maximum error is half the interval length. The procedure continues until the size of the maximum error is "acceptable."[3]

The disadvantages of the bisection method are (a) the slowness in converging to the root, (b) the need to know the interval containing the root before starting, and (c) the inability to determine the existence of or find other real roots in the starting interval.

The bisection method starts with two values of the independent variable, $x = L_0$ and $x = R_0$, which straddle a root. Since the function passes through zero at a root, $f(L_0)$ and $f(R_0)$ will have opposite signs. The following algorithm describes the remainder of the bisection method.

Let n be the iteration number. Then, for $n = 0, 1, 2, \ldots$, perform the following steps until sufficient accuracy is attained.

step 1: Set $m = \frac{1}{2}(L_n + R_n)$.

step 2: Calculate $f(m)$.

step 3: If $f(L_n)f(m) \le 0$, set $L_{n+1} = L_n$ and $R_{n+1} = m$. Otherwise, set $L_{n+1} = m$ and $R_{n+1} = R_n$.

step 4: $f(x)$ has at least one root in the interval $[L_{n+1}, R_{n+1}]$. The estimated value of that root, x^*, is

$$x^* \approx \tfrac{1}{2}(L_{n+1} + R_{n+1})$$

The maximum error is $\frac{1}{2}(R_{n+1} - L_{n+1})$.

Example 14.1

Use two iterations of the bisection method to find a root of

$$f(x) = x^3 - 2x - 7$$

Solution

The first step is to find L_0 and R_0, which are the values of x that straddle a root and have opposite signs. A table can be made and values of $f(x)$ calculated for random values of x.

x	−2	−1	0	+1	+2	+3
$f(x)$	−11	−6	−7	−8	−3	+14

Since $f(x)$ changes sign between $x = 2$ and $x = 3$, $L_0 = 2$ and $R_0 = 3$.

First iteration, $n = 0$:

$$m = \tfrac{1}{2}(L_n + R_n) = \left(\tfrac{1}{2}\right)(2 + 3) = 2.5$$

$$f(2.5) = x^3 - 2x - 7 = (2.5)^3 - (2)(2.5) - 7 = 3.625$$

[1]Most advanced hand-held calculators have "root finder" functions that use numerical methods to iteratively solve equations.
[2]The equation does not have to be a pure polynomial. The bisection method requires only that the equation be defined and determinable at all points in the interval.
[3]The bisection method is not a closed method. Unless the root actually falls on the midpoint of one iteration's interval, the method continues indefinitely. Eventually, the magnitude of the maximum error is small enough not to matter.

Since $f(2.5)$ is positive, a root must exist in the interval [2, 2.5]. Therefore, $L_1 = 2$ and $R_1 = 2.5$. At this point, the best estimate of the root is

$$x^* \approx \left(\tfrac{1}{2}\right)(2 + 2.5) = 2.25$$

The maximum error is $\left(\tfrac{1}{2}\right)(2.5 - 2) = 0.25$.

Second iteration, $n = 1$:

$$m = \left(\tfrac{1}{2}\right)(2 + 2.5) = 2.25$$

$$f(2.25) = (2.25)^3 - (2)(2.25) - 7 = -0.1094$$

Since $f(2.25)$ is negative, a root must exist in the interval [2.25, 2.5]. Therefore, $L_2 = 2.25$ and $R_2 = 2.5$. The best estimate of the root is

$$x^* \approx \tfrac{1}{2}(L_{n+1} + R_{n+1}) \approx \left(\tfrac{1}{2}\right)(2.25 + 2.5) = 2.375$$

The maximum error is $\left(\tfrac{1}{2}\right)(2.5 - 2.25) = 0.125$.

3. FINDING ROOTS: NEWTON'S METHOD

Many other methods have been developed to overcome one or more of the disadvantages of the bisection method. These methods have their own disadvantages.[4]

Newton's method is a particular form of *fixed-point iteration*. In this sense, "fixed point" is often used as a synonym for "root" or "zero." However, fixed-point iterations get their name from functions with the characteristic property $x = g(x)$ such that the limit of $g(x)$ is the fixed point (i.e., is the root).

All fixed-point techniques require a starting point. Preferably, the starting point will be close to the actual root.[5] And, while Newton's method converges quickly, it requires the function to be continuously differentiable.

Newton's method algorithm is simple. At each iteration ($n = 0, 1, 2,$ etc.), Eq. 14.1 estimates the root. The maximum error is determined by looking at how much the estimate changes after each iteration. If the change between the previous and current estimates (representing the magnitude of error in the estimate) is too large, the current estimate is used as the independent variable for the subsequent iteration.[6]

$$x_{n+1} = g(x_n) = x_n - \frac{f(x_n)}{f'(x_n)} \qquad 14.1$$

[4]The *regula falsi (false position) method* converges faster than the bisection method but is unable to specify a small interval containing the root. The *secant method* is prone to round-off errors and gives no indication of the remaining distance to the root.

[5]Theoretically, the only penalty for choosing a starting point too far away from the root will be a slower convergence to the root.

[6]Actually, the theory defining the maximum error is more definite than this. For example, for a large enough value of n, the error decreases approximately linearly. Therefore, the consecutive values of x_n converge linearly to the root as well.

Example 14.2

Solve Ex. 14.1 using two iterations of Newton's method. Use $x_0 = 2$.

Solution

The function and its first derivative are

$$f(x) = x^3 - 2x - 7$$

$$f'(x) = 3x^2 - 2$$

First iteration, $n = 0$:

$$x_0 = 2$$

$$f(x_0) = f(2) = (2)^3 - (2)(2) - 7 = -3$$

$$f'(x_0) = f'(2) = (3)(2)^2 - 2 = 10$$

$$x_1 = x_0 - \frac{f(x_0)}{f'(x_0)} = 2 - \frac{-3}{10} = 2.3$$

Second iteration, $n = 1$:

$$x_1 = 2.3$$

$$f(x_1) = (2.3)^3 - (2)(2.3) - 7 = 0.567$$

$$f'(x_1) = (3)(2.3)^2 - 2 = 13.87$$

$$x_2 = x_1 - \frac{f(x_1)}{f'(x_1)} = 2.3 - \frac{0.567}{13.87} = 2.259$$

4. NONLINEAR INTERPOLATION: LAGRANGIAN INTERPOLATING POLYNOMIAL

Interpolating between two points of known data is common in engineering. Primarily due to its simplicity and speed, straight-line interpolation is used most often. Even if more than two points on the curve are explicitly known, they are not used. Since straight-line interpolation ignores all but two of the points on the curve, it ignores the effects of curvature.

A more powerful technique that accounts for the curvature is the *Lagrangian interpolating polynomial*. This method uses an nth degree parabola (polynomial) as the interpolating curve.[7] This method requires that $f(x)$ be continuous and real-valued on the interval

[7]The Lagrangian interpolating polynomial reduces to straight-line interpolation if only two points are used.

$[x_0, x_n]$ and that $n + 1$ values of $f(x)$ are known corresponding to $x_0, x_1, x_2, \ldots, x_n$.

The procedure for calculating $f(x)$ at some intermediate point x^* starts by calculating the Lagrangian interpolating polynomial for each known point.

$$L_k(x^*) = \prod_{\substack{i=0 \\ i\neq k}}^{n} \frac{x^* - x_i}{x_k - x_i} \quad \text{14.2}$$

The value of $f(x)$ at x^* is calculated from Eq. 14.3.

$$f(x^*) = \sum_{k=0}^{n} f(x_k) L_k(x^*) \quad \text{14.3}$$

The Lagrangian interpolating polynomial has two primary disadvantages. The first is that a large number of additions and multiplications are needed.[8] The second is that the method does not indicate how many interpolating points should be (or should have been) used. Other interpolating methods have been developed that overcome these disadvantages.[9]

Example 14.3

A real-valued function has the following values.

$$f(1) = 3.5709$$

$$f(4) = 3.5727$$

$$f(6) = 3.5751$$

Use the Lagrangian interpolating polynomial to determine the value of the function at 3.5.

Solution

The procedure for applying Eq. 14.2, the Lagrangian interpolating polynomial, is illustrated in tabular form. Notice that the term corresponding to $i = k$ is omitted from the product.

$k = 0$: $\quad i = 0 \quad i = 1 \quad i = 2$

$$L_0(3.5) = \left(\frac{3.5-1}{1-1}\right)\left(\frac{3.5-4}{1-4}\right)\left(\frac{3.5-6}{1-6}\right) = 0.08333$$

(first factor crossed out)

$k = 1$:

$$L_1(3.5) = \left(\frac{3.5-1}{4-1}\right)\left(\frac{3.5-4}{4-4}\right)\left(\frac{3.5-6}{4-6}\right) = 1.04167$$

(second factor crossed out)

$k = 2$:

$$L_2(3.5) = \left(\frac{3.5-1}{6-1}\right)\left(\frac{3.5-4}{6-4}\right)\left(\frac{3.5-6}{6-6}\right) = -0.12500$$

(third factor crossed out)

Equation 14.3 is used to calculate the estimate.

$$\begin{aligned} f(3.5) &= (3.5709)(0.08333) + (3.5727)(1.04167) \\ &\quad + (3.5751)(-0.12500) \\ &= 3.57225 \end{aligned}$$

5. NONLINEAR INTERPOLATION: NEWTON'S INTERPOLATING POLYNOMIAL

Newton's form of the interpolating polynomial is more efficient than the Lagrangian method of interpolating between known points.[10] Given $n + 1$ known points for $f(x)$, the *Newton form of the interpolating polynomial* is

$$f(x^*) = \sum_{i=0}^{n} \left(f[x_0, x_1, \ldots, x_i] \prod_{j=0}^{i-1} (x^* - x_j) \right) \quad \text{14.4}$$

$f[x_0, x_1, \ldots, x_i]$ is known as the ith *divided difference*.

$$f[x_0, x_1, \ldots, x_i] = \sum_{k=0}^{i} \left(\frac{f(x_k)}{\begin{array}{c}(x_k - x_0)\cdots(x_k - x_{k-1}) \\ \times (x_k - x_{k+1})\cdots(x_k - x_i)\end{array}} \right) \quad \text{14.5}$$

It is necessary to define the following two terms.

$$f[x_0] = f(x_0) \quad \text{14.6}$$

$$\prod (x^* - x_j) = 1 \quad [i=0] \quad \text{14.7}$$

Example 14.4

Repeat Ex. 14.3 using Newton's form of the interpolating polynomial.

[8]As with the numerical methods for finding roots previously discussed, the number of calculations probably will not be an issue if the work is performed by a calculator or computer.

[9]Other common methods for performing interpolation include the *Newton form* and *divided difference table*.

[10]In this case, "efficiency" relates to the ease in adding new known points without having to repeat all previous calculations.

Solution

Since there are $n + 1 = 3$ data points, $n = 2$. Evaluate the terms for $i = 0$–2.

$i = 0$:

$$f[x_0]\prod_{j=0}^{-1}(x^* - x_j) = f(x_0)(1) = f(x_0)$$

$i = 1$:

$$f[x_0, x_1]\prod_{j=0}^{0}(x^* - x_j) = f[x_0, x_1](x^* - x_0)$$

$$f[x_0, x_1] = \frac{f(x_0)}{x_0 - x_1} + \frac{f(x_1)}{x_1 - x_0}$$

$i = 2$:

$$\begin{aligned} f[x_0, x_1, x_2]\prod_{j=0}^{1}(x^* - x_j) &= f[x_0, x_1, x_2](x^* - x_0) \\ &\quad \times (x^* - x_1) \end{aligned}$$

$$\begin{aligned} f[x_0, x_1, x_2] &= \frac{f(x_0)}{(x_0 - x_1)(x_0 - x_2)} \\ &\quad + \frac{f(x_1)}{(x_1 - x_0)(x_1 - x_2)} \\ &\quad + \frac{f(x_2)}{(x_2 - x_0)(x_2 - x_1)} \end{aligned}$$

Substitute known values.

$$\begin{aligned} f(3.5) &= 3.5709 + \left(\frac{3.5709}{1 - 4} + \frac{3.5727}{4 - 1}\right)(3.5 - 1) \\ &\quad + \left(\frac{3.5709}{(1 - 4)(1 - 6)} + \frac{3.5727}{(4 - 1)(4 - 6)} \right. \\ &\quad \left. + \frac{3.5751}{(6 - 1)(6 - 4)}\right) \\ &\quad \times (3.5 - 1)(3.5 - 4) \\ &= 3.57225 \end{aligned}$$

This answer is the same as that determined in Ex. 14.3.

15 Advanced Engineering Mathematics

1. Overview 15-1
2. Power Series Method: Homogeneous Linear Differential Equations 15-1
3. Extended Power Series Method: Homogeneous Linear Differential Equations 15-3
4. General Method for Solving Nonhomogeneous Differential Equations 15-3
5. Orthogonality of Functions 15-5
6. Laplace Transform: Initial and Final Value Theorems 15-5
7. Half-Range Expansions 15-6
8. Partial Differential Equations: Separation of Variables 15-7
9. Special Functions 15-9
10. Continuous-Time Systems: Fourier Series and Transforms 15-13
11. Continuous-Time Systems: Fourier Series Representation of a Periodic Signal 15-14
12. Continuous-Time Systems: Fourier Transform of an Aperiodic Signal 15-15
13. Convolution 15-16
14. Laplace Transforms 15-17
15. Discrete-Time Systems: Fourier Series and Transforms 15-17
16. Discrete-Time Systems: Fourier Series Representation of a Periodic Signal 15-17
17. Discrete-Time Systems: Fourier Transform of an Aperiodic Signal 15-18
18. z-Transforms 15-18
19. Transformation of Integrals 15-18
20. Line Integrals 15-19
21. Complex Numbers 15-21

1. OVERVIEW

This chapter emphasizes advanced engineering mathematics not covered in earlier chapters. A grasp of such material is necessary in order to understand various aspects of electrical engineering, especially at the design and theoretical levels. The province of the engineer in mathematics, however, resides in the recognition of the form of, and restrictions on, the solution.

2. POWER SERIES METHOD: HOMOGENEOUS LINEAR DIFFERENTIAL EQUATIONS

When the coefficients of a homogeneous linear differential equation are a function of x, the methods explained in Chap. 11 are no longer applicable.[1] Further, the solutions may be *nonelementary functions*.[2] Equations of this type include Bessel's equation and Legendre's equation.[3]

- Bessel Equation of Order ν

$$x^2y'' + xy' + (x^2 - \nu^2)y = 0 \qquad \textit{15.1}$$

- Legendre Equation of Order λ

$$(1 - x^2)y'' - 2xy' + \lambda(\lambda + 1)y = 0 \quad [-1 < x < 1] \qquad \textit{15.2}$$

The solutions of such equations are power series; the method of obtaining them is the *power series method*. A power series is of the following form.[4]

$$\sum_{m=0}^{\infty} c_m(x-a)^m = c_0 + c_1(x-a) + c_2(x-a)^2 + \cdots \qquad \textit{15.3}$$

The constants c_0, c_1, ... are the *coefficients* of the series. The constant a is called the *center*. The variable is x.

A common power series is the Maclaurin series.

$$\frac{1}{1-x} = \sum_{m=0}^{\infty} x^m = 1 + x + x^2 + \cdots \quad \left[\begin{array}{l}\text{If } |x| < 1, \text{ then this power series}\\ \text{is known as a geometric series.}\end{array}\right] \qquad \textit{15.4}$$

[1]Equations discussed in this chapter are *ordinary differential equations*, or ODEs, as are most equations in Chap. 11. The term "ordinary" distinguishes these equations from *partial differential equations*, or PDEs.

[2]*Elementary functions* are those that can be formed by a finite number of mathematical operations (addition, subtraction, multiplication, and division) from algebraic functions and exponential, logarithmic, and trigonometric functions. *Nonelementary functions* are those that cannot be so formed. Do not confuse nonelementary functions with transcendental functions. *Transcendental functions* are nonalgebraic functions, for example, exponential, logarithmic, and trigonometric.

[3]Bessel and Legendre equations and their associated polynomials occur in many engineering applications. As a single example, they are used in approximating the response of filter circuits.

[4]A power series does not include negative or fractional powers of x.

$$e^x = \sum_{m=0}^{\infty} \frac{x^m}{m!}$$

$$= 1 + x + \frac{x^2}{2!} + \frac{x^3}{3!} + \cdots \quad \text{15.5}$$

$$\cos x = \sum_{m=0}^{\infty} \frac{(-1)^m x^{2m}}{2m!}$$

$$= 1 - \frac{x^2}{2!} + \frac{x^4}{4!} - \cdots \quad \text{15.6}$$

$$\sin x = \sum_{m=0}^{\infty} \frac{(-1)^m x^{2m+1}}{(2m+1)!}$$

$$= x - \frac{x^3}{3!} + \frac{x^5}{5!} - \cdots \quad \text{15.7}$$

The steps to the power series method are as follows.

step 1: Given a differential equation, represent the given functions in the equation by a power series in powers of x (or $x - a$, if solutions of this type are desired). If the functions are polynomials, as is often the case in practical applications, nothing need be done in this first step.

step 2: Assume a solution in the form of a power series.

step 3: Substitute this series and the series obtained by termwise differentiation into the original equation.

step 4: Collect like powers of x and equate the sum of their coefficients to zero.

step 5: Determine the unknown coefficients in the assumed power series solution successively.

step 6: Substitute the values of the unknown coefficients into the assumed power series solution.

Example 15.1

Solve the following homogeneous, first-order linear ordinary differential equation with constant coefficients.[5]

$$y' + ky = 0$$

Solution

step 1: Represent the given functions in the equation by a power series in powers of x. Because the coefficients are constant (i.e., no functions are used in the equation), this step is not required.

step 2: Assume a solution in the form of a power series.

$$y(x) = c_0 + c_1 x + c_2 x^2 + \cdots = \sum_{m=0}^{\infty} c_m x^m$$

step 3: Substitute this series and the series obtained by termwise differentiation into the original equation.

[5]The solution to this differential equation was found using other methods in Sec. 11.2. The power series method is used on this simple differential equation to clarify the method.

Differentiation of $y(x)$ termwise results in the following equation.

$$y'(x) = 0 + c_1 + 2c_2 x + 3c_3 x^2 + \cdots$$

Substituting into the original equation gives

$$y' + ky = 0$$

$$(0 + c_1 + 2c_2 x + 3c_3 x^2 + \cdots) + k(c_0 + c_1 x + c_2 x^2 + \cdots) = 0$$

step 4: Collect like powers of x and equate the sum of their coefficients to zero.

$$(c_1 + kc_0) + (2c_2 + kc_1)x + (3c_3 + kc_2)x^2 + \cdots = 0$$

For this equation to be true, the sum of the coefficients of each power of x must be zero. That is,

$$c_1 + kc_0 = 0$$

$$2c_2 + kc_1 = 0$$

$$3c_3 + kc_2 = 0 \cdots$$

step 5: Determine the unknown coefficients in the assumed power series successively. That is, using the results of step 4, determine the coefficients in terms of c_0, which remains arbitrary.

$$c_1 = -kc_0$$

$$c_2 = \frac{-k}{2} c_1 = \frac{k^2}{2!} c_0$$

$$c_3 = \frac{-k}{3} c_2 = \frac{-k^3}{3!} c_0 \cdots$$

step 6: Substitute these values into the assumed power series solution from step 2.

$$y(x) = c_0 + c_1 x + c_2 x^2 + \cdots = \sum_{m=0}^{\infty} c_m x^m$$

$$= c_0 + (-kc_0)x + \left(\frac{k^2}{2!} c_0\right) x^2 + \left(\frac{-k^3}{3!} c_0\right) x^3 + \cdots$$

$$= c_0\left(1 - kx + \frac{k^2}{2!} x^2 - \frac{k^3}{3!} x^3 + \cdots\right)$$

For clarification, let $-kx = z$.

$$y(z) = c_0\left(1 + z + \frac{z^2}{2!} + \frac{z^3}{3!} + \cdots\right)$$

The terms in parentheses represent the Maclaurin series for e^z. The solution is

$$y(z) = c_0 e^z$$

or

$$y(x) = c_0 e^{-kx}$$

The initial conditions can be used to determine the value of the constant c_0.

3. EXTENDED POWER SERIES METHOD: HOMOGENEOUS LINEAR DIFFERENTIAL EQUATIONS

If a negative power of x exists in an equation—in other words, if the equation is not analytic at $x = 0$—the *extended power series method* is used. (In simpler terms, with a negative power of x, the x is in the denominator; the equation cannot be defined as $x = 0$ because this involves division by zero. This is also referenced as a *singularity*.) The extended power series method can be used on any differential equation of the following form.

$$y'' + \left(\frac{a(x)}{x}\right)y' + \left(\frac{b(x)}{x^2}\right)y = 0 \qquad 15.8$$

The Cauchy or Euler equation, Eq. 15.9, is of this form.[6]

$$x^2y'' + axy' + by = 0 \qquad 15.9$$

Rearranging to show that the equation is not analytic at $x=0$ (i.e., dividing by x^2 and allowing a and b to be functions of x) changes Eq. 15.9 into the form of Eq. 15.8.

Gauss' hypergeometric equation, Eq. 15.10, can also be solved using this method.[7]

$$x(1-x)y'' + \Big(c - (a+b+1)x\Big)y' - aby = 0 \qquad 15.10$$

In Eq. 15.8, it is assumed that $a(x)$ and $b(x)$ are analytic at $x=0$; therefore, at least one solution can be represented in the form[8]

$$\begin{aligned} y(x) &= x^r \sum_{m=0}^{\infty} c_m x^m \\ &= x^r(c_0 + c_1x + c_2x^2 + \cdots) \end{aligned} \qquad 15.11$$

[6]This form is equivalent to that given in Chap. 11. One should become accustomed to seeing various formats for the same-named equation.
[7]This equation finds application in probability and statistics.
[8]Using the form of the solution shown, the method is called the *Frobenius method.*

The exponent r may be any real or complex number and is chosen so that $c_0 \neq 0$.[9] Expanding $a(x)$ and $b(x)$ into a power series, differentiating Eq. 15.11, substituting into Eq. 15.8, and equating the sum of the coefficients of each power of x to zero yields the *indicial equation*, from the coefficients associated with x^r.

$$r^2 + (a_0 - 1)r + b_0 = 0 \qquad 15.12$$

Equation 15.12 will yield a basis of solutions. One of the solutions will be of the form of Eq. 15.11; the other form depends on the roots of the indicial equation. There are three possible cases depending on these roots.

case 1: distinct roots r_1 and r_2, which do not differ by an integer[10]

$$y_1(x) = x^{r_1}(c_0 + c_1x + c_2x^2 + \cdots)$$
$$y_2(x) = x^{r_2}(d_0 + d_1x + d_2x^2 + \cdots)$$

case 2: double root, r

$$y_1(x) = x^r(c_0 + c_1x + c_2x^2 + \cdots)$$
$$r = \frac{1-a_0}{2}$$
$$y_2(x) = y_1(x)\ln x + x^r \sum_{m=1}^{\infty} A_m x^m$$

The a_0 in the equation for $y_1(x)$ is from the indicial equation, Eq. 15.12.

case 3: distinct roots, r_1 and r_2, that differ by an integer, for example, 1, 2, 3, ...

$$y_1(x) = x^{r_1}(c_0 + c_1x + c_2x^2 + \cdots)$$
$$y_2(x) = k_p y_1(x)\ln x + x^{r_2} \sum_{m=0}^{\infty} c_m x^m \quad [x > 0]$$

4. GENERAL METHOD FOR SOLVING NONHOMOGENEOUS DIFFERENTIAL EQUATIONS

In Chap. 11, the method of undetermined coefficients was used to find a particular solution to nonhomogeneous equations. It was limited to simpler differential equations in which the coefficients $p(x)$ and $q(x)$ are constants and the forcing function $f(x)$ was of a particular form. The method of *variation of parameters* is completely general but more complicated, and it removes all such restrictions.

[9]Restraining r to be a nonnegative integer would make the entire expression a power series as covered in the Power Series Method sections of this chapter.
[10]This includes complex conjugate roots.

Consider the nonhomogeneous second-order linear differential equation, Eq. 15.13.[11]

$$y'' + p(x)y' + q(x)y = f(x) \qquad 15.13$$

The homogeneous portion of Eq. 15.13 has a solution of the following form.

$$y_h(x) = c_1 y_1(x) + c_2 y_2(x) \qquad 15.14$$

The coefficients c_1 and c_2 are selected to be functions $u(x)$ and $v(x)$ that make the following a particular solution to Eq. 15.13.[12]

$$y_p(x) = u(x)y_1(x) + v(x)y_2(x) \qquad 15.15$$

The parameters $u(x)$ and $v(x)$ are then determined in terms of $y_1(x)$ and $y_2(x)$ by rearranging the particular solution, Eq. 15.15. The result is substituted into the homogeneous solution, Eq. 15.14. The associated first and second derivatives are determined. The results are substituted into Eq. 15.13. The result is a particular solution of the following completely general form.

$$\begin{aligned} y_p(x) &= -y_1(x)\int \frac{y_2(x)f(x)}{W}\,dx \\ &\quad + y_2(x)\int \frac{y_1(x)f(x)}{W}\,dx \end{aligned} \qquad 15.16$$

The term W is called the Wronskian of y_1 and y_2 and is given by

$$W = y_1(x)y_2'(x) - y_1'(x)y_2(x) \qquad 15.17$$

The parameters $u(x)$ and $v(x)$ are

$$u(x) = -\int \frac{y_2(x)f(x)}{W}\,dx \qquad 15.18$$

$$v(x) = \int \frac{y_1(x)f(x)}{W}\,dx \qquad 15.19$$

The overall result is that the combination of Eq. 15.14 and Eq. 15.16 constitute a general solution to Eq. 15.13, that is, to any nonhomogeneous differential equation.

Example 15.2

Solve the following differential equation.

$$y'' + y = \sec x$$

[11]The functions p, q, and f are continuous on an open interval I.

[12]This is defined on the same open interval I mentioned in the previous footnote. Also, recall that since $j = 1\angle 90°$, $j^2 = 1\angle 180° = -1$, making $j = \sqrt{-1}$.

Solution

The homogeneous portion of the equation is

$$y'' + y = 0$$

The characteristic equation is[13]

$$\begin{aligned} r^2 + 0r + 1 &= 0 \\ r^2 &= -1 \\ r &= \pm\sqrt{-1} = \pm j \end{aligned}$$

The homogeneous solution is

$$y_h(x) = c_1 \cos x + c_2 \sin x \qquad \text{[I]}$$

From this,

$$y_1 = \cos x \text{ and } y_2 = \sin x$$

From Eq. 15.17, the Wronskian is

$$\begin{aligned} W\big(y_1(x), y_2(x)\big) &= y_1(x)y_2'(x) - y_1'(x)y_2(x) \\ &= \cos x \cos x - (-\sin x)\sin x \\ &= \cos^2 x + \sin^2 x = 1 \end{aligned}$$

Substitute into Eq. 15.16 to determine the particular solution.

$$\begin{aligned} y_p(x) &= -y_1(x)\int \frac{y_2(x)f(x)}{W}\,dx \\ &\quad + y_2(x)\int \frac{y_1(x)f(x)}{W}\,dx \\ &= -\cos x\int \frac{\sin x \sec x}{1}\,dx \\ &\quad + \sin x\int \frac{\cos x \sec x}{1}\,dx \end{aligned}$$

Substitute the definition of the secant and then integrate.

$$\begin{aligned} y_p(x) &= -\cos x\int \sin x\left(\frac{1}{\cos x}\right)dx \\ &\quad + \sin x\int \cos x\left(\frac{1}{\cos x}\right)dx \\ &= -\cos x\int \tan x\,dx + \sin x\int dx \\ &= -\cos x(-\ln|\cos x|) + (\sin x)(x) \\ &= \cos x \ln|\cos x| + x\sin x \end{aligned} \qquad \text{[II]}$$

[13]See Chap. 11.

Combine the homogeneous (or complementary) solution I with the particular solution II to obtain the general solution.

$$\begin{aligned} y(x) &= y_h(x) + y_p(x) \\ &= c_1\cos x + c_2\sin x + \cos x \ln|\cos x| + x\sin x \\ &= (c_1 + \ln|\cos x|)\cos x + (c_2 + x)\sin x \end{aligned}$$

5. ORTHOGONALITY OF FUNCTIONS

Let $g_m(x)$ and $g_n(x)$ be two real-valued functions defined on an interval $a \le x \le b$. Further, define the integral of the two as

$$(g_m, g_n) = \int_a^b g_m(x) g_n(x) dx \qquad 15.20$$

The functions g_m and g_n are *orthogonal* on the interval $a \le x \le b$ if the integral in Eq. 15.20 equals zero.[14] That is,

$$\begin{aligned} (g_m, g_n) &= \int_a^b g_m(x) g_n(x) dx \\ &= 0 \quad [m \neq n] \end{aligned} \qquad 15.21$$

The nonnegative square root of (g_m, g_m) is called the *norm* of g_m.

$$\|g_m\| = \sqrt{(g_m, g_m)} = \sqrt{\int_a^b g_m^2(x) dx} \qquad 15.22$$

6. LAPLACE TRANSFORM: INITIAL AND FINAL VALUE THEOREMS

Laplace transforms simplify the solution of nonhomogeneous differential equations by reducing many steps in the process to algebra. Additionally, the initial conditions are automatically included in the analysis.[15] Nevertheless, obtaining the inverse transform (i.e., moving from the s domain back into the time domain) may not be an easy task.

$$\begin{aligned} f(t) &= L^{-1}\{F(s)\} \\ &= \frac{1}{2\pi j} \int_{c-j\infty}^{c+j\infty} F(s) e^{st} ds \end{aligned} \qquad 15.23$$

Equation 15.23 can be used for this task, although the use of transform tables is more common. Further, the exact inverse transform may not be necessary. It may be more fundamental to understand how the equation, that is, the circuit, reacts at $t=0$ and $t=\infty$. To that end, the *initial value theorem*, Eq. 15.24, can be used.[16]

$$f(0) = \lim_{s\to\infty} sF(s) \qquad 15.24$$

The *final value theorem*, Eq. 15.25, can be used to determine the value of the function in the time domain as time approaches infinity.[17]

$$f(\infty) = \lim_{s\to 0} sF(s) \qquad 15.25$$

Example 15.3

The Laplace transform of a given voltage is shown. Evaluate the voltage at $t=0$.

$$V(s) = \frac{4s}{(s^2+9)(s^2+1)}$$

Expand the denominator and divide the numerator and denominator by s^2.

$$V(s) = \frac{4s}{s^4 + 10s^2 + 9} = \frac{\frac{4}{s}}{s^2 + 10 + \frac{9}{s^2}}$$

Apply the initial value theorem, Eq. 15.24.

$$\begin{aligned} v(0) &= \lim_{s\to\infty} sV(s) \\ &= \lim_{s\to\infty} s\left(\frac{\frac{4}{s}}{s^2 + 10 + \frac{9}{s^2}}\right) \\ &= \lim_{s\to\infty}\left(\frac{4}{s^2 + 10 + \frac{9}{s^2}}\right) \\ &= 0 \end{aligned}$$

The voltage function associated with the given voltage is

$$v(t) = \sin t \sin 2t$$

The value of $v(t)$ is indeed zero at $t=0$ as $\sin(0)=0$, confirming the answer.

Example 15.4

Given the voltage in the frequency domain (or s domain), as shown, what is the value of the voltage at $t=\infty$?[18]

$$V(s) = \frac{1}{s(s+1)^2}$$

[14] Both Legendre polynomials and Bessel functions are orthogonal. Important orthogonal sets of functions arise as solutions to linear second-order differential equations.

[15] The terms $f(0)$ and $f'(0)$ are required when transforming a second-order differential, thereby immediately placing the initial conditions into use.

[16] $f(0)$ is more technically $f(0^+)$, that is, the value at zero when approached from the positive side. This is based on the normally used Laplace transform being defined from zero to infinity.

[17] Technically, Eq. 15.25 should be $\lim_{t\to\infty} f(t) = \lim_{s\to 0} sF(s)$.

[18] Asking for the value of the voltage at $t=\infty$ is the equivalent of asking for the steady-state value.

Expand the denominator.

$$V(s) = \frac{1}{s(s+1)^2} = \frac{1}{s(s+1)(s+1)}$$
$$= \frac{1}{s(s^2+2s+1)} = \frac{1}{s^3+2s^2+s}$$

Apply the final value theorem, Eq. 15.25.

$$v(\infty) = \lim_{s\to\infty} sV(s)$$
$$= \lim_{s\to 0} s\left(\frac{1}{s^3+2s^2+s}\right)$$
$$= \lim_{s\to 0}\left(\frac{s}{s^3+2s^2+s}\right)$$
$$= \lim_{s\to 0}\left(\frac{1}{s^2+2s+s}\right) = 1$$

The voltage function associated with the given voltage is

$$v(t) = 1 - e^{-t} - te^{-t}$$

The value of $v(t)$ as $t \to \infty$ can be seen to be equal to 1 from this equation, confirming the solution.[19]

7. HALF-RANGE EXPANSIONS

In some engineering problems a function $f(t)$ defined over a finite interval may be more readily manipulated and a solution more easily obtained if the function is represented as a Fourier series, as will be shown in the Partial Differential Equations section of this chapter.

Given a function $f(t)$ defined on some interval $0 \le t \le l$ or $0 \le t \le T/2$ where T is the period,[20] the *even periodic extension* of $f(t)$ of period $T = 2l$ is given by the Fourier cosine series.

$$f(t) = a_0 + \sum a_n \cos\frac{n\pi t}{l} \qquad 15.26$$

The coefficients of the series are given by the following.

$$a_0 = \frac{1}{l}\int_0^l f(t)dt \qquad 15.27$$

$$a_n = \frac{2}{l}\int_0^l f(t)\cos\frac{n\pi t}{l}\,dt$$
$$[n = 1, 2, 3, \ldots] \qquad 15.28$$

[19]As t approaches ∞, the value of s approaches 0. In other words, as the function, or circuit, approaches a steady-state condition, the number of oscillations (frequency) approaches zero. The physical model represented by the mathematics matches what one would expect in the real world.

[20]The definition of the interval means the period is $T/2 = l$ or $T = 2l$.

The *odd periodic extension* of $f(t)$ on the same interval is given by the Fourier sine series.

$$f(t) = \sum_{n=1}^{\infty} b_n \sin\frac{n\pi t}{l} \qquad 15.29$$

The coefficients are

$$b_n = \frac{2}{l}\int_0^l f(t)\sin\frac{n\pi t}{l}\,dt$$
$$[n = 1, 2, 3, \ldots] \qquad 15.30$$

The series represented by Eq. 15.26 and Eq. 15.29 along with their associated coefficients are called the *half-range expansions* of the given function $f(t)$. Figure 15.1 shows a given function $f(t)$ and its half-range expansions.

Figure 15.1 *Periodic Extensions*

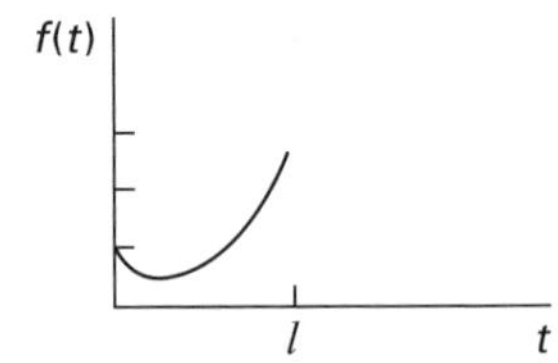

(a) given function *f*(*t*)

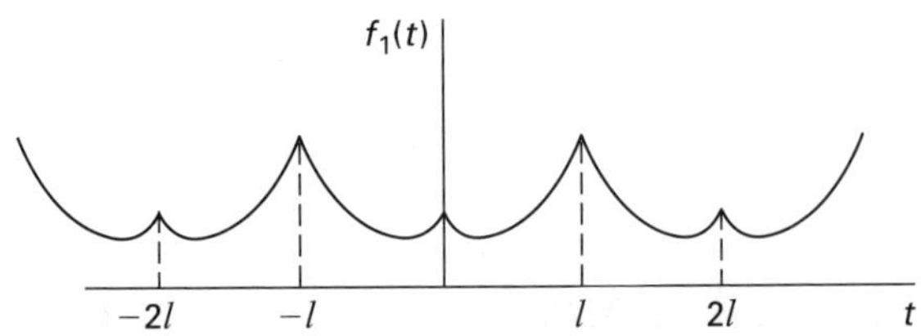

(b) even periodic extension of *f*(*t*)

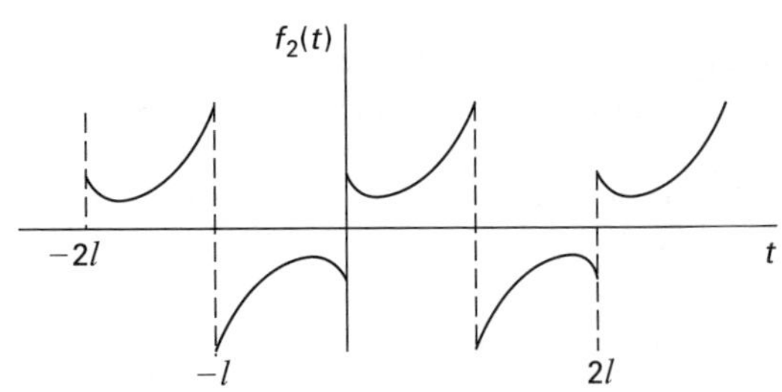

(c) odd periodic extension of *f*(*t*)

In the half-range expansion, $f_1(t) = f(t)$ on the interval $0 \le t \le 1$. Also, $f_2(t) = f(t)$ on this same interval.

Mentally summing the infinite series from the graphs in Fig. 15.1 leaves the original function $f(t)$ in place as was given mathematically in Eq. 15.29. This is especially apparent in part (c) of the figure, providing a clearer view for what is occurring mathematically.

8. PARTIAL DIFFERENTIAL EQUATIONS: SEPARATION OF VARIABLES

A *partial differential equation* involves one or more partial derivatives of an unknown function of two or more variables. As before, the order of the equation is the order of the highest derivative. The equation is *linear* if the dependent variable and its partial derivatives are of the first degree, that is, not raised to a power greater than one. The equation is defined as *homogeneous* if each term consists of a dependent variable or one of its derivatives.

Some important second-order partial differential equations follow. (In the following equations, c is a constant, t is time, and u represents the unknown function. All are shown in Cartesian coordinates. Additionally, the constant c is shown as c^2 to indicate a positive quantity in the physical world.)

One-dimensional wave equation:[21]

$$\frac{\partial^2 u}{\partial t^2} = c^2 \frac{\partial^2 u}{\partial x^2} \qquad 15.31$$

One-dimensional heat equation:

$$\frac{\partial u}{\partial t} = c^2 \frac{\partial^2 u}{\partial x^2} \qquad 15.32$$

Two-dimensional Laplace equation:

$$\frac{\partial^2 u}{\partial x^2} + \frac{\partial^2 u}{\partial y^2} = 0 \qquad 15.33$$

Two-dimensional Poisson equation:

$$\frac{\partial^2 u}{\partial x^2} + \frac{\partial^2 u}{\partial y^2} = f(x, y) \qquad 15.34$$

Three-dimensional Laplace equation:[22]

$$\frac{\partial^2 u}{\partial x^2} + \frac{\partial^2 u}{\partial y^2} + \frac{\partial^2 u}{\partial z^2} = 0 \qquad 15.35$$

The equations are solved based on the conditions at a boundary, that is, the *boundary conditions*. Those equations having time as one of the variables may already have the conditions at $t=0$ (the *initial conditions*) given or known, and they are solved as constrained by those conditions.

An important aspect of the solutions to linear partial differential equations is the following fundamental theorem. If u_1 and u_2 are any of the solutions of a linear homogeneous partial differential equation in some region, then another solution is

$$u = c_1 u_1 + c_2 u_2 \qquad 15.36$$

The terms c_1 and c_2 represent constants.

The following method of solving such problems is termed the *separation of variables*.[23]

step 1: Apply the separation of variables to obtain ordinary differential equations.[24]

step 2: Determine solutions that satisfy the boundary conditions.

step 3: Compose the solutions such that the initial conditions are satisfied.

Example 15.5

Solve the one-dimensional wave equation with the given boundary and initial conditions.

$$\frac{\partial^2 u}{\partial t^2} = c^2 \left(\frac{\partial^2 u}{\partial x^2} \right)$$

The boundary conditions are that the function $u(x, t)$ equals zero at $x=0$ and $x=l$.[25] This can be expressed in mathematical terms as

$$u(0, t) = 0 \qquad [I]$$

$$u(l, t) = 0 \qquad [II]$$

Further, this function must take on the values in conditions I and II for all t. Let the initial deflection be $f(x)$ and the initial velocity be $g(x)$.

$$u(x, 0) = f(x) \qquad [III]$$

$$\left. \frac{\partial u}{\partial t} \right|_{t=0} = g(x) \qquad [IV]$$

Solution

step 1: Apply the separation of variables to obtain ordinary differential equations. Separating variables yields an unknown function of the following form.

$$u(x, t) = F(x)G(t)$$

[21]A solution to this one-dimensional case will be shown. Solutions to the two-dimensional case in Cartesian coordinates involve double Fourier series. Solutions to the two-dimensional case in polar coordinates involve Bessel functions.

[22]The theory of solutions to this equation is called *potential theory*.

[23]The separation of variables method is also called the *product method*.

[24]Recall that ordinary differential equations are those that do not involve partial differentiation.

[25]A vibrating string secured at each end would provide such conditions, as would a confined electromagnetic wave.

From this function,

$$\frac{\partial^2 u(x,t)}{\partial t^2} = F(x)\ddot{G}(t)$$

$$\frac{\partial^2 u(x,t)}{\partial x^2} = F''(x)G(t)$$

The dots represent differentiation with respect to t. The primes represent differentiation with respect to x. Insert these results into the original wave equation.

$$\frac{\partial^2 u}{\partial t^2} = c^2\left(\frac{\partial^2 u}{\partial x^2}\right)$$

$$F(x)\ddot{G}(t) = c^2 F''(x)G(t)$$

$$\frac{F''(x)}{F(x)} = \frac{\ddot{G}(t)}{c^2 G(t)}$$

The variables t and x are now separated. Further, the result must be equal to a constant, say k. This can be deduced from the fact that changing the variable t changes only one side of the equation, and vice versa for the variable x.

$$\frac{F''(x)}{F(x)} = \frac{\ddot{G}(t)}{c^2 G(t)} = k$$

This leads to two linear ordinary differential equations, ODEs.

$$F''(x) - kF(x) = 0$$

$$\ddot{G}(t) - kc^2 G(t) = 0$$

step 2: Determine solutions based on the boundary conditions. Starting with the function $F(x)$, the boundary conditions in I and II were

$$u(0,t) = 0 = F(0)G(t)$$

$$u(l,t) = 0 = F(l)G(t)$$

Letting $G(t) \equiv 0$ gives $u(x,t) \equiv 0$, which is trivial. Therefore, $G(t) \neq 0$. This means that

$$F(0) = 0 \text{ and } F(l) = 0$$

If $k = 0$, the general solution of the ODE for the function $F(x)$ is of the form $F(x) = ax + b$.[26] Because $F(0) = 0$, b must be zero. Because $F(l) = 0$, a must be zero. Hence, $F(x) \equiv 0$ giving $u(x,t) \equiv 0$, all of which results in a trivial solution.

Let k be some positive value, $k = \mu^2$. The solution is now of the following form.

$$F(x) = Ae^{\mu x} + Be^{-\mu x}$$

But again, applying the boundary conditions makes $A = 0$ and $B = 0$. Therefore, $F(x) \equiv 0$ and $u(x,t) \equiv 0$.

To have a useful solution, k must be negative, say $k = -p^2$. This provides a solution of the following form.

$$F(x) = A\cos px + B\sin px$$

Now, $F(0) = 0$ makes $A = 0$. The second boundary condition yields the following.

$$F(l) = 0 = B\sin pl$$

Because $B = 0$ would give $F(x) \equiv 0$ and $u(x,t) \equiv 0$, $\sin pl$ must be equal to zero. That is,

$$pl = n\pi \text{ or } p = \frac{n\pi}{l} \quad [n = 0, 1, 2, \ldots]$$

Letting $B = 1$, one obtains infinitely many solutions of the following form.

$$F_n(x) = \sin\frac{n\pi}{l}x \quad [n = 1, 2, 3, \ldots]$$

Because k is restricted to negative values, that is, $k = -p^2$, the ODE for the function containing the t variable is

$$\ddot{G}(t) - kc^2 G(t) = 0$$

$$\ddot{G}(t) + p^2 c^2 G(t) = 0$$

$$\ddot{G}(t) + \left(\frac{n\pi}{l}\right)^2 c^2 G(t) = 0$$

$$\ddot{G}(t) + \lambda_n^2 G(t) = 0 \quad \left[\lambda_n = \left(\frac{n\pi}{l}\right)c\right]$$

The solution is therefore[27]

$$G_n(t) = B_n\cos\lambda_n t + B_n^*\sin\lambda_n t$$

Writing the function $u_n(x,t) = F_n(t)G_n(t)$ gives

$$u(x,t) = \sin\frac{n\pi}{l}x \times (B_n\cos\lambda_n t + B_n^*\sin\lambda_n t) \quad [n = 1, 2, 3, \ldots] \qquad \text{[V]}$$

[26]The characteristic equation results in a double root of zero. See Chap. 11 for the method used on homogeneous, second-order linear differential equations with constant coefficients.

[27]The symbols B_n and B_n^* are commonly used as b_n in representations of Fourier sine series. Do not confuse the symbol B_n^* with that of a complex number.

This function satisfies the boundary conditions given in Eqs. I and II. The functions are *eigenfunctions*, or characteristic functions. The values for λ_n are *eigenvalues*, or characteristic values. The function u_n represents harmonic motion of a frequency $\lambda_n/2\pi = cn/2l$ cycles per unit time. The *fundamental mode* occurs when $n=1$. Each mode n has $n-1$ nodes, that is, points that do not move.[28] Some fundamental nodes are shown in the following illustration.

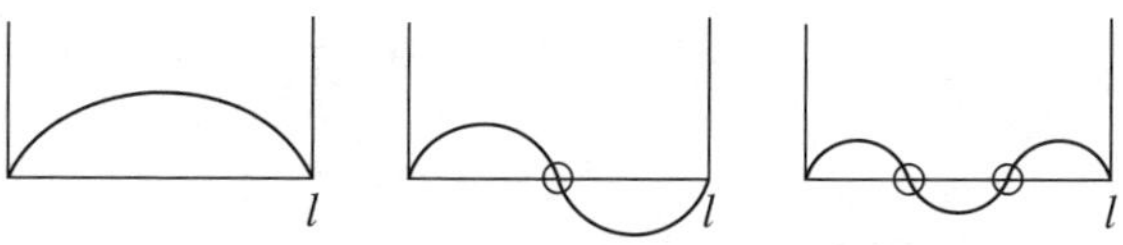

n = 1 n = 2 n = 3

step 3: Compose the solution such that the initial conditions are satisfied.

A single solution of the form of Eq. V will not, in general, satisfy both initial conditions III and IV. Based on the fundamental theorem given by Eq. 15.36, the sum of the solutions is a solution. Therefore, the following combination of solutions to Eq. V provides a solution to the one-dimensional wave equation.

$$u(x,t) = \sum_{n=1}^{\infty} u_n(x,t)$$

$$= \sum \sin \frac{n\pi}{l} x$$

$$(B_n \cos \lambda_n t + B_n^* \sin \lambda_n t) \quad \text{[VI]}$$

Applying the initial condition III to Eq. VI gives

$$u(x,0) = \sum_{n=1}^{\infty} B_n \sin \frac{n\pi}{l} x = f(x) \quad \text{[VII]}$$

To make Eq. VI satisfy initial condition III, B_n must be chosen as the half-range expansion of $f(x)$, that is, the Fourier sine series of $f(x)$.[29]

$$B_n = \frac{2}{l}\int_0^l f(x) \sin \frac{n\pi}{l} x \, dx$$

$$[n = 1, 2, 3, \ldots] \quad \text{[VIII]}$$

By making this choice for B_n, Eq. VII becomes valid and Eq. VI satisfies initial condition III.

One item now remains: make Eq. VI satisfy initial condition IV. Start by taking the derivative of Eq. VI with respect to time and applying condition IV. This gives the following equation.

$$\left.\frac{\partial u}{\partial t}\right|_{t=0}$$

$$= \sum_{n=1}^{\infty} \sin \frac{n\pi}{l} x \left. \begin{pmatrix} -B_n \lambda_n \sin \lambda_n t \\ + B_n^* \lambda_n \cos \lambda_n t \end{pmatrix} \right|_{t=0}$$

$$= \sum_{n=1}^{\infty} B_n^* \lambda_n \sin \frac{n\pi}{l} x = g(x) \quad \text{[IX]}$$

Using the same logic, and recalling that $\lambda_n = cn\pi/l$ gives

$$B_n^* \lambda_n = \frac{2}{l}\int_0^l g(x) \sin \frac{n\pi}{l} x \, dx$$

$$B_n^* = \frac{2}{cn\pi}\int_0^l g(x) \sin \frac{n\pi}{l} x \, dx \quad \text{[X]}$$

With B_n^* equal to the condition given in Eq. X, Eq. VI satisfies initial condition IV. All the conditions are now met.

Eq. VI is a solution to the one-dimensional wave equation with the coefficients B_n and B_n^* defined by Eq. VIII and Eq. IX.[30]

9. SPECIAL FUNCTIONS

Several different special functions of use occur as *named equations* with different forms and symbols. Some of particular use to the electrical engineer are discussed in this section.

Unit Step Function

Symbols: u_t; $S_k(t)$; $H(t-k)$

[28]This is not counting the constrained endpoints.

[29]The Fourier sine series results in an odd extension of $f(x)$. In other words, the original function $f(x)$ is extended in the coefficient B_n which, when summed, leaves only $f(x)$. See the graphs in Sec. 15.7 of this chapter for a physical picture of what is occurring.

[30]Also, the series represented in Eq. VI must converge, and the series obtained by differentiating Eq. VI twice (termwise) with respect to x and t must converge and sum to $\partial^2 u/\partial x^2$ and $\partial^2 u/\partial t^2$. Other solution methods are possible. One is called the *D'Alembert* solution. Another is using a variation of parameters and Laplace transforms. Also, Laplace transforms can change the partial differential equation into an ordinary differential equation that can be solved using standard methods.

This function may be used to represent what occurs when a switch is turned on.

$$u_t = \begin{cases} 0 & 0 < t < k \\ 1 & t > k \end{cases} \quad \text{15.37}$$

Equation 15.37 mathematically represents what is shown in Fig. 15.2. The symbol $H(t-k)$ is called the *Heaviside step function.*[31]

Figure 15.2 *Unit Step Function*

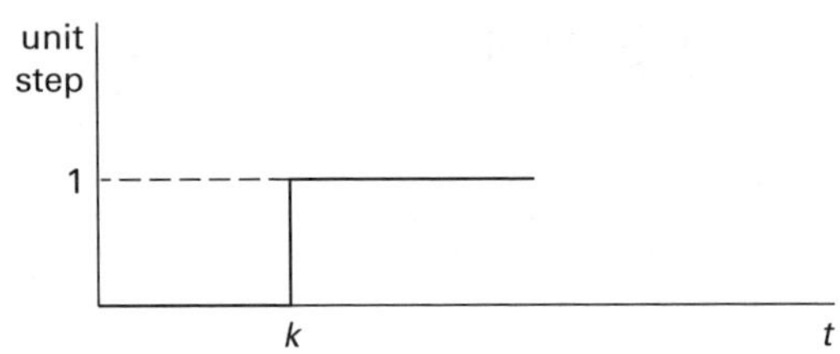

Unit Finite Impulse Function

$$I(h, t-t_0) = H(t-t_0) - H\big(t-(t_0+h)\big) \quad \text{15.38}$$

$$I(h, t-t_0) = \begin{cases} 0 & t < t_0 \text{ and } t > (t_0+h) \\ 1 & t_0 < t < (t_0+h) \end{cases} \quad \text{15.39}$$

The *unit finite impulse function* is shown graphically in Fig. 15.3.

Of special importance in Fig. 15.3 part (c) is that the unit finite impulse function, that is, the shaded region, represents an area equal to 1 for any $h > 0$. This area represents a rectangle with the base equal to $t_0 - (t_0+h) = h$ and the height equal to $1/h$; therefore, the area is $h(1/h) = 1$. The impulse is the integral of force $(1/h)$ over the interval of time (h).

Delta Function; Dirac Function; Impulse Function; Unit Impulse Function

Symbols: δt; $\delta(t-t_0)$

If the limit as $h \rightarrow 0$ is taken on the unit finite impulse function, the result is the *delta*, or *Dirac function.*[32] This function is sometimes called the *impulse function* or *unit*

Figure 15.3 *Unit Finite Impulse Function*

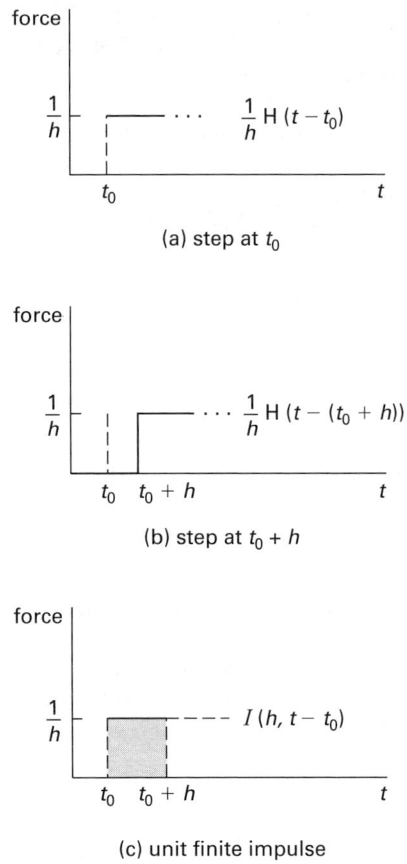

impulse function and not always differentiated from the unit finite impulse function.

$$\lim_{h \to 0} I(h, t-t_0) \equiv \delta(t-t_0) = \begin{cases} 0 & t \neq t_0 \\ \infty & t = t_0 \end{cases} \quad \text{15.40}$$

This is shown graphically in Fig. 15.4.[33]

Figure 15.4 *Delta Function*

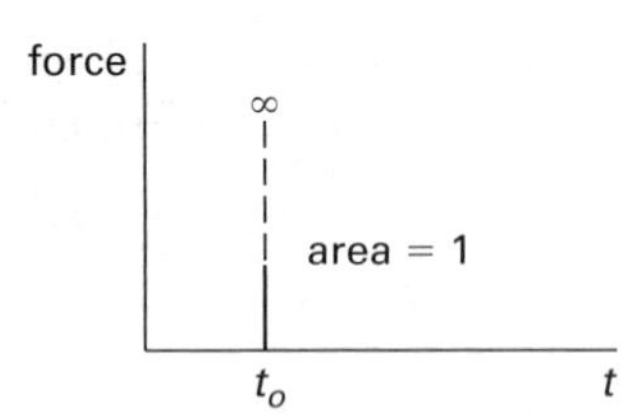

[31]Oliver Heaviside was an English electrical engineer whose treatment of the derivative process (represented by D) in algebraic terms was the spur to the development of the Laplace transform.

[32]Paul Dirac was an English physicist whose theory of negative energy holes predicted the existence of positrons.

[33]There is much more to the delta function. The mathematics shown is not strictly true. The delta function is not a proper function and is therefore shown as equivalent to (≡) the limit.

The delta function is strictly defined only in terms of integration. The delta function displays a property called *sifting* that is given by Eq. 15.41.[34]

$$\int_{-\infty}^{+\infty} \delta(t-t_0)f(t)\,dt = f(t_0) \qquad 15.41$$

Gamma Function

Symbols: $\Gamma(\alpha)$; $(\alpha-1)!$

Also known as the *factorial function*, the *gamma function* is defined by the following integral.[35]

$$\Gamma(\alpha) = \int_0^\infty e^{-t}t^{(\alpha-1)}\,dt \quad [\alpha > 0] \qquad 15.42$$

The gamma function is shown graphically in Fig. 15.5.

Figure 15.5 *Gamma Function*

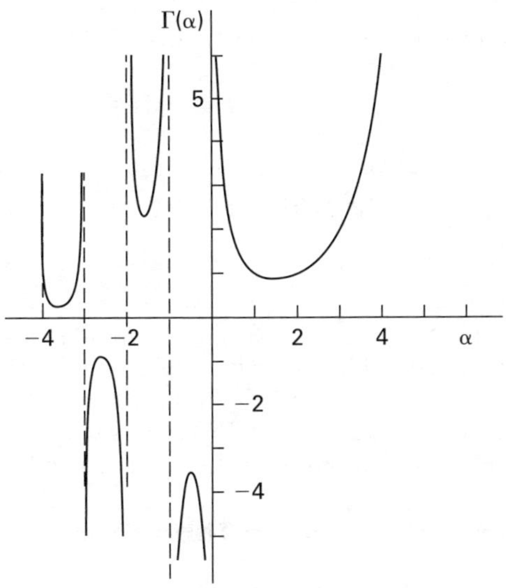

The important functional relation of the gamma function is

$$\Gamma(\alpha+1) = \alpha\Gamma(\alpha) \qquad 15.43$$

Example 15.6

Find $\Gamma(2)$.

Solution

Apply the definition given by Eq. 15.42.

$$\Gamma(\alpha) = \int_0^\infty e^{-t}t^{(\alpha-1)}\,dt \quad [\alpha > 1]$$

$$\Gamma(2) = \int_0^\infty e^{-t}t^{(2-1)}\,dt = \int_0^\infty e^{-t}t\,dt$$

From a mathematical handbook of integrals, the result of the previous equation is

$$\Gamma(2) = \frac{e^{-t}}{(-1)^2}(-t-1)\Big|_0^\infty = e^{-t}(-t-1)\Big|_0^\infty = \frac{-(t+1)}{e^t}\Big|_0^\infty$$

Use L'Hôpital's rule to determine the value at infinity.

$$\frac{-(t+1)}{e^t}\Big|_0^\infty = 0 - \left(\frac{-(0+1)}{e^0}\right) = 1$$

This result could also have been accomplished using the functional relation of Eq. 15.43 as shown.

$$\Gamma(\alpha+1) = \alpha\Gamma(\alpha)$$

$$\Gamma(2) = \Gamma(1+1) = 1\Gamma(1) = 1$$

Finally, the factorial symbol could have been used.[36]

$$\Gamma(\alpha) = (\alpha-1)!$$

$$\Gamma(2) = (2-1)! = 1! = 1$$

A table of gamma function values is given in App. 15.A.

Error Function and Complementary Error Function

The error function, also called the *probability integral*, was defined in Chap. 10 as

$$\text{erf}(x) = \frac{2}{\sqrt{\pi}}\int_0^x e^{-t^2}\,dt \qquad 15.44$$

$2/\sqrt{\pi}$ normalizes the function so that $\text{erf}(\infty) = 1$. By contrast, the *complementary error function* is defined as

$$\text{erfc}(x) = 1 - \text{erf}(x) = \frac{2}{\sqrt{\pi}}\int_x^\infty e^{-t^2}\,dt \qquad 15.45$$

The error function is shown graphically in Fig. 15.6.

[34]The impulse function is a powerful analytical tool. For example, electrical signals can be represented as weighted sums of shifted impulses—essentially this can be the interpretation of Eq. 15.41. Using this interpretation, the response of a *linear time invariant* (LTI) system can be analyzed for any arbitrary input, including *discrete* inputs, that is, those inputs that are not continuous. (Recall that the term "linear" indicates that superposition can be used; "time invariant" means a time shift in the input signal causes an identical time shift in the output signal.) Using the principle of superposition, one can then obtain the output of an electrical circuit to any input. Using the unit finite impulse function, one can do the same for continuous time systems.

[35]Also defined in terms of the gamma function are the *beta function* and *incomplete gamma functions*.

[36]An important value of gamma is $\Gamma(1/2) = \sqrt{\pi}$.

Figure 15.6 Error Function

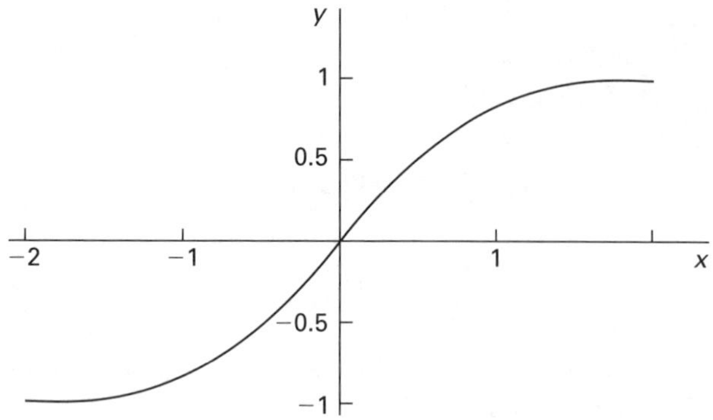

Bessel Functions of the First Kind

A Bessel differential equation of order ν is given by

$$x^2y'' + xy' + (x^2 - \nu^2)y = 0 \qquad 15.46$$

A particular solution is given by the Bessel function of the first kind of order ν. (This result is obtained using the power series method.)

$$J_\nu(x) = x^\nu \sum_{m=0}^{\infty} \frac{(-1)^m x^{2m}}{2^{2m+\nu} m!\Gamma(\nu + m + 1)} \qquad 15.47$$

Integer values of ν are often denoted by n. The functions J_0 and J_1 are important in many engineering problems. These functions are shown graphically in Fig. 15.7.

A table of values for $J_0(x)$ and $J_1(x)$ is given in App. 15.B.

Figure 15.7 Bessel Functions of the First Kind

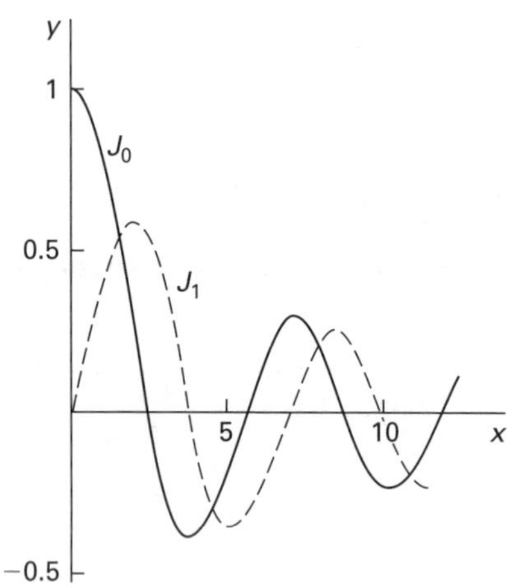

Bessel Functions of the Second Kind

If the value of ν or n is zero in a Bessel differential equation and one divides by the variable x, the result is

$$xy'' + y' + xy = 0 \qquad 15.48$$

The standard particular solution is known as the *Bessel function* of the second kind of order zero and is given by the following equation. This is also called *Neumann's function* of order zero.

$$Y_0(x) = \frac{2}{\pi}\left(\begin{array}{l} J_0(x)\left(\ln\frac{x}{2} + \gamma\right) \\ \quad + \sum\left(\frac{(-1)^{m-1}h_m}{2^{2m}(m!)^2}\right)x^{2m} \end{array}\right) \qquad 15.49$$

The constant γ is Euler's constant (also given as C_E, C, and $\ln\gamma$) which is defined as

$$\begin{aligned}\gamma &= \lim_{s\to\infty}\left(1 + \frac{1}{2} + \cdots + \frac{1}{s} - \ln s\right) \\ &\approx 0.577\end{aligned} \qquad 15.50$$

The variable h_m is defined as

$$h_m = 1 + \frac{1}{2} + \cdots + \frac{1}{m} \quad [m = 1, 2, 3 \ldots] \qquad 15.51$$

The second standard solution to Eq. 15.48 for all ν is given by

$$\begin{aligned}Y_\nu(x) &= \left(\frac{1}{\sin\nu\pi}\right) \\ &\quad \times \left(J_\nu(x)\cos\nu\pi - J_{-\nu}(x)\right)\end{aligned} \qquad 15.52$$

$$Y_n(x) = \lim_{\nu\to n} Y_\nu(x) \qquad 15.53$$

Equation 15.52 and Eq. 15.53 are known as *Bessel functions* of the second kind of order ν. These are also called *Neumann's functions* of order ν and are sometimes given as $N_\nu(x)$. They are also called *Weber's functions* in some texts. These functions are shown graphically in Fig. 15.8.

Bessel Functions of the Third Kind

Figure 15.8 Bessel Functions of the Second Kind

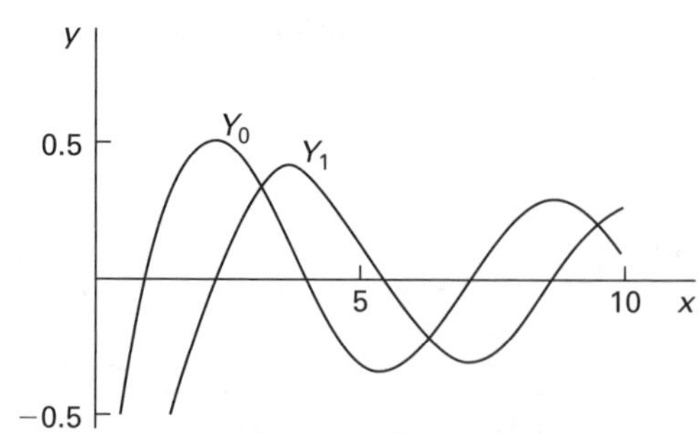

A general solution to Bessel's equation for all values of ν is given by the following formula.

$$y(x) = c_1 J_\nu(x) + c_2 Y_\nu(x) \qquad 15.54$$

Sometimes a solution needs to be complex for real values of x. In such cases, solutions take on the following form.

$$H_\nu^{(1)}(x) = J_\nu(x) + jY_\nu(x) \quad \text{15.55}$$

$$H_\nu^{(2)}(x) = J_\nu(x) - jY_\nu(x) \quad \text{15.56}$$

These linearly independent functions are called *Bessel functions* of the third kind of order ν. They are also called the *first and second Hankel functions* of order ν.

Hyperbolic Bessel Functions

Hyperbolic Bessel functions are given by Eq. 15.57.

$$I_m(x) = j^{-m}J_m(jx) \quad \text{15.57}$$

10. CONTINUOUS-TIME SYSTEMS: FOURIER SERIES AND TRANSFORMS

Fourier series are used to analyze periodic signals. Fourier integrals, also called Fourier transforms, are used to analyze aperiodic signals. The power of Fourier analysis is that if the input signal can be expressed in terms of periodic complex exponentials or sinusoids, then the output can be expressed in the same form, simplifying analysis. Further, representing otherwise intractable equations as a periodic Fourier series allows for analysis in ordinary and partial differential equations. This important concept is shown in the following example.

Example 15.7

Show that the response of a linear time-invariant system (LTI) to a complex exponential input is the same complex exponential with a change in amplitude.[37] In other words, show[38]

$$e^{st} \rightarrow H(s)e^{st}$$

The term $H(s)$ is the complex amplitude factor and in general will be a function of the complex variable s.

Solution

Assume an input signal of the form $x(t) = e^{st}$ provided to an LTI circuit with impulse response $h(t)$.[39]

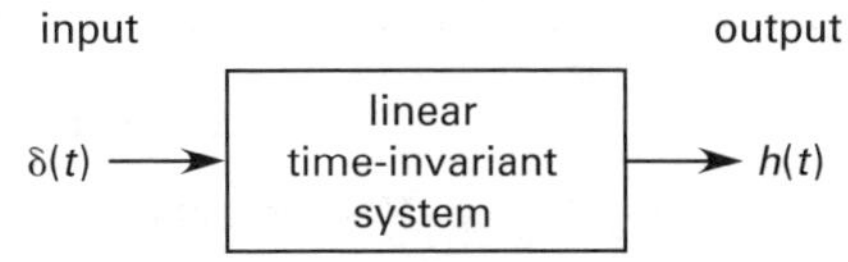

(a)
LTI system with unit impulse response $h(t)$

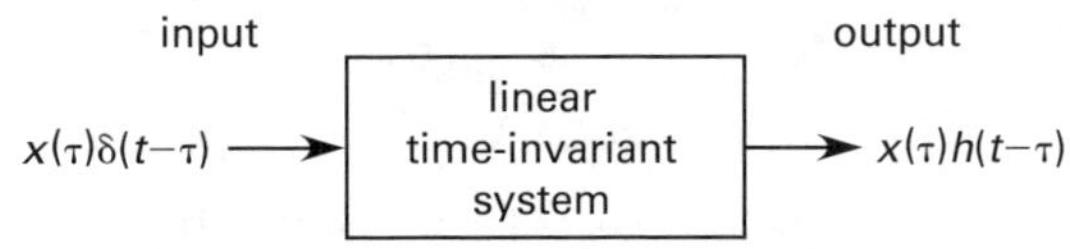

(b)
LTI system response to an impulse at time τ with amplitude $x(\tau)$

Writing the output $y(t)$ in terms of the convolution integral gives the following equation. (Recall that the convolution integral takes two functions defined in the s domain, in this case the impulse and the input signal $x(t)$, and gives the output in the time domain t directly without determining the inverse Laplace transform.)

$$y(t) = \int_{-\infty}^{+\infty} h(t-\tau)x(\tau)\,d\tau = \int_{-\infty}^{+\infty} h(\tau)e^{-s\tau}e^{st}\,d\tau$$

Because the term e^{st} does not affect the integral, it can be moved outside the integral.

$$y(t) = e^{st}\int_{-\infty}^{+\infty} h(\tau)e^{-s\tau}\,d\tau$$

This can be rewritten as

$$y(t) = e^{st}H(s) = H(s)e^{st}$$

$$H(s) = \int_{-\infty}^{+\infty} h(\tau)e^{-s\tau}\,d\tau$$

Any complex exponential is an eigenfunction of an LTI system. The constant $H(s)$ for a specified value s is the eigenvalue of the eigenfunction e^{st}.[40] In practical terms, what this means can best be illustrated as follows.

Consider an input signal of the form

$$x(t) = a_1e^{s_1t} + a_2e^{s_2t} + a_3e^{s_3t}$$

The response to each term separately is

$$a_1e^{s_1t} \rightarrow a_1H(s_1)e^{s_1t}$$

$$a_2e^{s_2t} \rightarrow a_2H(s_2)e^{s_2t}$$

$$a_3e^{s_3t} \rightarrow a_3H(s_3)e^{s_3t}$$

[37] "Linear" indicates that superposition applies. "Time invariant" indicates that a shift in the input signal $x(t)$ to $x(t-\tau)$ causes a shift in the output signal $y(t)$ to $y(t-\tau)$. When such a shift occurs in an equation, the equation is referred to as a *difference equation* and, as will be shown, represents a discrete time system.

[38] This is equivalent to showing the same thing for sinusoids, because from Euler's relation $e^{j\omega t} = \cos\omega t + j\sin\omega t$.

[39] The impulse response $h(t)$ occurs due to an input $\delta(t)$. This is simply a notation, which can vary considerably.

[40] The value $H(s)$ is determined by the electrical circuit and represents the circuit's response to an impulse $\delta(t-\tau)$. $H(s)$, with $s = j\omega$, is the Fourier transform of the response of the circuit to an impulse signal. Generally written as $H(\omega)$, it is termed the *frequency response* of the system.

Because, by the definition of "linear," superposition applies, the output $y(t)$ is

$$y(t) = a_1 H(s_1) e^{s_1 t} + a_2 H(s_2) e^{s_2 t} + a_3 H(s_3) e^{s_3 t}$$

In general terms,

$$\begin{gathered} \text{input} \rightarrow \text{output} \\ \sum_k a_k e^{s_k t} \rightarrow \sum_k a_k H(s_k) e^{s_k t} \end{gathered}$$

One final note before considering how this applies to electrical circuits in specific cases: In general, s is complex and of the form $s = a + j\omega$. If s is restricted to be equal to $j\omega$, that is, $s = j\omega$, then the exponentials are of the form $e^{j\omega t}$ and Fourier transforms are used. If s is allowed to be of the more general form $s = a + j\omega$, then Laplace transforms are used.

11. CONTINUOUS-TIME SYSTEMS: FOURIER SERIES REPRESENTATION OF A PERIODIC SIGNAL

A *continuous-time signal* is one in which the independent variable is continuous, that is, the signal is defined for a continuum of values. Figure 15.9 shows an example of such a signal.

Figure 15.9 *Continuous-Time Periodic Signal*

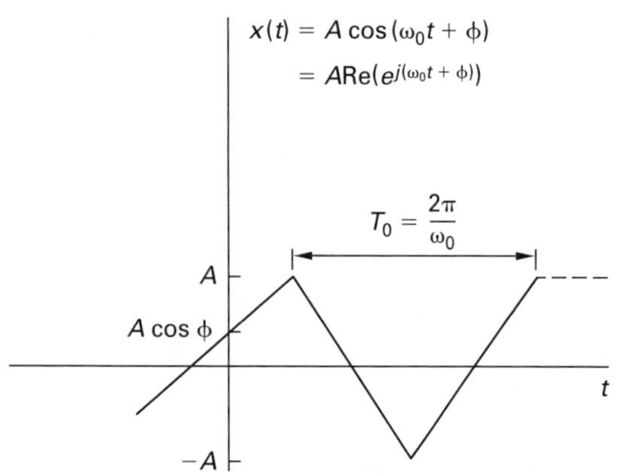

In Fig. 15.9, the values T_0 and ω_0 represent the *fundamental period* and *fundamental frequency*, respectively.

That is, they represent the minimum positive, nonzero value for which $x(t) = x(t + T)$ is true for a periodic function.

The Fourier series representation for any continuous time periodic signal is

$$x(t) = \sum_{k=-\infty}^{+\infty} a_k e^{jk\omega_0 t} \qquad 15.58$$

$$a_k = \frac{1}{T_0} \int_{T_0} x(t) e^{-jk\omega_0 t} dt \qquad 15.59$$

There are several important items:

1. There are other representations of a Fourier series. (See Sec. 10.15 for an example.) This complex exponential form is very useful in electrical analysis. It is called both the *complex Fourier series* and the *exponential Fourier series.*

2. The input term $x(t)$ is real and has the form of a cosine function.[41]

3. The value of $x(t)$ when $k=0$ represents the DC or constant value.

4. Fourier analysis is the determination of coefficients using the *analysis equation*, Eq. 15.59, to obtain the result.

5. Item (4) restated is the crux of this concept. The output $y(t)$ for a circuit is the input $x(t)$ represented as a Fourier series (see Eq. 15.58 and Eq. 15.59) multiplied by an amplitude factor $H(s)$. The amplitude factor $H(s)$ is determined by the circuit through which the given input (i.e., the signal) is processed.[42] The amplitude factor $H(s)$ is the circuit's response, $h(t-\tau)$, to an impulse, $\delta(t-\delta)$.[43] Some properties of the Fourier series for periodic signals are given in App. 15.C.

Example 15.8

Represent $x(t)$ from Fig. 15.9 as a Fourier series.

Solution

One could apply the analysis equation given in Eq. 15.59. However, in this simple case, it is easier to expand the equation in terms of complex exponentials and identify the Fourier series coefficients by inspection. Specifically, the cos $(\omega_0 t + \phi)$ function can be expressed as

$$\cos(\omega_0 t + \phi) = \tfrac{1}{2} e^{j(\omega_0 t + \phi)} + \tfrac{1}{2} e^{-j(\omega_0 t + \phi)}$$

[41]This is because the summation is from $-\infty$ to $+\infty$ and $a_k = a_{-k}$ (sometimes written $a_k = a^*_{-k}$), so the imaginary terms cancel. Hint: Use Euler's relation and trigonometric identities on the first few terms as a proof.

[42]This may seem rather circular. In different terms, if one represents the input as a Fourier series, that is, as Eq. 15.58, and then replaces the a_k with the Fourier coefficients (also called the *spectral coefficients*) as given by Eq. 15.59, the output is obtained by multiplying by the Fourier transform of the impulse input signal (given by $H(s)$ earlier in this section). The s is restricted to $s = j\omega$ when using the Fourier transform.

[43]When using the Fourier transform, the s is limited to $j\omega$ and $H(s)$ is the Fourier transform of $h(t-\tau)$. If s has both real and imaginary parts, that is, $s = a + j\omega$, $H(s)$ is the Laplace transform of $h(t-\tau)$.

In the first term $k = 1$, therefore

$$a_k = a_1 = \tfrac{1}{2}$$

In the second term $k = -1$, therefore

$$a_k = a_{-1} = \tfrac{1}{2}$$

For all other values of k, $a_k = 0$. The function in terms of a Fourier series can be represented as

$$x(t) = \sum_{-\infty}^{+\infty} a_k e^{jk(\omega_0 t+\phi)} = \tfrac{1}{2}\, e^{j(\omega_0 t+\phi)} + \tfrac{1}{2} e^{-j(\omega_0 t+\phi)}$$

The shift of $+\phi$ is the equivalent of multiplying by $e^{j\phi}$.

To confirm the result, use Euler's relation to expand the Fourier series and obtain the original equation. That is,

$$\begin{aligned} x(t) &= \tfrac{1}{2}\, e^{j(\omega_0 t+\phi)} + \tfrac{1}{2} e^{-j(\omega_0 t+\phi)} \\ &= \tfrac{1}{2}\Big(\cos(\omega_0 t+\phi) + j\sin(\omega_0 t+\phi)\Big) \\ &\quad + \tfrac{1}{2}\begin{pmatrix} \cos\Big(-(\omega_0 t+\phi)\Big) \\ + j\sin\Big(-(\omega_0 t+\phi)\Big) \end{pmatrix} \end{aligned}$$

Applying the trigonometric identities $\sin(-\alpha) = -\sin\alpha$ and $\cos(-\alpha) = \cos\alpha$ gives

$$\begin{aligned} x(t) &= \tfrac{1}{2}\Big(\cos(\omega_0 t+\phi) + j\sin(\omega_0 t+\phi)\Big) \\ &\quad + \tfrac{1}{2}\Big(\cos(\omega_0 t+\phi) - j\sin(\omega_0 t+\phi)\Big) \\ &= \cos(\omega_0 t+\phi) \end{aligned}$$

12. CONTINUOUS-TIME SYSTEMS: FOURIER TRANSFORM OF AN APERIODIC SIGNAL

Many signals of interest to an electrical engineer are not periodic, that is, they are *aperiodic*. The Fourier series representation can still be utilized if one considers the aperiodic signal as the limit of a periodic signal as the period becomes arbitrarily large.[44] The result is[45]

$$x(t) = \frac{1}{2\pi}\int_{-\infty}^{+\infty} X(\omega)\, e^{j\omega t}\, d\omega \qquad 15.60$$

$$X(\omega) = \int_{-\infty}^{+\infty} x(t)\, e^{-j\omega t}\, dt \qquad 15.61$$

[44]This is equivalent to saying $\omega_0 \to 0$ because $\omega_0 = 2\pi/T_0$. In other words, no periodicity exists.

[45]The derivation of most of the formulas used in this book will not be given unless they enhance the understanding of a fundamental principle. The electrical engineer is, after all, concerned with the results and how to utilize these results.

Equation 15.60 and Eq. 15.61 are referred to as a *Fourier transform pair*. Equation 15.61 is the *Fourier transform* or *Fourier integral*. Equation 15.60 is the *inverse Fourier transform* that is the *synthesis equation* of the pair.[46]

Some properties of the Fourier transform for aperiodic signals are given in App. 15.D. Some Fourier transform pairs are given in App. 15.E.

The arrangement for aperiodic signals is analogous to that for periodic functions. Equation 15.60 correlates with Eq. 15.58. The complex exponentials formed by the synthesis equations have amplitudes represented by the coefficients a_k given by Eq. 15.59 in the periodic case—representing a discrete set of harmonically related frequencies $k\omega_0$ with $k = 0,\ \pm 1,\ \pm 2, \ldots$ For aperiodic signals, Eq. 15.60 represents the linear combination of complex exponentials with amplitudes $(X(\omega)/2\pi)\,d\omega$, with $X(\omega)$ given by Eq. 15.61. Examples of Fourier transformations are shown in Fig. 15.10. (Obviously, the sine and cosine functions shown in the figure are not aperiodic, but rather periodic. The Fourier transform of a periodic signal with Fourier coefficients can be

Figure 15.10 *Fourier Transformations*

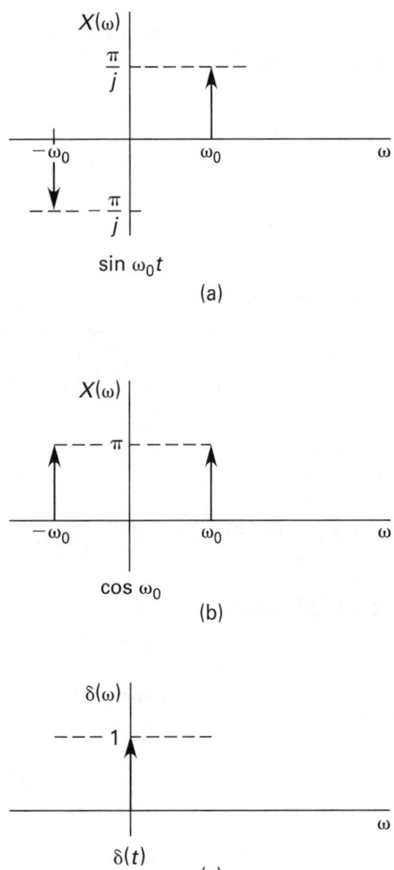

[46]This is because the inverse Fourier transform decomposes a signal into a linear combination of complex exponentials similar to the function of Eq. 15.58.

thought of as a train of impulses occurring at harmonically related frequencies for which the area of the impulse at the kth harmonic frequency ($k\omega_0$) is 2π times the kth Fourier series coefficient a_k. All the calculations in the aperiodic section apply.)

Fourier transforms are a useful tool for the analysis of continuous-time systems that are linear time-invariant. (The broad range of circuitry using resistors, capacitors and inductors are linear time-invariant.) The output signals of such circuits are related proportionally to the input signals by an amplitude factor.

In mathematical terms, an input signal $x(t)$ can be represented by an infinite number of pulses, summed as shown in Eq. 15.62.

$$x(t) = \int_{-\infty}^{+\infty} x(\tau)\delta(t-\delta)\,d\tau \qquad 15.62$$

The output signal is then given by the following equation.[47]

$$y(t) = \int_{-\infty}^{+\infty} x(\tau)h(t-\tau)\,d\tau \qquad 15.63$$

Equation 15.63 is of the form of a convolution integral (see Sec. 15.13). Using data from App. 15.D, Eq. 15.63 can be written as[48]

$$\begin{aligned} y(t) &= \int_{-\infty}^{+\infty} x(\tau)h(t-\tau)\,d\tau \\ &= \mathcal{F}^{-1}\big(X(\omega)H(\omega)\big) \end{aligned} \qquad 15.64$$

Herein lies the power of the Fourier transform: the often difficult convolution can be completely avoided. Convert the input signal $x(t)$ into the frequency domain as $X(\omega)$. Convert the impulse response $h(t)$ into $H(\omega)$. The convolution of $x(t)$ and $h(t)$ is simply the product of $X(\omega)$ and $H(\omega)$, and the output $y(t)$ is realized by applying the inverse Fourier transform to the result. In other words, a convolution in the time domain is multiplication in the frequency domain.[49]

13. CONVOLUTION

The convolution property of a given transform T is one that gives the inverse transform $T^{-1}\{X(\tau)H(\tau)\}$ directly in terms of the original functions $x(t)$ and $h(t)$.[50] This was evident in Eq. 15.64 where the inverse transform $T^{-1}\{X(\tau)H(\tau)\}$, that is, $y(t)$, was given in terms of the original functions $x(t)$ and $h(t)$.[51] A generalized form of the *convolution integral* is

$$\begin{aligned} f(t) * g(t) &= \int_a^b f(\tau)g(t-\tau)\,d\tau \\ &= T^{-1}\{F(s)G(s)\} \end{aligned} \qquad 15.65$$

When the convolution property is invoked for a Fourier transform, the limits of integration a and b become $-\infty$ and $+\infty$, respectively. Also, s becomes ω (or $j\omega$ in some texts) and $G(\omega)$ represents $H(\omega)$. ($H(\omega)$ is the Fourier transform of the system impulse response, or the frequency response, for LTI circuits as discussed in previous sections of this chapter.) When invoked for a Laplace transformation, a and b become 0 and $+\infty$, respectively, and s is generalized to real and imaginary components, $s=\sigma+j\omega$. (This essentially gives the properties of magnitude and phase.)

Convolution properties follow. The variable t has been dropped in Eq. 15.66 through Eq. 15.69 as convolution can be used with more than one variable.

$$f*g = g*f \quad \text{[commutative law]} \qquad 15.66$$

$$f*(g_1+g_2) = f*g_1 + f*g_2 \quad \text{[distributive law]} \qquad 15.67$$

$$(f*g)*v = f*(g*v) \quad \text{[associative law]} \qquad 15.68$$

$$f*0 = 0*f = 0 \qquad 15.69$$

Convolution is used to linearly filter a signal. It is used in electrical engineering (e.g., signal processing) because of its relationship to Fourier transforms. Specifically, the convolution of two functions in the time domain equals the multiplication of the transforms in the frequency domain, as shown in Eq. 15.65. In many problems, this property simplifies calculations. *Correlation* is similar to convolution in mathematical form, and is used to display the interdependence or association of two variables. It also is used in signal processing (e.g., image processing), probability and statistics, and areas of engineering

[47]Here no change in the input signal representation occurs. When using the Fourier series of $x(t)$ in the process of obtaining $y(t)$, such a change does occur. Though the transform method is more straightforward, the Fourier series representation has numerous applications and was therefore presented in depth in Sec. 15.11.

[48]Some texts write $X(\omega)$ and $H(\omega)$ as $X(j\omega)$ and $H(j\omega)$. The j is normally understood to be present.

[49]For reference, the transformation is from the domain of k-space, which is referred to as the *frequency domain* when ω replaces k.

[50]The dummy variable τ is used to represent the transform and can be any transform: Fourier, Fourier sine, Fourier cosine, Laplace and others.

[51]It is often easier to find the result using convolution rather than applying the inverse transform to $X(\omega)H(\omega)$.

requiring such applications, for example, communications. The autocorrelation function of a signal $f(t)$ is

$$\phi_{fg}(t) = \int_{-\infty}^{+\infty} f(t+\tau)g(\tau)\,d\tau \qquad 15.70$$

14. LAPLACE TRANSFORMS

The Laplace transform, whose properties and uses are given in Chap. 11, is repeated here for convenience.

$$\mathcal{L}(f(t)) = F(s) = \int_{0}^{\infty} e^{-st} f(t)\,dt \qquad 15.71$$

Equation 15.71 is more specifically termed the *unilateral Laplace transformation.* The primary use of this transform is in the solution of linear constant-coefficient differential equations with initial conditions. (For an example, see Sec. 11.13.)

By contrast, the *bilateral Laplace transform* of a general signal $f(t)$ is given by Eq. 15.72.

$$\mathcal{L}(f(t)) = F(s) = \int_{-\infty}^{+\infty} e^{-st} f(t)\,dt \qquad 15.72$$

The unilateral transform is more commonly encountered. The term unilateral is not often used. The differentiation between the two types is simply shown in the limits of integration. From the bilateral Laplace transform, the relationship between it and the Fourier transform is easily seen.

$$F(s)\big|_{s=j\omega} = \mathcal{F}\big(f(t)\big) \qquad 15.73$$

If s is not purely imaginary, the relationship is given mathematically as

$$F(\sigma + j\omega) = \int_{-\infty}^{+\infty} \left(f(t)e^{-\sigma t}\right) e^{-j\omega t}\,dt \qquad 15.74$$

In other words, the Laplace transform of a function with $s = \sigma + j\omega$ is the Fourier transform of $f(t)$ multiplied by the real exponential $e^{-\sigma t}$. Properties of the Laplace transform as well as some common transforms themselves are given in App. 15.D and App. 15.G. These appendices include the *region of convergence* (ROC or R in some texts), that is, the restrictions necessary to make the transform valid, which will become important when the frequency response of electrical circuits is considered.

15. DISCRETE-TIME SYSTEMS: FOURIER SERIES AND TRANSFORMS[52]

A *discrete-time signal* is one in which the independent variable is defined only at discrete times. Figure 15.11 shows an example of such a signal.

Figure 15.11 *Discrete-Time Periodic Signal*

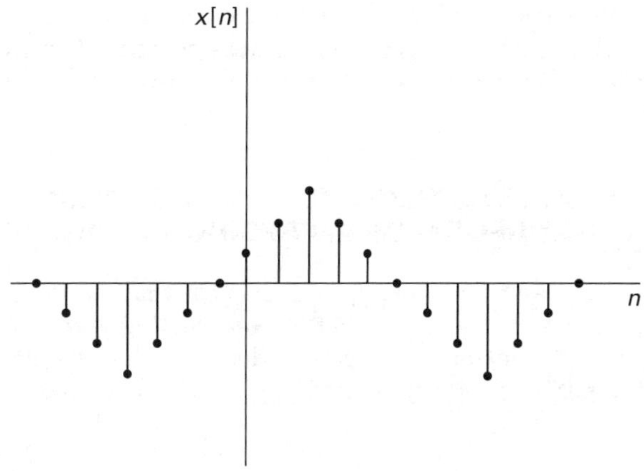

n replaces t as the variable in order to distinguish between the continuous-time and discrete-time cases. Additionally, brackets are used, $[n]$, to further distinguish discrete-time systems.

The response of an LTI system to inputs that can be represented as complex exponential sequences is analogous to continuous-time systems.[53] That is, in general terms,

$$x[n] = \sum_k a_k z_k^n \rightarrow y[n] = \sum_k a_k H[z_k] z_k^n \qquad 15.75$$

$$\text{input} \rightarrow \text{output}$$

Here $H[z_k]$, instead of being the Fourier or Laplace transform of the system impulse response $h(t)$, is the Fourier or the z-transform of the discrete time system impulse response $h[n]$. When $z = e^{j\Omega}$ with Ω real, $H[z_k]$ is the Fourier transform. When $z = e^{\sigma + j\Omega}$, $H[z_k]$ is the z-transform.[54]

16. DISCRETE-TIME SYSTEMS: FOURIER SERIES REPRESENTATION OF A PERIODIC SIGNAL

Similar to the continuous-time case, the Fourier series representation of a discrete-time periodic signal, of period N, is given by the following synthesis and analysis

[52]These systems, and their associated Fourier series and transforms, are sometimes called *finite* vice discrete.

[53]See the example in Sec. 15.10 for the form of the proof, as only the results will be presented here.

[54]This discrete transform, the z-transform, is discussed in Sec. 15.18. It is the discrete-time counterpart to the Laplace transform for continuous-time cases.

equations. (The symbol $\langle N \rangle$ indicates an interval of length N.)

$$k[n] = \sum_{k=\langle N \rangle} a_k e^{jk(2\pi/N)n} \qquad 15.76$$

$$a_k = \frac{1}{N} \sum_{n=\langle N \rangle} x[n] e^{-jk(2\pi/N)n} \qquad 15.77$$

The important difference in the Fourier series representation in this discrete-time case as compared to the continuous-time case is that it is finite (over the interval N).[55]

17. DISCRETE-TIME SYSTEMS: FOURIER TRANSFORM OF AN APERIODIC SIGNAL

Applying an analogous procedure to that used for continuous-time systems, that is, constructing a periodic signal from the aperiodic signal and then taking the limit as the period approaches infinity, gives the following.

$$x[n] = \frac{1}{2\pi} \int_{2\pi} x[\Omega] e^{j\Omega n} d\Omega \qquad 15.78$$

$$X[\Omega] = \sum_{n=-\infty}^{+\infty} x[n] e^{-j\Omega n} \qquad 15.79$$

The function given by Eq. 15.79 is called the *discrete-time Fourier transform* and the two equations together, Eq. 15.78 and Eq. 15.79, are termed the *Fourier transform pair.* As before, Eq. 15.78 is the synthesis equation and Eq. 15.79 is the analysis equation. The properties of the discrete-time Fourier series (used with periodic signals) are given in App. 15.H. Discrete-time Fourier transforms (used with aperiodic signals) are given in App. 15.I.

18. *z*-Transforms

The z-transform is the discrete-time counterpart to the Laplace transform. Also, it represents the generalization of the Fourier transform for discrete-time systems; that is, it is the transform to use when the value of s in a discrete-time system is $s = \sigma + j\Omega$.[56] The z-transform of a sequence is defined as[57]

$$X[z] = \sum_{n=-\infty}^{+\infty} x[n] z^{-n} \qquad 15.80$$

The variable z is complex. The relationship between the z-transform and the discrete-time Fourier transform is given by the following.

$$X[z]\big|_{z=e^{j\Omega}} = F[x[n]] \qquad 15.81$$

Some properties of the z-transform are given in App. 15.J. Some common z-transforms are given in App. 15.K.

19. TRANSFORMATION OF INTEGRALS

A *transform* takes on the following form.[58]

$$T\{f\} = \int_a^b f(k) k(x, \tau) dx = f(\tau) \qquad 15.82$$

Essentially one integrates out a variable, x in this case, and places the function in another *space* or *domain*, that of τ in this case. The symbol $T\{f\}$ indicates the transform of the function and $k(x, \tau)$ is referred to as the *kernal.* For the Fourier transform the kernal is $e^{j\omega_0}\tau$. For the Laplace transform, the kernal is $e^{-s\tau}$. Many others exist. A transform places the function in another *function space*, for example, k, ω, s, and so on.

By contrast, the *transformation of integrals* changes the way a physical system is viewed and has important consequences in electrical engineering. Such transformations follow.

Gauss' Divergence Theorem[59]

$$\iint_{A(V)} \mathbf{F} \cdot d\mathbf{A} = \iiint_{V(A)} \nabla \cdot \mathbf{F} dV \qquad 15.83$$

An arbitrary vector field is represented by $\mathbf{F}$. The symbol $A(V)$ indicates the surface area over the volume of concern. The $d\mathbf{A}$ indicates the normal unit differential surface area, and $V(A)$ is the volume bounded by the surface A. Equation 15.83 says the net flux through a closed surface equals the summation of the scaler sources inside; that is, it equals the divergence of the field integrated over the volume. The symbols

$$\iint_{A(V)} \text{ and } \iiint_{V(A)}$$

are sometimes represented as

$$\int_s \text{ or } \oint_S \text{ or } \iint_S \text{ or } \oint\!\!\oint_S$$

and

$$\int_V \text{ or } \iiint_V$$

[55]This finiteness is important because it is applied in the FFT or Fast Fourier Transform. The FFT algorithm is well-suited to computers allowing efficient analysis of signals using discrete methods, which prior to computers had been too time consuming.

[56]This means that s has both real and imaginary parts, that is, a magnitude and a phase.

[57]This is the *bilateral z-transform.* The *unilateral z-transform* is summed from $n = 0$ to $n = \infty$ much as the unilateral Laplace transform commonly used. Also, again similar to the unilateral Laplace transform, the unilateral z-transform is useful in solving for the output of circuits with a response given by constant coefficient difference equations with nonzero initial conditions.

[58]This is a general linear integral transformation; that is, superposition is applicable.

[59]More will be said on the electrical consequences and applications of this and other theorems later in this section.

Stokes' Theorem

Stokes' theorem is analogous to Gauss' theorem. It relates the curl of a vector field **F** inside a closed line, or contour, to the circulation—that is, the curl of **F** along that contour. See Fig. 15.12 for a physical picture of the theorem.[60]

Figure 15.12 Curl and Circulation of Vector Field

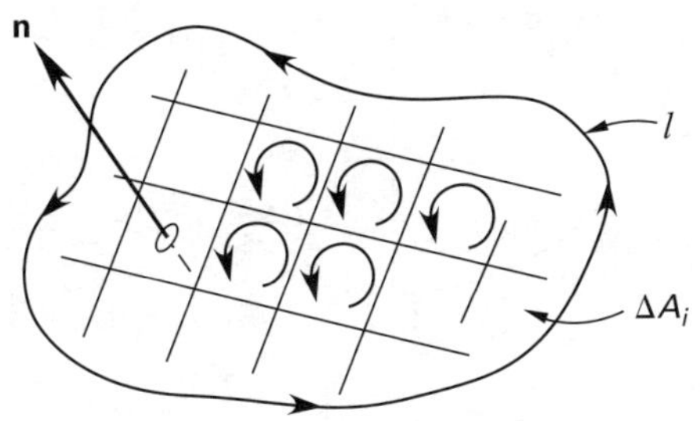

In mathematical terms, Stokes' theorem is

$$\oint_{l(A)} \mathbf{F}\cdot d\mathbf{l} = \iint_{A(l)} \nabla\times\mathbf{F}\cdot d\mathbf{A} \qquad 15.84$$

The symbols

$$\oint_{l(A)} \text{ and } \iint_{A(l)}$$

are sometimes represented as

$$\int_C \text{ and } \int_S$$

C indicates a *contour*, or *line*, *integral* and *S* indicates a *surface integral.*

Green's Theorem

Green's theorem is actually a special case of Stokes' theorem with the vector **F** in the same plane as the surface *A*. If $g(x, y)$ and $f(x, y)$ are functions defined in this plane, say x and y, then Green's theorem in mathematical terms is

$$\oint_{l(A)} (f(x,y)\,dx + g(x,y)\,dy) = \iint_{A(l)} \left(\frac{\partial g}{\partial x} - \frac{\partial f}{\partial y}\right) dx\,dy \qquad 15.85$$

This theorem transforms double integrals, that is, surface integrals, into line integrals. (This is similar to Gauss' theorem that transformed volume integrals, that is, triple integrals, into area integrals (double integrals).) Green's theorem is sometimes useful in simplifying the calculation of the integral.

20. LINE INTEGRALS

A line integral is a generalization of the concept of a definite integral. For example, a definite integral is given by the following.

$$\int_a^b f(x)\,dx \qquad 15.86$$

Let this integral apply to the curve in Fig. 15.13.

Figure 15.13 Simple Curve

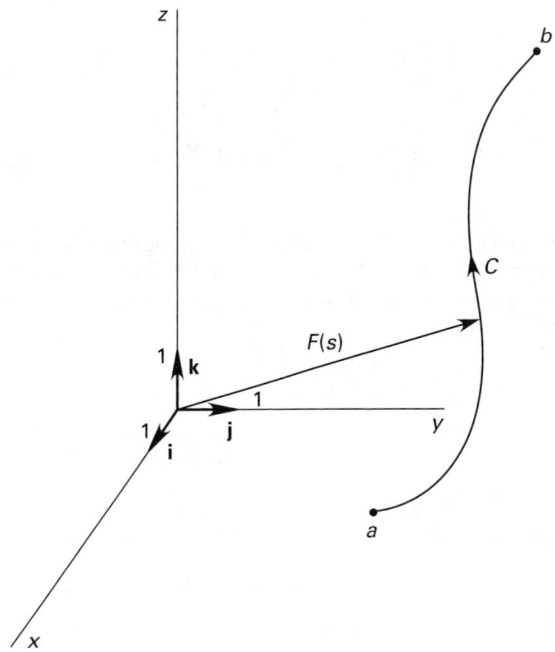

Because curve *C* is *simple*, that is, it has no double points, it may be represented in Cartesian coordinates as[61]

$$a \le s \le b$$
$$\mathbf{r}(s) = x(s)\mathbf{i} + y(s)\mathbf{j} + z(s)\mathbf{k} \qquad 15.87$$

The *line integral*[62] of *f* along *C* from *a* to *b* is

$$\int_C f(x,y,z)\,ds \qquad 15.88$$

The curve *C* is called the *path of integration.*

[60]A paddle wheel's rate of rotation can be thought of as measuring the curl of a turbulent fluid of velocity **F** with the curl in the direction of the paddle wheel axis, $\mathbf{n}dA = d\mathbf{A}$.

[61]Another way of defining a smooth curve is to say it has unique tangents at all points within the curve of the line.

[62]Probably a better name would be *curve integral* rather than line integral, but this term is not commonly used.

Figure 15.14 Closed Path

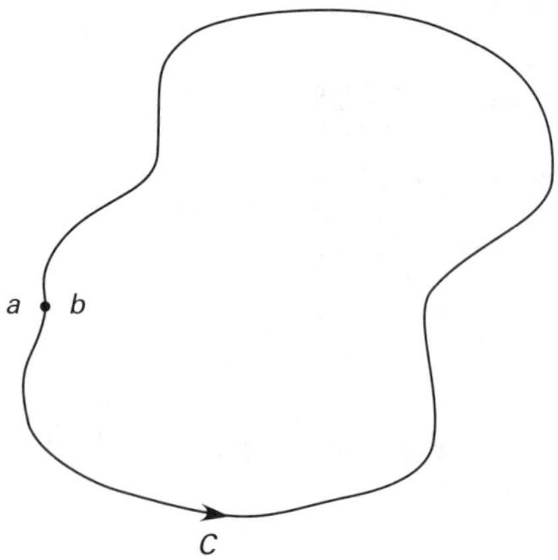

If the line integral is over a closed path as illustrated in Fig. 15.14, the integral would be written as

$$\oint_C f(x, y, z)\,ds \qquad 15.89$$

Line integrals are evaluated by changing them into definite integrals. The key is to represent the path of integration C in the desired form. One possible form is

$$\int_C f(x, y, z)\,ds = \int_a^b f\big(x(s), y(s), z(s)\big)\,ds \qquad 15.90$$

The value of $\mathbf{r}$ may also be represented as

$$\mathbf{r}(t) = x(t)\mathbf{i} + y(t)\mathbf{j} + z(t)\mathbf{k}$$
$$t_0 \le t \le t_1 \qquad 15.91$$

The variable t can be any parameter, not just time.

In this case, the line integral can be calculated using the following.

$$\int_C f(x, y, z)\,ds = \int_{t_0}^{t_i} f\big(x(t), y(t), z(t)\big)\frac{ds}{dt}\,dt \qquad 15.92$$

In this equation,

$$\frac{ds}{dt} = \sqrt{\dot{\mathbf{r}} \cdot \dot{\mathbf{r}}} = \sqrt{\dot{x}^2 + \dot{y}^2 + \dot{z}^2} \qquad 15.93$$

Example 15.9

Integrate the function $f(x, y) = 3xy^2$ over the portion of the circle from a to b as shown. The direction of C determines the positive direction and the integration path.

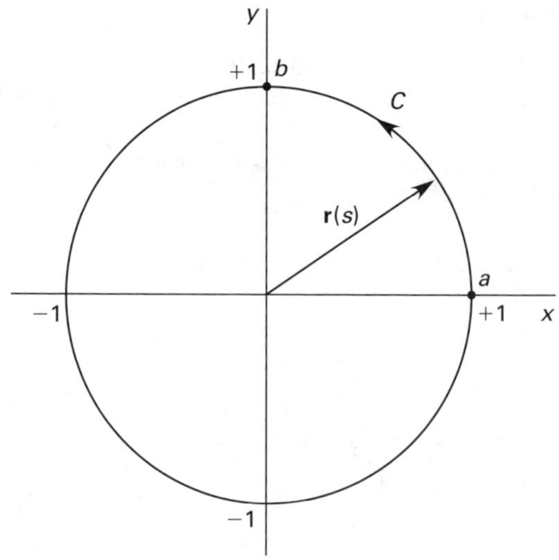

Solution

The value of $\mathbf{r}(s)$ can be seen as

$$\mathbf{r}(s) = \cos s\mathbf{i} + \sin\mathbf{j}$$

Using Eq. 15.87 gives

$$\int_C f(x, y)\,ds = \int_a^b f\big(x(s), y(s)\big)\,ds$$

$$\int_C 3xy^2\,ds = \int_0^{\pi/2} 3\cos s\sin^2 s\,ds = 3\int_0^{\pi/2} \cos s\sin^2 s\,ds$$

Using a change of variables, let $u = \sin s$. The integration becomes[63]

$$3\int_0^{\pi/2} \cos s\sin^2 s\,ds = 3\int_0^1 u^2\,du = 3\frac{u^3}{3}\bigg|_0^1 = 3\left(\tfrac{1}{3} - \tfrac{0}{3}\right) = 3\left(\tfrac{1}{3}\right) = 1$$

Example 15.10

Consider a particle upon which a variable force $\mathbf{p}$ acts along the curve C. The work done by the force $\mathbf{p}$ is given by the line integral

$$w = \int_C \mathbf{p}\cdot d\mathbf{r}$$

Show that this equals the gain in kinetic energy.

[63]At this point, one could choose to determine the integral from standard math tables.

Solution

First,

$$d\mathbf{r} = \frac{d\mathbf{r}}{dt}\,dt = \mathbf{v}\,dt$$

Substituting gives

$$w = \int_{t_0}^{t_1} \mathbf{p}\cdot\mathbf{v}\,dt$$

The values t_0 and t_1 are the initial and final values of dt. Newton's second law states that

$$\mathbf{p} = m\frac{d^2\mathbf{r}}{dt^2} = m\frac{d\mathbf{v}}{dt}$$

The variable m represents the mass. Substituting a second time gives

$$\begin{aligned} w &= \int_{t_0}^{t_1} \mathbf{p}\cdot\mathbf{v}\,dt = \int_{t_0}^{t_1}\left(m\frac{d\mathbf{v}}{dt}\right)\cdot\mathbf{v}\,dt \\ &= \int_{t_0}^{t_1} \frac{d}{dt}(m\mathbf{v}\cdot\mathbf{v})\,dt = \int_{t_0}^{t_1} dm|\mathbf{v}|^2 \\ &= \frac{m}{2}|\mathbf{v}|^2\Big|_{t_0}^{t_1} \end{aligned}$$

This is the desired result.

The value of a line integral of a given function depends not only on the endpoints but also on the shape of the integration path.

21. COMPLEX NUMBERS

Two additional important items not covered in Chap. 4 regarding complex numbers are now considered.

Let any complex number z be represented by

$$z = x + jy \qquad 15.94$$

The z is not bolded as in Chap. 4 to avoid confusion with vector variables. Although it is somewhat similar, it is a point and not a direction. Further, j is used instead of i to represent $\sqrt{-1}$ to avoid confusion with the variable for current.

The complex conjugate is

$$\overline{z} = z^* = x - jy \qquad 15.95$$

The real portion (Re or $\mathcal{R}$) of a complex number z is

$$\text{Re}(z) = x = \frac{1}{2}(z + z^*) \qquad 15.96$$

The imaginary part (Im or $\mathcal{I}$) of a complex numberz is

$$\text{Im}(z) = y = \frac{1}{2j}(z - z^*) \qquad 15.97$$

Also, the familiar formulas for real trigonometric functions hold for complex values.

Example 15.11

Express the Euler formula for complex values in terms of real functions. The Euler formula is

$$e^{j\theta} = \cos\theta + j\sin\theta$$

Solution

If θ is complex, and representing θ as z (which is standard in electrical engineering symbology), gives

$$e^{jz} = \cos z + j\sin z$$

Using a trigonometric identity, the individual terms are

$$\begin{aligned} \cos z &= \cos(x + jy) \\ &= \cos x\cos jy - \sin x\sin jy \\ \sin z &= \sin(x + jy) \\ &= \sin x\cos jy + \cos x\sin jy \end{aligned}$$

Because the familiar formulas apply, the hyperbolic cosine and sine for a complex number are

$$\begin{aligned} \cos z &= \frac{1}{2}(e^{jz} + e^{-jz}) \\ \sin z &= \frac{1}{2j}(e^{jz} - e^{-jz}) \end{aligned}$$

This allows the following representation.

$$\begin{aligned} \cos jy &= \tfrac{1}{2}(e^{j(jy)} + e^{-j(jy)}) \\ &= \tfrac{1}{2}(e^{-y} + e^{+y}) = \cosh y \end{aligned}$$

Also,

$$\begin{aligned} \sin jy &= \frac{1}{2j}\left(e^{j(jy)} - e^{-j(jy)}\right) \\ &= \frac{1}{2j}(e^{-y} - e^{+y}) = j\sinh y \end{aligned}$$

Substituting gives

$$\begin{aligned} \cos z &= \cos(x + jy) \\ &= \cos x\cosh y - j\sin x\sinh y \\ \sin z &= \sin(x + jy) \\ &= \sin x\cosh y + j\cos x\sinh y \end{aligned}$$

It is this real representation of the terms that is useful in the actual calculation of $\cos z$ and $\sin z$.

The theory of *complex analytic functions* can be useful in the study of potential theory in electrostatics. The development of such functions will not be presented but is mentioned to place them in the context of electrical engineering. The results of such calculations are shown as force field lines, or potential lines, in field theory.

Topic II: Basic Theory

Chapter

16. Electromagnetic Theory
17. Electronic Theory
18. Communication Theory
19. Acoustic and Transducer Theory

16 Electromagnetic Theory

Part 1: Fundamentals 16-2

Part 2: Electric and Magnetic Phenomena . . 16-3
1. Electromagnetic Effects 16-3
2. Conduction Effects 16-4
3. Dielectric Phenomena 16-4
4. Magnetic Phenomena 16-4
5. Thermoelectric Phenomena 16-5

Part 3: Electrostatics 16-5
6. Electric Charges 16-5
7. Coulomb's Law 16-6
8. Electric Fields 16-7
9. Permittivity and Susceptibility 16-9
10. Electric Flux 16-9
11. Electric Flux Density 16-10
12. Gauss' Law for Electrostatics 16-10
13. Capacitance and Elastance 16-11
14. Capacitors 16-11
15. Energy Density in an Electric Field 16-12

Part 4: Electrokinetics 16-12
16. Speed and Mobility of Charge Carriers 16-12
17. Current 16-13
18. Convection Current 16-13
19. Displacement Current 16-13
20. Conduction Current 16-14

Part 5: Magnetostatics 16-14
21. Magnetic Poles 16-14
22. Biot-Savart Law 16-15
23. Magnetic Fields 16-16
24. Permeability and Susceptibility 16-17
25. Magnetic Flux 16-17
26. Magnetic Field Strength 16-17
27. Gauss' Law for Magnetic Flux 16-18
28. Inductance and Reciprocal Inductance 16-18
29. Inductors 16-19
30. Energy Density in a Magnetic Field 16-19

Part 6: Magnetokinetics 16-20
31. Speed and Direction of Charge Carriers . . . 16-20
32. Voltage and the Magnetic Circuit 16-21
33. Magnetic Field-Induced Voltage 16-22
34. Eddy Currents 16-22
35. Displacement Current Magnetic Effect 16-22
36. Magnetic Hysteresis 16-22

Nomenclature

a	acceleration	m/s^2
a	unit vector	–
A, **A**	area	m^2
B, **B**	magnetic flux density	T or Wb/m^2
c	speed of light, 2.9979×10^8	m/s
C	capacitance	F
d	distance	m
D, **D**	electric flux density	C/m^2
e^-	charge on an electron, -1.6022×10^{-19}	C
E, **E**	electric field strength	V/m
F, **F**	force	N
G	conductance	S or A/V
G	universal gravitational constant, 6.672×10^{-11}	$N{\cdot}m^2/kg^2$
h	Planck's constant, 6.6256×10^{-34}	J·s
H, **H**	magnetic field strength	A/m
i	variable current	A
I, **I**	constant or rms current	A
J, **J**	current density	A/m^2
k	electrostatic proportionality constant, 8.987×10^9	$N{\cdot}m^2/C^2$
l, **l**	length	m
ℓ	azimuthal quantum number	–
L	inductance	H
m	mass	kg
m	spatial quantum number	–
m_s, s	spin quantum number	–
M, **M**	magnetization	A/m
M	mutual induction	H
n	principal quantum number	–
N	number of turns	–
p	pole strength	Wb or N/T
p^+	charge on a proton, 1.6022×10^{-19}	C
P	polarization	C/m^2
q, Q	charge	C
r, **r**	radius	m
R	ratio	–
R	resistance	Ω
$\mathcal{R}$	reluctance	A/Wb
s, **s**	surface area	m^2
S	elastance	1/F
t	thickness	m
t	time	s
u	energy density	J/m^3
U	energy	J
v	variable voltage	V
v, **v**	velocity[1]	m/s
V	constant or rms voltage	V
V_{volume}	volume	m^3

[1]The term "velocity" is often used in text where it would be more technically correct to use the term "speed." *Velocity* is a vector quantity with speed as its magnitude.

Symbols

Γ	reciprocal inductance	1/H
ϵ	permittivity	$C^2/N{\cdot}m^2$ or F/m
ϵ_0	free-space permittivity, 8.854×10^{-12}	$C^2/N{\cdot}m^2$ or F/m
θ	angle	degrees
Λ	flux linkage	Wb
μ	mobility	$m^2/V{\cdot}s$
μ	permeability	N/A^2 or H/m
μ_0	free-space permeability, 1.2566×10^{-6}	N/A^2
ρ	charge density	C/m^3
σ	conductivity	S·m
ϕ, Φ	magnetic flux	Wb
χ	susceptibility	–
ψ, Ψ	electric flux	C
ω	angular velocity	rad/s

Subscripts

1–2	from 1 to 2; or effect on 2 caused by the field of 1
ave	average
c	conduction
ci	cast iron
d	displacement drift
e	electric
l	line
m	magnetic
m	mean
r	relative

PART 1: FUNDAMENTALS

As a general rule, electrical effects emanate from static charges, while magnetic effects emanate from charges in motion. To be more precise, *electricity* is any manifestation of energy conversion of charge carriers that results in forces in the direction of motion of those charge carriers.[2] *Magnetism* is any manifestation of the kinetic energy of charge carriers that arises from forces or produces forces in the direction perpendicular to the motion of those charge carriers. The two were long thought to be separate phenomena, but were united by Maxwell's equations.[3] *Electromagnetism* deals with control of the average movement of charges, in generally linear elements, and usually at considerable power levels. For comparison, see the description of electronics in Sec. 17.1.

The charges of interest are the electron and the proton, designated as e^- and p^+. The mass, charge, and charge-to-mass ratios for the proton and electron are given in Table 16.1. The SI unit of charge is the coulomb (C). The charge of one electron is sometimes referred to as one *electrostatic unit* (esu). One C is approximately equal to 6.24×10^{18} esu. One mole of electrons is referred to as a *faraday.* One faraday is approximately equal to 96 500 C.

Table 16.1 *Properties of the Electron and Proton*

	electron	proton
mass at rest, kg	9.1096×10^{-31}	1.6726×10^{-27}
charge, C	-1.6022×10^{-19}	$+1.6022 \times 10^{-19}$
charge-to-mass ratio, C/kg	1.7588×10^{11}	9.5791×10^{7}

The *extranuclear structure* of an atom, that is, its electrical or electronic structure, determines the defining characteristics of that atom. Orbiting electrons are arranged in shells designated *K, L, M, N, O, P,* and *Q.* The inner shell electrons are tightly bound and interact only with high-energy particles, such as gamma rays. The outer shells in complex atoms—those with numerous neutrons and protons, and thus numerous electrons—tend to be loosely bound. It is these outer shells that determine the electrical and chemical properties of the elements. The energy of the orbiting electron is determined by four quantum numbers: the *principle quantum number* (n), the *azimuthal quantum number* (ℓ), the *spatial quantum number* (m), and the *spin quantum number* (m_s or s).[4,5]

When electrons are in close proximity, such as in a crystalline solid, nearby atoms affect their behavior and the electron's energy is no longer uniquely determined. Indeed, the single energy level of an electron in a free atom is spread into a band, or range, of energy levels.[6] The *conduction band* is the range of energy states in a solid in which electrons can move freely (that is, they can effect transitions between energy levels). In other words, the band is not full of *conduction electrons.* When every energy level is full, the material is known as an *insulator* or a *dielectric.*

[2]The forces produce displacement, velocity, or acceleration.

[3]Special Relativity also confirms the link between electricity and magnetism in that it shows that electric and magnetic fields appear or vanish depending upon the motion of the observer. Further, it was the Lorentz transformation that required invariance in Maxwell's equations; that is, electric and magnetic fields must remain in the same form for stationary and moving systems. This requirement is necessary because, without it, electrical and optical phenomena would differ depending on one's motion. It is important to understand that, while the experimental results do differ depending on the motion of the observer, the fundamental phenomena remain unchanged. For example, Maxwell's equations describing electromagnetism remain in the same form. The Lorentz transformation was the genesis of Einstein's Special Theory of Relativity.

[4]The azimuthal quantum number, ℓ, specifies the angular orbital momentum. Electrons whose value of ℓ is 0, 1, 2, and 3 are referred to as the *s*, *p*, *d*, and *f* electrons (for historical reasons related to the "picture" these electrons produced during experiments).

[5]The spin has two values, $\pm 1/2$, and is equal to $h/2\pi$ where h is Planck's constant.

[6]This band of energy levels allows what is referred to in some texts as an *electron cloud* to exist. The theory is by no means universal. Consult electrical conduction theory texts for more information.

Table 16.2 Common Conductors[a]

material	gauge[b]	diameter (in)	area[c] (circular mils)	nominal DC resistance (ohms/1000 ft at 68°F (20°C))
aluminum[d]	AWG 8	0.1285	16,510	1.030
copper[e]	AWG 8	0.1285	16,510	0.6533
steel[f]	BWG 8	0.165	27,239[g]	3.28[h]

[a]The English Engineering System is used in this table, as it represents the system most commonly used in this area of electrical engineering.
[b]AWG stands for American Wire Gauge, the usual standard for nonferrous wires, rods, and plates. BWG stands for Birmingham Wire Gauge, the usual standard for galvanized iron and steel wire.
[c]One circular mil is a unit having the area corresponding to the area of a circle with a diameter of 0.001 in.
[d]This information is for bare, solid, all-aluminum hard-drawn wire and is taken from ASTM (American Society of Testing Materials) Standard B230-55T.
[e]This value is considered a trade maximum. It depends on the specific process used when manufacturing the wire. ASTM requirements do exist for the resistance of copper for various processes.
[f]The type of steel referenced here is used for telephone and telegraph wire.
[g]The area in circular mils varies slightly depending upon the construction, that is, the number of strands and arrangement.
[h]This number is referenced to 1000 ft here for convenience in comparison with the other conductors. Transmission cables and telephone or telegraph wires are often referenced to longer lengths, as in ohms/mile.

PART 2: ELECTRIC AND MAGNETIC PHENOMENA

1. ELECTROMAGNETIC EFFECTS

Electromagnetic effects are caused by the dynamic behavior of elementary particles, that is, the electron and the proton, due to their mass and charge. These effects can differ greatly from free-space effects, depending upon the material in which the particles find themselves. There are three major classes of electromagnetic materials.

- *Conductors or semiconductors* are materials through which charges flow more or less easily. Some common conductors are given in Table 16.2. The properties of widely used semiconductors are given in Table 16.3. Good conductors have a conductivity range of 1×10^3 S/m to 6×10^3 S/m.
- *Dielectrics or insulators* are materials that inhibit the passages of charges. Some common insulators are given in Table 16.4. Insulators have a conductivity range of 10^{-18} S/m up to approximately 10^{-4} S/m. The determining factors for the conductivity of insulators are the application and the voltage stress (that is, the magnitude of the voltage being insulated).
- *Magnetic materials* are materials in which the motion of the charges produces perpendicular effects, that is, enhanced transverse effects. A very small portion of the elements in the periodic table exhibit such effects. These materials are given in Table 16.5.

Chapter 52 further discusses the properties of electrical materials. This information is necessary to understanding how electromagnetic and related theories can be applied to practical ends.

Table 16.3 Common Semiconductors

semiconductor	symbol	periodic table group
silicon	Si	IV
germanium	Ge	IV
gallium arsenide	GaAs	III and V
diamond and graphite	–	–
selenium	Se	VI
silicon carbide	SiC	IV
indium antimonide	InSb	III and V

Table 16.4 Common Insulators

insulating material
polyvinyl chloride (PVC)
butyl rubber
neoprene
plastics
ceramics
insulating gasses and liquids

Table 16.5 Magnetic Materials

magnetic material	symbol	atomic number
iron	Fe	26
cobalt	Co	27
nickel	Ni	28

2. CONDUCTION EFFECTS

Conduction effects occur in systems having mobile charges when an electric force is applied. The charges move in the direction of the applied force. As mentioned in Sec. 16.1, conduction effects result from the loosely bound outer shell electrons of atoms. In very strong conductors, electrons can be released and move under the application of small chemical, thermal, or other forces.

3. DIELECTRIC PHENOMENA

When a system contains bound charges, the electrons still move under the application of an electric force, but only to a limited extent. This process is called the *dielectric phenomenon*. It results in a current termed the *displacement current*, which is related to the *electric flux density*, **D**, in the dielectric material. In turn, the electric flux density, **D**, is proportional to the *electric field strength* or *intensity*, **E**.[7] Heat losses occur in dielectrics due to hysteresis effects at very high frequencies. Hysteresis effects occur in magnetic materials at much lower frequencies.

Dielectrics can sustain only a certain field strength before they break down and conduction occurs. This maximum field strength is called the *dielectric strength*, and its unit of measure is volts per meter of thickness of the material tested. The results vary, depending upon thickness and the environmental conditions of the test. For the same material, thin dielectrics break down at lower field strengths than thick dielectrics.

4. MAGNETIC PHENOMENA

Magnetic phenomena are caused by the directed motion of charges. The effects occur perpendicular to this motion. In magnetic materials, the phenomena are associated with the orientation of the orbiting electrons and, to some extent, the spin of the electrons. Magnetic phenomena occur in a relatively small number of materials whose outer shell orbits fill prior to the inner shells. This occurs because the outer shell orbit configuration is actually a lower energy configuration than the inner shell.

Several types of magnetism exist.

- *Ferromagnetism* is produced by the exchange of forces between atomic moments. This type of magnetism produces strongly magnetic materials with high permeability. Large clusters of atoms group together to form *magnetic domains*, each of which has the same atomic moment alignment within the domain. Ferromagnetism, named for iron, is the common magnetism one associates with horseshoes or toy magnets. See Fig. 16.1 for an example of a crystalline structure with domains and Fig. 16.2 for an example of magnetization curves. Figure 16.3 shows the spin arrangements of several types of magnetism.

- *Paramagnetism* is produced by the orbital or spin moments of the electrons, or both. The atomic alignment is minimal, that is, there is little or no domain formation, and as a result, paramagnetic materials are not strongly magnetic.

- *Antiferromagnetism* is produced by the exchange of forces between atomic moments. Antiferromagnetic materials have an antiparallel arrangement of equal spins. Their magnetism is of a similar strength to that of paramagnetic materials.

- *Diamagnetism* is produced by electron spins in antiparallel pairs in closed electron shells. Diamagnetic materials are weakly repulsive to external magnetic fields. These materials possess an internal magnetic field that opposes any externally applied magnetic field.

- *Ferrimagnetism* is produced by the moment that results from the combination of two antiferromagnetic lattices. Ferrimagnetic materials contain two kinds of magnetic ions, arranged antiparallel but with unequal spins.

Figure 16.1 Magnetic Domains

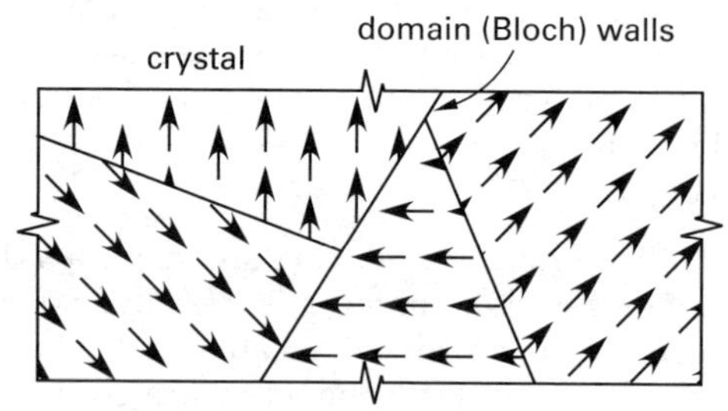

Figure 16.2 Magnetization Curves

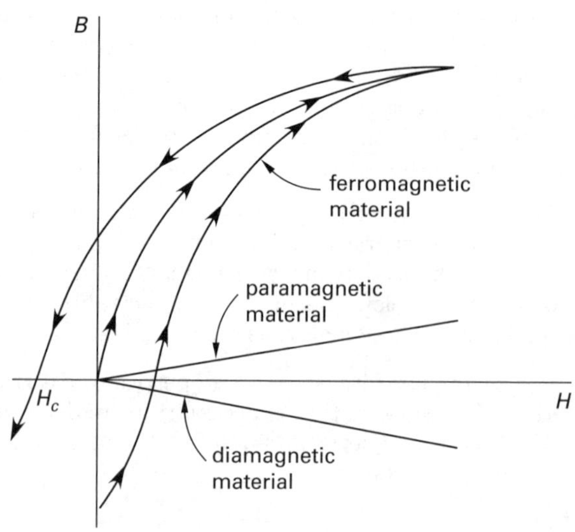

[7]Both of these are discussed more fully in Sec. 16.11.

Figure 16.3 *Magnetic Spin Arrangements*

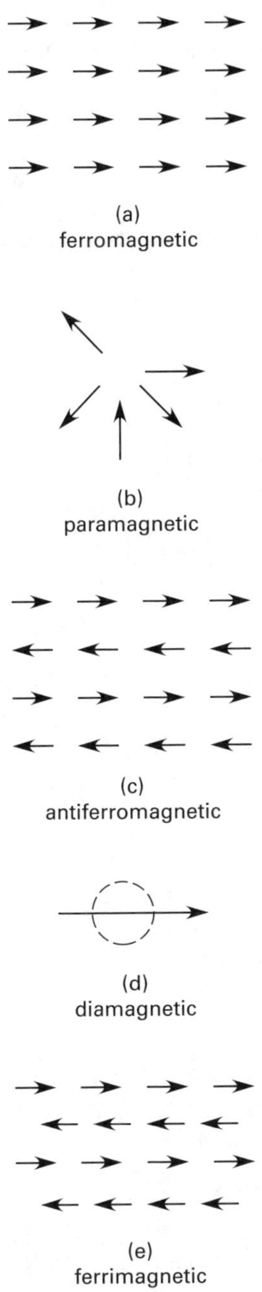

(Note: The circle in (d) represents the internal magnetic field.)

5. THERMOELECTRIC PHENOMENA

There are three significant relations between electricity and heat of importance to the electrical engineer. These are the Seebeck effect, the Thomson effect, and the Peltier effect. Each will be explained in terms of its applications. For now, the important point is that heat is capable of providing the energy to release and transport electrons.

PART 3: ELECTROSTATICS

As the name implies, electrostatics is the field of electrical engineering that deals with the electrical effects of stationary charges. An important concept in this field is that all electrical phenomena are expressible in terms of the masses and charges of the elementary particles involved.

6. ELECTRIC CHARGES

An electrically neutral material has equal numbers of positive and negative charge carriers. Equation 16.1 describes bodies in this unelectrified state.

$$Q^+ = Q^- = 0 \qquad 16.1$$

Only integer numbers of charge carriers exist. Therefore, the transport of an integer number n charge carriers to a body from another body means that

$$Q^+ = np^+ \text{ and } Q^- = ne^- \qquad 16.2$$

Like charges exhibit an electric repulsion. Unlike charges exhibit an electric attraction. An attraction also exists between the masses of any two charged particles. The force associated with the gravitational attraction is very small in comparison to the force of the electrical attraction. Therefore, the gravitational force can be ignored except when the charges are accelerated.

Static electricity formed by positive charges, such as that produced by rubbing silk on a glass rod, is called *vitreous static electricity.* When the static electricity is formed by negative charges, such as that produced by rubbing fur on a rubber, amber, or plastic rod, it is termed *resinous static electricity.*

Example 16.1

What is the gravitational force of attraction between two electrons located 1 cm apart?

Solution

Use Newton's Law of Gravitation.

$$F = \frac{Gm_1m_2}{r^2}$$

Newton's universal constant or *gravitational constant*, G, is equal to 6.672×10^{-11} N·m^2/kg^2. The mass of the electrons (taken from Table 16.1) and the 1 cm distance between the electrons are substituted into the equation, giving the following.

$$F = \frac{Gm_1m_2}{r^2} = \frac{Gm_e^2}{r^2}$$

$$= \frac{\left(6.672\times10^{-11}\ \frac{\text{N·m}^2}{\text{kg}^2}\right)(9.1096\times10^{-31}\ \text{kg})^2}{(1\ \text{cm})^2\left(\frac{1\ \text{m}}{100\ \text{cm}}\right)^2}$$

$$= 5.537\times10^{-67}\ \text{N}$$

7. COULOMB'S LAW

Coulomb's law states that the force between charges at rest in an isotropic medium is proportional to the product of the charges and inversely proportional to both the square of the distance between them and the dielectric coefficient of the medium. The force acts in a straight line between the centers of the charges.[8] Equation 16.3 describes Coulomb's law in mathematical terms.

$$\mathbf{F}_{1\text{-}2} = \frac{Q_1Q_2}{4\pi\epsilon r_{1\text{-}2}^2}\mathbf{a}_{r_{1\text{-}2}} \qquad 16.3$$

The vector $\mathbf{a}_{r_{1-2}}$ is a unit vector having a direction along a straight line connecting Q_1 and Q_2. The term ϵ represents the *dielectric coefficient*, more commonly called the *permittivity*. The 4π term occurs in many electric field calculations and, for this reason, is included in this *rationalized* form of Coulomb's law in order to cancel terms in integrals over spheres and make calculations easier. This is perhaps more readily seen when Coulomb's law is written in terms of the electric field, **E**.

$$\mathbf{F}_{1\text{-}2} = \mathbf{F}_2 = Q_2\mathbf{E}_1 \qquad 16.4$$

Coulomb's law is also stated in its Gaussian SI form as follows.

$$\mathbf{F}_{1\text{-}2} = \frac{kQ_1Q_2}{\epsilon_r r^2}\mathbf{a}_{1\text{-}2} \qquad 16.5$$

The term ϵ_r represents the *relative permittivity* of the dielectric. ($\epsilon_r = 1.0\ \text{C}^2/\text{N·m}^2$ for a vacuum.) The constant k has an approximate value of $8.987\times10^9\ \text{N·m}^2/\text{C}^2$. In Eq. 16.3 through Eq. 16.5 the force is measured in newtons, the distance in meters, and the charge in coulombs.

[8]In Eq. 16.3, the unit vector, and thus the force, is *from 1 to 2*. Don't be confused by this notation. The force direction is the direction charge 2 will move due to the electric field of charge 1. See Eq. 16.4. The notation often varies. The unit vector $\mathbf{a}_{r_{1-2}}$ (see Eq. 16.3) or $a_{1\text{-}2}$ (see Eq. 16.4) is sometimes written $\mathbf{a}_r$, $\mathbf{a}_{12}$, or even $\mathbf{a}_{21}$. The $\mathbf{a}_{21}$ notation is used as a reminder that, although the vector is *from 1 to 2*, it is determined mathematically by subracting the position elements of 2 from 1; for example, $x_2 - x_1$, $y_2 - y_1$, and $z_2 - z_1$. (The unit vector is then determined by dividing the distance between the two points.) At times, the notation 1-2 merely indicates the connection between the points and one must determine the correct direction from other information.

Example 16.2

What is the magnitude of the electrostatic force between two electrons 1 cm apart in a vacuum?

Solution

Take the magnitude of Eq. 16.5 and substitute the values for the relative permittivity and k given in this section. The relative permittivity has no units. As the term "relative" suggests, it is a comparison. This is discussed more fully in Sec. 16.9.

$$\mathbf{F}_{1\text{-}2} = k\left(\frac{Q_1Q_2}{\epsilon_r r^2}\right)\mathbf{a}_{1\text{-}2}$$

$$|\mathbf{F}_{1\text{-}2}| = F = k\left(\frac{Q_1Q_2}{\epsilon_r r^2}\right)$$

$$= \frac{\left(8.987\times10^9\ \frac{\text{N·m}^2}{\text{C}^2}\right)(-1.6022\times10^{-19}\ \text{C})^2}{(1)(1\ \text{cm})^2\left(\frac{1\ \text{m}}{100\ \text{cm}}\right)^2}$$

$$= 2.307\times10^{-24}\ \text{N}$$

Example 16.3

What is the ratio of the electrostatic force to the gravitational force between two electrons 1 cm apart in a vacuum?

Solution

As found in Ex. 16.2, the electrostatic repulsion is 2.307×10^{-24} N. As found in Ex. 16.1, the gravitational attraction is 5.537×10^{-67} N. The ratio of electrostatic force to gravitational force is

$$R = \frac{2.307\times10^{-24}\ \text{N}}{5.537\times10^{-67}\ \text{N}} = 4.166\times10^{42}$$

The electrostatic force that these elementary particles experience is 42 powers of ten greater than the gravitational force of attraction between them. Clearly, the gravitational attraction between such particles can be ignored in all but the most exacting calculations.

Coulomb's law implies that the principle of superposition applies for electric charges. This means that a system of charges is a linear system. The total resultant effect at any point within the system can be determined from the vector sum of the individual components.

Example 16.4

Three point charges in a vacuum are arranged in a straight line. Find the force on point charge A.

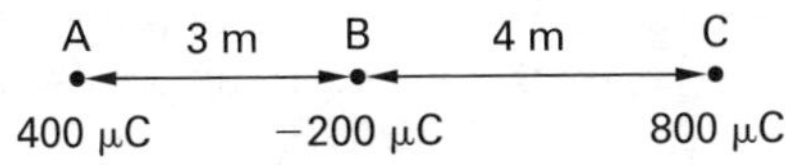

Solution

Because all three charges are in line, vector analysis is not needed. The force on point charge A is the sum of the individual forces from the other two point charges. Use the Gaussian form of Coulomb's law.

From Eq. 16.5,

$$\begin{aligned} F_{\text{A-B\&C}} &= F_{\text{B\&C-A}} \\ &= F_{\text{A-B}} + F_{\text{A-C}} = \frac{kQ_{\text{A}}Q_{\text{B}}}{r^2_{\text{A-B}}} + \frac{kQ_{\text{A}}Q_{\text{C}}}{r^2_{\text{A-C}}} \\ &= 8.987 \times 10^9\ \frac{\text{N·m}^2}{\text{C}^2} \\ &\quad \times \left(\frac{(400 \times 10^{-6}\ \text{C})(-200 \times 10^{-6}\ \text{C})}{(3\ \text{m})^2} + \frac{(400 \times 10^{-6}\ \text{C})(800 \times 10^{-6}\ \text{C})}{(7\ \text{m})^2} \right) \\ &= -79.9\ \text{N} + 58.7\ \text{N} \\ &= -21.2\ \text{N} \end{aligned}$$

Because the sign is negative, the force on A is toward B and C.

Example 16.5

Determine the magnitude and direction of the force on point charge A caused by point charges B and C.

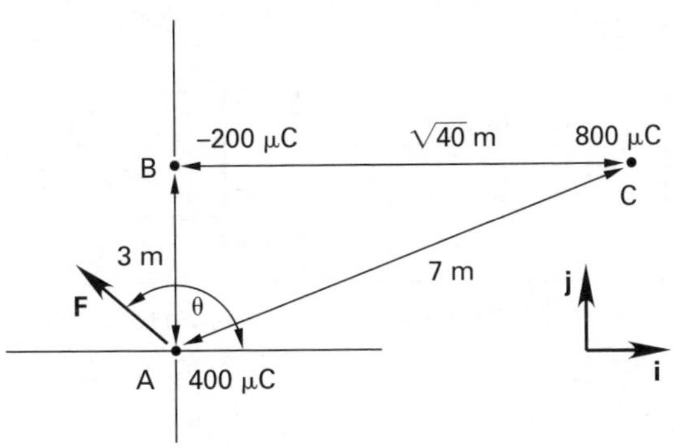

Solution

The charges and distances are the same as in Ex. 16.4.

$$F_{\text{B-A}} = F_{\text{A-B}} = -79.9\ \text{N} \quad \text{[attractive]}$$

$$F_{\text{C-A}} = F_{\text{A-C}} = -58.7\ \text{N} \quad \text{[repulsive]}$$

Determine the unit vectors for these forces. Let point A correspond to the origin. The unit vectors ($\mathbf{a}_{\text{B-A}}$ and $\mathbf{a}_{\text{C-A}}$) from B and C to A are determined as follows:

$$\begin{aligned} \mathbf{a}_{\text{B-A}} &= \frac{(x_{\text{A}} - x_{\text{B}})\mathbf{i} + (y_{\text{A}} - y_{\text{B}})\mathbf{j}}{\sqrt{(x_{\text{A}} - x_{\text{B}})^2 + (y_{\text{A}} - y_{\text{B}})^2}} \\ &= \frac{(0-0)\mathbf{i} + (0-3)\mathbf{j}}{\sqrt{(0-0)^2 + (0-3)^2}} \\ &= -\mathbf{j} \end{aligned}$$

$$\begin{aligned} \mathbf{a}_{\text{C-A}} &= \frac{(0-\sqrt{40})\mathbf{i} + (0-3)\mathbf{j}}{\sqrt{(0-\sqrt{40})^2 + (0-3)^2}} \\ &= -0.904\mathbf{i} - 0.429\mathbf{j} \end{aligned}$$

The total force is

$$\begin{aligned} F_{\text{B-A}}\mathbf{a}_{\text{B-A}} + F_{\text{C-A}}\mathbf{a}_{\text{C-A}} &= (-79.9\ \text{N})(-\mathbf{j}) + (58.7\ \text{N})(-0.904\mathbf{i} - 0.429\mathbf{j}) \\ &= -53.1\mathbf{i}\ \text{N} + 54.7\mathbf{j}\ \text{N} \end{aligned}$$

The magnitude of the force is

$$|\mathbf{F}_{\text{A-B\&C}}| = \sqrt{(-53.1\ \text{N})^2 + (54.7\ \text{N})^2} = 76.2\ \text{N}$$

The direction (counterclockwise angle with respect to the horizontal) is

$$\theta = 180° - \arctan\frac{54.7\ \text{N}}{53.1\ \text{N}} = 180° - 45.9° = 134.1°$$

8. ELECTRIC FIELDS

The *electric field*, **E**, with units of V/m, is the region in which an electric charge exerts a measurable force on another charge, above and beyond the gravitational force that the two charges exert on each other. It is a *vector force field* because a test charge placed in the field will experience a force in a given direction and with a specific magnitude. The direction of the field is along flux lines. These lines, ideally, represent the direction of the force on an infinitely small, isolated positive test charge.[9] The actual direction is represented by the unit vector **a**. Equation 16.6 represents the *electric field strength*, or *electric field intensity*, in a substance with permittivity ϵ at a distance r from a point charge Q.

$$\mathbf{E} = \frac{Q}{4\pi\epsilon r^2}\mathbf{a} \qquad 16.6$$

Coulomb forces obey the principle of linear superposition in free space and in many dielectrics. Therefore, the electric field strength or intensity acting on a charge Q

[9]The positive test charge is arbitrary, but universal. The charge would ideally have to be infinitely small in order to prevent its own electric field from distorting the original field.

located at point 1 as a result of charges $Q_1, Q_2, \ldots, Q_i$ at distances r_{i1} from point 1 can be represented as the following sum. The vector a_{ri1} is the unit vector with direction from Q_i to point 1.

$$\mathbf{E}_1 = \sum_{i=1}^{i} \frac{Q_i}{4\pi\epsilon r_{i1}^2} \mathbf{a}_{r_{i1}} \tag{16.7}$$

Additionally, the electric field intensity at any point p produced by charges $Q_1, Q_2, \ldots, Q_n$ at distances $r_1, r_2, \ldots, r_n$ from the point is the vector sum of the field strengths produced by the charges individually.

If one is dealing with charges distributed uniformly throughout space, the electric field strength at a distance r that is large when compared to dV_{volume} is given by Eq. 16.8.

$$\mathbf{E}_1 = \int_{V_{\text{volume}}} \frac{\rho\, dV_{\text{volume}}}{4\pi\epsilon r_{i1}^2} \mathbf{a}_{r_{i1}} \tag{16.8}$$

The unit vector $\mathbf{a}_{ri1}$ is directed from dv to point 1, at which $\mathbf{E}$ is evaluated. The unit vector, the distance, r, and the charge density, ρ, are usually functions of the variable of integration. The total charge, Q, in the volume V_{volume} that has the charge density ρ is

$$Q = \int_{V_{\text{volume}}} \rho\, dV_{\text{volume}} \tag{16.9}$$

The electric fields discussed to this point have been radial in nature. Not all fields are radial. As seen in Eq. 16.8, the electric field depends on the orientation of the surface or volume that contains the charge. A *uniform electric field*, commonly used in sensing instruments and other important applications, can be created between two flat plates, as shown in Fig. 16.4.

Figure 16.4 *Uniform Electric Field*

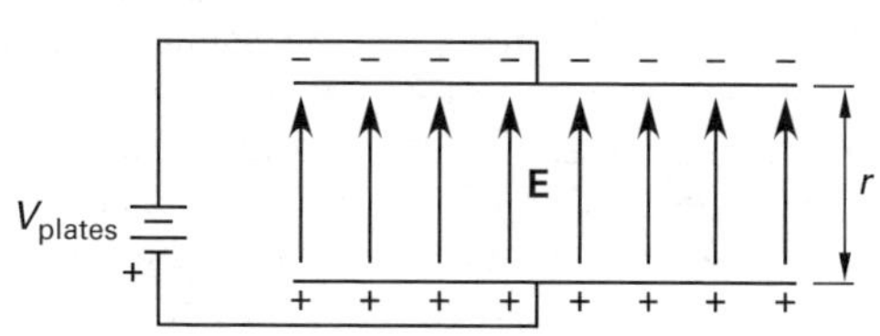

With a potential difference of V volts between the flat plates separated by distance r, the electric field strength is

$$E_{\text{uniform}} = \frac{V_{\text{plates}}}{r} \tag{16.10}$$

The field strengths, or intensities, for several configurations are given in Table 16.6.

An electric field is an abstraction that allows one to visualize how charges in different locations within the field will experience a force. A *line of electric force* is a curve drawn so that it has the direction of the force acting on the test charge, (i.e., it is in the direction of the electric field). A *tube of force* is created by drawing lines of force through the boundary of any closed curve. The lines form a tubular surface not cut by any other lines of force.

Table 16.6 *Electric Fields and Capacitance for Various Configurations*

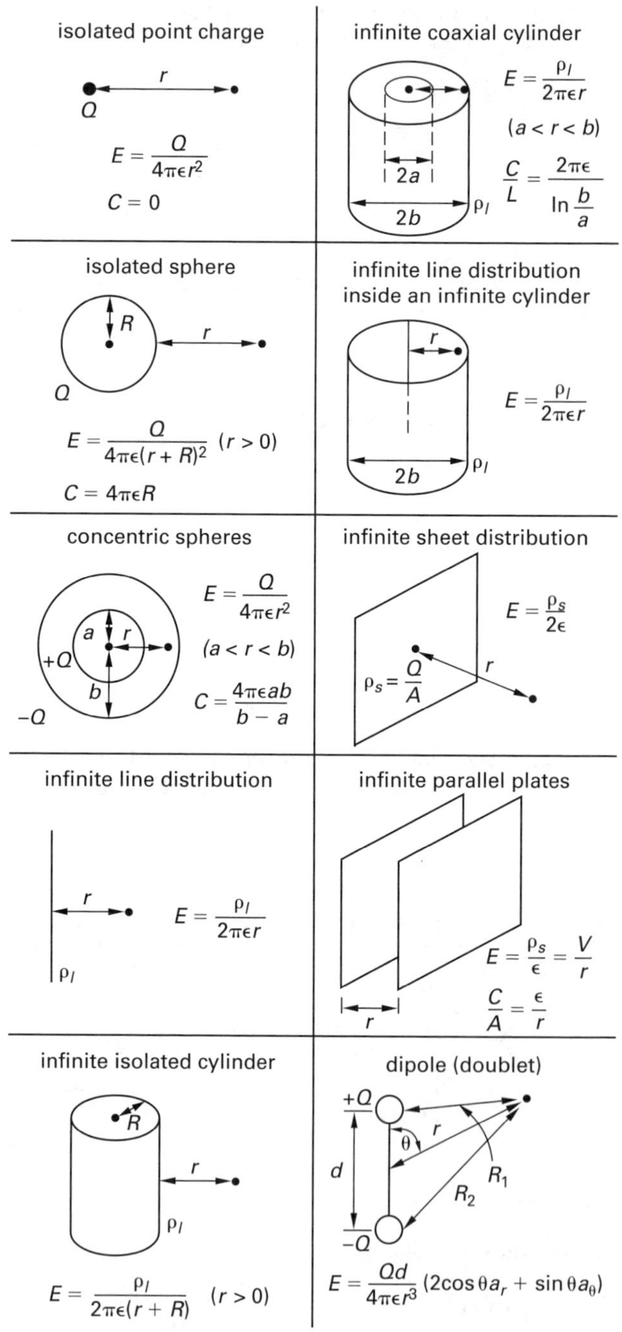

Reprinted with permission from *Core Engineering Concepts for Students and Professionals*, by Michael R. Lindeburg, PE, copyright 2010, by Professional Publications, Inc.

9. PERMITTIVITY AND SUSCEPTIBILITY

The forces between electric charges depend upon the environment in which they are located. The maximum coulombic force occurs in free space. In all other materials, the coulombic force is reduced. Another way of viewing this phenomenon is to say that the electric flux does not pass equally through all materials.[10] In fact, it cannot pass through conductive material at all and is diminished to varying degrees by various dielectrics.

The *permittivity of free space*, that is, a vacuum, ϵ_0, may be considered to be representative of the dielectric properties of free space. In actuality, it is a factor that provides consistent units for Coulomb's law when units of force, charge, and distance are arbitrary. The product of the permittivity of free space and the *relative permittivity* of a medium gives the *permittivity* of that medium.

$$\epsilon = \epsilon_0 \epsilon_r \qquad 16.11$$

In general, the relative permittivity is a complex number. Except for very high frequencies, the imaginary component, or quadrature component, can be ignored. The real component is often termed the *dielectric constant.* Because the value of ϵ_r depends on frequency as well as other factors, the term *dielectric coefficient* is more appropriate. The relative permittivity can also be viewed in terms of capacitance, as shown in Eq. 16.12.

$$\epsilon_r = \frac{C_{\text{with dielectric}}}{C_{\text{vacuum}}} \qquad 16.12$$

Typical relative permittivities are given in Table 16.7.

The permittivity can also be expressed as follows.

$$\epsilon = \epsilon_0(1 + \chi_e) \qquad 16.13$$

The term χ_e represents the electric susceptibility of a given dielectric. It is a numeric measure of the polarization or displacement of electrons in the atoms or molecules of the dielectric.

The units for permittivity are $C^2/N{\cdot}m^2$ or F/m. The permittivity of free space, or vacuum, ϵ_0, in SI units is 8.854×10^{-12} $C^2/N{\cdot}m^2$ (F/m).

Example 16.6

A certain capacitor is constructed of two square parallel plates with air as the dielectric. If a teflon insert is used in place of the air, by what factor does the capacitance increase?

Solution

The capacitance is directly proportional to the permittivity and is given by the following formula.

$$C = \frac{\epsilon A}{r} = \frac{\epsilon_0 \epsilon_r A}{r}$$

[10]The concept of flux is another abstraction used to help one visualize electric interactions. It will be discussed more fully in the Electric Flux section in this chapter.

Because the example asks for the factor by which the capacitance increases, not the actual value of the capacitance, one need merely compare the permittivity of teflon to that of air. Using data from Table 16.7,

$$\frac{\epsilon_{r,\text{teflon}}}{\epsilon_{r,\text{air}}} = \frac{2.0}{1.00059} = 2.0$$

One could use the upper bound of the permittivity for the teflon, in which case the factor by which the capacitance increases would be approximately 2.2.

Table 16.7 *Typical Relative Permittivities (20°C and 1 atm)*

material	ϵ_r	material	ϵ_r
acetone	21.3	mineral oil	2.24
air	1.00059	mylar	2.8–3.5
alcohol	16–31	olive oil	3.11
amber	2.9	paper	2.0–2.6
asbestos paper	2.7	paper (kraft)	3.5
asphalt	2.7	paraffin	1.9–2.5
bakelite	3.5–10	polyethylene	2.25
benzene	2.284	polystyrene	2.6
carbon dioxide	1.001	porcelain	5.7–6.8
carbon tetrachloride	2.238	quartz	5
		rock	≈ 5
castor oil	4.7	rubber	2.3–5.0
diamond	16.5	shellac	2.7–3.7
glass	5–10	silicon oil	2.2–2.7
glycerine	56.2	slate	6.6–7.4
hydrogen	1.003	sulfur	3.6–4.2
lucite	3.4	teflon	2.0–2.2
marble	8.3	vacuum	1.000
methanol	22	water	80.37
mica	2.5–8	wood	2.5–7.7

10. ELECTRIC FLUX

Flux is defined as the electric or magnetic lines of force in a region. The *electric flux*, Ψ, is a scalar field representing these imaginary lines of force. Flux lines leave or enter any surface at right angles to that surface, and the orientations of the electric field lines and the electric flux lines always coincide. By convention, the flux lines are directed outward from the positive charge and inward toward the negative charge, or to infinity if no negative charge exists in the area, as shown in Fig. 16.5. In Fig. 16.5, only a few representative lines of flux are shown. Also, if the charge on a hollow sphere is all the same sign, the flux outside this sphere is the same as the flux that would exist from a point charge at the center. Therefore, for the purposes of drawing flux lines and calculating forces, a hollow charged sphere can be replaced by a point charge at the center.

Figure 16.5 *Flux Lines*

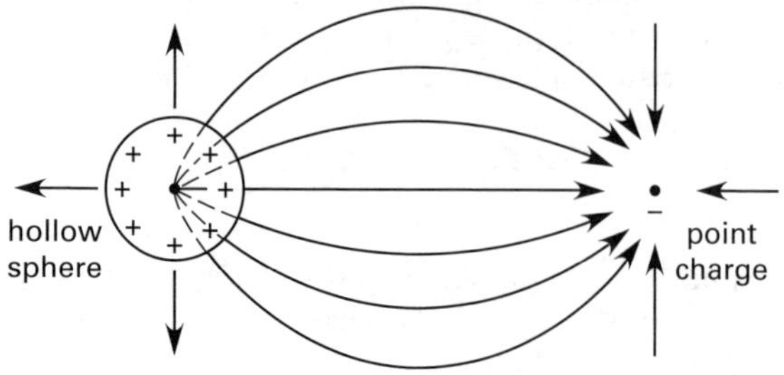

The flux, Ψ, is numerically equal to the charge. That is, by definition, one coulomb of electric charge gives rise to one coulomb of electric flux.

$$\Psi = Q \tag{16.14}$$

The electric flux is also called the *displacement flux.* This is because the flux is a quantity associated with the amount of bound charge, Q, displaced in a dielectric material that is subject to an electric field. This terminology can be understood by referring to Fig. 16.6.

Figure 16.6 *Displacement Flux Experiment*

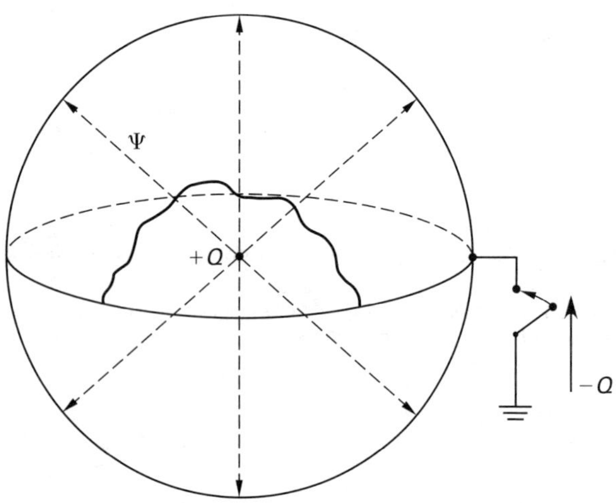

Early experimenters enclosed a fixed positive charge inside a spherical conducting shell as shown in Fig. 16.6. When the switch was closed, a charge of $-Q$, equal in magnitude to the enclosed positive charge, was found on the shell and was believed to be caused by the transient flow of negative charge induced by the flux from the $+Q$. In other words, the positive charge *displaced* the $-Q$ charge and forced it onto the surface. The scalar electric flux field, Ψ, is not directly measurable as is the electric field, $\mathbf{E}$. Instead it is inferred from such experiments.

11. ELECTRIC FLUX DENSITY

The *electric flux density,* $\mathbf{D}$, is also referred to as the *displacement density* or simply the *displacement.* It is also loosely referred to as the *vector flux density.* It is a vector quantity that, like the electric flux, Ψ, cannot be directly measured. It is given in Eq. 16.15. Unlike the electric field strength, $\mathbf{E}$, the electric flux density or displacement, $\mathbf{D}$, is independent of the medium.[11]

$$\mathbf{D} = \epsilon\mathbf{E} = \epsilon_0\epsilon_r\mathbf{E} \tag{16.15}$$

The magnitude of $\mathbf{D}$, or D, is also referred to as the *flux density.* It is the number of flux lines per unit area perpendicular to the flux. It is a scalar quantity with units of C/m^2.

$$D = |\mathbf{D}| = \frac{\psi}{A} = \frac{Q}{A} = \sigma \tag{16.16}$$

When the medium is nonisotropic, the permittivity, ϵ, is a tensor and the displacement, $\mathbf{D}$, and electric field strength, $\mathbf{E}$, are not necessarily in the same direction.

Many terms with differing units are used to refer to "flux density," such as ρ_l in units of C/m and ρ_V in units of C/m^3. This confusion is avoided in texts by referring to such terms as ρ_l and ρ_V as charge densities.[12] Care must be exercised when interpreting the phrase "flux density."

12. GAUSS' LAW FOR ELECTROSTATICS

Gauss' law for electrostatics is a necessary consequence of Coulomb's law. Gauss' law states that the amount of electric flux, Ψ, passing through any closed surface is proportional to (and in our rationalized system of SI units, equal to) the total charge, Q, contained within the surface. The surface is called a *Gaussian surface.* Gauss' law in a variety of forms is given in Eq. 16.17, Eq. 16.18, and Eq. 16.19. $d\mathbf{s}$ is an element of the surface area with magnitude ds and direction normal to $d\mathbf{s}$, and θ is the angle between the normal to the surface area and $\mathbf{D}$.

$$\begin{aligned}\Psi &= \iint D\,dA = \iint \sigma\,dA \\ &= Q \quad [d\mathbf{A} \text{ parallel to } \mathbf{D}]\end{aligned} \tag{16.17}$$

$$\begin{aligned}\Psi &= \iint \mathbf{D}\cdot d\mathbf{s} = \iint D\cos\theta\,ds \\ &= Q \quad [\text{arbitrary surface}]\end{aligned} \tag{16.18}$$

$$\Psi = \iiint \rho\,dv = Q \quad [\text{arbitrary volume}] \tag{16.19}$$

[11]This is because $\mathbf{E}$ is multiplied by the permittivity. See Eq. 16.6 by way of explanation, noting the permittivity in the denominator.

[12]This is consistent with the electric flux, Ψ, being numerically equal to the charge. Another way of considering this is to realize that a rationalized set of units (the SI system) that makes the flux equal to the charge is being used. If the system were not rationalized, the flux would merely be proportional to the charge.

Because the electric field strength or intensity, **E**, is related to the displacement or electric flux density, **D**, by the permittivity, Gauss' law as given in Eq. 16.17 can also be stated mathematically as

$$\Psi = \iint \epsilon E dA = Q \quad [d\mathbf{A} \text{ parallel to } \mathbf{E}] \qquad 16.20$$

Special Gaussian surfaces are used in highly symmetrical charge configurations to simplify calculations with Gauss' law. Such surfaces meet the following criteria.

1. The surface is closed.
2. At each point on the chosen surface, **D**, is either normal or tangential to the surface.
3. The flux density, or displacement, magnitude, D, is sectionally constant over the portion of the surface where **D** is normal.

   Requirement (1) is a general requirement of Gauss' law. Requirement (2) completely eliminates integrals and makes the calculation simple. Requirement (3) allows D to be removed from the integral.

Example 16.7

Given a point charge of constant value Q, what is the flux density, D, at a distance r from the charge?

Solution

First, arbitrarily construct a sphere around the point charge as shown.

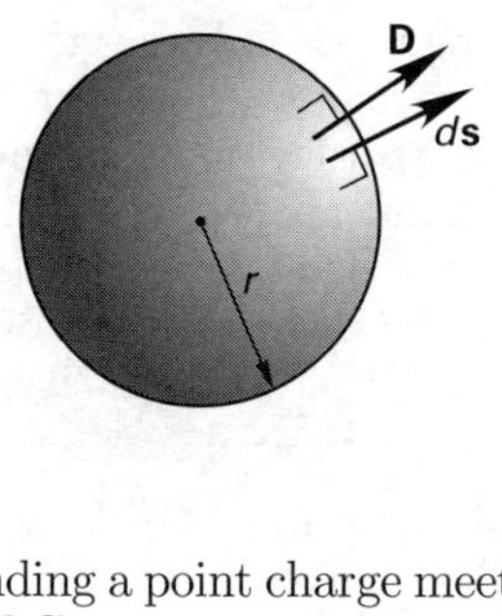

A sphere surrounding a point charge meets all the requirements of a special Gaussian surface. Using Eq. 16.18 gives the following.

$$\Psi = \iint \mathbf{D} \cdot d\mathbf{s} = \iint D \cos\theta ds = Q$$

Because the lines of flux emanate radially outward from the point charge, **D** and ds are parallel and $\cos\theta = 1$. Additionally, the magnitude of the flux density, D, is constant at a given r. This results in the following manipulations.

$$Q = \oint D \cos\theta ds$$
$$= \oint D(1) ds = D \oint ds$$

The surface area of a sphere is given by $4\pi r^2$.

$$Q = D \oint ds = D 4\pi r^2$$
$$D = \frac{Q}{4\pi r^2}$$

Using the same logic as in Ex. 16.7 on a uniform line charge of density ρ_l of length l at a distance of r gives the following equation for displacement magnitude.

$$D = \frac{\rho_l}{2\pi r} \qquad 16.21$$

13. CAPACITANCE AND ELASTANCE

If charges Q^+ and Q^- exist on conductors separated by free space or by a dielectric, the mutual attraction "stores" the charges, holding them in place. The charges produce a stress in the dielectric measured by the electric field strength, **E**, and a potential difference exists between the plates. *Capacitance* is the property of a system of conductors and dielectrics that permits this storage of charges.

If a potential difference is applied across two conductors, charges will build up, creating an electric field between the conductors. The amount of the charge, Q, is proportional to the applied voltage. The proportionality constant is called the *capacitance* and has units of *farads*, F. This is shown in Eq. 16.22. A farad is a large capacitance and very few capacitors are constructed with a value of even 1.0 F. Most circuit elements have capacitances in the range of μF (10^{-6}) or pF (10^{-12}).

$$Q = CV \qquad 16.22(a)$$

$$C = \frac{Q}{V} \qquad 16.22(b)$$

The *elastance*, S, is the reciprocal of the capacitance.

$$S = \frac{1}{C} = \frac{V}{Q} = \frac{\frac{U}{Q}}{Q} \qquad 16.23$$

As shown in Eq. 16.23, the elastance can also be defined as the amount of energy, U, required per charge, Q, to transfer a unit charge between two conductors separated by a dielectric.

14. CAPACITORS

A *capacitor* (once called a *condenser*) is a device that stores electric charge. It consists of conductors separated by a dielectric. A common type of capacitor consists of two parallel plates of equal area A separated by a distance r. This type of capacitor is called a *parallel plate capacitor* and its capacitance is determined by Eq. 16.24.

$$C = \frac{\epsilon A}{r} \tag{16.24}$$

Equation 16.24 ignores the effects of *fringing*, as do many calculations involving capacitors. An example of fringing is shown in Fig. 16.7. The formula for the capacitance of other configurations is given in Table 16.6.

Assuming the plates in a capacitor begin with the same potential ($V=0$), the total energy, U (in J), in the electric field of a capacitor with capacitance C charged to a potential V is given by Eq. 16.25.

$$U = \tfrac{1}{2}CV^2 = \tfrac{1}{2}VQ = \tfrac{1}{2}\frac{Q^2}{C} \tag{16.25}$$

For n capacitors in parallel, the total capacitance is

$$C_{\text{total}} = C_1 + C_2 + C_3 + \cdots + C_n \tag{16.26}$$

For n capacitors in series, the total capacitance is

$$\frac{1}{C_{\text{total}}} = \frac{1}{C_1} + \frac{1}{C_2} + \frac{1}{C_3} + \cdots + \frac{1}{C_n} \tag{16.27}$$

Figure 16.7 *Fringing*

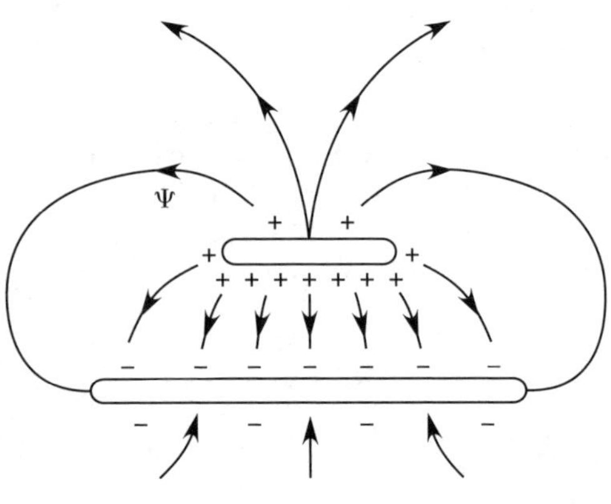

15. ENERGY DENSITY IN AN ELECTRIC FIELD

In a rectangular dielectric solid of surface area A and thickness t with a potential difference of V applied to either surface, the average energy (in J) is given by

$$\begin{aligned} U_{\text{ave}} &= \tfrac{1}{2}V_{\text{voltage}}Q = \tfrac{1}{2}(Et)(DA) = \tfrac{1}{2}EDV_{\text{volume}} \\ &= \tfrac{1}{2}\epsilon E^2 V_{\text{volume}} = \tfrac{1}{2}\frac{D^2}{\epsilon}V_{\text{volume}} \end{aligned} \tag{16.28}$$

Equation 16.28 in all its forms assumes that **D** and **E** are in the same direction. If not, the dot product must be used to determine the energy.

The average energy density at any point in this electric field is given in Eq. 16.29.

$$u_{\text{ave}} = \frac{U_{\text{ave}}}{V_{\text{volume}}} = \tfrac{1}{2}\epsilon E^2 = \tfrac{1}{2}\frac{D^2}{\epsilon} \tag{16.29}$$

PART 4: ELECTROKINETICS

16. SPEED AND MOBILITY OF CHARGE CARRIERS

A positive free charge moving in an electric field through a vacuum is shown in Fig. 16.8. Such a particle would experience a force given by the following.[13]

$$\mathbf{F} = Q\mathbf{E} = m\mathbf{a} \tag{16.30}$$

Figure 16.8 *Free Charge in the Electric Field of a Vacuum*

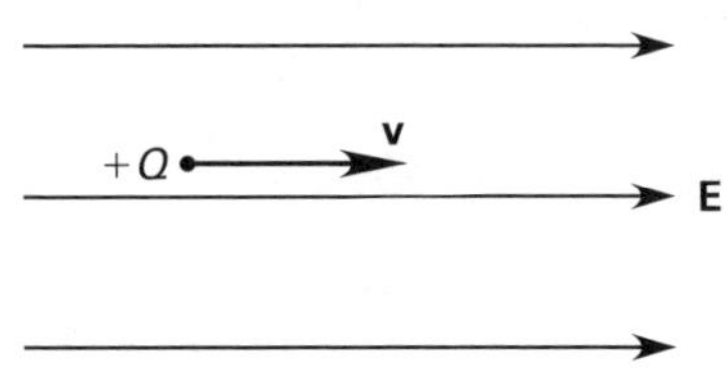

The force the particle experiences is unopposed and results in the constant acceleration of the particle as long as it remains in the electric field.

In metals, according to *electron-gas theory*, the electrons instead reach an average *drift velocity*, $\mathbf{v}_d$, because they undergo collisions as they move through the electric field. These collisions are between the electrons and the atoms within the crystalline structure of the metal and are caused by the thermal vibrations of the individual atoms.[14] The average drift velocity is

$$\begin{aligned} \mathbf{v}_d &= \int \mathbf{a}\,dt = \int \frac{\mathbf{F}}{m}\,dt \\ &= \frac{Q}{m}t_m\mathbf{E} = \mu\mathbf{E} \end{aligned} \tag{16.31}$$

The term t_m is the mean free time between collisions and m is the mass of the particle.[15] The mobility of the moving charge is given by μ. Because the particles in the metal undergo an increase in vibratory motion as the temperature increases, the mobility varies with the

[13]This is identical to Eq. 16.4 without the subscripts. Nominally, subscripts need only be used to avoid confusion when multiple fields or charges are involved.

[14]The thermal vibrations can be treated in a quantum mechanical fashion as discrete particles called *phonons*.

[15]The variation in the overall field within the metal is caused by the periodic variations established by the lattice atoms. The field variations can result in collisions, accounted for in some formulas by an *effective electron mass*, m_e^*. Only the thermal vibrations must be determined when such a mass is used.

temperature—and with the crystalline structure of the material. As the temperature increases, the mobility decreases, resulting in a lower drift velocity and a lower current. In circuit analysis this phenomenon is called *resistivity.* The resistivity is the reciprocal of the conductivity.

17. CURRENT[16]

An *electric current* is the movement of charges past a particular reference point or through a specified surface. The symbol I is generally used for constant currents while i is used for time-varying currents. Current is measured in *ampères* (A), commonly called amps. 1 A = 1 C/s. By convention, the current moves in the direction a positive charge would move (from the positive terminal to the negative terminal), that is, opposite to the flow of electrons. *Ohm's law* relates current to voltage and resistance as shown in Eq. 16.32.

$$V = IR \qquad 16.32$$

For simple DC circuits, this relationship is adequate and the current is defined by Eq. 16.33.[17]

$$I = \frac{dQ}{dt} \qquad 16.33$$

However, when the charges exist in a liquid or a gas, as in the field of electronic engineering, or when the charges are both positive and negative and have different characteristics, as in semiconductors, Eq. 16.33 is inadequate and one must be more specific in defining current. As a result, the *current density vector*, **J**, with units of A/m^2 is more frequently utilized in electromagnetic theory.

The *total current* is given by the sum of the convection current, the displacement current, and the conduction current. The convection and conduction currents are considered true currents and the displacement current a virtual current. Unless otherwise stated, the "current" will normally be the conduction current, but care must be used when determining what "current" is being referenced.

18. CONVECTION CURRENT[18]

In a material containing a volume V of mobile charges with a charge density of ρ moving across a surface area A with an average speed of v_{ave}, the *convection current* will be given by Eq. 16.34.

$$I = \rho A v_{ave} = \left(\frac{Q}{V}\right) A v_{ave} = \left(\frac{Q}{Al}\right) A \left(\frac{l}{t}\right) = \frac{Q}{t} \qquad 16.34$$

In terms of the current density vector, the convection current is given by Eq. 16.35 and can be visualized as shown in Fig. 16.9.

$$\mathbf{J} = \rho \mathbf{v}_d \qquad 16.35$$

Figure 16.9 *Convection Current*

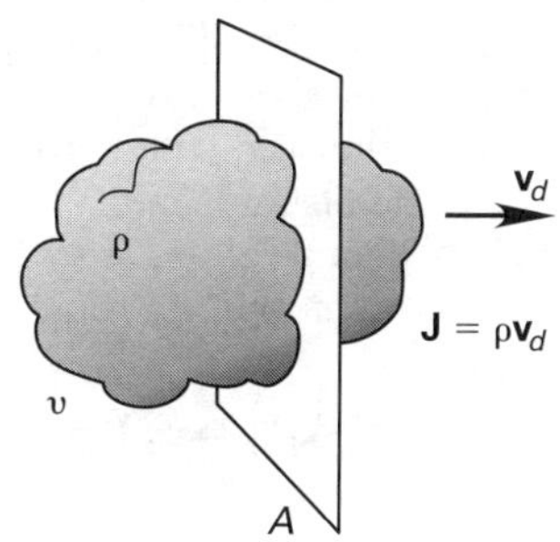

Convection, as the name implies, involves the diffusion of charges in a medium, as opposed to the movement of charges under the influence of an electric field. Although not of general use in electrical engineering, the concept of a convection current is sometimes useful in electromagnetic field theory.

19. DISPLACEMENT CURRENT[19]

When the electric field varies with time, a current can be made to flow in a dielectric material. That is, charges can be displaced within a dielectric by the electric fields of the charges moving outside the borders of the dielectric. This is because the electric fields of the charged particles extend well beyond the dimensions of the charges themselves. The current so created is called the *displacement current*, i_d, and is given by Eq. 16.36.

$$i_d = \frac{d\psi}{dt} \qquad 16.36$$

The displacement current in terms of the current density vector, **J**, is given by Eq. 16.37.

$$\mathbf{J}_d = \frac{\delta \mathbf{D}}{\delta t} \qquad 16.37$$

The displacement current, i_d, through any given surface can be obtained by integration of the normal component

[16]Compare the electric field, induced current, to the magnetic field, induced voltage, in Sec. 16.33.

[17]The use of the capital I indicates that the speed of the charges referred to is constant. If the speed is not constant then the current varies with time and i would be used as the appropriate symbol. In practical terms, this means that currents in DC circuits (unidirectional current flow) in which the current magnitude varies could be labeled i.

[18]Compare convection current with the movement of plasma (see Sec. 17.3).

[19]Compare displacement current to the displacement magnetic effect in Sec. 16.35.

Basic Theory

of the displacement current density vector, $\mathbf{J}_d$, over the surface as in Eq. 16.38.

$$i_d = \frac{d}{d_t}\int_A \mathbf{D}\cdot d\mathbf{A} \qquad 16.38$$

The displacement current is equal to zero if the electric flux is constant with respect to time. That is, to cause a current to "flow" through a dielectric without the benefit of a direct connection to a conductor (as in the case of a capacitor), the electric flux must be changing with respect to time.[20]

20. CONDUCTION CURRENT[21]

A *conduction current* occurs in the presence of an electric field within a conductor. The magnitude of this current is given by the number of mobile charges transferred per unit time within the conductor. The charges may be electrons (in metals and semiconductors), holes (in semiconductors), or ions (in gases and electrolytes). Conduction current is given by Eq. 16.39.

$$I_c = \sigma\left(\frac{A}{l}\right)V = GV \qquad 16.39$$

The symbol σ indicates the conductivity of the material and is a property of a unit volume of the material. The area of the conductor is given by A and the potential difference by V. The length represented by l is the length over which the potential is applied. The quantity G is the conductance of the specified conductor and depends on the dimensions of the material used.

The conduction current in terms of the conduction current density vector, $\mathbf{J}_c$, is given by Eq. 16.40.

$$\mathbf{J}_c = \rho\mathbf{v}_d \qquad 16.40$$

This is identical to Eq. 16.35. Applying Eq. 16.31, specifically $\mathbf{v}_d = \mu\mathbf{E}$, puts Eq. 16.40 in the more widely recognized form shown in Eq. 16.41.

$$\mathbf{J} = \sigma\mathbf{E} \qquad 16.41$$

Equation 16.41 is called the *point form of Ohm's law*.

PART 5: MAGNETOSTATICS

21. MAGNETIC POLES

Mobile charges set in motion by an electric field carry their own electric fields with them. The subsequent three-dimensional field can be resolved into longitudinal and transverse components, referenced to the direction of motion. The transverse fields produce magnetic forces between moving charges. This force also acts between the conductors in which the charges move. It is this force that constitutes the basis for motor and generator theory in electric power engineering. Moving charges represented by the current, i, and the magnetic field established by their motion are shown in Fig. 16.10(a). Another completely valid view of the same situation is shown in Fig. 16.10(b). If the charges move with a uniform motion, the field is referred to as *magnetostatic*. If the charges move nonuniformly, the field is referred to as *magnetokinetic*.

Figure 16.10 *Magnetic Fields of Moving Charges*

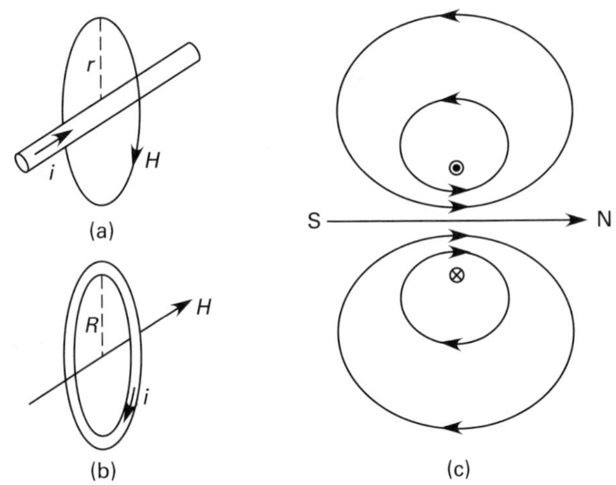

(The circled "x" indicates the current flows into the page. The circled "dot" indicates flow out of the page. That is, the dot indicates the head of an arrow and the "x" represents the tail. Assume positive current flow. This correlates with the *right-hand rule* that states that placing the thumb in the direction of the current flow and curling the fingers inward gives the direction of the magenetic field.)

A magnetic field can exist only with two equal and opposite poles, referred to as the *north pole* and the *south pole*.[22] The combination of the north and south pole linked together is referred to as a *dipole*. The magnetic *dipole moment*, m, depends on the pole strength, p, and the distance between the poles.

$$m = pd \qquad 16.42$$

The magnetic moment is a mathematical construct. Because the poles cannot be separated for measurement, the distance, d, is not measurable. Nevertheless, the magnetic moment is an important and fundamental quantity, and it can be measured. Magnetic moments are measured in terms of *Bohr magnetons*. In ferromagnetic material, a Bohr magneton is the moment

[20]The charges do not flow through the dielectric but are instead displaced within the dielectric. Also, this displacement ceases if the electric flux within the dielectric no longer changes.

[21]Compare conduction current with the principles of the magnetic circuit in Sec. 16.32.

[22]An electric charge can exist as a single charged object. Magnetic poles cannot exist as single north or south poles. (Mathematically there is no reason for the nonexistence of such a *magnetic monopole*, but none has yet been found.)

produced by one unpaired electron, which is equivalent to 9.24×10^{-27} A·m^2.

Figure 16.10(c) represents a loop of current creating a magnetic field. As the diameter of the current loop becomes infinitely small, the field approaches that of a magnetic dipole. A magnetic dipole field is shown in Fig. 16.11.

Figure 16.11 *Magnetic Dipole Field*

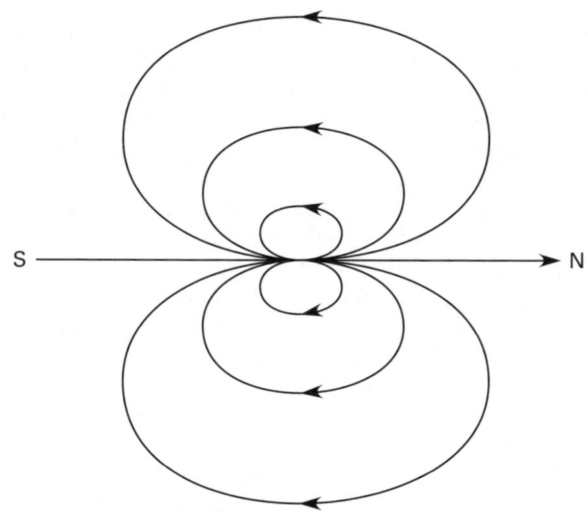

The infinitely small loop of current can be viewed as a single electron moving in an orbit. Hence, the magnetic moments of individual atoms and the usefulness of the Bohr magneton.[23]

For ease of presentation, and because in magnetic materials many groups of atoms tend to align themselves into domains, individual electrons or loops of current are not usually shown. Instead they are more often shown as in Fig. 16.12, which illustrates the standard convention that lines of *magnetic flux* are directed outward from the north pole (that is, the *magnetic source*), and inward at the south pole (that is, the *magnetic sink*).

Figure 16.12 *Typical Magnetic Field*

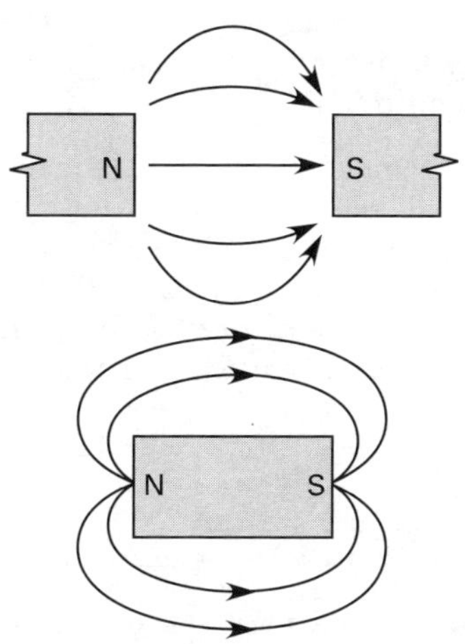

[23]It was Ampère who suggested over a hundred years ago that the magnetism observed in materials was caused by "molecular currents."

22. BIOT-SAVART LAW

The Biot-Savart law is basic to magnetostatics and magnetokinetics just as Coulomb's law is basic to electrostatics and electrokinetics. The law, in conjunction with Ampère's law, relates the force $\mathbf{F}_{1\text{-}2}$ between two current elements as in Eq. 16.43.

$$\mathbf{F}_{1\text{-}2} = \frac{\mu}{4\pi}\left(\frac{(I_2 d\mathbf{l}_2) \times (I_1 d\mathbf{l}_1) \times (\mathbf{r}_{1\text{-}2})}{r^2_{1\text{-}2}}\right) \qquad 16.43$$

The term μ is the magnetic permeability and $\mathbf{r}_{1\text{-}2}$ is a unit vector from the point represented by differential current element $d\mathbf{l}_1$ to $d\mathbf{l}_2$.[24] Both current elements are in the magnetic field of the other. For parallel currents Eq. 16.43 becomes

$$F_{1\text{-}2} = \frac{\mu}{4\pi r^2} I_1 dl_1 I_2 dl_2 \qquad 16.44$$

Because differential elements of current cannot be isolated, Eq. 16.43 and Eq. 16.44 are of use only when integrated over the entire path of the currents I_1 and I_2. The situation described by Eq. 16.43 and Eq. 16.44 is shown in Fig. 16.13.

Figure 16.13 *Magnetic Force Between Current Elements*

The magnetic force between two moving charges is given by Eq. 16.45. A specific case is shown in Fig. 16.13.

$$\mathbf{F}_{1\text{-}2} = \frac{\mu}{4\pi}\left(\frac{Q_1 Q_2}{r^2_{1-2}}\right)\mathbf{v}_2 \times (\mathbf{v}_1 \times \mathbf{r}_{1\text{-}2}) \qquad 16.45$$

If the charges are moving parallel to one another, the maximum magnetic force is

$$F = \frac{\mu}{4\pi r^2} Q_1 Q_2 \mathrm{v}^2 \qquad 16.46$$

[24]The magnetic permeability, μ, can only be used where one is operating over a linear region of the *BH* curve. If not, the magnetic permeability of free space, μ_0, should be used in its place. For hard ferromagnetic materials, **B** and **M** are directly related and μ need not be used.

Comparison of Magnetic and Electric Force

The maximum coulombic force between moving charges in free space can be deduced from Eq. 16.3 to be

$$F_e = \frac{Q_1 Q_2}{4\pi\epsilon_0 r^2} \quad 16.47$$

The maximum magnetic force between moving charges in free spaces can be deduced from Eq. 16.45 to be

$$F_m = \frac{\mu_0 Q_1 Q_2 \mathrm{v}^2}{4\pi r^2} \quad 16.48$$

Comparing the maximum magnetic force F_m to the maximum electric force F_e gives

$$\frac{F_m}{F_e} = \epsilon_0 \mu_0 \mathrm{v}^2 \quad 16.49$$

Consider Eq. 16.49. In electrical engineering practice the velocities of electrons are much less than the velocity of light, therefore the magnetic force is much weaker than the electric force. Additionally, the speed of light squared is fundamentally related to magnetic permeability and the electric permittivity as follows.

$$c^2 = \frac{1}{\mu_0 \epsilon_0} \quad 16.50$$

Finally, the magnetic force can be viewed as the electric force multiplied by the factor $(\mathrm{v}/c)^2$. That is, the magnetic force is a result of charges in relative motion.

$$F_m = F_e\left(\frac{\mathrm{v}}{c}\right)^2 \quad 16.51$$

The magnetic and electric forces on two reference charges moving in parallel are shown in Fig. 16.14.

Figure 16.14 *Magnetic and Electric Forces on Moving Charges*

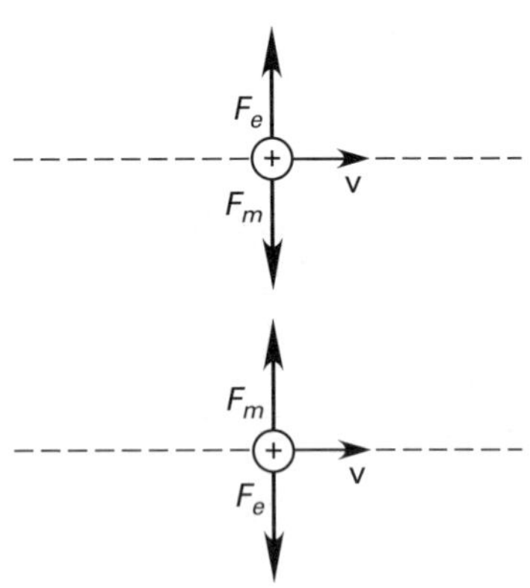

Magnetic Superposition

Magnetic forces occur as a result of the interaction of a pair of current-carrying conductors or the equivalent. One sets up the magnetic field and the other exhibits a force reaction to this same magnetic field. In free space and most nonmagnetic materials, the magnitude of the magnetic field is proportional to the total current that sets up the field as long as the permeability of the medium is constant. However, this is not true for iron and other ferromagnetic materials. Therefore, in general, the principle of linear superposition does not apply to current-carrying conductors in the vicinity of ferromagnetic materials.

23. MAGNETIC FIELDS

A magnetic field is a region in which a moving electric charge exerts a measurable force on another moving charge. It is a vector force field because a moving test charge placed in the field will experience a force in a given direction and with a specific magnitude. The direction of the field is along the flux lines, which are perpendicular to the current (that is, the moving charges creating the magnetic field). These lines, ideally, represent the direction of the force exerted on an infinitely small, isolated positive test charge moving with speed v. The actual direction is represented by the unit vector **a**. Equation 16.52 represents the *strength of the B-field*, or the *magnetic flux density*, in a substance with permeability μ at a distance r from current element $d\mathbf{l}_1$ that gives the direction of one turn, N, of the total current, I.[25] The denominator, $4\pi r^2$, represents the surface area of a sphere centered on the current element $d\mathbf{l}$ whose surface just touches the point in space where the **B**-field is being evaluated. Specific configurations may be evaluated with Eq. 16.52 by changing **B** to $d\mathbf{B}$ and l to dl and integrating.

$$\mathbf{B} = \frac{\mu I l}{4\pi r^2}\mathbf{a} \quad 16.52$$

B should not be called the magnetic field strength. Magnetic field strength is designated by **H**.[26] Although **H**, which will be defined in Sec. 16.26, is directly related to the current I (as **E** is directly related to the charge), the fundamental field is **B**, as **B** enters into the force relations (as **E** enters into the force relations in electrokinetics). **B** is analogous with **E**, and **H** is analogous with **D**. This is true though a review of the units in each case would seem to indicate otherwise.

The amount of *magnetic flux* in a magnetic field is symbolized by Φ or by the commonly used ϕ, both measured in webers (Wb). The magnetic flux density, **B**, is measured in teslas (T). One T is equivalent to 1 $\mathrm{Wb/m^2}$. **B** is also known as the *magnetic induction*, the *intensity of magnetization*, and the *dipole moment*

[25]The **B** field has two sources. One is the current, and the other is the magnetization, **M**.

[26]For historical reasons, the concept of the **B** field in terms of magnetic flux density developed first, prior to the concept of a field or field strength **H**. The names are opposites of those used when discussing electric quantities. Nevertheless, **B** is analogous to **E**, and **H** is analogous to **D**.

per unit volume.[27] **B** is given in terms of the flux in Eq. 16.53 and Eq. 16.54.

$$B = |\mathbf{B}| = \frac{\phi}{A} \quad 16.53$$

$$\mathbf{B} = \frac{\phi}{A}\mathbf{a} \quad 16.54$$

24. PERMEABILITY AND SUSCEPTIBILITY

The behavior and magnitude of magnetic forces between moving charges depend on the environment in which the moving charges are located. The largest magnetic forces occur in ferromagnetic materials. Another way of viewing this phenomenon is to say that magnetic flux does not pass equally through all materials. In fact, diamagnetic material repels magnets.

The *permeability of free space* (free space is a vacuum), μ_0, represents the magnetic properties of free space. However, it is better regarded as a constant relating the units in mechanical and electromagnetic systems. 1.2566×10^{-6} N/A^2 $= 4\pi \times 10^{-7}$ N/A^2 (H/m) is the value for μ_0 in the SI system. The product of the permeability of free space and the *relative permeability* of a medium gives the *permeability* of the medium.

$$\mu = \mu_0\mu_r \quad 16.55$$

Permeability is often defined in terms of the magnetic flux density and the magnetic field strength necessary to create it, as in Eq. 16.56.

$$\mu = \frac{B}{H} \quad 16.56$$

When **B** and **H** are not parallel, the permeability, μ, is a tensor. In general, when losses occur within a material, the permeability becomes a complex number and curves that show the variations in the real and imaginary terms are called the *magnetic spectrum* or *permeability spectrum* of the material.

The permeability can also be expressed as follows.

$$\mu = \mu_0(1 + \chi_m) \quad 16.57$$

The term χ_m is a dimensionless quantity called the *magnetic susceptibility.* Magnetic susceptibility is a measure of the alignment in a magnetic material. It is also the ratio of the magnetization, M, to the external applied magnetic field, H.

$$\chi = \frac{M}{H} \quad 16.58$$

For diamagnetic and paramagnetic materials, the susceptibility is essentially constant. For ferromagnetic materials, the susceptibility is nonlinear with respect to the applied field, as shown in Fig. 16.2.

25. MAGNETIC FLUX

The magnetic flux, Ψ_m, or more correctly, Φ, is defined as the surface integral of the normal component of the magnetic flux density, **B**, over an area, A, and is given by Eq. 16.59.

$$\Phi = \Psi_m = \iint \mathbf{B}\cdot d\mathbf{A} \quad 16.59$$

It is also useful to define the flux, ϕ, as the magnetic lines of force in a region.[28] In this case, the flux is given by Eq. 16.60, with N representing the number of complete turns of closely spaced conductors of length l carrying current I. The term NI in Eq. 16.60 is sometimes referred to as the *magnetomotive force* (mmf or preferably F_m).

$$\phi = \mu NIl = BA \quad 16.60$$

The orientation of the magnetic field lines and magnetic flux lines depends on the properties of the material, specifically the permeability. By convention, the flux lines exit the north pole as given by the right-hand rule.[29] The magnetic field lines always form closed loops. They do not terminate on anything but themselves (see Fig. 16.10 and Fig. 16.11). Since no magnetic charges exist, the magnetic flux on a closed sphere (or any closed surface) is equal to zero.

26. MAGNETIC FIELD STRENGTH

The *magnetic field strength,* **H**, in units of A/m, is derived from the magnetic flux density.

$$\mathbf{H} = \frac{1}{\mu}\mathbf{B} \quad 16.61$$

The magnetic field strength is a vector point function whose curl is the current density. Because it is proportional to the magnetic flux density in regions where no magnetized substances exist, and because the *cgs* system units define **H** in terms of lines of flux per unit area, **H** is referred to as the magnetic flux density in texts on magnetic theory. This explains its correlation with the electric flux density, **D**.[30] When using the SI system of

[27]The word "induction" stems from an earlier era. When an unmagnetized piece of iron was brought near a magnet, magnetic poles were said to be "induced" in the iron.

[28]The symbol for the flux, Ψ_m, does not necessarily have to change. It is shown here as commonly used, ϕ. The standard symbol for magnetic flux is Φ. The additional symbols are used to alert the reader to the range of symbols that may be encountered.

[29]In the case of straight wire, the thumb indicates the current direction and the fingers curl in the direction of the magnetic field. For a coil of wire, the fingers curl in the direction of the positive current flow and the thumb points in the direction of the magnetic field direction. The thumb points toward the north pole.

[30]There are other reasons that **H** is analogous to **D**. For instance, **D** is used to avoid direct reference to polarization charges and the polarization, **P**. It is then easier to work with **E** and **D**. **H** is used to avoid direct reference to the magnetization, **M**. It is then easier to work with **B** and **H**.

units, **H** should be referred to only as the magnetic field strength.

The magnetic field and inductance for various configurations are given in Table 16.8.

Example 16.8

A long wire with a radius of 0.001 m carries 30 A of current. What are the magnetic field strength and the magnetic flux density 0.01 m from the surface of the wire? (This is roughly equivalent to asking what are the magnetic field strength and flux density at the inside surface of a conduit containing a 10 AWG copper wire carrying the maximum current allowed by the *National Electrical Code.*)

Solution

From Table 16.8, the magnetic field strength is

$$H = \frac{I}{2\pi r} = \frac{30 \text{ A}}{2\pi(0.001 \text{ m} + 0.01 \text{ m})} = 434.06 \text{ A/m}$$

From Eq. 16.56, using the permeability of free space, the flux density is

$$\begin{aligned} B = \mu H &= \left(4\pi \times 10^{-7} \frac{\text{Wb}}{\text{A·m}}\right)\left(434.06 \frac{\text{A}}{\text{m}}\right) \\ &= 5.45 \times 10^{-4} \text{ T} \end{aligned}$$

27. GAUSS' LAW FOR MAGNETIC FLUX

A law similar to Gauss' law for electric flux can be written for magnetic flux. (See Sec. 16.12.) Such a law states that the amount of flux passing through any closed surface is equal to zero.

$$\Phi = \psi_m = \iint \mathbf{B} \cdot d\mathbf{A} = 0 \qquad 16.62$$

The physical significance of Eq. 16.62 is that magnetic flux lines form continuous closed loops. Additionally, this indicates that the divergence of any magnetic field is zero. In other words, no magnetic charge exists. The law refers to magnetic flux rather than magnetostatics as it is also applicable to time-varying magnetic fields (that is, magnetokinetics).

28. INDUCTANCE AND RECIPROCAL INDUCTANCE

When charges are in motion, a current is said to be flowing and a magnetic field is established. When the current varies with time, the generated magnetic field opposes the current change and in doing so "stores" energy in the magnetic field. *Inductance*, *L*, is the property of a system of conductors and circuits that permits this storage.

Table 16.8 *Magnetic Field and Inductance for Various Configurations*

straight infinite conductor

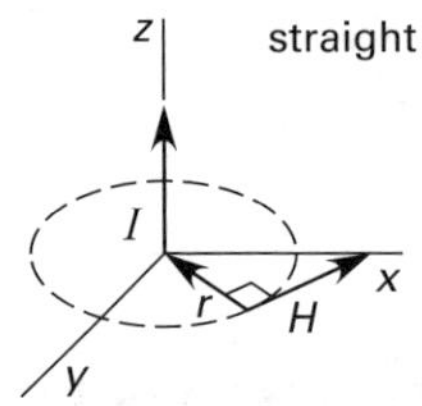

$$H = \frac{I}{2\pi r}$$

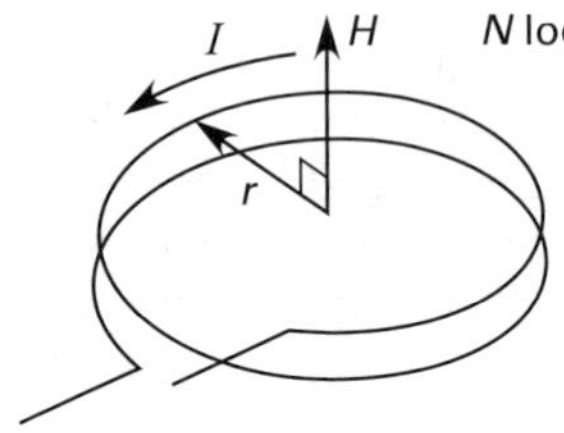

N loops

$$H = \frac{NI}{2r} \quad \text{[center of coil only]}$$

infinite cylindrical coil helix (solenoid)

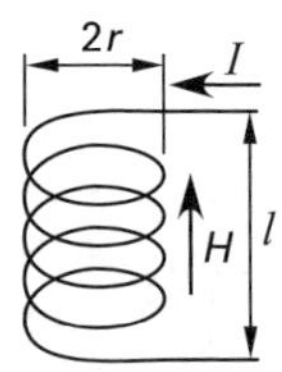

$$H = \frac{NI}{l} \quad [l >> r]$$

$$\frac{L}{l} = \mu\left(\frac{N}{l}\right)^2 A_{\text{coil}}$$

torus (toroidal coil)

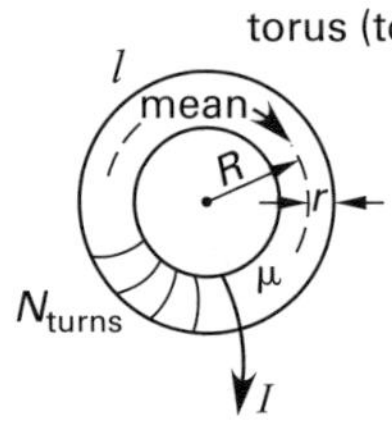

$$H = \frac{NI}{2\pi R} \quad [r << R]$$

$$L = \frac{\mu N^2 A_{\text{core}}}{l_{\text{mean}}} = \frac{\mu N^2 r^2}{2R}$$

coaxial cable (high frequencies)

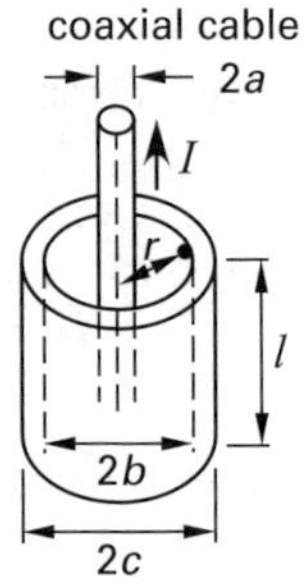

$$H = \frac{I}{2\pi r} \quad [a < r < b]$$

$$\frac{L}{l} = \frac{\mu}{2\pi} \ln \frac{b}{a}$$

parallel transmission lines (high frequencies)

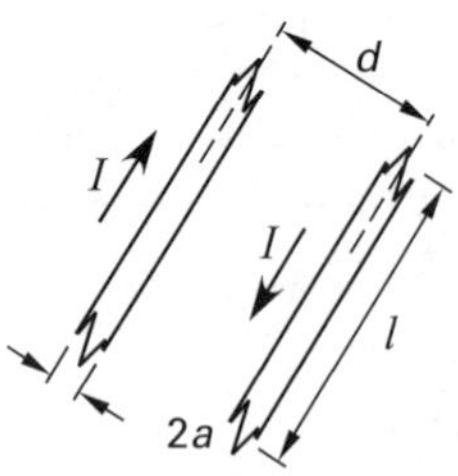

$$H = \frac{2I}{\pi d} \quad \left[\begin{matrix}\text{directly between} \\ \text{wires only}\end{matrix}\right]$$

$$\frac{L}{l} = \frac{\mu}{\pi} \ln \frac{d-a}{a} \quad [d >> a]$$

Reprinted with permission from *Core Engineering Concepts for Students and Professionals*, by Michael R. Lindeburg, PE, copyright 2010, by Professional Publications, Inc.

If a time-varying current is applied to a conductor, usually in the form of a coil of wire, the magnetic field builds up, producing a potential difference across the ends of the conductor. The amount of potential, V, is proportional to the magnitude of the rate of change of current. The proportionality constant is called the *inductance*, L, and is measured in *henrys*, H. This is shown in Eq. 16.63.[31]

$$\text{v} = L\frac{di}{dt} \qquad 16.63$$

The *reciprocal inductance*, Γ, a seldom used term, is the reciprocal of the inductance.

$$\Gamma = \frac{1}{L} \qquad 16.64$$

The type of inductance just described is also called *self-inductance*. It can also be described in terms of *flux linkage*, Λ (measured in Wb). Flux linkage is defined as the lines of flux that link the entire magnetic circuit. For example, in Fig. 16.15 the "linked flux" is the flux through the center of the solenoid that links the current flows of all the wire turns (the small flux loops around each turn of wire are ignored). Using this definition, the inductance would be given by Eq. 16.65.

$$L = \frac{\Lambda}{I} = \frac{N\phi}{I} \qquad 16.65$$

Mutual inductance, M, vital to transformer theory, is the term used to describe the fact that a changing current in one circuit induces an electromotive force in another. It is given by Eq. 16.66 where Λ represents the flux links in one winding and I is the current in the other winding.

$$M = \frac{\delta\Lambda}{\delta I} \qquad 16.66$$

Figure 16.15 *Magnetic Flux Links in a Solenoid*

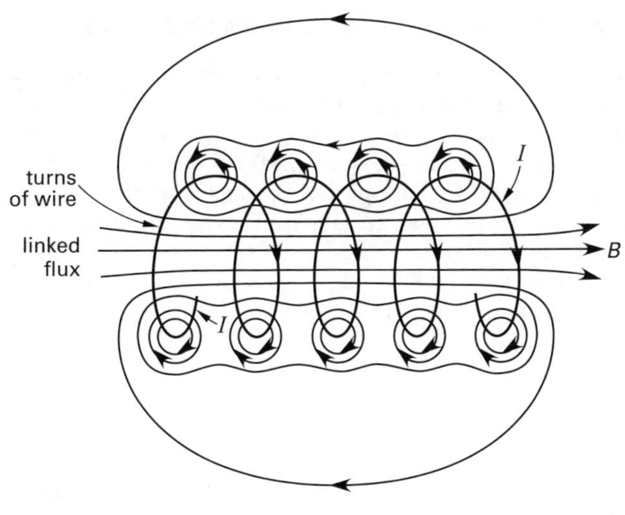

[31]This equation technically belongs in the Magnetokinetics section. It is often stated as the definition of inductance and is used here for that reason.

29. INDUCTORS

An *inductor* is a device that stores magnetic energy. It typically consists of coils of wire with a core material in the center, which can be air. A common type of inductor consists of N turns of wire with radius r and an iron core center with permeability μ. This type of inductor is called a *solenoid*, and its inductance is determined by Eq. 16.67.

$$L = \mu\frac{N^2 A_{\text{coil}}}{l} \qquad 16.67$$

The formula for the inductance of other configurations is given in Table 16.8.

The total energy, U (in joules), in the magnetic field of an inductor with inductance L and current I is

$$U = \tfrac{1}{2}LI^2 \qquad 16.68$$

For inductors in series, the total inductance is

$$L_{\text{total}} = L_1 + L_2 + L_3 + \cdots + L_n \qquad 16.69$$

The reciprocal of the total inductance for inductors in parallel is

$$\frac{1}{L_{\text{total}}} = \frac{1}{L_1} + \frac{1}{L_2} + \frac{1}{L_3} + \cdots + \frac{1}{L_n} \qquad 16.70$$

30. ENERGY DENSITY IN A MAGNETIC FIELD

The average energy (in joules) in a magnetic field is

$$U_{\text{ave}} = \tfrac{1}{2}\phi NI \qquad 16.71$$

The average energy density at any point in this magnetic field is

$$u_{\text{ave}} = \frac{U_{\text{ave}}}{V_{\text{volume}}} = \tfrac{1}{2}BH = \tfrac{1}{2}\mu H^2 = \tfrac{1}{2}\frac{B^2}{\mu} \qquad 16.72$$

The average energy stored in inductors that are mutually coupled is given by Eq. 16.73. Here M is the magnetization (with units of henrys) and I_1 and I_2 represent the currents in the two inductors.

$$U_{\text{ave}} = MI_1I_2 \qquad 16.73$$

Basic Theory

PART 6: MAGNETOKINETICS

31. SPEED AND DIRECTION OF CHARGE CARRIERS

A positive free charge moving with velocity, $\mathbf{v}$, in a magnetic field experiences a force given by Eq. 16.74.

$$\mathbf{F} = Q\mathbf{v} \times \mathbf{B} \qquad 16.74$$

The magnetic force can change a charge's direction but not its kinetic energy. This is in contrast to the force the charge experiences in an electric field, which does work on the charge and therefore changes its kinetic energy.[32]

If there are many charges flowing through a conductor in a magnetic field, that is, if a current-carrying conductor exists in an external magnetic field, the differential force equation is

$$\begin{aligned} d\mathbf{F} &= (dq)(\mathbf{v} \times \mathbf{B}) = (I\,dt)(\mathbf{v} \times \mathbf{B}) \\ &= I(d\mathbf{l} \times \mathbf{B}) \end{aligned} \qquad 16.75$$

In Eq. 16.75, $d\mathbf{l}$ is a unit length in the direction of conventional (i.e., positive) current flow I. If the conductor is straight and the magnetic field is constant along the length of the conductor, a version of *Ampère's law* results.

$$F = NIBl\sin\theta \qquad 16.76$$

N represents the number of turns of current-carrying conductor in the magnetic field and θ is the angle between the conductor (i.e., the current element $d\mathbf{l}$) and the magnetic field, $\mathbf{B}$. If these are at right angles, which is often the case, the force is represented by

$$F = NIBl \qquad 16.77$$

The situations in Eq. 16.75 through Eq. 16.77 are illustrated in Fig. 16.16.

Care must be taken to properly interpret directions in the figure. If an external magnetic force is applied in the direction shown, the conductor has a velocity in the same direction. Applying Eq. 16.74 then gives the direction of force on the charge carriers within the conductor—in this case, in the direction of the current shown. This is the basis of generator theory. If instead a current is applied to a loop of wire, of which the conductor represents one section, the velocity in Eq. 16.74 is in the direction of the charge carriers, that is, the direction of the current flow shown. In this case, the force given by Eq. 16.74 is in a direction opposite to the force arrow shown. This is the basis of motor theory.

[32]Consider that if a charge is not in motion with respect to a magnetic field, it cannot "see" the magnetic field and therefore experiences no change in kinetic energy. A charge stationary with respect to an electric field can "see" the field and therefore experiences a force that changes its kinetic energy.

Figure 16.16 Conductor in a Magnetic Field

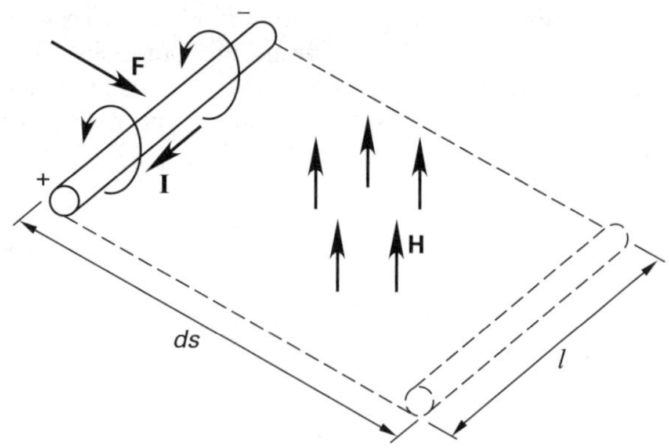

If a magnetic field, $\mathbf{B}$, is uniform within a region and a charge has an initial velocity perpendicular to this field, the resulting path of the particle is a circle of radius r. The magnitude of the force is determined by Eq. 16.74. Such a force would be directed toward the center of the circle. The centripetal acceleration is of a magnitude $\omega^2 r = \mathrm{v}^2/r$. Applying Newton's second law gives the following for the radius.

$$|Q|\,\mathrm{v}B = m\frac{\mathrm{v}^2}{r} \qquad 16.78$$

Rearranging Eq. 16.78 gives

$$r = \frac{m\mathrm{v}}{|Q|\,B} \qquad 16.79$$

When both electric and magnetic fields are in a given region at the same time, the force is given by the *Lorentz force equation*, Eq. 16.80.

$$\mathbf{F} = Q(\mathbf{E} + \mathbf{v} \times \mathbf{B}) \qquad 16.80$$

Electromagnetic Oscillations and Waves

A time-varying electric field produces time-varying currents that generate magnetic fields. If the electromagnetic variations are restricted to a small region, such as within a circuit, no spatial variations of the field occur and electrical oscillations are produced. If the electromagnetic variations are generated over large areas of space, compared to the wavelength of the oscillation, electromagnetic waves are generated that propagate through space with a velocity given by

$$\mathrm{v} = \frac{1}{\sqrt{\epsilon\mu}} \qquad 16.81$$

When a conduction current generates the electromagnetic field, the magnetic field strength, $\mathbf{H}$, is proportional to the electric field strength, $\mathbf{E}$, and an induction field is set up, keeping the energy within the circuit. If the electromagnetic field is generated by a displacement current, the magnetic field strength, **H**, is directly proportional to the rate of change of the electric field strength, $d\mathbf{E}/dt$. A radiation field is thereby created, and energy can leave the system (i.e., be radiated into space).

Electric and Magnetic Energy Conversion

Applying the law of conservation of energy to a lossless, energy-storing system in which charges move in directed nonuniform motion means the sum of the instantaneous values of electric and magnetic energy must remain constant. Changes in the electric energy necessarily mean changes in the magnetic energy as given in Eq. 16.82.

$$du_e = -du_m \qquad 16.82$$

32. VOLTAGE AND THE MAGNETIC CIRCUIT

An *electric potential* or *voltage* is the work done on a unit charge to bring it from some specified reference point to another point. The symbol V is generally used for constant voltages and v is used for time-varying systems. The unit is the *volt* with 1 V = 1 J/C. By convention, current flows from the positive terminal on a voltage source to the negative outside the source and from negative to positive inside the source. Also by convention, the current flows from the positive terminal to the negative terminal through a resistor.[33] This situation is illustrated in Fig. 16.17 along with an analogous *magnetic circuit.*

Figure 16.17 *Magnetic-Electric Circuit Analogy*

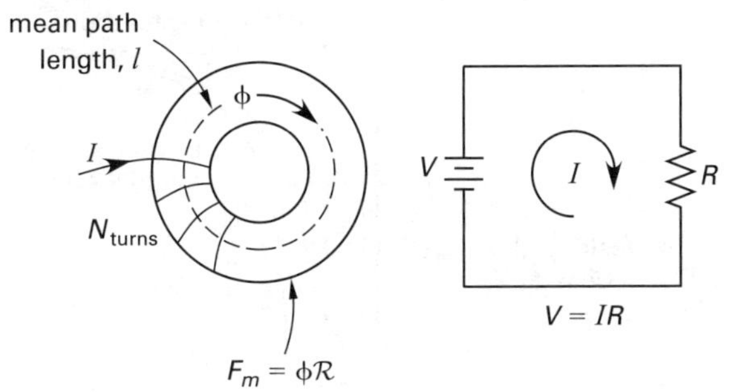

In a magnetic circuit, the flux density is proportional to the *magnetomotive force* (mmf), F_m, given in Eq. 16.83.

$$F_m = IN \qquad 16.83$$

The magnetomotive force is in units of amp-turns. Turns, like revolutions or radians, are for clarification only and can be disregarded in equation manipulations.

[33]These directions are based on conventional current flow and are arbitrary. As long as one remains consistent when choosing reference directions for both voltage and current, the correct results will be obtained.

The *reluctance*, $\mathcal{R}$, measured in A/Wb, is analogous to electrical resistance and is given by

$$\mathcal{R} = \frac{l}{\mu A} \qquad 16.84$$

The permeability depends on the flux density. Therefore, iterations between Eq. 16.84 and Eq. 16.83 may be needed if either the reluctance or flux density is an unknown.

The magnetic equation that correlates with Ohm's law in electric circuits is

$$F_m = Hl = \phi\mathcal{R} \qquad 16.85$$

Example 16.9

A 0.001 m slice is taken from a cast-iron toroidal coil ($\mu_r = 2000$) with a 0.35 m mean toroidal diameter and a 0.07 m core diameter. A steady unknown current flows through 300 turns of wire wrapped around the coil, producing a constant flux density in the core and air gap. What current is required to establish a flux of 10 mWb across the air gap?

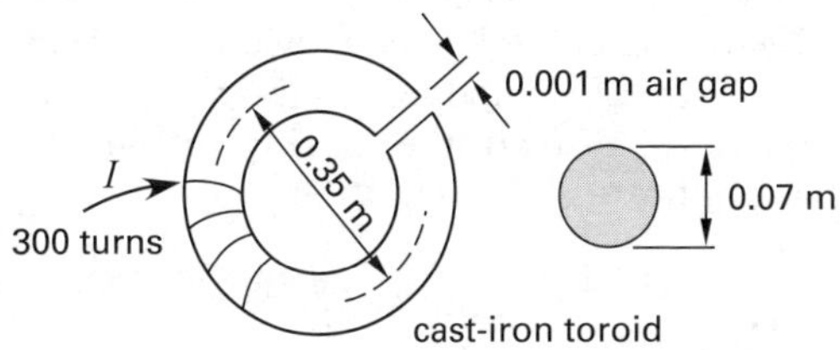

Solution

The mean path length through the cast iron is

$$l_{ci} = \pi(0.35 \text{ m}) - 0.001 \text{ m} = 1.0986 \text{ m}$$

The cross-sectional area of the flux path is

$$A = \frac{\pi}{4}d^2 = \frac{\pi}{4}(0.07 \text{ m})^2 = 0.003848 \text{ m}^2$$

The total reluctance is the sum of the reluctances of the cast iron and air paths. From Eq. 16.84,

$$\begin{aligned}\mathcal{R} &= \sum\frac{l}{\mu A} = \frac{1}{\mu_0 A}\sum\frac{1}{\mu_r} \\ &= \frac{1}{\left(4\pi \times 10^{-7} \frac{\text{Wb}}{\text{A·m}}\right)(0.003848 \text{ m}^2)} \\ &\quad \times \left(\frac{1.0986 \text{ m}}{2000} + \frac{0.001 \text{ m}}{1}\right) \\ &= 3.204 \times 10^5 \text{ A/Wb}\end{aligned}$$

The flux, ϕ, is given as 10^{-2} Wb. From Eq. 16.83, the magnetomotive force is

$$F_m = IN = I\ 300$$

From Eq. 16.83 and Eq. 16.85, the current is

$$I = \frac{\phi \mathcal{R}}{N}$$

$$= \frac{(10^{-2}\ \text{Wb})\left(3.203 \times 10^5\ \frac{\text{A}}{\text{Wb}}\right)}{300\ \text{turns}} = 10.68\ \text{A}$$

33. MAGNETIC FIELD-INDUCED VOLTAGE

When a conductor "cuts" lines of magnetic flux at a given rate, a voltage called the *induced electromotance*, or *electromotive force* (emf) is produced in the conductor, as illustrated in Fig. 16.16. The magnitude of such an *electromagnetic induction* is given in Eq. 16.86, which is the mathematical statement of *Faraday's law*. Faraday's law states that the induced voltage is proportional to the time rate of change of the magnetic flux linked with the circuit. The minus sign is mandated by *Lenz's law*, which states that the current flow in a conductor caused by an induced voltage moves in a direction that creates a magnetic flux that opposes the magnetic flux inducing the voltage. More succinctly, the direction of the induced emf is such as to oppose any change in current. This is a consequence of the law of conservation of energy. (The minus sign indicates this opposing nature of the induced voltage. It is not shown in all equations. Instead, the polarity of the voltage in the circuit indicates the opposition.)

$$V = -N\frac{d\phi}{dt} \tag{16.86}$$

Because the magnetic flux linkage is given by $\Lambda = N\phi$, and the induced voltage is the time rate of change of these linkages, the induced voltage can be written in terms of the velocity of the conductors.

$$V = N\frac{d\phi}{dt} = NBl\frac{ds}{dt} = NBl\text{v} \tag{16.87}$$

In both Eq. 16.86 and Eq. 16.87, N is the number of conductors, l is the length of the conductor in the magnetic field, and $d\mathbf{s}$ is the differential distance traveled by the conductor through the field (see Fig. 16.16). Equation 16.86 is the *flux-changing method* of generating a voltage. Equation 16.87 is the *flux-cutting method* of generating a voltage.

34. EDDY CURRENTS

Eddy currents, also known as *Foucault currents*, occur as a result of induced voltages in a conducting medium by a varying magnetic field. Because the currents so induced cannot leave the conducting medium, they "deflect" off the edges to form loops, or circulating currents. Such currents result in Joule heating of the conducting medium and can result in considerable losses if not adequately addressed in the design of the system.

35. DISPLACEMENT CURRENT MAGNETIC EFFECT

The displacement current explained in Sec. 16.19 occurs only in dielectrics and insulators. The conduction electrons in these materials are tightly bound and cannot move freely through the material. Nevertheless, the displacement current creates a magnetic field intensity, **H**, proportional to the time rate of change of the electric field intensity, **E**, producing the displacement current. (By contrast, in a conduction current, **H** and **E** are in time phase.) Because **H** is proportional to $d\mathbf{E}/dt$, and **E** is proportional to $d\mathbf{H}/dt$, it is possible to generate electromagnetic waves in nonconducting material.

36. MAGNETIC HYSTERESIS

Plotting the values of the magnetic flux density, **B**, for various values of the magnetic field strength, or magnetizing force, **H**, for a ferromagnetic material reveals a nonlinear relationship. This is illustrated in Fig. 16.18. The phenomenon causing values of **B** to lag behind **H** so that the fields differ in magnitude is called *hysteresis*. The loop traces a complete cycle of **BH** values and is termed the *hysteresis loop*.

Figure 16.18 *Magnetic Hysteresis*

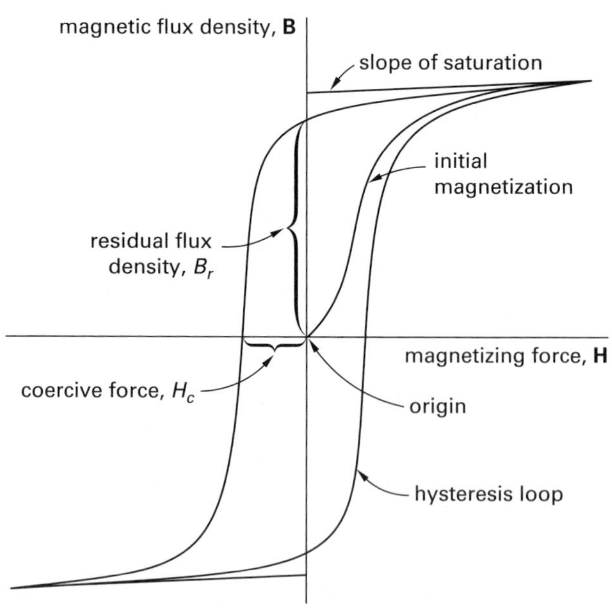

17 Electronic Theory

1. Fundamentals 17-1
2. Charges in a Vacuum 17-2
3. Charges in Liquids and Gases 17-2
4. Charges in Semiconductors 17-4

Nomenclature

A_0	constant	–
B, $\mathbf{B}$	magnetic flux density	T or Wb/m^2
C	constant	cm^{-6}
D	diffusion constant	m^2/s
e^-	electron charge, -1.6022×10^{-19}	C
E, $\mathbf{E}$	electric field strength	V/m or N/C
E	energy	J or eV
E_c	conduction-band energy edge	J or eV
E_v	valence band energy edge	J or eV
f	frequency	Hz
F, $\mathbf{F}$	force	N
h	Planck's constant, 6.6256×10^{-34}	J·s
J, $\mathbf{J}$	current density	A/m^2
k	Boltzmann's constant, 1.3807×10^{-23}	J/K
m	mass	kg
m_e	electron mass, 9.1094×10^{-31}	kg
n	electron carrier density	m^{-3}
N_c	effective conduction-band density state	cm^{-3}
N_v	effective valence-band density state	cm^{-3}
p	hole carrier density	m^{-3}
p^+	proton charge, 1.6022×10^{-19}	C
q	electric unit charge, 1.6022×10^{-19}	C
Q	charge	C
r	radius	m
T	temperature	K
U	energy	J
v, $\mathbf{v}$	velocity[1]	m/s
V	constant or rms voltage	V

Symbols

ϵ_0	free-space permittivity, 8.854×10^{-12}	C^2/N·m^2
λ	wavelength	m
μ	mobility	m^2/V·s
ν	frequency	rad/s
ρ	charge density	C/m^3
ρ_e	electron charge density	electrons/m^3
σ	conductivity	S/m
τ	mean lifetime	s
φ	work function[2]	J
ω	angular velocity	rad/s

Subscripts

0	initial or rest
d	drift
e	electron
g	gap
h	hole
i	intrinsic
n	electrons
p	holes
T	temperature
$\parallel$	parallel
$\perp$	perpendicular

1. FUNDAMENTALS

Electronic theory is the study of subatomic particles, such as electrons, in materials other than metals. It is concerned with the release, transport, control, and collection of charge carriers in a vacuum, in gases, and in semiconductors. While the charge carriers are usually electrons, they can be holes (a concept used to describe the absence of an electron), positive ions, or negative ions.

The fundamentals of electronic engineering are identical to those for electricity and magnetism. Because of this, many of the formulas used in this chapter will be repetitious. Their form and symbology may differ slightly, as often occurs in electrical engineering texts; this is purposefully done to expose readers to variations they may encounter. The principles and results are consistent. All the electromagnetic theory presented in Chap. 16 is applicable to electronic theory as well.

The primary difference between electronics and electromagnetism is in their applications. Electronic engineering allows for much greater control over the instantaneous, rather than the average, movement of charge carriers. Additionally, the control of these carriers can be extremely rapid. Electronic devices are generally nonlinear. That is, the output terminals can provide an amplified version of the current, voltage, or power supplied to the input terminals. The devices tend to be

[1]The term "velocity" is often used in text where it would be more technically correct to use the term "speed." Velocity is a vector quantity with speed as its magnitude.

[2]The units for the work function are often electron volts, eV. One eV is equivalent to 1.6022×10^{-19} J.

smaller than their electromagnetic counterparts and use considerably less energy. The field of electronic engineering is focused on *active electron devices*, those devices requiring an external power source in order to maintain their terminals at suitable operating voltages and currents. Electronics is generally more dependent on the material properties of a medium, such as composition and electronic structure, as well as the mode of electric conduction.

2. CHARGES IN A VACUUM

A charge Q in a vacuum experiencing an electric field intensity $\mathbf{E}$ (with units of V/m) will undergo an acceleration in the direction of the field. The force is

$$\mathbf{F}_{\parallel} = Q\mathbf{E} \qquad 17.1$$

If the charge is in a magnetic field of flux density $\mathbf{B}$ (with units of T) and velocity $\mathbf{v}$ in m/s, it will experience a transverse force of

$$\mathbf{F}_{\perp} = Q\mathbf{v} \times \mathbf{B} \qquad 17.2$$

The direction of motion of the particle will be a circle if the magnetic field is uniform and the particle enters it at right angles. The radius of the circle is

$$r = \frac{m\mathrm{v}}{QB} \qquad 17.3$$

If the charge is in an electromagnetic field, the total force is

$$\mathbf{F} = \mathbf{F}_{\parallel} + \mathbf{F}_{\perp} = Q(\mathbf{E} + (\mathbf{v} \times \mathbf{B})) \qquad 17.4$$

The energy of a charge Q in a vacuum, moving in an electric field $\mathbf{E}$ from point 1 to point 2, will experience a change in energy U (in units of J)[3] of

$$U = Q\int_{x_1}^{x_2} \mathbf{E}\cdot d\mathbf{x} = Q(V_1 - V_2) \qquad 17.5$$

Vacuum Tubes

A *vacuum tube* is a device in which electrons move through a vacuum or gaseous medium in a gas-tight envelope. In a basic vacuum tube, called a *triode*, the electrons are emitted from a heated *cathode* and move within the tube to the *plate* by an intervening *grid* that is used for amplification.

Tubes dominated electronic technology in the first half of the twentieth century as semiconductors do today. The release of electrons from the surface of a metal, which is the basis of the vacuum tube, remains an important area of study in electronic applications in space and in the construction of metal-semiconductor junctions.

[3]The magnetic field acts transverse to the direction of motion and can therefore only change the direction, not the energy.

Work Function

The *work function*, φ, is a measure of the energy required to remove an electron from the surface of a metal. It is usually given in terms of electron volts, eV. It is specifically the minimum energy necessary to move an electron from the *Fermi level* of a metal (an energy level below which all available energy levels are filled, and above which all are empty, at 0K) to infinity, that is, the vacuum level. This energy level must be reached in order to move electrons from semiconductor devices into the metal conductors that constitute the remainder of an electrical circuit. It is also the energy level of importance in the design of optical electronic devices.

In photosensitive electronic devices an incoming photon provides the energy to release an electron and make it available to the circuit, that is, free it so that it may move under the influence of an electric field. This phenomenon whereby a short wavelength photon interacts with an atom and releases an electron is called the *photoelectric effect*. The photon energy required to release an electron is

$$h\nu = \varphi + \tfrac{1}{2}m\mathrm{v}^2 \qquad 17.6$$

The term h is *Planck's constant* and ν is the *threshold frequency*. If the photon does not have the minimum energy required, no electron is released.

3. CHARGES IN LIQUIDS AND GASES

In any substance, such as a liquid or a gas, where the motion of a charged particle is hindered by surrounding particles, an average velocity called the *drift velocity*, $\mathbf{v}_d$, is established. The drift velocity is dependent on the electric field intensity, $\mathbf{E}$, and the *mobility*, μ, of the charge carrier as shown in Eq. 17.7.

$$\mathbf{v}_d = \mu\mathbf{E} \qquad 17.7$$

Mobility is measured in $\mathrm{m^2/V\cdot s}$. The concept of drift velocity in liquids and gases is illustrated in Fig. 17.1.

Conductivity and Resistivity

The mobility of the charge carriers is of special importance in electronic engineering, more so than in electrical engineering, although the principles are identical. The mobility is proportional to the conductivity.

$$\sigma = \rho\mu \qquad 17.8$$

The conductivity, σ, is measured in siemens per meter, S/m. The charge density is in units of $\mathrm{C/m^3}$. Metals have approximately 10^{28} atoms/m^3 with one to two electrons free to move, while silicon semiconductors have 10^{16} intrinsic charge carriers per cubic meter. Concentrations in gases vary widely with the type of gas and the conditions under which the gas is held.

Figure 17.1 *Drift Velocity in Liquids and Gases*

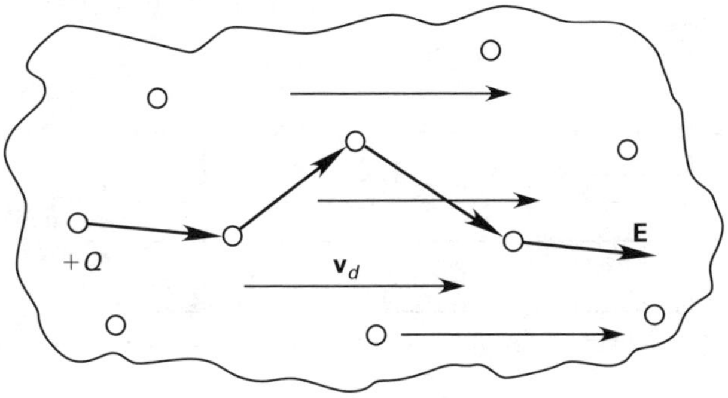

In metallic conductors, the conductivity is often defined by the point form of Ohm's law and given by

$$\sigma = \frac{|\mathbf{J}|}{|\mathbf{E}|} = \frac{J}{E} \qquad 17.9$$

Conductivity is more complex for liquids and gases. In general, both positive and negative ions may be present, some singly charged and some doubly charged, possibly with different masses. The conductivity must account for all these factors. If one assumes that the negative and positive ions are alike, the formula for conductivity in liquids and gases has only two terms. See Fig. 17.2 for the formula and a comparison to the formulas for conductors and semiconductors.

Example 17.1

A particular 10 AWG copper wire has a conductivity of 58×10^6 S/m. Assuming 10^{28} conduction electrons per cubic meter, what is the mobility of the electrons in the copper?

Solution

From Eq. 17.8, the conductivity is

$$\sigma = \rho\mu$$

Rearranging gives

$$\mu = \frac{\sigma}{\rho}$$

The conductivity is given. The charge density is

$$\begin{aligned}\rho &= \left(10^{28}\ \frac{\text{electron}}{\text{m}^3}\right)\left(-1.6022 \times 10^{-19}\ \frac{\text{C}}{\text{electron}}\right) \\ &= -1.6022 \times 10^{9}\ \text{C/m}^3\end{aligned}$$

Substituting gives

$$\begin{aligned}\mu &= \frac{\sigma}{\rho} = \frac{58 \times 10^6\ \frac{\text{S}}{\text{m}}}{-1.6022 \times 10^9\ \frac{\text{C}}{\text{m}^3}} \\ &= -3.6 \times 10^{-2}\ \frac{\text{S}\cdot\text{m}^2}{\text{C}} \quad (-3.6 \times 10^{-2}\ \text{m}^2/\text{V}\cdot\text{s})\end{aligned}$$

Figure 17.2 *Conductivity Formulas*

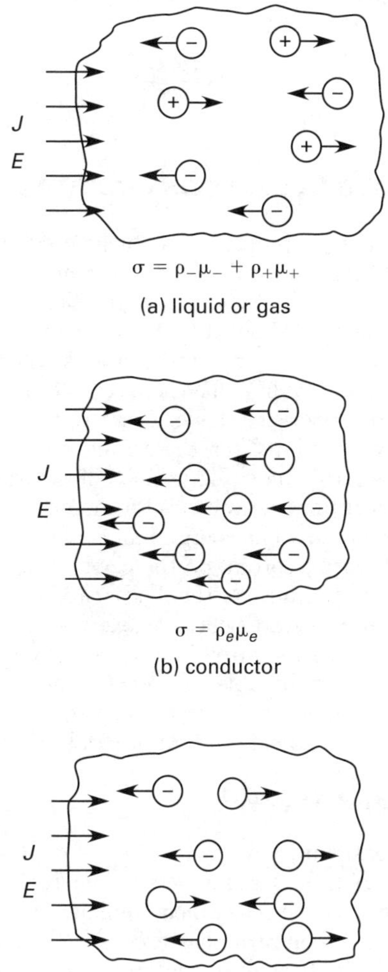

$\sigma = \rho_-\mu_- + \rho_+\mu_+$

(a) liquid or gas

$\sigma = \rho_e\mu_e$

(b) conductor

$\sigma = \rho_e\mu_e + \rho_h\mu_h$

(c) semiconductor

In metallic conductors, the charge carriers are electrons that move in a direction opposite that of the electric field. The charge density, ρ, and the mobility, μ, are both negative, leaving the conductivity positive. By convention, the electrons moving in one direction are treated as positive charges moving in the opposite direction, and ρ and μ are reported as positive.

Plasmas

Plasma, often viewed as a fourth state of matter, can be considered a highly ionized gas. Plasmas are highly conductive, with high concentrations of roughly equal numbers of positive and negative ions and a low voltage gradient. The negative charge carriers tend to be electrons that vibrate longitudinally about their mean position with a frequency of

$$f = \frac{1}{2\pi}\sqrt{\frac{\rho_e Q^2}{\epsilon_0 m_e}} \qquad 17.10$$

Let λ be the wavelength of oscillation of the frequency given by Eq. 17.10. Then the velocity of the electrons is

$$v = \lambda\sqrt{\frac{\rho_e Q^2}{\epsilon_0 m_e}} \qquad 17.11$$

4. CHARGES IN SEMICONDUCTORS

The conductivity is proportional to the concentration of free charge carriers (see Eq. 17.8). In a good conductor, this concentration is very large ($\approx 10^{28}$ electrons/m^3). In an insulator, it is very small ($\approx 10^7$ electrons/m^3). Semiconductor charge carrier concentrations vary between these approximate values. The semiconducting nature of these types of materials originates not so much from the number of carriers but from the arrangement of allowed energy states into bands. Figure 17.3 illustrates the general arrangement of energy levels for metals, semiconductors, and insulators. The energy bands and gaps are not to scale and vary according to the element or compound utilized. The importance of the figure can be summed up by the following statement: an electric field cannot impart energy to an electron in a completely filled band. The electron must be able to absorb energy and move into a new allowable energy level in order to move under the influence of the field or of a potential difference.

Energy-Band Theory[4]

Electrons exist in *core shells* (i.e., completely filled allowed energy levels) and in *valence shells* (i.e., allowed energy levels, some of which remain vacant). For example, the carbon atom has a core shell that is completely filled, the K shell (in electron configuration parlance, $1s^2$). The carbon atom's valence shell, the L shell, has only four of the possible eight electrons it could hold (in electron configuration parlance, $2s^2 2p^2$). When individual carbon atoms are brought together in a crystalline structure, the valence electrons are exposed to the potential energy changes of nearby atoms as illustrated in Fig. 17.4.

Because of this periodic potential, each energy level in the outermost shell splits into a number of discrete levels.[5] This is illustrated in Fig. 17.5. The periodic potential is usually accounted for in electronic equations by the use of an *effective mass* for electrons and holes, m_e^* and m_h^*.

At the atomic spacing d, atoms behave as individual atoms with discrete energy levels centered around the K and L shells. At atomic spacing c, the valence band splits, with two states in a lower band and six in an upper. At atomic spacing b, the bands overlap. Discrete energy levels still exist, but now an electron can move from low in the band to high in the band in small discrete steps of energy called *quanta*.[6] At atomic spacing a, the original L valence shell has been split into two distinct bands of

Figure 17.3 *Energy Band Diagrams*

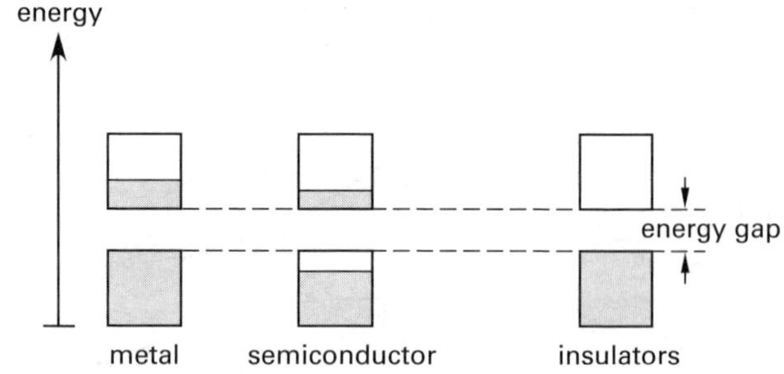

Figure 17.4 *Electron Potential Energy Variation in a One-Dimensional Crystal*

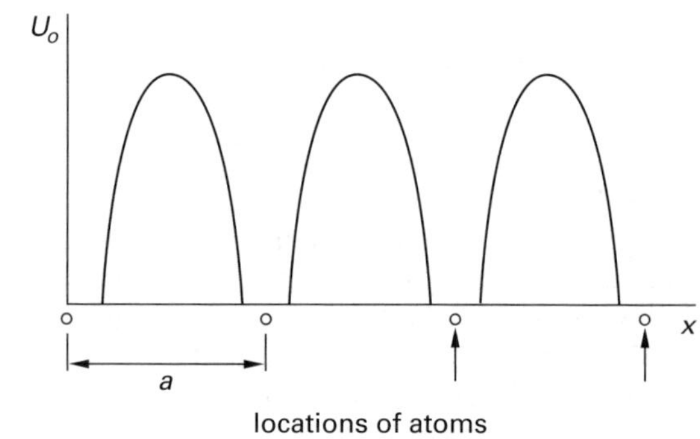

Figure 17.5 *Energy Bands of Carbon as a Function of Atomic Spacing*

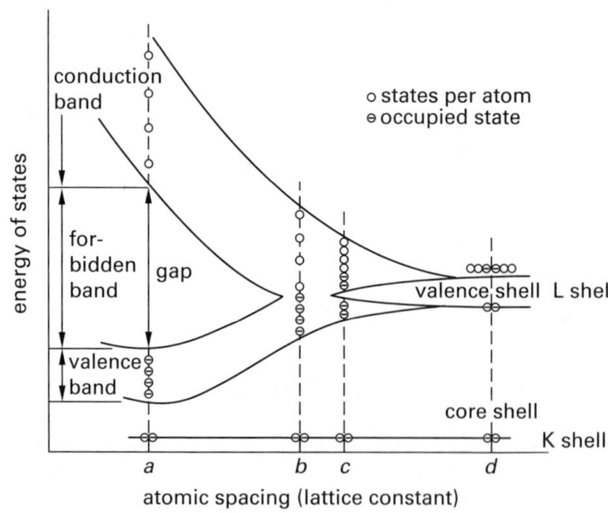

---

[4]Quantum theory plays the dominant role in explaining energy-band theory. Electrical engineering is primarily concerned with the results.

[5]In quantum mechanical terms, this occurs because the eigenfunctions of each atom obtained from a solution of Schrödinger's equation overlap those of neighboring atoms. This phenomenon is similar to the resonant double peak that occurs when two LC tuning circuits with identical resonant frequencies are brought close together.

[6]At atomic spacing c and d, the electrons can also move in discrete energy levels, but of several quanta.

energy. The lower retains the name *valence band* and the upper is called the *conduction band*. These bands are normally shown as in Fig. 17.6.

Figure 17.6 *Multiple Energy Bands*

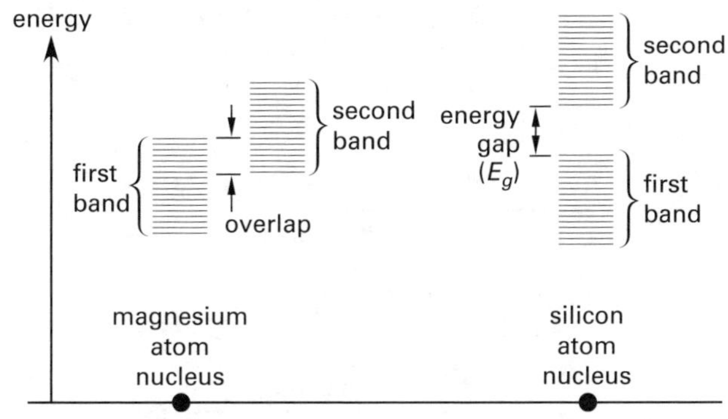

Intrinsic and Extrinsic Semiconductors

Silicon and germanium are the bases of most semiconductor devices. They each possess four valence electrons, leaving four unfilled electron positions. Their crystalline lattice structure uses covalent bonding, that is, a sharing of the electrons, to "complete" the outer shell of eight electrons. (The sharing of electrons does not actually fill the remaining positions but does bond the individual atoms into the crystalline structure.) This situation is illustrated symbolically in Fig. 17.7. In the figure, only silicon atoms are present. When a semiconductor is pure, it is called an *intrinsic semiconductor*.

The energy the structure possesses simply due to its temperature is sometimes adequate to break a covalent bond and free an electron. The amount of energy needed to free an electron is the necessary energy to overcome the energy gap, E_g, and raise the electron to the conduction band. In this manner, a *hole* is created in the valence band and a free electron in the conduction band, as illustrated in Fig. 17.8. The hole and the free electron are both *carriers*. That is, they both contribute significantly to the current upon the application of a voltage. This type of carrier generation is termed *thermal carrier generation*. The current so generated is the *intrinsic conduction*. The intrinsic carrier density, n_i, is given by Eq. 17.12.

$$n_i^2 = Ce^{-E_g/kT} \qquad 17.12$$

As Eq. 17.12 illustrates, the intrinsic concentration is a function of temperature. The constant, C, takes into account the effective masses of electrons and holes and is actually constant only at thermal equilibrium.[7] If temperature varies, the constant C can be replaced with A_0T^3, where A_0 is constant and T is the absolute temperature. The Boltzmann constant, k, is given in units consistent with energy gap units, 8.62×10^{-5} eV/K. The

[7]The constant C is the product of the *effective density states* N_c and N_v (see App. 17.A).

Figure 17.7 *Crystal Structure of Silicon*

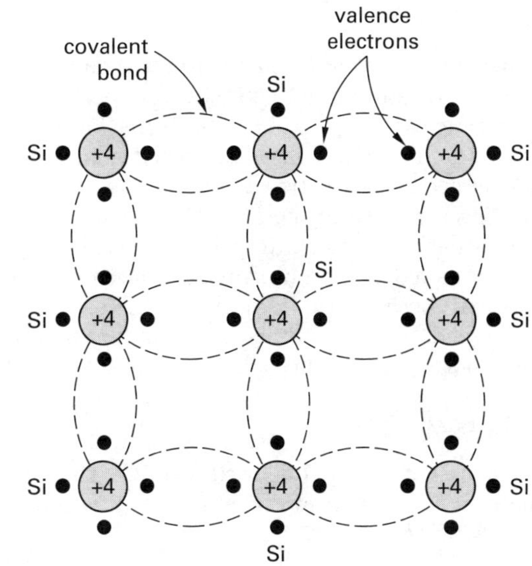

Figure 17.8 *Silicon Crystal with Broken Covalent Bond*

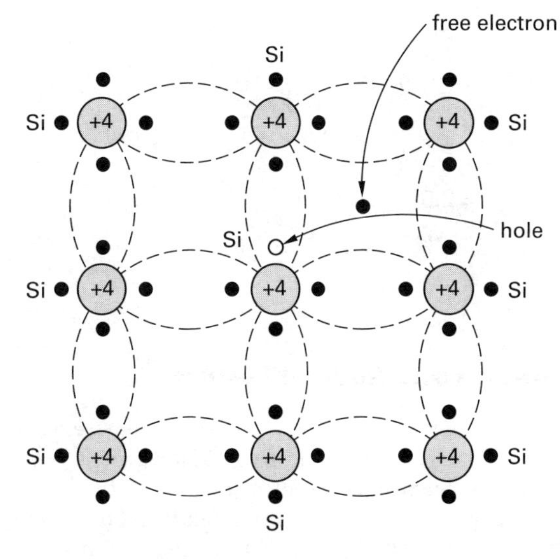

energy gap in silicon is approximately 1.1 eV, and in germanium is approximately 0.8 eV.

Silicon and germanium atoms are tetravalent (i.e., they have four electrons in the valence band). The conductivity of semiconductors composed of these elements can be increased by *doping*, that is, adding impurities to the mix. When a semiconductor is doped, it is called an *extrinsic semiconductor*.

The doping of silicon with pentavalent substitutional impurities such as phosphorus, arsenic, or antimony results in a higher electron density. This is because the fifth electron on the impurity atom is easily ionized, because its energy level is very close to the edge of the conduction-band energy, E_c. This situation is illustrated in Fig. 17.9(a). Such *dopants* are called *donors*. Any such donor mix is called *n-type*, because the *majority carriers* are electrons. Regions doped with donors are sometimes labeled n^+. The *minority carriers* in such materials are the holes.

The addition of trivalent impurity atoms such as boron, aluminum, and gallium attracts electrons and thereby creates vacancies (i.e., a lack of electrons—in essence, holes). This results in a higher hole density, because the energy level necessary to attract the electron (and create the hole) is very close to the edge of the valence band, E_v. This situation is illustrated in Fig. 17.9(b). Such dopants are called *acceptors*. Any such acceptor mix is called *p-type*, because the majority carriers are holes. Regions doped with acceptors are sometimes labeled p^+. The minority carriers in such materials are electrons.

Example 17.2

Given a constant, C, of 7.62×10^{38} cm^{-6}, what is the intrinsic carrier concentration in silicon at room temperature (300K)?

Solution

The intrinsic carrier concentration is given by Eq. 17.12.

$$\begin{aligned} n_i^2 &= Ce^{-E_g/kT} \\ &= (7.62 \times 10^{38}\ \text{cm}^{-6}) \\ &\quad \times \left(e^{-1.1\ \text{eV}/(8.62\times10^{-5}\ \text{eV/K})(300\text{K})}\right) \\ &= 2.56 \times 10^{20}\ \text{cm}^{-6} \\ n_i &= \sqrt{2.56 \times 10^{20}\ \text{cm}^{-6}} \\ &= 1.6 \times 10^{10}\ \text{cm}^{-3} \end{aligned}$$

Generation and Recombination

Thermal agitation continually generates new electron-hole pairs within a semiconductor.[8] However, electron-hole pairs disappear as a result of *recombination*. This is the process whereby free electrons fall into empty covalent bonds, filling the holes and reducing the number of charge carriers. The average electron or hole lives a certain mean time, τ_n for electrons and τ_p for holes.[9] These are important parameters in the design of semiconductor devices.

Law of Mass Action

Adding n-type impurities decreases the number of holes below the number existing in an intrinsic semiconductor; conversely, adding p-type impurities decreases the number of free electrons. Theoretical analysis indicates that, under thermal equilibrium, the product of free negative and positive (hole) concentrations remains constant regardless of the amount of doping. This is a form of the *law of mass action* and is given by

$$np = n_i^2 \qquad 17.13$$

The result of this law is that the doping of an intrinsic semiconductor increases the conductivity and results in a conductor that utilizes either predominately electrons (n-type) or predominately holes (p-type) as carriers. The intrinsic concentration is a function of temperature (see Eq. 17.12).

Figure 17.9 *Energy Band Diagrams of n-type and p-type Semiconductors*

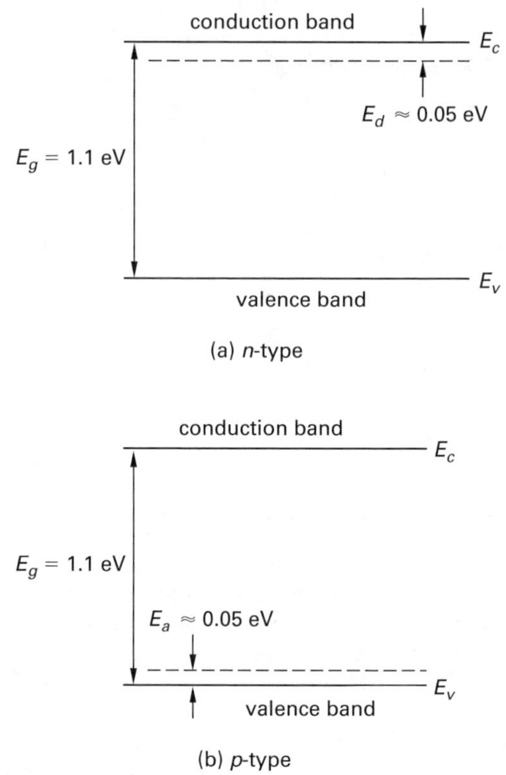

Diffusion

Whenever a concentration gradient of any kind exists, a flow will occur in an attempt to equalize the concentration. This phenomenon is called *diffusion* and is governed by a *diffusion coefficient*, D, measured in units of m^2/s. In a semiconductor, when an n-type region is placed in contact with a p-type region, a *diffusion current* flows to equalize the concentration of electrons and holes. This diffusion current has no analog in metals. The magnitude of the diffusion current density is

$$J = -qD\frac{d\rho}{dx} \qquad 17.14$$

The minus sign indicates that the current flow is in a direction opposite the concentration gradient. This is shown in Fig. 17.10.

[8]Any form of energy that can be absorbed by the semiconductor, such as illumination, radiation, and acoustic energy, can create electron-hole pairs.

[9]Do not confuse these terms with relaxation time that occurs for conduction and uses identical symbols.

Figure 17.10 *Diffusion Current Flow*

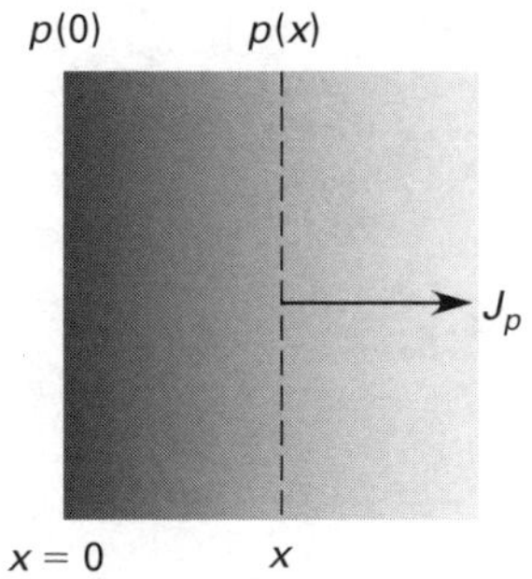

The diffusion and mobility of the electron and hole carriers are governed by the *Einstein equation*, Eq. 17.15.

$$V_T = \frac{D_p}{\mu_p} = \frac{D_n}{\mu_n} \qquad 17.15$$

The term V_T is the *volt-equivalent of temperature* and is defined by Eq. 17.16.

$$V_T = \frac{kT}{q} \qquad 17.16$$

Boltzmann's constant in Eq. 17.16 must be in units of J/K and the temperature, T, is in units of K. The unit of charge, q, is 1.6022×10^{-19} C.

The diffusion current will continue to flow until an electric field is established that opposes the flow. When this occurs, a region is created between the two semiconductor types that is devoid of free charge carriers. This region is called the *space-charge region*, because the region is space containing a charge but no free carriers. This region is also called the *depletion region* or *transition region*. The space-charge region is illustrated in Fig. 17.11 and forms the basis for the operation and control of all semiconductor devices.

A partial listing of the properties of silicon, germanium, and gallium arsenide is given in App. 17.A.

Example 17.3

What is the volt-equivalent of temperature at 300K?

Solution

From Eq. 17.16, the volt-equivalent of temperature is

$$\begin{aligned} V_T &= \frac{kT}{q} \\ &= \frac{\left(1.3807 \times 10^{-23}\ \frac{\text{J}}{\text{K}}\right)(300\text{K})}{1.6022 \times 10^{-19}\ \text{C}} \\ &= 2.58 \times 10^{-2}\ \text{V} \end{aligned}$$

Figure 17.11 *pn Junction Properties*

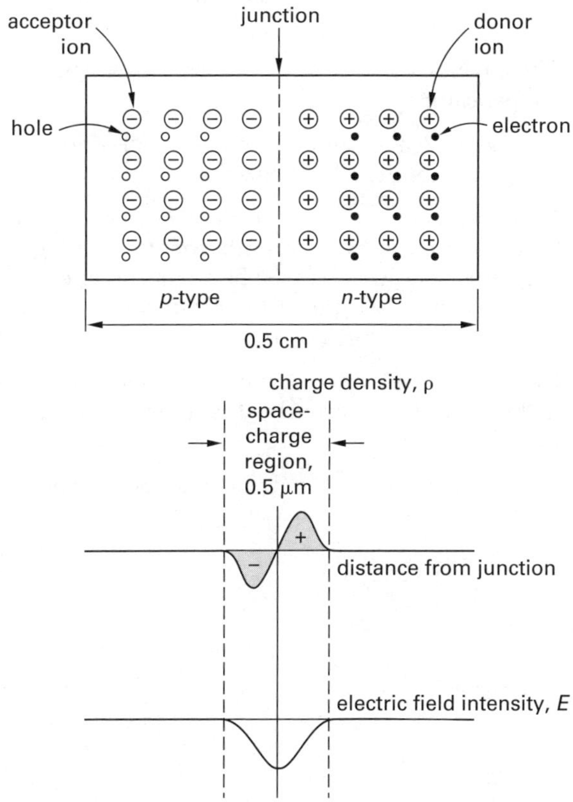

Example 17.4

If the mobility of electrons in silicon is 1.5 cm²/V·s at room temperature (300K), what is the value of the diffusion coefficient, D?

Solution

From Eq. 17.15, the Einstein relationship gives

$$V_T = \frac{D_n}{\mu_n}$$

Rearranging gives

$$\begin{aligned} D_n &= V_T \mu_n \\ &= (0.0258\ \text{V})\left(1.5\ \frac{\text{cm}^2}{\text{V·s}}\right)\left(\frac{1\ \text{m}}{100\ \text{cm}}\right)^2 \\ &= 3.87 \times 10^{-6}\ \text{m}^2/\text{s} \end{aligned}$$

Summary of Fundamental Semiconductor Principles

The principles of semiconductor behavior are summarized as follows.

- Two types of charge carriers exist: electrons and holes.
- A semiconductor may have donor impurities and, as a result, mobile charge carriers that are primarily electrons.

- A semiconductor may have acceptor impurities and, as a result, mobile charge carriers that are primarily holes.
- The intrinsic concentration of carriers is a function of temperature.
- At room temperature, essentially all donor and acceptor impurities are ionized.
- Current occurs as a result of two phenomena: (1) Diffusion current is by means of a concentration gradient, and (2) drift current is by means of an external electric field (identical to the conduction current in metals).
- Charge carriers are continuously generated thermally and simultaneously disappear, by means of recombination.
- Across an open circuited *pn* junction, a potential exists.

18 Communication Theory

1. Fundamentals 18-1
2. Channels and Bands 18-6
3. Coding 18-6
4. Noise 18-7
5. Modulation 18-8

Nomenclature

A	alphabet or message size	–
$c(t)$	carrier signal	–
f	frequency	Hz
$f(t)$	time domain function	–
$F(\omega)$	frequency domain function	–
$H(X)$	information entropy	–
i	variable current	A
I	constant or rms current	A
$I(x_i)$	information content of x_i	–
k	Boltzmann's constant, 1.3807×10^{-23}	J/K
n	number of carriers	–
$p(x_i)$	probability of element x_i	–
P	power	W
q	charge per carrier	C
r	ratio	dB
R	resistance	Ω
t	time	s
T	temperature	K
T_S	sampling period	s
v	variable voltage	V
V	constant or rms voltage	V

Symbols

Δt	rise time or time duration	s
$\Delta\Omega$	equivalent rectangular bandwidth	Hz
τ_d	delay time	s
ω	angular frequency	rad/s

Subscripts

c	center or carrier
I	information (signal)
m	message (signal)
n	noise
R	received or receiver
s	signal
S	sampling

1. FUNDAMENTALS

The purpose of any communication system is to deliver *information* represented as a set of data, $\{I\}$, over a given *channel*, C, to the desired *destination* or *receiver*, R. The information comes from a *source* and is usually transmitted as electric signals controlled by the sender. This is illustrated in Fig. 18.1.

Figure 18.1 *Communication System*

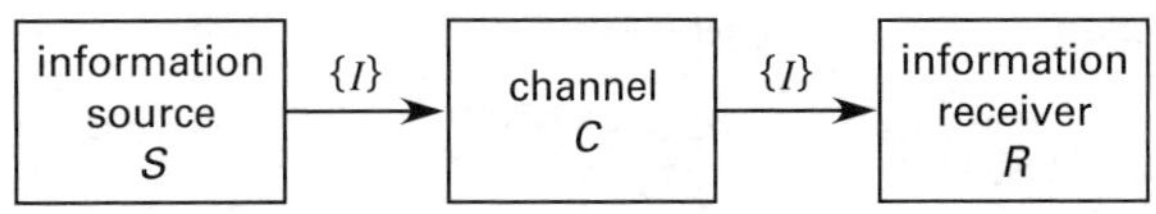

Ideally, the information received would match the information sent. However, because of channel limitations and noise, this may not be the case. Instead, a corrupted version of the information, $\{I^*\}$, may be received. The design objectives in any communication system are to minimize channel distortion and noise and to maximize the amount and rate of the information sent. As in most engineering problems, a tradeoff exists. Sending more information at a greater rate increases distortion and error.

Some forms of information are better suited for transmission over given channels than others. Therefore, encoders are used at the source to change the information set, $\{I\}$, into a more suitable form of information, $\{I_E\}$.[1] This is illustrated by Fig. 18.2.

Analog and Digital Signals

An *analog signal* operates with variables that are continuously measurable. An analog signal has an infinite number of values, although a given instrument may not have the resolution necessary to determine them. An example of such a signal is the human voice. The receiver, the human ear, does not always hear or resolve the entire signal, but the continuously varying values are in the voice waveform.

[1]The information is encoded in the sense that its format is changed to suit the medium of the channel, not for security purposes. Encoding for security purposes is termed *encrypting*.

Figure 18.2 *Communication System with Encoding and Decoding*

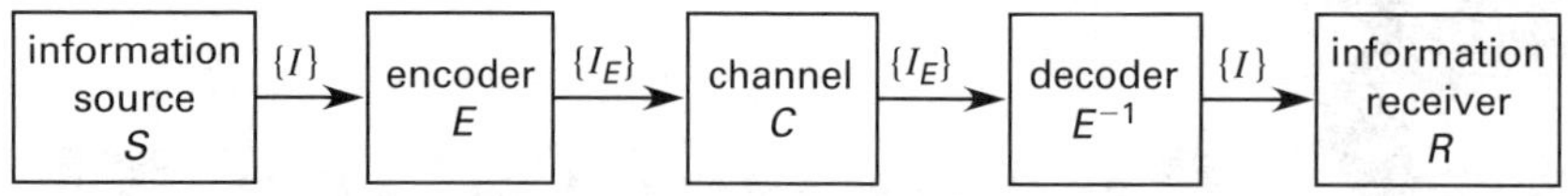

A *digital signal* operates with nominally discontinuous, or discrete, variables that can vary in frequency, amplitude, or polarity. Computer inputs and computer information processing utilize digital signals.

It is often necessary in electrical engineering to change an analog signal into a digital signal. To do so, the analog signal must be sampled. Figure 18.3 illustrates the sampling of a portion of an analog signal, part (a), in the time domain. The sampling is accomplished with impulse pulses, part (b), and results in a digital signal whose individual pulses vary in amplitude, part (c).

Figure 18.3 *Sampling*

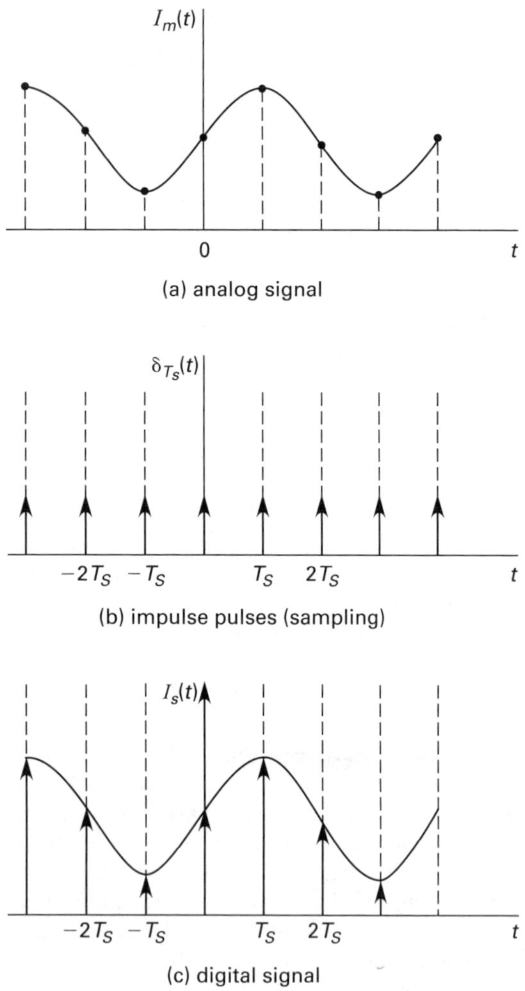

The *sampling theorem* is the basis for conversion of an analog signal to a digital signal. A signal that is sampled the minimum number of times while retaining all the properties of the original in the sampled signal is called an *ideal sampled signal*. The rate that achieves this ideal sampling is called the *Nyquist rate*, f_S, and is given by Eq. 18.1.[2]

$$f_S = \frac{1}{T_S} = 2f_I \qquad 18.1$$

The term f_I is the frequency of the original information signal (sometimes called the *message signal*). The *Nyquist interval*, T_S, is the sampling period corresponding to the Nyquist rate and is given by

$$T_S = \frac{1}{2f_I} \qquad 18.2$$

Example 18.1

In general, telephone line voice signals are limited to about 3300 Hz. What is the minimum sampling rate required in order to retain the original signal?

Solution

The minimum sampling rate is given by the Nyquist rate, Eq. 18.1. Applying this equation,

$$\begin{aligned} f_S &= 2f_I \\ &= (2)(3300\ \text{Hz}) = 6600\ \text{Hz} \end{aligned}$$

Example 18.2

What is the Nyquist interval for a wideband channel with a bandwidth of 4000 Hz?

Solution

The Nyquist interval is the period of sampling corresponding to the Nyquist rate. That is, it is the period corresponding to twice the information frequency. Using Eq. 18.2 and substituting gives

$$\begin{aligned} T_S &= \frac{1}{2f_I} = \left(\frac{1}{(2)(4000\ \text{Hz})}\right) \\ &= 1.250 \times 10^{-4}\ \text{s} \end{aligned}$$

[2]In actual practice the sampling rate is slightly higher than the Nyquist criteria to avoid *aliasing*, or *foldover*, which distorts the original signal. Also, sampling rates higher than the Nyquist rate are required to properly sample a sinusoidal signal because if one happened to sample when the sinusoid was equal to zero, all subsequent samples would be zero.

Basic Signal Theory

Consider the arbitrary nonnegative function, $f(t)$, shownin Fig. 18.4(a), which is the impulse response of a particular linear circuit. The time that $f(t)$ is appreciably different from zero is the *duration* of $f(t)$. When used in terms of the system, the duration is called the *rise time* or *response time* and is given the symbol Δt. The term τ_d is called the *delay time* of the system. The integral of the pulse represents the step-function response to the impulse $f(t)$ (see Fig. 18.4(b)).

Figure 18.4 *Signal Duration and Delay*

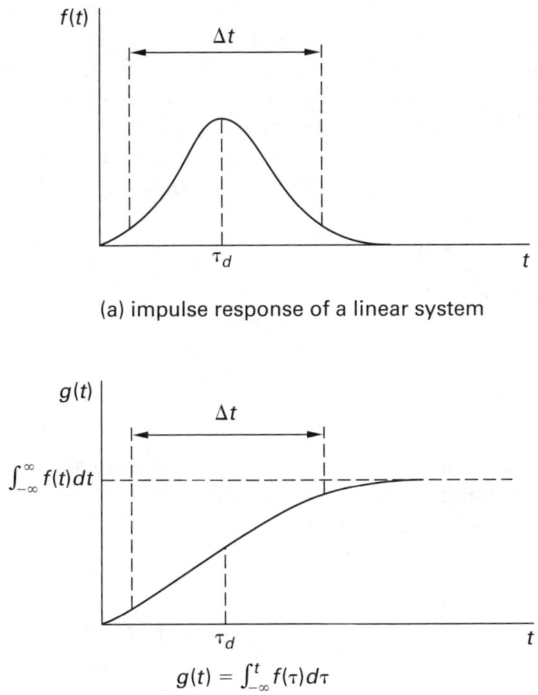

(a) impulse response of a linear system

(b) integral of the impulse response

Because the function $f(t)$ is nonnegative, a new nonnegative function with unit area can be defined. This function is

$$\frac{f(t)}{\int_{-\infty}^{+\infty} f(t)\,dt} \qquad 18.3$$

Equation 18.3 and its frequency domain counterpart, that is, its Fourier transform counterpart, are used to define an equivalent rectangular signal of bandwidth, $\Delta\Omega$, and height, $F(\omega)$. This is illustrated by the dashed lines in Fig. 18.5. The same process is followed in the time domain and yields an equivalent rectangular duration, Δt.

The *bandwidth* is defined by the frequency range over which $F(\omega)$ is appreciably greater than zero. The duration in the time domain is directly related to the bandwidth in the frequency domain. Furthermore, the two

Figure 18.5 *Equivalent Rectangular Bandwidth*

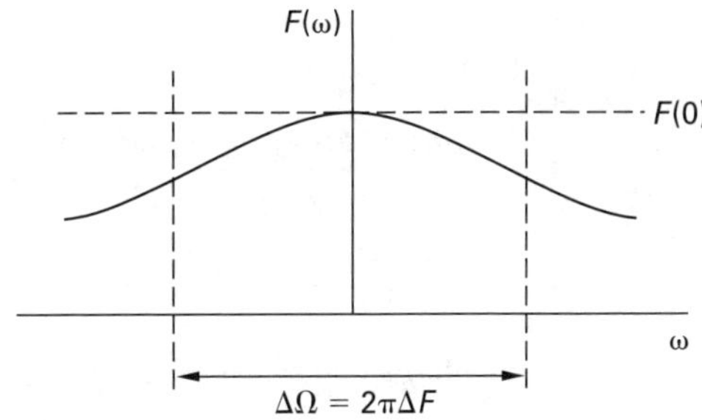

are related in a fundamental way by various *uncertainty relationships*. The primary uncertainty relationship is

$$\Delta t \Delta\Omega = 2\pi \qquad 18.4$$

The second moment uncertainty relationship, which relates the energy, is

$$\Delta t_2 \Delta\Omega_2 \geq \tfrac{1}{2} \qquad 18.5$$

Many other uncertainty relationships exist.

Signals, such as the function $f(t)$, are modified prior to being placed in a communication channel for transmission. Signals containing the desired information are placed on, or mixed with, a carrier signal and then transmitted to the receiver. The situation is illustrated in mathematical terms in Fig. 18.6.

Figure 18.6 *Basic Signal Processing: Functional Representation*

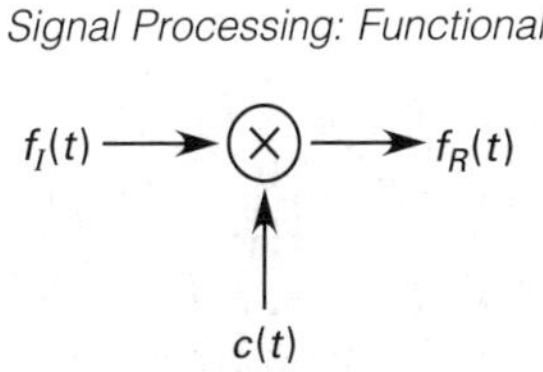

The processing indicated in the figure takes place in the information source, S, of Fig. 18.1 or in the encoder, E, of Fig. 18.2. Consider a specific signal. Let the information signal, $f_I(t)$, be given by Fig. 18.7(a). When mixed with a carrier signal, $c(t)$, of Fig. 18.7(b), the resultant signal, $f_R(t)$, is given as Fig. 18.7(c).[3]

If the amplitude of the carrier signal is sufficient, the resultant signal, $f_R(t)$, does not cross the x-axis and is represented as shown in Fig. 18.8.

The carrier is centered at zero on the frequency axis. One *sideband* is centered at ω_c and the other at $-\omega_c$. In each sideband, the frequencies above the center frequency $|\omega_c|$ are termed the *upper sideband* and those below constitute the *lower sideband*. If the information

[3]The specific signal shown in Fig. 18.7(c) represents $f_R(t) \approx f(t)\cos \omega_c t$ and is called a *suppressed-carrier amplitude modulation system*.

Figure 18.7 Basis Signal Processing: Graphical Representation

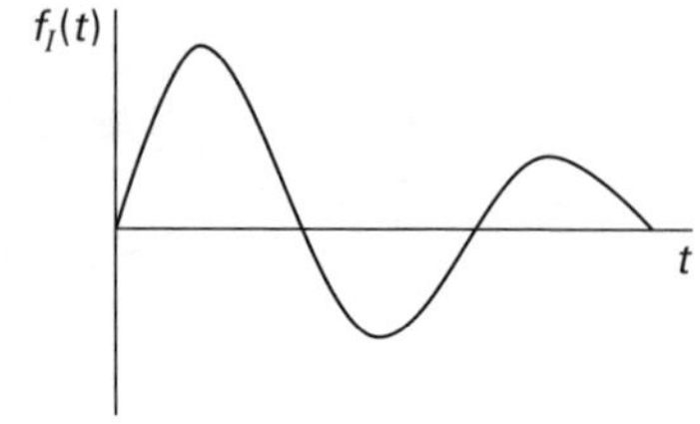

(a) information signal

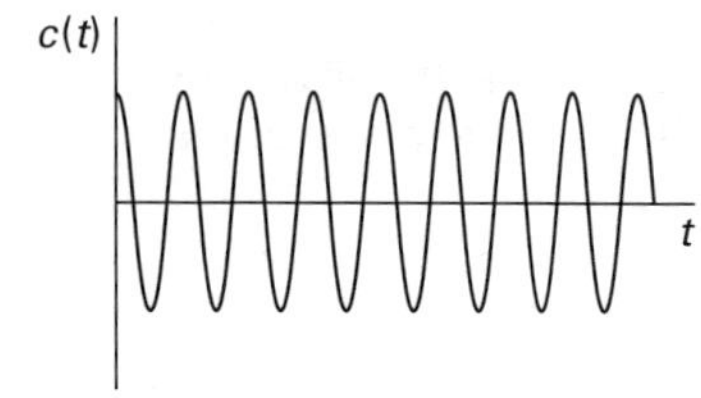

(b) carrier signal

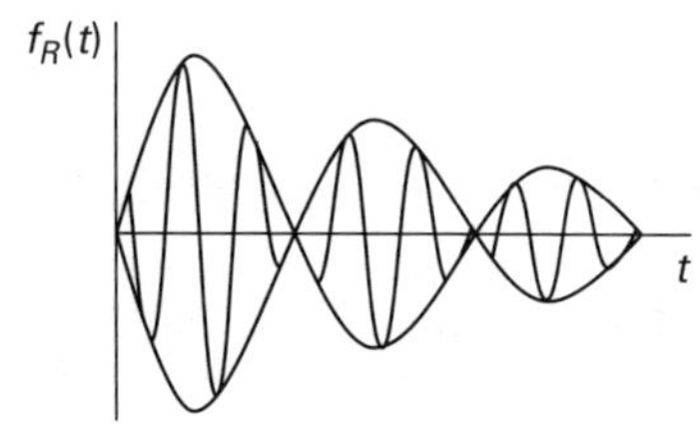

(c) resultant signal

Figure 18.8 Double Sideband AM Signal and Frequency Spectrum

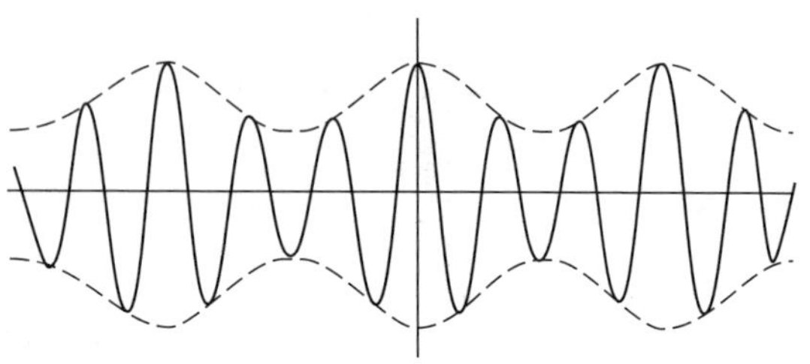

(a) double sideband AM signal

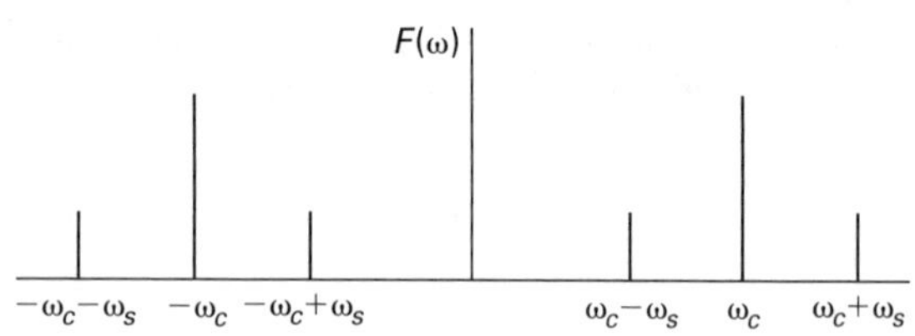

(b) frequency spectrum

signal is not sinusoidal, the frequency spectrum spreads as shown in Fig. 18.9.

Information signals, or message sets, are modified in order to transmit information over a given channel with

Figure 18.9 Spreading of Frequency Spectrum

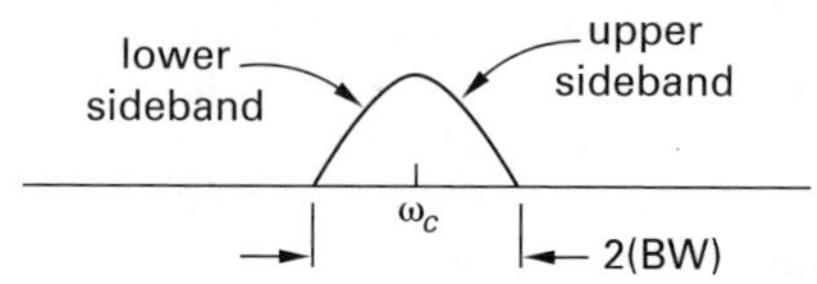

less error. Additionally, a carrier makes it easier to demodulate the signal at the receiver. However, the information desired, that is, the original signal, is completely contained within the upper or lower sideband of a single sideband. The transmitted carrier simply absorbs power and the second sideband unnecessarily expands the bandwidth. The former lowers the efficiency. The latter makes poor use of the frequency spectrum, meaning less information can be passed in a given bandwidth. Nevertheless, such a system is simple to construct and less expensive than comparable systems. In fact, this is the type of signal used for AM radio.

Example 18.3

If the bandwidth of a narrowband, or telegraph-grade, communication system is 300 Hz, what is the response time?

Solution

Assuming the bandwidth given is the equivalent rectangular bandwidth, the applicable formula is given by Eq. 18.4. Applying this equation,

$$\Delta t \Delta\Omega = 2\pi$$

$$\Delta t = \frac{2\pi}{\Delta\Omega} = \frac{2\pi}{300\ \text{Hz}} = 0.02\ \text{s}$$

Information Entropy

Information theory quantifies the communication process. To model a communication system, the information must be predictable. It must be possible for each piece of information to be represented by a *probability of occurrence*, $p(x_i)$. The more probable a particular piece of information, the less *self-information*, $I(x_i)$, a message contains. The information content is defined by Eq. 18.6.

$$I(x_i) = \log\frac{1}{p(x_i)} = -\log p(x_i) \qquad 18.6$$

The choice of the base for the logarithm is arbitrary, but is normally chosen as 2. In this case the unit of self-information, $I(x_i)$, is the common *binary digit* or *bit*. If base-e is used the unit is called the *nat*. (Base-e is convenient for its analytical properties when integration and differentiation are realized with electric circuitry.) Base-10 (rarely used) would result in the unit called a *hartley*.

Example 18.4

What is the information content of a message with only two possible values, zero and one, of equal probability?

Solution

Assume base 2. Using Eq. 18.6, and noting that equal probability means a 0.5 chance of a zero and a 0.5 chance of a one, gives

$$I(x_i) = \log_2 \frac{1}{p(x_i)} = \log_2 \frac{1}{0.5} = 1 \text{ bit}$$

(Unless otherwise told, assume base 2. Also, the bit is not a unit in the true sense.)

Example 18.5

The probability of occurrence of a given message in a computer system is 1.5259×10^{-5}. What is the information content of a message?

Solution

Use Eq. 18.6 and substitute the given information.

$$\begin{aligned} I(x_i) &= \log_2 \frac{1}{p(x_i)} \\ &= \log_2 \frac{1}{1.529 \times 10^{-5}} = 16 \text{ bits} \end{aligned}$$

One of the more important statistical properties of the information variable $I(x_i)$ is its mean or average value. In information theory, the average value of a message is called its *entropy* and is given the symbol $H(X)$. The entropy is defined by Eq. 18.7.

$$\begin{aligned} H(X) &= \sum_{i=1}^{A} p(x_i) I(x_i) \\ &= -\sum_{i=1}^{A} p(x_i) \log_2 p(x_i) \end{aligned} \qquad 18.7$$

The term A is the alphabet size, that is, the total number of elements in the message set. Properly interpreted, Eq. 18.7 is identical to entropy in statistical mechanics or thermodynamics. In statistical mechanics, entropy is a measure of the disorder of the system.

Similarly, in information theory, entropy is a measure of the uncertainty associated with a message. The units of entropy are the *bit*, the *nat*, and the *hartley* and are sometimes given as bits/message, nats/message, or hartleys/message.

The following properties emerge for a binary system.

- $H(X) \geq 0$.
- $H(X) = 0$ only for $p = 0$ or 1.
- $H(X)$ is a maximum at $p = 1/2$.

In the general case, for an alphabet size of A, the entropy has the following properties.

- $H(X) \geq 0$.
- $H(X) = 0$ if and only if all the probabilities are zero except for one, which is unity.
- $H(X) \leq \log_b A$.
- $H(X) = \log_b A$ if and only if all the probabilities are equal so that $p(x_i) = 1/A$ for all i.

Example 18.6

What is the entropy of a binary system with probabilities p and $1 - p$?

Solution

The entropy is given by Eq. 18.7. The alphabet size is two. That is, the value of the elements can be either a zero or a one in a binary system.

$$\begin{aligned} H(X) &= -\sum_{i=1}^{A} p(x_i) \log_2 p(x_i) \\ &= -\sum_{i=1}^{2} p(x_i) \log_2 p(x_i) \\ &= -\Big(p \log_2 p + (1-p) \log_2 (1-p)\Big) \\ &= -\Big((0.5) \log_2 0.5 + (1-0.5) \log_2 (1-0.5)\Big) \\ &= 1 \text{ bit/message} \end{aligned}$$

More enlightening than the mathematical answer in Ex. 18.6 is the plot of $H(X)$ versus the probability.

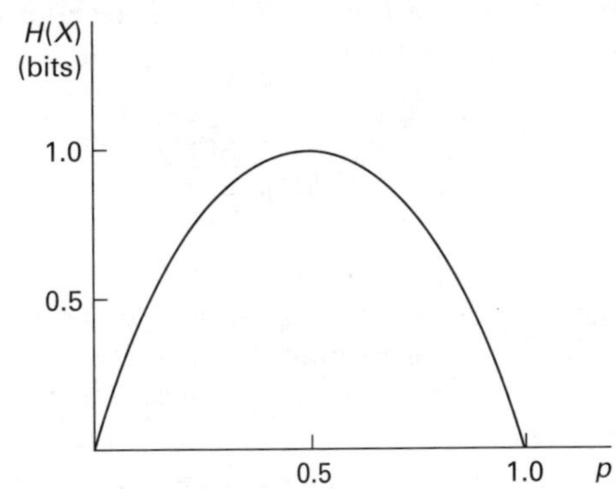

Example 18.7

A certain communication system consists of three possible messages with probabilities of 1/2, 1/4, and 1/4. What is the entropy of the message?

Solution

The information content of the probability 1/2 message is as follows.

From Eq. 18.6,

$$I(x_i) = \log_2 \frac{1}{p(x_i)} = \log_2 \frac{1}{0.5} = 1 \text{ bit}$$

The information content of the probability $1/4$ message is as follows.

From Eq. 18.6,

$$I(x_i) = \log_2 \frac{1}{p(x_i)} = \log_2 \frac{1}{0.5} = 2 \text{ bits}$$

From Eq. 18.7, the entropy is

$$\begin{aligned} H(X) &= -\sum_{i=1}^{A} p(x_i) \log_2 p(x_i) \\ &= \sum_{i=1}^{3} p(x_i) I(x_i) \\ &= \left(\tfrac{1}{2}\right)(1 \text{ bit}) + \left(\tfrac{1}{4}\right)(2 \text{ bits}) + \left(\tfrac{1}{4}\right)(2 \text{ bits}) \\ &= 1.5 \text{ bits/message} \end{aligned}$$

Decibels

The use of a unit known as the *decibel*, dB, is common in communications theory and electrical engineering in general. The decibel is a unit that describes the ratio of two powers or intensities. In acoustics, it also describes the ratio of a power to a reference power. In mathematical terms, the decibel is

$$r = 10 \log_{10} \frac{P_2}{P_1} \tag{18.8}$$

When the ratio is negative, a power loss is indicated. Because the decibel is a log function, decibel gain or loss in the successive stages of a communication system can be added algebraically.

When the ratio is taken with respect to 1 watt, that is, when $P_1 = 1$ W, the ratio is commonly called the *decibel watt* and abbreviated dBW. When the ratio is taken with respect to 1 milliwatt, that is, when $P_1 = 1 \times 10^{-3}$ W, the ratio is termed the *decibels above 1 milliwatt* and abbreviated dBm. Several other decibel ratios are used in electrical engineering and care must be taken to ensure that the correct reference power level is used.

The decibel is also used as a ratio of currents and voltages, though technically this should only be done if the impedances of the measured quantities are equal. This is often ignored, and the decibel for currents or voltages is given by Eq. 18.9

$$\begin{aligned} r &= 10 \log \left(\frac{I_2}{I_1}\right)^2 = 10 \log \left(\frac{V_2}{V_1}\right)^2 \\ &= 20 \log \frac{I_2}{I_1} = 20 \log \frac{V_2}{V_1} \end{aligned} \tag{18.9}$$

2. CHANNELS AND BANDS

As shown in Fig. 18.1, all communication systems use a channel. The *channel* is the medium that connects the information source to the destination, or receiver. The media may be metallic conductors such as copper, specialized cables such as coaxial cable or fiber optic cable, waveguides (devices that guide the propagation of electromagnetic waves using the physical construction of the guide), the atmosphere or, in the case of satellites, space itself.

Channels are classified according to bandwidth. Narrowband channels use bandwidths of 300 Hz or less. Voice-grade channels use 300 Hz to 4 kHz (the audible frequency range of the human voice is approximately 300–3400 Hz). Wideband channels use 4 kHz and greater bandwidths.[4]

Bands represent a somewhat arbitrary range of electromagnetic-wave frequencies between definite limits. Though somewhat arbitrary, the properties of the bands differ, creating different uses for each of the bands. For example, VLF (very low frequency) transmissions form surface waves that travel great distances and are able to penetrate water to a certain degree, making them useful for submarine communications. Standard AM radio uses the MF band while FM radio uses the VHF band. Television utilizes the VHF and UHF bands. Satellites, with their need for high data rates and minimal environmental interference, as well as their ability to utilize line-of-sight communication from Earth, function in the UHF and SHF bands. The bands are shown in Table 18.1. The electromagnetic spectrum is shown in App. 18.A.

3. CODING

In designing a communication system, one attempts to minimize the probability of information error, or bit error, and at the same time minimize the length of the message. Coding theorems guarantee the existence of codes that allow transmission within a channel's capacity at an arbitrarily small error rate. However, to do so, especially in the presence of noise, may require excessively long messages—that is, a low data rate—or raising the signal power above acceptable rates. To

[4]The channel capacity is determined by the type of channel and is given by the *Shannon coding theorem*. The results of this theorem are among the most important in information theory and communications engineering.

Table 18.1 Frequency Bands

symbol	frequency range	frequency band
VLF	3–30 kHz	very low
LF	30–300 kHz	low
MF	300–3000 kHz	medium
HF	3–30 MHz	high
VHF	30–300 MHz	very high
UHF	300–3000 MHz	ultra high
SHF	3–30 GHz	super high
EHF	30–300 GHz	extremely high

overcome this difficulty, *error-correcting codes* have been developed to correct some of the errors that occur during the transmission of an information signal through a noisy channel. A simple example is a parity check code. An example of an *even parity* check code for detecting single errors is shown in Table 18.2. This code determines the total number of ones or zeros in the message and assigns a one if the total is odd and a zero if the total is even.

Table 18.2 Even Parity Check Code

information	check digit	transmitted signal
000	0	0000
001	1	1001
010	1	1010
011	0	0011
100	1	1100
101	0	0101
110	0	0110
111	1	1111

4. NOISE

Noise or *interference* is any unwanted input to a communication system. Such input causes meaningless or erroneous information that must be ignored or removed. Noise results from random processes and can be classified in mathematical terms, but it is more useful to classify random processes as either *natural noise* or *artificial noise*. However, one mathematical classification is worthy of mention. *White noise* or *additive white Gaussian noise* is defined as a waveform in which all observations are jointly Gaussian and mutually independent. The power spectral density is constant with respect to frequency variation. That is, the noise is random and unaffected by changes in frequency. (Such noise cannot technically exist, but the concept is usefully applied to a system with a constant power spectral density over a frequency range wider than the bandwidth.)

Natural Sources of Noise

Thermal noise, or *Johnson noise*, is generated by the random motion of electrons in conductors and semiconductors. The mean squared value of the thermal noise voltage is given by Eq. 18.10.

$$\overline{v_n^2} = 4kTR(\text{BW}) \qquad 18.10$$

The bandwidth, BW, is the frequency bandwidth of the monitoring equipment. The resistance of the circuit or element is given by R in ohms. The temperature, T, is in kelvins. Because the voltage is squared, Eq. 18.10 is proportional to the average power generated by the noise. Thermal noise is constant from DC values to very high frequencies and is considered white noise.

Shot noise, sometimes called *Schottky noise*, is caused by electrons moving across a potential barrier in a random way. It was originally defined for vacuum tubes as the voltage developed because of random variations in the number and velocity of electrons emitted by a heated cathode. Current can be defined as

$$I = q\frac{dn}{dt} \qquad 18.11$$

q represents the charge per carrier and dn/dt is the number of carriers passing through the potential per second. Because of the random thermal velocities of the electrons, the value of dn/dt fluctuates. If the average current is given by I and the total current by $i(t)$, then the noise current is

$$i_n(t) = i(t) - I \qquad 18.12$$

Assuming the process is completely random, all frequencies are equally likely. The mean squared value of the shot noise current is

$$\overline{i_n^2} = 2qI(\text{BW}) \qquad 18.13$$

This noise causes sputtering or popping sounds in radio receivers and snow effects in televisions.

Defect noise describes a number of related phenomena exhibited by noise voltages across terminals of devices when DC currents pass through them. The magnitude depends on the surface imperfections of the terminals. Such noise is also called *flicker noise*, *contact noise*, *excess noise*, *current noise*, and *1/f noise*.

Example 18.8

What is the noise voltage of a 10,000 Ω resistor operating in a 4000 Hz frequency bandwidth at room temperature?

Basic Theory

Solution

Use Eq. 18.10. Room temperature is 300K.

$$\begin{aligned}\overline{v_n^2} &= 4kTR(\text{BW})\\ &= (4)\left(1.38\times 10^{-23}\ \frac{\text{J}}{\text{K}}\right)\\ &\quad\times (300\text{K})(10\times 10^3\ \Omega)(4000\ \text{Hz})\\ &= 6.62\times 10^{-13}\ \text{V}^2\end{aligned}$$

Solve for the voltage directly.

$$\begin{aligned}\overline{v_n^2} &= 6.62\times 10^{-13}\ \text{V}^2\\ v_n &= \sqrt{6.62\times 10^{-13}\ \text{V}^2}\\ &= 8.14\times 10^{-7}\ \text{V}\end{aligned}$$

Artificial Noise

The natural noise explained in Sec. 18.4 is fundamental in nature in that the proximate cause of the noise is the random, noncontinuous nature of electric charge under the influence of a potential. When a noise is generated by electrical or electronic equipment, which is in principle under human control, the noise is classified as *artificial* or *environmental noise.* Artificial noise may be grouped into three main categories.

Interference noise results when a radio station or television channel interferes with another. This may be caused by poor antenna design or unexpected atmospheric effects. This also includes *crosstalk* between communication links caused by multipath propagation or reflection.

Hum is a periodic signal arising from power lines. It occurs at the power supply frequency or its harmonics (60 Hz or 120 Hz in the U.S.). In transformer iron cores with loose laminations, the frequency of the hum is twice the supply frequency and is related to the alignment of magnetic domains in the material. Hum can be eliminated by proper construction, filtering, and shielding.

Impulse noise describes numerous phenomena. Impulse noises are often modeled as low-frequency shot processes, that is, as the superposition of a small number of large impulses. Sources may include the ignition noise from automobiles, corona noise from high-voltage transmission lines, switching noise, and lightning. Impulse noises are usually best handled with limiters or blankers that protect the system during the occurrence of the noise and return it to normal when the noisy period has passed.

5. MODULATION

Modulation is the process whereby some parameter of one wave is altered by some parameter of another wave. A particular case of modulation is illustrated in Fig. 18.6 and Fig. 18.7. Modulation is necessary because in many instances it may be inefficient or impossible to transmit a given spectrum through the required channel. In addition, modulation is used for frequency translation of a signal in order to maximize the efficiency of an antenna or move the signal to an assigned frequency band. Modulation makes signal processing, particularly the recovery of the signal information, easier. It can also be used to enhance *multiplexing*, a process whereby multiple signals are transmitted within the same channel. Figure 18.2 shows the block diagram for a communication system with the modulator replaced by an encoder. In fact, modulation is a form of encoding. In addition, the encoding covered in Sec. 18.3 would still occur alongside any modulation.

Signals can be modulated in amplitude or frequency. (AM stands for amplitude modulation and FM for frequency modulation.) Both sidebands or a single sideband can be transmitted. The original carrier wave may or may not be present.

19 Acoustic and Transducer Theory

1. Introduction 19-1
2. Sound Wave Propagation Velocity 19-2
3. Energy and Intensity of Sound Waves 19-3
4. Transduction Principles 19-3
5. Equivalent Circuit Elements 19-4

Nomenclature

B	bulk modulus	Pa
c	specific heat	J/kg·K
C	capacitance	F
E	modulus of elasticity	Pa
f	frequency	Hz
I	intensity	W/m^2
L	inductance	H
L_p	sound pressure level	dB
m	mass per area	kg/m^2
M	total mass	kg
MW	molecular weight (mass)[1]	kg/kmol
p	pressure	Pa
R	resistance	Ω
R	specific gas constant, 287	J/kg·K
R^*	universal gas constant, 8314.3	J/kmol·K
T	period	s
T	temperature	K
v	velocity	m/s
V	voltage	V

Symbols

β	sound intensity level	dB
γ	ratio of specific heats	–
γ	isentropic exponent	–
η	energy density	J/m^3
λ	wavelength	m
ρ	density	kg/m^3

Subscripts

0	reference
p	pressure
s	sound
v	volume

[1]This is also called the *kilogram molecular weight*. It is truly a mass, not a weight. The *gram molecular weight*, measured in g/mol·K, is also common. Care should be exercised when applying the units.

1. INTRODUCTION

Sound is an alteration of properties, such as pressure, particle displacement or velocity, or density, in an elastic medium. Sound is a wave phenomenon, like electromagnetic radiation, but with two important fundamental differences. First, sound is transmitted as a *longitudinal wave*, also known as a *compression wave*, that causes alternating compression and expansion (*rarefaction*), while electromagnetic waves are transverse waves. Second, sound waves require a medium through which to travel. Electromagnetic waves do not; they can travel through a vacuum.

For visualization purposes, sound waves are often represented as transverse waves as shown in Fig. 19.1. The *wavelength* is the distance between successive compressions or expansions.

Figure 19.1 *Longitudinal Waves*

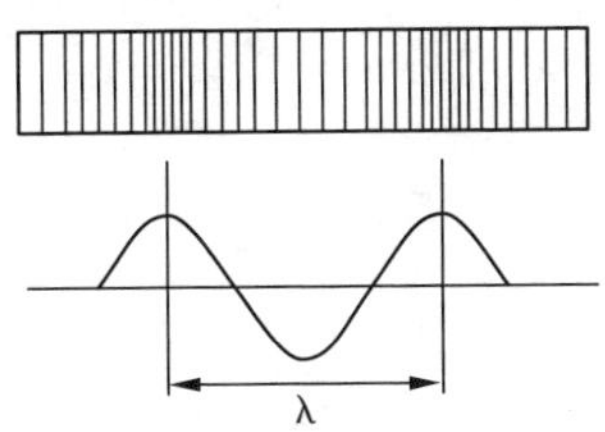

The usual relationships among wavelength, propagation velocity, frequency, and period remain valid.

$$\mathrm{v}_s = f\lambda \qquad 19.1$$

$$T = \frac{1}{f} \qquad 19.2$$

Sound waves traveling through a medium change the energy level of the medium. This energy is called *acoustic energy* and represents the difference between the total energy of the system and the system energy without the sound waves present. Electrical engineering is concerned with the manipulation and use of this energy. The basic interface between acoustic energy and electrical engineering is illustrated in Fig. 19.2.

A *transducer* provides the interface and is the device that provides a usable output in response to a specified *measurand.* The measurand is a physical quantity, property, or condition to be measured.[2] The transducer can be *self-generating,* in which case no external excitation power is required. Such transduction elements are termed *active elements,* whereas those that require excitation power are *passive elements.*

Figure 19.2 *Acoustic/Electrical Interface*

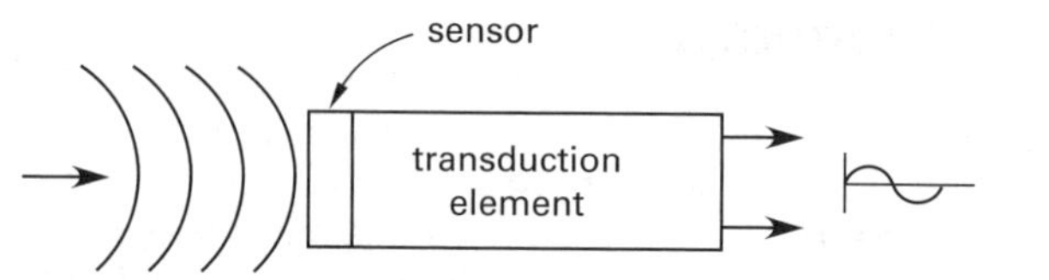

2. SOUND WAVE PROPAGATION VELOCITY

The propagation velocity of sound, v_s, is commonly called the *speed of sound* and *sonic velocity.* The sound velocity depends on the compressibility of the medium through which the sound wave is traveling. This occurs because the sound wave physically moves atoms in the medium, that is, places them closer together (compression) or farther apart (rarefaction), as it propagates. In solids, the compressibility is accounted for by the *elastic modulus, E.* In liquids, the elastic modulus is referred to as the *bulk modulus, B.* The speed of sound in solids and liquids is given by

$$v_s = \sqrt{\frac{E}{\rho}} \qquad 19.3$$

For ideal gases, the propagation velocity is

$$v_s = \sqrt{\frac{\gamma R^* T}{\text{MW}}} \qquad 19.4$$

In thermodynamics, the term γ is the ratio of the specific heat at a constant pressure to the specific heat at a constant volume, c_p/c_v, and is called the *isentropic exponent.*

The speed of sound in a few common materials is given in Table 19.1. Various values of the universal gas constant are given in Table 19.2. When the universal gas constant is divided by the molecular weight, that is, R^*/MW, it becomes the *specific gas constant, R.*

[2]These definitions are paraphrased from the American National Standard ANSI MC6.1.

Table 19.1 *Approximate Speeds of Sound (at one atmospheric pressure ≈ 10^5 Pa)*

material	speed of sound (m/s)
air	330 @ 0°C
aluminum	4990
carbon dioxide	260 @ 0°C
hydrogen	970 @ 0°C
steel	5150
water	1490 @ 20°C

Table 19.2 *Values of the Universal Gas Constant, R^**

units in SI and metric systems
8.3143 kJ/kmol·K
8314.3 J/kmol·K
0.08206 atm·L/mol·K
1.986 cal/mol·K
8.3143 J/mol·K
82.06 atm·cm^3/mol·K
0.08206 atm·m^3/kmol·K
8314.3 kg·m^2/s^2·kmol·K
8314.3 m^3·Pa/kmol·K
8.314 × 10^7 erg/mol·K

Example 19.1

What is the speed of sound in air at 293K, given an isentropic exponent of 1.4?

Solution

Use Eq. 19.4.

$$v_s = \sqrt{\frac{\gamma R^* T}{\text{MW}}}$$

The universal gas constant is 8.3143 J/mol·K. The molecular weight of air is approximately 29 g/mol. (This assumes 78% N_2, 21% O_2, and 1% Ar.) Substituting gives

$$\begin{aligned} v_s &= \sqrt{\frac{\gamma R^* T}{\text{MW}}} \\ &= \sqrt{\frac{(1.4)\left(8.3143\ \frac{\text{J}}{\text{mol·K}}\right)(293\text{K})}{\left(29\ \frac{\text{g}}{\text{mol}}\right)\left(\frac{1\ \text{kg}}{10^3\ \text{g}}\right)}} \\ &= 343\ \text{m/s} \end{aligned}$$

3. ENERGY AND INTENSITY OF SOUND WAVES

A sound wave, or acoustic energy, manifests itself in the displacement and motion of the medium through which it travels. One could measure the energy of the medium in terms of this displacement or motion. However, the most widely used and measurable property of sound waves is *sound pressure*, the fluctuation above and below the equilibrium pressure (or reference pressure), p_0, of the medium in which the sound wave travels. The energy density, η, is

$$\eta = \tfrac{1}{2}\frac{p_0^2}{\rho \mathrm{v}^2} \qquad 19.5$$

Like many other items in electrical engineering, the *intensity* of the sound wave, which is the power delivered per unit of wave front, is the value of concern. The intensity is

$$I = \eta \mathrm{v} = \tfrac{1}{2}\frac{p_0^2}{\rho \mathrm{v}} \qquad 19.6$$

The human ear can sense sound wave intensities from approximately 10^{-12} W/m^2 up to the pain threshold of 1 W/m^2. Because this encompasses a 12-decade range, the intensities are normally measured in decibels, dB, compared to a reference intensity, as shown in Eq. 19.7.

$$\beta = 10 \log \frac{I}{I_0} \qquad 19.7$$

Because the pressures that sound waves create are the normally measured quantities, the *sound-pressure level*, L_p, as defined in Eq. 19.8 is normally used in lieu of Eq. 19.7.

$$L_p = 20 \log \frac{p}{p_0} \qquad 19.8$$

The defined reference pressure for sounds in air is 20 μPa. This correlates with 0 dB, which is the threshold of human hearing.

Example 19.2

The density of air is approximately 1.3 kg/m^3. What pressure is expected at a transducer's sensor for a sound intensity level of 1 W/m^2 in air at 273K?

Solution

Use Eq. 19.6.

$$I = \tfrac{1}{2}\frac{p_0^2}{\rho \mathrm{v}}$$

From Table 19.1, the velocity (speed) of sound in air at 273K (0°C) is 330 m/s. The density of air is given. Rearranging to solve for the equilibrium pressure and substituting gives

$$I = \tfrac{1}{2}\frac{p_0^2}{\rho \mathrm{v}}$$

$$\begin{aligned} p_0 &= \sqrt{2I\rho \mathrm{v}} = \sqrt{\left(1\ \frac{\mathrm{W}}{\mathrm{m}^2}\right)(2)\left(1.3\ \frac{\mathrm{kg}}{\mathrm{m}^3}\right)\left(330\ \frac{\mathrm{m}}{\mathrm{s}}\right)} \\ &= 29.3\ \mathrm{Pa} \end{aligned}$$

The small change in pressure caused by a sound wave is superimposed on normal atmospheric pressure of approximately 10^5 Pa. Though small, this pressure results in a sensation of pain for most people. Any audio system design should use this pressure as an absolute maximum.

4. TRANSDUCTION PRINCIPLES

The transduction principle of a given transducer determines nearly all its other characteristics. There are three *self-generating* transduction types: photovoltaic, piezoelectric, and electromagnetic. All other types require the use of an external excitation power source.

In *photovoltaic transduction*, light is directed onto the junction of two dissimilar metals, generating a voltage. This type of transduction is used primarily in optical sensors. It can also be used with the measurand controlling a mechanical-displacement shutter that varies the intensity of the built-in light source. *Piezoelectric transduction* occurs because certain crystals generate an electrostatic charge or potential when placed in compression or tension or by bending forces. In *electromagnetic transduction*, the measurand is converted into a voltage by a change in magnetic flux that occurs when magnetic material moves relative to a coil with a ferrous core. These self-generating types of transduction are illustrated in Fig. 19.3.

In *capacitive transduction*, the measurand is converted into a change in capacitance. This can be realized by moving the electrodes of a given capacitor or changing the position of the dielectric. *Inductive transduction* is the change in self-inductance of a single coil. *Photoconductive transduction* converts the measurand into a conductance change within a semiconductor material caused by a change in the incident illumination. This is similar to photovoltaic transduction (see Fig. 19.4 to compare).

In *reluctive transduction*, the measurand is converted to an AC voltage change caused by a change in the reluctance path between two or more coils. Reluctance in a magnetic circuit is analogous to resistance in an electrical circuit. This principle also applies to differential transformers and inductance bridges. *Potentiometric transduction* converts the measurand into a position

Figure 19.3 *Self-Generating Transducers*

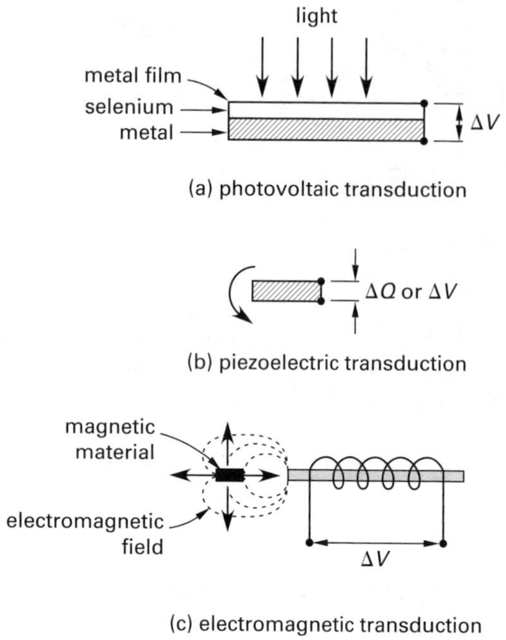

Figure 19.4 *Photoconductive Transduction*

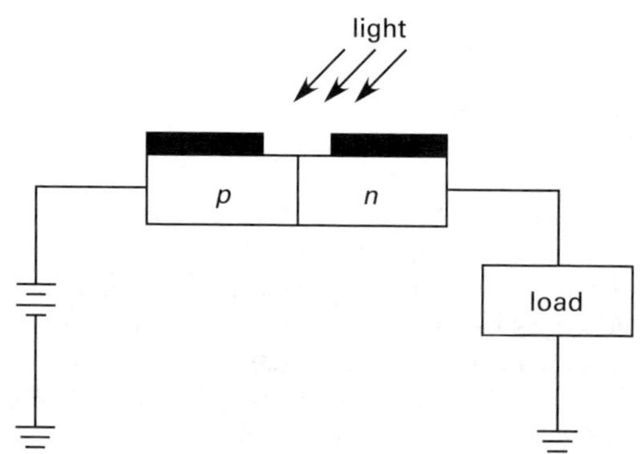

change of a moveable contact on a resistance element. This changes the end-to-end resistance relative to the wiper-arm position controlled by the measurand. *Resistive transduction* is similar to potentiometric transduction except that the change in resistance is brought about by heating or cooling, mechanical stresses, or other external forces. *Strain-gauge transduction* is a special case of resistive transduction generated by a resistance change, caused by a strain, in two of the four arms of a Wheatstone bridge. Here, however, the output is always given by the bridge output voltage.

5. EQUIVALENT CIRCUIT ELEMENTS

In electrical engineering, analogs have developed for the basic elements of acoustic and mechanical systems. These analogs are illustrated in Fig. 19.5.

Figure 19.5 *Basic Elements of Electrical, Acoustic, and Mechanical Systems*

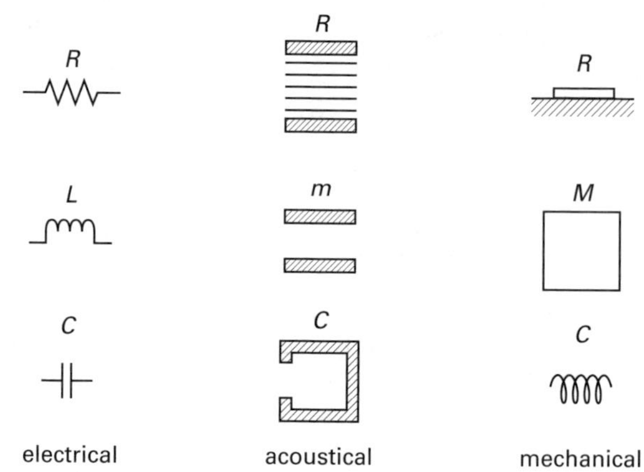

Resistance in acoustic systems can be defined as air friction or viscosity effects in narrow slots. In mechanical systems, resistance is analogous to friction. Inductance in acoustic systems is mass per unit area in a constriction or narrow opening, while in a mechanical system it is the mass itself. Capacitance in an acoustic system is the inverse stiffness of an enclosed volume of air when subjected to piston-like action. It is proportional to the volume enclosed. In mechanical systems, a spring is analogous to capacitance.

Topic III: Field Theory

Chapter

20. Electrostatics
21. Electrostatic Fields
22. Magnetostatics
23. Magnetostatic Fields
24. Electrodynamics
25. Maxwell's Equations

20 Electrostatics

Part 1: Electrostatic Fields 20-1
1. Point Charge 20-2
2. Dipole Charge 20-3
3. Coulomb's Law 20-3
4. Line Charge 20-4
5. Plane Charge 20-4
Part 2: Divergence 20-4
Part 3: Potential 20-5
Part 4: Work and Energy 20-6

Nomenclature

a	unit vector	–
d, **d**	distance	m
D, **D**	electric flux density	C/m^2
E, **E**	electric field strength	N/C or V/m
F, **F**	force	N
l, **l**, L	length	m
p, **p**	dipole moment	C·m
q	electric unit charge, 1.6022×10^{-19}	C
Q	charge	C
ΔQ	infinitesimal test charge	C
r, **r**, R	radius or distance	m
s, S	surface area	m^2
T, **T**	torque	N·m
U	energy	J
v	variable voltage	V
v	volume	m^3
V	constant or rms voltage	V
ΔV	potential difference	V
W	work	J

Symbols

ϵ	permittivity	$C^2/N{\cdot}m^2$ or F/m
ϵ_0	free-space permittivity, 8.854×10^{-12}	$C^2/N{\cdot}m^2$
ρ	charge density	C/m^3
ρ_l	line charge density	C/m
ρ_s	surface charge density	C/m^2
ρ_v	volume charge density	C/m^3
Ψ	electric flux	C

Subscripts

1–2	from 1 to 2; or effect on 2 caused by the field of 1
e	electric
l	line
r	radial
s	surface

PART 1: ELECTROSTATIC FIELDS

A *field* is an entity that acts as an intermediary in interactions between particles. It is distributed over part or all of space and is a function of space coordinates and time (if the field is not static). An electric field is fundamental in nature, causing a charged body to be attracted or repelled by another charged body. The electric field, **E**, represents the force and direction a positive test charge would experience if placed in the field. It is measured in N/C, which is equivalent to V/m. The electric field strength is given by[1]

$$\mathbf{E}_1 = \frac{\mathbf{F}_{1-2}}{\Delta Q_2} = \left(\frac{Q_1}{4\pi\epsilon r^2}\right)\mathbf{a} \qquad 20.1$$

The unit vector, **a**, points in the direction the positive test charge will move at the point of concern within the field, that is, along the lines of flux. The electric field is a vector force field in that a charge placed within the field experiences a force in a specific direction. The force field for a single charged particle is illustrated in Fig. 20.1.

Figure 20.1 *Electric Field of a Positive Charge*

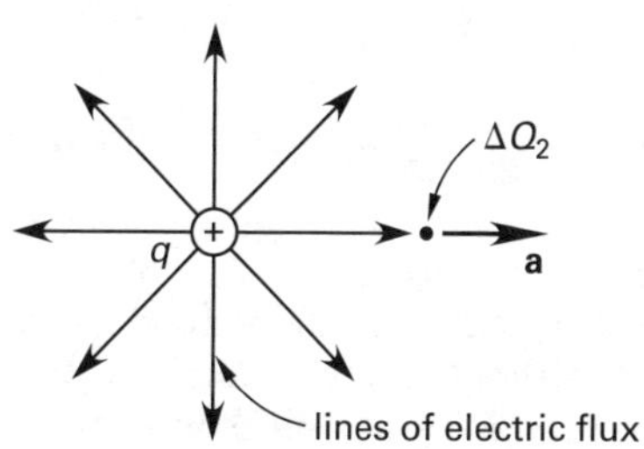

By convention, lines of flux are said to originate at positive charges and terminate at negative charges. Within an electric field, like charges repel and unlike charges attract. Therefore, a positive sign on vector **E** indicates repulsion and a negative sign indicates attraction.

[1]The electric field is defined in terms of the force per unit positive test charge. The rightmost equation is actually the realization of this definition for a point charge.

Field Theory

The force fields of multiple charges interact and the principle of superposition is applicable. The interaction of two unlike charges and of two like charges is illustrated in Fig. 20.2.

Figure 20.2 *Electric Field Interaction*

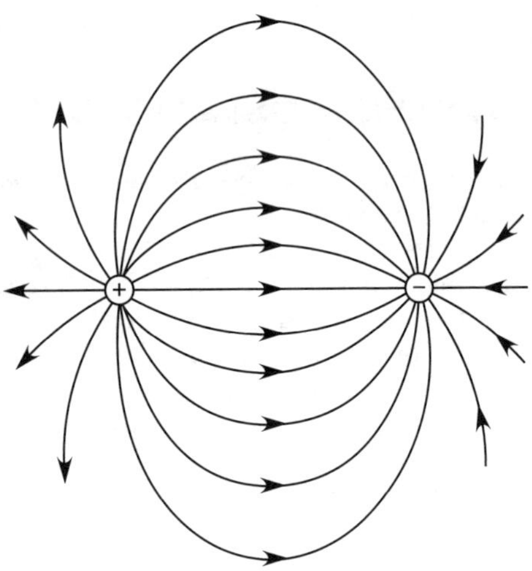

(a) unlike charges

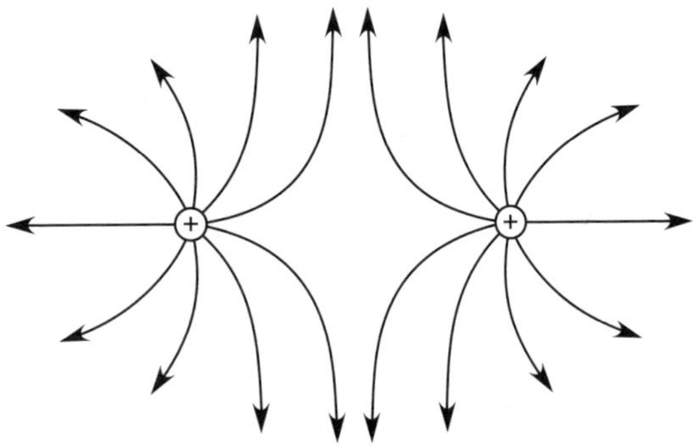

(b) like charges

Equation 20.1 was written in terms of the test charge. Normally an electric field is understood to be the field as it would exist without the test charge present. Further, subscripts are unnecessary in simple configurations. Therefore, in general, the force on a charged object (or distribution of charged objects) is

$$\mathbf{F} = Q\mathbf{E} \tag{20.2}$$

The lines of electric flux comprise a scalar field normally given the symbol Ψ. The flux is equal in magnitude to the charge enclosed. By definition, one coulomb of electric charge gives rise to one coulomb of electric flux. The electric flux density, $\mathbf{D}$, is a vector field with units of flux per unit area or C/m^2. This is not a measurable field. It is used in electrical engineering to remove the effect of the medium, as can be seen by its defining equation.

$$\mathbf{D} = \epsilon\mathbf{E} \tag{20.3}$$

1. POINT CHARGE

The electric field of a point charge was given in Eq. 20.1. It is repeated here with the unit vector specifically showing the radial spreading of the electric field. The electric field of a point charge follows the *inverse-square law* and is normally written in spherical coordinates.

$$\mathbf{E} = \left(\frac{Q}{4\pi\epsilon r^2}\right)\mathbf{a}_r \tag{20.4}$$

The potential, V, associated with a point charge is given by Eq. 20.5. The potential is a *work field* and is always a scalar. It represents the energy required to transport a test charge from one point to another in an electric field. The unit of potential (v or V) is the volt, V, which is equivalent to J/C.

$$V = \frac{Q}{4\pi\epsilon r} \tag{20.5}$$

Example 20.1

A positive charge is located at the origin of a Cartesian coordinate system. What is the electric field at point $(2, 2, 3)$ (distance in units of meters) with no dielectric present?

Solution

The situation is as shown.

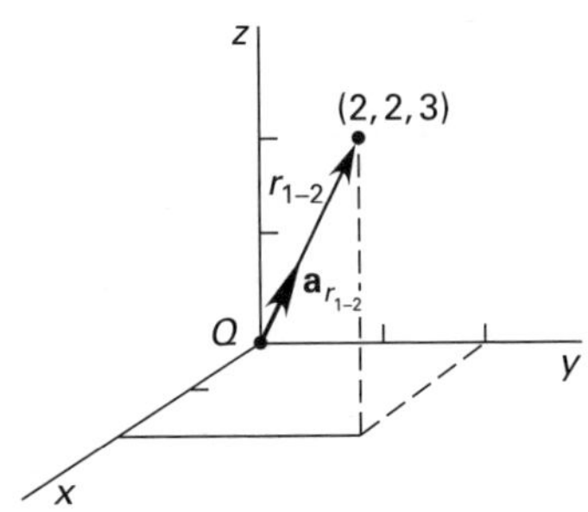

Using Eq. 20.1 and noting that point 1 of the equation is the origin and point 2 is the point (2, 2, 3) gives

$$\mathbf{E}_1 = \left(\frac{Q_1}{4\pi\epsilon r_{1-2}^2}\right)\mathbf{a}_{r_{1-2}}$$

Because no dielectric is present, $\epsilon = \epsilon_0 = 8.854 \times 10^{-12}$ C^2/N·m^2. The distance r_{1-2} is

$$r_{1-2} = \sqrt{(2\text{ m} - 0)^2 + (2\text{ m} - 0)^2 + (3\text{ m} - 0)^2} = 4.1\text{ m}$$

The vector $\mathbf{r}_{1-2} = 2\mathbf{a}_x + 2\mathbf{a}_y + 3\mathbf{a}_z$. Therefore the unit vector in the direction of $\mathbf{r}_{1-2}$ is given by

$$\begin{aligned}\mathbf{a}_{r_{1-2}} &= \frac{\mathbf{r}_{1-2}}{r_{1-2}} = \frac{2\mathbf{a}_x + 2\mathbf{a}_y + 3\mathbf{a}_z}{4.1\text{ m}} \\ &= 0.49\mathbf{a}_x + 0.49\mathbf{a}_y + 0.73\mathbf{a}_z\end{aligned}$$

The magnitude of the electric field is

$$\begin{aligned} E_1 &= \frac{Q_1}{4\pi\epsilon_0 r_{1-2}^2} \\ &= \frac{1.6022 \times 10^{-19}\ \text{C}}{4\pi\left(8.854 \times 10^{-12}\ \dfrac{\text{C}^2}{\text{N}\cdot\text{m}^2}\right)(4.1\ \text{m})^2} \\ &= 8.6 \times 10^{-11}\ \text{N/C} \end{aligned}$$

Therefore, the electric field is

$$\begin{aligned} \mathbf{E}_1 &= E_1 \mathbf{a}_{r_{1-2}} \\ &= \left(8.6 \times 10^{-11}\ \frac{\text{N}}{\text{C}}\right)(0.49\mathbf{a}_x + 0.49\mathbf{a}_y + 0.73\mathbf{a}_z) \end{aligned}$$

2. DIPOLE CHARGE

A *dipole* consists of equal and opposite point charges separated by a small distance d. Substances whose molecules are dipoles are referred to as *polar substances.* Water is the most significant of these substances. The two hydrogens are separated by 105° with the oxygen atom at the vertex. A dipole is illustrated in Fig. 20.3.

Figure 20.3 *Dipole*

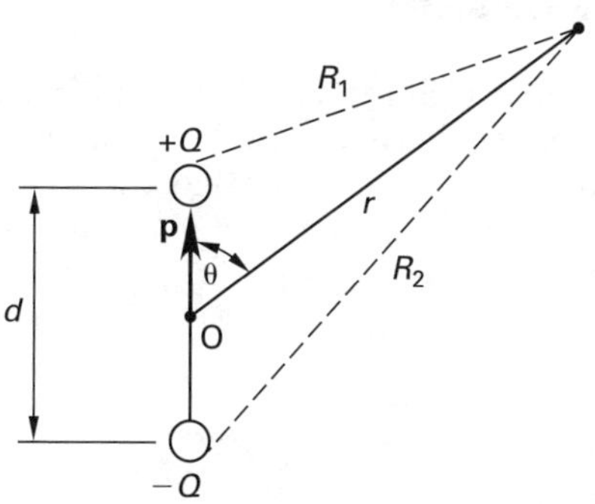

The *dipole moment* is defined by Eq. 20.6.[2]

$$\mathbf{p} = Q\mathbf{d} \qquad 20.6$$

The electric field of a dipole, given in spherical coordinates, is

$$\mathbf{E} = \left(\frac{Qd}{4\pi\epsilon r^3}\right)(2\cos\theta\mathbf{a}_r + \sin\theta\mathbf{a}_\theta) \qquad 20.7$$

The potential arising from two charges at a distance r from the dipole's center of symmetry is[3]

$$V = \frac{Q(R_2 - R_1)}{4\pi\epsilon R_1 R_2} \approx \frac{Qd\cos\theta}{4\pi\epsilon r^2} \qquad 20.8$$

A dipole in an electric field will experience a torque that tends to align the dipole moment, **p**, with the field, **E**. This is shown in Fig. 20.4. The torque is given by Eq. 20.9 and Eq. 20.10.

$$\mathbf{T} = \mathbf{p} \times \mathbf{E} \qquad 20.9$$

$$\text{T} = dF\sin\theta = dQE\sin\theta \qquad 20.10$$

Figure 20.4 *Dipole in an Electric Field*

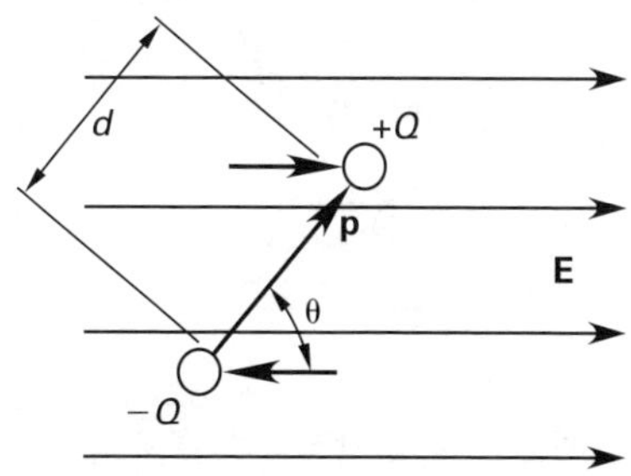

The minimum and maximum energy configurations for a dipole in the presence of an electric field are shown in Fig. 20.5.

Figure 20.5 *Dipole Energy Configurations*

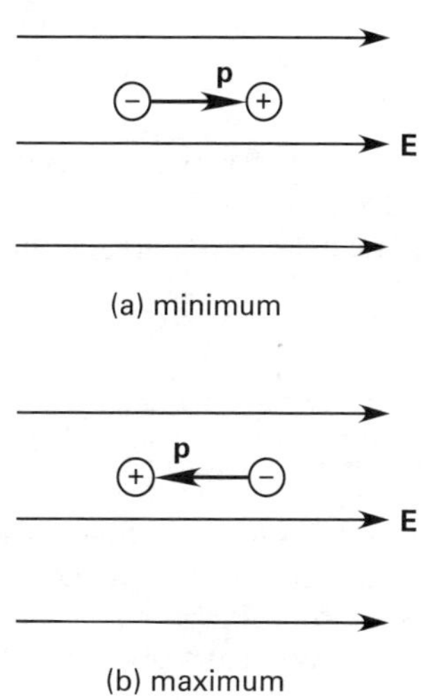

3. COULOMB'S LAW

So far, the equations in this chapter have described the effect of the two charges in the dipole on a third test charge within their electric field. The force between the

[2]The charge, q, used in the dipole moment equation is equal to the *unit positive charge*, often called simply the positive charge. This is done because a unit positive charge is the arbitrary reference. In any work (energy) equation involving a dipole, a 2 multiplies the result to account for both charges.

[3]When $r \gg d$, the distances R_1 and R_2 from the two dipole charges to the point (r, θ) are essentially the same.

two charges in the dipole, or between any two charges, is described by Coulomb's law and is defined by

$$\mathbf{F} = \left(\frac{Q_1 Q_2}{4\pi\epsilon r^2}\right)\mathbf{a} \qquad 20.11$$

4. LINE CHARGE

Another useful standard charge configuration is that of an infinite, straight-line distribution of charges. This distribution is illustrated in Fig. 20.6.

Figure 20.6 *Infinite Line Charge*

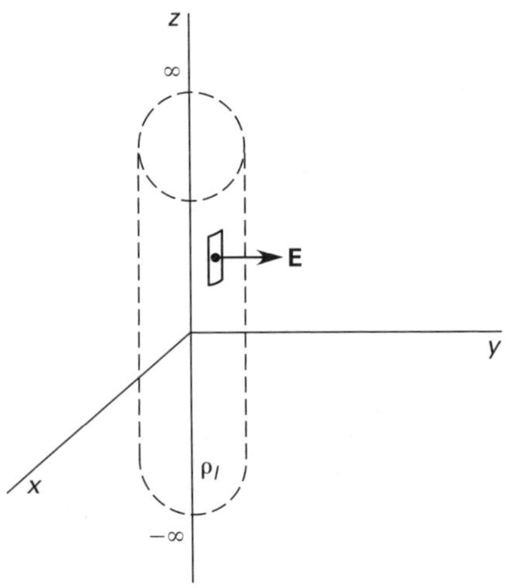

The electric field of such a distribution is given in cylindrical coordinates as

$$\mathbf{E} = \left(\frac{\rho_l}{2\pi\epsilon r}\right)\mathbf{a}_r \qquad 20.12$$

If the line is not an infinite line charge but instead exists over some finite length, L, then a line integral must be utilized. The electric field for a finite line charge density is

$$\mathbf{E} = \int_L \left(\frac{\rho_l \mathbf{a}_r}{4\pi\epsilon r^2}\right) dl \qquad 20.13$$

Example 20.2

An infinite uniform line charge in free space has a charge density, ρ_l, of 10 nC/m and lies along the x-axis of an arbitrary Cartesian coordinate system using meters as the unit of distance. What is the electric field strength, **E**, at point (2, 2, 3)?

Solution

The situation is as shown.

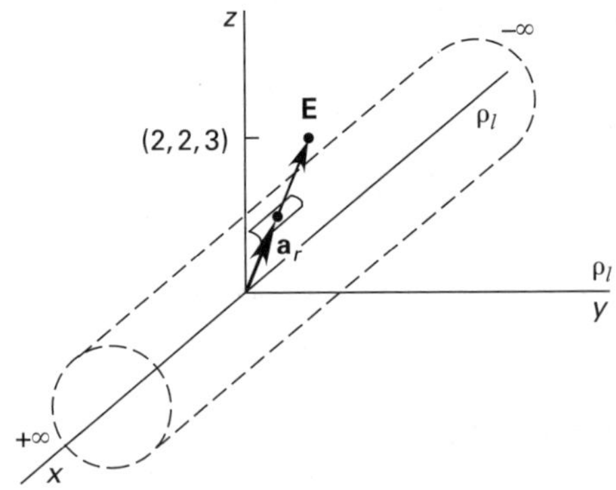

The field is constant with respect to x. Charges in the y-z plane must be determined in cylindrical coordinates.

$$r = \sqrt{(2\text{ m})^2 + (3\text{ m})^2} = 3.6\text{ m}$$

The electric field is

$$\begin{aligned}\mathbf{E} &= \left(\frac{\rho_l}{2\pi\epsilon_0 r}\right)\mathbf{a}_r \\ &= \left(\frac{10\times 10^{-9}\ \frac{\text{C}}{\text{m}}}{2\pi\left(8.854\times 10^{-12}\ \frac{\text{C}^2}{\text{N}\cdot\text{m}^2}\right)(3.6\text{ m})}\right)\mathbf{a}_r \\ &= (49.9\text{ N/C})\mathbf{a}_r \quad ((49.9\text{ V/m})\mathbf{a}_r)\end{aligned}$$

5. PLANE CHARGE

Consider electric charges distributed uniformly over an infinite plane with density ρ_s as shown in Fig. 20.7. Such a charge density produces an electric field given by Eq. 20.14. The term $\mathbf{a}_n$ represents the unit normal vector to the surface, S.

$$\mathbf{E} = \left(\frac{\rho_s}{2\epsilon}\right)\mathbf{a}_n \qquad 20.14$$

This type of charge configuration is also known as a *surface charge* or *sheet charge*. For a finite surface charge that exists in more than one plane as shown in Fig. 20.8, the electric field is given by the surface integral shown in Eq. 20.15.

$$\mathbf{E} = \int_S \left(\frac{\rho_s \mathbf{a}_r}{4\pi\epsilon r^2}\right) ds \qquad 20.15$$

PART 2: DIVERGENCE

The main indicators of a changing vector field are the *divergence* and the *curl*. The curl is best studied in terms of magnetic fields. The divergence is a scalar that bears a similarity to the derivative of a function. In terms of

Figure 20.7 *Infinite Surface Charge (viewed edge-on)*

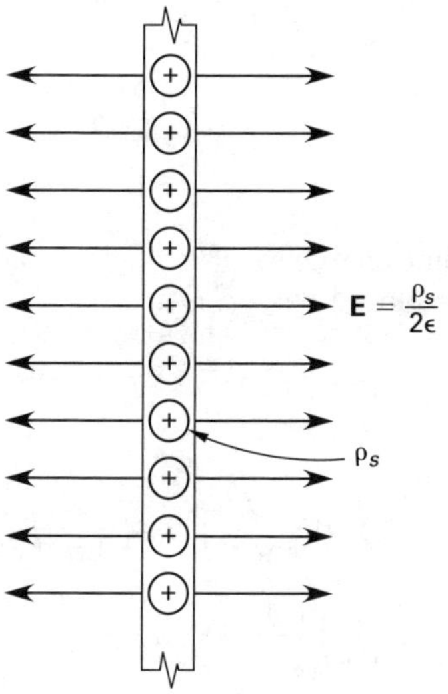

Figure 20.8 *Sheet Charge*

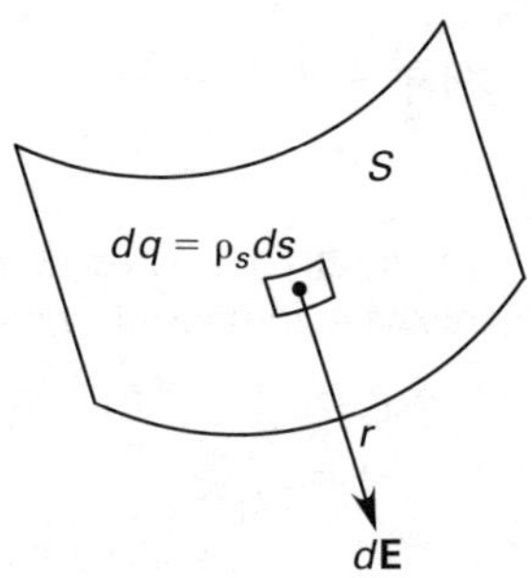

any vector field, $\mathbf{A}$, the *divergence theorem*, or *Gauss' divergence theorem*, is

$$\oint_S \mathbf{A}\cdot d\mathbf{s} = \int_V (\nabla\cdot\mathbf{A})dv \qquad 20.16$$

If the vector field is the electric flux density, $\mathbf{D}$, then the divergence theorem relates the flux emitted over a closed surface to the divergence of the flux density vector from the same volume, that is, the outward flow of the flux per unit volume. The term $\nabla\cdot\mathbf{D}$ is defined as the divergence of $\mathbf{D}$. The term ∇ is called the *del operator*.[4]

$$\begin{aligned}\text{div }\mathbf{D} &= \nabla\cdot\mathbf{D}\\ &= \left(\frac{\partial}{\partial x}\right)\mathbf{x} + \left(\frac{\partial}{\partial y}\right)\mathbf{y} + \left(\frac{\partial}{\partial z}\right)\mathbf{z}\cdot(D_x + D_y + D_z)\\ &= \frac{\partial D_x}{\partial x} + \frac{\partial D_y}{\partial y} + \frac{\partial D_z}{\partial z}\end{aligned} \qquad 20.17$$

[4]The del operator is defined only in the Cartesian coordinate system. It is used in the cylindrical and spherical coordinate systems, but care must be taken to ensure correct results.

From Gauss' law for electrostatics, the divergence of the electric flux density is

$$\text{div }\mathbf{D} = \nabla\cdot\mathbf{D} = \rho \qquad 20.18$$

Equation 20.18 is the differential form of Gauss' law given in integral form in Sec. 16.12.

Example 20.3

In the space charge region on the n-side of a pn junction, the electric flux density is

$$\mathbf{D} = 8x^2\mathbf{x} + 2\mathbf{y} + 2\mathbf{z}$$

What is the charge density in this region?

Solution

The charge density (in units of C/m^3) can be determined from Eq. 20.18.

$$\begin{aligned}\rho &= \nabla\cdot\mathbf{D} = \left(\frac{\partial D_x}{\partial x}\right)\mathbf{x} + \left(\frac{\partial D_y}{\partial D_y}\right)\mathbf{y} + \left(\frac{\partial D_z}{\partial D_z}\right)\mathbf{z}\\ &= \left(\frac{\partial(8x^2)}{\partial x}\right)\mathbf{x} + \left(\frac{\partial(2)}{\partial y}\right)\mathbf{y} + \left(\frac{\partial(2)}{\partial z}\right)\mathbf{z}\\ &= 16x + 0 + 0\\ \rho &= 16x\ \text{C/m}^3\end{aligned}$$

PART 3: POTENTIAL

The term "potential" was defined in Sec. 20.1 as the work that must be done against an electric field to bring a unit charge from an arbitrary reference point to the point in question. The reference point is normally infinity. For practical purposes, the surface of the earth or another large conductor is the reference point. The potential at a given point does not depend on the charge placed at the point, but rather on the charge or charge distribution that creates the electric field. The units of potential are volts (V). One volt represents one joule (J) of work expended moving one coulomb (C) of charge from one point to another (J/C). The potential is given in terms of the electric field as

$$\mathbf{E} = -\nabla V \qquad 20.19$$

The negative gradient of the potential is equal to the electric field. The minus sign is required, since positive charges moving in the direction of the electric field move from higher to lower potentials. That is, the potential decreases as the distance from the source of the electric field increases.

Because the potential is the work done per unit positive test charge in moving the charge from, say, point B to point A, the potential can be represented in integral form as follows.

$$V_{\mathrm{BA}} = \frac{W}{\Delta Q} = -\int_{\mathrm{B}}^{\mathrm{A}} \mathbf{E}\cdot d\mathbf{l} \qquad 20.20$$

The potential is at point A with respect to point B and is the work done in moving the positive test charge from B to A. The potential difference between two points is often represented as ΔV (not to be confused with the gradient of the potential, ∇V).

It is often useful to represent the potential in yet another way. Gauss' law was stated in Eq. 20.18 in terms of the electric flux density, **D**. Restating this equation in terms of the electric field, **E**, and assuming the permittivity is a constant, Gauss' law becomes[5]

$$\text{div}\ \mathbf{E} = \nabla\cdot\mathbf{E} = \frac{\rho}{\epsilon} \qquad 20.21$$

Substituting Eq. 20.19 into Eq. 20.21 gives the following useful result and introduces the Laplacian operator, ∇^2.

$$\nabla^2 v = -\frac{\rho}{\epsilon} \qquad 20.22$$

Equation 20.22 is called *Poisson's equation.* When there are no charges present, the right-hand side of Eq. 20.22 is zero and the equation is called *Laplace's equation.* The Laplacian operator for Cartesian coordinates is

$$\nabla^2 \equiv \frac{\partial^2}{\partial x^2} + \frac{\partial^2}{\partial y^2} + \frac{\partial^2}{\partial z^2} \qquad 20.23$$

Table 20.1 shows the relationships (proportionalities) between the electric field, the potential, and the distance, r, from the charge or charge distribution.

Table 20.1 *Electric Field and Potential versus Distance*

charge distribution		electric field		potential
point charge	$\propto$	$\frac{1}{r^2}$	$\propto$	$\frac{1}{r}$
line charge	$\propto$	$\frac{1}{r}$	$\propto$	$\ln r$
surface charge	$\propto$	constant	$\propto$	r

Example 20.4

Determine the equation for the potential of a point charge.

Solution

The potential is defined by Eq. 20.20 as

$$V_{\mathrm{BA}} = -\int_{\mathrm{B}}^{\mathrm{A}} \mathbf{E}\cdot d\mathbf{l}$$

Because a point charge's electric field is radial, the term $\mathbf{E}\cdot d\mathbf{l}$ can be replaced with $E_r dr$.

$$\begin{aligned} V_{\mathrm{BA}} &= -\int_{\mathrm{B}}^{\mathrm{A}} E_r dr \\ &= -\int_{\mathrm{B}}^{\mathrm{A}} \left(\frac{Q}{4\pi\epsilon r^2}\right) dr = \frac{-Q}{4\pi\epsilon}\int_{\mathrm{B}}^{\mathrm{A}} \frac{1}{r^2}\,dr \\ &= \left(\frac{Q}{4\pi\epsilon}\right)\left(\frac{1}{r_A} - \frac{1}{r_B}\right) \end{aligned}$$

The normal reference point is infinity.

$$V_{\mathrm{A}-\infty} = \left(\frac{Q}{4\pi\epsilon}\right)\left(\frac{1}{r_A} - \frac{1}{\infty}\right) = \frac{Q}{4\pi\epsilon r_A}$$

The subscripts are normally not shown, and free space is assumed unless otherwise indicated. Therefore,

$$V = \frac{Q}{4\pi\epsilon_0 r}$$

(Equation 20.19 could have been used as the starting point as well.)

PART 4: WORK AND ENERGY

Work is force moving an object over a distance. In an electric field, a positive charge will experience a force tending to move it in the direction of the electric field. Work, dW, is accomplished only by the force in the direction of motion. When a force is applied in a direction other than the direction of motion, the charge will change direction but no energy will be transferred to the charge. This is why magnetic fields change the direction of moving charges but transfer no energy.

Let the direction of motion be given by $d\mathbf{l}$. Let θ be the angle between the direction of motion, $d\mathbf{l}$ (the actual path of the charge), and the direction of the electric field, **E**. The situation for an arbitrary path is shown in Fig. 20.9.[6]

The differential work done by the electric field is

$$dW = -Q\mathbf{E}\cdot d\mathbf{l} = -QE\cos\theta\, dl \qquad 20.24$$

[5]The permittivity assumption is valid for linear, homogeneous, and isotropic mediums—a wide class of useful materials.

[6]External forces not shown would be required to make the positive charge follow the indicated path.

Figure 20.9 *Positive Charge in an Electric Field (arbitrary path)*

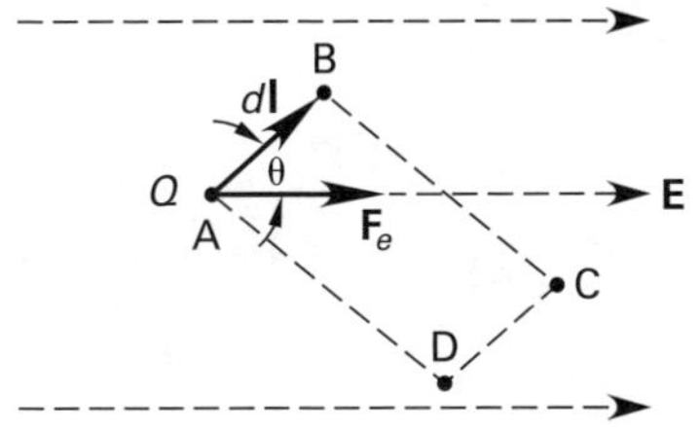

The total work done in moving a charge from point A to point B in an electric field is

$$\begin{aligned} W_{AB} &= -\int_A^B Q\mathbf{E}\cdot d\mathbf{l} \\ &= -\int_A^B QE\cos\theta dl \end{aligned} \qquad 20.25$$

The work is positive if an external force is required to move the charge (as in bringing two repulsive charges closer together or moving a charge against the force of the electric field). Work is negative if the electric field does the work; for example, allowing attracting charges to come together or allowing a positive charge to move in the direction of the electric field. Positive work indicates that the system requires an energy input and negative work indicates work output. No work is done in moving a charge perpendicular to a field, that is, across the lines of electric flux or anywhere along the lines of an equipotential surface.

When Eq. 20.25 is calculated for all the charges in a given charge configuration creating an electrostatic field, the result is the total energy, U, stored in the field. The energy is "stored" in the volume of the electrostatic field and is given by the following expressions.

$$\begin{aligned} U &= \tfrac{1}{2}\int \rho V dv = \tfrac{1}{2}\int \mathbf{D}\cdot\mathbf{E} dv \\ &= \tfrac{1}{2}\int \epsilon E^2 dv = \tfrac{1}{2}\int \frac{D^2}{\epsilon} dv \end{aligned} \qquad 20.26$$

An electrostatic field is a *conservative field.* No work is done in moving a charge around a closed path within an electrostatic field, such as path ABCDA in Fig. 20.9. Further, the work done in moving a charge between any two points within the field is independent of the path.

This means the integral of the electric field around a closed path equals zero as shown in Eq. 20.26.

$$W = -\oint Q\mathbf{E}\cdot d\mathbf{l} = 0 \qquad 20.27$$

The result given by Eq. 20.26 is true for electrostatic fields and must be modified for time varying fields.

Example 20.5

Neglecting fringing, what is the total energy stored in a parallel plate capacitor with a capacitance of 50 μF across 120 V?

Solution

Neglecting fringing, the capacitor's electric field is identical to the field of infinite parallel plates and is uniform. That is, from Table 16.6,

$$E = \frac{V}{r}$$

Substituting and noting that the field is constant and can be removed from inside the integral gives

$$U = \tfrac{1}{2}\int \epsilon E^2 dv$$

$$\begin{aligned} U &= \tfrac{1}{2}\epsilon E^2 \int dv = \tfrac{1}{2}\epsilon\left(\frac{V}{r}\right)^2 \int dv \\ &= \tfrac{1}{2}\frac{\epsilon V^2}{r^2}\int dv \end{aligned}$$

The capacitance is given by

$$C = \frac{\epsilon A}{r}$$

Substituting,

$$U = \tfrac{1}{2}\frac{\epsilon V^2}{r^2}\int dv = \tfrac{1}{2}\frac{CV^2}{rA}\int dv$$

The volume equals the area of the plates, A, multiplied by the distance between them, r.

$$U = \tfrac{1}{2}\left(\frac{CV^2}{rA}\right)(Ar) = \tfrac{1}{2}CV^2$$

Substituting the given information gives

$$\begin{aligned} U &= \tfrac{1}{2}(50\times 10^{-6}\text{ F})(120\text{ V})^2 \\ &= 0.36\text{ J} \end{aligned}$$

Field Theory

21 Electrostatic Fields

1. Polarization 21-1
2. Polarized Fields 21-2
3. Electric Displacement 21-3
4. Polarization in Dielectrics versus Conductors 21-4

Nomenclature

A	area	m^2
d, $\mathbf{d}$	distance	m
D, $\mathbf{D}$	electric flux density	C/m^2
E, $\mathbf{E}$	electric field strength	N/C (V/m)
f	frequency	Hz
J, $\mathbf{J}$	current density	A/m^2
p, $\mathbf{p}$	dipole moment	C·m
P, $\mathbf{P}$	polarization	C/m^2
q	electric unit charge, 1.6022×10^{-19}	C
Q	charge	C
t	time	s
v	volume	m^3
W	work	J

Symbols

ϵ	permittivity	$C^2/N{\cdot}m^2$ or F/m
ϵ_0	free-space permittivity, 8.854×10^{-12}	$C^2/N{\cdot}m^2$
θ	angle	degrees
ρ_{sp}	polarization surface charge density	C/m^2
σ	conductivity	S/m
χ	susceptibility	–
ω	angular frequency	rad/s

Subscripts

0	free space
c	conduction
d	displacement
D	density
e	electric
i	induced or internal
r	relative
sp	surface polarization

1. POLARIZATION

When an electric field is present in a dielectric, the bound electric charges tend to align in the lowest energy configuration possible. In doing so, they increase the flux density, $\mathbf{D}$. The alignment is referred to as the *polarization* of the dielectric and is given the symbol $\mathbf{P}$. The polarization is the total dipole moment, $\mathbf{p}$, per unit volume, v.

$$\mathbf{P} = \frac{\mathbf{p}}{v} \qquad 21.1$$

The flux density is

$$\mathbf{D} = \epsilon_0\mathbf{E} + \mathbf{P} \qquad 21.2$$

Polarization increases the flux density above the level that would exist under free-space conditions with the same electric field intensity (the polarization of free space equals zero). This occurs even when permanent dipole molecules are not initially present in the dielectric. This can be understood by considering that when an atom is placed within the influence of an electric field, the positive protons tend to move toward the direction of the electric field and the electrons tend to move toward the direction opposite the electric field. In this manner, a dipole is set up within the atom. See Fig. 21.1 for an illustration of this concept. The charge in such induced dipoles is usually not a complete unit charge (1.6022×10^{-19} C) but is instead some fraction thereof, because the charges are not completely separate. The dipole is analogous to spring energy in mechanical systems. The work done in the distortion of the atom is recoverable upon removal of the electric field.

Figure 21.1 *Induced Dipole Moment*

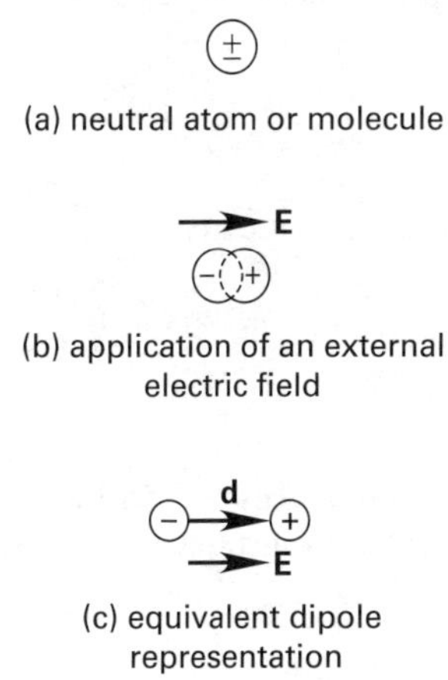

The electric field and the polarization can have different directions, though this is often not the case. In isotropic, linear material, $\mathbf{E}$ and $\mathbf{P}$ are parallel at each point.

$$\mathbf{P} = \chi_e\epsilon_0\mathbf{E} \qquad 21.3$$

The term *isotropic* indicates that the medium's properties are independent of direction. The term *linear* indicates that the permittivity, ϵ, is constant regardless of the magnitude of the charge present. The term χ_e is the electric susceptibility. This dimensionless term measures the ease of polarization in a material. For an isotropic linear material, Eq. 21.2 can be rewritten as[1]

$$\mathbf{D} = \epsilon_0(1+\chi_e)\mathbf{E} \qquad 21.4$$

The relative permittivity can be represented by Eq. 21.5.

$$\epsilon_r = 1+\chi_e \qquad 21.5$$

In free space, then, because no dielectrics are present, no polarization occurs. The electric flux density, $\mathbf{D}$, is

$$\mathbf{D} = \epsilon_0\mathbf{E} \qquad 21.6$$

When a dielectric is present, the flux density is increased and is given by Eq. 21.2. Importantly, Eq. 21.2 is often written without direct reference to polarization, but rather in terms of the relative permittivity. The relationship is shown explicitly in Eq. 21.7.

$$\mathbf{D} = \epsilon\mathbf{E} = \epsilon_0\epsilon_r\mathbf{E} = \epsilon_0\mathbf{E}+\mathbf{P} \qquad 21.7$$

Example 21.1

Water in vapor form has a fractional charge of approximately 5×10^{-20} C and a dipole distance of approximately 10^{-10} m. What is the maximum possible polarization for one mole of water molecules at standard temperature and pressure (STP)?

Solution

The maximum polarization is

$$P = \frac{p}{v}$$

Determine the dipole moment.

$$\begin{aligned} p &= qd \\ &= (5\times10^{-20}\text{ C})(10^{-10}\text{ m}) = 5\times10^{-30}\text{ C·m} \end{aligned}$$

At STP, the volume of 1 mol of an ideal gas is 22.4 L or 0.0224 m^3. The maximum polarization is

$$\begin{aligned} P &= \frac{p}{v} \\ &= \frac{5\times10^{-30}\text{ C·m}}{0.0224\text{ m}^3} \\ &= 2.2\times10^{-28}\text{ C/m}^2 \end{aligned}$$

[1] If the material is anisotropic or nonlinear, or both, χ_e is a tensor.

Example 21.2

What are the magnitudes of $\mathbf{P}$ and $\mathbf{D}$ for a dielectric in an electric field with a strength of $E = 38.8\times10^3$ V/m with a relative permittivity of 4.1 $C^2/N{\cdot}m^2$?

Solution

First,

$$P = \chi_e\epsilon_0E$$

The susceptibility is found from

$$\begin{aligned} \epsilon_r &= 1+\chi_e \\ \chi_e &= \epsilon_r - 1 = 4.1-1 = 3.1 \end{aligned}$$

So,

$$\begin{aligned} P &= \chi_e\epsilon_0E \\ &= (3.1)\left(8.854\times10^{-12}\ \frac{\text{C}^2}{\text{N·m}^2}\right)\left(38.8\times10^3\ \frac{\text{V}}{\text{m}}\right) \\ &= 1.06\times10^{-6}\text{ C/m}^2 \end{aligned}$$

Second,

$$\begin{aligned} D &= \epsilon_0\epsilon_rE \\ &= \left(8.854\times10^{-12}\ \frac{\text{C}^2}{\text{N·m}^2}\right)\left(4.1\ \frac{\text{C}^2}{\text{N·m}^2}\right) \\ &\quad\times\left(38.8\times10^3\ \frac{\text{V}}{\text{m}}\right) \\ &= 1.41\times10^{-6}\text{ C/m}^2 \end{aligned}$$

Example 21.3

An electric field strength is measured as 0.20×10^6 V/m. If the electric flux density is required to be 8×10^6 C/m^2 by the design, what is the minimum susceptibility required of the dielectric?

Solution

The flux density is given. Using this value,

$$\begin{aligned} D &= \epsilon_0(1+\chi_e)E \\ \chi_e &= \frac{D}{\epsilon_0E} - 1 \\ &= \frac{8\times10^{-6}\ \frac{\text{C}}{\text{m}^2}}{\left(8.854\times10^{-12}\ \frac{\text{C}^2}{\text{N·m}^2}\right)\left(0.20\times10^6\ \frac{\text{V}}{\text{m}}\right)} - 1 \\ &= 3.5 \end{aligned}$$

2. POLARIZED FIELDS

If a dipole is present, it can be aligned either partially or fully with the electric field. An example of such an alignment is illustrated in Fig. 21.2.

Figure 21.2 *Dipole Alignment About a Free Positive Charge*

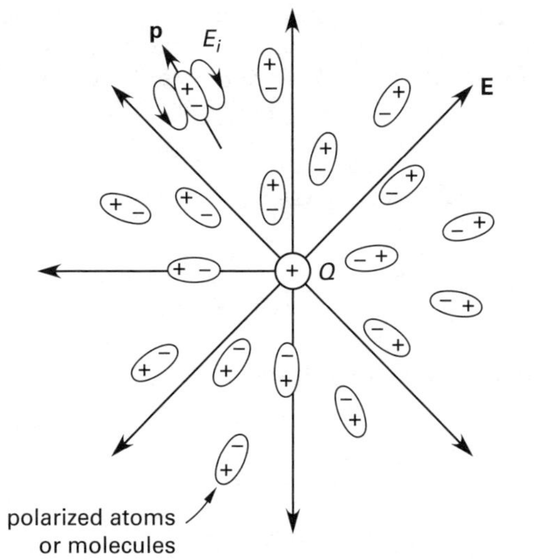

If a molecule contains a permanent dipole moment, it is called a *polar molecule*. Two common examples are water and salt (NaCl). If the molecule has no dipole moment in the absence of an electric field, it is called a *nonpolar molecule*. In this case, the dipole moment can be induced as explained in Sec. 21.1 and shown in Fig. 21.1. Yet a third type of dielectric exists that exhibits permanent polarization in the absence of an electric field but does so only after an induced polarization has taken place. Some waxes, for instance, are polarized when molten and exposed to an electric field. After cooling some of the induced polarization remains. This special case of aligned polar molecules occurs below a temperature called the *Curie point*. Such materials are analogous to permanent magnets and are called ferroelectrics. Ferroelectrics exhibit the properties of electrical hysteresis.

A dielectric with its dipole moments aligned produces an internal electric field, E_i, that opposes the original electric field, E_0, that created it. The energy expended in the alignment results in a dielectric, or dipole, electric field, E_d. This situation is illustrated in Fig. 21.3(a). The net result is that a surface charge appears at either end of the dielectric as shown in Fig. 21.3(b). The electric flux density is increased above the level that would exist in free space by an amount equal to the polarization as indicated by Eq. 21.2. This can be inferred from the units used to measure electric flux density, C/m^2. The charges per unit area have been increased above the level that exists in free space. An important point to remember is that the net charge in the dielectric remains zero. Also, the polarization surface charge is a real accumulation of charge. It does contribute to the internal and external electric fields. It does not, however, enter into boundary conditions when solving electrical engineering problems. Furthermore, the divergence of **D** depends only on the free charge density.

Figure 21.3 *Dielectric Electric Fields and Surface Charge*

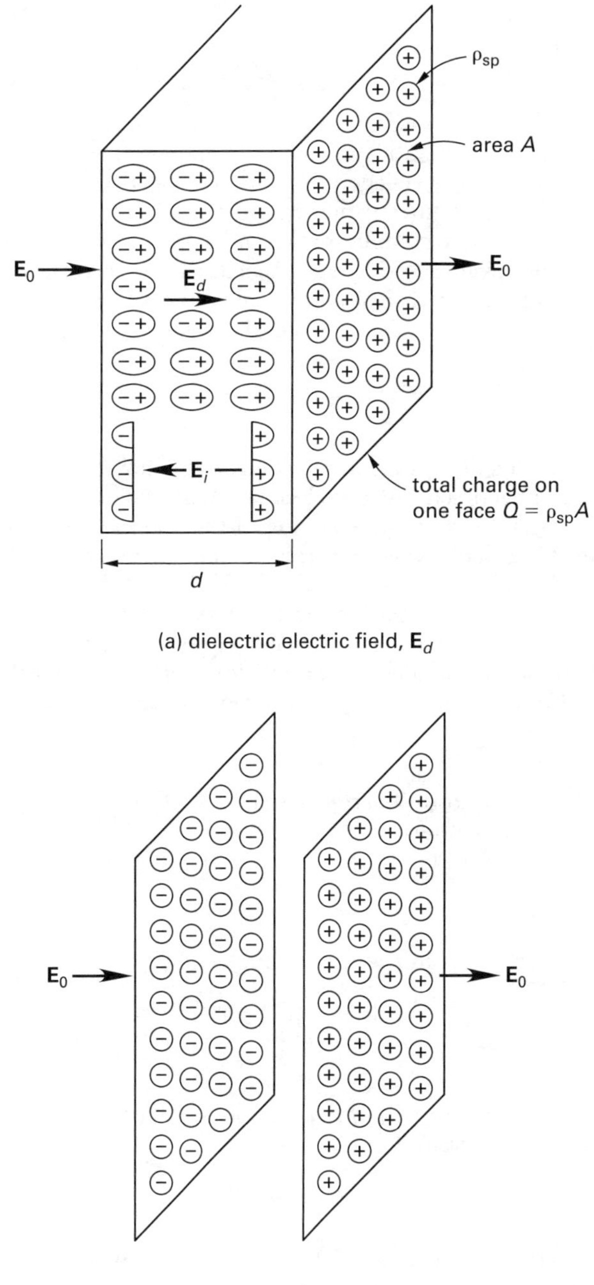

3. ELECTRIC DISPLACEMENT

The electric flux density, **D**, is also called the displacement density, for reasons now clear. It represents the displacement of charges within a dielectric. The motion of the charges within the dielectric solid is not translational, as it is for conduction current. Instead, the dipole moments rotate under the influence of the electric field (see Fig. 20.4). If the angle $\pi/2$ is chosen as the arbitrary zero energy reference, the work of rotating a dipole is

$$W = -\mathbf{p}\cdot\mathbf{E} \qquad 21.8$$

This motion of charges is the *displacement current.* In terms of the current density vector, **J**, the displacement current is defined by Eq. 21.9.

$$\mathbf{J}_D = \frac{\delta \mathbf{D}}{\delta t} \qquad 21.9$$

Equation 21.9 indicates that a time-varying electric field, and subsequently a time-varying electric flux density, results in current flow in a dielectric material that contains no free charges. In this manner, current is made to "flow" from one side of a capacitor to another. For a poor conductor, or *lossy dielectric*, the ratio of the conduction current density to the displacement current is

$$\frac{J_c}{J_d} = \frac{\sigma}{\omega\epsilon} \qquad 21.10$$

Displacement and displacement current define the electrical phenomena in a dielectric. While polarization is necessary for an understanding of the situation, it complicates calculations. For this reason, in practice with simple dielectrics, work is done directly with **D** and not **P** by using the relative permittivity as shown in Eq. 21.7. The relative permittivity is easily determined (for isotropic, linear dielectrics) and for this reason, can be used to account for all polarization effects.

Example 21.4

What is the total energy required to "flip" a dipole of unit positive charge a dipole distance of 10^{-10} m in an electric field of 0.20 MV/m?

Solution

The work done in rotating a dipole is

$$\begin{aligned} W &= -\mathbf{p}\cdot\mathbf{E} \\ &= -pE\cos\theta \end{aligned}$$

The zero reference point is $\theta = \pi/2$. The total energy required to rotate the dipole 90° is

$$|W_{90^\circ}| = pE$$

To flip the dipole (i.e., to rotate 180°) requires double that amount of energy.

$$|W_{180^\circ}| = 2pE$$

Substituting,

$$\begin{aligned} W_{\text{total}} &= 2pE = 2qdE \\ &= (2)(1.6022 \times 10^{-19}\ \text{C})(10^{-10}\ \text{m}) \\ &\quad \times \left(0.20 \times 10^6\ \frac{\text{V}}{\text{m}}\right) \\ &= 6.4 \times 10^{-24}\ \text{J} \end{aligned}$$

Example 21.5

What is the ratio of conduction current to displacement current in a conductor operating at 60 Hz with a conductance of 30×10^6 S/m and a ϵ_r of 1 $\text{C}^2/\text{N}\cdot\text{m}^2$?

Solution

The applicable formula is

$$\frac{J_c}{J_d} = \frac{\sigma}{\omega\epsilon}$$

Expanding and substituting,

$$\begin{aligned} \frac{J_c}{J_d} &= \frac{\sigma}{\omega\epsilon} = \frac{\sigma}{2\pi f \epsilon_0 \epsilon_r} \\ &= \frac{30 \times 10^6\ \frac{\text{S}}{\text{m}}}{2\pi(60\ \text{Hz})\left(8.854 \times 10^{-12}\ \frac{\text{C}^2}{\text{N}\cdot\text{m}^2}\right)\left(1\ \frac{\text{C}^2}{\text{N}\cdot\text{m}^2}\right)} \\ &= 9.0 \times 10^{15} \end{aligned}$$

The displacement current is a minor effect in conductors. Only at high frequencies need it be considered.

4. POLARIZATION IN DIELECTRICS VERSUS CONDUCTORS

In a metal conductor, free conduction charges are mobile and move to the surface of the metal under the influence of an electrostatic field. Because the charges are plentiful and mobile, the surface charge builds up until the induced electric field cancels the applied electric field. Consequently, the electric field in the interior of a metal is zero. That is, the polarization of a conductor by the external electric field is complete.

For dielectrics, on the other hand, the charges are bound and can only move a small fraction of an atomic diameter. The induced electric field, E_i, partially cancels the applied electric field. The result is an internal electric field in the dielectric that points in the same direction as the applied field (for isotropic, linear dielectrics). Furthermore, the induced charges on the surface of the dielectric are not equal in magnitude to the induced charges on metallic surfaces. (This could occur only if ϵ_r were infinite.) In short, a dielectric in an external electric field becomes polarized, but not to such an extent that the interior electric field goes to zero.

22 Magnetostatics

Part 1: The Magnetic Field 22-1
1. Biot-Savart Law 22-3
2. Force on a Moving Charged Particle 22-3
3. Force on Current Elements 22-4
4. Force on Distributed Current Elements . . . 22-5

Part 2: Lorentz Force Law 22-5

Part 3: Traditional Magnetism 22-5
5. Magnetic Pole 22-6
6. Magnetic Dipole 22-6
7. Coulomb's Law Equivalent: Force Between Magnetic Poles 22-8
8. Magnetic Potential 22-8
9. Magnetic Potential Energy 22-8

Part 4: Curl 22-8

Part 5: Magnetic Potential 22-9

Part 6: Magnetic Vector Potential 22-9

Part 7: Work and Energy 22-9

Nomenclature

a	unit vector	–
A	vector magnetic potential	Wb/m (T·m)
A	area	m^2
B, **B**	magnetic flux density	T (Wb/m^2)
c	speed of light (vacuum), 3.00×10^8	m/s
d, **d**	distance	m
E, **E**	electric field strength	V/m
F, **F**	force	N
H, **H**	magnetic field strength	A/m
I, **I**	constant or rms current	A
J, **J**	current density	A/m^2
K, **K**	sheet current	A
l, **l**	length	m
m, **m**	magnetic moment	$A{\cdot}m^2$ (J/T)
m	mass	kg
M	magnetization	A/m
n	unit vector	–
N	number of turns	–
p	pole strength	N/(A/m) or J/A or Wb
p_m	magnetic dipole moment	$A{\cdot}m^2$ (J/T)
p	electric dipole moment	C·m
q	electric unit charge, 1.6022×10^{-19}	C
q_m	magnetic charge	A·m or N/T
Q	charge	C
r, **r**	radius or distance	m
s, **s**	surface area	m^2
T	torque	N·m
U, **U**	magnetic potential	A
v, **v**	velocity[1]	m/s
V	volume	m^3
V	constant or rms voltage	V
V_m	magnetomotance	A
W	work	J

Symbols

ϵ	permittivity	$C^2/N{\cdot}m^2$ or F/m
θ	angle	degrees
μ	permeability	N/A^2 or H/m
μ_0	free-space permeability, 1.2566×10^{-6}	N/A^2 or H/m
ρ	charge density	C/m^3
Φ, ϕ	magnetic flux	Wb or N/(A/m) or J/A
ω	angular frequency	rad/s

Subscripts

0	initial
1–2	from 1 to 2; or effect on 2 caused by the field of 1
a	applied
e	electrical or electron
f	final
m	magnetic
p	pole
r	relative
V	per unit volume

PART 1: THE MAGNETIC FIELD

A *field* is an entity that acts as an intermediary in interactions between moving charges. It is distributed over part or all of space and is a function of space coordinates and time (if the field is not static). A *magnetic field* is elementary in nature. Charges moving through the field experience a force. The strength of the **B**-field is the electromotive force (V or J/C) produced when one turn of a loop of area A is linked in 1 sec.[2] It is measured in Wb/m^2, equivalent to $V{\cdot}s/m^2$. The strength of the **B**-field is

$$\mathbf{B} = \left(\frac{\mu I l}{4\pi r^2}\right)\mathbf{a} \qquad 22.1$$

[1]The term "velocity" is often used in text where it would be more technically correct to use the term "speed." Velocity is a vector quantity with speed as its magnitude.

[2]The term *linked* indicates that the magnetic flux is passed through the area of concern—in this case, the area of the loop.

The unit vector, **a**, points in the direction of the **B**-field, along the lines of flux. It is perpendicular to the direction of the force experienced by the positive moving charges (that is, the current). This occurs because a magnetic force always acts in a direction transverse to the movement of charges. The denominator in Eq. 22.1, $4\pi r^2$, represents the surface area of a sphere centered on the current element dl, whose surface just touches the point in space where the **B**-field is being evaluated. Specific configurations may be evaluated with Eq. 22.1 by changing **B** to $d\mathbf{B}$ and l to dl and integrating. The magnetic field is a vector force field, in that a moving charge within the field experiences a force in a specific direction. The force field for a single line of moving charges is illustrated in Fig. 22.1.

Figure 22.1 *Magnetic Field of a Positive Current*

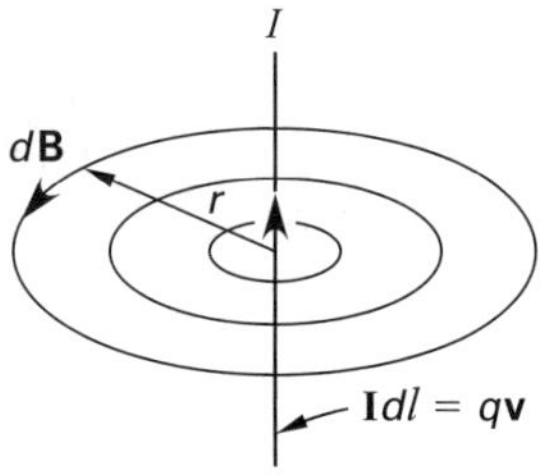

Lines of magnetic flux form continuous loops. By convention, these lines are said to emanate from the north pole of a magnetic dipole and enter the south pole. Like poles repel, and unlike poles attract.

The magnetic force fields of multiple currents interact, and the principle of superposition applies as long as the permeability of the medium is constant. Unfortunately, constant permeability is not exhibited by iron and other ferromagnetic materials, which are widely used around magnetic fields. Therefore, in general, the principle of linear superposition does not apply to current-carrying conductors in the vicinity of ferromagnetic materials.

Equation 22.1 is normally written in terms of the magnetic flux density as shown in Eq. 22.2. Both equations use identical units, Wb/m^2, which in the SI system is a tesla, T. The term **B** is most often called the magnetic flux density.[3]

$$B = \frac{\Phi}{A} \qquad 22.2$$

The force on a current element in a magnetic field is given by the Biot-Savart law in Eq. 16.43. This law can also be viewed in terms of the magnetic field produced by the current elements as explained in Sec. 22.2.

[3]The magnetic flux density, **B**, is the fundamental field in magnetics, and it appears in the force equations. Though the name is reminiscent of the electric flux density, **D**, this is a result of the historical development in the study of magnetics. **B** and **E** are the fundamental fields. **H** and **D** are constructs to aid in the solution of electrical engineering problems.

Magnetic flux, Φ, is defined as the surface integral of the normal component of the flux density, B, over the area.

$$\Phi = \int_A \mathbf{B}\cdot d\mathbf{A} \qquad 22.3$$

It is often more helpful to visualize the magnetic flux as the total quantity of lines of magnetic force set up around a current-carrying conductor. If there are N conductors of length l, the flux is

$$\Phi = \mu N I l = BA \qquad 22.4$$

Example 22.1

A wire carrying a current of 6 A is shown. The current is in the plane of the paper. Determine the magnetic flux density 0.5 m away from a 1 cm segment of the wire at the point indicated.

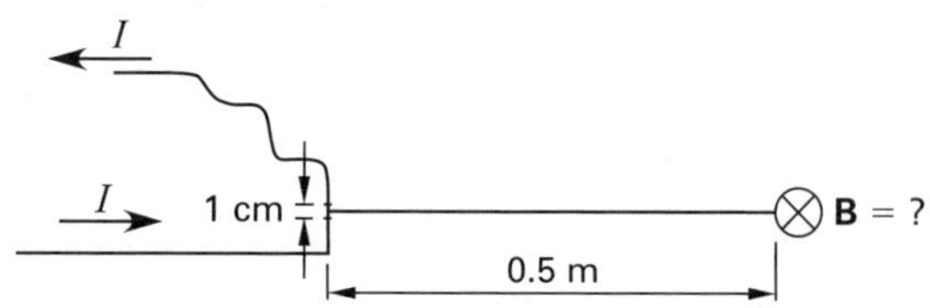

Solution

The magnetic flux density caused by the small segment, dl, 1 cm long is

$$\mathbf{B} = \left(\frac{\mu I l}{4\pi r^2}\right)\mathbf{a}$$

Because the segment is small compared to the distance, r, it is not necessary to integrate. (When to integrate, that is, when to use $d\mathbf{B}$ and dl, depends on the amount of error tolerable in the solution.) Substituting and assuming free-space conditions gives

$$\begin{aligned}\mathbf{B} &= \left(\frac{\mu I l}{4\pi r^2}\right)\mathbf{a} \\ &= \frac{\left(1.2566\times 10^{-6}\ \frac{\text{N}}{\text{A}^2}\right)(6\ \text{A})(1\ \text{cm})\left(0.01\ \frac{\text{m}}{\text{cm}}\right)\mathbf{a}}{4\pi(0.5\ \text{m})^2} \\ &= (2.4\times 10^{-8}\ \text{T})\ \mathbf{a}\end{aligned}$$

Using the right-hand rule, the direction of the **B**-field is into the paper, as shown.

Example 22.2

A loop of wire in an electric motor approximates a square with four conductors, each of length 0.15 m. Because no other condition is specified, free space is assumed. What is the total flux of the loop if the motor is rated to carry 15 A?

Solution

The applicable equation is

$$\Phi = \mu NIl = BA$$

Substituting known and given values results in the following.

$$\begin{aligned}\Phi = \mu NIl &= \left(1.2566 \times 10^{-6}\ \frac{\text{H}}{\text{m}}\right)(4)(15\ \text{A})(0.15\ \text{m}) \\ &= 1.1 \times 10^{-5}\ \text{Wb}\end{aligned}$$

1. BIOT-SAVART LAW

The Biot-Savart law is normally given as relating the force between two current elements as in Sec. 16.5. This is done to emphasize the similarity between electrostatics and magnetostatics and to exhibit the relationship between the Biot-Savart law and Coulomb's law in electromagnetics.

When considering **B** as the flux density, the Biot-Savart law is

$$d\mathbf{B} = \frac{\mu_0}{4\pi}\left(\frac{\mathbf{I} \times \mathbf{r}}{r^2}\right) dl \qquad 22.5$$

The term **r** is the unit vector pointing from the current element $\mathbf{I}dl$ to $d\mathbf{B}$.[4] The magnitude of Eq. 22.5 is given in Eq. 22.6.

$$dB = \frac{\mu_0}{4\pi}\left(\frac{Idl}{r^2}\right)\sin\theta \qquad 22.6$$

Comparing Eq. 22.6 with the equation for an electric field (see Eq. 16.6), the magnetic charge (if one existed) would have the form $q_m = Idl$.

Because the current element $\mathbf{I}dl$ is equivalent to the moving point charge $q\mathbf{v}$, the Biot-Savart law can be written as

$$\mathbf{B} = \frac{\mu_0}{4\pi}\left(\frac{q\mathbf{v} \times \mathbf{r}}{r^2}\right) \qquad 22.7$$

The cross product in Eq. 22.5 and Eq. 22.7 is accounted for when determining the direction of the magnetic field by using the *right-hand rule*. The right-hand rule states that when one's right thumb is pointed in the direction of the positive current flow, the fingers of the right hand curled inward toward the palm give the direction of the magnetic field. The rule can be verified by using the directions shown in Figure 22.1.

[4]Importantly, the direction of $\mathbf{I}dl$ can be ascribed to the current and written as shown, or it can be ascribed to the differential length and be written as $Id\mathbf{l}$ or $I\mathbf{dl}$.

2. FORCE ON A MOVING CHARGED PARTICLE

A magnetic field has no effect on stationary charged particles. If, however, the charge Q is moving with velocity **v** in an external magnetic field with flux density **B**, the force on the charge is

$$\mathbf{F} = Q\mathbf{v} \times \mathbf{B} \qquad 22.8$$

The magnetic field referred to in Eq. 22.8 is an external field, not the field created by the moving charged particle. The acceleration of the particle is given by Newton's second law.

The cross product gives the direction of the acceleration at an angle perpendicular to both the velocity **v** and the external magnetic field **B**. A charged particle in a uniform magnetic field will travel in a circular path with a radius and an angular velocity of

$$r = \frac{m\text{v}}{QB} \qquad 22.9$$

$$\omega = \frac{QB}{m} \qquad 22.10$$

Equation 22.8 is sometimes used to define the magnetic field. In doing so, the magnitude of Eq. 22.8 is written as

$$B = \frac{F}{q\text{v}} \qquad 22.11$$

Comparing Eq. 22.11 to Eq. 16.4 shows that just as the electric field can be defined in terms of the force per unit charge that is exerted at a point, the magnetic field can be defined as the force per unit moving charge that is exerted at a point.

Example 22.3

What is the maximum magnitude of the force on an electron moving at $0.1c$ in a uniform magnetic field of 5.0×10^{-5} T?

Solution

One-tenth the speed of light is the approximate maximum speed at which relativistic effects can be ignored. The electron is moving at $0.1c = 0.3 \times 10^8$ m/s. (The magnetic field strength is that of the Earth's.) The applicable equation is

$$\mathbf{F} = Q\mathbf{v} \times \mathbf{B}$$

Because the maximum magnitude is to be determined, the following adjustments can be made.

$$F = |Q|\text{v}B\sin\theta = |Q|\text{v}B(1) = |Q|\text{v}B$$

Field Theory

Substitute the given values.

$$\begin{aligned} F &= |Q|\mathrm{v}B \\ &= (1.6022 \times 10^{-19}\ \mathrm{C})\left(0.3 \times 10^{8}\ \frac{\mathrm{m}}{\mathrm{s}}\right)(5.0 \times 10^{-5}\ \mathrm{T}) \\ &= 2.4 \times 10^{-16}\ \mathrm{N} \end{aligned}$$

3. FORCE ON CURRENT ELEMENTS

A current element $\mathbf{I}dl$ within a magnetic field $\mathbf{B}$ experiences a force of

$$d\mathbf{F} = \mathbf{I}dl \times \mathbf{B} \qquad 22.12$$

Because the term $\mathbf{I}dl$ is a mathematical construct and cannot exist alone but must instead be part of a complete loop or circuit, the force on the entire loop can be obtained from integrating around the loop as shown in Eq. 22.13.

$$\mathbf{F} = \oint \mathbf{I}dl \times \mathbf{B} \qquad 22.13$$

For a straight conductor of length l, the total force on a current-carrying conductor is

$$\mathbf{F} = l\mathbf{I} \times \mathbf{B} \qquad 22.14$$

In terms of magnitude, assuming a uniform magnetic field, Eq. 22.14 can be written as

$$F = lIB\sin\theta \qquad 22.15$$

The direction of the magnetic force can be ascertained by the left-hand *FBI rule* explained in Fig. 22.2.

For a straight, infinitely long conductor, the magnitude of the magnetic field is

$$B = \frac{\mu I}{2\pi r} \qquad 22.16$$

Two straight, infinitely long conductors situated within the magnetic fields of one another experience a force of

$$\frac{\mathbf{F}}{l} = \mathbf{I} \times \mathbf{B} = \left(\frac{\mu I_1 I_2}{2\pi r}\right)\mathbf{a} \qquad 22.17$$

The relationship given by Eq. 22.17 is useful in defining current in terms of force and distance, enabling the measurement of electrical current in terms of familiar mechanical variables. The distance, r, is perpendicular to the wires.

Figure 22.2 *Magnetic Force Direction*

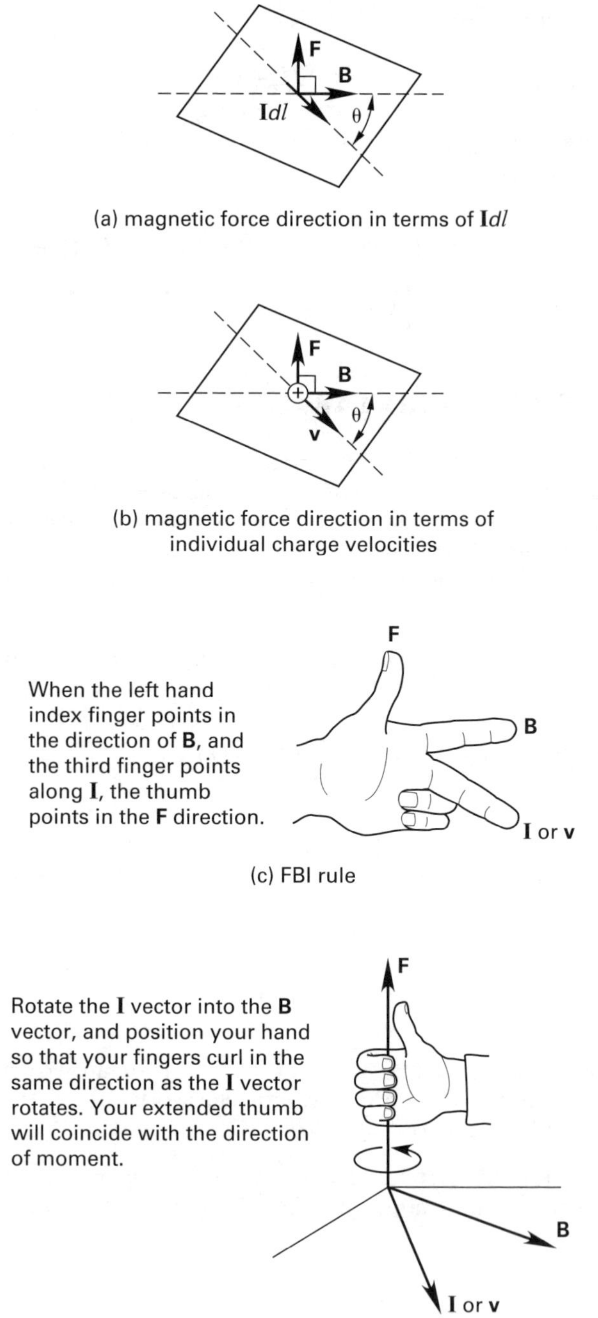

(a) magnetic force direction in terms of $\mathbf{I}dl$

(b) magnetic force direction in terms of individual charge velocities

(c) FBI rule

(d) right-hand rule

Example 22.4

The force on two parallel wires, approximated as infinite, is measured to be 30×10^{-3} N/m. The wires carry equal currents and repel one another. What are the magnitude and direction of the current in the two wires if the wires are 0.01 m apart?

Solution

The force per unit length is given. Assuming free space, the applicable equation is

$$\frac{\mathbf{F}}{l} = \mathbf{I} \times \mathbf{B} = \left(\frac{\mu I_1 I_2}{2\pi r}\right)\mathbf{a}$$

Determine the magnitude of the current by rearranging the equation to

$$I^2 = \frac{2\pi\left(\frac{F}{l}\right)r}{\mu_0} = \frac{2\pi\left(30 \times 10^{-3}\ \frac{\text{N}}{\text{m}}\right)(0.01\ \text{m})}{1.2566 \times 10^{-6}\ \frac{\text{N}}{\text{A}^2}}$$

$$= 1.5 \times 10^3\ \text{A}^2$$

$$I = \sqrt{1.5 \times 10^3\ \text{A}^2} = 38.7\ \text{A}$$

The direction can be determined in a number of ways, such as by using the right-hand rule or by determining if the magnetic fields aid or oppose each other (that is, if they "link"). Using the equation, and knowing that the force must be in a direction that causes repulsion, the currents must be as shown.

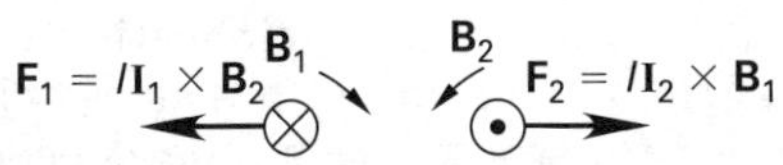

The magnitude of the current is 38.7 A flowing in opposite directions in the two wires.

4. FORCE ON DISTRIBUTED CURRENT ELEMENTS

Equation 22.8 and Eq. 22.12 can be generalized to the case of force per unit volume.[5] In doing so, the force can be written as

$$\mathbf{F}_V = \mathbf{J} \times \mathbf{B} \qquad 22.18$$

PART 2: LORENTZ FORCE LAW

When an electric field and a magnetic field are present at the same time, the force on a charge is given by the general force law, commonly called the *Lorentz force law*. The Lorentz force combines the electrical and magnetic force and is given in Eq. 22.19.

$$\mathbf{F} = \mathbf{F}_e + \mathbf{F}_m = Q(\mathbf{E} + \mathbf{v} \times \mathbf{B}) \qquad 22.19$$

[5]This is useful for understanding, among other things, plasmas and convection or diffusion currents in semiconductors.

Similarly, if a charge cloud of density ρ at each point exists in a volume, the Lorentz force is

$$d\mathbf{F} = \rho(\mathbf{E} + \mathbf{v} \times \mathbf{B})dV \qquad 22.20$$

PART 3: TRADITIONAL MAGNETISM

Traditional magnetism was based on concepts similar to those in electrostatics. As a result, the concept of a magnetic charge, q_m, was used, and magnets were described in terms of pole strength, p. The isolated magnetic charge, which does not exist in nature, is called a *monopole*. Nevertheless, the concept of an isolated magnetic charge at the end of a bar magnet is useful in understanding the effects of magnetism. Figure 22.3 illustrates this concept for a bar magnet (magnetic dipole) in an external magnetic field. The force on a magnetic charge is defined as the charge multiplied by the strength of the **B**-field as shown in Eq. 22.21. A magnetic field may, therefore, be defined in the same manner as an electric field; that is, the field represents the force exerted on an infinitesimal magnetic test charge at a point in space ($B = F/q_m$).

$$\mathbf{F} = q_m\mathbf{B} \qquad 22.21$$

Figure 22.3 *Magnetic Charge in a Magnetic Field (Dipole in a Magnetic Field)*

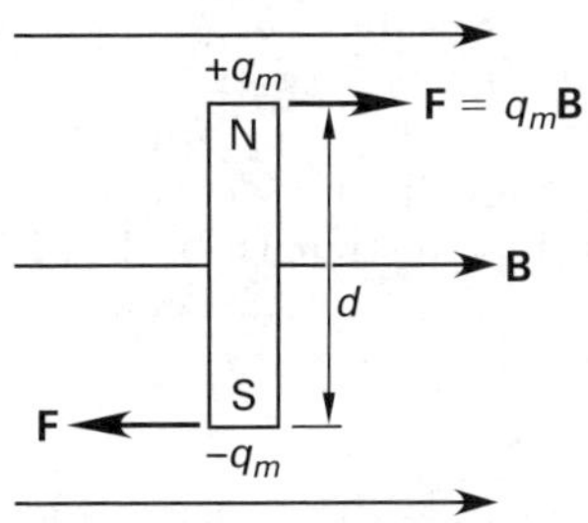

The torque exerted on the magnetic charge in an external magnetic field is

$$\mathbf{T} = q_m\mathbf{d} \times \mathbf{B} \qquad 22.22$$

The magnetic flux density, **B**, is also called the *intensity of magnetization* and the *dipole moment per unit volume*. For a bar magnet of pole strength p, length l, and cross-sectional area A, the magnetic flux density is

$$B = \frac{pl}{lA} = \frac{p}{A} = \frac{\mu m}{V} \qquad 22.23$$

The cgs system of units was the first to be used in the field of magnetism. This resulted in equations whose

Table 22.1 Magnetic Units

quantity	symbol	cgs units	SI units	conversion: cgs to SI
pole strength or flux	p or Φ or ϕ^a	maxwells[b]	Wb	10^8 maxwells = 1 Wb
flux density	B	gauss[c]	T^d	10^4 gauss = 1 T
field strength (intensity)	H	oersted	A/m[e]	$4\pi \times 10^{-3}$ oersted = 1 A/m
magnetization[f]	M	oersted	A/m[e]	$4\pi \times 10^{-3}$ oersted = 1 A/M
permeability	μ	gauss/oersted	H/m[g]	$\mu_{cgs}\,(4\pi \times 10^{-7}) = \mu_{mks}{}^h$

[a]The SI symbol for magnetic flux is Φ. The symbol p is used when referring to magnetic poles. The symbol ϕ is a generic flux symbol used in many texts.
[b]A *maxwell* is a line of force.
[c]A gauss is equivalent to lines/cm^2 or maxwells/cm^2.
[d]A tesla is equivalent to Wb/m^2.
[e]A/m is equivalent to N/Wb.
[f]The units used in this row assume the following equation for flux density: $B = \mu_0 H + \mu_0 M$. Other forms of this equation exist. Use care in determining the correct units or in comparing values from different references.
[g]H/m is equivalent to Wb/A·m and Ω·s/m.
[h]The value of μ_{cgs} in the cgs system is 1.

forms differ significantly from those in the SI system. To ensure consistency, exercise care when using equations in systems other than the SI system. A comparison of the cgs and SI system units for magnetism is given in Table 22.1.

5. MAGNETIC POLE

Defining *pole strength* (not to be confused with the magnetic field strength) provides a method of calculating the force between two permanent magnetic poles analogous to Coulomb's law. The number of *unit-poles* or just *poles*, p, is numerically equal to the flux, ϕ, emanating from the end of the magnet. (In the cgs system, the flux is numerically equal to $4\pi p$.) The area of the pole, A_p, and the flux density from the pole, B_p, are related to the pole strength by Eq. 22.24.

$$p = \phi = B_p A_p \qquad 22.24$$

The pole strength, as used here, and the magnetic charge used in Sec. 22.3 are related by Eq. 22.25.

$$q_m = \frac{p}{\mu} \qquad 22.25$$

In the cgs system, the value of the permeability of free space is 1. In free space, the magnetic charge and the pole strength can be used interchangeably.[6] In the SI system, this is not the case, and care must be exercised to ensure consistent units and terminology. In some SI texts, the pole strength is referred to as q_m.

The moving charge of modern magnetism in Eq. 22.11 is the magnetic charge of traditional magnetism in Eq. 22.21. Modern electrical engineering teaches that magnetism is caused by circulating currents (called *amperian currents*) and electron spins. The magnetic charge in such a context is equal to Idl. Equation 22.26 relates these concepts.

$$q_m = q\mathrm{v} = I\,dl \qquad 22.26$$

6. MAGNETIC DIPOLE

Individual magnetic poles do not exist. Magnetic dipoles do exist and are the result of circulating currents. A small loop of circulating current has an associated *magnetic moment*, **m**, whose magnitude depends on the current flow and the area encompassed by that flow as given in Eq. 22.27. The direction of the magnetic moment is perpendicular to the plane of the current loop and is shown as the unit vector **n**.

$$\mathbf{m} = IA\mathbf{n} \qquad 22.27$$

Two views of the magnetic moment are illustrated in Fig. 22.4. The first is associated with a loop of wire in a motor or generator (see Fig. 22.4(a) and Fig. 22.4(b)). The second might be associated with a circulating electron (see Fig. 22.4(c)).

The torque experienced by the magnetic moment is given by Eq. 22.28.

$$\mathbf{T} = \mathbf{m} \times \mathbf{B} \qquad 22.28$$

In traditional magnetism, the magnetic dipole was viewed as a bar magnet with a magnetic dipole moment $p_m = q_m d$. A comparison of the electric dipole moment, the modern view of the magnetic moment, and the traditional magnetic dipole moment is given in Fig. 22.5.

The magnetic moment in terms of current loops, magnetic charge, and pole strength is

$$m = IA = q_m d = \left(\frac{p}{\mu}\right) l \qquad 22.29$$

[6]The permeability, μ, in the cgs system is equivalent to the relative permeability, μ_r, in the mks system.

Figure 22.4 *Magnetic Moment*

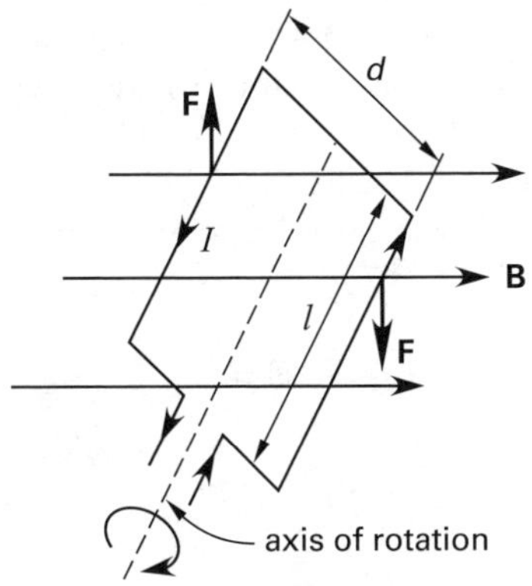

(a) loop of wire carrying current I and encompassing area $A = ld$

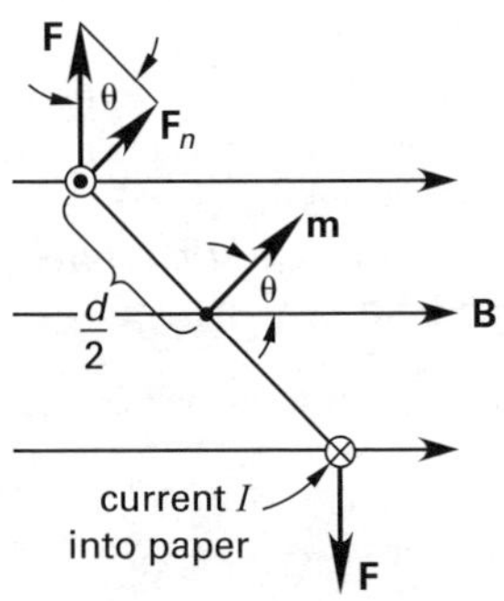

(b) cross section of loop showing the magnetic moment, **m**

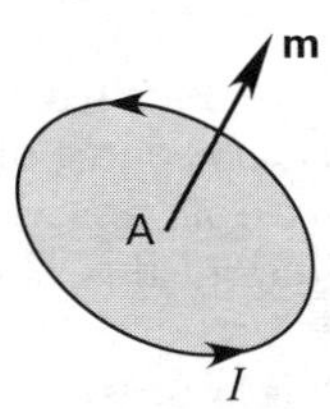

(c) generic view of a magnetic moment

Example 22.5

Determine the magnitude of the magnetic moment of a single electron at the first *Bohr radius* of 5.292×10^{-11} m if the current of the electron moving in the orbit is equivalent to approximately 1.05×10^{-3} A.

Solution

The magnitude of the magnetic moment, m, is

$$m = IA$$

Figure 22.5 *Dipole Moment Comparison*

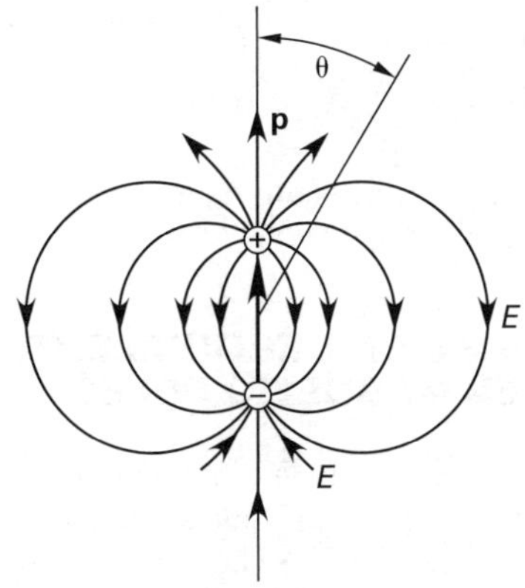

(a) electric dipole moment

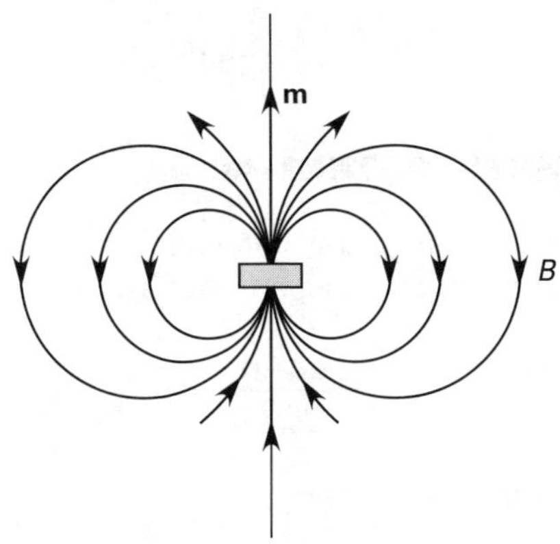

(b) magnetic dipole moment

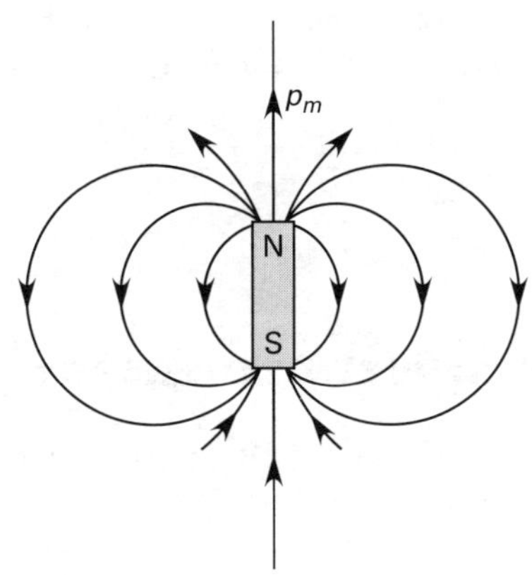

(c) traditional magnetic dipole moment

The area correlating to the Bohr radius is

$$\begin{aligned} A &= \pi r^2 = \pi (5.292 \times 10^{-11} \text{ m})^2 \\ &= 8.798 \times 10^{-21} \text{ m}^2 \end{aligned}$$

Substituting gives

$$m = IA = (1.05 \times 10^{-3}\ \text{A})(8.798 \times 10^{-21}\ \text{m}^2) = 9.24 \times 10^{-24}\ \text{A·m}^2$$

This is the approximate equivalent of one Bohr magneton, which equals $q_e h/4\pi m_e$.

7. COULOMB'S LAW EQUIVALENT: FORCE BETWEEN MAGNETIC POLES

The force between the pole ends of two magnets separated by a distance, r, written similarly to Coulomb's law, is given by Eq. 22.30. The orientation of the magnets is not important. Vector addition is required if more than two poles are present.

$$\mathbf{F}_{1-2} = p_2 \mathbf{H}_1 = \left(\frac{p_1 p_2}{4\pi\mu r^2}\right)\mathbf{a} \qquad 22.30$$

8. MAGNETIC POTENTIAL

The traditional *magnetic potential,* U_m (measured in amps), possessed by a magnet of pole strength p_A is

$$U_m = \frac{U_{\text{potential}}}{p_A} = \frac{-p_B}{4\pi\mu r} \qquad 22.31$$

The *magnetic potential gradient* is

$$H = \frac{-\Delta U_m}{\Delta r} \qquad 22.32$$

In modern magnetism, the magnetic potential gradient, called the *magnetic field strength,* is

$$H = \frac{dI}{dl} \qquad 22.33$$

9. MAGNETIC POTENTIAL ENERGY

The *work* required to change the separation distance, r, between two magnetic poles is

$$W = \int_{r_0}^{r_f} \mathbf{F}\cdot d\mathbf{r} = \left(\frac{-p_A p_B}{4\pi\mu}\right)\left(\frac{1}{r_f} - \frac{1}{r_0}\right) \qquad 22.34$$

The total *magnetic potential energy* in a system of two magnetic poles separated by a distance r is

$$U_{\text{potential}} = W_{\infty - r} = \int_{\infty}^{r} \mathbf{F}\cdot d\mathbf{r} = \frac{-p_A p_B}{4\pi\mu r} \qquad 22.35$$

PART 4: CURL

In Electrostatics, Sec. 20.3, the divergence of an electric field was found to equal the charge enclosed within the surface bounding the divergence. The applicable equation is repeated here for convenience. Equation 22.36 assumes the permittivity is constant (i.e., the medium is isotropic and linear).

$$\nabla\cdot\mathbf{E} = \frac{\rho}{\epsilon} \qquad 22.36$$

The curl of an electrostatic field is equal to zero.

$$\nabla\times\mathbf{E} = 0 \qquad 22.37$$

Equation 22.36 and Eq. 22.37 can be understood as follows: Any point in space surrounded by a closed surface with a net flux emanating from the volume must contain a charge, Q, within that volume. The charge is the cause of the net flux and is called the *scalar source* of the flux. No sideways variation in the field is present. That is, the field varies in the direction of the source. The electric field is an *irrotational field.* (An irrotational field is any force field in which $\oint \mathbf{F}\cdot dl = 0$.)

In magnetostatics, no magnetic charge exists. The divergence of the **B**-field is zero.

$$\nabla\cdot\mathbf{B} = 0 \qquad 22.38$$

The curl of the **B**-field in free space is

$$\nabla\times\mathbf{B} = \mu_0 \mathbf{J} \qquad 22.39$$

Equation 22.38 and Eq. 22.39 can be understood as follows: The current is the source of the flux in a magnetic field. The net flux from any volume, however, will be zero, because the lines of flux close upon themselves (no magnetic charges are ever enclosed). A sidewise variation exists and can be used to measure the source as given by Eq. 22.39. The current is a vector source. The magnetic field is called a *solenoidal field.* (A solenoidal field is any field in which the divergence is zero.)

In terms of the magnetic field strength, Eq. 22.39 is

$$\nabla\times\mathbf{H} = \mathbf{J} \qquad 22.40$$

In integral form, the results of Eq. 22.39 and Eq. 22.40 are known as *Ampère's law.* Ampère's law for steady currents states that the line integral of the magnetic field strength around a closed loop is equal to the current enclosed by the path. Ampère's law is

$$\oint \mathbf{H}\cdot d\mathbf{l} = \int_{\text{surface}} \mathbf{J}\cdot d\mathbf{s} = I_{\text{enclosed}} \qquad 22.41$$

PART 5: MAGNETIC POTENTIAL

The difference in the magnetic potential between any two points in a magnetic field is given by the product of the magnetic field strength and the length of the path.

$$\int_0^f \mathbf{H} \cdot d\mathbf{l} = \mathbf{U}_f - \mathbf{U}_0 \qquad 22.42$$

If the path is closed and no current is contained within it, no potential difference exists, as shown by Eq. 22.43.

$$\oint \mathbf{H} \cdot d\mathbf{l} = 0 \qquad 22.43$$

In a closed path that encloses N turns carrying current I, the potential is

$$\oint \mathbf{H} \cdot d\mathbf{l} = NI = V_m \qquad 22.44$$

The term NI, given the symbol V_m, is called the *magnetomotance*. The magnetic potential and the magnetomotance are specified in terms of amperes or ampere-turns.

PART 6: MAGNETIC VECTOR POTENTIAL

The electric field intensity, **E**, was first obtained from known charge configurations. The concept of a potential was then developed, and **E** was determined from the negative gradient of the potential. As a starting point in electrical engineering problems, the potentials on the boundary conductors are known. Then Laplace's equations are used to determine the potential V. Once V is known, **E** and **D** can be determined. Similarly, a *vector magnetic potential*, **A**, can be defined to function in the same manner as electric potential in electric circuits.

To ensure the vector magnetic potential satisfies known conditions in magnetic fields, the following conditions must be met.

$$\nabla \times \mathbf{A} = \mathbf{B} \qquad 22.45$$

$$\nabla \cdot \mathbf{A} = 0 \qquad 22.46$$

The condition required by Eq. 22.46 is called the *gauge condition*, known in magnetostatics as the *Coulomb gauge*. Magnetic vector potential is measured in Wb/m or T·m.

The vector magnetic potential for three standard current configurations—a current filament (see Eq. 22.47), a sheet current (see Eq. 22.48), and a volume current (see Eq. 22.49)—follow.

$$\mathbf{A} = \oint \frac{\mu I \, d\mathbf{l}}{4\pi r} \qquad 22.47$$

$$\mathbf{A} = \int_s \frac{\mu \mathbf{K} \, ds}{4\pi r} \qquad 22.48$$

$$\mathbf{A} = \int_V \frac{\mu \mathbf{J} \, dV}{4\pi r} \qquad 22.49$$

At infinity, the magnetic vector potential is assumed to be zero. Equation 22.47 through Eq. 22.49 cannot be applied if the current distributions are infinite.

PART 7: WORK AND ENERGY

Magnetic forces on moving charge particles and current-carrying conductors result from the magnetic field. Any forces ($\mathbf{F}_a$) used to counter these must be applied from a source external to the magnetic field and be equal and opposite to the force of the field. If motion of the charged particle or current-carrying conductor occurs, the work is given by

$$W = \int_0^f \mathbf{F}_a \cdot d\mathbf{l} \qquad 22.50$$

A positive result from the initial position on the path (l_0) to the final position (l_f) indicates that work was done by an external force on the system; that is, energy was expended against the field. Because the magnetic force is generally not a conservative force field, the entire path of integration must be specified.

23 Magnetostatic Fields

1. Magnetization 23-1
2. Magnetic Fields 23-3
3. Auxiliary Field **H** 23-5
4. Magnetization in Nonmagnetic and Magnetic Materials 23-5

Nomenclature

A	area	m^2
B, $\mathbf{B}$	magnetic flux density	T (Wb/m^2)
H, $\mathbf{H}$	magnetic field strength	A/m
I	constant or rms current	A
l	length	m
m, $\mathbf{m}$	magnetic moment	$A{\cdot}m^2$ (J/T)
M, $\mathbf{M}$	magnetization	A/m
N	number of turns	–
p_m	magnetic dipole moment	$A{\cdot}m^2$ (J/T)
$\mathbf{P}$	polarization	C/m^2
Q	charge	C
q_m	magnetic charge	A·m or N/T
r	radius	m
T	temperature	K
v	volume	m^3

Symbols

μ	permeability	N/A^2 (H/m)
μ_0	free-space permeability 1.2566×10^{-6}	N/A^2 (H/m)
χ	susceptibility	–

Subscripts

0	initial or free space
C	Curie
m	magnetic
r	relative

1. MAGNETIZATION

When a magnetic field is present in a magnetizable material, the magnetic moments tend to align themselves into the lowest energy configuration possible. In doing so, they increase the flux density, **B**. The alignment is referred to as *magnetization* of the material and is given the symbol **M**. The magnetization is the total magnetic moment, **m**, per unit volume, v.

$$\mathbf{M} = \frac{\mathbf{m}}{v} \qquad 23.1$$

The flux density is then given by the following general formula.[1]

$$\mathbf{B} = \mu_0\mathbf{H} + \mathbf{M} \qquad 23.2$$

Magnetization is technically defined as the magnetic polarization divided by the magnetic constant of the system used. In the SI system, the magnetic constant is the permeability of free space. Therefore, the equation becomes

$$\mathbf{B} = \mu_0(\mathbf{H} + \mathbf{M}) = \mu_0\mathbf{H} + \mu_0\mathbf{M} \qquad 23.3$$

The magnetic **B**-field has two sources. The first is the current, I, and the other is the magnetization, **M**. The current is the source of the magnetic field strength, **H**. The magnetic dipoles of individual atoms or molecules (which make up the larger domains) are the source of the magnetization, **M**.

Magnetization increases the flux density in magnetizable materials above the level that would exist under free-space conditions with the same magnetic field strength.[2] This occurs even when the magnetic moments of a material are not initially aligned. When an atom is placed within the influence of a magnetic field, its magnetic moment tends to align with the external magnetic field. The extent to which this occurs depends on the material. This is analogous to spring energy in mechanical systems. The work done in rotating the magnetic moment is recoverable upon removal of the magnetic field. The extent of the recovery again depends on the material.

The magnetic field and the magnetization can have different directions, though this is often not the case on a macroscopic level. In isotropic, linear material, **H** and **M** are parallel and related by

$$\mathbf{M} = \chi_m\mathbf{H} \qquad 23.4$$

The term *isotropic* indicates that the medium's properties are independent of direction. The term *linear* indicates that the permeability, μ, is constant regardless of

[1]Equation 23.2 is of a form similar to that used in electrostatic fields, but it cannot be used in the SI system as written. The study of magnetics developed mainly with the cgs system. In the cgs system $\mu_0 = 1$, but cgs was not a rationalized system, so $\mathbf{B} = \mathbf{H} + 4\pi\mathbf{M}$. (The terms **B**, **H**, and **M** are often parallel, and the equation is, therefore, written as $B = H + 4\pi M$.) The SI system is rationalized, but $\mu_0 = 4\pi \times 10^{-7}$ H/m.

[2]The magnetization, **M**, of free space is zero. A diamagnetic material's magnetization is negative and opposes the applied field. Diamagnetic materials are an exception to the statement in the text.

the magnitude of the current present. The term χ_m is the magnetic susceptibility. This dimensionless term measures the ease of magnetization in a material. For an isotropic linear material, Eq. 23.3 can be rewritten as

$$\mathbf{B} = \mu_0 (1 + \chi_m)\mathbf{H} \qquad 23.5$$

Importantly, for ferromagnetic materials (those materials with large permeabilities), Eq. 23.4 and Eq. 23.5 do not hold true. (By contrast, Eq. 23.3 is always valid.) Ferromagnetic materials are nonlinear and exhibit the property of hysteresis. This nonlinearity is easily seen on the *BH* curve in Fig. 23.1, which displays typical hysteresis loops.[3] Because ferromagnetic materials are so important in engineering applications, many permeabilities are defined in the various linear portions of the *BH* curve—and utilized only in those linear regions.

Figure 23.1 *Hysteresis Loops*

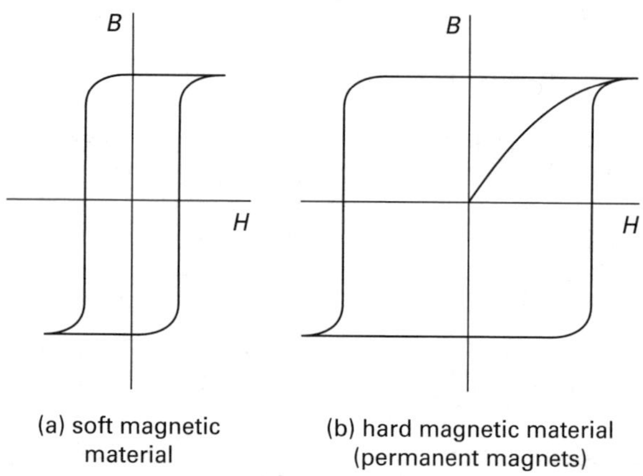

For those magnetic materials that are linear, and for ferromagnetic materials on linear portions of the *BH* curve, the relative permeability is

$$\mu_r = 1 + \chi_m \qquad 23.6$$

In free space, no magnetization occurs because no magnetic materials are present, and the magnetic flux density, **B**, is given as

$$\mathbf{B} = \mu_0 \mathbf{H} \qquad 23.7$$

If a magnetizable material is present, the flux density is increased and is given by Eq. 23.3. Importantly, Eq. 23.3 is often written without direct reference to the magnetization but rather in terms of the relative permeability. The relationship is shown explicitly in Eq. 23.8.

$$\mathbf{B} = \mu\mathbf{H} = \mu_0\mu_r\mathbf{H} = \mu_0\mathbf{H} + \mu_0\mathbf{M} \qquad 23.8$$

[3] *MH* curves are not commonly plotted, though they are identical in shape to *BH* curves.

Example 23.1

The dimensions of a small magnet used in instrumentation are 1 cm × 1 cm × 0.3 cm. The volume is uniformly magnetized and possesses a measured flux density of 3×10^{-2} T and a magnetic moment of 0.01 A·m². What is the magnetization, **M**, of the magnet?

Solution

The magnetization is determined from Eq. 23.1.

$$\mathbf{M} = \frac{\mathbf{m}}{v}$$

The volume is determined from the dimensions

$$v = (1\ \text{cm})(1\ \text{cm})(0.3\ \text{cm})\left(\frac{1\ \text{m}}{100\ \text{cm}}\right)^3 = 0.3 \times 10^{-6}\ \text{m}^3$$

Substituting gives

$$\mathbf{M} = \frac{\mathbf{m}}{v} = \frac{0.01\ \text{A·m}^2}{0.3 \times 10^{-6}\ \text{m}^3} = 3.3 \times 10^4\ \text{A/m}$$

Example 23.2

What is the magnetic field strength in the magnet described in Ex. 23.1?

Solution

The magnetic field strength, **H**, is related to the flux density and magnetization in all cases as follows.

$$\mathbf{B} = \mu_0(\mathbf{H} + \mathbf{M}) = \mu_0\mathbf{H} + \mu_0\mathbf{M}$$

Rearranging to solve for the desired quantity and substituting the given and calculated information gives

$$\mathbf{B} = \mu_0\mathbf{H} + \mu_0\mathbf{M}$$

Inside a magnet, the **B**-field and magnetization **M** are parallel, while the **H**-field is in opposition. Knowing this, one can change from vector quantities to magnitudes.

$$\begin{aligned} B &= \mu_0 H + \mu_0 M \\ H &= \frac{B - \mu_0 M}{\mu_0} \\ &= \frac{3 \times 10^{-2}\ \text{T} - \left(1.2566 \times 10^{-6}\ \frac{\text{H}}{\text{m}}\right)\left(3.3 \times 10^4\ \frac{\text{A}}{\text{m}}\right)}{1.2566 \times 10^{-6}\ \frac{\text{H}}{\text{m}}} \\ &= -9.1 \times 10^3\ \text{A/m} \end{aligned}$$

Consider the following.[4]

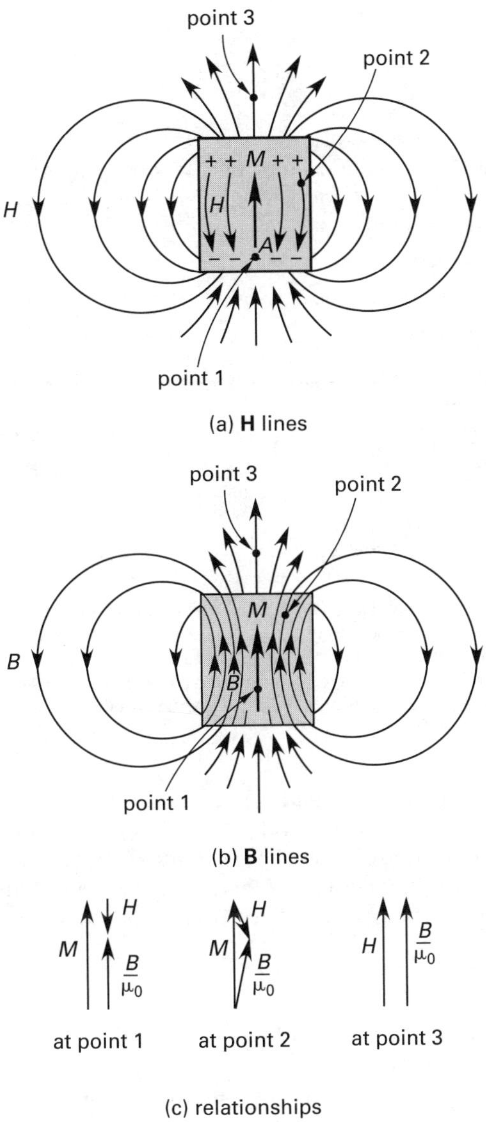

Outside the magnet, $\mathbf{B} = \mu_0\mathbf{H}$ and $\mathbf{M}$ equals zero. $\mathbf{B}$ and $\mathbf{H}$ are therefore parallel outside the magnet.

2. MAGNETIC FIELDS

All materials of concern to electronic engineers contain electrons in motion. In specific orbits, these electrons can be viewed as circulating currents. All materials then contain magnetic dipoles. Such dipoles can fully or partially align with a magnetic field. Parts (a) and (c) of Fig. 23.2 illustrate such an alignment.

[4]Considerably more explanation is possible but unnecessary for the majority of magnetic applications in electrical engineering.

Figure 23.2 *Permanent Magnet: Two Theoretical Views*

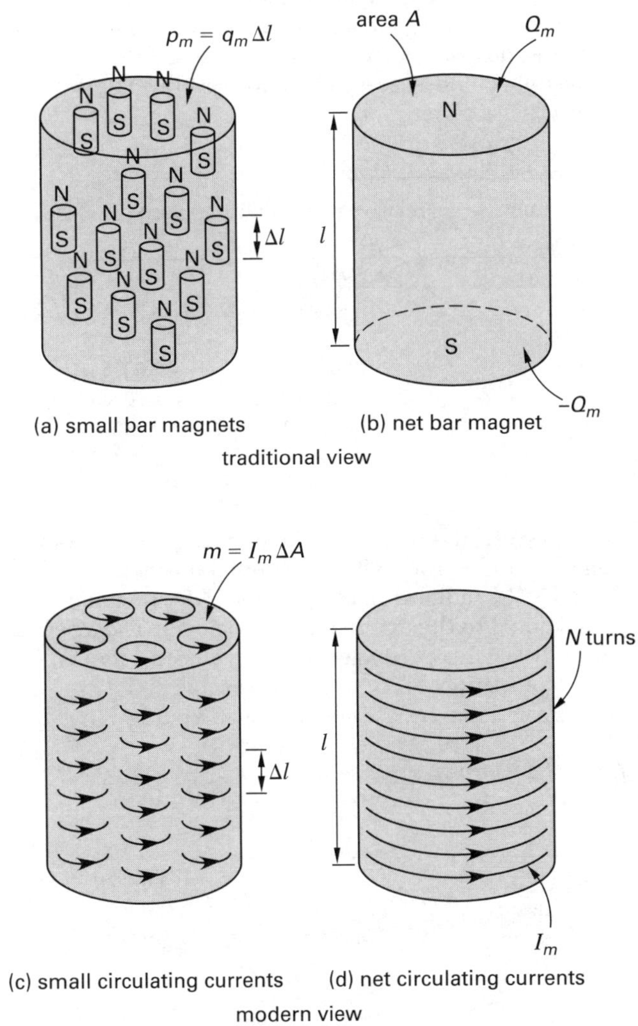

In addition to the movement of the electron in its orbit, the electron exhibits a property known as *spin*. The spin is the equivalent of the rotation of the electron about its axis. This rotation can be viewed as the equivalent of a circulating current and the source of an additional magnetic moment.

The magnetic moment of an atom may take two forms. First, the magnetic moments of all the electrons may be oriented such that they cancel, so that the magnet as a whole has no net magnetic moment. This condition leads to diamagnetism. Second, the cancellation of the electron moments, orbital and spin, may be partial, so that the atom is left with a net magnetic moment. Such an atom is referred to as a *magnetic atom*. This condition leads to several types of magnetism, including para-, ferro-, antiferro-, and ferrimagnetism.

Diamagnetism is exhibited by all materials and is caused by the orbital motion of the electrons. When an external magnetic field is applied, the magnetic moments of orbital pairs of electrons are unbalanced. Currents are induced in the circulating electrons in such a way as to create an internal magnetic field that opposes the external magnetic fields.[5] This is known as *antimagnetism*. No permanent magnetization, **M**, occurs. The induced magnetic field is too weak to be of use. The values of permeability and susceptibility for several diamagnetic materials are given in Table 23.1.

Table 23.1 *Diamagnetic Materials*

substance	relative permeability $\mu_r = 1 + \chi_m$	susceptibility χ_m
bismuth	0.99983	-1.66×10^{-4}
mercury	0.999968	-3.20×10^{-5}
gold	0.999964	-3.60×10^{-5}
silver	0.99998	-2.60×10^{-5}
lead	0.999983	-1.70×10^{-5}
copper	0.999991	-0.98×10^{-5}
water	0.999991	-0.88×10^{-5}

Paramagnetism is caused by the partial alignment of the spin axes of electrons with an external magnetic field. In this case the induced magnetization, **M**, is parallel and proportional to the external magnetic field. The internal field is weak. No permanent magnetization occurs. The values of permeability and susceptibility for several paramagnetic materials are given in Table 23.2.

Table 23.2 *Paramagnetic Materials*

substance	relative permeability $\mu_r = 1 + \chi_m$	susceptibility χ_m
air	1.00000036	3.6×10^{-7}
aluminum	1.000021	2.5×10^{-5}
palladium	1.00082	8.2×10^{-4}

The most important type of magnetism is ferromagnetism. Ferromagnetism is primarily caused by unpaired electron magnetic moments. The magnetic moments of individual atoms or molecules align themselves within small volumes called *domains*.[6] This effect occurs in iron, nickel, cobalt, and their alloys; in some manganese compounds; and in some rare earth elements. The domains can be completely aligned with weak external magnetic fields, resulting in very large magnetization, **M**. This effect is shown in Fig. 23.2. In part (a) of this figure, the individual magnetic moments (or groups of atoms in domains) are viewed as small bar magnets that have aligned with the external magnetic field (not shown). The net result is shown in part (b). Parts (a) and (b) illustrate the traditional view of magnetism. In the modern view, ferromagnetism is caused by circulating currents (both orbital and spin) as shown in part (c). Because of the cancellation of interior currents, the net result is a sheet of current on the surface of the cylinder as shown in part (d). The magnetization is significant enough in ferromagnetic materials that for all practical purposes $\mathbf{H} \ll \mathbf{M}$, so

$$\mathbf{B} \approx \mu\mathbf{M} \qquad 23.9$$

The internal magnetic field is very strong. Permanent magnetization can occur. The values of permeability and susceptibility for several ferromagnetic materials are given in Table 23.3.

Table 23.3 *Ferromagnetic Materials*

substance	relative permeability $\mu_r = 1 + \chi_m$
cobalt	250
nickel	600
commercial iron (0.2 impurity)	6000
high-purity iron (0.05 impurity)	2×10^5

(Note: The relative permeabilities are nonlinear for ferromagnetic materials. The value, however, will not change, regardless of the system of units used. Nevertheless, the numbers are for comparison of materials only. To have meaning, the *B* or *H* value at which the permeability was measured would need to be specified.)

Both paramagnetic and ferromagnetic materials are temperature sensitive. When ferromagnetic materials are heated above the *Curie temperature*, T_C, they become paramagnetic. The magnetization and temperature dependence of these three classes of magnetic materials are shown in Fig. 23.3.

Figure 23.3 *Magnetization Curves and Temperature Dependence*

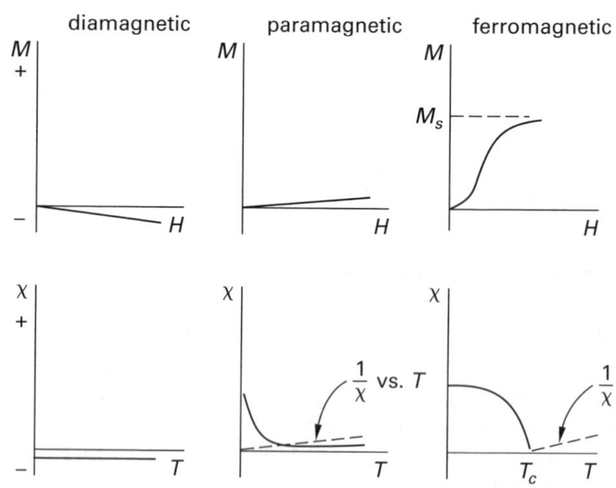

[5]This reaction, best known as *Lenz's law*, is the *counter-electromotive force* (emf) in motors and generators.

[6]Both paramagnetism and ferromagnetism occur as a result of unpaired electron moments. The difference is that ferromagnets exhibit domain formation caused by the quantum mechanical effect of *exchange forces*.

3. AUXILIARY FIELD H

The vector **H**, known as the magnetic field strength, is used to separate the sources of the **B**-field into those caused by external currents and those caused by the magnetization of the material itself. The magnetic field strength, **H**, represents the portion of the flux caused by external currents. The magnetization, **M**, represents the portion of the flux caused by the movement of magnetic dipoles within the material itself.

The magnetic field strength specifies the amount of force per unit flux produced by a conductor of N turns and length l carrying current I. After some manipulation, the magnitude of the magnetic field strength is given by

$$H = \frac{NI}{l} \qquad 23.10$$

Example 23.3

What is the magnetic field strength in a circular electrical device of radius 2.5 cm with 10 turns of wire carrying 30 A?

Solution

The magnetic field strength, H, which does not depend on the material present, is

$$H = \frac{NI}{l}$$

The path length is the circumference of a circle with a radius of 2.5 cm. That is,

$$l = 2\pi r = 2\pi(0.025\ \text{m}) = 0.16\ \text{m}$$

Substituting the given and the calculated gives

$$H = \frac{NI}{l} = \frac{(10)(30\ \text{A})}{0.16\ \text{m}} = 1.9 \times 10^3\ \text{A/m}$$

4. MAGNETIZATION IN NONMAGNETIC AND MAGNETIC MATERIALS

The use of the magnetic field strength, **H**, allows material media to be categorized into three groups. The magnetization vector, **M**, is avoided if possible (as is the polarization vector, **P**) to simplify calculations.

The first group consists of free space and nonmagnetic materials. In this group, the current is the only source of the magnetic field. As the magnetization is zero,

$$\mathbf{B} = \mu_0 \mathbf{H} \qquad 23.11$$

The second group consists of *soft ferromagnetic materials.* Such materials are suitable for linearization. That is, the current can be restricted to operate over a limited portion of the BH curve so that μ is constant. In this case, the governing equation is

$$\mathbf{B} = \mu\mathbf{H} = \mu_0\mu_r\mathbf{H} \qquad 23.12$$

The magnetization is embedded in the relative permeability term.

The third group consists of the *hard ferromagnetic materials.* These materials are very nonlinear and are suitable for use as permanent magnets. Because of the nonlinearity, the relative permeability term cannot be used. Instead, B and H are given by graphs that are experimentally determined and are typically as shown in Fig. 23.1. Because M is significantly larger than H in such materials, B is related directly to M as indicated in the magnetization curves of Fig. 23.3.

Example 23.4

What magnitude of the magnetic flux density in air is associated with a magnetic field strength of 50 A/m?

Solution

The relative permeability, μ_r, of air (1.00000036) very closely approximates that of free space (1). When air is specified, μ_r can be assumed to be equal to one. Use Eq. 23.11.

$$B = \mu_0 H = \left(1.2566 \times 10^{-6}\ \frac{\text{H}}{\text{m}}\right)\left(50\ \frac{\text{A}}{\text{m}}\right)$$
$$= 6.3 \times 10^{-5}\ \text{T}$$

Example 23.5

What is the flux density if commercial iron is substituted for air in Ex. 23.4?

Solution

Commercial iron has a relative permeability of 6000. From Ex. 23.4, $\mu_0 H = 6.3 \times 10^{-5}$ T. Using these values and Eq. 23.12 gives

$$B = \mu_0 H \mu_r = (6.3 \times 10^{-5}\ \text{T})(6000) = 0.38\ \text{T}$$

Field Theory

24 Electrodynamics

1. Electromotive Force 24-1
2. Faraday's Law of Electromagnetic Induction 24-2
3. Faraday's Law of Electrolysis 24-2
4. Potential 24-3
5. Energy and Momentum: Poynting's Vector 24-3
6. Electromagnetic Waves 24-3

Nomenclature

a	unit vector	–
A	area	m^2
AW	atomic weight	g/mol
B, **B**	magnetic flux density	T (Wb/m^2)
c	speed of light, 2.9979×10^8	m/s
E, **E**	electric field strength	V/m
$\mathcal{E}$	electromotive force	V
F	Faraday constant, 96 485	C/mol
F, **F**	force	N
H, **H**	magnetic field strength	A/m
i	variable current	A
I, **I**	constant or rms current	A
J, **J**	current density	A/m^2
l, **l**	length	m
L	self-inductance	H
m	mass	g
N	number of turns	–
P_a, $\mathbf{P}_a$	apparent power	VA
$\mathcal{P}$	Poynting vector	W/m^2
R	resistance	Ω
s	surface area	m^2
S	Poynting's vector	W/m^2
t	time	s
v	velocity	m/s
v, **v**	variable voltage	V
V, **V**	constant or rms voltage	V
V_m	magnetic potential	A or A-turns
W_e	equivalent weight	–
z	valence change	–

Symbols

ϵ	permittivity	C^2/N·m or F/m
μ	permeability	N/A^2 or H/m
μ_0	free-space permeability, 1.2566×10^{-6}	N/A^2 or H/m
σ	conductivity	S/m
ϕ	magnetic flux	Wb

Subscripts

ave	average
e	equivalent
g	grams
m	magnetic

1. ELECTROMOTIVE FORCE

It takes energy to drive a charge through a conducting medium. The source of the energy can be electromagnetic, chemical, or mechanical. Whatever the source, the net effect of the energy in any closed circuit is called the *electromotive force*, emf, or *electromotance*, and is given the symbol $\mathcal{E}$. It can be found from the line integral of the force over the path of the circuit.

$$\begin{aligned}\mathcal{E} &= \oint \mathbf{F}\cdot d\mathbf{l} = \oint \frac{\mathbf{J}}{\sigma}\cdot d\mathbf{l} \\ &= \oint \frac{I}{A\sigma}\,dl = I\oint \frac{1}{A\sigma}\,dl = IR\end{aligned} \quad \text{24.1}$$

The term R represents the total resistance in the external circuit. The force that moves the charges is the electric field, **E**. The electromotive force is measured in volts. In actuality, it is not a force, but the integral of the force per unit charge. The emf can be viewed as energy per unit charge, especially when considering that the unit for emf is the volt. The term electromotive force and the symbol $\mathcal{E}$ are reserved for the sources within an electric circuit, while v and V are used for the remaining components.[1]

Electromotive force can be generated by chemical means as in a battery, mechanically as in a piezoelectric crystal, by heat as in a thermocouple, or by many other methods. The two methods of considerable concern in electrical engineering are dynamic in nature.

The first is the *time-varying magnetic flux method.* The time-varying flux induces a voltage in any conductor present that depends on the number of turns, N, of the conductor and the *flux linkage*, $N\phi$ (the total number of flux lines cut). This is given by Eq. 24.2.

$$v = -N\frac{d\phi}{dt} \quad \text{24.2}$$

[1]Though this is generally true, in many instances the term *voltage* and its associated symbols (v, V) are used to indicate sources. It is understood that the voltage in those cases represents an electromotive force. Except where confusion may arise, this text will use v and V.

The minus sign is used to indicate that any current caused to flow by an induced electromotive force does so in a direction that generates a magnetic field opposing the flux that induced the emf. This is known as *Lenz's law.* The minus sign is not always shown. In circuit analysis, it is considered understood, though it is still used in equations written to obtain a numeric result. Equation 24.2 is often called *Faraday's law,* though in actuality it is the result of the law for time-varying fields.

Equation 24.2 holds true even when the time-varying magnetic flux in a given circuit element is changing because of the current in the element itself. Therefore, Eq. 24.2 can be written as follows.

$$v = -N\frac{d\phi}{di}\frac{di}{dt} = -L\frac{di}{dt} \qquad 24.3$$

The term L is known as *self-inductance.*

The second dynamic method of inducing an electromotive force is to maintain a constant flux and move the conductors. This is called the *flux-cutting method.* It results in an induced voltage, or electromotive force, of

$$V = -NBlv \qquad 24.4$$

2. FARADAY'S LAW OF ELECTROMAGNETIC INDUCTION

Faraday's law of electromagnetic induction states that the induced voltage is proportional to the rate of change of magnetic flux linked with the circuit. If the rate of change of magnetic flux linked is due to a time rate of change of the magnetic flux, that is, if the circuit (conductor) is stationary, then Faraday's law is

$$\begin{aligned} v &= \oint \mathbf{E}\cdot d\mathbf{l} = -\frac{d}{dt}\int_s \mathbf{B}\cdot ds \\ &= -\int_s \frac{\partial \mathbf{B}}{\partial t}\cdot ds = -\frac{d\phi}{dt} \end{aligned} \qquad 24.5$$

If the rate of change of the magnetic flux linked is due to the motion of the circuit (conductor) and the magnetic field is constant with respect to time, Faraday's law is

$$v = \oint \mathbf{E}\cdot d\mathbf{l} = \oint (\mathbf{v}\times\mathbf{B})\cdot d\mathbf{l} \qquad 24.6$$

The general case of Faraday's law when the magnetic flux is changing and the circuit (conductor) is in motion is

$$\begin{aligned} v &= \oint \mathbf{E}\cdot d\mathbf{l} = \oint (\mathbf{v}\times\mathbf{B})\cdot d\mathbf{l} \\ &= -\int_s \frac{\partial \mathbf{B}}{\partial t}\cdot ds \end{aligned} \qquad 24.7$$

Faraday's law applies to changing magnetic fields. In electrical engineering, the formulas of magnetostatics (Ampère's law, the Biot-Savart law, and others) are used to calculate the fields. While they give results that are only approximately correct, the errors are negligible. Even in the case where the current through a wire is interrupted by a fast-acting switch, or is simply cut, Ampère's law still holds true. Magnetostatic formulas can be used in conjunction with Faraday's law in *quasi-static conditions.* The magnetostatic equations must be abandoned only when the frequencies are extremely high, as in electromagnetic waves and radiation.

3. FARADAY'S LAW OF ELECTROLYSIS

Faraday's law of electrolysis states that the mass of substance deposited at one electrode or liberated at the other varies as in Eq. 24.8. F is the Faraday constant, equal to 96 485 C/mol.[2]

$$m_g = \frac{ItW_e}{F} \qquad 24.8$$

The term W_e is the equivalent weight of an element, the atomic weight in grams divided by the valence change that occurs during electrolysis, z.

$$W_e = \frac{\text{AW}}{z} \qquad 24.9$$

The atomic weight in the case of Faraday's law is given in grams. This is because one *faraday* of electricity will produce one gram of equivalent weight.

Example 24.1

The following electrolysis reaction can occur to copper.

$$\text{Cu}^{+2} + 2\text{e}^- \rightarrow \text{Cu}$$

What current is required to produce five grams of metallic copper (AW = 63.5) from a copper sulfate solution in 1 hr?

Solution

Because the change in the valence of copper is 2, the equivalent weight is

$$W_e = \frac{\text{AW}}{z} = \frac{63.5}{2} = 31.8$$

[2]An electron has a charge of 1.6022×10^{-19} C. A *faraday* is the charge associated with one mole of electrons. Therefore, a faraday is

$$\begin{aligned} F &= \left(1.6022\times10^{-19}\ \frac{\text{C}}{\text{electron}}\right)(6.022\times10^{23}\ \text{electrons}) \\ &= 96\,485\ \text{C/mol} \end{aligned}$$

Rearranging Eq. 24.8 to determine the current required gives

$$m_g = \frac{ItW_e}{F}$$

$$I = \frac{m_g F}{tW_e}$$

$$= \frac{(5\text{ g})\left(96\,485\ \frac{\text{C}}{\text{mol}}\right)}{(1\text{ hr})\left(3600\ \frac{\text{sec}}{\text{hr}}\right)(31.8)}$$

$$= 4.2\text{ A}$$

4. POTENTIAL

The electrostatic potential, or voltage, V, is a steady-state phenomenon that is well defined in space and is associated with a conservative electric field. The induced voltage, or potential, of Faraday's law, v, is time dependent and a multivalued function of position. It is associated with a nonconservative field, that of electromotive force.[3] When the potential is associated with a source of energy, the potential is called the electromotive force, or emf.

In magnetic circuits, the potential is called the *magnetomotive force*, mmf, or *magnetomotance*. Like its electric counterpart, mmf is actually not a force, but the line integral of the force around a path. In the magnetic case, it is the energy that moves the flux around the path. Hence, the mmf, or magnetic potential, is

$$\oint \mathbf{H}\cdot d\mathbf{l} = NI = V_m \qquad 24.10$$

5. ENERGY AND MOMENTUM: POYNTING'S VECTOR

Momentum conservation in electrodynamics occurs because the electric and magnetic fields themselves carry momentum. The conservation of energy in electrodynamics is given by *Poynting's theorem* as

$$\frac{dW}{dt} = -\frac{d}{dt}\int_V \tfrac{1}{2}(\epsilon E^2 + \mu H^2)\,dv - \oint_A (\mathbf{E}\times\mathbf{H})\cdot d\mathbf{A} \qquad 24.11$$

The first integral represents the stored energy in the fields. The second integral represents the rate at which energy is transferred out of volume V across the surface boundary A. Poynting's theorem says the work done on charges by the electromotive force is equal to the decrease in the energy stored in the field less the energy that flows out of the surface.

[3]A conservative field is one in which the work done in moving a charge depends only on its initial and final positions and not its path.

The cross product of **E** and **H** is Poynting's vector, **S** (see Fig. 24.1). It represents the instantaneous power density (in W/m^2) carried by an electromagnetic wave. The direction of **S** is perpendicular to both **E** and **H**, which is normal to the wave front as shown in Fig. 24.1. (In Fig. 24.1, **E** and **B** are shown as the primary fields, consistent with these two vectors representing the fundamental fields.) Poynting's vector is given in Eq. 24.12. In Eq. 24.12, **a** is a unit vector in the direction of propagation of the wave, that is, normal to the wave front.

$$\mathbf{S} = c\epsilon E^2\mathbf{a} = c\mu H^2\mathbf{a} = \mathbf{E}\times\mathbf{H} \qquad 24.12$$

The notation Re(x) stands for the real part of the complex number x. Poynting's vector can be used to solve any problem involving the flow of electric or magnetic energy. The only restriction is that the electric and magnetic fields be related—from the same source. Though Eq. 24.12 represents the instantaneous value of the power density, allowing **E** and **H** to vary dynamically, with a time dependence of $e^{j\omega t}$, leads to a time-average of Poynting's vector, $\mathcal{P}_{\text{ave}}$, given by

$$\mathcal{P}_{\text{ave}} = \tfrac{1}{2}\text{Re}(\mathbf{E}\times\mathbf{H}^*) \qquad 24.13$$

Figure 24.1 Poynting's Vector

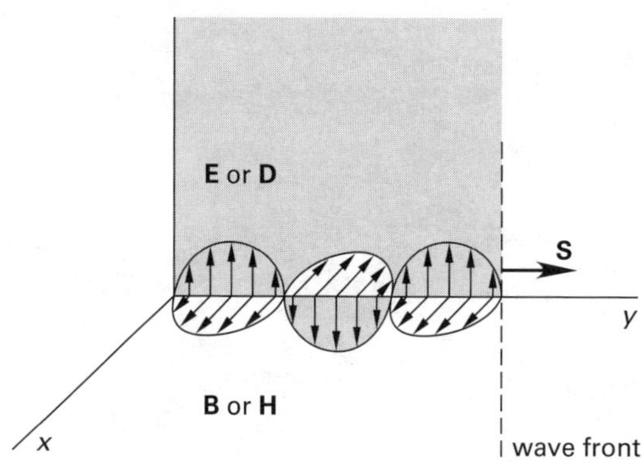

The complex conjugate of **H** is represented by **H***. This follows the results of complex power analysis in which the apparent power is $\mathbf{P}_a = \frac{1}{2}\mathbf{VI}^*$, where the active power is the real part of the apparent power, that is, $\mathcal{P} = \frac{1}{2}\text{Re}\mathbf{VI}^*$.

Poynting's vector, **S**, describes the flow of energy, just as the current density, **J**, describes the flow of current.

6. ELECTROMAGNETIC WAVES

Stationary charges produce an electric field. If the charges are moving with a uniform velocity, a second effect occurs called *magnetism*. If the charges are accelerating, they produce a radiating electromagnetic field capable of transporting energy. The radiating electromagnetic field can be treated as a wave.[4]

The treatment of electrical phenomena as waves dominates electrical engineering fields where the transmission of electromagnetic energy takes place over long distances (often without metallic conductors), such as communications, power transmission, and related areas. In other electrical engineering fields, it is usually more convenient to work with the results, that is, potentials, currents, and resistances.

[4]Electromagnetic waves undergo all of the processes commonly affecting waves, that is, reflection, refraction, diffraction, wave interference, dispersion, and scattering.

25 Maxwell's Equations

1. Maxwell's Equations 25-1
2. Electromagnetic Field Vectors 25-1
3. Comparison of Electric and Magnetic Equations 25-2

Nomenclature

A	area	m^2
$\mathbf{B}$	magnetic flux density	T or Wb/m^2
$\mathbf{D}$	electric flux density	C/m^2
$\mathbf{E}$	electric field strength	V/m
$\mathcal{E}$	electromotive force	V
F_m	magnetomotive force	A or amp-turns
$\mathbf{F}$	force	N
G	conductance	S
$\mathbf{H}$	magnetic field strength	A/m
I	constant or rms current	A
$\mathbf{J}$	current density	A/m^2
l, $\mathbf{l}$	length	m
L	inductance	H
$\mathbf{M}$	magnetization	A/m
P_m	permeance	H
$\mathbf{P}$	polarization	C/m^2
Q	charge	C
R	resistance	Ω
$\mathcal{R}$	reluctance	1/H
$\mathbf{s}$	surface area	m^2
t	time	s
T	torque	N·m
U	energy density	J/m^3
v	variable voltage	V
v, $\mathbf{v}$	velocity	m/s
V	constant or rms voltage	V
V	volume	m^3
V_m	magnetic potential	A or amp-turns
W	energy	J

Symbols

ϵ	permittivity	$C^2/N{\cdot}m^2$ or F/m
ϵ_0	free-space permittivity, 8.854×10^{-12}	$C^2/N{\cdot}m^2$
μ	permeability	N/A^2 or H/m
μ_0	free-space permeability, 1.2566×10^{-6}	N/A^2 or H/m
ρ	charge density	C/m^3
ρ	resistivity	Ω·m
σ	conductivity	S/m
ϕ	magnetic flux	Wb
χ	susceptibility	–

Subscripts

c	conduction
e	electrical
m	magnetic

Field Theory

1. MAXWELL'S EQUATIONS

Maxwell's equations are Gauss', Coulomb's, Faraday's, and Ampère's laws extended by the concept of a displacement current and the notion of a field. Though seldom used in the exact form presented in Table 25.1, they govern all of the electromagnetic phenomena in a medium that is stationary with respect to the coordinate system used.[1] They are valid for linear and nonlinear, isotropic and nonisotropic, homogeneous and nonhomogeneous media for frequencies ranging from zero to microwave frequencies, including many phenomena of light. The general set of Maxwell's equations is given in Table 25.1.

In free space, no charges exist ($\rho = 0$), and no conduction currents exist ($\mathbf{J}_c = 0$). Maxwell's equations then take on the forms given in Table 25.2.

The integral forms of Maxwell's equations are more fundamental in that they display the underlying physical laws. Maxwell's equations describe electromagnetic phenomena on a macroscopic (that is, nonquantum) level. Regions used in the integrals should be larger than atomic dimensions, and time intervals should be long enough to average atomic fluctuations.

2. ELECTROMAGNETIC FIELD VECTORS

The following vectors are electromagnetic field vectors: **E**, **D**, **B**, **H**, and **J**. These vectors define the properties of any medium in terms of the auxiliary equations summarized in Table 25.3.[2]

[1]If the medium is moving, the concepts of relativity must be invoked.
[2]These are auxiliary equations in the sense that they refine terms given in Maxwell's equations.

Table 25.1 *Maxwell's Equations*

integral form	point form	remarks
$\oint_s \mathbf{D} \cdot d\mathbf{s} = \int_V \rho \, dv$	$\nabla \cdot \mathbf{D} = \rho$	Gauss' law
$\oint_s \mathbf{B} \cdot d\mathbf{s} = 0$	$\nabla \cdot \mathbf{B} = 0$	nonexistence of magnetic monopoles
$\oint \mathbf{E} \cdot d\mathbf{l} = \int_s \left(\frac{-\partial \mathbf{B}}{\partial t}\right) \cdot d\mathbf{s}$	$\nabla \times \mathbf{E} = -\frac{\partial \mathbf{B}}{\partial t}$	Faraday's law
$\oint \mathbf{H} \cdot d\mathbf{l} = \int_s \left(\mathbf{J}_c + \frac{\partial \mathbf{D}}{\partial t}\right) \cdot d\mathbf{s}$	$\nabla \times \mathbf{H} = \mathbf{J}_c + \frac{\partial \mathbf{D}}{dt}$	Ampère's law

Table 25.2 *Maxwell's Equations: Free-Space Form*

integral form	point form
$\oint_s \mathbf{D} \cdot d\mathbf{s} = 0$	$\nabla \cdot \mathbf{D} = 0$
$\oint_s \mathbf{B} \cdot d\mathbf{s} = 0$	$\nabla \cdot \mathbf{B} = 0$
$\oint \mathbf{E} \cdot d\mathbf{l} = \int_s \left(\frac{-\partial \mathbf{B}}{\partial t}\right) \cdot d\mathbf{s}$	$\nabla \times \mathbf{E} = -\frac{\partial \mathbf{B}}{\partial t}$
$\oint \mathbf{H} \cdot d\mathbf{l} = \int_s \left(\frac{\partial \mathbf{D}}{\partial t}\right) \cdot d\mathbf{s}$	$\nabla \times \mathbf{H} = \frac{\partial \mathbf{D}}{dt}$

Table 25.3 *Electromagnetic Field Vector Equations*

$$\mathbf{D} = \epsilon \mathbf{E} = \epsilon_0 \mathbf{E} + \mathbf{P} = \epsilon_0 (1 + \chi_e) \mathbf{E}$$

$$\mathbf{B} = \mu \mathbf{H} = \mu_0 \mathbf{H} + \mu_0 \mathbf{M} = \mu_0 (1 + \chi_m) \mathbf{H}$$

$$\mathbf{J} = \sigma \mathbf{E} = \rho \mathrm{v}$$

Even though the continuity equation is embedded in Maxwell's equation (as Gauss' and Ampère's laws), for completeness, one should list it separately along with the Lorentz force equation. The *continuity equation* is

$$\nabla \cdot \mathbf{J} = -\frac{\partial \rho}{\partial t} \qquad 25.1$$

The Lorentz force equation is repeated here for convenience.

$$\mathbf{F} = Q(\mathbf{E} + \mathbf{v} \times \mathbf{B}) \qquad 25.2$$

The electromagnetic field vector equations define the electromagnetic properties of the medium. The field vector equations, along with the continuity equation and the Lorentz force equation, are used as the practical applications of Maxwell's equations to the relationships between **E**, **D**, **B**, **H**, and **J**.

3. COMPARISON OF ELECTRIC AND MAGNETIC EQUATIONS

A comparison of electric and magnetic phenomena in terms of circuits is given in Table 25.4.

A more extensive comparison of electric and magnetic equations and concepts is given in App. 25.A.

Table 25.4 *Electric and Magnetic Circuit Analogies*

electric	magnetic
emf $= V = IR$	mmf $= V_m = \phi \mathcal{R}$
current I	flux ϕ
emf $\mathcal{E}$ or V	mmf V_m
resistance $R = \rho l/A = l/\sigma A$	reluctance $\mathcal{R} = l/\mu A$
resistivity ρ	reluctivity $1/\mu$
conductance $G = 1/R$	permeance $P_m = \mu A/l$
conductivity $\sigma = 1/\rho$	permeability μ

Topic IV: Circuit Theory

Chapter

26. DC Circuit Fundamentals
27. AC Circuit Fundamentals
28. Transformers
29. Linear Circuit Analysis
30. Transient Analysis
31. Time Response
32. Frequency Response

26 DC Circuit Fundamentals

1. Voltage 26-1
2. Current 26-1
3. Resistance 26-2
4. Conductance 26-3
5. Ohm's Law 26-3
6. Power 26-3
7. Decibels 26-4
8. Energy Sources 26-4
9. Voltage Sources in Series and Parallel 26-4
10. Current Sources in Series and Parallel 26-5
11. Source Transformations 26-5
12. Maximum Energy Transfer 26-5
13. Voltage and Current Dividers 26-5
14. Kirchhoff's Laws 26-6
15. Kirchhoff's Voltage Law 26-6
16. Kirchhoff's Current Law 26-6
17. Series Circuits 26-7
18. Parallel Circuits 26-7
19. Analysis of Complicated Resistive Networks 26-7
20. Delta-Wye Transformations 26-8
21. Substitution Theorem 26-9
22. Reciprocity Theorem 26-9
23. Superposition Theorem 26-9
24. Thevenin's Theorem 26-10
25. Norton's Theorem 26-12
26. Loop-Current Method 26-12
27. Node-Voltage Method 26-13
28. Determination of Method 26-13
29. Practical Application: Batteries 26-13

Nomenclature

A	area	m^2
d	diameter	m
$\mathcal{E}$	electromotive force	V
G	conductance	S
i	variable current	A
I	constant or rms current	A
l	length	m
n	number of items	–
P	power	W
R	resistance	Ω
S	surface area	m^2
t	time	s
T	temperature	°C
v	variable voltage	V
V	constant or rms voltage	V
W	work	J

Symbols

α	thermal coefficient of resistivity	1/°C
ρ	resistivity	Ω·m
σ	conductivity	S/m

Subscripts

0	initial
cmil	circular mils
Cu	copper
e	equivalent
fl	full load
nl	no load
N	Norton
oc	open circuit
s	source
sc	short circuit
Th	Thevenin

1. VOLTAGE

Voltage is the energy, or work, per unit charge exerted in moving a charge from one position to another. One of the positions is a reference position and is given an arbitrary value of zero. Voltage is also referred to as the *potential difference* to distinguish it from the unit of potential difference, the volt. One volt, V, is the potential difference if one joule of energy moves one coulomb of charge from the reference position to another position. Therefore, 1 V = 1 J/C. Additional equivalent units are W/A, C/F, A/S, and Wb/s.

In practical terms, a potential of one volt is defined as the potential existing between two points of a conducting wire carrying a constant current, I, of one ampere when the power dissipated between these two points is one watt. When the voltage refers to a source of electrical energy, it is termed the *electromotive force* (emf) and sometimes given the symbol $\mathcal{E}$. In a *direct-current* (DC) *circuit*, the voltage, v, may vary in amplitude but not polarity. In many applications, the voltage magnitude, V, is constant as well. That is, it does not vary with time.

2. CURRENT

Current is the amount of charge transported past a given point per unit time. When current is constant or time-invariant, it is given the symbol I.[1] The unit of

[1]The symbol for current is derived from the French word *intensité*.

measure for current is the ampere (A), which is equivalent to coulombs per second (C/s). That is, one ampere is equal to the flow of one coulomb of charge past a plane surface, S, in one second. Though current is now known to be electron movement, it was originally viewed as flowing positive charges. This frame-of-reference current is called *conventional current* and is used in this and most other texts. When current refers to the actual flow of electrons, it is termed *electron current.* These concepts are illustrated in Fig. 26.1.

Figure 26.1 *Current Flow*

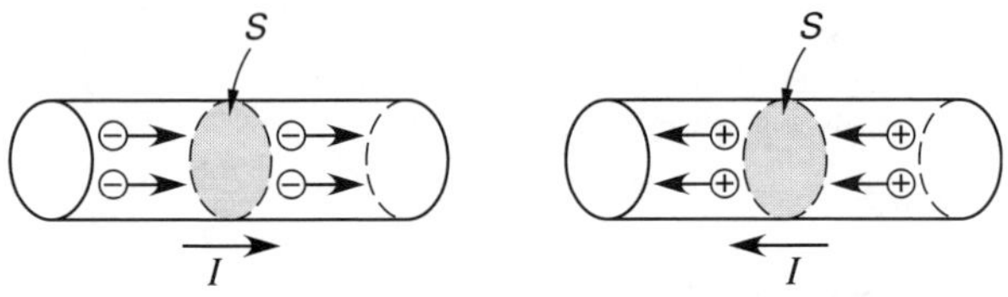

(a) electron current (b) conventional current

Current does not "flow," though this term is often used in conjunction with current. Additionally, the conservation of charge causes all circuit elements to be electrically neutral, meaning that no net positive or negative charge accumulates on any circuit element.

Example 26.1

A conductor has a constant current of 10 A. How many electrons pass a fixed plane in the conductor in 1 min?

Solution

First, using amperes, determine the charge flowing per minute.

$$10 \text{ A} = \left(10 \ \frac{\text{C}}{\text{s}}\right)\left(60 \ \frac{\text{s}}{\text{min}}\right) = 600 \text{ C/min}$$

Using the charge of an electron, compute the number of electrons flowing.

$$\frac{600 \ \dfrac{\text{C}}{\text{min}}}{1.6022 \times 10^{-19} \ \dfrac{\text{C}}{\text{electron}}} = 3.75 \times 10^{21} \text{ electrons/min}$$

3. RESISTANCE

Resistance, R, is the property of a circuit element that impedes current flow.[2] It is measured in ohms, Ω. A circuit with zero resistance is a *short circuit,* while a circuit with infinite resistance is an *open circuit.*

Resistors are often constructed from carbon compounds, ceramics, oxides, or coiled wire. Adjustable resistors are known as *potentiometers* and *rheostats.*

The resistance of any substance depends on its *resistivity* (ρ), its length, and its cross-sectional area. Assuming resistivity and cross-sectional area are constant, the resistance of a substance is

$$R = \frac{\rho l}{A} \qquad 26.1$$

Though SI system units are the standard, the units of area, A, are often put in the English Engineering System, especially in many tabulated values in the *National Electrical Code.* The area is given and tabulated for various conductors in *circular mils,* abbreviated cmil. One cmil is the area of a 0.001 in diameter circle. The concept of area in circular mils is represented by Eq. 26.2, Eq. 26.3, and Eq. 26.4.

$$A_{\text{cmil}} = \left(\frac{d_{\text{inches}}}{0.001}\right)^2 \qquad 26.2$$

$$A_{\text{in}^2} = 7.854 \times 10^{-7} \times A_{\text{cmil}} \qquad 26.3$$

$$A_{\text{cm}^2} = 5.067 \times 10^{-6} \times A_{\text{cmil}} \qquad 26.4$$

Resistivity is dependent on temperature. In most conductors, it increases with temperature, because electron movement through the lattice structure becomes increasingly difficult at higher temperatures. The variation of resistivity with temperature is specified by a *thermal coefficient of resistivity,* α, with typical units of 1/°C. The actual resistance or resistivity for a given temperature is calculated with Eq. 26.5 and Eq. 26.6.

$$R = R_0(1 + \alpha \Delta T) \qquad 26.5$$

$$\rho = \rho_0(1 + \alpha \Delta T) \qquad 26.6$$

The thermal coefficients for several common conducting materials are given in Table 26.1.

Example 26.2

What is the resistance of 305 m of 10 AWG (with an area of approximately 10,000 cmil) copper conductor with a resistivity of 2.82×10^{-6} Ω·cm?

Solution

From Eq. 26.1, and using the conversion of Eq. 26.4, the resistance is

$$\begin{aligned} R &= \frac{\rho l}{A} \\ &= \frac{(2.82 \times 10^{-6} \ \Omega\cdot\text{cm})(305 \text{ m})\left(100 \ \frac{\text{cm}}{\text{m}}\right)}{(10{,}000 \text{ cmil})\left(5.067 \times 10^{-6} \ \frac{\text{cm}^2}{\text{cmil}}\right)} \\ &= 1.70 \ \Omega \end{aligned}$$

[2]Resistance is the real part of *impedance,* which opposes changes in current flow.

Table 26.1 Approximate Temperature Coefficients of Resistance[a] and Percent Conductivities

material	$\alpha(1/°C)$	conductivity (%)
aluminum, 99.5% pure	0.00423	63.0
aluminum, 97.5% pure	0.00435	59.8
constantan[b]	0.00001	3.1
copper, IACS (annealed)	0.00402	100.0
copper, pure annealed	0.00428	102.1
copper, hard drawn	0.00402	97.8
gold, 99.9% pure	0.00377	72.6
iron, pure	0.00625	17.5
iron wire, EBB[c]	0.00463	16.2
iron wire, BB[d]	0.00463	13.5
manganin[e]	0.00000	3.41
nickel	0.00622	12.9
platinum, pure	0.00367	14.6
silver, pure annealed	0.00400	108.8
steel wire	0.00463	11.6
tin, pure	0.00440	12.2
zinc, very pure	0.00406	27.7

(Multiply 1/°C by 0.5556 to obtain 1/°F.)

[a]between 0°C and 100°C
[b]58% Cu, 41% Ni, 1% Mn
[c]Extra Best Best grade
[d]Best Best grade
[e]84% Cu, 4% Ni, 12% Mn

4. CONDUCTANCE

The reciprocal of resistivity is *conductivity*, σ, measured in siemens per meter (S/m), often given in siemens per centimeter. The reciprocal of resistance is *conductance*, G, measured in siemens (S). Siemens is the metric unit for the older unit known as the *mho* ($1/\Omega$).

$$\sigma = \frac{1}{\rho} \qquad 26.7$$

$$G = \frac{1}{R} \qquad 26.8$$

The *percent conductivity* is the ratio of a given substance's conductivity to the conductivity of standard IACS (International Annealed Copper Standard) copper, simply called *standard copper*. Alternatively, the percent conductivity is the ratio of standard copper's resistivity to the substance's resistivity.

$$\begin{aligned} \%\ \text{conductivity} &= \frac{\sigma}{\sigma_{\text{Cu}}} \times 100\% \\ &= \frac{\rho_{\text{Cu}}}{\rho} \times 100\% \end{aligned} \qquad 26.9$$

The standard resistivity of copper at 20°C is approximately

$$\begin{aligned} \rho_{\text{Cu,20°C}} &= 1.7241 \times 10^{-6}\ \Omega\cdot\text{cm} \\ &= 0.3403\ \Omega\cdot\text{cmil/cm} \end{aligned} \qquad 26.10$$

The percent conductivities for various substances are given in Table 26.1.

5. OHM'S LAW

Voltage, current, and resistance are related by Ohm's law.[3] The numerical result of Ohm's law is called the *voltage drop* or the *IR drop*.[4]

$$V = IR \qquad 26.11$$

Ohm's law presupposes a *linear circuit*, that is, a circuit consisting of linear elements and linear sources. In mathematical terms, this means that voltage plotted against current will result in a straight line with the slope represented by resistance. A *linear element* is a passive element, such as a resistor, whose performance can be portrayed by a linear voltage-current relationship. A *linear source* is one whose output is proportional to the first power of the voltage or current in the circuit. Many elements are linear or can be represented by equivalent linear circuits over some portion of their operation. Most sources, though not linear, can be represented as ideal sources with resistors in series or parallel to account for the nonlinearity.

6. POWER

If a steady current and voltage produce work, W, in time interval t, the electric power (energy conversion rate) is

$$\begin{aligned} P &= \frac{W}{t} = \frac{VIt}{t} \\ &= VI = \frac{V^2}{R} \end{aligned} \qquad 26.12$$

If the voltage and current vary with time, Eq. 26.12 still applies with these terms expressed as v and i. Equation 26.12 then represents the instantaneous power. The same is true for Eq. 26.13.

[3]This book uses the convention that uppercase letters represent fixed, maximum, effective values or direct current (DC) values; and that lowercase letters represent values that change with time, such as alternating current (AC) values. Direct current (DC) values can change in amplitude over time (but not polarity or direction), and for this reason are sometimes shown as lowercase letters.

[4]It is sometimes helpful to consider electrical problems and theories in terms of their mechanical analogs. The mechanical analogy to Ohm's law in terms of fluid flow is as follows: Voltage is the pressure, current is the flow, and resistance is the head loss caused by friction and restrictions within the system.

Circuit Theory

Equation 26.13 represents a form of power sometimes referred to as *I squared R* (I^2R) *losses*. Equation 26.13 is also the mathematical statement of *Joule's law of heating effect.*

$$P = I^2R \tag{26.13}$$

7. DECIBELS

Power changes in many circuits range over decades, that is, over several orders of magnitude. As a result, decibels are adopted to express such changes. The use of decibels also has an advantage in that the power gain (or loss) of cascaded stages in series is the sum of the individual stage decibel gains (or losses), making the mathematics easier. Strictly speaking, decibels refer to a power ratio with the denominator arbitrarily chosen as a particular value, P_0. The reference value changes, depending upon the specific area of electrical engineering usage. For example, communications and acoustics use different references.[5] Decibels have come into common usage for referring to voltage and current ratios as well, though an accurate comparison between circuits requires the two terminals being compared to have equivalent resistances.

$$\begin{aligned}\text{ratio (in dB)} &= 10\log_{10}\frac{P}{P_0}\\ &\approx 20\log_{10}\frac{V_2}{V_1}\\ &= 20\log_{10}\frac{I_2}{I_1}\end{aligned} \tag{26.14}$$

Time-varying quantities can also be in the ratio. Care must be taken to compare only instantaneous values, or effective (rms) or time-averaged values to values of the same type.

8. ENERGY SOURCES

Sources of electrical energy include friction between dissimilar substances, contact between dissimilar substances, thermoelectric action (for example, the Thomson, Peltier, and Seebeck effects), the Hall effect, electromagnetic induction, the photoelectric effect, and chemical action. These sources of energy manifest themselves by the potential (V) they generate. When sources of energy are processed through certain electronic circuits, a current source can be created.

An *ideal voltage source* supplies power at a constant voltage, regardless of the current drawn. An *ideal current source* supplies power in terms of a constant current, regardless of the voltage between its terminals.

[5]Communications uses many such reference values. One references all power changes to 1 kW. In audio acoustics, the reference is the power necessary to cause the minimum sound pressure level audible to the human ear at 2000 Hz, that is, a pressure of 20 μPa.

However, real sources have internal resistances that, at higher currents, reduce the available voltage. Consequently, a *real voltage source* cannot maintain a constant voltage when the currents are large. A *real current source* cannot maintain a constant current completely independent of the voltage at its terminals. Real and ideal voltage and current sources are shown in electrical schematic form in Fig. 26.2.

Figure 26.2 *Ideal and Real Energy Sources*

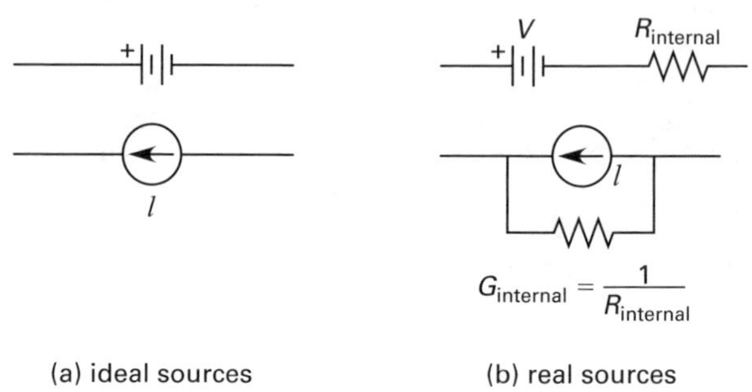

The change in the output of a voltage source is measured as its *regulation*, given by

$$\text{regulation} = \frac{V_{nl} - V_{fl}}{V_{fl}} \times 100\% \tag{26.15}$$

Independent sources deliver voltage and current at their rated values regardless of circuit parameters. *Dependent sources* deliver voltage and current at levels determined by a voltage or current somewhere else in the circuit.

9. VOLTAGE SOURCES IN SERIES AND PARALLEL

Voltage sources connected in series, as in Fig. 26.3(a), can be reduced to an equivalent circuit with a single voltage source and a single resistance, as in Fig. 26.3(b). The equivalent voltage and resistance are calculated per Eq. 26.16 and Eq. 26.17.

$$V_e = \sum V_i \tag{26.16}$$

$$R_e = \sum R_i \tag{26.17}$$

Figure 26.3 *Voltage Sources in Series and Parallel*

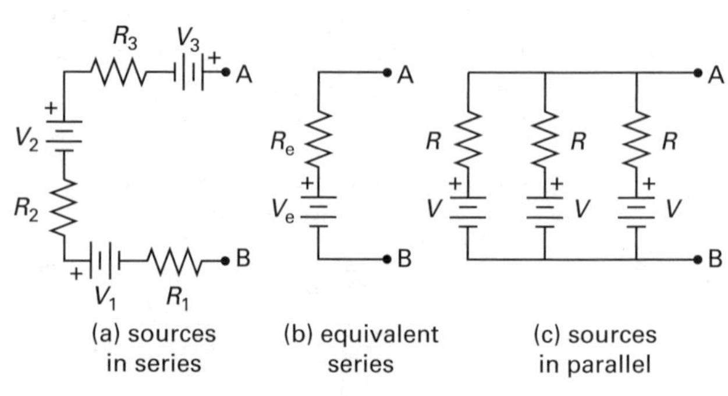

Millman's theorem states that n identical sources of voltage V and resistance R connected in parallel, as in Fig. 26.3(c), can be reduced to an equivalent circuit with a single voltage source and a single resistance, as in Fig. 26.3(b). The equivalent voltage and resistance are calculated per Eq. 26.18 and Eq. 28.19.

$$V_e = V \qquad 26.18$$

$$R_e = \frac{R}{n} \qquad 26.19$$

Nonidentical sources can be connected in parallel, but the lower-voltage sources may be "charged" by the higher-voltage sources. That is, the current may enter the positive terminal of the source. A loop current analysis is needed to determine the current direction and magnitude through the voltage sources.

10. CURRENT SOURCES IN SERIES AND PARALLEL

Currents sources may be placed in parallel. The circuit is then analyzed using any applicable method to determine the power delivered to the various elements. Current sources of differing magnitude cannot be placed in series without damaging one of them.

11. SOURCE TRANSFORMATIONS

A voltage source of V_s with an internal series resistance of R_s can be represented by an equivalent circuit with a current source of I_s with an internal parallel resistance of the same R_s, and vice versa. Consequently, if Eq. 26.20 is valid, the circuits in Fig. 26.4 are equivalent.

$$V_s = I_s R_s \qquad 26.20$$

Figure 26.4 *Equivalent Sources*

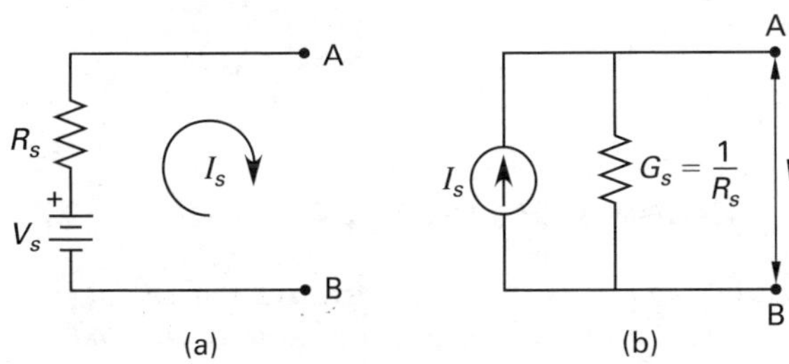

12. MAXIMUM ENERGY TRANSFER

The maximum energy transfer, that is, the maximum power transfer, from a voltage source is attained when the series source resistance (R_s) of Fig. 26.4(a) is reduced to the minimum possible value, with zero being the ideal case. This assumes the load resistance is fixed and the source resistance can be changed. Though this is the ideal situation, it is not often the case. Where the load resistance varies and the source resistance is fixed, maximum power transfer occurs when the load and source resistances are equal.

Similarly, the maximum power transfer from a current source occurs when the parallel source resistance (R_s) of Fig. 26.4(b) is increased to the maximum possible value. This assumes the load resistance is fixed and the source resistance can be changed. In the more common case, where the load resistance varies and the source resistance is fixed, the maximum power transfer occurs when the load and source resistances are equal.

The maximum power transfer occurs in any circuit when the load resistance equals the Norton or Thevenin equivalent resistance.

13. VOLTAGE AND CURRENT DIVIDERS

At times, a source voltage will not be at the required value for the operation of a given circuit. For example, the standard automobile battery voltage is 12 V, while some electronic circuitry uses 8 V for DC biasing. One method of obtaining the required voltage is by use of a circuit referred to as a *voltage divider*. A voltage divider is illustrated in Fig. 26.5(a). The voltage across resistor 2 is given by

$$V_2 = V_s\left(\frac{R_2}{R_1 + R_2}\right) \qquad 26.21$$

Figure 26.5 *Divider Circuits*

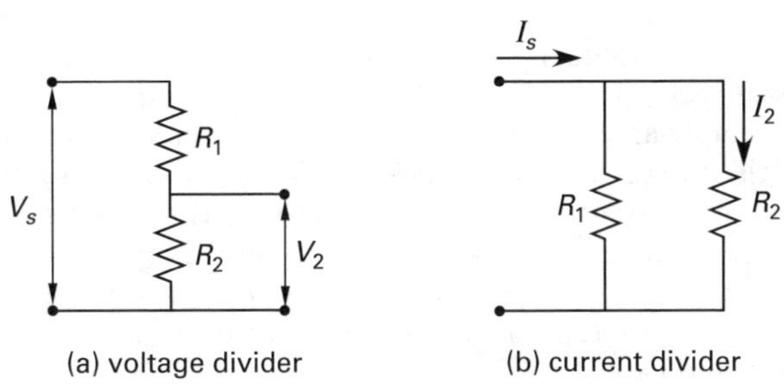

If the fraction of the source voltage, V_s, is found as a function of V_2, the result is known as the *gain* or the *voltage-ratio transfer function*. Rearranging Eq. 26.21 to show the specific relationship between R_1 and R_2 gives

$$\frac{V_2}{V_s} = \frac{1}{1 + \frac{R_1}{R_2}} \qquad 26.22$$

Circuit Theory

An analogous circuit called a *current divider* can be used to produce a specific current. A current divider circuit is shown in Fig. 26.5(b). The current through resistor 2 is

$$I_2 = I_s\left(\frac{R_1}{R_1 + R_2}\right) = I_s\left(\frac{G_2}{G_1 + G_2}\right) \qquad 26.23$$

14. KIRCHHOFF'S LAWS

A fundamental principle in the design and analysis of electrical circuits is that the dimensions of the circuit are small. That is, the circuit's dimensions are small when compared to the wavelengths of the electromagnetic quantities that pass within them. Only time variations need be considered, not spatial variations. Consequently, Maxwell's equations, which are partial integrodifferential equations, become ordinary integrodifferential equations that vary only with time.

The net result is that Gauss' law, in integral form, reduces to Kirchhoff's current law. This occurs because the net charge enclosed in any particular volume of the circuit is zero and, therefore, the sum of the currents at any point must be zero.

Another result is that Faraday's law of electromagnetic induction reduces to Kirchhoff's voltage law. This is because, with a "small" circuit, the surface integral of magnetic flux density is zero; the sum of the voltages around any closed loop must be zero.

15. KIRCHHOFF'S VOLTAGE LAW

Kirchhoff's voltage law (KVL) states that the algebraic sum of the voltages around any closed path within a circuit or network is zero. Some of the voltages may be sources, while others will be voltages caused by current in passive elements. Voltages through the passive elements are often called *voltage drops.* A *voltage rise* may also appear to occur across passive elements in the initial stages of the application of Kirchhoff's voltage law. For resistors, this is only because the direction of the current is arbitrarily chosen. For inductors and capacitors, a voltage rise indicates the release of magnetic or electrical energy. Stated in terms of the voltage rises and drops, KVL is

$$\sum_{\text{loop}} \text{voltage rises} = \sum_{\text{loop}} \text{voltage drops} \qquad 26.24$$

Kirchhoff's voltage law can also be stated as follows: The sum of the voltages around a closed loop must equal zero. The reference directions for voltage rises and drops used in this book are shown in Fig. 26.6.[6]

[6]This alternative statement of Kirchhoff's voltage law and the arbitrary directions chosen are sometimes useful in simplifying the writing of the mathematical equations used in applying KVL.

Figure 26.6 *Voltage Reference Directions*

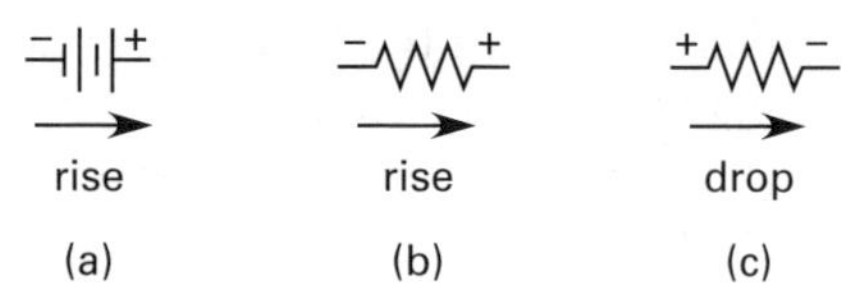

Example 26.3

Consider the following circuit. Determine the expression for V_{AB} across the load in the circuit in the figure.

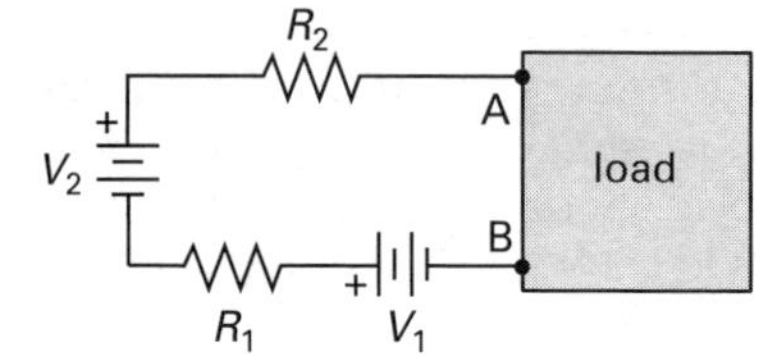

Solution

For consistency in all problems, apply KVL in a clockwise direction as shown. The unknown voltage is V_{AB}, which indicates that the voltage is arbitrarily referenced from A to B. That is, terminal A is assumed to be the high potential (+). Assign polarities to the resistors based on the assumed direction.

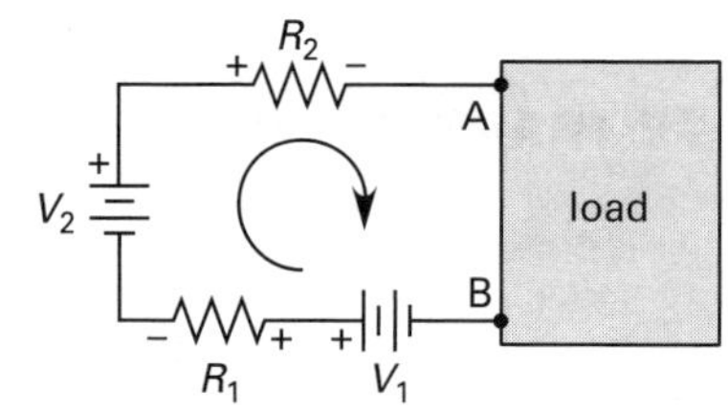

The KVL expression for the loop, clockwise starting at terminal B, is

$$V_1 - V_{R_1} + V_2 - V_{R_2} - V_{AB} = 0$$

Rearrange and solve for V_{AB}.

$$V_{AB} = V_1 - V_{R_1} + V_2 - V_{R_2}$$

16. KIRCHHOFF'S CURRENT LAW

Kirchhoff's current law (KCL) states that the algebraic sum of the currents at a node is zero. A *node* is a connection of two or more circuit elements. When the connection is between two elements, the node is a *simple node.* When the connection is between three or more elements, the node is referred to as a *principal node.* Stated in terms of the currents directed into and out of the node, KCL is

$$\sum_{\text{node}} \text{currents in} = \sum_{\text{node}} \text{currents out} \qquad 26.25$$

Kirchhoff's current law can also be stated as follows: The sum of the currents flowing out of a node must equal zero.[7] The reference directions for positive currents (currents "out") and negative currents (currents "in") used in this book are illustrated in Fig. 26.7.

Figure 26.7 *Current Reference Directions*

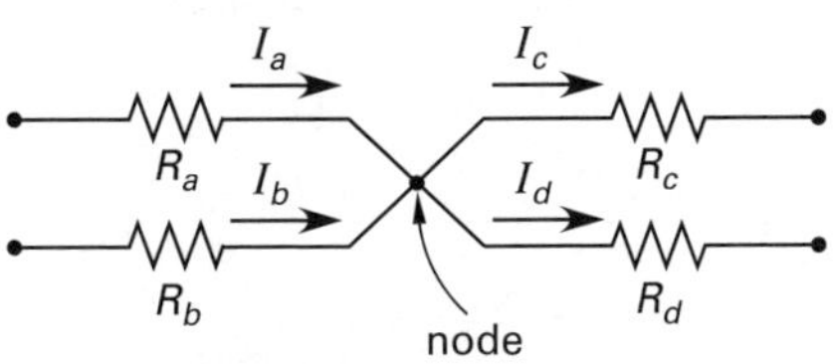

(a) I_a, I_b: negative currents or currents in
(b) I_c, I_d: positive currents or currents out

17. SERIES CIRCUITS

A simple series circuit, as shown in Fig. 26.8, has the following properties (generalized to any number of voltage sources and series resistances).

- Current is the same through all the circuit elements.

$$I = I_{R_1} = I_{R_2} = I_{R_3} \cdots = I_{R_n} \quad \text{26.26}$$

- The equivalent resistance is the sum of the individual resistances.

$$R_e = R_1 + R_2 + R_3 + \cdots + R_n \quad \text{26.27}$$

- The equivalent applied voltage is the sum of all the voltage sources, with the polarity considered.

$$V_e = \pm V_1 \pm V_2 \pm \cdots \pm V_n \quad \text{26.28}$$

- The sum of the voltage drops across all circuit elements is equal to the equivalent applied voltage (KVL).

$$V_e = IR_e \quad \text{26.29}$$

Figure 26.8 *Series Circuit*

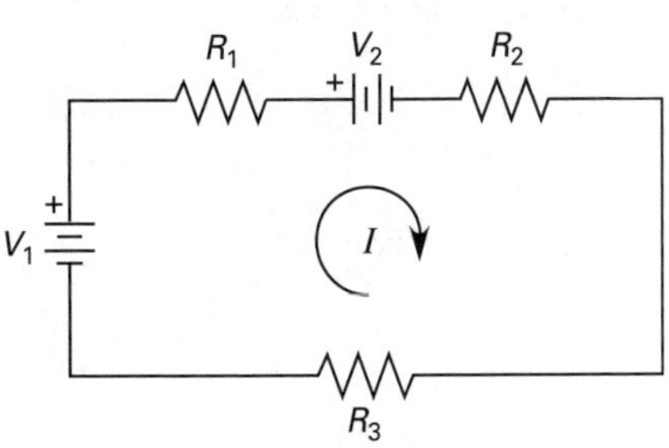

[7]This alternative statement of Kirchhoff's current law and the arbitrary directions chosen are sometimes useful in simplifying the writing of the mathematical equations used in applying KCL.

18. PARALLEL CIRCUITS

A simple parallel circuit with only one active source, as shown in Fig. 26.9, has the following properties (generalized to any number of resistors).

- The voltage across all legs of the circuit is the same.

$$\begin{aligned} V &= V_1 = V_2 = V_3 \cdots = V_n \\ &= I_1R_1 = I_2R_2 = I_3R_3 \cdots = V_n \end{aligned} \quad \text{26.30}$$

- The reciprocal equivalent resistance is the sum of the reciprocals of the individual resistances. The equivalent conductance is the sum of the individual conductances.

$$\frac{1}{R_e} = \frac{1}{R_1} + \frac{1}{R_2} + \frac{1}{R_3} + \cdots + \frac{1}{R_n} \quad \text{26.31(a)}$$

$$G_e = G_1 + G_2 + G_3 + \cdots + G_n \quad \text{26.31(b)}$$

- The total current is the sum of the currents in the individual legs of the circuit (KCL).

$$\begin{aligned} I &= I_1 + I_2 + I_3 + \cdots + I_n \\ &= \frac{V}{R_1} + \frac{V}{R_2} + \frac{V}{R_3} + \cdots + \frac{V}{R_n} \\ &= V(G_1 + G_2 + G_3 + \cdots + G_n) \end{aligned} \quad \text{26.32}$$

Figure 26.9 *Parallel Circuit*

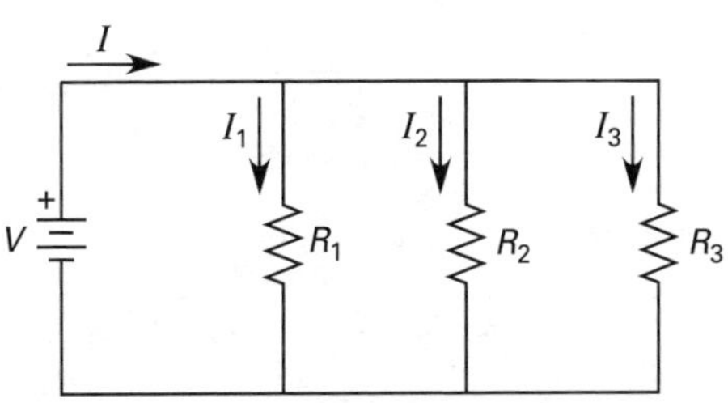

19. ANALYSIS OF COMPLICATED RESISTIVE NETWORKS

The following is a general method that can be used to determine the current flow and voltage drops within a complicated resistive circuit.

step 1: If the circuit is three-dimensional, draw a two-dimensional representation.

step 2: Combine series voltage sources.

step 3: Combine parallel current sources.

step 4: Combine series resistances.

step 5: Combine parallel resistances.

step 6: Repeat steps 2–5 as needed to obtain the current and/or voltage at the desired point, junction, or branch of the circuit. (Do not simplify the circuit beyond what is required to determine the desired quantities.)

step 7: If applicable, utilize the delta-wye transformation of Sec. 26.20.

20. DELTA-WYE TRANSFORMATIONS

Electrical resistances arranged in the shape of the Greek letter Δ (delta) or the English letter Y (wye) are known as *delta-wye configurations.* They are also called *pi-T configurations.* Two such circuits are shown in Fig. 26.10.

Figure 26.10 *Delta (Pi)-Wye (T) Configurations*

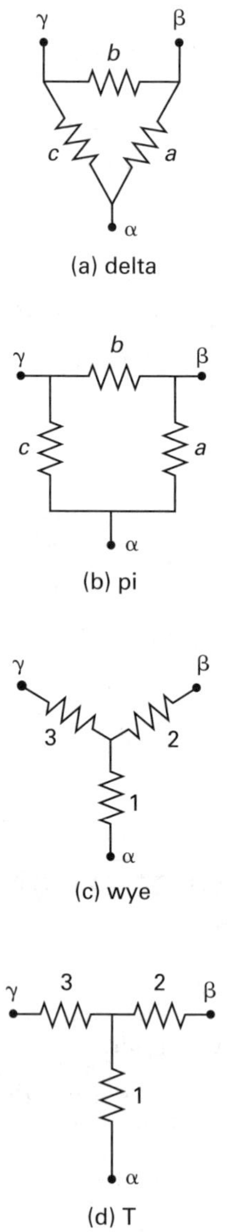

The equivalent resistances, which allow transformation between configurations, are

$$R_a = \frac{R_1R_2 + R_1R_3 + R_2R_3}{R_3} \qquad 26.33$$

$$R_b = \frac{R_1R_2 + R_1R_3 + R_2R_3}{R_1} \qquad 26.34$$

$$R_c = \frac{R_1R_2 + R_1R_3 + R_2R_3}{R_2} \qquad 26.35$$

$$R_1 = \frac{R_aR_c}{R_a + R_b + R_c} \qquad 26.36$$

$$R_2 = \frac{R_aR_b}{R_a + R_b + R_c} \qquad 26.37$$

$$R_3 = \frac{R_bR_c}{R_a + R_b + R_c} \qquad 26.38$$

Example 26.4

Simplify the circuit and determine the total current.

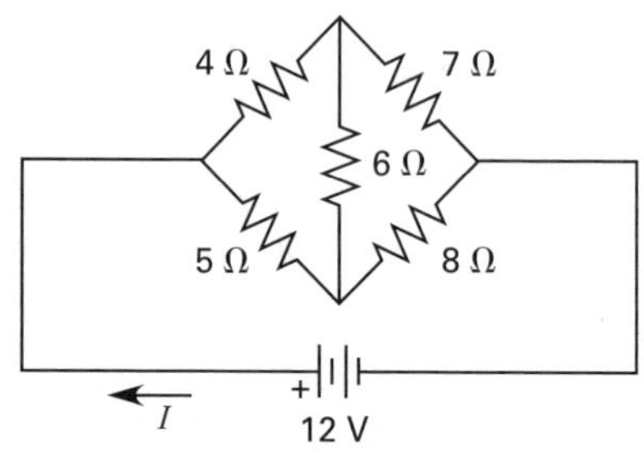

Solution

Convert the 4 Ω-5 Ω-6 Ω delta connection to wye form.

$$\begin{aligned} R_1 &= \frac{R_aR_c}{R_a + R_b + R_c} \\ &= \frac{(5\ \Omega)(4\ \Omega)}{5\ \Omega + 6\ \Omega + 4\ \Omega} = 1.3\ \Omega \\ R_2 &= \frac{R_aR_b}{R_a + R_b + R_c} \\ &= \frac{(5\ \Omega)(6\ \Omega)}{15\ \Omega} = 2.0\ \Omega \\ R_3 &= \frac{R_bR_c}{R_a + R_b + R_c} \\ &= \frac{(6\ \Omega)(4\ \Omega)}{15\ \Omega} = 1.7\ \Omega \end{aligned}$$

The transformed circuit is

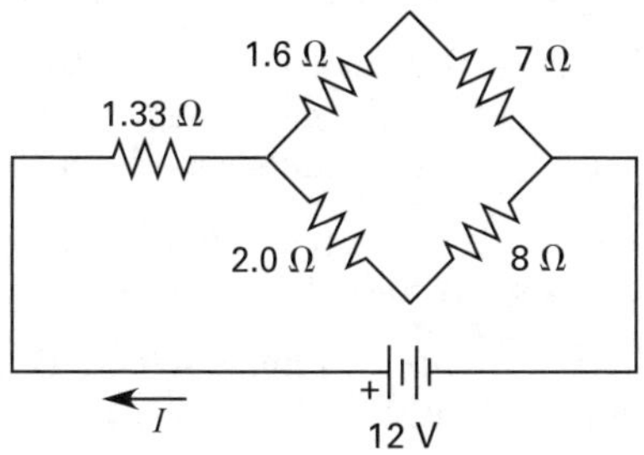

The total equivalent resistance is

$$R_e = 1.33\ \Omega + \frac{1}{\dfrac{1}{1.6\ \Omega + 7\ \Omega} + \dfrac{1}{2\ \Omega + 8\ \Omega}} = 5.95\ \Omega$$

The current is

$$I = \frac{V}{R_e} = \frac{12\ \text{V}}{5.95\ \Omega} = 2.02\ \text{A}$$

21. SUBSTITUTION THEOREM

The *substitution theorem*, also known as the *compensation theorem*, states that any branch in a circuit can be replaced by a substitute branch, as long as the branch voltage and current remain the same, without affecting voltages and currents in any other portion of the circuit.

22. RECIPROCITY THEOREM

In any linear, time-independent circuit with independent current and voltage sources, the ratio of the current in a short circuit in one part of the network to the output of a voltage source in another part of the network is constant—even when the voltage source and the short-circuit positions are interchanged. This principle of *reciprocity* is also applicable to the ratio of the current from a current source and the voltage across an open circuit. Reciprocity between a voltage source and short circuit is illustrated in Fig. 26.11.

Figure 26.11 *Reciprocal Measurements*

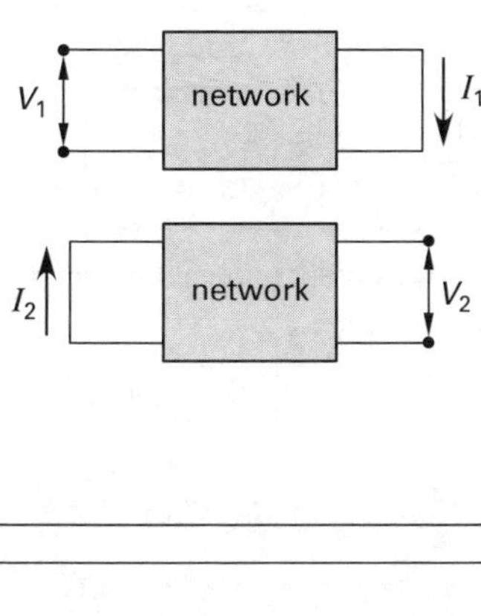

Alternatively, if the network is comprised of linear resistors, the ratio of the applied voltage to the current measured at any point is constant—even when the positions of the source and meter are interchanged. The ratio in both cases is

$$R_{\text{transfer}} = \frac{V_1}{I_1} = \frac{V_2}{I_2} \qquad \textit{26.39}$$

The ratio is called the *transfer resistance*. The term *transfer* is used when a given response is determined at a point in a network other than where the driving force is applied. In this case, the voltage is the driving force on one side of the circuit and the current is measured on the other side.

23. SUPERPOSITION THEOREM

The principle of *superposition* is that the response (that is, the voltage across or current through) of a linear circuit element in a network with multiple independent sources is equal to the response obtained if each source is considered individually and the results are summed. The steps involved in determining the desired quantity are as follows.

step 1: Replace all sources except one with their internal resistances.[8] Replace ideal current sources with open circuits. Replace ideal voltage sources with short circuits.

step 2: Compute the desired quantity (either voltage or current) for the element in question attributable to the single source.

step 3: Repeat steps 1 and 2 for each of the sources in turn.

step 4: Sum the calculated values obtained for the current or voltage obtained in step 2. The result is the actual value of the current or voltage in the element for the complete circuit.

Example 26.5

Determine the current through the center leg.

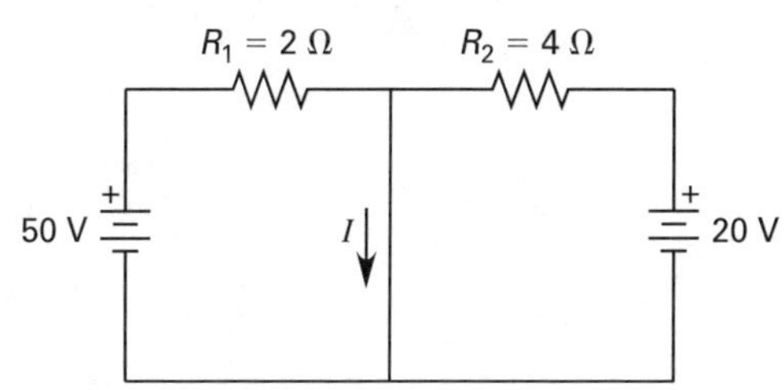

[8]As in all the DC theorems presented in this chapter, this principle also applies to AC circuits or networks. In the case of AC circuits or networks, the term *impedance* would be applicable instead of *resistance*.

Circuit Theory

Solution

First, work with the left (50 V) battery. Short out the right (20 V) battery.

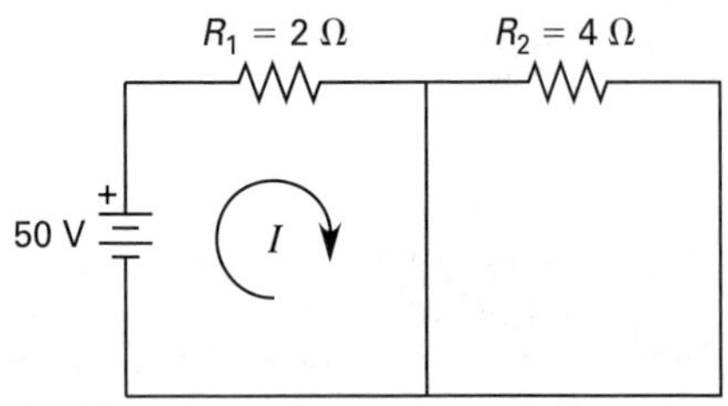

No current will flow through R_2. The equivalent resistance and current are

$$R_e = 2\ \Omega$$

$$I = \frac{V}{R_e} = \frac{50\ \text{V}}{2\ \Omega} = 25\ \text{A}$$

Next, work with the right (20 V) battery. Short out the left (50 V) battery.

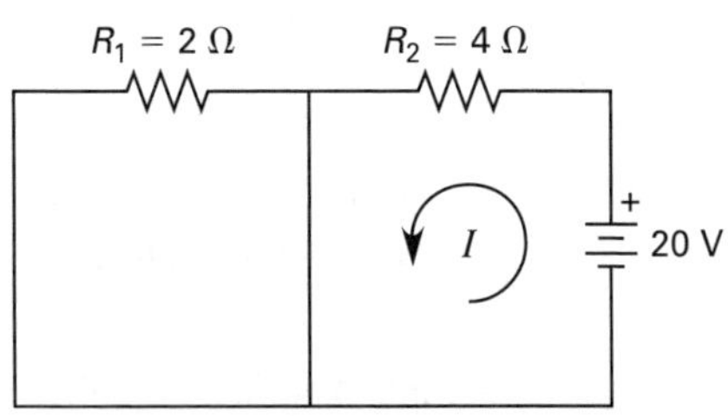

No current will flow through R_1. The equivalent resistance and current are

$$R_e = 4\ \Omega$$

$$I = \frac{V}{R_e} = \frac{20\ \text{V}}{4\ \Omega} = 5\ \text{A}$$

Considering both batteries, the total current flowing is $25\ \text{A} + 5\ \text{A} = 30\ \text{A}$.

24. THEVENIN'S THEOREM

Thevenin's theorem states that, insofar as the behavior of a linear circuit at its terminals is concerned, any such circuit can be replaced by a single voltage source, V_{Th}, in series with a single resistance, R_{Th}. The method for determining and utilizing the *Thevenin equivalent circuit*, with designations referring to Fig. 26.12, follows.

step 1: Separate the network that is to be changed into a Thevenin equivalent circuit from its load at two terminals, say A and B.

step 2: Determine the open-circuit voltage, V_{oc}, at terminals A and B.

step 3: Short circuit terminals A and B and determine the current, I_{sc}.

step 4: Calculate the Thevenin equivalent voltage and resistance from the following equations.

$$V_{\text{Th}} = V_{\text{oc}} \qquad 26.40$$

$$R_{\text{Th}} = \frac{V_{\text{oc}}}{I_{\text{sc}}} \qquad 26.41$$

step 5: Using the values calculated in Eq. 26.40 and Eq. 26.41, replace the network with the Thevenin equivalent. Reconnect the load at terminals A and B. Determine the desired electrical parameters in the load.

Steps 3 and 4 can be altered by using the following shortcut. Determine the Thevenin equivalent resistance by looking into terminals A and B toward the network with all the power sources altered. Specifically, change independent voltage sources into short circuits and independent current sources into open circuits, then calculate the resistance of the altered network. The resulting resistance is the Thevenin equivalent resistance.

Figure 26.12 *Thevenin Equivalent Circuit*

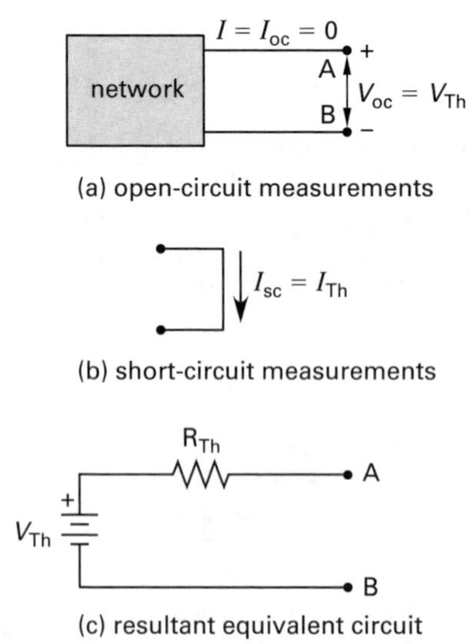

Example 26.6

Determine the Thevenin equivalent circuit that the load resistor, R_{load}, sees.

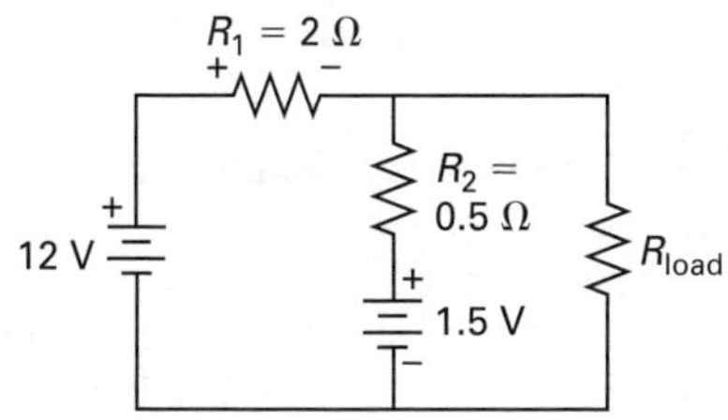

Solution

step 1: Separate the load resistor from the portion of the network to be changed.

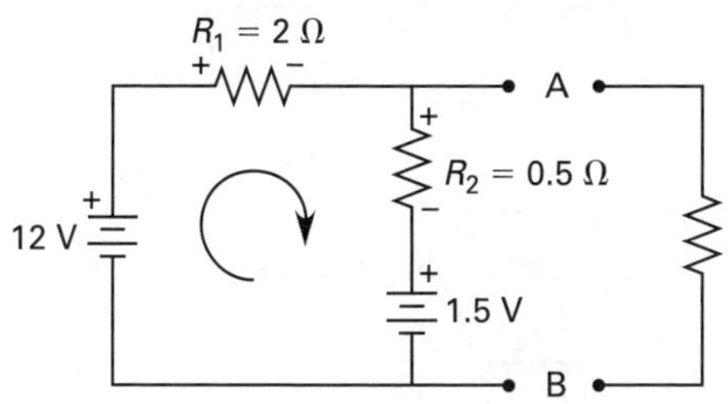

step 2: Apply Kirchhoff's voltage law to find the open-circuit voltage. Starting at terminal B, the equation is

$$12\ \text{V} - V_{R_1} - V_{R_2} - 1.5\ \text{V} = 0$$

$$V_{R_1} + V_{R_2} = 10.5\ \text{V}$$

For clarification, the circuit is redrawn using the result calculated.

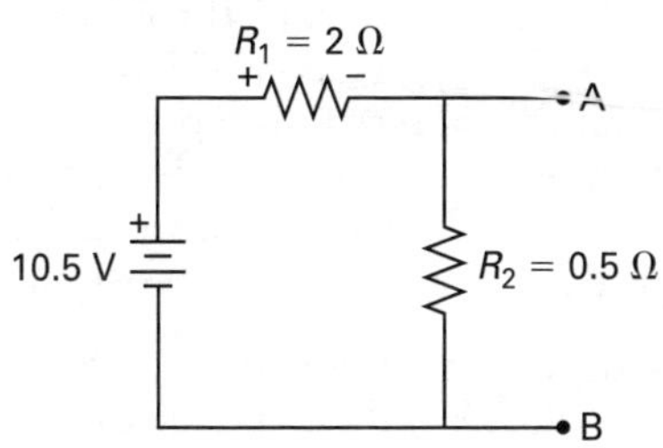

The voltage across R_2 is found using the voltage-divider concept.

$$V_{R_2} = (10.5\ \text{V})\left(\frac{R_2}{R_1 + R_2}\right)$$

$$= (10.5\ \text{V})\left(\frac{0.5\ \Omega}{2\ \Omega + 0.5\ \Omega}\right) = 2.1\ \text{V}$$

The circuit branch from A to B can now be illustrated as follows.

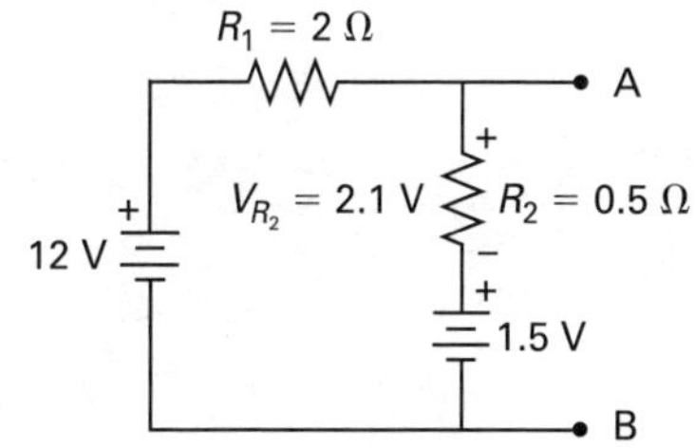

The open-circuit voltage is the voltage across R_2 and the 1.5 V source, so $V_{oc} = 3.6$ V. (This can be found by using KVL around the loop containing the terminals A and B.)

step 3: Short terminals A and B to determine the short-circuit current.

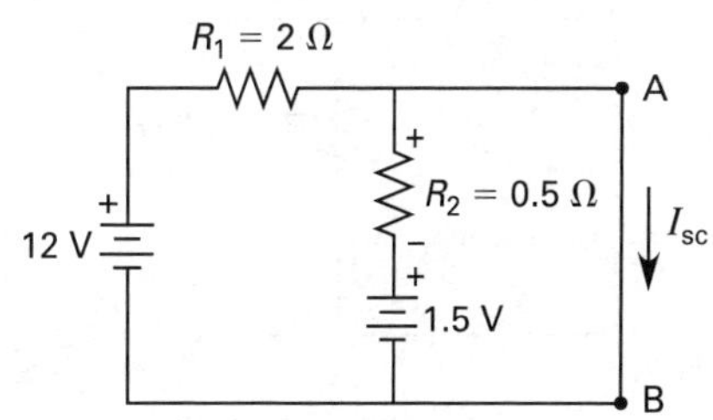

Using either Ohm's law or KVL around the outer loop, determine the short-circuit current contribution from the 12 V source.

$$I_{sc} = \frac{12\ \text{V}}{2\ \Omega} = 6\ \text{A}$$

Using either Ohm's law or KVL around the loop containing the 1.5 V source, R_2, and terminals A and B, determine the short-circuitcurrent contribution from the 1.5 V source.

$$I_{sc} = \frac{1.5\ \text{V}}{0.5\ \Omega} = 3\ \text{A}$$

The total short-circuit current, using the principle of superposition, is 6 A + 3 A = 9 A.

step 4: Using Eq. 26.40 and Eq. 26.41, calculate the Thevenin equivalent voltage and resistance.

$$V_{Th} = V_{oc} = 3.6\ \text{V}$$

$$R_{Th} = \frac{V_{oc}}{I_{sc}} = \frac{3.6\ \text{V}}{9\ \text{A}} = 0.4\ \Omega$$

step 5: Replace the network with the Thevenin equivalent.

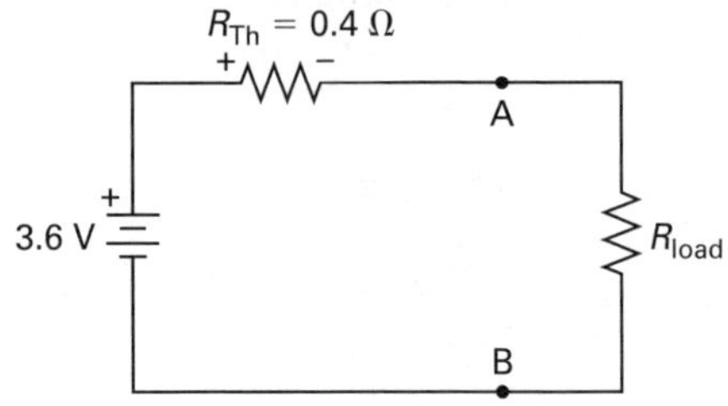

The method outlined is systematic. The shortcut for finding the Thevenin equivalent circuit mentioned earlier in this section would have simplified the solution. For example, short the voltage sources and determine the Thevenin equivalent resistance.

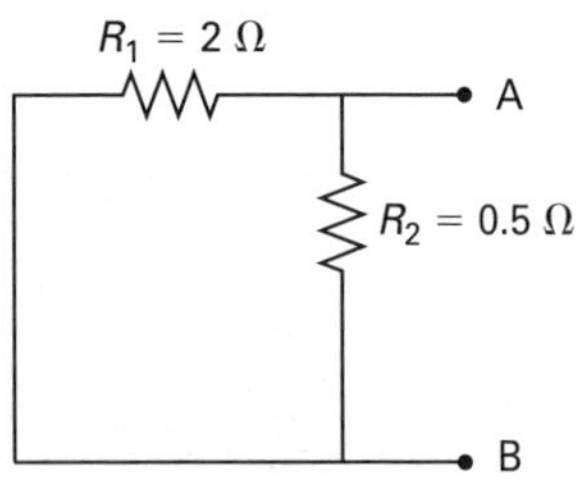

$$R_{Th} = \frac{R_1 R_2}{R_1 + R_2} = \frac{(2\ \Omega)(0.5\ \Omega)}{2\ \Omega + 0.5\ \Omega} = 0.4\ \Omega$$

The short-circuit current, found using superposition, is 9 A. The Thevenin equivalent voltage can then be found by combining Eq. 26.29 and Eq. 26.41.

$$V_{Th} = I_{Th} R_{Th}$$

Substituting the short-circuit current for I_{sc} and the calculated value for R_{Th} results in an identical answer, as required by Thevenin's theorem.

25. NORTON'S THEOREM

Norton's theorem states that, insofar as the behavior of a linear circuit at its terminals is concerned, any such circuit can be replaced by a single current source, I_N, in parallel with a single resistance, R_N. The method for determining and utilizing the *Norton equivalent circuit*, with designations referring to Fig. 26.13, follows.

step 1: Separate the network that is to be changed into a Norton equivalent circuit from its load at two terminals, say A and B.

step 2: Determine the open-circuit voltage, V_{oc}, at terminals A and B.

step 3: Short circuit terminals A and B and determine the current, I_{sc}.

step 4: Calculate the Norton equivalent current and resistance from the following equations.

$$I_N = I_{sc} \qquad 26.42$$

$$R_N = \frac{V_{oc}}{I_{sc}} \qquad 26.43$$

step 5: Using the values calculated in Eq. 26.42 and Eq. 26.43, replace the network with the Norton equivalent. Reconnect the load at terminals A and B. Determine the desired electrical parameters in the load.

Figure 26.13 *Norton Equivalent Circuit*

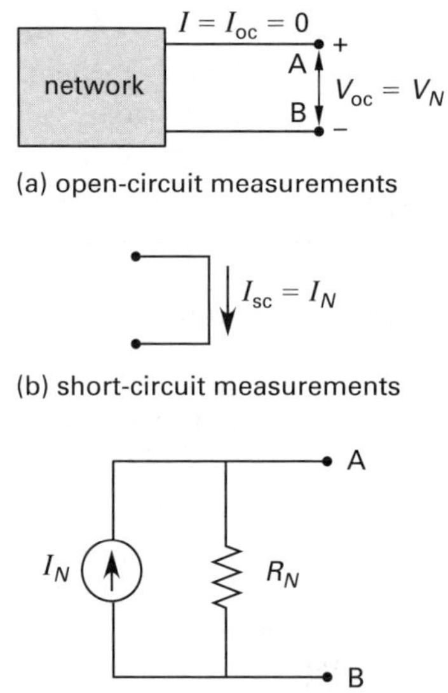

(c) resultant equivalent circuit

Steps 3 and 4 can be altered by using the following shortcut. Determine the Norton equivalent resistance by looking into terminals A and B toward the network with all the power sources altered. Specifically, change independent voltage sources into short circuits and independent current sources into open circuits, then calculate the resistance of the altered network. The resulting resistance is the Norton equivalent resistance.

The Norton equivalent resistance equals the Thevenin equivalent resistance for identical networks.

26. LOOP-CURRENT METHOD

The *loop-current method* is a systematic network-analysis procedure that uses currents as the unknowns. It is also called *mesh analysis* or the *Maxwell loop-current method*. The method uses Kirchhoff's voltage law and is performed on planar networks. It requires $n-1$ simultaneous equations for an n-loop system. The method's steps are as follows.

step 1: Select $n-1$ loops (one loop less than the total number of possible loops).

step 2: Assume current directions for the selected loops. (While this is arbitrary, clockwise directions will always be chosen in this text for consistency. Any incorrectly chosen current direction will result in a negative result when the simultaneous equations are solved in step 4.) Show the direction of the current with an arrow.

step 3: Write Kirchhoff's voltage law for each of the selected loops. Assign polarities based on the direction of the loop current. The voltage of a source is positive when the current flows out of the positive terminal, that is, from the negative terminal to the positive terminal inside the source. The selected direction of the loop current always results in a voltage drop in the resistors of the loop (see Fig. 26.6 (c)). Where two loop currents flow through an element, they are summed to determine the voltage drop in that element, using the direction of the current in the loop for which the equation is being written as the positive (i.e., correct) direction. (Any incorrect direction for the loop current will be indicated by a negative sign in the solution in step 4.)

step 4: Solve the $n-1$ equations (from step 3) for the unknown currents.

step 5: If required, determine the actual current in an element by summing the loop currents flowing through the element. (Sum the absolute values to obtain the correct magnitude. The correct direction is given by the loop current with the positive value.)

27. NODE-VOLTAGE METHOD

The *node-voltage method* is a systematic network-analysis procedure that uses voltages as the unknowns. The method uses Kirchhoff's current law. It requires $n-1$ equations for an n-principal node system (equations are not necessary at simple nodes, that is, nodes connecting only two circuit elements). The method's steps are as follows.

step 1: Simplify the circuit, if possible, by combining resistors in series or parallel or by combining current sources in parallel. Identify all nodes. (The minimum total number of equations required will be $n-1$, where n represents the number of principal nodes.)

step 2: Choose one node as the reference node, that is, the node that will be assumed to have ground potential (0 V). (To minimize the number of terms in the equations, pick the node with the largest number of circuit elements as the reference node.)

step 3: Write Kirchhoff's current law for each principal node except the reference node, which is assumed to have a zero potential.

step 4: Solve the $n-1$ equations (from step 3) to determine the unknown voltages.

step 5: If required, use the calculated node voltages to determine any branch current desired.

28. DETERMINATION OF METHOD

When analyzing electrical networks, the method used depends on the circuit elements and their configurations. The loop-current method using Kirchhoff's voltage law is used in circuits without current sources. The node-voltage method using Kirchhoff's current law is used in circuits without voltage sources. When both types of sources are present, one of the following two methods may be used.

method 1: Use each of Kirchhoff's laws, assigning voltages and currents as needed, and substitute any known quantity into the equations as written.

method 2: Use source transformation or source shifting to change the appearance of the circuit so that it contains only the desired sources—voltage sources when using KVL and current sources when using KCL. (Source shifting is manipulating the circuit so that each voltage source has a resistor in series and each current source has a resistor in parallel.)

Use the method that results in the least number of equations. The loop-current method produces $n-1$ equations, where n is the total number of loops. The node-voltage method produces $n-1$ equations, where n is the number of principal nodes. Count the number of loops and the number of principal nodes prior to writing the equations; whichever is least determines the method used.

Additional methods exist, some of which require less work. The advantage of the loop-current and node-voltage methods is that they are systematic and guarantee a solution.

29. PRACTICAL APPLICATION: BATTERIES

In general, a *battery* is defined as a direct-current voltage source made up of one or more units that convert chemical, thermal, nuclear, or solar energy into electrical energy. The most widely used battery type is one that converts the chemical energy contained in its active materials directly into electrical energy by means of an oxidation-reduction reaction.[9] A battery consists of two dissimilar metals, an anode and a cathode, immersed in an electrolyte. The cathode is reduced during the reaction. The transfer of charge is completed within the electrolyte by the flow of ions. This is shown for a single cell in Fig. 26.14. Regardless of the cell or device type, the flow of electrons is always from anode-to-cathode outside and cathode-to-anode inside the cell or device.

Figure 26.14 *Electrochemical Battery (Discharge)*

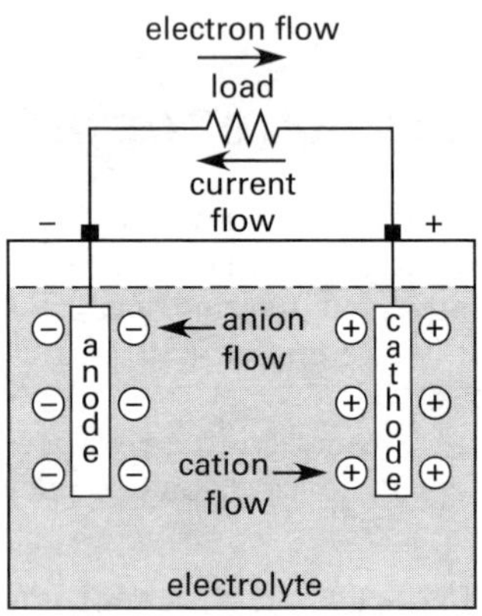

The battery terminals have no absolute voltage value. They have a value relative only to each other, and the terminals and the battery are said to *float.* Usually, one of the terminals is assigned as the reference and given the value of 0 V. This terminal is then the *reference* or *datum* and is said to be *grounded.* If the battery is allowed to float and the circuit is grounded elsewhere, all voltages are assumed to be measured with respect to this ground. The earth is generally regarded as being at 0 V and any circuit tied to earth by an electrical wire is grounded. Various symbols for grounds are shown in Fig. 26.15.

Figure 26.15 *Ground Symbols*

The voltage of a battery is determined by the number of cells used. The voltage of the cell is determined by the materials used, as this determines the half-cell oxidation potentials. The capacity of the battery is determined by the amount of materials used and is measured in ampere-hours (A·h). One gram-equivalent weight of material supplies 96,485 C or 26.805 A·h of electric charge.

A *primary battery* is one that uses an electrochemical reaction that is not effciently reversible. An example is the common flashlight battery. This type of battery is also called a *dry cell,* as the electrolyte is a moist paste instead of a liquid solution. A *secondary battery* is rechargeable and has a much higher energy density. An example is the lead-acid storage battery used in automobiles. Such batteries are treated as ideal voltage sources with a series resistance representing internal resistance. Using the specified voltage and resistance, any network containing a battery can be analyzed using the techniques described in this chapter.

[9]In *nonelectrochemical reactions,* the transfer of electrons takes place directly and only heat is involved.

27 AC Circuit Fundamentals

1. Fundamentals 27-1
2. Voltage 27-2
3. Current 27-3
4. Impedance 27-3
5. Admittance 27-3
6. Voltage Sources 27-3
7. Average Value 27-4
8. Root-Mean-Square Value 27-5
9. Phase Angles 27-7
10. Sinusoid 27-7
11. Phasors 27-7
12. Complex Representation 27-8
13. Resistors 27-9
14. Capacitors 27-9
15. Inductors 27-9
16. Combining Impedances 27-10
17. Ohm's Law 27-11
18. Power 27-11
19. Real Power and the Power Factor 27-11
20. Reactive Power 27-12
21. Apparent Power 27-12
22. Complex Power and the Power Triangle . . . 27-12
23. Maximum Power Transfer 27-13
24. AC Circuit Analysis 27-13

Nomenclature

a_n, b_n	Fourier coefficients	–
B	susceptance	S
C	capacitance	F
f	frequency	Hz
G	conductance	S
i	variable current	A
I, $\mathbf{I}$	constant or rms current	A
I^*, $\mathbf{I}^*$	complex conjugate of current	A
L	inductance	H
p	instantaneous power	W
pf	power factor	–
P, $\mathbf{P}$	power	W
Q, $\mathbf{Q}$	reactive power	VAR
R	resistance	Ω
S, $\mathbf{S}$	apparent or complex power	VA
t	time	s
T	period	s
v	variable voltage	V
V, $\mathbf{V}$	constant or rms voltage	V
X	reactance	Ω
Y, $\mathbf{Y}$	admittance	S
Z, $\mathbf{Z}$	impedance	Ω
$\mathbf{Z}$	complex number	–

Symbols

θ	phase angle	rad
ϕ	impedance angle or power angle	rad
ϕ	phase angle difference	rad
ϕ_{pf}	power factor angle	rad
φ	current angle	rad
ω	angular frequency	rad/s

Subscripts

ave	average
C	capacitative or capacitor
e	equivalent
eff	effective
i	imaginary
I	current
L	inductor
m	maximum
p	peak
pf	power factor
R	resistive or resistor
rms	root-mean-square
s	source
thr	threshold
Z	impedance

1. FUNDAMENTALS

Alternating waveforms have currents and voltages that vary with time in a symmetrical manner. Possible waveforms include square, sawtooth, and triangular along with many variations on these themes. However, for many applications in electrical engineering the variations are sinusoidal in time. In this book, unless otherwise specified, currents and voltages referred to are sinusoidal.[1] When sinusoidal, the waveform is nearly always referred to as AC, that is, *alternating current*, indicating that the current is produced by the application of a sinusoidal voltage. This means that the flow of electrons changes directions, unlike DC circuits, where the flow of electrons is unidirectional (though the magnitude can change in time). A circuit is said to be in a *steady-state* condition if the current and voltage time variation is purely constant

[1]Nearly all periodic functions can be represented as sinusoidal functions. The superposition theorem allows the effects of individual sinusoids to be summed to obtain an overall effect, which simplifies the mathematics necessary to obtain a result.

(DC) or purely sinusoidal (AC).[2] In this book, unless otherwise specified, the circuits are in a steady-state condition. Importantly, all the methods, basic definitions, and equations presented in Chap. 26 involving DC circuits are applicable to AC circuits. AC electrical parameters have both magnitudes and angles. Nevertheless, following common practice, phasor notation will be used only if necessary to avoid confusion.

2. VOLTAGE

Sinusoidal variables can be expressed in terms of sines or cosines without any loss of generality.[3] A sine waveform is often the standard. If this is the case, Eq. 27.1 gives the value of the instantaneous voltage as a function of time.

$$v(t) = V_m \sin(\omega t + \theta) \qquad 27.1$$

The *maximum value* of the sinusoid is given the symbol V_m and is also known as the *amplitude*. If $v(t)$ is not zero at $t = 0$, the sinusoid is *shifted* and a *phase angle*, θ, must be used, as shown in Fig. 27.1.[4] Also shown in Fig. 27.1 is the *period*, T, which is the time that elapses in one cycle of the sinusoid. The *cycle* is the smallest portion of the sinusoid that repeats.

Figure 27.1 Sinusoidal Waveform with Phase Angle

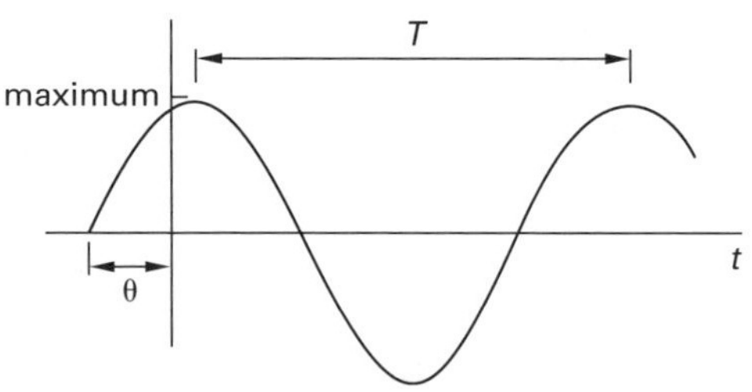

Because the horizontal axis of the voltage in Fig. 27.1 corresponds to time, not distance, the waveform does not have a wavelength. The frequency, f, of the sinusoid is the reciprocal of the period in hertz (Hz). The angular frequency, ω, in rad/s can also be used.

$$f = \frac{1}{T} = \frac{\omega}{2\pi} \qquad 27.2$$

$$\omega = 2\pi f = \frac{2\pi}{T} \qquad 27.3$$

An AC voltage waveform without a phase angle is plotted as a function of differing variables in Fig. 27.2.

[2]Steady-state AC may have a *DC offset.*

[3]The point at which time begins, that is, where $t = 0$, is of no consequence in steady-state AC circuit problems, as the signal repeats itself every cycle. Therefore, it makes no difference whether a sine or cosine waveform is used (though, if one exists, care must be taken to keep the phase angle correct).

[4]The term *phase* is not the same as *phase difference*, which is the difference between corresponding points on two sinusoids of the same frequency.

Figure 27.2 Sine Wave Plots

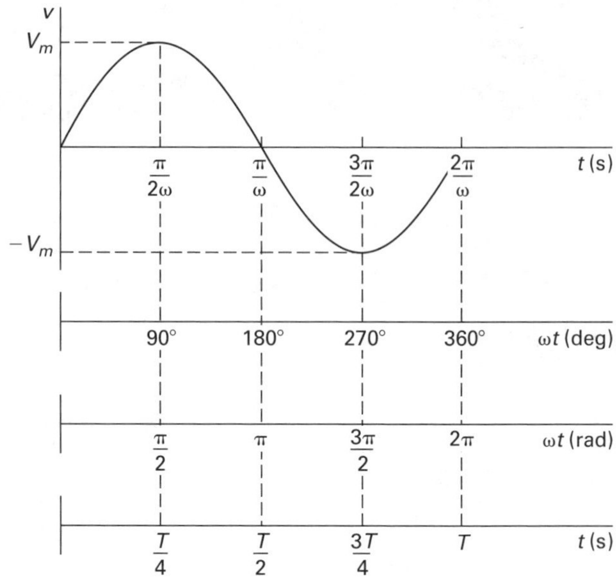

Exponentials, $e^{j\theta}$ and $e^{-j\theta}$, can be combined to produce $\sin\theta$ and $\cos\theta$ terms. As a result, sinusoids can be represented in the following equivalent forms.

- trigonometric: $V_m \sin(\omega t + \theta)$
- exponential: $V_m e^{j\theta}$
- polar or phasor: $V_m \angle\theta$
- rectangular: $V_r + jV_i$

The use of complex exponentials and phasor analysis allows sinusoidal functions to be more easily manipulated mathematically, especially when dealing with derivatives. Exponents are manipulated algebraically, and the resulting sinusoid is then recovered using Euler's relation. This avoids complicated trigonometric mathematics. The angles used in exponentials and with angular frequency, ω, must be in radians. From Euler's relation,

$$e^{j\theta} = \cos\theta + j\sin\theta \qquad 27.4$$

$$j = \sqrt{-1} = 1\angle\frac{\pi}{2} \qquad 27.5$$

Therefore, to regain the sinusoid from the exponential, use Euler's equation, keeping in mind two factors. First, either the real or the imaginary part contains the sinusoid desired, not both.[5] Second, the exponential should not be factored out of any equation until all the derivatives, or integrals, have been taken.

[5]The cosine function could be used as the reference sinusoid, and is in many texts. Over time, many standards have come to use the cosine as the preferred reference. In this text, the sine reference is used when graphing, since this allows the value at $t = 0$ to be zero. Otherwise, the cosine reference will be used. The desired form can be determined from the representation of the voltage (or current) source waveform as either a cosine or a sine function (see Eq. 27.1). Either way, only the phase angle changes.

3. CURRENT

Current is the net transfer of electric charge per unit time. When a circuit's driving force, the voltage, is sinusoidal, the resulting current is also (though it may differ by an amount called the *phase angle difference*). The equations and representations in Sec. 27.2, as well as any other representations of sinusoidal waveforms provided in this chapter, apply to current as well as to voltage.

4. IMPEDANCE

Electrical impedance, also known as *complex impedance*, is the total opposition a circuit presents to alternating current. It is equal to the ratio of the complex voltage to the complex current. Impedance, then, is a ratio of phasor quantities and is not itself a function of time. Relating voltage and current in this manner is analogous to Ohm's law, which for AC analysis is referred to as *extended Ohm's law*. Impedance is given the symbol **Z** and is measured in ohms. The three passive circuit elements (resistors, capacitors, and inductors), when used in an AC circuit, are assigned an angle, ϕ, known as the *impedance angle*. This angle corresponds to the phase difference angle produced when a sinusoidal voltage is applied across that element alone.

Impedance is a complex quantity with a magnitude and associated angle. It can be written in *phasor form*—also known as *polar form*—for example, $\mathbf{Z}\angle\phi$, or in *rectangular form* as the complex sum of its resistive (R) and reactive (X) components.

$$\mathbf{Z} \equiv R + jX \qquad 27.6$$

$$R = Z\cos\phi \quad \left[\begin{array}{c}\text{resistive or}\\ \text{real part}\end{array}\right] \qquad 27.7$$

$$X = Z\sin\phi \quad \left[\begin{array}{c}\text{reactive or}\\ \text{imaginary part}\end{array}\right] \qquad 27.8$$

The resistive and reactive components can be combined to form an *impedance triangle*. Such a triangle is shown in Fig. 27.3.

Figure 27.3 *Impedance Triangle*

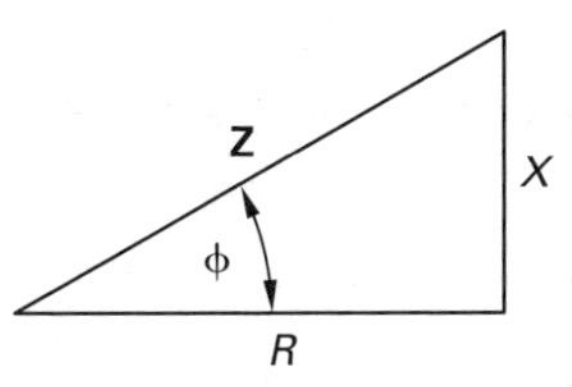

The characteristics of the passive elements in AC circuits, including impedance, are given in Table 27.1.

Table 27.1 *Characteristics of Resistors, Capacitors, and Inductors*

	resistor	capacitor	inductor
value	R (Ω)	C (F)	L (H)
reactance, X	0	$\frac{-1}{\omega C}$	ωL
rectangular impedance, **Z**	$R + j0$	$0 - \frac{j}{\omega C}$	$0 + j\omega L$
phasor impedance, **Z**	$R\angle 0°$	$\frac{1}{\omega C}\angle{-90°}$	$\omega L\angle 90°$
phase	in-phase	leading	lagging
rectangular admittance, **Y**	$\frac{1}{R} + j0$	$0 + j\omega C$	$0 - \frac{j}{\omega L}$
phasor admittance, **Y**	$\frac{1}{R}\angle 0°$	$\omega C\angle 90°$	$\frac{1}{\omega L}\angle{-90°}$

5. ADMITTANCE

The reciprocal of impedance is called *admittance*, **Y**. Admittance is useful in analyzing parallel circuits, as admittances can be added directly. The defining equation is

$$\mathbf{Y} = \frac{1}{\mathbf{Z}} = \frac{1}{Z}\angle{-\phi} \qquad 27.9$$

The reciprocal of the resistive portion of an impedance is known as *conductance*, G. The reciprocal of the reactive part is termed the *susceptance*, B.

$$G = \frac{1}{R} \qquad 27.10$$

$$B = \frac{1}{X} \qquad 27.11$$

Using these definitions and multiplying by a complex conjugate, admittance can be written in terms of resistance and reactance. Using the same method, impedance can be written in terms of conductance and susceptance. Equation 27.12 and Eq. 27.13 show this and can be used for conversion between admittance and impedance and vice versa.

$$\mathbf{Y} = G + jB = \frac{R}{R^2 + X^2} - j\left(\frac{X}{R^2 + X^2}\right) \qquad 27.12$$

$$\mathbf{Z} = R + jX = \frac{G}{G^2 + B^2} - j\left(\frac{B}{G^2 + B^2}\right) \qquad 27.13$$

6. VOLTAGE SOURCES

The energy for AC voltage sources comes primarily from electromagnetic induction. The concepts of ideal and real sources, as well as regulation, apply to AC sources (see Sec. 26.8). Independent sources deliver voltage and current at their rated values regardless of circuit parameters. Dependent sources, often termed *controlled sources*, deliver voltage and current at levels determined by a voltage or current somewhere else in the circuit. These types of sources occur in electronic circuitry and are also used to model electronic elements, such as transistors. The symbols used for AC voltage sources are shown in Fig. 27.4.

Figure 27.4 *AC Voltage Source Symbols*

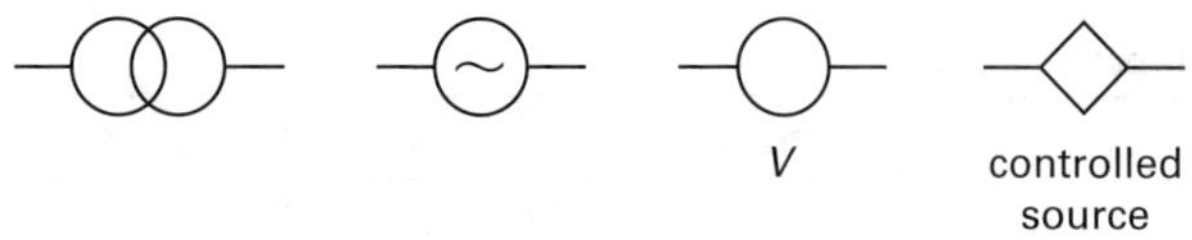

7. AVERAGE VALUE

In purely mathematical terms, the average value of a periodic waveform is the first term of a Fourier series representing the function; that is, it is the zero frequency or DC value. If a function, $f(t)$, repeats itself in a time period T, then the average value of the function is given by Eq. 27.14. In Eq. 27.14, t_1 is any convenient time for evaluating the integral, that is, the time that simplifies the integration. The integral itself is computed over the period.

$$f_{\text{ave}} = \frac{1}{T}\int_{t_1}^{t_1+T} f(t)\,dt \qquad 27.14$$

Integration can be interpreted as the area under the curve of the function. Equation 27.14 divides the net area of the waveform by the period T. This concept is illustrated in Fig. 27.5 and stated in mathematical terms by Eq. 27.15.

$$f_{\text{ave}} = \frac{\text{positive area} - \text{negative area}}{T} \qquad 27.15$$

For the function shown in Fig. 27.5, the area above the axis is called the positive area and the area below is called the negative area. The average value is the net area remaining after the negative area is subtracted from the positive area and the result is divided by the period.

Figure 27.5 *Average Value Areas Defined*

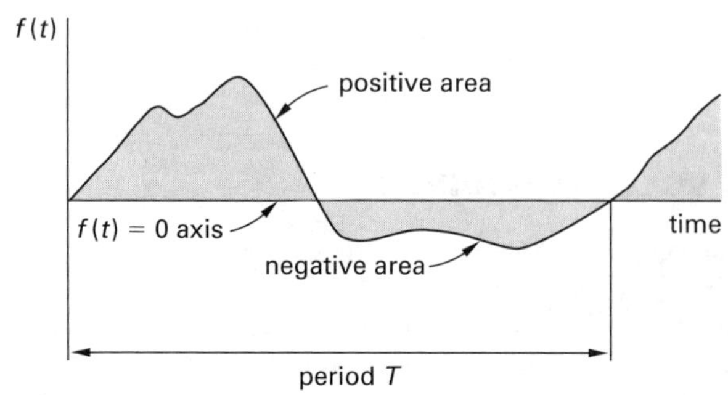

For any periodic voltage, Eq. 27.16 calculates the average value.

$$V_{\text{ave}} = \frac{1}{2\pi}\int_0^{2\pi} v(\theta)\,d\theta = \frac{1}{T}\int_0^{T} v(t)\,dt \qquad 27.16$$

Any waveform that is symmetrical with respect to the horizontal axis will result in a value of zero for Eq. 27.16. While mathematically correct, the electrical effects of such a voltage occur on both the positive and negative half-cycles. Therefore, the average is instead taken over only half a cycle. This is equivalent to determining the average of a *rectified waveform*, that is, the absolute value of the waveform. The average voltage for a rectified sinusoid is

$$\begin{aligned} V_{\text{ave}} &= \frac{1}{\pi}\int_0^{\pi} v(\theta)\,d\theta \\ &= \frac{2V_m}{\pi} \quad \text{[rectified sinusoid]} \end{aligned} \qquad 27.17$$

A DC current equal in value to the average value of a rectified AC current produces the same electrolytic effects, such as capacitor charging, plating operations, and ion formation. Nevertheless, not all AC effects can be accounted for using the average value. For instance, in a typical DC meter, the average DC current determines the response of the needle. An AC current sinusoid of equal average magnitude will not result in the same effect since the torque on each half-cycle is in opposite directions, resulting in a net zero effect. Unless the AC signal is rectified, the reading will be zero.

Example 27.1

What is the average value of the pure sinusoid shown?

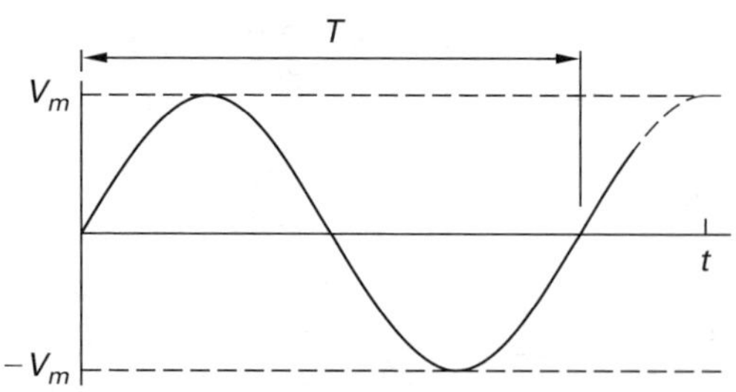

Solution

The sinusoid shown is a sine wave. Substituting into Eq. 27.14 with $t_1 = 0$ gives

$$\begin{aligned} f_{\text{ave}} &= \frac{1}{T}\int_{t_1}^{t_1+T} f(t)\,dt = \frac{1}{T}\int_0^{T} V_m \sin t\,dt \\ &= \left(\frac{V_m}{T}\right)(\cos t)\Big|_0^T = \left(\frac{V_m}{T}\right)(\cos T - \cos 0) \\ &= \left(\frac{V_m}{T}\right)(1-1) = 0 \end{aligned}$$

This result is as expected, because a pure sinusoid has equal positive and negative areas.

Example 27.2

Diodes are used to rectify AC waveforms. Real diodes will not pass current until a threshold voltage is reached. As a result, the output is the difference between the sinusoid and the threshold value. This is illustrated in the following figures.

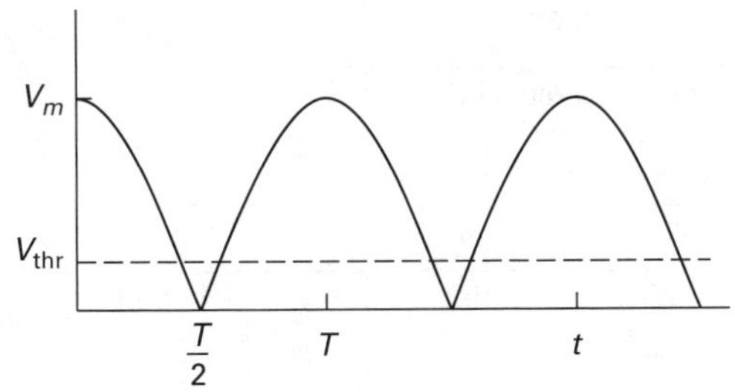

(a) ideal rectified waveform

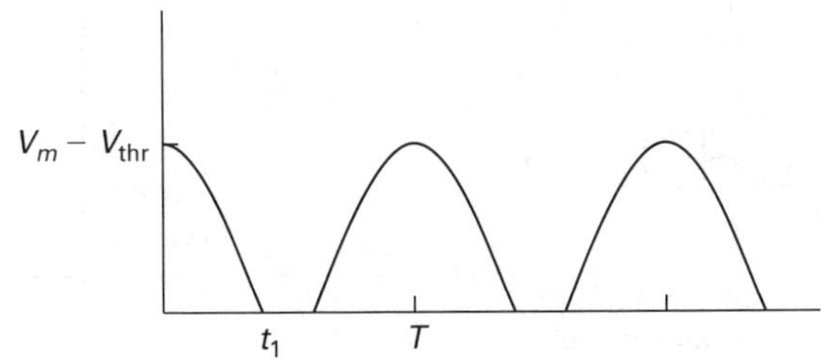

(b) net voltage

What is the expression for the average voltage on the output of a real rectifier diode with a sinusoidal voltage input?

Solution

Because of the symmetry, the average value can be found using half the average determined from Eq. 27.16 by integrating from zero to $T/2$, simplifying the calculation. The waveform is represented by

$$v(t) = \begin{cases} V_m \cos \frac{\pi}{T} t - V_{\text{thr}} & [\text{for } 0 < t < t_1] \\ 0 & [\text{for } t_1 < t < T/2] \end{cases}$$

At $t = t_1$, the following condition exists.

$$V_m \cos \frac{\pi}{T} t_1 = V_{\text{thr}}$$

The equation for one-half the average is

$$\begin{aligned} \tfrac{1}{2} v_{\text{ave}} &= \frac{1}{T} \int_0^{t_1} \left(V_m \cos \frac{\pi}{T} t - V_{\text{thr}} \right) dt \\ &= \left(\frac{1}{T}\right)\left(\frac{T}{\pi}\right) V_m \sin \frac{\pi}{T} t_1 - V_{\text{thr}} \left(\frac{t_1}{T}\right) \end{aligned}$$

Using the equation for the condition at $t = t_1$ and the trigonometric identity $\sin^2\theta + \cos^2\theta = 1$ results in the following equations.

$$\sin \frac{\pi}{T} t_1 = \sqrt{1 - \left(\frac{V_{\text{thr}}}{V_m}\right)^2}$$

$$\frac{t_1}{T} = \frac{1}{\pi} \arccos \frac{V_{\text{thr}}}{V_m}$$

Substituting into the equation for one-half the average and rearranging gives the following final result.

$$\begin{aligned} v_{\text{ave}} = \frac{2}{\pi} V_m \sqrt{1 - \left(\frac{V_{\text{thr}}}{V_m}\right)^2} \\ - \frac{2}{\pi} V_{\text{thr}} \arccos \frac{V_{\text{thr}}}{V_m} \end{aligned}$$

The arccos must be expressed in radians. The conclusion drawn is that real diode average values are more complex than values for ideal diode full-wave rectification. However, if ideal diodes are assumed, that is, if the threshold voltage is considered negligible, the following errors are generated.

- 1% error for $V_{\text{thr}}/V_m = 0.0064$
- 2% error for $V_{\text{thr}}/V_m = 0.0128$
- 5% error for $V_{\text{thr}}/V_m = 0.0321$
- 10% error for $V_{\text{thr}}/V_m = 0.0650$

For most practical applications, the ideal diode assumption results in an error of less than 10%.

8. ROOT-MEAN-SQUARE VALUE

In purely mathematical terms, the *effective*, or *root-mean-square* (rms) *value*, of a periodic waveform represented by a function, $f(t)$, which repeats itself in a time period, is given by Eq. 27.18. In Eq. 27.18, t_1 is any convenient time for evaluating the integral, that is, the time that simplifies the integration. The integral itself is computed over the period.

$$f_{\text{rms}}^2 = \frac{1}{T} \int_{t_1}^{t_1+T} f^2(t)\,dt \qquad 27.18$$

For any periodic voltage, Eq. 27.19 calculates the effective or rms value.

$$\begin{aligned} V = V_{\text{eff}} = V_{\text{rms}} &= \sqrt{\frac{1}{2\pi} \int_0^{2\pi} v^2(\theta)\,d\theta} \\ &= \sqrt{\frac{1}{T} \int_0^T v^2(t)\,dt} \qquad 27.19 \end{aligned}$$

Normally when a voltage or current variable, V or I, is left unsubscripted, it represents the effective (rms) value.[6] Rectification of the waveform is not necessary to calculate the effective value. (The squaring of the waveform ensures the net result of integration is something other than zero for a sinusoid.) For a sinusoidal waveform, $V = V_m/\sqrt{2} \approx 0.707\,V_m$. A DC current of magnitude I produces the same heating effect as an AC current of magnitude I_{eff}.

Table 27.2 gives the characteristics of various commonly encountered alternating waveforms. In this table, two additional terms are introduced. The *form factor* is given by

$$\text{FF} = \frac{V_{\text{eff}}}{V_{\text{ave}}} \qquad 27.20$$

The *crest factor*, CF, also known as the *peak factor* or *amplitude factor*, is

$$\text{CF} = \frac{V_m}{V_{\text{eff}}} \qquad 27.21$$

Nearly all periodic functions can be represented by a Fourier series. A Fourier series can be written as

$$f(t) = \tfrac{1}{2}a_0 + \sum_{n=1}^{\infty} a_n \cos\frac{2\pi n}{T}t + \sum_{n=1}^{\infty} b_n \sin\frac{2\pi n}{T}t \qquad 27.22$$

The average value is the first term, as mentioned in Sec. 27.7. The rms value is

$$f_{\text{rms}} = \sqrt{\left(\tfrac{1}{2}a_0\right)^2 + \tfrac{1}{2}\sum_{n=1}^{\infty}(a_n^2 + b_n^2)} \qquad 27.23$$

Table 27.2 *Characteristics of Alternating Waveforms*

waveform	$\frac{V_{\text{ave}}}{V_m}$	$\frac{V_{\text{rms}}}{V_m}$	FF	CF
sinusoid 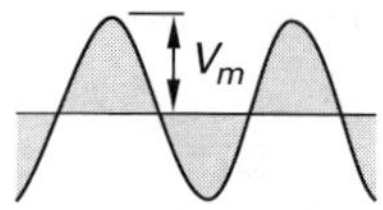	0	$\frac{1}{\sqrt{2}}$	–	$\sqrt{2}$
full-wave rectified sinusoid 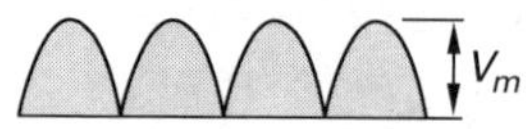	$\frac{2}{\pi}$	$\frac{1}{\sqrt{2}}$	$\frac{\pi}{2\sqrt{2}}$	$\sqrt{2}$
half-wave rectified sinusoid 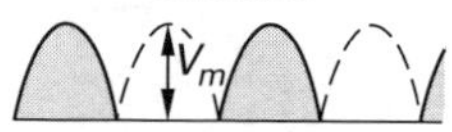	$\frac{1}{\pi}$	$\frac{1}{2}$	$\frac{\pi}{2}$	2
symmetrical square wave 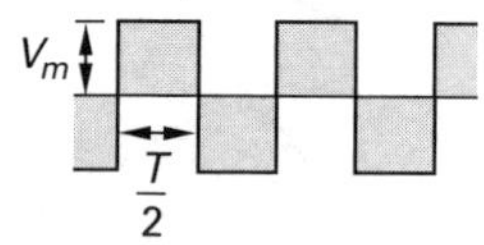	0	1	–	1
unsymmetrical square wave 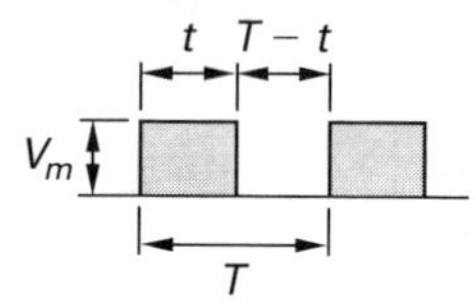	$\frac{t}{T}$	$\sqrt{\frac{t}{T}}$	$\sqrt{\frac{T}{t}}$	$\sqrt{\frac{T}{t}}$
sawtooth and symmetrical triangular 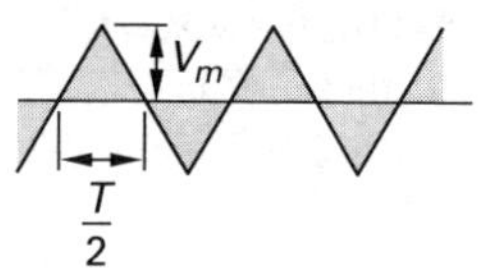	0	$\frac{1}{\sqrt{3}}$	–	$\sqrt{3}$
sawtooth and symmetrical triangular 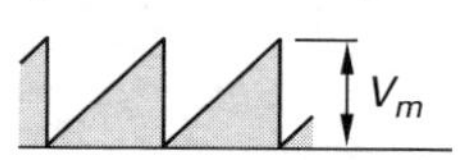	$\frac{1}{2}$	$\frac{1}{\sqrt{3}}$	$\frac{2}{\sqrt{3}}$	$\sqrt{3}$

Example 27.3

A peak sinusoidal voltage, V_p, of 170 V is connected across a 240 Ω resistor in a lightbulb. What is the power dissipated by the bulb?

Solution

From Table 27.2, the effective voltage is

$$V = \frac{V_p}{\sqrt{2}} = \frac{170\text{ V}}{\sqrt{2}} = 120.21\text{ V}$$

The power dissipated is

$$P = \frac{V^2}{R} = \frac{(120.21\text{ V})^2}{240\ \Omega} = 60.21\text{ W}$$

Example 27.4

What is the rms value of a constant 3 V signal?

Solution

Using Eq. 27.18, the rms value is

$$v_{\text{rms}}^2 = \frac{1}{T}\int_0^T (3\text{ V})^2 dt = \left(\frac{1}{T}\right)(9\text{ V})^2(T-0)$$

$$v_{\text{rms}} = \sqrt{(9\text{ V})^2} = 3\text{ V}$$

[6]The value of the standard voltage in the United States, reported as 115–120 V, is an effective value.

9. PHASE ANGLES

AC circuit inductors and capacitors have the ability to store energy in magnetic and electric fields, respectively. Consequently, while the same shape, the voltage and current waveforms differ by an amount called the *phase angle difference*, ϕ. Ordinarily, the voltage and current sinusoids do not peak at the same time. In a *leading circuit*, the phase angle difference is positive and the current peaks before the voltage (see Fig. 27.6). A leading circuit is termed a *capacitive circuit*. In a *lagging circuit*, the phase angle difference is negative and the current peaks after the voltage. A lagging circuit is termed an *inductive circuit*. These cases are represented mathematically as

$$v(t) = V_m \sin(\omega t + \theta) \quad \text{[reference]} \tag{27.24}$$

$$i(t) = I_m \sin(\omega t + \theta + \phi) \quad \text{[leading]} \tag{27.25}$$

$$i(t) = I_m \sin(\omega t + \theta - \phi) \quad \text{[lagging]} \tag{27.26}$$

Figure 27.6 *Leading Phase Angle Difference*

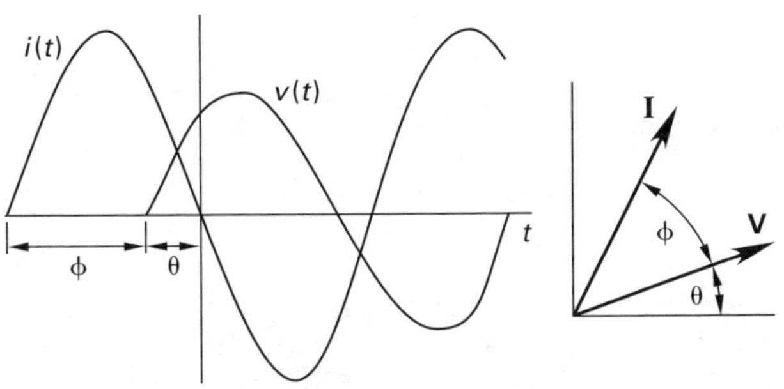

(a) time response

(b) phasor response

10. SINUSOID

The most important waveform in electrical engineering is the sine function, or a waveform that is sinusoidal in value with respect to time. A sinusoidal waveform has an amplitude (also known as a *magnitude* or *peak value*) that remains constant, such as V_m. The waveform, however, repeats or goes through cycles. From Eq. 27.1, the sinusoid can be characterized by three quantities: magnitude, frequency, and phase angle.[7] The major properties of the sinusoid are covered in Sec. 27.2.

A circuit processes or changes waveforms. This is called *signal processing* or *waveform processing*. *Signal analysis* is the determination of these waveforms. The sinusoid can be represented as a phasor and used to analyze electrical circuits only if the circuit is in a steady-state condition. The terms AC and DC imply steady-state conditions. For AC circuits, *sinusoidal steady-state* indicates that all voltages and currents within the circuit are sinusoids of the same frequency as the excitation, that is, the driving voltage.

Example 27.5

A reference voltage of $v(t) = 170 \sin \omega t$ is used during an experiment. A measurement of a second voltage, $v_2(t)$ occurs. The second voltage reaches its peak 2.5 ms before the reference voltage and has the same peak value 20 ms later. The peak value is 1.8 times the reference peak. What is the expression for $v_2(t)$?

Solution

Any sinusoidal voltage can be represented as in Eq. 27.1.

$$v(t) = V_m \sin(\omega t + \theta)$$

The peak value (equivalent to V_m) is 1.8 times the reference, or $1.8 \times 170\text{ V} = 306\text{ V}$. Calculate the angular frequency using Eq. 27.3 and the given period of 20 ms.

$$\omega = \frac{2\pi}{T} = \frac{2\pi}{20\text{ ms}} = 314\text{ rad/s}$$

The phase angle is determined by changing the time, 2.5 ms, into its corresponding angle, θ, as follows.

$$\theta = \omega t = \left(314\ \frac{\text{rad}}{\text{s}}\right)(2.5 \times 10^{-3}\text{ s}) = 0.79\text{ rad}$$

Converting 0.79 rad gives a result of 45°. Substituting the calculated values gives the result.

$$v_2(t) = V_m \sin(\omega t + \theta) = 306\ \sin(314t + 45^\circ)$$

The result mixes the use of radians (314) and degrees (45). This is often done for clarity. Ensure one or the other is converted prior to calculations. It is often best to deal with all angles in radians.

11. PHASORS

The addition of voltages and currents of the same frequency is simplified mathematically by treating them as phasors.[8] Using a method called the *Steinmetz algorithm*, the following two quantities, introduced in Sec. 27.2, are considered analogs.

$$v(t) = V_m \sin(\omega t + \theta) \text{ and } V = V_m \angle\theta \tag{27.27}$$

Both forms show the magnitude and the phase, but the *phasor*, $V_m \angle\theta$, does not show frequency. In the phasor form, the frequency is implied. The phasor form of Eq. 27.27 is shown in Fig. 27.7.

The phasor is actually a point, but is represented by an arrow of magnitude V_m at an angle θ with respect to a reference, normally taken to be $\theta = 0$. A *reference phasor* is one of known value, usually the driving voltage

[7]A phase angle of ±90° changes the sine function into a cosine function. Also, frequency refers in this case to the angular frequency, ω, in radians per second. The term frequency also refers to the term *f*, measured in hertz (Hz).

[8]For all circuit elements at the same frequency, the circuit must be in a steady-state condition.

Figure 27.7 *Phasor of Magnitude V_m and Angle θ*

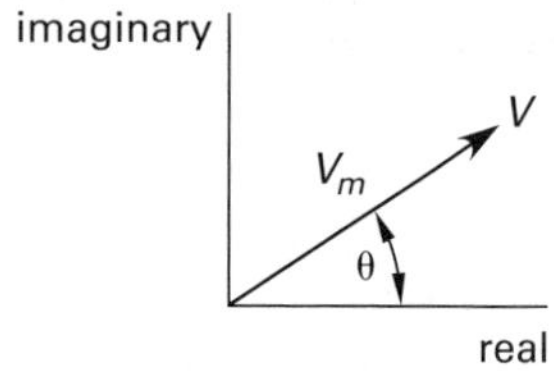

of a circuit (i.e., the voltage phasor). The reference phasor would be shown in the position of the real axis with θ equal to zero. Phasors are summed using phasor addition, commonly referred to as *vector addition.*[9] The Steinmetz algorithm is illustrated in Fig. 27.8.

Figure 27.8 *Steinmetz Algorithm Steps*

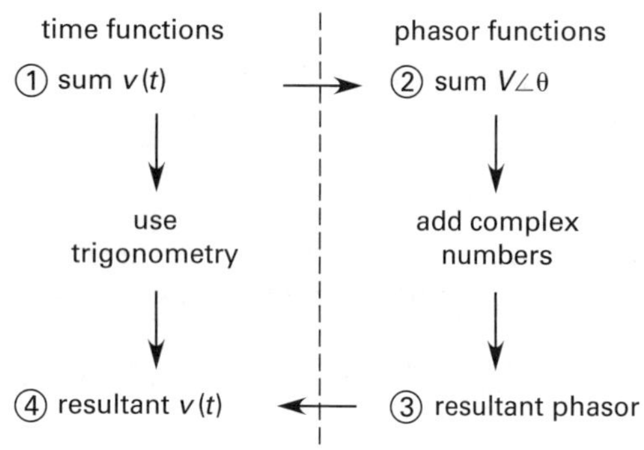

In electric circuits with sinusoidal waveforms, an alternate phasor representation is often used. The alternate representation is called the *effective value phasor notation.* In this notation, the rms values of the voltage and current are used instead of the peak values. The angles are also given in degrees. The phasor notation of Eq. 27.27 is then modified to

$$V = V_{\text{rms}}\angle\theta = \frac{V_m}{\sqrt{2}}\angle\theta = V\angle\theta \qquad 27.28$$

The effective value phasor notation represented by the first and second parts of Eq. 27.28 is used by NCEES. When the notation is not specified (for example, when a subscript is not used as in the third part of Eq. 27.28), assume the effective notation. Additionally, many texts do not show phasors in bold font, but instead let the angle symbol indicate that a phasor is being used. Care should be taken to determine the type of notation in use.

Ensure that angles substituted into exponential forms are in radians or errors in mathematical calculations will result.

[9]The term "vector addition" is not technically correct because phasors, with the exception of impedance and admittance phasors, rotate with time. The methods, however, are identical.

12. COMPLEX REPRESENTATION

Phasors are plotted in the complex plane. Figure 27.9 shows the relationships among the complex quantities introduced.

In Fig. 27.9, **Z** represents a complex number, not the impedance. Equation 27.1 is repeated here for convenience.

$$v(t) = V_m \sin(\omega t + \theta) \qquad 27.29$$

Figure 27.9 *Complex Quantities*

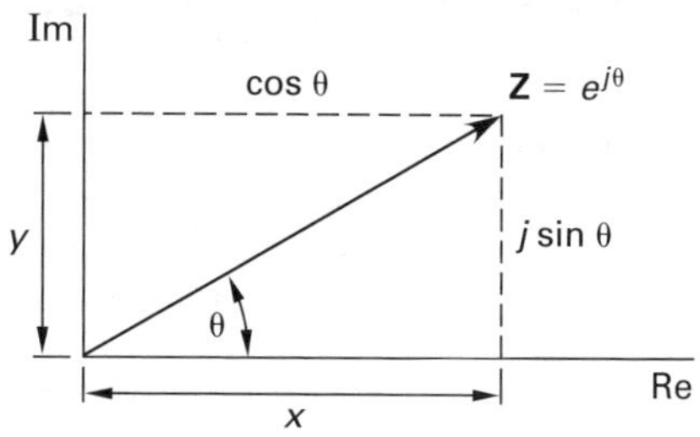

Using Euler's relation, Eq. 27.4, the voltage $v(t)$ can be represented as

$$v(t) = V_m e^{j(\omega t + \theta)} \qquad 27.30$$

Consequently, $v(t)$ is the imaginary part of Eq. 27.30. If the cosine function is used, $v(t)$ is the real part of Eq. 27.30. The fixed portion of Eq. 27.30 ($e^{j\theta}$) can be separated from the time-variable portion ($e^{j\omega t}$) of the function. This is shown in Fig. 27.10(a) and (b). The magnitude of $e^{j\omega t}$ remains equal to one, but the angle increases (rotates counterclockwise) linearly with time. All functions with an $e^{j\omega t}$ term are assumed to rotate counterclockwise with an angular velocity of ω in the complex plane. The voltage can be represented as in Fig. 27.10(c) and

$$v(t) = V_m e^{j\omega t} e^{j\theta} \qquad 27.31$$

If the rotating portion of Eq. 27.31 is assumed to exist, the voltage can be written

$$v(\theta) = V_m e^{j\omega t} e^{j\theta} \qquad 27.32$$

Changing Eq. 27.32 to the phasor form yields

$$\begin{aligned} \mathbf{V} &= V_m\angle\theta \quad \text{or} \\ \mathbf{V} &= V_{\text{rms}}\angle\theta = \frac{V_m}{\sqrt{2}}\angle\theta = V\angle\theta \end{aligned} \qquad 27.33$$

Figure 27.10 *Phasor Rotation in the Complex Plane*

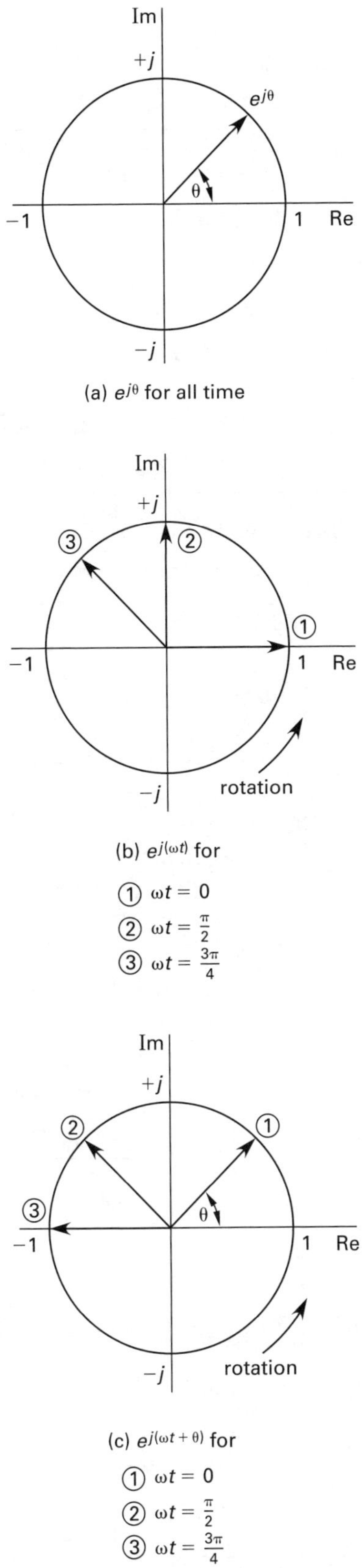

The properties of complex numbers, which can represent voltage, current, or impedance, are summarized in Table 27.3. The designations used in the table are illustrated in Fig. 27.9.

13. RESISTORS

Resistors oppose the movement of electrons. In an *ideal* or *pure resistor*, no inductance or capacitance exists. The magnitude of the impedance is the resistance, R, with units of ohms and an impedance angle of zero. Therefore, voltage and current are in phase in a purely resistive circuit.

$$\mathbf{Z}_R = R\angle 0 = R + j0 \qquad 27.34$$

14. CAPACITORS

Capacitors oppose the movement of electrons by storing energy in an electric field and using this energy to resist changes in voltage over time. Unlike a resistor, an *ideal* or *perfect capacitor* consumes no energy. Equation 27.35 gives the impedance of an ideal capacitor with capacitance C. The magnitude of the impedance is termed the *capacitive reactance*, X_C, with units of ohms and an impedance angle of $-\pi/2(-90°)$. Consequently, the current leads the voltage by 90° in a purely capacitive circuit.[10]

$$\mathbf{Z}_C = X_C\angle{-90°} = 0 + jX_C \qquad 27.35$$

$$X_C = \frac{-1}{\omega C} = \frac{-1}{2\pi f C} \qquad 27.36$$

15. INDUCTORS

Inductors oppose the movement of electrons by storing energy in a magnetic field and using this energy to resist changes in current over time. Unlike a resistor, an *ideal* or *perfect inductor* consumes no energy. Equation 27.37 gives the impedance of an ideal inductor with inductance L. The magnitude of the impedance is termed the *inductive reactance*, X_L, with units of ohms and an impedance angle of $\pi/2$ (90°). Consequently, the current lags the voltage by 90° in a purely inductive circuit.[11]

$$\mathbf{Z}_L = X_L\angle 90° = 0 + jX_L \qquad 27.37$$

$$X_L = \omega L = 2\pi f L \qquad 27.38$$

[10]The impedance angle for a capacitor is negative, but the current phase angle difference is positive—hence the term "leading." This occurs mathematically because the current is obtained by dividing the voltage by the impedance: $\mathbf{I} = \mathbf{V}/\mathbf{Z}$.

[11]The impedance angle for an inductor is positive, but the current phase angle difference is negative—hence the term "lagging." This occurs mathematically because the current is obtained by dividing the voltage by the impedance: $\mathbf{I} = \mathbf{V}/\mathbf{Z}$.

Table 27.3 Properties of Complex Numbers

	rectangular form	polar/exponential form
	$\mathbf{Z} = x + jy$	$\mathbf{Z} = \lvert\mathbf{Z}\rvert\angle\theta$ $\mathbf{Z} = \lvert\mathbf{Z}\rvert e^{j\theta} = \lvert\mathbf{Z}\rvert\cos\theta + j\lvert\mathbf{Z}\rvert\sin\theta$
relationship between forms	$x = \lvert\mathbf{Z}\rvert\cos\theta$ $y = \lvert\mathbf{Z}\rvert\sin\theta$	$\lvert\mathbf{Z}\rvert = \sqrt{x^2+y^2}$ $\theta = \tan^{-1}\frac{y}{x}$
complex conjugate	$\mathbf{Z}^* = x - jy$ $\mathbf{Z}\mathbf{Z}^* = (x^2+y^2) = \lvert z\rvert^2$	$\mathbf{Z}^* = \lvert\mathbf{Z}\rvert e^{-j\theta} = \lvert\mathbf{Z}\rvert\angle-\theta$ $\mathbf{Z}\mathbf{Z}^* = (\lvert\mathbf{Z}\rvert e^{j\theta})(\lvert\mathbf{Z}\rvert e^{-j\theta}) = \lvert\mathbf{Z}\rvert^2$
addition	$\mathbf{Z}_1 + \mathbf{Z}_2 = (x_1+x_2) + j(y_1+y_2)$	$\mathbf{Z}_1 + \mathbf{Z}_2 = (\lvert\mathbf{Z}_1\rvert\cos\theta_1 + \lvert\mathbf{Z}_2\rvert\cos\theta_2)$ $+ j(\lvert\mathbf{Z}_1\rvert\sin\theta_1 + \lvert\mathbf{Z}_2\rvert\sin\theta_2)$
multiplication	$\mathbf{Z}_1\mathbf{Z}_2 = (x_1x_2 - y_1y_2) + j(x_1y_2 + x_2y_1)$	$\mathbf{Z}_1\mathbf{Z}_2 = \lvert\mathbf{Z}_1\rvert\lvert\mathbf{Z}_2\rvert\angle\theta_1+\theta_2$
division	$\frac{\mathbf{Z}_1}{\mathbf{Z}_2} = \frac{(x_1x_2+y_1y_2)+j(x_2y_1-x_1y_2)}{\lvert\mathbf{Z}_2\rvert^2}$	$\frac{z_1}{z_2} = \frac{\lvert\mathbf{Z}_1\rvert}{\lvert\mathbf{Z}_2\rvert}\angle\theta_1-\theta_2$

16. COMBINING IMPEDANCES

Impedances in combination are like resistors: Impedances in series are added, while the reciprocals of impedances in parallel are added. For series circuits, the resistive and reactive parts of each impedance element are calculated separately and summed. For parallel circuits, the conductance and susceptance of each element are summed. The total impedance is found by a complex addition of the resistive (conductive) and reactive (susceptive) parts. It is convenient to perform the addition in rectangular form.

$$Z_e = \sum Z = \sqrt{\left(\sum R\right)^2 + \left(\sum X_L - \sum X_C\right)^2} \quad \text{[series]} \qquad 27.39$$

$$\frac{1}{Z_e} = \sum\frac{1}{Z} = Y_e = \sqrt{\left(\sum G\right)^2 + \left(\sum B_L - \sum B_C\right)^2} \quad \text{[parallel]} \qquad 27.40$$

The impedance of various series-connected circuit elements is given in App. 27.A. The impedance of various parallel-connected circuit elements is given in App. 27.B.

Example 27.6

Determine the impedance and admittance of the following circuits.

(a)

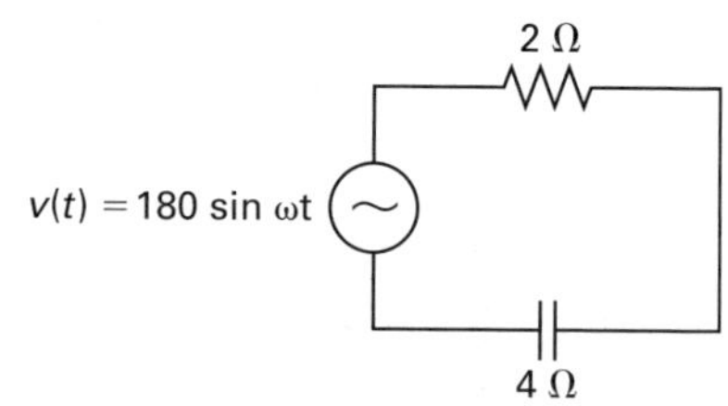

(b)

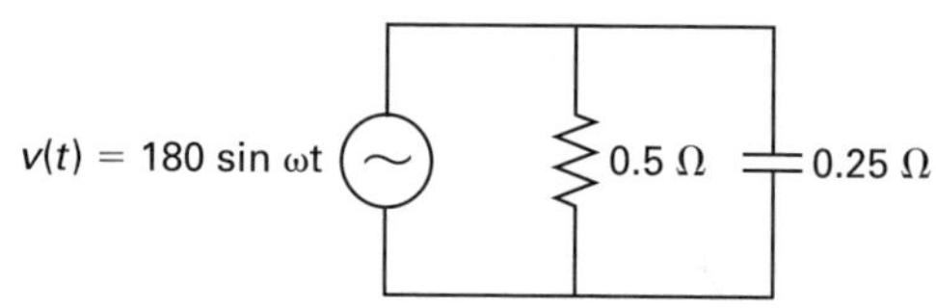

Solution

(a) From Eq. 27.39,

$$Z = \sqrt{R^2 + Z_C^2} = \sqrt{(2\ \Omega)^2 + (4\ \Omega)^2} = 4.47\ \Omega$$

$$\phi = \arctan\frac{X_C}{R} = \arctan\frac{-4\ \Omega}{2\ \Omega} = -63.4°$$

$$\mathbf{Z} = 4.47\ \Omega\angle-63.4°$$

From Eq. 27.9,

$$\mathbf{Y} = \frac{1}{\mathbf{Z}} = \frac{1}{4.47\ \Omega\angle-63.4°} = 0.224\ \text{S}\angle63.4°$$

(b) Because this is a parallel circuit, work with the admittances.

$$G = \frac{1}{R} = \frac{1}{0.5\ \Omega} = 2\ \text{S}$$
$$B_C = \frac{1}{X_C} = \frac{1}{0.25\ \Omega} = 4\ \text{S}$$
$$Y = \sqrt{G^2 + B_C^2} = \sqrt{(2\ \text{S})^2 + (4\ \text{S})^2} = 4.47\ \text{S}$$
$$\phi = \arctan\frac{B_C}{G} = \arctan\frac{4\ \text{S}}{2\ \text{S}} = 63.4^\circ$$
$$\mathbf{Y} = 4.47\ \text{S}\angle 63.4^\circ$$
$$\mathbf{Z} = \frac{1}{\mathbf{Y}} = \frac{1}{4.47\ \text{S}\angle 63.4^\circ} = 0.224\ \Omega\angle -63.4^\circ$$

17. OHM'S LAW

Ohm's law for AC circuits with linear circuit elements is similar to Ohm's law for DC circuits.[12] The difference is that all the terms are represented as phasors.[13]

$$\mathbf{V} = \mathbf{IZ} \qquad 27.41$$

$$V\angle\theta v = (I\angle\varphi_I)(Z\angle\phi_Z) \qquad 27.42$$

18. POWER

The instantaneous power, p, in a purely resistive circuit is

$$\begin{aligned} p_R(t) &= i(t)v(t) = (I_m \sin\omega t)(V_m \sin\omega t) \\ &= I_m V_m \sin^2\omega t \\ &= \tfrac{1}{2}I_m V_m - \tfrac{1}{2}I_m V_m \cos 2\omega t \end{aligned} \qquad 27.43$$

The second term of Eq. 27.43 integrates to zero over the period. Therefore, the *average power* dissipated is

$$\begin{aligned} P_R &= \tfrac{1}{2}I_m V_m = \frac{V_m^2}{2R} \\ &= I_{\text{rms}} V_{\text{rms}} = IV \end{aligned} \qquad 27.44$$

Equation 27.28 was used to change the maximum, or peak, values into the more usable rms, or effective, values.

In a purely capacitive circuit, the current leads the voltage by 90°. This allows the current term of Eq. 27.43 to be replaced by a cosine (because the sine and cosine differ by a 90° phase). The instantaneous power in a purely capacitive circuit is

$$\begin{aligned} p_C(t) &= i(t)v(t) \\ &= (I_m \cos\omega t)(V_m \sin\omega t) \\ &= I_m V_m \sin\omega t \cos\omega t \\ &= \tfrac{1}{2}I_m V_m \sin 2\omega t \end{aligned} \qquad 27.45$$

Because the $\sin 2\omega t$ term integrates to zero over the period, the average power is zero. Nevertheless, power is stored during the charging process and returned to the circuit during the discharging process.

$$P_C = 0 \qquad 27.46$$

Similarly, the instantaneous power in an inductor is

$$p_L(t) = -\tfrac{1}{2}I_m V_m \sin 2\omega t \qquad 27.47$$

Again, the $\sin 2\omega t$ integrates to zero over the period, and the average power is zero. Nevertheless, power is stored during the expansion of the magnetic field and returned to the circuit during the contraction of the magnetic field.

$$P_L = 0 \qquad 27.48$$

19. REAL POWER AND THE POWER FACTOR

In a circuit that contains all three circuit elements (resistors, capacitors, and inductors) or the effects of all three, the average power is calculated from Eq. 27.14 as

$$P_{\text{ave}} = \frac{1}{T}\int_0^T i(t)v(t)\,dt \qquad 27.49$$

Let the generic voltage and current waveforms be represented by Eq. 27.50 and Eq. 27.51. The *current angle* φ is equal to $\theta \pm \phi$.[14]

$$i(t) = I_m \sin(\omega t + \varphi) \qquad 27.50$$
$$v(t) = V_m \sin(\omega t + \theta) \qquad 27.51$$

Substituting Eq. 27.50 and Eq. 27.51 into Eq. 27.49 gives

$$P_{\text{ave}} = \left(\frac{I_m V_m}{2}\right)\cos\phi_{\text{pf}} = I_{\text{rms}} V_{\text{rms}} \cos\phi_{\text{pf}} \qquad 27.52$$

The angle $\phi_{\text{pf}} = |\theta - \varphi|$ is the *power factor angle.* It represents the difference between the voltage and current angles. This difference is the impedance angle, $\pm\phi$. Because the $\cos(-\phi) = \cos(+\phi)$, only the absolute value of the impedance angle is used. Because the absolute value is used in the equation, the terms *leading* (for a capacitive circuit) and *lagging* (for an inductive circuit)

[12] Ohm's law can be used on *nonlinear devices* (NLD) if the region analyzed is restricted to be approximately linear. When this condition is applied, the analysis is termed *small signal analysis.*

[13] In general, phasors are considered to rotate with time. A secondary definition of a phasor is any quantity that is a complex number. As a result, the impedance is also considered a phasor even though it does not change with time.

[14] Using a current angle simplifies the derivational mathematics (not shown) and clarifies the definition of the power factor angle given in this section.

must be used when describing the power factor. The power factor of a purely resistive circuit equals one; the power factor of a purely reactive circuit equals zero.

Equation 27.52 determines the magnitude of the product of the current and voltage in phase with one another, that is, of a resistive nature. Consequently, Eq. 27.52 determines the power consumed by the resistive elements of a circuit. This average power is called the *real power* or *true power*, and sometimes the *active power*. The term $\cos\phi_{pf}$, or $\cos|\phi|$, is called the *power factor* or *phase factor* and given the symbol pf. The power factor is also equal to the ratio of the real power to the apparent power (see Sec. 27.22). Often, no subscript is used on P when it represents the real power, nor are subscripts used on rms or effective values. The real power is then given by

$$P = IV\cos\phi_{pf} = IV\text{pf} \tag{27.53}$$

It is important to realize that the real power cannot be obtained by multiplying phasors. Doing so results in the addition of the voltage and current angles when the difference is required.

$$P \neq I\angle\varphi V\angle\theta = VI\angle\theta + \varphi \tag{27.54}$$

If phasor power is used, the difficulty can be alleviated by using Eq. 27.55.

$$P = \text{Re}(\mathbf{VI}^*) \tag{27.55}$$

Equation 27.55 is equivalent to Eq. 27.53.

20. REACTIVE POWER

The *reactive power*, Q, measured in units of VAR, is given by

$$Q = IV\sin\phi_{pf} \tag{27.56}$$

The reactive power, sometimes called the *wattless power*, is the product of the rms values of the current and voltage multiplied by the *quadrature* of the current. Sin ϕ_{pf} is called the *reactive factor*.[15] The reactive power represents the energy stored in the inductive and capacitive elements of a circuit.

21. APPARENT POWER

The apparent power, S, measured in voltamperes, is given by

$$S = IV \tag{27.57}$$

The apparent power is the product of the rms values of the voltage and current without regard to the angular relationship between them. The apparent power is representative of the combination of real and reactive power (see Sec. 27.22). As such, not all of the apparent power is dissipated or consumed. Nevertheless, electrical engineers must design systems to adequately handle this power as it exists in the system.

22. COMPLEX POWER AND THE POWER TRIANGLE

The real, reactive, and apparent powers can be related to one another as vectors. The *complex power vector*, **S**, is the vector sum of the *reactive power vector*, **Q**, and the *real power vector*, **P**. The magnitude of **S** is given by Eq. 27.57. The magnitude of **Q** is given by Eq. 27.56. The magnitude of **P** is given by Eq. 27.53. In general, $\mathbf{S} = \mathbf{I}^*\mathbf{V}$ where $\mathbf{I}^*$ is the complex conjugate of the current, that is, the current with the phase difference angle reversed. This convention is arbitrary, but emphasizes that the *power angle*, ϕ, associated with **S** is the same as the overall impedance angle, ϕ, whose magnitude equals the power factor angle, ϕ_{pf}.

The relationship between the magnitudes of these powers is

$$S^2 = P^2 + Q^2 \tag{27.58}$$

A drawing of the power vectors in a complex plane is shown in Fig. 27.11 for leading and lagging conditions. The following relationships are determined from the drawing.

$$P = S\cos\phi \tag{27.59}$$

$$Q = S\sin\phi \tag{27.60}$$

Figure 27.11 *Power Triangle*

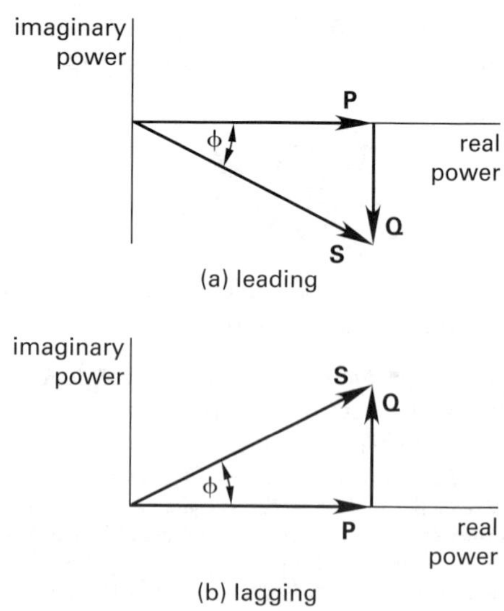

[15]The power factor angle, ϕ_{pf}, is used instead of the absolute value of the impedance angle (as in the power factor) because the sine function results in positive and negative values depending upon the difference between the voltage and current angles. If ϕ_{pf} were not used, $\pm\phi$ would have to be used.

Circuit Theory

Example 27.7

For the circuit shown, find the (a) apparent power, (b) real power, and (c) reactive power; and (d) draw the power triangle.

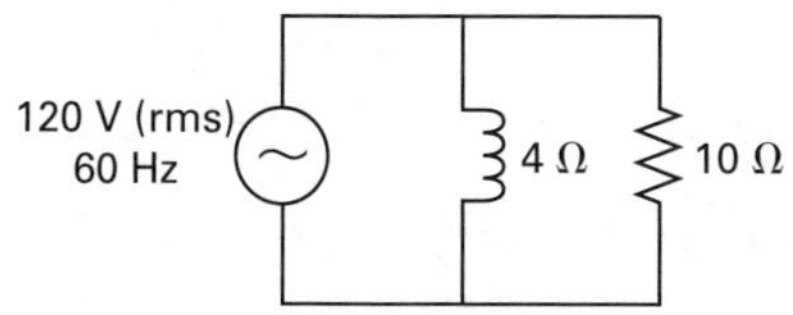

Solution

The equivalent impedance is

$$\frac{1}{Z} = \frac{1}{j4\ \Omega} + \frac{1}{10\ \Omega} = -j0.25 + 0.10\ \text{S}$$
$$Z = 3.714\ \Omega$$
$$\phi_Z = \arctan\frac{0.25\ \text{S}}{0.10\ \text{S}} = 68.2°$$

The total current is

$$I = \frac{V}{Z} = \frac{120\ \text{V}\angle 0°}{3.714\ \Omega\angle 68.2°} = 32.31\ \text{A}\angle -68.2°$$

(a) The apparent power is

$$S = I^*V = (32.31\ \text{A}\angle 68.2°)(120\ \text{V}\angle 0°)$$
$$= 3877\ \text{VA}\angle 68.2°$$

(The angle of apparent power is usually not reported.)

(b) The real power is

$$P = \frac{V_R^2}{R} = \frac{(120\ \text{V}\angle 0°)^2}{10\ \Omega} = 1440\ \text{W}$$

Alternatively, the real power can be calculated from Eq. 27.59.

$$P = S\cos\phi = (3877\ \text{VA})\cos 68.2° = 1440\ \text{W}$$

(c) The reactive power is

$$Q = \frac{V_L^2}{X_L} = \frac{(120\ \text{V})^2}{4\ \Omega} = 3600\ \text{VAR}$$

Alternatively, the reactive power can be calculated from Eq. 27.60.

$$Q = S\sin\phi = (3877\ \text{VA})\sin 68.2° = 3600\ \text{VAR}$$

(d) The real power is represented by the vector (in rectangular form) 1440 + j0. The reactive power is represented by the vector 0 + j3600. The apparent power is represented (in phasor form) by the vector 3877∠68.2°. The power triangle is

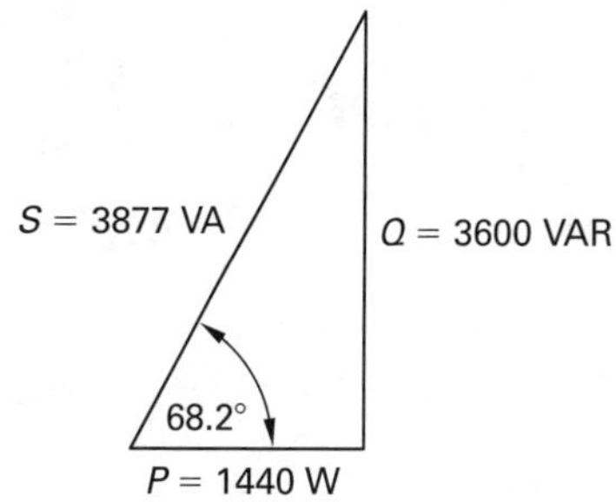

23. MAXIMUM POWER TRANSFER

Assuming a fixed primary impedance, the maximum power condition in an AC circuit is similar to that in a DC circuit. That is, the maximum power is transferred from the source to the load when the impedances match. This occurs when the circuit is in resonance. Resonance is discussed more fully in Chap. 30. The conditions for maximum power transfer, and therefore resonance, are given by

$$R_{\text{load}} = R_s \qquad 27.61$$

$$X_{\text{load}} = -X_s \qquad 27.62$$

24. AC CIRCUIT ANALYSIS

All of the methods and equations presented in Chap. 26, such as Ohm's and Kirchhoff's laws and loop-current and node-voltage methods can be used to analyze AC circuits as long as complex arithmetic is utilized.

28 Transformers

1. Fundamentals 28-1
2. Magnetic Coupling 28-1
3. Ideal Transformers 28-3
4. Impedance Matching 28-4
5. Real Transformers 28-5
6. Magnetic Hysteresis: BH Curves 28-6
7. Eddy Currents 28-6
8. Core Losses 28-6

Nomenclature

a	turns ratio	–
B	magnetic flux density	T
C	capacitance	F
f	frequency	Hz
H	magnetic field strength	A/m
i	variable current	A
I	constant or rms current	A
k	coupling coefficient	–
L	inductance or self-inductance	H
M	mutual inductance	H
N	number of items	–
O	origin	–
R	resistance	Ω
t	time	s
v	variable voltage	v
V	constant or rms voltage	V
V_s	effective voltage	V
X	reactance	Ω
Z	impedance	Ω

Symbols

μ	permeability	H/m
Φ	magnetic flux	Wb
ω	angular frequency	rad/s

Subscripts

0	initial
c	coercive or core
C	capacitative or capacitor
ep	effective primary
E	exciting
L	inductor
m	magnetizing, maximum, or mutual
p	primary
r	residual
ref	reflected
s	secondary

1. FUNDAMENTALS

A *transformer* is an electrical device consisting of two or more multiturn coils of wire in close proximity to link the magnetic field of one to the other. This linkage allows the transfer of electrical energy from one or more alternating circuits to one or more other alternating circuits by the process of magnetic induction. The transformer can be used to raise or lower the value of a capacitor, inductor, or resistor. It enables the efficient transmission of electrical energy at high voltage over great distances, and then is used to lower the voltage to safe values for industrial, commercial, and household use. The principle of the transformer is also the basic principle of induction motors, alternators, and synchronous motors. The transformer and these other devices are based on laws of magnetic induction.

2. MAGNETIC COUPLING

Any coil of wire, energized by a sinusoidal waveform, produces a magnetic flux, Φ, that is also sinusoidal.[1]

$$\Phi(t) = \Phi_m \sin \omega t \qquad 28.1$$

Faraday's law gives the voltage induced when a portion of the magnetic flux produced in one coil of wire with N_p turns passes through a second coil of wire with N_s turns. The minus sign is a result of Lenz's law, which states that the induced voltage, and therefore the induced flux, always acts to oppose the voltage (flux) that created it.

$$v_s(t) = -N_s\left(\frac{d\Phi(t)}{dt}\right) = -N_s\omega\Phi_m \cos \omega t \qquad 28.2$$

The effective voltage is obtained by dividing Eq. 28.2 by the square root of two. The net result is[2]

$$\begin{aligned} V_s(t) &= \frac{-N_s(2\pi f)\Phi_m \cos \omega t}{\sqrt{2}} \\ &= -4.44 N_s f \Phi_m \cos \omega t \end{aligned} \qquad 28.3$$

The magnetic flux that passes through both coils is called the *mutual flux*. The magnetic flux that couples only one coil is called the *leakage flux*. The term "leakage" is used because any flux that does not link the coils

[1]This ignores harmonic effects in the core of the transformer stemming from variable permeability. These harmonics must be accounted for if inductive interference with communications circuits is a concern.

[2]The capital V_s is used in Eq. 28.3 to indicate the effective voltage value, even though the voltage remains a function of time.

will not contribute to inducing a voltage in the secondary of the transformer. The fluxes are illustrated for a *core transformer* in Fig. 28.1. The magnetic flux Φ_{12} is the flux generated by coil number one that links coil two. The magnetic flux Φ_{11} is the flux generated by coil one that links only coil one. Similarly, the magnetic flux Φ_{21} is the flux generated by coil two that links coil one, and the magnetic flux Φ_{22} is the flux generated by coil two that links only coil two. Magnetic fluxes Φ_{11} and Φ_{22} result in the *self-inductance* of coils one (L_p) and two (L_s), normally called the *primary winding* and *secondary winding*, respectively.

Figure 28.1 *Fluxes in a Magnetically Coupled Circuit*

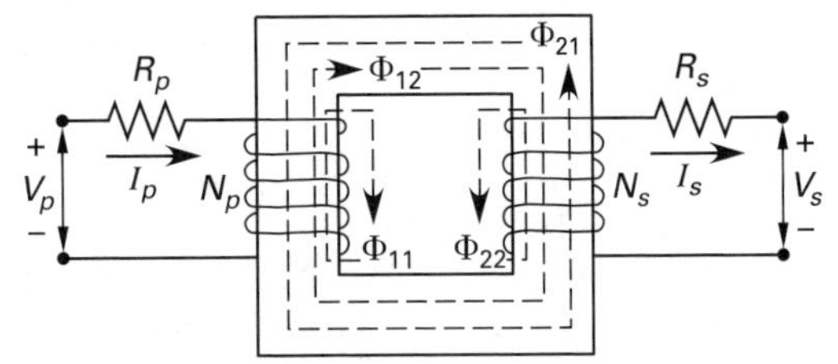

The *mutual inductance*, *M*, resulting from magnetic fluxes Φ_{12} and Φ_{21}, is defined as the ratio of the electromotive force induced in one coil (circuit) to the rate of change of current in the other coil (circuit). It is the proportionality constant between the induced voltage and the rate of current change.

$$v_p(t) = M\left(\frac{di_s(t)}{dt}\right) \qquad 28.4$$

$$v_s(t) = M\left(\frac{di_p(t)}{dt}\right) \qquad 28.5$$

Just as leakage flux is associated with self-inductance, mutual flux is associated with a *mutual reactance*, which is given by

$$X_m = \omega M \qquad 28.6$$

The induced voltage and generated flux associated with mutual reactance always oppose the flux and current that created them. For example, Φ_{12} in Fig. 28.1, generated by the current in the primary winding, results in a voltage in the secondary winding. This secondary voltage causes a current to flow that generates Φ_{21}, which opposes the original flux and generates a voltage drop of $-(I_s)j\omega M$ in the *primary*. By similar reasoning, a voltage drop of $-I_pj\omega M$ is generated in the *secondary* of the transformer as well. This is illustrated in Fig. 28.2(a), which is equivalent to Fig. 28.1, and in Eq. 28.7 and Eq. 28.8, which are obtained by applying KVL around the indicated loops.

$$\begin{aligned} V_p &= I_p(R_p + j\omega L_p) - I_sj\omega M \\ &= I_p(R_p + jX_p) - I_sjX_m \\ &= I_pZ_p - I_sjX_m \end{aligned} \qquad 28.7$$

$$\begin{aligned} V_s &= I_s(R_s + j\omega L_s) - I_pj\omega M \\ &= I_s(R_s + jX_s) - I_pjX_m \\ &= I_sZ_s - I_pjX_m \end{aligned} \qquad 28.8$$

Figure 28.2 *Transformer Models*

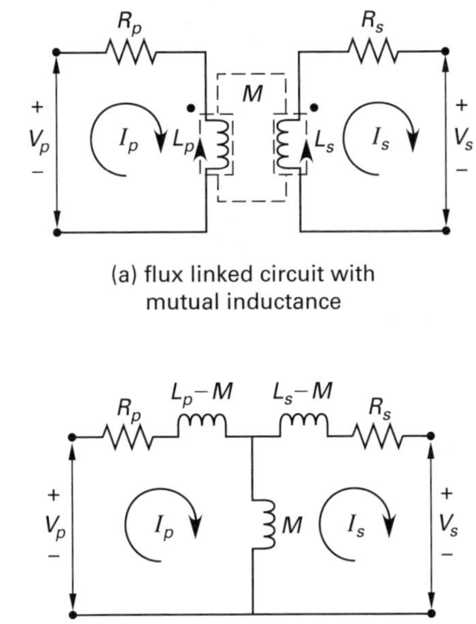

(a) flux linked circuit with mutual inductance

(b) equivalent conductively coupled circuit

The dots shown in Fig. 28.2(a) are used to determine the sign of the mutual inductance *when assuming directions of current flow* in analysis. When analyzing magnetically coupled circuits, the winding sense, or direction, can be determined using the right-hand rule (see Fig. 28.3). First, select a current direction on the primary side and place a dot where the current enters the winding (see Fig. 28.3(a)). Then use the right-hand rule to determine the current direction necessary for an opposing flux to be generated (see Fig. 28.3(b)). Place a dot where the current *leaves* this winding.[3] The two dotted terminals are positive simultaneously. With the instantaneous polarity determined, the need for the pictorial representation of the windings shown in Fig. 28.1 is eliminated. The coupled coils can be shown as in Fig. 28.3(c).

[3]This is equivalent to considering the primary of the transformer to be a load on the source and the secondary of the transformer to be the source for the secondary portion of the circuit.

Figure 28.3 *Dot Rule Determination*

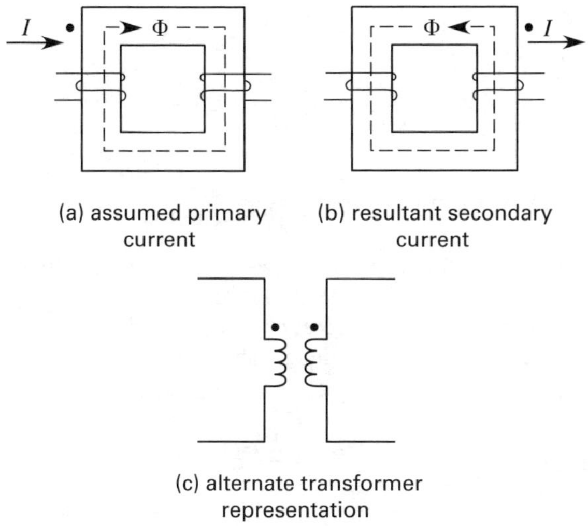

The *dot rule* is stated as follows.

- When the assumed currents both enter or both leave a pair of coupled coils by the dotted terminals, the sign of the mutual reactance, or M-terms, is the same as the L-terms in Eq. 28.7 and Eq. 28.8.
- When one current enters by a dotted terminal and the other current leaves by a dotted terminal, the sign of the M-terms is opposite the sign of the L-terms in Eq. 28.7 and Eq. 28.8.

This is shown in Fig. 28.4. The rule is used so that in the application of analysis laws, such as KVL, the resulting equations (see Eq. 28.7 and Eq. 28.8) will be consistent and the mathematical result will be valid for the circuit.

Figure 28.4 *Positive and Negative Mutual Inductances*

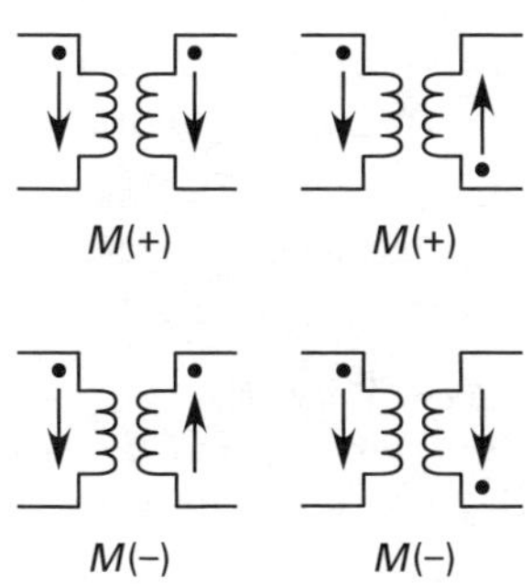

The amount of magnetic coupling is measured by the *coupling coefficient (coefficient of coupling)*, k, which is the fraction of the total flux that links both coils. The value of k varies from near zero for radio coils to near 1.0 for iron-core transformers.[4]

$$k = \frac{M}{\sqrt{L_p L_s}} = \frac{X_m}{\sqrt{X_p X_s}} \qquad 28.9$$

[4]In terms of the fluxes in Fig. 28.1, the coupling coefficient $k = 1 - \Phi_{11}/(\Phi_{11} + \Phi_{12})$. The high value of the coupling coefficient in iron cores is the primary reason for their use in transformers.

3. IDEAL TRANSFORMERS

Transformers, which are magnetically coupled circuits, are used to change voltages, match impedances, and isolate circuits. A primary winding current produces a magnetic flux, which induces a current in the secondary coil. To minimize flux leakage and improve efficiency, transformer coils are wound on magnetic permeable cores that improve the mutual inductance, M (see Sec. 28.2).[5] Two common designs, *core transformers* and *shell transformers*, are shown in Fig. 28.5.

Figure 28.5 *Core and Shell Transformers*

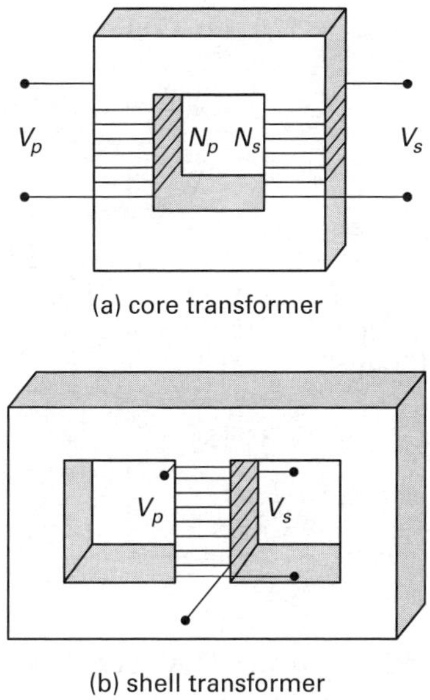

An *ideal transformer* has the following properties.

- The coupling coefficient is unity ($k=1$); that is, all the flux passes through each coil. There is no leakage flux. So, $|M| = |L_p| = |L_s|$, and $|X_m| = |X_p| = |X_s|$.
- The coils have no resistance; that is, $R_p = R_s = 0\ \Omega$. Combined with the first condition, this means that $Z_p = Z_s$ (see Fig. 28.2(b)).
- All the flux is contained inside a path of constant area. This allows the dimensions of the core to be used.
- Core *iron losses* are zero. This indicates that no heating losses exist due to eddy currents or hysteresis loop effects; that is $R_c = 0\,\Omega$ (see Sec. 28.5, Sec. 28.6, and Sec. 28.7).
- The core is infinitely permeable, so no *magnetizing current* (also called the *exciting current*) is required to create the flux in the core (see Sec. 28.5).

The last two conditions are technically the conditions for an *ideal core*. The transformer and the core can be

[5]An iron core transformer is indicated by one or more lines drawn between the coil symbols on a circuit diagram.

treated and modeled separately if necessary. The model for an ideal transformer with an ideal core is shown in Fig. 28.6.

Figure 28.6 *Ideal Transformer and Core*

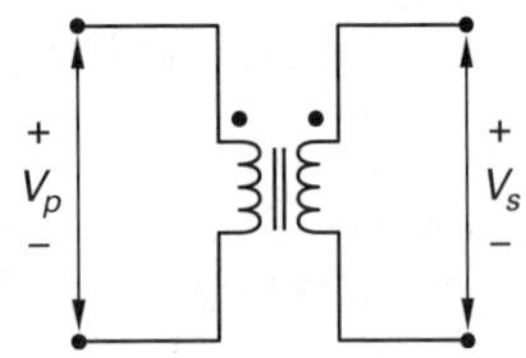

The ratio of the number of turns of primary to secondary windings is called the *turns ratio* (*ratio of transformation*), *a*.[6]

$$a = \frac{N_p}{N_s} \qquad \text{28.10}$$

In a lossless transformer, that is, a transformer with 100% efficiency, the power transfer from the primary side equals the power to the secondary side. Therefore, the relationship in Eq. 28.11 is valid.[7]

$$I_p V_p = I_s V_s \qquad \text{28.11}$$

$$a = \frac{N_p}{N_s} = \frac{V_p}{V_s} = \frac{I_s}{I_p} = \sqrt{\frac{Z_p}{Z_s}} \qquad \text{28.12}$$

If the turns ratio is greater than unity, the transformer decreases the voltage and is termed a *step-down transformer*. If the turns ratio is less than unity, the transformer increases the voltage and is termed a *step-up transformer*. The assumptions made for ideal transformers are very good approximations, especially for large iron-core transformers.

Example 28.1

An ideal step-down transformer has 200 primary turns and 50 secondary turns in its coils. 440 V are applied across the primary. The secondary load resistance is 5 Ω. What is the secondary voltage and current?

Solution

The turns ratio is

$$a = \frac{N_p}{N_s} = \frac{200}{50} = 4$$

The secondary voltage and current are

$$V_s = \frac{V_p}{a} = \frac{440 \text{ V}}{4} = 110 \text{ V}$$

$$I_s = \frac{V_s}{R_s} = \frac{110 \text{ V}}{5 \ \Omega} = 22 \text{ A}$$

4. IMPEDANCE MATCHING

Matching of impedances between source and load is accomplished to effect the maximum power transfer to the load. The transformer can impedance match. In Fig. 28.7, the impedance in the secondary is

$$Z_s = \frac{V_s}{I_s} \qquad \text{28.13}$$

Figure 28.7 *Impedance Matching*

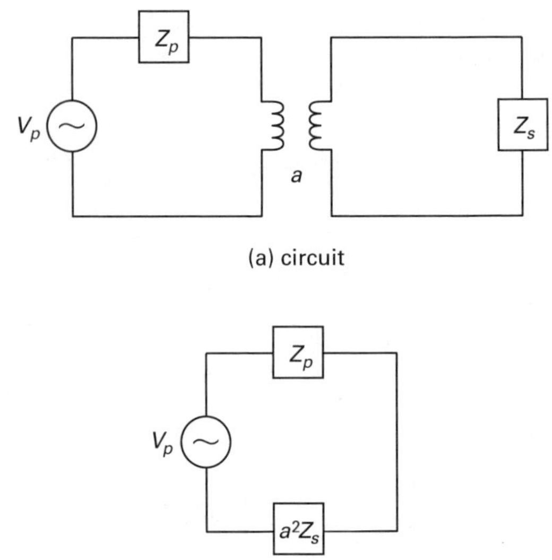

The impedance "seen" by the primary side, Z_{ep}, is termed the *effective primary impedance* or the *reflected impedance*, Z_{ref}. The reflected impedance is

$$Z_{ep} = Z_{ref} = \frac{V_p}{I_p} = Z_p + a^2 Z_s \qquad \text{28.14}$$

The turns ratio can be found from any of the relationships in Eq. 28.12, or specifically from

$$a = \sqrt{\frac{Z_p}{Z_s}} \qquad \text{28.15}$$

For a fixed primary impedance, the maximum power consumption in the load, that is, the maximum power transfer from the source, takes place when the secondary impedance is the conjugate of the primary impedance.

[6]The turns ratio, *a*, is alternatively defined as the ratio of secondary to primary turns, in which case the relationships of Eq. 28.12 are inverted. Unless specified otherwise, assume the ratio is defined as in Eq. 28.10.

[7]Phasor notation is not used, as all the elements are in phase in an ideal transformer. Further, this is not the formula for power. The 1/2 has been dropped as has the complex conjugate of the current (see Chap. 27).

Consequently, the conditions necessary to impedance match are

$$R_p = a^2 R_s \qquad 28.16$$

$$X_p = -a^2 X_s \qquad 28.17$$

Example 28.2

A 15 Ω speaker must be impedance matched to a 10 kΩ amplifier in order to produce the highest quality sound. What is the required turns ratio?

Solution

From Eq. 28.15,

$$a = \sqrt{\frac{Z_p}{Z_s}} = \sqrt{\frac{10 \times 10^3\ \Omega}{15\ \Omega}} = 25.8$$

Example 28.3

The primary of an ideal transformer with a turns ratio of 5 contains a 10 Ω resistor and an 80 mH inductor in series. What is the secondary reactance required to maximize the power transfer in this 60 Hz circuit?

Solution

Equation 28.17 gives the relationship between the reactances.

$$X_p = -a^2 X_s$$

$$X_s = \frac{X_p}{-a^2}$$

The primary reactance is found from

$$\begin{aligned} X_p &= X_L = \omega L = 2\pi f L \\ &= \frac{2\pi(60\ \text{Hz})(80\ \text{mH})}{1000\ \frac{\text{mH}}{\text{H}}} = 30.16\ \Omega \end{aligned}$$

Substituting the known and calculated values gives

$$X_s = \frac{X_p}{-a^2} = \frac{30.16\ \Omega}{-(5)^2} = -1.21\ \Omega$$

Example 28.4

For the reactance in Ex. 28.3, what is the value of the capacitance required?

Solution

From Ex. 28.3 and the definition of capacitive reactance,

$$X_s = X_C = -\frac{1}{\omega C}$$

$$\begin{aligned} C &= -\frac{1}{2\pi f X_C} = -\frac{1}{(2\pi)(60\ \text{Hz})(-1.21\ \Omega)} \\ &= 2.19 \times 10^{-3}\ \text{F} \end{aligned}$$

5. REAL TRANSFORMERS

Real transformers have flux leakage, resistance in the coils, and resistance and reactance associated with moving the domains within the core. A model of a real transformer is shown in Fig. 28.8.

Figure 28.8 *Real Transformer Equivalent Circuit*

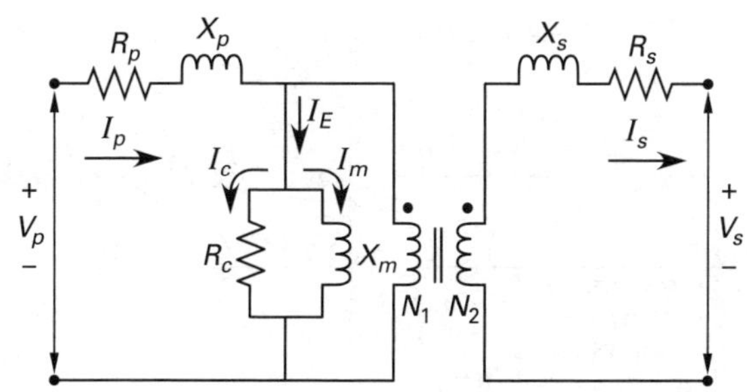

In the model of Fig. 28.8, all the core terms are shown on the primary side. This is arbitrary. Many other equivalent models are possible, though they will all contain these components. Components not yet defined follow.

- *Exciting current*, I_E: the total current needed to produce the magnetic flux in the core.
- *Mutual reactance*, X_m, and *magnetizing current*, I_m: the magnetizing current is the current necessary to generate the flux in the core. It is 90° out of phase with the core current.
- *Core resistance*, R_c, and *core current*, I_c: the core resistance represents iron losses (see Sec. 28.6 and Sec. 28.7). The core current is the current necessary to overcome those losses. This current flows in the primary windings but it is caused by resistive losses physically occurring in the core.

Each leakage flux corresponds to a "loss" of induced voltage in its coil. The loss is considered to be caused by an equivalent reactance, X_p or X_s. In terms of the flux, these reactances are

$$X_p = \omega L_p = \frac{V_{X_p}}{I_p} = \frac{4.44 f \Phi_p N_p}{I_p} \qquad 28.18$$

$$X_s = \omega L_s = \frac{V_{X_s}}{I_s} = \frac{4.44 f \Phi_s N_s}{I_s} \qquad 28.19$$

The turns ratio is related to the leakage inductances by

$$a = \sqrt{\frac{L_p}{L_s}} \qquad 28.20$$

6. MAGNETIC HYSTERESIS: BH CURVES

The core of a transformer, especially when specifically designed to increase the coupling, is made from iron. Iron is a ferromagnetic material. When a magnetic field, in this case produced by the coil of the transformer, is applied to a ferromagnetic material, a phenomenon known as *magnetic hysteresis* takes place. This phenomenon is shown in Fig. 28.9.

Figure 28.9 *Magnetic Hysteresis Loop*

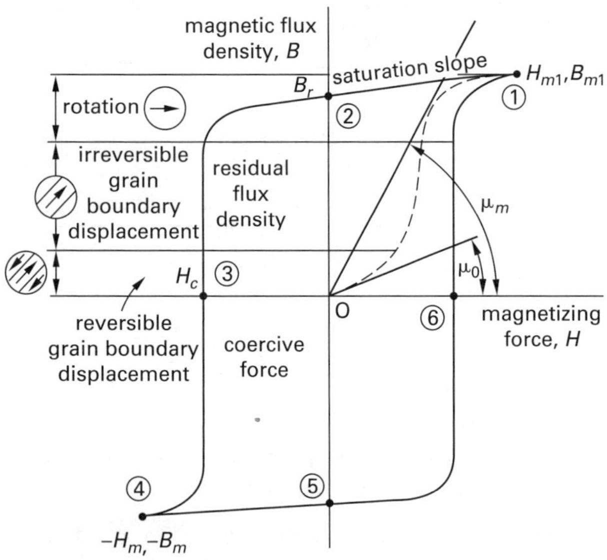

Figure 28.9 plots the magnetic flux density, B, produced for various values of the magnetizing force, H. The relationship is decidedly nonlinear.[8] When a material is first magnetized, the initial BH values occur along the curve from the origin, O, to point 1. This part of the curve, O1, is known as the *initial* or *normal magnetization curve*. The initial slope of O1 is called the *initial permeability*, μ_o (not to be confused with the permeability of free space). The slope of the line from the origin to a point tangent to the normal magnetization curve is called the *maximum permeability*, μ_m. The maximum value of B reached at point 1 is called the *saturation flux density*. If the magnetic field is then decreased, the flux density decreases as well, but not in proportion to the decrease in magnetizing field strength. When H equals zero, at point 2, the amount of magnetic flux density remaining in the core is called the *residual flux density*, B_r. In order to reduce the magnetic field in the core to zero, the magnetizing force must increase in reverse direction of that initially applied. The magnitude of the magnetizing force where B again becomes zero, at point 3, is called the *coercive force*, H_c.

As the magnetizing force continues to increase in the negative direction, the flux density again increases, but in opposite polarity, eventually reaching saturation in this direction at point 4. As the magnetic field is again removed, the magnitude of the magnetic flux density decreases to point 5. A positive value of H (equal to H_c) is required to drive the flux density to zero, at point 6.

As the magnetizing force is increased further, the curve from point 6 to point 1 is followed, completing the cycle but not returning to the origin. The domain, or grain boundary movement, that is the ultimate cause of this curve is shown for the positive portion of the cycle. The phenomenon whereby the values of B lag behind the values of H is called *hysteresis*. The entire loop is called the *hysteresis loop*. The area within the loop represents the energy loss that is due to heating from the movement of the magnetic domains.[9]

7. EDDY CURRENTS

Any conducting mass exposed to variations in magnetic flux contains *eddy*, or *Foucault*, *currents*. These currents are a result of magnetic induction caused by the relative motion caused by either variations in the magnetic field or the movement of the mass itself. Such currents result in Joule heating in the mass—in this case, in the core of the transformer. Such an energy loss, in a transformer or in the iron portions of electric machinery, is minimized by dividing the magnetically conducting material into small segments called *laminations*, coated with insulating material. This reduces the length of the path in which current exists.

8. CORE LOSSES

Core losses are defined as the rate of energy conversion into heat in a magnetic material caused by an alternating or pulsing magnetic field. This type of loss is also called the *excitation loss* or *iron loss* (see Sec. 28.5). Core losses are the combination of magnetic hysteresis and eddy current losses.

[8]To this point, all the relationships and equations have assumed linearity in the transformers.

[9]The portion of the energy used to create the flux comes from I_m (see Sec. 28.5). The portion of the energy that is lost as heat comes from I_c (see Sec. 28.5). The current I_c also supplies any eddy current losses (see Sec. 28.7).

29 Linear Circuit Analysis

1. Fundamentals 29-1
2. Ideal Independent Voltage Sources 29-2
3. Ideal Independent Current Sources 29-2
4. Ideal Resistors 29-2
5. Dependent Sources 29-3
6. Dependent Voltage Sources 29-3
7. Dependent Current Sources 29-3
8. Linear Circuit Elements 29-3
9. Resistance 29-3
10. Capacitance 29-5
11. Inductance 29-6
12. Mutual Inductance 29-7
13. Linear Source Models 29-7
14. Source Transformations 29-8
15. Series and Parallel Source Combination Rules 29-8
16. Redundant Impedances 29-8
17. Delta-Wye Transformations 29-9
18. Thevenin's Theorem 29-9
19. Norton's Theorem 29-9
20. Maximum Power Transfer Theorem 29-10
21. Superposition Theorem 29-10
22. Miller's Theorem 29-10
23. Kirchhoff's Laws 29-11
24. Kirchhoff's Voltage Law 29-11
25. Kirchhoff's Current Law 29-11
26. Loop Analysis 29-12
27. Node Analysis 29-12
28. Determination of Method 29-12
29. Voltage and Current Dividers 29-13
30. Steady-State and Transient Impedance Analysis 29-13
31. Two-Port Networks 29-14

Nomenclature

A	amplification factor	–
A	area	m^2
b	y-intercept	–
C	capacitance	F
f	frequency	Hz
i	variable current	A
I, $\mathbf{I}$	constant or rms current	A
l	length	m
L	inductance or self-inductance	H
m	slope	–
M	mutual inductance	H
n, N	number of items	–
p	instantaneous power	W
P	power	W
Q	charge	C
r	distance	m
R	resistance	Ω
t	time	s
T	temperature	°C
U	energy	J
v	instantaneous voltage	V
V, $\mathbf{V}$	constant or rms voltage	V
Y	admittance	S
Z, $\mathbf{Z}$	impedance	Ω

Symbols

α	thermal coefficient of resistance	1/°C
ϵ	permittivity	F/m
ϵ_0	free-space permittivity, 8.854×10^{-12}	F/m
ϵ_r	relative permittivity	–
ζ	constant	–
κ	arbitrary constant	–
μ	permeability	H/m
μ_0	free-space permeability, 1.2566×10^{-6}	H/m
ξ	constant	–
ρ	resistivity	Ω·cm
Ψ	magnetic flux	Wb
ω	angular frequency	rad/s

Subscripts

0	initial or free space (vacuum)
20	referenced to 20°C
C	capacitor
e	equivalent
fl	full load
int	internal or source
l	load
L	inductor
m	magnetizing
nl	no load
N	Norton
oc	open circuit
r	ratio or relative
R	resistive or resistor
s	source
sc	short circuit
Th	Thevenin

1. FUNDAMENTALS

Circuit analysis is fundamental to the practice of electrical engineering. Such analysis is based on Kirchhoff's two circuit laws and the values, as well as the fluctuations, of the *circuit variables*, that is, the voltages between terminals and the currents within the network. Circuit theory is nominally divided into determination of *equivalent circuits* and *mathematical analysis* of those equivalent circuits.[1] Equivalent circuits are composed of *lumped elements*. A lumped element is a single element

[1]In digital systems, the final equivalent circuit is a logic diagram.

representing a particular electrical property in the entire circuit, or in some portion of the circuit.[2] Equivalent circuit theory is sometimes called *network theory.* The term "network," while often used synonymously with "circuit," is usually reserved for more complicated arrangements of elements. In fact, "circuit" often designates a single loop of a network. Networks are often modeled using ideal elements.

2. IDEAL INDEPENDENT VOLTAGE SOURCES

An ideal independent voltage source maintains the voltage at its terminals regardless of the current flowing through the terminals. The symbols for an ideal independent voltage source and its characteristics are shown in Fig. 29.1. The subscript *s* generally indicates an independent source. The source value can vary with time, but its effective value does not change. That is, the magnitude of its variations in time is without regard to the current. The polarity assigned is arbitrary, but it is often assigned in the direction of positive power flow out of the source. This type of notation is called *source notation.*

Figure 29.1 *Ideal Voltage Source*

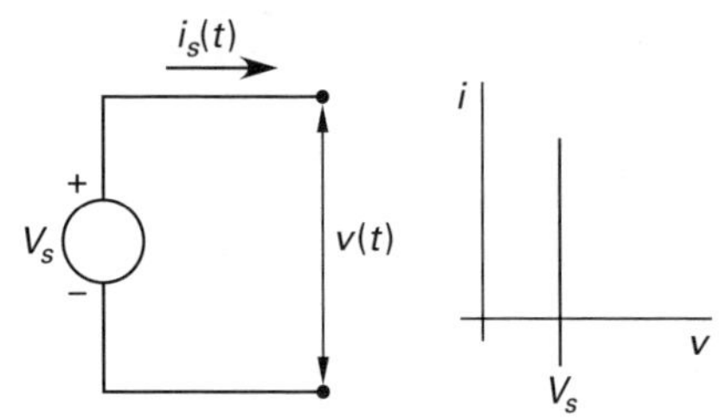

A *real voltage source* cannot maintain the voltage at its terminals when the current becomes too large (in either the positive or the negative direction). The equivalent circuit of a real voltage source is shown in Fig. 29.2. The decrease in voltage as the current increases is measured by the *voltage regulation* given in Eq. 29.1.

$$\text{regulation} = \frac{V_{\text{nl}} - V_{\text{fl}}}{V_{\text{fl}}} \times 100\% \tag{29.1}$$

Figure 29.2 *Real Voltage Source*

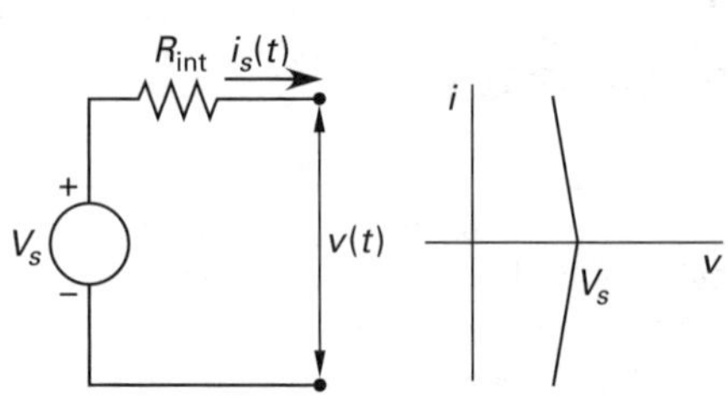

[2]Electrical properties are distributed throughout elements, though they are concentrated in certain areas. Analyzing such *distributed elements* is difficult. Because of this difficulty, they are represented by lumped parameters. This method is important in the analysis of electrical energy transmission.

3. IDEAL INDEPENDENT CURRENT SOURCES

An *ideal independent current source* maintains the current regardless of the voltage at its terminals. The symbols for an ideal independent current source and its characteristics are shown in Fig. 29.3, the subscript *s* indicating the independent source. The source value can vary with time, but its effective value does not change. That is, the magnitude of its variations in time is without regard to the voltage. The polarity is again arbitrary but is assigned using the source notation.

Figure 29.3 *Ideal Current Source*

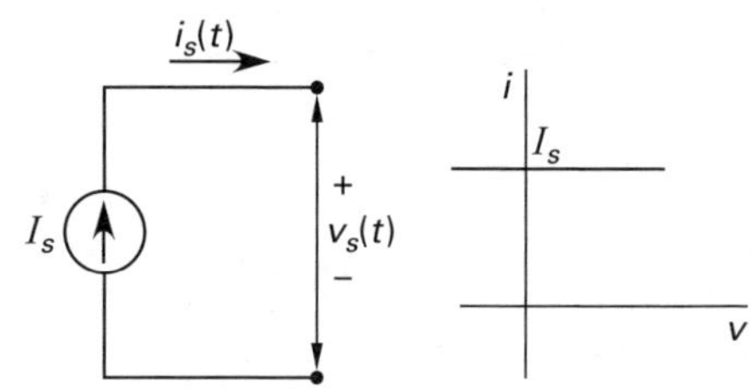

A *real current source* cannot maintain the current at its terminals when the voltage becomes too large (in either the positive or the negative direction). The equivalent circuit of a real current source is shown in Fig. 29.4.

Figure 29.4 *Real Current Source*

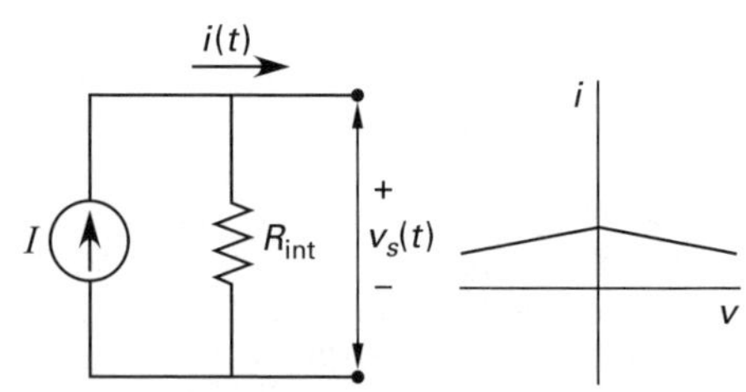

4. IDEAL RESISTORS

An *ideal resistor* maintains the same resistance at its terminals regardless of the power consumed. That is, it is a linear element whose voltage/current relationship is given by Ohm's law.

$$v = iR \tag{29.2}$$

The symbols for an ideal resistor are identical to those for a real resistor and are shown in Fig. 29.5, along with the characteristics of the ideal resistor. The polarity is assigned to make Eq. 29.2, Ohm's law, valid as written. This is sometimes called the *sink notation,* as it corresponds to positive power flow into the resistor.

A real resistor undergoes changes in resistance as the temperature changes. The resistance of metals and most alloys increases with temperature while that of carbon and electrolytes decreases. The resistance change is accounted for in circuit design by ensuring the design allows for proper circuit functioning over the expected temperature range (see Sec. 26.3).

Figure 29.5 *Resistor*

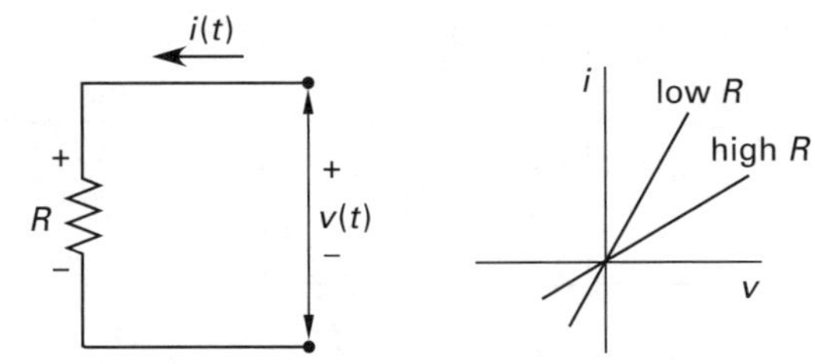

(a) ideal and real (b) ideal

5. DEPENDENT SOURCES

Any circuit that is designed to deliver more power to the load than is available at the signal frequency must use an *active device*. An active device is defined as a component capable of amplifying the current or voltage, such as a transistor. An active device uses a source of power other than the input signal in order to accomplish the modification. In order to analyze circuits containing active devices, the devices are replaced by an equivalent circuit of *passive linear elements*, that is, elements that are not sources of energy (such as resistors, inductors, and capacitors), and one or more dependent sources. Such sources may be treated as standard sources using all the linear circuit analysis techniques, with the exception of superposition. Superposition can only be used if the device is limited to operation over a linear region.

6. DEPENDENT VOLTAGE SOURCES

A *dependent voltage source* is one in which the output is controlled by a variable elsewhere in the circuit. For this reason, dependent voltage sources are sometimes called *controlled voltage sources*. The controlling variable can be another voltage, a current, or any other physical quantity, such as light intensity or temperature. The symbols for dependent voltage sources are shown in Fig. 29.6. A *voltage-controlled voltage source* is controlled by a voltage, v, present somewhere else in the circuit. A *current-controlled voltage source* is controlled by a current, i, present somewhere else in the circuit. The terms ζ and ξ are constants.

Figure 29.6 *Dependent Voltage Sources*

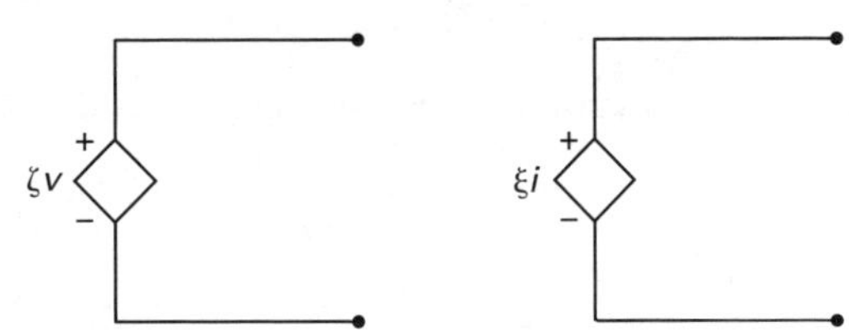

(a) voltage-controlled voltage source (b) current-controlled voltage source

7. DEPENDENT CURRENT SOURCES

A *dependent current source* is one in which the output is controlled by a variable elsewhere in the circuit. For this reason, dependent current sources are sometimes called *controlled current sources*. The controlling variable can be another current, voltage, or any other physical quantity, such as light intensity or temperature. The symbols for dependent current sources are shown in Fig. 29.7. A *voltage-controlled current source* is controlled by a voltage, v, present somewhere else in the circuit. A *current-controlled current source* is controlled by a current, i, present somewhere else in the circuit. The terms ζ and ξ are constants.

Figure 29.7 *Dependent Current Sources*

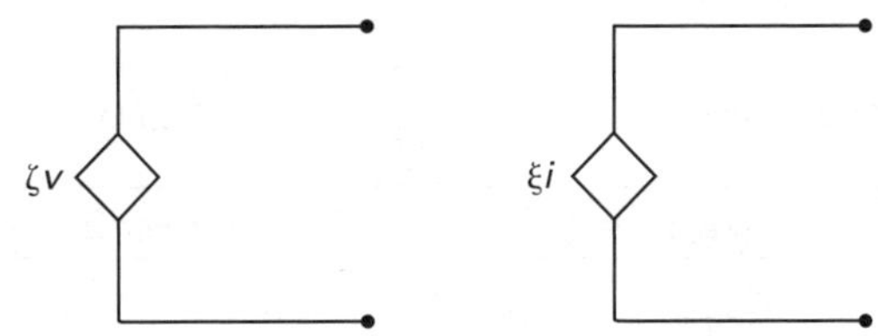

(a) voltage-controlled current source (b) current-controlled current source

8. LINEAR CIRCUIT ELEMENTS

The basic circuit elements are resistance, capacitance, and inductance. Related to the basic element of inductance are mutual inductance and ideal (linear) transformers. A *linear network* is one in which the resistance, capacitance, and inductance are constant with respect to current and voltage, and in which the current or voltage of sources is independent or directly proportional to the other currents and voltages, or their derivatives, in the network. That is, a linear network is one that consists of linear elements and linear sources.[3] Circuit elements or sources are linear if the superposition principle is valid for them, and vice versa. A summary of linear circuit element parameters is given in Table 29.1. The time domain and frequency domain behavior of linear circuit elements, along with their defining equations, are shown in Table 29.2.

9. RESISTANCE

Resistance is the opposition to current flow in a conductor. Technically, it is the opposition that a device or material offers to the flow of direct current. In AC circuits, it is the real part of the complex impedance. The resistance of a conductor depends on the resistivity, ρ, of the conducting material. The resistivity is a function of temperature that varies approximately linearly in accordance with Eq. 29.3. A slightly different version

[3]A linear source is one in which the output is proportional to the first power of a voltage or current in the circuit.

Table 29.1 *Linear Circuit Element Parameters*

circuit element	voltage	current	instantaneous power	average power	average energy stored
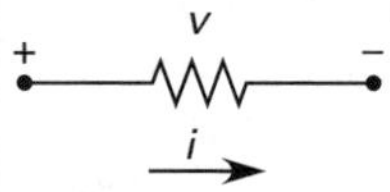	$v = iR$	$i = \frac{v}{R}$	$p = iv = i^2R = \frac{v^2}{R}$	$P = IV = I^2R = \frac{V^2}{R}$	–
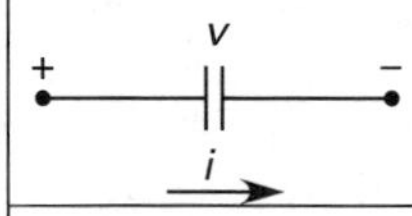	$v = \frac{1}{C}\int idt + \kappa$	$i = C\frac{dv}{dt}$	$p = iv = Cv\frac{dv}{dt}$	0	$U = \frac{1}{2}CV^2$
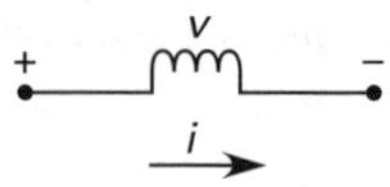	$v = L\frac{di}{dt}$	$i = \frac{1}{L}\int vdt + \kappa$	$p = iv = Li\frac{di}{dt}$	0	$U = \frac{1}{2}LI^2$

Table 29.2 *Linear Circuit Parameters, Time and Frequency Domain Representation*

parameter	defining equation	time domain	frequency domain[a]
resistance	$R = \frac{v}{i}$	$v = iR$	$V = IR$
capacitance	$C = \frac{Q}{V}$	$i = C\frac{dv}{dt}$	$I = j\omega CV$
self-inductance	$L = \frac{\Psi}{I}$	$v = L\frac{di}{dt}$	$V = j\omega LI$
mutual inductance[b]	$M_{12} = M_{21} = M = \frac{\Psi_{12}}{I_2} = \frac{\Psi_{21}}{I_1}$	$v_1 = L_1\frac{di_1}{dt} + M\frac{di_2}{dt}$ $v_2 = L_2\frac{di_2}{dt} + M\frac{di_1}{dt}$	$V_1 = j\omega(L_1I_1 + MI_2)$ $V_2 = j\omega(L_2I_2 + MI_1)$

[a]The voltages and currents in the frequency domain column are not shown as vectors; for example, **I** or **V**. They are, however, shown in phasor form. Across any single circuit element (or parameter), the phase angle is determined by the impedance and embodied by the j. If the $j\omega$ (C or L or M) were not shown and instead given as **Z**,the current and voltage would be shown as **I** and **V**, respectively.
[b]Currents I_1 and I_2 (in the mutual inductance row) are in phase. The angle between either I_1 or I_2 and V_1 or V_2 is determined by the impedance angle embodied by the j. Even in a real transformer, this relationship holds, since the equivalent circuit accounts for any phase difference with a magnetizing current, I_m.

of this equation was given in Sec. 26.3. Here, the resistivity is referenced to 20°C and is written ρ_{20}. For copper, ρ_{20} is approximately 1.8×10^{-8} Ω·m. The thermal coefficient of resistivity, also referred to as the *temperature coefficient of resistivity*, is referenced to 20°C and written α_{20}. For copper, α_{20} is approximately 3.9×10^{-3} °C^{-1}. The equation for resistivity at any temperature, T (in °C), then becomes[4]

$$\rho = \rho_{20}\big(1 + \alpha_{20}(T - 20)\big) \qquad 29.3$$

The resistance of a conductor of length l and cross-sectional area A is

$$R = \frac{\rho l}{A} \qquad 29.4$$

[4]Copper with a conductivity of 100% will experience a resistance change of approximately 10% for a 25°C (77°F) temperature change.

The defining equation for resistance is Ohm's law, or the extended Ohm's law for AC circuits, which relates the linear relationship between the voltage and the current.

$$V = IR \quad \text{or} \quad \mathbf{V} = \mathbf{I}\mathbf{Z}_R \qquad 29.5$$

The power dissipated by the resistor is[5]

$$P = IV = I^2R = \frac{V^2}{R} \qquad 29.6$$

The equivalent resistance of resistors in series is

$$R_e = R_1 + R_2 + \ldots + R_n \qquad 29.7$$

[5]When the term "power" is used, the average power is the referenced quantity. The instantaneous power depends on the time within the given cycle and is written p.

The equivalent resistance of resistors in parallel is

$$\frac{1}{R_e} = \frac{1}{R_1} + \frac{1}{R_2} + \ldots + \frac{1}{R_n} \qquad 29.8$$

10. CAPACITANCE

Capacitance is a measure of the ability to store charge. Technically, it is the opposition a device or material offers to changing voltage. In a DC circuit, that is, a circuit in which the voltage is constant, the capacitor appears as an open circuit (after any transient condition has subsided). In a DC circuit, if the voltage is zero, the capacitor appears as a short circuit. In AC circuits, it is the constant of proportionality between the changing voltage and current.

$$i = C\frac{dv}{dt} \qquad 29.9$$

Capacitance is a function of the insulating material's *permittivity*, ϵ, between the conducting portions of the capacitor. The permittivity is the product of the *free-space permittivity*, ϵ_0, and the *relative permittivity* or *dielectric constant*, ϵ_r. For free space, that is, a vacuum, the permittivity is 8.854×10^{-12} F/m. The dielectric constant varies, and for many linear materials it has values between 1 and 10.

$$\epsilon = \epsilon_r \epsilon_0 \qquad 29.10$$

The capacitance of a simple parallel plate capacitor is given by Eq. 29.11. The term r is the distance between the two parallel plates of equal area, A, and ϵ is the permittivity of the material between the plates.

$$C = \frac{\epsilon A}{r} \qquad 29.11$$

The defining equation for capacitance, which relates the amount of charge stored to the voltage across the capacitor terminals, is

$$C = \frac{Q}{V} \qquad 29.12$$

The (average) power dissipated by a capacitor is zero. The (average) energy stored in the electric field of a capacitor is

$$U = \tfrac{1}{2}CV^2 = \tfrac{1}{2}QV = \tfrac{1}{2}\left(\frac{Q^2}{C}\right) \qquad 29.13$$

The equivalent capacitance of capacitors in series is

$$\frac{1}{C_e} = \frac{1}{C_1} + \frac{1}{C_2} + \ldots + \frac{1}{C_n} \qquad 29.14$$

The equivalent capacitance of capacitors in parallel is

$$C_e = C_1 + C_2 + \ldots + C_n \qquad 29.15$$

From Eq. 29.9, the voltage cannot change instantaneously, as this would make the current and the power infinite, which clearly cannot be the case. Using this fact, the transient behavior of capacitors can be analyzed. Using this principle is the equivalent of treating a capacitor's arbitrary constant κ in Table 29.1 as the initial voltage, V_0.

Example 29.1

In the circuit shown, the switch has been open for an "extended time," which in such circuits is considered to be a minimum of five time constants. The capacitor has no charge on it at $t = 0$ s when the switch is closed. What is the current through the capacitor at the instant the switch is closed?

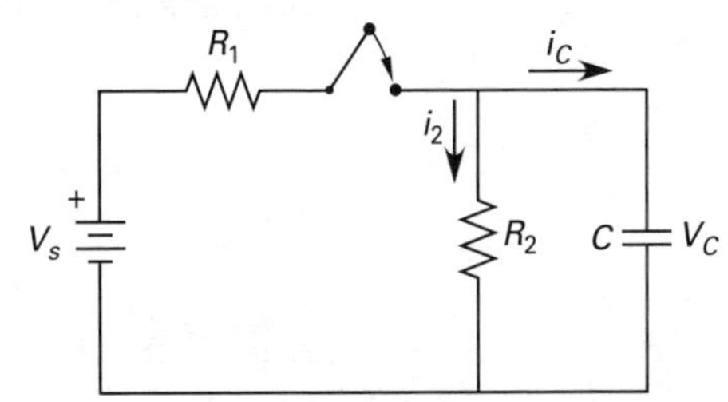

Solution

Because the voltage across the capacitor cannot change instantly, $v(0^-) = v(0^+)$, and the capacitor remains uncharged. The capacitor thus initially acts as a short circuit. All the current flows through the capacitor, bypassing R_2. The current is

$$i_C(0^+) = \frac{V_s}{R_1}$$

The initial voltage across the capacitor, $V_C(0^+)$, is zero.

Example 29.2

Using the circuit given in Ex. 29.1, what is the steady-state voltage across the capacitor?

Solution

The term "steady-state" is another way of saying "an extended period of time," which in such circuits is considered to be a minimum of five time constants. The capacitor will be fully charged and will act as an open circuit. Consequently, $dv/dt = 0$ and $i_C = 0$. All the current flows through R_2 and is given by

$$I_{R_2} = \frac{V_s}{R_1 + R_2}$$

Circuit Theory

Because they are in parallel, the voltage across the capacitor is the same as the voltage across R_2. Thus,

$$V_C(\infty) = V_{R_2} = I_2 R_2 = \left(\frac{V_s}{R_1 + R_2}\right) R_2 = V_s\left(\frac{R_2}{R_1 + R_2}\right)$$

The steady-state current through the capacitor, $I_C(\infty)$, is zero.

11. INDUCTANCE

Inductance is a measure of the ability to store magnetic energy. Technically, it is the opposition a device or material offers to changing current. In a DC circuit, that is, a circuit in which the current is constant, the inductor appears as a short circuit (after any transient condition has subsided). In a DC circuit, if the current is zero, the inductor appears as an open circuit. In AC circuits, it is the constant of proportionality between the changing current and voltage.

$$v = L\frac{di}{dt} \quad \text{29.16}$$

Inductance is a function of a medium's *permeability*, μ, between the magnetically linked portions of the inductor. The permeability is the product of the *free-space permeability*, μ_0, and the *relative permeability* or *permeability ratio*, μ_r. For free space, that is, a vacuum, the permeability is 1.2566×10^{-6} H/m. The permeability ratio varies. Depending on the material and the flux density, it can have values in the hundreds and sometimes thousands.

$$\mu = \mu_r \mu_0 \quad \text{29.17}$$

The inductance of a simple toroid shown in Fig. 29.8 is given by Eq. 29.18. The term l is the average flux path distance around the toroid of equal area, A, and μ is the permeability of the medium between the coil turns, N.

$$L = \frac{\mu N^2 A}{l} \quad \text{29.18}$$

The defining equation for inductance, which relates the amount of flux present to the current flowing through the inductor terminals, is

$$L = \frac{\Psi}{I} \quad \text{29.19}$$

The (average) power dissipated by an inductor is zero. The (average) energy stored in the magnetic field of an inductor is given by Eq. 29.20. Positive energy indicates

Figure 29.8 *Toroid Inductance Terms*

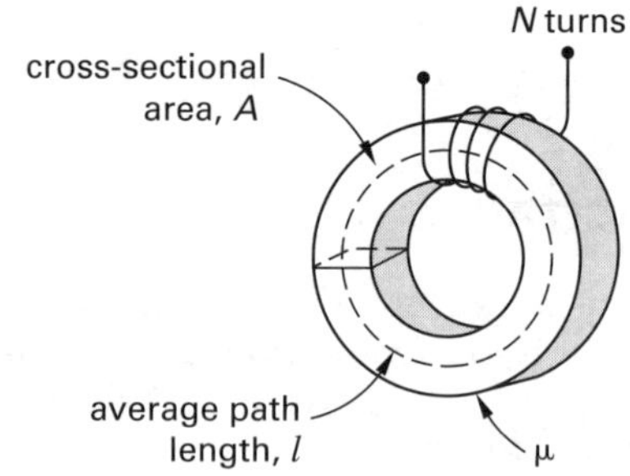

energy stored in the inductor. Negative energy indicates energy supplied to the circuit by the inductor.

$$U = \tfrac{1}{2}LI^2 = \tfrac{1}{2}\Psi I = \tfrac{1}{2}\left(\frac{\Psi^2}{L}\right) \quad \text{29.20}$$

The equivalent inductance of inductors in series is

$$L_e = L_1 + L_2 + \ldots + L_n \quad \text{29.21}$$

The equivalent inductance of inductors in parallel is

$$\frac{1}{L_e} = \frac{1}{L_1} + \frac{1}{L_2} + \ldots + \frac{1}{L_n} \quad \text{29.22}$$

From Eq. 29.16, the current cannot change instantaneously, as this would make the voltage and the power infinite, which clearly cannot be the case. Using this fact, the transient behavior of inductors can be analyzed. Using this principle is the equivalent of treating an inductor's arbitrary constant κ in Table 29.1 as the initial current, I_0.

Example 29.3

In the circuit shown, the switch has been open for an extended time, that is, greater than five time constants. The inductor has no associated magnetic field on it at $t = 0$ s when the switch is closed. What is the current through the inductor at the instant the switch is closed?

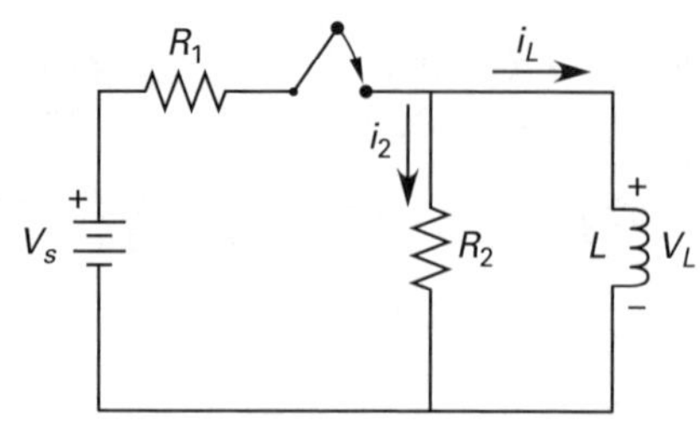

Solution

Because the current across the inductor cannot change instantaneously, $i(0^-) = i(0^+)$ and the inductor remains without a magnetic field. The inductor initially acts as an open circuit. All the current flows through R_2.

$$i_L(0^+) = 0$$

Example 29.4

Using the circuit given in Ex. 29.3, what is the initial voltage across the inductor?

Solution

All the current flows through R_2 and is given by

$$I_{R_2} = \frac{V_s}{R_1 + R_2}$$

Because they are in parallel, the voltage across the inductor is the same as the voltage across R_2. So,

$$\begin{aligned} V_L(0^+) &= V_{R_2} = I_2 R_2 \\ &= \left(\frac{V_s}{R_1 + R_2}\right) R_2 = V_s\left(\frac{R_2}{R_1 + R_2}\right) \end{aligned}$$

Example 29.5

Using the circuit given in Ex. 29.3, what is the steady-state value of the current through the inductor?

Solution

The term "steady-state" is another way of saying "an extended period of time," which in such circuits is considered to be a minimum of five time constants. The inductor's magnetic field will be at a maximum and the inductor acts as a short circuit. Consequently, $di/dt = 0$, and $v_L = 0$. All the current flows through the inductor, bypassing R_2, and, using Ohm's law, is

$$i_L(\infty) = \frac{V_s}{R_1}$$

12. MUTUAL INDUCTANCE

Mutual inductance is the ratio of the electromotive force induced in one circuit to the rate of change of current in the other circuit. Applicable equations are given in Table 29.2. The concept is explained in Chap. 28. The total energy stored in a system involving mutual inductance is

$$U = \tfrac{1}{2}L_1 I_1^2 + \tfrac{1}{2}L_2 I_2^2 + M I_1 I_2 \qquad 29.23$$

13. LINEAR SOURCE MODELS

Electric power sources tend to be nonlinear. In order to analyze such sources using circuit analysis, they are modeled as linear sources in combination with linear impedance. Such a source is classified as an *independent source.*

Dependent sources, whose output depends on a parameter elsewhere in the circuit, are associated with active devices, such as transistors. These are not actually sources of power but can be modeled as linear sources in combination with an impedance. The range of operation over which the model is valid must be selected to be approximately linear. Circuit analysis of this type is called *small-signal analysis.*

A third source type is the *transducer* that produces a current or voltage output from a nonelectric input. *Thermocouples* are transducers that produce voltages as functions of temperature. *Photodiodes* are transducers that produce currents as functions of incident illumination level. *Microphones* are transducers that produce voltages as functions of incident sound intensity.

All these sources can be modeled by linear approximation to measured values. Specific models for the electronic components are given in Chap. 36.

Example 29.6

The accompanying illustration shows a representative plot of battery voltage versus load current. (a) What is the linear expression of the battery for the range of operation from 8 A to 12 A? (b) Show the associated model.

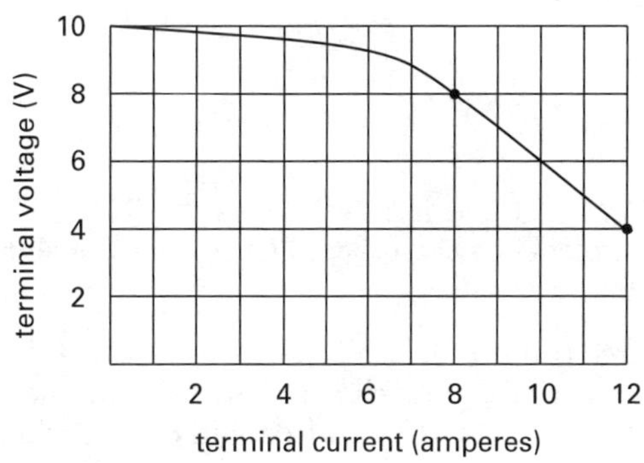

Solution

(a) Using the *two-point method*, equivalent to determining the parameters of a straight line in the formula $y = mx + b$, for the points (8, 8) and (4, 12) gives

$$\frac{I - 8\text{ A}}{A - 8\text{ V}} = \frac{12\text{ A} - 8\text{ A}}{4\text{ V} - 8\text{ V}} = -1\text{ S}$$

(b) Either the voltage or the current may be selected as the dependent variable.

$$V = -I + 16\text{ V} \qquad [\text{I}]$$

$$I = -V + 16\text{ A} \qquad [\text{II}]$$

Equation I is a loop voltage equation with the terminal voltage determined by two sources, one independent (the 16 V) and the other dependent on the current (the -1 S). Equation II is a node current equation with the terminal current determined by two sources, one independent (the 16 A) and the other dependent on the voltage (the -1 S). The resulting models are

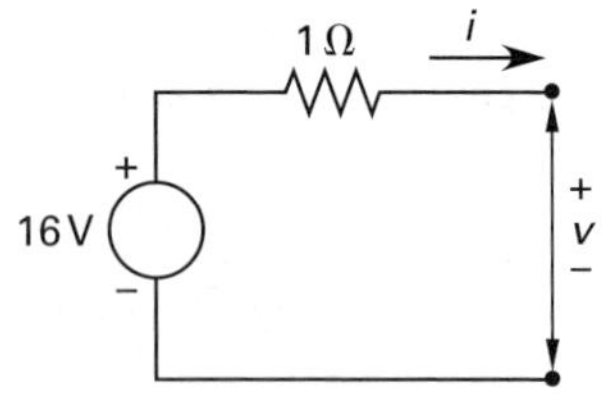

[I]

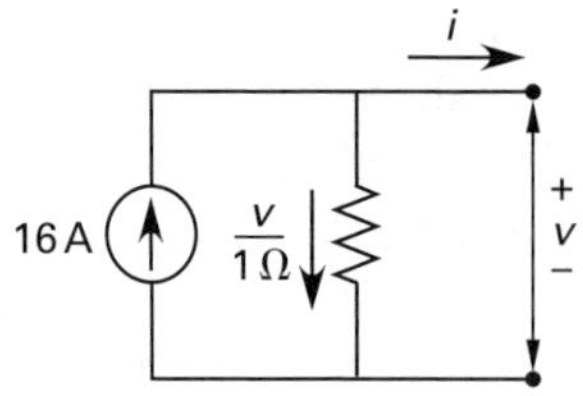

[II]

14. SOURCE TRANSFORMATIONS

An electrical source can be modeled as either a voltage source in series with an impedance or a current source in parallel with an impedance. The two models are equivalent in that they have the same terminal voltage/current relationships. Changing from one model to the other is called *source transformation* and is a *circuit reduction* technique. The models' parameters are determined by

$$v_{\text{Th}} = i_N Z_{\text{Th}} \qquad 29.24$$

$$i_N = \frac{v_{\text{Th}}}{Z_{\text{Th}}} \qquad 29.25$$

Using Eq. 29.24 and Eq. 29.25, it is possible to convert from one model to the other. The models are shown in Fig. 29.9.

The source transformation is not defined when Z equals zero or when Z approaches ∞. Also, both the open-circuit voltage and the short-circuit current direction must be the same in each model (see Sec. 26.11).

Figure 29.9 *Source Transformation Models*

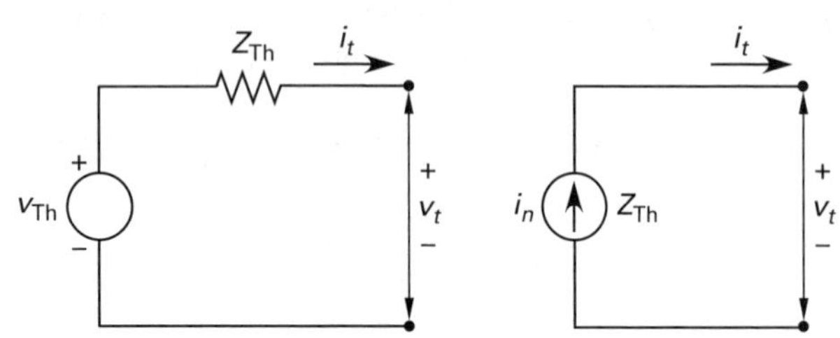

(a) equivalent voltage source

(b) equivalent current source

15. SERIES AND PARALLEL SOURCE COMBINATION RULES

One aspect of circuit reduction is the removal of sources that are not required. Section 29.14 referred to ideal current and voltage sources. These can be manipulated, combined, or eliminated as indicated by the following rules.

- An ideal voltage source cannot be placed in parallel with a second ideal voltage source, because the resulting voltage is indeterminate.
- An ideal current source cannot be placed in series with a second ideal current source, because the resulting current is indeterminate.
- An ideal voltage source in parallel with an ideal current source makes the current source redundant. The voltage is set by the voltage source, which absorbs all the current from the current source.
- An ideal current source in series with an ideal voltage source makes the voltage source redundant. The current is set by the current source, which is able to overcome any voltage of the voltage source.
- Ideal voltage sources in series can be combined algebraically using Kirchhoff's voltage law.
- Ideal current sources in parallel can be combined algebraically using Kirchhoff's current law.

The rules above can be used in an iterative manner to reduce a network to a single equivalent source and impedance. This technique is useful during phasor analysis or where a capacitance or inductance is present in a complicated network and the transient current or voltage is of interest. The iterative technique is also useful in the analysis of networks containing nonlinear or active devices.

16. REDUNDANT IMPEDANCES

Circuit reduction occurs when unnecessary impedances are removed according to the following rules.

- Impedance in parallel with an ideal voltage source may be removed. Theoretically, the ideal voltage supplies whatever energy is required to maintain the voltage, so the current flowing through the parallel impedance has no impact on the remainder of the circuit. Care must be exercised when calculating the source current to take into account the current through the parallel impedance.
- Impedance in series with an ideal current source may be removed. Theoretically, the ideal current source supplies whatever energy is required to maintain the current, so the energy loss in the series impedance has no impact on the remainder of the circuit. The voltage across the current source or current source/series impedance combination is determined by the remainder of the circuit.

17. DELTA-WYE TRANSFORMATIONS

Electrical impedances arranged in the shape of the Greek letter delta (Δ) or the English letter *Y* (wye) are known as *delta-wye configurations*, or *pi-T configurations*. A delta-wye configuration (see Fig. 29.10) can be useful in reducing circuits in power networks.

The equivalent impedances, which allow transformation between configurations, are

$$Z_a = \frac{Z_1Z_2 + Z_1Z_3 + Z_2Z_3}{Z_3} \quad 29.26$$

$$Z_b = \frac{Z_1Z_2 + Z_1Z_3 + Z_2Z_3}{Z_1} \quad 29.27$$

$$Z_c = \frac{Z_1Z_2 + Z_1Z_3 + Z_2Z_3}{Z_2} \quad 29.28$$

$$Z_1 = \frac{Z_aZ_c}{Z_a + Z_b + Z_c} \quad 29.29$$

$$Z_2 = \frac{Z_aZ_b}{Z_a + Z_b + Z_c} \quad 29.30$$

$$Z_3 = \frac{Z_bZ_c}{Z_a + Z_b + Z_c} \quad 29.31$$

Figure 29.10 *Delta (Pi)-Wye (T) Configurations*

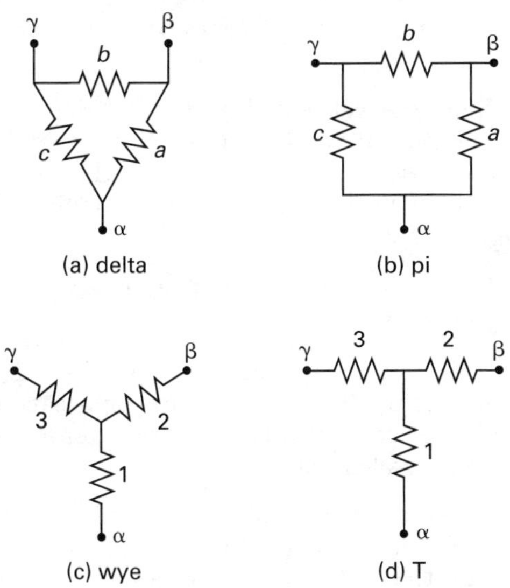

18. THEVENIN'S THEOREM

Thevenin's theorem states that, insofar as the behavior of a linear circuit at its terminals is concerned, any such circuit can be replaced by a single voltage source, $\mathbf{V}_{\text{Th}}$, in series with a single impedance, $\mathbf{Z}_{\text{Th}}$. The method for determining and utilizing the *Thevenin equivalent circuit*, with designations referring to Fig. 29.11, follows.

step 1: Separate the network that is to be changed into a Thevenin equivalent circuit from its load at two terminals, say A and B.

step 2: Determine the open-circuit voltage, $\mathbf{V}_{\text{oc}}$, at terminals A and B.

step 3: Short-circuit terminals A and B and determine the current, $\mathbf{I}_{\text{sc}}$.

step 4: Calculate the Thevenin equivalent voltage and resistance from the following equations.

$$\mathbf{V}_{\text{Th}} = \mathbf{V}_{\text{oc}} \quad 29.32$$

$$\mathbf{Z}_{\text{Th}} = \frac{\mathbf{V}_{\text{oc}}}{\mathbf{I}_{\text{sc}}} \quad 29.33$$

step 5: Using the values calculated in Eq. 29.32 and Eq. 29.33, replace the network with the Thevenin equivalent. Reconnect the load at terminals A and B. Determine the desired electrical parameters in the load.

Steps 3 and 4 can be altered by using the following shortcut. Determine the Thevenin equivalent impedance by looking into terminals A and B toward the network with all the power sources altered. Specifically, change independent voltage sources into short circuits and independent current sources into open circuits. Then calculate the resistance of the altered network. The resulting resistance is the Thevenin equivalent impedance.

Figure 29.11 *Thevenin Equivalent Circuit*

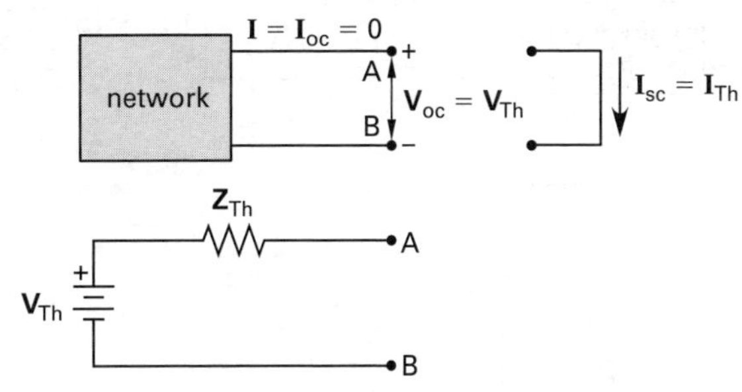

19. NORTON'S THEOREM

Norton's theorem states that, insofar as the behavior of a linear circuit at its terminals is concerned, any such circuit can be replaced by a single current source, $\mathbf{I}_N$, in parallel with a single impedance, $\mathbf{Z}_N$. The method for determining and utilizing the *Norton equivalent circuit*, with designations referring to Fig. 29.12, follows.

step 1: Separate the network that is to be changed into a Norton equivalent circuit from its load at two terminals, say A and B.

step 2: Determine the open-circuit voltage, $\mathbf{V}_{\text{oc}}$, at terminals A and B.

step 3: Short-circuit terminals A and B and determine the current, $\mathbf{I}_{\text{sc}}$.

step 4: Calculate the Norton equivalent current and resistance from the following equations.

$$\mathbf{I}_N = \mathbf{I}_{\text{sc}} \quad 29.34$$

$$\mathbf{Z}_N = \frac{\mathbf{V}_{\text{oc}}}{\mathbf{I}_{\text{sc}}} \quad 29.35$$

step 5: Using the values calculated in Eq. 29.34 and Eq. 29.35, replace the network with the Norton equivalent. Reconnect the load at terminals A and B. Determine the desired electrical parameters in the load.

Steps 2 and 4 can be altered by using the following shortcut. Determine the Norton equivalent impedance by looking into terminals A and B toward the network with all the power sources altered. Specifically, change independent voltage sources into short circuits and independent current sources into open circuits. Then calculate the resistance of the altered network. The resulting impedance is the Norton equivalent impedance.

The Norton equivalent impedance equals the Thevenin equivalent impedance for identical networks.

Figure 29.12 *Norton Equivalent Circuit*

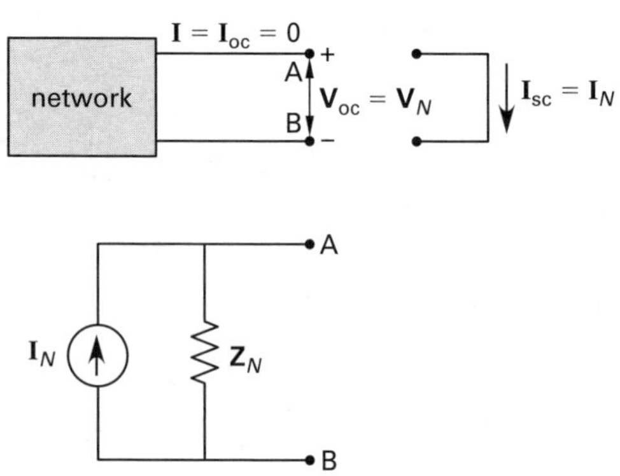

20. MAXIMUM POWER TRANSFER THEOREM

The maximum energy, that is, maximum power transfer from a voltage source, occurs when the series source impedance, $\mathbf{Z}_s$, is reduced to the minimum possible value, with zero being the ideal case. This assumes the load impedance is fixed and the source impedance can be changed. Though this is the ideal situation, it is not often the case. Where the load impedance varies and the source impedance is fixed, maximum power transfer occurs when the load and source impedances are complex conjugates. That is, $\mathbf{Z}_l = \mathbf{Z}_s^*$ or $R_{\text{load}} + jX_{\text{load}} = R_s - jX_s$.

Similarly, the maximum power transfer from a current source occurs when the parallel source impedance, $\mathbf{Z}_s$, is increased to the maximum possible value. This assumes the load impedance is fixed and the source impedance can be changed. In the more common case where the load impedance varies and the source impedance is fixed, the maximum power transfer occurs when the load and source impedances are complex conjugates.

The maximum power transfer in any circuit occurs when the load impedance equals the complex conjugate of the Norton or Thevenin equivalent impedance.

21. SUPERPOSITION THEOREM

The principle of *superposition* is that the response of (that is, the voltage across or current through) a linear circuit element in a network with multiple independent sources is equal to the response obtained if each source is considered individually and the results summed. The steps involved in determining the desired quantity follow.

step 1: Replace all sources except one by their internal resistances. Ideal current sources are replaced by open circuits. Ideal voltage sources are replaced by short circuits.

step 2: Compute the desired quantity, either voltage or current caused by the single source, for the element in question.

step 3: Repeat steps 1 and 2 for each of the sources in turn.

step 4: Sum the calculated values obtained for the current or voltage obtained in step 2. The result is the actual value of the current or voltage in the element for the complete circuit.

Superposition is not valid for circuits in which the following conditions exist.

- The capacitors have an initial charge (i.e., an initial voltage) not equal to zero. This principle can be used if the charge is treated as a separate voltage source and the equivalent circuit analyzed.
- The inductors have an initial magnetic field (i.e., an initial current) not equal to zero. This principle can be used if the energy in the magnetic field is treated as a separate current source and this equivalent circuit analyzed.
- Dependent sources are used.

22. MILLER'S THEOREM

A standard two-port network is shown in Fig. 29.13(a). *Miller's theorem* states that if an admittance, Y, is connected between the input and output terminals of a two-port network, as shown in Fig. 29.13(b), the output voltage is linearly dependent on the input voltage. Further, the circuit can be transformed as shown in

Fig. 29.13(c). The transformation equations, with A as a constant that must be determined by independent means, are

$$Y_1 = Y(1 - A) \tag{29.36}$$

$$Y_2 = Y\left(1 - \frac{1}{A}\right) \tag{29.37}$$

Similar formulas can be developed from Eq. 29.36 and Eq. 29.37 for impedances Z_1 and Z_2, though the theorem is not often used in such a manner. Miller's theorem is useful in transistor high-frequency amplifiers, among other applications.

Figure 29.13 *Miller's Theorem*

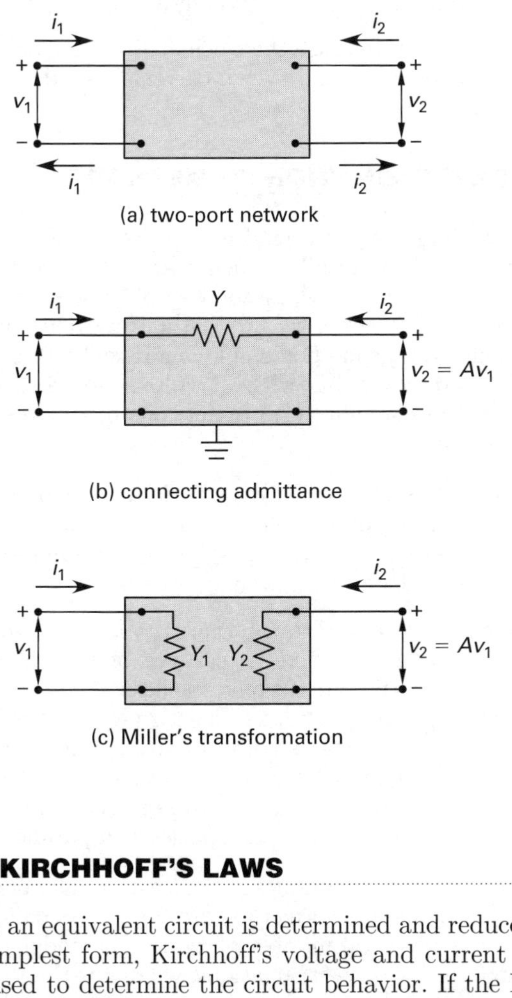

(a) two-port network

(b) connecting admittance

(c) Miller's transformation

23. KIRCHHOFF'S LAWS

Once an equivalent circuit is determined and reduced to its simplest form, Kirchhoff's voltage and current laws are used to determine the circuit behavior. If the KVL and KCL equations are written in terms of instantaneous quantities, $v(t)$ and $i(t)$, a differential equation results with an order equal to the number of independent energy-storing devices in the circuit. The solution of this equation results in a steady-state component and a transient component, which decays with time. If the equation is written in terms of phasor quantities, **V** and **I**, an algebraic equation results. The solution of this equation alone results in a steady-state component. For an equation with an order higher than two, the most efficient solution method is computer-aided design software.

24. KIRCHHOFF'S VOLTAGE LAW

Kirchhoff's voltage law (KVL) states that the algebraic sum of the phasor voltages around any closed path within a circuit or network is zero. Stated in terms of the voltage rises and drops, the sum of the phasor voltage rises equals the sum of the phasor voltage drops around any closed path within a circuit or network. The method for applying KVL follows.

step 1: Identify the loop.

step 2: Pick a loop direction.[6]

step 3: Assign the loop current in the direction picked in step 2.

step 4: Assign voltage polarities consistent with the loop current direction in step 3.

step 5: Apply KVL to the loop using Ohm's law to express the voltages across each circuit element.

step 6: Solve the equation for the desired quantity.

25. KIRCHHOFF'S CURRENT LAW

Kirchhoff's current law (KCL) states that the algebraic sum of the phasor currents at any node is zero. Stated in terms of the currents out of and into a node, the sum of the currents directed out of a node must equal the sum of the currents directed into the same node. The method for applying KCL follows.

step 1: Identify the nodes and pick a reference or datum node.

step 2: Label the node-to-datum voltage for each unknown node.

step 3: Pick a current direction for each path at every node.[7]

step 4: Apply KCL to the nodes using Ohm's law to express the currents through each circuit branch.

step 5: Solve the equations for the desired quantity.

[6]Clockwise is the direction chosen as the standard in this text. Any direction is allowed. The resulting mathematics determines the correct direction of current flow and voltage polarities. Consistency is the key to preventing calculation errors.

[7]Current flow out of the node is the direction chosen as the standard for positive current in this text. Any direction is allowed. The resulting mathematics determines the correct direction of current flow and voltage polarities. Consistency is the key to preventing calculation errors.

26. LOOP ANALYSIS

The *loop-current* method is a systematic network-analysis procedure that uses currents as the unknowns. It is also called *mesh analysis* or the *Maxwell loop-current* method. The method uses Kirchhoff's voltage law and is performed on planar networks. It requires $n-1$ simultaneous equations for an n-loop system. The method's steps are as follows.

step 1: Select $n-1$ loops, that is, one loop less than the total number of possible loops.

step 2: Assign current directions for the selected loops. (While this is arbitrary, clockwise directions will always be chosen in this text for consistency.) Any incorrectly chosen current direction will cause a negative result when the simultaneous equations are solved in step 4. Show the direction of the current with an arrow.

step 3: Write Kirchhoff's voltage law for each of the selected loops. Assign polarities based on the direction of the loop current. The voltage of a source is positive when the current flows out of the positive terminal, that is, from the negative terminal to the positive terminal inside the source. The selected direction of the loop current always results in a voltage drop in the resistors of the loop (see Fig. 26.6). Where two loop currents flow through an element, they are summed to determine the voltage drop in that element, using the direction of the current in the loop for which the equation is being written as the positive (i.e., correct) direction. Any incorrect direction for the loop current will be indicated by a negative sign in the solution in step 4.

step 4: Solve the $n-1$ equations from step 3 for the unknown currents.

step 5: If required, determine the actual current in an element by summing the loop currents flowing through the element. Sum the absolute values to obtain the correct magnitude. The correct direction is given by the loop current with the positive value.

27. NODE ANALYSIS

The *node-voltage* method is a systematic network-analysis procedure that uses voltages as the unknowns. The method uses Kirchhoff's current law. It requires $n-1$ equations for an n-principal node system (equations are not necessary at simple nodes, that is, nodes connecting only two circuit elements). The method's steps are as follows.

step 1: Simplify the circuit, if possible, by combining resistors in series or parallel or by combining current sources in parallel. Identify all nodes. The minimum number of equations required will be $n-1$ where n represents the number of principal nodes.

step 2: Choose one node as the reference node, that is, the node that will be assumed to have ground potential (0 V). To minimize the number of terms in the equations, select the node with the largest number of circuit elements to serve as the reference node.

step 3: Write Kirchhoff's current law for each principal node except the reference node, which is assumed to have a zero potential.

step 4: Solve the $n-1$ equations from step 3 to determine the unknown voltages.

step 5: If required, use the calculated node voltages to determine any branch current desired.

28. DETERMINATION OF METHOD

The method used to analyze an electrical network depends on the circuit elements and their configurations. The loop-current method employing Kirchhoff's voltage law is used in circuits without current sources. The node-voltage method employing Kirchhoff's current law is used in circuits without voltage sources. When both types of sources are present, one of two methods may be used.

method 1: Use each of Kirchhoff's laws, assigning voltages and currents as needed, and substitute any known quantity into the equations as written.

method 2: Use source transformation or source shifting to change the appearance of the circuit so that it contains only the desired sources, that is, voltage sources when using KVL and current sources when using KCL. *Source shifting* is manipulating the circuit so that each voltage source has a resistor in series and each current source has a resistor in parallel.

Use the method that results in the least number of equations. The loop-current method produces $n-1$ equations where n is the total number of loops. The node-voltage method produces $n-1$ equations where n is the number of principal nodes. Count the number of loops and the number of principal nodes prior to writing the equations. Whichever is least determines the method used.

Additional methods exist, some of which require less work. The advantage of the loop-current and node-voltage methods is that they are systematic and guarantee a solution.

29. VOLTAGE AND CURRENT DIVIDERS

At times, a source voltage will not be at the required value for the operation of a given circuit. For example, the household voltage of 120 V is too high to properly bias electronic circuitry. One method of obtaining the required voltage without using a transformer is to use a *voltage divider*. Figure 29.14(a) illustrates a voltage divider. The voltage across impedance 2 is

$$\mathbf{V}_2 = \mathbf{V}_s \left(\frac{\mathbf{Z}_2}{\mathbf{Z}_1 + \mathbf{Z}_2} \right) \qquad 29.38$$

If the fraction of the source voltage, $\mathbf{V}_s$, is found as a function of $\mathbf{V}_2$, the result is known as the *gain* or the *voltage-ratio transfer function*. Rearranging Eq. 29.38 to show the specific relationship between $\mathbf{Z}_1$ and $\mathbf{Z}_2$ gives

$$\frac{\mathbf{V}_2}{\mathbf{V}_s} = \frac{1}{1 + \frac{\mathbf{Z}_1}{\mathbf{Z}_2}} \qquad 29.39$$

A specific current can be obtained through an analogous circuit called a *current divider*, shown in Fig. 29.14(b). The current through impedance 2 is

$$\mathbf{I}_2 = \mathbf{I}_s \left(\frac{\mathbf{Z}_1}{\mathbf{Z}_1 + \mathbf{Z}_2} \right) = \mathbf{I}_s \left(\frac{\mathbf{G}_2}{\mathbf{G}_1 + \mathbf{G}_2} \right) \qquad 29.40$$

Figure 29.14 *Divider Circuits*

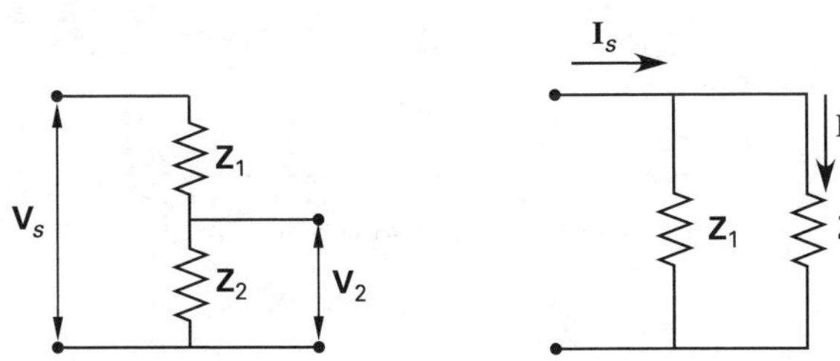

30. STEADY-STATE AND TRANSIENT IMPEDANCE ANALYSIS

When the electrical parameters of a circuit do not change with time, the circuit is said to be in a *steady-state* condition. In DC circuits, this condition exists when the magnitude of the parameter is constant. In AC circuits, this condition exists when the frequency is constant. A *transient* is a temporary phenomenon occurring prior to a network reaching steady state.

DC steady-state impedance analysis is based on constant voltages and currents and, therefore, time derivatives of zero. The DC impedances are

- Resistance:

$$Z_R|_{\text{DC}} = R \qquad 29.41$$

- Inductance:

$$v = L\frac{di}{dt} = L(0) = 0 \qquad 29.42$$

$$Z = \frac{v}{i} = \frac{0}{i} = 0 \qquad 29.43$$

$$Z_L|_{\text{DC}} = 0 \quad \text{[short circuit]} \qquad 29.44$$

- Capacitance:

$$i = C\frac{dv}{dt} = C(0) = 0 \qquad 29.45$$

$$Z = \frac{v}{i} = \frac{v}{0} \rightarrow \infty \qquad 29.46$$

$$Z_C|_{\text{DC}} = \infty \quad \text{[open circuit]} \qquad 29.47$$

AC steady-state impedance analysis is based on the phasor form where $df(t)/dt = j\omega f(t)$. The AC steady-state impedances are

- Resistance:

$$Z_R|_{\text{AC}} = R \qquad 29.48$$

- Inductance:

$$v = L\frac{di}{dt} = Lj\omega i \qquad 29.49$$

$$Z = \frac{v}{i} = \frac{Lj\omega i}{i} = j\omega L \qquad 29.50$$

$$Z_L|_{\text{AC}} = j\omega L \qquad 29.51$$

- Capacitance:

$$i = C\frac{dv}{dt} = Cj\omega v \qquad 29.52$$

$$Z = \frac{v}{i} = \frac{v}{Cj\omega v} = \frac{1}{j\omega C} \qquad 29.53$$

$$Z_C|_{\text{AC}} = \frac{1}{j\omega C} \qquad 29.54$$

Transient impedance analysis is based on the phasor form with the complex variable s substituted for $j\omega$. The variable $s = \sigma + j\omega$ and is the same as the Laplace transform variable. The derivative is $df(t)/dt = sf(t)$. The transient impedances are

- Resistance:

$$Z_R = R \qquad 29.55$$

- Inductance:

$$Z_L = sL \qquad 29.56$$

Circuit Theory

- Capacitance:

$$Z_C = \frac{1}{sC} \tag{29.57}$$

Transient impedances are useful in the stability analysis as well as during transients.

31. TWO-PORT NETWORKS

An electric circuit or network is often used to connect a source to a load, modifying the source energy or information in a given manner as required or desired by the load. If the circuit is such that the current flow into one terminal is equal to the current flow out of a second terminal, the terminal pair is called a *port*. A given network may have any number of ports, the most common being the *two-port network* shown in Fig. 29.15. Four variables exist in this representation: v_1, v_2, i_1, and i_2. The subscript 1 indicates the *input port*, and the subscript 2 indicates the *output port*.

Figure 29.15 *Two-Port Network*

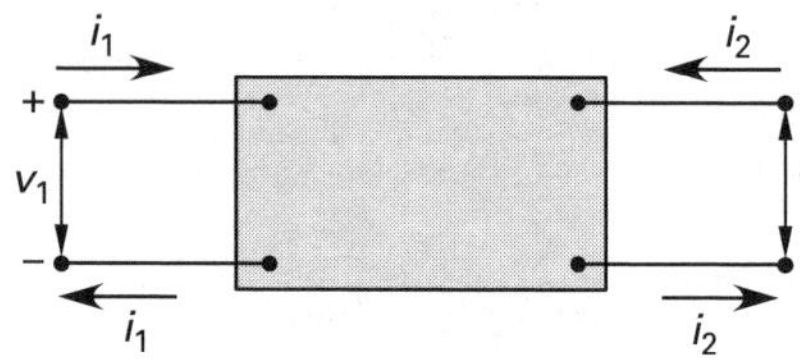

Two-port networks can represent either three- or four-terminal devices, such as transistors or transformers (see Fig. 29.16). The electrical properties of the device within the network are described by a set of parameters identified by double subscripts, the first representing the row and the second the column of a matrix. The *parameter type* is determined by the selection of independent variables for which to solve.

Figure 29.16 *Transformer and Transistor Two-Port Networks*

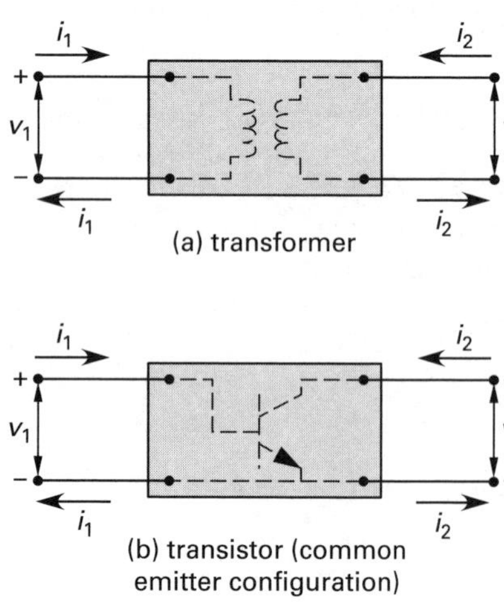

(a) transformer

(b) transistor (common emitter configuration)

Open-circuit impedance parameters, or z parameters, occur when the two currents are selected as the independent variables. The z parameter model and *deriving equations* for the individual parameters are shown in Fig. 29.17. The applicable matrix and resulting equations follow.

$$\begin{bmatrix} v_1 \\ v_2 \end{bmatrix} = \begin{bmatrix} z_{11} & z_{12} \\ z_{21} & z_{22} \end{bmatrix} \begin{bmatrix} i_1 \\ i_2 \end{bmatrix} \tag{29.58}$$

$$v_1 = z_{11} i_1 + z_{12} i_2 \tag{29.59}$$

$$v_2 = z_{21} i_1 + z_{22} i_2 \tag{29.60}$$

Figure 29.17 *Impedance Model Parameters*

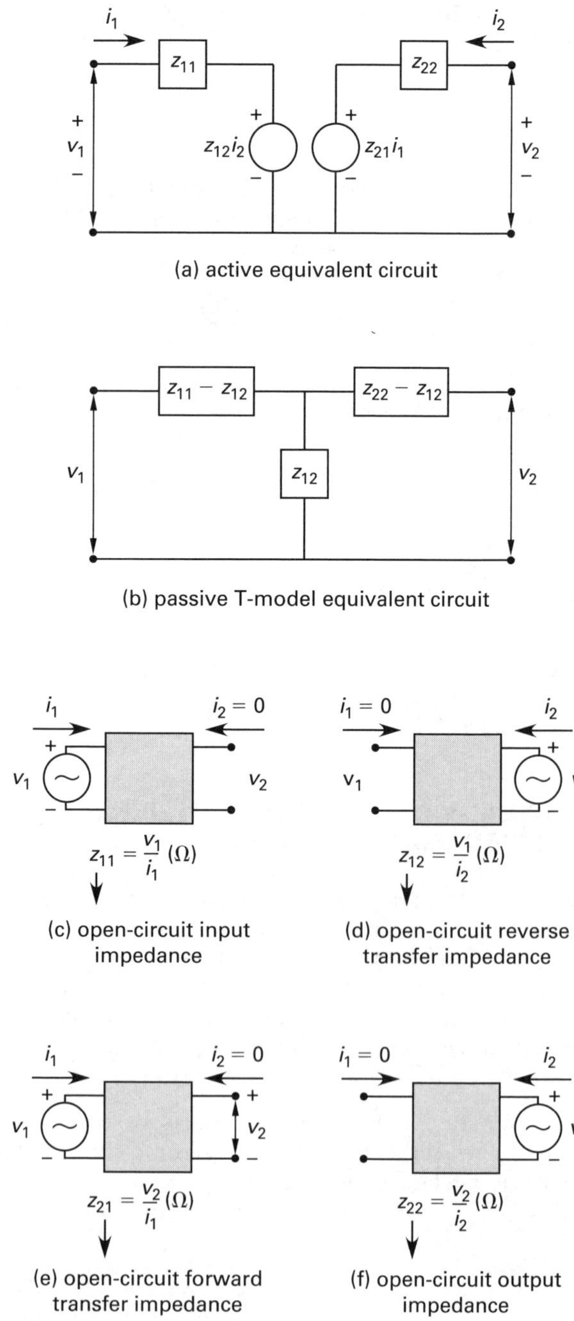

(a) active equivalent circuit

(b) passive T-model equivalent circuit

(c) open-circuit input impedance

(d) open-circuit reverse transfer impedance

(e) open-circuit forward transfer impedance

(f) open-circuit output impedance

Short-circuit admittance parameters, or y parameters, occur when the two voltages are selected as the independent variables. The y parameter model and *deriving equations* for the individual parameters are shown in Fig. 29.18. The applicable matrix and resulting equations follow.

$$\begin{bmatrix} i_1 \\ i_2 \end{bmatrix} = \begin{bmatrix} y_{11} & y_{12} \\ y_{21} & y_{22} \end{bmatrix} \begin{bmatrix} v_1 \\ v_2 \end{bmatrix} \quad 29.61$$

$$i_1 = y_{11}v_1 + y_{12}v_2 \quad 29.62$$

$$i_2 = y_{21}v_1 + y_{22}v_2 \quad 29.63$$

Figure 29.18 *Admittance Model Parameters*

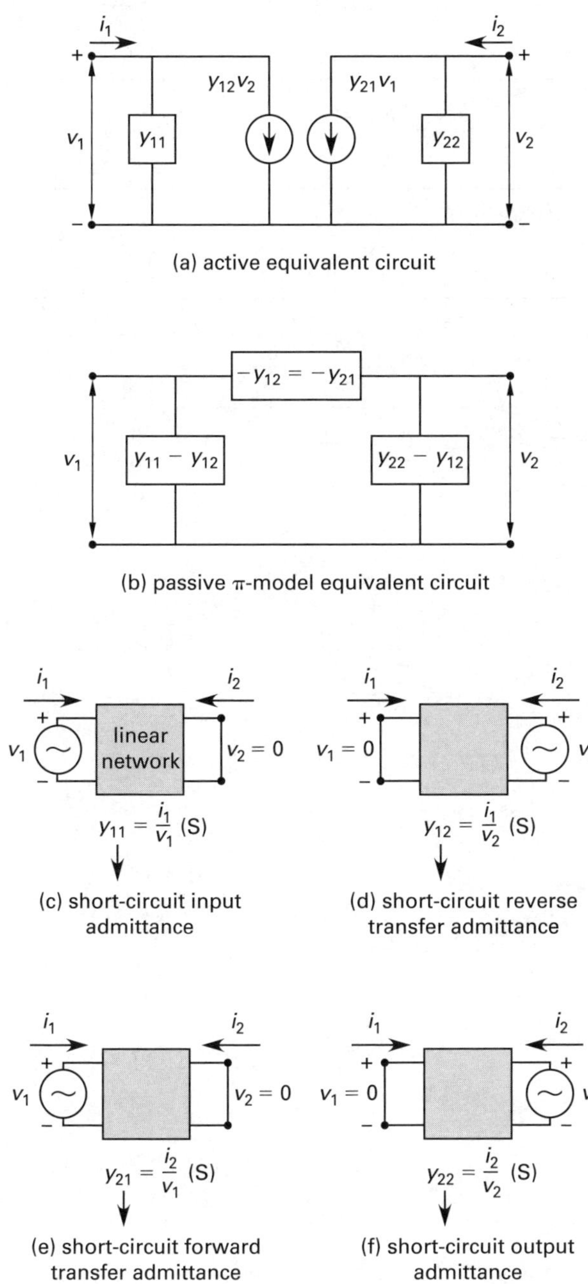

Hybrid parameters, or h parameters, occur when one voltage and one current are selected as the independent variables. Figure 29.19 shows the h parameter model and *deriving equations* for the individual parameters. The applicable matrix and resulting equations follow.

$$\begin{bmatrix} v_1 \\ i_2 \end{bmatrix} = \begin{bmatrix} h_{11} & h_{12} \\ h_{21} & h_{22} \end{bmatrix} \begin{bmatrix} i_1 \\ v_2 \end{bmatrix} \quad 29.64$$

$$v_1 = h_{11}i_1 + h_{12}v_2 \quad 29.65$$

$$i_2 = h_{21}i_1 + h_{22}v_2 \quad 29.66$$

Figure 29.19 *Hybrid Model Parameters*

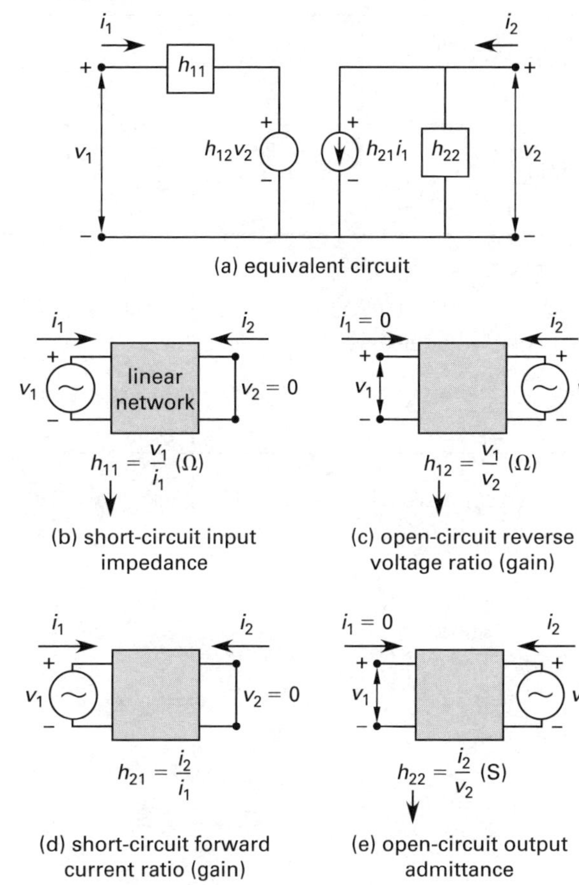

Inverse hybrid parameters, or g parameters, also occur when one voltage and one current are picked as the independent variables. The applicable matrix and resulting equations follow.

$$\begin{bmatrix} i_1 \\ v_2 \end{bmatrix} = \begin{bmatrix} g_{11} & g_{12} \\ g_{21} & g_{22} \end{bmatrix} \begin{bmatrix} v_1 \\ i_2 \end{bmatrix} \quad 29.67$$

$$i_1 = g_{11}v_1 + g_{12}i_2 \quad 29.68$$

$$v_2 = g_{21}v_1 + g_{22}i_2 \quad 29.69$$

Other parameter representations are possible. Some examples include the *transmission line* or *chain parameters*, and the *inverse transmission line parameters*. The choice of parameter type depends on numerous factors, including whether or not all the parameters exist or are defined for a given network, the mathematical convenience of using a certain set of parameters, and the accuracy of sensitivity of the parameters when considered as part of the overall circuit to which it is connected. The parameters and their deriving equations are summarized in Table 29.3. The formulas for conversion between the parameter types are given in Sec. 26.3.

Table 29.3 *Two-Port Network Parameters*

representation		deriving equations		
impedance $\begin{bmatrix} z_{11} & z_{12} \\ z_{21} & z_{22} \end{bmatrix}$	$z_{11} = \frac{V_1}{I_1}$ $I_2 = 0$	$z_{12} = \frac{V_1}{I_2}$ $I_1 = 0$	$z_{21} = \frac{V_2}{I_1}$ $I_2 = 0$	$z_{22} = \frac{V_2}{I_2}$ $I_1 = 0$
admittance $\begin{bmatrix} y_{11} & y_{12} \\ y_{21} & y_{22} \end{bmatrix}$	$y_{11} = \frac{I_1}{V_1}$ $V_2 = 0$	$y_{12} = \frac{I_1}{V_2}$ $V_1 = 0$	$y_{21} = \frac{I_2}{V_1}$ $V_2 = 0$	$y_{22} = \frac{I_2}{V_2}$ $V_1 = 0$
hybrid $\begin{bmatrix} h_{11} & h_{12} \\ h_{21} & h_{22} \end{bmatrix}$	$h_{11} = \frac{V_1}{I_1}$ $V_2 = 0$	$h_{12} = \frac{V_1}{V_2}$ $I_1 = 0$	$h_{21} = \frac{I_2}{I_1}$ $V_2 = 0$	$h_{22} = \frac{I_2}{V_2}$ $I_1 = 0$
inverse hybrid $\begin{bmatrix} g_{11} & g_{12} \\ g_{21} & g_{22} \end{bmatrix}$	$g_{11} = \frac{I_1}{V_1}$ $I_2 = 0$	$g_{12} = \frac{I_1}{I_2}$ $V_1 = 0$	$g_{21} = \frac{V_2}{V_1}$ $I_2 = 0$	$g_{22} = \frac{V_2}{I_2}$ $V_1 = 0$
transmission or chain $\begin{bmatrix} A & B \\ C & D \end{bmatrix}$	$A = \frac{V_1}{V_2}$ $I_2 = 0$	$B = \frac{-V_1}{I_2}$ $V_2 = 0$	$C = \frac{I_1}{V_2}$ $I_2 = 0$	$D = \frac{-I_1}{I_2}$ $V_2 = 0$
inverse transmission $\begin{bmatrix} \alpha & \beta \\ \gamma & \delta \end{bmatrix}$	$\alpha = \frac{v_2}{v_1}$ $i_1 = 0$	$\beta = \frac{v_2}{i_1}$ $v_1 = 0$	$\gamma = \frac{-i_2}{v_1}$ $i_1 = 0$	$\delta = \frac{i_2}{i_1}$ $v_i = 0$

30 Transient Analysis

1. Fundamentals 30-1
2. Resistor-Capacitor Circuits: Natural Response 30-2
3. Resistor-Capacitor Circuits: Forced Response 30-3
4. Resistor-Inductor Circuits: Natural Response 30-4
5. Resistor-Inductor Circuits: Forced Response 30-4
6. RC and RL Circuits: Solution Method 30-5
7. Rise Time 30-6
8. Damped Oscillations: Ringing 30-7
9. Sustained Oscillations: Resonance 30-8
10. Resonant Circuits 30-8
11. Series Resonance 30-9
12. Parallel Resonance 30-10

Nomenclature

a	constant	–
A	natural response coefficient	–
b	constant	–
c	constant	–
C	capacitance	F
f	function	–
G	conductance	S
i	variable current	A
I	constant or rms current	A
L	self-inductance	H
N	number	–
P	power	W
q	instantaneous charge	C
Q	quality factor	–
R	resistance	Ω
s	Laplace transform variable	–
t	time	s
v	variable voltage	V
V	constant or rms voltage	V
X	reactance	Ω
Y	admittance	S
Z	impedance	Ω

Symbols

κ	integration constant	–
σ	damping, real part of s	–
τ	time constant	s
ϕ	angle	rad
ϕ	phase difference angle	rad
ω	frequency	rad/s
ω_0	zero resistance frequency	rad/s

Subscripts

0	initial
bat	battery
C	capacitor
d	delay
L	inductor
m	maximum
r	rise
R	resistor
s	source
ss	steady-state
tr	transient

1. FUNDAMENTALS

Whenever a network or circuit undergoes a change, the currents and voltages experience a transitional period during which their properties shift from their former values to their new steady-state values. This time period is called a *transient*. The determination of a circuit's behavior during this period is *transient analysis*. There are three classes of transient problems, all based on the nature of the energy source. These are *DC switching transients*, *AC switching transients*, and *pulse transients*. *Switching transients* occur as a result of a change in topology of the circuit, that is, the physical elements within the circuit change (become connected or disconnected) as a result of the operation of a switch. The switch can be a physical switch or an electrical one, such as a transistor operating in the switching mode. A circuit's response to a switching transient is known as a *step response*.

Pulse transients involve a change in the current or voltage waveform, not in the topology of a circuit. A circuit's response to a pulse transient is known as an *impulse response*.

The energy storage elements in electric circuits are the capacitor and the inductor. When one of these is present, the mathematical representation of the circuit is a first-order differential equation of the form

$$f(t) = b\left(\frac{dx}{dt}\right) + cx \qquad 30.1$$

Either the current or the voltage can be represented as the dependent variable, that is, $f(t)$, in Eq. 30.1. For capacitors, the voltage is used because the energy storage is a function of voltage and a true differential equation results (the capacitor current is an integral equation). For inductors, similar reasoning holds and

Circuit Theory

the current is considered the dependent variable, $f(t)$. The solution to Eq. 30.1 is of the general form

$$x(t) = \kappa + Ae^{-t/\tau} \quad 30.2$$

See Table 30.1 for the applicable equations for both the inductor and capacitor.

The term A in Eq. 30.2 is the natural response coefficient and is set to match the conditions at some known time in the transient, typically at the onset. The term κ is value-dependent on the forcing function, $f(t)$. The term τ is called the *time constant* and is equal to

$$\tau = \frac{b}{c} \quad 30.3$$

The time constant for an RC circuit is

$$\tau = RC \quad 30.4$$

The time constant for an RL circuit is

$$\tau = \frac{L}{R} \quad 30.5$$

There are two parts to Eq. 30.2, or any solution to a first-order differential equation: the homogeneous (or complementary) solution and the particular solution. The homogeneous solution, that is, the solution where $f(t) = 0$, is called the *natural response* of the circuit and is represented by the exponential term in Eq. 30.2. The decay behavior of a general exponential is shown in Fig. 30.1. The particular solution is called the *forced response*, since it depends on the forcing function, $f(t)$, and is represented by κ in Eq. 30.2.

Figure 30.1 *Exponential Behavior*

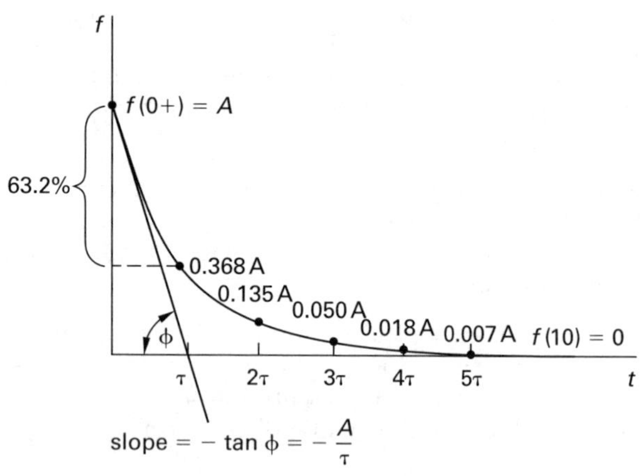

When two independent energy storage elements are present, the mathematical representation of the circuit is a second-order differential equation of the form

$$f(t) = a\left(\frac{d^2x}{dt^2}\right) + b\left(\frac{dx}{dt}\right) + cx \quad 30.6$$

There are three forms of the solution, depending on the magnitudes of the constants a, b, and c (see Chap. 31).

2. RESISTOR-CAPACITOR CIRCUITS: NATURAL RESPONSE

Consider the generic resistor-capacitor *source-free circuit* shown in Fig. 30.2. When the charged capacitor, C, is connected to the resistor, R, via a complete electrical path, the capacitor will discharge in an attempt to resist the change in voltage. In the process, the capacitor's stored electrical energy is dissipated in the resistor until no further energy remains and no current flows. This gradual decrease is the transient. The voltage follows the form of Fig. 30.1.

Figure 30.2 *Resistor-Capacitor Circuit: Natural Response*

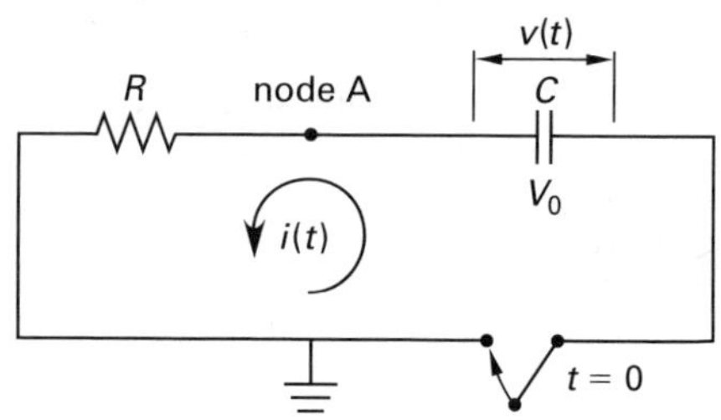

Example 30.1

Determine the formula for the voltage in the circuit of Fig. 30.2 from $t=0$ onward.

Solution

Write Kirchhoff's current law (KCL) for the simple node A between the resistor and the capacitor. The reference node is located at the switch, opposite node A. In keeping with the convention of this book, both currents are assumed to flow out of the node, giving

$$i_C + i_R = 0$$

Substitute the expression for current flow through a capacitor and the Ohm's law expression for current flow in a resistor.

$$C\left(\frac{dv}{dt}\right) + \frac{v}{R} = 0$$

$$\frac{dv}{dt} + \left(\frac{1}{RC}\right)v = 0$$

This is a homogeneous first-order linear differential equation. The solution could be written directly from the information in the mathematics chapters or by knowing that only an exponential form of v can be linearly combined with its derivative. In order to show the process, separate the variables and integrate both sides.

$$\frac{dv}{v} = -\left(\frac{1}{RC}\right)dt$$
$$\int \frac{1}{v}\,dv = \int -\left(\frac{1}{RC}\right)dt$$
$$\ln v = -\frac{1}{RC}t + \kappa$$
$$v(t) = e^{-(1/RC)t+\kappa} = Ae^{-t/RC} = Ae^{-t/\tau}$$

This is the natural response of the system. To determine the constant A, the value of $v(t)$ at some time must be known. Because the initial voltage is V_0, the value of the constant A is

$$v(0) = V_0 = Ae^{-0/\tau} = A$$

The final solution is

$$v(t) = V_0 e^{-t/RC}$$

3. RESISTOR-CAPACITOR CIRCUITS: FORCED RESPONSE

Consider the generic resistor-capacitor circuit shown in Fig. 30.3. When the charged capacitor, C, is connected to the resistor, R, via a complete electrical path, the capacitor will discharge or charge, depending on the magnitude of V_{bat}, in an attempt to resist the change in voltage. In the process, the capacitor's stored electrical energy is either dissipated or enhanced until no further energy change occurs and no current flows. This gradual decrease or increase is the transient. After the passage of time equal to five time constants, 5τ, the final value is within less than 1% of its steady-state value and the transient is considered complete. The voltage follows the form of Fig. 30.1, or its inverse, with the exception that the final steady-state voltage is the voltage of the driving force.

Figure 30.3 *Resistor-Capacitor Circuit: Forced Response*

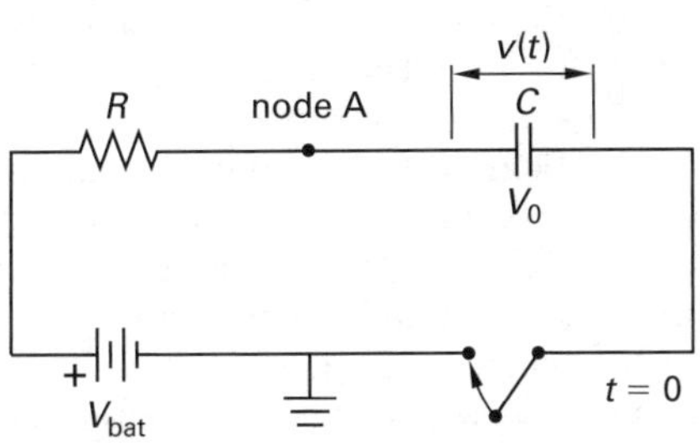

Example 30.2

Determine the formula for the voltage in the circuit of Fig. 30.3 from $t = 0$ onward.

Solution

Write KCL for the simple node A between the resistor and the capacitor. The node between the negative terminal of the battery and the capacitor, opposite node A, is the reference node. Both currents are assumed to flow out of the node in keeping with the convention of this text, giving

$$i_C + i_R = 0$$

Substitute the expression for current flow through a capacitor and the Ohm's law expression for current flow in a resistor.

$$C\left(\frac{dv}{dt}\right) + \frac{v - V_{\text{bat}}}{R} = 0$$
$$\frac{dv}{dt} + \left(\frac{1}{RC}\right)(v - V_{\text{bat}}) = 0$$
$$\frac{dv}{dt} + \left(\frac{1}{RC}\right)v = \left(\frac{1}{RC}\right)V_{\text{bat}}$$

This is a nonhomogeneous first-order linear differential equation. The known constant to the right of the equal sign is called the *forcing* or *driving function.* Such an equation is solved in two steps. The first step is to determine the homogeneous, or complementary, solution, that is, the solution with the forcing function equal to zero. In Ex. 30.1 this was determined to be

$$v_{\text{tr}}(t) = Ae^{-t/RC}$$

This is the natural response of the system and represents the transient portion of the solution. The next step is to determine the particular solution, or forced response, of the system. Because the particular solution is the steady-state solution, the expectation is that the solution would be of the same form as the forcing function. Assuming a constant, κ, as the particular solution and substituting this into the KCL expression for v gives

$$v_{\text{ss}} = \kappa$$
$$\frac{dv}{dt} + \left(\frac{1}{RC}\right)v = \left(\frac{1}{RC}\right)V_{\text{bat}}$$
$$\frac{dv_{\text{ss}}}{dt} + \left(\frac{1}{RC}\right)v_{\text{ss}} = \left(\frac{1}{RC}\right)V_{\text{bat}}$$
$$\frac{d\kappa}{dt} + \left(\frac{1}{RC}\right)\kappa = \left(\frac{1}{RC}\right)V_{\text{bat}}$$
$$0 + \left(\frac{1}{RC}\right)\kappa = \left(\frac{1}{RC}\right)V_{\text{bat}}$$
$$\kappa = V_{\text{bat}}$$

The particular solution is

$$v(\infty) = v_{\text{ss}} = V_{\text{bat}}$$

The total solution is the combination of the complementary (homogeneous) and particular (nonhomogeneous) solutions, that is, the combination of the transient and steady-state solutions. The total solution is

$$v(t) = V_{\text{bat}} + Ae^{-t/RC}$$

The order of the transient and steady-state solutions is unimportant. The exponential term (the transient term) is often listed to the right in an equation.

The natural response coefficient A is again determined by the initial condition $v(0) = V_0$.

$$v(0) = V_0 = V_{\text{bat}} + Ae^{-0/RC} = V_{\text{bat}} + A$$
$$A = V_0 - V_{\text{bat}}$$

The final solution is

$$v(t) = V_{\text{bat}} + (V_0 - V_{\text{bat}})e^{-t/RC}$$

The voltage $v(t)$ is the voltage at node A, that is, the voltage across the capacitor.

4. RESISTOR-INDUCTOR CIRCUITS: NATURAL RESPONSE

Consider the generic resistor-inductor source-free circuit shown in Fig. 30.4. When the inductor, L, whose magnetic field is at a maximum due to initial current, I_0, is connected to the resistor, R, via a complete electrical path, the inductor will discharge in an attempt to resist the change in current. In the process, the inductor's stored magnetic energy is dissipated in the resistor until no further energy remains and no current flows. This gradual decrease is the transient. After the passage of time equal to five time constants, 5τ, the final value is within less than 1% of its steady-state value and the transient is considered complete. The current follows the form of Fig. 30.1.

Figure 30.4 *Resistor-Inductor Circuit: Natural Response*

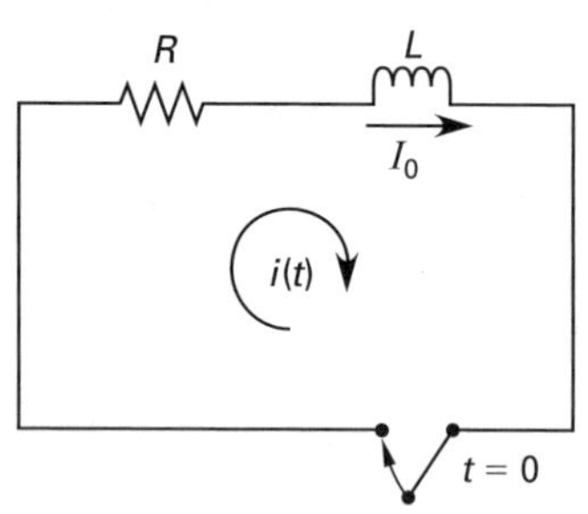

Example 30.3

Determine the formula for the current in the circuit of Fig. 30.4 from $t = 0$ onward.

Solution

Write Kirchhoff's voltage law (KVL) for the loop containing the resistor and the inductor. In keeping with the convention of this book, the reference direction is clockwise, giving

$$v_L + v_R = 0$$

Substitute the expression for voltage across an inductor and the Ohm's law expression for voltage drop in a resistor.

$$L\left(\frac{di}{dt}\right) + iR = 0$$
$$\frac{di}{dt} + \left(\frac{R}{L}\right)i = 0$$

This is a homogeneous first-order linear differential equation. The solution could be written directly from the information in the mathematics chapters or by knowing that only an exponential form of i can be linearly combined with its derivative. The process is similar to that in Ex. 30.1.

$$\frac{di}{i} = -\left(\frac{R}{L}\right)dt$$
$$\int\left(\frac{1}{i}\right)di = \int -\left(\frac{R}{L}\right)dt$$
$$\ln i = -\left(\frac{R}{L}\right)t + \kappa$$
$$i(t) = e^{-(R/L)t+\kappa} = Ae^{-t/(L/R)} = Ae^{-t/\tau}$$

This is the natural response of the system. To determine the constant A, the value of $i(t)$ at some time must be known. Because the initial current is I_0, the value of the constant A is

$$i(0) = I_0 = Ae^{-0/\tau} = A$$

The final solution is

$$i(t) = I_0e^{-t/(L/R)}$$

5. RESISTOR-INDUCTOR CIRCUITS: FORCED RESPONSE

Consider the generic resistor-inductor circuit shown in Fig. 30.5. When the inductor, L, whose magnetic field is at a maximum due to initial current, I_0, is connected to the resistor, R, via a complete electrical path, the inductor's magnetic field will collapse or expand in an attempt to resist the change in current. In the process, the inductor's stored magnetic energy is either dissipated or enhanced until no further energy change occurs and no further current change occurs. This gradual decrease or increase is the transient. After the passage of time equal to five time constants, 5τ, the final value is within less than 1% of its steady-state value and the transient is considered complete. The voltage follows the form of Fig. 30.1 or its inverse, with the exception that the final steady-state current is the current resulting from the driving force and is determined by the remainder of the circuit.

Figure 30.5 *Resistor-Inductor Circuit: Forced Response*

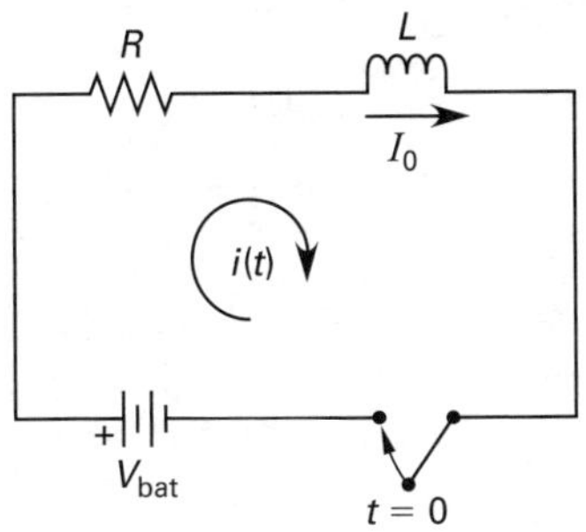

Example 30.4

Determine the formula for the current in the circuit of Fig. 30.5 from $t = 0$ onward.

Solution

Write KVL for the loop containing the resistor and the inductor. In keeping with the convention of this book, the reference direction is clockwise, giving

$$v_L + v_R - V_{bat} = 0$$

Substitute the expression for voltage across an inductor and the Ohm's law expression for voltage drop in a resistor.

$$L\left(\frac{di}{dt}\right) + iR - V_{bat} = 0$$

$$\frac{di}{dt} + \left(\frac{R}{L}\right)i = \left(\frac{1}{L}\right)V_{bat}$$

$$\frac{di}{dt} + \left(\frac{1}{\frac{L}{R}}\right)i = \left(\frac{1}{L}\right)V_{bat}$$

This is a nonhomogeneous first-order linear differential equation. The known constant to the right of the equal sign is called the *forcing* or *driving function.* Such an equation is solved in two steps. The first step is to determine the particular solution, or forced response, of the system. Because the particular solution is the steady-state solution, the expectation is that the solution will take the same form as the forcing function. Assuming a constant, κ, as the particular solution and substituting this into the KVL expression for i gives

$$i_{ss} = \kappa$$

$$\frac{di}{dt} + \left(\frac{1}{\frac{L}{R}}\right)i = \left(\frac{1}{L}\right)V_{bat}$$

$$\frac{di_{ss}}{dt} + \left(\frac{1}{\frac{L}{R}}\right)i_{ss} = \left(\frac{1}{L}\right)V_{bat}$$

$$\frac{d\kappa}{dt} + \left(\frac{1}{\frac{L}{R}}\right)\kappa = \left(\frac{1}{L}\right)V_{bat}$$

$$0 + \left(\frac{1}{\frac{L}{R}}\right)\kappa = \left(\frac{1}{L}\right)V_{bat}$$

$$\kappa = \left(\frac{1}{R}\right)V_{bat}$$

The particular solution is

$$i(\infty) = i_{ss} = \left(\frac{1}{R}\right)V_{bat}$$

The next step is to determine the homogeneous, or complementary, solution, that is, the solution with the forcing function equal to zero. In Ex. 30.3 this was determined to be

$$i_{tr}(t) = Ae^{-t/(L/R)}$$

This is the natural response of the system and represents the transient portion of the solution. The total solution is the combination of the particular (nonhomogeneous) and the complementary (homogeneous) solution, that is, the combination of the steady-state and transient solutions. The total solution is

$$i(t) = \left(\frac{1}{R}\right)V_{bat} + Ae^{-t/(L/R)}$$

The natural response coefficient A is again determined by the initial condition $i(0) = I_0$. So,

$$i(0) = I_0 = \left(\frac{1}{R}\right)V_{bat} + Ae^{-0/(L/R)} = \left(\frac{1}{R}\right)V_{bat} + A$$

$$A = I_0 - \left(\frac{1}{R}\right)V_{bat}$$

The final solution is

$$i(t) = \left(\frac{1}{R}\right)V_{bat} + \left(I_0 - \left(\frac{1}{R}\right)V_{bat}\right)e^{-t/(L/R)}$$

The current $i(t)$ is the current through the inductor.

6. RC AND RL CIRCUITS: SOLUTION METHOD

The method presented in Sec. 30.2 through Sec. 30.5 is valid for any circuit that can be represented by a single equivalent capacitor or inductor and a single equivalent resistor. The solutions were obtained for DC transients.

Table 30.1 is a summary of DC transient responses.[1] AC transients are handled in the same manner except that the forcing function, that is, the source, is represented as a sinusoid. Phasor methods, or sinusoidal methods, are then used to solve the resulting differential equation. If the switch closure were instead treated as a pulse from the source and represented by the unit step function, $u(t)$, the voltage applied at $t = 0^+$ would have been $V_{\text{bat}}u(t)$ and the equations would have been handled in the same manner. The steps for solving single resistor and single capacitor/inductor circuits follow.

step 1: Find the steady-state response from the particular solution of the nonhomogeneous equation. The solution will be of the same form as the forcing function.[2] Regardless of transient type, the result will be a steady-state value, κ.

step 2: Determine the transient response from the homogeneous equation or from the known parameters of the time constant. The transient response will take the form $Ae^{-t/\tau}$.

step 3: Sum the steady-state and transient solutions, that is, the particular and complementary solutions, to obtain the total solution. This will take the form $\kappa + Ae^{-t/\tau}$.

step 4: Determine the initial value from the circuit initial conditions. Use this information to calculate the constant A.

step 5: Write the final solution.

This method is used for resistor-capacitor circuits, which are primarily used in electronics. Inductors tend to be large and to change value with both temperature and time. They do not lend themselves well to miniaturization. Capacitors are more easily manufactured on integrated circuits. Inductors are important in power circuits that handle large amounts of current or high voltages and in transformers.

Table 30.1 *Transient Response*

type of circuit	response
series RC, charging $\tau = RC$ $e^{-N} = e^{-t/\tau} = e^{-t/RC}$ 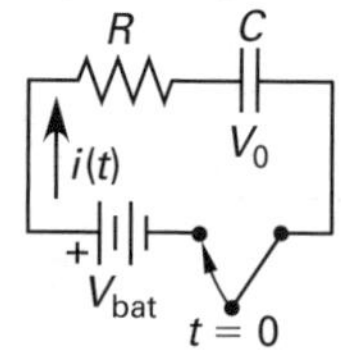	$V_{\text{bat}} = v_R(t) + v_C(t)$ $i(t) = \left(\dfrac{V_{\text{bat}} - V_0}{R}\right)e^{-N}$ $v_R(t) = i(t)R$ $= (V_{\text{bat}} - V_0)e^{-N}$ $v_C(t) = V_0 + (V_{\text{bat}} - V_0) \times (1 - e^{-N})$ $Q_C(t) = C\begin{pmatrix} V_0 + (V_{\text{bat}} - V_0) \\ \times(1 - e^{-N}) \end{pmatrix}$
series RC, discharging $\tau = RC$ $e^{-N} = e^{-t/\tau} = e^{-t/RC}$ 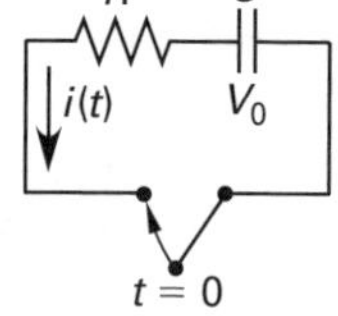	$0 = v_R(t) + v_C(t)$ $i(t) = \left(\dfrac{V_0}{R}\right)e^{-N}$ $v_R(t) = -V_0e^{-N}$ $v_C(t) = V_0e^{-N}$ $Q_C(t) = CV_0e^{-N}$
series RL, charging $\tau = \dfrac{L}{R}$ $e^{-N} = e^{-t/\tau} = e^{-t/(L/R)}$ 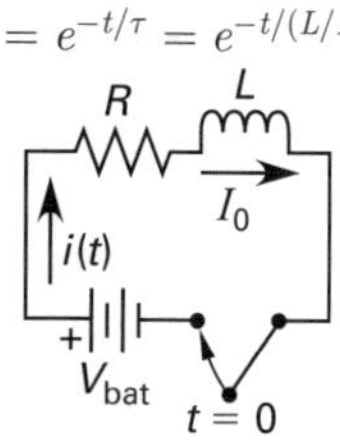	$V_{\text{bat}} = v_R(t) + v_L(t)$ $i(t) = I_0e^{-N} + \left(\dfrac{V_{\text{bat}}}{R}\right)(1 - e^{-N})$ $v_R(t) = i(t)R$ $= I_0Re^{-N} + V_{\text{bat}}(1 - e^{-N})$ $v_L(t) = (V_{\text{bat}} - I_0R)e^{-N}$
series RL, discharging $\tau = \dfrac{L}{R}$ $e^{-N} = e^{-t/\tau} = e^{-t/(L/R)}$ 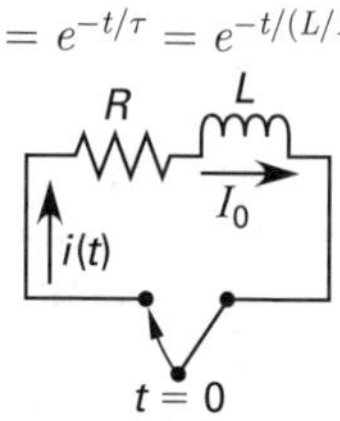	$0 = v_R(t) + v_L(t)$ $i(t) = I_0e^{-N}$ $v_R(t) = I_0Re^{-N}$ $v_L(t) = -I_0Re^{-N}$

7. RISE TIME

When the input of a capacitive or inductive circuit like the one in Fig. 30.6(a) undergoes a step change, the circuit response is called a *step response*. The step response is illustrated in Fig. 30.6(b). The speed of the response may be quantified in several ways. The time constant, τ, is one possibility. The shorter the time constant, the less time it takes for the circuit to reach a specified value. The value of the time constant is dependent on the circuit. A second possibility is the *rise time*, which is a *defined quantity* applicable to responses in general regardless of the circuit. Let t_{10} be the time a response has reached 10% of its final steady-state value. Let t_{90} be the time a response has reached 90% of its final steady-state value. The rise time, t_r, is then defined as

$$t_r = t_{90} - t_{10} \qquad 30.7$$

[1]The final expression of the response given in the table appears to differ from that given in step 3 of the solution method. The table uses the common form $1 - e^{-t/\tau}$. The forms are equivalent and can be interchanged. Both contain a steady-state and transient component. The steady-state term κ can be found in the $1 - e^{-t/\tau}$ form by letting $t \rightarrow \infty$.

[2]If the excitation is constant, the form of the particular solution is a constant and can be determined by simply open-circuiting the terminals of the capacitor and determining the voltage present. This voltage will be the steady-state voltage to which the capacitor will be driven.

Figure 30.6 *Rise Time*

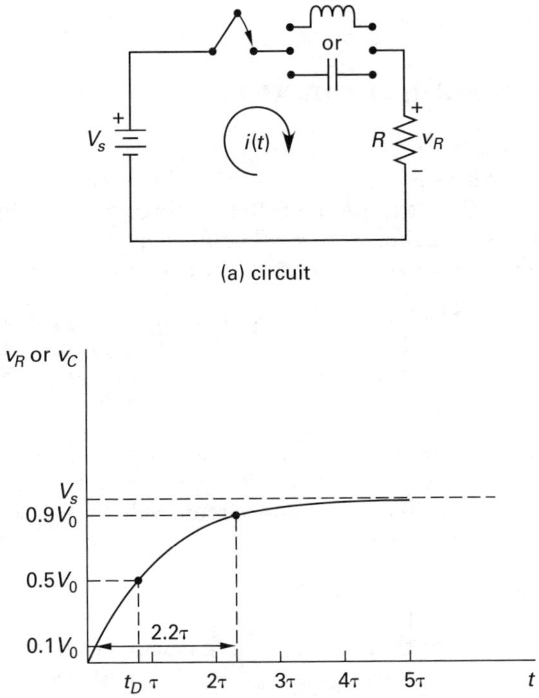

The rise time is related to the time constant by Eq. 30.8.

$$t_r = 2.2\tau \qquad 30.8$$

Another defined quantity is the *time delay*, t_d, which is the time for a response to reach 50% of its final steady-state value. The rise and delay times and their relationship to the time constant are shown graphically in Fig. 30.6(b).

8. DAMPED OSCILLATIONS: RINGING

When energized, some circuits undergo voltage or current oscillations that decrease in magnitude over time. Such oscillation is called *ringing*. An example of such a circuit and its response to a switch closure at $t = 0$ is shown in Fig. 30.7. The capacitor is initially charged to V_0 and the inductor has no initial magnetic field. Writing KVL for the circuit yields

$$\frac{1}{C}\int_0^t i\,dt + iR + L\left(\frac{di}{dt}\right) = V_0 \qquad 30.9$$

The current is the rate of change of the charge on the capacitor, that is, $i = dq/dt$. Rearranging mathematically gives

$$\frac{d^2q}{dt^2} + \left(\frac{R}{L}\right)\left(\frac{dq}{dt}\right) + \left(\frac{1}{LC}\right)q = \frac{q_0}{LC} \qquad 30.10$$

Figure 30.7 *Ringing Circuit*

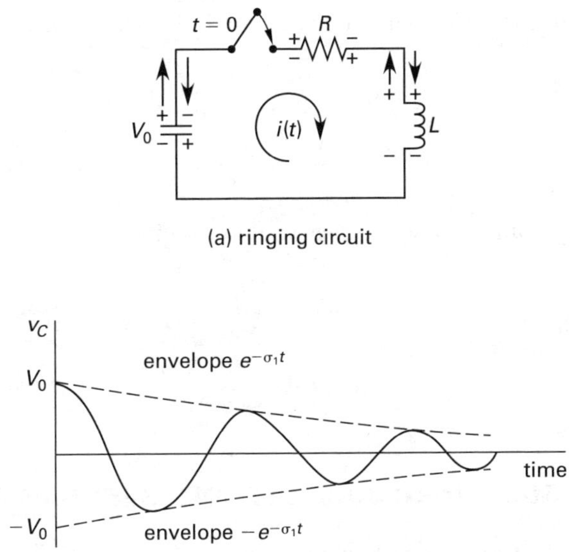

Representing this nonhomogeneous second-order differential equation in the *s* domain, appropriate for Laplace transform analysis, results in the following characteristic equation.[3]

$$s^2 + \left(\frac{R}{L}\right)s + \frac{1}{LC} = 0 \qquad 30.11$$

The roots of Eq. 30.11, found from the quadratic equation, take one of three possible forms: (1) real and unequal, (2) real and equal, or (3) complex conjugates.

Ringing occurs for case 3 with roots given by

$$s_1, s_2 = -\sigma_1 \pm j\omega_1 \qquad 30.12$$

The solution to Eq. 30.10 for the ringing circuit is

$$v_C(t) = V_0 e^{-\sigma_1 t}\cos\omega_1 t \qquad 30.13$$

The term σ_1 is called the *damping*. The exponential, $e^{-\sigma_1 t}$, is termed the *envelope* of the waveform. The frequency of the oscillation is given by ω_1. These quantities are illustrated in Fig. 30.7(b). The damping and frequency are found in terms of the circuit parameters from Eq. 30.14 and Eq. 30.15.

$$\sigma_1 = \frac{R}{2L} \qquad 30.14$$

$$\omega_1 = \sqrt{\frac{1}{LC} - \left(\frac{R}{2L}\right)^2} \qquad 30.15$$

[3]A second-order equation should be expected because the circuit contains two independent energy storage devices.

The damping and frequency are also defined in terms of the *quality factor*, Q, of the inductor coil. If ω_0 is the frequency of the circuit when $R = 0$, the quality factor is

$$Q = \frac{\omega_0 L}{R} = \frac{1}{\omega_0 CR} = \frac{X_L}{R} = \frac{X_C}{R} \qquad 30.16$$

The damping and frequency can be defined as

$$\sigma_1 = \frac{\omega_0}{2Q} \qquad 30.17$$

$$\omega_1 = \omega_0 \sqrt{1 - \left(\frac{1}{2Q}\right)^2} \qquad 30.18$$

9. SUSTAINED OSCILLATIONS: RESONANCE

Capacitors and inductors within the same circuit exchange energy. In the ideal, that is, lossless, circuit shown in Fig. 30.8, this exchange can continue indefinitely and is known as a *sustained oscillation*.[4]

Figure 30.8 *Resonant Circuit*

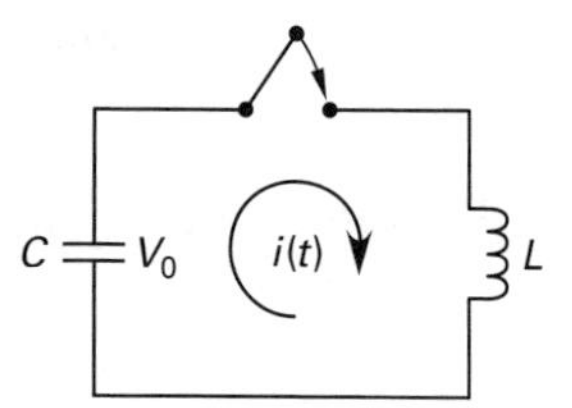

The characteristic equation for the circuit in Fig. 30.8 is

$$s^2 + \frac{1}{LC} = 0 \qquad 30.19$$

The solution is

$$s_1, s_2 = \pm j\sqrt{\frac{1}{LC}} = \pm j\omega_0 \qquad 30.20$$

No damping term, σ_1, exists and the oscillations can continue unfettered. This unimpeded flow of energy naturally occurs at the frequency ω_0 and is called *resonance*.

In a real circuit, resistive components are present and will yield a damping term, σ_1, in the solution to the circuit equation. The result is that an external source of energy is required to make up the loss. Resonance is then defined as a phenomenon of an AC circuit whereby relatively large currents occur near certain frequencies, and a relatively unimpeded oscillation of energy from potential to kinetic occurs. The frequency at which this occurs is called the *resonant frequency*. At the resonant frequency the capacitive reactance equals the inductive reactance, that is, $X_C = X_L$, the current phase angle difference, ϕ, is zero ($\phi = 0$), and the power factor, pf, equals one (pf = 1).

[4]This type of circuit is more than academic and can be realized, or nearly so, with superconducting inductors.

10. RESONANT CIRCUITS

A *resonant circuit* has a zero current phase angle difference. This is equivalent to saying the circuit is purely resistive (i.e., the power factor is equal to one) in its response to an AC voltage. The frequency at which the circuit becomes purely resistive is the *resonant frequency*.

For frequencies below the resonant frequency, a series *RLC* circuit will be capacitive (leading) in nature; above the resonant frequency, the circuit will be inductive (lagging) in nature.

For frequencies below the resonant frequency, a parallel *GLC* circuit will be inductive (lagging) in nature; above the resonant frequency, the circuit will be capacitive (leading) in nature.

Circuits can become resonant in two ways. If the frequency of the applied voltage is fixed, the elements must be adjusted so that the capacitive reactance cancels the inductive reactance (i.e., $X_L - X_C = 0$). If the circuit elements are fixed, the frequency must be adjusted.

As Fig. 30.9 and Fig. 30.10 illustrate, a circuit approaches resonant behavior gradually. ω_1 and ω_2 are the *half-power points* (*70% points*, or *3 dB points*) because at those frequencies, the power dissipated in the resistor is half of the power dissipated at the resonant frequency.

$$Z_{\omega 1} = \sqrt{2}R \qquad 30.21$$

$$I_{\omega 1} = \frac{V}{Z_{\omega 1}} = \frac{V}{\sqrt{2}R} = \frac{I_0}{\sqrt{2}} \qquad 30.22$$

$$P_{\omega 1} = I^2 R = \left(\frac{I_0}{\sqrt{2}}\right)^2 R = \tfrac{1}{2}P_0 \qquad 30.23$$

Figure 30.9 *Series Resonance (Band-Pass Filter)*

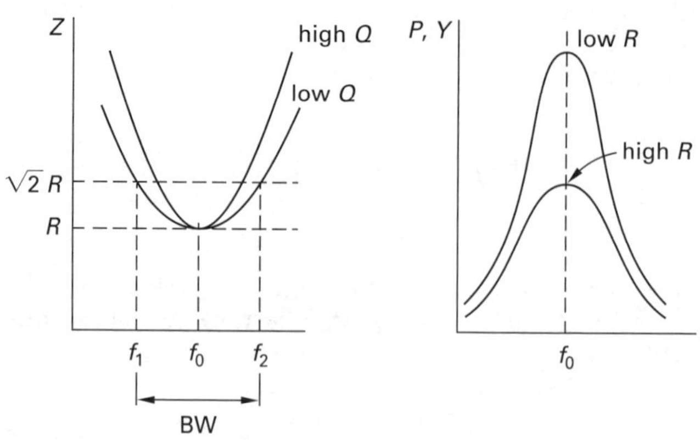

Figure 30.10 *Parallel Resonance (Band-Reject Filter)*

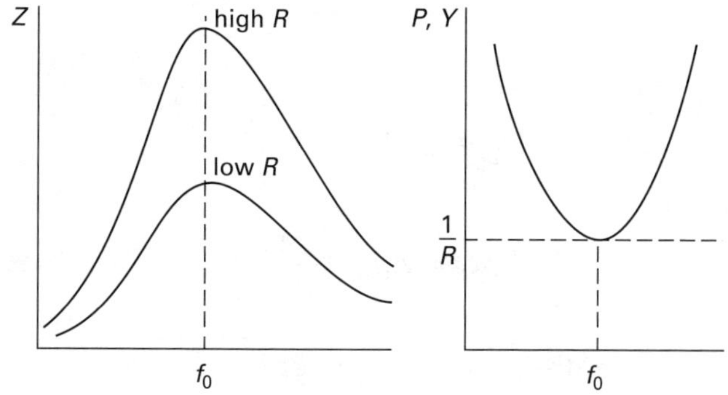

The frequency difference between the half-power points is the *bandwidth*, BW, a measure of circuit selectivity. The smaller the bandwidth, the more selective the circuit.

$$\text{BW} = f_2 - f_1 \qquad 30.24$$

The *quality factor*, Q, for a circuit is a dimensionless ratio that compares the reactive energy stored in an inductor each cycle to the resistive energy dissipated.[5] The effect the quality factor has on the frequency characteristic is illustrated by Fig. 30.9.

$$\begin{aligned} Q &= 2\pi\left(\frac{\text{maximum energy stored per cycle}}{\text{energy dissipated per cycle}}\right) \\ &= \frac{f_0}{(\text{BW})_{\text{Hz}}} = \frac{\omega_0}{(\text{BW})_{\text{rad/s}}} \\ &= \frac{f_0}{f_2 - f_1} = \frac{\omega_0}{\omega_2 - \omega_1} \quad \left[\begin{array}{l}\text{parallel} \\ \text{or series}\end{array}\right] \end{aligned} \qquad 30.25$$

Then, the energy stored in the inductor of a series RLC circuit each cycle is

$$U = \frac{I_m^2 L}{2} = I^2 L = Q\left(\frac{I^2 R}{2\pi f_0}\right) \qquad 30.26$$

The relationships between the half-power points and quality factor are

$$\begin{aligned} f_1, f_2 &= f_0\left(\sqrt{1 + \frac{1}{4Q^2}} \mp \frac{1}{2Q}\right) \\ &\approx f_0 \mp \frac{f_0}{2Q} = f_0 \mp \frac{\text{BW}}{2} \end{aligned} \qquad 30.27$$

Various resonant circuit formulas are tabulated in App. 30.A.

[5]The name *figure of merit* refers to the quality factor calculated from the inductance and internal resistance of a coil.

11. SERIES RESONANCE

In a resonant series RLC circuit,

- impedance is minimum
- impedance equals resistance
- current and voltage are in phase
- current is maximum
- power dissipation is maximum

The total impedance (in rectangular form) of a series RLC circuit is $R + j(X_L - X_C)$. At the resonant frequency, $\omega_0 = 2\pi f_0$,

$$X_L = X_C \quad \text{[at resonance]} \qquad 30.28$$

$$\omega_0 L = \frac{1}{\omega_0 C} \qquad 30.29$$

$$\omega_0 = 2\pi f_0 = \frac{1}{\sqrt{LC}} \qquad 30.30$$

The power dissipation in the resistor is

$$\begin{aligned} P &= \tfrac{1}{2} I_m^2 R = \frac{V_m^2}{2R} \\ &= I^2 R \\ &= \frac{V^2}{R} \end{aligned} \qquad 30.31$$

The quality factor for a series RLC circuit is

$$\begin{aligned} Q &= \frac{X}{R} = \frac{\omega_0 L}{R} \\ &= \frac{1}{\omega_0 RC} = \frac{1}{R}\sqrt{\frac{L}{C}} \\ &= \frac{\omega_0}{(\text{BW})_{\text{rad/s}}} = \frac{f_0}{(\text{BW})_{\text{Hz}}} \\ &= G\omega_0 L = \frac{G}{\omega_0 C} \end{aligned} \qquad 30.32$$

Example 30.5

A series RLC circuit is connected across a sinusoidal voltage with peak of 20 V. (a) What is the resonant frequency in rad/s? (b) What are the half-power points in rad/s? (c) What is the peak current at resonance? (d) What is the peak voltage across each component at resonance?

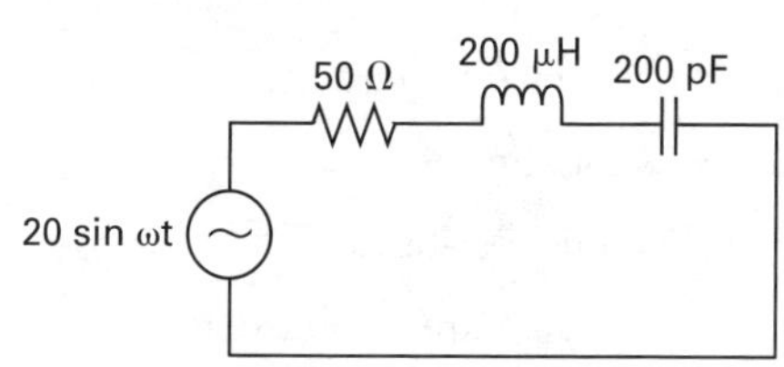

Circuit Theory

Solution

(a) Equation 30.30 gives the resonant frequency.

$$\omega_0 = \frac{1}{\sqrt{LC}} = \frac{1}{\sqrt{(200 \times 10^{-6}\ \text{H})(200 \times 10^{-12}\ \text{F})}}$$
$$= 5 \times 10^6\ \text{rad/s}$$

(b) The half-power points are

$$\omega_1, \omega_2 = \omega_0 \mp \frac{\text{BW}}{2} = \omega_0 \mp \frac{\omega_0}{2Q} = \omega_0 \mp \frac{R}{2L}$$
$$= 5 \times 10^6\ \frac{\text{rad}}{\text{s}} \mp \frac{50\ \Omega}{(2)(200 \times 10^{-6}\ \text{H})}$$
$$= 5.125 \times 10^6\ \text{rad/s},\ 4.875 \times 10^6\ \text{rad/s}$$

(c) The total impedance is the resistance at resonance. The peak resonant current is

$$I_0 = \frac{V_m}{Z_0} = \frac{V_m}{R}$$
$$= \frac{20\ \text{V}\angle 0^\circ}{50\ \Omega}$$
$$= 0.4\ \text{A}\angle 0^\circ$$

(d) The peak voltages across the components are

$$V_R = I_0 R = (0.4\ \text{A}\angle 0^\circ)(50\ \Omega) = 20\ \text{V}\angle 0^\circ$$
$$V_L = I_0 X_L = I_0 j\omega_0 L$$
$$= j(0.4\ \text{A}\angle 0^\circ)\left(5 \times 10^6\ \frac{\text{rad}}{\text{s}}\right)(200 \times 10^{-6}\ \text{H})$$
$$= 400\ \text{V}\angle 90^\circ$$
$$V_C = I_0 X_C = \frac{I_0}{j\omega_0 C}$$
$$= \frac{0.4\ \text{A}\angle 0^\circ}{j\left(5 \times 10^6\ \frac{\text{rad}}{\text{s}}\right)(200 \times 10^{-12}\ \text{F})}$$
$$= 400\ \text{V}\angle -90^\circ$$

12. PARALLEL RESONANCE

In a resonant parallel *GLC* circuit,

- impedance is maximum
- impedance equals resistance
- current and voltage are in phase
- current is minimum
- power dissipation is minimum

The total admittance (in rectangular form) of a parallel *GLC* circuit is $G + j(B_C - B_L)$. At resonance,

$$X_L = X_C \qquad 30.33$$

$$\omega_0 L = \frac{1}{\omega_0 C} \qquad 30.34$$

$$\omega_0 = 2\pi f_0 = \frac{1}{\sqrt{LC}} \qquad 30.35$$

The power dissipation in the resistor is

$$P = \tfrac{1}{2} I_m^2 R = \frac{V_m^2}{2R}$$
$$= I^2 R$$
$$= \frac{V^2}{R} \qquad 30.36$$

The quality factor for a parallel *RLC* circuit is

$$Q = \frac{R}{X} = \omega_0 RC = \frac{R}{\omega_0 L}$$
$$= R\sqrt{\frac{C}{L}} = \frac{\omega_0}{(\text{BW})_{\text{rad/s}}} = \frac{f_0}{(\text{BW})_{\text{Hz}}}$$
$$= \frac{\omega_0 C}{G} = \frac{1}{G\omega_0 L} \qquad 30.37$$

Example 30.6

A parallel *RLC* circuit containing a 10 Ω resistor has a resonant frequency of 1 MHz and a bandwidth of 10 kHz. To what should the resistor be changed in order to increase the bandwidth to 20 kHz without changing the resonant frequency?

Solution

From Eq. 30.25 and Eq. 30.37,

$$Q_{\text{old}} = \frac{f_0}{\text{BW}} = 2\pi f_0 RC$$
$$C = \frac{1}{2\pi R(\text{BW})} = \frac{1}{(2\pi)(10\ \Omega)(10 \times 10^3\ \text{Hz})}$$
$$= 1 \times 10^{-5}/2\pi\ \text{F}$$

The new quality factor is

$$Q_{\text{new}} = \frac{f_0}{\text{BW}}$$
$$= \frac{10^6\ \text{Hz}}{20 \times 10^3\ \text{Hz}}$$
$$= 50$$

From Eq. 30.37, the required resistance is

$$
\begin{aligned}
R &= \frac{Q}{2\pi f_0 C} \\
&= \frac{50}{(2\pi)(10^6\ \text{Hz})\left(\dfrac{1\times 10^{-5}}{2\pi}\ \text{F}\right)} \\
&= 5\ \Omega
\end{aligned}
$$

31 Time Response

1. Fundamentals 31-1
2. First-Order Analysis 31-2
3. First-Order Analysis: Switching Transients . . 31-3
4. First-Order Analysis: Pulse Transients 31-6
5. Second-Order Analysis 31-7
6. Second-Order Analysis: Overdamped 31-9
7. Second-Order Analysis: Critically Damped . . 31-11
8. Second-Order Analysis: Underdamped 31-12
9. Second-Order Analysis: Pulse Transients . . 31-13
10. Higher-Order Circuits 31-14
11. Complex Frequency 31-14
12. Laplace Transform Analysis 31-15
13. Capacitance in the s Domain 31-16
14. Inductance in the s Domain 31-17
15. Laplace Transform Analysis: First- and Second-Order Systems 31-17

Nomenclature

a, A	constant	–
A	natural response coefficient	–
b, B	constant	–
c	constant	–
C	capacitance	F
i	variable current	A
I, $\mathbf{I}$	constant or rms current	A
L	self-inductance	H
Q	quality factor	–
R	resistance	Ω
s, $\mathbf{s}$	complex frequency	Hz
s	Laplace transform variable	–
t	time	s
v	variable voltage	V
V, $\mathbf{V}$	constant or rms voltage	V
$\mathbf{Z}$	impedance	Ω

Symbols

α	damping	Hz
β	oscillation frequency	rad/s
ζ	damping ratio	–
θ	phase angle	rad
κ	steady-state constant	–
σ	neper frequency or napier frequency	Np/s or Hz
τ	time constant	s
ω	angular frequency	rad/s or Hz
ω_0	resonance frequency	rad/s

Subscripts

0	initial or resonant
C	capacitor
DC	direct current
L	inductor
L/C	inductor or capacitor
n	natural
p	peak
s	source
so	switch open
ss	steady-state
tr	transient
Th	Thevenin

1. FUNDAMENTALS

Time domain analysis is analysis that takes place with time as the governing variable. *Sinusoidal analysis* is analysis that occurs when the current and voltage functions are manipulated in their trigonometric form. *Frequency domain analysis* deals with Fourier or Laplace transforms of time functions, which are then functions of frequency. The Fourier transform occurs in k-space which, when k is replaced by ω, is called the frequency domain. The Laplace transform occurs in s-space, also considered the frequency domain. The variable $\mathbf{s}$ is called the *complex frequency* and is given by Eq. 31.1. (In Eq. 31.3, the variable $\mathbf{s}$ is shown in bold to emphasize its vector properties, important during pole-zero analysis. It is commonly shown as unbolded and will be shown as such in the remainder of this book.)

$$\mathbf{s} = \sigma + j\omega \qquad 31.1$$

The variable σ is termed the *neper frequency* and is assigned units of Np/s or Hz. The variable σ represents the ratio of two currents, voltages, or other analogous quantities. The number of nepers is the natural log of this ratio. The term "napier" is sometimes used in place of neper.[1] The Laplace transform and the complex frequency s are used to represent exponential and damped sinusoidal motion in the same manner that ordinary frequency expresses simple harmonic motion. That is, using the variable s allows constants, exponential decay,

[1]This occurs since the natural logarithm is alternately called the *Naperian logarithm.*

sinusoids, and damped sinusoids (sinusoids with an exponential envelope) to be represented in the same form, Ae^{st} (see Sec. 31.11). *Sinusoidal analysis* is also considered any analysis using the assumption that the forcing function, the source, is sinusoidal in nature. Sinusoidal analysis in Chap. 27 through Chap. 30 assumed a single frequency. When the frequencies are allowed to vary, the circuit output is called the *frequency analysis* or *frequency response.*

2. FIRST-ORDER ANALYSIS

First-order analysis is performed upon circuits with one energy storage element—that is, one equivalent capacitor or inductor. In circuits containing numerous sources and resistors, the Thevenin or Norton equivalent is taken as seen from the terminals of the single capacitor or inductor. This allows the methods of Chap. 30 to be applied, modified so that the Thevenin voltage, v_{Th}, is used in place of the source voltage and the Thevenin current, i_{Th}, is used as the current through the capacitor and inductor. Figure 31.1(a) represents a generic circuit with a single capacitor or inductor. The source can be either AC or DC. Figure 31.1(b) shows the Thevenin equivalent of the circuit with the switch closed. Figure 31.1(c) shows the new Thevenin equivalent with the switch open. Part (c) of this figure is a convenient way of showing that if the source is eliminated (or changed to a DC source), the resulting currents and voltages will be exponentials (the natural response) and will share a new time constant and possibly new initial currents and voltages.

The differential equation for first-order circuits takes on the general form

$$f(t) = b\frac{dx}{dt} + cx \qquad 31.2$$

The solutions are of the form

$$x(t) = \kappa + Ae^{-t/\tau} \qquad 31.3$$

The time constant $\tau = b/c$. κ is a constant, the steady-state component of the solution. The method for determining the complete solution is as follows.

step 1: Determine the Thevenin or Norton equivalent of the circuit as seen from the terminals of the capacitor or inductor.

step 2: Find the steady-state response from the particular solution of the nonhomogeneous equation. The solution will be of the same form as the forcing function.[2] Regardless of transient type, the result will be a steady-state value, κ.

step 3: Determine the transient response from the homogeneous equation or from the known parameters of the time constant. The transient response will be of the form $Ae^{-t/\tau}$.

step 4: Sum the steady-state and transient solutions, that is, the complementary and particular solutions, to obtain the total solution.[3] This will be of the form $\kappa + Ae^{-t/\tau}$.

step 5: Determine the initial value from the circuit initial conditions. Use this information to calculate the constant A.

step 6: Write the final solution.

Figure 31.1 *First-Order Circuit*

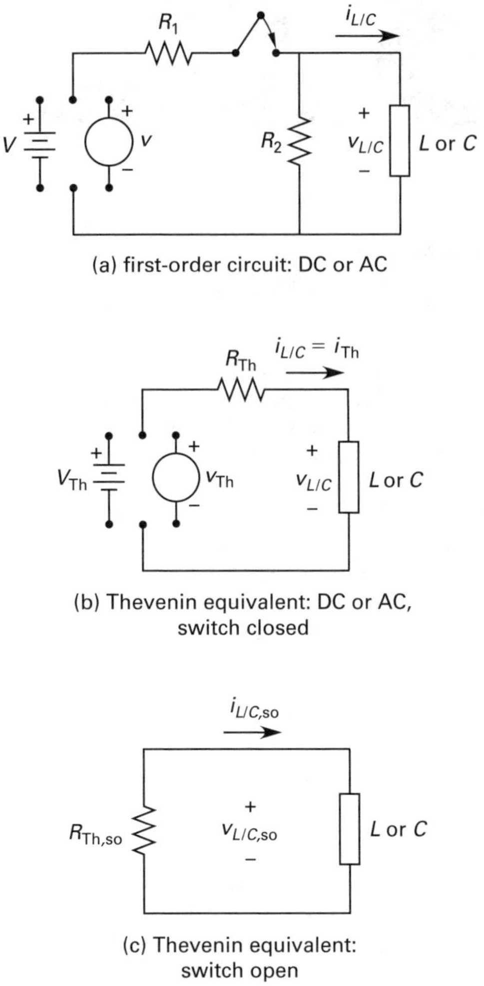

A capacitive first-order circuit generally uses voltage as the dependent variable. Applying KVL to Fig. 31.1(b) gives

$$v_{\text{Th}} = i_{\text{Th}}R_{\text{Th}} + v_C \qquad 31.4$$

[2]If the excitation is constant, the form of the particular solution is a constant and can be determined by simply open-circuiting the terminals of the capacitor and determining the voltage present. This voltage will be the steady-state voltage to which the capacitor will be driven.

[3]The final expression often varies between texts, with some using $1 - e^{-t/\tau}$. The forms are equivalent and can be interchanged. Both contain a steady-state and transient component. The steady-state term κ can be found in the $1 - e^{-t/\tau}$ form by letting $t \rightarrow \infty$.

The Thevenin current is the current flowing through the capacitor, $i_{\text{Th}} = i_C = C(dv_C/dt)$. Substituting gives

$$v_{\text{Th}} = R_{\text{Th}}C\left(\frac{dv_C}{dt}\right) + v_C \qquad 31.5$$

Equation 31.5 is in the general form of a first-order differential equation shown in Eq. 31.2. The solution is

$$v_C(t) = V_{C,\text{ss}} + Ae^{-t/\tau} \qquad 31.6$$

An inductive first-order circuit generally uses current as the dependent variable. Applying KVL to Fig. 31.1(b) gives

$$v_{\text{Th}} = i_{\text{Th}}R_{\text{Th}} + v_L \qquad 31.7$$

The Thevenin current is flowing through the inductor, $i_{\text{Th}} = i_L$. Substituting for the inductor voltage and rearranging gives

$$v_{\text{Th}} = L\left(\frac{di_L}{dt}\right) + R_{\text{Th}}i_L \qquad 31.8$$

Equation 31.8 is in the general form of a first-order differential equation shown in Eq. 31.2. The solution is

$$i_L(t) = I_{L,\text{ss}} + Ae^{-t/\tau} \qquad 31.9$$

3. FIRST-ORDER ANALYSIS: SWITCHING TRANSIENTS

First-order switching transients result from the operation of a switch. The switching mechanism can be an actual switch, relay contacts, or a nonlinear device such as a transistor operating in the saturation or cutoff region. A sudden change in the source results in sudden changes in the voltages and currents within a circuit. A jump in the capacitor voltage requires an impulse current. A jump in the inductor current requires an impulse voltage. Because the conditions in a capacitor or inductor cannot change instantaneously, the post-switched conditions of capacitance, C, or self-inductance, L, are derived from the pre-switched conditions.

Example 31.1[4]

For the network shown, $R_1 = 20\ \Omega$, $R_2 = 10\ \Omega$, $V_{\text{DC}} = 12$ V, and $L = 0.20$ H. The switch has been open for a long time and is closed at $t = 0$. Determine the steady-state response of the network.

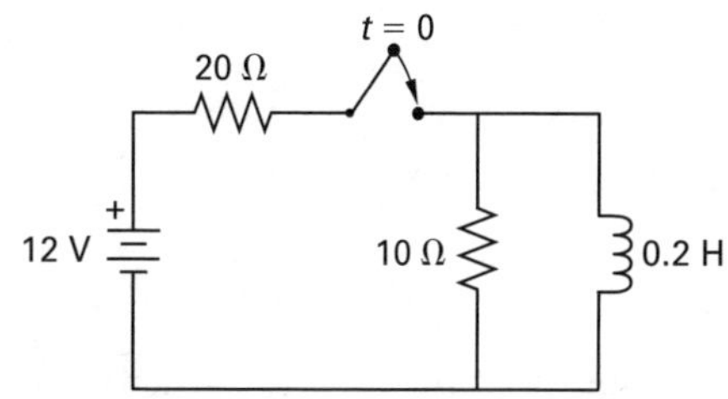

Solution

Using the voltage divider concept, the Thevenin voltage is 4 V. The Thevenin resistance is calculated as seen from the terminals of the inductor, as shown in Fig. 31.1(b). Therefore,

$$\begin{aligned} R_{\text{Th}} &= R_1 \| R_2 = \frac{R_1R_2}{R_1+R_2} = \frac{(20\ \Omega)(10\ \Omega)}{20\ \Omega + 10\ \Omega} \\ &= 6.67\ \Omega \end{aligned}$$

The steady-state condition is determined from Eq. 31.8, noting that steady-state current is constant $(di/dt = 0)$.

$$\begin{aligned} v_{\text{Th}} &= L\left(\frac{di_L}{dt}\right) + R_{\text{Th}}i_L \\ &= (0.2\ \text{H})(0) + (6.67\ \Omega)i_L = 4\ \text{V} \\ i_{L,\text{ss}} &= \frac{4\ \text{V}}{6.67\ \Omega} = 0.60\ \text{A} \end{aligned}$$

The inductor is a short circuit for steady-state DC conditions, as expected.

Example 31.2

Determine the transient response of the circuit in Ex. 31.1.

Solution

The transient response is of the form $Ae^{-t/\tau}$. The time constant is

$$\tau = \frac{L}{R_{\text{Th}}} = \frac{(0.20\ \text{H})\left(1000\ \frac{\text{s}}{\text{ms}}\right)}{6.67\ \Omega} = 30\ \text{ms}$$

The steady-state current was found to be 0.60 A in Ex. 31.1. Using Eq. 31.9, the total current is

$$i_L(t) = I_{L,\text{ss}} + Ae^{-t/\tau} = 0.60\ \text{A} + Ae^{-t/(30\times10^{-3}\ \text{s})}$$

Because the switch was open for a long time, steady-state conditions are assumed to exist and the initial current in the inductor is zero. That is, $i(0^-) = i(0^+) = 0$. Substituting this information to find the constant A gives

$$\begin{aligned} i_L(0) &= 0.60\ \text{A} + Ae^{-0/(30\times10^{-3}\ \text{s})} = 0.60\ \text{A} + A = 0 \\ A &= -0.60\ \text{A} \end{aligned}$$

[4]Since transient conditions, and a possible AC source, are used in some examples, lowercase letters will be used to designate the voltage and current.

Circuit Theory

The transient response is

$$i_{\text{tr}}(t) = (-0.60\ \text{A})e^{-t/(30\times10^{-3}\ \text{s})}$$

Example 31.3

What is the current through the inductor in Ex. 31.1 120 ms after the switch closes?

Solution

The total solution is the combination of the steady-state and transient responses. From Ex. 31.1 and Ex. 31.2,

$$i_L(t) = 0.60\ \text{A} - (0.60\ \text{A})e^{-t/(30\times10^{-3}\ \text{s})}$$

At 120 ms, or 4τ, after the switch closes, the current is

$$\begin{aligned} i_L(120\ \text{ms}) &= 0.60\ \text{A} - (0.60\ \text{A})e^{(-120\times10^{-3}\ \text{s})/(30\times10^{-3}\ \text{s})} \\ &= 0.60\ \text{A} - (0.60\ \text{A})e^{-4} = 0.59\ \text{A} \end{aligned}$$

Example 31.4

For the network shown, let $R_1 = 10\ \Omega$, $R_2 = 10\ \Omega$, $V_{\text{DC}} = 10$ V, and $C = 50\ \mu$F. The switch has been open for an extended time and is closed at $t = 0$. Determine the steady-state capacitance response.

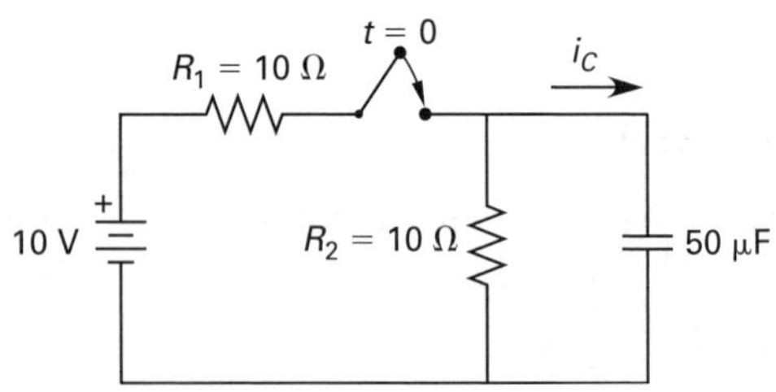

Solution

Using the voltage divider concept, the Thevenin voltage is 5 V. The Thevenin resistance is calculated as seen from the terminals of the capacitor.

$$\begin{aligned} R_{\text{Th}} = R_1 \| R_2 &= \frac{R_1R_2}{R_1+R_2} = \frac{(10\ \Omega)(10\ \Omega)}{10\ \Omega + 10\ \Omega} \\ &= 5\ \Omega \end{aligned}$$

The steady-state condition is determined from Eq. 31.5, noting that the voltage is constant ($dv/dt = 0$).

$$\begin{aligned} v_{\text{Th}} &= R_{\text{Th}}C\left(\frac{dv_C}{dt}\right) + v_C \\ &= (5\ \Omega)(50\times10^{-6}\ \text{F})(0) + v_C \\ &= 5\ \text{V} \\ v_{C,\text{ss}} &= 5\ \text{V} \end{aligned}$$

The capacitor is an open circuit for steady-state DC conditions, as expected.

Example 31.5

For the network in Ex. 31.4, the switch opens three time constants into the transient. Determine the response of the network from this time onward.

Solution

After the switch opens, $v_{\text{Th}} = 0$ (see Fig. 31.1(c)). Therefore, the steady-state voltage will be zero. The transient response must be of the form $Ae^{-t/\tau}$.

$$v_C(t) = Ae^{-t/\tau} = Ae^{-t/R_{\text{Th}}C}$$

Once the switch is open, the Thevenin resistance is

$$R_{\text{Th}} = R_2 = 10\ \Omega$$

The time constant is

$$\tau = R_{\text{Th}}C = (10\ \Omega)(50\times10^{-6}\ \text{F})\left(10^6\ \frac{\mu\text{s}}{\text{s}}\right) = 500\ \mu\text{s}$$

The total solution is the combination of the steady-state and transient responses, as given in Eq. 31.6.

$$v_C(t) = V_{C,\text{ss}} + Ae^{-t/\tau} = 0\ \text{V} + Ae^{-t/(500\times10^{-6}\ \text{s})}$$

In order to determine the constant A, the voltage condition must be known at a certain time. Because the capacitor voltage cannot change instantaneously, $v_C(0^-) = v_C(0^+)$. The switch was opened three time constants into the transient determined by the conditions in Ex. 31.4. The transient response of the circuit in Ex. 31.4 is

$$\begin{aligned} v_C(t) &= Ae^{-t/R_{\text{Th}}C} = Ae^{-t/((5\ \Omega)(50\times10^{-6}\ \text{F}))} \\ &= Ae^{-t/250\times10^{-6}\ \text{s}} \end{aligned}$$

The total solution for the circuit in Ex. 31.4, as given by Eq. 31.6, and noting that the initial voltage is zero, is

$$\begin{aligned} v_C(t) &= V_{C,\text{ss}} + Ae^{-t/\tau} = 5\ \text{V} + Ae^{-t/(250\times10^{-6}\ \text{s})} \\ v_C(0) &= 5\ \text{V} + Ae^{-0/(250\times10^{-6}\ \text{s})} = 5\ \text{V} + A = 0\ \text{V} \\ A &= -5\ \text{V} \\ v_C(t) &= 5\ \text{V} - 5e^{-t/(250\times10^{-6}\ \text{s})}\ \text{V} \end{aligned}$$

At three time constants into the initial transient,

$$v_C(3\tau) = 5\ \text{V} - 5e^{-((3)(250\times10^{-6}\ \text{s}))/(250\times10^{-6}\ \text{s})}\ \text{V} = 4.75\ \text{V}$$

Use 4.75 V as the initial voltage for the network in Ex. 31.5 to determine the value of A.

$$v_C(t) = V_{C,\text{ss}} + Ae^{-t/\tau} = 0\text{ V} + Ae^{-t/(500\times10^{-6}\text{ s})}$$
$$v_C(0^-) = v_C(0^+) = 4.75\text{ V}$$
$$4.75\text{ V} = 0\text{ V} + Ae^{-0/(500\times10^{-6}\text{ s})} = A$$

Therefore, the response of the network from the time the switch opens is

$$v_C(t) = V_{C,\text{ss}} + Ae^{-t/\tau} = 0\text{ V} + 4.75e^{-t/(500\times10^{-6}\text{ s})}\text{ V}$$
$$= 4.75e^{-t/(500\times10^{-6}\text{ s})}\text{ V}$$

Example 31.6

For the network shown, $R_1 = 10\ \Omega$, $R_2 = 10\ \Omega$, $v = 170\text{ V}\sin(377t + 30°)$, and $L = 0.15$ H. Determine the steady-state current with the switch closed.

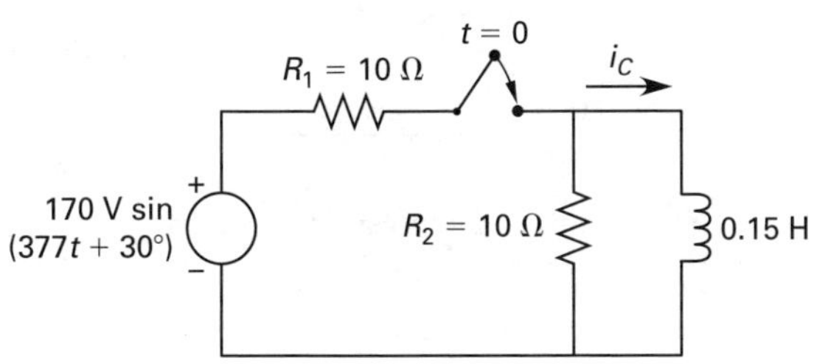

Solution

Using Eq. 31.8 and noting that multiplying by $j\omega$ is equivalent to taking the derivative, gives

$$v_{\text{Th}} = L\left(\frac{di_L}{dt}\right) + R_{\text{Th}}i_L = Lj\omega i_L + R_{\text{Th}}i_L$$
$$i_L = \frac{v_{\text{Th}}}{j\omega L + R_{\text{Th}}}$$

The source is given as sinusoidal. Analysis is the same for sinusoidal sources, but the mathematics is more cumbersome. Instead, use the phasor representation of the sinusoid source and Eq. 31.8.

$$i_L = \frac{v_{\text{Th}}}{R_{\text{Th}} + j\omega L}$$
$$\mathbf{I}_L = \frac{\mathbf{V}_{\text{Th}}}{\mathbf{Z}} = \frac{\mathbf{V}_{\text{Th}}}{\mathbf{Z}_{R_{\text{Th}}} + \mathbf{Z}_L}$$

Write the source voltage in phasor form as 120 V $\angle 30°$. Using the voltage-divider concept, the Thevenin voltage is one-half this value. The impedances of the resistor and inductor are

$$\mathbf{Z}_{R_{\text{Th}}} = 5\ \Omega\angle 0°$$
$$\mathbf{Z}_L = j\omega L = j\left(377\ \frac{\text{rad}}{\text{s}}\right)(0.15\text{ H}) = 56.55\ \Omega\angle 90°$$
$$\mathbf{Z}_{\text{Th}} = \mathbf{Z}_{R_{\text{Th}}} + \mathbf{Z}_L = (5 + j0) + (0 + j56.55)$$
$$= 5 + j56.55 = 56.77\ \Omega\angle 84.9°$$

Substitute to determine the steady-state current.

$$\mathbf{I}_L = \frac{\mathbf{V}_{\text{Th}}}{\mathbf{Z}} = \frac{\mathbf{V}_{\text{Th}}}{\mathbf{Z}_{R_{\text{Th}}} + \mathbf{Z}_L} = \frac{60\text{ V}\angle 30°}{56.77\ \Omega\angle 84.9°}$$
$$= 1.06\text{ A}\angle{-55°}$$

Convert to the time (sinusoidal) form.

$$\mathbf{I}_L = 1.06\text{ A}\angle{-55°} = 1.5\text{ A}\sin(377t - 55°)$$

Effective or rms values are used with phasor quantities. Peak values are used with sinusoidal quantities.[5] Radians and degrees are mixed in the equations strictly for clarity. Radians are necessary for calculations. Transient calculation using sinusoids follows a similar pattern, with phasor analysis being an efficient calculational method. During transients using sinusoids, the exact time of the transient must be specified.

Example 31.7

For the network shown, $R_1 = 10\ \Omega$, $R_2 = 10\ \Omega$, $v = 636\text{ V}\cos(2513t + 60°)$, and $C = 1\ \mu$F. Determine the steady-state capacitance voltage with the switch closed.

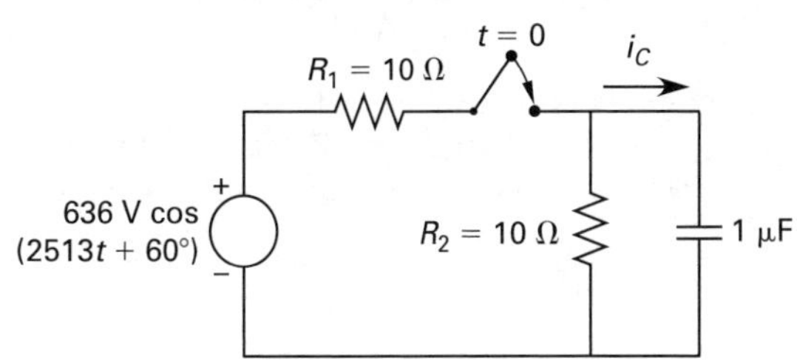

Solution

Using Eq. 31.5 gives

$$v_{\text{Th}} = R_{\text{Th}}C\left(\frac{dv_C}{dt}\right) + v_C$$

Represent the source voltage in phasor form as 450 V$\angle 60°$. Using the voltage divider concept, the Thevenin voltage is one-half this value. Recall that, in the

[5]The representation of phasors as rms quantities and sinusoidal variables as peak quantities is standard in this text. Care should be taken to determine how phasors are being represented in a given document.

frequency domain, multiplying by $j\omega$ is equivalent to taking the derivative. Combining the two gives

$$v_{\text{Th}} = R_{\text{Th}}C\left(\frac{dv_C}{dt}\right) + v_C$$

$$\mathbf{V}_{\text{Th}} = \mathbf{Z}_{R_{\text{Th}}}Cj\omega\mathbf{V}_C + \mathbf{V}_C = (j\omega\mathbf{Z}_{R_{\text{Th}}}C + 1)\mathbf{V}_C$$

$$\mathbf{V}_C = \frac{\mathbf{V}_{\text{Th}}}{j\omega\mathbf{Z}_{R_{\text{Th}}}C + 1}$$

$$= \frac{225\ \text{V}\angle 60^\circ}{j\left(2513\ \frac{\text{rad}}{\text{s}}\right)(5\ \Omega)(1\times 10^{-6}\ \text{F}) + 1}$$

$$= \frac{225\ \text{V}\angle 60^\circ}{1.01\ \Omega\angle 0.72^\circ} = 225\ \text{V}\angle 59.28^\circ$$

Transforming into the time domain gives

$$v_{C,\text{ss}}(t) = 318\ \text{V}\cos(2513t + 59^\circ)$$

At the angular frequency given, equivalent to 400 Hz, the capacitance is essentially an open circuit and has little impact on the network. The voltage source in this example was given as the cosine function to emphasize that the choice is arbitrary and does not affect the analysis.

Circuit Theory

4. FIRST-ORDER ANALYSIS: PULSE TRANSIENTS

A *pulse* is generally defined as a variation in a quantity that is normally constant. When the variation occurs at a single point in time, the pulse is called a *step pulse*. If the magnitude of the step pulse equals one, it is termed a *unit step*. A pulse has a finite duration that is normally brief compared to the time scale of interest. When the duration of the pulse is so short that it can be thought of as infinitesimal, it is called an *impulse*. Figure 31.2 shows a *unit pulse* of each type.

The unit step is defined by Eq. 31.10. The unit impulse is the derivative of the unit step function. The relationships between various forcing functions are shown in Table 31.1.

$$u(t-t_1) = 0 \quad [t < t_1]$$
$$u(t-t_1) = 1 \quad [t < t_1] \qquad 31.10$$

Table 31.1 Operations on Forcing Functions

	function when	
function	differentiated	integrated
unit impulse	–	unit step
unit step	unit impulse	unit ramp
unit ramp	unit step	unit parabola
unit parabola	unit ramp	(third degree)
unit exponential	unit exponential	unit exponential
unit sinusoid	unit sinusoid	unit sinusoid

Figure 31.2 Pulse Transients

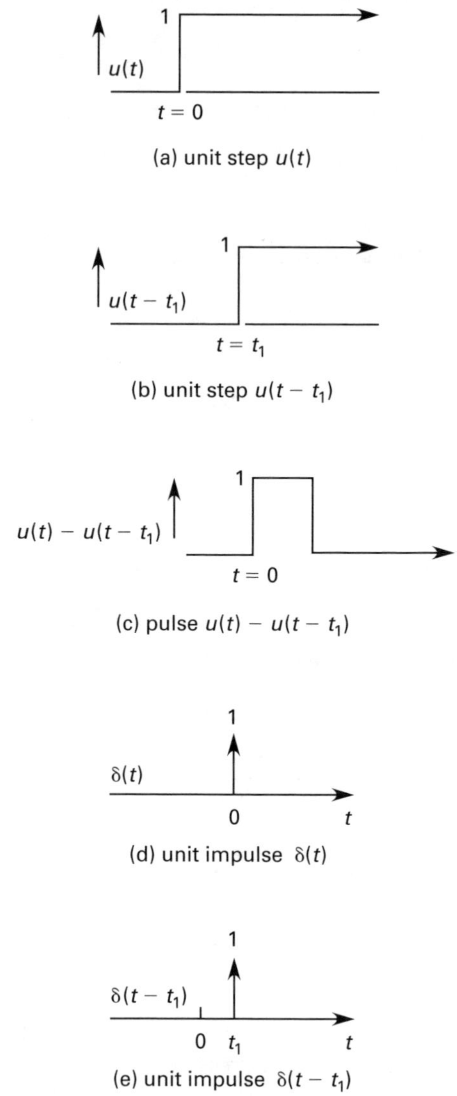

The response of a circuit to a unit step is called a *step response*. The source voltage magnitude, in this case the Thevenin voltage magnitude, is one for a unit step. For a first-order linear circuit like the one shown in Fig. 31.3, the response is determined in terms of the *state variables* of the energy storage device, that is, inductive current or capacitive voltage.[6] If other electrical quantities are required, they can be found from any of the circuit relations. The response equations are

$$v_C(t) = u(t)(1 - e^{-t/\tau}) \qquad 31.11$$

$$i_L(t) = u(t)\left(\frac{1}{R_{\text{Th}}}\right)(1 - e^{-t/\tau}) \qquad 31.12$$

The time constant, τ, for the capacitive circuit is $R_{\text{Th}}C$ and for the inductive circuit is L/R_{Th}.

[6]A state variable is a variable sufficient to specify the condition of the network and define its future behavior.

Figure 31.3 *First-Order Circuit*

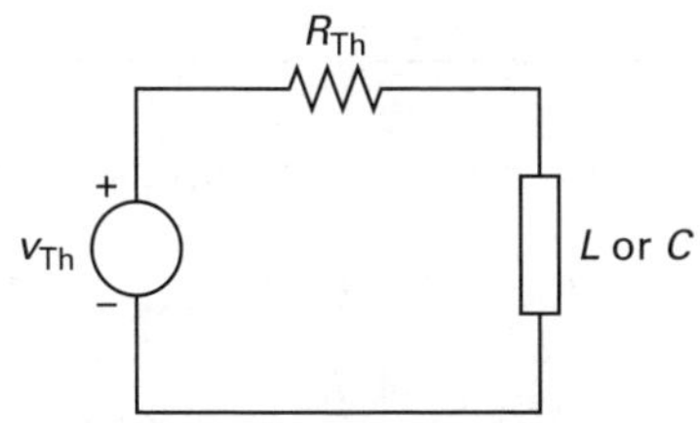

The step and impulse responses for *RC* and *RL* circuits are shown in Table 31.2. Of importance, the magnitude of the voltage and current sources for a unit step response is one.[7]

Example 31.8

Determine the capacitance voltage in the circuit shown.

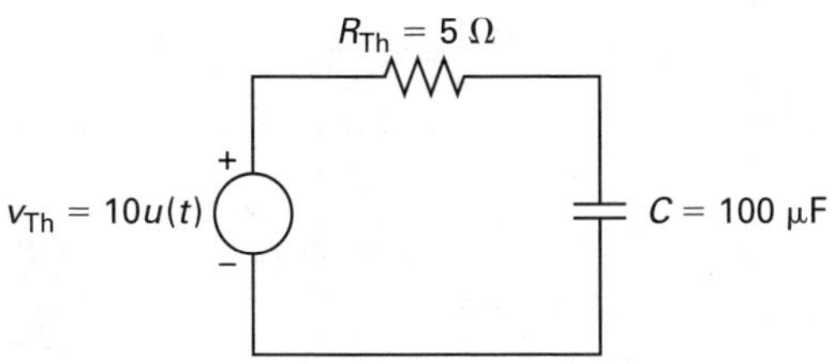

Solution

From Eq. 31.11 or Table 31.2, the form of the capacitance voltage is

$$v_C(t) = u(t)(1 - e^{-t/\tau})$$

The time constant is

$$\tau = R_{\rm Th}C = (5\ \Omega)(100 \times 10^{-6}\ \text{F}) = 500 \times 10^{-6}\ \text{s}$$

The unit step has a magnitude of one. A factor of 10 is required to scale the unit step to the actual value of the Thevenin voltage. The final solution is

$$v_C(t) = 10u(t)\left(1 - e^{-t/(500\times 10^{-6}\ \text{s})}\right)$$

Example 31.9

If the source in Ex. 31.8 outputs two 10 V pulses of 0.010 s duration 0.020 s apart, what is the capacitance voltage?

[7]The magnitude of one for a unit step is the reason for the apparent nonappearance of the current and voltage in the equations.

Solution

From Eq. 31.10 and Fig. 31.2, the voltage source is represented as follows. The brackets in Eq. 31.10 enclose each pulse for clarification.

$$\begin{aligned} v_{\rm Th}(t) = &\left(10u(t) - 10u(t - 0.010\ \text{s})\right) \\ &+ \left(10u(t - 0.030\ \text{s}) - 10u(t - 0.040\ \text{s})\right) \end{aligned}$$

Using Eq. 31.11 or Table 31.2, scaling by a factor of 10, and applying the principle of superposition gives the total response.

$$\begin{aligned} v_C(t) = &\ 10u(t)\left(1 - e^{-t/(500\times 10^{-6}\ \text{s})}\right) \\ &- 10u(t - 0.010\ \text{s})\left(1 - e^{(-t-0.010\ \text{s})/(500\times 10^{-6}\ \text{s})}\right) \\ &+ 10u(t - 0.030\ \text{s})\left(1 - e^{(-t-0.030\ \text{s})/(500\times 10^{-6}\ \text{s})}\right) \\ &- 10u(t - 0.040\ \text{s})\left(1 - e^{(-t-0.040\ \text{s})/(500\times 10^{-6}\ \text{s})}\right) \end{aligned}$$

5. SECOND-ORDER ANALYSIS

Second-order analysis involves circuits with one equivalent capacitor and one equivalent inductor, that is, two energy storage elements. As in first-order analysis, the method for determining the desired quantity is mathematical analysis in the time domain. Frequency domain analysis, which is also performed in first-order networks, significantly simplifies second-order analysis and is accomplished by transforming either the applicable differential equation or the circuit itself into the *s* domain.

Laplace transform analysis is then used to determine the desired quantity. (Fourier transform analysis could also be used, but the Laplace transform has the advantage of representing constants, sinusoids, and exponential functions in a single variable, *s*.) Figure 31.4 represents a generic circuit with a single capacitor and inductor.

Figure 31.4 *Second-Order Circuits*

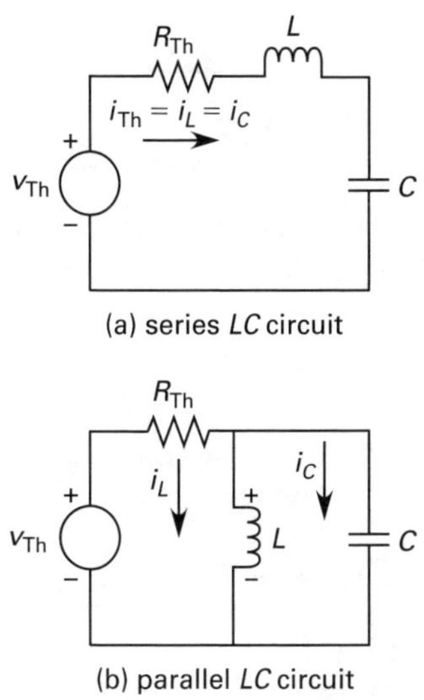

Table 31.2 Step and Impulse Responses for First-Order Circuits

circuit	unit step response	unit impulse response
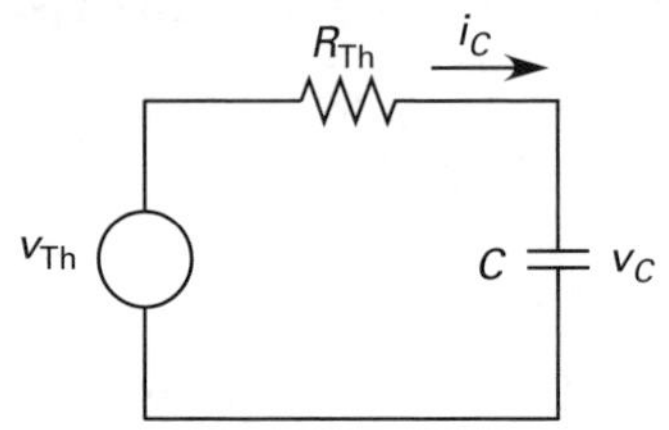	$v_{\text{Th}}(t) = u(t)$ $v_C(t) = u(t)\left(1 - e^{-t/(R_{\text{Th}}C)}\right)$ $i_C(t) = u(t)\left(\frac{1}{R_{\text{Th}}}\right)\left(e^{-t/(R_{\text{Th}}C)}\right)$	$v_{\text{Th}}(t) = \delta(t)$ $h_v(t) = u(t)\left(\frac{1}{R_{\text{Th}}C}\right)\left(e^{-t/(R_{\text{Th}}C)}\right)$ $h_i(t) = u(t)\left(-\frac{1}{R_{\text{Th}}^2 C}\right)\left(e^{-t/(R_{\text{Th}}C)}\right) + \delta(t)\left(\frac{1}{R_{\text{Th}}}\right)$
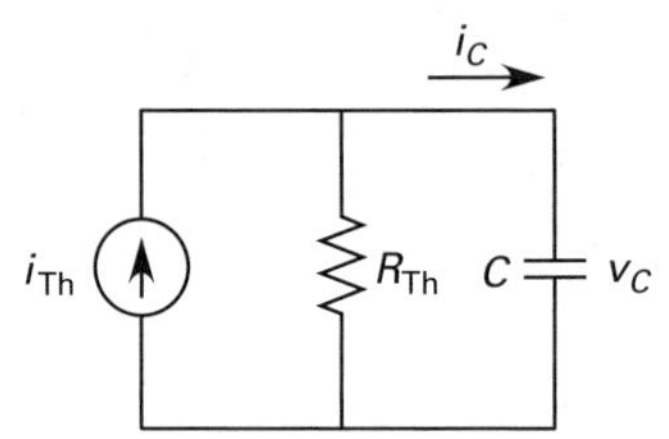	$i_{\text{Th}}(t) = u(t)$ $v_C(t) = u(t)R_{\text{Th}}\left(1 - e^{-t/(R_{\text{Th}}C)}\right)$ $i_C(t) = u(t) = u(t)\left(e^{-t/(R_{\text{Th}}C)}\right)$	$i_{\text{Th}}(t) = \delta(t)$ $h_v(t) = u(t)\left(\frac{1}{C}\right)\left(e^{-t/(R_{\text{Th}}C)}\right)$ $h_i(t) = u(t)\left(-\frac{1}{R_{\text{Th}}C}\right)\left(e^{-t/(R_{\text{Th}}C)}\right) + \delta(t)$
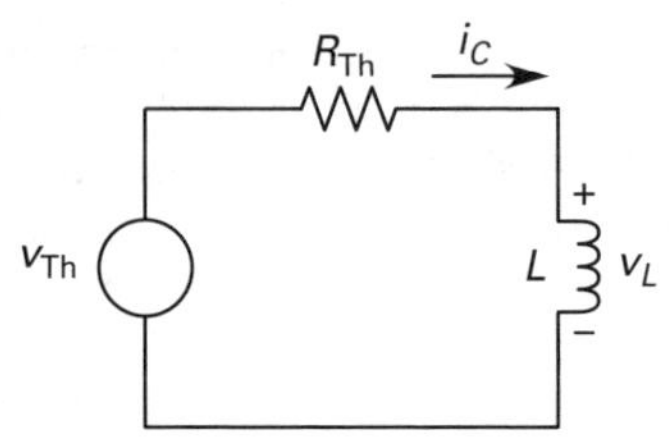	$v_{\text{Th}}(t) = u(t)$ $v_L(t) = u(t)\left(e^{-t/(L/R_{\text{Th}})}\right)$ $i_L(t) = u(t)\left(\frac{1}{R_{\text{Th}}}\right)\left(1 - e^{-t/(L/R_{\text{Th}})}\right)$	$v_{\text{Th}}(t) = \delta(t)$ $h_v(t) = u(t)\left(\frac{R_{\text{Th}}}{L}\right)\left(e^{-t/(L/R_{\text{Th}})}\right) + \delta(t)$ $h_i(t) = u(t)\left(\frac{1}{L}\right)\left(e^{-t/(L/R_{\text{Th}})}\right)$
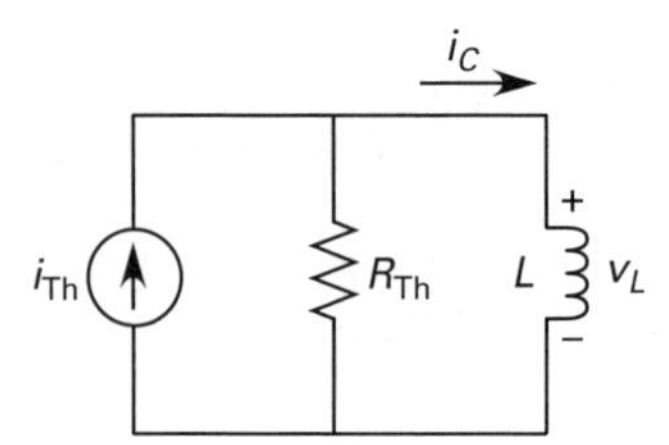	$i_{\text{Th}}(t) = u(t)$ $v_L(t) = u(t)R\left(e^{-t/(L/R_{\text{Th}})}\right)$ $i_L(t) = u(t)\left(1 - e^{-t/(L/R_{\text{Th}})}\right)$	$i_{\text{Th}}(t) = \delta(t)$ $h_v(t) = u(t)\left(-\frac{R_{\text{Th}}^2}{L}\right)\left(e^{-t/(L/R_{\text{Th}})}\right) + R\delta(t)$ $h_i(t) = u(t)\left(\frac{R_{\text{Th}}}{L}\right)\left(e^{-t/(L/R_{\text{Th}})}\right)$

The differential equation for second-order circuits takes on the general form

$$f(t) = a\frac{d^2x}{dt^2} + b\frac{dx}{dt} + cx \qquad 31.13$$

The associated characteristic equation, written with the roots shown in the s domain, is

$$as^2 + bs + c = 0 \qquad 31.14$$

Equation 31.13 can be written in terms of any circuit variable. It is normally written in terms of capacitance voltage or inductance current for reasons identical to those in first-order circuits. The solutions take on one of three forms, depending upon the magnitude of the constants (see Sec. 31.5 through Sec. 31.7). The general response of the three forms is illustrated in Fig. 31.5. *Overdamping* in a linear system is defined as damping over that is required for critical damping. *Critical damping* in a linear system is damping on the threshold between oscillatory and exponential behavior. *Underdamping* is damping that results in oscillatory behavior.

The method for determining the complete solution to second-order networks follows.

step 1: Determine the second-order equation for the circuit to be analyzed by using electrical engineering

Figure 31.5 Second-Order Circuit Response to a Step Change

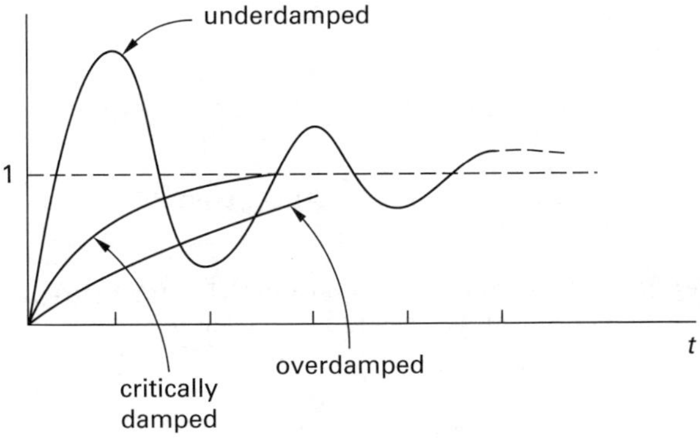

relationships and laws such as Ohm's law, KVL, KCL, and others.

step 2: Write the characteristic equation associated with the second-order equation. Using the constants of this equation, a, b, and c, determine the form of the solution.

step 3: Write the solution form.

step 4: If required, determine the initial values from the circuit's initial conditions.[8] Determine the steady-state values, x_{ss} and x'_{ss}.

step 5: Calculate the roots of the characteristic equation.

step 6: Use the information in steps 4 and 5 to calculate the constants in the final solution, A and B. (This involves the solution form in step 3 and its derivative or the direct use of the generic equations for the constants.)

step 7: Write the final solution from the information found in steps 4, 5, and 6.

Steps 4, 5, and 6 are equivalent to performing the following.

step 4′: Find the steady-state response from the particular solution of the nonhomogeneous equation. The solution will be of the same form as the forcing function. Regardless of transient type, the result will be a steady-state value, κ.

step 5′: Determine the transient response from the homogeneous equation. Regardless of the transient type, the transient response will be of the form $Ae^{s_1t} + Be^{s_2t}$. For the critical damped case $s_1 = s_2$, and for the underdamped or oscillatory case the variable s will take on the values $\alpha \pm j\beta$.

step 6′: Sum the steady-state and transient solutions, that is, the complementary and particular solutions, to obtain the total solution. This will be of the form $\kappa + Ae^{s_1t} + Be^{s_2t}$. Calculate the constants in the final solution, A and B.

6. SECOND-ORDER ANALYSIS: OVERDAMPED

If the two roots of Eq. 31.14 are real and different from one another, equivalent to the condition $b^2 > 4ac$, the solution is of the form

$$x(t) = \kappa + Ae^{s_1t} + Be^{s_2t} \qquad 31.15$$

The first term, κ, is a constant, that is, the steady-state component of the solution. This is the particular solution and results from the effect of an energy source or forcing function. The exponential terms represent the natural response of the circuit.[9] The roots of the characteristic equation are

$$s_1 = \frac{-b + \sqrt{b^2 - 4ac}}{2a} \qquad 31.16$$

$$s_2 = \frac{-b - \sqrt{b^2 - 4ac}}{2a} \qquad 31.17$$

The coefficients A and B in Eq. 31.15 are found from the initial and steady-state values of x and dx/dt. Because capacitor voltage and inductor current cannot change instantaneously, knowing the conditions immediately prior to any switching transient means knowing the conditions immediately after. By isolating the two energy storage elements and determining the Thevenin equivalent, any second-order circuit can be represented as in Fig. 31.4. In the series RLC circuits of Fig. 31.4(a), the results of applying the initial conditions are

$$v_C(0^+) = V_0 \qquad 31.18$$

$$\frac{dv_C(0^+)}{dt} = \frac{I_0}{C} \qquad 31.19$$

These conditions are a result of the common current path, and $i_C = C(dv_C/dt)$.

For the parallel RLC circuits in Fig. 31.4(b), the results of applying the initial conditions are

$$i_L(0^+) = I_0 \qquad 31.20$$

$$\frac{di_L(0^+)}{dt} = \frac{v_C(0^+)}{L} \qquad 31.21$$

These conditions are a result of the common voltage, and $v_L = L(di_L/dt)$.

[8]Initial conditions in a circuit are the initial values of the state variable function and its derivative. Two known conditions are required for the solution of a second-order equation.

[9]The natural response of an electric circuit is the response that would occur without a forcing function, that is, without an energy source. This portion of the solution goes by numerous names including natural, zero energy, homogeneous, and complementary.

Circuit Theory

The constants A and B are determined by substituting Eq. 31.18 through Eq. 31.21 into Eq. 31.15 and its derivative (multiplying by s is the equivalent of taking the derivative) and then solving for the specific circuit. Equation 31.18 through Eq. 31.21 can be used if network initial conditions are given and circuit element parameters, for example, L and C, are known. The results of such a substitution, in generic terms, are

$$x(0^+) = A + B + x_{ss} \qquad 31.22$$

$$x'(0^+) = s_1 A + s_2 B + x'_{ss} \qquad 31.23$$

Solving Eq. 31.22 and Eq. 31.23 simultaneously gives the values of the constants.

$$A = \tfrac{1}{2}\left(1 + \frac{b}{\sqrt{b^2 - 4ac}}\right)(x(0^+) - x_{ss}) + \left(\frac{a}{\sqrt{b^2 - 4ac}}\right)(x'(0^+) - x'_{ss}) \qquad 31.24$$

$$B = \tfrac{1}{2}\left(1 - \frac{b}{\sqrt{b^2 - 4ac}}\right)(x(0^+) - x_{ss}) - \left(\frac{a}{\sqrt{b^2 - 4ac}}\right)(x'(0^+) - x'_{ss}) \qquad 31.25$$

Example 31.10

Determine the second-order equation for the circuit in Fig. 31.4(a).

Solution

Using KVL and rearranging the result gives

$$v_{\text{Th}} = iR_{\text{Th}} + L\left(\frac{di}{dt}\right) + \frac{1}{C}\int i dt$$

Differentiating gives

$$\frac{dv_{\text{Th}}}{dt} = R_{\text{Th}}\left(\frac{di}{dt}\right) + L\left(\frac{d^2 i}{dt^2}\right) + \left(\frac{1}{C}\right)i$$

Rearrange to put into the form of Eq. 31.13.

$$\frac{dv_{\text{Th}}}{dt} = L\left(\frac{d^2 i}{dt^2}\right) + R_{\text{Th}}\left(\frac{di}{dt}\right) + \left(\frac{1}{C}\right)i$$

Example 31.11

A certain network is described by the following differential equation.[10]

$$24 \text{ V} = 2\left(\frac{d^2 v}{dt^2}\right) + 5\left(\frac{dv}{dt}\right) + 2v$$

[10]Dividing the equation by two shows that 12 is the value of the forcing function. Further, the 12 actually represents a step change of the source (Thevenin) voltage at $t = 0$. After the transient ends, the final value of the voltage across the capacitor will be 12 V.

The initial conditions are $v = 0$ and $dv/dt = 0$. Determine the expression for the voltage for $t \geq 0$.

Solution

Use the method for solving second-order networks outlined in Sec. 31.5.

step 1: Because the second-order equation is given, step 1 is complete.

step 2: The characteristic equation using constants a, b, and c from the second-order equation is

$$2s^2 + 5s + 2 = 0$$

From this, $a = 2$, $b = 5$, $c = 2$, and $b^2 > 4ac$. The network is overdamped.

step 3: The solution form is

$$x(t) = \kappa + Ae^{s_1 t} + Be^{s_2 t}$$

step 4: The initial conditions are given. The forcing function is a constant 24 V. At steady-state, dv/dt and d^2v/dt^2 equal zero. The steady-state voltages are[11]

$$\begin{aligned} 24 \text{ V} &= 2\left(\frac{d^2 v_{ss}}{dt^2}\right) + 5\left(\frac{dv_{ss}}{dt}\right) + 2v_{ss} \\ &= 0 + 0 + 2v_{ss} \\ v_{ss} &= \tfrac{24}{2} = 12 \text{ V} \end{aligned}$$

The steady-state voltage in this case is not a function of time. Therefore,

$$v'_{ss} = 0 \text{ V}$$

step 5: Using Eq. 31.16 and Eq. 31.17, the roots of the characteristic equation are

$$\begin{aligned} s_1 &= \frac{-b + \sqrt{b^2 - 4ac}}{2a} \\ &= \frac{-5 + \sqrt{(5)^2 - (4)(2)(2)}}{(2)(2)} = -\tfrac{1}{2} \\ s_2 &= \frac{-b - \sqrt{b^2 - 4ac}}{2a} \\ &= \frac{-(5) - \sqrt{5^2 - (4)(2)(2)}}{(2)(2)} = -2 \end{aligned}$$

[11]Units are not shown where they might cause confusion with the variables used, such as the voltage v and volts V. Units should be shown in all final solutions.

step 6: Use Eq. 31.24 and Eq. 31.25 to determine the constants A and B.

$$\begin{aligned} A &= \left(\tfrac{1}{2}\right)\left(1+\frac{b}{\sqrt{b^2-4ac}}\right)(v(0^+)-v_{ss}) \\ &\quad + \left(\frac{a}{\sqrt{b^2-4ac}}\right)(v'(0^+)-v'_{ss}) \\ &= \left(\tfrac{1}{2}\right)\left(1+\frac{5}{\sqrt{(5)^2-(4)(2)(2)}}\right)(0-12) \\ &\quad + \left(\frac{2}{\sqrt{(5)^2-(4)(2)(2)}}\right)(0-0) \\ &= -16 \end{aligned}$$

$$\begin{aligned} B &= \left(\tfrac{1}{2}\right)\left(1-\frac{b}{\sqrt{b^2-4ac}}\right)(v(0^+)-v_{ss}) \\ &\quad - \left(\frac{a}{\sqrt{b^2-4ac}}\right)(v'(0^+)-v'_{ss}) \\ &= \left(\tfrac{1}{2}\right)\left(1-\frac{5}{\sqrt{(5)^2-(4)(2)(2)}}\right)(0-12) \\ &\quad - \left(\frac{2}{\sqrt{(5)^2-(4)(2)(2)}}\right)(0-0) \\ &= 4 \end{aligned}$$

step 7: Write the final solution.

$$\begin{aligned} v(t) &= \kappa + Ae^{s_1 t} + Be^{s_2 t} \\ &= 12 - 16e^{(-1/2)t} + 4e^{-2t} \end{aligned}$$

7. SECOND-ORDER ANALYSIS: CRITICALLY DAMPED

If the two roots of Eq. 31.14 are real and the same, equivalent to the condition $b^2 = 4ac$, the solution is of the form

$$x(t) = \kappa + Ae^{st} + Bte^{st} \qquad \text{31.26}$$

The first term, κ, is a constant, that is, the steady-state component of the solution. The exponential terms represent the natural response of the circuit. The root of the characteristic equation is

$$s = \frac{-b}{2a} \qquad \text{31.27}$$

The coefficients A and B in Eq. 31.26 are found from the initial and steady-state values of x and dx/dt. By the same reasoning used in Sec. 31.6, the initial conditions for a series RLC circuit remain unchanged and are repeated here for convenience.

$$v_C(0^+) = V_0 \qquad \text{31.28}$$

$$\frac{dv_C(0^+)}{dt} = \frac{I_0}{C} \qquad \text{31.29}$$

Similarly, the initial conditions for parallel RLC circuits are

$$i_L(0^+) = I_0 \qquad \text{31.30}$$

$$\frac{di_L(0^+)}{dt} = \frac{v_C(0^+)}{L} \qquad \text{31.31}$$

The constants A and B are determined by substituting Eq. 31.28 through Eq. 31.31 into Eq. 31.26 and its derivative (multiplying by s is the equivalent of taking the derivative) and then solving for the specific circuit. Equation 31.28 through Eq. 31.31 can be used if network initial conditions are given and circuit element parameters such as L and C are known. The results of such a substitution, in generic terms, are

$$x(0^+) = A + x_{ss} \qquad \text{31.32}$$

$$x'(0^+) = sA + B + x'_{ss} \qquad \text{31.33}$$

Solving Eq. 31.32 and Eq. 31.33 simultaneously gives the values of the constants.

$$A = x(0^+) - x_{ss} \qquad \text{31.34}$$

$$B = x'(0^+) - x'_{ss} + \left(\frac{b}{2a}\right)A \qquad \text{31.35}$$

Example 31.12

Consider the series RLC circuit shown, with $v_{Th} = 10$ V, $R = 80\ \Omega$, $L = 80 \times 10^{-3}$ H, and $C = 50 \times 10^{-6}$ F. The capacitor is initially charged to 5 V and the initial current in the circuit is 187.5×10^{-3} A. Determine the behavior of the circuit, in terms of the capacitance voltage, for $t \geq 0$.

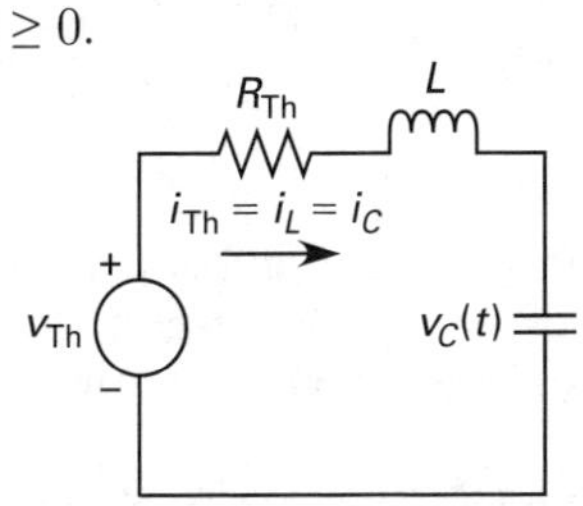

Circuit Theory

Solution

Using KVL and writing the resulting equation in terms of the capacitance voltage,

$$10\ \text{V} = iR_{\text{Th}} + L\left(\frac{di}{dt}\right) + v_C$$

Substituting $i = C(dv/dt)$,

$$10\ \text{V} = C\left(\frac{dv_C}{dt}\right)R_{\text{Th}} + LC\left(\frac{d^2v_C}{dt^2}\right) + v_C$$

$$10\ \text{V} = LC\left(\frac{d^2v_C}{dt^2}\right) + R_{\text{Th}}C\left(\frac{dv_C}{dt}\right) + v_C$$

The characteristic equation follows.

$$LCs^2 + RCs + 1 = 0$$

Determine the solution form from the relationship between the coeffcients of the characteristic equation.

$$\begin{aligned} b^2 &= (RC)^2 = \left((80\ \Omega)(50 \times 10^{-6}\ \text{F})\right)^2 = 16 \times 10^{-6} \\ 4ac &= (4)(LC)(1) \\ &= (4)(80 \times 10^{-3}\ \text{H})(50 \times 10^{-6}\ \text{F})(1) = 16 \times 10^{-6} \\ b^2 &= 4ac \end{aligned}$$

The circuit is critically damped. The solution is of the form given by Eq. 31.26.

$$v_C(t) = \kappa + Ae^{st} + Bte^{st}$$

The initial conditions are given. The steady-state conditions are found from the results of the KVL analysis.

$$\begin{aligned} 10 &= LC\left(\frac{d^2v_C}{dt^2}\right) + R_{\text{Th}}C\left(\frac{dv_C}{dt}\right) + v_C \\ &= LC(0) + R_{\text{Th}}C(0) + v_C \\ v_{C,\text{ss}} &= 10 \\ v'_{C,\text{ss}} &= 0 \end{aligned}$$

The roots of the characteristic equation are found from Eq. 31.27 or actual calculation to be[12]

$$\begin{aligned} s &= \frac{-b}{2a} = \frac{-R_{\text{Th}}C}{2LC} \\ &= \frac{-R_{\text{Th}}}{2L} = \frac{-(80\ \Omega)}{(2)(80 \times 10^{-3}\ \text{H})} = -500 \end{aligned}$$

Because the initial conditions and circuit parameters are known, instead of using the generic equations (see Eq. 31.32 through Eq. 31.35), use the solution form directly with the initial conditions from Eq. 31.28 and Eq. 31.29.

$$\begin{aligned} v_C(t) &= \kappa + Ae^{st} + Bte^{st} \\ v_C(0^+) &= 10 + Ae^{s(0)} + B(0)\,e^{s(0)} \\ 5 &= 10 + A \\ A &= -5 \end{aligned}$$

The second initial condition results in the following value for B.

$$\begin{aligned} v_C(t) &= k + Ae^{st} + Bte^{st} \\ \frac{dv_C(0^+)}{dt} = \frac{I_0}{C} &= 0 + Ase^{st} + B(e^{st} + tse^{st}) \\ \frac{187.5 \times 10^{-3}\ \text{A}}{50 \times 10^{-6}\ \text{F}} &= 0 + (-5)(-500)\left(e^{(-500)(0)}\right) \\ &\quad + B\left(e^{(-500)(0)} + (0)(-500) \times \left(e^{(-500)(0)}\right)\right) \\ 3750 &= 2500 + B \\ B &= 1250 \end{aligned}$$

The final solution is

$$\begin{aligned} v_C(t) &= \kappa + Ae^{st} + Bte^{st} \\ &= 10 - 5e^{-500t} + 1250te^{-500t} \end{aligned}$$

8. SECOND-ORDER ANALYSIS: UNDERDAMPED

If the two roots of Eq. 31.14 are complex conjugates, equivalent to the condition $b^2 < 4ac$, the solution is of the form[13]

$$x(t) = \kappa + e^{\alpha t}\left(A_e e^{j\beta t} + B_e e^{-j\beta t}\right) \qquad 31.36$$

This can also be represented in the sinusoidal form as

$$x(t) = \kappa + e^{\alpha t}\left(A\cos\beta t + B\sin\beta t\right) \qquad 31.37$$

The first term, κ, is a constant, that is, the steady-state component of the solution. The exponential terms represent the natural response of the circuit. Equation 31.36 is the ringing response, or damped oscillation, shown in Fig. 30.7(b). The sinusoidal response is enveloped by the exponential, since α is negative.

[12]Units are not commonly shown for s. They are inverse seconds.

[13]The subscript e on the constant is a reminder that the constants in Eq. 31.36 and Eq. 31.37 differ.

The variable s used in the overdamped and critically damped cases has been expanded to clarify this behavior and is represented by Eq. 31.38 and Eq. 31.39.[14]

$$s_1 = \alpha + j\beta \qquad 31.38$$

$$s_2 = \alpha - j\beta \qquad 31.39$$

The roots of the characteristic equation are

$$\alpha = \frac{-b}{2a} \qquad 31.40$$

$$\beta = \frac{\sqrt{4ac - b^2}}{2a} \qquad 31.41$$

The coeffcients A_e and B_e in Eq. 31.36, or A and B in Eq. 31.37, are found from the initial and steady-state values of x and dx/dt. By the same reasoning used in Sec. 31.6, the initial conditions for a series RLC circuit remain unchanged and are repeated here for convenience.

$$v_C(0^+) = V_0 \qquad 31.42$$

$$\frac{dv_C(0^+)}{dt} = \frac{I_0}{C} \qquad 31.43$$

Similarly, the initial conditions for parallel RLC circuits are

$$i_L(0^+) = I_0 \qquad 31.44$$

$$\frac{di_L(0^+)}{dt} = \frac{v_C(0^+)}{L} \qquad 31.45$$

The constants are determined by substituting Eq. 31.42 through Eq. 31.45 into Eq. 31.36 or Eq. 31.37 and their derivatives (multiplying by $j\beta$ is the equivalent of taking the derivative) and then solving for the specific circuit. Equation 31.42 through Eq. 31.45 can be used if network initial conditions are given and circuit element parameters such as L and C are known. The results of such a substitution, in generic terms, are

$$x(0^+) = A + x_{\text{ss}} \qquad 31.46$$

$$x'(0^+) = -\alpha A + \beta B + x'_{\text{ss}} \qquad 31.47$$

Solving Eq. 31.46 and Eq. 31.47 simultaneously gives the values of the constants.

$$A = x(0^+) - x_{\text{ss}} \qquad 31.48$$

$$B = \left(\frac{\alpha}{\beta}\right)A + \left(\frac{1}{\beta}\right)(x'(0^+) - x'_{\text{ss}}) \qquad 31.49$$

[14]The terminology of α and β is often used in transient analysis. The variable s was split for the underdamped case to clearly show the sinusoids enveloped by the exponentials.

9. SECOND-ORDER ANALYSIS: PULSE TRANSIENTS

The output of a second-order circuit responding to a step change is illustrated in Fig. 31.5. The electrical parameters defining the response vary depending upon the type of analysis required (see Sec. 31.10). The response of a second-order network with zero initial conditions, that is, $x(0) = 0$ and $x'(0)$, to a step change has a wide range of applications. Because a unit step change can be scaled to any specific pulse, it forms the baseline input. Consider Eq. 31.13 with zero initial conditions and the unit step as the input, that is, $f(t) = 1$. The steady-state value of the state variable is

$$x_{\text{ss}} = \frac{f(t)}{c} = \frac{1}{c} \qquad 31.50$$

The equations in Sec. 31.6 through Sec. 31.8 are then modified using Eq. 31.50 and the zero initial conditions requirement as follows.[15]

For an overdamped circuit ($b^2 > 4ac$),

$$x(t) = \frac{1}{c} + Ae^{s_1 t} + Be^{s_2 t} \qquad 31.51$$

$$s_1 = \frac{-b + \sqrt{b^2 - 4ac}}{2a} \qquad 31.52$$

$$s_2 = \frac{-b - \sqrt{b^2 - 4ac}}{2a} \qquad 31.53$$

$$A = -\left(\frac{1}{2c}\right)\left(1 + \frac{b}{\sqrt{b^2 - 4ac}}\right) \qquad 31.54$$

$$B = -\left(\frac{1}{2c}\right)\left(1 - \frac{b}{\sqrt{b^2 - 4ac}}\right) \qquad 31.55$$

For a critically damped circuit ($b^2 = 4ac$),

$$x(t) = \frac{1}{c} + Ae^{st} + Bte^{st} \qquad 31.56$$

$$s = \frac{-b}{2a} \qquad 31.57$$

$$A = -\frac{1}{c} \qquad 31.58$$

$$B = -\frac{b}{2ac} \qquad 31.59$$

For an underdamped circuit ($b^2 < 4ac$),

$$x(t) = \frac{1}{c} + e^{\alpha t}(A\cos\beta t + B\sin\beta t) \qquad 31.60$$

$$s_1 = \alpha + j\beta \qquad 31.61$$

$$s_2 = \alpha - j\beta \qquad 31.62$$

[15]Because x_{ss} is a constant, x'_{ss}. This condition is also incorporated.

Circuit Theory

$$\alpha = \frac{-b}{2a} \quad 31.63$$

$$\beta = \frac{\sqrt{4ac - b^2}}{2a} \quad 31.64$$

$$A = -\frac{1}{c} \quad 31.65$$

$$B = -\frac{\alpha}{\beta c} \quad 31.66$$

10. HIGHER-ORDER CIRCUITS

A circuit with two or more independent energy storage elements is considered a *higher-order circuit.* Second-order systems dominate the behavior of third- and higher-order systems and can be used to approximate the behavior of such networks. Higher-order analysis takes place primarily in the s domain, using the concept of a complex frequency, s. The characteristic equation derived from the network differential equation is described with different terminology depending upon the type of electrical analysis performed. In purely mathematical terms, the characteristic equation is

$$as^2 + bs + c = 0 \quad 31.67$$

Control systems analysis uses the form given by Eq. 31.68. The *natural frequency*, ω_n, is sometimes written simply as ω. The term ζ is called the *damping ratio.*

$$s^2 + 2\zeta\omega_n s + \omega_n^2 = 0 \quad 31.68$$

Resonance analysis uses the form given by Eq. 31.69. The resonant frequency is symbolized by ω_0 and the quality factor by Q.

$$s^2 + \left(\frac{\omega_0}{Q}\right)s + \omega_0^2 = 0 \quad 31.69$$

Transient analysis uses the form given by Eq. 31.70. The *damping* is given the symbol α while the *frequency of oscillation* is β.

$$(s+\alpha)^2 + \beta^2 = 0 \quad 31.70$$

In the transient analysis of Sec. 31.6 through Sec. 31.8, the three cases could have been represented as $\alpha > \omega_0$ (overdamped), $\alpha = \omega_0$ (critically damped), and $\alpha < \omega_0$ (underdamped). The parameters are related as follows.

$$\omega_0^2 = \omega_n^2 = \alpha^2 + \beta^2 \quad 31.71$$

$$\zeta = \frac{1}{2Q} \quad 31.72$$

$$\alpha = \frac{\omega_0}{2Q} \quad 31.73$$

These characteristic equations are the denominators of transfer functions, which are ratios from a network's output point to a network's input point. Their roots, called *poles*, will aid in determining the frequency response of a network without solving the original differential equation. The roots of the characteristic equation determine the natural response of the network, that is, the response without a forcing function (without input). The transfer function and the characteristic equation describe the response of the network independent of input. Any input can then be attached and the output derived by knowing the circuit's transfer function and natural response.[16]

11. COMPLEX FREQUENCY

Electrical circuits can respond to input with a resulting constant output, A, and exponential output, $e^{\alpha t}$, or a sinusoid, $\sin \omega t$.[17] The complex frequency, $\mathbf{s}$, is used to combine the three possibilities and represent each in terms of a single exponential, e^{st}.[18] The complex frequency is

$$\mathbf{s} = \sigma + j\omega \quad 31.74$$

Consider a signal containing all three possibilities represented by Eq. 31.75, including a phase shift. In such a case, the peak value can be represented by the constant A. (The peak value is used to allow the constant to remain unchanged when in the exponential or sinusoidal form. In the sinusoidal form the peak value must be used to properly represent the function. The form $A\angle\theta$ normally uses an rms value for A, while the exponential form uses a peak value. Care must be taken to ensure the desired form is used.) Such a signal is given by

$$A_p e^{\sigma t}\sin(\omega t + \theta) \quad 31.75$$

Representing the sinusoid in exponential terms, using the sine term, and dropping the Im prefix as is common practice, allows Eq. 31.75 to be represented as

$$A_p e^{j\theta} e^{(\sigma + j\omega)t} = A_p e^{j\theta} e^{st} \quad 31.76$$

The e^{st} term will be common to all elements in a network, just as $e^{j\omega t}$ was common to all elements. Only the magnitude A and the angle θ of a given signal need to be carried in calculations. If desired, the e^{st} term is recovered upon completion of analysis, giving the exact form of the response. The variable σ is termed the *neper frequency* and assigned units of Np/s or Hz. The neper frequency σ is always zero or negative in an actual circuit. In Eq. 31.75, for example, if σ were positive, the exponential term would increase without bound, resulting in an infinite response. Clearly, this is not possible. Nevertheless, the σ is considered to carry its sign (in other words, the negative is internal), though the sign is sometimes shown explicitly to clarify the

[16]The transfer function gives the steady-state response. The poles provide the transient response.

[17]Even aperiodic signals can be represented in terms of sinusoids. Consequently, these three possibilities represent the full range.

[18]It is common practice to write the complex frequency $\mathbf{s}$ simply as s, regardless of its actual value, which can be real or complex. This book will follow this practice.

decaying exponential aspect. The variable ω is the angular frequency and is assigned units of rad/s or Hz. A negative value of ω merely implies that the sinusoid carries a negative sign, because $\sin(-\omega t) = -\sin\omega t$.

If $\sigma = 0$ and $\omega = 0$, the signal or input is a constant. If $\sigma < 0$ and $\omega = 0$, the signal represented is an exponential decay. If $\sigma < 0$ and $\omega \neq 0$, the signal is a damped sinusoid.[19] If $\sigma = 0$ and $\omega \neq 0$, the signal is a sinusoid. Table 31.3 shows examples of each of the signal response types based on Eq. 31.76 along with their values in the s domain. Table 31.4 gives the generalized impedances in the s domain. Table 31.4 also exhibits the similarity of form between the frequency domain and the s domain. Working in the s domain is always possible and simplifies the mathematics. If only a sinusoidal response is desired, upon completion of circuit manipulation, s can be replaced with $j\omega$ and the result will be in the correct form for any computations. Figure 31.6 illustrates how signal response terms plot.

Table 31.3 *Example Signal Representation in the s Domain*

time function	$A_p e^{j\theta}$ or $A_p\angle\theta^a$	s^b
50	$50\angle 0°$	$0 + j0$
$50e^{-100t}$	$50\angle 0°$	$-100 + j0$ Np/s
$50\sin(200t - 60°)$	$50\angle -60°$	$0 + j200$ rad/s
$50e^{-100t}\sin(-200t + 60°)$	$50\angle 60°$	$-100 - j200\ \text{s}^{-1}$

[a]The format $A\angle\theta$ used in this text normally indicates rms values. Here the peak value is used. If the rms format is used, the values in this column would be divided by $\sqrt{2}$.
[b]A negative frequency $(-\omega)$ value merely implies that the sinusoid carries a negative sign, because $\sin(-\omega t) = -\sin\omega t$ and s would then have a negative j term.

Table 31.4 *Generalized Impedance in the s Domain*

impedance type	frequency domain value	s domain value
Z_R	R	R
Z_L	$j\omega L$	sL
Z_C	$\dfrac{1}{j\omega C}$	$\dfrac{1}{sC}$

12. LAPLACE TRANSFORM ANALYSIS

The Laplace transform method of analysis moves the state variables, that is, the capacitive voltage and the inductive current, from the time domain into the s domain. In the s domain, many forms of excitation, such as the constant, the exponential, and the sinusoid, can

Figure 31.6 *Signal Response in the s Domain*

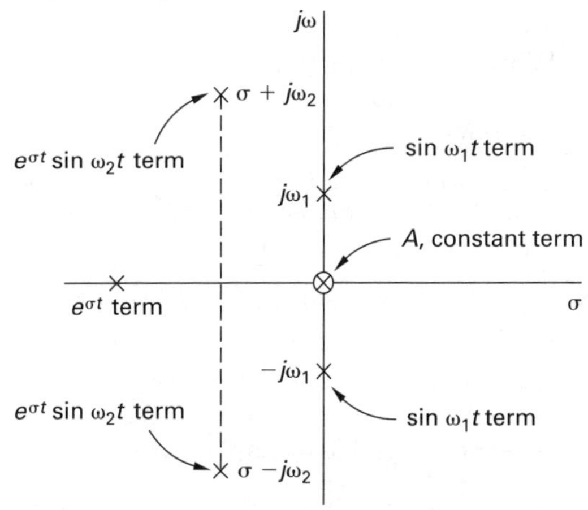

be united and easily handled mathematically. The Laplace transform is given by[20]

$$\mathcal{L}\{f(t)\} = \int_{0+}^{\infty} f(t)e^{-st}dt \qquad 31.77$$

The calculation of Eq. 31.77 seldom needs to be accomplished. Numerous tables of Laplace transforms exist (see App. 11.A). Using the data in a Laplace transform table, the properties given by Eq. 31.78 and Eq. 31.79 are used to transform a differential equation representing the behavior of an electrical network.

$$\mathcal{L}\{f(t)\} = F(s) \qquad 31.78$$

$$\mathcal{L}\left\{\frac{d^2f(t)}{dt^2}\right\} = s^2F(s) - sf(0^+) - f'(0^+) \qquad 31.79$$

$$\mathcal{L}\left\{\frac{df(t)}{dt}\right\} = sF(s) - f(0^+) \qquad 31.80$$

Because the energy in a capacitor or inductor cannot change instantaneously, the conditions at $t = 0$ are those at $t = 0^+$. Once the transformation is complete, the inverse Laplace transform can be taken using Eq. 31.81.

$$\mathcal{L}^{-1}\{F(s)\} = f(t) = \frac{1}{2\pi j}\int_{\sigma-j\infty}^{\sigma+j\infty} F(s)e^{st}dt \qquad 31.81$$

The calculation of Eq. 31.81 seldom needs to be accomplished. The same tables used for Laplace transforms contain the inverses. The path of integration of Eq. 31.81 is a straight line parallel to the $j\omega$ axis. All the poles of $F(s)$ lie to the left of this path.

[19]The condition for a damped sinusoid is sometimes given as $\sigma \neq 0$ and $\omega \neq 0$. This is true as long as the value of σ is restricted to negative values. (In real circuits, σ is either zero or negative.) If the negative sign is not carried with the σ but is instead shown in the exponential as $e^{-(\sigma+j\omega)t}$, the condition $\sigma > 0$ and $\omega \neq 0$ represents the damped sinusoid.

[20]The term s is the complex frequency or the *complex frequency variable* and could be shown as s. Also, taking the Laplace transform of a physical quantity introduces an additional unit of time. For example, $I(s)$ has units of A·s or C. Because the inverse transform removes the extra time unit, such units will be omitted in the s domain and $I(s)$ is still referred to as the current. The same is done for the voltage, $V(s)$.

Circuit Theory

The *classical method* for analyzing electrical networks using Laplace transform analysis follows.

step 1: Write the differential equation describing the circuit behavior from known circuit relationships and laws.

step 2: Transform the differential equation, term by term, into the s domain using Laplace transform tables (e.g., App. 11.A).

step 3: Solve for the transformed variable, $V(s)$ or $I(s)$, in terms of s.

step 4: Use partial fraction expansion, or any other desired method, to rearrange the terms into those recognizable in Laplace transform tables.

step 5: Equate the coeffcients of the powers of s. Solve for the constants.

step 6: Obtain the inverse Laplace transform.

The classical method is useful when the desired quantity is one of the state variables, capacitor voltage or inductor current. If the desired variables are other than the state variables, the *circuit transformation method* is the most direct.

The circuit transformation method follows.

step 1: Transform each individual circuit element into the s domain. Initial condition sources will exist for energy storage elements. For the capacitor, an initial condition voltage source will be added in series. For the inductor, an initial condition current source will be added in parallel.

step 2: If the desired quantity is one of the variables of an energy storage device (i.e., a state variable), obtain the equivalent Thevenin or Norton form of the circuit as seen from the terminals of the capacitor or inductor.

step 3: Write the differential equation, describing the circuit behavior from known circuit relationships and laws.

step 4: Solve for the desired variable in terms of s.

step 5: Use partial fraction expansion, or any other desired method, to rearrange the terms into those recognizable in Laplace transform tables.

step 6: Equate the coeffcients of the powers of s. Solve for the constants.

step 7: Obtain the inverse Laplace transform.

Regardless of the method used, the transformed equations obtained in step 3 are arranged using partial fraction expansion in order to be inversely transformed. This is necessary for transient analysis of first- and second-order systems, because the initial conditions impact the transient. Because the initial conditions do not impact the steady-state value, in sinusoidal analysis, usually called frequency analysis or frequency response, the equations of step 3 reduce to phasor form, so steps 4 through 7 can be skipped and the response analyzed using graphical means (see Chap. 32).

13. CAPACITANCE IN THE s DOMAIN

Capacitors oppose rates of change in voltage. In capacitive circuits, voltage is the state variable and is preferred in system equations. Because the capacitor voltage cannot change instantaneously, $v_C(0+) = v_C(0)$. With this information, the capacitive voltage, $v_C(t)$, and the capacitive current, $C(dv_C(t))$, are transformed by Eq. 31.82 and Eq. 31.83.

$$\mathcal{L}\{v_C(t)\} = V_C(s) \qquad \text{31.82}$$

$$\mathcal{L}\{i_C(t)\} = \mathcal{L}\left\{C\frac{dv_C(t)}{dt}\right\}$$
$$= sCV_C(s) - Cv_C(0) \qquad \text{31.83}$$

The transformed capacitance must exhibit the properties at its terminals given by Eq. 31.83. Such a model is shown in Fig. 31.7. Using the extended Ohm's law on the current source $sCV_C(s)$, that is, taking the voltage across the source's terminals, $V_C(s)$, and dividing by the current, $sCV_C(s)$, gives the model of Fig. 31.8. Taking the Thevenin equivalent of the model shown in Fig. 31.8 results in a third model given in Fig. 31.9 that shows the initial voltage explicitly. Any of the models is valid; Fig. 31.9 is perhaps the most intuitive.

Figure 31.7 *Capacitor Two-Source Model*

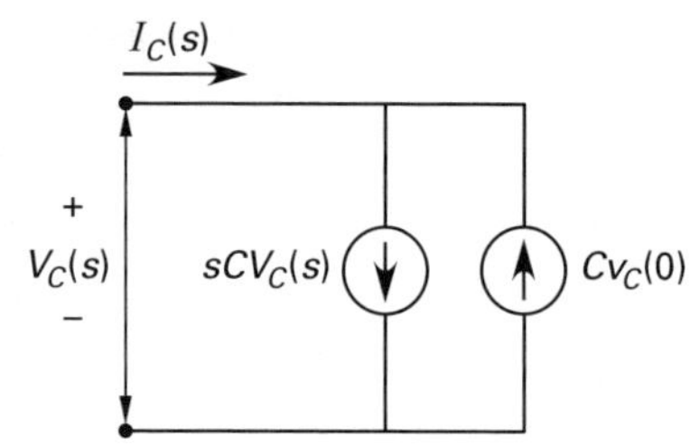

Figure 31.8 *Capacitor Impedance and Initial Current Source Model*

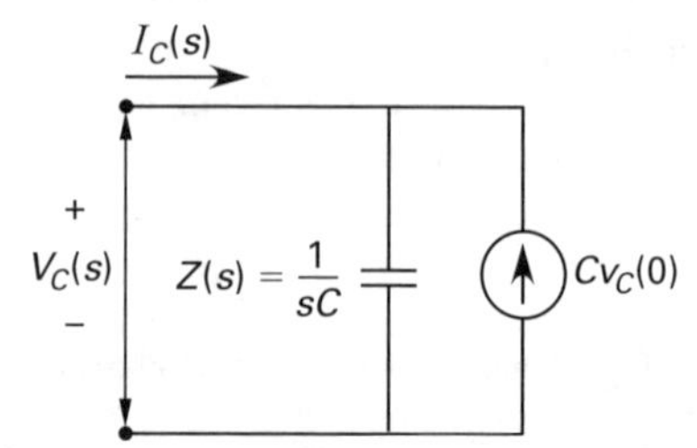

Figure 31.9 Capacitor Impedance and Initial Voltage Source Model

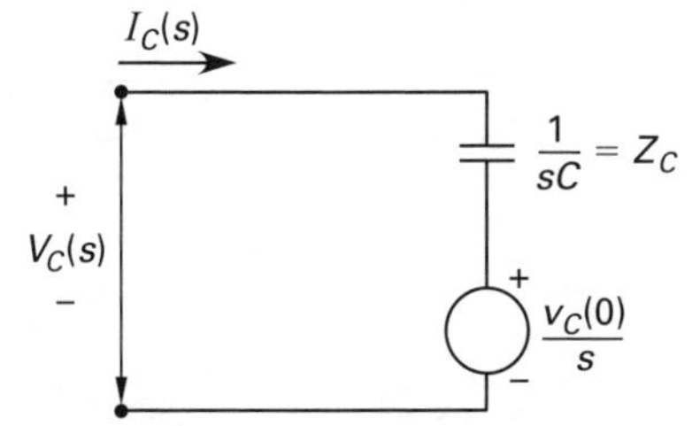

14. INDUCTANCE IN THE *s* DOMAIN

Inductors oppose rates of change in current. In inductive circuits, current is the state variable and it is preferred in system equations. Because the inductor current cannot change instantaneously, $i_L(0+) = i_L(0)$. With this information, the inductive current, $i_L(t)$, and the inductive voltage, $L(di_L(t)/dt)$, are transformed by Eq. 31.84 and Eq. 31.85.

$$\mathcal{L}\{i_L(t)\} = I_L(s) \quad \text{31.84}$$

$$\mathcal{L}\{v_L(t)\} = \mathcal{L}\left\{L\frac{di_L(t)}{dt}\right\}$$

$$= sLI_L(s) - Li_L(0) \quad \text{31.85}$$

The transformed inductance must exhibit the properties at its terminals given by Eq. 31.85. Such a model is shown in Fig. 31.10. Using the extended Ohm's law on the voltage source $sLI_L(s)$, that is, dividing by the current through the source's terminals, $I_L(s)$, gives the model shown in Fig. 31.11. Taking the Norton equivalent of the model in Fig. 31.11 results in a third model given in Fig. 31.12, which shows the initial current explicitly. Any of the models is valid; Fig. 31.12 is perhaps the most intuitive.

15. LAPLACE TRANSFORM ANALYSIS: FIRST- AND SECOND-ORDER SYSTEMS

Either of the methods in Sec. 31.12 may be applied. If variables other than the capacitor voltage or inductor current are required, the circuit transformation method is the most direct method. The Thevenin form is preferred for capacitive circuits because the final capacitive voltage must equal the Thevenin voltage, thereby acting as a check on the solution. The Norton equivalent is preferred for inductive circuits, because the final inductive current must equal the Norton equivalent current, thereby acting as a check on the solution. Any of the network analysis techniques is applicable when working in the *s* domain.

Figure 31.10 Inductor Two-Source Model

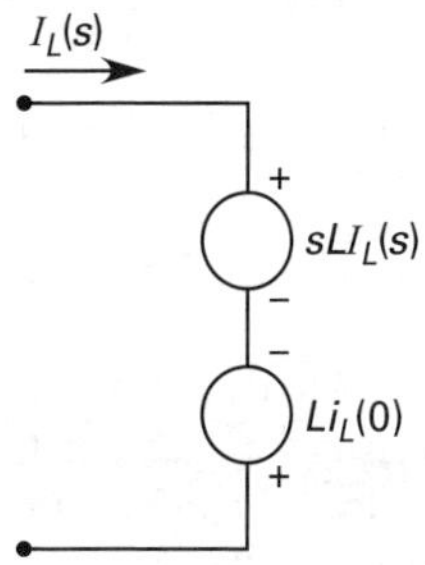

Figure 31.11 Inductor Impedance and Initial Voltage Source Model

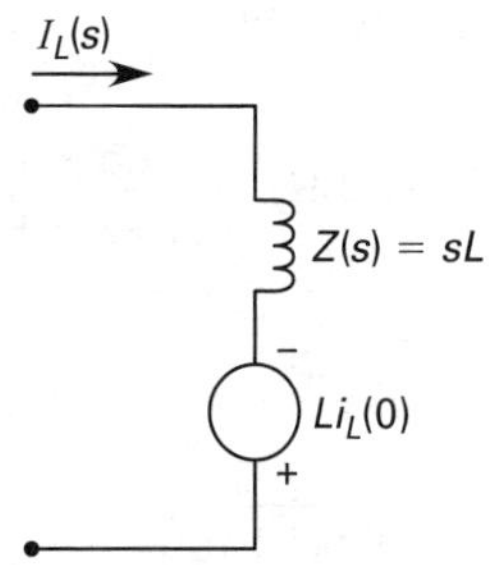

Figure 31.12 Inductor Impedance and Initial Current Source Model

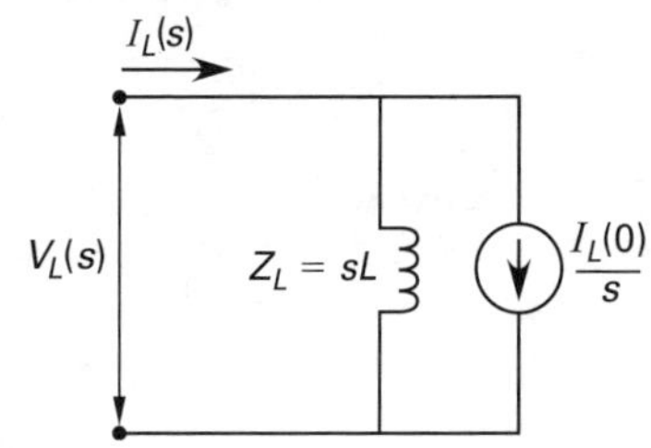

Example 31.13

In the circuit shown, the capacitor has an initial charge of 4.5×10^{-3} C. The switch closes at $t = 0$, applying a steady 120 V source. Determine the current using the Laplace transform method.

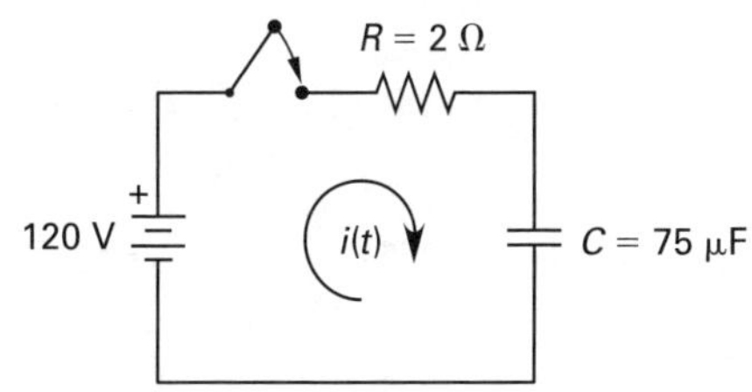

Solution

The classical approach will be used. Using KVL in the direction shown gives

$$V_s - i(t)R - \frac{1}{C}\int i(t)\,dt = 0$$

Though not stated explicitly, the behavior of the circuit from $t = 0$ onward is the desired result. So,

$$V_s = i(t)R + v_C(0^+) + \frac{1}{C}\int_0^\infty i(t)\,dt$$

The initial capacitor voltage can be determined from the initial charge, using $C = Q/V$. The constant-source voltage can be considered a unit step at $t = 0$. The initial capacitor voltage is a constant as well. Both transform to the s domain using $1/s$ (see App. 11.A). Integration equates to multiplication by $1/s$. Substituting and transforming gives

$$V_s = i(t)R + \frac{Q_0}{C} + \frac{1}{C}\int_0^\infty i(t)\,dt$$

$$\frac{120\ \text{V}}{s} = I(s)(2\ \Omega) + \frac{4.5\times 10^{-3}\ \text{C}}{(75\times 10^{-6}\ \text{F})(s)} + \frac{I(s)}{(75\times 10^{-6}\ \text{F})(s)}$$

$$I(s) = \frac{30\ \text{A}}{s + 6.65\times 10^3}$$

Performing the inverse Laplace transform on $I(s)$ gives

$$i(t) = 30\ \text{A}e^{-6.65\times 10^3 \frac{1}{s}t}$$

32 Frequency Response

1. Fundamentals 32-1
2. Transfer Function 32-2
3. Steady-State Response 32-3
4. Transient Response 32-4
5. Magnitude and Phase Plots 32-5
6. Bode Plot Principles: Magnitude Plot 32-7
7. Bode Plot Principles: Phase Plot 32-10
8. Bode Plot Methods 32-11

Nomenclature

A	factor or constant	–
B	constant	–
D	denominator magnitude	–
G	gain	dB
i	instantaneous current	A
I, $\mathbf{I}$	effective or DC current	A
K	constant	–
N	numerator magnitude	–
$\mathbf{p}$	pole	Hz
$r(t)$	response function, time domain	–
$R(s)$	response function, s domain	–
s, $\mathbf{s}$	complex frequency	Hz
$T(s)$, $\mathbf{T}(\mathbf{s})$	transfer or network function	–
v	instantaneous voltage	V
V, $\mathbf{V}$	effective or DC voltage	V
X	constant	–
Y	constant	–
$\mathbf{z}$	zero	Hz
$\mathbf{Z}$	impedance	Ω

Symbols

α	angle	rad
β	angle	rad
θ	angle	rad
σ	neper frequency or napier frequency	Np/s or Hz
σ_0	break or corner frequency	Np/s or Hz
ω	angular frequency	rad/s

Subscripts

d	number or denominator
G	gain
n	number or numerator
net	network
p	pole
z	zero

1. FUNDAMENTALS

In a network with capacitors, or inductors, or both, transients occur.[1] Transients are represented by mathematical circuit models consisting of differential equations. The solutions to these equations depend on the circuit initial conditions.[2] The methods outlined in Chap. 31 can be used to obtain the solutions, with the Laplace transform method changing differential equation manipulation to algebraic manipulation and the phasor method simplifying sinusoidal mathematics.

Transient analysis requires the complete analysis of a circuit, because both the transient and steady-state components are relevant. The transient portion (natural response) of any output eventually settles on the steady-state (forced response) value determined by the forcing function, that is, the *signal*.[3] Steady-state analysis of a circuit is independent of the circuit initial conditions. Solutions to problems in steady-state analysis can be determined from the circuit characteristics alone. These characteristics are contained within a *transfer function*, given in Eq. 32.1.

$$T(s) = \mathcal{L}\left\{\frac{r(t)}{f(t)}\right\} \qquad 32.1$$

The circuit's input signal or forcing function is $f(t)$, and the response to this function is $r(t)$. The transfer function is given the symbol $T(s)$.[4] The transfer function is evaluated in the frequency domain rather than in the time domain. This type of analysis is called *frequency response analysis*. When the signal is sinusoidal, that is, $s = j\omega$, the term *sinusoidal steady-state analysis* is used. Unlike phasor analysis, in which a fixed frequency is assumed, frequency response analysis allows the frequency behavior over a range of values to be examined. This allows the effects of *variable frequency inputs* and *spectral inputs*, which have multiple frequencies, to be analyzed.

If the time response to a sinusoidal signal (that is, $s = j\omega$) is required, the signal magnitude is multiplied

[1]In purely resistive circuits, the response is considered instantaneous.
[2]The arbitrary constant occurring in integration of electrical parameters represents the initial condition.
[3]The natural response is determined from the homogenous solution (i.e., the forcing function is equal to zero) of the differential equation. The forced response is from the particular solution whose form must be that of the forcing function.
[4]The term $T(s)$ is actually the Laplace transform of the transform function. The symbolism for each of the terms varies widely, but Laplace transforms are normally shown as uppercase while time functions are shown as lowercase.

by the transfer function magnitude, $|F(s)||T(s)|$, which gives the magnitude of the output or response, $|R(s)|$. The angle is determined by adding the signal angle to the transfer function angle, $\angle F(s) + \angle T(s)$, which gives the angle of the response, $\angle R(s)$. The result is a phasor that can be changed into the sinusoidal form using the known value of $j\omega$. The transfer function can be used to determine the output for a variety of inputs since it contains the circuit's characteristics.

Frequency response analysis is used to simplify calculations and increase understanding compared to equivalent analysis in the time domain. The main principle is that a sinusoidal input produces a sinusoidal output.

2. TRANSFER FUNCTION[5]

The term *transfer function* refers to the relationship of one electrical parameter in a network to a second electrical parameter elsewhere in the network.[6] The *network function*, $\mathbf{T}_{\text{net}}(\mathbf{s})$, refines this definition to be the ratio of the complex amplitude of an exponential output, $\mathbf{Y}(\mathbf{s})$, to the complex amplitude of an exponential input, $\mathbf{X}(\mathbf{s})$. If $x(t)$ is Xe^{st} and $y(t)$ is Ye^{st}, and if the linear circuit is made up of lumped elements, the network function can be derived from an input-output differential equation as a rational function of $\mathbf{s}$, and can be written in the following general form.[7]

$$\mathbf{T}_{\text{net}}(\mathbf{s}) = \frac{\mathbf{Y}(\mathbf{s})}{\mathbf{X}(\mathbf{s})} = A\left(\frac{(\mathbf{s}-\mathbf{z}_1)(\mathbf{s}-\mathbf{z}_2)\cdots(\mathbf{s}-\mathbf{z}_n)}{(\mathbf{s}-\mathbf{p}_1)(\mathbf{s}-\mathbf{p}_2)\cdots(\mathbf{s}-\mathbf{p}_d)}\right) \qquad 32.2$$

The complex constants represented as $\mathbf{z}_y (y = 1, 2, \ldots, n)$ are the *zeros* of $\mathbf{T}_{\text{net}}(\mathbf{s})$, and are plotted in the $\mathbf{s}$ domain as circles. The complex constants represented as $\mathbf{p}_x$ $(x = 1, 2, \ldots, d)$ are the *poles* of $\mathbf{T}_{\text{net}}(\mathbf{s})$ and are plotted in the $\mathbf{s}$ domain as X's. Equation 32.2 can be interpreted as the ratio of the response in one part of the $\mathbf{s}$ domain network to the excitation in another part of the network. When $\mathbf{s} = \mathbf{z}_y$, the response of the network will be zero regardless of the excitation. When $\mathbf{s} = \mathbf{p}_x$, the response will be infinite regardless of the excitation. Equation 32.2 is independent of the forcing function. That is, the equation contains the characteristics of the network itself. The response to some source (input) is found by substituting the value of $\mathbf{s} = \sigma_s + j\omega_s$ for the source into Eq. 32.2. The properties of the circuit elements themselves are held in the complex zeros and poles, $\mathbf{z}$ or $\mathbf{p}$.

Example 32.1

Determine the network function of the circuit shown in terms of a current output, $\mathbf{I}(\mathbf{s})$, for a given voltage input, $\mathbf{V}(\mathbf{s})$.

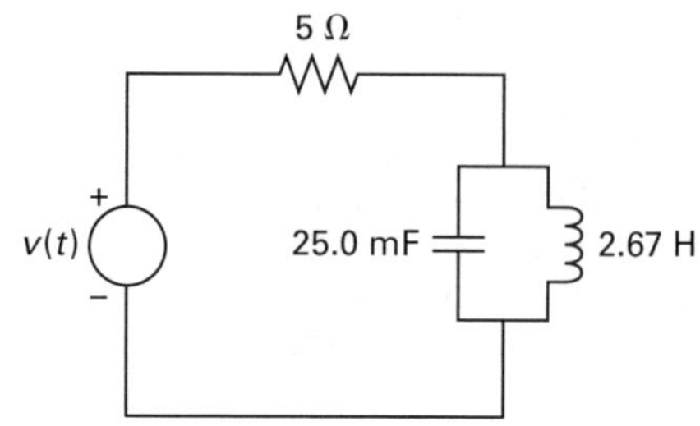

Solution

The network function is

$$\mathbf{T}_{\text{net}}(\mathbf{s}) = \frac{\mathbf{I}(\mathbf{s})}{\mathbf{V}(\mathbf{s})} = \frac{1}{\mathbf{Z}(\mathbf{s})}$$

Because the calculation will be accomplished in the $\mathbf{s}$ domain, transform the circuit. The result is

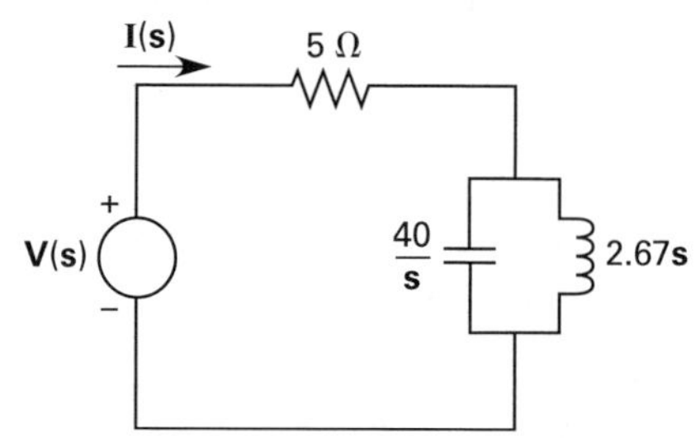

The impedance consists of a resistor in series with the parallel combination of a capacitor and an inductor.

$$\mathbf{Z}(\mathbf{s}) = \left(5 + \frac{\left(\frac{40}{\mathbf{s}}\right)(2.67\mathbf{s})}{\frac{40}{\mathbf{s}} + 2.67\mathbf{s}}\right)\ \Omega$$

After manipulating mathematically, the result is

$$\mathbf{Z}(\mathbf{s}) = \left(\frac{(5)(\mathbf{s}^2 + 8\mathbf{s} + 15)}{\mathbf{s}^2 + 15}\right)\Omega = \left(\frac{(5)(\mathbf{s}+3)(\mathbf{s}+5)}{\mathbf{s}^2 + 15}\right)\Omega$$

The network function is

$$\mathbf{T}_{\text{net}}(\mathbf{s}) = \frac{1}{\mathbf{Z}(\mathbf{s})} = 0.2\left(\frac{\mathbf{s}^2 + 15}{(\mathbf{s}+3)(\mathbf{s}+5)}\right)\Omega$$

[5]Functions of s are not normally shown in bold, though they are complex, to avoid confusion with phasors. They are shown as complex in this section to clarify several points regarding their properties.

[6]When the function is unitless—that is, when it is the ratio of voltages, the ratio of currents, and so on—it is called the *transfer ratio*.

[7]This equation is often shown with positive signs. In this case, the zeros are the $-\mathbf{z}$ values. The minus signs are shown here to clarify how the $\mathbf{s}$ and $\mathbf{z}$ values would be combined in the $\mathbf{s}$ domain (see Sec. 32.3).

Example 32.2

Plot zeros and poles of the network function in Ex. 32.1.

Solution

The zeros are determined by the roots in the numerator as $\mathbf{s} = \pm j\sqrt{15}$, complex conjugate pairs. The poles are determined by the roots in the denominator as $\mathbf{s} = -3$ and $\mathbf{s} = -5$. Plotting gives the following result.

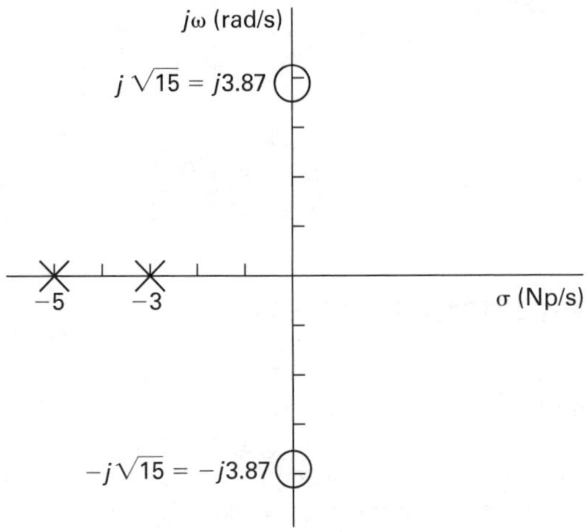

The plot is a complete representation of the network function.

3. STEADY-STATE RESPONSE[8]

The network function is a ratio of phasor quantities, such as $\mathbf{V}_2/\mathbf{V}_1$, $\mathbf{I}_2/\mathbf{I}_1$, $\mathbf{V}_2/\mathbf{I}_1$, and so on. The variables can also be impedances or admittances. As for any phasor quantity, the network function can be expressed as a magnitude at an angle, N or $D\angle\theta$. Each of the zero terms in the numerator and pole terms in the denominator in Eq. 32.2 can be represented as in Eq. 32.3 and Eq. 32.4, respectively.

$$(\mathbf{s} - \mathbf{z}_n) = N_n\angle\alpha_n \qquad 32.3$$

$$(\mathbf{s} - \mathbf{p}_d) = D_d\angle\beta_d \qquad 32.4$$

Substituting Eq. 32.3 and Eq. 32.4 into the generic network equation, Eq. 32.2, provides a mathematical picture of the steady-state response phasors.

$$\begin{aligned}\mathbf{T}_{\text{net}}(\mathbf{s}) &= A\left(\frac{(N_1\angle\alpha_1)(N_2\angle\alpha_2)\cdots(N_n\angle\alpha_n)}{(D_1\angle\beta_1)(D_2\angle\beta_2)\cdots(D_d\angle\beta_d)}\right)\\ &= A\left(\frac{N_1N_2\cdots N_n}{D_1D_2\cdots D_d}\right)\angle\left(\begin{matrix}\alpha_1+\alpha_2+\cdots\\ +\alpha_n\end{matrix}\right)\\ &\quad -(\beta_1+\beta_2+\cdots+\beta_d)\end{aligned} \qquad 32.5$$

Equation 32.3 and Eq. 32.4 considered together indicate that the steady-state response of a network to a forcing function input of $\mathbf{s} = \sigma + j\omega$ is determined by the lengths of the vectors from the zeros to the $\mathbf{s}$ and the vectors from the poles to the $\mathbf{s}$. In addition, the angle is determined by the angles these vectors make with the σ axis.[9] The angles α and β are always between 0° and 90°. Equation 32.5 is a geometric means of determining the network function from the locations of the zeros and poles, without knowing the analytical expression. If the poles and zeros are known, the network function can be written to within a factor of A.

Example 32.3

For the circuit in Ex. 32.1 and the response given in Ex. 32.2, determine the current output for a voltage excitation input of $v = e^{st}$ V where $\mathbf{s} = \sqrt{15}$ Np/s.

Solution

Place the input on the pole-zero plot. Draw the vectors from the zeros to the input $\mathbf{s}$. Draw the vectors from the poles to the input $\mathbf{s}$.

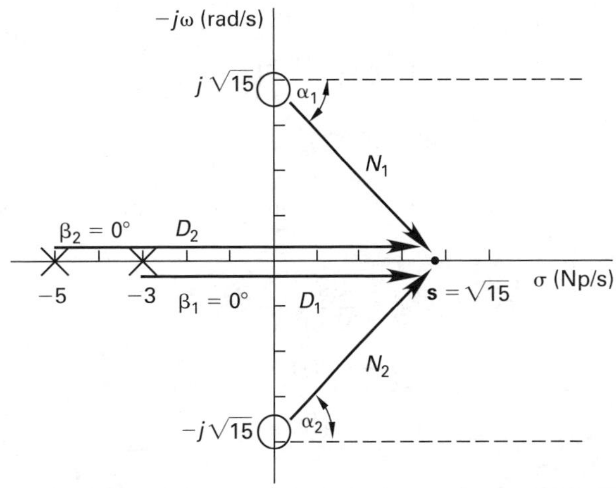

The magnitudes and angles are then calculated. For the zeros,

$$\begin{aligned}N_1 = N_2 &= \sqrt{\left(\sqrt{15}\ \frac{\text{Np}}{\text{s}}\right)^2 + \left(\sqrt{15}\ \frac{\text{Np}}{\text{s}}\right)^2}\\ &= \sqrt{30}\ \frac{\text{Np}}{\text{s}}\end{aligned}$$

$$\alpha_1 = \tan^{-1}\left(\frac{-\sqrt{15}\ \frac{\text{Np}}{\text{s}}}{\sqrt{15}\ \frac{\text{Np}}{\text{s}}}\right) = -45°$$

$$\alpha_2 = \tan^{-1}\frac{\sqrt{15}\ \frac{\text{Np}}{\text{s}}}{\sqrt{15}\ \frac{\text{Np}}{\text{s}}} = 45°$$

$$\alpha_1 + \alpha_2 = 0°$$

[8]Functions of $\mathbf{s}$ are shown as complex in this section to clarify several points regarding their properties, such as the phasor addition that takes place in Ex. 32.3.

[9]The vectors represent what the circuit must do to respond to the forcing function. Recall that the steady-state output of a circuit always takes the form of the forcing or input function.

Circuit Theory

For the poles,

$$D_1 = 3\ \frac{\text{Np}}{\text{s}} + \sqrt{15\ \frac{\text{Np}}{\text{s}}} = 6.87\ \frac{\text{Np}}{\text{s}}$$

$$D_2 = 5\ \frac{\text{Np}}{\text{s}} + \sqrt{15\ \frac{\text{Np}}{\text{s}}} = 8.87\ \frac{\text{Np}}{\text{s}}$$

$$\beta_1 = 0°$$

$$\beta_2 = 0°$$

$$\beta_1 + \beta_2 = 0°$$

The factor A was determined in Ex. 32.1. The network function is calculated as

$$\begin{aligned}\mathbf{T}_{\text{net}}(\mathbf{s}) &= A\left(\frac{N_1 N_2 \cdots N_n}{D_1 D_2 \cdots D_d}\right)\angle(\alpha_1 + \alpha_2 + \cdots + \alpha_n) \\ &\quad - (\beta_1 + \beta_2 + \cdots + \beta_d)\end{aligned}$$

$$\begin{aligned}\mathbf{T}_{\text{net}}(\sqrt{15}) &= (0.2)\left(\frac{(\sqrt{30})(\sqrt{30})}{\left(6.87\ \frac{\text{Np}}{\text{s}}\right)\left(8.87\ \frac{\text{Np}}{\text{s}}\right)}\right)\angle(0° + 0°) \\ &= 98.46 \times 10^{-3}\angle 0°\end{aligned}$$

The response for the given input can now be determined.

$$\mathbf{T}_{\text{net}}(\mathbf{s}) = \frac{\mathbf{I}(\mathbf{s})}{\mathbf{V}(\mathbf{s})}$$

$$\begin{aligned}i(t) &= t_{\text{net}}(t)v(t) \\ &= 98.46 \times 10^{-3} e^{\sqrt{15}t}\end{aligned}$$

The result indicates that as time increases the current becomes infinite, clearly an impossibility. This is consistent with the condition in Sec. 31.10 that σ takes on zero or negative values only.

Example 32.4

Write the expression for the transfer function in the pole-zero plot shown.

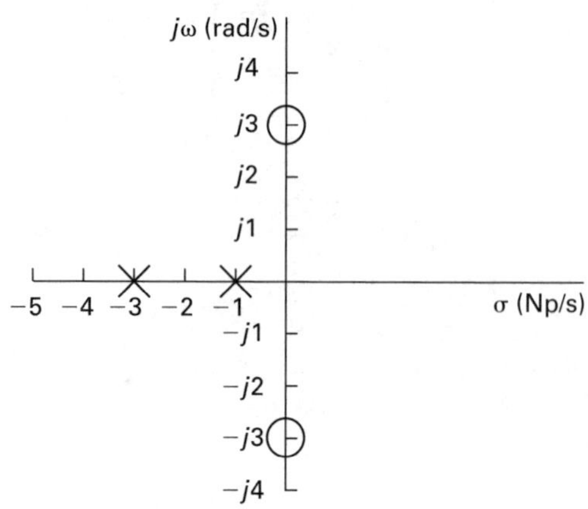

Solution

The zeros are located at $\pm j3$. Consequently, the numerator is

$$\text{s}^2 + 9$$

The poles are located at -1 and -3. Consequently, the denominator is

$$(\text{s} + 3)(\text{s} + 1)$$

The transfer function is

$$\mathbf{T}_{\text{net}}(\mathbf{s}) = A\left(\frac{\text{s}^2 + 9}{(\text{s}+3)(\text{s}+1)}\right)$$

4. TRANSIENT RESPONSE

The network function uses the complex frequency to determine the steady-state response. Nevertheless, the circuit characteristics, including the natural response, are embedded in the input equation in the denominator of the network function. Specifically, the transient response is characterized by the poles of the network function.

The real portion of the complex number representing a pole contains the neper frequency, σ, of the circuit for that pole. The *neper frequency*, also called the napier frequency, determines the exponential terms in a solution.[10] The exponential terms represent the transient response of a circuit (see Chap. 31). Near a pole, then, the natural response of the circuit elements impacts the overall output of a network, while far away from a pole little effect occurs.

Example 32.5

What is the mathematical form of the transient response for the network function in Ex. 32.4?

Solution

The network function in Ex. 32.4 was

$$T_{\text{net}}(s) = A\left(\frac{s^2 + 9}{(s+3)(s+1)}\right)$$

In keeping with common practice, the s is no longer indicated as a complex number, though it still is. Because each term is $(s - p)$, the poles occur at $(s-(-p))$. That is, the poles are at -3 and -1. The expanded denominator is

$$s^2 + 4s + 3$$

[10]The term "napier" is most commonly known for its association with natural logarithms. It is used here associated with the natural response.

The coeffcients $a=1$, $b=4$, and $c=3$ indicate that $b^2>4ac$. The transient response is of the form

$$\text{transient response} = Ae^{s_1 t} + Be^{s_2 t}$$

The transient response could be either the current or voltage response, depending upon the ratio represented by the network function. The initial conditions, not included in the network function, would be required to determine the constants A and B.

5. MAGNITUDE AND PHASE PLOTS

A plot of the magnitude and phase of a transfer function in the **s** domain provides information regarding the response of that function as frequency varies.[11] Magnitude and phase plots are fundamental to constructing Bode plots.

Consider a transfer function with a single zero at the origin. As only the steady-state response to a sinusoid is of concern, let $s=j\omega$. The transfer function is

$$T(s) = s = s+0 \quad 32.6$$

$$T(j\omega) = j\omega \quad 32.7$$

The frequency is allowed to vary as illustrated in Fig. 32.1(a). The magnitude is plotted in Fig. 32.1(b) from $|T(j\omega)|=\omega$. The phase angle variation with frequency is shown in Fig. 32.1(c) from $\alpha=\tan^{-1}\omega/0$ or directly from j, which is $1\angle 90°$.

Consider a transfer function with a single pole at the origin. As only the steady-state response to a sinusoid is of concern, let $s=j\omega$. The transfer function is

$$T(s) = \frac{1}{s} = \frac{1}{s+0} \quad 32.8$$

$$T(j\omega) = \frac{1}{j\omega} \quad 32.9$$

The frequency is allowed to vary as illustrated in Fig. 32.2(a). The magnitude is plotted in Fig. 32.2(b) from $|T(j\omega)|=1/|\omega|$. The phase angle variation with frequency is shown in Fig. 32.2(c) from $\beta=-\tan^{-1}\omega/0$ or directly from $1/j$, which is $-j$ or $1\angle -90°$.

Consider a transfer function with a single zero but located on the negative real axis at $-\sigma_n$.[12] Again, only the steady-state response to a sinusoid is of concern, so $s=j\omega$. The transfer function is

$$T(s) = s+\sigma_n \quad 32.10$$

$$T(j\omega) = j\omega+\sigma_n \quad 32.11$$

Figure 32.1 *Single Zero at the Origin*

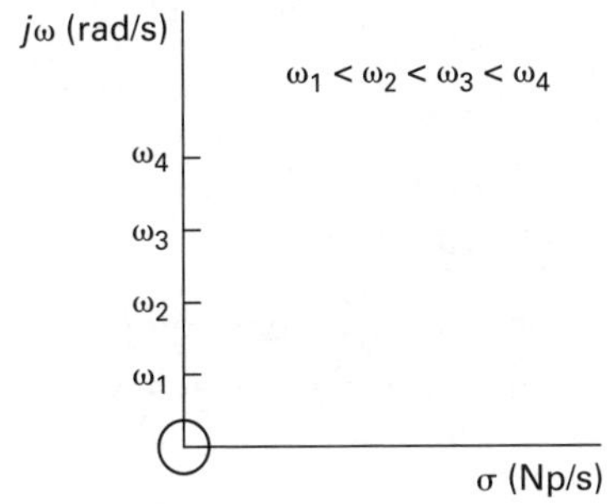

(a) zero at origin, frequency increasing

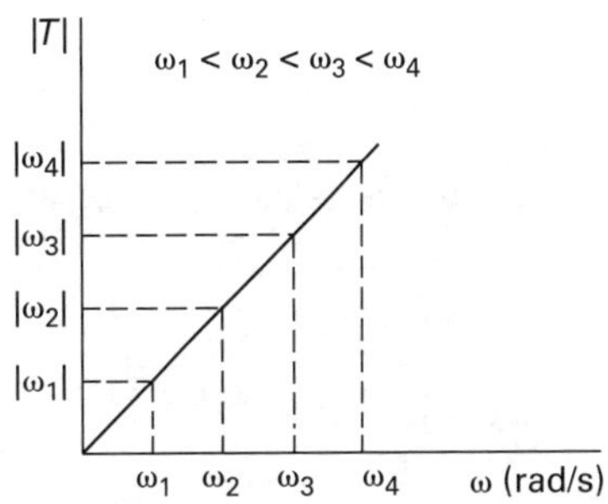

(b) transfer function magnitude, frequency increasing

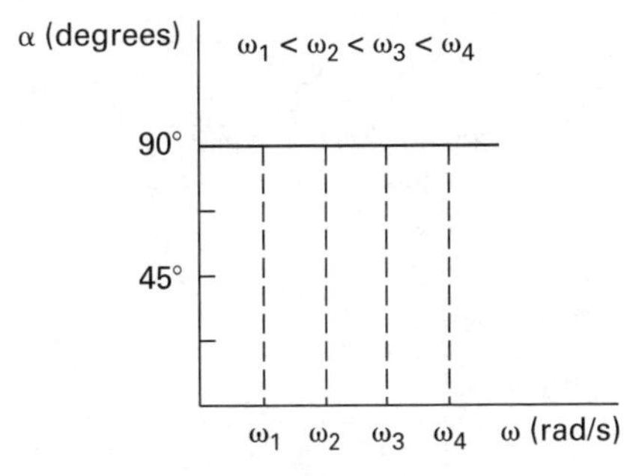

(c) transfer function angle, frequency increasing

The frequency is allowed to vary as illustrated in Fig. 32.3(a).[13] The magnitude is plotted in Fig. 32.3(b) from $|T(j\omega)|=\sqrt{\omega^2+\sigma_n^2}$. The phase angle variation with frequency is shown in Fig. 32.3(c) from $\alpha=\tan^{-1}\omega/\sigma_n$.

[11]Because the methods apply to all functions in the **s** domain, not merely network functions, the term "transfer function" and the symbol $T(s)$ will be used instead of $\mathbf{T}_{\text{net}}(\mathbf{s})$.

[12]All practical devices will have zeros to the left of the imaginary axis. The restriction of being on the negative real axis will be removed for Bode plots but is used here to clarify the concept of a *break* or *corner frequency* that occurs at zeros and poles, that is, where the real portion of the zero or pole is σ_n or σ_p. The neper frequency, σ, represents the exponential solution or the natural response of the circuit elements that impacts the steady-state response in the area of a zero or a pole.

[13]The figure is not to scale. The relationships between drawings are approximate, but individual drawings show their true shapes.

Consider a transfer function with a single pole but located on the negative real axis at $-\sigma_n$. Again, only the steady-state response to a sinusoid is of concern, so $s = j\omega$. The transfer function is

$$T(s) = \frac{1}{s + \sigma_d} \qquad 32.12$$

$$T(j\omega) = \frac{1}{j\omega + \sigma_d} \qquad 32.13$$

Figure 32.2 *Single Pole at Origin*

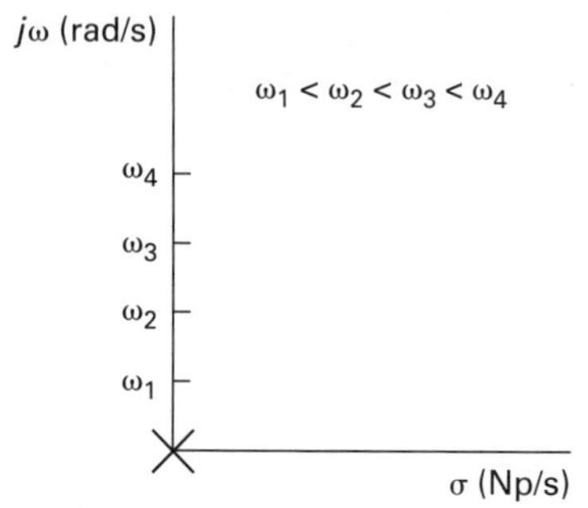

(a) pole at origin, frequency increasing

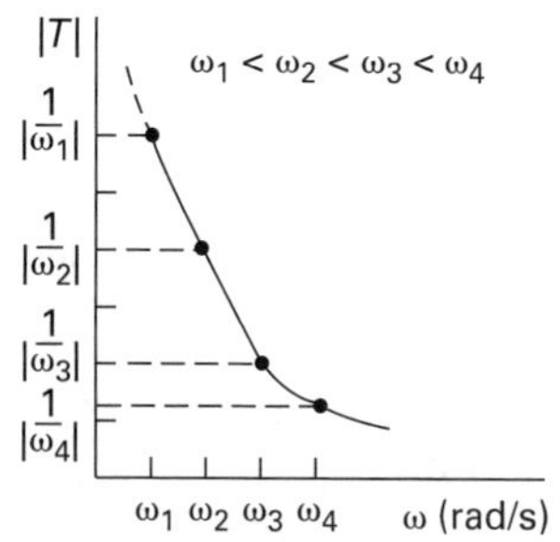

(b) transfer function magnitude, frequency increasing

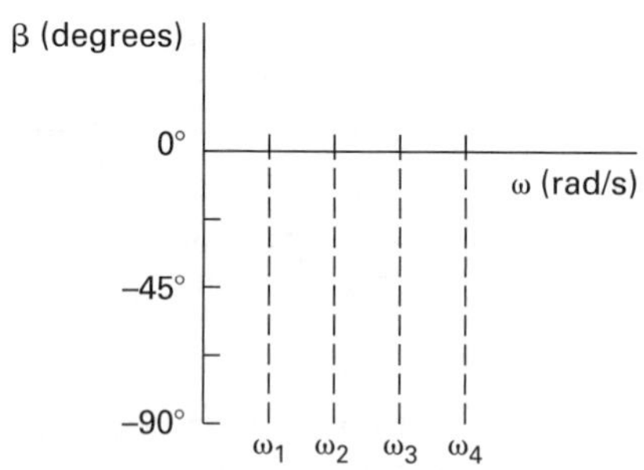

(c) transfer function angle, frequency increasing

Figure 32.3 *Single Zero on Negative Real Axis ($-\sigma_n$)*

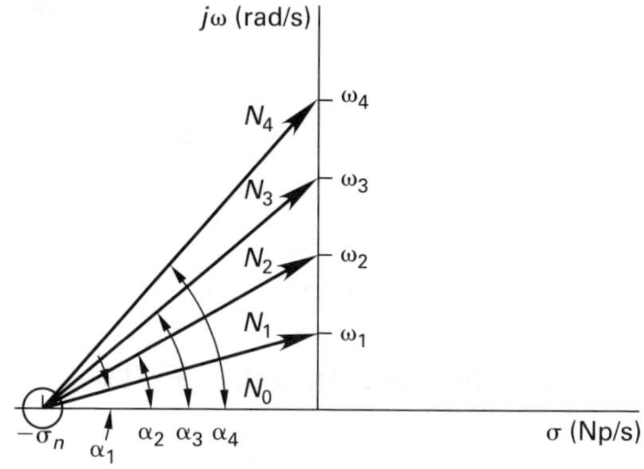

(a) zero on negative real axis, frequency increasing

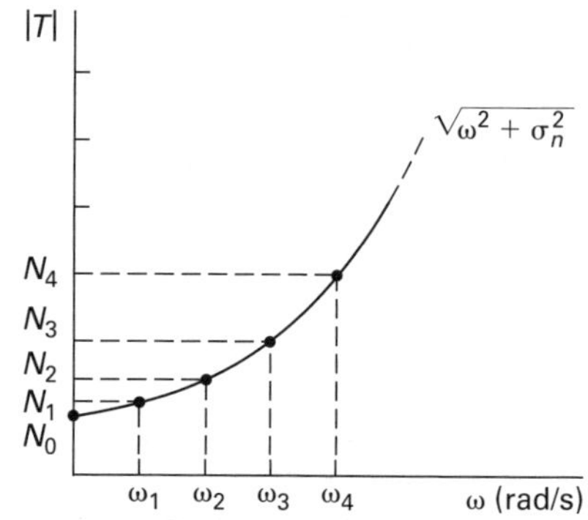

(b) transfer function magnitude, frequency increasing

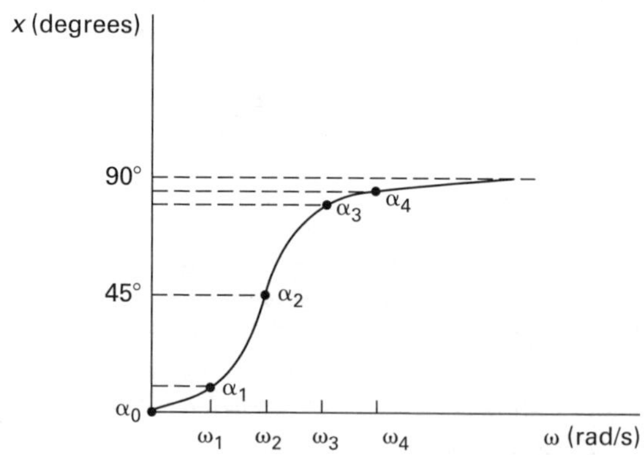

(c) transfer function angle, frequency increasing

The frequency is allowed to vary as illustrated in Fig. 32.4(a).[14] The magnitude is plotted in Fig. 32.4(b) from $|T(j\omega)| = 1/\sqrt{\omega^2 + \sigma_d^2}$. The phase angle variation with frequency is shown in Fig. 32.4(c) from $\beta = -\tan^{-1}\omega/\sigma_d$.

[14]The figure is not to scale. The relationships between drawings are approximate, but individual drawings show their true shapes.

Figure 32.4 *Single Pole on Negative Real Axis ($-\sigma_p$) Plot*

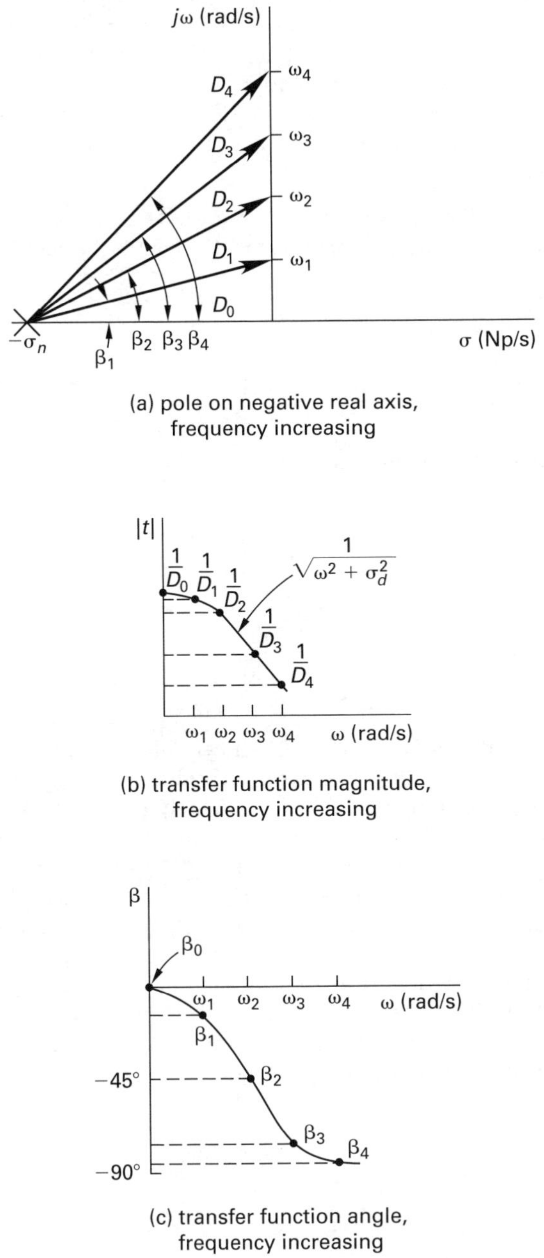

When $\sigma = \omega$, the magnitude of the transfer function for a zero is $\sqrt{\omega^2 + \sigma_n^2} = \sqrt{2}\sigma_n$ and the phase angle is $\alpha = +45°$. When $\sigma = \omega$, the magnitude of the transfer function for a pole is $1/(\sqrt{2}\sigma_d)$ and the angle is $\beta = -45°$. When the transfer function contains numerous zeros and poles, the magnitude is obtained by multiplying the magnitudes of the individual transfer functions, and the phase is obtained by summing the zero angles and subtracting the sum of the pole angles. Equation 32.5 is such a combination, with each $N\angle\alpha$ and $D\angle\beta$ representing the magnitude of one transfer function. The multiplication and division can be replaced by addition and subtraction using logarithmic quantities, guided by Eq. 32.14 and Eq. 32.15.

$$\log xy = \log x + \log y \qquad 32.14$$

$$\log \frac{x}{y} = \log x - \log y \qquad 32.15$$

The magnitude of the transfer function, Eq. 32.5, can be written as

$$\log|T(s)| = \log A + \log N_1 + \log N_2$$

$$+ \cdots + \log N_n - \log D_1$$

$$- \log D_2 - \cdots - \log D_d \qquad 32.16$$

6. BODE PLOT PRINCIPLES: MAGNITUDE PLOT

A *Bode plot* or *Bode diagram* is a plot of the gain or phase of an electrical device or network against the frequency. The units of the plot are normally magnitude in decibels versus logarithm of the frequency. The logarithm is useful for algebraic manipulation (see Eq. 32.14 and Eq. 32.15) and the wide frequency range of response. The *decibel*, in units dB, is defined as 10 times the logarithmic ratio of two powers. Nevertheless, it is used for transfer functions in general. Because the I^2 and V^2 are proportional to power, the magnitudes of transfer functions containing the voltage and current are[15]

$$G = 10 \log \left|\frac{I_2}{I_1}\right|^2 = 20 \log \left|\frac{I_2}{I_1}\right| \qquad 32.17$$

$$G = 10 \log \left|\frac{V_2}{V_1}\right|^2 = 20 \log \left|\frac{V_2}{V_1}\right| \qquad 32.18$$

In general terms, any transfer function's magnitude, called the *gain* (G), is

$$G = 20 \log|T(s)| \qquad 32.19$$

The transfer function for a single zero at the origin is given by Eq. 32.6. Normalizing this with the factor σ_0 gives

$$T(s) = \sigma_0\left(\frac{s}{\sigma_0}\right) \qquad 32.20$$

[15]To be used in the ratio, the currents or voltages should act on the same impedance. Standard practice ignores this technicality.

Circuit Theory

Using Eq. 32.19 and Eq. 32.20, $s = j\omega$, and the properties of logarithms from Eq. 32.14 and Eq. 32.15 gives

$$\begin{aligned} G &= 20 \log \left| \sigma_0 \left(\frac{j\omega}{\sigma_0} \right) \right| \\ &= 20 \log \sigma_0 + 20 \log \frac{\omega}{\sigma_0} \end{aligned} \quad \textit{32.21}$$

The resulting Bode plot is shown in Fig. 32.5. The normalizing factor is the frequency where the gain (magnitude) has a value of 0 dB, which gives an actual normalizing factor of one. A zero at the origin results in a line of positive slope passing through $\omega = 1$ or $\log \omega = 0$.[16] The slope is 6 dB/octave or 20 dB/decade. An *octave* is an interval between any two frequencies with a ratio of two to one; for example, $\omega_2 = 2\omega_1$. On the log scale the magnitude change between octaves is

$$\begin{aligned} \Delta G &= 20 \log \omega_2 - 20 \log \omega_1 \\ &= 20 \log \frac{\omega_2}{\omega_1} \\ &= 20 \log \frac{2\omega_1}{\omega_1} \\ &= 20 \log 2 \\ &= 6.02 \approx 6 \text{ dB/octave} \end{aligned} \quad \textit{32.22}$$

A decade of frequency change alters the magnitude by

$$\begin{aligned} \Delta G &= 20 \log \omega_2 - 20 \log \omega_1 \\ &= 20 \log \frac{\omega_2}{\omega_1} \\ &= 20 \log \frac{10\omega_1}{\omega_1} \\ &= 20 \log 10 \\ &= 20 \text{ dB/decade} \end{aligned} \quad \textit{32.23}$$

Figure 32.5 *Bode Plot: Single Zero at Origin*

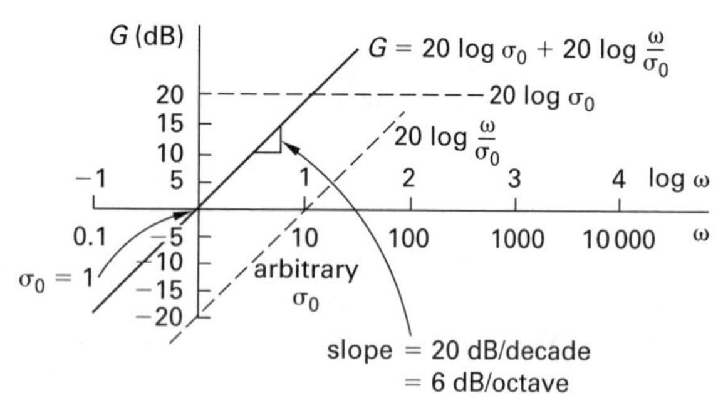

The two slopes are equivalent. The constant $20 \log \sigma_0$ does not change the shape of the plot. The constant can be accounted for in one of two ways. First, it can shift the plot upward, as was done in this case. Second, the value of 0 dB could be changed to a new value that is equal to the constant, after plotting the other factors of the gain. The angle for a zero at the origin remains a constant 90°.

The transfer function for a single pole at the origin is given by Eq. 32.8. Normalizing this with the factor σ_0 gives

$$T(s) = \frac{1}{\sigma_0 \left(\dfrac{s}{\sigma_0} \right)} \quad \textit{32.24}$$

With $s = j\omega$, the gain is

$$\begin{aligned} G &= 20 \log \left| \frac{1}{\sigma_0 \left(\dfrac{j\omega}{\sigma_0} \right)} \right| \\ &= -20 \log \sigma_0 - 20 \log \frac{\omega}{\sigma_0} \end{aligned} \quad \textit{32.25}$$

The resulting Bode plot is shown in Fig. 32.6. The normalizing factor is the frequency where the gain (magnitude) has a value of 0 dB, which gives an actual normalizing factor of one. A pole at the origin results in a line of negative slope passing through $\omega = 1$ or $\log \omega = 0$.[17] The slope is −6 dB per octave or −20 dB per decade. The constant $-20 \log \sigma_0$ does not change theshape of the plot. In this case, the constant is used to shift the plot downward. Another option is to change the gain coordinates so that the zero value correlates with the value of the constant, after plotting the other terms of the gain. The angle for a pole at the origin remains a constant −90°.

Figure 32.6 *Bode Plot: Single Pole at Origin*

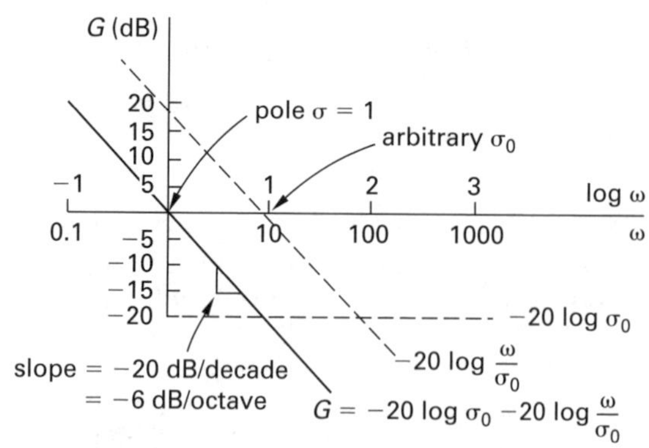

[16]Considering Eq. 32.21 and Fig. 32.5, regardless of where the example normalizing factor, σ_0, is located (in this case, it was at $\sigma_0 = 10$), the final result remains unchanged: a +20 dB/decade line through $\omega = 1$ or $\log \omega = 0$ with an actual σ_0 of 1.

[17]Considering Eq. 32.25 and Fig. 32.6, regardless of where the example normalizing factor, σ_0, is located (in this case, it was at $\sigma_0 = 10$), the final result remains unchanged: a −20 dB per decade line through $\omega = 1$ or $\log \omega = 0$ with an actual σ_0 of 1.

The transfer function for a single zero on the negative real axis at $-\sigma_0$ in the s plane is given by Eq. 32.10. Normalizing this equation gives

$$T(s) = \sigma_0\left(1 + \frac{s}{\sigma_0}\right) \quad 32.26$$

With $s = j\omega$, the gain is

$$\begin{aligned} G &= 20 \log\left|\sigma_0\left(1 + \frac{j\omega}{\sigma_0}\right)\right| \\ &= 20 \log\sigma_0 + 20 \log\left|\left(1 + \frac{j\omega}{\sigma_0}\right)\right| \end{aligned} \quad 32.27$$

The resulting Bode plot is shown in Fig. 32.7. The constant $20 \log \sigma_0$ does not change the shape of the plot but rather shifts it upward or changes the value of the 0 dB to that of the constant (not shown). The normalizing factor is the frequency where the gain (magnitude) has a value of 0 dB. To determine the gain on either side of the normalizing factor, which will now be called the *break frequency* or *corner frequency*, consider the behavior of the asymptotes.[18] At low frequencies, ignoring the constant term of Eq. 32.27 because it only changes the plot's position, the gain is

$$G \approx 20 \log\left|\left(1 + \frac{0}{\sigma_0}\right)\right| = 0 \text{ dB} \quad [\omega \ll \sigma_0] \quad 32.28$$

Figure 32.7 *Bode Plot: Single Zero on Negative Real Axis*

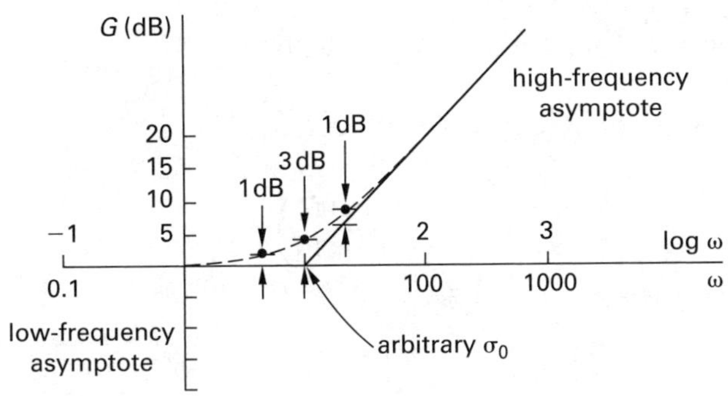

The low-frequency asymptote is approximated by drawing a line from the break frequency horizontally to the left. At high frequencies, ignoring the constant term of Eq. 32.27, the gain is

$$\begin{aligned} G &\approx 20 \log\left|\left(\frac{j\omega}{\sigma_0}\right)\right| \\ &= 20 \log \frac{\omega}{\sigma_0} \quad [\omega \gg \sigma_0] \end{aligned} \quad 32.29$$

The high-frequency asymptote is approximated by a straight line with a slope of +20 dB/decade. That is, the curve "breaks upward" at the break frequency. At the break frequency $\omega = \sigma_0$, the magnitude of the gain is

$$\begin{aligned} G_0 &= 20 \log\left|\left(1 + \frac{j\omega}{\sigma_0}\right)\right| \\ &= 20 \log\left(\sqrt{(1)^2 + (1)^2}\right) \\ &= 20 \log \sqrt{2} = 3 \text{ dB} \end{aligned} \quad 32.30$$

At $\frac{1}{2}\sigma_0$ and $2\sigma_0$, the gain differs by 1 dB from the approximating asymptotes. The exact value of the frequency for a given gain, G_ω, can be determined by rearranging Eq. 32.27 (ignoring the constant factor). The result is

$$\omega = \sigma_0\sqrt{(10)^{G_\omega/10} - 1} \quad 32.31$$

While Eq. 32.31 could be used to determine exact values, the asymptotes provide reasonable accuracy and adequate information for most electrical engineering applications. Use of the asymptotic approximations results in an *idealized Bode plot.* Computer programs are available to accurately plot the response of a given circuit should such accuracy become necessary. The angle for a zero on the negative real axis of the s domain is determined from Eq. 32.26 to be $\tan^{-1}\omega/\sigma_0$. At the break frequency, this angle is 45°.

The transfer function for a single pole on the negative real axis at $-\sigma_0$ in the s plane is given by Eq. 32.12. Normalizing the equation and letting $s = j\omega$ results in a gain of

$$\begin{aligned} G &= 20 \log\left|\left(\frac{1}{\sigma_0\left(1 + \frac{j\omega}{\sigma_0}\right)}\right)\right| \\ &= -20 \log\sigma_0 - 20 \log\left|\left(1 + \frac{j\omega}{\sigma_0}\right)\right| \end{aligned} \quad 32.32$$

The plot for the pole is identical to that for a zero, with two exceptions: it has a negative slope and the plot breaks downward at the break frequency as shown in Fig. 32.8.

Figure 32.8 *Bode Plot: Single Pole on Negative Real Axis*

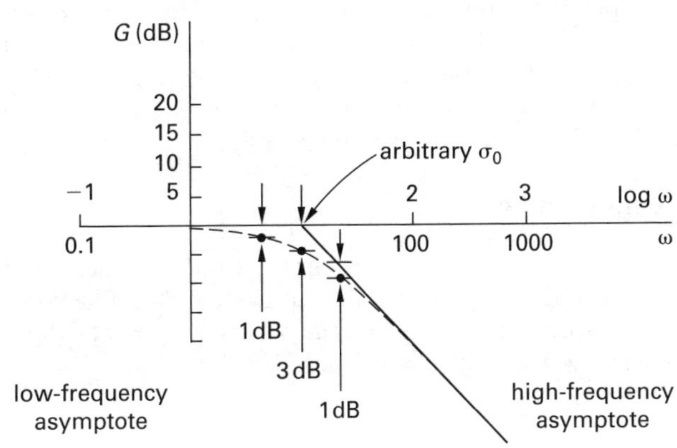

[18]The break frequency or corner frequency corresponds to the zero or pole location.

Circuit Theory

The transfer function for a circuit with numerous zeros and poles can be generalized as

$$T(s) = A\left(\frac{(s+\sigma_{z1})(s+\sigma_{z2}) \cdots \times (s+\sigma_{zn})s^y}{(s+\sigma_{p1})(s+\sigma_{p2}) \cdots \times (s+\sigma_{pd})s^x}\right) \quad 32.33$$

The term s^y represents zeros at the origin, with y equal to the number of zeros. Zeros at the origin of the s domain have a normalizing factor of one. The term s^x represents poles at the origin, with x equal to the number of poles. Poles at the origin of the s domain also have a normalizing factor of one. Changing the form to that useful for Bode plotting,

$$\begin{aligned} G &= 20 \log|T(s)| \\ &= 20 \log|T(j\omega)| \\ &= 20 \log\left(A\left(\frac{\sigma_{z1}\sigma_{z2}\cdots\sigma_{zn}}{\sigma_{p1}\sigma_{p2}\cdots\sigma_{pd}}\right)\right) \\ &\quad + 20 \log\left|\left(1+\frac{j\omega}{\sigma_{z1}}\right)\right| + 20 \log\left|\left(1+\frac{j\omega}{\sigma_{z2}}\right)\right| + \cdots \\ &\quad + 20 \log\left|\left(1+\frac{j\omega}{\sigma_{zn}}\right)\right| + y20 \log|j\omega| \\ &\quad - 20 \log\left|\left(1+\frac{j\omega}{\sigma_{p1}}\right)\right| - 20 \log\left|\left(1+\frac{j\omega}{\sigma_{p2}}\right)\right| - \cdots \\ &\quad - 20 \log\left|\left(1+\frac{j\omega}{\sigma_{pd}}\right)\right| - x20 \log|j\omega| \end{aligned} \quad 32.34$$

The normalization used in Eq. 32.34 results in each of the terms being plotted from the break frequency on the $G=0$ dB line. The break frequency (normalizing factor) for the single zeros and poles at the origin is $\omega=1$. (The zeros and poles plot as straight lines and do not actually "break" up or down at the origin. The zero or pole at zero frequency does not get plotted on a Bode plot, because the value is $-\infty$ and $+\infty$.)

The entire plotting process can be reversed. For example, if the output of a circuit is determined experimentally, the resulting Bode plot can be estimated and the transfer function derived from the plot. Then, using the value of the asymptotic plot at some point, the value of A in Eq. 32.33 can be determined.

Example 32.6

The figure shows a certain circuit's output, measured and plotted as a dashed line on a Bode plot. An asymptotic response is estimated and plotted as a solid line. Determine the transfer function.

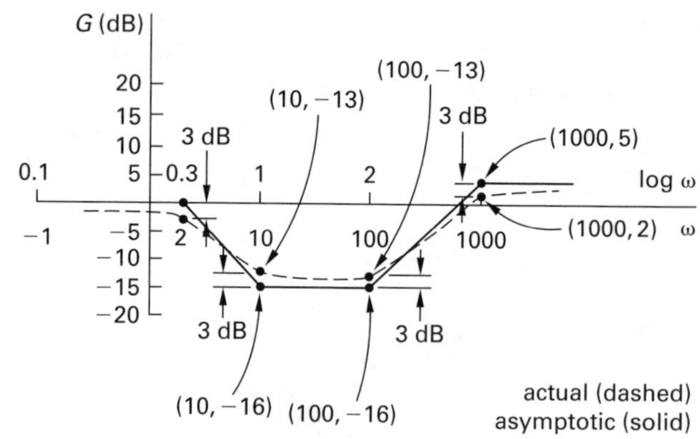

Solution

The transfer function must be in the form of Eq. 32.33. From the plot, the zeros and poles are the points where $G=0$ dB on the frequency axis where the breaks occur. Consequently, the breaks upward at $\omega=10$ and $\omega=100$ indicate zeros. The breaks downward at $\omega=2$ and $\omega=1000$ indicate poles. The transfer function is

$$T(s) = A\left(\frac{(s+10)(s+100)}{(s+2)(s+1000)}\right)$$

Determine A from the value of the asymptotic plot when $s=0$. At $s=0$, the plot indicates a magnitude of 0 dB.

$$G = 20 \log|T(0)| = 0 \text{ dB}$$
$$|T(0)| = \text{antilog } 0 = 1$$

Substituting $T(0)=1$ and $s=0$ gives

$$\begin{aligned} T(s) &= A\left(\frac{(s+10)(s+100)}{(s+2)(s+1000)}\right) \\ 1 &= A\left(\frac{(10)(100)}{(2)(1000)}\right) \\ A &= 2 \end{aligned}$$

The final solution is

$$T(s) = (2)\left(\frac{(s+10)(s+100)}{(s+2)(s+1000)}\right)$$

7. BODE PLOT PRINCIPLES: PHASE PLOT

The phase of a zero at the origin, such as $s+0$ in the numerator of Eq. 32.33, provides an angle of $+90°$. The phase of a pole at the origin, such as $s+0$ in the denominator of Eq. 32.33, provides an angle of $-90°$. For zeros or poles on the negative real axis of the s domain, with $s=j\omega$, the angles for each term in Eq. 32.33 are determined from Eq. 32.35 and Eq. 32.36, respectively.

$$\theta_z = \tan^{-1}\frac{\omega}{\sigma_{zn}} \quad \text{[radians]} \quad 32.35$$

$$\theta_p = -\tan^{-1}\frac{\omega}{\sigma_{pd}} \quad \text{[radians]} \quad 32.36$$

Circuit Theory

At the break frequency, $\omega = \sigma$ for either zeros or poles, and the phase angle is $+45°$ for zeros and $-45°$ for poles. Equation 32.35 and Eq. 32.36 could be used directly to plot the phase angle as a function of frequency. An *idealized Bode phase plot* that is similar to that used for magnitude can be plotted with approximations. Consider Eq. 32.35 and Eq. 32.36 at a frequency $\omega = 0.1\sigma$. The absolute value of the zero or pole angle at this frequency is

$$\begin{aligned} |\theta| &= \left| \pm \tan^{-1} \frac{0.1\sigma}{\sigma} \right| = 0.09 \text{ rad} \\ &= 5° \approx 0 \end{aligned} \qquad 32.37$$

This is analogous to the low-frequency asymptote for the magnitude of Eq. 32.28. Consider a frequency one decade higher than the break frequency, $\omega = 10\sigma$, for either a zero or a pole. The absolute value of the zero or pole angle at this frequency is

$$\begin{aligned} |\theta| &= \left| \pm \tan^{-1} \frac{10\sigma}{\sigma} \right| \\ &= 1.47 \text{ rad} = 84.3° \approx 90° \end{aligned} \qquad 32.38$$

This is analogous to the high-frequency asymptote for the magnitude of Eq. 32.29. At the break frequency, $\omega = \sigma$ and the angle is 45° for either a zero or a pole. The result is that, from one decade below the break frequency to one decade above, the phase angle for a zero or pole changes from an absolute value of 0° to 90° at a rate of 45° per decade. This approximation is illustrated in Fig. 32.9 for an arbitrary break frequency, that is, arbitrary zero or pole. The approximations used in the idealized phase plot are accurate to within 6°. The phase angles for the entire transfer function must be added as in Eq. 32.5 to obtain the resultant angle. This is done on the Bode plot in a manner analogous to adding the magnitudes.

Figure 32.9 *Bode Phase Plot for a Zero or Pole*

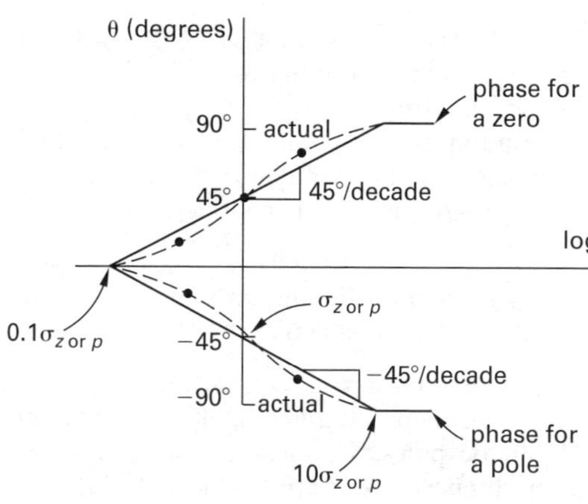

8. BODE PLOT METHODS

The general form of the transfer function of a circuit is[19]

$$T(s) = A\left(\frac{(s+z_1)(s+z_2) \cdots \times (s+z_n)(s)^y}{(s+p_1)(s+p_2) \cdots \times (s+p_d)(s)^x}\right) \qquad 32.39$$

The break frequencies in Eq. 32.39 are the zeros and the poles.[20] The break frequency, better termed the *normalizing factor*, for the zeros at the origin (s^y) and poles at the origin (s^x) is $\omega = 1$. As the sinusoidal response is of interest, $s = j\omega$. Using the normalized notation changes Eq. 32.39 to Eq. 32.40.

$$T(j\omega) = A\left(\frac{z_1 z_2 \cdots z_n}{p_1 p_2 \cdots p_d}\right) \times \left(\frac{\left(1+\frac{j\omega}{z_1}\right)\left(1+\frac{j\omega}{z_2}\right) \cdots \times \left(1+\frac{j\omega}{z_n}\right)(j\omega)^y}{\left(1+\frac{j\omega}{p_1}\right)\left(1+\frac{j\omega}{p_2}\right) \cdots \times \left(1+\frac{j\omega}{p_d}\right)(j\omega)^x}\right) \qquad 32.40$$

Combining the constant term A with the values of the zeros and poles gives a new constant, K.[21]

$$T(j\omega) = K\left(\frac{\left(1+\frac{j\omega}{z_1}\right)\left(1+\frac{j\omega}{z_2}\right) \cdots \times \left(1+\frac{j\omega}{z_n}\right)(j\omega)^y}{\left(1+\frac{j\omega}{p_1}\right)\left(1+\frac{j\omega}{p_2}\right) \cdots \times \left(1+\frac{j\omega}{p_d}\right)(j\omega)^x}\right) \qquad 32.41$$

The constant K is

$$K = A\frac{\prod_1^n z_n}{\prod_1^d p_d} \qquad 32.42$$

[19]Transfer functions are often called the *gain* of a circuit. This chapter uses gain to indicate $20\log|T(s)|$.

[20]Using the zero and pole notation removes the restriction of the break frequency occurring only on the negative real axis of the s domain. The zeros and poles are located at $-z$ and $-p$, respectively. This is the reason the positive sign is used.

[21]The constant K is a function of $j\omega$ because of the zeros and poles included.

Circuit Theory

The gain as used in Bode plots is derived from Eq. 32.41 to be

$$G = 20\log K + \sum_{n=1}^{n} 20\log\left|\left(1+\frac{j\omega}{z_n}\right)\right| + 20y\log\omega - \sum_{d=1}^{d} 20\log\left|1+\frac{j\omega}{p_d}\right| - 20x\log\omega \quad 32.43$$

The angle of the gain, θ_G, in degrees is

$$\theta_G = \sum_{n=1}^{n}\tan^{-1}\frac{\omega}{z_n} + 90y - \sum_{d=1}^{d}\tan^{-1}\frac{\omega}{p_d} - 90x \quad 32.44$$

The overall method for determining the frequency response of a circuit utilizing Bode plots follows.

step 1: Determine the transfer function for the circuit of interest and arrange it in the form of Eq. 32.39 or Eq. 32.41. The break frequencies for each term are the zeros and the poles. Recall that the break frequency for a zero or pole at the origin of the s domain is $\omega = 1$.

step 2: Plot the break frequency points on the axis of a log frequency scale. That is, plot the points where the magnitude, in decibels, of the transfer function is zero.

step 3: For each zero at the s domain origin (s^y terms), plot a straight line with a slope of +20 dB per decade through the origin of the Bode plot axis, that is, through $\omega = 1$ ($\log\omega = 0$) and $20\log|T(j\omega)| = 0$.

step 4: For each pole at the s domain origin (s^x terms), plot a straight line with a slope of −20 dB per decade through the origin of the Bode plot axis, that is, through $\omega = 1$ ($\log\omega = 0$) and $20\log|T(j\omega)| = 0$.

step 5: For all other zeros, draw two asymptotic straight lines from the associated break frequency. The first is of zero slope and extends to lower frequencies. The other has a slope of +20 dB per decade (or 6 dB per octave) and extends to higher frequencies.

step 6: For all other poles, draw two asymptotic straight lines from the associated break frequency. The first is of zero slope and extends to lower frequencies. The other has a slope of −20 dB per decade (or −6 dB per octave) and extends to higher frequencies.

step 7: Sum the asymptotic lines drawn. The result is the idealized Bode plot for the transfer function.

step 8: If greater accuracy is desired, at each break frequency plot a point displaced 3 dB from the intersection of the asymptotic lines. The point is +3 dB from lines that break upward and −3 dB from lines that break downward. At differences of plus or minus one octave from the break frequency, plot a point +1 dB from lines that break upward and −1 dB from lines that break downward. Additional points may be plotted using Eq. 32.31. Join the points in a smooth curve with the ends asymptotic to the idealized Bode plot lines. This smooth curve is the actual plot.

step 9: Determine the value of K from the value of the asymptotic or actual curve at some frequency or by using Eq. 32.42.[22] Using this value of $20\log K$, either shift the plot upward or downward by the amount calculated, or change the 0 dB point on the axis to the value of $20\log K$. In either case, the shape of the Bode plot remains unchanged.

The overall method for determining the phase response of a circuit utilizing Bode plots follows.

step 1: Determine the transfer function for the circuit of interest and arrange it in the form of Eq. 32.39 or Eq. 32.41. The break frequencies for each term are the zeros and the poles. Recall that the break frequency for a zero or pole at the origin of the s domain is $\omega = 1$.

step 2: Plot the break frequency points on the axis of a log frequency scale. That is, plot the points where the angle $\theta = 0°$.

step 3: For each zero at the s domain origin (s^y terms), plot a constant angle of +90°.

step 4: For each pole at the s domain origin (s^x terms), plot a constant angle of −90°.

step 5: For all other zeros, plot a point at the break frequency corresponding to +45°. At one decade below the break frequency, plot a point corresponding to 0°. At one decade above the break frequency, plot a point corresponding to +90°. Connect the points. The slope will be +45° per decade.

step 6: For all other poles, plot a point at the break frequency corresponding to −45°. At one decade below the break frequency, plot a point corresponding to 0°. At one decade above the break frequency, plot a point corresponding to −90°. Connect the points. The slope will be −45° per decade.

step 7: Sum the asymptotic lines drawn. The result is an idealized Bode phase plot.

step 8: If greater accuracy is required, additional points may be drawn using Eq. 32.35 for zeros and Eq. 32.36 for poles. Connecting the points with a smooth curve results in the actual plot.

[22] A pole or zero at the origin in the s domain has a pole or zero value in Eq. 32.42 of one.

Circuit Theory

Example 32.7

A certain circuit has the following transfer function.

$$T(s) = \frac{10}{s(s+10)}$$

Draw the associated Bode plot.

Solution

Unless otherwise specified, an idealized Bode plot is assumed. Following the steps in Sec. 32.8 results in the following plot.

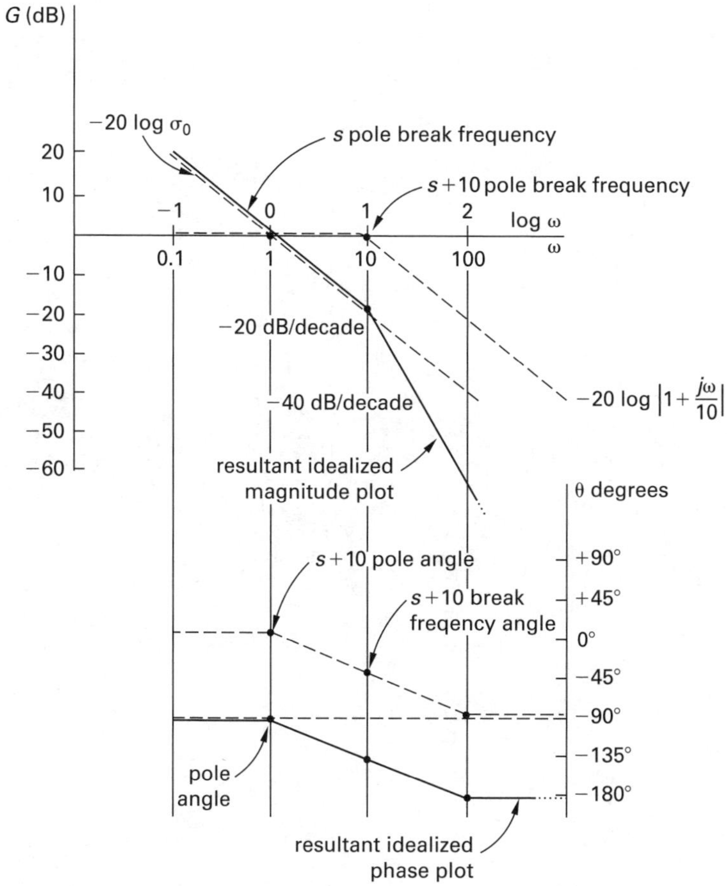

Topic V: Power Generation

Chapter

33. Batteries, Fuel Cells, and Power Supplies

Power Generation

33 Batteries, Fuel Cells, and Power Supplies

1. Battery Fundamentals 33-1
2. Battery Theory 33-1
3. Battery Types and Capacity 33-3
4. Fuel Cells 33-4
5. Power Supplies 33-4

Nomenclature

C	capacity	A·h
$\mathcal{E}$	electromotive force	V
I	effective or DC current	A
P	power	W
Q	overcharge	A·h
R	resistance	Ω
V	effective or DC voltage	V
V	volume	L

Subscripts

bat	battery
cell	per cell
H_2	hydrogen
int	internal
s	source
t	terminal

1. BATTERY FUNDAMENTALS

A *battery* is a direct-current voltage source (made up of one or more units) that converts chemical, thermal, nuclear, or solar energy into electrical energy. Chemical conversion is used in the majority of batteries and will be the focus of this section. A *cell* is a single unit of a battery.[1] A *primary cell* undergoes chemical change in such a manner that one or both electrodes are unusable following discharge. A *secondary cell* undergoes chemical change in a reversible manner and can be recharged following operation. *Shelf life* is the time a battery can be stored without use before its capacity drops to a value that makes it uneconomical to store further. The *cycle life* is the number of times a battery can be discharged or charged before the output voltage drops to a prescribed value. A *cycle* is a predetermined number of amp-hours (A·h) of discharge or charge. The advantages of a battery as a source of electrical energy are its portability, storage capability, and nonpolluting operation. The disadvantages are its general energy inefficiency and low power capability, both of which are changing as battery technology advances.[2]

[1]The first storage cell was invented by Italian professor Alessandro Volta in 1800.

[2]*Energy efficiency* is the ratio of the discharging energy (in J) to the charging energy (in J). Values between 50% and 70% are common.

2. BATTERY THEORY

An *electrolyte* is a compound that, when dissolved in certain solvents (usually water), will conduct electricity. When placed in solution, an electrolyte dissociates into ions as shown in Fig. 33.1(a). Placing two dissimilar electrodes in the solution results in a migration of the ions according to each ion's affinity. The result is a potential difference between the two electrodes as shown in Fig. 33.1(b). The magnitude of the potential difference depends on the materials used for the electrodes and is determined by their specific electronic structures. Some typical combinations are shown in Table 33.1.

Table 33.1 *Cell Voltages*

battery type/name	anode (negative electrode)[a]	cathode (positive electrode)[b]	nominal cell voltage (V)
primary/Leclanche	Zn	MnO_2	1.2
primary/ mercury-zinc	Zn	HgO	1.2
primary/ silver-zinc	Zn	AgO	1.6
secondary/ lead-acid	Pb	PbO_2	1.8
secondary/ nickel-cadmium	Cd	NiOOH	1.2
secondary/ silver-zinc	Zn	AgO	1.5
secondary/ nickel-metal hydride	H_2·(metals)	NiOOH	1.2

[a]The anode is positive with respect to the battery, which results in a negatively charged electrode.
[b]The cathode is negative with respect to the battery, which results in a positively charged electrode.

When an external load is attached to the terminals, a current flows from the positive electrode (cathode) to the negative electrode (anode) in the external circuit as shown in Fig. 33.1(c). In the internal circuit, current flows from the anode to the cathode. Positive ions (cations) migrate toward the cathode (positive electrode), where they capture incoming electrons and become neutral atoms, transforming the material on the cathode. Negative ions (anions) migrate toward the anode where they lose their electrons by combining with the active material of the anode. When the material on either plate is consumed, the chemical reaction stops and the current ceases as shown in Fig. 33.1(d).

Power Generation

Figure 33.1 Battery Theory

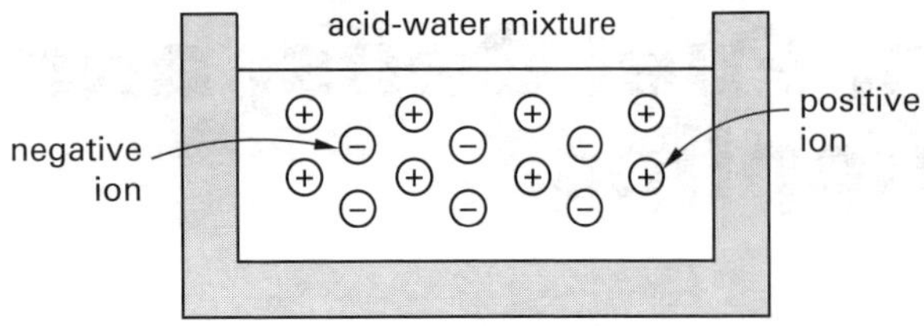

(a) electrolyte dissociation

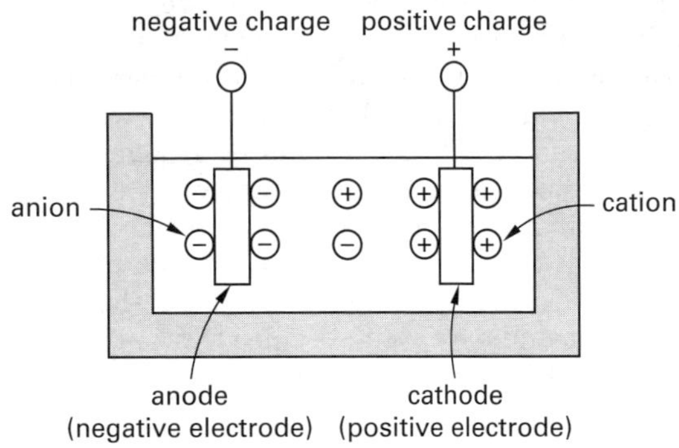

(b) potential due to electrode affinity

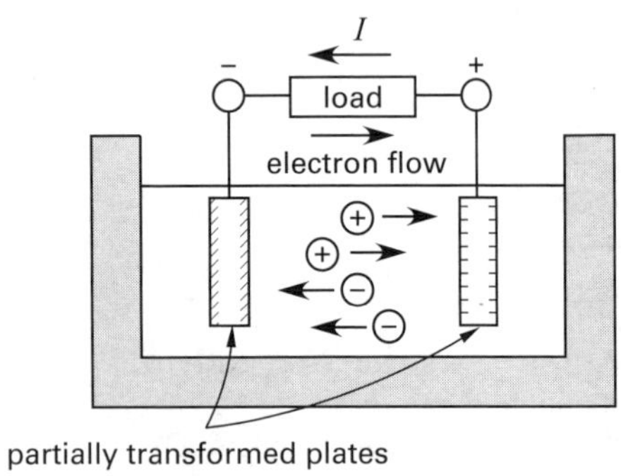

(c) current flow

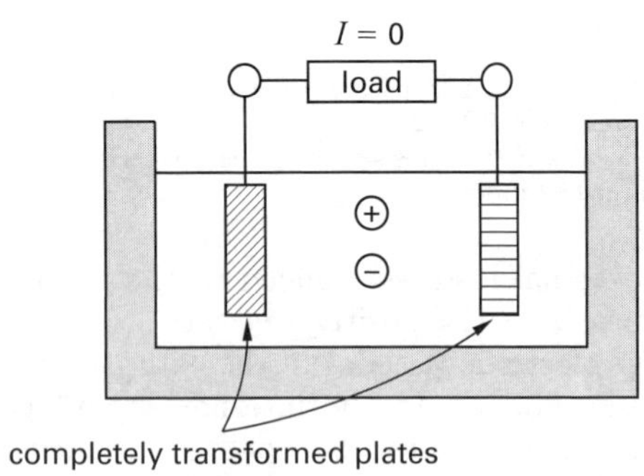

(d) cell discharged (electrode chemical change complete)

Care should be taken when referring to positive and negative potential on a battery. The anode is positive with respect to the internal circuit, that is, the battery itself. The cathode is negative with respect to the internal circuit, that is, the battery itself. In electrical circuit schematics showing a battery, the power flows out of the positively charged electrode (the cathode) during discharge, because it is positive with respect to the external circuit. The positive, or conventional, current flows from the anode to the cathode within the battery and from the cathode to the anode in the external circuit (that is, from the negative electrode to the positive electrode within the battery and from the positive to the negative in the external circuit). The cathode (positive electrode) is the one labeled as positive on an actual battery terminal. The anode (negative electrode) is alternatively defined as the terminal at which conventional current enters a battery.

All batteries possess an internal resistance to current flow that results in a lower output voltage as the current increases. The equivalent circuit for a battery is shown in Fig. 33.2.

Figure 33.2 Battery Equivalent Circuit

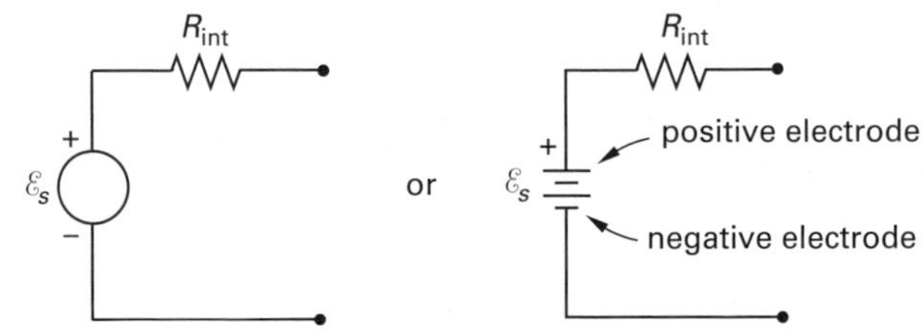

Example 33.1

A 1.5 V battery with an internal resistance of 0.3 Ω is connected to a 1 Ω load. What is the load current?

Solution

The circuit is that of Fig. 33.2 with a 1 Ω resistor attached to the terminals. Using Ohm's law, the current is

$$\begin{aligned} V &= IR \\ I &= \frac{V}{R} = \frac{V}{R_{\text{int}} + R_{\text{load}}} \\ &= \frac{1.5\ \text{V}}{0.3\ \Omega + 1\ \Omega} = 1.15\ \text{A} \end{aligned}$$

Example 33.2

For the battery and load in Ex. 33.1, what is the power consumed by the internal resistance?

Solution

The power is determined by

$$\begin{aligned} P &= I^2R \\ &= (1.15\ \text{A})^2(0.3\ \Omega) = 0.397\ \text{W} \end{aligned}$$

Example 33.3

For the battery and load in Ex. 33.1, what is the terminal voltage?

Solution

Using KVL around the loop of Fig. 33.2 gives

$$\mathcal{E}_s - IR_{\text{int}} - V_t = 0$$

Using the current from Ex. 33.1 and solving for the terminal voltage gives

$$\begin{aligned} V_t &= \mathcal{E}_s - IR_{\text{int}} \\ &= 1.5\ \text{V} - (1.15\ \text{A})(0.3\ \Omega) = 1.15\ \text{V} \end{aligned}$$

3. BATTERY TYPES AND CAPACITY

Primary batteries are those that cannot be recharged. Primary batteries that contain no free or liquid electrolyte are called *dry cell batteries.* A *reserve battery* is one in which an essential component is withheld, making the battery inert and permitting long-term storage. A *secondary battery* is one designed so that it can be recharged electrically following discharge. Charging is accomplished by passing a current through the battery in the direction opposite that of the discharge current.

Batteries are also typed according to cell sizes established by the American National Standards Institute (ANSI). Some examples of typical sizes are given in Table 33.2.

Table 33.2 *Cell Size*

cell size	starting drain current (mA)	service capacity (h)
AAA	2	290
	10	45
	20	17
AA	3	350
	15	40
	30	15
B	5	420
	25	65
	50	25
C	5	430
	25	100
	50	40
D	10	500
	50	105
	100	45

As seen in Table 33.2, service capacity of a battery varies. As the current withdrawn is increased, the capacity drops. This occurs primarily because of the battery's internal resistance and the effect of *polarization.* Polarization is the phenomenon whereby, when a cell discharges current, hydrogen is given off at one of the terminals. The hydrogen bubbles act to insulate the electrode, lowering the chemical reaction rate and the current.

Hydrogen production also occurs during battery charging. For a battery that is fully charged, the approximate hydrogen production is

$$V_{H_2} = 0.42\left(\frac{V_{\text{bat}}Q}{V_{\text{cell}}}\right) \qquad 33.1$$

The term V_{H_2} is the volume of hydrogen produced in liters. Each liter of hydrogen produced correlates to the electrolysis of 0.8 mL of water. The term V_{bat} is the total battery voltage while V_{cell} is the individual cell voltage. The term Q is the amount of overcharge in amp-hours (A·h).

The service capacity combined with the current withdrawn determines the *battery capacity.* Because the capacity varies with current, it is normally measured in amp-hours (A·h) over a standard period. Standard time periods include one, five, eight, and ten hours.

Example 33.4

A 12 V lead-acid battery with six 2 V cells is left on a charger for 24 h following completion of the charge. Assuming a charging current of 3 A, how much hydrogen is produced?

Solution

The volume of hydrogen produced is

$$V_{H_2} = 0.42\left(\frac{V_{\text{bat}}Q}{V_{\text{cell}}}\right)$$

The battery voltage is given. The cell voltage is 2 V. The amount of overcharge, Q, is

$$Q = (3\ \text{A})(24\ \text{h}) = 72\ \text{A·h}$$

Substituting gives

$$\begin{aligned} V_{H_2} &= 0.42\left(\frac{V_{\text{bat}}Q}{V_{\text{cell}}}\right) = (0.42)\left(\frac{(12\ \text{V})(72\ \text{A·h})}{2\ \text{V}}\right) \\ &= 181\ \text{L} \end{aligned}$$

Example 33.5

How much does the capacity of a D-cell battery drop when the drain current is increased from 10 mA to 100 mA?

Solution

From Table 33.2, the capacity in A·h at 10 mA is

$$C = (10 \times 10^{-3}\ \text{A})(500\ \text{h}) = 5000 \times 10^{-3}\ \text{A·h}$$

From Table 33.2, the capacity in A·h at 100 mA is

$$C = (100 \times 10^{-3}\ \text{A})(45\ \text{h}) = 4500 \times 10^{-3}\ \text{A·h}$$

The decrease in capacity is

$$1 - \frac{4500 \times 10^{-3}\ \text{A·h}}{5000 \times 10^{-3}\ \text{A·h}} = 0.1 \quad (10\%)$$

4. FUEL CELLS

A *fuel cell* is a device that converts chemical energy directly into electrical energy. A fuel cell differs from a battery in that the fuel and oxidant must be continuously supplied and, at least in theory, no heat is produced as a result of the reaction. A fuel cell uses both liquid and gaseous fuels, such as hydrogen, hydrazine, and coal gas. The oxidant is oxygen or air. A hydrogen-oxygen fuel cell is shown in Fig. 33.3.

Figure 33.3 *Hydrogen-Oxygen Fuel Cell*

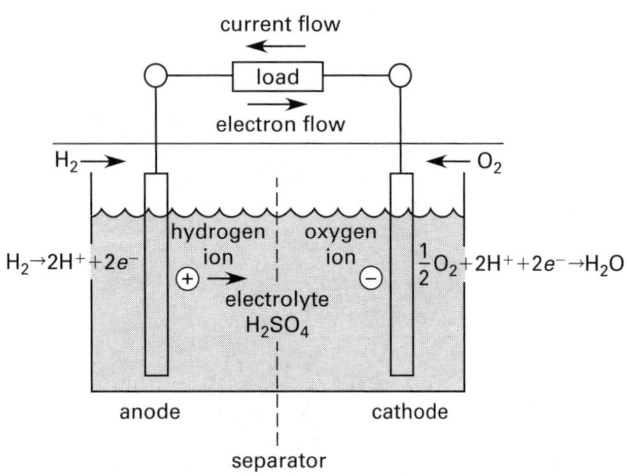

A fuel, in this case hydrogen, is supplied continuously to an electrode, where it undergoes a catalytic reaction producing electrons and a fuel ion ($2H^+$). The electrons travel through the external circuit, supplying power to the load. The fuel ion migrates through the electrolyte and gas separation barrier to the oxygen electrode. The incoming oxygen reacts with the cathode to produce negative oxygen ions that diffuse in the electrolyte. At the oxidizing electrode's surface, the oxygen, hydrogen ions, and electrons combine through a catalytic reaction to form water. The net result is hydrogen and oxygen combining to form water and produce electrical energy.

Fuel cells are of interest because of their high efficiency, minimal environmental impact, and ability to be constructed in a modular fashion.

5. POWER SUPPLIES

A *power supply* is any source of electrical energy, although the term generally refers to a battery or power line used to supply an electronic circuit with the proper electric voltages and currents for operation. An example is shown in Fig. 33.4.

The purpose of the power supply is to take a given input and shape the output such that the electronic circuitry can operate accurately and without short- or long-term damage. The principles of the component parts are covered in Chap. 28, Chap. 40, and Chap. 36.

Figure 33.4 *Power Supply*

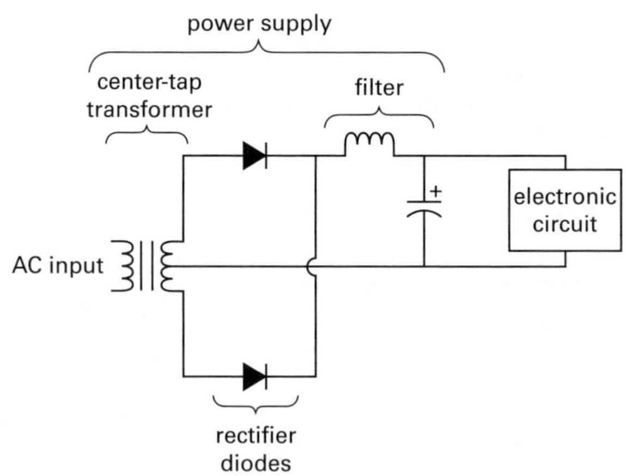

Topic VI: Protection and Safety

Chapter

34. Protection and Safety

34 Protection and Safety

1. Power System Structure 34-1
2. Power System Grounding 34-1
3. Power System Configurations 34-2
4. Relays 34-3
5. Relay Reliability 34-3
6. Zones of Protection 34-4
7. Relay Speed 34-4
8. Protection System Elements 34-4
9. Transformers 34-4
10. Relay Types 34-5
11. Universal Relay Equation 34-6
12. Solid-State Relays 34-7
13. Protective Devices 34-7

Nomenclature

C	capacitance	F
E	voltage	V
I	current	A
pf	power factor	–
R	resistance	Ω
V	constant or rms voltage	V
X	reactance	Ω
Z	impedance	Ω

Symbols

θ	torque angle during normal conditions	degrees
τ	time delay	s
τ	torque	N·m
ϕ	impedance angle or torque angle during referenced condition	degrees

Subscripts

f	fault
p	phase or pickup
r	ratio or relay

1. POWER SYSTEM STRUCTURE

The protection and safety of power systems are determined by the structures of the systems. Most power systems consist of the three elements shown in Fig. 34.1. The *control system* uses equipment to maintain a nominal voltage and frequency, as well as maintain the power system's economic operation and security in interconnected networks. The control system operates continuously on the power system to adjust operating parameters. The *protection system* operates circuit breakers and other devices, such as relays, that change the structure of the power system to protect it from abnormal situations, such as electrical faults. The protection system operates when required and is faster than the control system. The advance of computer-based electronic protection systems blurs the line between protection and control, incorporating both elements into one system; however, the functions remain the same.

Figure 34.1 *Power System Structure*

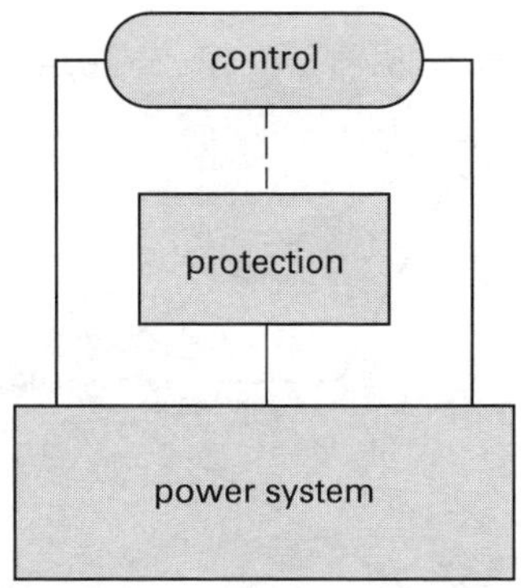

2. POWER SYSTEM GROUNDING

A power system's *grounding scheme* influences the fault current levels. As a result, the grounding scheme impacts the protective system *relaying*. (See Sec. 34.5.) In a true ungrounded system, there is zero fault current. In a real system, capacitive coupling of the feeders to ground results in a ground path if a fault occurs. The ungrounded system with a ground fault is shown in Fig. 34.2.

If the capacitance is large enough, the capacitive ground fault becomes self-sustaining and must be interrupted by circuit breakers. *Protective relays* (devices used to make or break one or more connections in an electric circuit) must sense the low level of ground-fault current before a ground fault can damage the power system. To do so, impedance is inserted in the dashed box, as illustrated in Fig. 34.2. The impedance ensures adequate current flow that can then be sensed by the protective system, which can then react before the fault causes excessive damage.

Ungrounded systems provide for high system reliability because one ground does not cause protective systems to react. A second ground fault, or the capacitive ground fault, is required to cause breakers to open and equipment to be removed from service. A disadvantage of ungrounded systems is that when a ground fault occurs on one phase, as shown in Fig. 34.2, the other two phase voltages become $\sqrt{3}$ larger than their normal values. The phasor representation of a grounded phase is shown in Fig. 34.3. Phase A voltage changes the reference on

Figure 34.2 *Ungrounded System with Ground Fault*

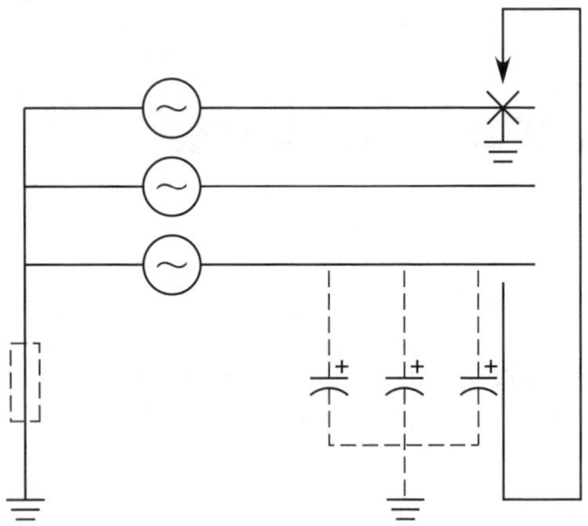

Figure 34.3 *Phasor Diagram of Grounded Phase*

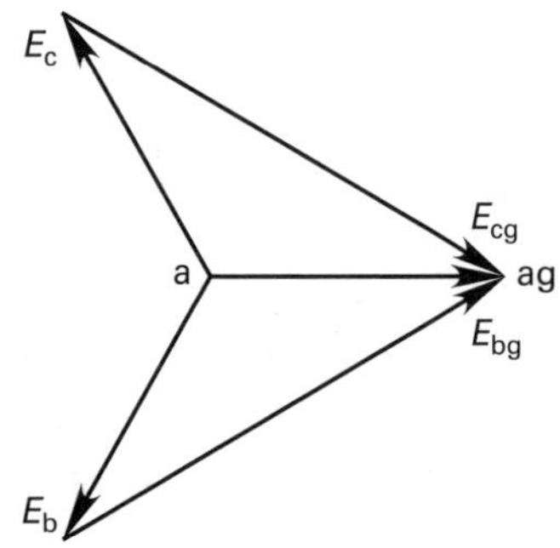

the ground resulting in increased voltage differences on phases B and C to ground. Insulation levels on the conductors must account for this possible change.

On low voltage systems, the insulation level is usually set based on lightning-induced phenomena, which are generally larger than the fault-induced voltage increase. On high voltage systems (e.g., those with voltages greater than 100 kV), fault-induced voltages are more critical. As a result, on high voltage systems, a solidly grounded neutral is used. Such grounded systems are called *effectively grounded*. These systems have high fault currents and the impedance in the dashed box shown in Fig. 34.2 is set to limit the fault current to a value that circuit breakers can safely interrupt.[1]

The system in Fig. 34.2 is called an ungrounded system even though the wye connection may be grounded to allow detection of faults. (The conductors in the distribution system are *ungrounded*.) Such a ground is called a *functional earth*. A functional earth is defined as a ground that serves a purpose other than protecting against electrical shock. The conductors supply various loads that are usually encased in metal frames. These frames must be grounded for the safety of personnel. The ground used to provide this safety is called a *protective earth* (PE).[2]

3. POWER SYSTEM CONFIGURATIONS

The configuration of buses making up a power system will determine the available protection system options. A *radial system* is a serially connected system with a single source, as shown in Fig. 34.4. Such a system is normally limited to distribution systems (generally operating at voltages of 100 kV or less). It has the advantage of being economical to build and the disadvantage of being less reliable since a problem at any point in the network affects all downstream loads. Radial systems are also easier to protect because a fault current can only flow from the source to the fault, making it less complex to calculate. Because the system is usually remote from the source, the fault current is approximately independent of the generating capacity.

Figure 34.4 *Radial Bus Configuration*

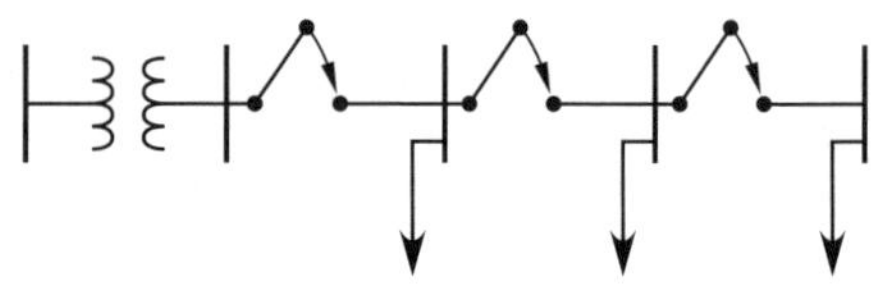

A *network system*, as shown in Fig. 34.5, is normally used for transmission and sub-transmission systems (generally operating at voltages of 100–200 kV or higher). It has increased the reliability of power delivery to loads, and the loss of a single source of power has minimal impact on reliability. However, network systems are more complicated to protect than radial systems. A fault current can occur from multiple sources and from multiple directions. Because the source may be close to the fault, the fault current may vary widely depending on location.

Figure 34.5 *Network Bus Configuration*

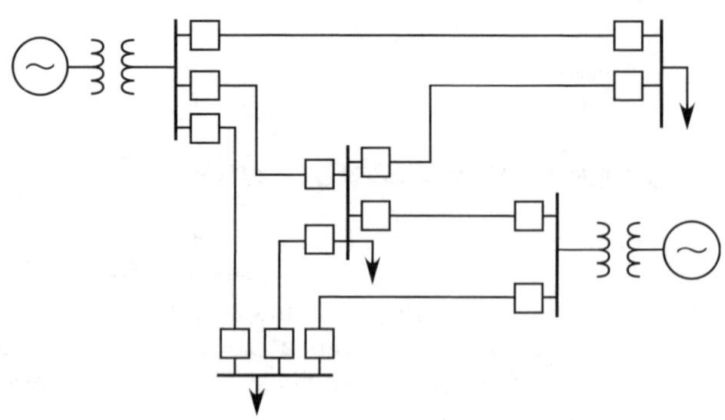

[1]In lower voltage networks, an alternative to ungrounded system is the *ground fault neutralizer* (GFN) known as the *Petersen coil*.

[2]International standards call an ungrounded system an IT system. The "I" means isolation and the "T" is for terra (or Earth). When the neutral is effectively grounded along with the protective earth, the system is called a *TN system*. The "T" means terra and the "N" means neutral. When the neutral is grounded, but is physically separate from the protective earth ground, the system is called a TT system, for terra-terra.

4. RELAYS

Relays are designed to determine electrical circuit operating systems and to trip circuit breaks when a fault is detected. They can be electromagnetic or electronic. The various protective functions available on a relay are given by standardized numbers in ANSI/IEEE Standard C37.2, examples of which are given in Table 34.1. Prefixes to the numbers indicate its location. For example, RE stands for *remote*. Numerical prefixes indicate an association with a given unit. For example, 101 is a master element associated with unit 1 of a multiunit plant. Suffix letters indicate auxiliary devices and actuating quantities. For example, CL indicates a device that is energized when the main device is closed, while a suffix of C stands for *current*, P for *power*, VAR for *reactive power*, and so on.

The goal of any protective system is a quick and accurate diagnosis of an anomaly, a fast response time to the indicated issue, and a minimal disturbance of the entire power system.

5. RELAY RELIABILITY

The purpose of *relaying* is to remove any element that is functioning abnormally from a given power system. In general, the protective system does not prevent equipment damage. For the most part, relays operate after some damage has already occurred. Their purpose is to prevent further damage, minimize danger to personnel, and stabilize the system.

A relay's *reliability* is the probability that it will perform its intended function under a given set of conditions. The conditions can be environmental or electrical. Additionally, the assumption is made that maintenance is accomplished as specified.

A relay can be unreliable in two manners. First, a relay can fail to operate when intended. Secondly, a relay can operate when not expected. Therefore, the definition of protective systems relays must be expanded to include the requirement that they are also dependable and secure. A relay is *dependable* if there is a measure of certainty that it will operate for designed faults. A relay is *secure* if there is a measure of certainty that it will not

Table 34.1 *ANSI/IEEE Power System Device Function Numbers*

number	title	function
1	master element	places equipment in or out of operation
2	time delay or closing relay	provides desired amount of time delay
3	interlocking relay	allows or stops a given operating sequence
6	starting circuit breaker	connects device to source of voltage
11	multifunction device	combination of functions in one device
12	overspeed device	operates on machine overspeed
14	underspeed device	operates on machine underspeed
21	distance relay	functions when admittance, impedance, or reactance increases or decreases beyond a certain value[a]
24	volts per hertz relay	operates on a given V/Hz ratio[b]
27	undervoltage relay	operates on machine undervoltage
32	directional power relay	operates on power flow in a given direction[c]
46	phase-balance relay	operates when negative phase sequence exceeds a given value[d]
52	AC circuit breaker	operates a circuit breaker
58	rectification relay	operates when a rectifier fails to conduct or block correctly
59	overvoltage relay	operates on overvoltage condition
64	ground detector relay	operates upon failure of insulation to ground
72	DC circuit breaker	operates a circuit breaker
87	differential protection relay	operates upon a quantitative difference between two or more currents or other electrical quantities

[a]A *distance relay* is useful in determining if a fault exists in a given distribution system.
[b]In variable speed devices, the volts per hertz (V/Hz) ratio is adjusted to maintain the current to a machine constant. A constant torque results in a constant speed.
[c]Reverse power on a generator can damage the machine. This relay detects the angular relationship between the current and voltage, which determines the direction of power flow.
[d]This relays checks for negative phase-sequence currents above a given value. Negative phase-sequence currents indicate unbalance phases meaning that a fault exists in the system.

operate incorrectly. As a system becomes more dependable, it generally becomes less secure.

6. ZONES OF PROTECTION

The reliability requirement that relays are dependable (i.e., they operate only as intended) is further defined by the zones in which they operate. A *zone* is a region of the power system and can be closed or open. A *closed zone*, also known as a *differential*, *unit*, or *absolutely selective zone* is one where all power apparatuses entering the zone are monitored at the entry points. In an *open zone*, also known as an *unrestricted*, *non-unit*, or *relatively selective zone*, monitoring does not occur, and the zone is defined by level of the fault current.

Current transformers (CTs) whose input is utilized by the relays monitoring the zones, are the proximate boundaries of the zones.[3] An example of zones is shown in Fig. 34.6.

Figure 34.6 *Closed and Open Zones*

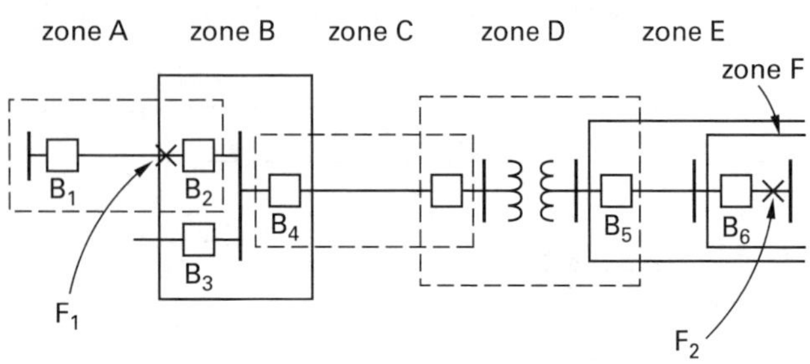

Coordination in protective systems is designed to meet the general goal of having the breaker, or other protective device, closest to the fault open first, and the breaker farthest from the fault open last. In Fig. 34.6, zones A, B, C, and D are closed zones. Zones E and F are open zones. Fault 1 occurs in two zones: A and B. As a result, breakers 1 through 4 will open, even though breakers 3 and 4 do not isolate the fault. Fault 2 occurs in an open zone. Breaker 6 is programmed to open first. If it fails to do so, breaker 5 will open after a specified delay. The setup in Fig. 34.6 represents a form of coordination.

7. RELAY SPEED

A relay senses the operating parameters of a given system. When those parameters or waveforms are distorted due to a fault, the relay must analyze the distortion to determine the relevant information and make a decision, based on internal logic parameters, regarding the fault. The time it takes for the relay to make the decision and the security of that decision is an inverse relationship. This inverse-time operating characteristic applies to all protection relays.

An *instantaneous relay* operates as soon as the logic circuitry makes the decision. No intentional delay is inserted after the decision. A *time-delay relay* has an intentional time delay inserted after the decision is made, such as would be done for backup protection. A *high-speed relay* is defined as one that operates in a specified time, such as 50 ms, or three cycles of a 60 Hz system, as given in various electrical standards. An *ultra high-speed relay* generally indicates a relay that operates in 4 ms or less, or approximately one-quarter of a cycle in a 60 Hz system, though it is not defined by standards.

8. PROTECTION SYSTEM ELEMENTS

A protection system is comprised of three main elements: transducers, relays, and breakers. *Transducers* sense the desired signal and convert it into a form usable by the relays. Current and voltage instrument transformers are examples of transducers. *Relays* monitor the input from the transducers and provide an output to operate breakers in the event of a fault. *Breakers* are designed to interrupt power and isolate sections of the power system. They come in a variety of types and interrupting mediums, such as air, oil, gas, solid dielectric, and SF_6.

When a fault occurs in a system, the AC voltage available may not be adequate to allow protective system operation. To ensure the necessary trips occur, battery backups are used to power protective system components. An *uninterruptable power supply* (UPS) is often used. A UPS ensures normal system power is provided to components unless that power is interrupted or inadequate. When such an interruption occurs, the batteries in the UPS automatically, and with minimal interruption time, provide the necessary backup power.

Two separate battery systems are sometimes used. One system powers electromechanical relays, and the other system powers solid-state relays. Electromechanical relays can produce severe transients on battery leads and potentially impact sensitive solid-state devices, which is why they are separated.

9. TRANSFORMERS

The symbol for a *current transformer* (CT) is shown in Fig. 34.7. Current transformers are used as instrument transformers (also known as *metering transformers*) or as transducers for protection circuits (i.e., relaying transformers).[4] They have low power, and ideally, high accuracy. The accuracy is measured in terms of the transformer's ability to lower the current; in other words, they transform the current without changing the expected magnitude or the phase relationship. In Fig. 34.7, the dots imply that the current leaving the

[3]Neighboring zones always overlap to ensure system protection. Faults in an overlap region can disconnect both regions, possibly even more of the system, so overlaps are as small as practical.

[4]Metering transformers are designed to operate most accurately between no-load and full-load. Relaying transformers are designed to operate accurately at expected fault conditions.

secondary dot and flowing to the load is in phase with the current entering the primary dot.[5]

Figure 34.7 *Current Transformer*

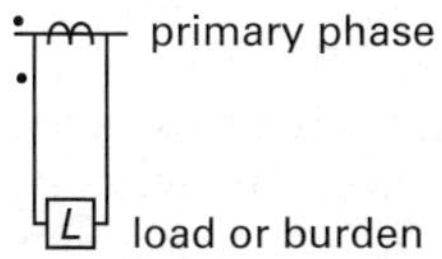

The *burden* is the amount of power drawn from the circuit connecting the secondary terminals of instrument transformers, usually given as apparent power in voltamperes. The term burden distinguishes power for operation of the device itself from power sensed, which is being monitored and is referred to as the load. As long as the burden is within the rating of the CT, only acceptably small errors are introduced in the phase relationship between the primary and secondary currents, and the current transformation ratio is exact. For example, a CT with a primary current angle of 30° with respect to the voltage and a ratio of 100:1 will transform a 100 A current at 30° degrees to a 5 A current at 30°. When the burden is too high, or when non-ideal conditions exist, a CT will exhibit both *phase angle errors* and *transformation errors*. These errors vary with the loading of the transformer and the power factor.

Voltage transformers are also used for metering or relaying. They have low power and, ideally, high accuracy. Accuracy is measured in terms of the ability to transform the voltage without changing the expected magnitude and phase relationship. The change in voltage output from no-load to full-load is called the *voltage regulation*.[6] The standard connections for transformers are shown in Fig. 34.8(b) and Fig. 34.8(c). The polarities are called *additive* or *subtractive*. Additive and subtractive polarities are a way to avoid the dot convention. Figure 34.8(a) shows a voltage transformer with positive polarity using the dot convention. Figure 34.8(b) shows additive polarity, and Fig. 34.8(c) shows subtractive polarity. The "H" connection represents the primary phase. The "X" connection represents the secondary phase.

10. RELAY TYPES

Generally when faults occur, currents increase, voltages drop, harmonics develop, and phase relationships shift. Relays, both analog and digital, can be made to sense and respond to these changes. A relay can be comprised of multiple types embedded in a single device. An explanation of some of these types follows.

Figure 34.8 *Voltage Transformer*

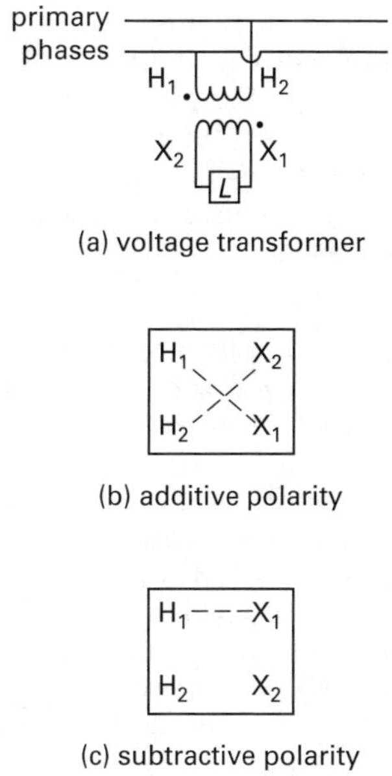

Level detection relays, also called *magnitude* and *overcurrent relays*, operate when a fault exceeds a given value.[7] The *pickup setting* is the setting at which the relay operates. The properties of a typical level detection relay are shown in Fig. 34.9. The x-axis is a normalized value; that is, a ratio of the fault current to the pickup current.

Figure 34.9 *Level Detector Characteristics*

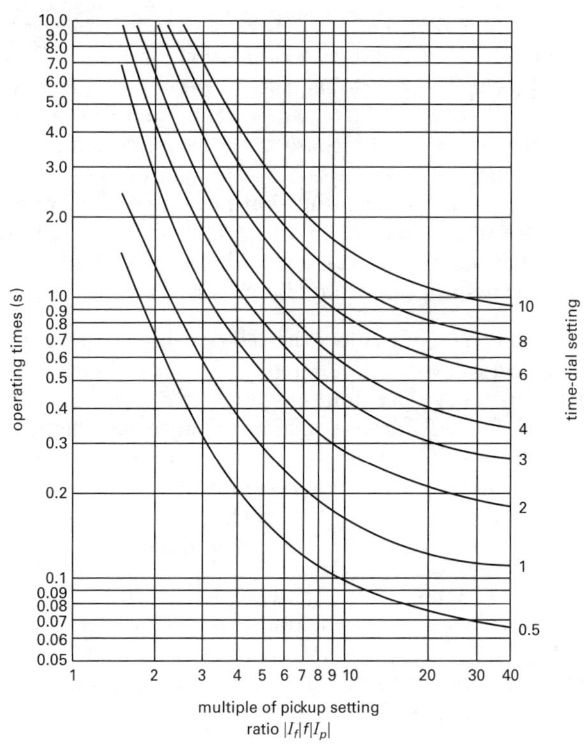

[5]In terms of voltage polarity, both dots are positive at the same time.

[6]The phase change can be eliminated with filtering circuits.

[7]A *fuse* is a simple level detector.

The time dial setting and the pickup setting are selected to give the desired time response in the event of a fault. The time for the relay to function is found by first dividing the fault current by the pickup setting to obtain the multiple of the pickup setting. Using this value, starting at the bottom axis, trace a straight line to the value of the time-dial setting selected (see the right axis). From this point, trace a line to the left axis to obtain the operating time. The characteristic curves are plotted on log-log scale.

Differential comparison relays operate when currents at the input and output of a zone are not identical. This method is very sensitive to changes and is effective in isolating faults. A sample scheme is shown in Fig. 34.10. The disadvantage to this scheme is that currents at each end of the zone can be separated by a significant distance, requiring an increased equipment cost for wiring and imposing possible time delays.

Figure 34.10 *Differential Relay Scheme*

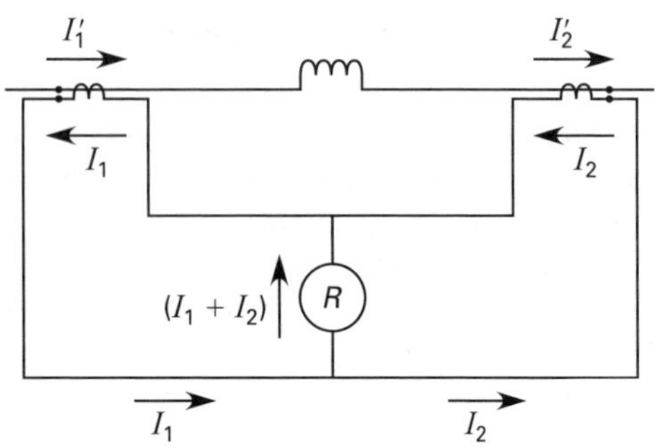

A *phase angle relay* compares the current and voltage to determine, for example, the direction of power flow. When power flow is normal, the impedance angle of the fault circuit, ϕ, is 37° for a load power factor of 0.8. When power flow reverses, the phase angle becomes 180° ± 37°. If a fault occurs, the phase angle becomes either $-\phi$ or $180° - \phi$. For power transmission networks, the impedance is primarily inductive and ϕ is near 90°. A phase angle relay detecting a 90° impedance angle can react under the assumption that a fault has occurred.

A *distance relay* measures the relationship between the voltage and current at a single point. Assuming standard impedance per unit distance, the relay can be adjusted for any desired distance. Since the distance relay depends on the ratio of the voltage to the current, it is also called a *ratio relay* or *impedance relay*. If the relay is not bidirectional, but rather measures faults in only one direction, it is called a *mho relay*.

Other relay types include the *pilot relay*, which operates with information received from a remote location, and the *harmonic relay*, which operates on the harmonic content of the transmission system. *Frequency relays* operate on changes in frequency.

11. UNIVERSAL RELAY EQUATION

Equation 34.1 is called the *universal relay equation*. Developed from analog relay devices, it can be used with *solid-state relays* or programmed into *computer relays*. The term θ is the torque angle in an analog system during normal conditions and is related directly to the impedance angle. The term ϕ is the torque angle in an analog system at the referenced (potentially, the fault) condition, which is also related to the impedance angle. By manipulating the constants in either an analog or digital protective system, a variety of relays can be realized. Figure 34.11 shows the relay types that can occur.[8]

$$\begin{aligned}\tau &= \tau_{\text{magnetic}} - \tau_{\text{spring}} \\ &= C_1 I^2 + C_2 V^2 + C_3 I V \sin(\theta - \phi) - \tau_{\text{spring}}\end{aligned} \qquad 34.1$$

Figure 34.11 *Universal Relay Equation Characteristics*

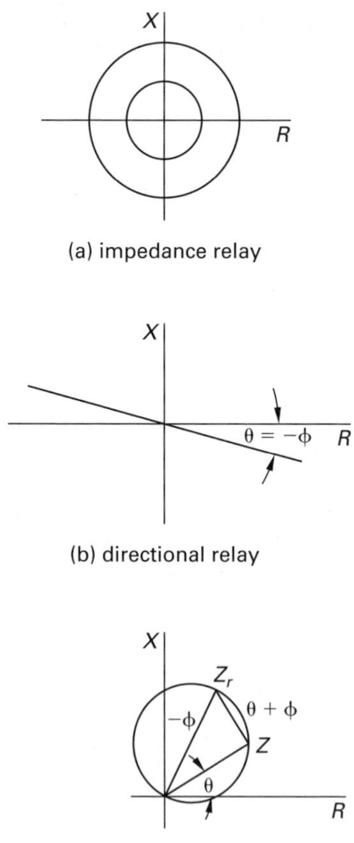

While the *R*-*X* plane used in Fig. 34.11 appears to use negative resistance, this is not the case. A relay can be located in any number of locations within a system. The power flows, under normal conditions, toward the load. If a fault occurs upstream of the relay's location, power

[8]The term Z_r is the impedance of the transmission line at the maximum response of the relay; that is, the maximum distance from the relay that will result in a response. The angle $(\theta + \phi)$ between a line connecting Z_r and the actual impedance, Z, when less than 90°, means the point lies inside the circle and a trip will result. When the angle is greater than 90°, the point lies outside the circle and the trip will be blocked.

flow reverses, and the power will then flow from the load to the fault point. Recall that motors exposed to a fault in the system become generators feeding power to the fault from their inertial energy. Transformers do the same from the inductive energy stored in their magnetic fields.

The various zones of the *R-X* plane are shown in Fig. 34.11. Current and voltage in the system can be plotted on the plane to indicate phase relationships. Impedance values plot as points. Changes to the voltage-current relationships can then be analyzed as conditions shift due to faults.

Figure 34.12 *R-X Diagram*

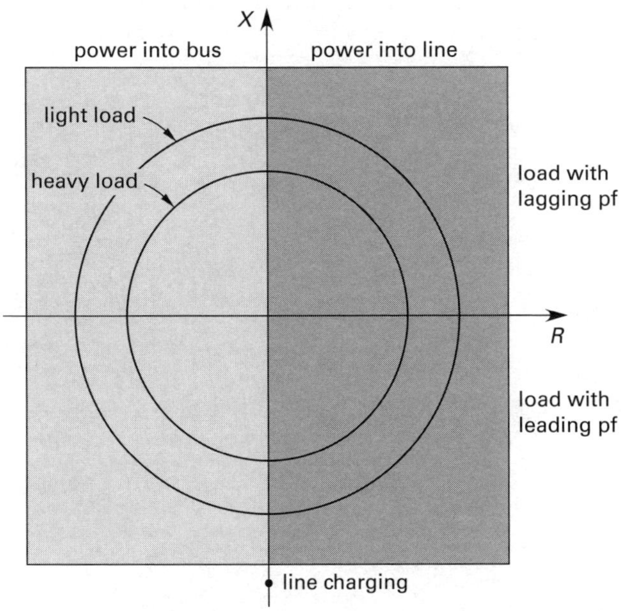

12. SOLID-STATE RELAYS

Many relays are electronic, or *solid-state*. Such relays offer numerous advantages: rugged design, ease of adjustment and programmability, accurate response, and high reliability with little or no maintenance. A solid-state level relay is shown in Fig. 34.13. A level relay is also known as an *overcurrent relay*. The signal sensed is passed through the resistor, *R*, rectified by the bridge network, amplified and filtered by the *R-C* circuit, and then compared to a reference voltage by an operational amplifier, *A*. If the signal exceeds the reference value, the output of the op amp increases, resulting in a protective signal after a self-imposed time delay occurs.

13. PROTECTIVE DEVICES

A wide variety of protective devices are available. Table 34.2 lists these devices in order of complexity and price.

Fuses are simple and inexpensive and common in protective systems. They come in a series of standard sizes, each with a time-current characteristic curve.

Figure 34.13 *Solid-State Overcurrent Relay*

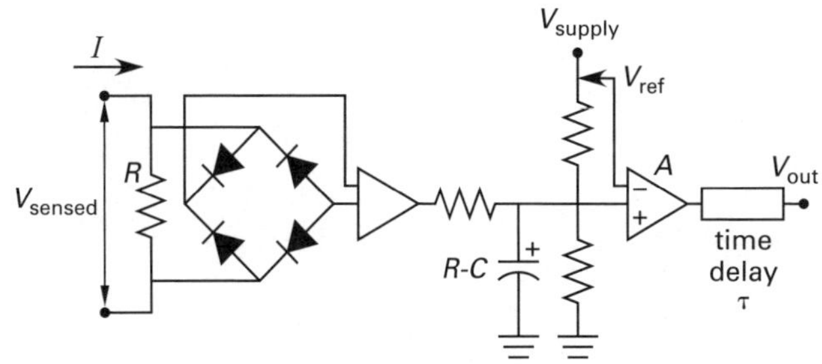

Table 34.2 *Protective Devices*

protective device	cost and complexity
fuse	low
sectionizer	↑
recloser	
instantaneous overcurrent	
inverse time-delay	
overcurrent	
directional overcurrent	
distance	↓
pilot	high

A *sectionalizer* is a self-contained device used in conjunction with other devices, such as reclosers and circuit breakers, to permanently isolate a faulted portion of a distribution system. A sectionalizer does not interrupt a fault current. Instead, it counts the number of times a fault is seen (i.e., sensed) and opens when the count reaches a pre-programmed level. Only then is the circuit deenergized, usually by a recloser or circuit breaker. Sectionalizers are an economical way to isolate faults without the additional step of coordinating trip levels.

A *recloser* senses a fault and interrupts power to a system. After a preset time, the recloser attempts to reenergize the system by closing. If other protective devices have isolated the fault, the recloser remains closed.

An *overcurrent device* operates when a fault exceeds a given value.

Pilot protection refers to a communication channel between the ends of a long transmission line that enables the system to instantaneously clear faults over 100% of the line. Such communication channels include the power line itself using a *power line carrier*, *microwaves*, *fiber optics*, or *communication cables*.

Non-pilot protection refers to a system that responds only to the parameters at one end of a transmission line. If a fault occurs on the line away from the protection, a time-delay is incurred automatically and the fault is not interrupted instantaneously at both ends of the line. Such a system is also called a *graded*, *relatively selective*, or *non-unit* system.

Topic VII: Power System Management

Chapter

35. Power System Management

35 Power System Management

1. Energy and Power Units 35-1
2. Energy Growth Rate 35-1
3. Electric Power Economics 35-2
4. Energy Management System 35-2
5. Information Management Subsystem 35-2

Nomenclature

a	per unit growth rate	–
E	energy	J
P	power	W
t	time	s

Subscripts

0	initial
d	doubling

1. ENERGY AND POWER UNITS

Energy is the work that can be done by a force over a distance. It can result from electrical, chemical, mechanical, nuclear, and other processes. Electrically, energy is the work required to move a positive charge through an electric field. Chemically, the rearrangement of electron positions and energy levels releases energy in the form of heat. Mechanically, kinetic or potential energy results in forces that move objects over a distance. Nuclear energy is obtained from the conversion of mass directly to energy. All these sources, and others, can be used to create electrical energy through the appropriate power conversion system.

Work and energy have the same units. In colloquial terms, the energy of a body is its capacity to do work. The unit of energy in electrical engineering is the watt-second (W·s), which is equivalent to a joule (J). A more common unit is the kilowatt hour (kWh). The relationship between the two as well as between other units is given in Table 35.1.

Power is the rate at which work is done, or energy per unit time. Common power units are the watt, where 1 W equals 1 J/s, and horsepower, where 1 hp equals 745.7 W.

$$P = \frac{E}{t} \qquad 35.1$$

2. ENERGY GROWTH RATE

Accommodation for future growth in energy usage requires knowledge of past growth and assumptions regarding that growth in the future. Energy usage, as a general rule, grows at a rate proportional to the amount of energy that is currently being used, E. The per unit growth rate is given by a.

Table 35.1 *Energy Equivalence*

energy	equivalent
1 W·s	1 J
1 kWh	3.6×10^6 J
1 cal	4.186 J
1 Btu	1.055×10^3 J
1 Btu	0.252×10^3 cal
1 quad	10^{15} Btu*

*Sometimes a quad will be defined as 10^{18} Btu.

$$\frac{dE}{dt} = aE \qquad 35.2$$

The solution to Eq. 35.2 is given in Eq. 35.3, where E_0 is the energy usage at some initial time.

$$E = E_0 e^{at} \qquad 35.3$$

The *doubling time*, or the time it takes for energy usage to double, is given by Eq. 35.4.

$$t_d = \frac{\ln 2}{a} = \frac{0.693}{a} \qquad 35.4$$

Example 35.1

Worldwide energy usage is expected to increase from 495 quad to 739 quad between 2007 and 2035, according to the Reference Case Projections published by the U.S. Energy Information Administration. What per unit growth rate is the Energy Information Administration predicting?

Solution

Use Eq. 35.3 to calculate the per unit growth rate over the 28 year period.

$$E = E_0 e^{at}$$

The future energy usage ratio is

$$\frac{E}{E_0} = e^{at}$$

Power System Management

Take the natural log of both sides and solve for a.

$$t = 2035 - 2007 = 28 \text{ yr}$$

$$\ln \frac{E}{E_0} = \ln e^{at} = at$$

$$a = \frac{\ln \frac{E}{E_0}}{t} = \frac{\ln \frac{678 \text{ quad}}{472 \text{ quad}}}{28 \text{ yr}}$$

$$= 0.0129 \text{ yr}^{-1} \quad (1.29\% \text{ per year})$$

3. ELECTRIC POWER ECONOMICS

The cost of electricity can be separated into three general categories of utility operations: generation, transmission, and distribution. These can be further separated into three major components: fuel, equipment, and labor.

The fuel costs depend on the fuel selected. Major sources of energy include fossil fuels (e.g., oil, coal, and natural gas); nuclear fuel (e.g., uranium and plutonium); hydroelectric power; geothermal steam; fuel cells; batteries; solar; waves; and wind. Fossil fuels develop their power from chemical conversion, nuclear from mass to energy conversion; hydroelectric from potential to kinetic energy conversion; geothermal from convection heat transfer; fuel cells and batteries from chemical conversion; solar from radiation to electric conversion; wave and wind from kinetic to mechanical conversion.

To compare electrical projects, the equipment (capital costs) must be combined with annual costs for fuel and labor. The principles involved are covered in Chap. 55.

4. ENERGY MANAGEMENT SYSTEM

An *energy management system* (EMS) is the system of components used to control a utility network. The task of an EMS is to control power generation and scheduling, network analysis, and the training of operators. Such a system includes everything from the transducers that sense the network's condition, to the computers at the control center displaying the information. This is illustrated in Fig. 35.1.

Figure 35.1 *Energy Management System Block Diagram*

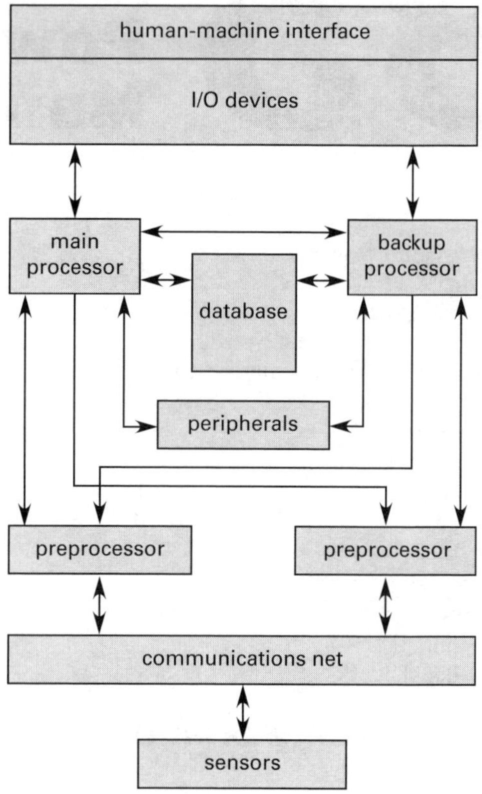

The *Supervisory Control and Data Acquisition* (SCADA) subsystem of the EMS provides for three critical functions: data acquisition, alarm display and control, and supervisory control. The supervisory processors communicate with the attached electronic devices using a serial communications protocol called *Modbus*®.

Data acquisition consists of collection, processing, monitoring, calculations, and configuration control. Collection includes gathering information from remote sensors. Processing converts the raw data—often analog—into *engineering units* (eu) that an operator or computer can utilize for calculation purposes. The system continuously monitors the information and uses the calculations to provide alarms or automatic protective actions. Configuration control is the function that shifts transducers or monitoring points, or removes such sensors or portions of the system when errors occur.

The SCADA system can also be used for *scheduling*. This is the process whereby the system evaluates the load, load forecast, fuel cost, and the usage of each generation machine. Then, the system adjusts the loading on each machine to ensure the most efficient, cost-effective usage.

5. INFORMATION MANAGEMENT SUBSYSTEM

The *information management subsystem* (IMS) consists of the computer network and supporting devices that gather, store, and display the information used by the energy management system, EMS. Advances in computers have made such systems powerful tools for fast and accurate monitoring of electrical systems. Software languages vary, but Fortran retains a high usage rate. New applications, especially those with *graphics user interfaces* (GUI), are usually written in C or C++.

Topic VIII: Measurement and Instrumentation

Chapter

36. Measurement and Instrumentation

36 Measurement and Instrumentation

1. Fundamentals . . . 36-1
2. Signal Representation . . . 36-2
3. Measurement Circuit Types . . . 36-3
4. Permanent Magnet Moving-Coil Instruments . . . 36-4
5. DC Voltmeters . . . 36-4
6. DC Ammeters . . . 36-5
7. Moving-Iron Instruments . . . 36-6
8. Electrodynamometer Instruments . . . 36-6
9. Power Measurements . . . 36-7
10. Power Factor Measurements . . . 36-7
11. Electronic Instrumentation . . . 36-8
12. Insulation and Ground Testing . . . 36-8

Nomenclature

a, b	Fourier series coefficient	–
B	magnetic flux density	T
C	compensating winding	–
C	capacitance	F
d	deflection	various
f	frequency	Hz
$f(t)$	function of time	–
H	magnetic field strength	A/m
I	effective or DC current	A
k	coupling coefficient	–
κ	Boltzmann's constant, 1.3807×10^{-23}	J/K
n, N	number of turns	–
P	power	W
Q	quality factor	–
R	resistance	Ω
S	apparent power	VA
t	time	s
T	temperature	K
T	period	s
T	torque	N·m
v	instantaneous voltage	V
V	effective or DC voltage	V

Symbols

θ	angular displacement	rad
ϕ	angle	rad

Subscripts

0	initial
ave	average
ext	external
fs	full scale
H	higher
L	lower
n	number
rms	root-mean-square

1. FUNDAMENTALS

An *instrument* is a device for measuring, and sometimes for controlling or recording, a value under observation. *Instrumentation* is the use of such devices for measurement, detection, observation, computation, communication, or control. Instrument outputs are given in terms of some reference value. A *reference standard* is an accepted measurement base.[1] *Precision* is the range of values of a set of measurements. An instrument's precision is a measure of the reproducibility of observations. The *accuracy* is a statement of the limits that bound the departure of the measured value from the true quantity. Accuracy includes the imprecision of the instrument and any accumulated errors from the national reference standards in the measurement itself. The accuracy of meter movements is given in terms of the percentage of error relative to the full-scale value, regardless of where the error occurs. That is, the error may occur in the low end of the indicating range, but the accuracy is nevertheless given as a percentage of the highest indicated value.

Errors can be separated into two general categories: systematic and random. A *systematic error* is an error that is reproducible and the result of some physical effect, such as a weakened magnetic field, altered resistance value, repeatable signal distortion, and the like. A *random error* is a deviation in the output caused by chance. Random errors are accidental, inconsistent in magnitude and sign, and not reproducible. The effects of errors are analyzed using statistical mathematics.

The *resolution* is the smallest increment that can be distinguished by the instrumentation system. The resolution is determined by the amount of deflection per unit input and is limited by the amount of signal beyond the noise. *Noise* is any portion of a signal that does not convey useful information, such as unwanted currents and voltages in an electrical device or system. Random disturbances caused by thermal agitation of electrons are a significant contributor to noise in electrical

[1]United States national reference standards are maintained by the National Institute of Standards and Technology, formerly the National Bureau of Standards (NBS).

systems and are called *Johnson noise*. A measure of the Johnson noise in resistors is[2]

$$V_{\text{noise}} = \sqrt{4\kappa TR\Delta f} = \sqrt{4\kappa TR(\text{BW})} \qquad 36.1$$

The term V_{noise} is the rms equivalent voltage generated by the thermal agitation. Boltzmann's constant is κ, the absolute temperature is T, the resistance is R, and the bandwidth over which the noise occurs is Δf or BW.

Environmental factors influence the accuracy of instrumentation systems. Stray magnetic fields can combine with instruments' magnetic fields to alter the output. Ambient temperature differences between readings can result in differences caused by mechanical changes in the instrumentation or changes in resistance in the circuitry. Electrostatic charges influence the position of indicating needles. For example, the rubbing of an instrument face can move a needle with constant sensing input. Mechanical shock, vibration, and even mounting position influence the output of instrumentation and must be accounted for or protected against in order to obtain accurate indications.

Most measurement circuits are not exposed to the full voltage or current of the system monitored. The measurement circuits are isolated from the system by *instrument transformers*. An instrument transformer extends the range measured by a given meter, isolates the meter from the line, and allows the instrument circuit to be grounded. When the primary winding is in series with the line, the instrument transformer is called a *current transformer*. When the primary winding is in parallel with the line, the instrument transformer is called a *potential transformer*. Instrument transformers are tested in the same manner as power transformers.

Example 36.1

What is the expected thermal agitation noise at 20°C in a digital voltmeter using a 10 kΩ resistor and designed to operate within the frequency range of 0 Hz to 1000 Hz?

Solution

Ignoring any other resistance in the voltmeter, which will be small compared to the 10 kΩ resistance, and noting that 20°C is equal to 293K, the Johnson noise voltage is

$$\begin{aligned} V_{\text{noise}} &= \sqrt{4\kappa TR\Delta f} \\ &= \sqrt{\begin{array}{c}(4)\left(1.3807\times 10^{-23}\ \frac{\text{J}}{\text{K}}\right)(293\text{K}) \\ \times\ (10\times 10^{3}\ \Omega)(1000\ \text{Hz})\end{array}} \\ &= 4.02\times 10^{-7}\ \text{V} \end{aligned}$$

Such a small voltage level is of significance in only the most sensitive of measurements.

2. SIGNAL REPRESENTATION

DC signals are represented by their constant values. AC signals can be represented in a number of ways. The average value of a generic AC signal is

$$f_{\text{ave}} = \frac{1}{T}\int_{t_1}^{t_1+T} f(t)\,dt \qquad 36.2$$

The average value of a sinusoidal function is zero, so a separate method of measuring AC sinusoidal signals is required in order to use their values in electrical calculations. The root-mean-square (rms) is that method. The rms or effective value of a generic function is

$$f_{\text{rms}} = \sqrt{\frac{1}{T}\int_{t_1}^{t_1+T} f^2(t)\,dt} \qquad 36.3$$

Because nearly all periodic functions can be represented by a Fourier series, many important nonsinusoidal signals are changed into such series for further manipulation. A Fourier series is given by

$$\begin{aligned} f(t) = \tfrac{1}{2}a_0 &+ \sum_{n=1}^{\infty} a_n \cos\left(\frac{2\pi n}{T}\right)t \\ &+ \sum_{n=1}^{\infty} b_n \sin\left(\frac{2\pi n}{T}\right)t \end{aligned} \qquad 36.4$$

The average value of a function given by a Fourier series is

$$f_{\text{ave}} = \tfrac{1}{2}a_0 \qquad 36.5$$

The rms value of a function given by a Fourier series is

$$f_{\text{rms}} = \sqrt{\left(\tfrac{1}{2}a_0\right)^2 + \tfrac{1}{2}\sum_{n=1}^{\infty}(a_n^2 + b_n^2)} \qquad 36.6$$

For a sinusoid, the average and rms values are given in Eq. 36.7 through Eq. 36.9. (Equation 36.7 is the average of a rectified sinusoid.)

$$V_{\text{ave}} = \frac{2}{\pi}V_p \qquad 36.7$$

$$V_{\text{rms}} = \frac{1}{\sqrt{2}}V_p \qquad 36.8$$

$$V_{\text{rms}} = 1.11\,V_{\text{ave}} \qquad 36.9$$

[2]Random disturbances in mechanical systems are explained by the mechanism of *Brownian motion*. In magnetic systems, the *Barkhausen effect* explains such disturbances.

Example 36.2

A voltage waveform is represented by the truncated Fourier series,

$$v(t) = 5 + 3\cos 100t + 0.8\cos 300t + 4\sin 100t + 0.5\sin 200t$$

Determine the average value of this voltage.

Solution

The average value is the zero frequency term as determined by Eq. 36.5.

$$a_0 = (2)(5 \text{ V}) = 10 \text{ V}$$
$$f_{\text{ave}} = \tfrac{1}{2}a_0 = \left(\tfrac{1}{2}\right)(10 \text{ V}) = 5 \text{ V}$$

Example 36.3

For the voltage waveform given in Ex. 36.2, what is the rms voltage?

Solution

The rms value is given by Eq. 36.6.

$$f_{\text{rms}} = \sqrt{\left(\tfrac{1}{2}a_0\right)^2 + \tfrac{1}{2}\sum_{n=1}^{\infty}(a_n^2 + b_n^2)}$$
$$= \sqrt{(5.0)^2 + \left(\frac{1}{2}\right)\left((3)^2 + (0.8)^2 + (4)^2 + (0.5)^2\right)}$$
$$= 6.16 \text{ V}$$

3. MEASUREMENT CIRCUIT TYPES

Electrical parameters can be measured with *electromechanical circuits.* Such instruments use the measured parameter to generate a magnetic field that interacts with a separate magnetic field to produce the rotation of a restrained needle. Because of mechanical inertia, the instruments do not function for AC in the range of 5 Hz to 20 Hz. Because of inductive effects, the instruments do not function well above a maximum of 100 Hz to 2500 Hz, depending on the design. Because of these restrictions, the electromechanical instruments are primarily used in DC systems.

Substitution circuits compare an unknown electrical quantity with a reference. *Bridge circuits* are a simple means of determining impedance by such substitution. A *Wheatstone bridge* is shown in Fig. 36.1.

Figure 36.1 *Wheatstone Bridge*

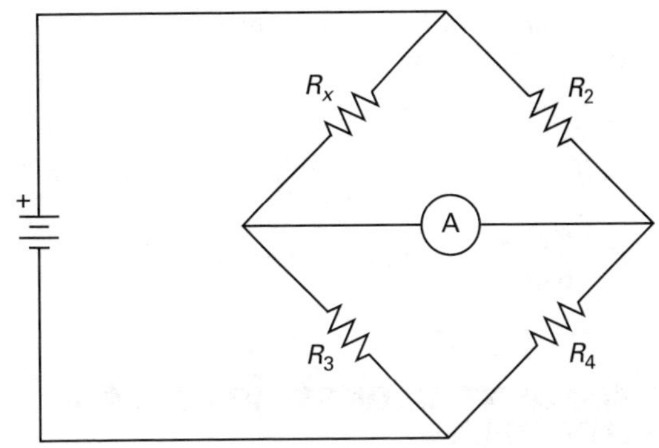

Balance exists in the bridge when no current flows through the ammeter. When balanced, the relationship between the resistors is

$$\frac{R_x}{R_3} = \frac{R_2}{R_4} \qquad \text{36.10}$$

Typically, R_3 and R_4 are fixed resistors that can be selected to vary their ratio from 10^{-3} to 10^3. Resistor R_2 is adjustable in steps and used to balance the bridge with the unknown resistor R_x. When balance is achieved, the unknown resistance is given by Eq. 36.10. Many other types of bridges are used to measure such things as inductance, capacitance, frequency, angles, and the like.[3]

Analog circuits use the mathematical relationships among electrical parameters to determine the desired parameters. Designs vary, but the instruments are used to determine such parameters as impedance, quality factor of coils (Q), the coefficient of coupling (k), magnetic parameters (H and B), frequency (f), and the like.

Digital circuits use digital processing of analog signals and analog-to-digital converters to display the results.[4] Digital instruments offer ease of use, quick response, and accurate results.

Transducer circuits are used in conjunction with transducers themselves to provide a signal of sufficient strength for use in follow-on systems.[5]

Example 36.4

A Wheatstone bridge with an R_3/R_4 ratio of 500 is balanced when the adjustable resistor has a value of 10 Ω. What is the unknown resistance?

[3] Some of the named bridges are the Maxwell-Wien, Anderson, Owen, Hay, Campbell, and Schering.
[4] The digital processing is often accomplished with the use of operational amplifiers, which are covered in Chap. 40.
[5] Chapter 19 provides background on transducer principles.

Solution

At a balanced condition, the resistance is given by Eq. 36.10.

$$\frac{R_x}{R_3} = \frac{R_2}{R_4}$$

$$R_x = \left(\frac{R_3}{R_4}\right)R_2 = (500)(10\ \Omega) = 5000\ \Omega \quad (5\ \text{k}\Omega)$$

4. PERMANENT MAGNET MOVING-COIL INSTRUMENTS

A permanent magnet moving-coil instrument is shown in Fig. 36.2. Such an instrument is also called a *d'Arsonval movement* or a *galvanometer*. A current is passed through the coil on the rotor. The rotor is located within a magnetic field created by the permanent magnets. Because a current-carrying conductor within a magnetic field will experience a torque, the coil rotates. The torque produced is proportional to the current through the coil and the number of turns of wire in the coil ($T_{\text{coil}} \propto IN$). The spring is restrained in such a manner as to provide an opposing torque proportional to the angular displacement ($T_{\text{spring}} \propto \theta$). The pointer comes to rest when the torques are equal in magnitude.

Figure 36.2 *Permanent Magnet Moving-Coil Mechanism (d'Arsonval)*

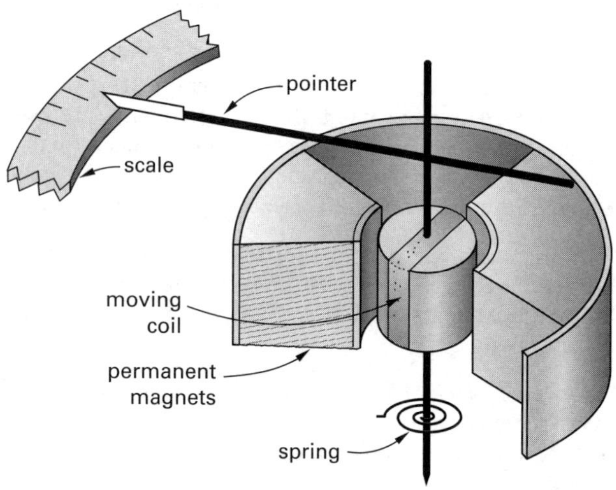

The current necessary to cause full-scale deflection of a specific instrument is given the symbol I_{fs}. The usual range of currents is 50 μA to 10 mA. In ammeter applications, the voltage at which full-scale current flow occurs is usually in the range of voltages from 50 mV to 100 mV, with a nominal value of 50 mV used as the standard. Because this is usually more than required by the coil for I_{fs} to flow, a *swamping resistor* is placed in series with the coil.

Because the permanent magnet's field is unidirectional, a change in the direction of the current in the coil reverses the torque, so the unmodified instrument is useful for detecting DC quantities only.[6] It can be made to detect AC quantities with the addition of a rectifier circuit on the coil input. When measuring AC rectified quantities, the d'Arsonval movement responds to the average voltage or current. In view of Eq. 36.9, adjusting the scale by a factor of 1.11 allows the rms value to be read directly.

The d'Arsonval movement can be designed to detect AC quantities with the use of a thermocouple. The hot junction of a thermocouple is placed in thermal, but not electrical, contact with a resistance carrying the current to be measured. The cold junction is connected to a heat sink that maintains ambient temperature. The measured current heats the wire via the resistance and subsequently heats one end of the thermocouple. A potential difference is developed between the ends of the dissimilar metals of the thermocouple proportional to the temperature difference ($I \propto \Delta T$), which is fed as input to the d'Arsonval meter movement. Because the heating effect is proportional to the square of the current, using a square-root factor on the scale output allows the rms value to be read directly. Such a *thermoelement instrument* can function up to several hundred megahertz.

5. DC VOLTMETERS

A d'Arsonval meter movement configured to perform as a DC voltmeter is shown in Fig. 36.3.

Figure 36.3 *DC Voltmeter*

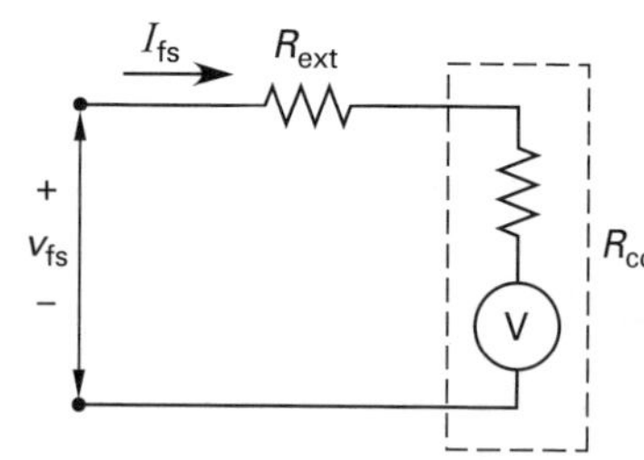

The external resistance is used to limit the current to the full-scale value, I_{fs}, at the desired full-scale voltage, V_{fs}. The electrical relationships in the voltmeter are given by Eq. 36.11.

$$\frac{1}{I_{\text{fs}}} = \frac{R_{\text{ext}} + R_{\text{coil}}}{V_{\text{fs}}} \qquad 36.11$$

The quantity $1/I_{\text{fs}}$ is fixed for a given instrument and is called the *sensitivity*. The sensitivity is measured in ohms per volt (Ω/V).

[6]Specially designed d'Arsonval movements can respond to currents up to 100 Hz.

Example 36.5

Common DC voltmeters have a sensitivity of 1000 Ω/V. Given such a meter with a coil resistance of 200 Ω and a 1% accuracy, determine the external resistance necessary to use a full-scale voltage of 10 V.

Solution

The external resistance is found from Eq. 36.11 by noting that the sensitivity is $1/I_{fs}$.

$$\frac{1}{I_{fs}} = \frac{R_{ext} + R_{coil}}{V_{fs}} = 1000\ \Omega/\text{V}$$

$$\begin{aligned} R_{ext} &= V_{fs}\left(1000\ \frac{\Omega}{\text{V}}\right) - R_{coil} \\ &= (10\ \text{V})\left(1000\ \frac{\Omega}{\text{V}}\right) - 200\ \Omega \\ &= 9800\ \Omega \quad (9.8\ \text{k}\Omega) \end{aligned}$$

Example 36.6

An overall accuracy of 2% is desired for the meter in Ex. 36.5. What is the maximum error possible for the external resistance?

Solution

The maximum error introduced by the external resistance will occur at full-scale voltage. The maximum error from the movement itself is given at the full-scale voltage as 1%, regardless of where the error actually occurs. The external resistance must be a 9.8 kΩ resistor with a 1% accuracy.

Example 36.7

The manufacturer's data for a DC voltmeter lists the sensitivity as 20,000 Ω/V. What current is required to deflect the movement to the 100% value?

Solution

The 100% value is the full-scale deflection. The full-scale deflection current is found from the definition of the sensitivity.

$$\text{sensitivity} = \frac{1}{I_{fs}}$$

$$I_{fs} = \frac{1}{20{,}000\ \frac{\Omega}{\text{V}}} = 5 \times 10^{-6}\ \text{A} \quad (50\ \mu\text{A})$$

6. DC AMMETERS

A d'Arsonval meter movement configured to perform as a DC ammeter is shown in Fig. 36.4.

The swamping resistance is used to limit the current to the full-scale value, I_{fs}, at the desired full-scale voltage, V_{fs}. For ammeters, the standard full-scale voltage is 50 mV.[7] The electrical relationships in the ammeter are given by Eq. 36.12.

Figure 36.4 DC Ammeter

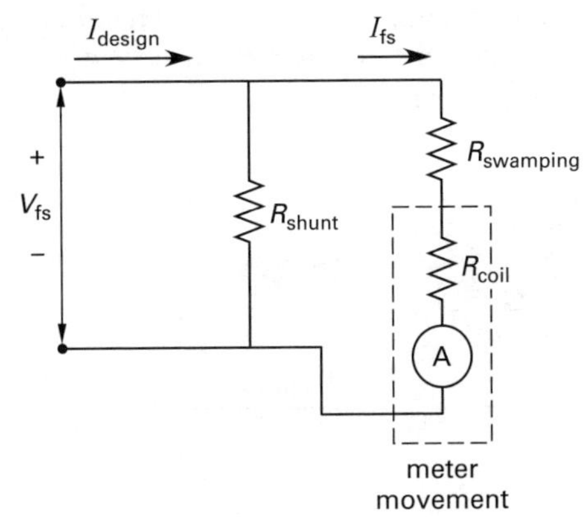

$$I_{design} = \frac{V_{fs}}{R_{shunt}} + I_{fs} \qquad 36.12$$

Example 36.8

A d'Arsonval movement is used as an ammeter with coil and swamping resistors totaling 200 Ω. The full-scale current value, I_{fs}, is 0.5 μA. If no shunt resistor is used, what percentage of the full-scale current will flow through the meter movement?

Solution

Without a shunt resistor, and noting that the standard full-scale standard voltage is 50 mV, the full-scale current is

$$\begin{aligned} I_{fs} &= \frac{V_{fs}}{R_{swamping} + R_{coil}} \\ &= \frac{50 \times 10^{-3}\ \text{V}}{200\ \Omega} = 2.5 \times 10^{-4}\ \text{A} \end{aligned}$$

The actual full-scale current should be 0.5 μA. The percentage of full-scale current flowing without the shunt resistor is

$$\%I = \left(\frac{2.5 \times 10^{-4}\ \text{A}}{0.5 \times 10^{-6}\ \text{A}}\right) \times 100\% = 50{,}000\%$$

If the full-scale current had not been specified as 0.5 μA, the calculated current of 2.5×10^{-4} A would have been the correct value for proper operation.

[7]The standard ammeter is designed to withstand a 50 mV voltage across the shunt with I_{fs} flowing in the movement at the desired design current flow.

Measurement/ Instrumentation

Example 36.9

What value of shunt resistance is required for the ammeter in Ex. 36.8 to function properly with a 1 A full-scale deflection?

Solution

Use Eq. 36.12.

$$I_{\text{design}} = \frac{V_{\text{fs}}}{R_{\text{shunt}}} + I_{\text{fs}}$$

$$R_{\text{shunt}} = \frac{V_{\text{fs}}}{I_{\text{design}} - I_{\text{fs}}} = \frac{50 \times 10^{-3}\ \text{V}}{1\ \text{A} - (0.5 \times 10^{-6}\ \text{A})} = 0.05\ \Omega$$

7. MOVING-IRON INSTRUMENTS

A *moving-iron instrument* is shown in Fig. 36.5. The instrument shown is called a *radial-vane mechanism*. A current is passed through the coil. A stationary vane and a moving vane are located within the resulting magnetic field and are similarly magnetized. Because the induced magnetic field is the same for both vanes, they repel one another. The moving vane is attached to a pointer and restraining spring. The angular displacement of the moving vane is proportional to the square of the current and the number of turns of wire in the coil ($T_{\text{coil}} \propto I^2N$). The spring is restrained in such a manner as to provide an opposing torque proportional to the angular displacement ($T_{\text{spring}} \propto \theta$). The pointer comes to rest when the torques are equal in magnitude.

Figure 36.5 *Radial-vane Mechanism*

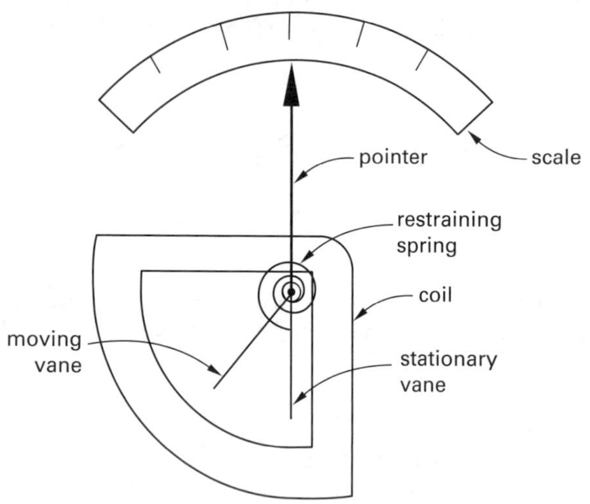

The vanes repel one another regardless of the polarity of the current or its changing value. As a result, the instrument can be used to measure AC quantities. Because the torque is proportional to the square of the current, using a square-root factor on the scale output allows the rms value to be read directly. The instrument can measure AC quantities up to approximately 500 Hz. With special compensation for eddy currents and skin effects, this range can be extended to 10 kHz.

Most moving-iron instruments can also measure DC quantities. They are less useful for DC quantities because significant remnant magnetization occurs over time.

Moving-iron instruments can be used for voltmeters or ammeters using setups similar to those of d'Arsonval instruments. The sensitivity and power consumption of moving-iron instruments is between that of d'Arsonval movements and electrodynamometer movements.

8. ELECTRODYNAMOMETER INSTRUMENTS

An *electrodynamometer instrument* is shown in Fig. 36.6(a). A current is passed through a moving coil and two stationary coils. The magnetic field of the moving coil interacts with the magnetic fields of the stationary coils to produce a torque. The torque produced is proportional to the current and the number of turns of wire in the moving coil as well as the current and number of turns in the stationary coils ($T \propto I^2N$). The spring is restrained in a manner that provides an opposing torque proportional to the angular displacement ($T_{\text{spring}} \propto \theta$). The pointer comes to rest when the torques are equal in magnitude.

Figure 36.6 *Electrodynamometer*

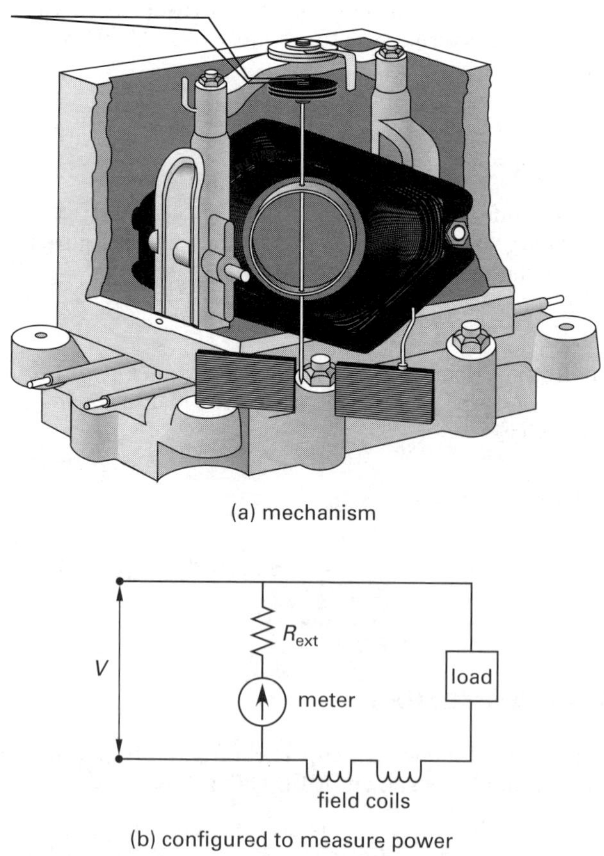

Because the torque is proportional to the square of the current, using a square-root factor on the scale output allows the rms value to be read directly.[8] The instrument can measure AC quantities in the range of 25–2500 Hz. The lower limit is caused by the instrument's attempt to follow the instantaneous values of the AC quantities. The upper limit is caused by inductive reactance.

Though currents and voltages can be measured, the dynamometer is most commonly used to measure power. When the moving coil is placed across a load, the current in the moving coil will be proportional to the voltage across the load. When the stationary coils are placed in series with the load, the current in the stationary coils will be proportional to the current in the load. The combined magnetic field is then proportional to the instantaneous product of the voltage and the current—that is, the power. One possible configuration for measuring power is shown in Fig. 36.6(b). Corrections for the power consumption of the coils are required to ensure an accurate result.

The electrodynamometer can also be used to measure DC quantities because average DC values are equivalent to rms AC values.

The instruments can be used for voltmeters or ammeters using setups similar to those of d'Arsonval instruments. The sensitivity is higher than either d'Arsonval or moving-iron instruments, but the power consumption is higher as well.

9. POWER MEASUREMENTS

Many different types of instruments are used to measure power. All require compensation for the power they consume, inductance of the measuring coils, and phase angle corrections for any instrument transformers used. Three-phase systems require a minimum of two wattmeters to measure power. The power meter most commonly used in residential applications is the induction-type watt-hour meter shown in Fig. 36.7.

The parallel coil voltage, V, produces a magnetic flux at its pole tip that lags the applied voltage by nearly 90°, because of its high inductance and the large number of turns used in the coil. The fluxes in the series coils, I_1 and I_2, are nearly in phase with the current. All three poles induce currents in a circular disk located between the poles.[9] A torque, proportional to the power, is generated in the disk and causes it to rotate.[10] The disk speed is made proportional to the driving torque by a braking torque that is also proportional to the disk speed. This breaking torque (countertorque) arises from the interaction of the rotating disk with a permanent magnetic field (not shown) that induces eddy currents in the disk. The compensating winding, C, with adjustable resistor, R, is used to ensure that the flux from the parallel pole, V, is exactly 90° out of phase. The net result is a rotating disk speed proportional to the power delivered to the load.

Figure 36.7 Induction-Type Watt-Hour Meter

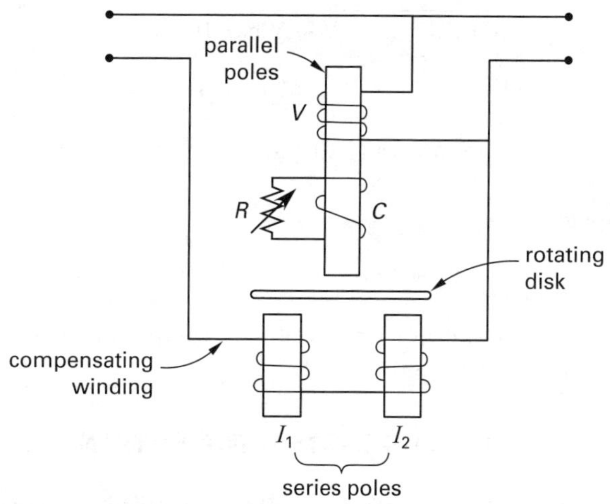

10. POWER FACTOR MEASUREMENTS

The power factor, pf, is given by the ratio of the true power to the apparent power.

$$\text{pf} = \frac{P}{S} \qquad 36.13$$

For a balanced three-phase, three-wire system, results obtained using the *two-wattmeter method* can be used to determine the angle between the current and the voltage, and therefore the power factor. Let P_H be the larger power, and let P_L be the power measured by the other wattmeter. The desired angle is determined from

$$\tan\phi = \frac{\sqrt{3}(P_H - P_L)}{P_H + P_L} \qquad 36.14$$

Power factor meters can also be constructed from electrodynamometers with two moving coils. Both coils carry the series current of the load, but one does so in series with a resistor and the other does so in series with an inductor. Consequently, the two coils' currents are nearly in quadrature, that is, 90° apart. At unity power factor, the torque on the coil in series with the resistor will be at a maximum. At zero power factor, the torque on the coil in series with the inductor will be at a maximum. The angular displacement of the needle measures the power factor of the load.

Example 36.10

Using the two-wattmeter method, the power of a three-phase circuit is measured as 1074 kW (1074×10^3 W) and 426 kW (426×10^3 W). What is the power factor?

[8]This is true when the same current is flowing in all coils. That is, the device is configured so that the coils are in series.

[9]This is the same principle as that of the induction motor.

[10]The torque is proportional to the power only if the fluxes between the series coils and the parallel coil are in quadrature at unity power factor.

Solution

The angle necessary to determine the power factor comes from Eq. 36.14.

$$\tan\phi = \frac{\sqrt{3}(P_H - P_L)}{P_H + P_L}$$
$$= \frac{\sqrt{3}(1074 \times 10^3 \text{ W} - 426 \times 10^3 \text{ W})}{1074 \times 10^3 \text{ W} + 426 \times 10^3 \text{ W}} = 0.748$$

The angle between the current and the voltage is

$$\phi = \tan^{-1} 0.748 = 36.8°$$

The power factor is

$$\text{pf} = \cos\phi = \cos 36.8° = 0.8$$

Given the nature of most systems, the power factor is 0.8 lagging.

11. ELECTRONIC INSTRUMENTATION

The advent of improved electronics has advanced instrumentation accuracy and rapidity. Digital instruments combine the analog circuit and sensor outputs with digital processing. Combinations of the instruments previously discussed can be used to create a *multimeter*, a device for measuring multiple electric quantities. When such an instrument remains in the analog realm, it is called a VOM meter (volt-ohm-milliamp meter). When the parameter being sensed is the primary analog portion of the circuit and the remainder is digital, the instrument is called a *digital multimeter, DMM* (or *DVOM*).

Such DMMs, especially when combined with a microprocessor, provide the following advantages and properties. Because they have a single digital readout, parallax errors are eliminated. Circuitry can be added that allows auto-ranging, a feature that searches for and selects the correct range for the parameter being measured. Autopolarity is a feature that can display whether a voltage agrees or disagrees with the red (positive) and black (negative) leads. Sample and hold features circuitry that captures the reading and continues to display it after the leads are removed from the *element under test* (*EUT*). Sample and hold circuitry is also used to digitize a waveform so the digital form can be stored. The microprocessor can be programmed to provide for self-calibration, data conversion, data processing, linearization, and process monitoring and control. Meters can also have interfaces to personal computers to allow for automatic downloading and extraction of monitored parameters. Such interfaces include IrDA (Infrared Data Association), RS-232 (Recommended Standard 232), USB (Universal Serial Bus), and IEEE-488 (commonly called the GPIB or General Purpose Interface Bus).

A Texas Instruments block diagram of a digital multimeter is shown in Fig. 36.8. A Keithley Instruments block diagram is shown in Fig. 36.9. A highly accurate digital electrometer using a microcomputer is shown in Fig. 36.10.

The operation principles of the individual electronic components shown in Fig. 36.8, Fig. 36.9, and Fig. 36.10 are discussed in detail in Chap. 40.

12. INSULATION AND GROUND TESTING

Insulation testing is conducted to ensure equipment can operate properly and personnel are protected. There are a variety of standards, such as those of IEEE or NEMA, for insulation testing. A successful insulation test will yield a high resistance value, and generally, 1 MΩ indicates satisfactory resistance. Figure 36.11 illustrates a simple, low-voltage insulation test adequate for most checks. Using a digital multimeter, the leads are attached as shown. Care must be taken not to penetrate the insulation with the positive lead, and the resistance must be checked.

If the resistance is expected to be above 1 MΩ, a more accurate insulation test will be required, and can be accomplished by raising the voltage at which the test occurs. The voltage can be increased using a *megger*. (See Fig. 36.12.) The megger operates at a higher voltage than the digital multimeter and provides more accurate resistance readings. A hand-crank generator in simpler models powers the circuit. The generator provides the power to coils A and B, which are situated on a rotor, C, all of which can pivot in response to the interaction of the magnetic fields of the coils and the permanent magnet, M. When the circuit leads, labeled earth and line, are not connected and the hand-crank is turned, current flows through coil B, generating a magnetic field. This magnetic field aligns itself with the permanent magnetic field of M, pivoting the rotor counterclockwise such that the pointer points to infinity. When a circuit is under test, the earth and line connections are attached to the wire and insulation. As the generator turns, current flows through both coils A and B. Coil A is a deflecting coil and generates a torque that turns the rotor clockwise, causing the needle to point to a position that represents equal torques between coils A and B. Since the current and the torque generated by coil A are dependent on the external resistance of the insulation under test, this provides a reading of that resistance.

Hand-crank meggers operate on the order of 500 V. They often come equipped with a slip clutch that enables the generator to turn at a constant speed, providing consistent readings. A circuit under test should not be exposed to voltage levels that can damage the insulation. Depending on the standard and the test, this voltage level varies from one to three times the rated voltage.

If increased accuracy is desired, electronic meggers can be used. Such a megger is shown in Fig. 36.13. This is called the *leakage method* or *loss of charge method*. *R* represents the insulation resistance to be checked. The battery is the power supply and can be set to whatever voltage the

Figure 36.8 *Digital Multimeter Overview*

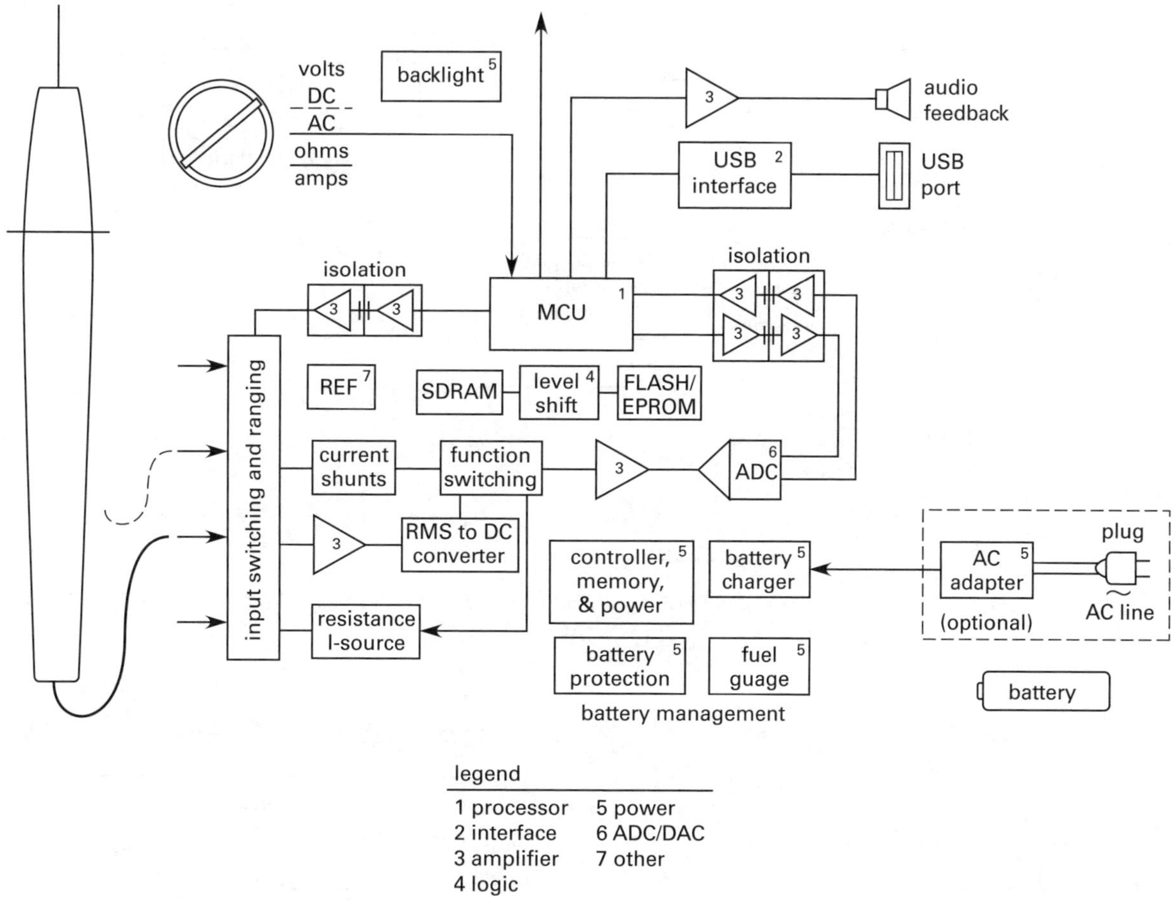

Reprinted with permission from http://www.ti.com, copyright © 1995-2010 by Texas Instruments.

Figure 36.9 *Keithley Instruments Block Diagram*

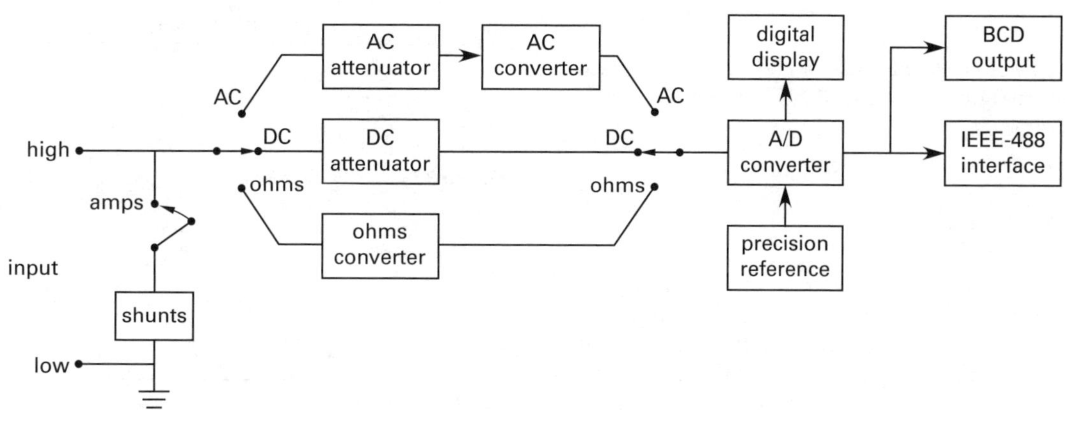

Reprinted with permission from *Electronics Engineers' Handbook*, Fourth Edition, by Donald Christiansen et al., copyright © 1997 by McGraw-Hill Companies, Inc.

meter is capable of generating. When the resistance is placed under test, switch A closes long enough to charge the capacitor, C, and then reopens. Switch B then immediately closes and the deflection, d_1, of the meter is noted. The process is repeated, allowing a period of time to pass before switch B is closed. The deflection, d_2, is then noted. The deflections differ by an amount equal to the charge loss on the capacitor due to the current flow

Figure 36.10 *Typical Digital Electrometer*

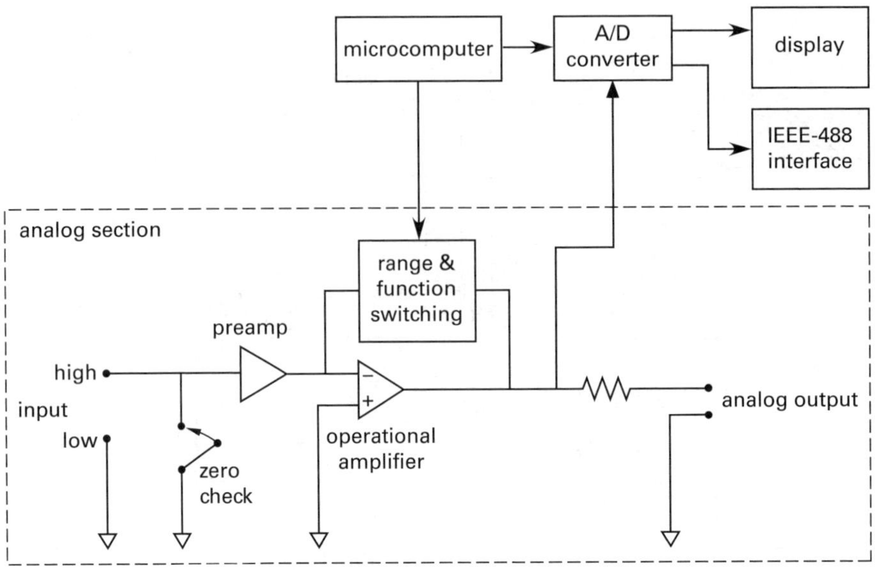

Figure 36.11 *Insulation Testing*

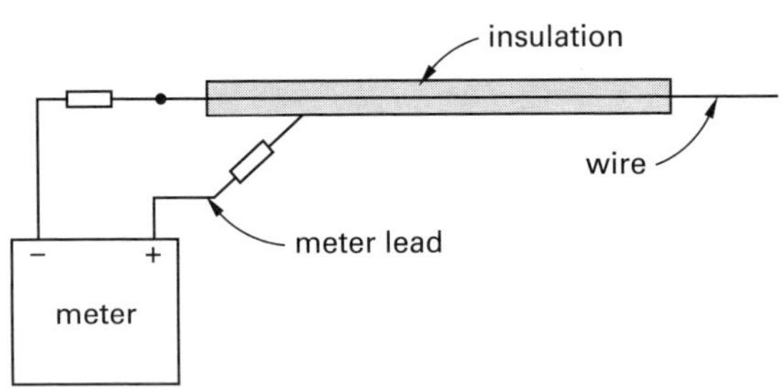

through the resistance, R. The loss can be used to calculate the insulation resistance per Eq. 36.15, with the capacitance in microfarads.

$$R = \frac{t}{2.303C\log_{10}\left(\frac{d_1}{d_2}\right)} \quad \text{[in the units of M}\Omega\text{]} \qquad 36.15$$

Ground testing indicates testing to determine if a conducting connection exists between an electrical circuit or equipment and the ground or some conducting body. If the connection is desired, as in a safety ground, then the result desired is a low resistance connection. If the connection is to have a specific value, as in impedance connections made on wye generators or transformers, then the desired value is the result. If a ground connection is not meant to exist, then a high value is desired. Standards exist for ground impedance values, and for checking said values. For building structures, such guidance can be found in the *National Electrical Code*, which is covered in Chap. 53.

Figure 36.12 *Simplified Megger Schematic*

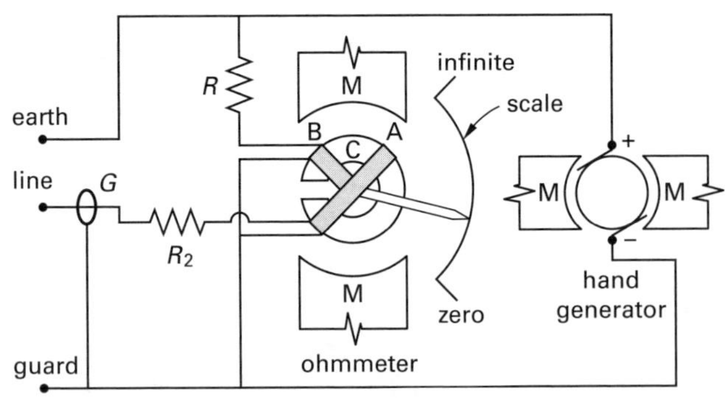

Figure 36.13 *Leakage Method for Measuring Insulation Resistance*

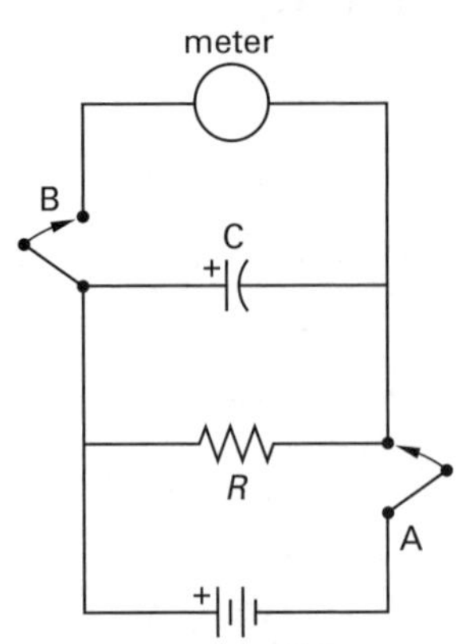

Measurement/Instrumentation

Topic IX: Electronics

Chapter

37. Electronics Fundamentals
38. Junction Transistors
39. Field Effect Transistors
40. Electrical and Electronic Devices

37 Electronics Fundamentals

1. Overview 37-1
2. Semiconductor Materials 37-2
3. Device Performance Characteristics 37-4
4. Bias 37-4
5. Amplifiers 37-4
6. Amplifier Classification 37-4
7. Load Line and Quiescent Point Concept . . . 37-5
8. *pn* Junctions 37-7
9. Diode Performance Characteristics 37-9
10. Diode Load Line 37-11
11. Diode Piecewise Linear Model 37-11
12. Diode Applications and Circuits 37-12
13. Schottky Diodes 37-12
14. Zener Diodes 37-13
15. Tunnel Diodes 37-13
16. Photodiodes and Light-Emitting Diodes . . . 37-14
17. Silicon-Controlled Rectifiers 37-14

Nomenclature

A	constant	$K^{-3}\cdot m^{-3}$
A	gain	–
c	speed of light, 2.9979×10^8	m/s
C	capacitance	F
D	diffusion constant	m^2/s
E	electric field strength	V/m
E	energy	J or eV
h	Planck's constant, 6.6256×10^{-34}	J·s
i	instantaneous or variable current	A
I	effective or DC current	A
n	electron concentration	cm^{-3}
n, N	concentration	m^{-3}
p	hole concentration	cm^{-3}
q	electric unit charge, 1.6022×10^{-19}	C
r, R	resistance	Ω
T	temperature	K
v	instantaneous or variable voltage	V
V	effective or DC voltage	V

Symbols

β	CE forward current ratio or gain	–
η	empirical constant	–
κ	Boltzmann's constant, 1.3807×10^{-23}	J/K
λ	wavelength	m
μ	mobility	$m^2/V\cdot s$
σ	conductivity	S/m

Subscripts

0	at 0K
0	barrier or source value
ac	alternating current or small signal
A	acceptor
be, BE	base to emitter
B	DC or biased base
BB	base supply or base biasing
BO	breakover
BQ	base to quiescent point
c	conduction band or collector
C	collector
CB	collector to base
CC	collector supply
CE	collector to emitter
CO	cutoff
d	diffusion
D	donor, diode, or drain
e, E	emitter
f	forward (AC or instantaneous)
F	forward (DC component) or total
G	gap or gate
H	hold
i	intrinsic or input (AC or instantaneous)
I	current
L	load
m	maximum
n	electrons or n-type
p	holes or p-type
pn	p to n
r	reverse (AC or instantaneous)
R	reverse (DC component or total)
s	saturation
S	source
t	transition
T	thermal
v	valence band
V	constant or effective voltage
Z	zener
ZK	zener knee

1. OVERVIEW

Electronics involves charge motion through materials other than metals, such as vacuums, gases, or semiconductors. The focus in this chapter will be on semiconductor materials. Because an understanding of the electron structure is vital to understanding electronics, a periodic table of the elements is given in App. 37.A.

An *electronic component* is one able to amplify, control, or switch voltages or currents without mechanical or other nonelectrical commands. The charge in metals is carried by the electron, with a charge of -1.6022×10^{-19} C. In semiconductor materials, the charge is carried both by the electron and by the absence of the electron in a covalent bond, which is referred to as a *hole*[1] with a charge of $+1.6022 \times 10^{-19}$ C. The concentration of electrons, n, and holes, p, is given by the *mass action law*, Eq. 37.1.

$$n_i^2 = np \qquad 37.1$$

The term n_i is the concentration of carriers in a pure (*intrinsic*) semiconductor.[2] In intrinsic semiconductor materials, the number of electrons equals the number of holes, and the mass action law is stated as in Eq. 37.1. Nevertheless, the mass action law applies for intrinsic and extrinsic semiconductor materials. *Extrinsic* semiconductor materials are those that have had impurities deliberately added to modify their properties, normally their conductivity. For extrinsic semiconductors, the mass action law is probably better understood as indicating that the product np remains constant regardless of position in the semiconductor or doping level. Carriers, either n or p, in a semiconductor are constantly generated by thermal creation of electron-hole pairs and constantly eliminated by recombination of electron-hole pairs. The carriers generated are caused to diffuse by concentration gradients in the semiconductor, a phenomenon that does not occur in metals.

Semiconductor devices are inherently nonlinear. Nevertheless, they are commonly analyzed over ranges in which their behavior is approximately linear. Such an analysis is called *small-signal analysis*. The DC or effective value of the electrical parameters in the models used for analysis will be represented by uppercase letters. AC or instantaneous values will be represented by lowercase letters. Equivalent parameters in the models will use lowercase letters. For convenience, the lowercase letter t is omitted from functions of time. For example, $v(t)$ is written simply as v. The subscripts on currents indicate the terminals into which current flows. Subscripts on voltages indicate the terminals across which the voltage appears. Subscripts indicating biasing voltages are capitalized. For example, $V_{\rm BB}$ is the base biasing voltage while $v_{\rm be}$ is the instantaneous voltage signal applied to the base-emitter junction. A list of common designations is given in App. 37.B. *Ultra large-scale integration* (ULSI) is a term used to describe circuits with 1,000,000 or more components per chip.

[1]The concept of holes explains the conduction of electricity without free electrons. The hole is considered to behave as a free positive charge—quantum mechanics justifies such an interpretation. (The *Hall voltage* is experimental confirmation.) The calculation of total charge motion in semiconductors is simplified as a result. The same equations used for electron movement can be used for hole movement with a change of sign and a change of values for some terms, such as mobility.

[2]Intrinsic means natural. Pure silicon is silicon in its natural state, with no doping, though it will have naturally occurring impurities. The terms pure and intrinsic are used interchangeably.

Electronic components are often connected in an array known as an *integrated circuit* (IC). An integrated circuit is a collection of active and passive components on a single semiconductor substrate (*chip*) that function as a complete electronic circuit. *Small-scale integration* (SSI) involves the use of less than 100 components per chip. *Medium-scale integration* (MSI) involves between 100 and 999 components per chip. *Large-scale integration* involves between 1000 and 9999 components per chip. *Very large-scale integration* involves more than 10,000 components per chip.

The nonlinearity of electronic devices makes them attractive for use as amplifiers. An *amplifier* is a device capable of increasing the amplitude or power of a physical quantity without distorting the wave shape of the quantity. The amplifying properties of transistors are covered in this chapter. Topics peculiar to amplifiers and amplifier types are covered in Chap. 40 and Chap. 41. The application of the electronic components to a variety of circuits is extremely broad. A few examples are provided in the end sections of this chapter. Additionally, some electronic circuits are used often enough to be designed into *standard modular devices*, that is, off-the-shelf ICs with all the necessary components to accomplish a given circuit task. An understanding of the basic electrical and electronic principles for the components used in such devices will enable an understanding of the overall circuit functions of the standard modular devices and electrical/electronic circuits in general.

Example 37.1

Silicon has an intrinsic carrier concentration of 1.6×10^{10} cm^{-3}. What is the concentration of holes?

Solution

Regardless of whether the material is intrinsic or extrinsic, the law of mass action applies. However, in intrinsic materials, the concentration of electrons and the concentration of holes are equal. Equation 37.1 can be used as follows.

$$n_i^2 = np = p^2$$

$$n_i = p = 1.6 \times 10^{10}\ \text{cm}^{-3}$$

2. SEMICONDUCTOR MATERIALS

The most common semiconductor materials are silicon (Si) and germanium (Ge). Both are in group IVA of the periodic table and contain four valence electrons. A *valence electron* is one in the outermost shell of an atom. Both materials use covalent bonding to fill the outer shell of eight when they create a crystal lattice. Semiconductor materials ($N \approx 10^{10}$ to 10^{13} electrons/m^3) are slightly more conductive than insulators ($N \approx 10^7$ electrons/m^3) but less conductive than metals ($N \approx 10^{28}$ electrons/m^3). The conductivity, σ, of semiconductor materials can be made to vary from approximately 10^{-7} to 10^5 S/m.

The semiconductor crystal lattice has many defects (i.e., free electrons and their corresponding holes). The formation of free electrons and holes is driven by the temperature and is called *thermal carrier generation.* The density of these electron-hole pairs in intrinsic materials is given by

$$n_i^2 = A_0 T^3 e^{-E_{G0}/\kappa T} \qquad 37.2$$

The term A_0 is a constant independent of temperature and related to the density states at the bottom edge of the conduction band and the top edge of the valence band. T is the absolute temperature. E_{G0} is the energy gap at 0K, that is, the energy between the conduction band and the valence band. This is the energy required to break the covalent bond. The energy gap for silicon at 0K is approximately 1.21 eV. The energy gap for germanium is approximately 0.78 eV. The term κ is Boltzmann's constant, with values of 1.3807×10^{-23} J/K and 8.621×10^{-5} eV/K.[3] At a given temperature at thermal equilibrium, Eq. 37.2 becomes

$$n_i^2 = N_c N_v e^{-E_G/\kappa T} \qquad 37.3$$

The terms N_c and N_v are the effective density states in the conduction band and valence band, respectively. The energy gap for silicon at room temperature (300K) is 1.12 eV. The energy gap for germanium at room temperature is 0.80 eV.

When minor amounts of impurities called *dopants* are added, the materials are termed *extrinsic semiconductors.*[4] If the impurities added are from group IIIA, with three valence electrons, an additional hole is created in the lattice. These dopants are called *acceptors*, and semiconductors with such impurities are called *p-types.* The *majority carriers* in *p*-type semiconductors are holes, and the *minority carriers* are electrons. The majority of the charge movement takes place in the valence band. Typical dopants are indium (In) and gallium (Ga). If the impurities added are from group VA, with five valence electrons, an additional electron is provided to the lattice.[5] These dopants are called *donors*, and semiconductors with such impurities are called *n-types.* The majority carriers in *n*-type semiconductors are electrons, and the minority carriers are holes. The majority of charge movement takes place in the conduction band. Typical dopants are phosphorus (P), arsenic (As), and antimony (Sb).

The conductivity of the semiconductor is determined by the carriers. The mobility of electrons is higher than that of holes.[6] The total conductivity in any semiconductor is a combination of the movement of the electrons and holes and is given by

$$\sigma = q(n\mu_n + p\mu_p) \qquad 37.4$$

The law of mass action, Eq. 37.1, applies to extrinsic semiconductors. With the addition of dopants, the law of electrical neutrality given by Eq. 37.5 also applies. The concentration of acceptor atoms is N_A, each contributing one positive charge to the lattice. (Do not confuse N_A with Avogadro's number.) The concentration of donor atoms is N_D, each contributing one negative charge. Since neutrality is maintained, the result is[7]

$$N_A + n = N_D + p \qquad 37.5$$

In a *p*-type material, the concentration of donors is zero, ($N_D = 0$). Additionally, the concentration of holes is much greater than the number of electrons ($p \gg n$). For a *p*-type material, Eq. 37.5 can be rewritten as $N_A \approx p$. Using the law of mass action, Eq. 37.1, the concentration of electrons in a *p*-type material is

$$n = \frac{n_i^2}{N_A} \qquad 37.6$$

In an *n*-type material, the concentration of acceptors is zero ($N_A = 0$). Additionally, the concentration of electrons is much greater than the number of holes ($n \gg p$). For an *n*-type material, Eq. 37.5 can be rewritten as $N_D \approx n$. Using the law of mass action, Eq. 37.1, the concentration of holes in an *n*-type material is

$$p = \frac{n_i^2}{N_D} \qquad 37.7$$

Example 37.2

The energy gap for silicon at room temperature (300K) is 1.12 eV. The energy density states are $N_c = 2.8 \times 10^{19}$ cm^{-3} and $N_v = 1.02 \times 10^{19}$ cm^{-3}. What is the intrinsic carrier concentration?

Solution

The intrinsic carrier concentration, assuming thermal equilibrium, is given by Eq. 37.3.

$$\begin{aligned} n_i^2 &= N_c N_v e^{-E_G/\kappa T} \\ &= (2.8 \times 10^{19}\ \text{cm}^{-3}) \\ &\quad \times (1.02 \times 10^{19}\ \text{cm}^{-3}) e^{-1.12\,\text{eV}/(8.621\times 10^{-5}\text{eV/K})(300\text{K})} \\ &= 4.452 \times 10^{19}\ \text{cm}^{-6} \\ n_i &= \sqrt{4.452 \times 10^{19}\ \text{cm}^{-6}} = 6.67 \times 10^9\ \text{cm}^{-3} \\ &\approx 0.7 \times 10^{10}\ \text{cm}^{-3} \end{aligned}$$

This number differs slightly from the number determined experimentally and more exact calculations (1.6×10^{10} cm^{-3}).

[3]The symbology for Boltzmann's constant often varies with the units of the energy gap. If eV is used as the energy unit, Boltzmann's constant may be seen as either κ or $\overline{\kappa}$ to distinguish between eV/K and J/K.

[4]A minor amount of dopant material is on the order of 10 parts per billion.

[5]The additional electron exists because the outer shell octet is satisfied by the covalent bonding of the first four electrons in the impurity.

[6]The mobility of electrons is a factor of 2.5 higher in silicon and about 2.1 higher in germanium.

[7]Neutrality is maintained because each atom of an added impurity removes one intrinsic atom.

3. DEVICE PERFORMANCE CHARACTERISTICS

Electronic components function on some variation of *pn* junction principles. Amplifiers function on some variation of transistor principles, that is, *pnp* or *npn* junction principles. Most electronic components can be modeled as two-port devices with two variables—current and voltage—for each port. The relationship between the variables depends on the type of device. The relationship can be expressed mathematically, as for MOSFETs (*metal-oxide semiconductor field-effect transistors*), modeled in equivalent circuits, as with transistor *h*-parameters, or described graphically, as with BJT characteristic curves.

Semiconductor devices are inherently nonlinear. The performance of such devices is analyzed in a linear fashion over small portions of the characteristic curve(s). Such analysis is called *small-signal analysis.* A small signal is one that is much less than the average, that is, steady-state, value for the device, which usually is the biasing value. The models used in each linear portion of the characteristic curve are termed *piecewise linear models.* If the operation is outside the linear region or if the input signal is large compared to the average value, the device distorts the input signal. Such operation is called *nonlinear operation.*

The characteristic curve for an ideal transistor operating as a current-amplifying device is shown in Fig. 37.1. (The shape of the curve is the same for a diode, that is, a *pn* junction.)

Figure 37.1 *Typical Semiconductor Performance*

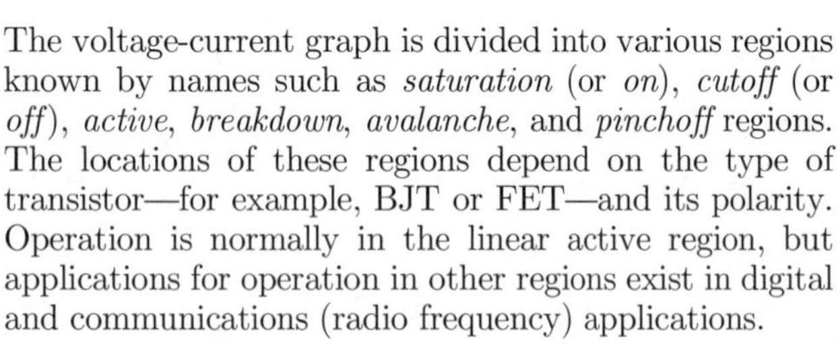

The voltage-current graph is divided into various regions known by names such as *saturation* (or *on*), *cutoff* (or *off*), *active*, *breakdown*, *avalanche*, and *pinchoff* regions. The locations of these regions depend on the type of transistor—for example, BJT or FET—and its polarity. Operation is normally in the linear active region, but applications for operation in other regions exist in digital and communications (radio frequency) applications.

4. BIAS

Bias is a DC voltage applied to a semiconductor junction to establish the operating point, also called the *quiescent* (*no-signal*) *point. Biasing* establishes the operating point with no input signal.[8] Biasing, then, is the process of establishing the DC voltages and currents (the bias) at the device's terminals when the input signal is zero (or nearly so).

For *pn* junctions, *forward bias* (or *on condition*) is the application of a positive voltage to the *p*-type material or, equivalently, the flow of current from the *p*-type to the *n*-type material. In a small semiconductor device, forward bias results in current in the milliampere range.

Reverse bias (or *off condition*) is the application of a negative voltage to the *p*-type material, or equivalently, the flow of current from the *n*-type to the *p*-type material. In a small semiconductor device, reverse bias results in current in the nanoampere range.

Self-biasing is the use of the amplifier's output voltage, rather than a separate power source, as the supply for the input bias voltage. This negative feedback control regulates the output current and voltage against variations in transistor parameters.

5. AMPLIFIERS

An *amplifier* produces an output signal from the input signal. The input and output signals can be either voltage or current. The output can be either smaller or larger (the usual case) than the input in magnitude. While most amplifiers merely scale the input voltage or current upward, the amplification process can include a sign change, a phase change, or a complete phase shift of 180°.[9] The ratio of the output to the input is known as the *gain* or *amplification factor*, *A*. A *voltage amplification factor*, A_V, and *current amplification factor*, A_I or β, can be calculated for an amplifier.

Figure 37.2 illustrates a simplified current amplifier with current amplification factor β. The additional current leaving the amplifier is provided by the bias battery, V_2.

$$i_{\text{out}} = \beta i_{\text{in}} \qquad 37.8$$

A capacitor, *C*, is placed in the output terminal to force all DC current to travel through the *load resistor*, R_L. Kirchhoff's voltage law for loop abcd is

$$V_2 = i_{\text{out}} R_L + V_{\text{ac}} = \beta i_{\text{in}} R_L + V_{\text{ac}} \qquad 37.9$$

If there is no input signal (i.e., $i_{\text{in}} = 0$), then $i_{\text{out}} = 0$ and the entire battery voltage appears across terminals ac ($V_{\text{ac}} = V_2$). If the voltage across terminals ac is zero, then the entire battery voltage appears across R_L so that $i_{\text{out}} = V_2/R_L$.

6. AMPLIFIER CLASSIFICATION

Amplifiers are classified on the basis of how much input is translated into output. A sinusoidal input signal is

[8]Bias is used both as a verb and a noun in electronics.

[9]An *inverting amplifier* is one for which $v_{\text{out}} = -A_V v_{\text{in}}$. For a sinusoidal input, this is equivalent to a phase shift of 180° (i.e., $V_{\text{out}} = A_V V_{\text{in}} \angle -180°$).

Figure 37.2 *General Amplifier*

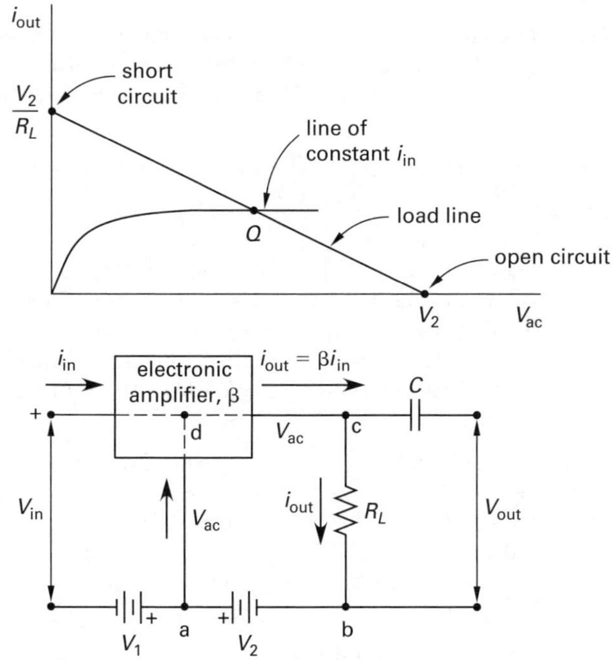

assumed. The output of an amplifier depends on the bias setting, which in turn establishes the quiescent point.

A *Class A amplifier*, as shown in Fig. 37.3, has a quiescent point in the center of the active region of the operating characteristics. Class A amplifiers have the greatest linearity and the least distortion. Load current flows throughout the full input signal cycle. Because the load resistance of a properly designed amplifier will equal the Thevenin equivalent source resistance, the maximum power conversion efficiency of an ideal Class A amplifier is 50%.

Figure 37.3 *Class A Amplifier*

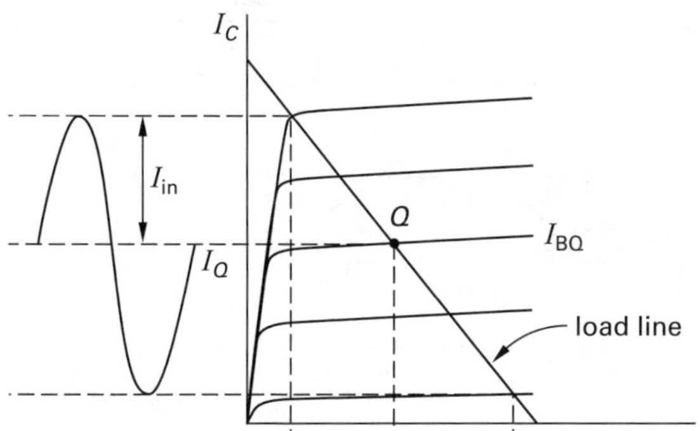

For *Class B amplifiers*, as shown in Fig. 37.4, the quiescent point is established at the cutoff point. A load current flows only if the signal drives the amplifier into its active region, and the circuit acts like an amplifying half-wave rectifier. Class B amplifiers are usually combined in pairs, each amplifying the signal in its respective half of the input cycle. This is known as *push-pull operation.* The output waveform will be sinusoidal except for the small amount of crossover distortion that occurs as the signal processing transfers from one amplifier to the other. The maximum power conversion efficiency, of an ideal Class B push-pull amplifier is approximately 78%.

Figure 37.4 *Class B Amplifier*

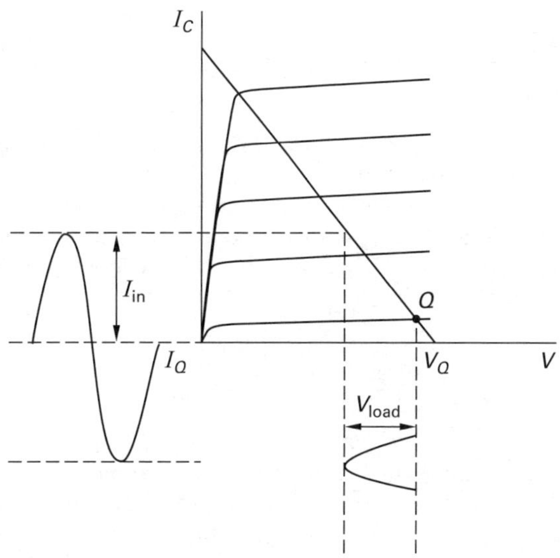

The intermediate *Class AB amplifier* has a quiescent point somewhat above cutoff but where a portion of the input signal still produces no load current. The output current flows for more than half of the input cycle. AB amplifiers are also used in push-pull circuits.

Class C amplifiers, as shown in Fig. 37.5, have quiescent points well into the cutoff region. Load current flows during less than one-half of the input cycle. For a purely resistive load, the output would be decidedly nonsinusoidal. However, if the input frequency is constant, as in radio frequency (rf) power circuits, the load can be a parallel LRC tank circuit tuned to be resonant at the signal frequency. The LRC circuit stores electrical energy, converting the output signal to a sinusoid. The power conversion effciency of an ideal Class C amplifier is 100%.

7. LOAD LINE AND QUIESCENT POINT CONCEPT

The i_{in}-v_{out} curves of Sec. 37.5 and Sec. 37.6 illustrate how amplification occurs. The two known points, (v_{out}, i_{out}) = (V_2, 0) and (v_{out}, i_{out}) = (0, V_2/R_L), are plotted

Figure 37.5 *Class C Amplifier (Resistive Load)*

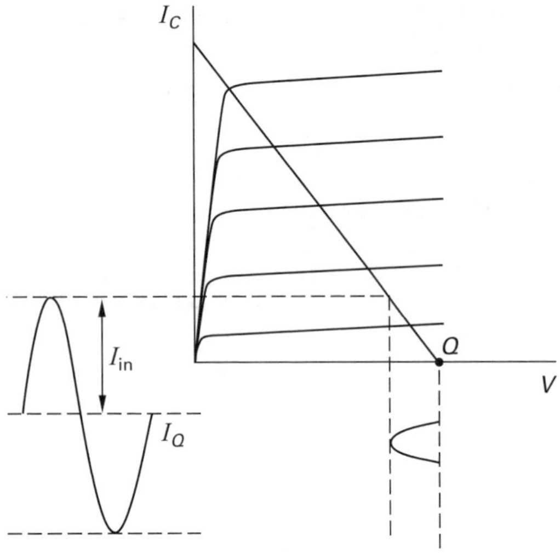

on the voltage-current characteristic curve. The straight *load line* is drawn between them. The change in output voltage (the horizontal axis) caused by a change in input voltage (parallel to the load line) can be determined. Equation 37.10 gives the voltage *gain* (*amplification factor*).[10,11]

$$A_V = \frac{\partial v_{out}}{\partial v_{in}} \approx \frac{\Delta v_{out}}{\Delta v_{in}} \qquad 37.10$$

Usually, a nominal current (the *quiescent current*) flows in the abcd circuit even when there is no signal. The point on the load line corresponding to this current is the *quiescent point* (*Q-point* or *operating point*). It is common to represent the quiescent parameters with uppercase letters (sometimes with a subscript Q) and to write instantaneous values in terms of small changes to the quiescent conditions.

$$v_{in} = V_Q + \Delta v_{in} \qquad 37.11$$

$$v_{out} = V_{out} + \Delta v_{out} \qquad 37.12$$

$$i_{out} = I_{out} + \Delta i_{out} \qquad 37.13$$

Because it is a straight line, the load line can also be drawn if the quiescent point and any other point, usually $(V_{BB}, 0)$, are known.

The ideal voltage amplifier has an infinite *input impedance* (so that all of v_{in} appears across the amplifier and no current or power is drawn from the source) and zero *output impedance* so that all of the output current flows through the load resistor.

Determination of the load line for a generic transistor amplifier is accomplished through the following steps.

step 1: For the configuration provided, label the x-axis on the *output characteristic curves* with the appropriate voltage. (For a BJT, this is V_{CE} or V_{CB}. For a FET, this is V_D.)[12]

step 2: Label the y-axis as the output current. (For a BJT, this is I_C. For a FET, this is I_D.)

step 3: Redraw the circuit with all three terminals of the transistor open. Label the terminals. (For a BJT, these are base, emitter, and collector. For a FET, these are gate, source, and drain.) Label the current directions all pointing inward, toward the amplifier. (For a BJT, these are I_B, I_E, and I_C. For a FET, these are I_D and I_S.)

step 4: Perform KVL analysis in the output loop. (For a BJT, this is the collector loop. For a FET, this is the drain loop.) The transistor voltage determined is a point on the x-axis with the output current equal to zero. Plot the point.

step 5: Redraw the circuit with all three terminals of the transistor shorted. Label as in step 3.

step 6: Use Ohm's law, or another appropriate method, in the output loop to determine the current. (For the BJT, this is the collector current. For the FET, this is the drain current.) The transistor current determined is a point on the y-axis with the applicable voltage in step 1 equal to zero. Plot the point.

step 7: Draw a straight line between the two points. This is the DC load line.[13]

Example 37.3

Consider the following common emitter (CE).

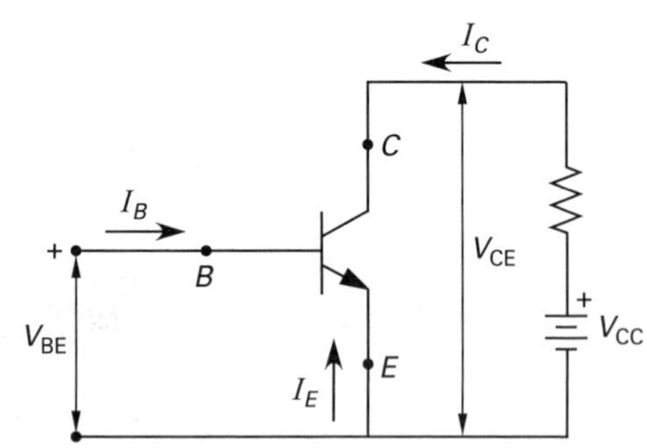

[10]Gain can be increased by increasing the load resistance, but a larger biasing battery, V_2, is required. The choice of battery size depends on the amplifier circuit devices, size considerations, and economic constraints.

[11]A *high-gain amplifier* has a gain in the tens of hundreds of thousands.

[12]The output characteristic curves for the BJT are also called the *collector characteristics* or the *static characteristics*. Figure 37.2 is an example.

[13]AC load lines are determined in the same manner, but active components, that is, inductors and capacitors, are accounted for in the analysis.

If the load resistance is 500 Ω and the collector supply voltage, V_{CC}, is 10 V, determine and draw the load line.

Solution

The x- and y-axes are drawn as shown.

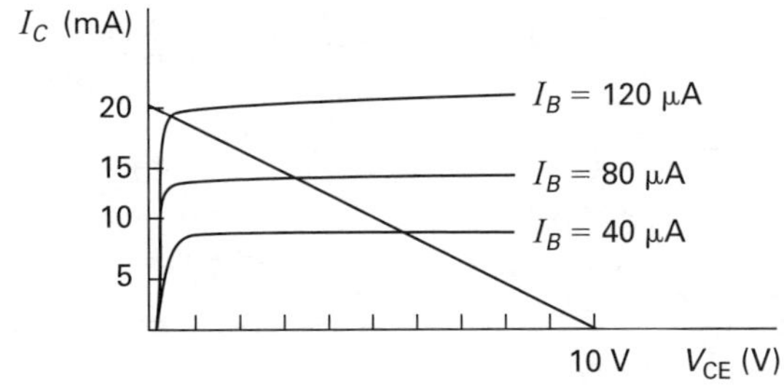

Redrawing the circuit and labeling gives

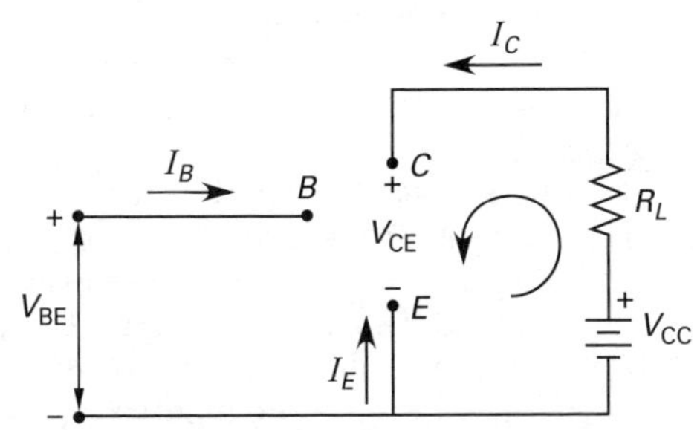

Write KVL around the indicated loop.

$$V_{CC} - I_C R_L - V_{CE} = 0$$

With the terminals open-circuited, $I_C = 0$. Substituting and rearranging gives

$$V_{CE} = V_{CC} = 10 \text{ V}$$

Plot this point (10, 0) on the x-axis. Redraw and label the circuit with the terminals shorted.

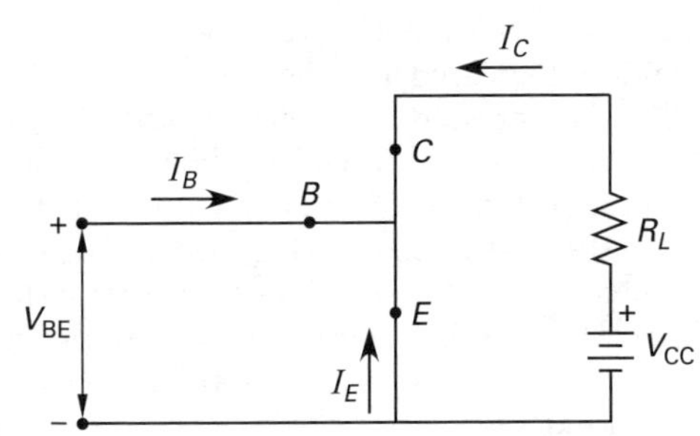

Using Ohm's law and the given values,

$$I_C = \frac{V_{CC}}{R_L} = \frac{10 \text{ V}}{500 \ \Omega} = 0.20 \text{ A} \quad (20 \text{ mA})$$

Plot this point on the y-axis. Draw a straight line between the two points. The load line is shown superimposed on the output characteristic curves in the first drawing.

8. *pn* JUNCTIONS

The *pn junction* forms the basis of diode and transistor operation. The *pn* junction is constructed of a p-type material (the anode) and an n-type material (the cathode) bonded together as shown in Fig. 37.6(a). Some of the acceptor atoms are shown as ions with a minus sign because after an impurity atom accepts an electron, it becomes negatively charged. Some of the donor atoms are shown as ions with a plus sign because after an impurity atom gives up an electron it is positively charged. (Overall, the law of charge neutrality holds, and the *pn* junction is neutral.)

The concentration gradient across the junction causes holes to diffuse to the right, and electrons to diffuse to the left. As a result, the concentration of holes on the p-side near the junction is depleted and a negative charge exists. The concentration of electrons on the n-sidenear the junction is depleted as well, and a positive charge exists. The result of this diffusion is shown in Fig. 37.6(b), the shape of which is determined by the level of doping.[14] The diffusion process continues until the electrostatic field set up by the charge separation is such that no further charge motion is possible. The net electric field intensity is shown in Fig. 37.6(c). The net result is a small region in which no mobile charge carriers exist. This region is called the *space-charge region*, *depletion region*, or *transition region*. Holes have a potential barrier they must overcome to move from left to right just as electrons have an energy barrier they must overcome to move from right to left, as shown in Fig. 37.6(d) and Fig. 37.6(e).

The flow of carriers caused by the concentration gradient is called the *diffusion current*, $I_{\text{diffusion}}$, also called the *recombination current* or the *injection current*. The flow of carriers caused by the established electric field is called the *drift current*, I_s, also called the *saturation current*, *thermal current*, or *reverse saturation current*.[15] The movement of carriers caused by recombination (diffusion) and drift (saturation) is continual, though at thermal equilibrium, without any applied voltage, the net current is zero.

$$I_{\text{junction}} = I_{\text{diffusion}} + I_s = 0 \quad \text{[algebraic sum]} \qquad 37.14$$

[14]A *step-graded junction* and a *linearly graded junction* are two possible types.

[15]Numerous symbols are used, among them I_o and I_{co}.

Figure 37.6 *pn Junction Characteristics*

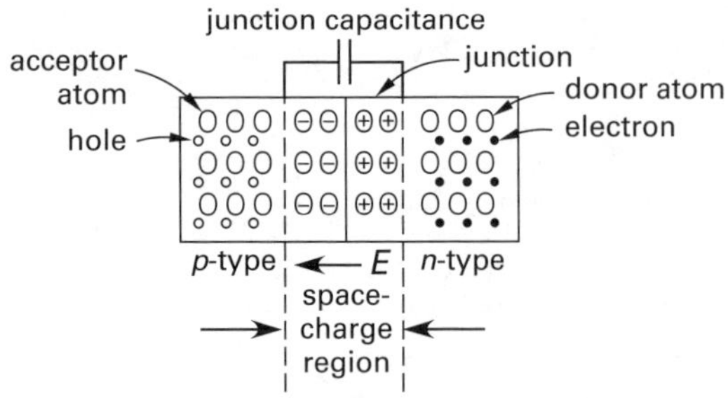

(a) *pn* junction

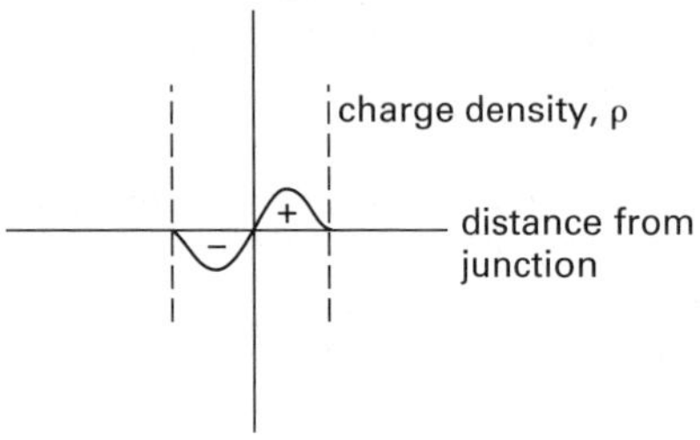

(b) charge density

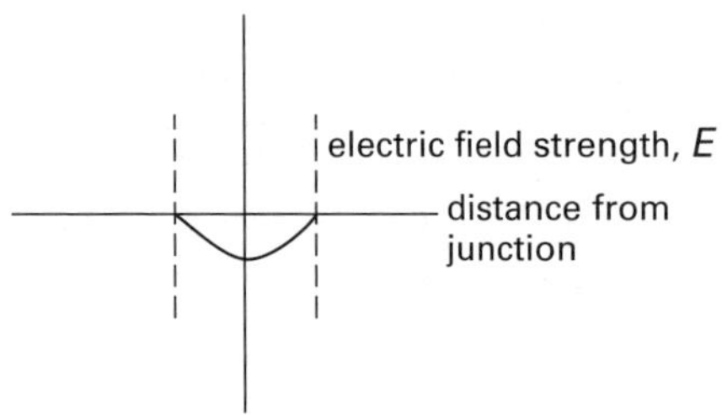

(c) electric field intensity

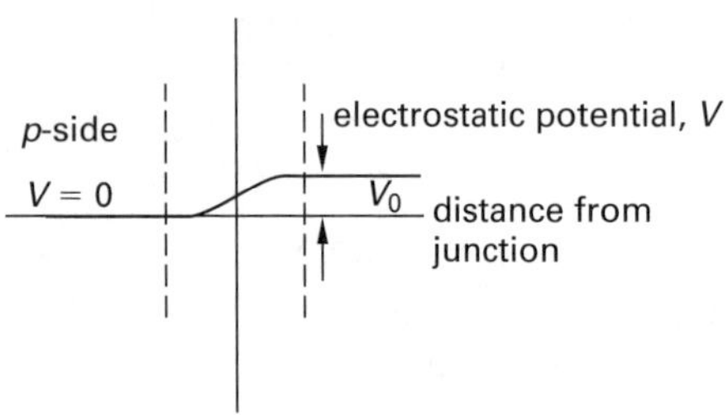

(d) electrostatic potential

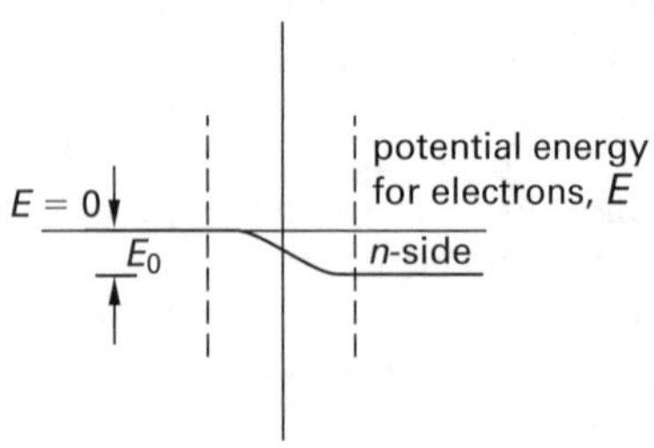

(e) potential energy

Figure 37.7 *pn Junction Carrier Movement*

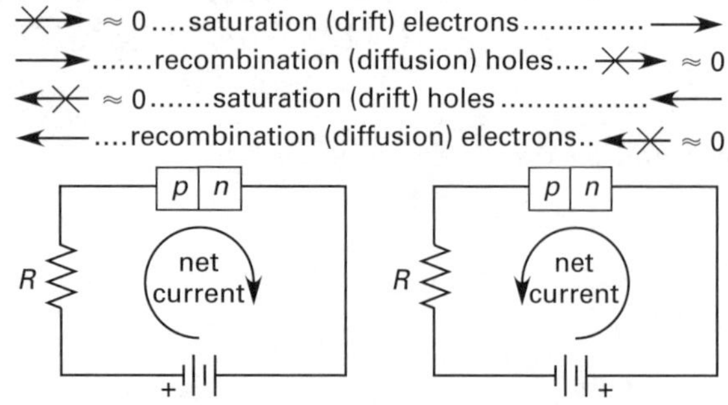

(a) forward bias

(b) reverse bias

A summary of the movement of the carriers is shown in Fig. 37.7.

The width of the space-charge region at equilibrium is shown in Fig. 37.8(a).

When an external voltage is applied, that is, a biasing voltage, the width of the space-charge region and the barrier height change. When the *p*-type material is connected to a positive potential, that is, forward biased, holes are repelled across the junction into the *n*-type material, and electrons are repelled across the junction into the *p*-type material. The width of the space-charge region is reduced and the barrier height for p to n current flow is reduced as shown in Fig. 37.8(b). A DC forward bias voltage, V_F, of approximately 0.5 to 0.7 V for silicon and 0.2 to 0.3 V for germanium is required to overcome the barrier voltage. Once the barrier is overcome, the junction current increases significantly from an increase in the diffusion current. That is, holes cross the junction into the *n*-type material, where they are considered injected minority carriers. Electrons cross the junction into the *p*-type material, where they too are injected minority carriers. Because hole movement in one direction and electron movement in the opposite direction constitute a current in the same direction, the total current is the sum of the hole and electron minority currents.

When the *p*-type material is connected to a negative potential, that is, reverse biased, holes and electrons move away from the junction. The width of the space-charge region is increased and the barrier height for p to n current flow is increased as shown in Fig. 37.8(c). The process nominally stops when the holes in the *p*-type material are depleted. However, a few holes in the *n*-type material are thermally generated, as there are electrons in the *p*-type material. These minority carriers thermally diffuse into the depletion region and are swept across by the electric field. The effect is constant for a given temperature and independent of the reverse bias. This is the reverse saturation current, I_s.[16] The reverse saturation

[16]The reverse saturation current also accounts for any current leakage across the surface of the semiconductor.

Figure 37.8 *pn Junction Space-Charge Region*

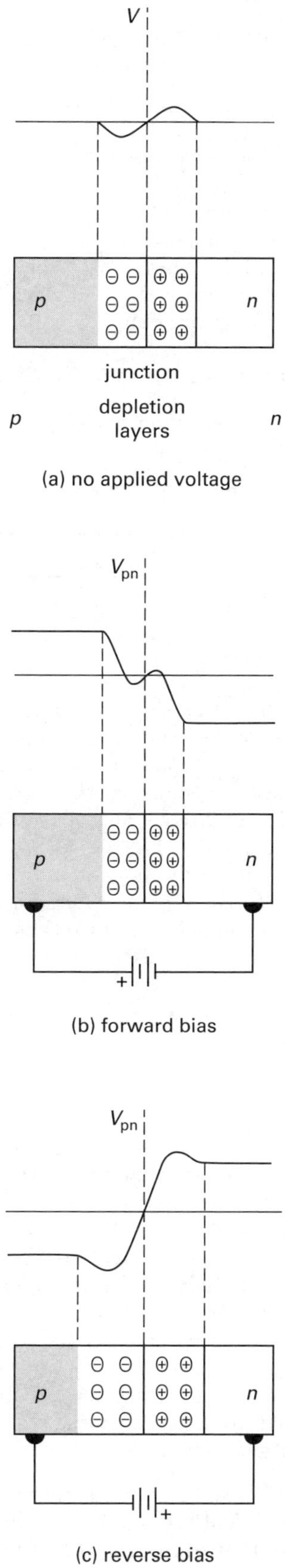

current is small ($\approx 10^{-9}$ A). An ideal *pn* junction, excluding the breakdown region, is governed by[17]

$$I_{\text{pn}} = I_s\left(e^{qV_{\text{pn}}/\kappa T} - 1\right) \qquad 37.15$$

Breakdown is a large, abrupt change in current for a small change in voltage. When a *pn* junction is reverse biased, the saturation current is small up to a certain reverse voltage, where it changes dramatically. Two mechanisms can cause this change. The first is *avalanche breakdown*. Avalanche occurs when thermally generated minority carriers are swept through the space-charge region and collide with ions. If they possess enough energy to break a covalent bond, an electron-hole pair is created. The same effect may occur for these newly generated carriers, resulting in an avalanche effect. Avalanche breakdown occurs in lightly doped materials at greater than 6 V reverse bias. The second breakdown mechanism is *zener breakdown*. Zener breakdown occurs through the disruption of covalent bonds caused by the strength of the electric field near the junction. No collisions are involved. The additional carriers created by the breaking covalent bonds increase the reverse current. Zener breakdown occurs in highly doped materials at less than 6 V reverse bias.[18]

9. DIODE PERFORMANCE CHARACTERISTICS

A *diode* is a two-electrode device. The diode is designed to pass current in one direction only. An *ideal diode*, approximated by a *pn* junction, has a zero voltage drop, that is, no forward resistance, and acts as a short circuit when forward biased (on). When reversed biased (off), the resistance is infinite and the device acts as an open circuit. Diode construction and theory is that of a *pn* junction. (See Sec. 37.8.)

The characteristics and symbol for a typical real semiconductor diode are shown in Fig. 37.9. The *reverse bias voltage* is any voltage below which the current is small, that is, less than 1% of the maximum rated current. The *peak inverse* (*reverse*) *voltage*, PIV or PRV, is the maximum reverse bias the diode can withstand without damage. The forward current is also limited by heating effects. Maximum forward current and peak inverse voltage for silicon diode rectifiers are approximately 600 A and 1000 V, respectively.

The ideal *pn* junction current was given in Eq. 37.15. For practical junctions (*diodes* or *rectifiers*), the equation becomes

$$I = I_s\left(e^{qV/\eta\kappa T} - 1\right) = I_s\left(e^{V/\eta V_T} - 1\right) \qquad 37.16$$

[17]The subscript "pn" is used here for clarification. Standard diode voltage and current directions are defined in Sec. 37.9, after which the subscript is no longer used.

[18]The name zener is commonly used regardless of the breakdown mechanism.

Figure 37.9 *Semiconductor Diode Characteristics and Symbol*

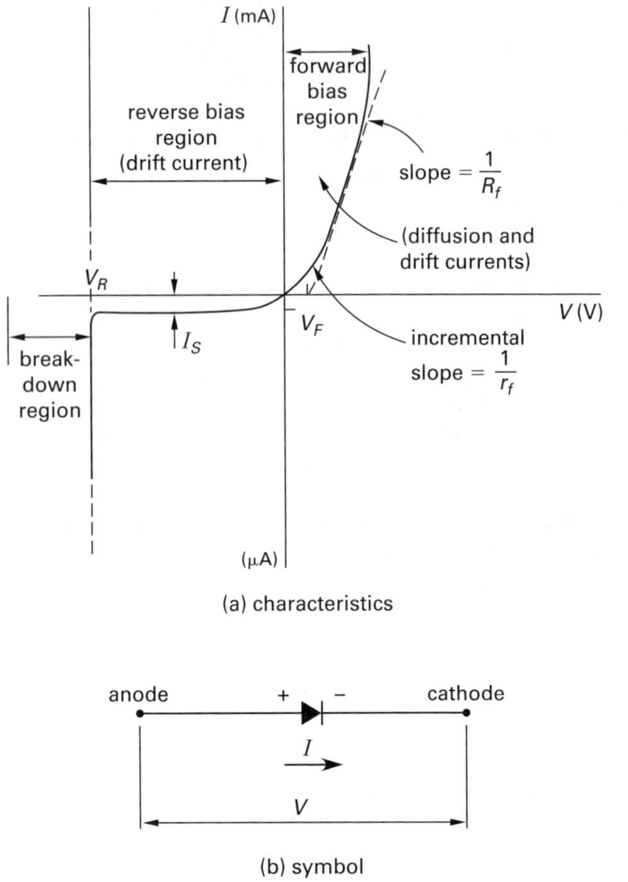

Equation 37.16, which is based on the *Fermi-Dirac probability function*, is valid for all but the breakdown region (see Fig. 37.9). The term η is determined experimentally. For discrete silicon diodes, $\eta = 2$. For germanium diodes, $\eta = 1$. The saturation current, taken from any value of I with a small reverse bias (for example, between 0 and -1 V) gives $I_s \approx 10^{-9}$ A for silicon and 10^{-6} A for germanium. The term V_T represents the *voltage equivalent of temperature* and is related to the diffusion occurring at the junction.[19]

$$V_T = \frac{\kappa T}{q} = \frac{D_p}{\mu_p} = \frac{D_n}{\mu_n} \qquad 37.17$$

Boltzmann's constant is given by κ. The absolute temperature is T. An electron charge is represented by q. Diffusion constants, measured in m^2/s, are given the symbol D. The mobility, measured in $m^2/V{\cdot}s$, is given by the symbol μ.

The voltage equivalent of temperature is also known as the *thermal voltage*. The value of V_T is often quoted at *room temperature*, which can vary from 293K (20°C) to 300K (27°C) depending upon the reference used. The temperature has the effect of doubling the saturation current every 10°C. So,

$$\frac{I_{s2}}{I_{s1}} = (2)^{T_2 - T_1/10°\text{C}} \qquad 37.18$$

A *real diode* can be modeled as shown in Fig. 37.10. The real diode model is composed of an ideal diode, a resistor, R_f, and a voltage source, V_F. The voltage source accounts for the barrier voltage. Typically, for silicon $V_F = 0.7$ V, and for germanium $V_F = 0.2$ V.

Figure 37.10 *Diode Equivalent Circuit*

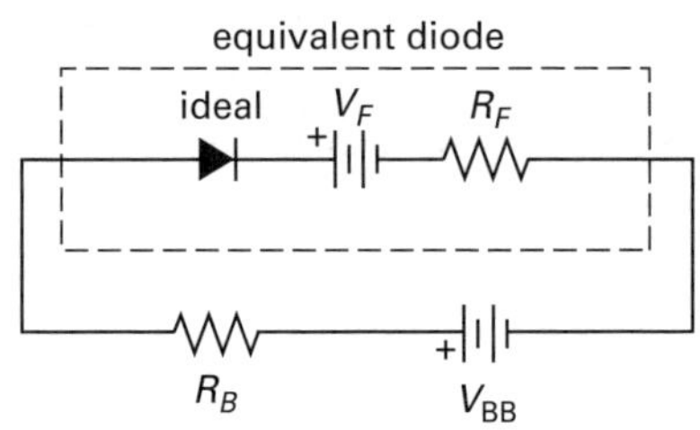

The resistance, R_f, is the slope of a line approximating the linear portion of the characteristic curve as shown in Fig. 37.9. R_f is the resistance for an ideal diode. It is not the static (average) resistance of the diode, as the average resistance is not a constant. The static (average) resistance is calculated as $R_{\text{static}} = V_D/I_D$. When the diode is forward biased by more than a few tenths of a volt, R_f is equal to the *dynamic forward resistance*, r_f. Disregarding lead contact resistance (less than 2 Ω), the dynamic forward resistance is

$$R_f = r_f = \frac{\eta V_T}{I_D} \qquad 37.19$$

A *dynamic reverse resistance*, r_r, also exists and is the inverse of the slope at a point in the reverse bias region.[20] Because the reverse current is very small, the resistance is often considered infinite. Capacitances associated with the junction are also ignored in most models of the diode. The *diffusion capacitance*, C_d, is associated with the charge stored during forward biased operation. The *transition capacitance*, C_t, is associated with the space-charge region width and is the primary capacitance of concern during reverse bias operation. (Diodes designed to take advantage of this voltage-sensitive capacitance are called *varactors*).

The model of Fig. 37.10 assumes that (1) the reverse bias current is suffciently small that the diode acts as an open circuit in that direction, (2) the reverse bias

[19]This is also called the *Einstein relationship*.

[20]Specifying the reverse current, I_{co}, is equivalent to specifying the reverse resistance.

voltage does not exceed the breakdown voltage, and (3) the switching time is instantaneous. The *switching time* is the transient that occurs from the time interval of the application of a voltage to forward (reverse) bias and the achievement of the actual condition. The switching time depends upon the speed of movement of minority carriers near the junction and the junction capacitance.

Example 37.4

What is the thermal voltage at a room temperature of 300K?

Solution

The thermal voltage is given by Eq. 37.17 as

$$V_T = \frac{\kappa T}{q} = \frac{\left(1.3807 \times 10^{-23}\ \frac{\text{J}}{\text{K}}\right)(300\text{K})}{1.6022 \times 10^{-19}\ \text{C}} = 0.026\ \text{V}$$

10. DIODE LOAD LINE

Figure 37.11 shows a forward-biased real diode in a simple circuit. V_{BB} is the *bias battery* (hence the subscripts), and R_B is a current-limiting resistor. R_f and V_f are equivalent diode parameters, not discrete components. If R_f is known in the vicinity of the operating point, the diode current, I_D, is found from Kirchhoff's voltage law. The diode voltage is found from Ohm's law: $V_D = I_D R_f$.

$$V_{BB} - V_f = I_D(R_B + R_f) \qquad 37.20$$

The diode current and voltage drop can also be found graphically from the *diode characteristic curve* (see Fig. 37.11) and the *load line*, a straight line representing the locus of points satisfying Eq. 37.20.[21] The load line is defined by two points. If the diode current is zero, all of the battery voltage appears across the diode (point $(V_{BB}, 0)$). If the voltage drop across the diode is zero, all of the voltage appears across the current-limiting resistor (point $(0, V_{BB}/R_B)$). (Because the diode characteristic curve implicitly includes the effects of V_f and R_f, these terms should be omitted.) The no-signal *operating point*, also known as the *quiescent point*, is the intersection of the diode characteristic curve and the load line.

The *static load line* is derived assuming there is no signal (i.e., $v_{in} = 0$). With a signal, the *dynamic load line* shifts left or right while keeping the same slope. This is equivalent to solving Eq. 37.20 with an additional voltage source.

Figure 37.11 Diode Load Line

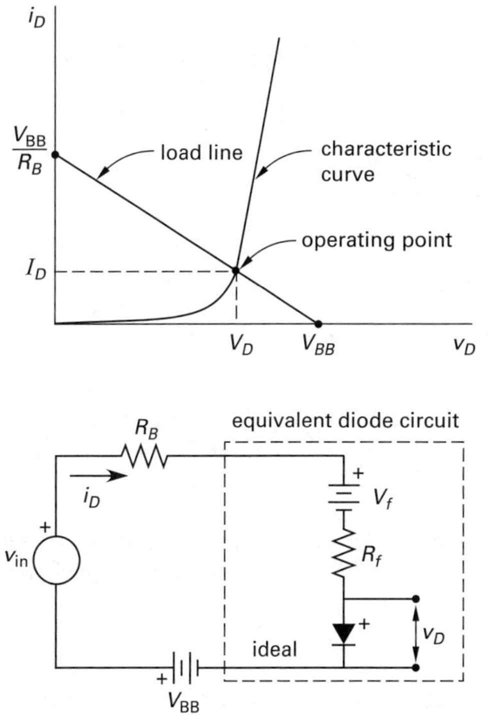

$$i_D = I_D + \Delta i_D = \frac{V_{BB} - V_f + v_{in}}{R_B + R_f} \qquad 37.21$$

Although presented in a slightly different manner, the method for determining the load line is similar to that given in Sec. 37.7. That is,

step 1: Open-circuit the electronic component's equivalent circuit and determine the point $(x, 0)$.

step 2: Short-circuit the electronic component's equivalent circuit and determine the point $(0, y)$.

step 3: Connect the two points.

The load line superimposed on the characteristics curve is then used to determine the operating point (Q-point) of the overall circuit.

11. DIODE PIECEWISE LINEAR MODEL

If the voltage applied to a diode varies over an extensive range, a piecewise linear model may be used.[22] For the real diode characteristic shown in Fig. 37.9, three regions are evident from V_R to V_F.

[21] Notice that the horizontal axis voltage is the voltage across the diode (modeled as a resistor).

[22] The model in Sec. 37.10 is for the forward bias region only and assumes no reverse current flow.

(1) *Reverse breakdown region,* $V_D < V_R$

$$I_D = \frac{V_D + V_R}{R_r} \quad [V_D \text{ and } V_R \text{ are negative}] \qquad 37.22$$

(2) *Off region,* $V_R < V_D < V_F$

$$I_D = 0 \qquad 37.23$$

(3) *Forward bias region,* $V_D > V_F$

$$I_D = \frac{V_D - V_F}{R_f} \qquad 37.24$$

Using ideal diodes, the real diode can be represented by the model in Fig. 37.12 over the entire range of operation.

Figure 37.12 *Piecewise Linear Model*

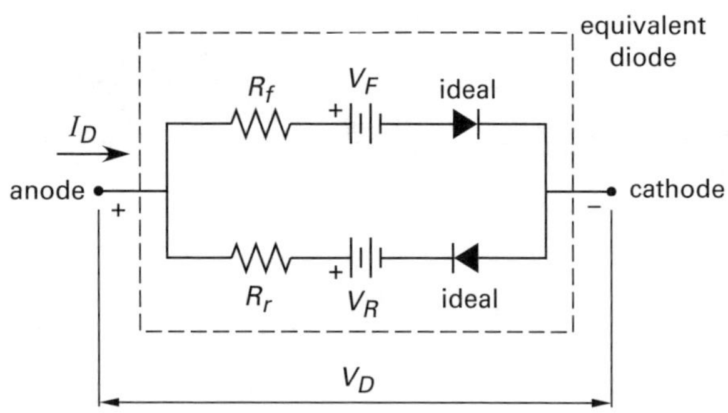

12. DIODE APPLICATIONS AND CIRCUITS

Diodes are readily integrated into *rectifier, clipping,* and *clamping circuits.* A clipping circuit cuts the peaks off of waveforms; a clamping circuit shifts the DC (average) component of the signal. Figure 37.13 illustrates the response to a sinusoid with peak voltage V_m for several simple circuits.

Figure 37.13 *Output from Simple Diode Circuits*

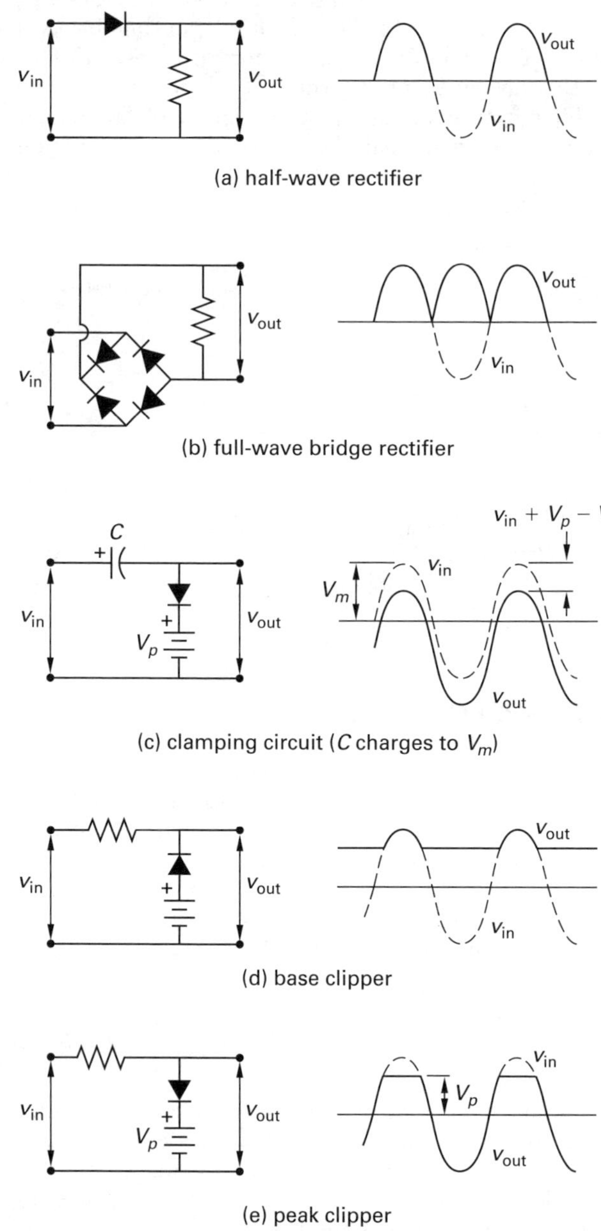

13. SCHOTTKY DIODES

A *Schottky diode,* also called a *barrier diode* or a *hot-carrier diode,* is a diode constructed with a metal semiconductor contact as shown in Fig. 37.14(a). The Schottky diode symbol is shown in Fig. 37.14(b). The metal semiconductor rectifying junction is similar to the *pn* junction, but the physical mechanisms are somewhat different.

In the forward direction, electrons from the lightly doped semiconductor cross into the metal anode, where electrons are plentiful.[23] The electrons in the metal are majority carriers, whereas in a *p*-type material they are minority carriers. Being majority carriers, they are indistinguishable from other carriers and are not stored near the junction. This means that no minority carrier population exists to move when switching occurs from forward to reverse (on to off) bias. Consequently, switching times are extremely short (approximately 10^{-12} s).

The Schottky symbol is used on any electronic component designed with a *Schottky contact,* that is, a rectifying, rather than an ohmic, contact as shown in Fig. 37.14(c).

[23]The electrons injected into the metal are above the *Fermi energy level,* determined by the Fermi-Dirac distribution of electron energies, and are called *hot carriers.*

In the figure, the Schottky contact lies between the base and the collector.

Figure 37.14 Schottky Diode

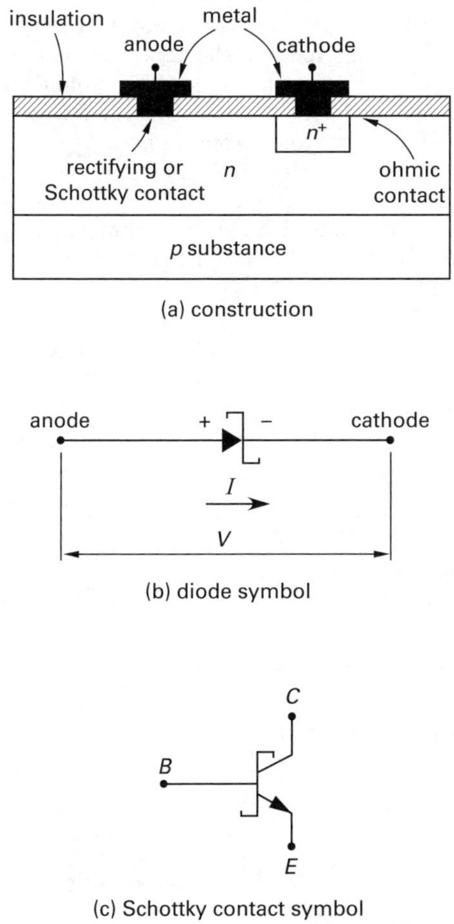

14. ZENER DIODES

A *zener diode* is a diode specifically designed to operate within the breakdown region. The construction and theory of a zener diode is similar to that of a *pn* junction, with the design allowing for greater heat dissipation capabilities. The characteristics and symbol are shown in Fig. 37.15.

When a reverse voltage, known as the *zener voltage*, V_Z (which is negative with respect to the anode), is applied, a reverse saturation current, I_Z, flows. The voltage-currentrelationship is nearly linear. Current flows until the reverse saturation current drops to I_{ZK} near the knee of the characteristic curve. This minimum current is the *keep-alive current*, that is, the minimum current for which the output characteristic is linear.

Because the current is large, an external resistor must be used to limit the current to within the power dissipation capability of the diode. Zener diodes are used as voltage regulating and protection devices. The regulated zener voltage varies with the temperature. The *temperature coefficient* is

$$\text{tc} = \frac{\Delta V_Z}{V_Z \Delta T} \times 100\% \qquad 37.25$$

Figure 37.15 Zener Diode

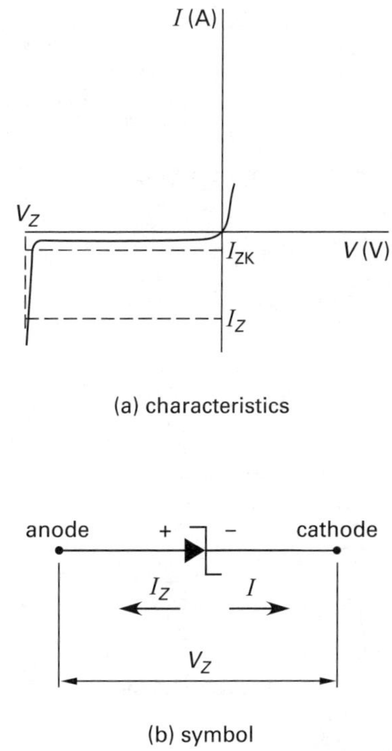

15. TUNNEL DIODES

A *tunnel diode* (*Esaki diode*) is a two-terminal device with an extremely thin potential barrier to electron flow, so that the output characteristic is dominated by the quantum-mechanical tunneling process. To make a thin potential barrier (50 Å to 100 Å), both regions of the diode are heavily doped. The amount of tunneling is limited by the electrons available in the *n*-type material or by the available empty energy states in the *p*-type material to which they can tunnel. The characteristics are shown in Fig. 37.16(a).

Point 1 in Fig. 37.16(a) corresponds to a biasing level that allows for maximum tunneling. (At this point, the minimum electron energy in the *n*-type material conduction band equals the Fermi level in the *p*-type material's valence band.) Between points 1 and 2, increases in bias voltage result in a decrease in current as available energy states are filled and the total amount of tunneling drops. This is an area of negative resistance, or negative conductance, and is the primary use for tunnel diodes. If the doping levels are slightly reduced from their typical values of $5 \times 10^{19}\ \text{cm}^{-3}$, the forward tunneling current becomes negligible and the output

Figure 37.16 Tunnel Diode

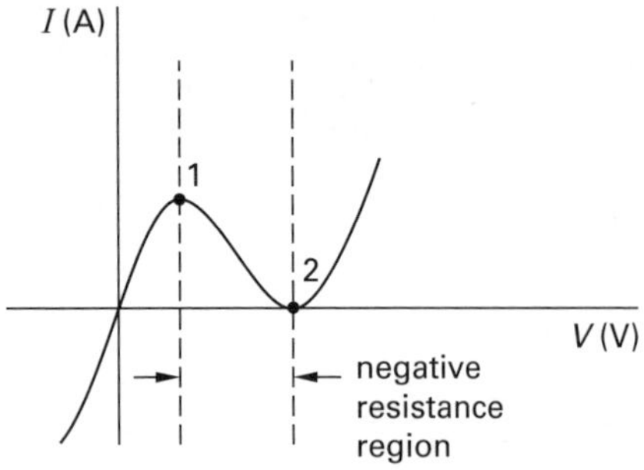

(a) tunnel diode characteristics

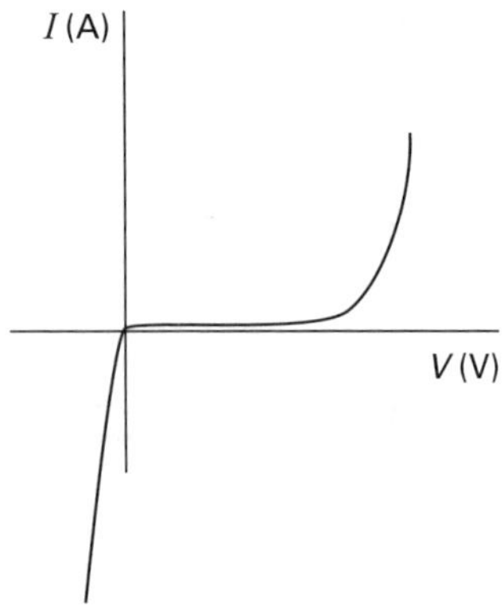

(b) backward diode

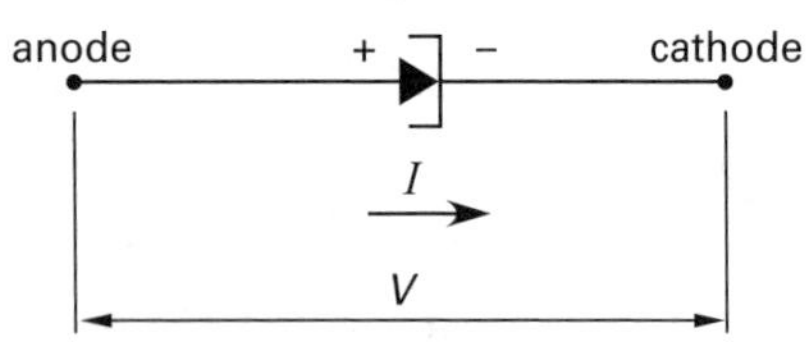

(c) symbol

characteristic becomes that of a *backward diode*, Fig. 37.16(b). The symbol for a tunnel diode is shown in Fig. 37.16(c).

16. PHOTODIODES AND LIGHT-EMITTING DIODES

If a semiconductor junction is constructed so that it is exposed to light, the incoming photons generate electron-hole pairs. When these carriers are swept from the junction by the electric field, they constitute a *photocurrent*, which is seen as an increase in the reverse saturation current. The holes generated move to the *p*-type material and the electrons move to the *n*-type material in response to the electric field that is established whenever *p*- and *n*-type semiconductors are joined (see Fig. 37.6(a)). Such devices are used as light sensors and are called *photodiodes*. When the device is designed without a biasing source, it becomes a *solar cell.*

When forward biased, diodes inject carriers across the junction that are above thermal equilibrium. When the carriers recombine, they emit photons from the *pn* junction area.[24] The photon is caused by the recombination of electron-hole pairs. The wavelength depends on the energy band gap and on the material used. Gallium arsenide (GaAs) and other binary compounds are commonly used. When the photon is in the infrared region, the mechanism is called *electroluminescence* and the diodes are called *electroluminescent diodes*. If the photons are in the visible region, the devices are called *light-emitting diodes* (LEDs). The emitted wavelength, λ, is given by Eq. 37.26. h is Planck's constant, with a value of 6.6256×10^{-34} J·s.

$$\lambda = \frac{hc}{E_G} \qquad 37.26$$

Example 37.5

The manufacturer's data sheet for a gallium arsenide diode shows a band gap energy of 1.43 eV. Will such a band gap result in emitted photons within the wavelength of visible light?

Solution

The photon emitted wavelength is given by

$$\begin{aligned}\lambda &= \frac{hc}{E_G} \\ &= \frac{(6.6256 \times 10^{-34}\ \text{J·s})\left(2.9979 \times 10^{8}\ \frac{\text{m}}{\text{s}}\right)}{(1.43\ \text{eV})\left(1.6022 \times 10^{-19}\ \frac{\text{J}}{\text{eV}}\right)} \\ &= 8.67 \times 10^{-7}\ \text{m}\end{aligned}$$

The wavelength of visible light is from approximately 8×10^{-7} to 4×10^{-7} m. Consequently, GaAs is a good choice for a light-emitting diode.

17. SILICON-CONTROLLED RECTIFIERS

A four-layer *pnpn* device with an anode, cathode, and gate terminal is called a *silicon-controlled rectifier* (SCR) or *thyristor*.[25] A conceptual construction is shown in Fig. 37.17(a).

When the SCR is reverse biased, that is, when the anode is negative with respect to the cathode, the characteristics are similar to a reverse-biased *pn* junction as shown in Fig. 37.17(b). When the SCR is forward biased, that is, when the anode is positive with respect to the cathode, four distinct regions of operation are evident.

[24]They can also release the energy as heat, or *phonons*.

[25]A *thyristor* is defined as a transistor with thyratron-like characteristics. That is, as collector current is increased to a critical value, the alpha (common base current gain) rises above unity and results in a high-speed triggering action.

Figure 37.17 *Silicon-Controlled Rectifier*

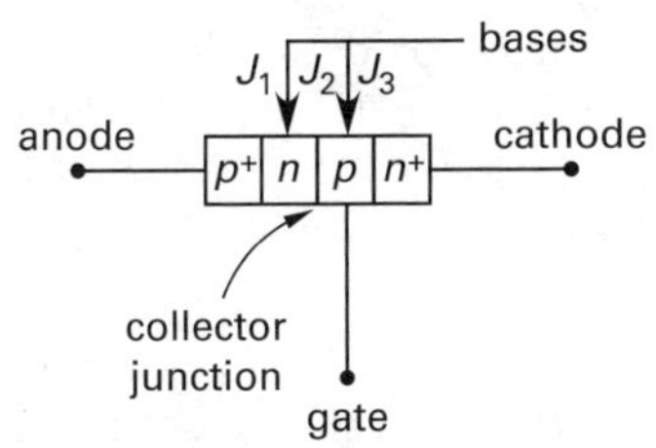

(a) conceptual construction

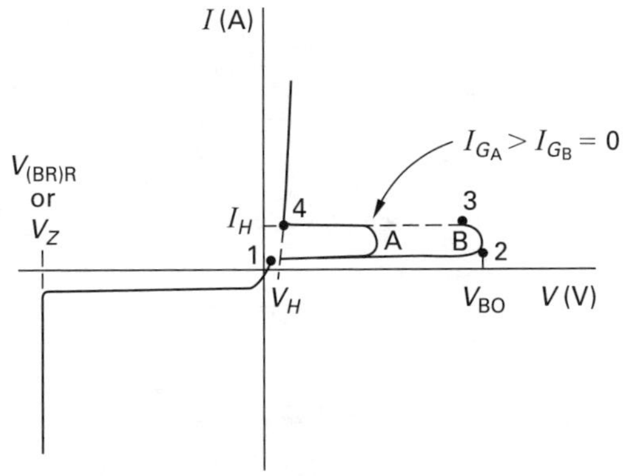

(b) characteristics

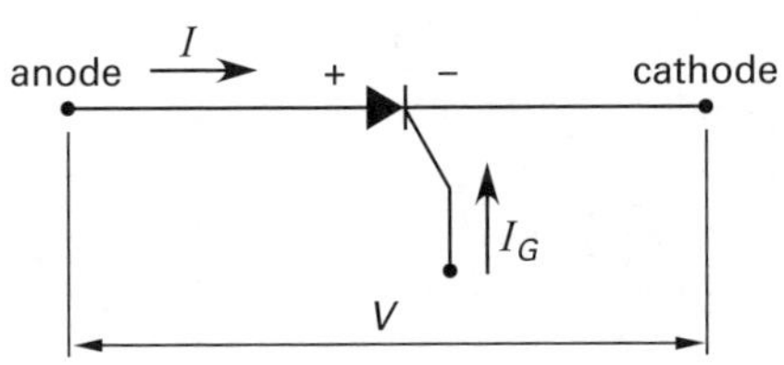

(c) symbol

From point 1 to point 2, junctions J_1 and J_3 are forward biased. Junction J_2 is reverse biased. The external voltage appears primarily across the reverse-biased junctions. The device continues to operate similarly to a reverse-biased *pn* junction. This is called the *off* or *high-impedance region.*

From point 2 to point 3, the current increases slowly to the *breakover voltage*, V_{BO}. At this point, junction J_2 undergoes breakdown and the current increases sharply.

From point 3 to point 4, the current increases as the voltage decreases. This region is called the *negative resistance region.*

From point 4, junction J_2 is forward biased. The voltage across the device is essentially that of a forward-biased *pn* junction (approximately 0.7 V). If the current through the diode is reduced by the external circuit, the diode remains on until the current falls below the *hold current*, I_H, or the *hold voltage*, V_H. Below this point, the diode switches off, to the high-impedance state.

The gate functions to increase the current at the collector junction of the *npn* transistor, J_2, which is an integral part of the SCR. By increasing the current through the reverse-biased junction, J_2, the anode current is increased. This increases the gain of the two transistors, resulting in a lowering of the forward breakover voltage. This is seen in the difference between paths A and B in Fig. 37.17(b). Consequently, for a given anode to cathode voltage, the gate can be used to turn the SCR on. Once on, however, the SCR must be reverse biased to turn off. A thyristor designed to be turned off by the gate is called a *gate turn-off thyristor.*

SCRs are used in power applications to allow small voltages and currents to control much larger electrical quantities. The symbol for an SCR is shown in Fig. 37.17(c).

Other devices with multiple *pn* junctions using the same principles are the *diac* and *triac*, shown in Fig. 37.18 and Fig. 37.19, respectively.

Most electronic components are unable to handle large amounts of current. When properly designed for power dissipation, semiconductors handling large amounts of power are called *power semiconductors.* Such devices constitute a branch of electronics called *power electronics.* Silicon-controlled rectifiers and Schottky diodes are

Figure 37.18 *Diac*

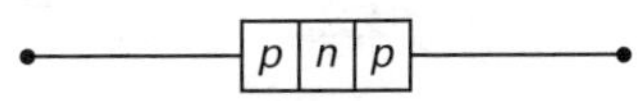

(a) conceptual construction

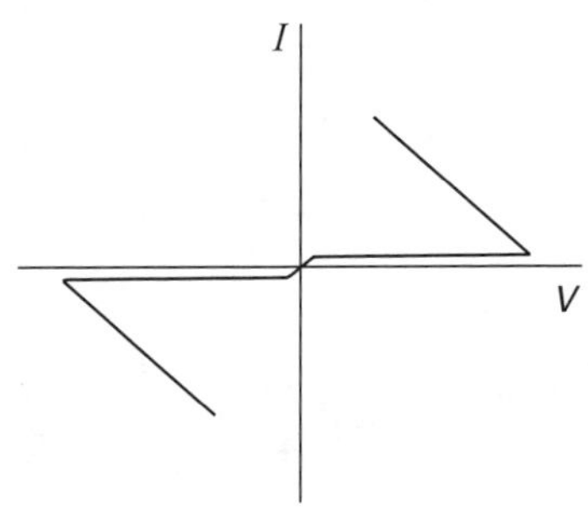

(b) characteristics

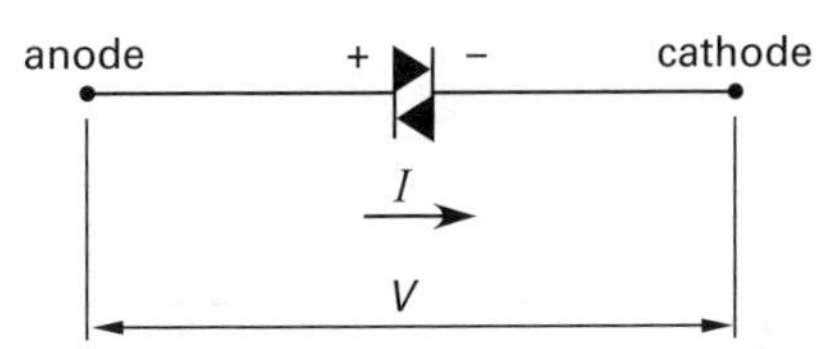

(c) symbol

Figure 37.19 *Triac*

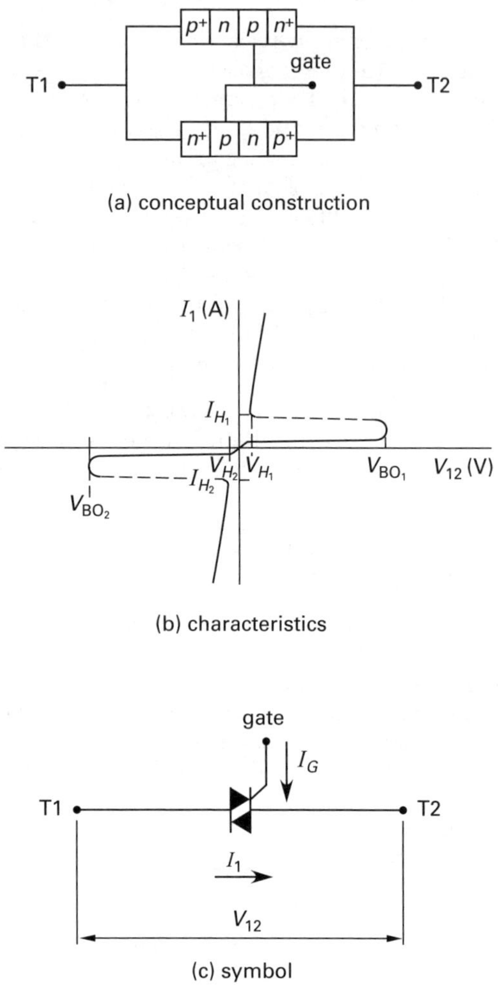

traditional power semiconductors. Newer designs include the *high-power bipolar junction transistor* (HPBT), power metal-oxide semiconductor field-effect transistor (MOSFET), *gate turn-off thyristor* (GTO), and *insulated gate bipolar transistor* (IGBT), sometimes called a *conductivity-modulated field-effect transistor* (COMFET).

Power semiconductors are classified as either trigger or control devices. *Trigger devices*, such as GTOs, start conduction by some trigger input and then behave as diodes. *Control devices* are normally BJTs and FETs used in full-range amplifiers.

38 Junction Transistors

1. Transistor Fundamentals 38-2
2. BJT Transistor Performance Characteristics 38-3
3. BJT Transistor Parameters 38-3
4. BJT Transistor Configurations 38-4
5. BJT Biasing Circuits 38-5
6. BJT Load Line 38-6
7. Amplifier Gain and Power 38-7
8. Cascaded Amplifiers 38-7
9. Equivalent Circuit Representation and Models 38-7
10. Approximate Transistor Models 38-8
11. Hybrid-π Model 38-8
12. Transistor Circuit Linear Analysis 38-9
13. Transistor Circuit High-Frequency Analysis 38-12
14. Unijunction Transistors 38-12
15. Darlington Transistors 38-12

Nomenclature

A	gain	–
C	capacitance	F
E	energy	J or eV
g	transconductance	S
g_m	mutual conductance or transconductance	S
h_f	forward transfer voltage ratio with output shorted	–
h_{fe}	CE small signal (AC) forward current transfer ratio or gain	–
h_i	input impedance with output shorted	Ω
h_{ie}	CE small signal (AC) input impedance	Ω
$h_{\text{i,o,f,r and e,c,b}}$	hybrid parameter[1]	–
$h_{\text{I,O,F,R and E,C,B}}$	hybrid parameter[2]	–
h_o	output admittance with input open	S
h_{oe}	CE small signal (AC) open-circuit output admittance	S
h_r	reverse transfer voltage ratio with input open	Ω
h_{re}	CE small signal (AC) reverse voltage transfer ratio or gain	–
i	instantaneous or variable current	A
I	effective or DC current	A
I_{CBO}	DC collector cutoff current, emitter open[3]	A
n	electron concentration	m^{-3} or cm^{-3}
n, N	concentration	m^{-3} or cm^{-3}
p	hole concentration	m^{-3} or cm^{-3}
P	power dissipation	W
Q	quiescent point	–
r	small-signal resistance[4]	Ω
$r_{\text{bb}'}$	base spreading resistance	Ω
$r_{\text{b}'\text{c}}$	feedback resistance	Ω
$r_{\text{b}'\text{e}}$	base input resistance	Ω
r_{ce}	output resistance	Ω
R	resistance	Ω
S	apparent power	VA
v	instantaneous or variable voltage	V
V	effective or DC voltage	V

Symbols

α	CB forward current ratio or gain	–
β	CE forward current ratio or gain	–
ω	angular frequency	rad/s

Subscripts

0	at 0K
0	barrier or source value
ac	alternating current or small signal
b	AC or small signal base
b′c	from internal base to collector
b′e	from internal base to emitter
bc, BC	base to collector
be, BE	base to emitter
B	DC or biased base
BB	base supply or base biasing
c	conduction band or collector
cb, CB	collector to base
ce, CE	collector to emitter
cf	corner frequency or cutoff frequency
co	cutoff
C	collector
CBO	collector cutoff current
CC	collector supply
CQ	Q-point collector current

[1]The first subscript refers to the AC or small signal input (i), output (o), forward (f), or reverse (r) value. The second subscript refers to the configuration, that is, e for common emitter (CE), c for common collector (CC), b for common base (CB).

[2]The first subscript refers to the DC or static input (I), output (O), forward (F), or reverse (R) value. The second subscript refers to the configuration, that is, E for common emitter (CE), C for common collector (CC), B for common base (CB).

[3]This is essentially the reverse-saturation current between the collector and the base.

[4]This is also known as the *dynamic resistance* or the *instantaneous resistance*. Any resistance represented with a lowercase letter is the small-signal value.

d	diffusion
DC	direct current or static
e	AC or small signal emitter
eb, EB	emitter to base
ec, EC	emitter to collector
E	DC or biased emitter
EE	emitter to supply voltage
f	forward (AC or instantaneous)
i	input (AC or instantaneous)
I	current or input (DC or total)
max	maximum
o	output (AC or instantaneous)
P	power
r	reverse (AC or instantaneous)
R	resistance
t	transition
T	thermal
V	constant or effective voltage

1. TRANSISTOR FUNDAMENTALS

A *transistor* is an active device comprised of semiconductor material with three electrical contacts (two rectifying and one ohmic). Two major types of transistors exist: *bipolar junction transistors* (BJTs) and *field-effect transistors* (FETs). A BJT uses the base current to control the flow of charges from the emitter to the collector. Field-effect transistors use an electric field, that is, voltage, established at the gate (equivalent to the base) to control the flow of charges from the source to the drain (equivalent to the emitter and collector). The controlling variable in a BJT is the current at the base; in a FET it is the voltage at the gate. An *npn* bipolar junction transistor (BJT) is shown in Fig. 38.1(a).

The transistor has three operating regions: *cutoff*, *saturation*, and *active* or *linear*. In the cutoff region, both the base-emitter junction and the collector-base junction are reverse biased. In the saturation region, both the base-emitter junction and the collector-base junction are forward biased. Small amounts of current injected into the base control the movement of charges in the output (from emitter to collector) in a manner similar in theory to that of an SCR, specifically, the triac (see Sec. 37.17). When operating from the cutoff to the saturation region, the transistor is a switch.

In the active, or linear, region, the base-emitter junction is forward biased and the collector-base junction is reverse biased. When so biased, the electron potential energy is of the shape shown in Fig. 38.1(b). Small amounts of current injected into the base lower the potential barrier across the emitter-base junction. As carriers move from the emitter to the base, most are swept across the narrow base by the relatively large electric field established by the reverse-biased collector-base junction. (See the large space-charge, or depletion region, shown in Fig. 38.1(b).) When operating in the active (linear) region, the transistor is an amplifier.

The symbol for BJTs is shown in Fig. 38.1(c). Positive currents are defined as those flowing into the associated terminals. Actual current flows follow the arrows.[5] Majority carriers are always injected into the base by the *emitter* (E). The thin controlling center of a transistor is the *base* (B). The majority carriers are gathered by the *collector* (C).

Figure 38.1 *Bipolar Junction Transistor*

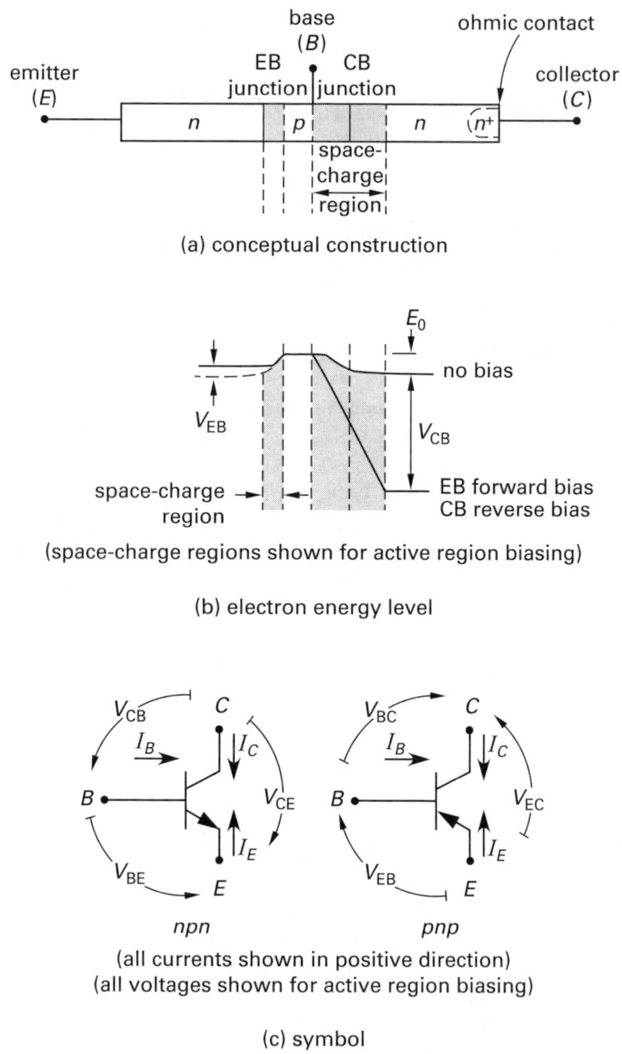

When operated over all three ranges—cutoff, saturation, and active—the transistor is modeled in a piecewise linear fashion. This is *large-signal analysis*. Operated as a switch, between the cutoff and saturation regions, the transistor is modeled as an open circuit and a short circuit. This is *digital circuit operation*. Operated as an amplifier, in the active region, the transistor is modeled as equivalent parameters, usually *h*-parameters. This is *analog circuit operation*. Analog analysis occurs in two phases. First, the operating point, or Q-point, is determined. Determining the Q-point is a combination of determining the load line for the biasing used and setting the base current by using the values of the biasing components (see Sec. 38.5 and Sec. 38.6). This is often a

[5] That is, conventional current flow follows the arrows. Electrons flow opposite to the arrows.

reiterative process. (For a FET, the items determined are the load line and gate voltage.) Second, the incremental AC performance is determined using *small-signal analysis*, that is, using *h*-parameters to model the transistor and replacing independent sources with the appropriate models (see Sec. 38.9). Real sources are replaced with internal resistance. Ideal voltage sources are replaced with short circuits. Ideal current sources are replaced with open circuits.

2. BJT TRANSISTOR PERFORMANCE CHARACTERISTICS

When the base-emitter junction is forward biased and the collector-base junction is reverse biased, the transistor is said to be operating in the *active region*.

When the base-emitter junction is not forward biased (as when V_{BB} is zero), the base current will be nearly zero and the transistor acts like a simple switch. This is known as being *off* or *open*, and operating in the *cutoff mode*.[6] Also, the collector and emitter currents are zero when the transistor operates in the *cutoff region*. However, a very small input voltage (in place of V_{BB}) will forward bias the base-emitter junction, which, because I_B is so low, instantly forces the transistor into its saturation region. This results in a large collector current.

When the collector-emitter voltage is very low (usually about 0.3 V for silicon and 0.1 V for germanium), the transistor operates in its *saturation region*. Regardless of the collector current, the transistor operates as a closed switch (i.e., a short circuit between the collector and emitter). This is known as being *on* or *closed*.

$$V_{CE} \approx 0 \quad \text{[saturation]} \qquad 38.1$$

The BJT operating regions are shown in Fig. 38.2. The output characteristics for the common emitter and common base configurations are shown in Fig. 38.3. (A sample maximum power curve is shown as a dashed line. The *Q*-point must be to the left and below the power curve.)

Figure 38.2 *BJT Operating Regions*

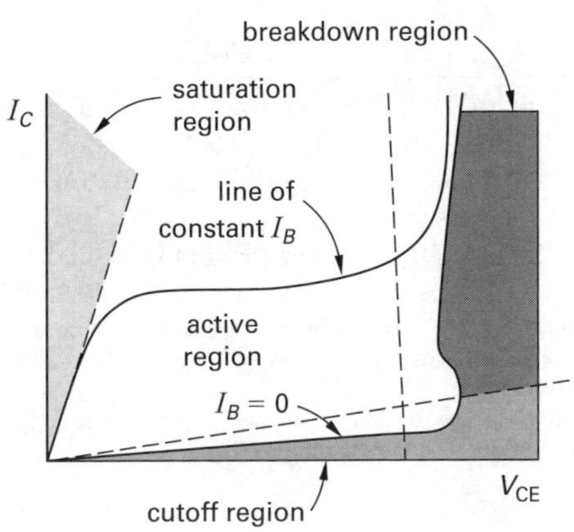

[6]Except for digital and switching applications, this condition usually results from improper selection of circuit resistances.

Figure 38.3 *BJT Output Characteristics*

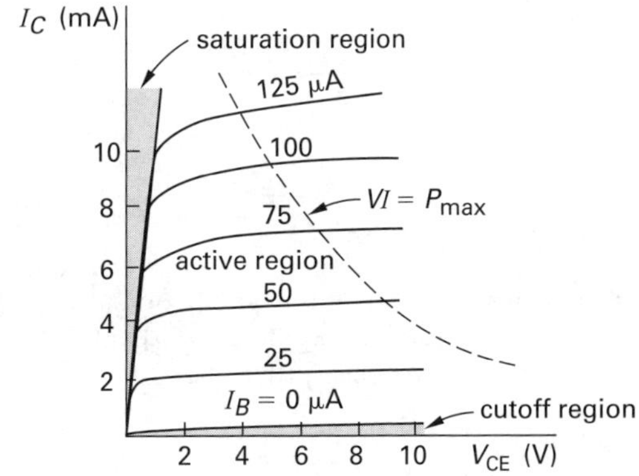

(a) common emitter

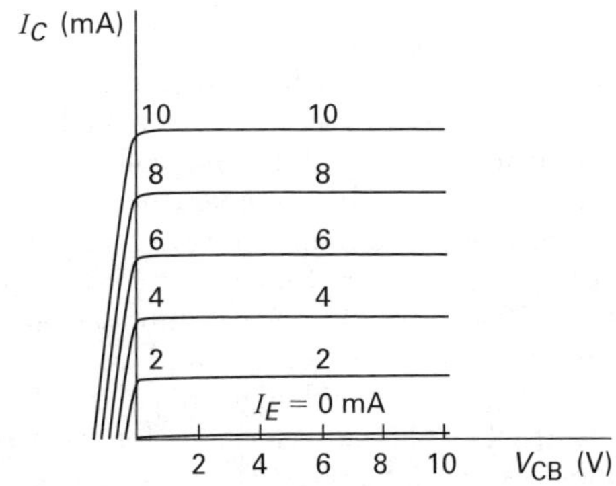

(b) common base

3. BJT TRANSISTOR PARAMETERS

Equation 38.2 is Kirchhoff's current law, taking the transistor as a node. Usually, the collector current is proportional to, and two or three orders of magnitude larger than, the base current, I_B. Because of this, a small change in base current of, for example, 1 mA, can produce a change in collector current of 100 mA. The *current (amplification) ratio*, β_{DC}, is the ratio of collector-base currents.

$$I_E = I_C + I_B \qquad 38.2$$

$$\beta_{DC} = \frac{I_C}{I_B} = \frac{\alpha_{DC}}{1 - \alpha_{DC}} \qquad 38.3$$

$$\alpha_{DC} = \frac{I_C}{I_E} = \frac{\beta_{DC}}{1 + \beta_{DC}} \qquad 38.4$$

Both α_{DC} and β_{DC} are for DC signals only. The corresponding values for small signals are designated α_{ac} and β_{ac}, respectively, and are calculated from differentials. (The difference between β_{ac} and β_{DC} is very small, and the two are not usually distinguished.)

$$\beta_{\text{ac}} = \frac{\Delta I_C}{\Delta I_B} = \frac{i_C}{i_B} \quad 38.5$$

$$\alpha_{\text{ac}} = \frac{\Delta I_C}{\Delta I_E} = \frac{i_C}{i_E} \quad 38.6$$

Thermal (saturation) current is small but always present and can be included in Eq. 38.2. I_{CBO} is the thermal current at the collector-base junction.

$$I_C = I_E - I_B \quad 38.7$$

$$I_C = \alpha I_E - I_{\text{CBO}} \approx \alpha I_E \quad 38.8$$

Transistors are manufactured from silicon and germanium, although silicon transistors have a higher temperature operating range. While the collector cutoff current is very small at room temperature, it doubles every 10°C, rendering germanium transistors useless around 100°C. Silicon transistors remain useful up to approximately 200°C. While germanium has a lower collector-emitter saturation voltage and may outperform silicon in high-speed and high-frequency devices, silicon is nevertheless the material used for most semiconductor devices and integrated circuit systems.

Figure 38.4 illustrates a family of curves for various base currents. The DC amplification factor, β_{DC}, can be found (for a wide range of V_{CE} values) by taking a point on any line in the active (horizontal line) region and calculating the ratio of the coordinates, I_C/I_B. The small-signal amplification factor, β_{ac}, is calculated as the difference in the I_C between two I_B lines divided by the differences in I_B.

Figure 38.4 *Small Signal Terms*

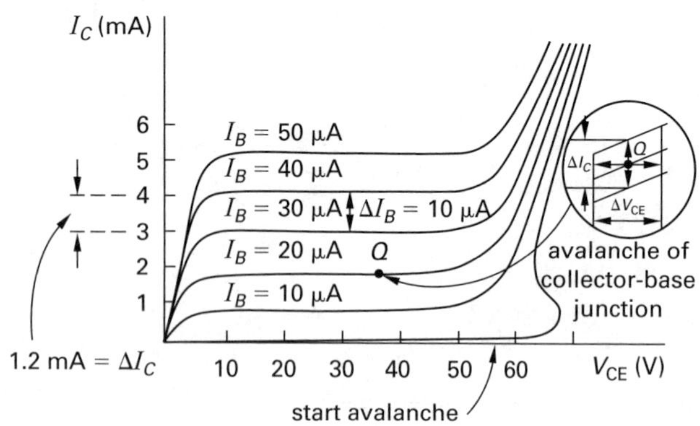

Many of the parameters used have multiple symbols. When given in manufacturers' data sheets or as equivalent parameters, h-parameter symbols are more common. The equivalences are given in Eq. 38.9 through Eq. 38.12.

$$\alpha_{\text{DC}} = h_{\text{FB}} \quad 38.9$$

$$\alpha_{\text{ac}} = h_{\text{fb}} \quad 38.10$$

$$\beta_{\text{DC}} = h_{\text{FE}} \quad 38.11$$

$$\beta_{\text{ac}} = h_{\text{fe}} \quad 38.12$$

4. BJT TRANSISTOR CONFIGURATIONS

There are six ways a BJT transistor can be connected in a circuit, depending on which leads serve as input and output. Only three configurations have significant practical use, however. The terminal not used for either input or output is referred to as the *common terminal.* For example, in a *common emitter* circuit, the base receives the input signal and the output signal is at the collector. For the *common collector* (also known as an *emitter follower*) circuit, the input is to the base and the output is from the emitter. For a *common base* circuit, the input is to the emitter and the output is from the collector. The various transistor configurations are shown in Fig. 38.5.[7] A summary of the relative properties of the configurations is given in Table 38.1.

The common emitter configuration is arguably the most versatile and useful. It is the only configuration with a voltage and current gain that is greater than unity. The common collector is widely used as a buffer stage. That is, it is used between a high-impedance source and a low-impedance load. The common base has the fewest applications, though it is sometimes used to match a low-impedance source with a high-impedance load, or to act as a noninverting amplifier with a voltage gain greater than unity.

Table 38.1 *Comparison of Transistor Configuration Properties*

symbol[a]	common emitter[a]	common collector[a]	common base
A_V	high (−100)	low (< 1)	high (+100)
A_I	high (−50)	high (+50)	low (< 1)
R_{in}	medium (1000 Ω)	high (> 100 kΩ)	low (< 100 Ω)
R_{out}	high (≈ 50 kΩ)[b]	low (< 100 Ω)	high (≈ 2 MΩ)[b]

[a]Approximate values in parentheses are based on a source and load resistance of 3 kΩ, an h_{ie} of approximately 1 kΩ, and an h_{fe} of approximately 50.
[b]The output resistance is often assumed to be infinite for this configuration, simplifying the transistor model.

[7]Bypass capacitors are not shown.

Figure 38.5 *Transistor Configurations*

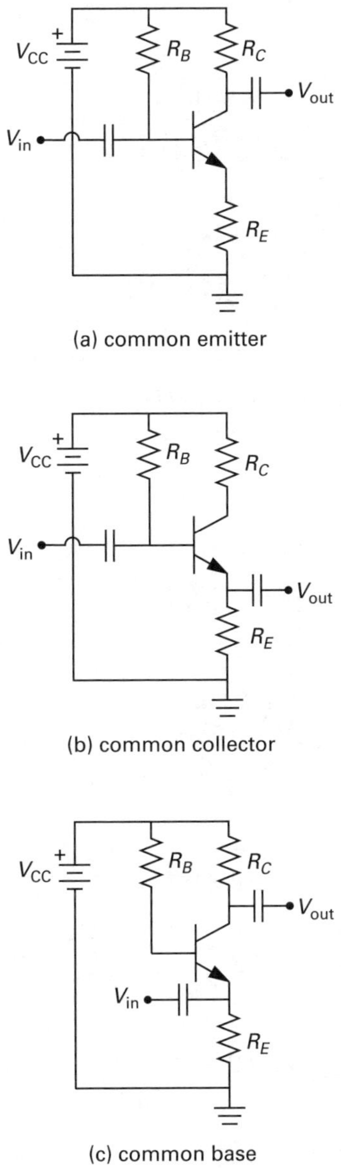

5. BJT BIASING CIRCUITS

To maximize the transistor's operating range when large swing-signals are expected, the quiescent point should be approximately centered in the active region. The purpose of biasing is to establish the base current and, in conjunction with the load line, the quiescent point. Figure 38.6 illustrates several typical biasing methods: fixed bias, fixed bias with feedback, self-bias, voltage-divider bias, multiple-battery bias, and switching circuit (cutoff) bias. All of the methods can be used with all three common-lead configurations and with both *npn* (shown) and *pnp* transistors.[8] It is common to omit the bias battery (shown in Fig. 38.6(a) in dashed lines) in transistor circuits.

The base current can be found by writing Kirchhoff's voltage law around the input loop, including the bias battery, the external resistances, and the V_{BE} barrier voltage that opposes the bias battery. Because the base is thin, there is negligible resistance from the base to the emitter, and V_{BE} (being less than 1 V) may be omitted as well. For the case of *fixed bias with feedback* (*fixed bias with emitter resistance*) illustrated in Fig. 38.6(b), the base current is found from

$$V_{CC} = I_B R_B + V_{BE} + I_E R_E \quad [v_{in} = 0] \tag{38.13}$$

$$\begin{aligned} I_B &= \frac{V_{CC} - V_{BE} - I_E R_E}{R_B} \quad [v_{in} = 0] \\ &= \frac{V_{CC} - V_{BE}}{R_B + \frac{\beta}{\alpha} R_E} \\ &= \frac{V_{CC} - V_{BE}}{R_B + (1+\beta) R_E} \quad [v_{in} = 0] \end{aligned} \tag{38.14}$$

The first step in designing a fixed-bias amplifier with emitter resistance is choosing the quiescent collector current, I_{CQ}. R_E is selected so that the voltage across R_E is approximately three to five times the intrinsic V_{BE} voltage (i.e., 0.3 V for germanium and 0.7 V for silicon).

$$R_E \approx \frac{3 V_{BE}}{I_{CQ}} \tag{38.15}$$

Once R_E is found, R_B can be calculated from Eq. 38.13.

The *bias stability*, with respect to any quantity M, is given by Eq. 38.16. Variable M commonly represents temperature (T), current amplification (β), collector-base cutoff (I_{CBO}), and base-emitter voltage (V_{BE}).

$$S_M = \frac{\frac{\Delta I_C}{I_{CQ}}}{\frac{\Delta M}{M}} \tag{38.16}$$

The collector-base cutoff current—essentially, the reverse saturation current—doubles with every 10°C rise in temperature, as was mentioned in Sec. 37.9 and Sec. 38.3. The thermal stability of a transistor, then, is affected by this current. The effect is self-reinforcing: as the temperature increases, the saturation current increases, which further increases the temperature. This phenomenon is called *thermal runaway*. The emitter resistor used in the biasing circuits of Fig. 38.6 helps stabilize the transistor against this trend. As the current rises, the voltage drop across the emitter resistor rises in a direction that opposes the forward biased base-emitter junction. This decreases the base current, and so the collector current increases less than it would without the self-biasing resistor R_E.

[8]The polarity of the DC supply voltages must be reversed to convert the circuits shown for use with *pnp* transistors.

Electronics

Figure 38.6 *DC Biasing Methods*

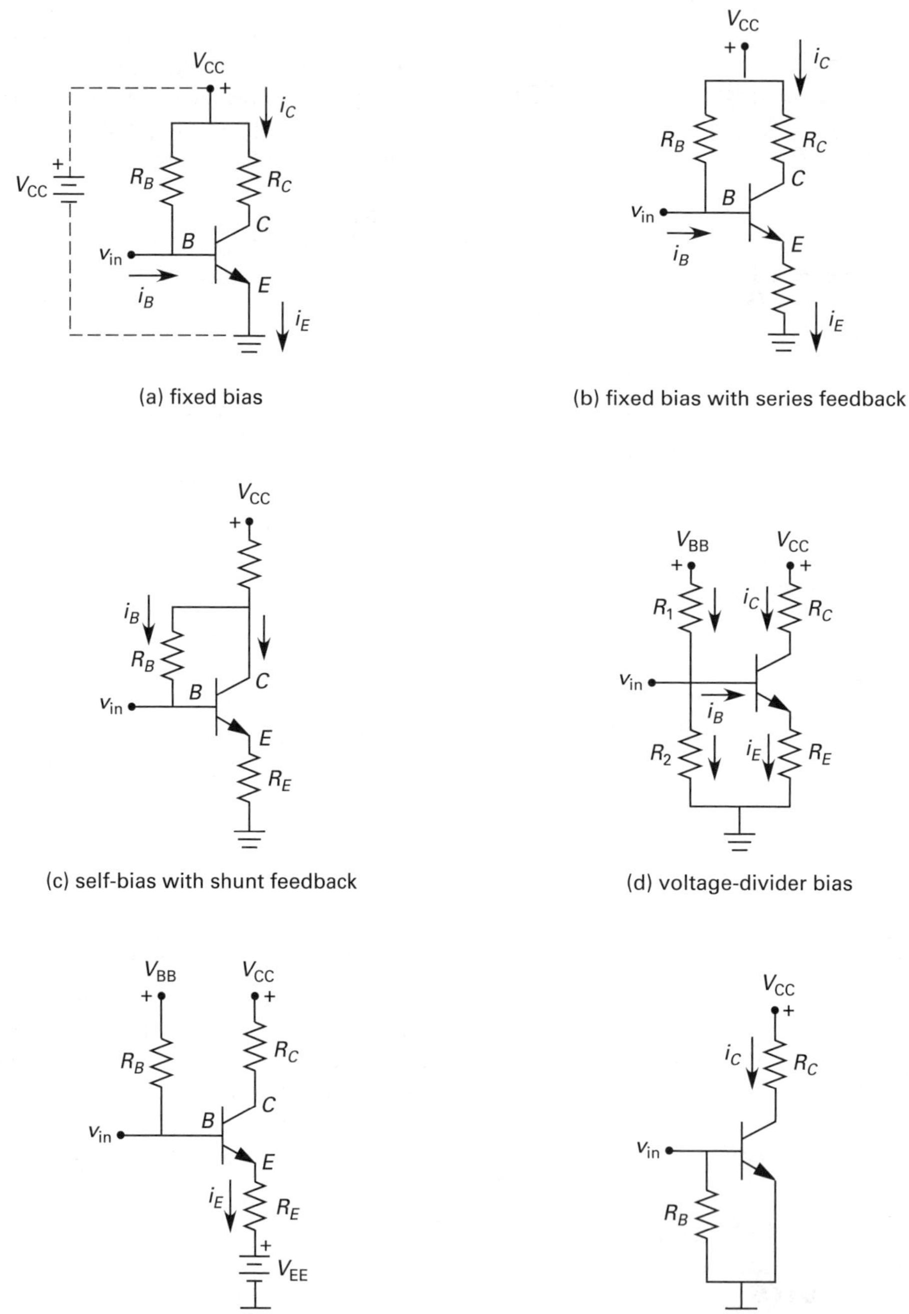

(a) fixed bias

(b) fixed bias with series feedback

(c) self-bias with shunt feedback

(d) voltage-divider bias

(e) multiple-battery bias

(f) cutoff bias

6. BJT LOAD LINE

Figure 38.7 illustrates part of a simple common emitter transistor amplifier circuit. The bias battery and emitter and collector resistances define the *load line*. If the emitter-collector junction could act as a short circuit (i.e., V_{CE} equals zero), the collector current would be $V_{\mathrm{CC}}/(R_C + R_E)$. If the signal is large enough, it can completely oppose the battery-induced current, in which case the net collector current is zero and the full bias battery voltage appears across the emitter-collector junction. The intersection of the load line and the base current curve defines the *quiescent point*. The load line, base current, and quiescent point are all illustrated in Fig. 38.7.

Determination of the load line for a BJT transistor is accomplished by the following steps.

step 1: For the configuration provided, label the x-axis on the *output characteristic curves* with the appropriate voltage, V_{CE} or V_{CB}.

step 2: Label the y-axis as the output current, I_C.

step 3: Redraw the circuit with all three terminals of the transistor open. Label the terminals as the base, emitter, and collector. Label the current directions all pointing inward, that is, toward the transistor, I_B, I_E, and I_C.

step 4: Perform KVL analysis in the collector loop. The transistor voltage determined is a point on the x-axis with the output current equal to zero. Plot the point.

step 5: Redraw the circuit with all three terminals of the transistor shorted. Label as in step 3.

step 6: Use Ohm's law, or another appropriate method, in the collector loop to determine the current. The transistor current determined is a point on the y-axis with the applicable voltage in step 1 equal to zero. Plot the point.

step 7: Draw a straight line between the two points. This is the DC load line. The DC load line is used to determine the biasing and Q-point. The AC load line is determined in the same manner, but with active components, that is, inductors and capacitors, included. The AC load line is used to determine the transistor's response to small signals.

Figure 38.7 *Common Emitter Load Line and Quiescent Point*

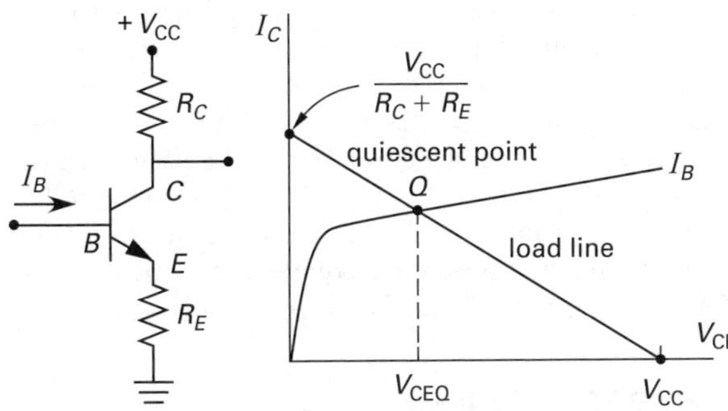

7. AMPLIFIER GAIN AND POWER

The *voltage-*, *current-*, *resistance-*, and *power-gain* are

$$A_V = \frac{\Delta V_{\text{out}}}{\Delta V_{\text{in}}} = \frac{v_{\text{out}}}{v_{\text{in}}} = \beta A_R \qquad 38.17$$

$$A_I = \frac{\Delta I_{\text{out}}}{\Delta I_{\text{in}}} = \frac{i_{\text{out}}}{i_{\text{in}}} = \beta \qquad 38.18$$

$$A_R = \frac{Z_{\text{out}}}{Z_{\text{in}}} = \frac{A_V}{\beta} \qquad 38.19$$

$$A_P = \frac{P_{\text{out}}}{P_{\text{in}}} = \beta^2 A_R = A_I A_V \qquad 38.20$$

The collector power dissipation, P_C, should not exceed the rated value. (This restriction applies to all points on the load line.) See Fig. 38.3(a) for a sample power restriction.

$$P_C = \tfrac{1}{2} I_C V_{\text{CE}} \quad \text{[rms values]} \qquad 38.21$$

8. CASCADED AMPLIFIERS

Several amplifiers arranged so that the output of one is the input to the next are said to be *cascaded amplifiers.*[9] When each *amplifier stage* is properly coupled to the following, the overall gain is

$$A_{\text{total}} = A_{V,1} A_{V,2} A_{V,3} \cdots \qquad 38.22$$

Capacitors are used in amplifier circuits to isolate stages and pass small signals. This is known as *capacitive coupling.*[10] A capacitor appears to a steady (DC) voltage as an open circuit. However, it appears to a small (AC) voltage as a short circuit, and so input and output signals pass through, leaving the DC portion behind.

9. EQUIVALENT CIRCUIT REPRESENTATION AND MODELS

A transistor can be modeled as any of the equivalent two-port networks described in Sec. 29.31. The different transistor configurations are shown as two-port networks in Fig. 38.8.

Figure 38.8 *Transistors as Two-Port Networks*

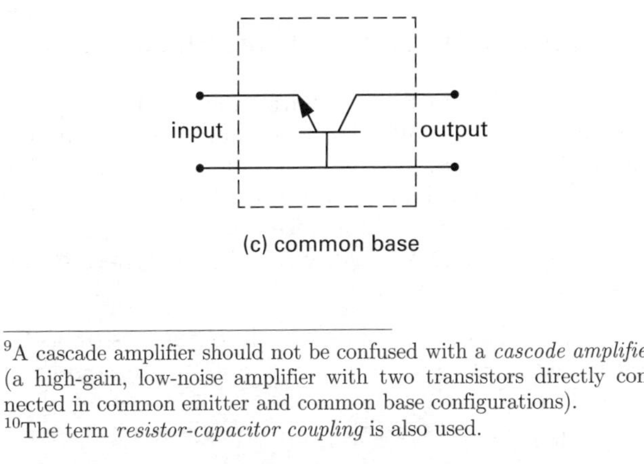

(a) common emitter (b) common collector (c) common base

[9]A cascade amplifier should not be confused with a *cascode amplifier* (a high-gain, low-noise amplifier with two transistors directly connected in common emitter and common base configurations).

[10]The term *resistor-capacitor coupling* is also used.

The equivalent parameters most often used are the hybrid parameters, also called the *h-parameters*. The h-parameters are defined for any two-port network in terms of Eq. 38.23 through Eq. 38.25.

$$\begin{bmatrix} v_1 \\ i_2 \end{bmatrix} = \begin{bmatrix} h_{11} & h_{12} \\ h_{21} & h_{22} \end{bmatrix} \begin{bmatrix} i_1 \\ v_2 \end{bmatrix} \qquad 38.23$$

$$v_1 = h_{11} i_1 + h_{12} v_2 \qquad 38.24$$

$$i_2 = h_{21} i_1 + h_{22} v_2 \qquad 38.25$$

Each of the h-parameters in Eq. 38.24 and Eq. 38.25 is defined specifically for transistor models in the following way.

- h_i input impedance with output shorted (Ω)
- h_r reverse transfer voltage ratio with input open (dimensionless)
- h_f forward transfer current ratio with output shorted (dimensionless)
- h_o output admittance with input open (S)

Using the h-parameters as so defined, Eq. 38.26 and Eq. 38.27 become the governing equations for *small-signal circuit models*, also called *AC incremental models*.

$$v_i = h_i i_{\text{in}} + h_r v_{\text{out}} \qquad 38.26$$

$$i_o = h_f i_{\text{in}} + h_o v_{\text{out}} \qquad 38.27$$

Normally the h-parameters are given with two subscripts. The first is defined as in Eq. 38.26 and in Eq. 38.27. The second subscript indicates the configuration, such as common emitter (e or E), common collector (c or C), or common base (b or B). The h-parameters are normally specified for the common emitter configuration only. Table 38.2 shows the equivalence between the parameters for the different transistor configurations. The defining equations for the small-signal models are given in Table 38.3. Typical values for a widely used *npn* transistor, the 2N2222A, are given in Table 38.4.

Table 38.2 *Equivalent Circuit Parameters*

symbol	common emitter	common collector	common base
h_{11}, h_{ie}	h_{ie}	h_{ic}	$\frac{h_{ib}}{1+h_{fb}}$
h_{12}, h_{re}	h_{re}	$1 - h_{rc}$	$\frac{h_{ib}h_{ob}}{1+h_{fb}} - h_{rb}$
h_{21}, h_{fe}	h_{fe}	$-1 - h_{fc}$	$\frac{-h_{fb}}{1+h_{fb}}$
h_{22}, h_{oe}	h_{oe}	h_{oc}	$\frac{h_{ob}}{1+h_{fb}}$
h_{11}, h_{ib}	$\frac{h_{ie}}{1+h_{fe}}$	$\frac{-h_{ic}}{h_{fc}}$	h_{ib}
h_{12}, h_{rb}	$\frac{h_{ie}h_{oe}}{1+h_{fe}} - h_{re}$	$h_{rc} - \frac{h_{ic}h_{oc}}{h_{fc}} - 1$	h_{rb}
h_{21}, h_{fb}	$\frac{-h_{fe}}{1+h_{fe}}$	$\frac{-1-h_{fc}}{h_{fc}}$	h_{fb}
h_{22}, h_{ob}	$\frac{h_{oe}}{1+h_{fe}}$	$\frac{-h_{oc}}{h_{fc}}$	h_{ob}
h_{11}, h_{ic}	h_{ie}	h_{ic}	$\frac{h_{ib}}{1+h_{fb}}$
h_{12}, h_{rc}	$1 - h_{re}$	h_{rc}	1
h_{21}, h_{fc}	$-1 - h_{fe}$	h_{fc}	$\frac{-1}{1+h_{fb}}$
h_{22}, h_{oc}	h_{oe}	h_{oc}	$\frac{h_{ob}}{1+h_{fb}}$

10. APPROXIMATE TRANSISTOR MODELS

The models used in Sec. 38.9 are exact. In many practical applications, sufficiently accurate results can be obtained with simplified models. The values of h_r and h_o are very small. That is, the reverse transfer voltage ratio and the output admittance are insignificant. The simplified models of Table 38.5 are obtained by ignoring these two parameters.

The simplified models assume that the output resistance is infinite; that is, $1/h_o$ is approximately infinite. (See Table 38.1.) The simplified models further assume that the reverse voltage source, $h_r v$, is negligible. The reverse voltage accounts for the narrowing of the base width as the collector-base junction reverse bias increases. As the effectiveness of the base current in controlling the output is minimized, gain decreases. This explains the opposing voltage at the base-emitter junction. This phenomenon is called the *Early effect* (see Fig. 38.1(a)). The simplified models can also be shown with the input impedance, h_i, replaced with a voltage source equal to the barrier voltage (0.7 V for silicon and 0.3 V for germanium).

11. HYBRID-π MODEL

Though the h-parameter model is common, another significant model is the *hybrid-π* or *Giacoletto* model. The hybrid-π model is shown in Fig. 38.9. The terms used in the model follow.

- $r_{bb'}$ *base spreading resistance*, or the *interbase resistance*, or the *small-signal base bulk resistance* (Ω)
- $r_{b'e}$ *small-signal base input resistance* (Ω)
- $r_{b'c}$ *small-signal feedback resistance* (Ω) This resistance accounts for the Early effect.
- r_{ce} *small-signal output resistance* (Ω)
- g_m *transistor transconductance* (S)

Table 38.3 *BJT Equivalent Circuits*

common connection	equivalent circuit	network equations
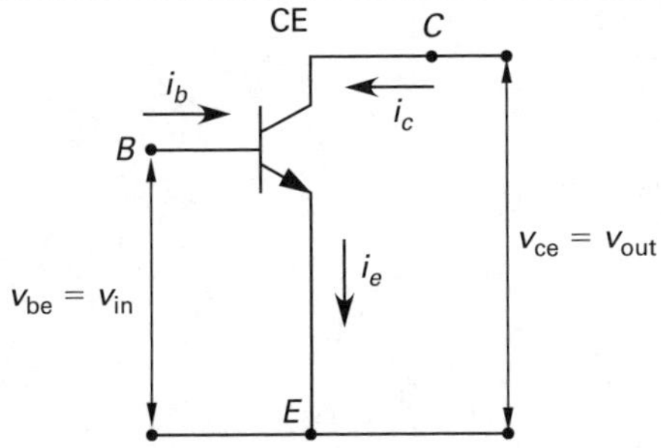	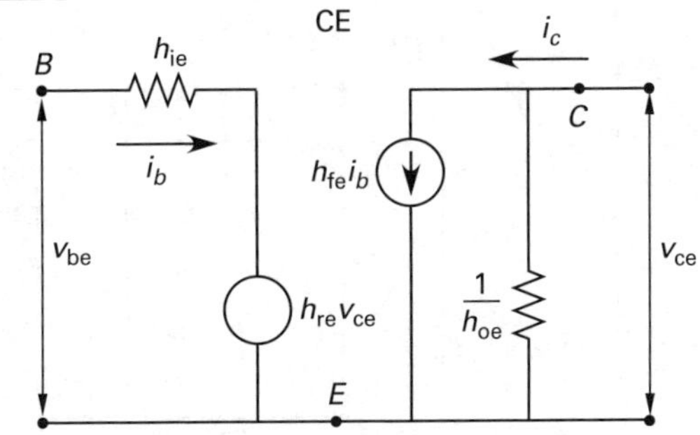	common emitter $v_{be} = h_{ie} i_b + h_{re} v_{ce}$ $i_c = h_{fe} i_b + h_{oe} v_{ce}$
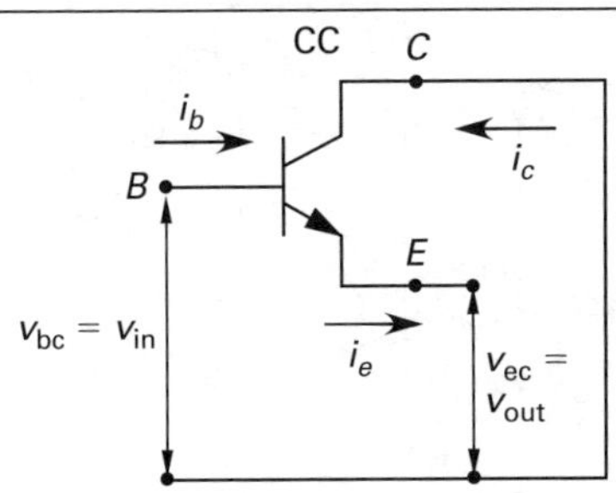	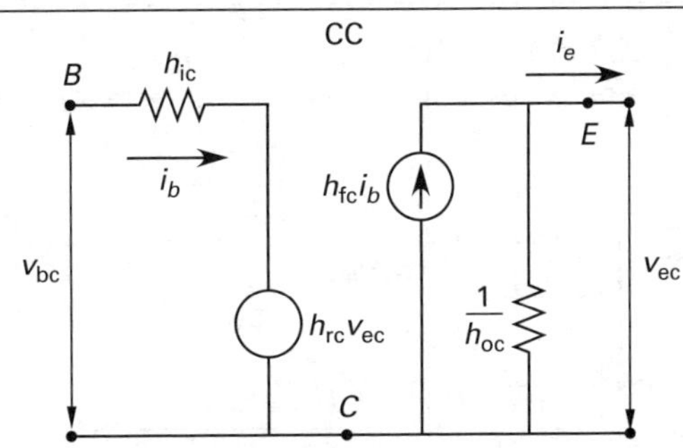	common collector $v_{bc} = h_{ic} i_b + h_{rc} v_{ec}$ $i_e = h_{fc} i_b + h_{oc} v_{ec}$
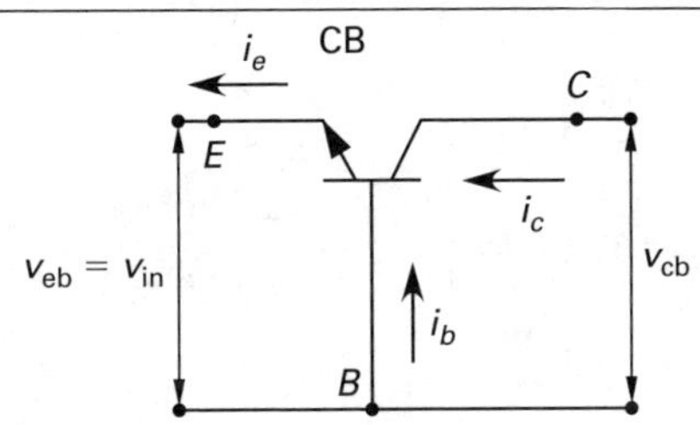	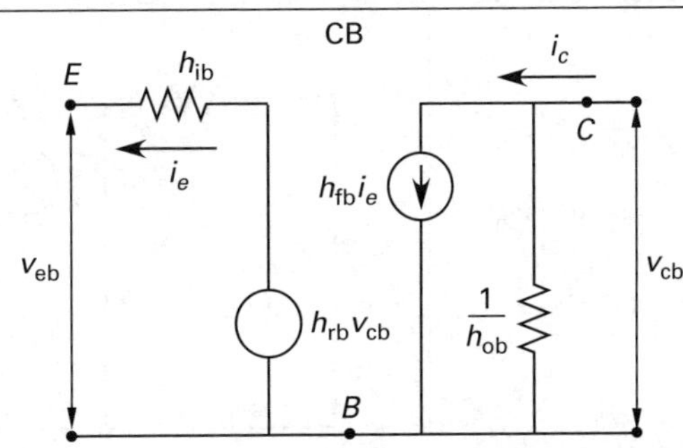	common base $v_{eb} = h_{ib} i_e + h_{rb} v_{cb}$ $i_c = h_{fb} i_e + h_{ob} v_{cb}$

Table 38.4 *Typical h-Parameter Values*

h-parameter	range of values	
h_{ie}	$0.25 \times 10^3\ \Omega$ to $8.0 \times 10^3\ \Omega$	
h_{re}	$4.0 \times 10^{-4}\ \Omega$ to $8.0 \times 10^{-4}\ \Omega$	[max values]
h_{fe}	50 to 375	
h_{oe}	5.0×10^{-6} S to 200×10^{-6} S	

Figure 38.9 *Hybrid-π Model*

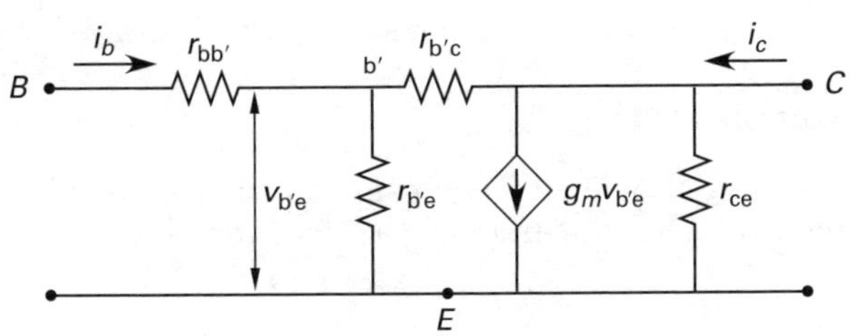

If the common emitter h-parameters are known, the hybrid-π parameters can be calculated from Eq. 38.28 through Eq. 38.32, in the order given.

$$g_m = \frac{|I_C|}{V_T} \qquad 38.28$$

$$r_{b'e} = \frac{h_{fe}}{g_m} \qquad 38.29$$

$$r_{bb'} = h_{ie} - r_{b'e} \qquad 38.30$$

$$r_{b'c} = \frac{r_{b'e}}{h_{re}} \qquad 38.31$$

$$g_{ce} = h_{oe} - \frac{1 + h_{fe}}{r_{b'c}} \qquad 38.32$$

12. TRANSISTOR CIRCUIT LINEAR ANALYSIS

Because the small-signal, low-frequency response of a transistor is linear, it can be obtained analytically rather

Table 38.5 *BJT Simplified Equivalent Circuits*

common connection	equivalent circuit	network equations
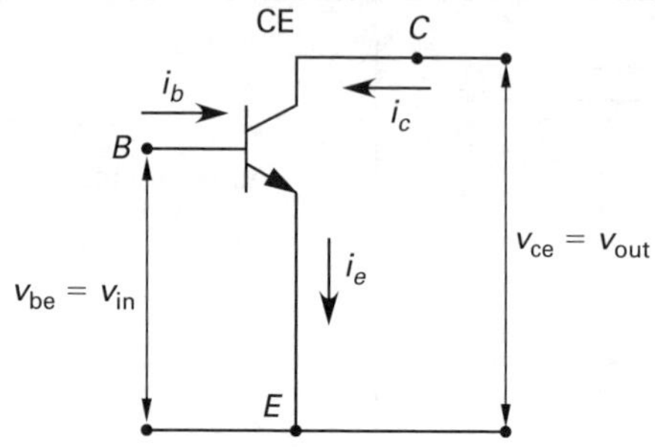	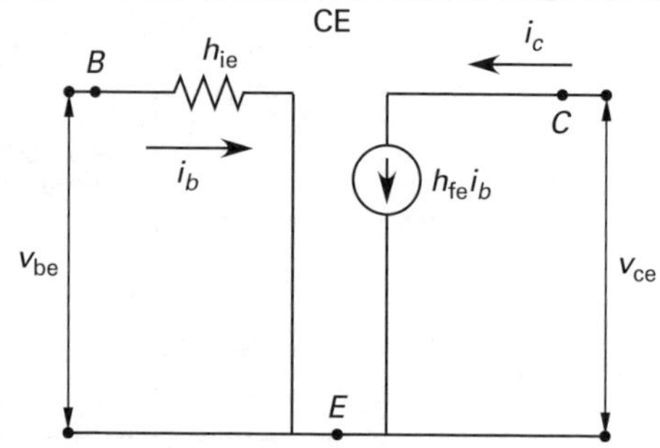	common emitter* $v_{be} = h_{ie} i_b \approx 0.7\text{ V}$ $i_c = h_{fe} i_b$
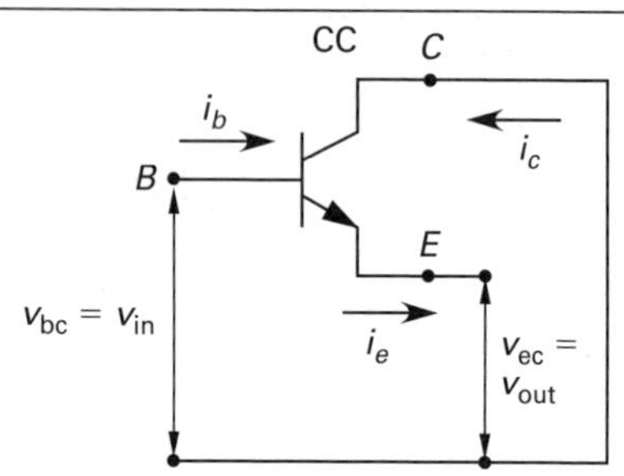	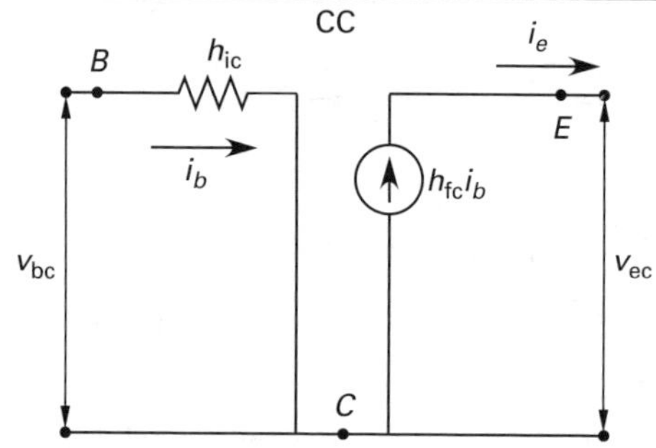	common collector $v_{bc} = h_{ic} i_b$ $i_e = h_{fc} i_b$
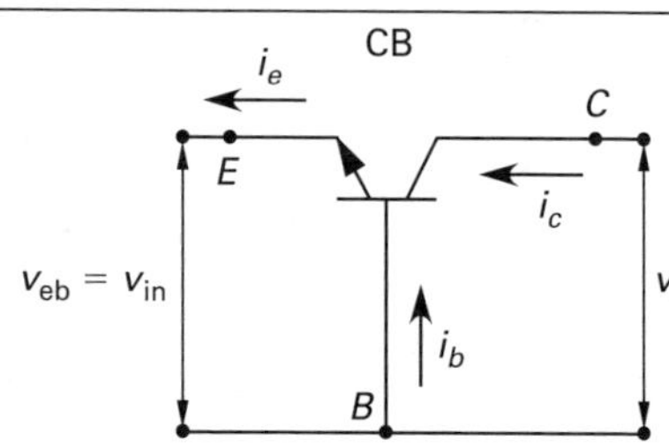	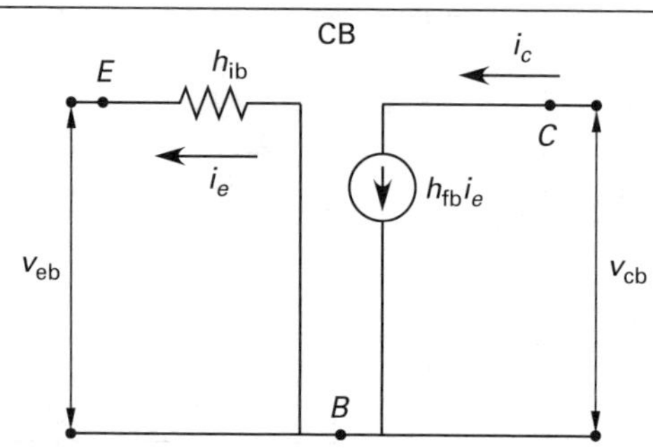	common base* $v_{eb} = h_{ib} i_e \approx 0.7\text{ V}$ $i_c = h_{fb} i_e$

*Germanium transistors are *pnp* types. $|v_{be}| = |v_{eb}| = 0.3$ V for germanium.

than graphically by using the models in Sec. 38.9.[11] The procedure for small-signal, or AC incremental, analysis follows.

step 1: Draw the circuit diagram. Include all external components, such as resistors, capacitors, and sources from the network.

step 2: Label the points for the base, emitter, and collector but do not draw the transistor. Maintain the points in the same relative position as in the original circuit.

step 3: Replace the transistor by the desired model. (Several models exist. Exact models were given in Sec. 38.9. Approximate models are given in Sec. 38.10. The hybrid-π model is given in Sec. 38.11.)

step 4: Because small-signal analysis is to be accomplished, only slight changes around the quiescent point are of interest. Therefore, replace each *independent* source by its internal resistance. An ideal voltage source is replaced with a short circuit. An ideal current source is replaced with an open circuit. (If biasing analysis were performed, that is, large-signal analysis, the sources would remain as drawn.) Additionally, replace inductors with open circuits, and capacitors with short circuits.

step 5: Solve for the desired parameter(s) in the resultant circuit using Kirchhoff's current and voltage laws.

Example 38.1

A transistor is used in a common emitter amplifier circuit as shown. Assume the inductor has infinite

[11]The *h*-parameters must be considered complex functions of the frequency to analyze high-frequency circuits, so the *h*-parameter model is not used.

impedance and the capacitors have zero impedance to AC signals. The transistor h-parameters are

$$h_{\text{ie}} = 750\ \Omega$$
$$h_{\text{oe}} = 9.09 \times 10^{-5}\ \text{S}$$
$$h_{\text{fe}} = 184$$
$$h_{\text{re}} = 1.25 \times 10^{-4}$$

(a) If a quiescent point is wanted at $V_{\text{CE}} = 11$ V and $I_C = 10 \times 10^{-3}$ A, what should the emitter resistance, R_E, be?

For parts (b) through (g), assume that $R_E = 800\ \Omega$. (b) Draw the DC load line. (c) If the base current (i_B) is 80×10^{-6} A, what is the collector current (i_C)? (d) What is the AC circuit voltage gain? (e) What is the AC circuit current gain? (f) What is the circuit's input impedance? (g) What is the output impedance? (h) Given $v_{\text{in}} = 0.25 \sin 1400t$ V, what is v_{out}? (i) What is the purpose of the inductor, L_C?

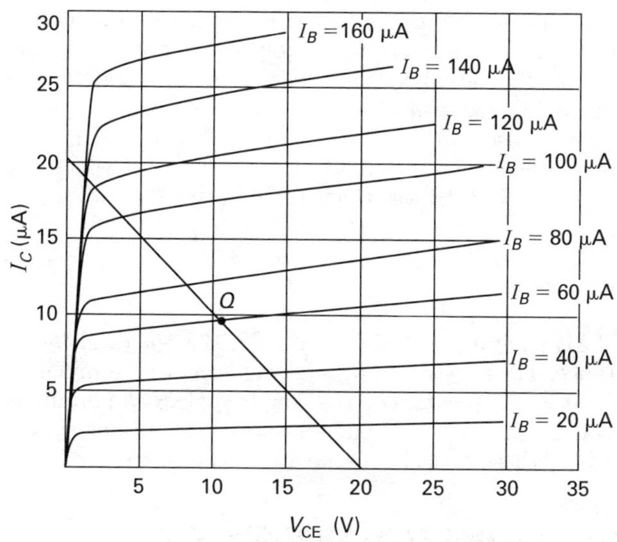

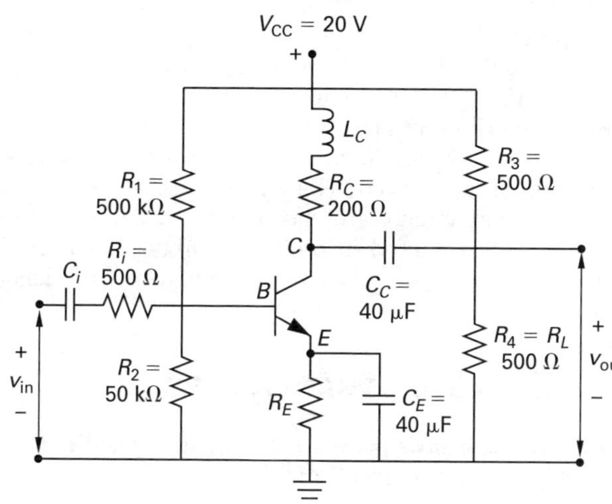

Solution

(a) Write the voltage drop in the common emitter circuit. Disregard the inductor (which passes DC signals). Use h_{fe} for h_{FE} because they are essentially the same and both are large. Kirchhoff's voltage law is

$$\alpha = \frac{\beta}{1+\beta} \approx \frac{h_{\text{fe}}}{1+h_{\text{fe}}} = \frac{184}{1+184} \approx 1$$

$$\begin{aligned} V_{\text{CC}} &= I_C R_C + I_E R_E + V_{\text{CE}} \\ &= I_C R_C + \left(\frac{I_C}{\alpha}\right) R_E + V_{\text{CE}} \\ &\approx I_C (R_C + R_E) + V_{\text{CE}} \end{aligned}$$

$$\begin{aligned} R_E &= \frac{V_{\text{CC}} - V_{\text{CE}}}{I_C} - R_C \\ &= \frac{20\ \text{V} - 11\ \text{V}}{10 \times 10^{-3}\ \text{A}} - 200\ \Omega \\ &\approx 700\ \Omega \end{aligned}$$

(b) If $I_C = 0$, then $V_{\text{CE}} = V_{\text{CC}}$. This is one point on the load line. If $V_{\text{CE}} = 0$, then

$$I_C = \frac{V_{\text{CC}}}{R_C + R_E} = \frac{20\ \text{V}}{200\ \Omega + 800\ \Omega} = 0.02\ \text{A}$$

These two points define the DC load line.

(c) From Eq. 38.3,

$$\begin{aligned} i_C &= \beta i_B = h_{\text{fe}} i_B \\ &= (184)(80 \times 10^{-6}\ \text{A}) = 0.0147\ \text{A} \quad (14.7\ \text{mA}) \end{aligned}$$

(d) To determine the AC circuit voltage gain, it is necessary to simplify the circuit. The bias battery V_{CC} is shorted out. (This is because the battery merely shifts the signal without affecting the signal swing.) Therefore, many of the resistors connect directly to ground. The inductor has infinite impedance, so R_C is disconnected. Both capacitors act as short circuits, so R_E is bypassed.

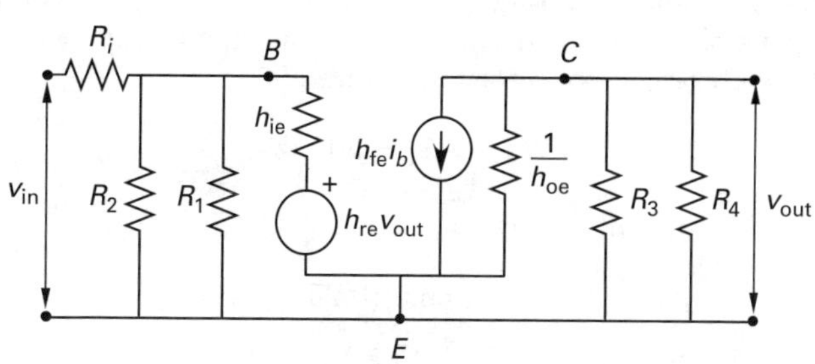

To simplify the circuit further, recognize that R_1, R_2, and $1/h_{\text{oe}}$ are very large and can be treated as infinite impedances.

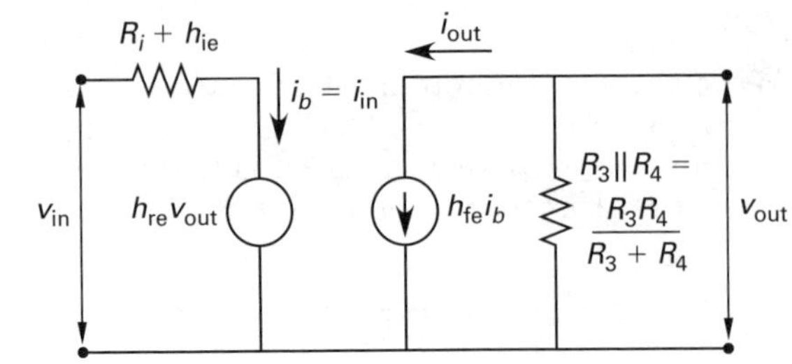

Continuing with the simplified analysis,

$$\begin{aligned} v_{out} &= h_{fe}i_b\left(\frac{R_3R_4}{R_3 + R_4}\right) \\ &= (184i_b)\left(\frac{(500\ \Omega)(500\ \Omega)}{500\ \Omega + 500\ \Omega}\right) = (46{,}000\ \Omega)i_b \end{aligned}$$

$$\begin{aligned} v_{in} &= i_b(R_i + h_{ie}) + h_{re}v_{out} \\ &= i_b(500\ \Omega + 750\ \Omega) + (1.25 \times 10^{-4})(46{,}000\ \Omega)i_b \\ &= (1256\ \Omega)i_b \end{aligned}$$

To perform an exact analysis, h_{oe} and h_{re} must be considered. Convert R_i to its Norton equivalent resistance and place it in parallel with R_1, R_2, and h_{ie}. Use the current-divider concept to calculate i_b.

The voltage gain is

$$A_V = \frac{v_{out}}{v_{in}} \approx \frac{(46{,}000\ \Omega)i_b}{(1256\ \Omega)i_b} = 36.6$$

(e) The current gain is

$$A_I = \frac{i_{out}}{i_{in}} \approx \frac{h_{fe}i_b}{i_b} = 184$$

(f) The input impedance (resistance) is

$$R_{in} = \frac{v_{in}}{i_{in}} \approx \frac{(1256\ \Omega)i_b}{i_b} = 1256\ \Omega$$

(g) The output impedance is effectively the Thevenin equivalent resistance of the output circuit. The load resistance (R_4 in this instance) is removed. The independent source voltage (v_{in}) is shorted, which effectively opens the controlled source $h_{fe}i_b$. The remaining resistance between collector and ground is

$$R_{out} = R_3 = 500\ \Omega$$

(h) The output voltage is

$$\begin{aligned} v_{out} = A_V v_{in} &= (36.6)(0.25\ \text{V})\sin 1400t \\ &= (9.15\ \text{V})\sin 1400t \end{aligned}$$

(i) There are several possible uses for the inductor. It might be included to limit voltage extremes, such as high-voltage spikes, which could damage the transistor when the amplifier is turned on or off. Alternatively, it might prevent AC current from being drawn across R_C, and, in so doing, hold V_C at a constant value.

13. TRANSISTOR CIRCUIT HIGH-FREQUENCY ANALYSIS

The *h*-parameter models are useful at low frequencies. For high frequencies, the hybrid-π model is modified to include the effects of various capacitances as shown in Fig. 38.10.

Figure 38.10 *High-Frequency Hybrid-π Model*

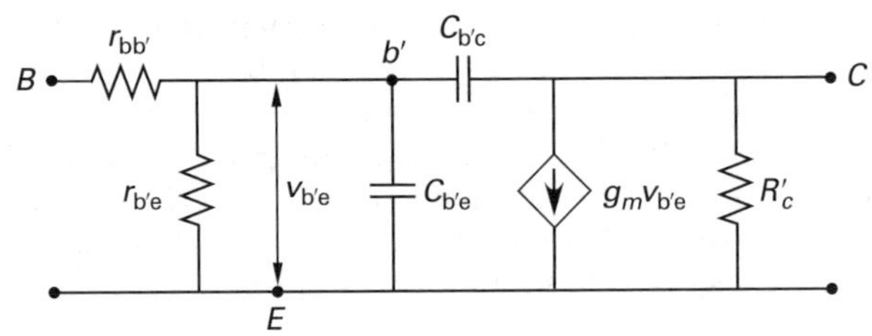

The capacitance, $C_{b'e}$, is the sum of the diffusion capacitance, C_d, and the transition capacitance, C_t (see Sec. 37.9). The capacitance, $C_{b'c}$, is the transition capacitance for the collector junction. The high-frequency model is used whenever the frequency exceeds the *corner frequency*, ω_{cf}, that is, the frequency at which the resistance is 3 dB less than its DC value.[12]

$$\omega_{cf} = \frac{1}{RC} \quad \text{38.33}$$

The capacitance term, C, in Eq. 38.33 is the total capacitance. The resistance, R, is as seen from the terminals of the total capacitance. (In low-frequency *h*-parameter models, the bypass capacitors are typically not shown. It is to these capacitors that Eq. 38.33 refers.)

14. UNIJUNCTION TRANSISTORS

A *unijunction transistor* is an electronic device with three contacts, two ohmic and one to a *pn* junction, as shown in Fig. 38.11(a). The output characteristics are illustrated in Fig. 38.11(b), which clearly shows a negative resistance region. The UJT operates in two states: on (low resistance) and off (high resistance). The symbol and equivalent circuit are shown in Fig. 38.11(c) and (d). The device is used in oscillator circuits, as a *UJT relaxation oscillator*, and in pulse generation and delay circuits.

15. DARLINGTON TRANSISTORS

A *Darlington transistor*, more commonly called a *Darlington pair*, is a current amplifier consisting of two

[12]When the frequency is $10\omega_{cf}$, the impedance is essentially the capacitive reactance. The resistance is considered insignificant at that point.

Figure 38.11 *Unijunction Transistor*

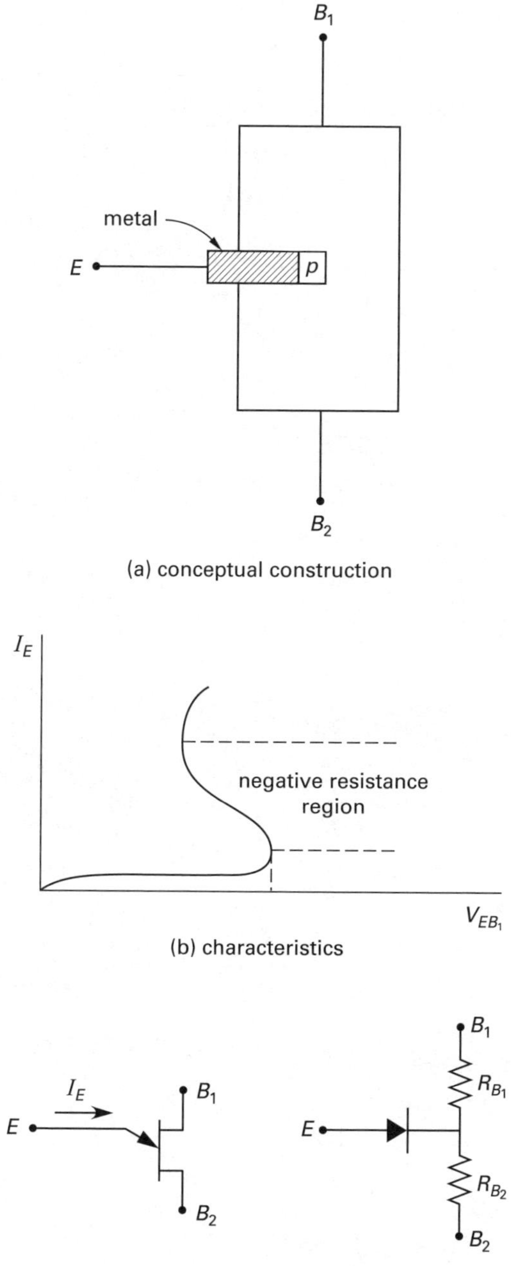

separate transistors treated as one. A Darlington transistor has a high input impedance and a significant current gain. For example, the forward current transfer gain, h_{fe}, can be as high as 30,000. A Darlington transistor is illustrated in Fig. 38.12.

Figure 38.12 *Darlington Transistor*

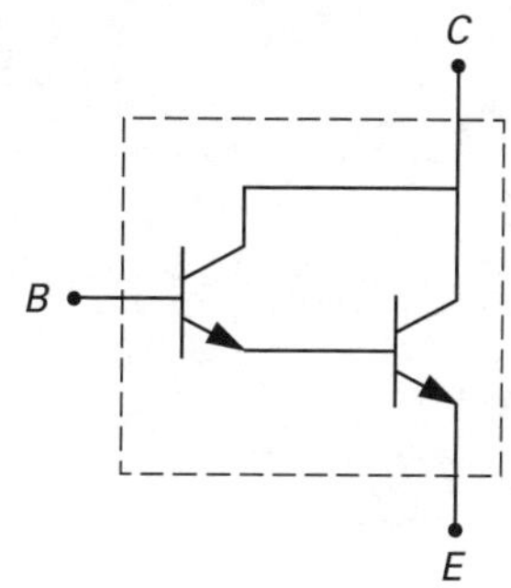

39 Field Effect Transistors

1. FET Fundamentals 39-1
2. JFET Characteristics 39-2
3. JFET Biasing 39-3
4. FET Models 39-4
5. MOSFET Characteristics 39-4
6. MOSFET Biasing 39-5

Nomenclature

A	gain	–
C	capacitance	F
g_m	mutual conductance or transconductance	S
i	instantaneous or variable current	A
I	effective or DC current	A
I_{DSS}	zero-gate-voltage drain current	A
n	electron concentration	m^{-3}
p	hole concentration	m^{-3}
r	small-signal resistance[1]	Ω
R	resistance	Ω
v	instantaneous or variable voltage	V
V	effective or DC voltage	V
V_{DD}	drain supply voltage	V

Subscripts

CC	collector supply
d	dynamic or diffusion
ds, DS	drain to source
D	donor, diode, or drain
DSQ	drain to source at Q-point
DSS	drain to source saturation
gs, GS	gate to source
G	gap, gate
GD	gate to drain
GSQ	gate to source at Q-point
i	intrinsic or input (AC or instantaneous)
iss	common-source short-circuit input
L	load
m	mutual
oss	common-source short-circuit output
P	pinchoff or power
$P0$	pinchoff at 0 V
s	saturation
S	source
SD	source to drain
SG	source to gate
V	constant or effective voltage

[1]This is also known as the *dynamic resistance* or the *instantaneous resistance*. Any resistance represented with a lowercase letter is the small-signal value.

1. FET FUNDAMENTALS

Field-effect transistors (FETs) are bidirectional devices constructed of an n-type channel surrounded by a p-type gate, or vice versa.[2] The connections are named in a different manner than BJTs in order to distinguish the two. For a FET (BJT), the connections are the *gate* (base), *source* (emitter), and *drain* (collector). In a BJT, the base current controls the overall operation of the transistor. In a FET, the gate voltage and the corresponding electric field control the overall operation of the transistor. The various configurations and the concepts regarding biasing, load lines, and amplifier operation described for BJTs also apply to FETs.

There are two major types of FETs: the *junction field-effect transistor* (JFET) and the *metal-oxide semiconductor field-effect transistor* (MOSFET). Both types are made in n- and p-channel types, with the n-channel types more common. The fundamental difference between the JFET and the MOSFET is that the latter can operate in the enhancement mode (see Sec. 39.5).[3]

Variations in construction result in different names and properties. An n-channel MOSFET is sometimes referred to as an NMOS, and the p-channel MOSFET is sometimes called a PMOS. A high-power MOSFET is called an HMOS. A MOSFET with increased drain current capacity caused by its V-type structure is called a VMOS. A double-diffused MOSFET, which has replaced the VMOS except in high-frequency applications, is called a DMOS. A bipolar transistor using an insulated gate is called an *insulated gate bipolar transistor* (IGBT). An SCR using an insulated gate for control is called a *MOS-controlled thyristor* (MCT). A MOSFET with p-channel and n-channel devices on the same chip is called a *complementary MOSFET* (CMOS) and is used primarily because of its low power dissipation.

When the conventional MOSFET oxide layer is replaced by a double layer of nitride and oxide, which allows electrons to tunnel from the gate, the device is called a *metal-nitride-oxide-silicon* (MNOS) transistor. The MNOS device is used as a read-only memory (ROM) and MOS transistors are also used as memory devices. The MOS transistor is used to control the transfer of the charge from one location to the next. As a result, it is also called a *charge-transfer device* (CTD). When the charge transfer is

[2]Bidirectional indicates that the current flow direction depends on the potential between the source and the drain.
[3]MOSFETs are also much more susceptible to damage from *electrostatic discharges* (ESD) due to the thin oxide layer at the gate.

accomplished on the circuit level by discrete MOS transistors and capacitors, it is called a *bucket-brigade device* (BBD). When the charge transfer takes place on the device level, it is called a *charge-coupled device* (CCD).

2. JFET CHARACTERISTICS

Junction field-effect transistor (JFET) construction, characteristics, and symbols are shown in Fig. 39.1.

A bidirectional channel for current flow exists between the source, S, and the drain, D. The current flow is controlled by the reverse biased *pn* junction at the gate, G, with the depletion width providing the control.[4] For example, when the gate is unbiased, a normal depletion zone is set up at the gate *pn* junction, as shown in Fig. 39.2(a) for an *n*-channel JFET.

Once biased, the depletion layer shifts, and the JFET operates in a relatively linear fashion in the *ohmic region*, also called the *triode region*. The ohmic depletion layer is shown in Fig. 39.2(b). The ohmic region is shown in Fig. 39.1(b). As the drain-source voltage increases, the depletion layer widens until it encompasses the drain terminal. At this point, the current is pinched off, and the JFET operates in a relatively constant manner in the *pinchoff region*. The pinchoff region depletion layer is shown in Fig. 39.2(c). The pinchoff region is shown in Fig. 39.1(b). As the drain-source voltage continues to increase, avalanche breakdown occurs and the voltage remains approximately constant as the current increases dramatically. This is known as the *breakdown* or *avalanche region* (not shown).

For a fixed value of V_{GS}, the drain-source voltage separating the resistive and pinchoff regions is the *pinchoff voltage*. As Fig. 39.1(b) shows, there is a value for V_{GS} for which no drain current flows. This is also referred to as the pinchoff voltage but is designated $V_{GS(off)}$. The drain current corresponding to the horizontal part of a curve (for a given value of V_{GS}) is the *saturation current*, represented by I_{DSS}.

The term "pinchoff voltage" and the symbol V_P are ambiguous, as the actual pinchoff voltage in a circuit depends on the gate-source voltage, V_{GS}. When V_{GS} is zero, the pinchoff voltage is represented unambiguously by V_{P0} (where the zero refers the value of V_{GS}). For other values of V_{GS},

$$V_P = V_{P0} + V_{GS} \qquad 39.1$$

Some manufacturers do not adhere to this convention when reporting the pinchoff voltage for their JFETs. They may give a value for V_{P0} and refer to it as V_P. The absence of a value for V_{GS} implies that the value given is actually V_{P0}.

[4]The source "emits" the *n*- or *p*-channel carriers, by convention. The reverse bias is therefore with respect to the gate. FETs are bidirectional, nevertheless.

Figure 39.1 *Junction Field-Effect Transistor*

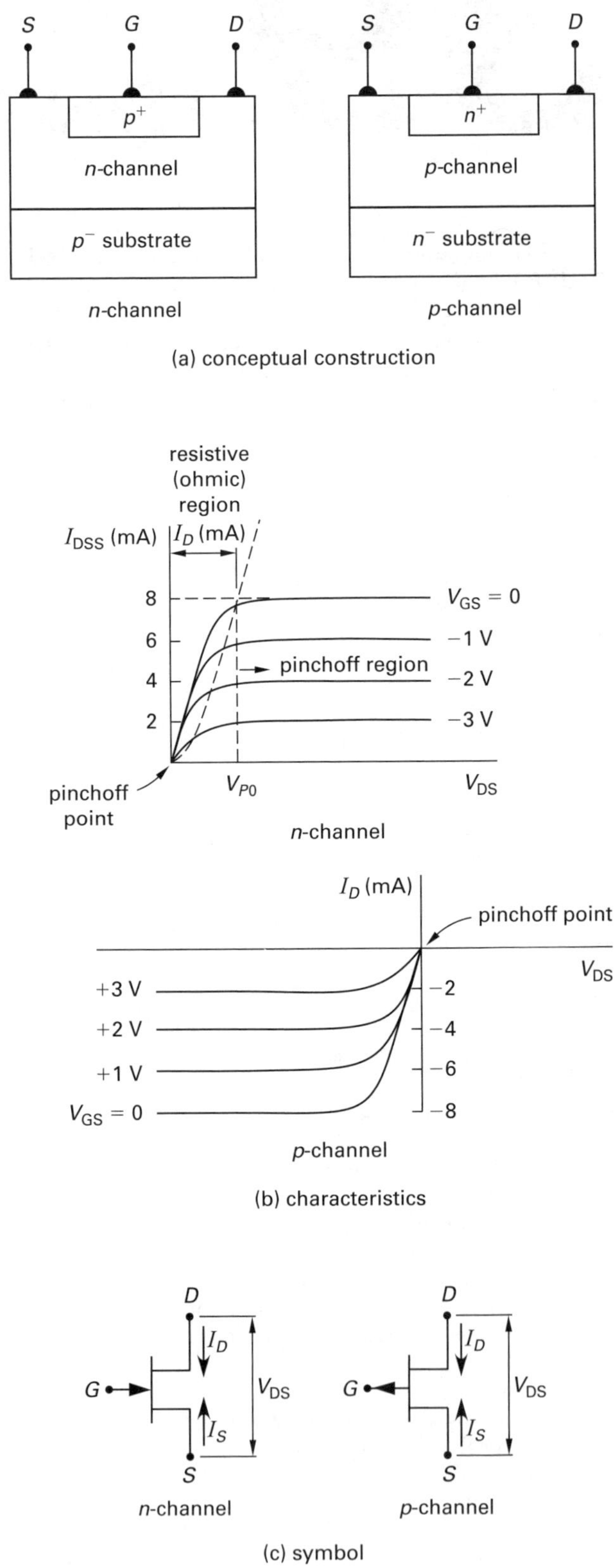

Figure 39.2 *Pinchoff Theory*

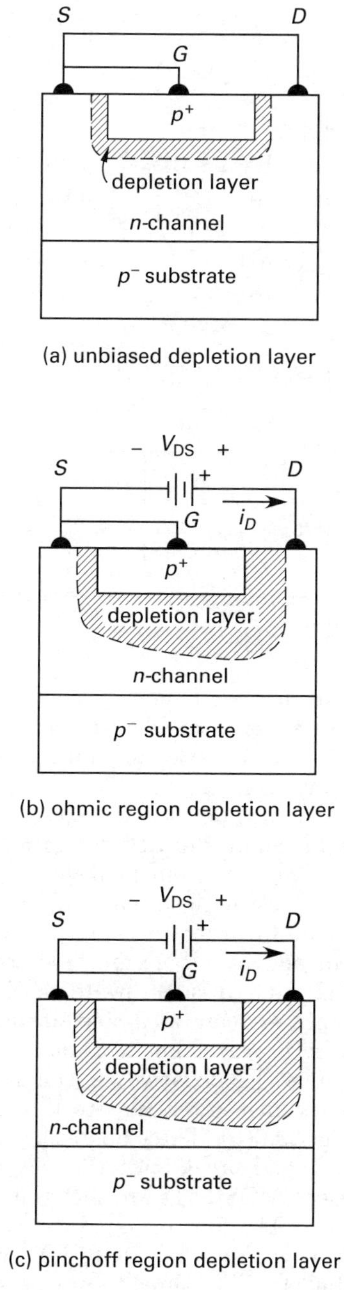

Some manufacturers do not provide characteristic curves, choosing instead to indicate only I_{DSS} and $V_{GS(off)}$. The characteristic curves can be derived, as necessary, from *Shockley's equation.*

$$I_D = I_{DSS}\left(1 - \frac{V_{GS}}{V_P}\right)^2 \qquad 39.2$$

The *transconductance*, g_m, is defined for small-signal analysis by Eq. 39.3.

$$\begin{aligned} g_m &= \frac{\Delta I_D}{\Delta V_{GS}} = \frac{i_D}{v_{GS}} \\ &= \left(\frac{-2I_{DSS}}{V_P}\right)\left(1 - \frac{V_{GS}}{V_P}\right) = g_{mo}\left(1 - \frac{V_{GS}}{V_P}\right) \\ &\approx \frac{A_V}{R_{out}} \end{aligned} \qquad 39.3$$

The drain-source resistance can be obtained from the slope of the V_{GS} characteristic in Fig. 39.1(b).

$$r_d = r_{DS} = \frac{\Delta V_{DS}}{\Delta I_D} = \frac{v_{DS}}{i_D} \qquad 39.4$$

3. JFET BIASING

JFETs operate with a reverse biased gate-source junction. The quiescent point is established by choosing V_{GSQ}. I_D is then determined by Shockley's equation (see Eq. 39.2).

Figure 39.3 shows a JFET self-biasing circuit. Because the gate current is negligible, the load line equation is

$$\begin{aligned} V_{DD} &= I_S(R_D + R_S) + V_{DS} \\ &= I_D(R_D + R_S) + V_{DS} \end{aligned} \qquad 39.5$$

Figure 39.3 *Self-Biasing JFET Circuit*

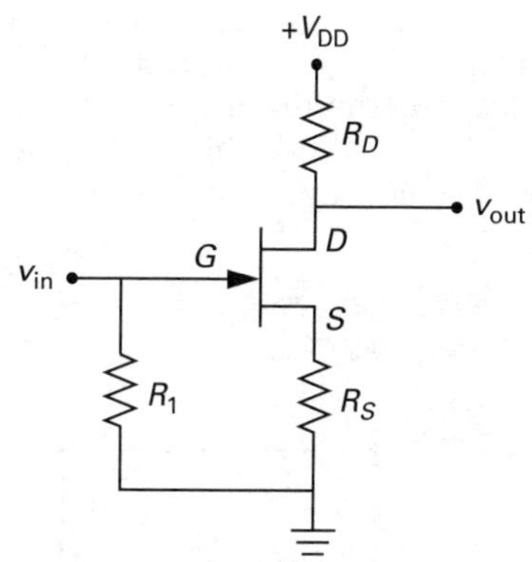

The load line for any JFET is determined using the procedures outlined in Sec. 37.7.

At the quiescent point, $V_{in} = 0$. From Kirchhoff's voltage law, around the input loop,

$$V_{GS} = -I_S R_S = -I_D R_S \qquad 39.6$$

Example 39.1

A JFET with the characteristics shown operates as a small-signal amplifier. The supply voltage is 24 V. A quiescent bias source current of 5 mA (0.005 A) is desired at a bias voltage of $V_{DS} = 15$ V. Design a self-biasing circuit similar to Fig. 39.3.

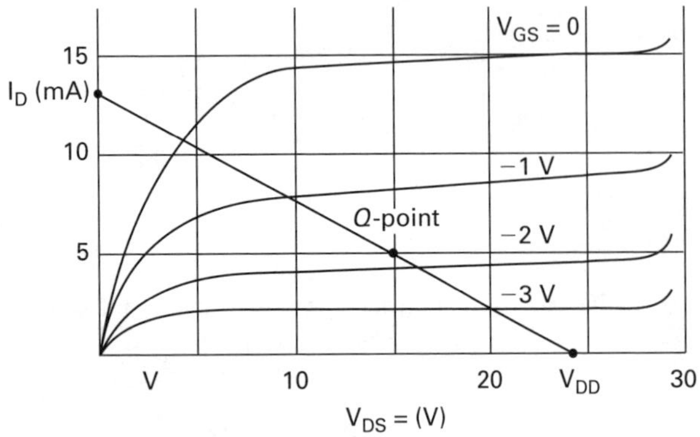

Solution

Because the gate draws negligible current, $I_D = I_S$. At the quiescent point, $V_{GS} = -1.75$ V. From Eq. 39.6,

$$R_S = \frac{-V_{GS}}{I_D} = \frac{-(-1.75 \text{ V})}{0.005 \text{ A}} = 350 \ \Omega$$

From Eq. 39.5,

$$\begin{aligned} R_D &= \frac{V_{DD} - V_{DS}}{I_S} - R_S = \frac{24 \text{ V} - 15 \text{ V}}{0.005 \text{ A}} - 350 \ \Omega \\ &= 1450 \ \Omega \end{aligned}$$

4. FET MODELS

A field-effect transistor can be modeled as shown in Fig. 39.4. The model is valid for both the JFET and the MOSFET. Because the drain resistance is very high (see the typical values in Table 39.1), the model can be simplified further by removing r_{DS}, that is, by assuming $r_{DS} \approx \infty$.

Figure 39.4 *FET Equivalent Circuit*

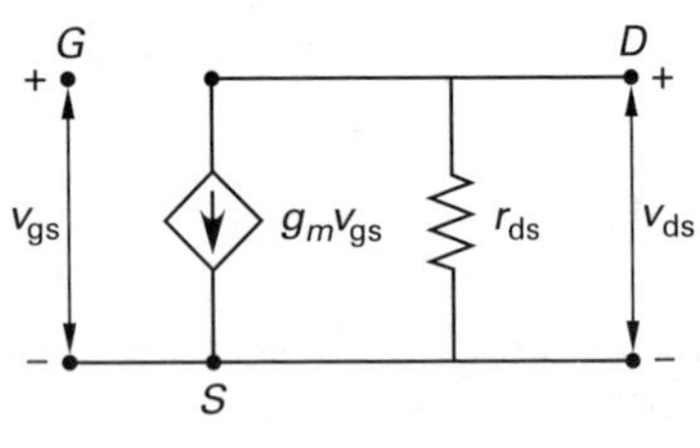

At high frequencies the various capacitances associated with the FET must be accounted for, as shown in the model in Fig. 39.5.

5. MOSFET CHARACTERISTICS

The *metal-oxide semiconductor field-effect transistor* (MOSFET) is constructed as either a *depletion device* or an *enhancement device* as shown in Fig. 39.6(a). In

Table 39.1 *Typical FET Parameter Values*

parameter	JFET	MOSFET
g_m	0.1×10^{-3} S to 10×10^{-3} S	0.1×10^{-3} S to 20×10^{-3} S or greater
r_{ds}	$0.1 \times 10^6 \ \Omega$ to $1 \times 10^6 \ \Omega$	$1 \times 10^3 \ \Omega$ to $50 \times 10^3 \ \Omega$
C_{ds}	0.1×10^{-12} F to 1×10^{-12} F	0.1×10^{-12} F to 1×10^{-12} F
C_{gd}, C_{gs}	1×10^{-12} F to 10×10^{-12} F	1×10^{-12} F to 10×10^{-12} F

Figure 39.5 *FET High-Frequency Model*

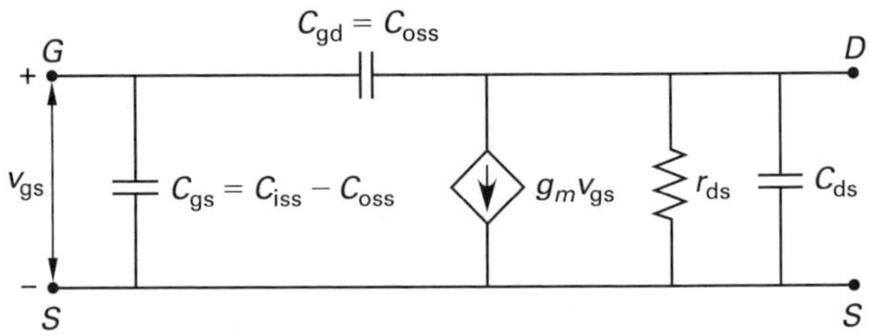

the depletion device, a channel for current exists with the gate voltage at zero. In order to control the current, a voltage applied to the gate must push channel majority carriers away from the gate, pinching off the current. For an *n*-channel device, the gate voltage must be negative. For a *p*-channel device, the gate voltage must be positive.[5] In an enhancement device, a channel for current does not exist with the gate voltage at zero, but must be created. In order to create the channel and control the current, a voltage must be applied to the gate to pull majority carriers toward the gate, increasing the current. For an *n*-channel device, the gate voltage must be positive. For a *p*-channel device, the gate voltage must be negative. A summary of these characteristics, which are similar to those for JFETs, is shown in Fig. 39.6(b). Three operating regions exist: ohmic, pinchoff, and breakdown. The various symbols used to represent MOSFETs are shown in Fig. 39.6(c) and Fig. 39.6(d). The four-terminal devices shown are IEEE standard symbols for MOSFETs with passive substrates (bulk). The three-terminal symbols are shorthand notation for active substrates (i.e., the bulk is connected to the source). The arrow of the bulk terminal of the four-terminal devices indicates the polarity of the bulk-drain, bulk-source, and bulk-channel diodes. The arrow points toward the channel for an NMOS device and away for a PMOS device. The arrow on the three-terminal devices indicates the source terminal and points in the direction of conventional (i.e., positive) current flow. The MOSFET model is identical to that for the JFET (see Sec. 39.4).

[5]The gate voltage is referenced to the source by convention, since the source is considered the emitter of majority carriers into the channel.

Figure 39.6 *Metal-Oxide Semiconductor Field-Effect Transistor*

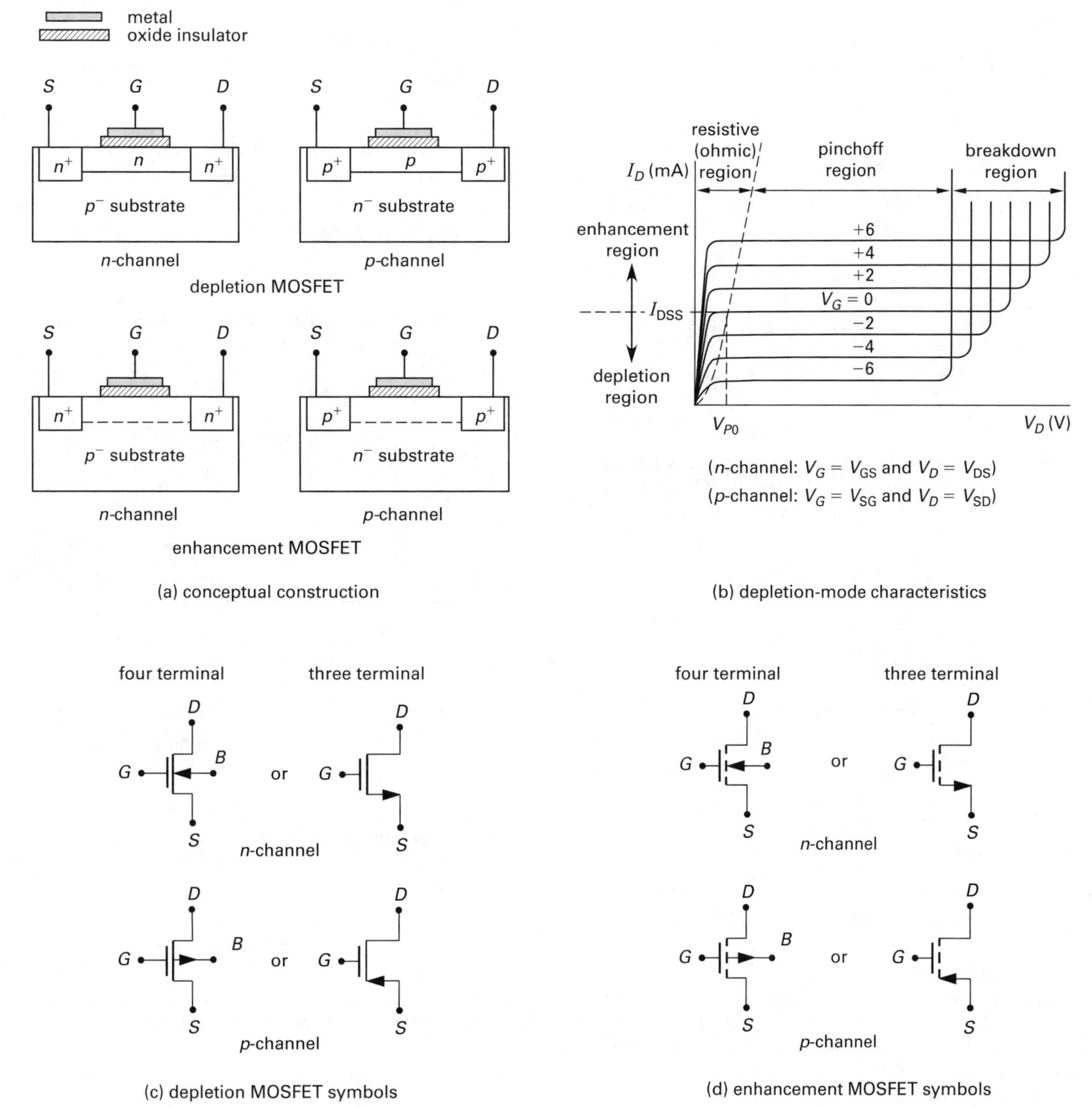

6. MOSFET BIASING

A typical MOSFET biasing circuit is shown in Fig. 39.7. Because MOSFETs do not have a *pn* junction between the gate and channel, biasing can be either forward or reverse. The polarity of the gate-source bias voltage depends on whether the transistor is to operate in the depletion mode or enhancement mode. Because no gate current flows, the resistor R_G is used merely to control the input impedance. When R_G is very large, the input impedance is essentially R_2.

Together, R_1 and R_2 form a voltage divider. Because the gate current is zero, the voltage divider is unloaded. Therefore, the gate voltage is

$$V_G = V_{\text{DD}}\left(\frac{R_2}{R_1 + R_2}\right) = V_{\text{GS}} + I_S R_S \qquad 39.7$$

The load line equation is found from Kirchhoff's voltage law and is the same as for the JFET.

$$V_{\text{DD}} = I_S(R_D + R_S) + V_{\text{DS}} \qquad 39.8$$

***Figure 39.7** Typical MOSFET Biasing Circuit*

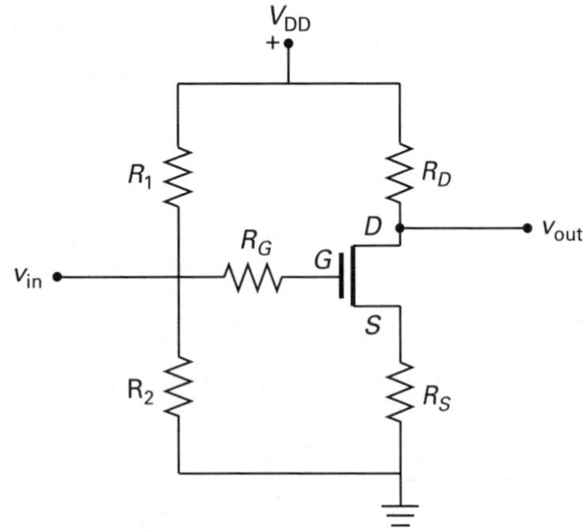

Example 39.2

A MOSFET is used in an amplifier as shown. All capacitors have zero impedance to AC signals. The no-signal drain current (I_D) is 20×10^{-3} A. The performance of the transistor is defined by

$$I_D = (14 + V_{GS})^2 \quad \text{[in A]}$$

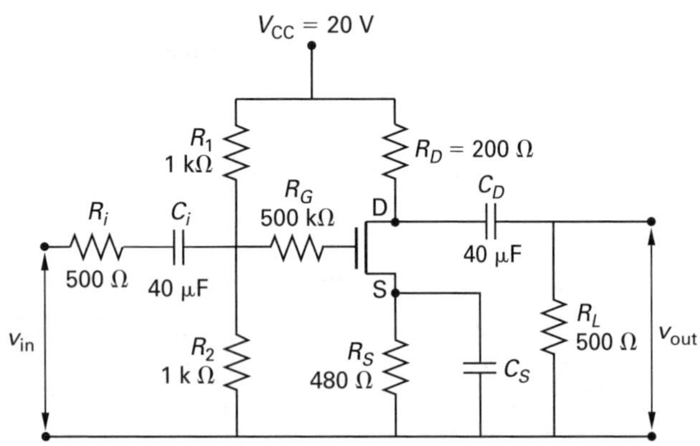

(a) What is the V_D (potential) at point D with no signal? (b) What is V_S with no signal? (c) What is V_{DSQ}? (d) If $V_G = 0$ V, what is V_{GS}? (e) What is the input impedance? (f) What is the voltage gain? (g) What is the output impedance?

Solution

(a) $V_D = V_{CC} - I_D R_D$

$$= 20 \text{ V} - (20 \times 10^{-3} \text{ A})(200 \ \Omega) = 16 \text{ V}$$

(b) The voltage drop between the source and the ground is through R_S. Because the gate draws negligible current, the drain and source currents are the same.

$$V_S = I_S R_S = (20 \times 10^{-3} \text{ A})(480 \ \Omega) = 9.6 \text{ V}$$

(c) The quiescent voltage drop across the drain-source junction is

$$V_{DSQ} = V_D - V_S = 16 \text{ V} - 9.6 \text{ V} = 6.4 \text{ V}$$

(d) If the gate is at zero potential and the source is at 9.6 V potential, then

$$V_{GS} = V_G - V_S = 0 \text{ V} - 9.6 \text{ V} = -9.6 \text{ V}$$

(e) To simplify the circuit, short out the bias battery, V_{CC}. Consider the capacitors as short circuits to AC signals.

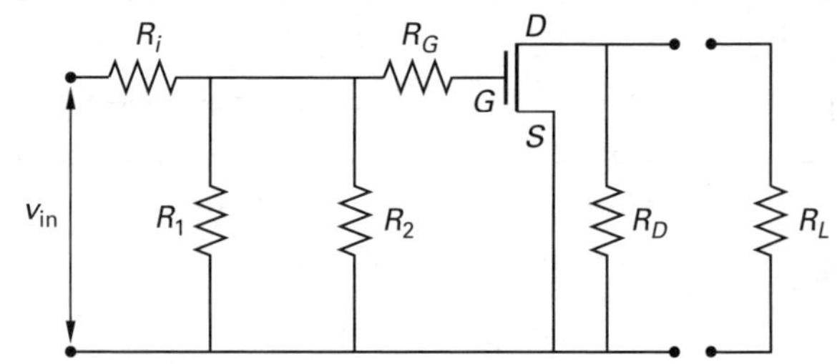

Because R_G is so large, it effectively is an open circuit. The input impedance (resistance) is

$$R_{in} = R_i + R_1 \| R_2 = 500 \ \Omega + \frac{(1000 \ \Omega)(1000 \ \Omega)}{1000 \ \Omega + 1000 \ \Omega}$$
$$= 1000 \ \Omega$$

(f) A FET has properties similar to a vacuum tube. The voltage gain is normally calculated from the transconductance.

$$A_V \approx g_m R_{out}$$

The transconductance is not known, but it can be calculated from Eq. 39.3 and the performance equation.

$$g_m = \frac{dI_D}{dV_{GS}} = \frac{d(14 + V_{GS})^2 \times 10^{-3}}{dV_{GS}}$$
$$= (2)(14 + V_{GS}) \times 10^{-3}$$

Because V_{GS} is -9.6 V at the quiescent point,

$$g_m = (2)(14 - 9.6 \text{ V}) \times 10^{-3} = 8.8 \times 10^{-3} \text{ S}$$

The resistance is a parallel combination of R_D, R_L, and r_d. The drain-source resistance, r_d, is normally very large and, because it was not given in this problem, is disregarded. Then,

$$\begin{aligned} R &= R_D \| R_L = \frac{R_D\ R_L}{R_D + R_L} \\ &= \frac{(200\ \Omega)(500\ \Omega)}{200\ \Omega + 500\ \Omega} = 143\ \Omega \\ A_V &= g_m R = (8.8 \times 10^{-3}\ \text{S})(143\ \Omega) = 1.26 \end{aligned}$$

This is a small gain. If advantage is not being taken of other properties possessed by the circuit, the resistances should be adjusted to increase the voltage gain.

(g) Proceeding as in the solution to Ex. 38.1, part (g), the output impedance is $R_D = 200\ \Omega$.

40 Electrical and Electronic Devices

Part 1: Introduction 40-2

Part 2: Amplifiers 40-2
1. Fundamentals 40-2
2. Ideal Operational Amplifiers 40-5
3. Operational Amplifier Limits 40-8
4. Amplifier Noise 40-8

Part 3: Pulse Circuits: Waveform Shaping and Logic 40-9
5. Pulse Circuit Fundamentals 40-9
6. Clamping Circuits 40-9
7. Zener Voltage Regulator Circuit: Ideal 40-13
8. Zener Voltage Regulator Circuit: Practical 40-14
9. Operational Amplifier Regulator Circuit . . . 40-16
10. Transistor Switch Fundamentals 40-17
11. JFET Switches 40-18
12. CMOS Switches 40-18
13. Active Waveform Shaping 40-19
14. Logic Families 40-19
15. Logic Gates 40-19
16. Simplification of Binary Variables 40-20
17. Logic Circuit Fan-Out 40-20
18. Logic Circuit Delays 40-23
19. Resistor-Transistor Logic (RTL) 40-24
20. Diode-Transistor Logic (DTL) 40-24
21. Transistor-Transistor Logic (TTL or T^2L) 40-24
22. Emitter-Coupled Logic (ECL) 40-26
23. MOS Logic 40-26
24. CMOS Logic 40-26
25. Multivibrators 40-27

Part 4: Devices and Circuits 40-28
26. Circuit: Phase-Locked Loop (PLL) 40-28

Nomenclature

A	gain	–
C	capacitance	F
D	diffusion constant	m^2/s
f	frequency	Hz
F_m	figure of merit	rad/s
G	gain	–
h_{fe}	CE small-signal (AC) forward current transfer ratio or gain	–
h_{FE}	CE static input (DC) forward current transfer ratio or gain	–
H	high	V
i	instantaneous or variable current	A
I	effective or DC current	A
I_{CBO}	reverse saturation current	A
I_{EBO}	DC emitter cutoff current, collector open	A
I_{ZK}	keep-alive current	A
κ	Boltzmann's constant, 1.3805×10^{-23}	J/K
N	fan-out (number of loads)	–
P	power	W
q	electric unit charge, 1.6022×10^{-19}	C
R	resistance	Ω
S_R	slew rate	V/s
t	time	s
T	temperature	K
v	instantaneous or variable voltage	V
V	effective or DC voltage	V
V_T	voltage equivalent of temperature	V
Z	impedance	Ω

Symbols

α	CB forward current ratio or gain	–
δ	feedback factor	–
ω	angular frequency	rad/s

Subscripts

0	resonance
amp	amplifier
A, B	input A, input B
b, B	base
BE	base to controller
c	AC or small signal collector
c	control or collector
cl	current limiting
cm	common mode
co	cutoff
C	collector
C	DC or biased collector
CC	collector supply
CE	collector to emitter

d	delay
dm	differential mode
D	diode
DC	direct current
DD	drain supply
e	AC or small signal emitter
E	DC or biased emitter
EBO	emitter to base, collector open
EE	emitter supply
f	fall, feedback, or forward (AC or instantaneous)
fl	full load
F	forward (DC component or total)
GS	gate to source
i	current
L	load
m	merit
max	maximum
min	minimum
n	noise
nl	no load
p	power
r	rise
ref	reference
R	resistance
s	signal, source, storage, supply, or input
T	thermal
u	unity
v	instantaneous voltage
V	voltage
Z	zener
ZK	zener knee
ZM	zener at maximum rated current

PART 1: INTRODUCTION

An understanding of basic electronics and associated circuits is necessary for the electronics engineer who will apply this knowledge to fabricate more complex circuits from the basic building blocks. A power electrical engineer makes use of such knowledge because of the widespread use of electronic devices in instrumentation, illumination controls, battery charging automatic circuits, power electronics, variable speed drives, protective relays, transmission line monitoring, and control circuits. A computer engineer, while primarily focused on utilizing the computer to accomplish various tasks, must have an appreciation of the underlying hardware in order to take full advantage of the available capabilities.

This chapter develops the principles of the operational amplifier, various pulse circuits used for waveform shaping and logic circuits, and specific circuits with widespread applications. These principles are essential to an electrical engineer, as they provide a basis for understanding electronics and associated circuits.

PART 2: AMPLIFIERS

1. FUNDAMENTALS

An *amplifier* is any device capable of increasing the magnitude or power level of a physical quantity. The prime concern of most amplifiers is gain, A. The gain can be a voltage (A_V), current (A_i), or power gain (A_p or G_p). The input impedance (Z_{in}) and the output impedance (Z_{out}) are also parameters of concern. The *bandwidth* is the range of frequencies for which the amplifier will respond. The bandwidth (BW) is normally defined as the difference between the upper and lower frequencies at the *3 dB half-power points*, also called the *3 dB downpoints*, as shown in Fig. 40.1.

Figure 40.1 Bandwidth

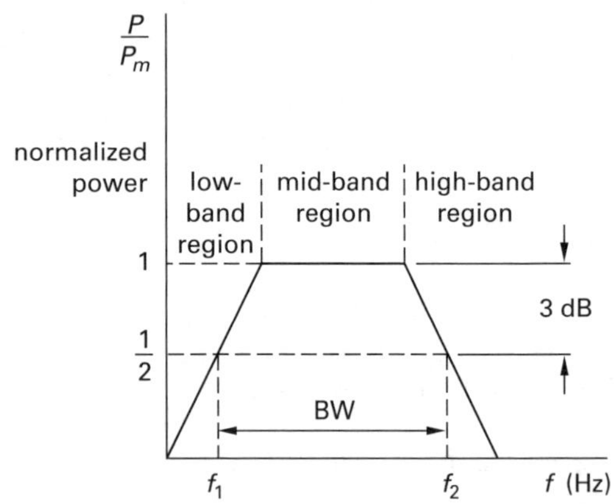

The amplifier in Fig. 40.1. is a *band-pass* amplifier. There are also *low-pass*, *high-pass*, and *notch* amplifiers, though they are more commonly called *filters*. The gain-bandwidth product is used as a *figure of merit*, F_m, measured in rad/s, for band-pass and high-pass amplifiers. (Bandwidth is sometimes measured in Hz instead of rad/s, in which case the figure of merit is also given in Hz.) The *gain-bandwidth product* (GBW) for a band-pass amplifier is

$$\text{GBW} = F_m = A_{\text{ref}}(\text{BW}) \qquad 40.1$$

The term A_{ref} is a reference gain, which is either the maximum gain or the gain at the frequency at which the gain is purely real or purely imaginary. For a lowpass amplifier, a high-frequency cutoff, ω_0, at the 3 dB downpoint is defined and the figure of merit is[1]

$$\text{GBW} = F_m = A_{\text{ref}}\omega_0 \tag{40.2}$$

The transistors described in Chap. 36, when biased in the linear region, are amplifiers. The small-signal models given are for the *mid-band region* of Fig. 40.1. In this region, the reactive elements act as short circuits and so are not included in any models.[2] In the *low-band region,* the impedance of the reactive elements, normally the coupling capacitors and emitter bypass capacitors, is significant.[3] Capacitors must be added to the small-signal models in the low-band region, and the gain decreases.[4] In the *high-band region,* the impedance caused by the capacitance of the device itself becomes significant, and the gain decreases. In this manner, the device capacitances are added to the models (see Sec. 38.13 and Sec. 39.4).

Individual transistors amplify. When used in conjunction with other transistors and electronic devices, different types of amplifiers can be constructed. An amplifier with a wide range of applications is the *operational amplifier* (op amp). It takes its name from the multiple configurations with which mathematical operations may be performed. Originally designed as the major component of an analog computer, the op amp is a high-gain DC (direct-coupled or direct current) amplifier with high stability and immunity to oscillation, which uses large amounts of negative feedback. The symbols used for the op amp are shown in Fig. 40.2.

Figure 40.2 *Operational Amplifier Symbols*

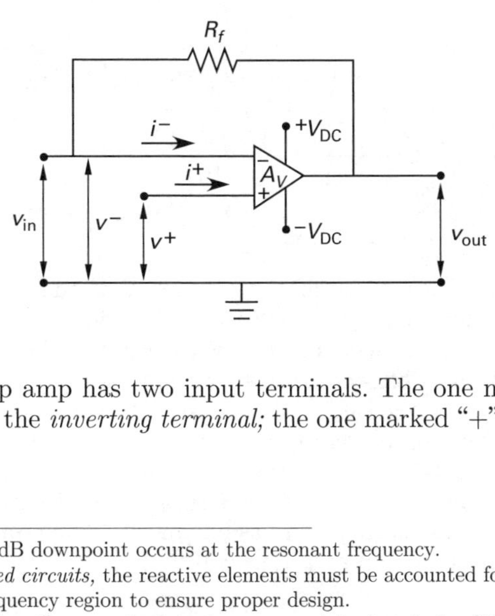

The op amp has two input terminals. The one marked "–" is the *inverting terminal;* the one marked "+" is the *noninverting terminal.* The fundamental op amp formula is

$$v_{\text{out}} = A_V(v^+ - v^-) \tag{40.3}$$

The voltages marked $+V_{\text{DC}}$ and $-V_{\text{DC}}$ are the power supplies to the op amp, and are often called the *rail voltages.* The range of v_{out} is between these two voltages.[5] In fact, the output voltage should stay 3 V from either value to avoid distortion of the output signal at rated values. The range of the input signal, derived from Eq. 40.3 and this distortion restriction, is

$$|v^+ - v^-| < \frac{V_{\text{DC}} - 3\text{ V}}{A_V} \tag{40.4}$$

When the op amp input is within the range of Eq. 40.4, the operation is linear. Typical values of op amp voltage gain, A_V, range from 10^5 to 10^8. Power supply voltages, $\pm V_{\text{DC}}$, are in the range of $\pm$10–15 V. Typical input impedances are greater than $10^5\ \Omega$, while output impedances are very low. The internal circuits of the op amp consist of the items in the block diagram of Fig. 40.3.

Figure 40.3 *Operational Amplifier Internal Configuration*

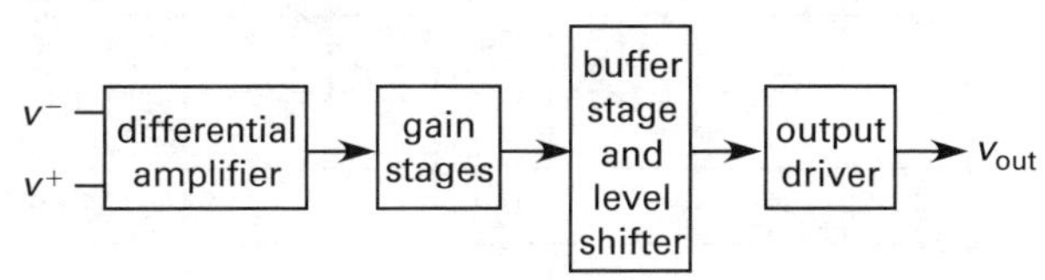

The *differential amplifier* constitutes the first stage of the op amp. A differential amplifier is shown in Fig. 40.4. The resulting output current, which is a combination of the emitter currents, is shown in Fig. 40.5. The linear region for operation restricts the voltage difference at the op amp input, as in Eq. 40.4, though here the difference is expressed in terms of the thermal voltage, V_T. Ideally, the differential amplifier amplifies only the difference between the signals. In practice, both the *differential-mode signal,* v_{dm}, and the *common-mode signal,* v_{cm}, are amplified.

$$v_{\text{dm}} = v^+ - v^- \tag{40.5}$$

$$v_{\text{cm}} = \tfrac{1}{2}(v^+ + v^-) \tag{40.6}$$

To measure how well the differential amplifier amplifies the difference signal and not the common signal, the concept of the *common-mode rejection ratio* (CMRR) is used. The CMRR is defined as

$$\text{CMRR} = \left|\frac{A_{\text{dm}}}{A_{\text{cm}}}\right| \tag{40.7}$$

[1]The 3 dB downpoint occurs at the resonant frequency.

[2]In *tuned circuits,* the reactive elements must be accounted for in the mid-frequency region to ensure proper design.

[3]"Significant" means the impedance seen at the terminals of the capacitor is of the same magnitude as the impedance of the capacitor itself.

[4]These models are not shown in Chap. 38. They would consist of the small-signal low-frequency model with coupling and emitter bypass capacitors added—the same circuit that would be used to determine the AC load line.

[5]Although the gain is large, the output cannot be greater than the available voltage. The rail voltages represent $\pm\infty$ for the output.

Figure 40.4 *Differential Amplifier*

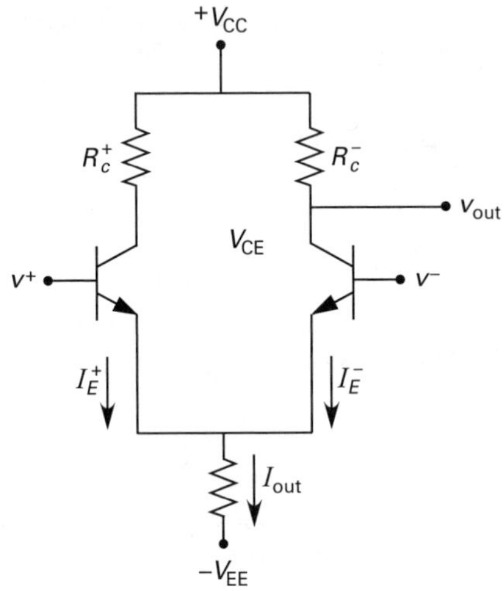

Figure 40.5 *Differential Amplifier Emitter Currents (Ideal Current Source Biasing)*

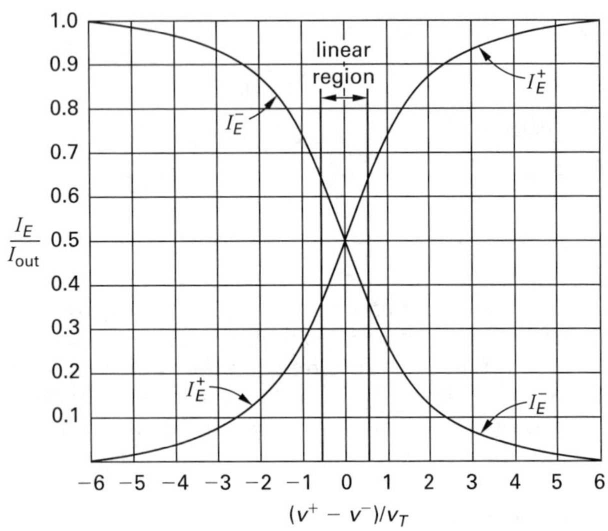

The output voltage in terms of the CMRR is

$$v_{\text{out}} = A_{\text{dm}} v_{\text{dm}} \left(1 + \left(\frac{1}{\text{CMRR}} \right) \left(\frac{v_{\text{cm}}}{v_{\text{dm}}} \right) \right) \qquad 40.8$$

The gain stages in Fig. 40.3 supply the required amplification and represent cascaded amplifiers, for example, cascaded CE amplifiers. The buffer stage has a high input resistance to prevent it from loading the output driver, for example, an *emitter follower* (CC configuration). Circuitry associated with the buffer shifts the DC bias level so that the output is zero when the input is zero. The output driver provides the necessary large-signal current or voltage gain. For example, for a large current gain, the common collector configuration could be used (see Sec. 38.4).[6]

Op amps are designed to closely approximate ideal op amp behavior. Because of this and because of their versatility of use, they are widely used in place of discrete transistor amplifiers. Op amps are used in linear systems such as voltage-to-current converters, DC instrumentation, voltage followers, and filters; and in nonlinear systems such as AC/DC converters, peak detectors, sample-and-hold systems, analog multipliers, and analog-to-digital and digital-to-analog converters, among many others. A low-frequency op amp equivalent circuit is shown in Fig. 40.6.

Figure 40.6 *Operational Amplifier Equivalent Circuit*

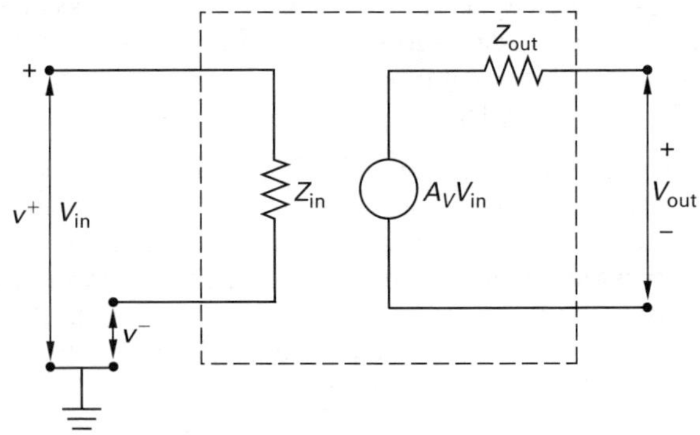

Example 40.1

An op amp has a voltage gain of 10^8. The DC power supply is ± 15 V. What is the maximum input voltage difference for linear operation?

Solution

The input voltage difference is given by Eq. 40.4.

$$|v^+ - v^-| < \frac{V_{\text{DC}} - 3}{A_V} = \frac{15 \text{ V} - 3 \text{ V}}{10^8}$$
$$= 0.12 \times 10^{-6} \text{ V} \quad (0.12\ \mu\text{V})$$

The difference between the terminals of the op amp must be kept within 0.12 μV to remain within the linear region. (This is accomplished with negative feedback.) This voltage difference is so small that the terminals are essentially at the same voltage. One of the assumptions for an ideal op amp is that the voltages are the same. Consequently, if one of the terminals is connected to ground, the other terminal is also at ground potential. This situation is referred to as having a *virtual ground* at the ungrounded op amp terminal.

[6]The common collector (CC) configuration is also called the *emitter follower*. Voltage gain is approximately one. This means that a change in base voltage appears as an equal change across the load at the emitter. Consequently, the emitter follows the input signal.

Example 40.2

The input impedance for the op amp in Ex. 40.1 is $10^5\ \Omega$. What is the input current?

Solution

The voltage difference was calculated as 0.12 μV. The current between the terminals is given by Ohm's law.

$$\begin{aligned} I &= \frac{V}{R} = \frac{0.12 \times 10^{-6}\ \text{V}}{10^5\ \Omega} \\ &= 1.2 \times 10^{-12}\ \text{A} \quad (1.2\ \text{pA}) \end{aligned}$$

This current is so low as to be negligible. An ideal op amp implies that the input current is zero.

Example 40.3

If the gain given for Ex. 40.1 is the differential-mode gain and the CMRR is 10,000, what is the output voltage when the op amp is at the edge of linear operation and $v^+ = 50\ \mu$V?

Solution

The op amp is at the edge of linear operation for the maximum voltage difference allowed by Eq. 40.4, which was calculated in Ex. 40.7 as 0.12 μV. The output voltage is given by Eq. 40.8.

$$v_{\text{out}} = A_{\text{dm}} v_{\text{dm}} \left(1 + \left(\frac{1}{\text{CMRR}}\right)\left(\frac{v_{\text{cm}}}{v_{\text{dm}}}\right)\right)$$

The differential-mode voltage was calculated as 0.12 μV. The common-mode voltage is unknown. From Eq. 40.4 and Eq. 40.5,

$$\begin{aligned} v_{\text{dm}} &= v^+ - v^- = 0.12\ \mu\text{V} = 50\ \mu\text{V} - v^- \\ v^- &= 50\ \mu\text{V} - 0.12\ \mu\text{V} = 49.88\ \mu\text{V} \\ v_{\text{cm}} &= \tfrac{1}{2}(v^+ + v^-) \\ &= \tfrac{1}{2}(50\ \mu\text{V} + 49.88\ \mu\text{V}) \approx 50\ \mu\text{V} \end{aligned}$$

Substituting the calculated and given values gives

$$\begin{aligned} v_{\text{out}} &= A_{\text{dm}} v_{\text{dm}} \left(1 + \left(\frac{1}{\text{CMRR}}\right)\left(\frac{v_{\text{cm}}}{v_{\text{dm}}}\right)\right) \\ &= (10^8)(0.12 \times 10^{-6}\ \text{V})\left(1 + \left(\frac{1}{10{,}000}\right)\left(\frac{50\ \mu\text{V}}{0.12\ \mu\text{V}}\right)\right) \\ &= 12.5\ \text{V} \end{aligned}$$

This output voltage is within 3 V of the power supply voltage. Distortion could be caused by the nonlinear operation. In this case, because $A_{\text{dm}} v_{\text{dm}} = 12$ V, the negative feedback must be increased.

2. IDEAL OPERATIONAL AMPLIFIERS

An *ideal operational amplifier* exhibits the following properties.

- $Z_{\text{in}} = \infty$
- $Z_{\text{out}} = 0$
- $A_V = \infty$
- $\text{BW} = \infty$

Two other properties not always mentioned are (1) $V_{\text{out}} = 0$ when $v^+ = v^-$ regardless of the magnitude of the input voltage (normally v^-), and (2) characteristics do not drift with temperature.[7] Because the actual op amp so closely approximates these conditions, it is possible to use ideal op amp analysis for most calculations. Only when high-frequency behavior is desired or circuit limitations have been calculated do the assumptions need to be discarded. A typical ideal op amp configuration with *voltage shunt feedback* is shown in Fig. 40.7.

Figure 40.7 *Ideal Operational Amplifier Typical Configuration*

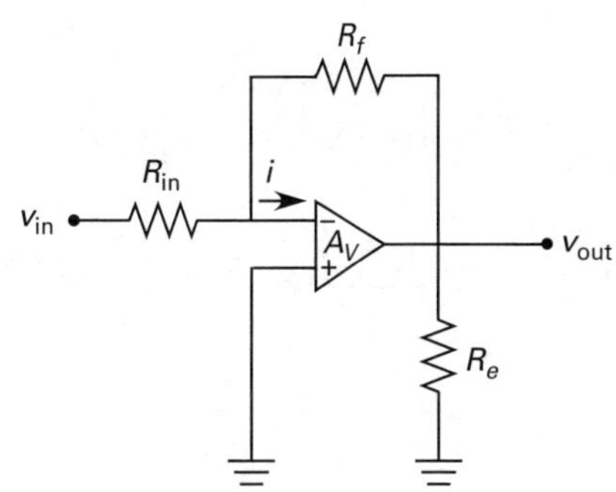

The assumptions regarding the properties of the ideal op amp result in the following practical results during analysis.

- The current to each input is zero.
- The voltage between the two input terminals is zero.
- The op amp is operating in the linear range.

The voltage difference of zero between the two terminals is called a *virtual short circuit*, or, because the positive terminal is often grounded, a *virtual ground.* The term "virtual" is used because, although the feedback from the output is used to keep the voltage difference between the terminals at zero, no current actually flows into the short circuit.

The procedure for analyzing an ideal op amp circuit follows.

step 1: Draw the circuit and label all the nodes, voltages, and currents of interest.

[7]Property (1) is for the op amp itself, *without feedback.* When negative feedback is used, the output voltage is whatever value maintains $v^+ = v^-$.

step 2: Write Kirchoff's current law (KCL) at the op amp input node (normally the inverting terminal).[8]

step 3: Simplify the resulting equation using the ideal op amp assumptions. Specifically, the current into the op amp is zero and the voltage difference between the terminals is zero. The known voltage at one terminal must be the voltage at the other (e.g., if the positive terminal is at ground, the negative terminal voltage is zero).

step 4: Solve for the desired quantity or expression.

Using the procedure given, the relationships or values for any op amp circuit parameter can be found. The relationship between the output voltage and the input voltage for a variety of op amp circuits is shown in Fig. 40.8.

Example 40.4

Consider the circuit shown. The op amp is considered ideal with a 1 μV input signal. What is the current through the feedback resistor?

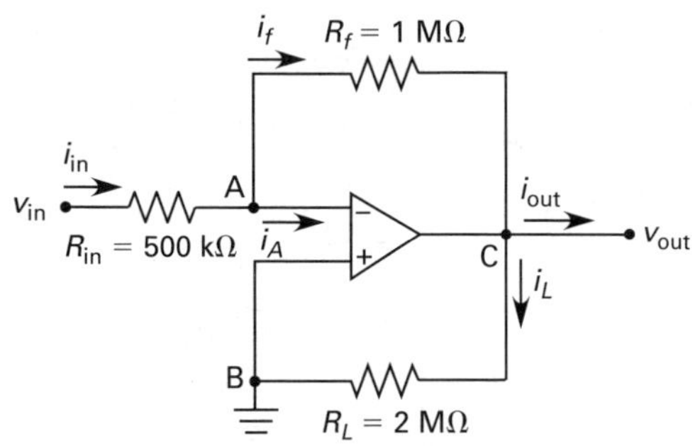

Solution

The circuit is already drawn. Writing KCL (assuming that currents out of the node are positive) at the input terminal of the op amp, that is, node A, gives[9]

$$-i_{in} + i_A + i_f = 0$$

$$i_f = i_{in} - i_A = \frac{v_{in} - V_A}{R_{in}} - i_A$$

Using the ideal op amp assumptions, the input current to the op amp is zero, that is, $i_A = 0$. Also, the voltage difference between the terminals is zero. Because the positive terminal is connected to ground, $v^+ = 0 = v^- = V_A$. Substituting gives

$$\begin{aligned} i_f &= i_{in} - i_A = \frac{v_{in} - V_A}{R_{in}} - i_A \\ &= \frac{1 \times 10^{-6}\ \text{V} - 0\ \text{V}}{500 \times 10^3\ \Omega} - 0\ \text{A} \\ &= 2 \times 10^{-12}\ \text{A} \end{aligned}$$

The negative sign indicates that the current direction is opposite that shown. Because i_A is zero, the input current is equal to 2×10^{-12} A.

Example 40.5

For the circuit in Ex. 40.4, determine the voltage gain.

Solution

The voltage gain is

$$A_V = \frac{v_{out}}{v_{in}}$$

The input voltage is known. The output voltage can be calculated in a number of ways. Using a portion of KCL at node A (or Ohm's law) gives

$$i_f = \frac{v_A - v_{out}}{R_f} = \frac{0\ \text{V} - v_{out}}{1 \times 10^6\ \Omega} = 2 \times 10^{-12}\ \text{A}$$

$$\begin{aligned} v_{out} &= (-1)(2 \times 10^{-12}\ \text{A})(1 \times 10^6\ \Omega) \\ &= -2 \times 10^{-6}\ \text{V} \end{aligned}$$

Care should be taken with the negative signs. The feedback current i_f is positive because the direction shown in the example figure is used.[10] Substituting gives

$$A_V = \frac{v_{out}}{v_{in}} = \frac{-2 \times 10^{-6}\ \text{V}}{1 \times 10^{-6}\ \text{V}} = -2$$

The gain is negative because the input occurs at the inverting terminal. The gain is determined by the resistors (since $|i_f| = |i_{in}|$), as can be seen by rewriting the gain formula.

$$A_V = \frac{v_{out}}{v_{in}} = \frac{-i_f R_f}{i_{in} R_{in}} = \frac{-R_f}{R_{in}}$$

Example 40.6

For the circuit in Ex. 40.4, determine the current through the load resistor.

[8]Methods other than KCL may be used, but this systematic method normally includes all the desired quantities and minimizes errors.

[9]The current into the node, i_{in}, is shown as negative to be consistent with the assumption that all currents into the node are negative and those out are positive. Ohm's law is written as shown to be consistent with the direction of the input current given.

[10]Recall that in this text positive currents flow out of the node and negative currents flow in (see Sec. 26.16).

Figure 40.8 Operational Amplifier Circuits

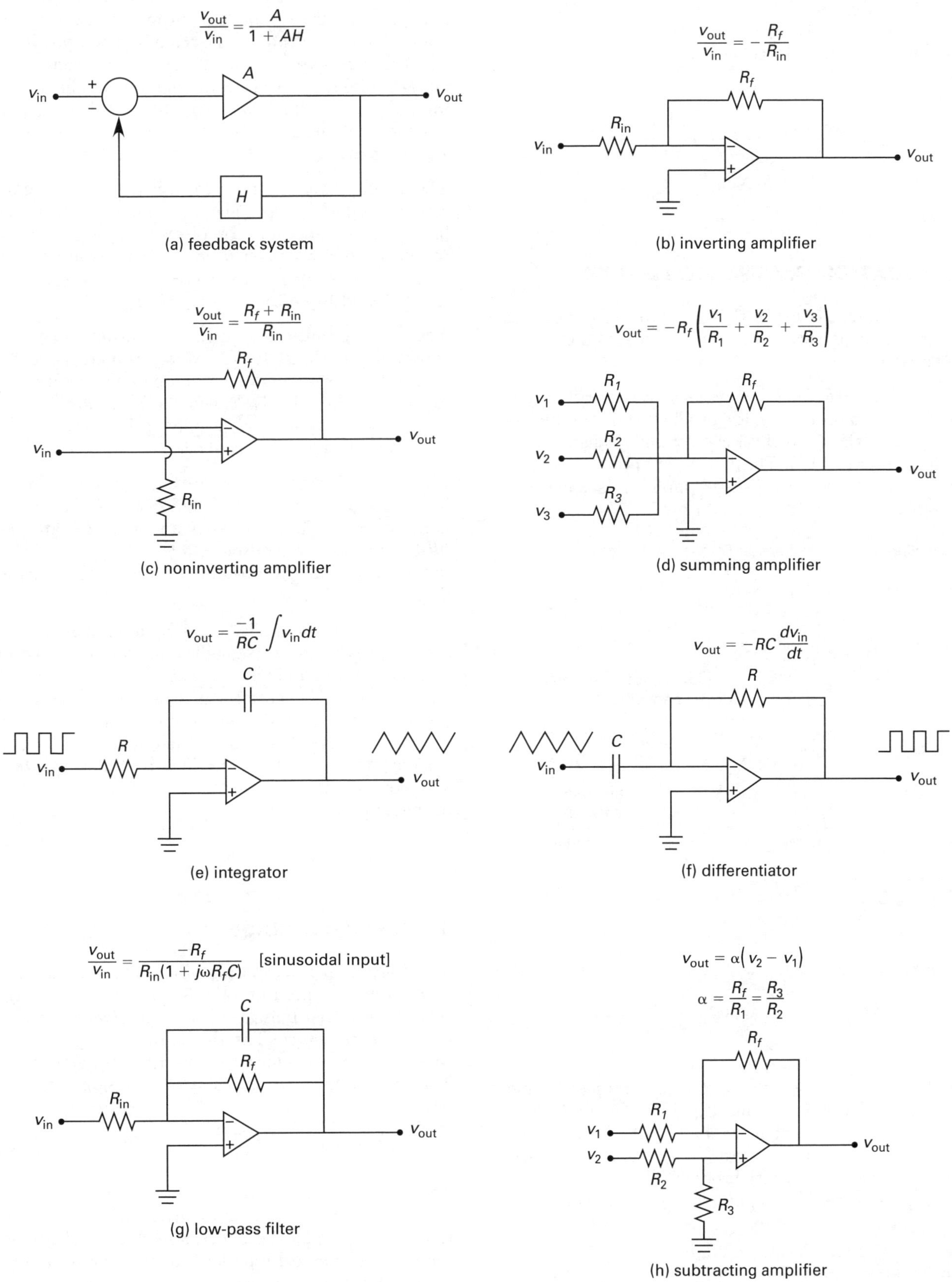

Solution

Several possible methods exist for determining the solution. Because v_{out} is known and the direction of i_L is given, using Ohm's law at node C provides the answer directly.

$$i_L = \frac{v_{\text{out}} - v_{\text{B}}}{R_L} = \frac{-2\times 10^{-6}\ \text{V} - 0\ \text{V}}{2\times 10^{6}\ \Omega}$$
$$= -1\times 10^{-12}\ \text{A}$$

3. OPERATIONAL AMPLIFIER LIMITS

A real op amp has limits that must be accounted for during design or use. Those limits are summarized in this section.

Op amps have a finite bandwidth. Specifically, they have a finite gain-bandwidth product. The product of the gain and the bandwidth is essentially constant (see Eq. 40.1 and Eq. 40.2). The first-order approximation of the frequency response of an amplifier is shown in Fig. 40.9.

Figure 40.9 *Operational Amplifier Frequency Response*

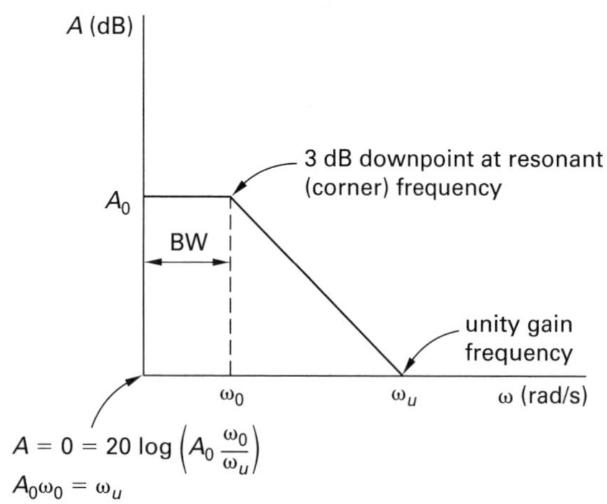

The *unity gain* is defined as

$$|A| = A_0 \frac{\omega_0}{\omega_u} = 1 \qquad 40.9$$

The term ω_u is called the *unity gain frequency* and replaces $A_0\omega_0$ in Eq. 40.1 and Eq. 40.2. The unity gain frequency is used to define the bandwidth for a real op amp. As the frequency increases the gain decreases, keeping the gain bandwidth product (GBW), also called the figure of merit (F_m), approximately constant.[11]

[11]The frequency response of any amplifier circuit can be found by transforming the external electrical components into the *s* domain (see Sec. 31.10). Then obtain the gain equation using the procedures of Sec. 40.2. Finally, plot the response on a Bode plot (see Chap. 32).

The output voltage is limited to the range supply voltages. Further, to ensure operation in the linear region and prevent distortion, the output voltage should be approximately 3 V from the maximum supply voltage. This limit is expressed in terms of the gain and the voltage difference that can be used at the terminal inputs (see Eq. 40.4 and Fig. 40.5). When operating outside the linear region, the op amp is considered *saturated.*

The op amp amplifies both the difference signal and the common signal. Amplifying the common signal results in problems, especially for precision applications. The common-mode rejection ratio (CMRR) is a measure of the op amp's ability to overcome this obstacle (see Eq. 40.7 and Eq. 40.8).

The gain is finite. As such, the output is less than expected (see Prob. 4). The voltage difference is determined from the fundamental op amp equation (see Eq. 40.3), which is also called the *constraint equation.* Nevertheless, the gain is usually so high that this is a minor problem at low frequencies.

The input currents are small but not zero when the input is zero. A voltage difference does exist when the input is zero. These currents and voltages are called *offset currents* and *offset voltages.* They are compensated for within the op amp by the level shifter (see Sec. 40.1 and Fig. 40.3).

If a large signal is applied to an op amp, driving it into saturation, the output signal is limited to the maximum current that the amplifier can supply. This means that an incoming exponential signal would become a ramp signal on the output, and a sinusoidal signal would become a sawtooth signal on the output. The slopes of the output ramps are determined by the *slew rate,* S_R. The slew rate is determined by the internal capacitance of the amplifier and is

$$S_R = \left.\frac{dv}{dt}\right|_{\text{max}} \approx \frac{I_{\text{max}}}{C} \qquad 40.10$$

4. AMPLIFIER NOISE

There are numerous sources of noise in amplifiers. The *thermal noise* is produced by the random thermal agitation of electrons. It is also called the *Johnson noise* or the *resistance noise.* The thermal noise is proportional to the root-mean-square of the noise voltage squared. The thermal noise is also called the *noise power,* P_n.

$$P_n = \frac{V^2_{\text{noise,rms}}}{4R} = \kappa T(\text{BW}) \qquad 40.11$$

The *shot noise* is due to the random variations in the velocity of the electrons in a current flow. It is proportional to $2qI$. *Low-frequency noise,* also called *flicker noise,* is the noise from slow changes, or drift, in the system parameters, caused by temperature, component aging, and so on. *Interference noise* is electromagnetic induction from sources outside the amplifier.

The total noise in a given circuit is compared to the incoming signal, or power, to ensure the ability of the circuit to discern between the two. The measure of this ability is the signal-to-noise ratio (SNR).

$$\text{SNR} = 10\log\frac{P_s}{P_n} = 20\log\frac{V_s}{V_n} \quad 40.12$$

Because the resistance is common to both the signal and the noise, the units of the signal power, P_s, and the noise power, P_n, can be expressed in watts or simply volts squared (in which case the noise power equals $4R\kappa T(\text{BW})$).[12] A signal-to-noise ratio of approximately 10 dB is required for the op amp output to reflect the signal.

PART 3: PULSE CIRCUITS: WAVEFORM SHAPING AND LOGIC

5. PULSE CIRCUIT FUNDAMENTALS

A *pulse circuit* is any active electrical network designed to respond to discrete pulses of current or voltage. Such circuits abound in computer and digital systems. Examples are waveform shaping circuits and logic circuits. An *active circuit*, such as a transistor circuit, is a circuit that is capable of amplifying a current or voltage signal. Waveform shaping is primarily accomplished with active nonlinear circuits. Active nonlinear elements are described in Chap. 38, though when used as switches, they are considered pulsed circuits—hence their coverage here. *Passive linear elements*, those elements that are not sources of energy (for example, resistors, capacitors, and inductors) are encountered in the design of pulse, waveform shaping, and logic circuits. Passive linear circuit elements are described in Chap. 29 through Chap. 31. *Passive nonlinear elements*, primarily diodes, are widely used for waveform shaping elements and elements of logic circuits. The delta (Dirac) function and the step function, described in Chap. 15, are used to mathematically describe the response of pulse circuits. *Pulse generators* are circuits that are designed with high-speed switches, capable of generating very short-duration pulses for physically testing the response of a network to a delta-like input.[13] The ability to recognize the overall function of a pulse, waveform, or logic circuit is a practical skill required in electrical engineering. Detailed analysis of such circuits involves the application of knowledge of the individual components and requires significant time and specialization.[14]

[12]When ratios are involved, such as the signal-to-noise ratio, the only term remaining is the rms noise voltage squared, V^2. This is why the square of the voltage is sometimes referred to as the *noise power*.

[13]An *RC differentiator* with a switching transistor to control the application of the power is a pulse generator. The *RC* differentiator is an *RC* circuit with the output taken across the capacitor.

[14]This chapter, and indeed this text, focuses on the general skills required in electrical engineering. The tools for detailed analysis are presented, but the overall principles and methods will be the focus.

6. CLAMPING CIRCUITS

A diode represents a purely resistive passive nonlinear element. The diode can be used as a simple means of limiting voltage to a desired value. When the voltage is limited in some manner, it is *clamped.* Two types of clamping exist: limiting and clipping. A *limiter* is an electronic circuit designed to prevent a waveform from exceeding a predetermined value. A *clipper* is an electronic circuit designed to maintain a waveform at a predetermined value when the input is above a certain level. The two types of circuits are shown in Fig. 40.10 for an ideal diode, that is, a diode with no resistance and a zero voltage drop. The ideal circuits can be used for many practical calculations and for determining overall circuit response.

Figure 40.10 *Ideal Diode Limiter and Clipper Circuits*

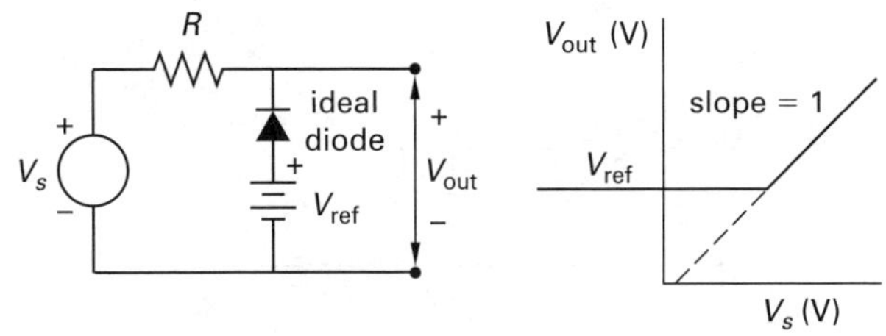

(a) shunt diode limiter circuit

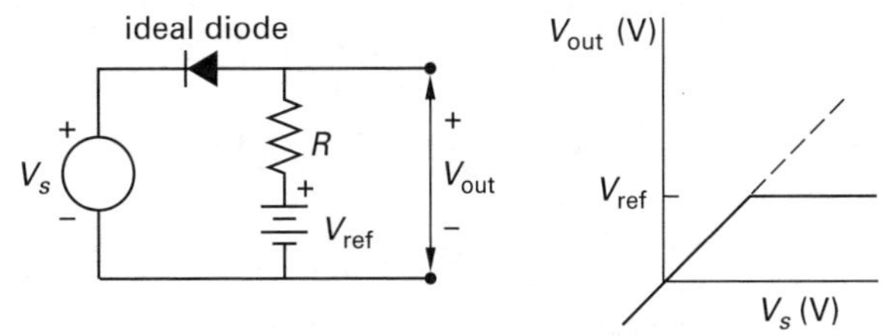

(b) series diode clipper circuit

The diode is not an ideal device. The forward threshold voltage drop, V_F, for silicon diodes is approximately 0.7 V, while that for germanium diodes is approximately 0.2 V. Additionally, diodes exhibit some forward resistance, R_f, which accounts for the small but finite slope of the forward characteristic. (The reverse resistance or zener resistance is not a concern, because limiter and clipper circuits do not operate in the breakdown region.) The limiter and clipper circuits for an actual diode are shown in Fig. 40.11. Actual circuits are more complex than ideal circuits but produce more accurate results.

Analysis of the circuit in Fig. 40.11(a) yields Eq. 40.13 and Eq. 40.14.

$$v_{out} = \frac{R(V_{ref} - V_F)}{R + R_f} + \frac{R_f V_{in}}{R_f + R} \quad [V_{in} < V_{ref} - V_F] \quad 40.13$$

$$V_{out} = V_{in} \quad [V_{in} > V_{ref} - V_F] \quad 40.14$$

Figure 40.11 *Real Diode Limiter and Clipper Circuits*

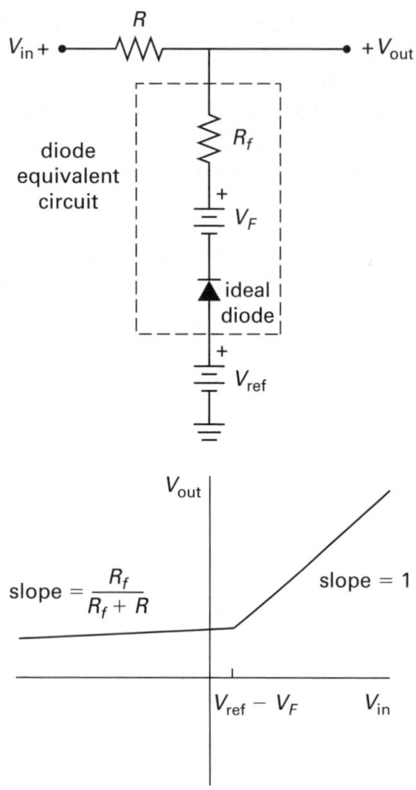

(a) limiter circuit

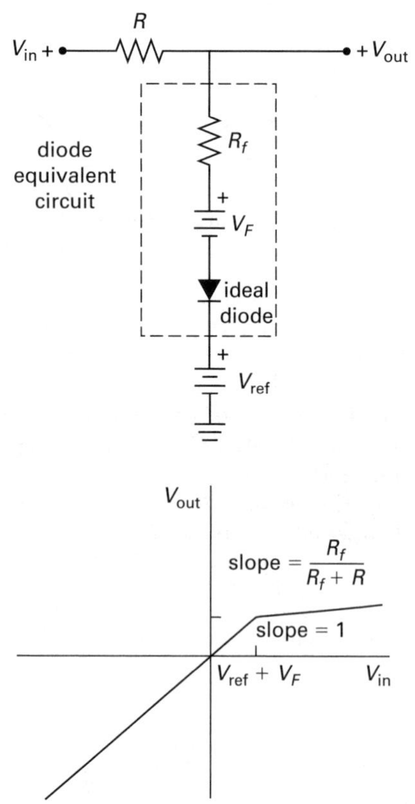

(b) clipper circuit

Equation 40.13 represents the forward-biased case. Equation 40.14 represents the reverse-biased case, ignoring the small reverse saturation current. Analysis of the circuit in Fig. 40.11(b) yields Eq. 40.15 and Eq. 40.16.

$$v_{\text{out}} = \frac{R_f V_{\text{in}}}{R_f + R} + \frac{R(V_{\text{ref}} + V_F)}{R + R_f} \quad [V_{\text{in}} > V_{\text{ref}} + V_F] \qquad 40.15$$

$$V_{\text{out}} = V_{\text{in}} \quad [V_{\text{in}} < V_{\text{ref}} + V_F] \qquad 40.16$$

Equation 40.15 represents the forward-biased case. Equation 40.16 represents the reverse-biased case, ignoring the small reverse saturation current.

The ideal limiter or clipper circuit would have R_f equal to zero and the effect of the threshold voltage would be negligible. This type of circuit can be realized using a *precision diode.* A precision diode circuit, equivalent circuit, and characteristics are shown in Fig. 40.12. A precision diode is used to rectify very small signals: those less than 0.7 V for silicon or 0.2 V for germanium. In fact, the precision diode rectifies microvolt-level signals.

In Fig. 40.12(a), a diode is placed between the operational amplifier output and the feedback point of a voltage follower. For negative voltage inputs, $-v_{\text{in}}$, the op amp output voltage, V_{amp}, is negative and the diode is reverse biased.[15] The output voltage is then equal to zero because no current flows through the output resistor, R_{out}. For positive voltage inputs, $+v_{\text{in}}$, the op amp output voltage, V_{amp}, is equal to $A_V V_{\text{in}}$ until the diode becomes forward biased, that is, until the diode voltage, V_D, exceeds V_F. This forward-biased condition, which occurs when $A_V V_{\text{in}} = V_D$ or $V_{\text{in}} = V_D / A_V$, is shown in Fig. 40.12(b). From the forward-biased condition onward, the output voltage across R_{out} is the input voltage magnified by the op amp as shown in Fig. 40.12(c). The approximate transfer equation for the precision diode of Fig. 40.12(a) is

$$V_{\text{out}} = V_{\text{in}} - \frac{V_F}{A_V} \qquad 40.17$$

Precision diode circuits sensitive to negative voltages are shown in Fig. 40.13.

Example 40.7

Consider the precision diode in Fig. 40.12(b) with a 12 V power supply and a voltage gain of 10^6. What is the maximum voltage difference that can be amplified without distortion or saturation?

[15] For proper operation, the diode must have a reverse breakdown voltage greater in magnitude than the maximum negative voltage of the op amp, that is, greater in magnitude than the negative supply voltage (not shown in the figure).

Figure 40.12 *Precision Diode Circuit*

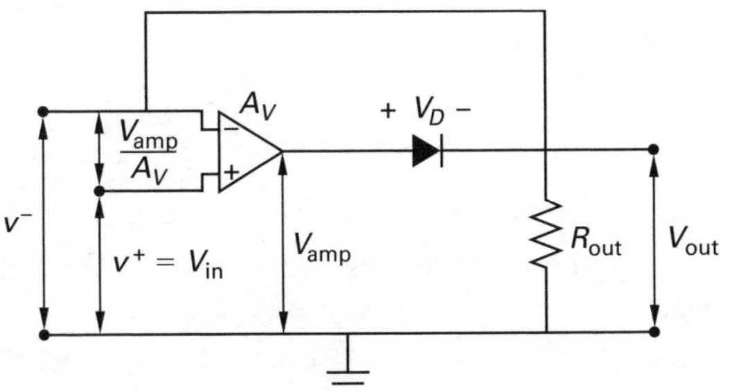

(a) precision diode

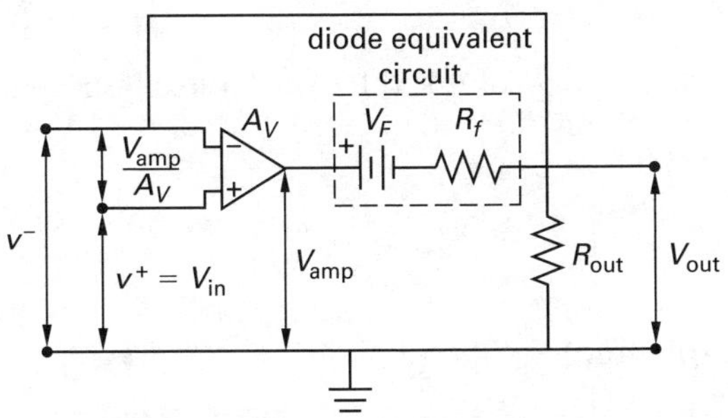

(b) forward-biased precision diode

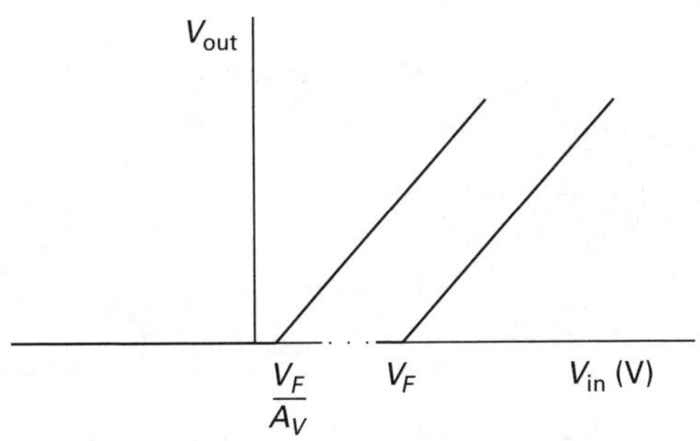

(c) characteristics

Figure 40.13 *Precision Diode Circuits Sensitive to Negative Voltage Input*

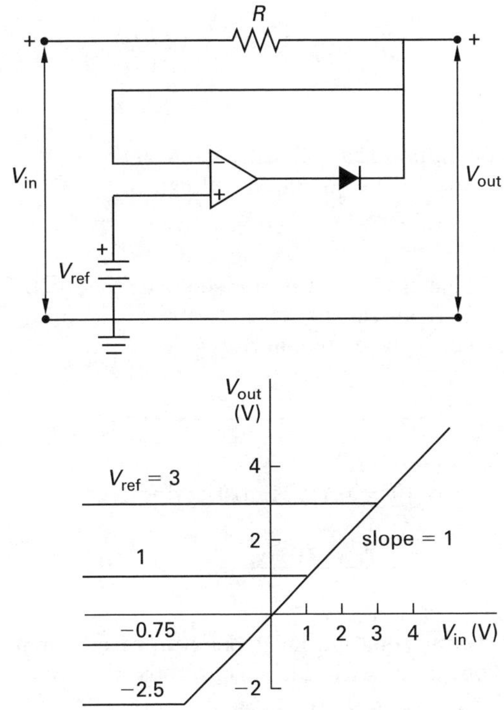

(a) precision limiter/clipper circuit

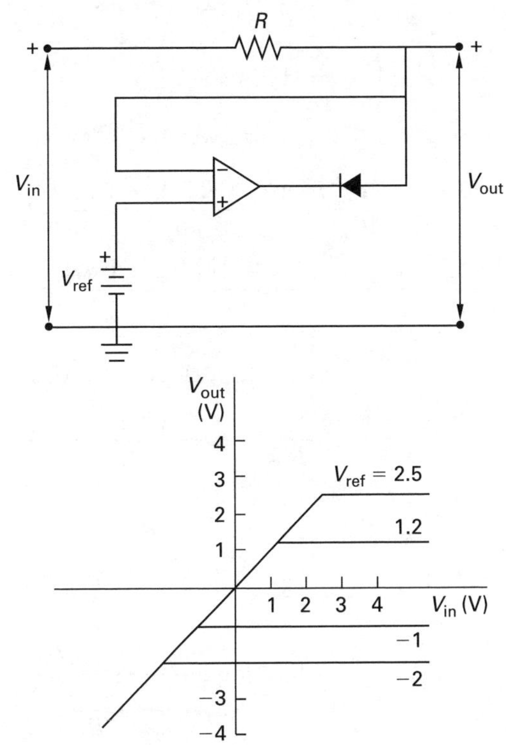

(b) precision limiter/clipper circuit sensitive to negative voltages

Solution

The maximum voltage difference that can be amplified is determined by the constraint equation.

$$v^+ - v^- = \frac{V_{amp}}{A_V}$$

The constraint equation is Eq. 40.17 with the diode removed, which changes V_F to V_{amp}. The maximum voltage of the amplifier is 9 V (3 V less than the maximum power supply voltage) to prevent distortion or saturation. So,

$$v^+ - v^- = \frac{V_{\text{amp}}}{A_V} = \frac{9\text{ V}}{10^6} \quad (9\ \mu\text{V})$$

Example 40.8

Use the op amp in Ex. 40.7 with a silicon diode. What is the output voltage if the input voltage is 9 μV?

Solution

With the silicon diode, the threshold voltage is 0.7 V. Writing KCL at the negative terminal (ignoring the small current into the op amp) gives

$$\frac{v^- - 0\text{ V}}{R_{\text{out}}} + \frac{v^- + V_F - V_{\text{amp}}}{R_f} = 0$$

$v^- = V_{\text{out}}$, and $V_F = 0.7$ V. Substituting gives

$$\frac{V_{\text{out}}}{R_{\text{out}}} + \frac{V_{\text{out}} + 0.7\text{ V} - V_{\text{amp}}}{R_f} = 0$$

The constraint equation must be used, rather than the assumption used with ideal op amps that $v^+ = v^-$, because the voltage response to an input is the unknown. That is, the transfer function must be determined. The constraint equation is

$$v^+ - v^- = V_{\text{in}} - V_{\text{out}} = \frac{V_{\text{amp}}}{A_V}$$

$$V_{\text{amp}} = A_V(V_{\text{in}} - V_{\text{out}})$$

Substituting and rearranging gives the following result.

$$\frac{V_{\text{out}}}{R_{\text{out}}} + \frac{V_{\text{out}} + 0.7\text{ V} - V_{\text{amp}}}{R_f} = 0$$

$$\frac{V_{\text{out}}}{R_{\text{out}}} + \frac{V_{\text{out}} + 0.7\text{ V} - A_V(V_{\text{in}} - V_{\text{out}})}{R_f} = 0$$

$$\frac{V_{\text{out}}}{R_{\text{out}}} = \frac{\begin{array}{c}V_{\text{out}} + 0.7\text{ V}\\ -A_V(V_{\text{in}} - V_{\text{out}})\end{array}}{R_f}$$

$$\frac{V_{\text{out}}}{R_{\text{out}}} + \frac{V_{\text{out}}}{R_f} + \frac{A_V V_{\text{out}}}{R_f} = \frac{A_V V_{\text{in}} - 0.7\text{ V}}{R_f}$$

$$V_{\text{out}}\left(\frac{1}{R_{\text{out}}} + \frac{1}{R_f} + \frac{A_V}{R_f}\right) = \frac{A_V V_{\text{in}} - 0.7\text{ V}}{R_f}$$

$$V_{\text{out}} = \frac{A_V V_{\text{in}} - 0.7\text{ V}}{R_f\left(\frac{1}{R_{\text{out}}} + \frac{1}{R_f} + \frac{A_V}{R_f}\right)}$$

$$= \frac{V_{\text{in}} - \frac{0.7\text{ V}}{A_V}}{\frac{R_f}{A_V R_{\text{out}}} + \frac{1}{A_V} + 1}$$

The gain is very large. The transfer function reduces to Eq. 40.17, with 0.7 V substituted for V_F.

$$V_{\text{out}} = V_{\text{in}} - \frac{0.7\text{ V}}{A_V}$$

Substituting the input voltage of 9 μV and the gain of 10^6 gives

$$V_{\text{out}} = V_{\text{in}} - \frac{0.7\text{ V}}{A_V} = 9\times 10^{-6}\text{ V} - \frac{0.7\text{ V}}{10^6}$$
$$= 8.3\times 10^{-6}\text{ V} \quad (8.3\ \mu\text{V})$$

The offset voltage of 0.7 V for silicon has been reduced by the precision diode to 0.7×10^{-6} V, or 0.7 μV, which is essentially zero.

Example 40.9

Design a circuit that has the following output characteristic.

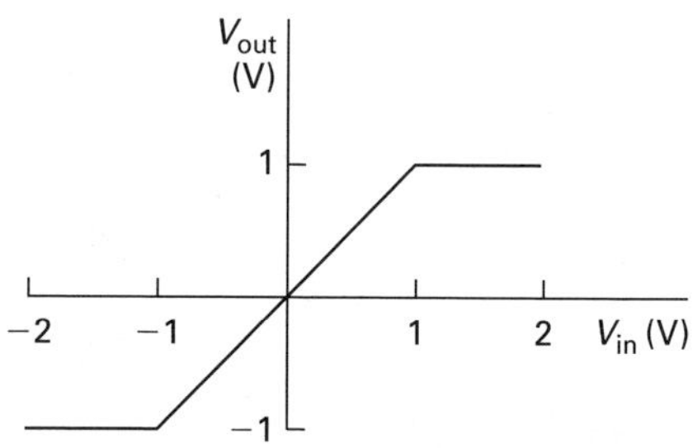

This circuit linearly transfers any signal of less than 1.0 V in magnitude, but clips all signals greater than 1.0 V in magnitude. The power supply voltage available is ±12.0 V.

Solution

A precision diode circuit set up as voltage followers exhibits the characteristics given. For the positive voltages, a clipper circuit similar to that in Fig. 40.13(b) can be used with the reference voltage at the noninverting terminal set to 1.0 V by a voltage divider. For the negative voltages, a limiter circuit similar to that in Fig. 40.13(a) can be used with the reference voltage at the noninverting terminal set to −1.0 V by a voltage divider. The resulting circuit follows.

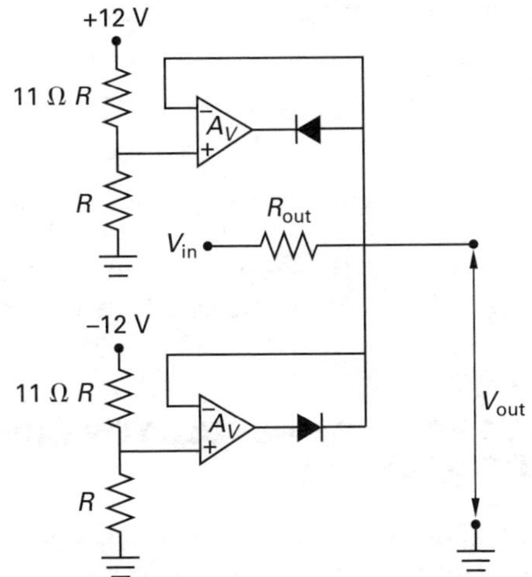

7. ZENER VOLTAGE REGULATOR CIRCUIT: IDEAL

Zener diodes operating in the breakdown region are used as voltage regulators. That is, they are used to control the load voltage (output voltage) within a narrow set of values in circuits for which the supply voltage (input voltage) varies and the load current ranges from zero to a maximum value.

An ideal zener diode voltage regulator circuit is shown in Fig. 40.14(a) with the model for the ideal zener illustrated in Fig. 40.14(b). The zener will maintain a constant voltage as long as the zener current, I_Z, is positive as shown in Fig. 40.14(c).

The design restrictions can be understood from Eq. 40.18 and Eq. 40.19, or derived from Ohm's law and the application of KCL to Fig. 40.14(b).

$$I_s = \frac{V_s - V_Z}{R_s} \tag{40.18}$$

$$I_Z = I_s - I_L \tag{40.19}$$

As the load current, I_L, varies, the supply current given by Eq. 40.18 remains constant, but, according to KCL as stated in Eq. 40.19, the zener current will vary to compensate. As the load current increases the zener diode current decreases. However, to keep the ideal diode forward biased, that is, to keep the zener diode in breakdown, the zener current must be zero or positive.[16] That is, the load voltage, which equals the zener voltage, must be equal to V_{ZM}. To minimize the total power consumption, the value of R_s selected should be as small as possible while still compensating for changes in supply voltage and load current. The combination of these requirements gives

Figure 40.14 *Ideal Zener Diode Voltage Regulator*

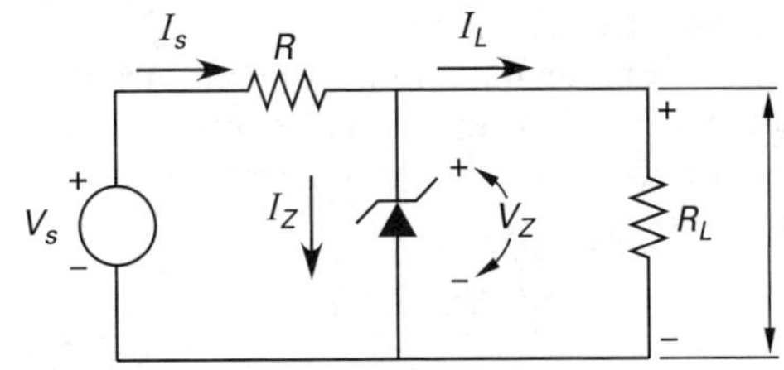

(a) regulator circuit

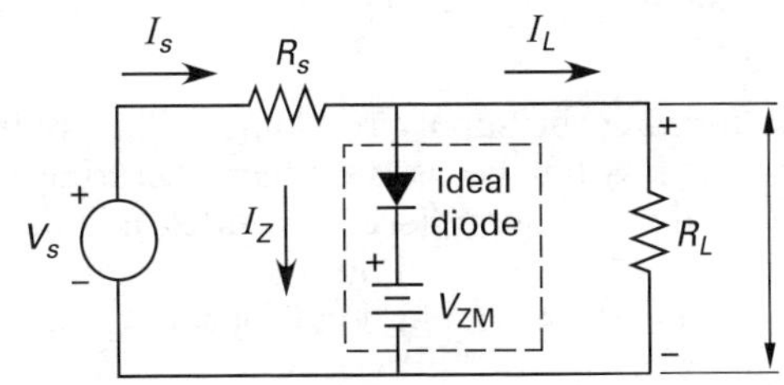

(b) regulator circuit with ideal zener model

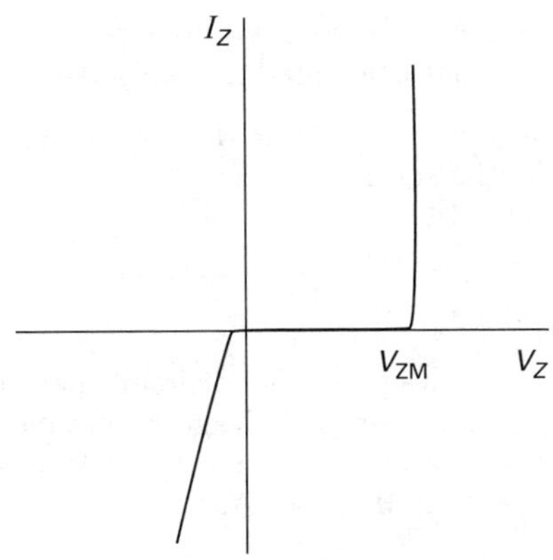

(c) ideal zener characteristics

$$R_s = \frac{V_{\text{in,min}} - V_{\text{ZM}}}{I_{L,\text{max}}} \tag{40.20}$$

$$\begin{aligned} I_{Z,\text{max}} &= \frac{V_{\text{in,max}} - V_{\text{ZM}}}{R_s} \\ &= \left(\frac{V_{\text{in,max}} - V_{\text{ZM}}}{V_{\text{in,min}} - V_{\text{ZM}}}\right) I_{L,\text{max}} \end{aligned} \tag{40.21}$$

Equation 40.20 represents the requirement that at the maximum possible load current, no negative current flows in the ideal diode, with an equal sign used to

[16]In a practical zener diode, the current must be greater than the keep-alive current, I_{ZK}.

minimize the value of R_s.[17] The phrase "no negative current" means that the minimum supply voltage, $V_{\text{in,min}}$, is greater than or equal to V_{ZM}. Equation 40.20 represents the maximum current drawn by the zener, setting the power requirements for the diode and the supply resistor, given by Eq. 40.22 and Eq. 40.23, respectively.

$$P_D = I_{Z,\text{max}} V_{\text{ZM}} \quad \text{40.22}$$

$$P_{R_s} = I^2_{Z,\text{max}} R_s \quad \text{40.23}$$

An ideal zener diode voltage regulator circuit is designed as follows.

step 1: Determine the supply resistance that minimizes power loss but keeps the zener operating in the breakdown region (see Eq. 40.20). This step accounts for the minimum supply (input) voltage and the maximum load (output) current.

step 2: Determine the maximum zener current drawn with the supply resistance calculated in step 1 (see Eq. 40.20). This step accounts for the maximum supply (input) voltage and the minimum load (output) current.

step 3: Determine the required diode power rating with the current calculated in step 2 (see Eq. 40.22).

step 4: Determine the required resistor power rating with the current calculated in step 2 (see Eq. 40.23).

Example 40.10

Design a zener voltage regulator circuit, patterned after Fig. 40.14(a). The output voltage requirement is 5 V, while the input varies from 10 V to 15 V and the load current ranges from 0 A to 5 A.

Solution

Following the design steps, the supply resistance is

$$R_s = \frac{V_{\text{in,max}} - V_{\text{ZM}}}{I_{L,\text{max}}} = \frac{10\ \text{V} - 5\ \text{V}}{5\ \text{A}} = 1\ \Omega$$

The maximum zener current drawn is

$$I_{Z,\text{max}} = \frac{V_{\text{in,max}} - V_{\text{ZM}}}{R_s} = \frac{15\ \text{V} - 5\ \text{V}}{1\ \Omega} = 10\ \text{A}$$

The diode power rating is

$$P_D = I_{Z,\text{max}} V_{\text{ZM}} = (10\ \text{A})(5\ \text{V}) = 50\ \text{W}$$

The resistor power rating must be at least

$$P_{R_s} = I^2_{Z,\text{max}} R_s = (10\ \text{A})^2(1\ \Omega) = 100\ \text{W}$$

Ideal zener diode regulator calculations can be used as first approximations of actual circuits. The inherent assumptions are that the keep-alive current, I_{ZK}, equals zero and that the zener resistance, R_Z, is negligible (see Sec. 40.8).

8. ZENER VOLTAGE REGULATOR CIRCUIT: PRACTICAL

A practical (equivalent) zener diode voltage regulator circuit, illustrated in Fig. 40.15, differs from the ideal case in a number of ways. First, the voltages that can be used are standardized. Appendix 40.A lists some of the standard zener diodes and their respective zener impedances, Z_Z, and keep-alive currents, I_{ZK}. Second, equivalent zener diodes exhibit some resistance, R_Z, as shown in the equivalent model of Fig. 40.15(b). This resistance, R_Z, can cause the zener voltage to vary as much as 20% from the stated value at rated current. The effect is significant for zener diodes rated for 8 V or less, as the zener effect dominates. Third, the zener diode requires a small but finite current called the keep-alive current, I_{ZK}, to remain in the breakdown region. The zener current must be kept above the keep-alive current for the voltage-regulating effect to occur.

The defining equations for the equivalent circuit are derived from Fig. 40.15(b).

$$V_{\text{in}} - V_L = (I_L + I_Z)R_s \quad \text{40.24}$$

$$V_L = V_{\text{ZM}} + I_Z R_Z \quad \text{40.25}$$

Equation 40.24 and Eq. 40.25 reduce to the ideal conditions when I_Z equals zero. Because the zener current must be kept above the keep-alive current, Eq. 40.24 and Eq. 40.25 must be solved simultaneously, which results in Eq. 40.26. (Equation 40.26 is similar to Eq. 40.20 for the ideal case, in that it is used to determine the source resistance with the input voltage at its minimum value, and the load current at its maximum value.)

$$V_L = \frac{V_{\text{ZM}} R_s + V_{\text{in}} R_Z}{R_s + R_Z} - I_L\left(\frac{R_s R_Z}{R_s + R_Z}\right) \quad \text{40.26}$$

The zener current is maximum when the input voltage, V_{in}, is maximum and the load current, I_L, is minimum.

$$\begin{aligned} I_{Z,\text{max}} &= \frac{V_{\text{ZM}} R_s + V_{\text{in,max}} R_Z}{R_Z(R_s + R_Z)} - \frac{V_{\text{ZM}}}{R_Z} \\ &= \frac{V_{\text{in,max}}}{R_s} - \frac{V_{\text{ZM}} R_s + V_{\text{in,max}} R_Z}{R_s(R_s + R_Z)} \end{aligned} \quad \text{40.27}$$

Equation 40.27 represents the maximum current drawn by the zener and sets the power requirements for the

[17] The actual requirement is that the supply resistance be less than or equal to the right-hand side of Eq. 40.20.

Figure 40.15 *Equivalent Zener Diode Voltage Regulator*

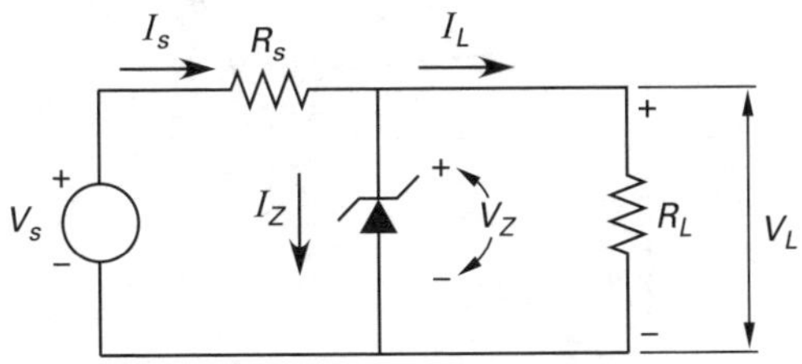

(a) regulator circuit

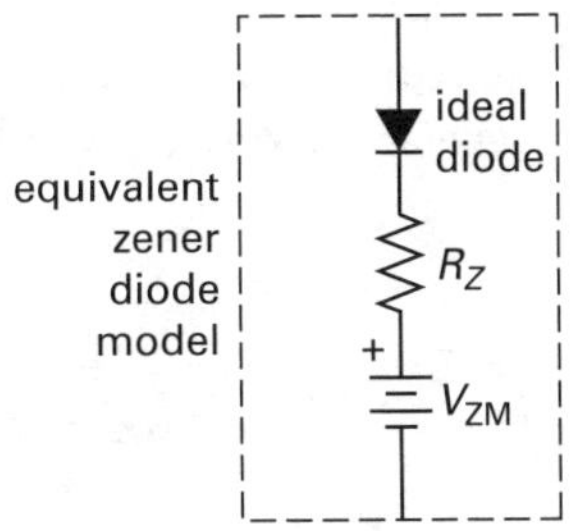

(b) regulator circuit with equivalent zener model

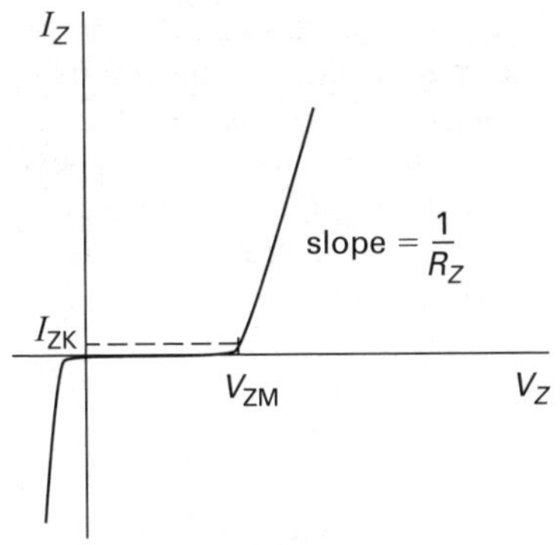

(c) characteristics

diode and the supply resistor, given by Eq. 40.28 and Eq. 40.29, respectively.

$$P_D = I_{Z,\text{max}} V_{\text{ZM}} + I_{Z,\text{max}}^2 R_Z \qquad 40.28$$

$$P_{R_s} = I_{Z,\text{max}}^2 R_s \qquad 40.29$$

If the load current, I_L, is allowed to exceed the maximum level, the zener current, I_Z, will decrease below the keep-alive level, I_{ZK}, and the zener diode will cease to regulate. Equation 40.26 and Eq. 40.27 will no longer be valid. The load voltage will be given by Eq. 40.30, which is derived from Eq. 40.24.

$$V_L = V_{\text{in}} - I_L R_s \qquad 40.30$$

A practical zener diode voltage regulator circuit is designed as follows.

step 1: Determine the supply resistance that minimizes power loss but keeps the zener operating in the breakdown region. Use Eq. 40.25 with the zener current, I_Z, equal to the keep-alive current, I_{ZK}, to determine the load voltage minimum that maintains the keep-alive current. Then use Eq. 40.26 with the calculated value of the minimum load voltage, the minimum input voltage, and the maximum load current to determine the source resistance. This step accounts for the minimum supply (input) voltage and the maximum load (output) current.

step 2: Determine the maximum zener current drawn with the supply resistance calculated in step 1 (see Eq. 40.27). This step accounts for the maximum supply (input) voltage and the minimum load (output) current.

step 3: Determine the required diode power rating with the current calculated in step 2 (see Eq. 40.28).

step 4: Determine the required resistor power rating with the current calculated in step 2 (see Eq. 40.29).

step 5: If desired, determine temperature effects on the zener diode at maximum expected temperature changes, and recalculate the parameters in the above steps to determine the circuit functions correctly (see Sec. 37.14).

step 6: If desired, determine the voltage regulation of the circuit. The regulation is calculated between the maximum input voltage at no load current and the minimum input voltage at maximum current.

Example 40.11

Determine the source resistance for a zener voltage regulator circuit, patterned after Fig. 40.15(b). The output voltage requirement is 5.1 V, while the input varies from 10 V to 15 V, and the load current ranges from 0 A to 5 A. The keep-alive current, I_{ZK}, equals 5 mA, and the zener resistance, R_Z, equals 0.12 Ω. (The parameters are identical to those used in Ex. 40.10 with the exception that the voltage output is 5.1 V, which is one of the standard zener voltages from App. 40.A.)

Solution

The supply resistance is determined from Eq. 40.26 by substituting the value of the minimum load voltage from Eq. 40.25. (See step 1 of the procedure in this section.) The minimum load voltage is

$$\begin{aligned} V_{L,\text{min}} &= V_{\text{ZM}} + I_{\text{ZK}} R_Z \\ &= 5.1 \text{ V} + (5 \times 10^{-3} \text{ A})(0.12 \text{ Ω}) \\ &= 5.1006 \text{ V} \end{aligned}$$

Substituting into Eq. 40.26, using the minimum input voltage and maximum load current, gives

$$V_L = \frac{V_{ZM} R_s + V_{\text{in,min}} R_Z}{R_s + R_Z} - I_{L,\text{max}}\left(\frac{R_s R_Z}{R_s + R_Z}\right)$$

$$5.1006\ \text{V} = \frac{(5.1\ \text{V})(R_s) + (10\ \text{V})(0.12\ \Omega)}{R_s + 0.12\ \Omega} - (5\ \text{A})\left(\frac{(R_s)(0.12\ \Omega)}{R_s + 0.12\ \Omega}\right)$$

$$R_s = 0.98\ \Omega$$

Using a practical zener diode circuit instead of an ideal diode circuit requires a minimal source resistance change.

Example 40.12

Determine the voltage regulation for the circuit in Ex. 40.11.

Solution

The voltage regulation (step 6 of the procedure in this section) is determined by obtaining the equation for the load voltage and current. The value of R_s for the circuit to operate in the breakdown region is 0.98 Ω (from Ex. 40.11). Substituting this value into Eq. 40.26 gives the relationship between the load current and voltage.

$$\begin{aligned} V_L &= \frac{V_{ZM} R_s + V_{\text{in}} R_Z}{R_s + R_Z} - I_L\left(\frac{R_s R_Z}{R_s + R_Z}\right) \\ &= \frac{(5.1\ \text{V})(0.98\ \Omega) + (V_{\text{in}})(0.12\ \Omega)}{0.98\ \Omega + 0.12\ \Omega} - I_L\left(\frac{(0.98\ \Omega)(0.12\ \Omega)}{0.98\ \Omega + 0.12\ \Omega}\right) \\ &= 4.54\ \text{V} + 0.109\ V_{\text{in}} - 0.107 I_L \end{aligned}$$

For an input voltage of 10 V and no-load current, the load voltage is 5.63 V. At the maximum load current of 5 A, the load voltage is 5.10 V. (If the load current goes above the maximum of 5 A, the diode ceases to regulate and the load voltage is determined by Eq. 40.30.) Repeating the calculation for an input voltage of 15 V gives the values of 6.18 V and 5.64 V for no-load current and maximum load current, respectively. Plotting the results gives the following.

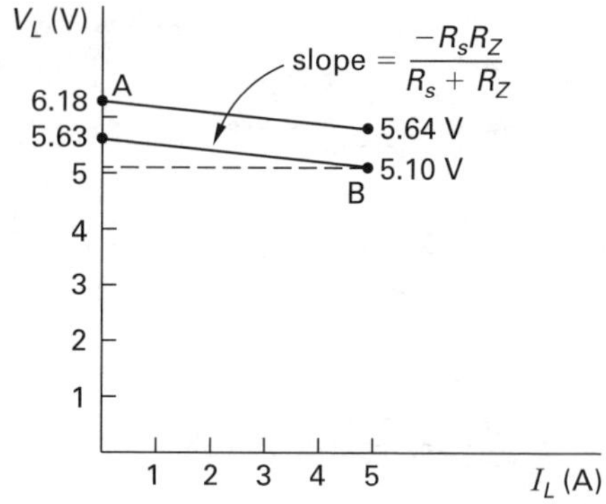

The regulation is calculated between the maximum input voltage at no-load current (point A) and the minimum input voltage at maximum current (point B).

$$\begin{aligned} \text{VR} &= \frac{V_{\text{nl}} - V_{\text{fl}}}{V_{\text{fl}}} \times 100\% \\ &= \frac{6.18\ \text{V} - 5.10\ \text{V}}{5.10\ \text{V}} \times 100\% = 21.2\% \end{aligned}$$

9. OPERATIONAL AMPLIFIER REGULATOR CIRCUIT

Zener diodes provide voltage regulation at low-current demands. Better regulation is achieved using a precision diode, but a precision diode also lowers the operational amplifier current output. Using the zener diode as a reference for input to an op amp and passing the output to the base of a transistor improves the regulation and increases the current capability. Such an op amp regulator is shown in Fig. 40.16.

Figure 40.16 *Operational Amplifier Voltage Regulator*

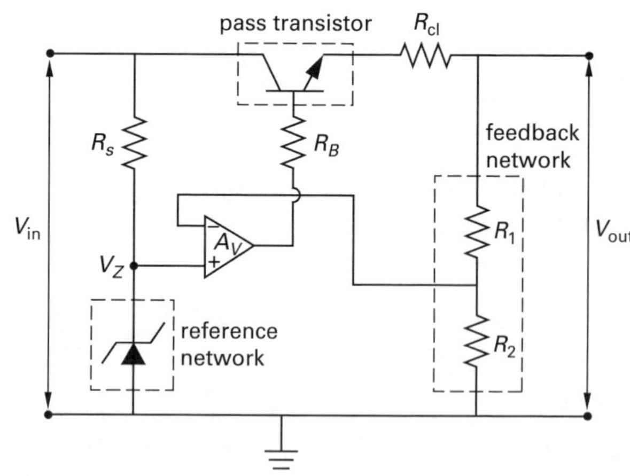

The common base transistor is called the *pass transistor* and is used to supply a current gain, α. The zener provides the reference voltage, V_Z, at the noninverting terminal of the op amp. The input to the op amp is

$V_Z - \delta V_{out}$ where delta is called the *feedback factor* and is given by[18]

$$\delta = \frac{R_2}{R_1 + R_2} \qquad 40.31$$

The output voltage is approximately

$$V_{out} \approx \frac{V_Z}{\delta} \approx V_Z\left(\frac{A_V}{1 + \delta A_V}\right) \qquad 40.32$$

The supply resistance, R_s, is set to ensure that the minimum keep-alive current flows in the zener diode.

$$R_s = \frac{V_{in,min} - V_{ZM}}{I_{ZK}} - R_Z \qquad 40.33$$

The equivalent circuit for the zener diode, which shows R_Z, is shown in Fig. 40.15(b). The minimum and maximum zener voltages are

$$V_{Z,min} = V_{ZM} + I_{ZK}R_Z \qquad 40.34$$

$$V_{Z,max} = \frac{V_{in,max}R_Z + V_{ZM}R_s}{R_s + R_Z} \qquad 40.35$$

The method of design follows the same basic steps given in Sec. 40.8.

10. TRANSISTOR SWITCH FUNDAMENTALS

The bipolar junction transistor (BJT) is an active nonlinear device that, when used between its saturation and cutoff regions, becomes an electronic switch (see Sec. 38.2). A transistor circuit set up to operate as a switch is shown in Fig. 40.17(a). The pulse input that controls the switch is shown in Fig. 40.17(b). The pulses can be considered to represent logic conditions.[19] Combinations of such switches can be used to build logic circuits. The characteristics of the switch, the two operating points, and approximate models are shown in Fig. 40.17(c).

When the transistor switch is in the on state, that is, saturated, the collector current is given by KVL in the output loop as

$$I_{C,sat} = \frac{V_{CC} - V_{CE,sat}}{R_L} \qquad 40.36$$

The collector-emitter voltage, V_{CE}, is typically 0.1 V for germanium diodes and 0.2 V for silicon diodes and can be ignored for first-order calculations—hence the short-circuit model of Fig. 40.17(c). The value of the load resistance must be determined so that when in saturation the condition of Eq. 40.37 is satisfied.[20]

Figure 40.17 Transistor Switch

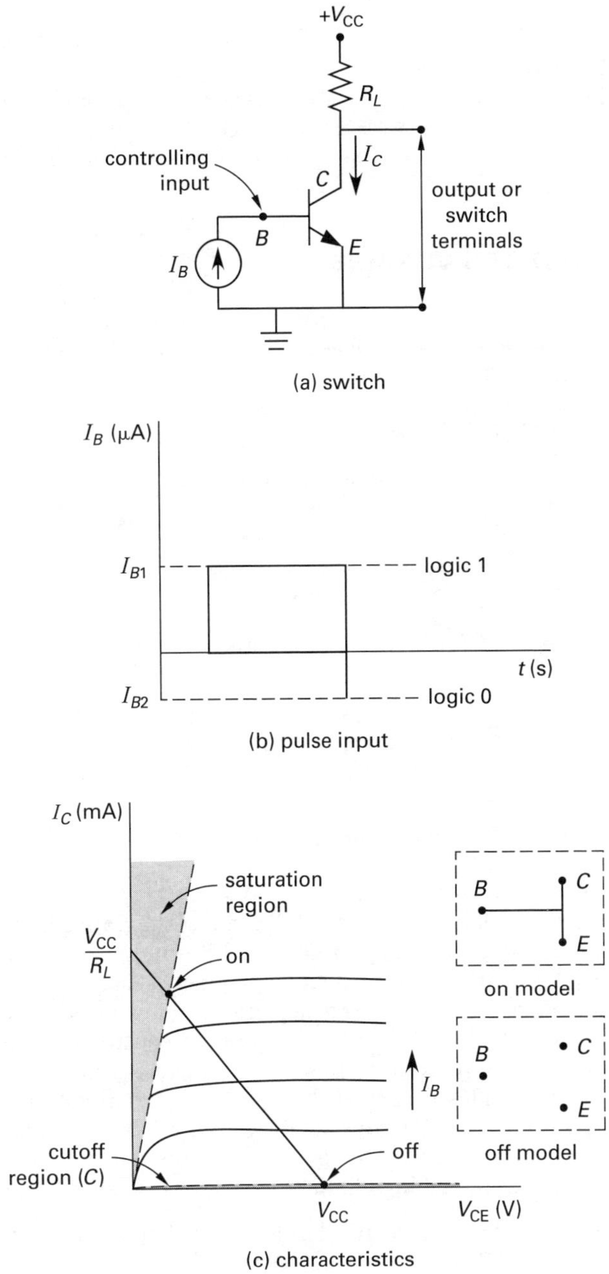

$$I_{B1} \geq \frac{I_{C,sat}}{h_{FE}} \approx \frac{V_{CC}}{h_{FE}R_L} \qquad 40.37$$

The base current calculated by Eq. 40.37 is the minimum current required to drive the transistor into saturation, ignoring the small collector-emitter voltage. When the transistor is in saturation (on), the base-emitter and collector-base junctions are both forward biased. When

[18]The *feedback factor* is sometimes symbolized β. This symbol is avoided here to prevent confusion with the CE current gain, β.

[19]Normally the logic conditions are voltage inputs. For transistor-transistor-logic (TTL) circuits, 0.0 V to 0.8 V is logic 0 and approximately 3.0 V is logic 1.

[20]At low frequencies, $h_{FE} \approx h_{fe}$, as is the case for all of the small-signal parameters.

the transistor is in cutoff (off), the base-emitter and collector-base junctions are both reverse biased.

Field-effect transistors (FETs) are also used as switches. The gate provides the controlling input. In the on condition, unlike the BJT, the properties of the FET are more like that of a resistor. In the off condition, small leakage currents flow, similar to reverse saturation currents in the BJT.

11. JFET SWITCHES

Junction field-effect transistors used as switches are constructed from depletion-mode *p*-channel types as shown in Fig. 40.18.

Figure 40.18 *JFET Switch*

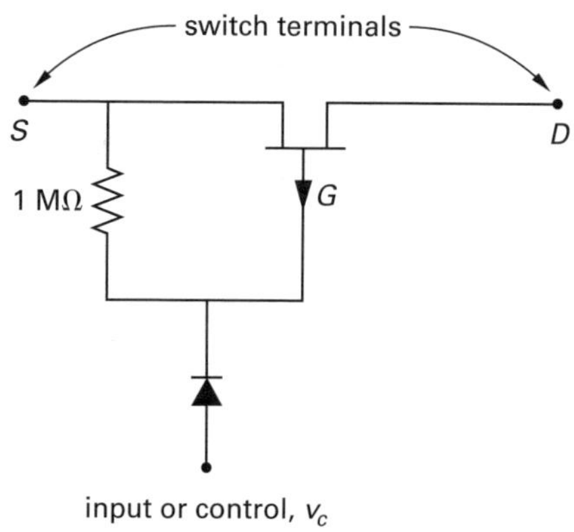

When the control voltage, v_c, is zero, the gate-source voltage, V_{GS}, is equal to zero and the switch is on. This occurs because the gate-channel *pn* junction controls the transistor by varying the depletion region caused by reverse bias (see Sec. 39.2). A positive control voltage results in a positive gate-source voltage, turning the switch off. This positive voltage is the pinchoff voltage, V_{GS}(off). The diode prevents forward bias (i.e., negative voltage) on the gate-channel *pn* junction, which would destroy the switching characteristics of the JFET.[21]

The advantage of the JFET is its low resistance while in the on condition. Disadvantages include complexity and price. Further, the JFET is restricted to switching voltages a few volts less than the control voltages, because anything larger could turn the switch on. Also, if the switching voltage becomes too large while in the on condition, the switch can turn off. That is, the JFET moves from the ohmic region to the pinchoff region (see Sec. 39.2).

12. CMOS SWITCHES

When a *p*-channel MOS device and an *n*-channel MOS device are fabricated on the same chip, the device is called a *complementary MOSFET* (CMOS). A CMOS device used as an *inverter* or digital switch is illustrated in Fig. 40.19 along with the approximate output characteristics.

Figure 40.19 *CMOS Inverter*

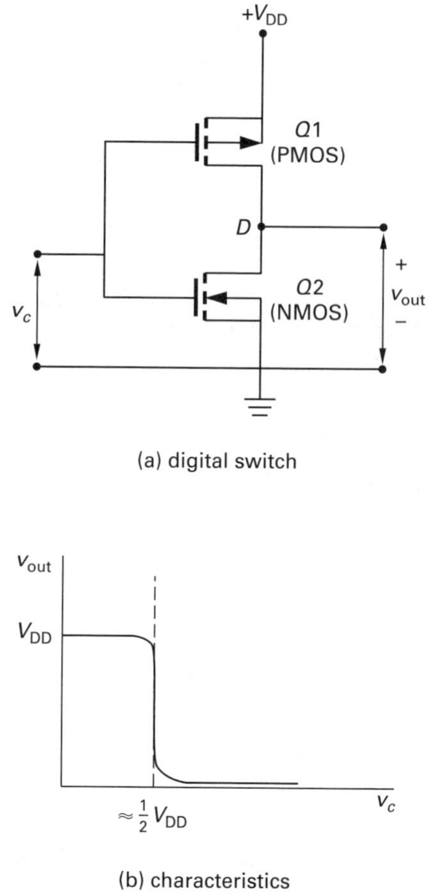

The control or input voltage, v_c, varies from 0 V (logic 0) to $+V_{DD}$ (logic 1). When the control voltage is 0 V, the gate-source voltage of $Q1$ is negative and $Q1$ is on. (The negative voltage attracts holes, enhancing the channel and allowing current flow.) The gate-source voltage of $Q2$ is zero and $Q2$ is off. (A positive voltage is required to attract electrons into the channel.) Consequently, $Q1$ can be modeled as a short circuit from source to drain, resulting in $v_{out} = +V_{DD}$ (logic 1). When the control voltage is $+V_{DD}$, the gate-source voltage of $Q1$ is zero and $Q1$ is off. The gate-source voltage of $Q2$ is positive and $Q2$ is on. Consequently, $Q2$ can be modeled as a short circuit from source to drain, resulting in $v_{out} = 0$ V (logic 0). Current flows only during the switching transient. The static current flow and static power are zero, so the power requirements of CMOS circuits are extremely low.

An analog switch is created when the inverter is combined with a second CMOS device as shown in Fig. 40.20.

[21]The current from the gate would add to that flowing from source to drain. The device would no longer function strictly as a switch controlling external current.

Figure 40.20 CMOS Switch

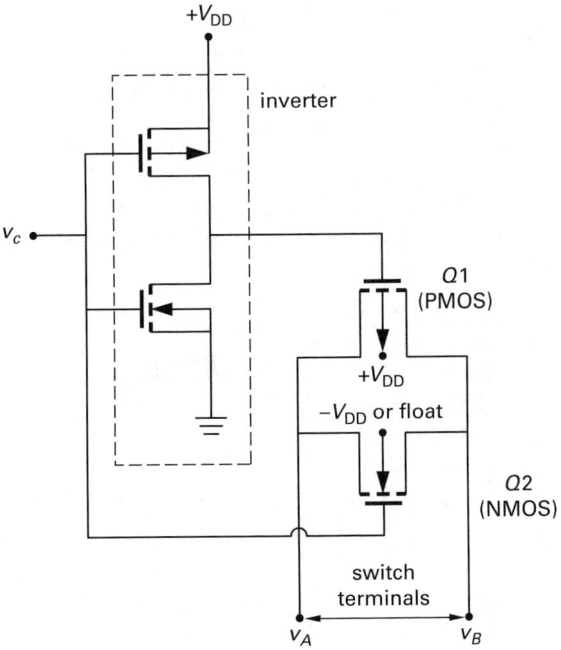

When the control voltage, v_c, is 0 V, both parallel switch channels ($Q1$ and $Q2$) are off. $Q1$ is off because $+V_{DD}$ is applied to its gate by the inverter. $Q2$ is off because 0 V is applied to its gate by the control voltage. In the off condition, restrictions on the switching terminals' voltage exist. For instance, $Q1$ will turn on if both terminals are more positive than $+V_{DD}$—hence the upper limit on the voltage. $Q2$ will turn on if both terminals are negative by an amount equal to V_{GS}—hence a lower limit on the voltage. When the control voltage, v_c, is $+V_{DD}$, $Q1$ will turn on when the switching voltage is positive enough to make V_{GS} negative enough to create the pÂ-channel. $Q2$ will turn on when the switching voltage is negative or within a few volts of V_{DD}.

13. ACTIVE WAVEFORM SHAPING

Waveform shaping occurs in many circuits using nonlinear active devices, with and without storage elements (capacitors and inductors). These devices can be discrete or integrated, although integrated circuits rely primarily on capacitors rather than inductors. No single simple theory describes the behavior of all these elements. The analysis is generally done by breaking the circuit into a sequence of linear problems and combining the results.

Examples of such circuits include operational amplifiers with capacitors and inductors (integrators and differentiators, filters), waveform generators (sweep, square, triangular), oscillator circuits (monostable, bistable, astable), and so on.

14. LOGIC FAMILIES

Integrated circuits are classified into families based on the circuit elements they contain and the manner in which these elements are combined. *Passive logic circuits* have no transistors (i.e., they have only resistors and diodes), while *active logic circuits* do have transistors.

One of the earliest logic families is *diode-transistor logic* (DTL). *High-threshold logic* (HTL) is similar to DTL but uses higher voltages and better diodes to produce greater noise immunity. *Variable-threshold logic* (VTL) is a variation of DTL with adjustable noise immunity. *Direct-coupled transistor logic* (DCTL) is an unsophisticated connection of transistors with poor immunity to noise and low logic levels. Because there are no resistors, it has low power dissipation. *Resistor-transistor logic* (RTL) is similar to DCTL with the addition of resistors to limit current hogging. *Resistor-capacitor transistor logic* (RCTL) is a low-cost compromise between speed and power. *Transistor-transistor logic* (TTL or T^2L) is the fastest of all logic families using transistors operating in the saturated region. The fastest of all logic families using bipolar junction transistors is *emitter-coupled logic* (ECL), also known as *current-mode logic* (CML). ECL transistors do not operate in the saturation region. ECL circuits consume more power than TTL.

Metal-oxide semiconductor (MOS) circuitry contains MOSFETs. It requires few resistors and has high packing density. *Complementary construction* uses both n-channel and p-channel transistors in the same circuit. NMOS logic uses n-channel MOSFETs, while PMOS uses p-channel MOSFETs. (NMOS is approximately six times slower than PMOS.) *Complementary metal-oxide semiconductor* (CMOS), also known as *complementary transistor logic* (CTL), circuits have high packing density and reliability. They are capable of higher switching speeds than MOS.

Circuitry based on bipolar transistors switches faster, offers greater current drive, and is more suitable for analog applications than CMOS circuitry.[22] On the other hand, CMOS consumes less power, dissipates less heat, and has a higher packing density. Bipolar and CMOS circuitry are combined in biCMOS integrated circuits.

15. LOGIC GATES

A *gate* performs a logical operation on one or more inputs. The inputs (labeled A, B, C, etc.) and output are limited to the values of zero or one. A listing of the output value for all possible input values is known as a *truth table*. Table 40.1 combines the symbols, names, and truth tables for the most common gates. For more information on logical operations, see Chap. 41.

[22]CMOS is equivalent to TTL in speed but slower than ECL.

Table 40.1 *Logic Gates*

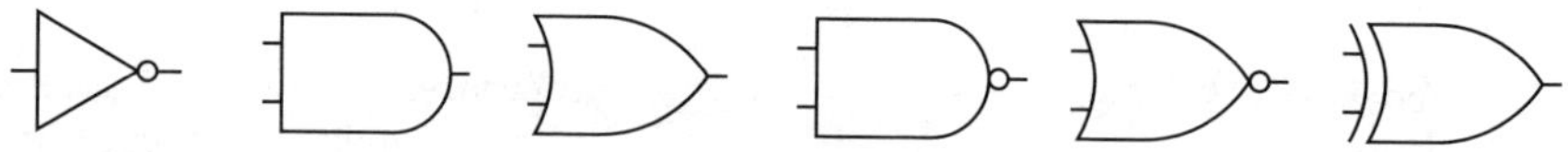

inputs		not	and	or	nand	nor	exclusive or
A	B	$-A$ or $\overline{A}$	AB	$A+B$	$\overline{AB}$	$\overline{A+B}$	$A\oplus B$
0	0	1	0	0	1	1	0
0	1	1	0	1	1	0	1
1	0	0	0	1	1	0	1
1	1	0	1	1	0	0	0

16. SIMPLIFICATION OF BINARY VARIABLES

The rules of *Boolean algebra* are used to write and simplify expressions of binary variables (i.e., variables constrained to two values). The basic laws governing Boolean variables are as follows.

- *commutative:* $A+B=B+A$
  $A\cdot B=B\cdot A$
- *associative:* $A+(B+C)=(A+B)+C$
  $A\cdot(B\cdot C)=(A\cdot B)\cdot C$
- *distributive:* $A\cdot(B+C)=(A\cdot B)+(A\cdot C)$
  $A+(B\cdot C)=(A+B)\cdot(A+C)$
- *absorptive:* $A+(A\cdot B)=A$
  $A\cdot(A+B)=A$

De Morgan's theorems are

$$\overline{A+B}=\overline{A}\cdot\overline{B}$$
$$\overline{A\cdot B}=\overline{A}+\overline{B}$$

The following basic identities are used to simplify Boolean expressions.

$$0+0=0 \quad 0\cdot 0=0 \quad A+0=A$$
$$0+1=1 \quad 0\cdot 1=0 \quad A+1=1$$
$$1+0=1 \quad 1\cdot 0=0 \quad A+A=A$$
$$1+1=1 \quad 1\cdot 1=1 \quad A+\overline{A}=1$$

$$A\cdot 0=0 \quad -0=1$$
$$A\cdot 1=A \quad -1=0$$
$$A\cdot A=A \quad -\overline{A}=A$$
$$A\cdot\overline{A}=0$$

Switching algebra, a form of Boolean algebra, is discussed in Sec. 41.5.

Example 40.13

Simplify and write the truth tables for the following network of logic gates.

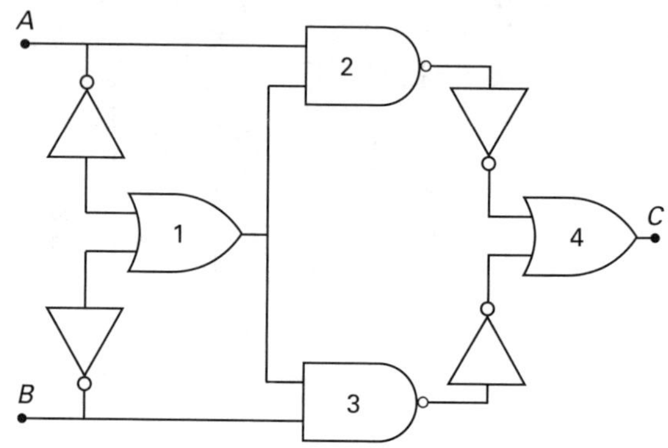

Solution

Determine the inputs and output of each gate in turn.

gate	inputs	output
1	$\overline{A}$, $\overline{B}$	$(\overline{A}+\overline{B})$
2	A, $(\overline{A}+\overline{B})$	$\overline{A\cdot(\overline{A}+\overline{B})}$
3	B, $(\overline{A}+\overline{B})$	$\overline{B\cdot(\overline{A}+\overline{B})}$
4	$A\cdot(\overline{A}+\overline{B})$, $B\cdot(\overline{A}+\overline{B})$	$A\cdot(\overline{A}+\overline{B})+B\cdot(\overline{A}+\overline{B})$

Simplify the output of gate 4.

$$A\cdot(\overline{A}+\overline{B})+B\cdot(\overline{A}+\overline{B}) \quad \text{[original]}$$
$$A\cdot\overline{A}+A\cdot\overline{B}+B\cdot\overline{A}+B\cdot\overline{B} \quad \text{[distributive]}$$
$$A\cdot\overline{B}+B\cdot\overline{A} \quad [\text{since } A\cdot\overline{A}=0]$$
$$A\oplus B \quad \text{[definition]}$$

The truth table is

A	B	C
0	0	0
0	1	1
1	0	1
1	1	0

17. LOGIC CIRCUIT FAN-OUT

Fan-out is the number of parallel loads that can be driven from one output node of a logic circuit. Because logic circuits are routinely cascaded in order to achieve the desired outcome, a single transistor's output must provide multiple inputs for subsequent stages.

Consider the NOT gate shown in Fig. 40.21, which is a common emitter-configured transistor with the output taken from the collector. The transistor is designed to operate as a switch between the saturation and cutoff regions. When the input is high (several volts), the base-emitter junction is forward biased and the base current is sufficient to saturate the transistor. The output is then low, nearly zero if the small collector-emitter voltage (0.2 V for silicon transistors and 0.1 V for germanium) is ignored. The input has been inverted, that is, changed from high to low—hence the designation as a NOT gate.[23]

Figure 40.21 *NOT Gate*

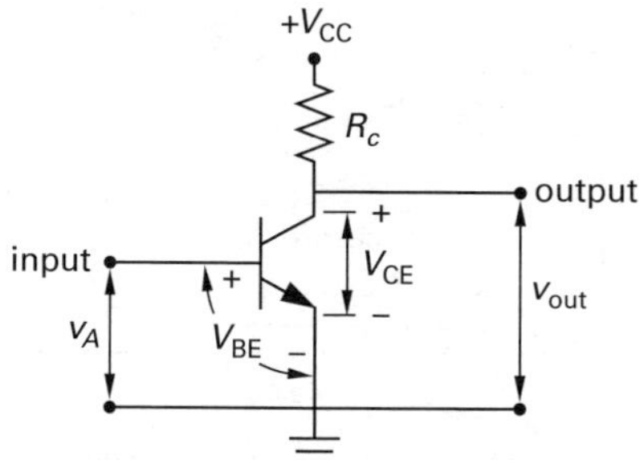

(a) configuration

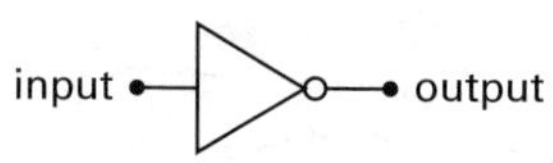

(b) logic symbol

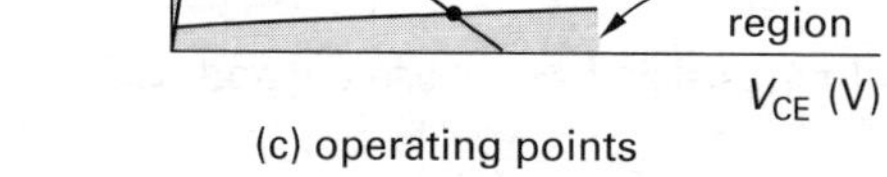

(c) operating points

v_A	v_{out}
L	H
H	L

(d) voltage truth table

[23]The logic described here and used throughout this chapter is *positive logic*, that is, logic where the high voltage is logic condition 1 and the low voltage is logic condition 0. *Negative logic* is the reverse. *Dynamic* or *pulse logic* is a system where the presence of a pulse is logic condition 1 and the absence of a pulse is logic condition 0.

In practical applications, this transistor does not operate in isolation, but instead is cascaded with other logic devices. The base current drawn while in saturation is *sourced* from previous stages. When in cutoff, the reverse saturation current, I_{CBO}, flowing through the collector-base junction must be *sunk* by previous stages. This situation is shown in Fig. 40.22.

Figure 40.22 *Fan-Out Unit Loads*

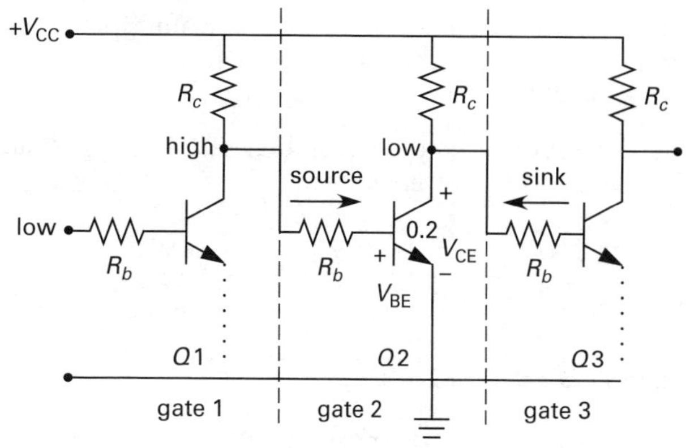

The number of stages a logic gate can source, that is, the current a stage can supply when its collector voltage is high, is called the fan-out. In Fig. 40.22, transistor $Q1$ is sourcing the current to $Q2$. The number of stages that a logic gate can sink, that is, the current the collector (or output circuit) can accept without removing the transistor from the saturated state, is also called the fan-out. In Fig. 40.22, transistor $Q2$ is sinking the current from $Q3$. A given logic family's fan-out specification is stated in terms of *unit loads*. A *unit source current* is the input current drawn *by* a single gate when the voltage is high. A *unit sink current* is the current drawn *from* the input of a single gate when the input voltage is low. The fan-out specification is the minimum number of unit loads, source or sink, the gate can provide.

Gate 1 in Fig. 40.22 sources a unit load to $Q2$. $Q1$ is cut off and appears as an open circuit.[24] The unit source current is the base current to $Q2$ given by

$$I_{B2} = \frac{V_{CC} - V_{BE2}}{R_c + R_b} \quad \textit{40.38}$$

$Q2$ is saturated, which causes $Q3$ to be cut off. The reverse saturation current, I_{CBO}, from the collector-base junction of $Q3$ is drawn to the collector of $Q2$. (This sink current raises the voltage at the collector of $Q2$. If too much current is drawn, the voltage rise can reverse-bias the collector-base junction of $Q2$, bringing it out of

[24]Because the threshold voltage is approximately 0.7 V for a silicon base-emitter junction, $Q1$ will be cut off at voltages of 0.6 V or less.

saturation.) The unit sink current is I_{CBO3}. The collector current is

$$I_{C2} = \frac{V_{CC} - V_{CE,sat}}{R_c} + I_{CBO3} \qquad 40.39$$

When a gate's function is to maintain multiple loads, the source and sink transistors can be modeled as shown in Fig. 40.23 in order to determine the fan-out.

For the sourcing condition of Fig. 40.23(a), applying KCL at the collector node gives

$$\frac{V_{CC} - V_{CE}}{R_c} - I_C = N\left(\frac{V_{CE} - V_{BE}}{R_b}\right) \qquad 40.40$$

Figure 40.23 *Fan-Out Circuts*

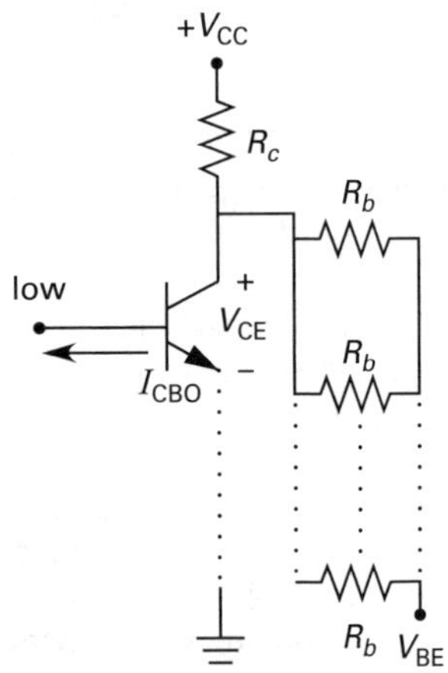

(a) source loads

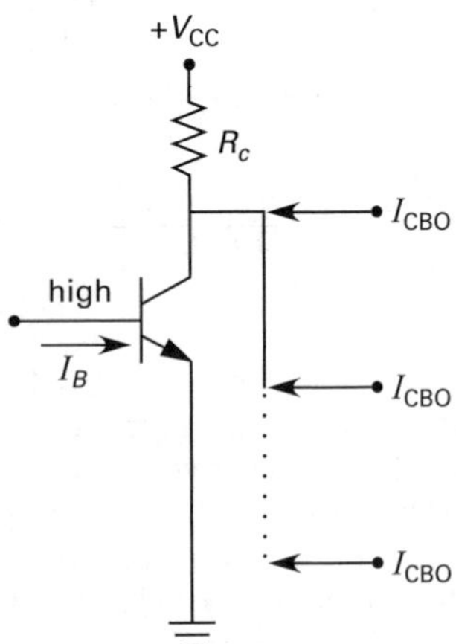

(b) sink loads

The forward base-emitter voltage required to hold the load transistors in saturation is typically ≈ 0.7 V, and the reverse saturation is I_{CBO}. The current in the collector resistor, R_c, includes the reverse saturation current, I_{CBO}, of the cutoff transistor.[25]

The load line equation without sourced loads is shown as a dashed line in Fig. 40.24(a). The load line for a transistor sourcing multiple loads is shown as a solid line in Fig. 40.24(a) and is given by Eq. 40.40 rearranged as

$$I_C + \left(\frac{1}{R_c} + \frac{N}{R_b}\right)V_{CE} = \frac{V_{CC}}{R_c} + \left(\frac{N}{R_b}\right)V_{BE} \qquad 40.41$$

Figure 40.24 *Load Lines*

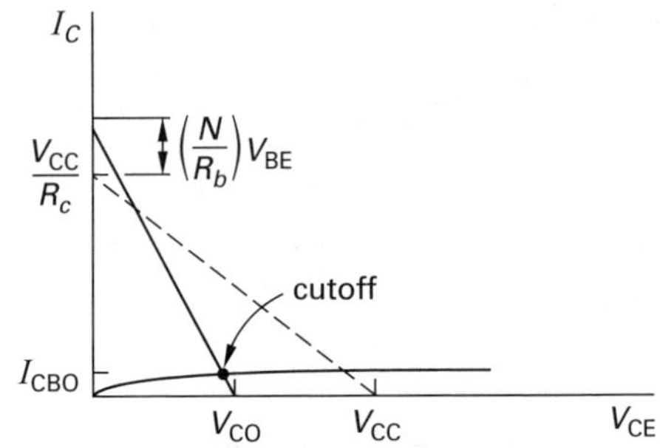

(a) cutoff: multiple source loads

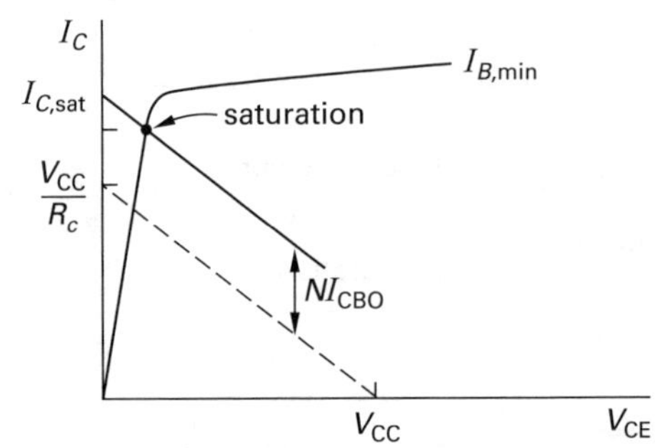

(b) saturated: multiple sink loads

Each load transistor has a base current given by

$$I_B = \frac{V_{CE} - V_{BE}}{R_b} \qquad 40.42$$

Substituting the restriction of Eq. 40.42 into the load line equation, Eq. 40.41, gives the sourcing equation for this type of logic gate.

$$I_C + \left(N + \frac{R_b}{R_c}\right)I_B = \frac{V_{CC} - V_{BE}}{R_c} \qquad 40.43$$

[25]The collector current is not shown in Eq. 40.40 as the reverse saturation current because this equation is used in a general manner to determine the load line.

For the sink condition of Fig. 40.23(b), applying KCL at the collector node gives

$$I_{C,\mathrm{sat}} = \frac{V_{\mathrm{CC}} - V_{\mathrm{CE,sat}}}{R_c} + NI_{\mathrm{CBO}} \qquad 40.44$$

The load lines for the sink condition are taken directly from Eq. 40.44 and are shown in Fig. 40.24(b). Once the minimum collector saturation current is determined by Eq. 40.44 for the sink condition, which depends on the sink fan-out (N), the minimum base current required to maintain the transistor in saturation can be graphically determined from the characteristic curve as shown in Fig. 40.24(b). The base current curve that exists at the point where the load line curve for multiple sinking loads intersects the calculated collector saturated current determines the minimum base current.[26]

Once the equations for the source condition, Eq. 40.43, and sink condition, Eq. 40.44, are determined, the collector current, I_C, is set as the reverse saturation current, I_{CBO}, in the source equation and both equations are solved for the fan-out, N. The *source fan-out* is

$$N + \frac{R_b}{R_c} \le \frac{V_{\mathrm{CC}} - V_{\mathrm{BE}} - I_{\mathrm{CBO}}R_c}{I_{B,\mathrm{min}}R_c} \qquad 40.45$$

The *sink fan-out* is

$$N \le \frac{I_{C,\mathrm{sat}}R_c + V_{\mathrm{CE,sat}} - V_{\mathrm{CC}}}{I_{\mathrm{CBO}}R_c} \qquad 40.46$$

The procedure for determining the fan-out for other types of logic is similar.

Example 40.14

Given a transistor with the characteristic shown, $R_c = 1000\ \Omega$ and $V_{\mathrm{CC}} = 5$ V, determine the base current necessary for a fan-out of five.

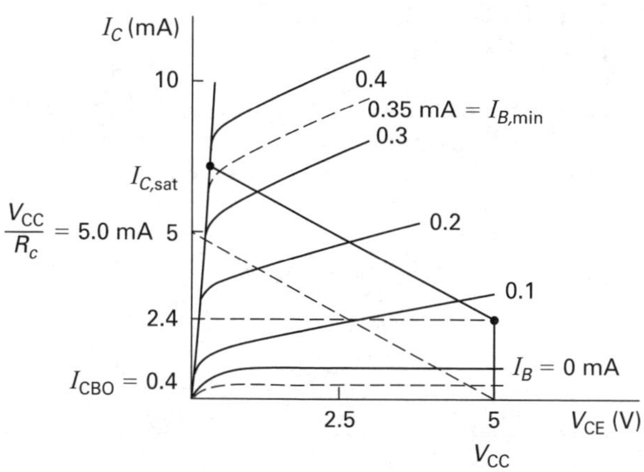

[26]The minimum base current, $I_{B,\mathrm{min}}$, is the base current that a sourcing gate must supply as well as the minimum base current it requires when operating in the sink condition.

Solution

Because $I_{\mathrm{CBO}} = 0.4$ mA, five load gates require $5 \times 0.4\ \mathrm{mA} = 2$ mA excess current. The corresponding load line is shown solid. The resulting saturation current requires a base current of about 0.35 mA. (Setting the value at 0.4 mA gives a margin of error for temperature effects and variations between transistors.)

Example 40.15

Determine the maximum base resistance, R_b, that will allow sourcing five gates from the transistor gate in Ex. 40.14.

Solution

Using Eq. 40.45 with $V_{\mathrm{BE}} = 0.7$ V and $I_{B,\mathrm{min}}$ set at 0.4 mA,

$$N + \frac{R_b}{R_c} \le \frac{V_{\mathrm{CC}} - V_{\mathrm{BE}} - I_{\mathrm{CBO}}R_c}{I_{B,\mathrm{min}}R_c}$$

$$5 + \frac{R_b}{1000\ \Omega} \le \frac{5\ \mathrm{V} - 0.7\ \mathrm{V} - (0.0004\ \mathrm{A})(1000\ \Omega)}{(0.0004\ \mathrm{A})(1000\ \Omega)}$$

$$R_{b,\mathrm{max}} = 4750\ \Omega$$

18. LOGIC CIRCUIT DELAYS

Logic circuits do not switch instantly from cutoff to saturation, or vice versa, upon application of an input pulse. Various delays are inherent, as shown in Fig. 40.25.

When the input voltage level (pulse) rises, a finite time, called the *delay time*, t_d, is required for the base-emitter junction of the transistor in the inverter circuit to reach the turn-on level. The time for the input voltage to rise enough for the collector output voltage to fall to the saturation level is called the *rise time*, t_r.

With the transistor in saturation, both the base-emitter and collector-base junctions are forward biased. Excess charge is stored in the base region, that is, the space-charge region is small. When the input voltage level (pulse) drops, a finite time is required to remove the excess charges in the base. This delay is called the *storage time*, t_s. Once the excess charge is removed, the collector current falls. This time is called the *fall time*, t_f.

The minimum time for a transistor to pass a pulse, that is, the time from cutoff to saturation and back to cutoff, or vice versa, is the sum of all the delays. The maximum frequency at which a transistor can operate is the reciprocal of the sum of these delays.

The sum of the delays is called the *propagation delay* for a given logic gate. Approximate fan-out and propagation delay values for various transistor logic families, listed left to right in the approximate order of development, are given in Table 40.2.

Figure 40.25 *Logic Circuit Delays*

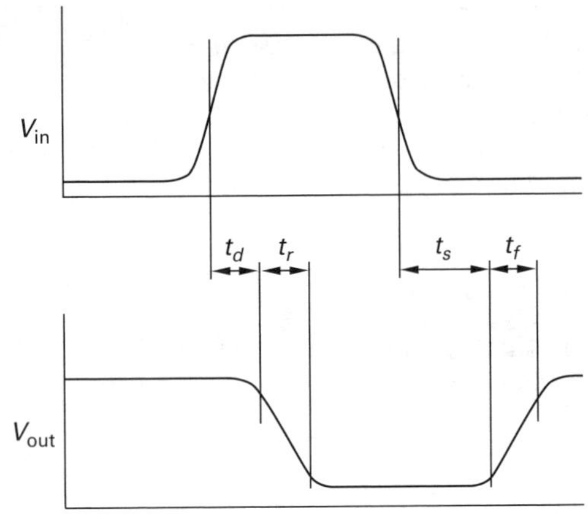

Figure 40.26 *Resistor-Transistor Logic*

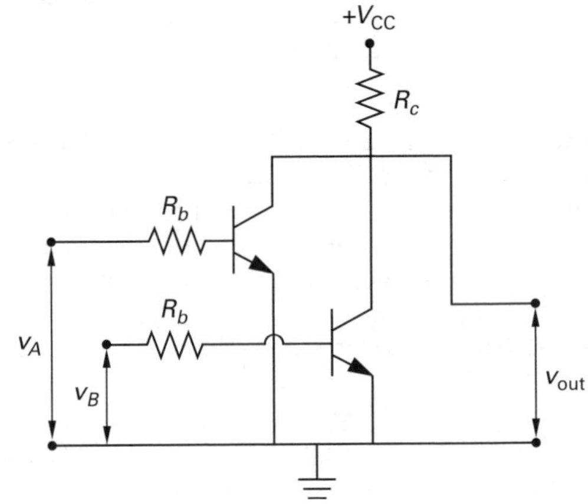

(a) basic gate: NOR

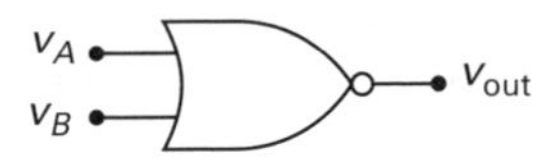

(b) logic symbol

V_A	V_B	V_{out}
L	L	H
L	H	L
H	L	L
H	H	L

(c) voltage truth table

19. RESISTOR-TRANSISTOR LOGIC (RTL)

The basic gate of the resistor-transistor logic (RTL) family is the NOR gate as shown in Fig. 40.26. The inverter used to describe fan-out in Sec. 40.17 is the basic building block. Although a *fan-in* of two is shown in Fig. 40.26, multiple inputs are possible. Logic 1 (high voltage) at any input saturates the associated transistor, causing the output to be connected to ground across the small collector-emitter voltage, resulting in a logic 0 (low voltage) output.

20. DIODE-TRANSISTOR LOGIC (DTL)

The basic gate of the diode-transistor logic (DTL) family is the NAND gate shown in Fig. 40.27. A fan-in of two is shown, but multiple inputs are possible. When either input is logic 0 (low voltage), the associated diode, D, conducts, resulting in a low voltage at node B. The base current to the transistor will be logic 0 because any current flow is blocked by diodes $D1$ and $D2$. (Two diodes are used to ensure that the transistor will not turn on with one input low.) The transistor is cut off and the output is high ($+V_{CC}$, logic 1). When both inputs are logic 1 (high voltage), neither of the diodes conduct and the voltage at node B rises toward $+V_{CC}$. Diodes $D1$ and $D2$ conduct, and the transistor saturates, causing V_{out} to be low ($V_{CE} \approx 0$ V, logic 0).

DTL uses less current than RTL when sourcing loads, but more current when sinking loads. Zero current is required to source a load using DTL logic because of the diodes. *High-threshold logic* (HTL) is similar but uses higher voltages and better diodes to improve noise performance.

21. TRANSISTOR-TRANSISTOR LOGIC (TTL OR T²L)

The basic gate of the transistor-transistor logic (TTL) family is the NAND gate as shown in Fig. 40.28. The basic TTL NAND gate can be considered a modification of the DTL NAND gate. That is, the input diodes have been replaced by a multi-emitter transistor's base-emitterjunctions. A fan-in of two is shown, but multiple

Table 40.2 *Logic Family Data*

parameter	RTL	DTL	HTL	TTL	ECL	MOS	CMOS
basic gate[a]	NOR	NAND	NAND	NAND	OR-NOR	NAND	NOR OR NAND
fan-out[b]	5	8	10	10	25	20	> 50
propagation delay[c]	12	30	90	10	0.1	10	0.1

[a]Positive logic is assumed when determining the basic gate.
[b]Worst-case condition.
[c]Approximate values in nanoseconds.

Figure 40.27 *Diode-Transistor Logic*

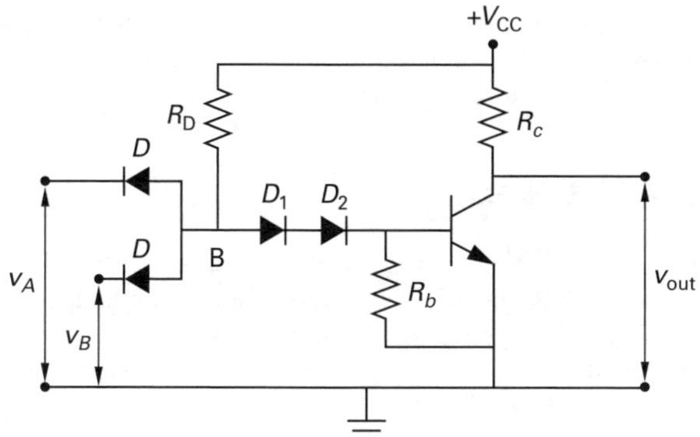

(a) basic gate: NAND

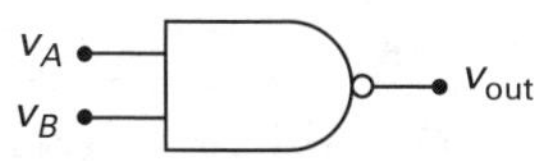

(b) logic symbol

v_A	v_B	v_{out}
L	L	H
L	H	H
H	L	H
H	H	L

(c) voltage truth table

Figure 40.28 *Transistor-Transistor Logic*

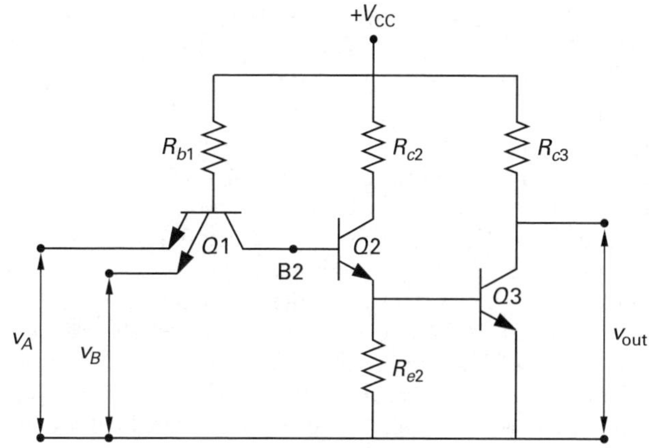

(a) basic gate: NAND

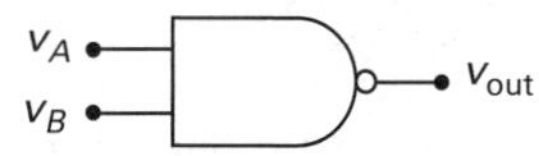

(b) logic symbol

v_A	v_B	v_{out}
L	L	H
H	L	H
L	H	H
H	H	L

(c) voltage truth table

inputs are possible. When at least one input is logic 0 (low voltage), the emitter of $Q1$ is forward biased. Assuming $Q2$ is cut off, the current supplied to the collector of $Q1$ is the reverse saturation current, I_{EBO}, of the emitter-base junction of $Q2$. Because this current is small, $I_{B1} > I_{EBO}/h_{FE}$ and $Q1$ is in saturation. The voltage at node B2 is then equal to the input voltage, v_A or v_B, summed with $V_{CE,sat}$, that is, about 0.4 V to 0.8 V. This voltage is not enough to forward bias $Q2$ and $Q3$, because about 0.7 V apiece is required. $Q3$ is cut off and the output is high ($+V_{CC}$, logic 1). When all the inputs are at logic 1 (high voltage), the emitter-base junction of $Q1$ is reverse biased. $Q1$ is cut off and the base-collector junction is forward biased by approximately $+V_{CC}$. This voltage is high enough to forward bias the base-collector junction of $Q1$ and the base-emitter junctions of $Q2$ and $Q3$, about 2.1 V. So, the output is low ($V_{CE3} \approx 0$ V, logic 0).

TTL logic is the fastest-saturating logic gate. The fall time of TTL circuits is improved by reducing the collector current in the saturated state. This is accomplished by using a *totem-pole configuration*, also called *active pull-up*, which uses an additional transistor and diode in series with R_{C3}.[27] Additionally, the collector of $Q3$ can be left open, in which case it is called the *open-collector TTL gate*. The open-collector TTL gate is used with *wired logic* or *collector logic*, because the collector outputs are wired to the next gate's output without intervening circuitry. This allows the logic voltage level to be changed. For example, in Fig. 40.28, the driving voltage is $+V_{CC}$. If the collector of $Q3$ is open, that is, not connected to $+V_{CC}$ via R_{c3}, then a power source can be attached to the collector of $Q3$, providing a different voltage logic, say V'_{CC}. (One can connect a +5 V logic circuit to a +12 V logic circuit without damaging the components.)

A type of logic with operational similarities to the open-collector logic is *tri-state logic*. In this case, the output of the transistor can be logic level 0, level 1, or level Z. The Z-level is a high-impedance level at which the output is completely disconnected from the rest of the circuit. This is accomplished by adding an enable (inhibit) circuit to the output, similar to the circuit used on clocked flip-flops to control sequencing (see Chap. 52). The high-impedance level allows multiple circuits to connect to the same output line or bus so that several circuits can timeshare a bus. Only the active circuit controls the bus. The others are in the Z-level state and have no impact on the voltage level, or signal, on the bus. Circuits with this tri-state logic are called tri-state buffers.

[27]This configuration is called the totem-pole because the additional transistor sits on top of $Q3$.

22. EMITTER-COUPLED LOGIC (ECL)

The basic gate of the emitter-coupled logic (ECL) family is the OR-NOR gate shown in Fig. 40.29. The basic building block of the family is the difference amplifier ($Q2$ and $Q3$ in Fig. 40.29). The transistors in this logic family do not saturate. Instead, they remain in the active region as the input voltages, v_A and v_B, operating above and below (logic 1 and logic 0) the reference voltage, V_{ref}. Because the emitter current through R_e remains approximately constant, when both inputs are logic 0 (i.e., below the reference voltage) the emitter currents from $Q1$ and $Q2$ are minimal and the emitter current of $Q3$ is maximum. This causes the voltage output at v_{OR} to be low (logic 0) and the voltage output at v_{NOR} to be high (logic 1). When either or both of the inputs are logic 1 (i.e., above the reference voltage) the emitter currents of $Q1$ and $Q2$ are maximum, and the emitter current of $Q3$ is minimal. This causes the voltage output at v_{OR} to be high (logic 1), and the voltage output at v_{NOR} to be low (logic 0).

ECL logic improves on the propagation delay of TTL by not driving the transistors into saturation. This minimizes the switching time, because charge carriers do not have to be cleared from the base junction. However, the power requirements of ECL are greater than TTL (see Table 40.2). Emitter-coupled logic is also called *current-mode logic* (CML).

Figure 40.29 *Emitter-Coupled Logic*

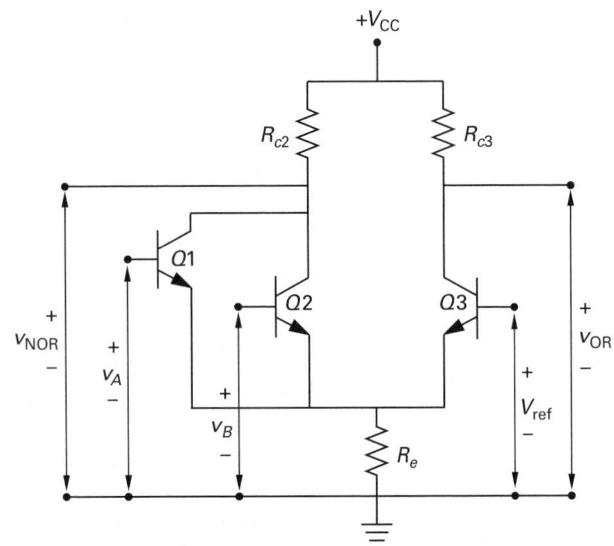

(a) basic gate: OR–NOR

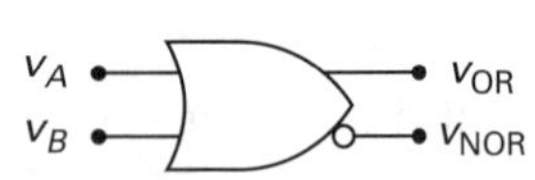

(b) logic symbol

v_A	v_B	v_{OR}	v_{NOR}
L	L	L	H
H	L	H	L
L	H	H	L
H	H	H	L

(c) voltage truth table

23. MOS LOGIC

The basic gate of the MOS logic family is the NAND gate as shown in Fig. 40.30.[28] The basic building block of the family is the inverter ($Q1$ and $Q2$ in Fig. 40.30). A fan-in of two is shown, but multiple inputs are possible. $Q1$ is always biased on by the drain power supply, $+V_{\text{DD}}$. When either input is logic 0 (low voltage), the associated transistor, $Q2$ or $Q3$, is off. No current flows and v_{out} is high ($+V_{\text{DD}}$, logic 1). When both of the inputs are logic 1 (high voltage), both transistors are on. Current flows and the output voltage is low ($v_{\text{out}} \approx 0$ V, logic 0).

MOS logic is simple and requires no external resistors or capacitors, making it well-suited for realization in integrated circuit form. The NMOS positive logic shown in Fig. 40.30(a) draws power in only one state. The PMOS positive logic (NOR) draws power in three states. In both cases, the power drawn is less than for other logic families, allowing much higher device densities. NMOS is faster, because electron mobility is higher than hole mobility. PMOS is less expensive. Advances have made the propagation delays comparable to those for other logic families (see Table 40.2).

24. CMOS LOGIC

The basic gate of the CMOS logic family may be either the NOR gate or the NAND gate shown in Fig. 40.31. (Both are positive logic gates.) A fan-in of two is shown, but multiple inputs are possible. For the NOR gate in Fig. 40.31(a), when both of the inputs are logic 0 (voltage low), $Q1$ and $Q2$ are on, and $Q3$ and $Q4$ are off. The output is caused to be high ($\approx +V_{\text{DD}}$, logic 1) by the connection to the drain supply via the conducting transistors $Q1$ and $Q2$. When either or both of the inputs are logic 1 (voltage high), one or both of the upper transistors ($Q1$ and/or $Q2$) is off. One or both of the lower transistors ($Q3$ and/or $Q4$) is on, connecting the output to ground via the conducting channel. The output is low (≈ 0 V, logic 0). For the NAND gate, a logic 0 (low voltage) on one or both inputs causes one or both upper transistors ($Q1$ and/or $Q2$) to conduct, connecting the output to the drain supply voltage. The output is high ($\approx +V_{\text{DD}}$, logic 1). If both of the inputs are logic 1 (high voltage), the lower series transistors, $Q3$ and $Q4$, turn on and connect the output to ground. The output is low (≈ 0 V, logic 0).

CMOS logic requires only a single supply voltage and is relatively easy to fabricate. The only input current that flows is that required to charge the gate-channel

[28]This is the basic gate for positive logic NMOS. For positive logic PMOS, the basic gate is the NOR gate.

Figure 40.30 *MOS Logic*

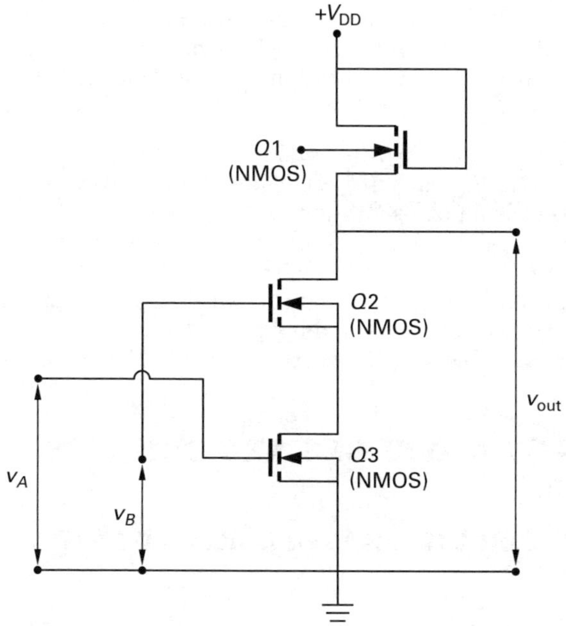

(a) basic gate: NAND

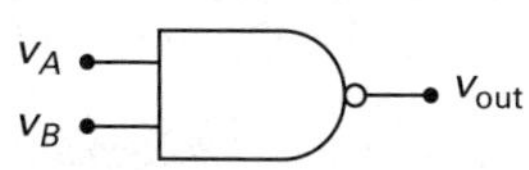

(b) logic symbol

v_A	v_B	v_{out}
L	L	H
L	H	H
H	L	H
H	H	L

(c) voltage truth table

capacitances and any leakage through the off transistor. The power supply consumption is extremely low. Because of the low power consumption, the fan-out is very high. Speeds are comparable to that for ECL (see Table 40.2). Consequently, CMOS has the best overall properties of any logic family.

25. MULTIVIBRATORS

Multivibrators are pulse generator circuits that produce rectangular-wave outputs. There are three types.

The *astable multivibrator*, also called the free-running multivibrator, has no particular stable state and generates a continuous flow of pulses without input, as shown

Figure 40.31 *CMOS Logic*

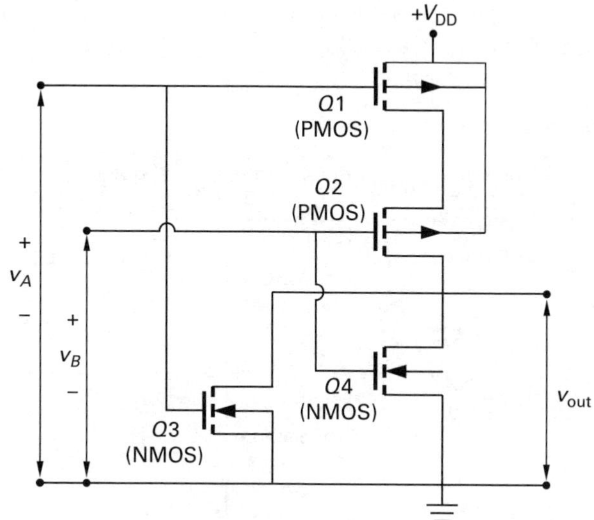

(a) basic gate: NOR

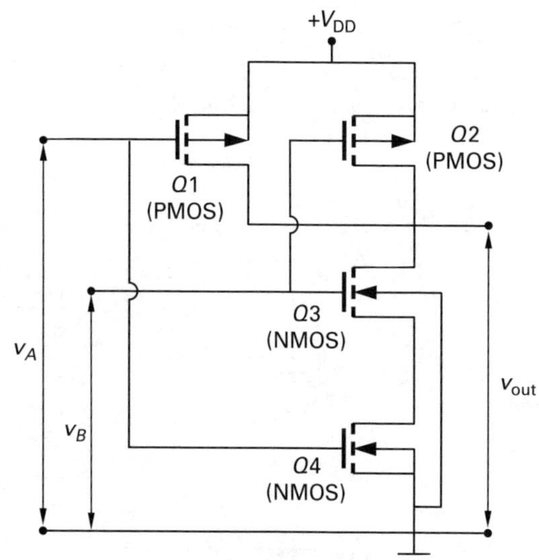

(b) basic gate: NAND

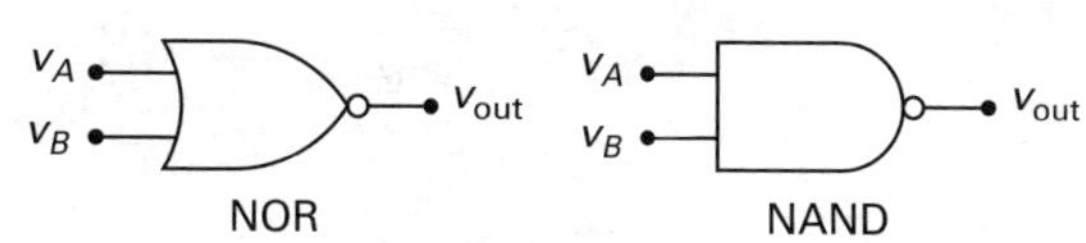

(c) logic symbol

v_A	v_B	v_{out}
L	L	H
L	H	L
H	L	L
H	H	L

NOR

v_A	v_B	v_{out}
L	L	H
L	H	H
H	L	H
H	H	L

NAND

(d) voltage truth tables

in Fig. 40.32(a). The astable multivibrator may be used as a computer or digital system clock. The operational amplifier circuitry realization of an astable multivibrator is shown in Fig. 40.33(a).

Figure 40.32 *Multivibrator Types*

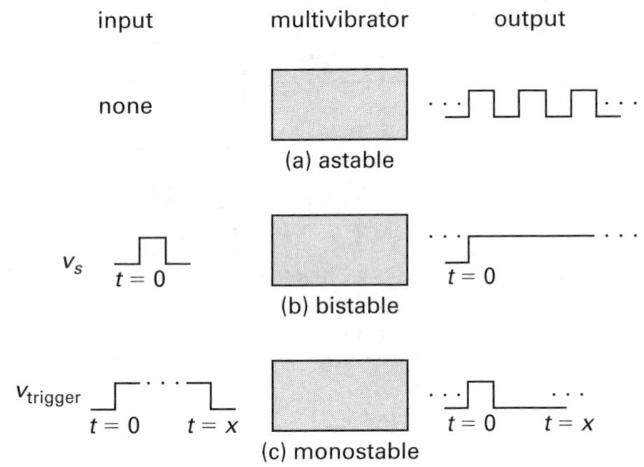

Figure 40.33 *Multivibrator [Op Amp] Circuits*

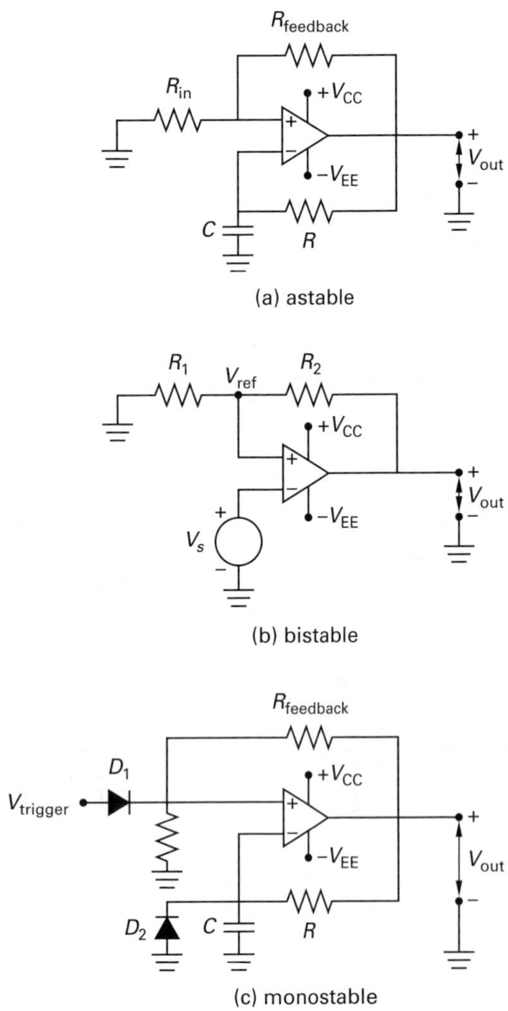

The *bistable multivibrator* has two stable states and generates either a LOW or HIGH output as shown in Fig. 40.32(b). The bistable multivibrator is also called a flip-flop, and is used as a logic device. The operational amplifier circuitry realization of a bistable multivibrator is shown in Fig. 40.33(b).

The *monostable multivibrator* has a single stable state as shown in Fig. 40.32(c). The monostable multivibrator is also called a one-shot multivibrator, and is used to generate a single pulse of known duration. This occurs because the duration of input pulse is unrelated to the duration of the output pulse, which is set by the circuitry. The operational amplifier circuitry realization of a monostable multivibrator is shown in Fig. 40.33(c).

PART 4: DEVICES AND CIRCUITS

26. CIRCUIT: PHASE-LOCKED LOOP (PLL)

A *phase-locked loop* (PLL) is a closed-loop control circuit for providing and outputting frequency that is exactly in phase with an incoming frequency.[29] A basic PLL is shown in Fig. 40.34. The circuit consists of a phase frequency detector, a loop (low-pass) filter, a voltage controlled oscillator (VCO), and a feedback path. Both digital and analog PLLs have in common the phase detector, the VCO, and the feedback path.

Figure 40.34 *Phase-Locked Loop Functional Block Diagram*

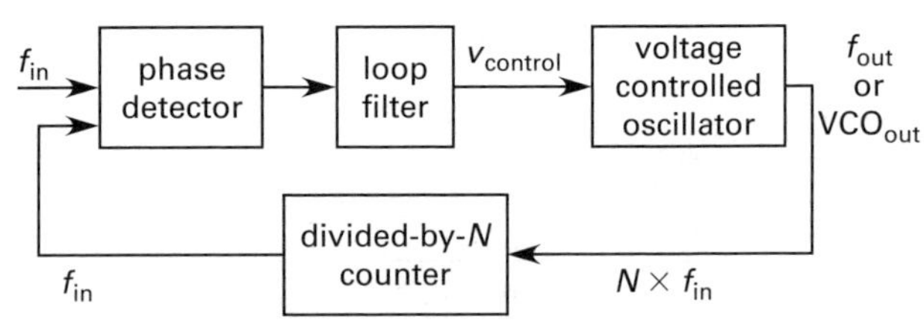

The phase detector compares the input frequency with the feedback frequency and outputs an error (or corrective) signal to the oscillator (via the loop filter) that then responds to drive the phase difference between the two inputs to zero. An example is an exclusive OR gate or a more complex state machine called a *phase frequency detector* that compares zero crossing points. Both such detectors are digital. Both use the low pass filter to smooth the output.

An example analog phase detector is the *four-quadrant multiplier* or *balance mixer* that multiplies the input reference signal with the oscillator feedback signal and generates a low-frequency signal with amplitude

[29]A mechanical analogy is tuning a piano. A tuning fork is used to provide a reference frequency. The piano string is adjusted until the beat frequency is inaudible. Another example is the manual transmission in an automobile. When being shifted, gears on the engine side and wheel side are turning at different rates. The rates must be matched before the shift is completed or a grinding sound is heard.

proportional to the phase difference (also called the *phase error*) between the input and the oscillator. The mixer also generates a higher frequency signal at twice the oscillator frequency, but this is eliminated by the low pass filter.

The *voltage controlled oscillator* uses the filter output and the controlling voltage input to generate an output frequency that is a multiple of the input frequency. The result, once fed back through the divide-by-N circuitry, drives the phase detector and filter to change the VCO voltage input until the frequencies are aligned at the output.

A digital VCO may consist of a clock source (crystal oscillator or multivibrator circuit), two counters, and a digital comparator. (Such digital circuitry is covered in Chap. 43.) An analog VCO is shown in Fig. 40.35. The VCO uses the LRC *tank circuit* or *antiresonant circuit* that sets the frequency of the oscillator circuit. (See Chap. 30 for a discussion of resonant circuits.)

Figure 40.35 Voltage Controlled Oscillator (VCO) Circuit

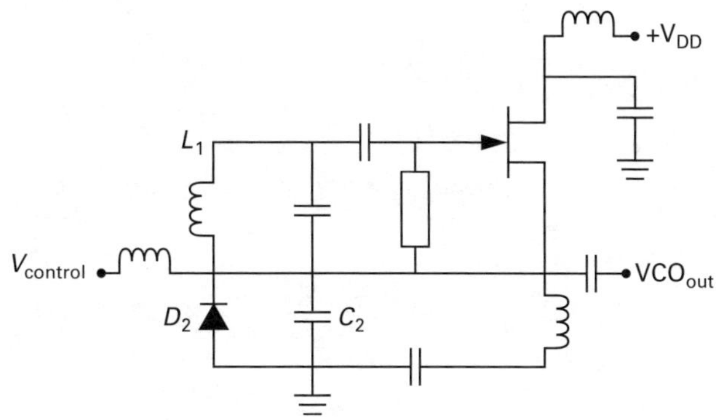

Topic X: Digital Systems

Chapter

41. Digital Logic
42. Logic Network Design
43. Synchronous Sequential Networks
44. Digital Systems

41 Digital Logic

1. Digital Information Representation 41-1
2. Computer Number Systems 41-1
3. Electronic Logic Device Levels and Limits . . 41-1
4. Fundamental Logic Operations 41-2
5. Switching Algebra 41-3
6. Minterms and Maxterms 41-4
7. Canonical Representation of Logic Functions 41-5
8. Canonical Realization of Logic Functions: SOP 41-6
9. Canonical Realization of Logic Functions: POS 41-6

Nomenclature

$f(x)$	scalar function of x	–
$F(X)$	vector function of X	–
m	minterm	–
M	maxterm	–
n	number	–
N	total number of possible values	–

1. DIGITAL INFORMATION REPRESENTATION

Computers process information in digital form. The simplest representation of digital information is that of a single binary variable called a *scalar*. The notation for a scalar is x_i. A *binary scalar* takes on a value of zero or one, and represents one bit of information. The smallest information storage unit is termed a *cell*.

At the electronic device level, a zero or one is defined in terms of a voltage output specific to the logic family chosen. For transistor-transistor logic (TTL), the range of values of the zero and one, called the *state* of the cell, is shown in Fig. 41.1. In TTL, zero is represented by the voltage range 0–0.8 V, while one is represented by the voltage range 2.0–5.0 V.

Figure 41.1 *Definition of Binary States for TTL*

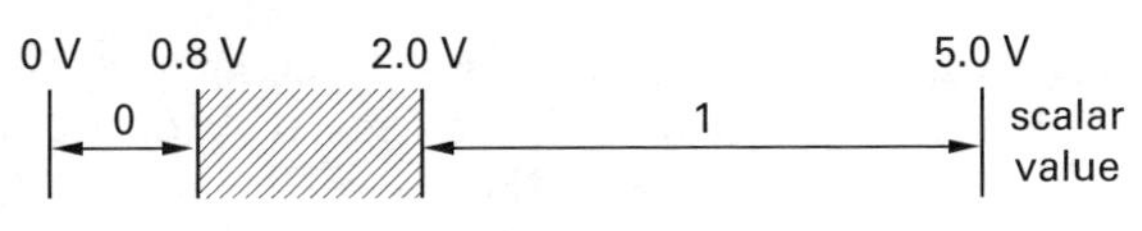

A grouping of n binary variables is called an *n-tuple* and is represented by $[x_1, x_2, x_2, \ldots, x_n]$. The value of n is called the *word length*. The cells can be listed and numbered left to right or right to left. Left to right is the normal notation. An exception occurs when representing a binary number, in which case the cells are listed right to left, from zero to $n-1$, corresponding to $2^0, 2^1, \ldots, 2^{n-1}$. Because each binary variable in the n-tuple takes on only one of two values, the total number of values that can be represented, N, is

$$N = 2^n \qquad 41.1$$

An ordered collection of cells is called a *register*. The information within each register is encoded to represent numerical, logical, or character information. The variables within the register are called *state variables* because they define the value (state) of the register (vector). Particular values within a register are called *states of the register*. For example, a 2-tuple register has four states: $[0, 0]$, $[0, 1]$, $[1, 0]$, and $[1, 1]$.

Example 41.1

A certain computer uses a 32-bit word length. What is the maximum number of possible state variables?

Solution

The maximum number of state variables is given by Equation 41.1.

$$N = 2^n = 2^{32} = 4{,}294{,}967{,}296$$

2. COMPUTER NUMBER SYSTEMS

Several types of number systems other than base 10 find application in computer systems. These include the binary, octal, and hexadecimal number systems. These systems and their manipulation are described in Chap. 13.

3. ELECTRONIC LOGIC DEVICE LEVELS AND LIMITS

Logic one is considered to indicate a true condition. Logic zero indicates a false condition. Logic one is normally associated with the higher voltage in an electronic circuit, and logic zero is associated with the lower voltage. This type of association is called *positive logic*. When the low-voltage condition represents the true statement, the logic is called *negative logic*. In order to avoid confusion, truth tables are sometimes given in terms of high (H) and low (L) voltages, rather than one and zero.

An alternative is to use *assertion levels*. When the assertion (true) condition is the high-voltage level, the term *active high* is used. This corresponds to positive logic. When the assertion (true) condition is the low-voltage level, the term *active low* is used.

In practice, the type of device and its limits need to be considered. Depending on the circuit, the high- and low-voltage levels will be represented by a range of values. Also, the fan-in and fan-out capabilities limit the number of variable inputs and the number of logic gates that can be attached to the output. Additionally, the propagation delays must be accounted for to ensure that levels of logic have time to respond, providing the proper output before the next set of variables arrives. The propagation delays, therefore, determine the operating speed of the logic network. Timing diagrams are used to map these delays and ensure proper operation.

4. FUNDAMENTAL LOGIC OPERATIONS

Information to be processed is combined with, and compared to, other information. The logic that does this in such a manner that the output depends only on the value of the inputs is called *combinational logic.*[1] *Logic variables* and *logic constants* have one of two possible values, zero or one. By convention, zero is considered "false," and one is considered "true."

Logic functions are commonly represented in *truth tables*. Because an n-tuple can take on 2^n values, a truth table has 2^n rows. Each of these rows has an associated output that can be either zero or one. Combining these two facts gives the number of scalar functions that can be generated from n logic variables. A logic operator involving two operands is called a *binary operator*. For a two-variable input, that is, a 2-tuple, there are 16 possible functions.

$$\text{no. of scalar functions} = (2)^{2^n} \qquad 41.2$$

Not all the scalar functions are useful. In fact, only six are of primary interest. Further, all logic functions can be represented in terms of only two of these, NAND and NOR. These often-used functions are called *logic operators*. A logic operator involving a single variable or operand is called a *unary operation*. A unary function is shown in Table 41.1.

Table 41.1 *Unary Function*

x	$f_1(x) = 0$	$f_2(x) = x$	$f_3(x) = \overline{x}$	$f_4(x) = 1$
0	0	0	1	1
1	0	1	0	1

The function $f_2(x)$ is called the *identity function*. The function $f_3(x)$ is called the NOT, or negation, function. The symbol and truth table for the NOT function are shown in Fig. 41.2.

Figure 41.2 *NOT Logic*

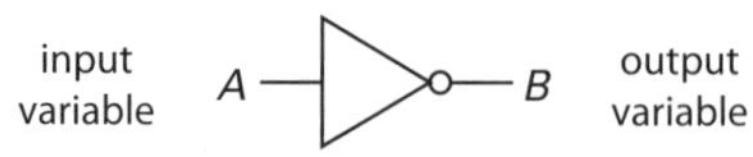

(a) symbol

A	B
0	1
1	0

(b) truth table

$$B = \overline{A}$$

(c) equation

Six logic operators (or gates) are used so often that they have been given their own symbols: AND, OR, XOR, NAND, NOR, and XNOR (or coincidence). The X indicates the word "exclusive," while N indicates the word "not." The logic operator AND is indicated by "·" or by "∧." The dot is commonly not used when all the variables are represented by single letters. (For example, $A \cdot B$ can be written as AB.) The logic operator OR is indicated by "+" or "∨." The small circle on the end of any gate symbol, or on the input or output, indicates the NOT function. The standard symbol for the NOT operation is $\overline{x}$. Alternate symbols for the NOT operation include $-x$ or x'. The NOT operator applied over an expression has the effect of enclosing the expression in parentheses. For example,

$$\overline{x_1 x_2} = \overline{(x_1 x_2)}$$

A circle around the AND or OR symbol indicates "exclusive" operation, ⊗ or ⊕. The fundamental logic operations and their properties are shown in Table 41.2.[2]

When combinations of logic operations are to be analyzed, the following *order of precedence* is applicable. If in doubt when writing an expression, use parentheses to establish the proper order of precedence.

step 1: Evaluate those items within parentheses first.

step 2: Evaluate logic expressions from left to right, applying all instances of the NOT operations first, the AND operations second, and the OR operations last.

[1]When the output has a memory, that is, when the output is used as feedback to the input, the logic is called *sequential*.

[2]Argument inputs and results are commonly shown using capital letters. They can be scalars or vectors. Scalars have only two values. Vectors represent information and may have numerous values. The properties and laws applicable to logic variables are also applicable to logic vectors. Capital letters are sometimes reserved for logic vectors in educational texts. The symbols of A, B, C, or X, Y, Z are used for inputs and outputs with no special significance attached.

Table 41.2 Logic Operators

operator	symbol	truth table			equation
		A	B	C	
AND		0	0	0	$A \bullet B = C$
		0	1	0	
		1	0	0	
		1	1	1	
		A	B	C	
OR	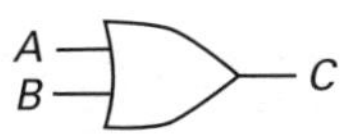	0	0	0	$A + B = C$
		0	1	1	
		1	0	1	
		1	1	1	
		A	B	C	
XOR	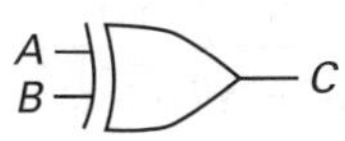	0	0	0	$A \oplus B = C$
		0	1	1	
		1	0	1	
		1	1	0	
		A	B	C	
NAND	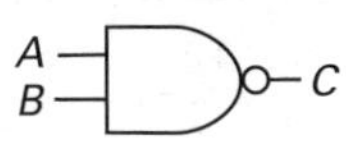	0	0	1	$\overline{A \bullet B} = C$
		0	1	1	
		1	0	1	
		1	1	0	
		A	B	C	
NOR	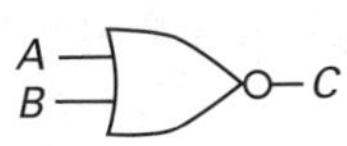	0	0	1	$\overline{A + B} = C$
		0	1	0	
		1	0	0	
		1	1	0	
		A	B	C	
XNOR or coincidence	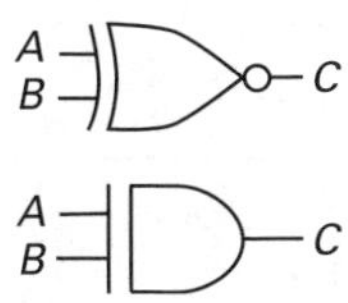	0	0	1	$A \odot B = C$ or $A \otimes B = C$
		0	1	0	
		1	0	0	
		1	1	1	

Two functions, f and g, are *logically equivalent* when they are defined for the same arguments and

$$f(x_n, \ldots, x_1) = g(x_n, \ldots, x_1) \qquad 41.3$$

Equation 41.3 must be valid for all combinations on the n-tuple. Logical equivalence is easily verified by using a truth table for both functions and ensuring that the results are identical. Additionally, the rules of *switching algebra*, usually referred to as *Boolean algebra*, can be used to transform one of the functions into the other, which proves equivalence.

Example 41.2

If $A = 1$ and $B = 0$, what is the output of the following expression?

$$\overline{\left(\overline{(AB)}\,\overline{(AB)}\right)}$$

Solution

Applying the rules of precedence, the expression is reduced as follows.

$$\begin{aligned}\overline{\left(\overline{(AB)}\,\overline{(AB)}\right)} &= \overline{\left(\overline{(1\cdot 0)}\,\overline{(1\cdot 0)}\right)} \\ &= \overline{\left(\overline{(0)}\,\overline{(0)}\right)} \\ &= \overline{(1\cdot 1)} \\ &= \overline{1} \\ &= 0\end{aligned}$$

5. SWITCHING ALGEBRA

Switching algebra is a particular form of Boolean algebra, though the distinction is seldom made. (Section 40.16 has more information on Boolean algebra.) Three basic postulates define switching algebra.

- *postulate 1*: A Boolean variable has two possible values, zero and one, and these values are exclusive. That is,

$$\text{if } x = 0 \text{ then } x \neq 1, \text{and}$$
$$\text{if } x = 1 \text{ then } x \neq 0$$

- *postulate 2*: The NOT operation is defined as

$$\overline{0} = 1 \text{ and } \overline{1} = 0$$

- *postulate 3*: The logic operations AND and OR are defined as in Table 41.2.

Using these postulates, the special properties, laws, and theorems of switching algebra are developed. The properties, laws, and theorems of switching algebra are summarized in Table 41.3.

De Morgan's theorem, stated in Table 41.3, has many applications and is used when constructing logic networks from NAND and NOR logic. NAND and NOR are *universal operations*, sometimes called *complete sets.* The term "complete set" indicates that any logic expression can be represented solely by NAND logic, or solely by NOR logic. Because any logic expression is a combination of NOT, AND, and OR operations, proving that NAND logic can represent any of these three proves universality. This proof is shown in Eq. 41.4 through Eq. 41.6 for the NOT, AND, and OR operations, respectively.

Table 41.3 *Basic Properties of Switching Algebra (Boolean algebra)*

name	property	dual
special properties: 0	$0 + A = A$	$0 \cdot A = 0$
special properties: 1	$1 + A = 1$	$1 \cdot A = A$
idempotence law	$A + A = A$	$A \cdot A = A$
complementation law	$A + \overline{A} = 1$	$A \cdot \overline{A} = 0$
involution	$(\overline{A}) = A$	
commutative law	$A + B = B + A$	$A \cdot B = B \cdot A$
associative law	$A + (B + C) = (A + B) + C$	$A \cdot (B \cdot C) = (A \cdot B) \cdot C$
distributive law	$A \cdot (B + C) = (A \cdot B) + (A \cdot C)$	$A + (B \cdot C) = (A + B)(A + C)$
absorption law	$A + (A \cdot B) = A$	$A \cdot (A + B) = A$
	$A + (\overline{A} \cdot B) = A + B$	$A \cdot (\overline{A} + B) = A \cdot B$
De Morgan's theorem	$\overline{(A + B)} = \overline{A} \cdot \overline{B}$	$\overline{(A \cdot B)} = \overline{A} + \overline{B}$

$$\overline{(A \cdot A)} = \overline{A} + \overline{A} = \overline{A} \qquad \textit{41.4}$$

$$\overline{\overline{(A \cdot B)} \cdot \overline{(A \cdot B)}} = (A \cdot B) + (A \cdot B) = A \cdot B \qquad \textit{41.5}$$

$$\overline{\overline{(A \cdot A)} \cdot \overline{(B \cdot B)}} = (A \cdot A) + (B \cdot B) = A + B \qquad \textit{41.6}$$

Using the dual of Eq. 41.4 through Eq. 41.6 proves the universality of the NOR operation. The dual of each of the items in Table 41.2 is obtained using the *principle of duality*. The dual of an expression is obtained by applying the following steps, without changing the variables involved.

step 1: If present, change zeros to ones and ones to zeros.

step 2: Change the original OR operations to AND.

step 3: Change the original AND operations to OR.

6. MINTERMS AND MAXTERMS

A *product term* is a function defined by a logical AND set of terms with either a variable, x_i, or its negation, $\overline{x_i}$. A product term where all the variables appear, but only once, is called a *minterm*. The name occurs because the minterms take on a value of one for only one of the 2^n combinations of an n-tuple, that is, for the input variables, and a value of zero for all other combinations. Expressed in terms of true/false notation, the minterms give the rows of a truth table in which the function is true. Minterms for a three-variable function are shown in Table 41.4.

A *sum term* is a function defined by a logical OR set of terms with either a variable x_i or its negation $\overline{x_i}$. A sum term where all the variables appear, but only once, is called a *maxterm*. The name occurs because the maxterms take on a value of zero for only one of the 2^n combinations of an n-tuple, that is, for the input variables, and a value of one for all other combinations. Expressed in terms of true/false notation, the maxterms give the rows of a truth table in which the function is false. Maxterms for a three-variable function are shown in Table 41.4.

Table 41.4 *Minterms and Maxterms*

decimal row number	binary input combinations ABC	minterm (product term)	maxterm (sum term)
0	000	$m_0 = \overline{A}\,\overline{B}\,\overline{C}$	$M_0 = A + B + C$
1	001	$m_1 = \overline{A}\,\overline{B}C$	$M_1 = A + B + \overline{C}$
2	010	$m_2 = \overline{A}B\overline{C}$	$M_2 = A + \overline{B} + C$
3	011	$m_3 = \overline{A}BC$	$M_3 = A + \overline{B} + \overline{C}$
4	100	$m_4 = A\overline{B}\,\overline{C}$	$M_4 = \overline{A} + B + C$
5	101	$m_5 = A\overline{B}C$	$M_5 = \overline{A} + B + \overline{C}$
6	110	$m_6 = AB\overline{C}$	$M_6 = \overline{A} + \overline{B} + C$
7	111	$m_7 = ABC$	$M_7 = \overline{A} + \overline{B} + \overline{C}$

Example 41.3

Consider the truth table for the two-variable function $F(A, B)$ shown. List the minterms.

A	B	$F(A, B)$
0	0	0
0	1	1
1	0	1
1	1	0

Solution

The minterms occur where the value of the function is true (that is, one). So, the minterms are

$$m_1 = \overline{A}B$$

$$m_2 = A\overline{B}$$

Example 41.4

For the function in Ex. 41.3, list the maxterms.

Solution

The maxterms occur where the value of the function is false (that is, zero). Therefore, the maxterms are

$$M_0 = A + B$$
$$M_3 = \overline{A} + \overline{B}$$

7. CANONICAL REPRESENTATION OF LOGIC FUNCTIONS

Any arbitrary function described by a truth table can be represented using minterms and maxterms. For example, the *minterm form* of the three-variable function in Table 41.4 is

$$F(A,B,C) = \sum_{i=0}^{7} m_i \qquad 41.7$$

When Eq. 41.7 is expanded, it is referred to as the *canonical sum-of-product form* (SOP) of the function. For example,

$$\begin{aligned} F(A,B,C) &= m_0 + m_1 + m_2 + m_3 + m_4 + m_5 \\ &\quad + m_6 + m_7 \\ &= \overline{A}\,\overline{B}\,\overline{C} + \overline{A}\,\overline{B}\,C + \overline{A}\,B\,\overline{C} + \overline{A}\,B\,C \\ &\quad + A\,\overline{B}\,\overline{C} + A\,\overline{B}\,C + A\,B\,\overline{C} + A\,B\,C \end{aligned} \qquad 41.8$$

The terms are called *minterms* because they have a value of one for only one of the 2^i possible values of Eq. 41.7. (These are the terms of a function that are true; that is, logic one.)

The *maxterm form* of the three-variable function in Table 41.4 is

$$F(A,B,C) = \prod_{i=0}^{7} M_i \qquad 41.9$$

When Eq. 41.9 is expanded, it is referred to as the *canonical product-of-sum form* (POS) of the function. For example,

$$\begin{aligned} F(A,B,C) &= M_0M_1M_2M_3M_4M_5M_6M_7 \\ &= (A+B+C)(A+B+\overline{C}) \\ &\quad \times (A+\overline{B}+C)(A+\overline{B}+\overline{C}) \\ &\quad \times (\overline{A}+B+C)(\overline{A}+B+\overline{C}) \\ &\quad \times (\overline{A}+\overline{B}+C)(\overline{A}+\overline{B}+\overline{C}) \end{aligned} \qquad 41.10$$

The terms are called *maxterms* because they have a value of zero for only one of the 2^i possible values of Eq. 41.9. (These are the terms of a function that are false; that is, logic zero.)

The term "canonical" in this context means a simple or significant form of an equation or function. Canonical forms contain all the input variables in each term. SOP and POS forms may not be the simplest expressions with regard to realizing these functions in terms of logic gates and may require minimization.

Example 41.5

Write the function in Ex. 41.3 in SOP form.

Solution

The minterms were determined in Ex. 41.3. The SOP form, using the format of Eq. 41.8 is

$$F(A,B) = \overline{A}B + A\overline{B}$$

Example 41.6

Write the function of Ex. 41.3 in POS form.

Solution

The maxterms were determined in Ex. 41.4. The POS form, using the format of Eq. 41.10 is

$$F(A,B) = (A+B) + (\overline{A}+\overline{B})$$

Example 41.7

Consider the truth table shown. What are the (a) SOP and (b) POS forms of the function?

X	Y	Z	F(X, Y, Z)
0	0	0	0
0	0	1	1
0	1	0	1
0	1	1	0
1	0	0	1
1	0	1	0
1	1	0	0
1	1	1	1

Solution

(a) Minterms (logic value one) exist in rows 1, 2, 4, and 7. The SOP form is

$$\begin{aligned} F(X,Y,Z) &= m_1 + m_2 + m_4 + m_7 \\ &= \overline{X}\,\overline{Y}\,Z + \overline{X}\,Y\,\overline{Z} + X\,\overline{Y}\,\overline{Z} + X\,Y\,Z \end{aligned}$$

(b) Maxterms (logic value zero) exist in rows 0, 3, 5, and 6. The POS form is

$$\begin{aligned} F(X,Y,Z) &= M_0M_3M_5M_6 \\ &= (X+Y+Z)(X+\overline{Y}+\overline{Z}) \\ &\quad \times (\overline{X}+Y+\overline{Z})(\overline{X}+\overline{Y}+Z) \end{aligned}$$

8. CANONICAL REALIZATION OF LOGIC FUNCTIONS: SOP

Realization is the process of creating a logic diagram from the truth table for a given function. Since any scalar function can be represented in either canonical sum-of-product or product-of-sum form,[3] any function can be realized by a two-level logic circuit. The connection points are shown with a dot. Undotted line crossings indicate nonconnecting crossings. It is normally assumed that the input variables and their complements are available.

The SOP form of the function in Ex. 41.7 indicates a two-level AND-OR realization, as shown in Fig. 41.3.

Figure 41.3 *SOP Realization*

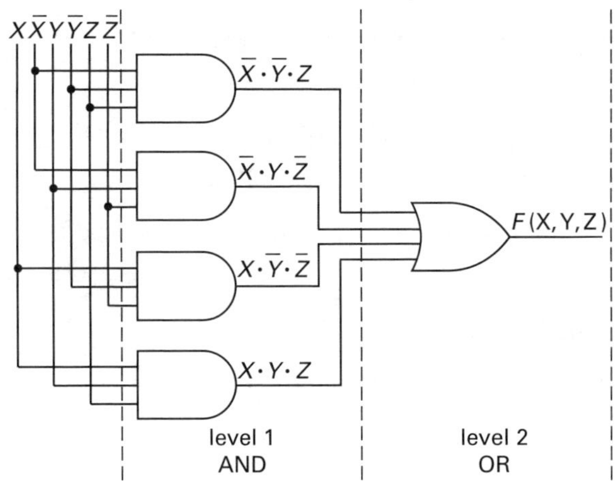

In Fig. 41.3, the AND gates are in logic level one. The OR gate is in logic level two.

The same realization can occur using only NAND gates. Consider the equivalence shown in Fig. 41.4.

Figure 41.4 *NAND Equivalence*

X	Y	Z	$\overline{X\cdot Y\cdot Z}$	$\bar{X}+\bar{Y}+\bar{Z}$
0	0	0	1	1
0	0	1	1	1
0	1	0	1	1
0	1	1	1	1
1	0	0	1	1
1	0	1	1	1
1	1	0	1	1
1	1	1	0	0

(a) truth table

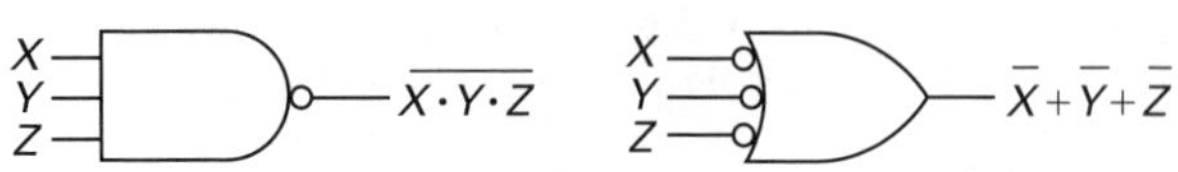

(b) symbols

The NAND gate can be represented in two ways: the standard way or as an OR gate with the inputs inverted. Using this equivalence, the SOP form of the function in Ex. 41.7 can be realized with NAND gates only, as shown in Fig. 41.5.

The realization in Fig. 41.5 is AND-OR. This can be seen by considering the inversions represented by the bubbles (small circles) on the output of the level-one logic to cancel with the inversions on the input to the level-two logic.[4]

Figure 41.5 *SOP NAND-NAND Realization*

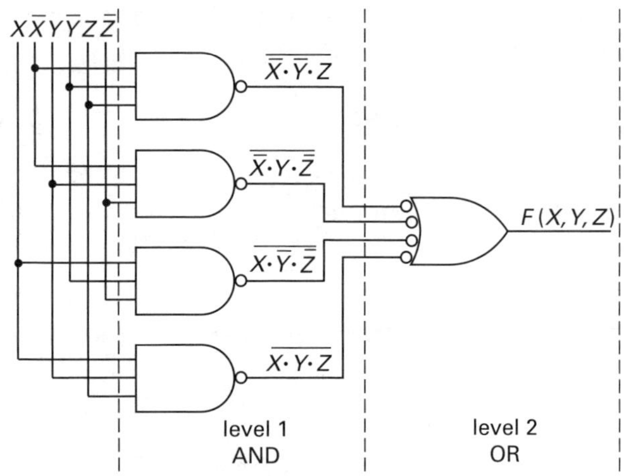

9. CANONICAL REALIZATION OF LOGIC FUNCTIONS: POS

The POS form of the function in Ex. 41.7 indicates a two-level OR-AND realization, as shown in Fig. 41.6.

Figure 41.6 *POS Realization*

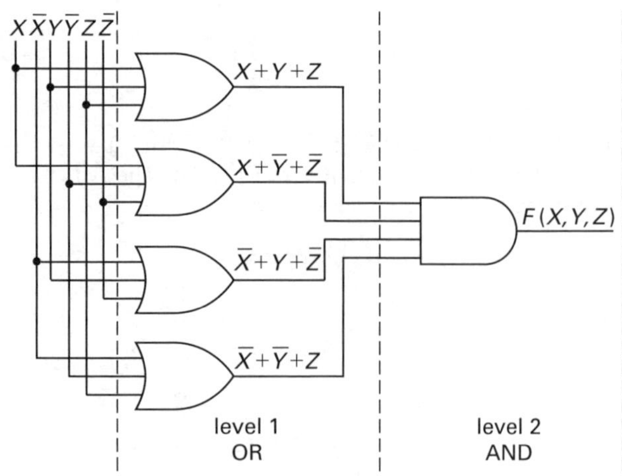

[3]Canonical forms of logic realization are useful in the design of read-only memory (ROM) and programmable logic array (PLA) circuits.

[4]Figure 41.5 is not a minimum realization because the function is the XOR of *X*, *Y*, and *Z* and could be realized with two two-input XOR gates.

In Fig. 41.6, the OR gates are in logic level one. The AND gate is in logic level two.

The same realization can occur using only NOR gates. Consider the equivalence shown in Fig. 41.7.

The NOR gate can be represented in two ways: the standard way or as an AND gate with the inputs inverted. Using this equivalence, the POS form of the function in Ex. 41.7 can be realized with NOR gates only, as shown in Fig. 41.8.

Figure 41.7 *NOR Equivalence*

X	Y	Z	$\overline{X+Y+Z}$	$\overline{X}\cdot\overline{Y}\cdot\overline{Z}$
0	0	0	1	1
0	0	1	0	0
0	1	0	0	0
0	1	1	0	0
1	0	0	0	0
1	0	1	0	0
1	1	0	0	0
1	1	1	0	0

(a) truth table

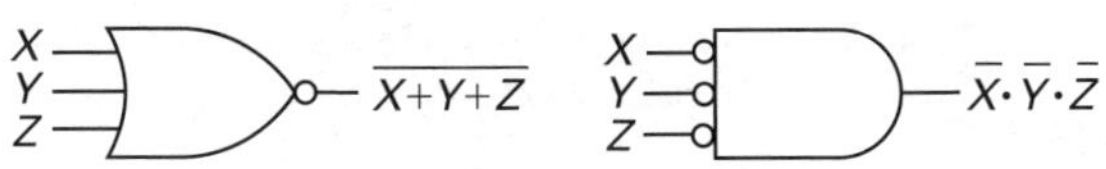

(b) symbols

The realization in Fig. 41.8 is OR-AND. This can be understood by considering the inversions represented by the bubbles on the output of the level-one logic to cancel with the inversions on the input to the level-two logic.[5]

Figure 41.8 *POS NOR-NOR Realization*

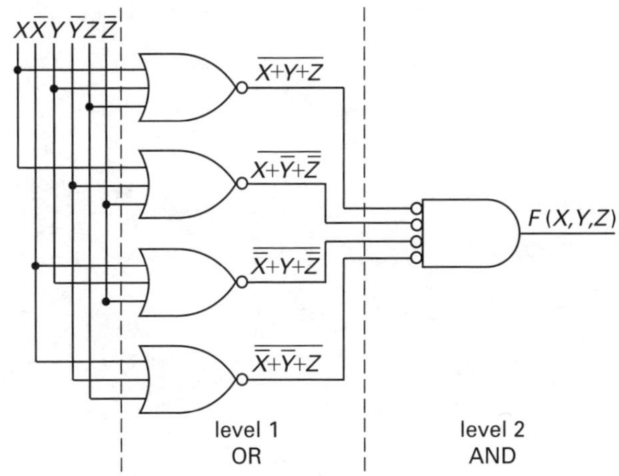

[5]Figure 41.8 is also not a minimum realization because the function is the XOR of X, Y, and Z and could be realized with two two-input XOR gates.

42 Logic Network Design

1. Combinational Logic Design Process 42-1
2. Operations on Digital Information 42-1
3. Minimization Principle 42-2
4. Karnaugh Map Fundamentals 42-2
5. Karnaugh Map Minimization 42-3
6. Veitch Diagrams 42-5
7. NAND/NOR Function Representation 42-8
8. Programmable Logic Arrays 42-8
9. Encoders and Decoders 42-9
10. Demultiplexers 42-9
11. Multiplexers 42-10
12. Arithmetic Logic Devices 42-10

Nomenclature

a, b, A, B, C, D	variable equal to zero or one	–
c	carry	–
d	don't care condition	–
$f(x)$	logic-level function of x	–
$F(X)$	system-level function of X	–
K	number of output lines	–
m	minterm	–
M	maxterm	–
N	number of input lines	–
Q	output signal	–
s	sum	–
S	control signal	–
x, X	input variable equal to zero or one	–
y, Y	output variable equal to zero or one	–

1. COMBINATIONAL LOGIC DESIGN PROCESS

The logic network design process starts with a set of requirements. The designer translates the requirements into a set of specifications that describe a combinational logic network, which must then be realized. Each design process is unique. Nevertheless, the general steps are as follows.

step 1: Using the set of requirements provided, identify and name all the inputs supplied to the network and all the outputs required of the network.

step 2: Establish the exact relationships between the inputs and the outputs. The relationships should be expressed in analytical form, that is, in a mathematical equation or truth table.

step 3: Code each input and output into digital form. The code may be fixed by the system requirements or open to the designer to interpret.

step 4: Develop the logic expressions that must be realized by the combinational logic network using the information in steps 2 and 3. Sometimes the expressions may be written directly from the relationships determined in step 2. If not, analytical techniques such as Boolean algebra or truth table techniques (described in this chapter) may be used to obtain the expressions.

step 5: Determine the logic circuit elements that will be used to realize the network. Applying analytical minimization techniques, reduce the complexity of the logic expressions to the desired level.[1]

step 6: Using the logic expressions from step 5, realize the network. Test the network to ensure that it meets the requirements of step 1.

2. OPERATIONS ON DIGITAL INFORMATION

Logic expressions (functions) at the logic level are symbolized using $f(x)$, while those at the system level use $F(X)$. At the logic level, design concerns include the coding of the information processed and the logic expressions necessary to realize the scalar functions. At the system level, design concerns include the type of information dealt with, its range of values, and the mathematical expressions required to obtain the desired output. The difference is illustrated in Fig. 42.1.

A scalar or vector's *domain* represents its range of values. Scalars have only two values, but those values can have meaning, such as true or false. Vectors have a wide range of values. Examples of each are given in Table 42.1.

Operands are any of the quantities entering into or resulting from an operation. Operands represent the information being processed. The representation of operands is illustrated in Fig. 42.2.

Vector *logic operations* use the same logic as scalar logic operations. *Relational operations* on vectors include greater than/less than, greater than or equal to/less

[1]The time spent reducing the complexity is inversely proportional to the number of circuits to be produced. The greater the number of circuits to be produced, the more economical it becomes to reduce the complexity.

than or equal to, and equal to/not equal to.[2] *Mixed-mode operations* allow the combination of a scalar with a vector.[3]

Figure 42.1 *Representation of Processing Tasks*

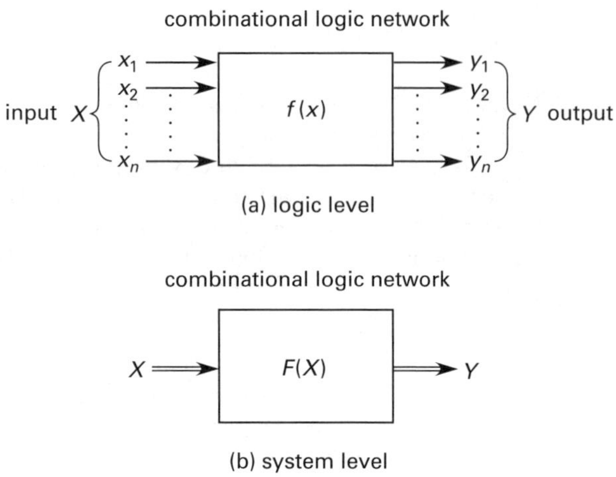

Figure 42.2 *Representation of Operands*

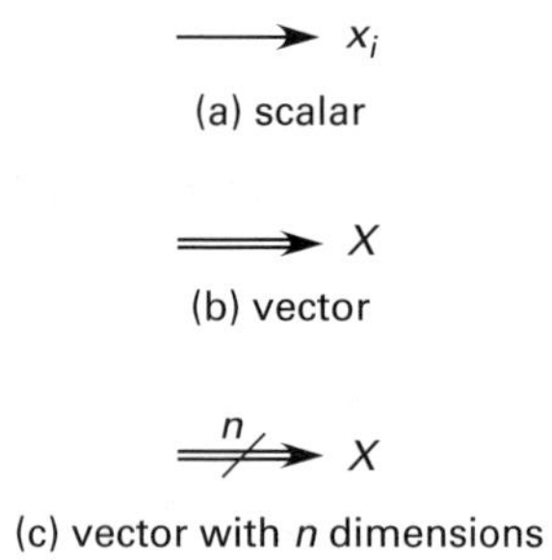

Table 42.1 *Scalar and Vector Examples*

identifier	domain	description
x	{0, 1}	scalar with a value of zero or one
f	{NOT READY, READY}	scalar representing two logic conditions
INST	{ADD, SUB, MUL, DIV}	vector representing four concepts—mathematical operations
SIGNAL	$\{-255, \ldots, 0, \ldots + 255\}$	vector representing decimal integers

[2]Coincidence logic is the simplest form of the "equal to" comparison.
[3]Mixed-mode operations are used by instruction registers to control the output of a combinational logic network.

3. MINIMIZATION PRINCIPLE

The complexity of a logic function depends on the Boolean expression representing the function. The complexity determines the number of gates necessary to realize the function and thus the circuit size, speed, and power consumption. Consequently, methods of minimization are used to reduce this complexity. The basis of minimization of logic functions is the *theorem of logical adjacency* given by either Eq. 42.1 or Eq. 42.2. The theorem applies to variables and functions.

$$A \cdot B + A \cdot \overline{B} = A \quad \text{[sum of products]} \qquad 42.1$$

$$(A + B) \cdot (A + \overline{B}) = A \quad \text{[product of sums]} \qquad 42.2$$

The logical adjacency theorem implies that SOP (see Eq. 42.1) or POS (see Eq. 42.2) functions can be minimized through the inspection of the logical adjacency of minterms and maxterms. *Adjacency* exists when minterms (or maxterms) differ by one value.

Example 42.1

A given logic function is defined by

$$F(A, B, C) = \prod_{i=2,4,5} M_i$$

Using the theorem of logical adjacency, simplify the function.

Solution

Maxterms exist where the value of the function is zero. Expanding the given function into canonical POS form gives

$$\begin{aligned} F(A, B, C) &= M_2 M_4 M_5 \\ &= (A + \overline{B} + C)(\overline{A} + B + C)(\overline{A} + B + \overline{C}) \end{aligned}$$

The maxterms are equal to the following binary values.

$$M_2 = 010$$
$$M_4 = 100$$
$$M_5 = 101$$

Comparing either the functional expression of variables or the binary representation, maxterms 4 and 5 differ by only one variable. The variable is C, in the 2^0 position in binary form. This variable can be eliminated, giving the simplified expression

$$F(A, B, C) = (A + \overline{B} + C)(\overline{A} + B)$$

4. KARNAUGH MAP FUNDAMENTALS

A *Karnaugh map* is a truth table that has been rearranged to form a geometrical pattern of functional relationships for determining gating configurations. The map allows gating requirements to be recognized in their simplest form through the principle of logical adjacency. Karnaugh two-, three-, and four-input maps are shown in Fig. 42.3 along with their associated truth tables.

Karnaugh maps are constructed with the inputs along the top and side, with the side having the least significant binary digits. The numbering scheme differs from that used in the truth table and is labeled "adjacency ordering" in Fig. 42.3. The numbering scheme is called *unit distance coding*, because adjacent numbers differ by only one bit; adjacent inputs differ at most by X or $\overline{X}$.

The output is within the box. In Fig. 42.3, the output is represented as a scalar function associated with the binary input, and the box is numbered in the lower right with the decimal number of the corresponding row of the truth table. These decimal numbers equate to the binary numbers in the truth table, and represent the locations of the minterms and maxterms. The actual output of a logical scalar expression will be either zero, one, or *d*. The *d* indicates a "don't care" condition. A "don't care" condition is used when the output is not defined for the given inputs. When such a condition exists the function is said to be *incompletely specified.* The *d* can be assigned a zero or one value as necessary to further the minimization process.

Five-input and six-input Karnaugh maps can be constructed, but the increasing complexity makes the adjacencies difficult to perceive. In practice, with five or more inputs, minimization is done with a computer. The *tabular method* or *Quine-McClusky method* lends itself well to computer applications.

5. KARNAUGH MAP MINIMIZATION

The sum-of-product (SOP) and product-of-sum (POS) expressions for a given function can be obtained from a Karnaugh map. The SOP is obtained by grouping the minterms, that is, the blocks where the function output equals one. The POS is obtained by grouping the maxterms, that is, the blocks where the function output equals zero.[4]

The rows and columns of a Karnaugh map are numbered so that only one variable changes from block to block. Rows are adjacent, with the top row adjacent to the bottom row as well. Columns are adjacent, with the leftmost column adjacent to the rightmost column as well. Combining adjacent rows and columns eliminates the variable that changes from one row to the next or from one column to the next. That is, the variable that appears in negated form in one of the minterms (or maxterms) and unnegated form in the other minterm (or maxterm) is eliminated (see Eq. 42.1 and Eq. 42.2).

[4]The resulting expressions may not be absolute minimizations. Nevertheless, the expressions are useful as starting points, and the techniques are useful in the design of flip-flop logic for sequential circuits.

Figure 42.3 *Karnaugh Maps and Associated Truth Tables*

F: B \ A	0	1
0	$f(0,0)_0$	$f(1,0)_2$
1	$f(0,1)_1$	$f(1,1)_3$

input / output

A	B	F(A,B)
0	0	f(0,0)
0	1	f(0,1)
1	0	f(1,0)
1	1	f(1,1)

(a) two-input

adjacency ordering

F: C \ AB	00	01	11	10
0	$f(0,0,0)_0$	$f(0,1,0)_2$	$f(1,1,0)_6$	$f(1,0,0)_4$
1	$f(0,0,1)_1$	$f(0,1,1)_3$	$f(1,1,1)_7$	$f(1,0,1)_5$

input / output

A	B	C	F(A,B,C)
0	0	0	f(0,0,0)
0	0	1	f(0,0,1)
0	1	0	f(0,1,0)
0	1	1	f(0,1,1)
1	0	0	f(1,0,0)
1	0	1	f(1,0,1)
1	1	0	f(1,1,0)
1	1	1	f(1,1,1)

(b) three-input

adjacency ordering

F: CD \ AB	00	01	11	10
00	$f(0,0,0,0)_0$	$f(0,1,0,0)_4$	$f(1,1,0,0)_{12}$	$f(1,0,0,0)_8$
01	$f(0,0,0,1)_1$	$f(0,1,0,1)_5$	$f(1,1,0,1)_{13}$	$f(1,0,0,1)_9$
11	$f(0,0,1,1)_3$	$f(0,1,1,1)_7$	$f(1,1,1,1)_{15}$	$f(1,0,1,1)_{11}$
10	$f(0,0,1,0)_2$	$f(0,1,1,0)_6$	$f(1,1,1,0)_{14}$	$f(1,0,1,0)_{10}$

adjacency ordering

input / output

A	B	C	D	F(A,B,C,D)
0	0	0	0	f(0,0,0,0)
0	0	0	1	f(0,0,0,1)
0	0	1	0	f(0,0,1,0)
0	0	1	1	f(0,0,1,1)
0	1	0	0	f(0,1,0,0)
0	1	0	1	f(0,1,0,1)
0	1	1	0	f(0,1,1,0)
0	1	1	1	f(0,1,1,1)
1	0	0	0	f(1,0,0,0)
1	0	0	1	f(1,0,0,1)
1	0	1	0	f(1,0,1,0)
1	0	1	1	f(1,0,1,1)
1	1	0	0	f(1,1,0,0)
1	1	0	1	f(1,1,0,1)
1	1	1	0	f(1,1,1,0)
1	1	1	1	f(1,1,1,1)

(c) four-input

When two minterms (or maxterms) are combined, one variable is eliminated. When four minterms (or maxterms) are combined, two variables are eliminated. The combination of eight minterms (or maxterms) eliminates three variables. Some of the typical, but less transparent, groupings of adjacent blocks are shown in Fig. 42.4.

Figure 42.4 *Typical Adjacent Minterm Groupings*

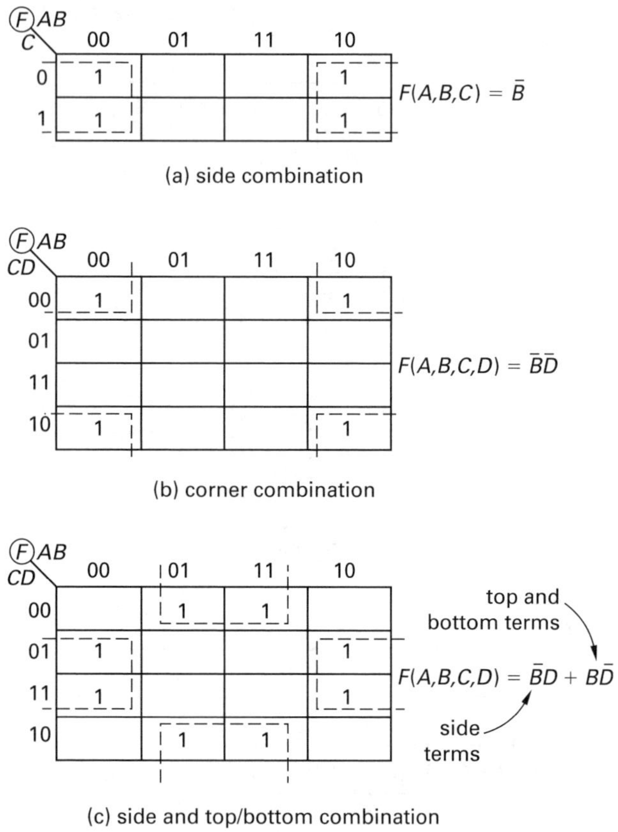

The general rules for combining or grouping minterms (or maxterms) are as follows.

- Each one-square (zero-square) must be accounted for in at least one grouping.
- Any grouping should be as large as possible.
- Groupings are made in terms of 1, 2, 4, 8, …, 2^n minterms (maxterms).
- A one-square grouping should not be used if it can be combined with another one-square. A two-square grouping should not be used if it can be combined into a four-square, and so on.
- Groupings should be used in such a manner as to minimize the total number of groups.

The minterm formulation (grouping ones) is used to find the SOP form. The maxterm formulation (grouping zeros) is used to find the POS form. Use the form with the smaller number of groups, because this generally leads to the minimum number of logic gate inputs. The general procedure for finding the minimal Karnaugh map follows.

step 1: Identify and circle all of the one-squares (zero-squares) that cannot be combined with any other squares.

step 2: Identify and circle all of the one-squares (zero-squares) that can combine with only one other square. Form groupings of two squares with these. If a square can be combined in any other manner, leave it uncombined at this step.

step 3: Identify and circle all of the one-squares (zero-squares) that can combine in groups of four only. Form groupings of four squares with these. If a square can be combined in any other manner, leave it uncombined at this step. Do not use the squares combined in step 2.

step 4: Identify and circle all of the one-squares (zero-squares) that can combine in groups of eight only. Form groupings of eight squares with these. If a square can be combined in any other manner, leave it uncombined at this step. Do not use the squares combined in steps 2 or 3.

step 5: Any square not assigned to a grouping is now arbitrarily placed in the largest group possible that includes the most uncovered squares. Add the minimum number of squares to cover all the remaining one-squares (zero-squares). Use squares with the don't care condition, *d*, to increase the grouping size.

step 6: Check the resulting expression's output against the original truth table. The results must match.

Example 42.2

A given three-variable function has the following truth table. Draw the Karnaugh map, and determine the SOP form of the function.

A	B	C	$F(A, B, C)$
0	0	0	0
0	0	1	0
0	1	0	0
0	1	1	1
1	0	0	0
1	0	1	1
1	1	0	0
1	1	1	1

Solution

The Karnaugh map is

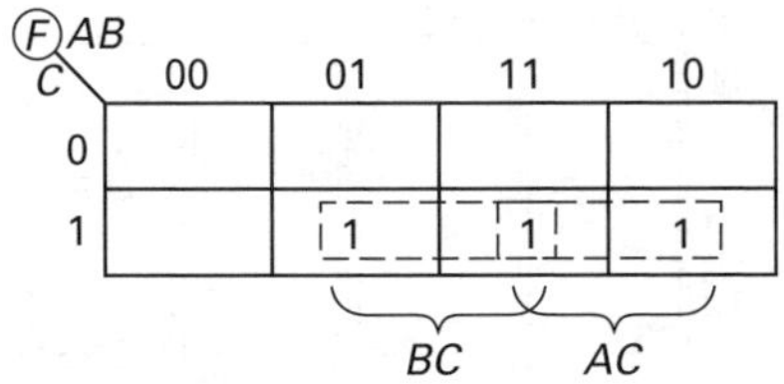

Step 2 has been applied to the map as indicated. The leftmost grouping (m_3 and m_7) contains A and $\overline{A}$, making it redundant. The rightmost grouping (m_7 and m_5) contains B and $\overline{B}$, making it redundant. The resulting expression is

$$F(A, B, C) = BC + AC$$

To spot-check, consider the input value of 011 for A, B, and C. Substituting this into the previous expression results in a value of one, which matches the truth table and the Karnaugh map.

Example 42.3

Realize the function in Ex. 42.2 with NAND-NAND logic.

Solution

Assuming all the variables and their complements are available, the circuit can be realized in NAND-NAND form directly from the SOP equation. The result is

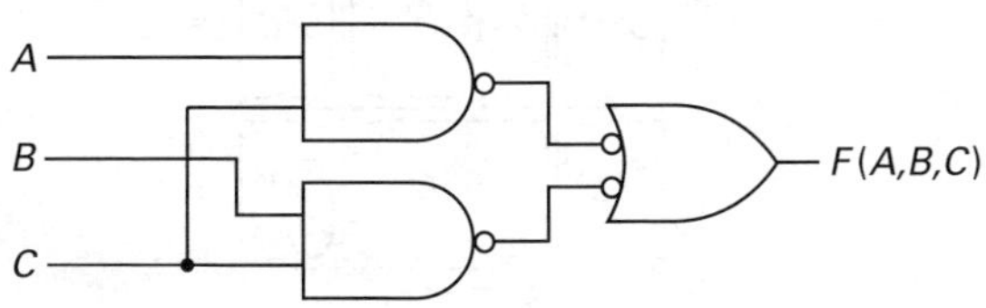

6. VEITCH DIAGRAMS

Adjacencies indicate which variable is to be eliminated during minimization. The variable that changes, that is, the one contained in a grouping in both its negated and unnegated form, is eliminated using the theorem of logical adjacency (see Eq. 42.1 and Eq. 42.2). On the Karnaugh map this variable is indicated by its values of zero and one. To improve the usefulness of the map, the variables are listed by name rather than by value. The variable name is given in the true condition. For example, when a general variable $X = 0$, the variable is false, and the true condition is $\overline{X}$. The map, when using named variables, is called a *Veitch diagram.* Three- and four-input Veitch diagrams are shown in Fig. 42.5.

The Veitch diagram is a Karnaugh map that shows when the variable is true. When the variable is not true, the complement of the variable is true. The Veitch diagram allows for the entry of zeros and ones immediately from a logic expression. Conversely, it allows the writing of logic expressions directly from the diagram.

When a map is constructed with the minterm information (values) and the Veitch diagram information (names), it is called a *Karnaugh-Veitch* (K-V) *map.* The groupings, also called *encirclements*, for a three-input K-V map are shown in Fig. 42.6. The groupings for a four-input K-V map are shown in Fig. 42.7. The groupings in both figures are by minterms. The groupings by maxterms are found using the zeros.

Example 42.4

Design the NAND-NAND logic to realize segment a of a seven-segment display for a calculator.

Solution

Step 1 of the design process in Sec. 42.1 is to identify and name all the inputs supplied to the network and all the outputs required of the network, using the set of requirements provided. The inputs will be binary numbers. The output will be a display such as

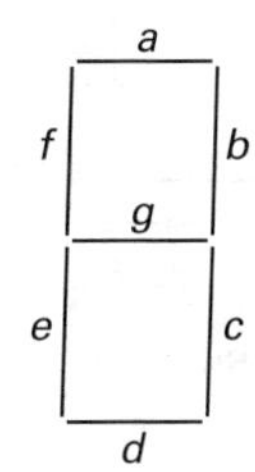

Step 2 is to establish the exact relationships between the inputs and outputs. Using a truth table for this purpose results in the following. The capital X indicates a don't care condition, because using d might cause confusion with the segment d.

inputs				segments							decimal
A	*B*	*C*	*D*	*a*	*b*	*c*	*d*	*e*	*f*	*g*	number
0	0	0	0	1	1	1	1	1	1	0	0
0	0	0	1	0	1	1	0	0	0	0	1
0	0	1	0	1	1	0	1	1	0	1	2
0	0	1	1	1	1	1	1	0	0	1	3
0	1	0	0	0	1	1	0	0	1	1	4
0	1	0	1	1	0	1	1	0	1	1	5
0	1	1	0	1	0	1	1	1	1	1	6
0	1	1	1	1	1	1	0	0	0	0	7
1	0	0	0	1	1	1	1	1	1	1	8
1	0	0	1	1	1	1	1	0	1	1	9
1	0	1	0	*X*	*X*	*X*	*X*	*X*	*X*	*X*	not used
1	0	1	1	*X*	*X*	*X*	*X*	*X*	*X*	*X*	not used
1	1	0	0	*X*	*X*	*X*	*X*	*X*	*X*	*X*	not used
1	1	0	1	*X*	*X*	*X*	*X*	*X*	*X*	*X*	not used
1	1	1	0	*X*	*X*	*X*	*X*	*X*	*X*	*X*	not used
1	1	1	1	*X*	*X*	*X*	*X*	*X*	*X*	*X*	not used

Digital Systems

Figure 42.5 *Veitch Diagram Fundamentals*

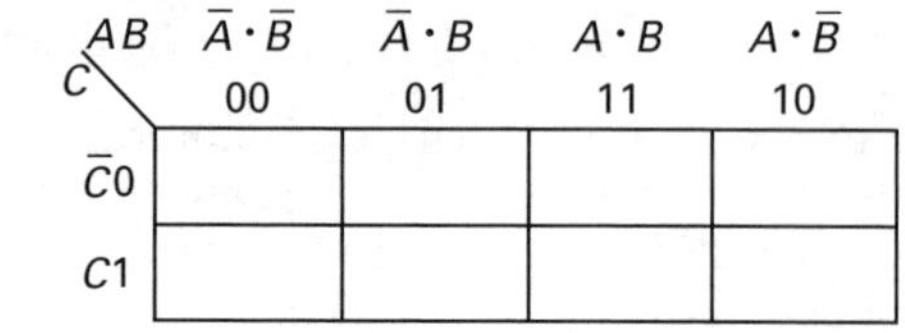

(a) three-input named variables

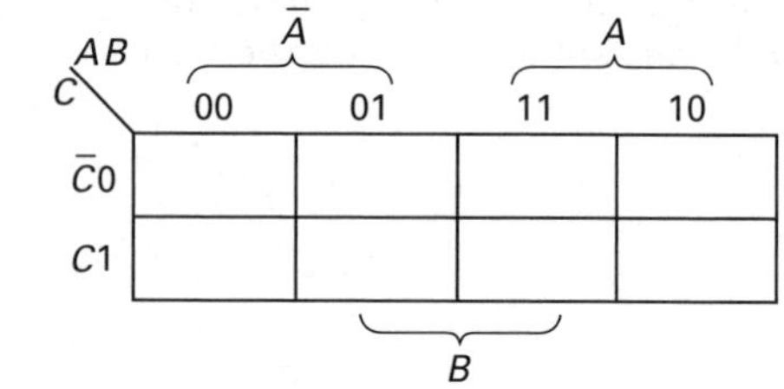

(b) three-input Veitch diagram

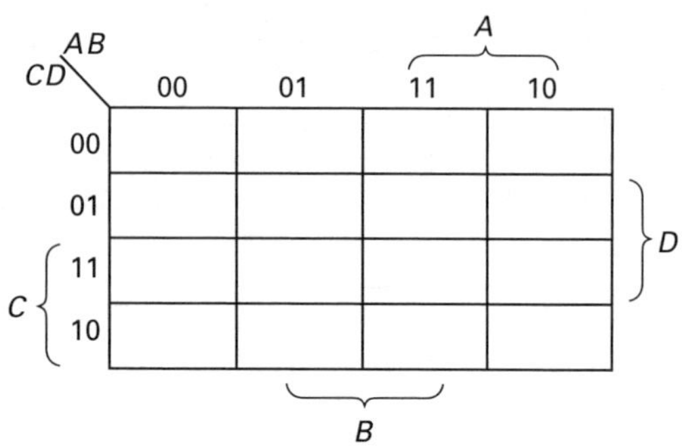

(c) four-input Veitch diagram

Figure 42.6 *Three-Input K-V Map: Minterms*

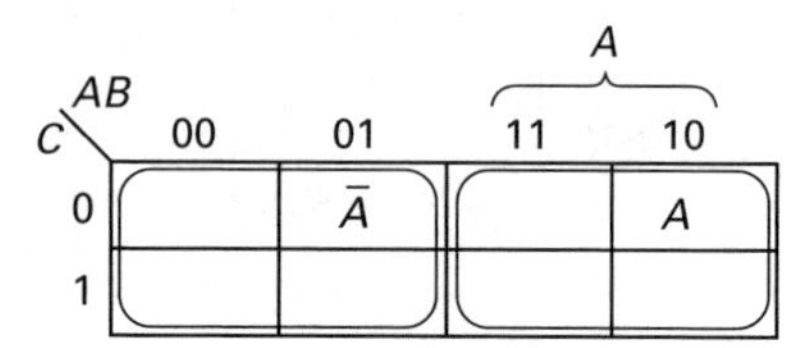

(a) A and $\bar{A}$

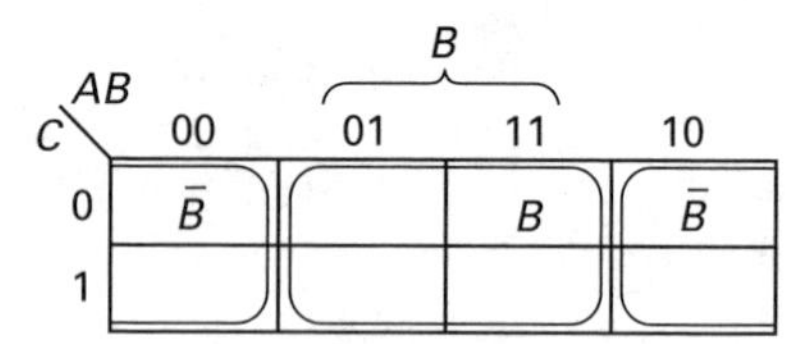

(b) B and $\bar{B}$

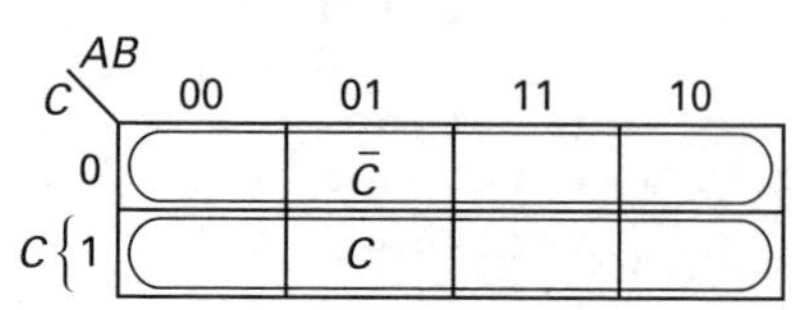

(c) C and $\bar{C}$

Figure 42.7 *Four-Input K-V Map: Minterms*

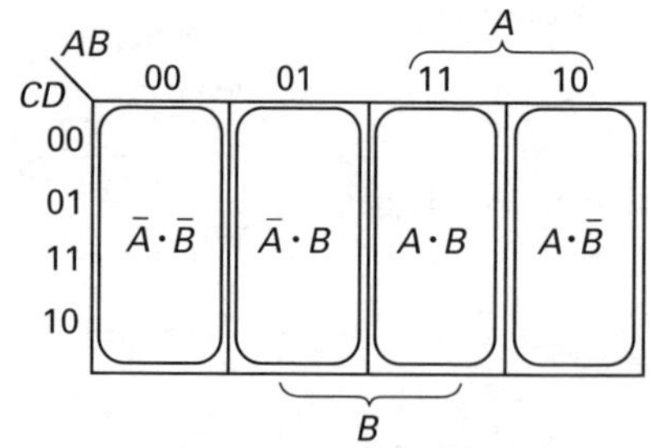

(a) A and B only

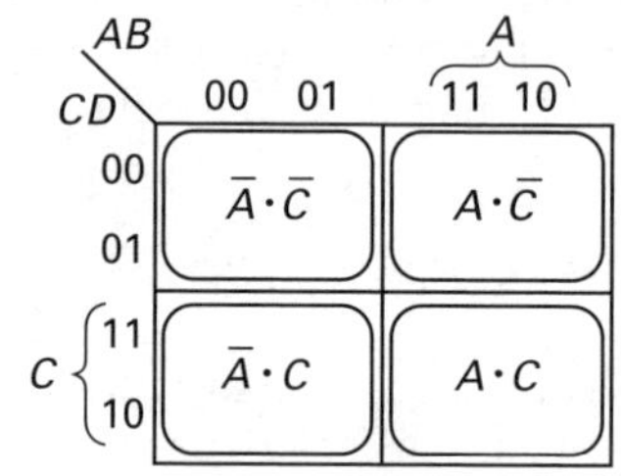

(b) A and C only

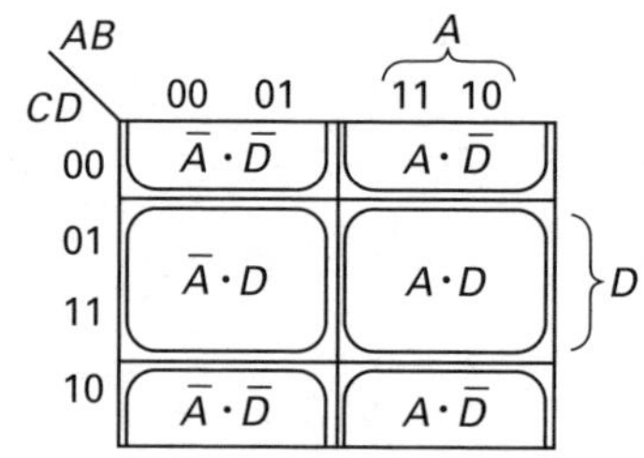

(c) A and D only

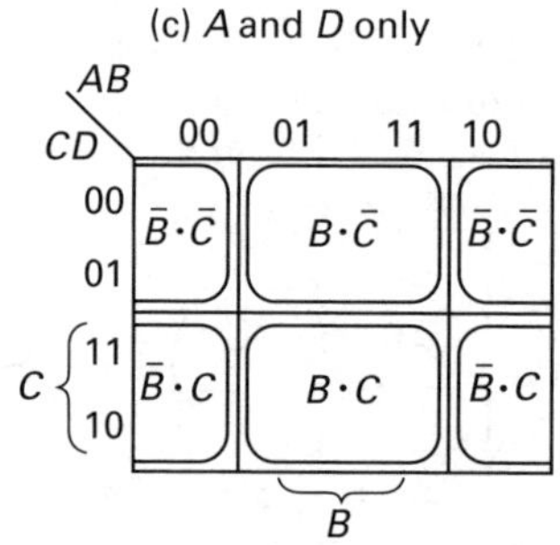

(d) B and C only

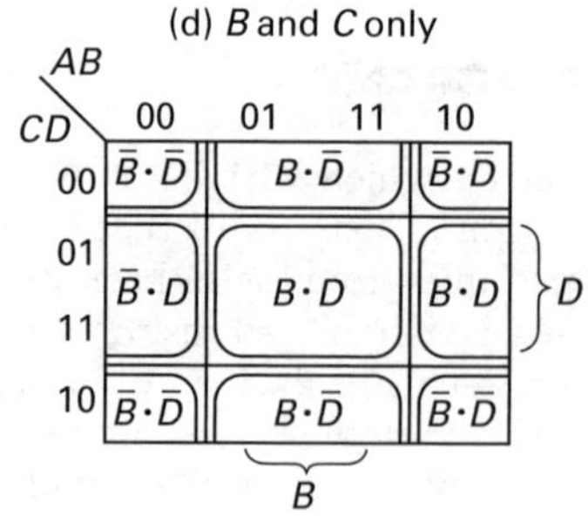

(e) B and D only

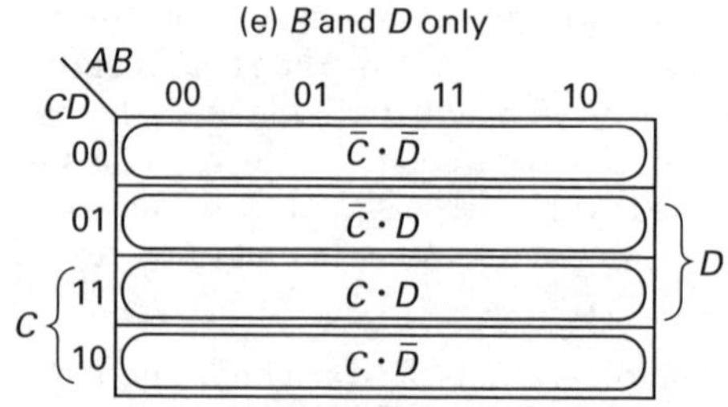

(f) C and D only

Step 3 is to code the inputs and outputs. In this case the ones and zeros are used to indicate the on-off condition of the particular segment of the display. The ones and zeros also correlate with the voltage necessary to drive the logic circuit.

Step 4 is to develop the logic expression. The Karnaugh map for any of the segments with the don't care conditions indicated is

CD \ AB	00	01	11	10
00	0	4	X_{12}	8
01	1	5	X_{13}	9
11	3	7	X_{15}	X_{11}
10	2	6	X_{14}	X_{10}

Placing the output values for segment a in the truth table, using the Veitch format, and grouping the terms to give the minimal map (see Sec. 42.5) gives

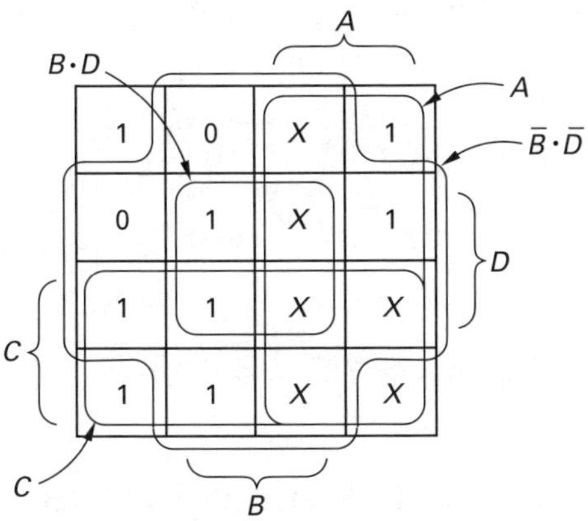

Step 5 is to determine the logic circuit elements that will be used to realize the network. In this case, NAND-NAND will be used. Applying analytical minimization techniques, reduce the complexity of the logic expressions to the desired level.

Minterm 10, m_{10}, is a "don't care" condition. Treating m_{10} as one allows the four corners, m_0, m_2, m_8, and m_{10}, to be combined to give $\overline{B}\,\overline{D}$. Minterms 13 and 15 are also treated as ones. This allows m_5 and m_7 to be combined with m_{13} and m_{15} to yield BD. Minterms 11 and 14 are also treated as ones. This, along with the "don't care" ones of m_{10} and m_{15}, allows the bottom two rows to be combined, yielding C. The remaining uncovered minterm is m_{12}. Making minterm 12 a "don't care" one allows the right half of the map to be encircled, yielding A. The resulting SOP expression is

$$f(\text{segment } a) = \overline{B}\,\overline{D} + BD + C + A$$

All ones were used for the "don't care" conditions. This does not have to be the case. In addition, multiple coverage of minterms occurred in the groupings. This is useful if it reduces the complexity.

The NAND-NAND expression is obtained by complementing the SOP form, applying De Morgan's law, and complementing the result (see Sec. 42.7). The result is

$$f(\text{segment } a) = \overline{(\overline{\overline{B}\,\overline{D}})(\overline{BD})\,(\overline{C})\,(\overline{A})}$$

Step 6 is to use the logic expression to realize the network and to test the network. This expression is realized with the following NAND-NAND network.

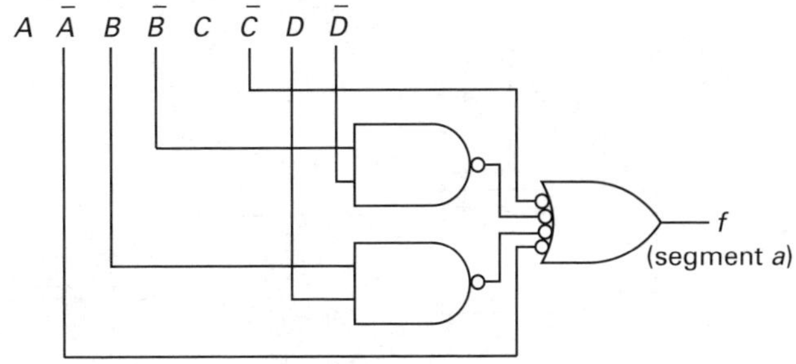

A check of the expression against the values in the truth table shows that the expression as written realizes the desired output.

Example 42.5

Using NOR-NOR logic, realize segment a of the seven-segment display in Ex. 42.4.

Solution

The NOR-NOR realization requires the use of maxterms. The maxterms occur where the output of the function is zero. The Veitch diagram for segment a is shown in the following illustration.

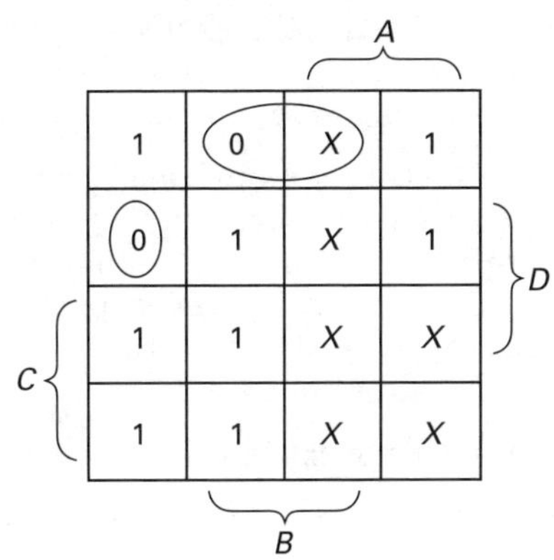

Maxterm 1 cannot be combined with any other zero. Recall that a maxterm is a sum term in which the variables take on the value of zero. So,

$$M_1 = A + B + C + \overline{D}$$

Making the "don't care" condition of maxterm 12 a zero allows M_4 and M_{12} to be combined, yielding $\overline{B} + C + D$. Combining this with M_1 yields the following POS expression.

$$f(\text{segment } a) = (A + B + C + \overline{D})(\overline{B} + C + D)$$

The NOR-NOR expression is obtained by complementing the POS form, applying De Morgan's law, and complementing the result (see Sec. 42.7). The result is

$$f(\text{segment } a) = \overline{\Big((A+B+C+\overline{D}) + (\overline{B}+C+D)\Big)}$$

The expression is realized using the following NOR-NOR network.

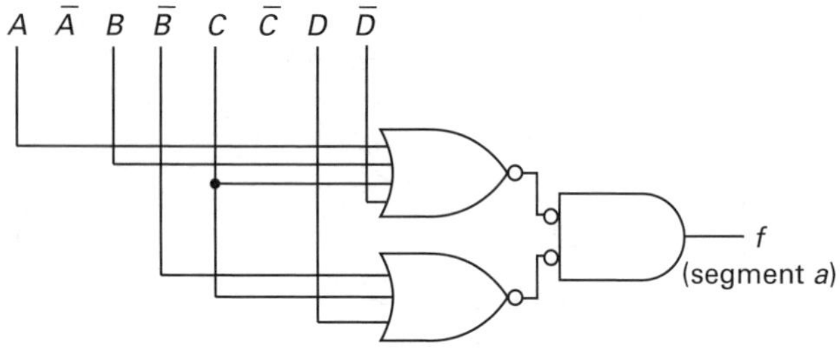

In Ex. 42.4 and Ex. 42.5, the conversion of the expression for the function is not required. The method shown in Sec. 41.6 and Sec. 41.7 can be used directly. The conversion directly shows the NAND-NAND or NOR-NOR relationship.

7. NAND/NOR FUNCTION REPRESENTATION

Functions in their SOP or POS form can be used to realize NAND-NAND only or NOR-NOR only logic as described in Sec. 41.6 and Sec. 41.7. Once a Karnaugh or Karnaugh-Veitch map has been minimized using the procedures of Sec. 42.5 and the resulting SOP or POS expression is found, the expression itself can be manipulated to specifically show the NAND-NAND or NOR-NOR relationship. The procedure follows.

step 1: For a NAND-only expression, group the minterms (output value one) in a map to form the sum-of-product (SOP) expression. For a NOR-only expression, group the maxterms (output value zero) in a map to form the product-of-sum (POS) expression.

step 2: Complement the expression from step 1.

step 3: Apply De Morgan's law to the expression in step 2.

step 4: Complement the result.

Example 42.6

Use a K-V map to find the minimum NAND-only expression for the following three-input function.

$$F(A,B,C) = \sum m(3,4,7) + d(6)$$

The term $d(6)$ indicates a "don't care" condition in minterm position 6, that is, in 110 binary.

Solution

The K-V map, with grouping, is

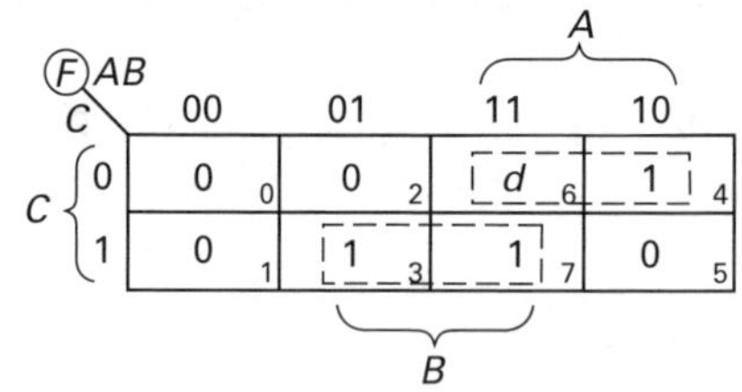

The minterm 6 "don't care" was arbitrarily given a value of 1. The resulting expression is

$$F(A,B,C) = A\overline{C} + BC$$

Complementing the result gives

$$\overline{F(A,B,C)} = \overline{(A\overline{C} + BC)}$$

Applying De Morgan's law results in the following expression.

$$\overline{F(A,B,C)} = \overline{(A\overline{C})}\;\overline{(BC)}$$

Complementing the result puts the expression in NAND-only form.

$$F(A,B,C) = \overline{\overline{(A\overline{C})}\,\overline{(BC)}}$$

8. PROGRAMMABLE LOGIC ARRAYS

The most direct method for realizing a network with multiple inputs and multiple outputs is to generate the canonical expressions that represent the function. This is accomplished with a *switching matrix*, an example of which is shown in Fig. 42.8.

The switching matrix shown in Fig. 42.8 realizes two-minterm expressions but could easily be connected for additional minterms. Consequently, a switching matrix can realize any combinational logic network. A device called a *programmable logic array* (PLA) is used to realize switching matrices. An unprogrammed PLA, along with a PLA programmed to realize the switching matrix in Fig. 42.8, is shown in Fig. 42.9.

Programmable logic arrays are manufactured in standardized sizes. The AND and OR connections are "burned in" by a variety of methods, either by the manufacturer or in the field. In some cases, the connections are reversible and can be "burned out."[5]

[5]A *read-only memory* (ROM) is a switching matrix that can be realized with a PLA. The ROM requires additional circuitry to address the desired portion of the ROM (that is, a decoder) and transfer the output to the specified location.

Figure 42.8 *Switching Matrix*

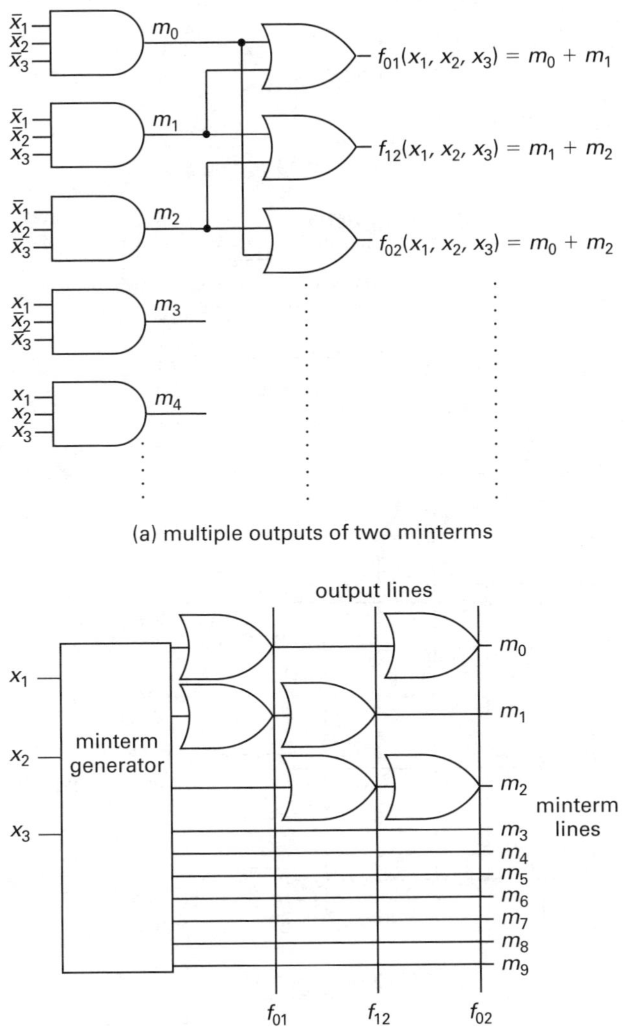

(a) multiple outputs of two minterms

(b) simplified representation

Figure 42.9 *Programmable Logic Array*

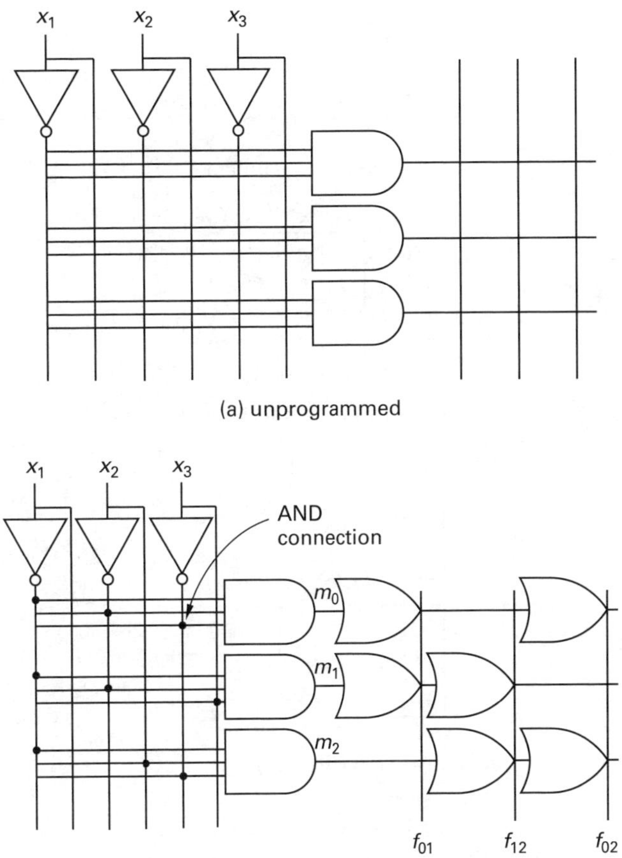

(a) unprogrammed

(b) programmed

9. ENCODERS AND DECODERS

An *encoder* is a network or system in which 2^N inputs, or fewer, are placed on N or fewer output lines. A *decoder* is a network or system that uses N input lines to select one of 2^N or less outputs. Decoders are referred to as N-line-to-K-line decoders. N represents the number of input lines and K is the number of output lines.

A simplified priority encoder is shown in Fig. 42.10. The output of the encoder provides information about the status of the highest priority line that is asserted. That is, the output is decimal 2 if the decimal 1 and 2 lines are asserted, decimal 3 if 3 and any combination below the 3 is asserted, and so on. In practice, an enable line is required to have following circuits discern when the output is valid. Encoders are used throughout computer systems. For example, encoders are available that encode ten input lines to four output lines, which would allow a decimal input to be coded as a binary coded decimal (BCD) number for processing.

A decoder is shown in Fig. 42.11. The decoder shown has a binary input that is used to select a minterm output. The decoder is an array of AND gates as used in a PLA. A decoder is sometimes referred to as a demultiplexer. Decoders are used throughout computer systems. For example, decoders are used to pick the desired information line given a specific address input in read-only memory (ROM). Also, instructions are decoded to enable the specific hardware required during central processing unit (CPU) operation.

10. DEMULTIPLEXERS

A *demultiplexer* is a device used to separate two or more signals previously combined by a compatible multiplexer and transmitted over a single channel or line. The demultiplexer is a *data selector*. A commonly used abbreviation is DEMUX or DMUX. A demultiplexer similar to the decoder in Sec. 42.9 is illustrated in Fig. 42.12. The demultiplexer is analogous to a multiposition switch that sends the input signal to the desired output line. Demultiplexers are used in the design of CPUs.

Figure 42.10 *Encoder*

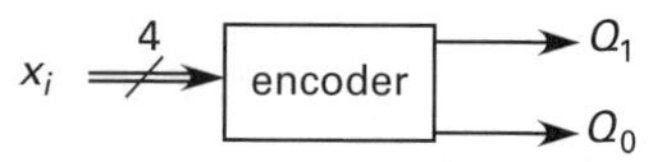

(a) priority encoder block diagram

input				output	
x_3	x_2	x_1	x_0	Q_1	Q_0
0	0	0	0	*d*	*d*
0	0	0	1	0	0
0	0	1	0	0	1
0	1	0	0	1	0
1	0	0	0	1	1

(b) priority encoder truth table

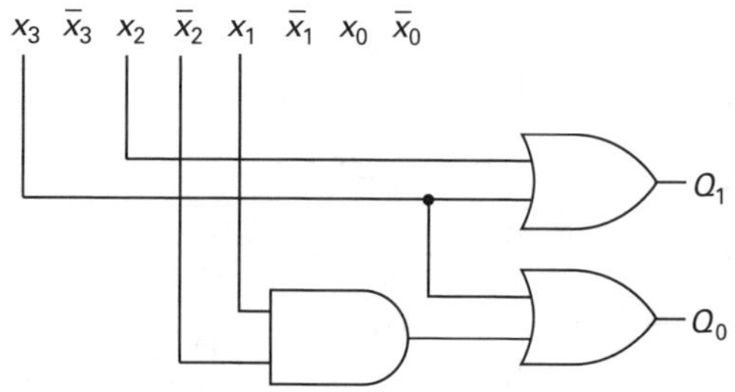

(c) AND-OR realization

Figure 42.11 *Decoder*

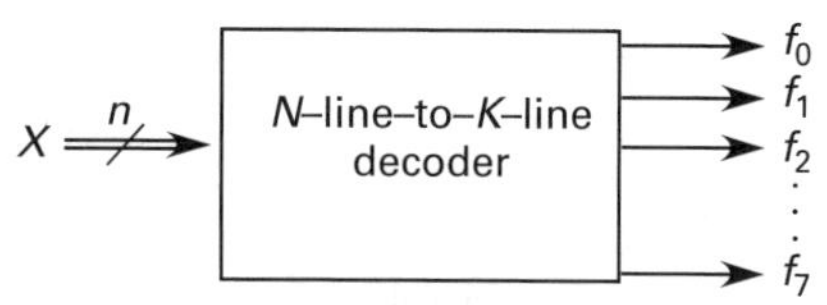

(a) decoder block diagram

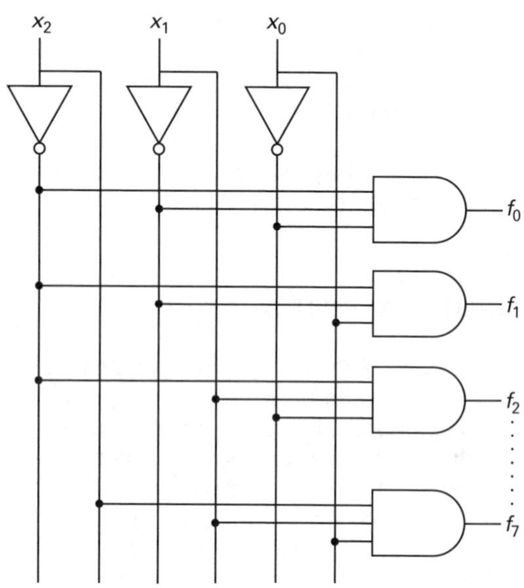

(b) 3-line-to-8-line decoder

Figure 42.12 *Demultiplexer*

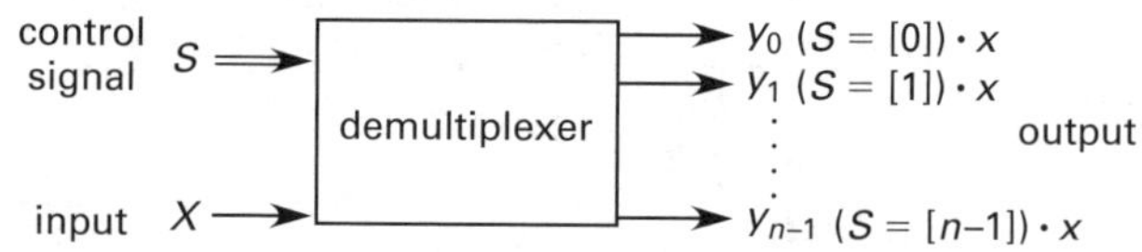

(a) demultiplexer block diagram

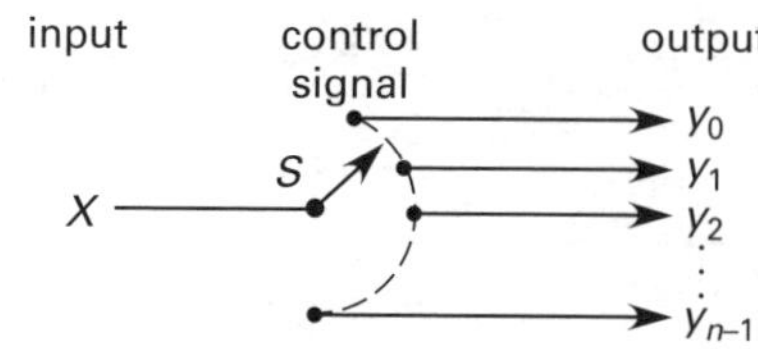

(b) switch analogy

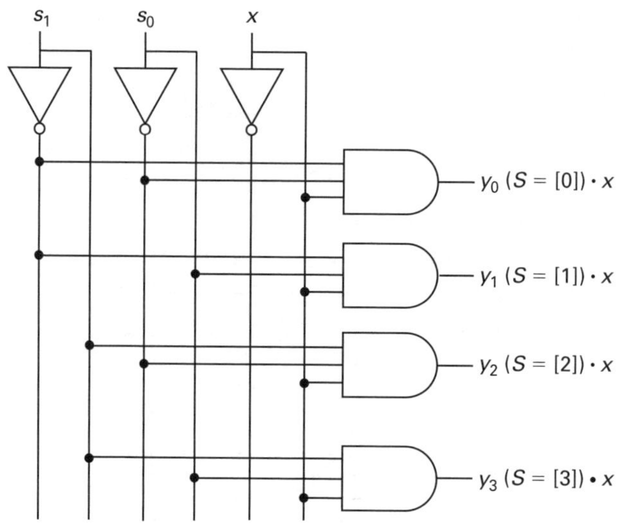

(c) 1:4 demultiplexer

11. MULTIPLEXERS

A *multiplexer* is a device for combining two or more signals for transmission over a single channel or line.[6] The multiplexer is a data selector. A commonly used abbreviation is MUX. A multiplexer compatible with the demultiplexer in Sec. 42.10 is shown in Fig. 42.13. An OR gate is used on the output to electronically isolate the AND gates from one another.

12. ARITHMETIC LOGIC DEVICES

The arithmetic logic unit (ALU) uses combinational logic devices to accomplish mathematical and logic operations. Binary addition is accomplished with the following rules.

- $0+0=$ sum 0 carry 0
- $0+1=$ sum 1 carry 0
- $1+0=$ sum 1 carry 0
- $1+1=$ sum 0 carry 1

[6]Multiplexer is also spelled multiplexor.

Figure 42.13 Multiplexer

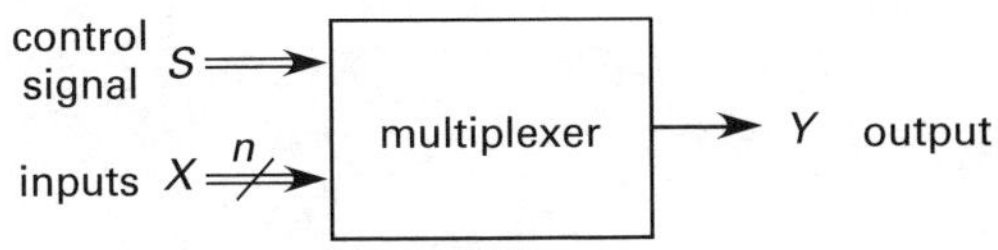

(a) multiplexer block diagram

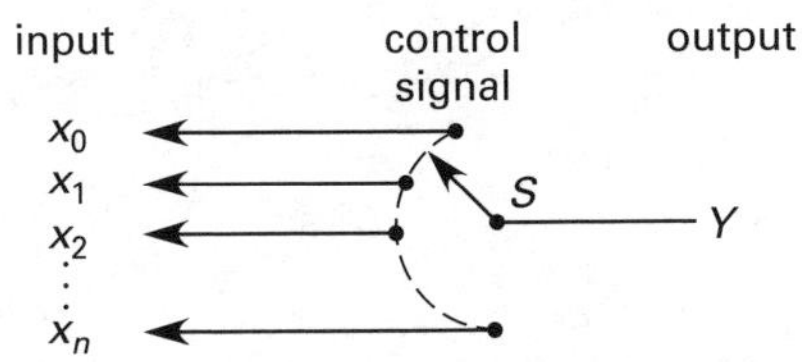

(b) switch analogy

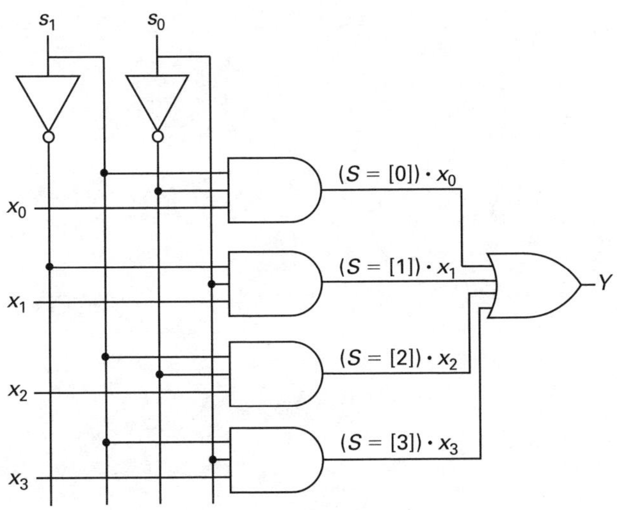

(c) 4:1 multiplexer

Given these rules, adder truth tables are constructed as shown in Fig. 42.14. From the truth tables, the necessary combinational logic circuits are realized. For the *half-adder*, the sum is XOR logic and the carry is AND logic.

To add an n-bit number, adders are cascaded to form a *ripple-carry adder* (see Fig. 42.15). The term *ovr* at the stage $n-1$ indicates an overflow condition, which occurs if the value is outside the range of numbers that can be stored.

Figure 42.14 Adders

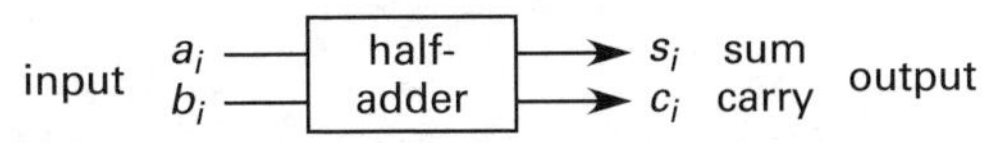

(a) half-adder block diagram

b_i	a_i	s_i	c_i
0	0	0	0
0	1	1	0
1	0	1	0
1	1	0	1

(b) half-adder truth table

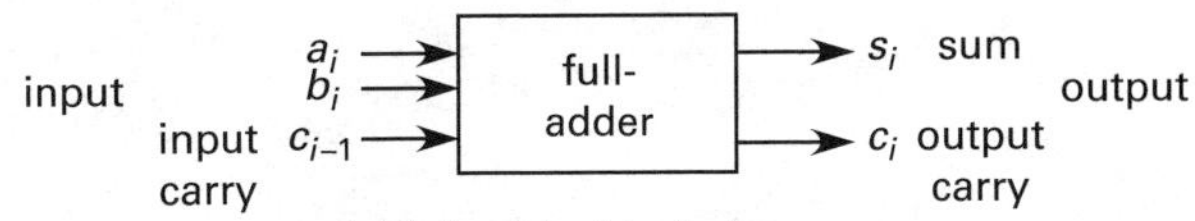

(c) full-adder block diagram

b_i	a_i	c_{i-1}	s_i	c_i
0	0	0	0	0
0	0	1	1	0
0	1	0	1	0
0	1	1	0	1
1	0	0	1	0
1	0	1	0	1
1	1	0	0	1
1	1	1	1	1

(d) full-adder truth table

Subtraction is accomplished using one's complement arithmetic. That is, negative numbers are changed to their one's complement and added.

Figure 42.15 Ripple-Carry Adder

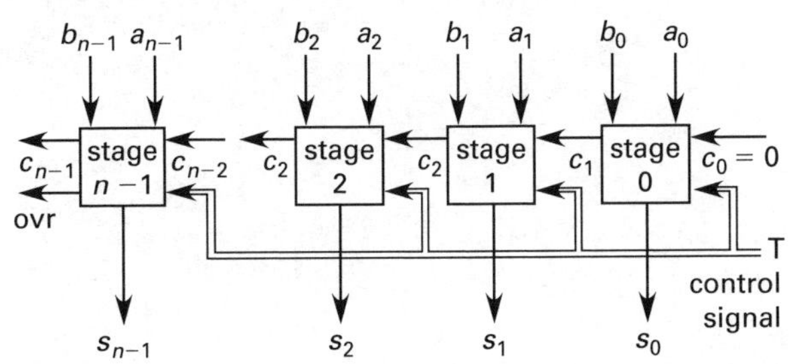

43 Synchronous Sequential Networks

1. Fundamentals . . . 43-1
2. Clocks and Timing . . . 43-1
3. *S-R* Flip-Flops . . . 43-2
4. *J-K* Flip-Flops . . . 43-2
5. *T* Flip-Flops . . . 43-3
6. *D* Flip-Flops . . . 43-3
7. Flip-Flop Transition Tables . . . 43-3
8. Counters . . . 43-4
9. Synchronous Binary Counter . . . 43-5
10. Adjacent-State Binary Counter . . . 43-6
11. Counter Analysis Procedure . . . 43-6
12. Sequence Detection . . . 43-8

Nomenclature

b	binary input or output	–
C	clock input	–
D	delay	–
I	next state	–
J	set condition	–
K	reset condition	–
Q	output or present state	–
$\overline{Q}$	complement of present state	–
Q^+	next state	–
R	reset condition	–
S	set condition or state	–
T	toggle condition	–
X	input	–
Y	output	–
Z	output	–

1. FUNDAMENTALS

Networks with memory are called *sequential networks.* A sequential network comprises *sequential logic elements* with at least one input and output and one internal state variable, and is designed so that the output signal depends on the past and present states of the inputs. If the state of the sequential network can change only when a transfer pulse or clock pulse is applied, the network is called a *synchronous sequential network.* The general form of a synchronous sequential network is shown in Fig. 43.1.

The combination logic network principles are described in Chap. 42. The *register* is the computer hardware for storing one machine word. Registers comprise a group of flip-flops, which are bistable multivibrators designed with logic circuits. (Flip-flops are the most commonly used memory devices in sequential networks.) The general sequential network is defined by three types of equations: the *output equations*, the *next-state equations*, and the *flip-flop control equations.* The equations are obtained from *state tables,* also called *transition tables,* that list the inputs, present state, and next state of the circuit. State tables represented in graphical form are called *transition diagrams* or *state diagrams.*

Figure 43.1 *General Synchronous Sequential Network*

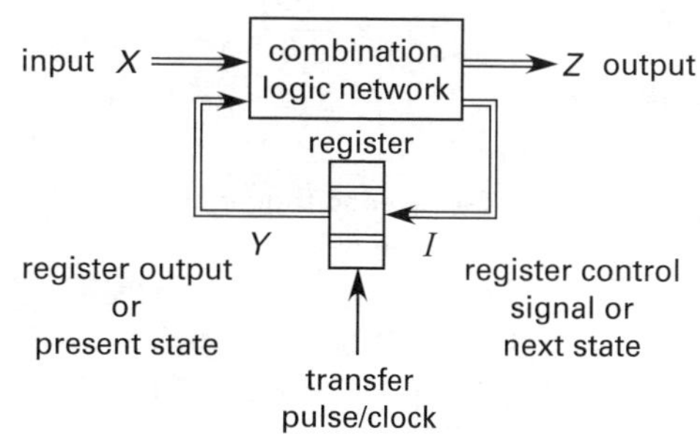

The general steps in designing a synchronous sequential network are as follows.

step 1: Describe the network specifications.

step 2: Determine the state table required.

step 3: Minimize the state table.

step 4: Make the state assignments; that is, encode the information in binary form in order to transform the state table into a transition table.

step 5: Realize the network.

These steps conform to the general steps in the design of combinational logic networks (see Sec. 42.1) though the methods differ. Only steps 3 and 5 are algorithmic, with step 3 being of lesser importance with improved integrated circuit technology.[1]

2. CLOCKS AND TIMING

Flip-flops, which are also called *latches*, can be designed to operate only when a clock pulse is present. When this occurs, the flip-flop (FF) is *triggered* by the clock. The flip-flop is triggered either on the rising edge or on the falling edge of the clock pulse. The symbols are shown in Fig. 43.2. The clock logic is realized with an AND gate, and can be used with any flip-flop. Any state table is

[1]The focus of this chapter will be on the construction and use of state and transition tables. Counters, which are used in numerous information processing tasks, will be the design examples.

Figure 43.2 Flip-Flop Clocks

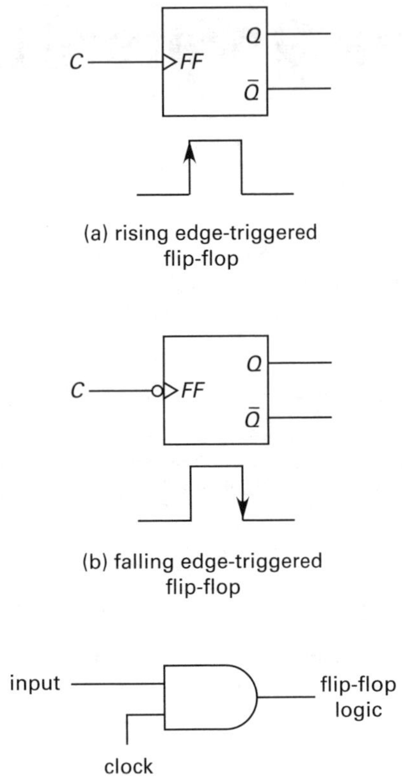

(a) rising edge-triggered flip-flop

(b) falling edge-triggered flip-flop

(c) clock logic realization

modified by simply adding a clock input that must be asserted to allow operation of the flip-flop.

Clocks are used to prevent the flip-flop from changing with spurious level changes in the inputs. A flip-flop without a clock is called a *transparent flip-flop*. Removing this transparency with a clock permits better synchronization and timing. Another method of removing this transparency is to use two clocked *S-R* flip-flops in series. (*S-R* flip-flops can be used to construct any of the other flip-flops, so this method works for any flip-flop.) The first flip-flop operates on the rising edge of the clock signal, and the second flip-flop operates on the falling edge of the clock signal, preventing unwanted transitions.

Additionally, if the data or control inputs to a flip-flop are changing at the instant of the clock pulse, a problem known as *metastability* may occur. In the metastable case, the flip-flop does not settle in to a stable state (0 or 1, LOW or HIGH) in the normally allotted time. Metastability is avoided by holding the information stable before and after the clock pulse for a set period of time, called the *setup time* for the data line and the *hold time* for the control line.

Additional inputs to any flip-flop are the *preset* and *clear signals*. Both these signals operate *asynchronously*; that is, they change the state of the flip-flop immediately, regardless of the state of the clock signal. The preset signal changes the output (the Q line) to a state of one. The clear signal changes the output, Q, to a value of zero. These control lines are used when initially starting a computer or digital system and at other appropriate times.

As with any electronic circuit, flip-flops have propagation delays and timing considerations. The delays generally vary for low-to-high and high-to-low transitions and for the type of transition, such as clock, preset, clear, and so on. These propagation delays determine the maximum operating frequency of the flip-flop. The *setup time* is the minimum time inputs must be stable before a clock transition can be triggered. The *hold time* is the minimum time the inputs must remain stable after the clock transition. Additionally, if the data or control inputs to a flip-flop are changing at the instant of the clock pulse, a problem known as *metastability* may occur. In the metastable case, the flip-flop does not settle in to a stable state (0 or 1, LOW or HIGH) in the normally allotted time. Metastability is avoided by holding the information stable before and after the clock pulse for a set period of time, called the *setup time* for the data line and the *hold time* for the control line.

3. *S-R* FLIP-FLOPS

The *S-R* flip-flop, also called the *S-C* flip-flop, is shown in Fig. 43.3(a). The S indicates the set condition, and the R indicates the reset or clear condition. The state table, in place of a truth table, is shown in Fig. 43.3(b). The present state is indicated by the Q. The next state is indicated by the Q^+. The condition that exists when both S and R have the value one is indeterminate and must not be allowed to occur. The realization of the flip-flop is shown in Fig. 43.3(c).

The *next-state equation*, which is also called the *characteristic equation* or *transition equation*, for the *S-R* flip-flop is

$$Q^+ = S + \overline{R}Q \qquad 43.1$$

This equation is valid as long as the indeterminate state is disallowed. That is, $S = R = 1$ cannot occur. The restriction on Eq. 43.1, then, is $S \cdot R = 0$.

4. *J-K* FLIP-FLOPS

The *J-K* flip-flop takes its initials from the inventors. The *J-K* flip-flop is identical to the *S-R* flip-flop without the restriction against allowing both inputs to have the value one. The *J-K* flip-flop is shown in Fig. 43.4(a). The J indicates the set condition, and the K indicates the reset or clear condition. The state table, in place of a truth table, is shown in Fig. 43.4(b). The present state is indicated by the Q. The next state is indicated by the Q^+. When both J and K have the value one, the output toggles. Figure 43.4(c) shows the realization of the flip-flop.

Figure 43.3 S-R Flip-Flop

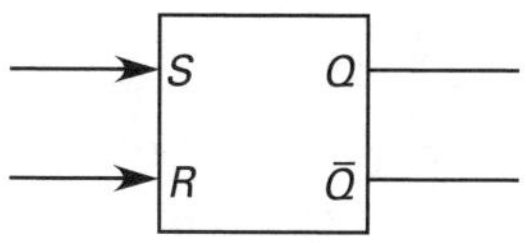

(a) symbol

S	R	Q	Q^+	remarks
0	0	0	0	hold
0	0	1	1	hold
0	1	0	0	reset
0	1	1	0	reset
1	0	0	1	set
1	0	1	1	set
1	1	0	indeterminate	disallowed condition
1	1	1	indeterminate	disallowed condition

(b) state table

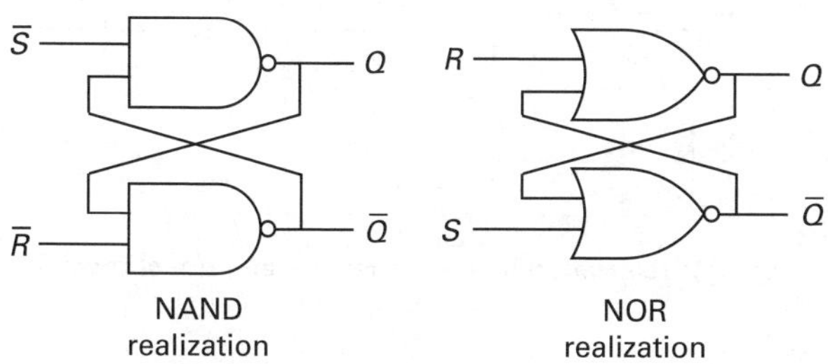

(c) logic realization

Figure 43.4 J-K Flip-Flop

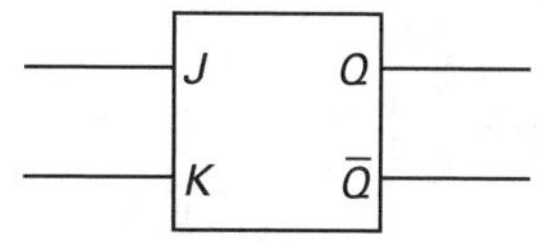

(a) symbol

J	K	Q	Q^+	remarks
0	0	0	0	hold
0	0	1	1	hold
0	1	0	0	reset
0	1	1	0	reset
1	0	0	1	set
1	0	1	1	set
1	1	0	1	toggle
1	1	1	0	toggle

(b) state table

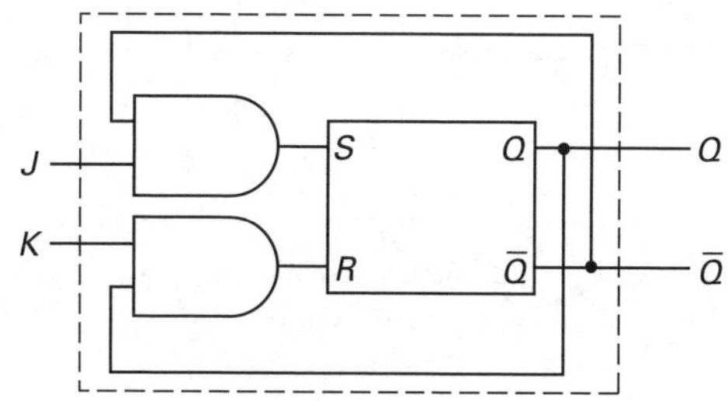

(c) logic realization

The next-state equation for the *J-K* flip-flop is

$$Q^+ = J\overline{Q} + \overline{K}Q \qquad 43.2$$

This equation is valid as long as the input signals do not change during the clock pulse.

5. *T* FLIP-FLOPS

The *T* flip-flop is shown in Fig. 43.5(a). The *T* indicates the toggle condition. The state table, in place of a truth table, is shown in Fig. 43.5(b). The present state is indicated by the Q. The next state is indicated by the Q^+. When T has the value one, the output toggles. The realization of the flip-flop is shown in Fig. 43.5(c).

The next-state equation for the *T* flip-flop is

$$Q^+ = T \oplus Q = T\overline{Q} + \overline{T}Q \qquad 43.3$$

This equation is valid as long as the clock pulse is high, which is no longer than the propagation time through the gates. This ensures that the flip-flop changes state only once per clock pulse.

6. *D* FLIP-FLOPS

The *D* flip-flop, also called the *delay* or *data flip-flop*, is shown in Fig. 43.6(a). The *D* indicates the delay in the output from the application of the clock pulse. The output follows the input but does so after the clock pulse is removed. That is, the input information is delayed one clock period. The *D* flip-flop is so named because it is used in the accurate transfer of data from one source to another. The state table, in place of a truth table, is shown in Fig. 43.6(b). The present state is indicated by the Q. The next state is indicated by the Q^+. The realization of the flip-flop is shown in Fig. 43.6(c). The delay aspect of the flip-flop is illustrated in the timing diagram of Fig. 43.6(d).

The next-state equation for the *D* flip-flop is

$$Q^+ = D \qquad 43.4$$

The *D* flip-flop is always a clocked device.

7. FLIP-FLOP TRANSITION TABLES[2]

In sequential networks, the transitions from the present state to the next state are controlled to satisfy design requirements. Four possible transitions are possible: zero to zero, zero to one, one to zero, and one to one. Though zero to zero and one to one do not appear to be

[2]Truth tables show inputs and outputs and are derived from output equations. Transition tables show inputs, current-state values, and next-state values and are derived from next-state equations. Excitation tables show known state transitions and the control signals necessary to produce those transitions. They are essentially rearranged transition tables and are used to generate the control equations.

Figure 43.5 *T Flip-Flop*

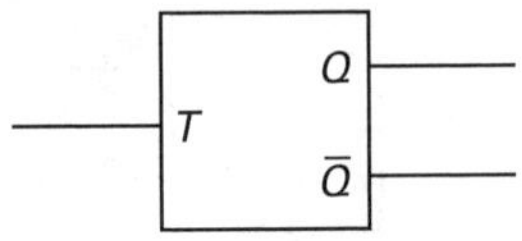

(a) symbol

T	Q	Q^+	remarks
0	0	0	hold
0	1	1	hold
1	0	1	toggle
1	1	0	toggle

(b) state table

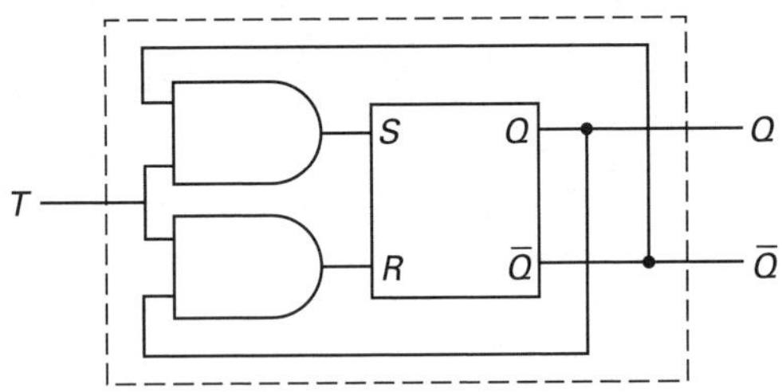

(c) logic realization

Figure 43.6 *D Flip-Flop*

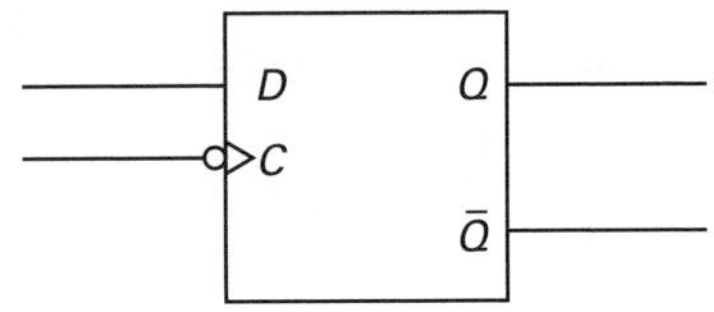

(a) symbol

D	Q	Q^+	remarks
0	0	0	The output follows
0	1	0	the input, *D*,
1	0	1	regardless of the
1	1	1	present state, *Q*.

(b) state table

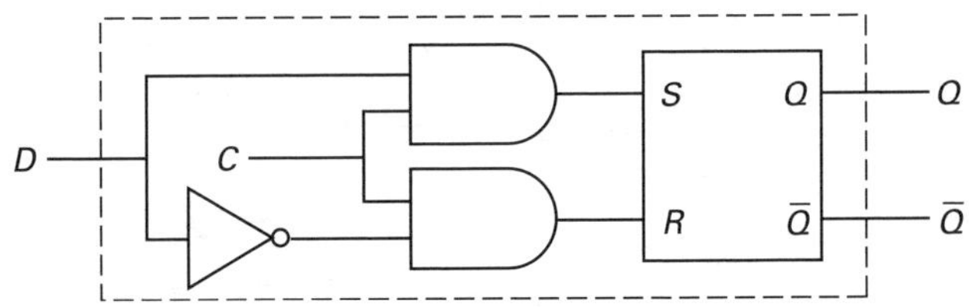

(c) logic realization (edge trigger circuitry not shown)

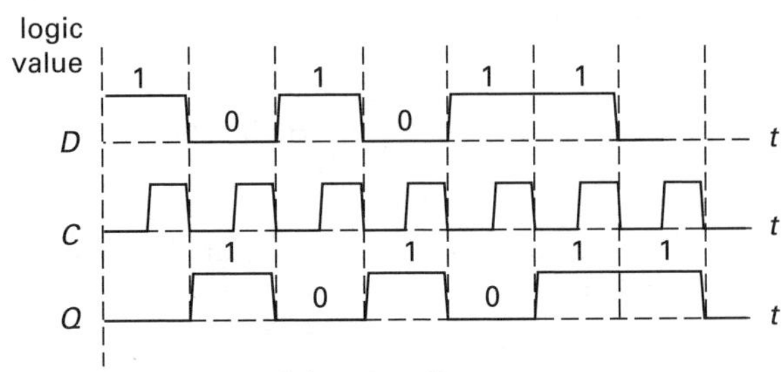

(d) timing diagram

transitions, they must be accounted for in logic design. The transitions for the principal flip-flops are summarized in Table 43.1.

8. COUNTERS

Sequential networks that function without any input signals are called *autonomous networks*. Such networks generate periodic control signals for use in information processing tasks. The general network is identical to that in Fig. 43.1 without the input signal, X. Autonomous networks are generalized counters.

Counters are sets of flip-flops, normally of the same type and functioning on the same edge of the clock signal. The count is determined by the number of states in a given sequence, limited to 2^n, where n is the number of flip-flops in the counter. When a counter starts at zero, counts to $2^n - 1$, and resets to zero, it is called a *binary up counter*, *modulo* 2^n *counter*, or *ripple counter*. An example is shown in Fig. 43.7.

The ripple counter in Fig. 43.7 is composed of *J*-*K* flip-flops with both inputs at value one, resulting in toggle action. The flip-flops start at clear ($Q=0$) and operate on the falling edge of the clock. The initial clock signal, arbitrarily labeled zero, is applied to the stage 0 flip-flop (FF), which is triggered on the falling edge. On clock signal 1, the stage 0 FF switches from high to low ($Q=1$ to $Q=0$), and the stage 1 FF is triggered from the output. On clock signal 2, the stage 0 FF switches from low to high and the stage 2 FF remains unchanged. On clock signal 3, the stage 0 FF switches from high to low,

Table 43.1 *Flip-Flop Transition Table*

present state Q		next state Q^+	S	R	J	K	T	D
0	⟶	0	0	X	0	X	0	0
0	⟶	1	1	0	1	X	1	1
1	⟶	0	0	1	X	1	1	0
1	⟶	1	X	0	X	0	0	1

$$Q^+ = S + \overline{R} \cdot Q$$
$$Q^+ = J \cdot \overline{Q} + \overline{K} \cdot Q$$
$$Q^+ = T \oplus Q$$
$$Q^+ = D$$

Figure 43.7 *Ripple Counter*

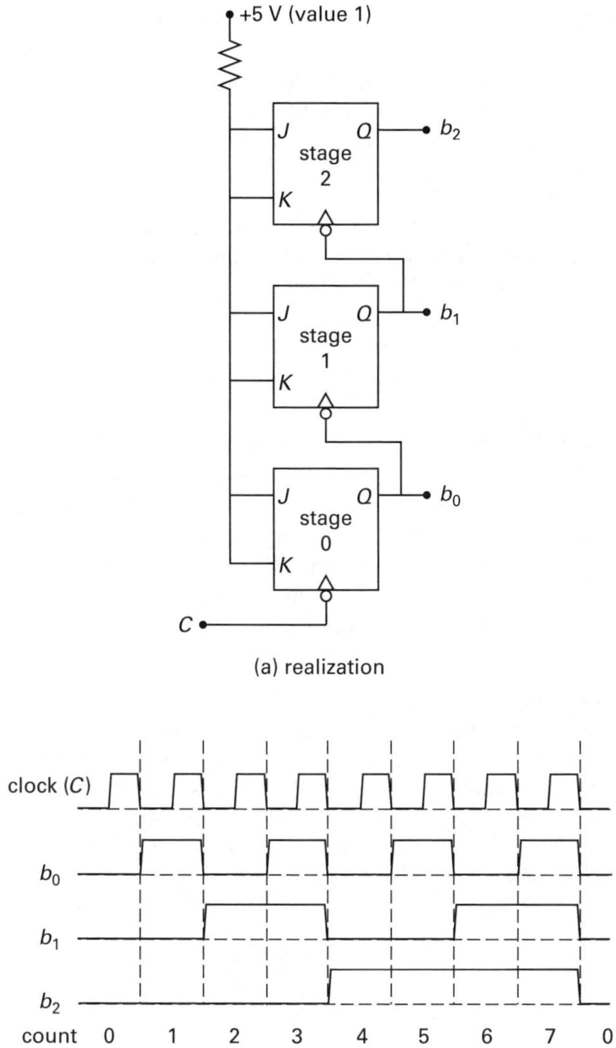

(a) realization

(b) timing diagram

Figure 43.8 *Synchronous Binary Counter*

present state $Q_2Q_1Q_0$	next state $Q_2^+Q_1^+Q_0^+$	required T inputs $T_2T_1T_0$
000	001	001
001	010	011
010	011	001
011	100	111
100	101	001
101	110	011
110	111	001
111	000	111

(a) excitation table

$$T_2 = Q_0 \cdot Q_1$$
$$T_1 = Q_0$$
$$T_0 = 1$$

(b) control equations

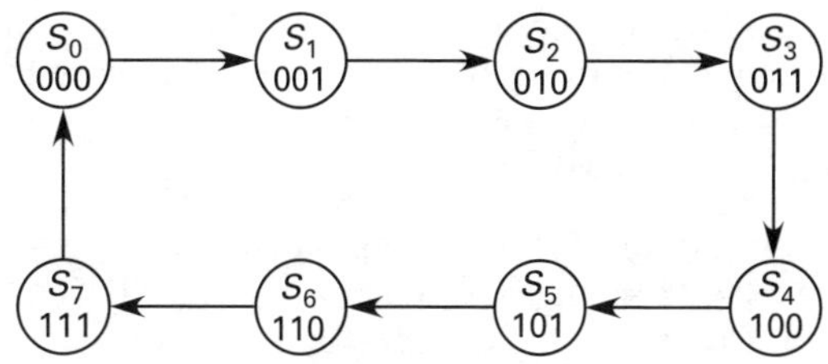

(c) state diagram

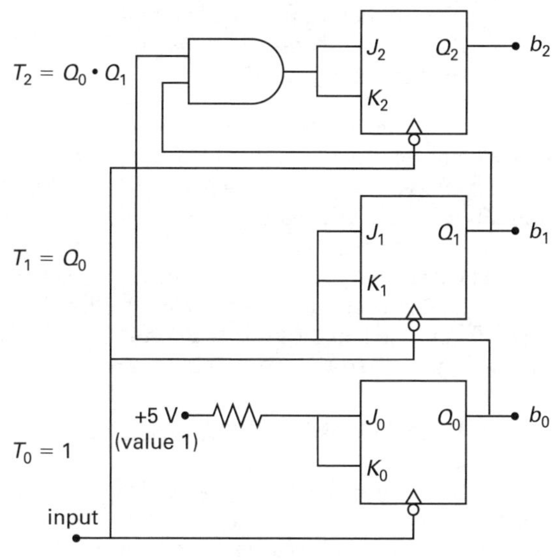

(d) realization

as does the stage 1 FF. The stage 2 FF is triggered from the output of the stage 1 FF. The process continues as shown through decimal 7 and then resets.

A ripple counter is *asynchronous* because the flip-flops are not actually triggered by the clock. Because the count ripples through the circuit, the propagation delay is the fundamental limitation of this type of counter. Counters using common clocking are required for count lengths more than a few bits long.

9. SYNCHRONOUS BINARY COUNTER

A counter is *synchronous* when all the flip-flops change states at the same time. When this occurs, a state table, referred to as an excitation table, can be created in which the *state* is determined by the Q outputs of all the flip-flops. An example of a synchronous binary counter is shown in Fig. 43.8.

The state table is constructed by listing the desired state sequence in order in the present-state column, as in Fig. 43.8(a). The next-state condition is also listed, along with the required inputs necessary to cause the transition. The logic necessary to generate the required inputs is normally determined using a Karnaugh map or Veitch diagram. For this counter, the logic can be determined by inspection and the resulting control equations are shown in Fig. 43.8(b). State diagrams, as shown in Fig. 43.8(c), are useful tools to visualize the transitions and aid in the design. The states are indicated by the S's and generally do not need to have the same subscripts as the counts. The realization of the network using *J-K* flip-flops configured as toggles is shown in Fig. 43.8(d).

A problem with this type of design is that more than one flip-flop can change state at the same time. As a consequence of differences in transition speeds, a false value can appear during the transition process. A false value can be avoided by allowing only one state variable to change during a clock period. This is equivalent to ensuring that state assignments are logically adjacent; that is, the next state is logically adjacent to the present state.[3]

10. ADJACENT-STATE BINARY COUNTER

A binary counter uses all 2^n states. Adjacent states can be assigned to all present states, and when done, the state variables will no longer correspond to the correct count. Logic circuitry must then be used to correct the output. An example of an adjacent-state binary counter is shown in Fig. 43.9.

An excitation table for the synchronous counter described in Sec. 43.9, rearranged to ensure logical adjacency, is shown in Fig. 43.9(a). The states are subscripted for the correct count and no longer match the binary count of the present state. The next state and the required inputs for D flip-flops are shown. In Fig. 43.9(b), Karnaugh maps are used to determine the control and output equations. The state diagram is given in Fig. 43.9(c). The circuit is realized in Fig. 43.9(d).

Redundancy exists in the circuit diagram because $Q_1 \cdot \overline{Q}_0$ is common to D_1 and D_2. Additionally, when using D flip-flops, the last column of Fig. 43.9(a) can be eliminated because the value of D is always the next-state value for the corresponding state variable. Even for this relatively simple circuit, numerous logic lines clutter the realization. Because of this, feedback lines are normally omitted, and the state variables themselves are shown as inputs to the various gate inputs.

11. COUNTER ANALYSIS PROCEDURE

Though each design situation will be unique, the general procedure in Sec. 43.1 can be used in the analysis or design of counters. The network specifications (step 1) are given information; they comprise the requirements of the system. The state table size (step 2) is determined by the system requirements as well. Minimization of the state table (step 3) is accomplished through the removal of redundant or unnecessary states from the table. The state assignments (step 4) are made using heuristic and advanced analytical techniques necessary for the large number of possibilities (see Sec. 43.12 for general rules). Once in binary form, the next-state equations for the chosen flip-flop (see Table 43.1) are used to fill in the next-state column of the state table. With the present and next states known, Karnaugh maps, Veitch diagrams, or other appropriate methods are used to obtain the control and output equations. The circuit is obtained (step 5) by constructing the logic elements necessary to realize the control and output equations for the given flip-flop. A check should also be done to ensure that the circuit realizes the requirements of step 1 and does not have *hang-up states*. Hang-up states, also called *lockout states*, are two or more unused states that have no next state in the desired sequence. A circuit check and search for hang-up states are efficiently accomplished using a state diagram and timing diagram.[4]

The following counter circuit analysis illustrates the principles involved. The focus for analysis, however, is determining the state table and developing the state diagram and timing diagram. Consider the counter in Fig. 43.10 with the control equations indicated. Because the system is already realized, all the design steps are complete.

Complete the present-state column of a state table, starting at zero, with the total number of states appropriate for the number of flip-flops used. The present-state column of Table 43.2 is the result. The states are listed in logical adjacency order (see Sec. 43.10 and Sec. 42.4). From the counter circuit in Fig. 43.10 and the next-state equations from Table 43.1, the next-state equations can be written as

$$\begin{aligned} Q_2^+ &= J_2 \cdot \overline{Q}_2 + \overline{K}_2 \cdot Q_2 \\ &= \overline{Q}_2 \cdot Q_1 \cdot \overline{Q}_0 + \overline{Q}_2 \cdot Q_2 \\ &= \overline{Q}_2 \cdot Q_1 \cdot \overline{Q}_0 \end{aligned} \quad 43.5$$

$$\begin{aligned} Q_1^+ &= S_1 + \overline{R}_1 \cdot Q_1 \\ &= Q_0 + (\overline{Q}_2 + Q_0) \cdot Q_1 \\ &= Q_0 + \overline{Q}_2 \cdot Q_1 \end{aligned} \quad 43.6$$

$$Q_0^+ = D_0 = \overline{Q}_1 \quad 43.7$$

Equation 43.5 through Eq. 43.7 generate the next-state column of Table 43.2.

Examining Table 43.2 shows that the counter resets at state S_4. Therefore, this is a five-state counter. Three states are unused, which indicates that hang-up states are possible. Checking Table 43.2 shows that all the unused states return to the main sequence. This is easily visualized with the state diagram shown in Fig. 43.11. A final check of the circuit is made to ensure proper operation using the timing diagram in Fig. 43.12.

[3]Ensuring the next state is logically adjacent to the present state is not always possible, but it is a desirable design goal.

[4]State diagrams are often useful in the beginning of design problems to aid in determining the inputs to the state table and to visualize the system requirements.

Figure 43.9 *Adjacent-State Binary Counter*

state	present state $Q_2Q_1Q_0$	next state $Q_2^+Q_1^+Q_0^+$	outputs $b_2b_1b_0$	D inputs $D_2D_1D_0$
S_0	000	001	000	001
S_1	001	011	001	011
S_2	011	010	010	010
S_3	010	110	011	110
S_4	110	111	100	111
S_5	111	101	101	101
S_6	101	100	110	100
S_7	100	000	111	000

(a) excitation table

D_2

Q_2Q_1 \ Q_0	0	1
00	0	0
01	1	0
11	1	1
10	0	1

$D_2 = Q_2 \cdot Q_0 + Q_1 \cdot \overline{Q}_0$

b_2

Q_2Q_1 \ Q_0	0	1
00	0	0
01	0	0
11	1	1
10	1	1

$b_2 = Q_2$

D_1

Q_2Q_1 \ Q_0	0	1
00	0	1
01	1	1
11	1	0
10	0	0

$D_1 = \overline{Q}_2 \cdot Q_0 + Q_1 \cdot \overline{Q}_0$

b_1

Q_2Q_1 \ Q_0	0	1
00	0	0
01	1	1
11	0	0
10	1	1

$b_1 = Q_2 \oplus Q_1$

D_0

Q_2Q_1 \ Q_0	0	1
00	1	1
01	0	0
11	1	1
10	0	0

$D_0 = \overline{Q_2 \oplus Q_1}$

b_0

Q_2Q_1 \ Q_0	0	1
00	0	1
01	1	0
11	0	1
10	1	0

$b_0 = Q_2 \oplus Q_1 \oplus Q_0$

(b) control and output equations

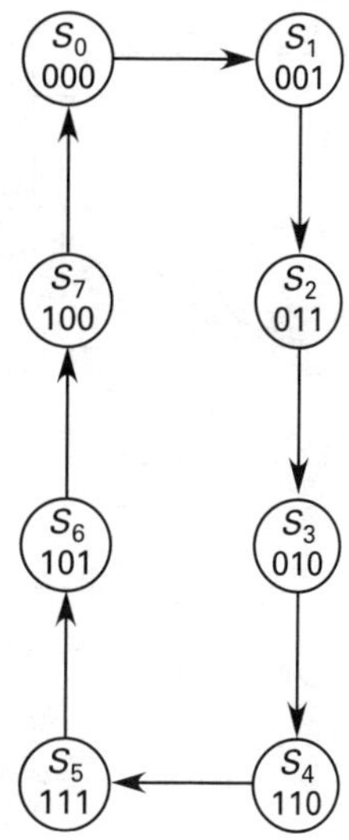

(c) state diagram

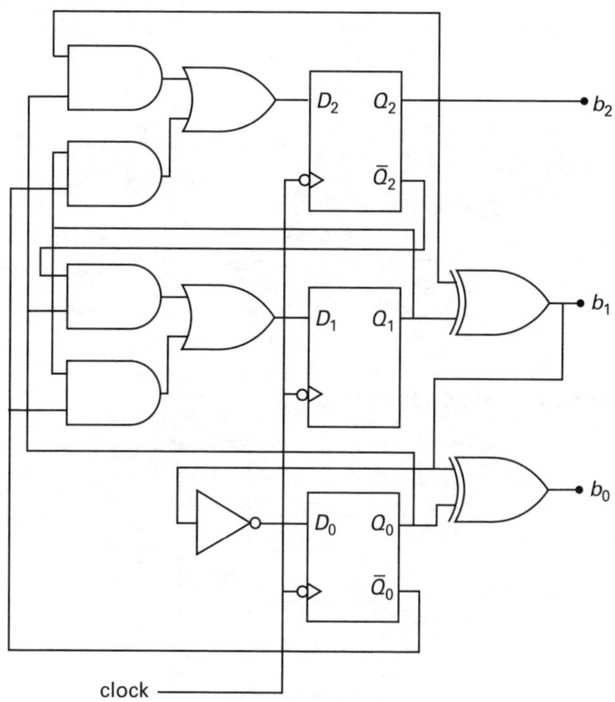

(d) realization

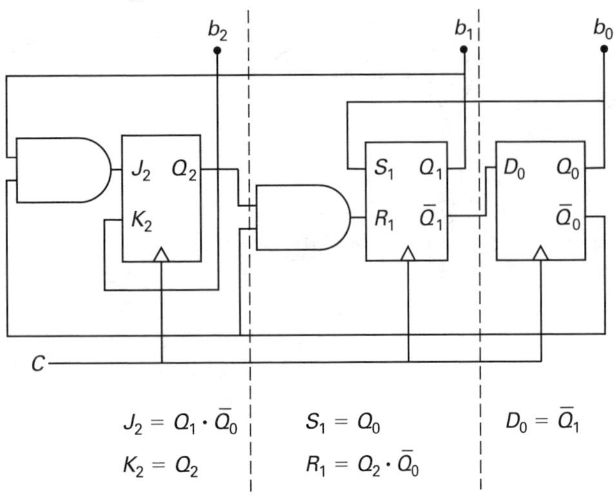

Figure 43.10 *Sample Counter*

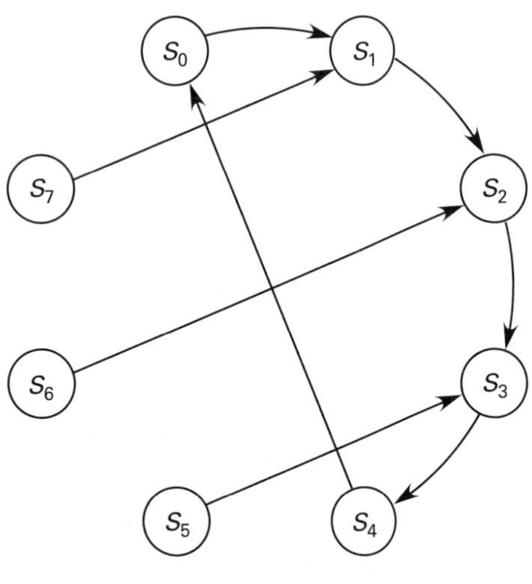

Figure 43.11 *State Diagram for Sample Counter*

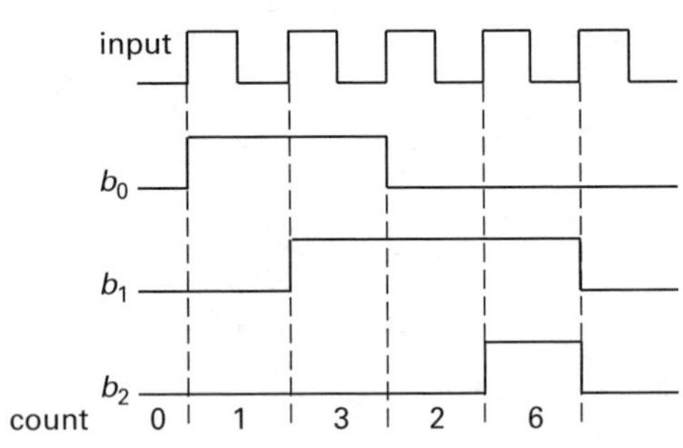

Figure 43.12 *Timing Diagram for Sample Counter*

12. SEQUENCE DETECTION

In general, a sequential circuit that operates in a predefined sequence with the inputs monitored using a clock on the same frequency as the outputs (i.e., with synchronous outputs) is called a *state machine*. If the state machine's next-state condition is determined by the current states only, it is called a *Moore machine*. If

Table 43.2 *State Table for Sample Counter*

state	present state $Q_2Q_1Q_0$	next state $Q_2^+Q_1^+Q_0^+$
S_0	0 0 0	0 0 1
S_1	0 0 1	0 1 1
S_2	0 1 1	0 1 0
S_3	0 1 0	1 1 0
S_4	1 1 0	0 0 0
S_5	1 1 1	0 1 0
S_6	1 0 1	0 1 1
S_7	1 0 0	0 0 1

the state machine's next-state condition is determined by the current state and the circuit inputs, it is called a *Mealy machine*. A state machine that determines when a specific input sequence has occurred is called a *sequence detector*. The design of the detector follows the general guidelines in Sec. 43.1 and Sec. 43.11.

As an example, consider the design of a sequence detector used to determine when 1001 occurs. The state diagram is shown in Fig. 43.13.

Figure 43.13 *State Diagram for Sequence 1001 Detector*

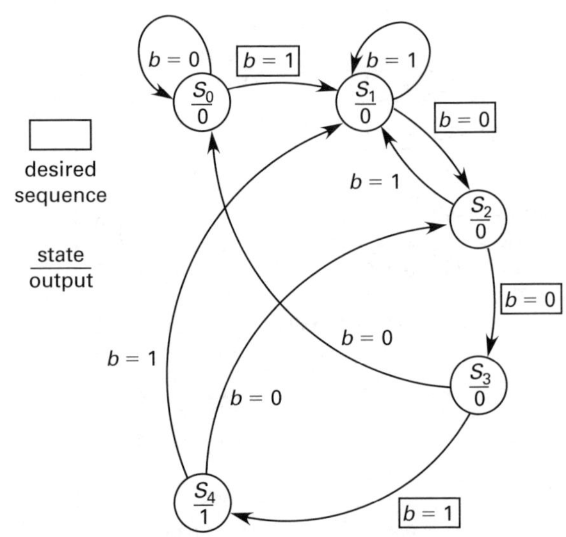

The detector starts in state S_0, which is the condition indicating that the previous input was not in the correct sequence. This is called the *start-up state*. The detector remains in S_0 as long as zeros are obtained. This is the reason for the loop labeled $b = 0$ at S_0. When a one is received, the detector moves to state S_1, which is the condition indicating the start of the sequence.

The detector remains in S_1 as long as ones are received. This is the reason for the loop labeled $b = 1$ at S_1. When a zero is received, the detector moves to state S_2, which indicates that a one followed by a zero has been received. If an input of one is now received, the detector must start over at state S_1. If a zero is received, the detector moves to state S_3, which is the condition indicating that the sequence 100 has been received. Either way, the detector remains in S_2 for one clock cycle only.

The detector also remains in S_3 for only one clock cycle. When a zero is received, the detector starts over at state S_0. If an input of one is now received, the detector moves to S_4, which indicates the sequence 1001 has been received. At S_4, the detector output must change to one to indicate the detection of the sequence. This is indicated in the state diagram underneath the state variable within the circle that indicates the state. At S_4 either a zero or a one input moves the detector out of the state. For a zero, the detector moves to S_2 since this indicates that a one followed by a zero has been received. For a one, the detector moves to S_1 since this indicates that a one has been received. The data is shown in Table 43.3.

Table 43.3 *State Table for Sequence 1001 Detector*

present state	next state for input b: $b=0$	$b=1$	present output
S_0	S_0	S_1	0
S_1	S_2	S_1	0
S_2	S_3	S_1	0
S_3	S_0	S_4	0
S_4	S_2	S_1	1

The state assignments must now be made. This process is complicated and a large number of possibilities exist. The following general rules can be used when assigning binary coding to the present state values to maximize logical adjacency and minimize logic gates.

- *Rule 1:* States that have the same next states for the same input should be given adjacent assignments.
- *Rule 2:* States that are the next state of the same state should be given adjacent assignments.
- *Rule 3:* States that have the same outputs for a given input should be given adjacent assignments.

Using these rules, the adjacency weighting for the sequence 1001 detector is shown in Table 43.4. The matrix lists the states and the number(s) of the rule(s) indicating that the states should be adjacent.

The number(s) of rule(s) applicable sets the priority of the adjacency assignments. Assuming equal weight among the rules, S_1 and S_2 are the first priority for adjacency. S_1 and S_0 are the second priority, followed by all the states with two rules satisfied. The lowest priorities are S_2, S_3, and S_2, S_4. Not all the groupings are possible. It is common to set S_0 to the zero state (in this case, 000). Given this, one efficient grouping is shown in the present-state column of Table 43.5.

Table 43.4 *Adjacency Weighting for Sequence 1001 Detector*

	S_4	S_3	S_2	S_1
S_0	1, 2	1, 3	1, 3	1, 2, 3
S_1	1, 1	2, 3	1, 2, 2, 3	
S_2	1	3		
S_3				

Using the next-state equations of Table 43.1, the next-state column of Table 43.5 is filled out. The output equation is simply $Y = Q_2$. The control equations, that is, the flip-flop input equations, are determined from the Karnaugh maps derived from the state table as shown in Fig. 43.14.

Using the control equations, the network is realized as shown in Fig. 43.15.

Figure 43.14 *Control Equations for Sequence 1001 Detector*

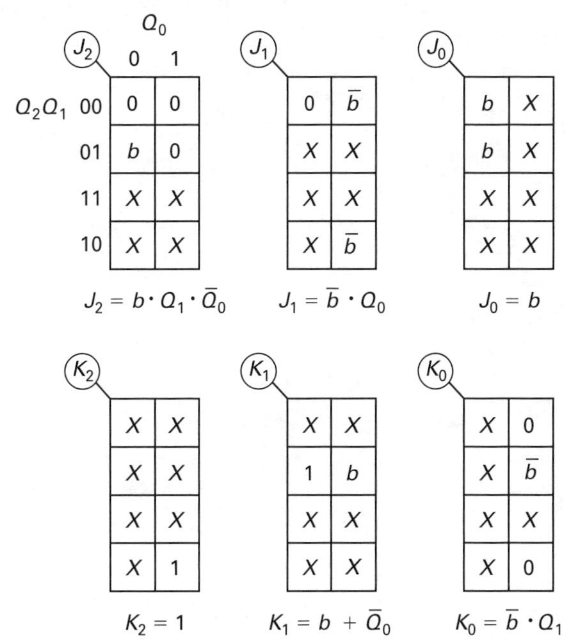

Table 43.5 *Expanded State Table for Sequence 1001 Detector*

state	present state	next state $b=0$	$b=1$	present output	flip-flop inputs					
S	$Q_2Q_1Q_0$	$Q_2^+Q_1^+Q_0^+$	$Q_2^+Q_1^+Q_0^+$	Y	J_2	K_2	J_1	K_1	J_0	K_0
S_0	0 0 0	0 0 0	0 0 1	0	0	X	0	X	b	X
S_1	0 0 1	0 1 1	0 0 1	0	0	X	$\overline{b}$	X	X	0
S_2	0 1 1	0 1 0	0 0 1	0	0	X	X	b	X	$\overline{b}$
S_3	0 1 0	0 0 0	1 0 1	0	b	X	X	1	b	X
S_4	1 0 1	0 1 1	0 0 1	1	X	1	$\overline{b}$	X	X	0

Figure 43.15 *Realization of Sequence 1001 Detector*

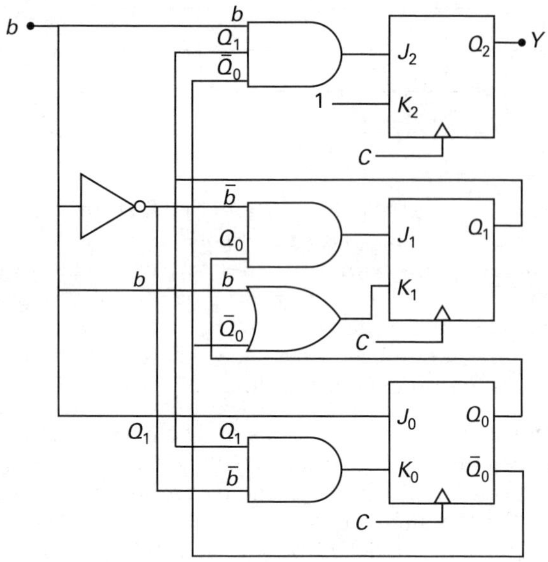

44 Digital Systems

1. Digital Systems Defined 44-1
2. Interface Fundamentals 44-1
3. Data Conversion Basics 44-2
4. Digital-to-Analog Conversion 44-2
5. Ladder-Type D/A Converter 44-3
6. Multiplying-Type D/A Converter 44-5
7. Counting A/D Converter 44-5
8. Parallel-Comparator A/D Converter 44-6
9. Dual-Slope A/D Converter 44-7
10. Protocols 44-7

Nomenclature

b	bit (zero or one)	–
C	capacitance	F
I	current	A
n	number of bits	–
N	n-bit number	–
R	resistance	Ω
t	time	s
V	effective or DC voltage	V

Subscripts

a	analog
d	digital
fb	feedback
LSB	least significant bit
max	maximum
ref	reference
s	sampled

1. DIGITAL SYSTEMS DEFINED

A digital system is defined as any of the levels of operation for a digital computer. General digital system architecture is shown in Fig. 44.1.

Figure 44.1 General Digital System Architecture

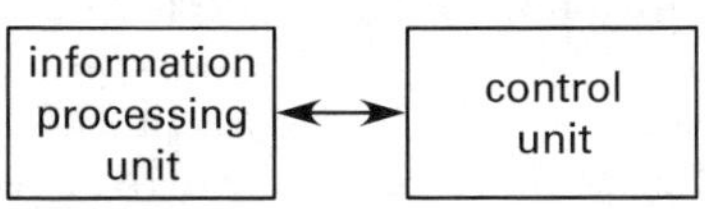

An essential difference is that the *information processing unit* in a digital system is designed to execute a specific sequence of operations in a fixed program. A *central processing unit* (CPU) in a computer system is capable of executing a varying sequence of operations without either the CPU or the control unit having to be redesigned for each sequence.

A digital system comprises the electronic elements in Chap. 37 through Chap. 40, arranged for logic operations, as described in Chap. 40, and is designed to accomplish logic tasks as described in Chap. 41 and Chap. 42.

2. INTERFACE FUNDAMENTALS

Digital systems process information at high rates. Output devices, such as printers, and input devices, such as humans, process information at considerably lower rates. The process of matching the flow of information between portions of a digital system is called the *interface problem.* The general solution to interface issues is shown in Fig. 44.2.

Incoming or outgoing information is temporarily stored in a *buffer*, which is a synchronizing element, that is, a register for the temporary storage of information. The *transfer pulse* is used to control the rate of information flow. A *flag* is used to indicate the occurrence of a given condition, which the input/output device (that is, the encoder) and the digital system use to determine when the next transfer pulse can be issued. Using a flag to control the interchange of information between the devices is called *handshaking.*

Two sets of standards are suggested by the interface model of Fig. 44.2. The first is that of *standard user interfaces.* Such standards determine the implementation of the interface and are *technical standards.* Technical standards for interfaces are usually de facto standards determined by dominance in the marketplace or consensus within the industry. The second is that of *standards for user interfaces.* Such standards determine the human-computer interaction and are *ergonomic standards.* Ergonomic standards are usually determined by standards-making bodies and do not involve guidance for implementation or specific programming requirements.

The input/output (I/O) interface is called the *surface* of the software. Interaction takes place at this surface using command names, shortcut keys, symbols, voice input, or *direct manipulation*—which is an action of the user on an object on the screen, such as pointing, clicking, moving, or changing said object. This type of interaction occurs through a *dialogue interface.* If a user interacts by changing data, rearranging or configuring electronic tools, or generating macros or commands, the interface is called a *tool* or *application interface.* The interface between a musical instrument and a computer occurs via a *musical instrument digital interface* (MIDI). A MIDI keyboard stores information about notes, note values, key pressure, control signals,

Figure 44.2 *Hardware Interface Overview*

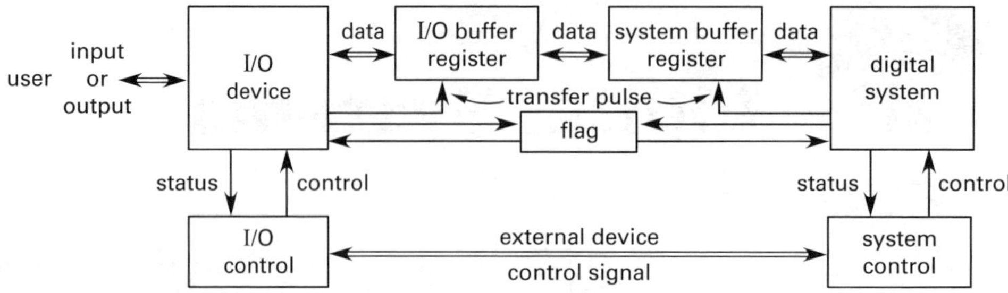

and so on, that is passed through the interface and can be reproduced by the system.

3. DATA CONVERSION BASICS

Hardware devices used with digital systems often operate on analog data. Input sensors often obtain data about the physical variable measured in terms of an analog voltage but analyzed by digital systems. Control systems on digital system input and output devices accept digital data but require analog voltages to function. Conversions are necessary. The conversion from digital signal voltage to analog voltage is called *digital-to-analog* (D/A) *conversion*. Similarly, the conversion from analog voltage to digital signal voltage is called *analog-to-digital* (A/D) *conversion*.

Analog signals contain a continuous range of values. Digital signals contain discrete values. The accuracy of an analog signal is ideally infinite. The digital signal approximates the analog signal, introducing error, though the error is extremely small in modern digital systems. *Quantization* is the process of approximating a continuous range of values by a smaller set of discrete values. The difference between the actual analog value and the discrete value is the *quantization error*. The accuracy of the digital signal is dependent on the *quantization resolution* of the A/D converter, that is, the number of bits in the output.[1]

Digital information is considered to be binary and weighted according to its position as a power of two. The value of a positive n-bit binary number N is

$$N = b_{n-1}2^{n-1} + b_{n-2}2^{n-2} + \cdots + b_2 2^2 + b_1 2^1 + b_0 2^0 \qquad 44.1$$

The binary bits, represented by b, have a value of zero or one. When coded in binary, N is[2]

$$N = b_{n-1}b_{n-2}\cdots b_2 b_1 b_0 \qquad 44.2$$

[1]For example, compact disc (CD) audio is sampled at 44.1 kHz and quantized with 16 bits (2 bytes). Therefore, the digital signal can take on one of 65,536 (that is, 2^{16}) values.

[2]The b occurs in parallel fashion or serial fashion with a storage register.

[3]The value represented by Eq. 44.3 is also the most steps that can occur in a digital system controlled by an n-bit circuit.

The digit b_{n-1} is the most significant bit (MSB) while b_0 is the least significant bit (LSB). The maximum value that can be represented by the code in Eq. 44.2 is[3]

$$\text{maximum value} = 2(\text{MSB}) - \text{LSB} = 2^n - 1 \qquad 44.3$$

4. DIGITAL-TO-ANALOG CONVERSION

A digital signal is considered to be low (digital zero) when the voltage is between 0.0 V and 0.8 V. A digital signal is high (digital one) when the voltage is between 2.0 V and 5.0 V.[4] As a result, these signals cannot be used directly for the conversion to analog information. Instead, the digital signal drives an electronic switch that connects the converter to a steady 0 V if the voltage is less than 0.8 V and a steady +5 V if the voltage is above 2.0 V. An example of a D/A converter using binary weighted resistors is shown in Fig. 44.3. The electronic switch, designated S in Fig. 44.3, can be any one of a variety of electronic devices as described in Chap. 40. Fig. 44.4 shows one possible configuration.

Example 44.1

What is the expression for V_{out} for the D/A converter in Fig. 44.3 in terms of the binary input?

Solution

Figure 44.3 is an 8-bit converter with a range of 10 V. The inputs to the electronic switches are the data bus lines. When a data line is high logic 1, the switch is in the upper position and connected to the +5 V supply. When a data line is low logic 0, the switch is in the lower position and connected to ground; that is, to 0 V. Superposition applies, so the techniques and assumptions for an ideal operational amplifier described in Chap. 45 determine the output of each binary bit. Then all outputs can be summed for the final solution. The output voltage for the MSB is determined by writing KVL at the inverting terminal.

$$\frac{0\text{ V} - V_7}{10\text{ k}\Omega} + \frac{0\text{ V} - V_{\text{out}}}{10\text{ k}\Omega} = 0$$

$$V_{\text{out}} = -\left(\frac{10\text{ k}\Omega}{10\text{ k}\Omega}\right)V_7$$

[4]The digital voltages mentioned are for TTL logic (see Chap. 41).

Repeating the procedure for each bit and summing gives the following.

$$\begin{aligned}V_{\text{out}} = &-\left(\frac{10\ \text{k}\Omega}{10\ \text{k}\Omega}\right)V_7 - \left(\frac{10\ \text{k}\Omega}{20\ \text{k}\Omega}\right)V_6\\ &-\left(\frac{10\ \text{k}\Omega}{40\ \text{k}\Omega}\right)V_5 - \left(\frac{10\ \text{k}\Omega}{80\ \text{k}\Omega}\right)V_4\\ &-\left(\frac{10\ \text{k}\Omega}{160\ \text{k}\Omega}\right)V_3 - \left(\frac{10\ \text{k}\Omega}{320\ \text{k}\Omega}\right)V_2\\ &-\left(\frac{10\ \text{k}\Omega}{640\ \text{k}\Omega}\right)V_1 - \left(\frac{10\ \text{k}\Omega}{1280\ \text{k}\Omega}\right)V_0\end{aligned}$$

When the binary input is one, the voltage is +5 V. When the binary input is zero, the voltage is 0 V. The output equation can be written as

$$\begin{aligned}V_{\text{out}} = &-5b_7 - 2.5b_6 - 1.25b_5 - 0.625b_4 - 0.3125b_3\\ &- 0.15625b_2 - 0.078125b_1 - 0.0390625b_0\end{aligned}$$

Figure 44.3 *D/A Converter: Binary Weighted Resistors*

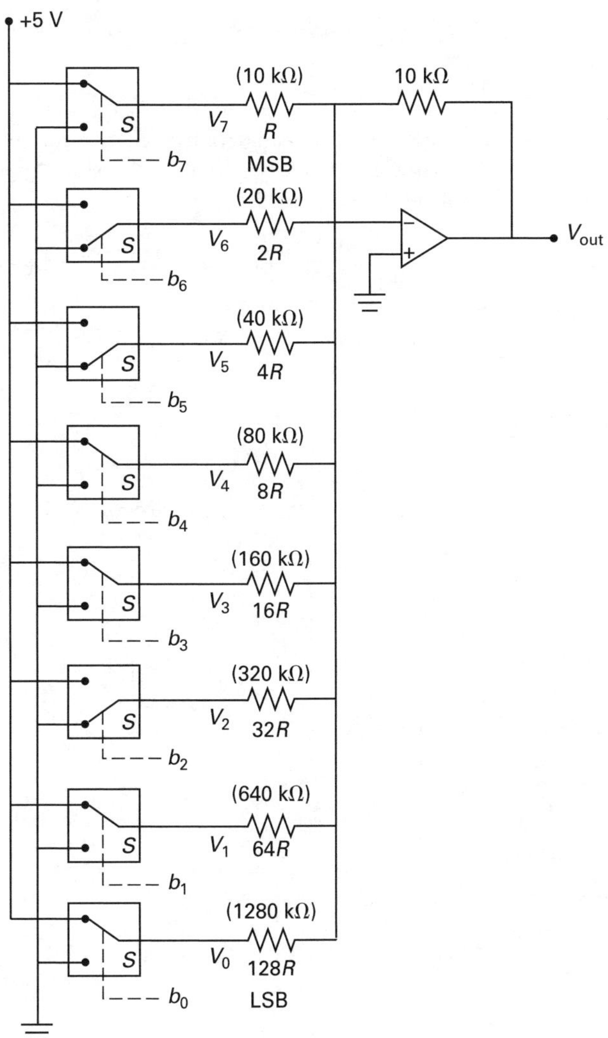

Figure 44.4 *Digitally Controlled Switch*

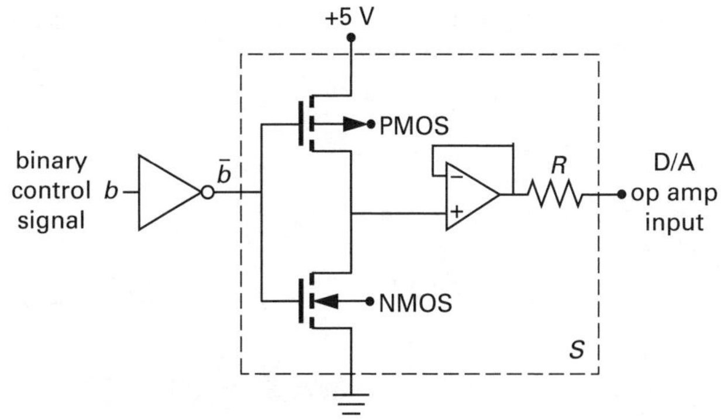

Example 44.2

What is the accuracy of the conversion for the D/A converter in Fig. 44.3?

Solution

The accuracy is always one-half the value of the least significant bit. Using the term for the LSB from the expression for V_{out} in Ex. 44.1 gives

$$\begin{aligned}\text{accuracy} &= \tfrac{1}{2}\,|V_{\text{LSB}}|\\ &= \left(\tfrac{1}{2}\right)(|-0.0390625|)\\ &= 0.01953125\ \text{V}\\ &= 19.5\ \text{mV}\end{aligned}$$

5. LADDER-TYPE D/A CONVERTER

A *ladder-type D/A converter* is based on a cell concept. Each succeeding cell has twice the weight of the previous one—hence the term "ladder."[5] Fig. 44.5(a) shows the design of the first cell, which corresponds to the LSB. A typical cell is shown in Fig. 44.5(b). Either the reference voltage, V_{ref}, or 0 V is applied, depending upon the electronic switch position.

Figure 44.5 *Cells of a Ladder-Type D/A Converter*

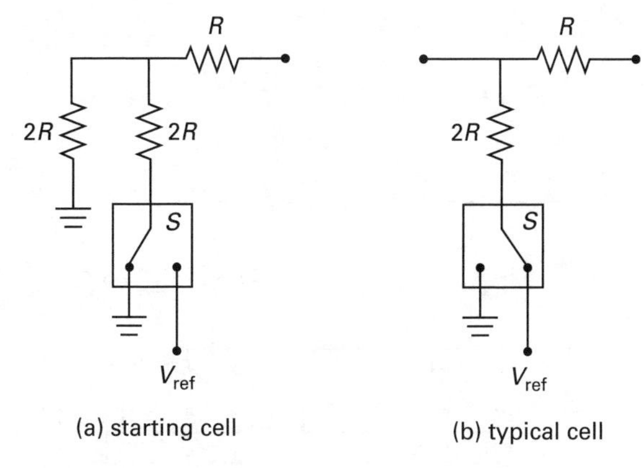

[5]The ladder-type D/A converter uses twice the number of resistors for a given number of bits as the binary weighted D/A converter, but the resistors used have only two values, R and $2R$.

Figure 44.6 shows the Thevenin equivalents of the starting cell, with the output taken as V_0; the succeeding cell, with the output taken as V_1; and the combination of the starting and the succeeding cells.

***Figure 44.6** D/A Converter Cells' Thevenin Equivalents*

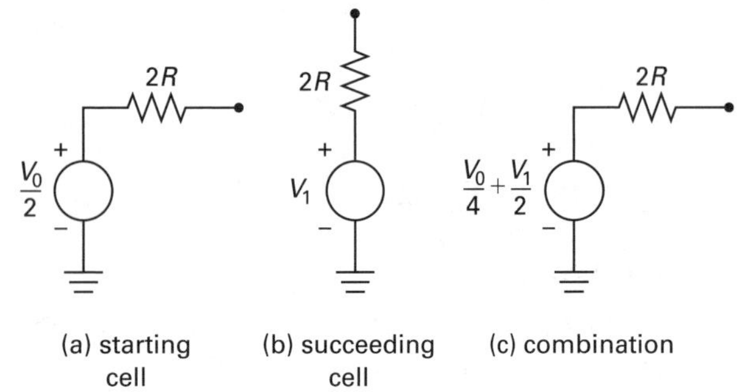

A ladder-type D/A converter with seven cells, identical to the number of cells in the converter in Sec. 44.4, has an output of

$$V_{\text{out}} = \frac{V_7}{2} + \frac{V_6}{4} + \frac{V_5}{8} + \frac{V_4}{16} + \frac{V_3}{32} + \frac{V_2}{64} + \frac{V_1}{128} + \frac{V_0}{256} \qquad 44.4$$

The voltages in Eq. 44.4 are either V_{ref} or 0 V, depending on the electronic switch settings for the particular bit in the ladder. The converter is completed by connection to the noninverting terminal of an operational amplifier, which provides buffering, as shown in Fig. 44.7.

When any switch in Fig. 44.7 closes, a transient occurs. Because of the capacitance from the nodes to the ground, a propagation delay occurs that is longer for the switches physically located further from the noninverting terminal. That is, the delay for S_0 (LSB) is longer than the delay for S_7 (MSB). These transients and delays are avoided by using an *inverting-type D/A converter*, which is shown in Fig. 44.8.

By inspection of Fig. 44.8, the current is

$$I = \frac{V_{\text{ref}}}{2R}\left(b_7 + \tfrac{1}{2}b_6 + \tfrac{1}{4}b_5 + \tfrac{1}{8}b_4 + \tfrac{1}{16}b_3 + \tfrac{1}{32}b_2 + \tfrac{1}{64}b_1 + \tfrac{1}{128}b_0\right) \qquad 44.5$$

Using ideal operational amplifier assumptions, all the current in Eq. 44.5 flows through the feedback resistor and none flows into the inverting terminal, making the output voltage

$$\begin{aligned} V_{\text{out}} &= -2RI \\ &= -V_{\text{ref}}\left(b_7 + \tfrac{1}{2}b_6 + \tfrac{1}{4}b_5 + \tfrac{1}{8}b_4 + \tfrac{1}{16}b_3 + \tfrac{1}{32}b_2 + \tfrac{1}{64}b_1 + \tfrac{1}{128}b_0\right) \end{aligned} \qquad 44.6$$

The maximum output occurs when all the binary input values are high (logic 1). In this condition, the magnitude of the maximum output for inverting ladder-type conductors of n-bit size is

$$|V_{\text{out,max}}| = V_{\text{ref}}\left(2 - \frac{1}{2^{n-1}}\right) \qquad 44.7$$

***Figure 44.7** Ladder-Type D/A Converter*

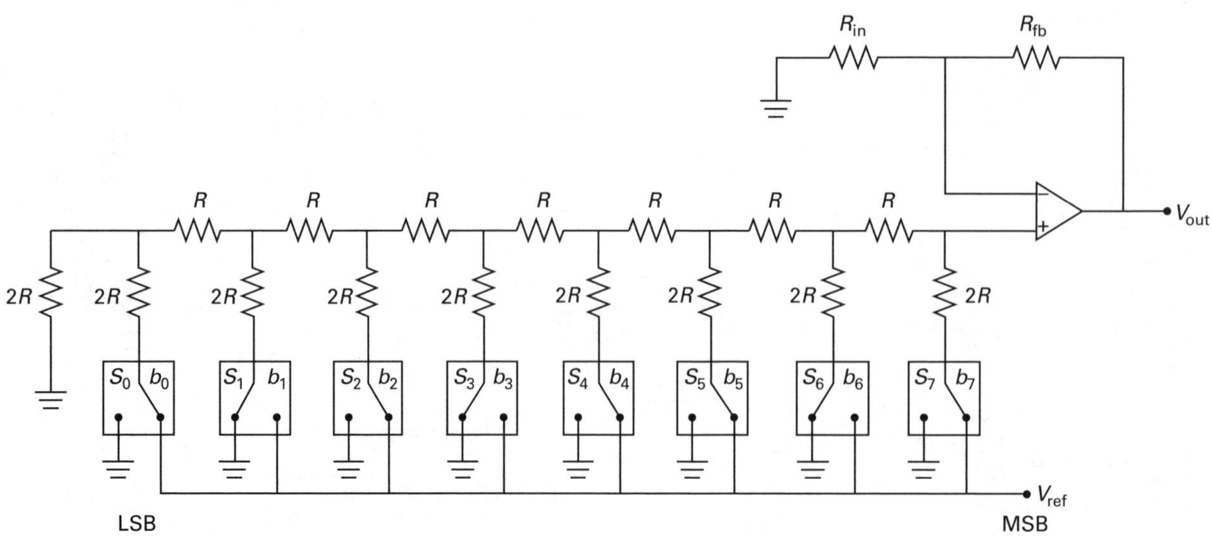

Figure 44.8 *Inverting-Type D/A Converter*

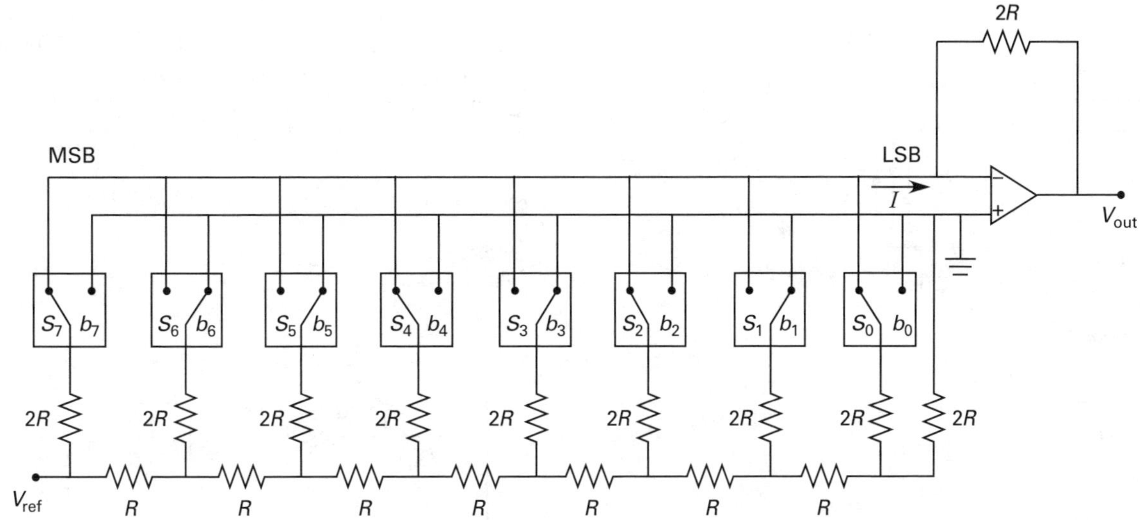

Example 44.3

What is the maximum output voltage for an 8-bit inverting ladder-type D/A converter with a reference voltage of 5 V?

Solution

For an 8-bit converter, $n=8$. Substituting into Eq. 44.7,

$$
\begin{aligned}
|V_{\text{out,max}}| &= V_{\text{ref}}\left(2-\frac{1}{2^{n-1}}\right) = (5\text{ V})\left(2-\frac{1}{(2)^{8-1}}\right)\\
&= (5\text{ V})\left(2-\frac{1}{128}\right)\\
&= 9.96\text{ V}
\end{aligned}
$$

6. MULTIPLYING-TYPE D/A CONVERTER

If the reference voltage of a noninverting ladder-type D/A converter (such as the one shown in Fig. 44.7) is replaced by an analog voltage signal, the output becomes a product of the analog signal and the binary input controlling the electronic switches. Such a device is called a *multiplying-type D/A converter.* This arrangement is also called a *programmable attenuator* because the output voltage is a fraction of the analog input with the attenuation controlled by the digital system or computer supplying the binary signal.

The *inverting-type D/A converter* (also called an *inverting ladder-type D/A converter*) of Fig. 44.8 can also be used as a programmable attenuator by changing the feedback resistance from $2R$ to R. The binary weighted D/A converter of Fig. 44.3 can also be used as a programmable attenuator by changing the feedback resistance from 10 kΩ to 5 kΩ. In both cases, the incoming voltage, V_{ref} or +5 V, is replaced by the analog input voltage that is to be attenuated.

7. COUNTING A/D CONVERTER

An analog-to-digital converter can be constructed as shown in Fig. 44.9(a).

The system shown in Fig. 44.9(a) is a *counting A/D converter.* The *sample-and-hold circuit* is used to avoid having the input change while conversion is occurring. The clock and comparator are inputs to the AND gate. The *comparator* is a high-gain differential amplifier whose output is high as long as the sampled analog signal is greater than the counter output, V_d.[6] When the comparator output is high, the AND gate passes the clock pulses to the binary counter. The binary counter sequences through binary numbers that comprise the output of the circuit. The digital output is also fed to a D/A converter whose output is attached to the inverting terminal of the comparator. When the magnitude of the D/A output, V_d, is above the sampled voltage, V_s, the comparator output changes to logic low and the clock signal is blocked from passing through the AND gate. The binary counter ceases to count. The digital output is proportional to the magnitude of the analog voltage sampled. The counter is then cleared in conjunction with another sample of the analog voltage, and the process repeats. The output of the circuit is illustrated in Fig. 44.9(b).

If an *up-down counter* is used, the clear signal and AND gate are not required. The counter counts up to the level of voltage sampled, forcing the comparator output low. The counter then counts down, but only by one LSB, at which point the comparator output goes high and the process repeats. The digital output ranges ±1 LSB around the correct value. Such a circuit is called a *tracking* or *servo converter.*

[6]The output of the comparator is designed for compatibility with the logic system used (i.e., 0 V for logic level low, and +5 V for logic level high in TTL systems).

Figure 44.9 *Counting A/D Converter*

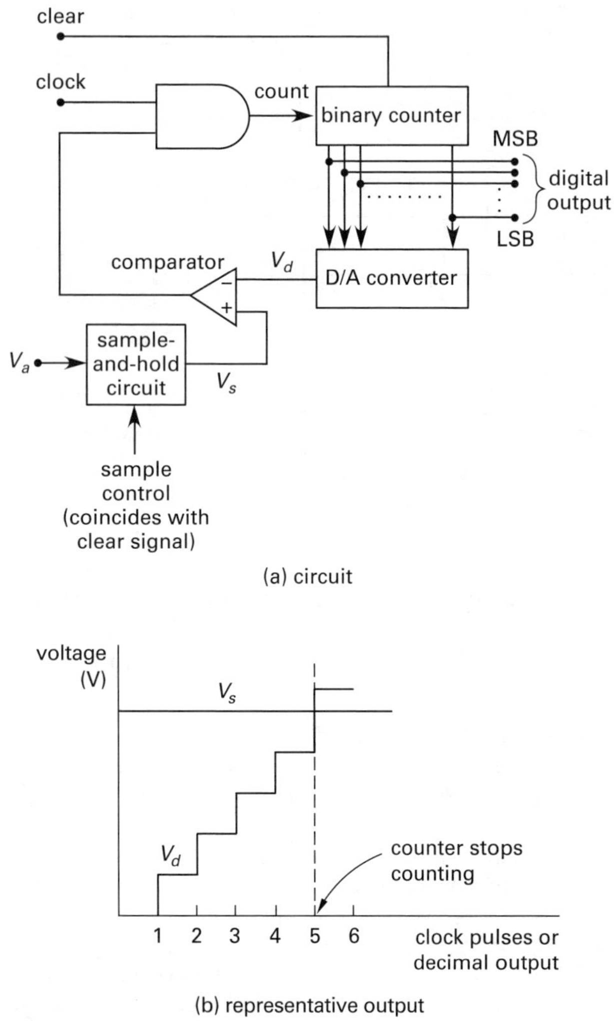

The *successive-approximation converter* is a variation of the tracking converter that replaces the up-down counter with a programmer. The programmer sets the MSB to logic level one and all other bits to zero. The comparator then checks the D/A converter output against the sampled analog signal. If the D/A converter output is larger, the one is removed from the MSB position and tried in the next position. This continues until the D/A converter output is smaller than the sampled analog signal, at which point the one remains in that bit position. For such an n-bit converter, n clock periods are required to obtain the digital output while 2^n pulses may be required for a counting-type A/D converter.

8. PARALLEL-COMPARATOR A/D CONVERTER

Another type of A/D converter is the *parallel-comparator*, which is the fastest of all converters. An example is shown in Fig. 44.10.

Figure 44.10 is a 3-bit A/D converter. The analog signal is applied simultaneously to parallel comparators with equally spaced thresholds. The signal is bracketed. That is, the output of the comparators is logic level zero for all comparators above the analog voltage input, and logic level one for all comparators below the analog voltage input. The signals on lines 1–7 are then priority-encoded to provide a 3-bit output. The truth table for the converter is given in Table 44.1.

Figure 44.10 *Parallel-Comparator A/D Converter*

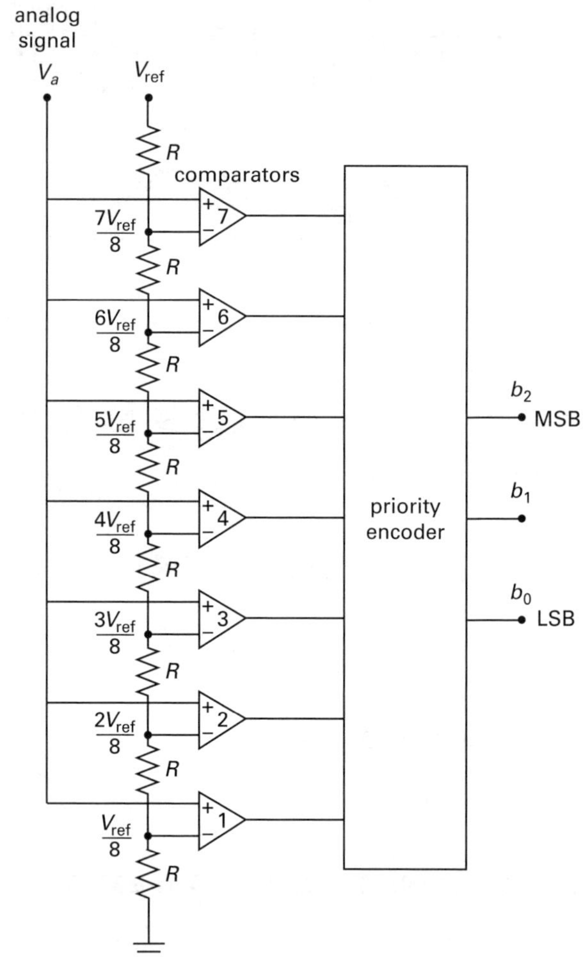

Table 44.1 *Parallel-Comparator A/D Converter Truth Table*

input							output		
7	6	5	4	3	2	1	b_2	b_1	b_0
0	0	0	0	0	0	0	0	0	0
0	0	0	0	0	0	1	0	0	1
0	0	0	0	0	1	1	0	1	0
0	0	0	0	1	1	1	0	1	1
0	0	0	1	1	1	1	1	0	0
0	0	1	1	1	1	1	1	0	1
0	1	1	1	1	1	1	1	1	0
1	1	1	1	1	1	1	1	1	1

9. DUAL-SLOPE A/D CONVERTER

The *dual-slope* or *ratiometric A/D converter* is shown in Fig. 44.11.

Figure 44.11 *Dual-Slope A/D Converter*

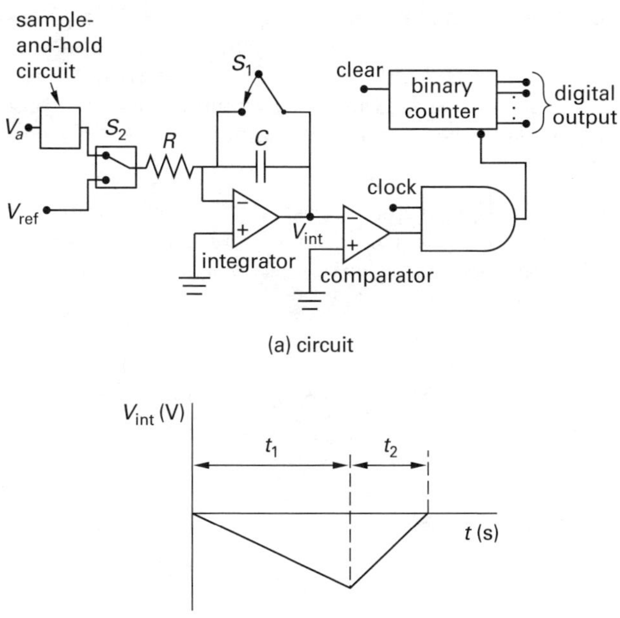

(b) representative integrator output

The positive input analog voltage is sampled and applied through electronic switch S_2 to the inverting terminal of an integrator. At the moment of sampling, switch S_1 closes to discharge the feedback capacitor. The counter is cleared at the same time, which is designated t_0. Because the output of the integrator is negative, the comparator output will be high. The clock signal is passed by the AND gate to the binary counter, which begins counting. The counter eventually reaches a state with all bits in logic level one, that is, high. The next clock pulse clears the counter and connects S_2 to the reference voltage. The process is triggered by the transition of the MSB from logic level one to logic level zero. At the time of connection to the reference voltage, designated t_1, the capacitor is charged to $-V_at_1/RC$. The integrator commences integration of the negative reference voltage with the output set by Eq. 44.8.

$$\begin{aligned} V_{\text{int}} &= \frac{-V_at_1}{RC} - \frac{V_{\text{ref}}(t-t_1)}{RC} \\ &= \frac{-V_at_1}{RC} + \frac{|V_{\text{ref}}|(t-t_1)}{RC} \end{aligned} \qquad 44.8$$

The reference voltage is negative in order to remove the charge buildup on the bypass capacitor.

As long as the integrator output remains negative, the comparator output is high and the clock pulse is passed to the counter. Once the integrator output becomes positive, the comparator output shifts to low, blocking the clock pulse and stopping the count. The termination of the count is designated t_2. The voltage at t_2 is found by setting the output voltage in Eq. 44.8 to zero.

$$V_a = \frac{|V_{\text{ref}}|(t_2-t_1)}{t_1} \qquad 44.9$$

The fraction $(t_2-t_1)/t_1$ is the ratio of the counter output at the time the comparator voltage reaches zero (which depends on the magnitude of the analog voltage sampled) to the full count. This ratio gives the ratiometric A/D converter its name. The final count is proportional to the analog voltage.

$$\frac{\text{final count}}{\text{full count}} = \frac{V_a}{V_{\text{ref}}} \qquad 44.10$$

The integrator output for a representative analog input is illustrated in Fig. 44.11(b). From this illustration, the reason for the "dual-slope" terminology may be seen.

10. PROTOCOLS

Any bus passing digital information is normally shared by several components. In order for each of these components to function properly, they must communicate in a standard manner, by a shared set of rules. These shared rules are called *protocols.*

The *asynchronous protocol* uses a handshaking technique to coordinate between components. For example, device A requests information. The information is found by device B. The address of the information is placed on the address bus, and an *acknowledge signal* is sent. Device A sees the acknowledge signal and stops making the request. It reads the address and waits for the data. Device B sends a *data ready signal* when the transfer is prepared. Device A acknowledges the data ready signal and begins transfer of the data. The process continues with the two components in lockstep responding to one another. The entire process has a self-timing nature. This *asynchronous transfer mode* (ATM) is insensitive to noise because of self-timing. Additionally, technological improvements, such as increases in speed of components, are easily handled because each component is keyed to the other and not to a clock. Furthermore, clock skew or timing is not an issue, allowing asynchronous busses to be physically longer than synchronous busses. The major disadvantage of asynchronous protocols is the significant overhead of such a mode. *Overhead* is the time a computer system spends performing computations (such as timing between components) that do not directly contribute to the progress of any user-assigned tasks. Asynchronous designs are used when it is necessary to accommodate numerous devices with a wide performance range and the ability to upgrade the system is important.

Synchronous protocol uses an explicit clock signal to control the flow of information on busses. The components respond to either the leading edge or trailing edge of the clock. The clock period is determined in a manner that will allow for propagation of all signals and allow the system to settle. Allowance must also be made for clock skew. *Clock skew* is a propagation delay of the clock signal throughout the digital system caused by differences in transmission lines, loading, and so on. Overall, clock frequency is limited by the speed of the logic elements used and the length of the bus (transmission) lines. Synchronous protocol is fast, requires fewer bus lines, and is easier to implement and test. The major disadvantage is that it is less flexible than asynchronous protocol, because all devices must operate at the same clock rate. Additionally, the clock rate is fixed and cannot be raised in a single component to take advantage of technological advances. Synchronous designs are used when it is necessary to connect a small number of highly coupled devices and speed is of paramount importance. For example, processor and memory subsystems are most often connected via synchronous buses.

The *split-transaction protocol* transmits information at the beginning and end of a transaction only. For this reason, only a request-and-reply transaction is transmitted. This protocol works only if more than one bus master is present and the memory system is sophisticated enough to track the overlapping transactions. Such a system has increased bus bandwidth over other protocols. The major disadvantage is that such a protocol is expensive to implement because each transaction must be identified and tracked. Such a protocol is used in systems with shared memories. Split-transaction protocol is also called *connect-disconnect*, *packet-switched*, or *pipelined protocol.*

Numerous other protocols exist and are implemented for a variety of purposes and applications, such as the *burst* or *diagram block-transfer protocol*, *cache coherence protocols*, and others.

Topic XI: Communications

Chapter

45. High-Frequency Transmission
46. Antenna Theory
47. Communication Links
48. Signal Formats
49. Signal Multiplexing
50. Communications Systems and Channels

45 High-Frequency Transmission

1. Fundamentals 45-1
2. High-Frequency Transmission Lines 45-2
3. Smith Chart Fundamentals 45-3
4. Smith Chart: Input Impedance 45-5
5. Smith Chart: Compensation Reactance 45-6
6. Impedance Determination using VSWR Measurements 45-7

Nomenclature

b	normalized susceptance	–
B	susceptance	S
C	capacitance	F
g	normalized conductance	–
G	conductance	S
I	effective or DC current	A
I	rms phasor current	A
l	length	m
L	inductance	H
n	integer	–
r	normalized resistance	–
R	resistance	Ω
v	velocity	m/s
V	effective or DC voltage	V
V	rms phasor voltage	V
x	distance variable	m
x	normalized reactance	–
X	reactance	Ω
y	normalized admittance	–
Y	admittance	S
z	normalized impedance	–
Z	impedance	Ω

Symbols

β	phase constant	rad/m
γ	propagation constant	rad/m
Γ	reflection coefficient	–
λ	wavelength	m
ϕ	angle	rad
ω	angular frequency	rad/s

Subscripts

0	characteristic
c	compensating
C	capacitor
cp	compensation point
g	generator
i	imaginary
l	per unit length
L	inductor
max	maximum
min	minimum
r	real
R	receiving end
R	resistance
S	sending end
unk	unknown

1. FUNDAMENTALS

Transmission lines are three-dimensional, being large in one direction and small in the other two. Power transmission lines operating at 60 Hz have a wavelength of 5000 km and a transverse dimension of about a meter, or less. Even the longitudinal dimension of a transmission line, which is rarely greater than 500 km, is a fraction of the wavelength. Coaxial lines are used as the frequency increases up to approximately 1 GHz. A *coaxial cable* or *line* is one in which the conductor is centered inside and insulated from an outer metal tube that acts as a second conductor. Above 1 GHz, the losses become prohibitive and waveguides are used. A *waveguide* is a device that constrains the propagation of an electromagnetic wave along a path predetermined by the physical construction of the guide. Waveguides are used up to approximately 100 GHz, where the losses become prohibitive. Above approximately 100 GHz, optical guides are used that operate at frequencies of approximately 2×10^{14} Hz with minimal losses.

Transmission lines are designed to transfer power from a source to a load. The distance between the source and load must be small compared to the wavelength in order to prevent the power from being radiated into space, which occurs when the wavelength is on the order of the conductor separating the source and load. Conductors specifically designed with dimensions comparable to the wavelength of the frequency are called *antennas.* High-frequency transmission lines and waveguides are used to connect transmitters (sources) and receivers (loads). The principles of high-frequency transmission must be understood in computer and digital system design to maximize power transfer and minimize power radiation.

Communications

2. HIGH-FREQUENCY TRANSMISSION LINES

At high frequencies, approximately 1 MHz and higher, wavelengths are shorter and even a few feet of line are treated as a long transmission line. The resistance is normally negligible, specifically, $R \ll \omega L$ and $G \ll \omega C$. Setting the resistance, R, equal to zero simplifies the governing equations. The impedance per unit length is[1]

$$Z_l = jX_{L,l} = j\omega L_l \quad (45.1)$$

The admittance per unit length is

$$\begin{aligned} Y_l &= jB_{C,l} \\ &= \frac{j}{X_{C,l}} = -\frac{1}{jX_{C,l}} = j\omega C_l \end{aligned} \quad (45.2)$$

The propagation constant, γ, in units of radians per meter (mile) is

$$\begin{aligned} \gamma &= \sqrt{Y_l Z_l} \\ &= \sqrt{(j\omega C_l)(j\omega L_l)} \\ &= j\omega\sqrt{L_l C_l} \\ &= j\beta \end{aligned} \quad (45.3)$$

The term β in Eq. 45.3 is the phase constant. The phase velocity, $\mathrm{v}_{\text{phase}}$, and the wavelength are

$$\mathrm{v}_{\text{phase}} = \frac{1}{\sqrt{L_l C_l}} \quad (45.4)$$

$$\lambda = \frac{2\pi}{\beta} = \frac{1}{f\sqrt{L_l C_l}} \quad (45.5)$$

The characteristic impedance, Z_0, is[2]

$$Z_0 = \sqrt{\frac{Z_l}{Y_l}} = \sqrt{\frac{j\omega L_l}{j\omega C_l}} = \sqrt{\frac{L_l}{C_l}} \quad (45.6)$$

The equations for long power transmission lines, Eq. 45.7 through Eq. 45.10, are applicable to high-frequency lines when the values from Eq. 45.1, Eq. 45.2, Eq. 45.3, and Eq. 45.6 are substituted.

$$\begin{aligned} V &= \tfrac{1}{2}V_R(e^{\gamma x} + e^{-\gamma x}) \\ &\quad + \tfrac{1}{2}I_R Z_0(e^{\gamma x} - e^{-\gamma x}) \\ &= V_R \cosh\gamma x + I_R Z_0 \sinh\gamma x \end{aligned} \quad (45.7)$$

$$\begin{aligned} I &= \tfrac{1}{2}I_R(e^{\gamma x} + e^{-\gamma x}) \\ &\quad + \tfrac{1}{2}\left(\frac{V_R}{Z_0}\right)(e^{\gamma x} - e^{-\gamma x}) \\ &= I_R \cosh\gamma x + \left(\frac{V_R}{Z_0}\right)\sinh\gamma x \end{aligned} \quad (45.8)$$

$$V_S = V_R\cosh\gamma l + I_R Z_0 \sinh\gamma l \quad (45.9)$$

$$I_S = I_R\cosh\gamma l + \frac{V_R}{Z_0}\sinh\gamma l \quad (45.10)$$

At high frequencies, however, circuits are designed to minimize reflected waves. This is accomplished by using capacitive compensation to make the load impedance appear to be the characteristic impedance. The expressions for the sending voltage and current are specified in terms of the reflection coefficient, Γ. The voltage traveling in the positive x direction toward the receiving end is associated with

$$V^+ = \tfrac{1}{2}V_R + \tfrac{1}{2}Z_0 I_R \quad (45.11)$$

Equation 45.11 is one of the constants from the solution of the wave equation, Eq. 45.12. The other constant is V^-. Both can be determined from the solution to the wave equation given by Eq. 45.7. The positive x direction is defined as the direction of power flow, that is, from the sending end to the receiving end.

$$\frac{\partial^2 V}{\partial x^2} = \gamma^2 V \quad (45.12)$$

The terms $\cosh j\beta x = \cos\beta x$ and $\sinh j\beta x = j\sin\beta x$, when combined with Eq. 45.11, allow the sending-end voltage and current to be written in terms of the reflection coefficient.

$$\begin{aligned} V_S &= V^+(e^{j\beta l} + \Gamma e^{-j\beta l}) \\ &= V^+\Big((1+\Gamma)\cos\beta l + j(1-\Gamma)\sin\beta l\Big) \end{aligned} \quad (45.13)$$

$$\begin{aligned} I_S &= \left(\frac{V^+}{Z_0}\right)(e^{j\beta l} - \Gamma e^{-j\beta l}) \\ &= \left(\frac{V^+}{Z_0}\right)\Big((1-\Gamma)\cos\beta l + j(1+\Gamma)\sin\beta l\Big) \end{aligned} \quad (45.14)$$

The input impedance is V_S/I_S and, in terms of the reflection coefficient, is given by

$$Z_{\text{in}} = Z_0\left(\frac{(1+\Gamma)\cos\beta l + j(1-\Gamma)\sin\beta l}{(1-\Gamma)\cos\beta l + j(1+\Gamma)\sin\beta l}\right) \quad (45.15)$$

The reflection coefficient is defined in Eq. 45.16. Using this definition, the input impedance can be written in terms of the load impedance, as in Eq. 45.17

$$\Gamma_L = \frac{V_{\text{reflected}}}{V_{\text{incident}}} = \frac{Z_{\text{load}} - Z_0}{Z_{\text{load}} + Z_0} \quad (45.16)$$

[1]The per-unit-length impedance and admittance are sometimes written with lowercase letters, z and y, to emphasize their per-unit qualities. In this chapter, the lowercase letters are used for normalized quantities.

[2]The characteristic impedance is the ratio of the voltage to the current on an infinite transmission line. It can be called the *characteristic resistance*, R_0, because the line is lossless. That is, Z_l and Y_l are purely reactive, making Z_0 purely resistive.

$$Z_{in} = Z_0 \left(\frac{Z_{load} \cos \beta l + jZ_0 \sin \beta l}{Z_0 \cos \beta l + jZ_{load} \sin \beta l} \right) \qquad 45.17$$

The voltage at any point on a transmission line is a function of the distance from the load because there are two waves traveling on any line, one incident to the load and one reflected from the load. The standing wave ratios defined by Eq. 45.18 and Eq. 45.19 measure the difference between the maximum rms voltage (current) and the minimum rms voltage (current) of the two waves. (The standing wave ratio is a physical variable that can be directly measured.) Where the magnitudes of the waves add, the voltage (current) is at a maximum. Where the magnitudes of the waves subtract, the voltage (current) is at a minimum. The reflection coefficient is defined in terms of the standing wave ratios by Eq. 45.20.

$$\text{VSWR} = \frac{V_{max}}{V_{min}} \qquad 45.18$$

$$\text{ISWR} = \frac{I_{max}}{I_{min}} \qquad 45.19$$

$$\Gamma = \frac{\text{SWR} - 1}{\text{SWR} + 1} \qquad 45.20$$

Rearranging Eq. 45.20 to define the voltage standing wave ratio, VSWR, in terms of the reflection coefficient gives

$$\text{VSWR} = \frac{|V^+| + |V^-|}{|V^+| - |V^-|} = \frac{1 + |\Gamma|}{1 - |\Gamma|} \qquad 45.21$$

The VSWR is related to the minimum and maximum impedance by Eq. 45.22 and Eq. 45.23.

$$Z_{max} = Z_0(\text{VSWR}) \qquad 45.22$$

$$Z_{min} = \frac{Z_0}{\text{VSWR}} \qquad 45.23$$

Example 45.1

A certain transmission line has a characteristic impedance of 50 Ω and a terminating resistance of 100 Ω. What is the reflection coefficient?

Solution

The reflection coefficient is defined by Eq. 45.16 as

$$\Gamma_l = \frac{V_{reflected}}{V_{incident}} = \frac{V_{reflected}}{V_{load}} = \frac{Z_{load} - Z_0}{Z_{load} + Z_0}$$

Substitute the given values.

$$\Gamma_l = \frac{Z_{load} - Z_0}{Z_{load} + Z_0} = \frac{100\ \Omega - 50\ \Omega}{100\ \Omega + 50\ \Omega} = 0.333$$

Example 45.2

For the transmission line described in Ex. 45.1, what is the input impedance if the line is an even number of wavelengths long?

Solution

The transmission line is restricted by the problem statement to be an even number of wavelengths for the frequency carried. In mathematical terms, with n as an integer, this is stated as

$$l = n\lambda = \frac{2\pi n}{\beta}$$

Equation 45.5 is used to change the wavelength, λ, into a function of the phase constant, β. Rearranging gives

$$\beta l = 2\pi n$$

This wavelength restriction causes the sinusoidal terms of the input impedance equation, Eq. 45.17, to be

$$\cos \beta l = 1$$

$$\sin \beta l = 0$$

Substituting these values into Eq. 45.17 gives

$$Z_{in} = Z_0 \left(\frac{Z_{load}}{Z_0} \right) = Z_{load} = 100\ \Omega$$

Example 45.3

For the transmission line in Ex. 45.1, what is the voltage standing wave ratio?

Solution

The VSWR can be found by substituting the calculated value of the reflection coefficient (determined in Ex. 45.1) into Eq. 45.21.

$$\text{VSWR} = \frac{1 + |\Gamma|}{1 - |\Gamma|} = \frac{1 + 0.333}{1 - 0.333} = 2$$

3. SMITH CHART FUNDAMENTALS

Electronic calculators and computer-aided design (CAD) programs have reduced the importance of graphical methods for the solution of transmission line problems.[3] Nevertheless, the Smith chart remains useful for explaining steady-state situations on transmission lines. A *Smith chart* is a special polar diagram with constant reflection coefficient circles, constant standing wave ratio circles, constant resistance circles, constant reactance arcs (portions of a circle), and radius lines

[3]A Smith chart is a graphical method for the solution of transmission line equations. A *bounce diagram*, also called a *time-distance plot*, is a graphical method for the solution of multiple reflections on transmission lines.

Communications

representing constant line-angle loci. A complete Smith chart is given in App. 45.A.[4] The Smith chart is essentially a polar representation of the reflection coefficient in terms of the normalized resistance and reactance. The normalization occurs with respect to the characteristic impedance, Z_0.

$$z = \frac{Z}{Z_0} = \frac{R + jX}{Z_0} = \frac{1 + \Gamma}{1 - \Gamma} \qquad 45.24$$

The normalized impedance (z), impedance (Z), resistance, reactance, and reflection coefficients are all functions of distance, normally measured with the load as the origin.[5] In the complex Γ plane, the curves of constant reflection coefficient magnitude, $|\Gamma|$, are circles as shown in Fig. 45.1(a). Constant standing wave ratios are also circles positioned exactly as shown in Fig. 45.1(a). Neither the constant reflection coefficient nor the constant standing wave ratio circles are shown on the final Smith chart. Instead, radial scales, shown in App. 45.A next to the chart, are used to determine their value.

The constant resistance circles are placed on the chart in the positions shown in Fig. 45.1(b). The origin of the chart is located at the intersection of the $r = 1$ circle and the line representing $x = 0$. The outermost circle is $r = 0$. The arcs of constant reactance are shown in Fig. 45.1(c). The constant resistance circles of Fig. 45.1(b) and the constant reactance arcs of Fig. 45.1(c) are combined to create the final Smith chart in App. 45.A. Because the circles of constant $|\Gamma|$ are not included, the value of $|\Gamma|$ is read from the left-hand external scale in App. 45.A. Because the circles of constant VSWR are not included, the value of VSWR is read from the right-hand external scale. Both values correspond to a point (r, x). There are three circumferential scales. The outer two measure fractions of a wavelength toward the generator (clockwise) or toward the load (counterclockwise). The full circumference represents one-half the wavelength. The inner circumferential scale measures the reflection coefficient angle, $\phi_L = \phi_R - \beta x$.[6]

Any circle drawn around the origin is called an *impedance locus circle*. At the origin, the impedance locus circle has a radius of zero, which is called the *matched condition*. At this point the total reactance is zero, the load impedance matches the characteristic impedance, making the normalized impedance equal to one, and the

Figure 45.1 *Smith Chart Components*

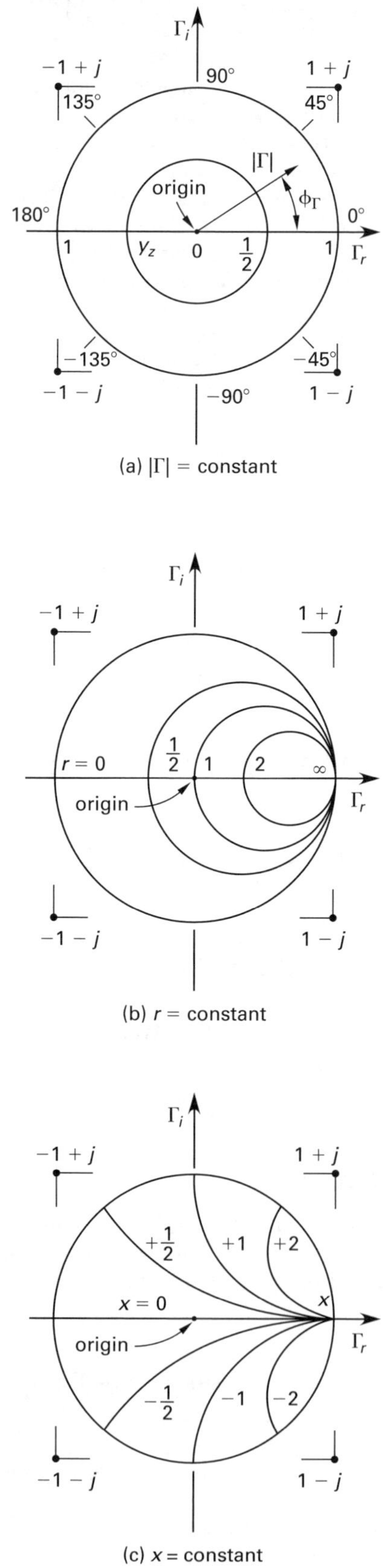

[4]The constant reflection coefficient circles and constant standing wave ratio circles are not plotted on the Smith chart. Instead, radial scales, shown in App. 45.A, are used to determine the value of the respective items as if a circle were drawn from the center of the Smith chart at a radius given by the scale.

[5]Do not confuse the distance x with the normalized reactance x. In some books, the distance is given the symbol d or the normalized reactance is given the symbol χ to avoid confusion. Here, the individual terms in Eq. 45.24 are not listed as functions of distance, to avoid the problem in this specific instance.

[6]The x used here is distance. When the total distance is from the load (receiving end) to the generator (sending end), the symbol l is used instead.

reflection coefficient equals zero. Typical values for other electrical conditions are shown in Table 45.1.

Table 45.1 *Smith Chart Electrical Conditions*

electrical condition	reflection coefficient (Γ)	normalized resistance (r)	normalized reactance (x)
open circuit	$1\angle 0°$	∞ (arbitrary)	arbitrary (∞)
short circuit	$1\angle 180°$	0	0
pure reactance	$1\angle \pm 90°$	0	± 1
matched line (pure resistance)	0	1	0

The Smith chart in App. 45.A and the components shown in Fig. 45.1 were arranged to show the real portion of the reflection coefficient along the standard x-axis and the imaginary component along the standard y-axis, subscripted as r and i, respectively. It is also common to rotate the chart 90° counterclockwise to place the $r=0$ and $x=0$ position, that is, the intersection of the zero resistance circle and the zero reactance line, at the top of the chart as shown in Fig. 45.2.

The Smith chart can be used for normalized admittances, y, by considering the r-circles as conductance circles (g-circles) and the x-arcs as susceptance arcs (b-arcs). The angle for the reflection coefficient, Γ, for a given admittance, y, is given by $180° + \phi_\Gamma$, and the point $y = 0 + j0$ represents an open-circuit condition.

The input impedance, compensating reactance, compensating position, reflection coefficient, and standing wave ratio can be graphically determined using the Smith chart.[7] The steps vary and are presented in examples in Sec. 45.4 and Sec. 45.5.

Figure 45.2 *Alternate Smith Chart Positioning*

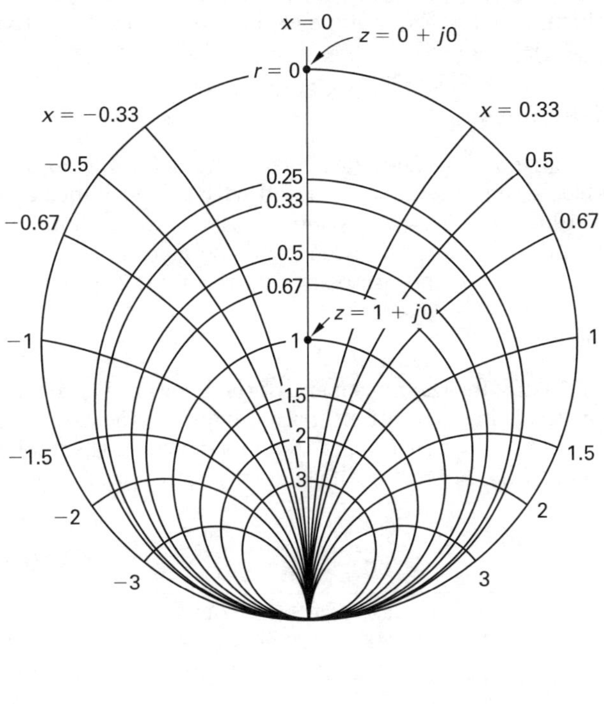

4. SMITH CHART: INPUT IMPEDANCE

The general procedure in finding the input impedance is to first determine the normalized load impedance, z_L, using Eq. 45.24. Plot the point z_L, representing the normalized resistance and reactance, at the intersection of the r circle and the x arc. Draw a circle centered at $1 + j0$ (the origin) through the value $r + jx$ corresponding to z_L. This is the impedance locus circle. Draw a radial line from the origin through z_L. Determine the angle βl and the normalized impedance a distance l from the load by following the impedance locus circle clockwise (toward the generator or sending end) through an angle of $2\beta l$.

Example 45.4

A transmission line with a length, βl, of 6 radians (344°) and a characteristic impedance, z_0, of 50 Ω is terminated with a load of $75 + j25$ Ω. What is the input impedance (i.e., the impedance seen by the generator)?

Solution

step 1: Calculate the normalized load impedance.

$$z_L = \frac{Z_L}{Z_0} = \frac{R_L + jX_L}{Z_0} = \frac{75 + j25\ \Omega}{50\ \Omega} = 1.5 + j0.5$$

step 2: Plot point $z_L = 1.5 + j0.5$ on the Smith chart.

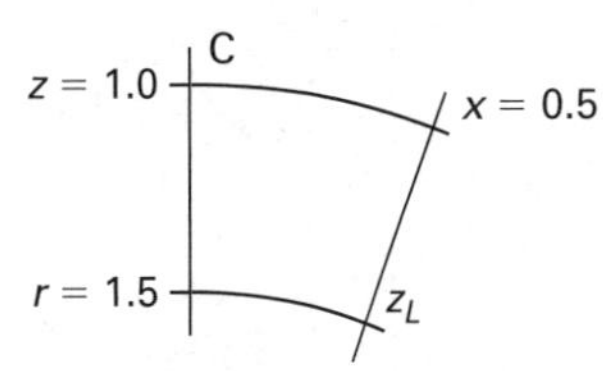

step 3: Draw a circle centered at $1.0 + j0$ (point C), that is, the origin, which passes through point z_L. This is the impedance locus circle.

[7]Though the reflection coefficient and standing wave ratio can be determined graphically, App. 45.A substitutes radial scales for the graphical procedure.

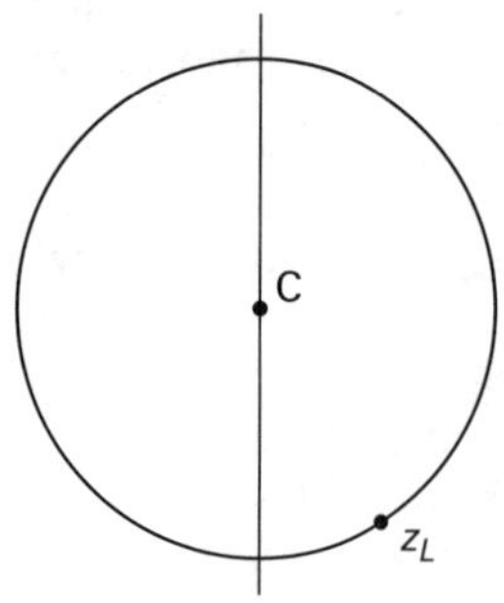

step 4: Extend a line from point C (the origin) through point z_L to the edge of the Smith chart, point L (load point).

step 5: Calculate the angular separation between the source and load points.

$$2\beta l = (2)(344°) = 688°$$

Because this is more than one revolution, subtract the full revolution.

$$688° - 360° = 328°$$

Move clockwise 328° from the load to the generator point.

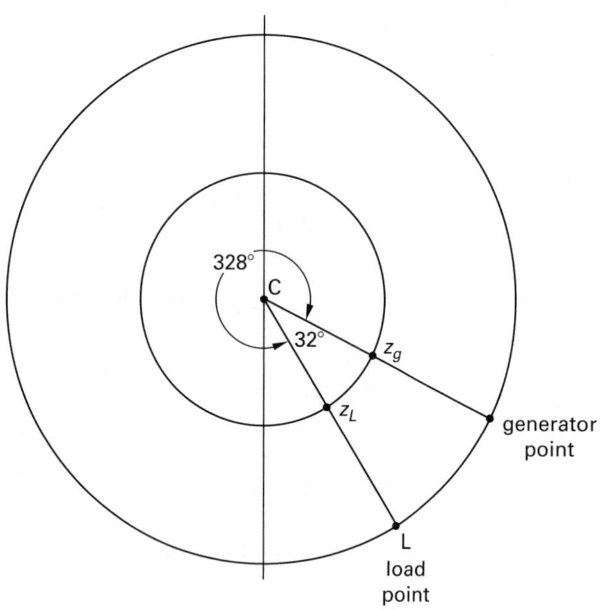

step 6: Read the impedance seen from the generator; that is, the input impedance at the intersection of the generator line and the impedance circle, point z_g.

$$z_g = 1.1 + j0.6$$

step 7: Calculate the input load impedance.

$$\begin{aligned} Z_g &= Z_0 z_g \\ &= (50\ \Omega)(1.1 + j0.6) \\ &= 55 + j30\ \Omega \end{aligned}$$

5. SMITH CHART: COMPENSATION REACTANCE

Compensation is used to make the total reactance of a given transmission line zero. Compensation is also called *impedance matching.* At low frequencies, the matching is accomplished with lumped reactive components (inductors and capacitors). At high frequencies, in order to minimize dissipation losses, a length of open- or short-circuited line is used. Compensation with a single line is called *single-stub matching.* Compensation with two lines is called *double-stub matching.*

The reactance required for compensation can be determined graphically on the Smith chart. The impedance locus circle is set up as before (see Ex. 45.4, steps 1 through 3). The line is traversed toward the source (clockwise) to the intersection of the impedance locus circle and the $r=1$ circle. The intersection is called the *compensation point.* At the compensation point, the impedance is

$$z_{\text{cp}} = 1 + jx \qquad 45.25$$

The compensation reactance is $-x$, which results in a total normalized impedance of $z=1$. Because capacitance is the most desirable compensating reactance for a variety of reasons, the compensation point is always chosen on the positive reactance side of the Smith chart. The positive reactance side of the chart is where $x > 0$, which makes the compensating reactance negative ($x < 0$).[8]

Example 45.5

A transmission line that has a characteristic impedance, Z_0, of 50 Ω is terminated with a load impedance of $100 + j100$ Ω. (a) Where in the transmission line should a compensating reactance be placed? (b) What is the value of the compensating reactance?

Solution

(a) Asking for the placement of the reactance is equivalent to asking at what electrical angle (βl) the capacitor should be placed to make the impedance at that point equal to Z_0. Following the suggested solution outline,

[8]The positive reactance side of the Smith chart is the top if positioned as in App. 45.A, and to the right if positioned as in the examples of this chapter.

step 1: Calculate the normalized load impedance.

$$z_L = \frac{Z_L}{Z_0} = \frac{100 + j100\ \Omega}{50\ \Omega} = 2 + j2$$

step 2: Plot the point z_L on the Smith chart.

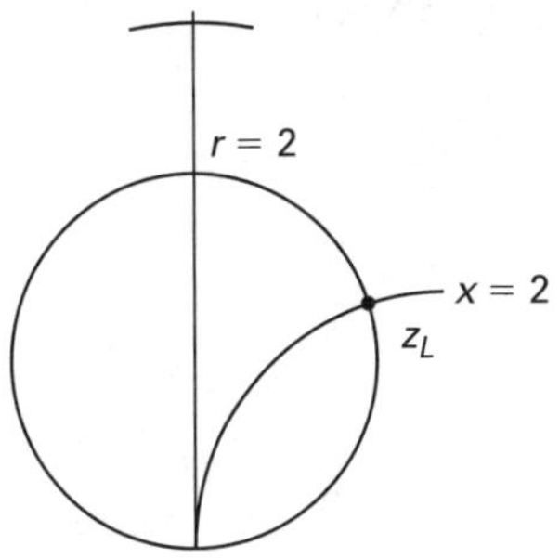

step 3: Draw a circle centered at $1 + j0$, that is, the origin, which passes through point z_L. This is the impedance locus circle.

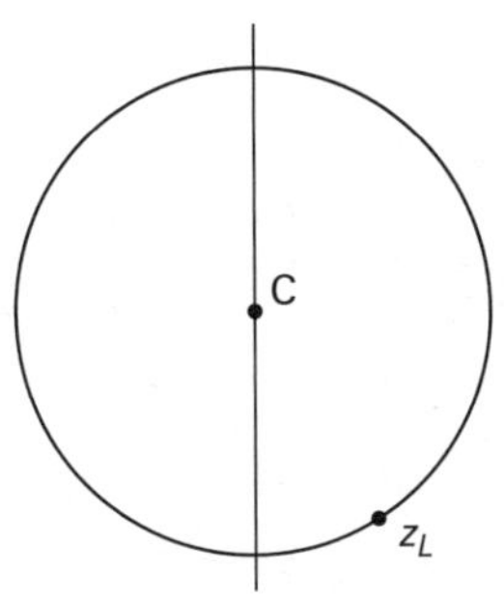

step 4: Extend a line from point C (the origin) through the load point z_L. From the load point, follow the impedance locus circle clockwise, toward the source to point CP (the compensation point), which is the intersection of the impedance locus circle and the $r=1$ circle on the positive reactance side of the Smith chart. (The angle traversed is $2\beta l$.)

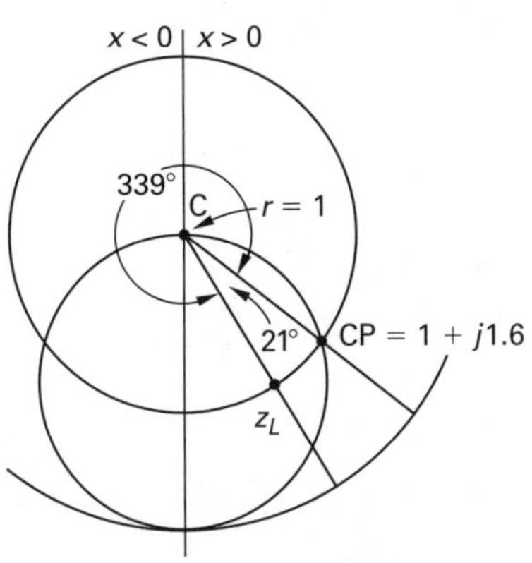

step 5: Measure the angular separation between the compensation and load points. In this case, the value is 21°. Because the electrical angle is always measured from the load toward the source, that is, clockwise, the angular separation is actually $360° - 21° = 339°$.

$$2\beta l = 339°$$

The electrical angle (βl) is

$$\beta l = \frac{339°}{2} = 169.5°$$

The compensation point is βl electrical degrees or $169.5°/360°$ wavelengths (0.471λ) from the load toward the source. The reactance should be placed a distance $l = 0.471\lambda$ from the load toward the source. The distance, l, can also be found using Eq. 45.5.

$$\lambda = \frac{2\pi}{\beta}$$

$$\beta\lambda = 2\pi$$

$$\beta l\lambda = 2\pi l$$

$$l = \left(\frac{\beta l}{2\pi}\right)\lambda = \left(\frac{169.5°}{360°}\right)\lambda = 0.471\lambda$$

step 6: Read the normalized impedance at the compensation point, z_{cp}.

$$z_{cp} = 1 + j1.6$$

(b) The normalized compensating reactance, x_c, must therefore be -1.6. The actual reactance is

$$X_c = Z_0 x_c = (50\ \Omega)(-1.6) = -80\ \Omega$$

6. IMPEDANCE DETERMINATION USING VSWR MEASUREMENTS

The voltage standing wave ratio (VSWR) can be measured using a *slotted line* in which a probe can be placed to determine the maximum and minimum voltages. The distance between a minimum and a maximum is $\lambda/4$ (one-quarter wavelength). The distance between two minimums or two maximums is $\lambda/2$ (one-half wavelength). The first minimum of the VSWR from the load is given the symbol l_{min}. With the load shorted, the distance from the short toward the generator of the first VSWR minimum determines the half-wavelength, which is the position where $\beta l = \pi$.[9] Using the measured VSWR, l_{min}, and the half-wavelength, an unknown impedance on a transmission line can be determined.

The voltage standing wave minimum occurs at the point of minimum impedance. Equation 45.17 can be written

[9]A short circuit is a minimum and is located at $0 + j0$ (see Table 45.1). A distance of one-half wavelength on the Smith chart is two circumferences (see App. 45.A), which places one at $0 + j0$ again—that is, at a minimum. The term βl is the argument of the sinusoidal functions that describe the voltage and will be at a minimum every π radians, or one-half wavelength.

for an unknown impedance, Z_{unk}, at a minimum input impedance as

$$\frac{Z_{\text{in,min}}}{Z_0} = \frac{\frac{Z_{\text{unk}}}{Z_0} + j\tan\beta l}{1 + j\left(\frac{Z_{\text{unk}}}{Z_0}\right)\tan\beta l} \qquad 45.26$$

Equation 45.23 is rearranged to explicitly show the relationship between the minimum impedance and the VSWR as

$$\frac{Z_{\text{min}}}{Z_0} = \frac{1}{\text{VSWR}} = \frac{Z_{\text{in,min}}}{Z_0} \qquad 45.27$$

Equation 45.26 and Eq. 45.27 are combined to determine an expression for the unknown impedance in terms of the VSWR, the minimum input impedance, and the minimum distance from the load at which the first minimum of the VSWR occurs.

$$Z_{\text{unk}} = Z_0\left(\frac{1 - j(\text{VSWR})\tan\beta l}{\text{VSWR} - j\tan\beta l}\right) \qquad 45.28$$

Communications

46 Antenna Theory

1. Antenna Fundamentals 46-1
2. Wave Propagation Fundamentals 46-2
3. Free-Space Wave Characteristics 46-3
4. Poynting's Vector and Power 46-3
5. Polarization 46-3
6. Interference 46-3
7. Reflection 46-4
8. Refraction 46-4
9. Diffraction and Scattering 46-4
10. Free-Space Basic Transmission Loss 46-5
11. Elemental Antenna 46-5
12. Elemental Antenna: Radiated Power 46-6
13. Elemental Antenna: Directivity 46-6
14. Elemental Antenna: Radiation Resistance 46-6
15. Elemental Antenna: Ohmic Resistance 46-6
16. Elemental Antenna: Efficiency 46-6
17. Elemental Antenna: Gain 46-7
18. Short Dipole Antenna 46-7
19. Dipole Antenna 46-8
20. Half-Wave Antenna 46-9

Nomenclature

A	vector magnetic potential	Wb/m
B	bandwidth	Hz
c	speed of light, 2.9979×10^8	m/s
D	directivity	–
D	electric flux density	c/m^2
E, **E**	electric field strength	V/m
f	frequency	Hz
G	gain	dB
h	antenna height (length)	m
H, **H**	magnetic field strength	A/m
I	rms phasor current	A
l	length	m
L	loss	dB
n	absolute index of refraction	–
p	power density	W/m^2
P	power or average power density	W or W/m^2
r	radial distance	m
R	resistance	Ω
S	Poynting's vector	W/m^2
t	time	s
T	temperature	K
v	velocity	m/s
dz	differential length of elemental antenna	m
Z_0	characteristic impedance	Ω

Symbols

α	angle	rad
β	phase constant	rad/m
δ	skin depth	m
ϵ	permittivity	F/m
ϵ_0	free-space permittivity, 8.854×10^{-12}	F/m
η	efficiency	%
θ	angle or polar angle	rad
κ	Boltzmann's constant, 1.3805×10^{-23}	J/K
λ	wavelength	m
μ	permeability	H/m
μ_0	free-space permeability, 1.2566×10^{-6}	H/m
ϕ	azimuthal angle	rad
ω	angular frequency	rad/s

Subscripts

0	free space
ave	average
b	basic
c	critical
g	gain (as a ratio)
G	gain (in dB)
i	incident
l	per unit length
max	maximum
n	noise
r	radiated, reflected, relative, radial, or radius
rad	radiation
rec	received or receiving
sys	system
trans	transmission or transmitting
θ	polar
ϕ	azimuthal

1. ANTENNA FUNDAMENTALS

An *antenna* is a device used for radiating or receiving electromagnetic waves. Antenna dimensions are comparable to the wavelength of the signal.

An electromagnetic wave is generated by the current within the antenna. The wave can be generated by an electric line source, an electric current loop, or a *Huygens' source*, which is a source with constant electric and magnetic current sources of equal length at the same point.

A group of components, such as antennae, arranged to vary the transmission or reception of electromagnetic radiation (waves) with direction is called an *array*.

Electromagnetic waves can be generated by a *simple array*, that is, a line of radiating elements (antennae); a *uniform linear array*, that is, a simple array with each element fed uniformly in signal amplitude; a *circular array*; and a *planar array*.

An antenna's shape determines its overall radiating pattern. A *wire antenna* may be positioned with the wire over or near a ground plane, V dipole, bent dipole, loop, and others. A *waveguide antenna* is constructed of a waveguide opening onto a conducting ground plane.[1] The waveguide itself is rectangular or circular and usually opens onto a finite ground plane. When the waveguide feeds a horn shape in a manner similar to the wire antenna, the result is called a *horn antenna*. There are sectoral, pyramid, conical, and corrugated horns. A *reflector antenna* is one whose input signal is reflected from a *dish*. Some examples of reflector antennae are shown in Fig. 46.1. Waveguide, horn, and reflector antennae are examples of aperture antennae. An *aperture* is an opening through which waves can pass. An *aperture antenna* is one in which the beam width is determined by the dimensions of the lens, horn, or reflector. Other antennae include the *surface wave* and *microstrip* types.

Figure 46.1 *Reflector Antennae*

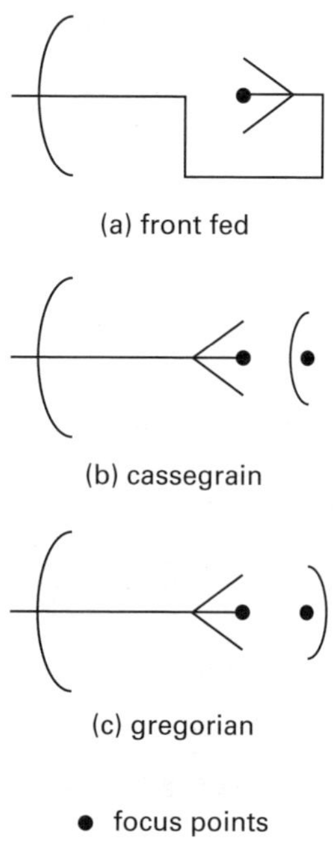

The radiated power of an antenna is the power transferred from the source current to space. When power is radiated uniformly in all directions, the antenna is *isotropic*.[2] The ratio of the isotropic power to the radiated power is called the *directivity* of the antenna and is a measure of the antenna's *focus*. The *antenna gain* is a measure of the effectiveness of a directional antenna compared to an *omnidirectional* or *isotropic antenna*. The gain is dependent on the angle from the antenna.[3] The gain and the directivity are related by the *antenna efficiency*, which relates the radiated power to the input power. The efficiency, in turn, depends on the *radiation resistance*, which is the total radiated power of an antenna divided by the square of the effective current measured at the point where the power is supplied to the antenna.[4] The efficiency also depends on the *ohmic resistance* of the antenna, primarily a function of the skin depth at high frequencies.

The *antenna polarization* is the orientation of the electric field lines in the radiated electromagnetic wave. That is, polarization characterizes the orientation of the field vector in its travel. The polarization usually refers to the electric field vector. When the electric field vector, **E**, and the magnetic field vector, **H**, are orthogonal to each other and in a plane transverse to the direction of propagation, the wave is *elliptically polarized*. When the rectangular components are equal and their phases differ by an odd integral of $\pi/2$, the wave is *circularly polarized*. A wave is *right-hand circularly polarized* (RHCP) if its rotation is clockwise when viewed in the direction of propagation, and *left-hand circularly polarized* (LHCP) if the rotation is counterclockwise.[5] When the electric field vector at a fixed point in space remains pointing in a fixed direction, varying only in magnitude, the wave is *linearly polarized*. This is also called *plane polarization*.[6]

The *antenna temperature* is the temperature a blackbody enclosure in thermal equilibrium would produce if it surrounded the antenna. The *noise power* is

$$P_n = \kappa TB \qquad 46.1$$

The term κ is Boltzmann's constant, T is the temperature in degrees kelvin, and B is the bandwidth in hertz. All noise sources that contribute to the noise power can therefore be considered to have an equivalent temperature.

2. WAVE PROPAGATION FUNDAMENTALS

The antenna type determines the radiation properties. The radiation pattern is obtained from the source currents creating the magnetic field or from the magnetic field about a line, area, or volume, all of which depend on the antenna type. Regardless, the magnetic field is

[1]This type of antenna is found in reflector antennae or flush-mounted antennae on aircraft and spacecraft.

[2]The power radiated in this manner is called the *equivalent isotropically radiated power* (EIRP).

[3]The ability of an array to direct a signal in a given direction is called *signal steering* and is accomplished by delaying the signal to the various elements of the array.

[4]The radiation resistance is termed the *radiation impedance* if *radiation reactance* is taken into account.

[5]This definition comes from the International Radio Regulations, Geneva, 1990. The definitions used in some texts differ.

[6]Plane polarization is similar to elliptical polarization, though without the rotation of the electric field vector about the direction of the propagation axis.

described by the permeability and the magnetic vector potential, **A**.

$$\mathbf{H} = \frac{1}{\mu}\nabla \times \mathbf{A} \qquad 46.2$$

The propagation of this magnetic field depends on numerous factors. One of the primary factors is the frequency of the radiated wave. Consequently, most discussion involving propagation is focused on certain frequency bands. For example, very low-frequency waves propagate thousands of kilometers over the surface of the earth with little attenuation. As frequency rises, the ground path is limited to short distances because of excessive losses. At medium to high frequencies, though, ionospheric reflections allow wave propagation, that is, radio communication, over long distances. Above about 30 MHz, the ionospheric reflections are not dependable, and line-of-sight propagation must be used. At very high frequencies, little interaction takes place with the environment and the wave propagates in a relatively straight line, making for short-distance communications unless the transmitter (and/or receiver) is located in space to increase the line-of-sight area.[7] At frequencies above approximately 10 GHz, absorption of the wave by air and water vapor as well as attenuation by rain can occur. In free space, that is, in a vacuum, electromagnetic waves propagate unhindered. A line-of-sight condition in a nonionized atmosphere approximates free space.[8]

3. FREE-SPACE WAVE CHARACTERISTICS

An electromagnetic wave has a wavelength, which is defined as the distance the wave propagates in one period.

$$\lambda = \frac{2\pi}{\beta} = \frac{2\pi}{\omega\sqrt{\mu_0\epsilon_0}} = \frac{c}{f} \qquad 46.3$$

The velocity of an electromagnetic wave in free space is

$$\mathrm{v}_0 = \frac{1}{\sqrt{\mu_0\epsilon_0}} = c \approx 3 \times 10^8 \text{ m/s} \qquad 46.4$$

The *characteristic impedance of free space* is

$$\begin{aligned} Z_0 &= \sqrt{\frac{\mu_0}{\epsilon_0}} = 120\pi\ \Omega = 377\ \Omega \\ &= 4\pi\mathrm{v}_0 \times 10^{-7}\ \Omega \end{aligned} \qquad 46.5$$

[7]The area "seen" by a satellite on the Earth's surface is called the *footprint*. Satellites use line-of-sight frequencies. When in geosynchronous orbit, the satellite has a footprint of approximately 40% of the earth's surface. Three satellites are necessary for complete coverage.

[8]Refraction effects in the atmosphere may still cause bending, reflection, scattering, and fading. Nevertheless, the assumption of free space in Earth-based communication calculations can be used for initial estimates.

4. POYNTING'S VECTOR AND POWER

Poynting's vector, **S**, is the instantaneous power density carried by an electromagnetic wave given by

$$\mathbf{S} = \epsilon c E^2 \mathbf{a} = \mathbf{E} \times \mathbf{H} \quad [\text{in W/m}^2] \qquad 46.6$$

The vector **a** represents a unit vector in the direction of propagation. Poynting's vector is illustrated in Fig. 46.2.

The real part of Poynting's vector is the average power density in the direction of transmission given by[9]

$$P_{\text{ave}} = \tfrac{1}{2}\text{Re}(\mathbf{E} \times \mathbf{H}) = \frac{E^2}{2Z_0} \quad [\text{in W/m}^2] \qquad 46.7$$

Figure 46.2 Poynting's Vector

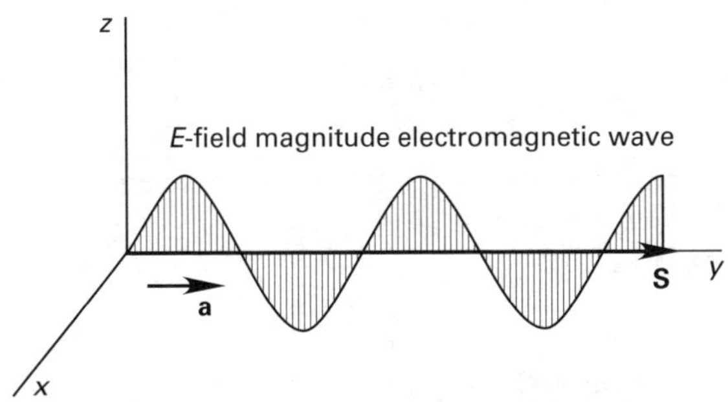

5. POLARIZATION

Polarization is a phenomenon exhibited by electromagnetic waves in which the direction of the electric field is constant or varies in a definite way. Polarization is illustrated in Fig. 46.3.

Figure 46.3 Polarization

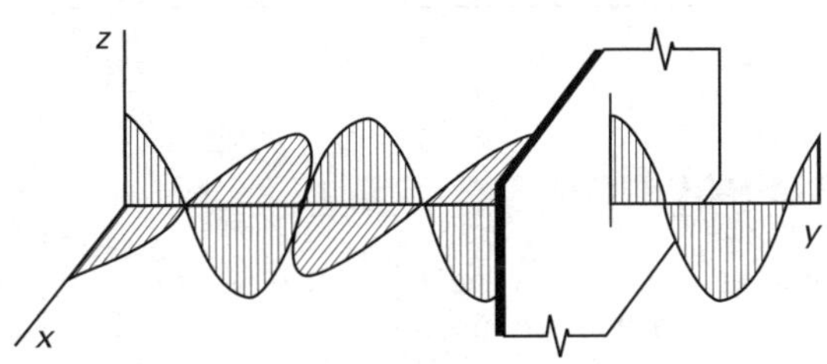

6. INTERFERENCE

Interference is the variation with distance or time of the amplitude of a wave that results from the algebraic or vector addition of two or more waves with identical or nearly identical frequencies. Interference is also called

[9]Equation 46.5 and Eq. 46.6 are consistent if the relative permittivity in Eq. 46.6 is equal to one.

Communications

cancellation when the waves subtract, and *reinforcement* or *superposition* when they add to one another.[10] In communication systems, interference refers to any undesired energy that interferes with the reception of the desired signals.

7. REFLECTION

Reflection is the return of waves or particles from a surface upon which they are incident. Reflection occurs at or near the boundary between mediums, such as between the air and the ionosphere or the air and the ground. The concept of reflection is shown in Fig. 46.4. At the *critical incident angle*, θ_c, an optically transparent surface becomes totally reflecting. Total internal reflection is the concept upon which *optical fibers* (also known as *light pipes*) function.[11]

$$\sin\theta_c = \frac{1}{n} \tag{46.8}$$

The critical angle depends on the *absolute index of refraction*, *n*, given by[12]

$$n = \sqrt{\mu_r \epsilon_r} \tag{46.9}$$

Any portion of the wave that enters a surface such as that in Fig. 46.4 would be called the *transmitted wave*. Transmitted waves are *refracted*.

Figure 46.4 *Reflection from a Surface*

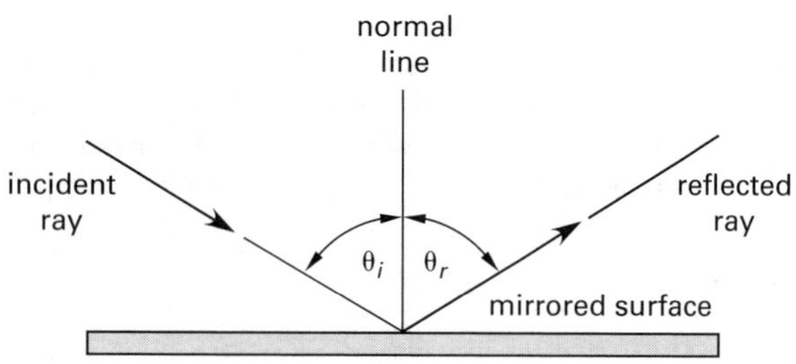

8. REFRACTION

Refraction is the bending of an electromagnetic wave as it passes from one medium to another. Refraction results from the change in the direction of the lines of force of an electric or magnetic field at the boundary of two mediums with differing permeabilities or permittivities. That is, refraction is the change in direction of wave propagation caused by a difference in wave velocity in the two mediums. Refraction is illustrated in Fig. 46.5(a).

Figure 46.5 *Refraction*

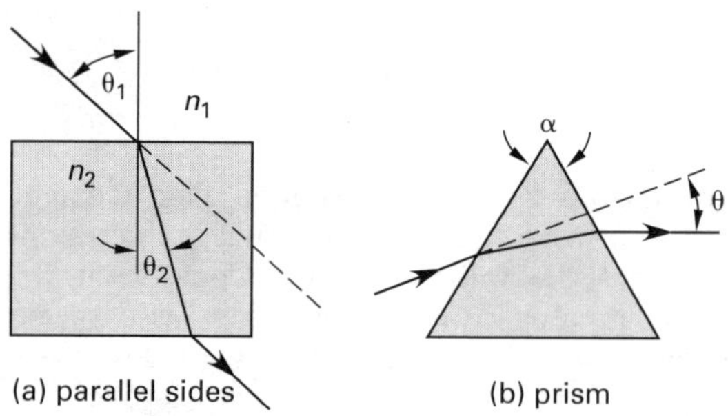

Snell's law, Eq. 46.10, relates the incident and reflected angles. Electromagnetic waves tend to bend *toward the normal* when entering a denser material. A vacuum (and, for practical purposes, air) has an index of refraction, *n*, equal to one ($n_1 = 1$).

$$n_{\text{relative}} = \frac{n_2}{n_1} = \frac{\sin\theta_1}{\sin\theta_2} \tag{46.10}$$

When the refraction takes place through a prism-shaped medium as in Fig. 46.5(b), the *angle of refraction* (*angle of deviation*), θ, is governed by

$$n_{\text{relative}} = \frac{\sin\frac{\alpha+\theta}{2}}{\sin\frac{\alpha}{2}} \tag{46.11}$$

9. DIFFRACTION AND SCATTERING

Diffraction is any redistribution in space of the intensity of a wave that results from the presence of an object, or, more technically, large-scale irregularities. That is, diffraction is the change in path of a wave when it passes around objects (or through narrow slits). Diffraction is illustrated in Fig. 46.6.

Scattering is the diffusion of a wave in a random manner. Scattering results from small-scale irregularities, such as small changes in the index of refraction in the upper atmosphere.

Figure 46.6 *Diffraction*

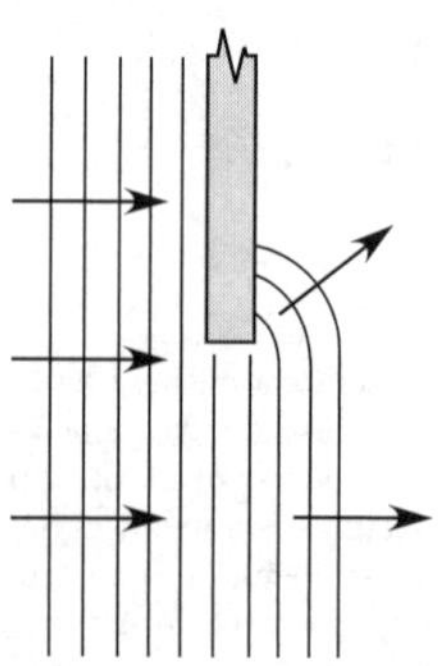

[10]The term "interference" is often generically used to mean the subtraction of the waves only.
[11]Equation 46.8 is used specifically for a wave incident upon a surface to free space, which air approximates well.
[12]The *r* in Eq. 46.9 indicates relative, not reflected, values.

10. FREE-SPACE BASIC TRANSMISSION LOSS

The gain, G, and loss, L, in a communication system are usually given in decibels (dB). The use of decibels allows losses for various conditions to be placed in tabular form for a variety of environmental conditions. The total power of a particular system at any given point is then easily determined from the algebraic combination of the gains and losses.

The system loss, L_{sys}, in a communications link is the difference between the transmitting antenna power density, p_{trans}, and the power density available at the receiving antenna, p_{rec}.

$$\begin{aligned} L_{\text{sys}} &= 10 \log \frac{p_{\text{trans}}}{p_{\text{rec}}} \\ &= p_{\text{trans}} - p_{\text{rec}} \quad \text{[in dB]} \end{aligned} \qquad 46.12$$

The *basic transmission loss*, L_b, is the loss within a given system assuming isotropic antennae and isotropic antenna polarization identical to that of the real antennae, and disregarding obstacles close to the antennae. The *free-space basic transmission loss*, L_{b0}, is the loss with isotropic antennae in a perfectly dielectric, homogenous, unlimited isotropic environment. The power density in an isotropic environment is spread over the surface of a sphere with area $4\pi r^2$, where r is the radial distance. The effective absorbing area of an isotropic receiving antenna is $\lambda^2/4\pi$. Combining the concepts gives an isotropic receiving antenna a *power density* of

$$\begin{aligned} p_{\text{rec}} &= \left(\frac{\lambda^2}{4\pi}\right)\left(\frac{1}{4\pi r^2}\right) p_{\text{trans}} \\ &= \left(\frac{\lambda}{4\pi r}\right)^2 p_{\text{trans}} \end{aligned} \qquad 46.13$$

The inverse of the term $(\lambda/4\pi r)^2$ of Eq. 46.13 represents the free-space basic transmission loss and is expressed in decibels as

$$L_{b0} = 20 \log \frac{4\pi r}{\lambda} \qquad 46.14$$

The term $(4\pi r/\lambda)^2$ is called the *path loss*. It is not a loss per se, but represents the decrease in power caused by the spherical spreading of the radiated energy. (The square of the term is reflected in the value of 20 in front of the logarithm, rather than 10.)

11. ELEMENTAL ANTENNA

An antenna differs from a transmission line in that its dimensions are comparable to the wavelength of the signal carried. The relationship between an element of antenna current and the resulting radiated electromagnetic field is based on retarded magnetic potential, with phasor notation used to account for the time variation of the current. An elemental antenna is a small section of uniform electric current, Idz, as shown in Fig. 46.7.

The expression for the magnetic field strength or intensity, $\mathbf{H}$, of an elemental antenna of length $dz = l$, using spherical coordinates, is derived from Eq. 46.2 as

$$\mathbf{H}_\phi = \frac{\mathrm{I}l}{4\pi} \sin\theta \left(j\frac{\beta}{r} + \frac{1}{r^2} \right) e^{-j\beta r} \quad \text{[in A/m]} \qquad 46.15$$

Figure 46.7 Elemental Antenna

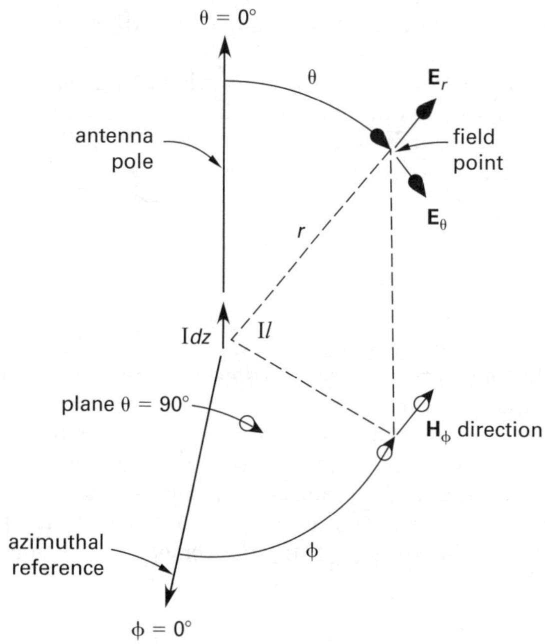

The phase constant in Eq. 46.15, β, from Eq. 46.3, is $\omega\sqrt{\mu_0\epsilon_0}$. The term ω is the radian frequency of the signal, μ_0 is the free-space permeability, and ϵ_0 is the free-space permittivity. In Fig. 46.7, θ is the polar angle, ϕ is the azimuthal angle, dz is the differential length of the elemental antenna, and r is the distance from the element Il to the indicated point in the field. The current term, I, is the rms phasor current and therefore has the time dependence embedded. The magnetic field, $\mathbf{H}$, is exclusively in the azimuthal direction, that is, perpendicular to r and parallel to the $\theta = 90°$ plane.

The components of the electric field, $\mathbf{E}$, are found using Maxwell's equation relationship (see Chap. 25).

$$\nabla \times \mathbf{H} = \text{curl } \mathbf{H} = \frac{\partial \mathbf{D}}{\partial t} = j\omega\epsilon_0 \mathbf{E} \qquad 46.16$$

The radial component of the electric field is

$$E_r = \frac{\mathrm{I}l}{2\pi} \cos\theta \left(\frac{1}{cr^2} + \frac{1}{j\beta r^3} \right) e^{-j\beta r} \quad \text{[in V/m]} \qquad 46.17$$

Communications

The speed of light, c, is equal to $1/\sqrt{\mu_0\epsilon_0}$ from Eq. 46.4. The polar component is

$$E_\theta = \frac{\text{I}l}{4\pi}\sin\theta\left(\frac{j\beta}{r}+\frac{1}{r^2}+\frac{1}{j\beta r^3}\right)e^{-j\beta r} \quad \text{[in V/m]} \qquad 46.18$$

12. ELEMENTAL ANTENNA: RADIATED POWER

From Poynting's theorem (see Sec. 46.4), the power density is obtained. The radiated power is in the r-direction only, and, because the cross product is involved, includes only $E_\theta H_\phi$. Additionally, at large distances from the antenna, only the $1/r$ components are significant contributors to the power density. The radiated power density is

$$p_r = \left(\frac{\text{I}l}{4\pi}\right)^2\sqrt{\frac{\mu_0}{\epsilon_0}}\sin^2\theta\left(\frac{j\beta}{r}\right)^2 e^{-2j\beta r} \quad [\text{in W/m}^2] \qquad 46.19$$

The total average power density, P_r, is obtained from Eq. 46.19 by integrating over a sphere of radius r. Because the power density is not a function of the azimuthal angle, ϕ, the surface area element is $(2\pi r \sin\theta)rd\theta$, which is integrated from $\theta=0$ to $\theta=\pi$. The term $e^{-2j\beta r}$ is a time delay that vanishes during the integration, because the integration data is the average value. The average radiated power in an elemental antenna of length l is

$$P_r = \left(\frac{(\text{I}l)^2}{6\pi}\right)\sqrt{\frac{\mu_0}{\epsilon_0}}\beta^2 \quad [\text{in W/m}^2] \qquad 46.20$$

13. ELEMENTAL ANTENNA: DIRECTIVITY

The *directivity* is the value of the directive gain of an antenna in the direction of its maximum value. The directivity is a measure of an antenna's *focus*. Directivity, D, is mathematically defined as the ratio of the isotropic power to the radiated power.

$$D_g = \frac{P_{\text{isotropic}}}{P_r} \qquad 46.21$$

Maximum power density is calculated from Eq. 46.19 at $\theta=90°$ and is

$$P_{r,\text{max}} = \left(\frac{\text{I}l}{4\pi}\right)^2\sqrt{\frac{\mu_0}{\epsilon_0}}\left(\frac{j\beta}{r}\right)^2 e^{-2j\beta r} \quad [\text{in W/m}^2] \qquad 46.22$$

An isotropic antenna radiates power equally in all directions, forming a sphere with a surface area of $4\pi r^2$. Multiplying Eq. 46.22 by this surface area and removing the time delay exponential term gives

$$P_{\text{isotropic}} = \frac{(\text{I}l)^2}{4\pi}\sqrt{\frac{\mu_0}{\epsilon_0}}\beta^2 \quad [\text{in W/m}^2] \qquad 46.23$$

The directivity of the elemental antenna is given by the ratio of Eq. 46.23 and Eq. 46.20.

$$D_g = \frac{P_{\text{isotropic}}}{P_r} = 1.5 \qquad 46.24$$

In terms of decibels, the directivity of the elemental antenna is

$$D_G = 10\log D_g = 1.76 \text{ dB} \qquad 46.25$$

14. ELEMENTAL ANTENNA: RADIATION RESISTANCE

The *radiation resistance*, R_{rad}, is mathematically defined as the ratio of the total radiated power to the square of the rms input current of an antenna.

$$R_{\text{rad}} = \frac{P_r}{\text{I}_{\text{in}}^2} \qquad 46.26$$

The radiation resistance of an elemental antenna is given by the ratio of Eq. 46.20 and the input current.

$$R_{\text{rad}} = \frac{l^2\beta^2}{6\pi}\sqrt{\frac{\mu_0}{\epsilon_0}} \qquad 46.27$$

15. ELEMENTAL ANTENNA: OHMIC RESISTANCE

The magnetic field calculations of Sec. 46.11 are based on an infinitesimally thin antenna, with the field calculations far from the antenna.[13] Antenna wire has a finite diameter and operates at a sufficiently high frequency that the skin effect dominates (see Sec. 37.3). The ohmic resistance (see Sec. 37.4) is given by

$$R_{\text{ohmic}} = \frac{R_l}{1-\left(1-\frac{\delta}{r}\right)^2} \qquad 46.28$$

16. ELEMENTAL ANTENNA: EFFICIENCY

The *antenna efficiency* is the ratio of the amount of power radiated into space to the antenna input power.

[13] *Far field* is the term used when describing an electromagnetic field at a significant distance from the antenna such that local effects can be ignored.

The efficiency, η, is determined from the radiation resistance and ohmic resistance as

$$\eta = \frac{R_{\text{rad}}}{R_{\text{ohmic}} + R_{\text{rad}}} \times 100\% \qquad 46.29$$

17. ELEMENTAL ANTENNA: GAIN

The *antenna gain* is the ratio of isotropic power to input power. That is, gain compares the maximum isotropically radiated power to the actual input power. Gain is a measure of the effectiveness of a directional antenna compared to a nondirectional antenna and is

$$G = \eta D_g \qquad 46.30$$

18. SHORT DIPOLE ANTENNA

A *dipole antenna* has two equal branches of height h, as shown in Fig. 46.8. In a *short dipole antenna*, the height (length), h, is much less than the wavelength of the signal.

Figure 46.8 *Short Dipole Antenna*

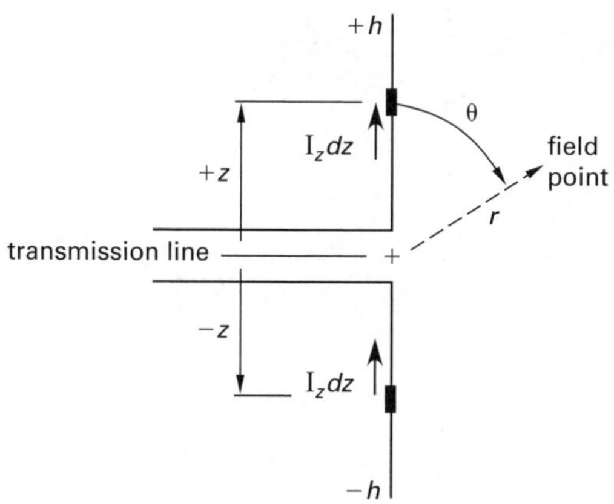

Two identical current elements, Idz, exist in each branch as standing waves, linearly diminishing to zero at $z = +h$ and $z = -h$. The current is described by

$$\begin{aligned} I_z &= I_{\text{in}}\left(1 - \frac{z}{h}\right) \quad [\text{for } 0 < z < +h] \\ &= I_{\text{in}}\left(1 + \frac{z}{h}\right) \quad [\text{for } -h < z < 0] \end{aligned} \qquad 46.31$$

The input current, I_{in}, is the rms phasor current leaving the lower branch and entering the upper branch. Because z is small compared to the wavelength, the current can be assumed to exist at the origin only. Therefore, Eq. 46.15 applies with l as dz.

$$H_\phi = \frac{\int_{-h}^{+h} I dz}{4\pi} \sin\theta\left(j\frac{\beta}{r} + \frac{1}{r^2}\right)\left(e^{-j\beta r}\right) \quad [\text{in A/m}] \qquad 46.32$$

Using the definition of the current from Eq. 46.31 in the integration of Eq. 46.32 gives the resulting magnetic field.

$$H_\phi = \frac{I_{\text{in}} h}{4\pi} \sin\theta\left(\frac{j\beta}{r} + \frac{1}{r^2}\right) e^{-j\beta r} \quad [\text{in A/m}] \qquad 46.33$$

This differs from the elemental antenna only in having the term $I_{\text{in}}h$ replace Il. The radiated *power density* is

$$p_r = \left(\frac{I_{\text{in}} h}{4\pi}\right)^2 \sqrt{\frac{\mu_0}{\epsilon_0}} \sin^2\theta \left(\frac{j\beta}{r}\right)^2 e^{-2j\beta r} \quad [\text{in W/m}^2] \qquad 46.34$$

The total average power radiated, over a sphere of radius r, is calculated as in Sec. 46.12 as

$$P_r = \frac{(I_{\text{in}} h)^2}{6\pi} \sqrt{\frac{\mu_0}{\epsilon_0}} \beta^2 \quad [\text{in W/m}^2] \qquad 46.35$$

The derivation of the directivity is identical to that in Sec. 46.13 with h replacing l in Eq. 46.22 and Eq. 46.23. The resulting *directivity* of the short dipole antenna is

$$D_g = \frac{P_{\text{isotropic}}}{P_r} = 1.5 \qquad 46.36$$

Therefore, the directivity of a short dipole antenna is identical to that of an elemental antenna.[14]

The radiation resistance, as defined in Sec. 46.14, for the short dipole antenna is

$$R_{\text{rad}} = \frac{P_r}{I_{\text{in}}^2} = \frac{h^2\beta^2}{6\pi}\sqrt{\frac{\mu_0}{\epsilon_0}} \qquad 46.37$$

Comparing Eq. 46.37 with Eq. 46.27, a similar expression, indicates that radiation resistance is proportional to h^2 for a short dipole antenna and l^2; that is, $(dz)^2$ for the elemental antenna. The length of the short dipole antenna is $2h$, while the length of the elemental antenna is l, that is, dz. Therefore, in determining the radiation resistance (see Eq. 46.37), the effective length of the short dipole antenna is one-half the actual length. This effect is caused by the nonuniform current distribution in the short dipole antenna.

The ohmic resistance experiences the same effect from the nonuniform current distribution. The *ohmic power* of a short dipole antenna is

$$P_{\text{ohmic}} = \int_{-h}^{+h} I_z^2 R_l dz = I_{\text{in}}^2 R_l \tfrac{2}{3} h \qquad 46.38$$

[14]This result should not be surprising because the antenna current is concentrated at the origin.

Communications

The *ohmic resistance* is

$$R_{\text{ohmic}} = \frac{P_{\text{ohmic}}}{\text{I}_{\text{ohmic}}^2} = \left(\frac{2h}{3}\right) R_l \qquad 46.39$$

Therefore, in determining the ohmic resistance (see Eq. 46.39), the effective length of the short dipole antenna is one-third the actual length.

The formulas for the short dipole antenna efficiency and the gain are identical to those for the elemental antenna; Eq. 46.29 and Eq. 46.30, respectively. Nevertheless, the short dipole antenna efficiency will be lower than that of the elemental antenna, even though the ohmic resistance is one-third the value and is in the denominator of Eq. 46.29, which by itself tends to make the efficiency larger. The efficiency is lower because the effective length of the short dipole is one-half that of the elemental antenna and this length (h) is squared in the calculation of the radiation resistance (see Eq. 46.37), which dominates the change in efficiency (see Eq. 46.29).

19. DIPOLE ANTENNA

When the antenna length (height) is comparable to the wavelength of the signal carried, the assumptions made regarding the antenna current being an infinitesimal element of uniform current (in an elemental antenna) or concentrated at the origin (in a short dipole antenna) are no longer valid. Specifically, the distance to the point of interest in the field from the elemental currents, Idz, in each portion of the dipole must be accounted for separately. The concept is illustrated in Fig. 46.9(a).

Because the antenna height (length), h, is much less than the radial distance to the point in the field, the distances from the two current elements are approximated as

$$r_1 \approx r - z\cos\theta \qquad 46.40$$

$$r_2 \approx r + z\cos\theta \qquad 46.41$$

The assumed current standing wave distribution is that of Fig. 46.9(b). The maximum rms current phasor is shown as I_{max}. The currents can be represented mathematically as

$$\begin{aligned}\text{I}_z &= \text{I}_{\text{max}} \sin\beta\,(h - z) \quad [\text{for } 0 < z < +h] \\ &= \text{I}_{\text{max}} \sin\beta\,(h + z) \quad [\text{for } -h < z < 0]\end{aligned} \qquad 46.42$$

The angle, θ, is assumed to be the same over the length of the antenna because the radial distance, r, is much greater than the antenna height, h. Additionally, this assumption means the differences in r_1 and r_2 from the actual distance r only occur in the exponential term of Eq. 46.15. Incorporating the radial distances in Eq. 46.40 and Eq. 46.41, using the current of Eq. 46.42, and considering only the significant contributors to the magnetic field and power density (i.e., assuming a $1/r$ dependence) gives

$$\begin{aligned}H_\phi &= j\left(\frac{\text{I}_{\text{max}}}{2\pi r\sin\theta}\right) e^{-j\beta r} \\ &\quad \times \cos(\beta h\cos\theta) - \cos\beta h \quad [\text{in A}/m]\end{aligned} \qquad 46.43$$

The polar component of the electric field is

$$E_\theta = \sqrt{\frac{\mu_0}{\epsilon_0}}H_\phi = 120\pi H_\phi \qquad 46.44$$

Figure 46.9 *Dipole Antenna*

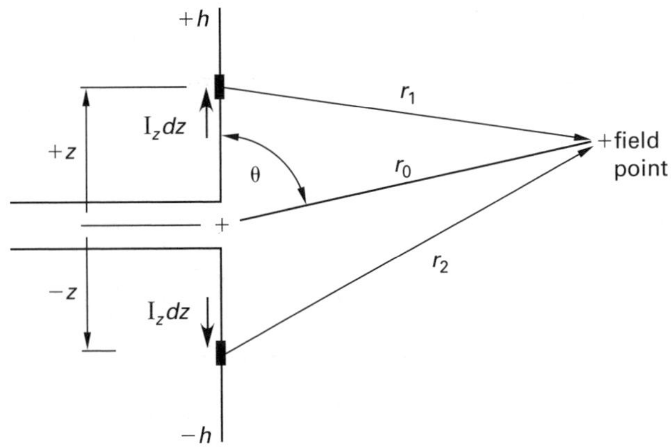

(a) antenna with length comparable to wavelength

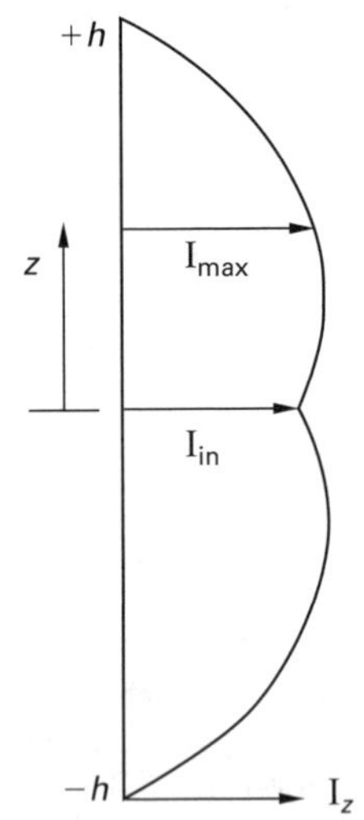

(b) assumed current standing wave distribution

The *power density* is determined from $E_\theta H_\phi = 120\pi H_\phi^2$.

$$p_r(r,\theta) = (120\pi)(-1)\left(\frac{\mathrm{I}_{\max}}{2\pi r \sin\theta}\right)^2 e^{-2j\beta r} \times \left(\cos(\beta h \cos\theta) - \cos\beta h\right)^2 \qquad 46.45$$

In order to determine the parameters of interest (the radiated power, directivity, radiation resistance, and so on), Eq. 46.45 is integrated from $\theta = 0$ to $\theta = \pi$. The results contain cosine and sine integrals (see Sec. 10.17), which are evaluated by computation or by using tables. The calculations are simplified if multiples of wavelengths are used, such as one-fourth or one-half wavelengths.

Antennae designed to be a one-fourth multiple of the signal wavelength are called *quarter-wave antennae.* Antennae designed as a one-half multiple of the signal wavelength are called *half-wave antennae*, or *half-wave dipoles.*

20. HALF-WAVE ANTENNA

A half-wave antenna is constructed so that the antenna height is a multiple of one-half the signal wavelength. Specifically, $2h = \lambda/2$. The radiated power for such an antenna is determined by integrating Eq. 46.45 over a sphere of differential volume $(2\pi r \sin\theta) r d\theta$. With $2h = \lambda/2$, the $\cos\beta h$ term equals zero (see Eq. 44.5). The average radiated power, with the average suppressing the exponential time-delay term, is

$$P_r = \int_0^\pi 120\pi\left(\frac{\mathrm{I}_{\max}}{2\pi r \sin\theta}\right)^2 \cos^2\left(\frac{\pi}{2}\cos\theta\right) \times (2\pi r \sin\theta) r d\theta \quad \text{[in W]} \qquad 46.46$$

Equation 46.46 simplifies to

$$P_r = 60\mathrm{I}_{\max}^2 \int_0^\pi \left(\frac{\cos^2\left(\frac{\pi}{2}\cos\theta\right)}{\sin\theta}\right) d\theta \quad \text{[in W]} \qquad 46.47$$

Using numerical techniques, Eq. 46.47 is evaluated to

$$P_r = 73.08\mathrm{I}_{\max}^2 \quad \text{[in W]} \qquad 46.48$$

From Eq. 46.45, the maximum power density occurs at $\theta = \pi/2$, which results in

$$p_{r,\max} = \frac{120\pi\mathrm{I}_{\max}^2}{(2\pi r)^2} \quad \text{[in W/m}^2\text{]} \qquad 46.49$$

The *isotropic power* is found from Eq. 46.49 by multiplying by the surface of a sphere at radial distance r from the antenna, that is, by $4\pi r^2$.

$$P_{\text{isotropic}} = 120\mathrm{I}_{\max}^2 \quad \text{[in W]} \qquad 46.50$$

The *directivity* for a half-wave antenna is the ratio of Eq. 46.50 to Eq. 46.48,

$$D_g = \frac{P_{\text{isotropic}}}{P_r} = \frac{120\mathrm{I}_{\max}^2}{73.08\mathrm{I}_{\max}^2} = 1.64 \qquad 46.51$$

The *radiation resistance* is

$$R_{\text{rad}} = \frac{P_r}{\mathrm{I}_{\max}^2} = \frac{73.08\mathrm{I}_{\max}^2}{\mathrm{I}_{\max}^2} = 73.08\ \Omega \qquad 46.52$$

The *ohmic power* is

$$P_{\text{ohmic}} = \int_{-h}^{+h} \mathrm{I}_z^2 R_l dz \qquad 46.53$$

The *ohmic resistance* can be shown to be

$$R_{\text{ohmic}} = hR_l \qquad 46.54$$

The formulas for the half-wave antenna efficiency and gain are identical to those for the elemental antenna, Eq. 46.29 and Eq. 46.30, respectively.

47 Communication Links

1. Fundamentals 47-1
2. Power Distribution 47-1
3. Antenna Apertures 47-2
4. Directivity 47-3
5. Beam Solid Angle Ω 47-3
6. Communication Link Transmission 47-3

Nomenclature

A	aperture or area	m^2
D	directivity	–
E	electric field strength	V/m
G	gain	– or dB
I	effective or DC current	A
l	length	m
L	loss	dB
p	power density	W/m^2
P	power	W
r	radial distance	m
R	resistance	Ω
V	effective or DC voltage	V
X	reactance	Ω
Z_0	characteristic impedance	Ω

Symbols

ϵ_0	free-space permittivity 8.854×10^{-12}	F/m
η	efficiency	–
λ	wavelength	m
μ_0	free-space permeability 1.2566×10^{-6}	H/m
Ω	solid angle	sr

Subscripts

amp	amplifier
eff	effective
g	gain (as a ratio)
max	maximum
p	path
r	radiated
rad	radiation
rec	received
R	receiver
scat	scattered
T	transmitter
ϕ	free space

1. FUNDAMENTALS

Communication, in the context of electrical engineering, is the science and technology that collects information from a source, transforms it into electrical currents or fields, transmits it via a medium (also known as a *communication channel*), and reconstitutes the information at the receiver. Common media include conducting wire, coaxial cable, fiber-optic material, waveguides, the atmosphere, and space.

Communication systems include computer networks as well as telephony, cable, optical, wireless, microwave, and satellite systems. These are broadly classified as broadcast, voice, or data communication systems. A *communication link* is a general term for the combination of a transmitting antenna, a receiving antenna, and the medium or channel through which they are connected, either by air or space (vacuum).[1] *Telecommunications* is a general term meaning communications over long distances.

Communications

2. POWER DISTRIBUTION

If the receiving antenna is positioned so that it obtains the maximum power from the transmitting antenna, the power density radiated from the transmitting antenna available at the receiving antenna is

$$p_r = \left(\frac{D_T}{4\pi r^2}\right) P_T \quad [\text{in W/m}^2] \qquad 47.1$$

The term p_r is the radiated power density at the radial distance r, P_T is the transmitted power, and D_T is the transmitting antenna directivity gain (see Sec. 46.13).

In free space, the *power density* is the electric field strength divided by the characteristic impedance, Z_0.[2]

$$p_r = \frac{E^2}{Z_0} = \frac{E^2}{\sqrt{\frac{\mu_0}{\epsilon_0}}} = \frac{E^2}{120\pi} \quad [\text{in W/m}^2] \qquad 47.2$$

Because the receiving antenna is oriented in the direction of the electric field lines, for maximum power reception, the voltage received is $El = V_{\text{rec}}$, where the effective antenna length is l. The receiving antenna has a generic input impedance of $R_{\text{in}} + jX_{\text{in}}$. The antenna

[1]Any development of equations in this chapter is accomplished using parameters for dipole antennae (see Chap. 46). Identical parameters exist for other types of antennae, though the exact equations differ. Where the results are applicable to all antennae, this is specifically noted.

[2]The characteristic impedance of free space is also called the *intrinsic impedance*.

supplies an amplifier with a generic impedance of $R_{amp} + jX_{amp}$. The current in the receiving antenna is

$$I_{rec} = \frac{V_{rec}}{R_{in} + R_{amp} + j(X_{in} + X_{amp})} = \frac{El}{R_{in} + R_{amp} + j(X_{in} + X_{amp})} \quad 47.3$$

Three resistances are involved. Two of them—the radiation resistance, R_{rad}, and the ohmic resistance, R_{ohmic}—are components of the input resistance, R_{in} (see Sec. 46.14 and Sec. 46.15). The third is the amplifier resistance, R_{amp}. The power is given by I^2R. Using the power and three resistances along with Eq. 47.2 and Eq. 47.3 gives the distribution of the power in the receiving antenna.

The power lost in the radiation resistance is reradiated. This power is considered to be *scattered* and is given by

$$P_{scat} = \frac{120\pi p_r l^2 R_{rad}}{(R_{in} + R_{amp})^2 + (X_{in} + X_{amp})^2} \quad 47.4$$

The power lost in the ohmic resistance of the antenna is

$$P_{loss} = \frac{120\pi p_r l^2 R_{ohmic}}{(R_{in} + R_{amp})^2 + (X_{in} + X_{amp})^2} \quad 47.5$$

The power used by the amplifier is

$$P_{amp} = \frac{V_{rec}^2 R_{amp}}{(R_{in} + R_{amp})^2 + (X_{in} + X_{amp})^2} = \frac{120\pi p_r l^2 R_{amp}}{(R_{in} + R_{amp})^2 + (X_{in} + X_{amp})^2} \quad 47.6$$

The effectiveness of a receiving antenna is improved by making the amplifier impedance (that is, the receiver impedance) the complex conjugate of the antenna input impedance. Doing so eliminates the reactance term in the denominator of Eq. 47.3. Half of the received power is then transmitted to the amplifier (receiver), and the remainder becomes ohmic and scattering losses.

3. ANTENNA APERTURES

An antenna aperture is an opening through which radio waves can pass. Antennae have a physical aperture based on their structure, measured in m^2. Because there are losses, the physical aperture does not deliver all of the power from the power flux incident upon it. The situation is illustrated in Fig. 47.1.

Figure 47.1 *Communication Link*

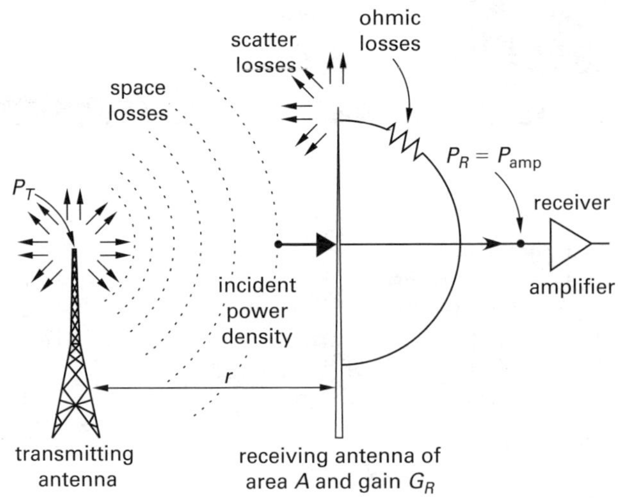

(Note: The power density is sometimes called the *flux density*.)

Using Eq. 47.4 through Eq. 47.6, scattering, loss, and effective apertures are defined as

$$A_{scat} = \frac{120\pi l^2 R_{rad}}{(R_{in} + R_{amp})^2 + (X_{in} + X_{amp})^2} \quad 47.7$$

$$A_{loss} = \frac{120\pi l^2 R_{ohmic}}{(R_{in} + R_{amp})^2 + (X_{in} + X_{amp})^2} \quad 47.8$$

$$A_{eff} = \frac{120\pi l^2 R_{amp}}{(R_{in} + R_{amp})^2 + (X_{in} + X_{amp})^2} \quad 47.9$$

The development of the apertures in Eq. 47.7 through Eq. 47.9 is based on the parameters for a linear antenna, such as the dipole antennae described in Chap. 46. Nevertheless, the results are general and the equations have counterparts for all types of antennae.

The relationships in Eq. 47.10 through Eq. 47.13 are valid regardless of the antenna type.

$$P_{scat} = A_{scat} p_r \quad 47.10$$

$$P_{loss} = A_{loss} p_r \quad 47.11$$

$$P_{amp} = A_{eff} p_r \quad 47.12$$

$$p_{r,total} = (A_{scat} + A_{loss} + A_{eff}) p_r = A_{total} p_r \quad 47.13$$

The *amplifier power* (sometimes called the *receiver power*, P_R) uses the effective aperture and is based on the received voltage (i.e., the received electric field strength). Therefore, the amplifier power and effective aperture account for the space losses shown in Fig. 47.1 and can be related to the directivity of the antenna.

4. DIRECTIVITY

The directivity is the directive gain of an antenna. Mathematically, it is defined as the ratio of the isotropic power to the radiated power, which is equivalent to the ratio of the maximum power density to the average power density. With this information, it can be shown that for any antenna the directivity is[3]

$$D_g = \left(\frac{4\pi}{\lambda^2}\right) A_{\text{eff}} \qquad 47.14$$

5. BEAM SOLID ANGLE Ω

The *beam angle*, also called the *beam width*, is the angle, measured in the horizontal plane, between the two directions where the intensity of the electromagnetic beam is one-half its maximum value. The beam angle is measured in radians. The concept is illustrated in Fig. 47.2.

Figure 47.2 *Beam Angle*

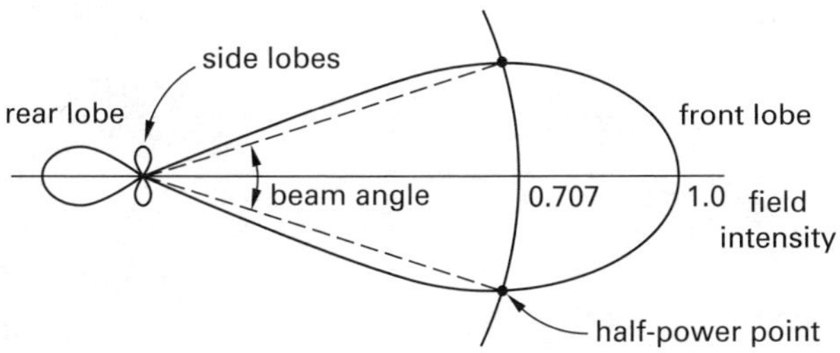

A *beam solid angle*, also referred to as the beam angle, is defined as the region where the antenna power density is one-half or more of the maximum power density.[4] The beam angle is alternatively defined as the solid angle that contains the total radiated power if the power density is uniformly equal to the maximum power density.

$$P_r = p_{r,\text{max}} \Omega r^2 \qquad 47.15$$

The directivity is then defined as the ratio of a sphere solid angle to the beam solid angle.

$$D_g = \frac{4\pi}{\Omega} \qquad 47.16$$

6. COMMUNICATION LINK TRANSMISSION

The power density available at the receiving antenna is given by Eq. 47.1. The relationship between the power density and the effective aperture is given by Eq. 47.12. Combining these concepts, the power at the receiving amplifier, assuming a lossless transmitting antenna, is

$$P_R = \left(\frac{D_T A_R}{4\pi r^2}\right) P_T \qquad 47.17$$

The receiving antenna power is P_R, which is the power at the amplifier input, P_{amp} (see Fig. 47.1). The term D_T is the directivity of the transmitting antenna given by Eq. 47.14, with the subscript g dropped for convenience. The maximum effective receiving antenna aperture is A_R, though it can be written $A_{R,\text{eff,max}}$. Incorporating Eq. 47.14 into Eq. 47.17, the receiving antenna power is written as

$$P_R = \left(\frac{A_R A_T}{\lambda^2 r^2}\right) P_T \qquad 47.18$$

Alternatively, the receiving antenna power is written as

$$P_R = \left(\frac{\lambda}{4\pi r}\right)^2 D_R D_T P_T \qquad 47.19$$

The receiving antenna power as written in Eq. 47.18 and Eq. 47.19 accounts for the path loss, L_p, which is the square of the inverse of the term in parentheses in Eq. 47.19 (see Sec. 46.10). The directivity terms, which contain the effective aperture, account for I^2R type losses from the receiving antenna itself to the associated amplifier. The total efficiency of the antenna, however, accounts for all the losses from the incident wavefront to the antenna output. These losses include the *illumination efficiency* or *aperture taper efficiency*, which is determined by the energy distribution of the feed across the aperture, and other losses such as spillover, blockage, phase errors, diffraction effects, polarization, and mismatch losses. The total efficiency relates the gain to the directivity (see Sec. 46.17) as follows.

$$G = \eta D \qquad 47.20$$

The receiving antenna power must be written in terms of the gain when the efficiencies are significant (low), giving

$$P_R = \left(\frac{\lambda}{4\pi r}\right)^2 G_R G_T P_T \qquad 47.21$$

If the efficiencies are important, Eq. 47.17 through Eq. 47.19 must be modified by factors of the antenna efficiencies as well. Parabolic reflector antennae have efficiencies in the range of 50–75%. Smaller antennae generally have lower efficiencies. Cassegrain antennae generally have higher efficiencies. Horn antennae have efficiencies of approximately 90%.

[3]The maximum value of the effective aperture is used to determine the directivity.

[4]See Sec. 7.16 for an illustration of the solid angle. The nomenclature used in Chap. 7 (ω) represents an area. The nomenclature used here (Ω) represents a region. The units are identical, and the symbols are used interchangeably.

Communications

Equation 47.21 is known as the *Friss transmission formula.* Equation 47.18 and Eq. 47.19 are also considered versions of this equation. In word form, the formula is

$$\text{power received} = \frac{\text{EIRP} \times \text{gain of receiving antenna}}{\text{path loss}} \qquad 47.22$$

The term EIRP stands for the *effective isotropically radiated power* and is the decibel value of $G_T P_T$. The path loss is the inverse of the term in parentheses in Eq. 47.19, and in decibels it is given the symbol L_p, which is equivalent mathematically to Eq. 46.14. The Friss formula then, in decibel terms, is

$$P_R = \text{EIRP} + G_R - L_p \quad \text{[in dBW]} \qquad 47.23$$

The individual terms are

$$\text{EIRP} = 10 \log G_T P_T \quad \text{[in dBW]} \qquad 47.24$$

$$G_R = 10 \log \eta \frac{4\pi A_R}{\lambda^2} \quad \text{[in dBW]} \qquad 47.25$$

$$L_p = 20 \log \frac{4\pi r}{\lambda} \quad \text{[in dBW]} \qquad 47.26$$

Using the Friss formula in the form of Eq. 47.23 allows multiple losses to be accounted for using algebraic manipulation, simplifying calculations and allowing the use of tabular data. Such additional losses might include losses in the transmitting antenna, losses from mispointing the antenna, attenuation from atmospheric effects (rain), and so on.

When using communication link equations, care should be taken to ensure a complete understanding of the terms. For example, in some texts the receiving antenna power, P_R, is the power incident upon the antenna and not the power at the input of the amplifier. Also, if the effective aperture, A_{eff}, includes all the antenna efficiencies, not just those associated with I^2R losses, Eq. 47.14 is equal to the gain, G, rather than the directivity, D (see Eq. 47.20).

48 Signal Formats

1. Fundamentals . . . 48-1
2. Analog Modulation . . . 48-2
3. Amplitude Modulation . . . 48-2
4. Single-Sideband Amplitude Modulation . . . 48-3
5. Vestigial Sideband Amplitude Modulation . . . 48-3
6. Quadrature Multiplexing Amplitude Modulation . . . 48-3
7. Amplitude Modulation: Bandwidth and Power . . . 48-4
8. Angle Modulation . . . 48-4
9. Frequency Modulation . . . 48-4
10. Phase Modulation . . . 48-4
11. Angle Modulation: Bandwidth and Power . . . 48-5
12. Digital Modulation . . . 48-5
13. Baseband Digital-Data Signals . . . 48-6
14. Inversion and Static . . . 48-6
15. Intersymbol Interference . . . 48-7
16. Pulse Amplitude Modulation . . . 48-7
17. Pulse Code Modulation . . . 48-7
18. Quadrature Amplitude Modulation . . . 48-8
19. Digital Modulation: Bandwidth and Power . . . 48-8
20. Digital Error Probability . . . 48-9
21. Forward Error Correcting . . . 48-9

Nomenclature

a	amplitude of modulating signal	V
A	amplitude	V
D	dimensionality	–
E	energy	J
f	frequency	Hz
$\mathcal{F}$	Fourier transform	–
$I(t)$	in-phase component	V
k	proportionality constant	–
k_f	frequency modulator constant	rad/s·V
k_p	phase modulator constant	rad/V
m	modulation index	–
M	map total number	–
N	noise power	W
N_0	noise spectral density	W/Hz
P	power	W
$\mathcal{P}(e)$	error probability, symbol error probability	–
$\mathcal{P}_b(e)$	bit error probability	–
Q	number of bits	–
$Q(t)$	quadrature component	V
R	binary digit rate	digits/s
$s(t)$	signal (time function)	V
S	average signal power	W
$S(\omega)$	signal (frequency function)	V
t	time	s
T	time period	s

Symbols

α	real integer variable	–
β	imaginary integer variable	–
θ	angle, phase angle	rad
$\Delta\theta$	peak phase deviation	rad
ϕ	signal angle	rad
ω	angular frequency	rad/s
$\Delta\omega$	peak frequency deviation	rad/s

Subscripts

0	characteristic or initial
AM	amplitude modulation
ave	average
b, bit	per bit
c	carrier
f	frequency
FM	frequency modulation
max	maximum
mod	modulated or modulation
p	phase
s	per second
x	x component
y	y component

1. FUNDAMENTALS

The purpose of a communication system is to deliver information from a source to a receiver via a channel. The information signals are normally confined to a maximum frequency. The maximum frequency can range from a few hertz for control systems to a few megahertz for television video signals. The bandwidth (BW) allotted for a given signal is usually on the order of the maximum frequency as well. That is, BW $\approx f_{\text{max}}$. The net result is that most communication baseband signal spectrums appear low-pass in character. Such a representation is shown in Fig. 48.1.

Figure 48.1 Typical Baseband Signal Spectrum

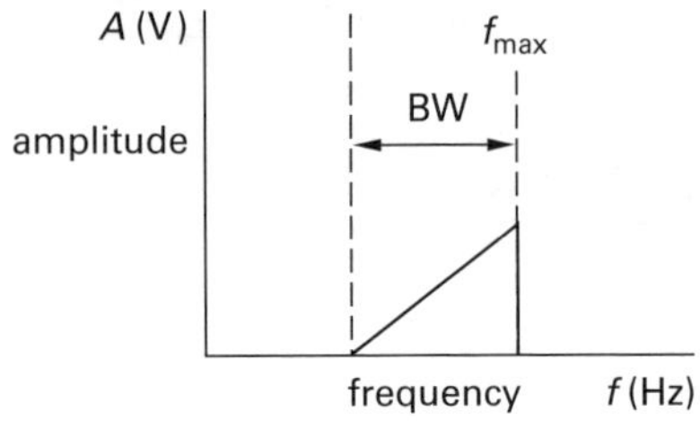

The spectrum shown is the band of frequencies that the system uses to modulate a given signal. The *baseband* is the band of frequencies used to modulate a transmitted carrier. This concept is illustrated in Fig. 48.2. By contrast, a *baseband system* is a communication system in which information is transmitted in an unmodulated band of frequencies. In Fig. 48.2(a), removal of the modulator and carrier would create a baseband system.

Figure 48.2 Baseband Signal

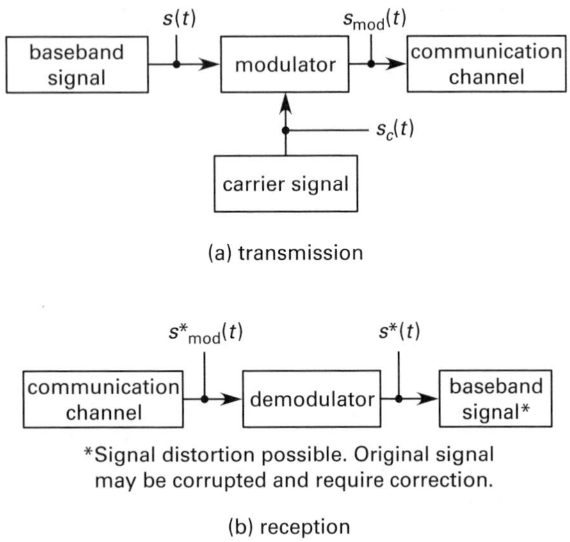

A carrier signal can be represented by the general equation

$$s_c(t) = A\cos(\omega_c t + \theta) = A\cos(2\pi f_c t + \theta) \qquad 48.1$$

The term A is the amplitude of the signal, normally a voltage parameter. The carrier frequency is given as ω_c or f_c, and θ is the phase angle, often assumed to be zero unless specifically noted otherwise. The information signal, that is, the baseband signal, can manipulate the carrier by changing the amplitude, the frequency, or the phase angle. The manipulation is called *modulation*, defined as the process or the result of the process by which a parameter in one wave is varied in accordance with the parameter in another. The possible manipulations are *amplitude modulation* (AM), *frequency modulation* (FM), and *phase modulation* (PM) as illustrated in Fig. 48.3.

Figure 48.3 Modulation Types

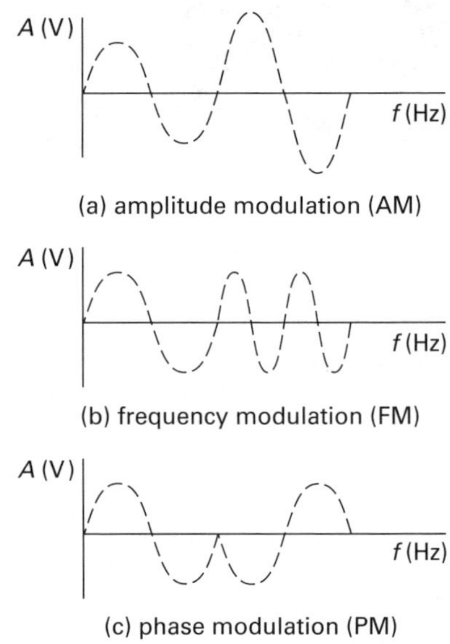

The modulations can be carried out in an analog or a digital fashion. Signals are modulated for a variety of reasons.

- Communication through the channel using the baseband signal alone may be inefficient or impossible.
- Translation to higher frequencies via modulation means smaller wavelengths and more reasonably sized antennae for a given amount of power. (To radiate large amounts of power, an antenna must be large compared to the wavelength.)
- The baseband signal may require frequency translation to an assigned band.
- Several signals may require transmission through the same channel over the same time range. Modulation techniques allow the signals to share the channel but remain recoverable at the receiver as separate signals. Such techniques are called *multiplexing.*

2. ANALOG MODULATION

Analog modulation is the modulation of a continuous complex exponential signal. The modulation is accomplished by varying the amplitude, the frequency, or the phase in a manner proportional to the physical phenomenon, the signal.

3. AMPLITUDE MODULATION

In amplitude modulation (AM), the angle, θ, of Eq. 48.1 is constant and usually assumed to be equal to zero. The amplitude term to be modulated, A, can be represented as the product of a modulator proportionality constant, k, and an information signal, $s(t)$.[1] The general modulated signal, which is a combination of the baseband signal (information signal) and the carrier signal, $s_c(t)$, is then represented as

$$s_{\mathrm{mod}}(t) = ks(t)\cos\omega_c t = \mathrm{Re}\big(ks(t)e^{j\omega_c t}\big) \qquad 48.2$$

Taking the Fourier transform of Eq. 48.2 yields the character of the amplitude modulation.

$$\begin{aligned} S_{\mathrm{mod}}(\omega) &= \mathcal{F}\Big(\tfrac{1}{2}ks(t)e^{j\omega_c t} + \tfrac{1}{2}ks(t)e^{-j\omega_c t}\Big) \\ &= \tfrac{1}{2}S(\omega-\omega_c) + \tfrac{1}{2}S(\omega+\omega_c) \end{aligned} \qquad 48.3$$

An analysis of Eq. 48.3 indicates that the amplitude modulated signal has been shifted (translated) in frequency to above and below the original carrier by $\pm\omega_c$. The signal is said to have two sidebands, an upper and a lower. The original carrier wave is no longer present and the bandwidth, BW, has doubled. This type of signal is called a *double-sideband suppressed-carrier* (DSB-SC).

Receivers using the DSB-SC signal must be designed to maintain synchronization. As a result, they are expensive. To eliminate this issue and lower the receiver cost, the carrier signal can be transmitted with the information signal. Doing so, however, lowers the efficiency because more power (not containing signal information) is used. The signal is

$$s_{\mathrm{mod}}(t) = ks(t)\cos\omega_c t + A\cos\omega_c t \qquad 48.4$$

The amplitude, A, in Eq. 48.4 is the pure carrier wave, and is not modified. That is, it is not proportional to the information signal. If A is large enough, the envelope of the modulated waveform is proportional to the information signal and demodulation becomes simply envelope detection using a diode and a low-pass filter. Commercial broadcast stations (radio stations) use this type of signal. Consequently, the generic term "AM" refers to this *double-sideband large-carrier* (DSB-LC) waveform, which is also called *double-sideband transmitted-carrier* (DSB-TC). This basic AM signal is described in Sec. 18.1, and illustrated in Fig. 18.8.

When the information signal is a single frequency, $s(t) = a\cos\omega_{\mathrm{mod}}t$, Eq. 48.4 can be rearranged as

$$\begin{aligned} s_{\mathrm{mod}}(t) &= ks(t)\cos\omega_c t + A\cos\omega_c t \\ &= A\left(\frac{ks(t)}{A} + 1\right)\cos\omega_c t \\ &= A\left(\left(\frac{ka}{A}\right)\cos\omega_{\mathrm{mod}}t + 1\right)\cos\omega_c t \\ &= A(m\cos\omega_{\mathrm{mod}}t + 1)\cos\omega_c t \end{aligned} \qquad 48.5$$

[1]When a high-frequency signal is simply switched on and off slowly, the term *continuous wave* (CW) *modulation* is sometimes used.

The dimensionless scaling factor in Eq. 48.5, m, is called the amplitude *modulation index*, *index of modulation*, or *modulation factor*, and is defined mathematically as[2]

$$m_{\mathrm{AM}} = \frac{ka}{A} = \frac{\Delta A}{A} \qquad 48.6$$

The modulation index for an AM signal is the ratio of the maximum amplitude deviation of the modulated carrier to the amplitude of the carrier. The modulation index is sometimes expressed as a percentage.

4. SINGLE-SIDEBAND AMPLITUDE MODULATION

The double-sideband signals of Sec. 48.3 double the bandwidth. Further, the same information is carried in the upper and lower sidebands, using power but not providing additional information. Elimination of one of the sidebands results in a signal called a *single-sideband* (SSB) signal. Single-sideband radios find application in marine communication.

5. VESTIGIAL SIDEBAND AMPLITUDE MODULATION

The filtering requirements for an SSB signal are stringent. An ideal bandpass filter with an infinitely sharp cutoff or an equivalent technique is required. Attenuating some, but not all, of one sideband minimizes the bandwidth and lowers the power without the strict filtering requirements. This type of modulation results in a signal called a *vestigial sideband* (VSB). Vestigial sideband amplitude modulation is used to transmit the video portion of commercial television signals.

6. QUADRATURE MULTIPLEXING AMPLITUDE MODULATION

In Eq. 48.2, the assumption is made that the information signal used for modulation, $s(t)$, is real. If this assumption is removed, the general modulated signal is represented as[3]

$$\begin{aligned} s_{\mathrm{mod}}(t) &= \mathrm{Re}\Big(\big(s_x(t) - js_y(t)\big)e^{j\omega_c t}\Big) \\ &= s_x(t)\cos\omega_c t + s_y(t)\sin\omega_c t \\ &= I(t) + Q(t) \end{aligned} \qquad 48.7$$

The in-phase component of the modulated signal is represented by $I(t)$, while the quadrature component is $Q(t)$. Both of the components are real. Because the cosine and sine functions are orthogonal, two signals

[2]The term "index" is sometimes reserved for frequency modulation of a sinusoidal signal.

[3]The real portion of the product of two complex values, $\mathrm{Re}(z_1z_2)$, is $\mathrm{Re}(z_1)\mathrm{Re}(z_2) - \mathrm{Im}(z_1)\mathrm{Im}(z_2)$. The $-j$ comes from the inductive nature of the signal.

can be transmitted simultaneously. Although this is modulation, the process is referred to as *quadrature multiplexing* (QM).

7. AMPLITUDE MODULATION: BANDWIDTH AND POWER

Amplitude modulation (AM) is a frequency translation. If the modulating signal, that is, the baseband signal, has low-pass characteristics and a bandwidth of BW, the allowable change in frequency of the modulated signal is

$$\text{BW} \le \Delta f \le 2(\text{BW}) \quad \text{48.8}$$

The lower limit corresponds to the single-sideband signal, the upper limit to the double-sideband signal.

The power in an AM signal is used by the carrier and the modulating signal. The average power is

$$\begin{aligned} P &= \frac{A^2}{2}\left((ks(t))^2_{\text{ave}} + 1\right) \\ &= P_{\text{signal}} + P_{\text{carrier}} \end{aligned} \quad \text{48.9}$$

The first term of Eq. 48.9 is the signal power and the second is the carrier power. The signal power carries the information and is thus useful power. The carrier power lowers the efficiency. The percent efficiency is

$$\eta = \frac{P_{\text{signal}}}{P_{\text{signal}} + P_{\text{carrier}}} \times 100\% \quad \text{48.10}$$

In double-sideband suppressed-carrier modulation, no power is used for a carrier and half the power exists in each sideband. The sidebands carry identical information, however, and the information-bearing power is one-half the total. In single-sideband suppressed-carrier modulation, all the power used carries information. Though they are more efficient, SSB-SC signals are complicated to generate and receive. Further, the use of carriers simplifies demodulation.

8. ANGLE MODULATION

The entire term in parentheses in Eq. 48.1 determines the angle of the carrier. That is, $\phi(t) = \omega_c(t) + \theta$. The phase angle can be manipulated in direct proportion to the information signal, $s(t)$, by

$$\phi(t) = \omega_c t + k_p s(t) + \theta_0 \quad \text{48.11}$$

The phase angle manipulation in Eq. 48.11 is called *phase modulation* (PM).

Defining the instantaneous frequency, ω_i, as the rate of change of the phase angle provides a second method of modulating the phase angle. The frequency is manipulated in Eq. 48.12.

$$\omega_i(t) = \omega_c + k_f s(t) \quad \text{48.12}$$

The phase angle manipulation in Eq. 48.12 is called *frequency modulation* (FM).

9. FREQUENCY MODULATION

Frequency modulation (FM) changes the carrier wave frequency in proportion to the instantaneous value of the modulating wave. If the modulating wave is a single-frequency sinusoid, $s(t) = a\cos\omega_{\text{mod}}t$, whose peak amplitude produces a maximum frequency deviation of $\Delta\omega$, a modulated FM signal is represented as[4]

$$\begin{aligned} s_{\text{mod}}(t) &= A\cos\left(\omega_c t + \left(\frac{\Delta\omega}{\omega_{\text{mod}}}\right)\sin\omega_{\text{mod}}t\right) \\ &= A\cos(\omega_c t + m_{\text{FM}}\sin\omega_{\text{mod}}t) \end{aligned} \quad \text{48.13}$$

The dimensionless ratio factor in Eq. 48.13 is called the *frequency modulation index*, *index of modulation*, or *modulation factor*, and is defined mathematically as

$$m_{\text{FM}} = \frac{\Delta\omega}{\omega_{\text{mod}}} \quad \text{48.14}$$

The modulation index for an FM signal is the ratio of the maximum frequency deviation of the modulated carrier to the modulating frequency. The modulation index is sometimes expressed as a percentage. The maximum frequency deviation is

$$\Delta\omega = ak_f \quad \text{48.15}$$

The term a is the amplitude of the modulating signal, that is, the baseband or information signal. The *frequency modulator constant*, k_f, has units of radians per second per volt.

The audio portion of television broadcasts is transmitted using frequency modulation. Additionally, FM radio stations use frequency modulation.

10. PHASE MODULATION

Phase modulation (PM) changes the carrier wave angle in proportion to the instantaneous value of the modulating wave. If the modulating wave is a single-frequency sinusoid, $s(t) = a\cos\omega_{\text{mod}}t$, whose peak amplitude

[4]Analysis of FM and PM signals requires the use of Fourier transforms, Bessel functions, and, alternatively, Fast Fourier Transforms (FFTs).

produces a maximum frequency deviation of $\Delta\omega$, a modulated PM signal is represented as

$$\begin{aligned} s_{\text{mod}}(t) &= A\cos(\omega_c t + (\Delta\theta)\cos\omega_{\text{mod}} t + \theta_0) \\ &= A\cos\left(\omega_c t + \left(\frac{\Delta\omega}{\omega_{\text{mod}}}\right)\cos\omega_{\text{mod}} t + \theta_0\right) \end{aligned} \quad 48.16$$

Comparing Eq. 48.16 and Eq. 48.13 illustrates the difference between PM and FM. Phase-modulated signals are proportional to the modulating signal, $s(t)$.[5] Frequency-modulated signals are proportional to the integral of the modulating signal for an input signal cosine function (hence the sine term in Eq. 48.13). The dimensionless ratio factor in Eq. 48.16, $\Delta\theta$, is mathematically identical to the frequency modulation index and is defined as

$$\Delta\theta = \frac{\Delta\omega}{\omega_{\text{mod}}} \quad 48.17$$

The term $\Delta\theta$ is the peak phase deviation. The peak phase deviation is related to the maximum frequency deviation by

$$\Delta\omega = ak_f\omega_{\text{mod}} = \Delta\theta\omega_{\text{mod}} \quad 48.18$$

In PM, then, the peak frequency deviation, $\Delta\omega$, is proportional to the amplitude of the modulating signal, a, and the modulating signal frequency, ω_{mod}. For this reason, phase modulation is less desirable for use if the frequency band is limited, although it has advantages over FM in the ease with which demodulation occurs.

11. ANGLE MODULATION: BANDWIDTH AND POWER

Angle modulation bandwidth analysis requires the use of Bessel functions of the first kind. Using the properties of these functions, an approximation for the bandwidth of an FM or PM signal is given by *Carson's rule*, which is

$$\begin{aligned} \text{BW} &\approx 2\omega_{\text{mod}}(m+1) \\ &= 2(\Delta\omega + \omega_{\text{mod}}) \\ &\approx 2(\Delta\omega + \omega_{\text{max}}) \end{aligned} \quad 48.19$$

The modulation index, m, is given by Eq. 48.14 for an FM signal or the mathematical equivalent in Eq. 48.17 for a PM signal. The first two equalities in Eq. 48.19 estimate the bandwidth when the modulating signal is a single-frequency sinusoid.[6] The bandwidth of an FM signal modulated by something other than a single-frequency sinusoid is most accurately estimated using the maximum modulating frequency, ω_{max}, instead of the actual modulating frequency as shown in the third part of Eq. 48.19.

When $\Delta\omega$ is larger than ω_{max}, the signal is known as *wideband*. When $\Delta\omega$ is smaller than ω_{max}, the signal is known as *narrowband*.

The calculation of power for FM or PM signals is more complicated than for AM signals and requires the use of advanced engineering mathematics.

12. DIGITAL MODULATION

Digital modulation is the mapping of binary symbols into signals, with the mapping method selected for a given communication channel and its characteristics.[7] Digital modulation is also defined as a method of placing digital information on a channel in discrete phase or frequency states controlled (i.e., modulated) by the digital information signal.

Three basic types of digital modulation involve modifications to the amplitude, phase, and frequency of a carrier. Those methods modifying the amplitude are called *pulse amplitude modulation* (PAM) *techniques*. Methods that map digital signals to changes in phase or frequency are called *pulse code modulation* (PCM) *techniques*. The term PCM derives from the digital signal being coded or *keyed* to certain phases or frequencies.

Digital information is represented as logic zeros and ones. *Binary modulation* is the mapping of the zeros and ones into two different waveforms. *Multilevel modulation* is the mapping of n binary digits onto $M = 2^n$ different waveforms.

The theoretical foundation for digital modulation is much the same as that for analog modulation, though discrete mathematics is used. An additional design issue in digital modulation is quantization. *Quantization* is the process whereby a continuum of wave values (a continuous time signal) is represented as a finite number of assigned (quantized) levels (a discrete time signal). Quantization is accomplished by sampling the analog signal at various intervals.

Sampling results in *quantization noise*, also known as *quantumization distortion* or *quantumization noise*. Such noise is the inherent distortion introduced in the process of conversion of the analog signal (with a range of values) into a digital signal (with quantized values).[8] In Fig. 48.4, the quantization noise is the difference between the analog signal and the digital output. This difference is added to the signal as noise.

Assuming uniform quantization error and uniform signal distribution across the quantization levels in an ideal

[5]The modulating signal is alternatively referred to as the information, input, or baseband signal. Sometimes the modulating wave is simply called the signal, as opposed to the carrier.

[6]A single-frequency sinusoid modulating signal is assumed throughout this chapter.

[7]The terms *mapping*, *coding*, and *keying* are used interchangeably in many instances.

[8]Consider the opposite process, converting a digital signal to an analog waveform. The error introduced in such a process is called *aliasing*.

Communications

Figure 48.4 Quantization Noise

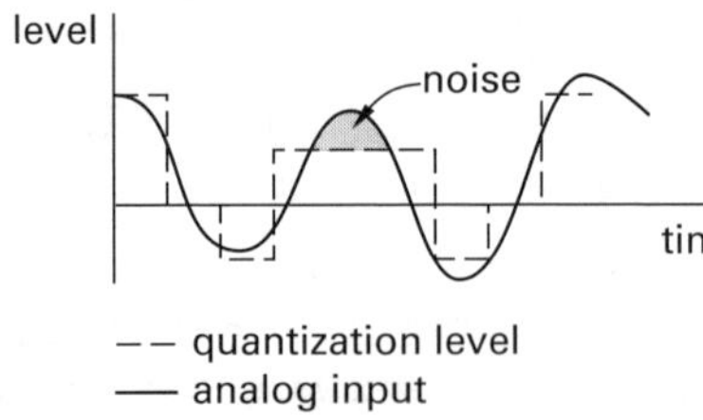

analog-to-digital converter, the signal-to-quantization-noise ratio (SQNR) is calculated from Eq. 48.20. Full amplitude sawtooth and triangular waveform signals also meet the requirements for use of Eq. 48.20. Q is the number of bits in the converter.

$$\text{SQNR} = 20\log_{10} 2^Q \approx 6.02Q \quad \text{[dB]} \qquad \textit{48.20}$$

When the signal is a full amplitude sine wave, the SQNR is calculated by Eq. 48.21.

$$\text{SQNR} \approx 1.761 + 6.02Q \quad \text{[dB]} \qquad \textit{48.21}$$

The actual quantized levels are called *codes*, and the process of representing or assigning the levels is termed *coding*. The basics of both sampling and coding are described in Chap. 18. For coding to be accurate, the transmitter and receiver in any communication must be synchronized. Synchronization is the process of maintaining one operation in step with another. For digital data, this means knowing or sensing the bit intervals (see Sec. 48.13). A *coherent* system is one in which the receiver is aware of the bit interval start and stop times.

13. BASEBAND DIGITAL-DATA SIGNALS

Some baseband digital signal formats used to transmit digital data are illustrated in Fig. 48.5. The time frames shown are called *bit intervals*. Each bit interval represents a single piece of information. Times t_1 through t_5 are bit intervals for the sample digital signal transmitted, 10110.

The data signal in Fig. 48.5(a) is *unipolar* in that it has only one polarity, in this case positive. The data signal in Fig. 48.5(b) is *bipolar* with the positive and negative voltage maximums constituting the poles. The bipolar signal shown in Fig. 48.5(b) is called *polar nonreturn-to-zero* (NRZ).[9] In a *return-to-zero* (RZ) *signal*, the data pulse returns to the logic 0 state *before* the next bit interval, as shown in Fig. 48.5(c). The bandwidth of an RZ signal is therefore not as wide as an NRZ signal.

If the channel is not DC coupled, NRZ and RZ signals cannot be used because the average value of an NRZ or RZ signal is less than one. Therefore, if DC is not transmitted, the value of logic 1 is never available.

[9]The NRZ signal is also called the NRZ-L, where the L stands for logic, meaning the voltage level stands for the logic level. Both positive logic and negative logic are termed NRZ-L.

Figure 48.5 Baseband Digital Signals

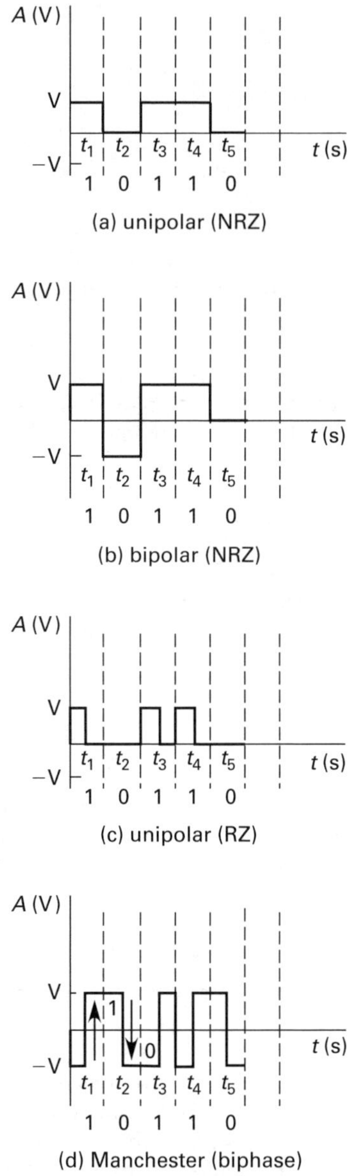

(Times t_1 through t_5 are bit intervals for the sample digital signal transmitted, 10110.)

A data signal that allows signal transmission for an AC-coupled channel is called *alternate-mark-inversion* (AMI). AMI is identical to unipolar RZ with every other nonzero pulse inverted, which also uses the polarity change to differentiate the one and the zero.

14. INVERSION AND STATIC

NRZ signals have two specific problems to overcome, inversion and static. *Inversion*, which can have many causes, is a condition in which the logic levels are reversed (inverted) in the channel. This results in the interpretation of the +V signals as −V signals, and vice versa. So, every bit is in error. *Static* in this instance means "without motion or change."[10] That is, static occurs when no change in logic level occurs from one bit interval to the next. Both conditions result in erroneous data, inversion by directly altering the information, and static by potential loss of timing information (synchronization) between bit intervals.

The inversion problem is overcome by using the NRZ-M or NRZ-S signals.[11] In the NRZ-M system, logic level one is represented by a change in the signal in successive bit intervals while a zero is represented by no change. Inversion has no impact. The NRZ-S signal is the reverse of the NRZ-M system in that logic level zero is represented by a change and one by no change.

The static problem remains in both the NRZ-M and NRZ-S signals in that consecutive bits may still remain unchanged. That is, a series of ones or zeros results in potential synchronization difficulties. The *Manchester signal* is a *biphase signal* that solves synchronization difficulties by representing a logic 1 by a change from low to high voltage and a logic 0 by a change from high to low. The Manchester signal is shown in Fig. 48.5(d). The inversion problem, however, remains.

A combination of *differential encoding* such as NRZ-M and NRZ-S with *biphase encoding* such as the Manchester signal solves both inversion and synchronization problems. Numerous combinations and signal types exist and are known either as *signal formats* or *codes*.

15. INTERSYMBOL INTERFERENCE

Another issue in digital system design is the interference of one pulse of information with another across bit intervals. Because the frequency response characteristic of any channel is not ideal, the pulse being sensed to create the logic 0 or 1 may be wider than the bit interval, with a pulse amplitude that varies. These two conditions are called *delay distortion* and *amplitude distortion*, respectively. The resulting excess energy from one pulse bit interval, also called a *keying interval*, interferes with the reception of the information signal in successive intervals. The interference is called *intersymbol interference* (ISI). ISI is avoided through use of an appropriate low-pass filter. By selecting a filter such that the pulse waveform represents the $(\sin x)/x$ shape, the waveform crosses the time axis at the bit intervals, and no interference is obtained. A zero ISI situation is shown in Fig. 48.6. The technique for selecting the filter to produce this result uses the *Nyquist criterion*.

In many cases, zero interference is not achieved and distortion of the signal exists at the receiver. An *equalizer* attempts to correct the received signal for any amplitude-frequency or phase-frequency differences from the selected channel signal format. The process of removing distortion is called *channel equalization*.

[10]Static is also used to mean the distortion of a signal by noninformation-bearing electromagnetic sources or environmental effects.

[11]The M stands for "mark" and the S for "set," from telegraph terminology.

Figure 48.6 *Zero Intersymbol Interference*

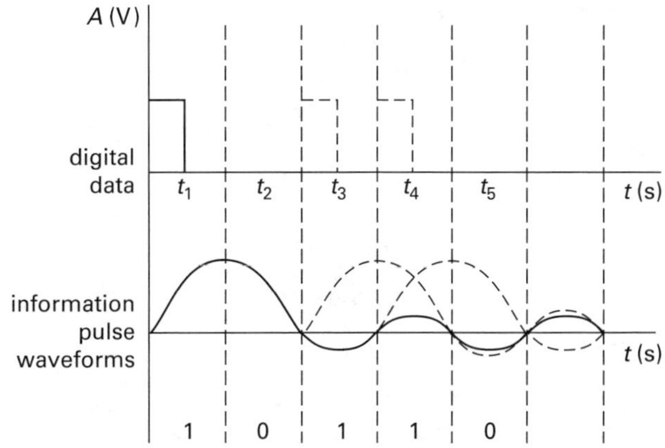

16. PULSE AMPLITUDE MODULATION

Pulse amplitude modulation (PAM) is a linear modulation scheme in which a signal is modified by a set of discrete amplitude values, A. The amplitudes may take on any number of set amplitude values. If only two amplitude values are used, with one being $+A$ and the other zero, the phrase *asymmetric amplitude-shift keying* (AASK) is used to describe the modulation (see Fig. 48.5(a) and (c)). If the two amplitude values are $+A$ and $-A$, the term *amplitude-shift keying* (ASK) describes the modulation (see Fig. 48.5(b) and (d)). The constellation of possible values for a 4-bit PAM system is shown in Fig. 48.7(a). From the figure, PAM is one-dimensional modulation.

17. PULSE CODE MODULATION

Pulse code modulation (PCM) is modulation in which a sampled analog signal is divided into quantized levels represented by binary codes. The binary codes represent the amplitude, phase, and/or frequency of the signal. PCM comprises three operations: sampling, which generates a PAM signal; quantization, which divides the amplitude of the signal into discrete levels; and encoding, which assigns binary codes to the levels.[12]

A simple one-dimensional PCM technique is to turn the carrier signal on and off. If the carrier signal is simply turned on and off, the term *binary on-off keying* (BOOK) is used. A two-dimensional linear modulation scheme that changes the phase by a set number of values is called *phase-shift keying* (PSK). An example is shown in Fig. 48.7(b). If only two phases are used, the phase shift is 180° and the term *binary phase-shift keying* (BPSK) is applied. When the frequency of the carrier is changed by a set number of values, the term *frequency-shift keying* (FSK) is applied. If only two

[12]Digital data cannot be transmitted on an antenna of reasonable size. Similar to analog data, frequency translation must occur. PCM is the method of translation (see Sec. 48.1).

Communications

Figure 48.7 *Digital Modulation Constellations*

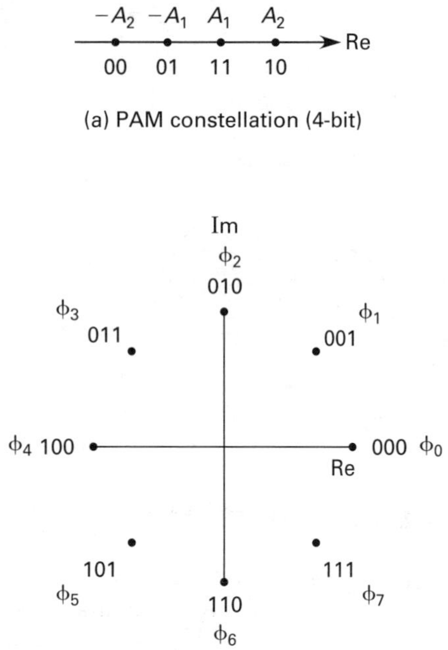

(a) PAM constellation (4-bit)

(b) PSK constellation (8-bit)

frequencies are used, the modulation is called *binary frequency-shift keying* (BFSK). Frequency modulation is normally multidimensional.

The term *binary* is used when a signal is modulated to only two possible values. The term *M-ary* is used for signals mapped to multiple modulation codes.[13]

18. QUADRATURE AMPLITUDE MODULATION

A two-dimensional linear modulation scheme that modifies the amplitude and phase is called *quadrature amplitude modulation* (QAM) and is shown in Fig. 48.8.[14]

19. DIGITAL MODULATION: BANDWIDTH AND POWER

The bandwidth of a digital signal can be defined in several ways. A familiar definition is the *half-power bandwidth*, which is the interval between the two frequencies at which the power spectrum is 3 dB below its peak value. The *equivalent noise bandwidth* is equivalent to a rectangle with a height equal to the maximum power spectrum density and an area equal to one-half the power of the modulated signal. There are numerous other bandwidth definitions.

[13]A *modem*, that is, a *modulator-demodulator*, is an interface device that uses PCM.

[14]The phase, that is, the angle associated with the digital points in QAM, determines the value of the imaginary portion and subsequently impacts the amplitude because $|\alpha + j\beta| = \alpha^2 + \beta^2$.

Figure 48.8 *QAM Constellation (16-bit)*

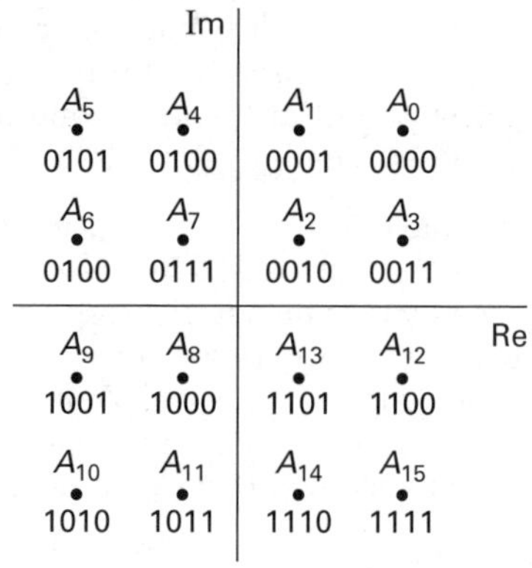

In order to compare modulation schemes, the *Shannon bandwidth* is used.

$$\text{BW} = \frac{D}{2T} \tag{48.22}$$

The variable D is called the *dimensionality* of the signal set. That is, it is the number of orthonormal signals within a given mapping (coding) scheme. The time duration over which the signals are defined is T.

The power is normally considered in terms of its ratio with the channel noise. The power is the product of the binary digit rate, R_s, in digits per second, and the bit energy, E_{bit}.

$$P = \frac{E_{\text{ave}}}{T} = E_{\text{bit}} R_s \tag{48.23}$$

The bit energy and digit rate are

$$E_{\text{bit}} = \frac{E_{\text{ave}}}{\log_2 M} \tag{48.24}$$

$$R_s = \frac{\log_2 M}{T} \tag{48.25}$$

The term M is the number of waveforms upon which the binary digits are mapped and is given by $M = 2^n$ where n is the number of binary digits available.

Assuming white Gaussian noise with a power spectral density of $N_0/2$, the *signal-to-noise ratio* (SNR) is

$$\text{SNR} = \frac{S}{N} = \frac{P}{N_0(\text{BW})} = \left(\frac{E_{\text{bit}}}{N_0}\right)\left(\frac{R_s}{\text{BW}}\right) \tag{48.26}$$

The first term in Eq. 48.26 is the energy transmitted per bit divided by white Gaussian noise power spectral density. The second term, the bit rate divided by the bandwidth, is called the *bandwidth efficiency*.

20. DIGITAL ERROR PROBABILITY

The performance of a given digital modulation scheme can be measured by the *symbol error probability*, $\mathcal{P}(e)$, which is the total probability that a given waveform is detected incorrectly. The probability of corruption for a single bit is called the *bit error probability*, $\mathcal{P}_b(e)$, also called the *bit error rate* (BER). Because each symbol carries $\log_2 M$ bits, a signal symbol error affects, at a minimum, one bit, and at a maximum, $\log_2 M$ bits.[15]

The overall relationship is

$$\frac{\mathcal{P}(e)}{\log_2 M} \leq \mathcal{P}_{\text{bit}}(e) \leq \mathcal{P}(e) \qquad 48.27$$

The equation for the probability of error, $\mathcal{P}(e)$, varies with the digital modulation scheme. Table 48.1 gives the equations for the major techniques. Appendix 48.A and App. 48.B list values for the *error function*, erf(x), also called the *probability function*, from which the *complementary error function*, erfc(x), mentioned in the table can be found.

$$\text{erfc}(x) = \frac{2}{\sqrt{\pi}}\int_x^\infty e^{-x^2}\,dx = 1 - \text{erf}(x). \qquad 48.28$$

21. FORWARD ERROR CORRECTING

Once detected, errors can be corrected by retransmission of the original signal. Redundancy in the transmitted signal ensures errors will be detectable.[16] Retransmission of a signal is not always feasible. The collection of techniques designed to detect errors and correct those errors without retransmission is called *forward error correcting* (FEC). Two general types of FEC codes are *block codes* and *convolution codes*.

Communications

Table 48.1 *Digital Modulation Error Probabilities, $\mathcal{P}(e)$*

amplitude-shift keying (ASK)*	phase-shift keying (PSK)	frequency-shift keying (FSK)
$\left(1-\frac{1}{M}\right)\text{erfc}\left(\sqrt{\left(\frac{3\log_2 M}{M^2-1}\right)\left(\frac{E_{\text{bit}}}{N_0}\right)}\right)$	$\approx \text{erfc}\left(\sqrt{\log_2 M\left(\frac{E_{\text{bit}}}{N_0}\right)}\sin\frac{\pi}{M}\right)$	$\frac{1}{2}\text{erfc}\left(\sqrt{\log_2 M\left(\frac{E_{\text{bit}}}{2N_0}\right)}\right)$ lower bound $\left(\frac{M-1}{2}\right)\text{erfc}\left(\sqrt{\log_2 M\left(\frac{E_{\text{bit}}}{2N_0}\right)}\right)$ upper bound

*Amplitude-shift keying is also known as pulse amplitude modulation (PAM).

[15]For example, a 4-ary scheme ($M=4$) contains $\log_2 4 = 2$ bits; that is, 00, 01, 10, and 11. If one of these symbols is in error (corrupted), at most two bits are affected.

[16]Claude Shannon proved in 1948 that arbitrary reliability is possible as long as a fixed portion of the transmitted signal is redundant.

49 Signal Multiplexing

1. Fundamentals 49-1
2. Frequency Multiplexing 49-1
3. Time Multiplexing 49-1
4. Space Multiplexing 49-2
5. Spread-Spectrum Multiplexing 49-2

Nomenclature

$s(t)$	signal (time function)	V
t	time	s

Symbols

ω	angular frequency	rad/s

Subscripts

n	number

1. FUNDAMENTALS

The signal formats described in Chap. 47, for the most part, are capable of carrying one information signal.[1] Once at the receiver, the information may have processing limitations unrelated to the channel. For example, a human listening to a voice signal will hear the signal as a continuous sound even if that signal is broken up into small digital pieces. A computer processing the information coming from a receiver is limited in its rate of information intake. In order to maximize the efficiency of a channel, several information signals can be sent. Sending multiple signals over a single channel in a manner that still allows for individual signal recognition at the receiver requires *multiplexing*. For multiplexing to be successful, the signals must not overlap. Mathematically, this means that the signals must be orthogonal.

2. FREQUENCY MULTIPLEXING

Frequency multiplexing or *frequency-division multiplexing* (FDM) is the process of sending two or more signals over a common channel using a different frequency band for each. The process is shown in Fig. 49.1.

[1]Where the carrier is split into upper and lower sidebands, different information can be sent in each sideband.

Figure 49.1 *Frequency Multiplexing*

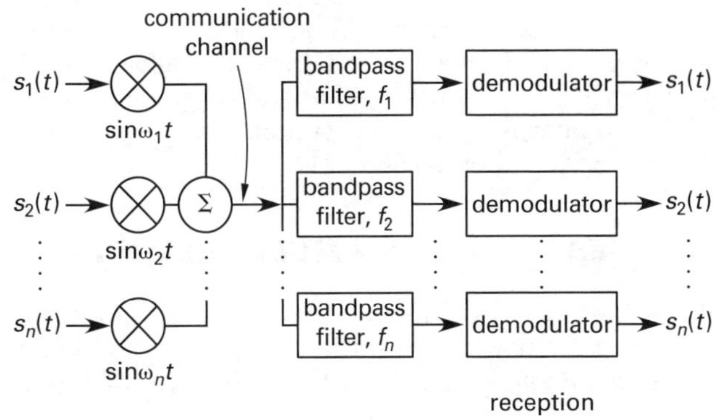

The original signals are combined with a sinusoid of the selected frequency. By selecting sinusoids of varying frequencies, the signals can be stacked in frequency without overlapping. To ensure that the signals remain separate, a *guard slot*, which is a frequency range designed to carry no signal, is included between the utilized frequencies to prevent inadvertent overlap of signals.

3. TIME MULTIPLEXING

The time domain and the frequency domain are mathematically related. Furthermore, they have complete duality with each other. Therefore, if a signal can be frequency multiplexed, it can be time multiplexed. The process is shown in Fig. 49.2.

Figure 49.2 *Time Multiplexing*

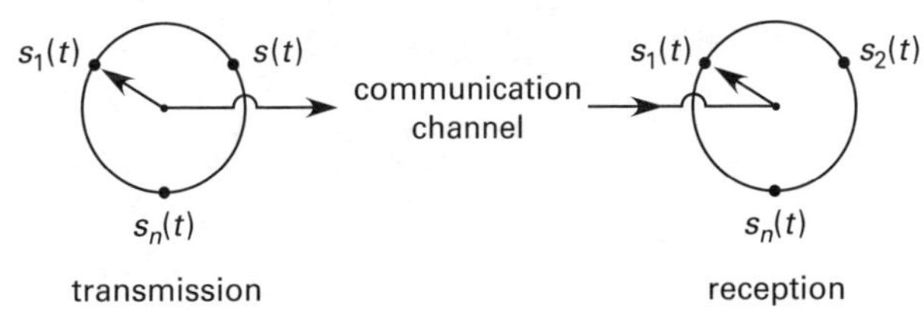

Time multiplexing or *time-division multiplexing* (TDM) is the process of sending two or more signals over a common channel using successive time intervals for the different signals.

4. SPACE MULTIPLEXING

Space multiplexing or *space-division multiplexing* (SDM) is the process of sending two or more signals over a common channel using separate geographical areas for the different signals. A variety of means are used to accomplish the separation. Lowering the power to deliberately minimize the signal coverage areas, as is done with cellular radio/phone systems, allows signals to use the same frequency or time assignments. Spot beam antennae focus signals into specific coverage areas, as is done in some satellite systems, allowing multiple signals to share frequency or time assignments. *Polarization multiplexing*, a type of space multiplexing, allows signals to share the same frequency, time, and geographical areas while obtaining orthogonal separation through the use of different polarization planes.

5. SPREAD-SPECTRUM MULTIPLEXING

When the transmission signal bandwidth is deliberately wide with the power spread thinly over the entire band, it is called a *spread-spectrum transmission*. By keeping the power near or below the noise level, narrow-band radios can transmit within the band without interference. Spread-spectrum transmission also minimizes jamming potential, improves privacy, and enhances security. Numerous techniques have been devised for multiplexing over this type of channel.

Code-division multiple access (CDMA) is the process by which each of the multiple signals on a single channel is assigned a pseudorandom noise code independent of the others (orthogonal). For example, in *frequency hopping* a carrier's frequency moves around the allowed bandwidth as determined by a pseudonoise sequence generator feeding a frequency hopper (controller). As long as the appropriate frequency hopper is receiving the signal, the original information can be retrieved.

Time-division multiple access (TDMA) is the process by which multiple signals transmitted over a single channel are each assigned a separate time slot. As a result, the information signals are transmitted in bursts allotted to specific transmitters.

Other types of multiple-access techniques exist.

50 Communication Systems and Channels

1. Systems and Channels Overview 50-1
2. Telephony 50-2
3. Broadcast Radio 50-2
4. Broadcast Television 50-2
5. Satellite Systems 50-3
6. Wireless Communication 50-3
7. Optical Communications 50-3
8. Electromagnetic Compatibility 50-4

Nomenclature

c	speed of light, 2.9979×10^8	m/s
C	channel capacity	bits/s
d	horizon distance	m
f	frequency	Hz
h	antenna height	m
i	detected current	A
N	noise power	W
N_0	noise spectral density	W/Hz
P	power	W
S	average signal power	W

Symbols

λ	wavelength	m
ρ	responsivity	A/W

1. SYSTEMS AND CHANNELS OVERVIEW

A *communication system* is a device in which electrical pulses from a transmission point are accurately reproduced at a reception point. A *communication channel* is the medium that connects the transmission point to the reception point.[1]

A communication channel may consist of copper wire, air, coaxial cable, a waveguide, space, fiber-optic material, or other media. A communication system may be telegraph, telephone (including wireless, cellular, and satellite), radio and television (broadcast and cable), satellite, local- and wide-area computer networks, and so on. In practice, the terms are used interchangeably, with some systems taking their names from the channels or even the frequencies used, such as fiber-optic and microwave systems.[2]

[1]A communication channel is also the band of frequencies assigned or allocated for a particular purpose. For example, a standard radio broadcasting channel is 10 kHz wide, while a television channel is 6 MHz wide.

[2]The term "microwave" dates to the 1940s when an electromagnetic wave of less than 1 m was difficult to utilize, and was therefore deemed to be a "micro-wave."

The channel capacity of an analog system is determined by the maximum frequency deviation. The capacity, C, of a digital channel in bits per second (bps) is determined by the signal-to-noise ratio and the bandwidth as shown in Eq. 50.1.

$$\begin{aligned} C &= (\mathrm{BW})\log_2\left(1+\frac{S}{N}\right) \\ &= (\mathrm{BW})\log_2\left(1+\frac{S}{N_0(\mathrm{BW})}\right) \end{aligned} \qquad 50.1$$

Equation 50.1 should be understood in the context of digital bandwidth and power (see Sec. 48.19). The value of the channel capacity as the bandwidth approaches infinity is called the *Shannon limit* and is[3]

$$C = 1.44\left(\frac{S}{N_0}\right) \qquad 50.2$$

The higher the frequency of the channel, the greater the bandwidth and capacity. *Microwave channels* involve frequencies of approximately 1–100 GHz, corresponding to 30 cm and 0.3 cm wavelengths. Microwave channels are treated as high-frequency transmission lines (see Chap. 45). Microwave signals are line-of-sight signals. An antenna has a limited coverage area. Considering the curvature of the Earth, the radio horizon, d, in kilometers, depends on the antenna height, h, in meters, as shown in Eq. 50.3.

$$d_{\mathrm{km}} = 4.1\sqrt{h_{\mathrm{m}}} \qquad 50.3$$

Optical channels use electromagnetic waves of approximately 10^{-6} m, correlating to 3×10^{14} Hz. Specially designed fiber-optic cables and connections must be used for the transmission of optical wavelength signals.

The information transferred is generally classified as voice, video, facsimile, data, or a combination of these. The term *broadcast* refers to radio or television. *Telecommunications* refers to communications over long distances but is generally used in reference to mobile radio, wireless, and cellular voice communication or local-area and wide-area computer networks.

In the United States, authority to control telecommunications comes from the *Code of Federal Regulations*, Title 47, Chap. 73, and is assigned to the Federal Communications Commission (FCC). Specific rules governing the frequency assignments of such communications are created by the National Telecommunications and Information Agency (NTIA) and published as the

[3]This result is a direct consequence of the *Shannon coding theorem*.

Manual of Regulations and Procedures for Federal Radio Frequency Management. This manual is commonly referred to as the "Redbook" because of the color of its cover.

2. TELEPHONY

The *telephone* is a system for converting sound waves into current variations for transmission to distant points. It is meant primarily for voice communication, although its uses are more numerous now. Because it was designed for human speech, the channel or band used is limited to between 0.3 kHz and 3.4 kHz. Bandpass filter imperfections and trunk signaling, which is done out of the normal band, require the voice channel to be standardized as a 4 kHz channel. The general concept of a telephone network is illustrated in Fig. 50.1.

Figure 50.1 *Public Switched Telephone Network*

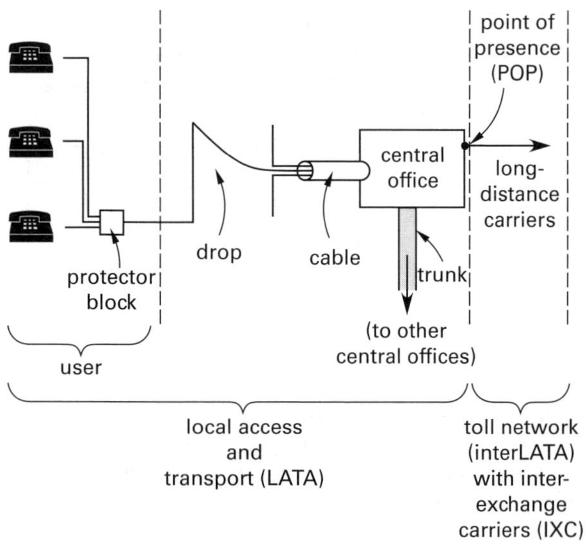

The *public switched telephone network* (PSTN) represented in Fig. 50.1 starts at the user or customer.[4] Multiuser lines are connected using twisted copper wire to a single point called a protector block. The *protector block* provides overvoltage protection to the network. Many user lines are then bundled in a cable that is buried underground or strung between telephone poles and run to the central office. The central office provides the first stage of switching, either to other central offices via a *trunk line* or to long-distance carriers. The interface point between the long-distance carriers and the central offices is called the *point of presence* (POP). The network up to the POP is called the local access and transport area (LATA). Beyond the POP, *interexchange carriers*, such as AT&T™, operate. The network described here is a generic system often referred to as *plain old telephone service* (POTS).

[4]Upgraded systems that are digital end to end are called *integrated services digital networks* (ISDNs).

3. BROADCAST RADIO

Broadcast radio refers to AM-band, shortwave, and FM-band transmissions meant for free reception by the public. Radio in the United States is regulated by the Federal Communications Commission (FCC). The rules governing broadcasting are published in the *Code of Federal Regulation* (CFR), specifically in Title 47 and various chapters. The chapters are referred to as *FCC rules*. For example, the AM bandwidth is covered in 47 CFR 73.14, that is, Title 47, Chap. 73.14.

The standard AM broadcast channel is from 535 kHz to 1705 kHz, with the carriers assigned center frequencies of 540–1700 kHz in 10 kHz steps. A total of 117 stations can technically exist in the same area (FCC Rule 73.14). The stations operate at anywhere from 0.1 kW to 50 kW depending on the type and class of channel. A *clear channel* is a station assigned to serve a wide area. Clear channels are so called because they are legally protected from interference. *Regional channels* serve population centers. *Local channels* serve small communities. Within the various channel designations there are a variety of classifications.

Above the standard AM band, shortwave radio is used. The band varies from 5950 kHz to 26,100 kHz. These stations are intended to be received by the general public in another country.

The standard FM broadcast channel is from 88 MHz to 108 MHz. One hundred numbered channels (stations) are assigned with bandwidths of 200 kHz each (FCC Rule 73.201). The stations operate between 0.25 kW and 100 kW depending on the classification. The classification depends on the *effective radiated power* (ERP) and the *height above average terrain* (HAAT) as compared to a class contour.

Radio receivers mix the incoming signal frequency with an *intermediate frequency* (IF), which then generates the sum and difference of the two. This process is called *heterodyning*.[5] The receivers that are designed to convert all incoming carrier signals to common intermediate-frequency (IF) values using heterodyning are called *superheterodyne receivers*. The IF signal is 455 kHz for AM and 10.7 MHz for FM.

4. BROADCAST TELEVISION

Television (TV) broadcast channels are designated by channel number and service level. *Low-band VHF*, consisting of channels 2 through 6, ranges from 54 MHz to 88 MHz. *High-band VHF*, consisting of channels 7 through 13, ranges from 174 MHz to 216 MHz. UHF signals consist of channels 14 through 69, which range from 470 MHz to 806 MHz. Each channel is

[5]Heterodyning allows a single bandpass filter to be used with all the incoming frequencies shifted to within the IF band.

6 MHz wide. The television channels operate from 100 W to 5000 kW.[6]

High-definition television (HDTV) is a digital system designed to fit within the 6 MHz bandwidth of standard broadcasting channels.

5. SATELLITE SYSTEMS

Satellites provide a line-of-sight communications platform with wide area coverage. Satellites are generally placed in three orbit categories, depending on the application. *Low Earth orbit* (LEO) is between 500 km (310 mi) and 900 km (560 mi) above the surface of the Earth. *Medium Earth orbit* (MEO) is between 5000 and 12 000 km (3110 and 7460 mi). *Geostationary Earth orbit* (GEO) is at approximately 36 000 km (22,370 mi) above the surface of the Earth. The term *geosynchronous orbit* is used also, though a geostationary orbit is a geosynchronous orbit that lies in the Earth's equatorial plane. That is, a geostationary orbit is a special geosynchronous orbit with a zero inclination.

A *geostationary satellite* is one that orbits the Earth from west to east at the GEO altitude in such a manner that it makes one revolution in 24 hr. That is, it is synchronous with the Earth's rotation. This means a GEO satellite remains fixed over a given surface position. The satellite *footprint* (the area that is covered by a communications satellite) in GEO is fixed as well. Three satellites properly positioned in GEO can provide communication coverage of the entire surface of the Earth.

The frequency bands used vary from 1 GHz to 30 GHz. The band designations have historical significance only and are given in Table 50.1. The transmission of satellite communications is similar to that of antennas, as described in Chap. 46.

Table 50.1 *Satellite and Radio Communication Frequency Bands*

band	frequency range (GHz)
L	1–2
S	2–4
C	4–8
X	8–12.5
Ku	12.5–18
K	18–26.5
Ka	26.5–40

6. WIRELESS COMMUNICATION

Wireless communication employs an antenna for radiating and receiving information using electromagnetic waves. Wireless communication encompasses a broad spectrum of systems.

Wireless telephones that have short ranges are usually called *cordless*. *Cellular telephones* (cell phones) achieve long-range capability by utilizing antennae that provide local coverage in an area called a *cell*. The system antenna in the cell containing the user's telephone controls the communication process. When the user moves into another cell, the communication process is passed to that cell.[7]

Wireless radio is termed *portable radio* if the transceiver can be moved but must be stationary while in use. *Mobile radio* can be operated while in motion.

Wireless systems can be totally Earth-based or utilize satellite resources. If Earth-based, the term *terrestrial radio system* or *radio system* is used. If at least one component is based in space, the term *satellite radio system* or *satellite system* is applied. Systems are also classified by the location of the user—for example, land, maritime, aeronautical, space, and deep space.

Because wireless systems now transmit a variety of information types, not merely voice, the multiservice networks are sometimes called *personal communication networks* (PCNs) or *personal communication systems* (PCSs).

7. OPTICAL COMMUNICATIONS

Optical communications refers to the transmission of information using signals with carrier waves oscillating in the optical (i.e., visible) frequency range. (See App. 18.A for an overview of the electromagnetic spectrum.) For historical reasons, optical oscillations are referred to by their wavelengths rather than their frequencies. The two are related by Eq. 50.4.

$$\lambda = \frac{c}{f} \qquad 50.4$$

A frequency of 3×10^{14} Hz represents the approximate beginning of the optical band. This correlates with a wavelength of 10^{-6} m, which is called a *micrometer*. This is the range of interest for optical communications.

Optical communication systems are noted for their wide bandwidth and low loss transmission. This means that such systems can carry vast amounts of information when compared to other modes of communication. While atmospheric transmission is possible, the majority of optical communications use glass fiber as the wave guide.

The schemes developed for optical information transmission include *wavelength-division multiplexing* (WDM and *optical frequency-division multiplexing* (OFDM). The topologies used are star- and ring-networks.

The major components of a point-to-point optical fiber transmission network are shown in Fig. 50.2.

The generic optical system is shown in Fig. 50.2(a). Like other communications systems, it consists of a transmitter, a transmission path (in this case the fiber), and a receiver. The internal block diagram of an optical transmitter is shown in Fig. 50.2(b). The *modulator* converts

[6]Radio transmissions are referred to as "stations" while television transmissions are called "channels." This is because the FCC assigns actual channel numbers to each frequency band.

[7]Cellular phones are a form of mobile radio with the signal patched into the telephone system network.

an electrical input signal, the voltage, to an optical output. The output of the modulator drives the light source, normally a laser diode or *light-emitting diode* (LED). The laser is the coherent source of light for the system. (*Laser* is an acronym for light amplification by stimulated emission of radiation.) The signal is then coupled to the fiber transmission line.

Figure 50.2 *Optical Transmission System*

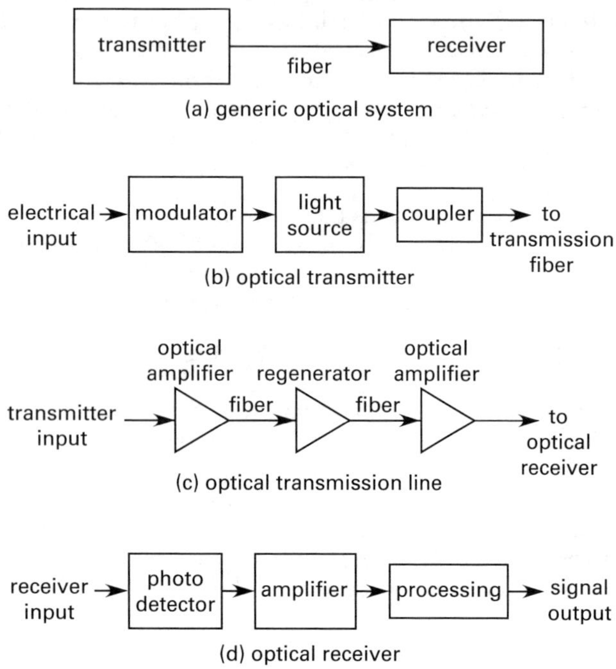

The optical transmission line, shown in Fig. 50.2(c), transmits the signal to the receiver, amplifies the signal when required (in the *optical amplifier*) and regenerates the signal (in the *regenerator*) to improve quality and compensate for a variety of effects that degrade the signal.

The optical receiver, shown in Fig. 50.2(d), senses the incoming signal using photodiodes, amplifies the signal, then converts the optical signal to an electrical signal and passes the information to the output. The output of the photo diode is given by Eq. 50.5.

$$i = \rho P \qquad 50.5$$

The detected current is given by i; P is the incident optical power; and ρ is the *responsivity* of the diode, usually on the order of 0.5 mA/mW.

Transmitters suffer from the analog effect called *chirp*, which is a phase shift associated with the switching of the modulator or laser source, and from the digital effect called *jitter*, which is a random and systematic deviation of the arrival times of pulses. Both are measured by an *eye pattern mask test* standardized in the *synchronous optical network* (SONET), and *synchronous digital hierarchy* (SDH) optical systems standards developed by Bell Labs in 1994. The eye pattern mask test is shown in Fig. 50.3. The "eye" is the pattern displayed on the oscilloscope of the optical signal in the transmission line. The shape of the "mask" determines the quality of the signal.

Figure 50.3 *Optical Eye Pattern Mask Test Display*

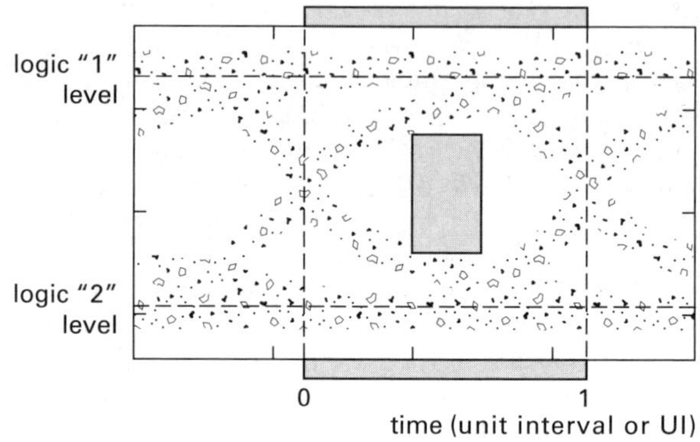

Fiber optic lines suffer from attenuation caused by absorption and scattering. There is also *dispersion*, caused by the differing speed of light in various materials. Such differences can result in a distorted signal at the receiver. Dispersion is generally of two types. The first is called *material, intramodal*, or *chromatic dispersion*, and is based on the type of fiber used. The other is multimode dispersion, also called intermodal dispersion, which is caused by the lack of complete monochromatic signal in all light sources.

Receivers suffer from the sensitivity limits of the photodiodes and the normal electronic component noise (see Chap. 18 and Chap. 40).

8. ELECTROMAGNETIC COMPATIBILITY

Electromagnetic compatibility (EMC) is the study of electrical and electronic components and systems to ensure functionality in the intended electromagnetic environment at design levels of efficiency. That is, EMC's purpose is to ensure that electrical and electronic equipment operates without interfering with equipment (including itself), without endangering life, and without allowing for unauthorized release of information or for unprotected use.

There are four basic EMC (interference) issues. *Interference* is any undesired energy that tends to interfere with the reception of the desired signals. The four EMC issues are shown in Fig. 50.4.

One can have a noisy component of power supply as shown in Fig. 50.4(a), resulting in *radiated emissions*, which are RF emissions of energy external to the equipment. One can have a noisy component of power supply as illustrated in Fig. 50.4(c), resulting in *conducted emissions*, which are emissions that remain in or on the conducting path to the equipment. Conversely, emissions from other equipment may interfere with the operation of one's susceptible equipment as shown in Fig. 50.4(b) and Fig. 50.4(d). This is known as *radiated susceptibility* or *conducted susceptibility*.

As Fig. 50.4 suggests, there are three general ways of minimizing interference.

- Suppress the emission of the interference at its source.
- Suppress the reception of the interference at its receiver.
- Minimize the coupling (radiated or conducted) efficiency.

Figure 50.4 *Basic EMC Issues*

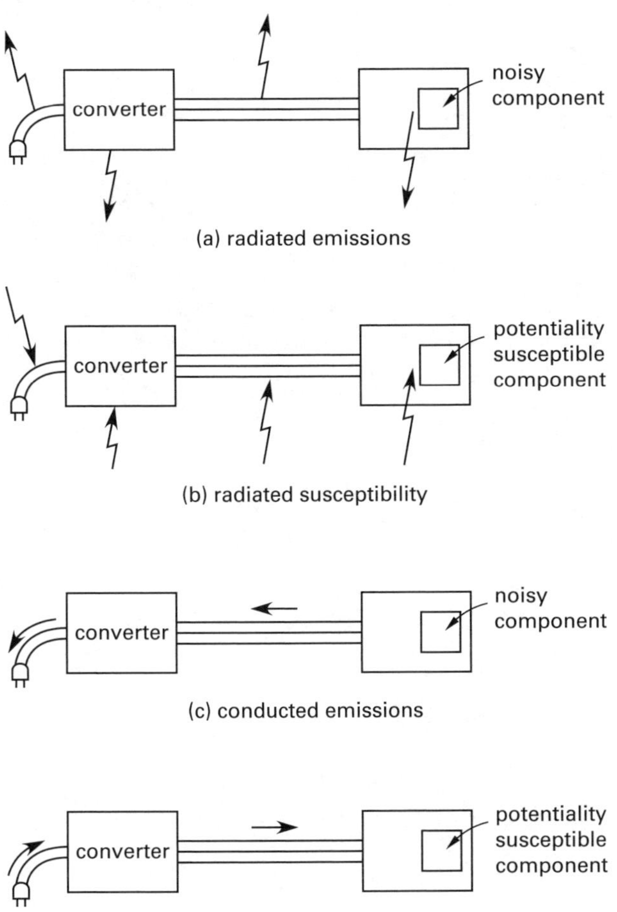

(a) radiated emissions

(b) radiated susceptibility

(c) conducted emissions

(d) conducted susceptibility

In addition to interference, there are other components of EMC that must be considered based on the design and requirements of the system. Some of these are shown in Fig. 50.5.

Electrostatic discharge (ESD) is the direct transfer of static charge buildup. ESD can be devastating to integrated circuits, resulting in a complete malfunction. The principle is shown in Fig. 50.5(a) and is familiar to anyone who has shuffled their feet across a carpet only to be shocked when next touching a metallic door handle.

Figure 50.5 *Aspects of Electromagnetic Compatability (EMC)*

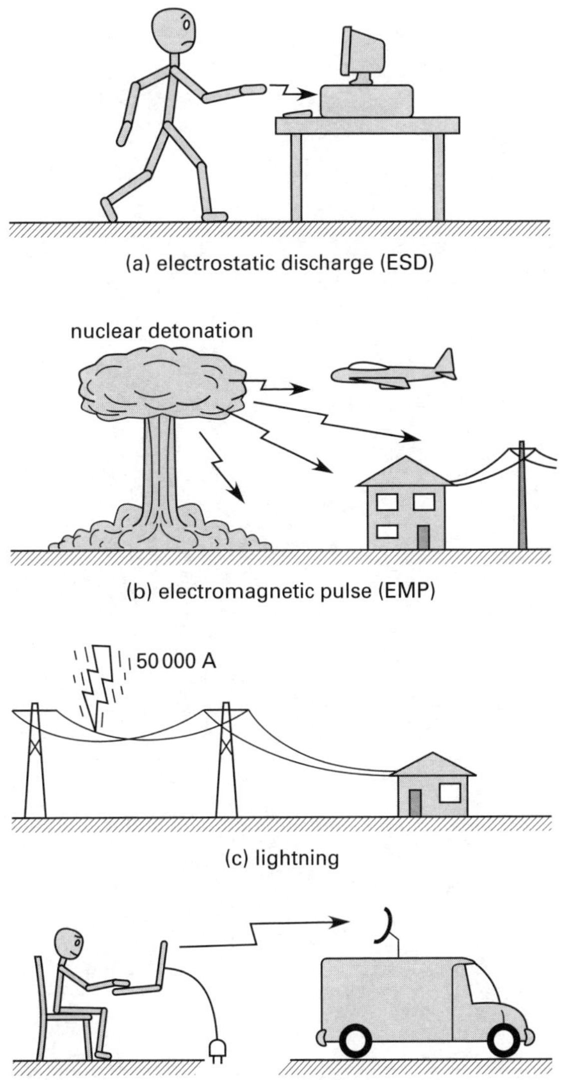

(a) electrostatic discharge (ESD)

(b) electromagnetic pulse (EMP)

(c) lightning

(d) TEMPEST (secure communication and data processing)

Luckily the electromagnetic pulse (EMP) shown in Fig. 50.5(b) is not as familiar. (The processes on the Sun can result in an electromagnetic interference but are not the same phenomenon as EMP.) An EMP can interfere with the operation of any electrical or electronic equipment by inducing voltages, and thereby currents, in electrical systems.

Lightning is an electrical discharge that can result in interference to power systems and nearby electrical systems as shown in Fig. 50.5(c).

Finally, EMC may be concerned with the emission of electromagnetic energy when such energy contains information that others are not authorized to view, as shown in Fig. 50.5(d). The military has developed requirements to ensure just such an event does not occur. The system is called TEMPEST. Because commercial concerns also desire to protect information, the term is often used to indicate the entire field of *emissions security* or *emanations security* (EMSEC).

Topic XII: Control Systems

Chapter

51. Analysis of Control Systems

51 Analysis of Control Systems

1. Types of Response 51-2
2. Graphical Solution 51-2
3. Classical Solution Method 51-3
4. Feedback Theory 51-4
5. Sensitivity 51-5
6. Block Diagram Algebra 51-6
7. Predicting System Time Response 51-7
8. Predicting Time Response from a Related Response 51-7
9. Initial and Final Values 51-7
10. Special Cases of Steady-State Response 51-8
11. Poles and Zeros 51-8
12. Predicting System Time Response from Response Pole-Zero Diagrams 51-9
13. Frequency Response 51-10
14. Gain Characteristic 51-10
15. Phase Characteristic 51-11
16. Stability 51-11
17. Bode Plots 51-12
18. Root-Locus Diagrams 51-12
19. Hurwitz Test 51-13
20. Routh Criterion 51-13
21. Nyquist Analysis 51-13
22. Application to Control Systems 51-13
23. Control System Types/Modes 51-14
24. State Model Representation 51-14

Nomenclature

$\mathbf{A}$	system matrix	–
$B(s)$	feedback signal	–
$\mathbf{B}$	control matrix	–
C	capacitance	F
C_1, C_2	constant	–
$C(s)$	control output	–
$\mathbf{C}$	ouput matrix	–
$e(t)$	error	–
$E(s)$	error, $\mathcal{L}(e(t))$	–
$f(t)$	forcing function	–
$F(s)$	forcing function, $\mathcal{L}(f(t))$	–
$G(s)$	forward transfer function	–
h	step height	–
$H(s)$	reverse transfer function	–
$i(t)$	current	A
$I(s)$	current, $\mathcal{L}(i(t))$	A
k, K	constant	–
K	gain	–
K	scale factor	–
L	inductance	H
L	length	–
M	fraction overshoot	–
M	gain	–
$M(s)$	manipulated signal	–
n	degrees of freedom, order of the system, or system type	–
N	Nyquist's number	–
$p(t)$	arbitrary function	–
P	number of poles	–
$P(s)$	arbitrary function, $\mathcal{L}(p(t))$	–
Q	quality factor	–
r	real value (root)	–
$r(t)$	time response	–
R	resistance	Ω
$R(s)$	response function, $\mathcal{L}(r(t))$, or reference input	1/s
s	Laplace transform variable or complex frequency variable	–
S	sensitivity	–
t	time	s
$T(s)$	transfer function, $\mathcal{L}(T(t))$	–
$T(t)$	transfer function	–
u	input variable	–
$U(s)$	disturbance	–
$\mathbf{U}$	control vector	–
v	variable voltage	V
V	constant or rms voltage	V
x	state variable	–
$\mathbf{X}$	state vector	–
y	output variable	–
$Y(s)$	output	–
$\mathbf{Y}$	output vector	–
Z	number of zeros	–

Symbols

α	pole-angle	rad
β	zero-angle	rad
ζ	damping ratio	–
τ	time constant	s
ω	natural frequency	rad/s

Subscripts

d	damped or delay
f	forced or feedback
i	input
n	natural
o	output
p	peak or pole-line
r	rise
s	settling
z	zero-line

1. TYPES OF RESPONSE

Natural response (also referred to as *initial condition response, homogeneous response,* and *unforced response*) is the manner in which a system behaves when energy is applied and then subsequently removed. The system is left alone and allowed to do what it would do naturally, without the application of further disturbing forces. In the absence of friction, natural response is characterized by a sinusoidal response function. The response function will contain exponentially decaying sinusoids otherwise.

Forced response is the behavior of a system that is acted upon by a force that is applied periodically. Forced response in the absence of friction is characterized by sinusoidal terms having the same frequency as the forcing function.

Natural and forced responses are present simultaneously in forced systems. The sum of the two responses is the *total response.*[1] This is the reason that differential equations are solved by adding a particular solution to the homogeneous solution. The homogeneous solution corresponds to the natural response; the particular solution corresponds to the forced response.

$$\text{total response} = \text{natural response} + \text{forced response} \quad 51.1$$

Because the influence of decaying functions disappears after a few cycles, natural response is sometimes referred to as *transient response.* Once the transient response effects have died out, the total response consists entirely of forced terms. This is the *steady-state response.*

2. GRAPHICAL SOLUTION

Graphical solutions are available in some cases, particularly those with homogeneous, step, and sinusoidal inputs. When the system equation is a homogeneous second-order linear differential equation with constant coefficients (see Sec. 11.6) in the form of Eq. 51.2, the natural time response can be determined from Fig. 51.1. The term ω is the natural frequency, ω_n, defined in Sec. 31.10, and ζ is the damping ratio.

$$x'' + 2\zeta\omega x' + \omega^2 x = 0 \quad 51.2$$

When the system equation is a second-order linear differential equation with constant coefficients and the forcing function is a step of height h (as in Eq. 51.3), the time response can be determined from Fig. 51.2.

$$x'' + 2\zeta\omega x' + \omega^2 x = \omega^2 h \quad 51.3$$

Figure 51.2 illustrates that a system responding to a step will eventually settle to the steady-state position of the step but that when damping is low ($\zeta < 1$), there will be *overshoot.* Figure 51.3 illustrates this and other parameters of second-order response to a step input. The settling time depends on the tolerance (i.e., the separation of the actual and steady-state responses). The *time delay,* t_d in Fig. 51.3, is the time needed to reach 50% of the steady-state value. The *time constant,* τ, is the time needed to reach approximately 63% of the steady-state value.

Figure 51.1 *Natural Response*

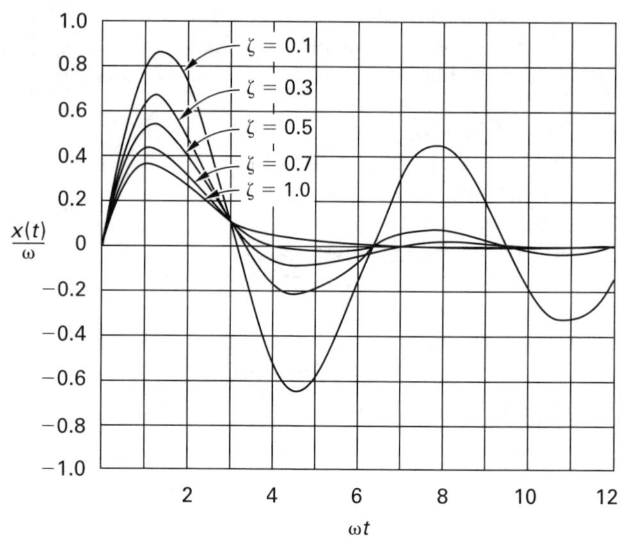

Figure 51.2 *Response to a Unit Step*

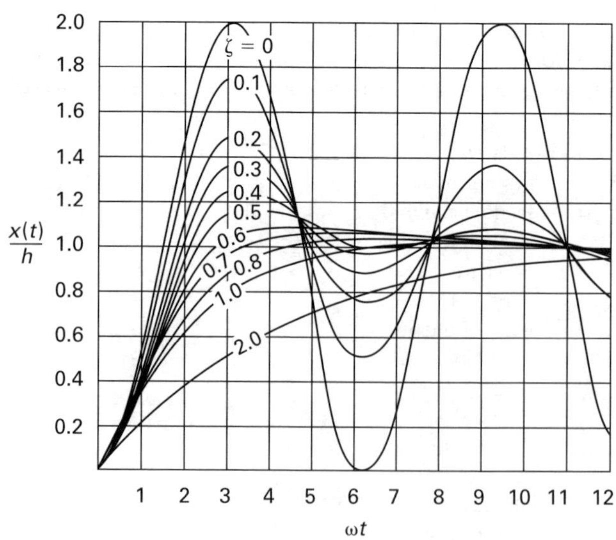

The damped frequency is

$$\omega_d = \omega\sqrt{1 - \zeta^2} \quad 51.4$$

The rise time is

$$t_r = \frac{\pi - \arccos\zeta}{\omega_d} \quad 51.5$$

[1]This response is a function of time, and, therefore, is referred to as *time response* to distinguish it from *frequency response.* (See Sec. 51.13.)

The peak time is

$$t_p = \frac{\pi}{\omega_d} \qquad 51.6$$

The peak gain (fraction overshoot) is

$$M_p = e^{(-\pi\zeta)/\sqrt{1-\zeta^2}} \qquad 51.7$$

The settling time is

$$t_s = \frac{3.91}{\zeta\omega_n} \quad \text{[2\% criterion]} \qquad 51.8$$

$$= \frac{3.00}{\zeta\omega_n} \quad \text{[5\% criterion]} \qquad 51.9$$

The time constant is

$$\tau = \frac{1}{\zeta\omega_n} \qquad 51.10$$

Figure 51.3 *Second-Order Step Time Response Parameters*

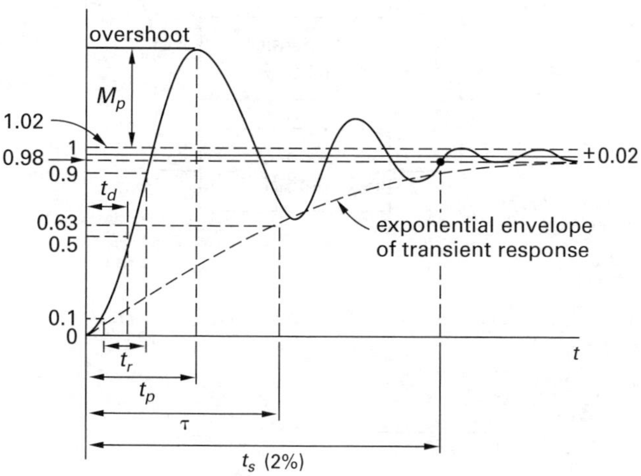

3. CLASSICAL SOLUTION METHOD

A system model can be thought of as a block where the *input signal* is the forcing function, $f(t)$, and the *output signal* is the response function, $r(t)$. The standard analytical method of determining the response of a system from its system equation uses Laplace transforms. Accordingly, the classical solution approach does not determine the output signal or response function directly but derives the *transfer function*, $T(s)$, instead.[2] The transfer function is also known as the *rational function.*

$$T(s) = \frac{\mathcal{L}r(t)}{\mathcal{L}f(t)} \qquad 51.11$$

[2]Strictly speaking, $r(t)/f(t)$ is the *transfer function* and $\mathcal{L}r(t)/\mathcal{L}f(t)$ is the *transform of the transfer function.* However, the distinction is seldom made.

Transfer functions in the s-domain are Laplace transformations of the corresponding time-domain functions. Transforming $r(t)/f(t)$ into $T(s)$ involves more than a simple change of variables. The s symbol can be thought of as a derivative operator; similarly, the integration operator is represented as $1/s$. By convention, Laplace transforms are represented by uppercase letters while operand functions are represented by lowercase letters.

Example 51.1

The following system equation in integrodifferential equation form has been derived for the output current from an electrical network. Convert the system equation to the s-domain.

$$i(t) = \frac{1}{L}\int (v_1 - v_2)dt + C\left(\frac{d(v_2 - v_1)}{dt}\right)$$

Solution

Replace all derivative operators by s; replace all integration operators by $1/s$. Replace time voltage, $v(t)$, with s-domain voltage, $V(s)$.

$$I(s) = \frac{V_1}{sL} - \frac{V_2}{sL} + sCV_2 - sCV_1$$

Example 51.2

Determine the voltage transfer function for the system shown and draw its black box representation.

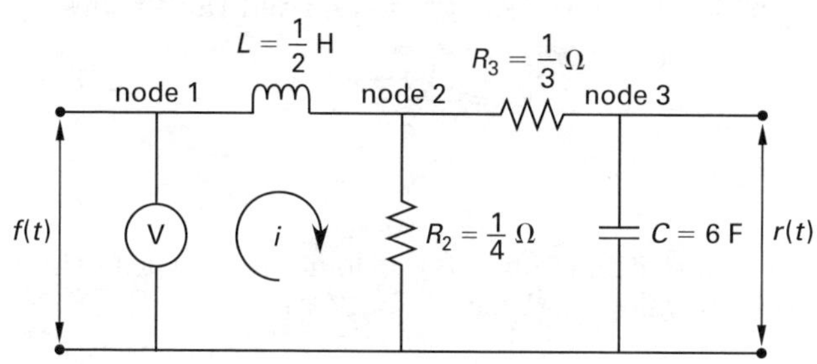

Solution

First, write the system equations. There are two loop currents, so two simultaneous equations are needed. One system equation can be written for each of the three nodes, so there will be one redundant system equation. (At this point, it is not obvious which of the two system equations can be used to derive the transfer function.)

At node 1,

$$i(t) = \frac{1}{L}\int (v_1 - v_2)dt = 2\int (v_1 - v_2)dt$$

At node 2,

$$\begin{aligned} 0 &= \frac{1}{L}\int (v_2 - v_1)\,dt + \frac{v_2}{R_2} + \frac{v_2 - v_3}{R_3} \\ &= 2\int (v_2 - v_1)\,dt + 4v_2 + 3(v_2 - v_3) \end{aligned}$$

At node 3,

$$\begin{aligned} 0 &= C\left(\frac{dv_3}{dt}\right) + \frac{v_3 - v_2}{R_3} \\ &= 6\left(\frac{dv_3}{dt}\right) + 3(v_3 - v_2) \end{aligned}$$

Next, convert the system equations to the s-domain by substituting s for derivative operations and $1/s$ for integration operations.

For node 1,

$$I(s) = \frac{2(V_1 - V_2)}{s}$$

For node 2,

$$0 = \frac{2(V_2 - V_1)}{s} + 4V_2 + 3(V_2 - V_3)$$

For node 3,

$$0 = 6sV_3 + 3(V_3 - V_2)$$

The transfer function is the ratio of the output to input voltages, V_3/V_1. It does not depend on V_2. The equation for node 1 cannot be used unless $i(t)$ is known, which it (generally) is not. V_2 is eliminated from the equations for nodes 2 and 3. From the second and third nodes,

$$\begin{aligned} V_2 &= \frac{2V_1 + 3sV_3}{2 + 7s} \\ &= V_3(1 + 2s) \end{aligned}$$

The transfer function, $T(s)$, is found by equating these two expressions and solving for V_3/V_1.

$$T(s) = \frac{V_3}{V_1} = \frac{1}{7s^2 + 4s + 1}$$

The black box representation of this system is

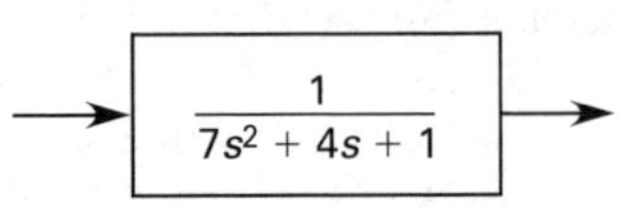

4. FEEDBACK THEORY

The output signal is returned as input in a feedback loop (feedback system). A basic feedback system (see Fig. 51.4) consists of two black box units (a *dynamic unit* and a *feedback unit*), a pick-off point (take-off point), and a summing point (*comparator* or *summer*). The summing point is assumed to perform positive addition unless a minus sign is present. The incoming signal, v_i, is combined with the feedback signal, v_f, to give the *error* (*error signal*), e. Whether addition or subtraction is used in Eq. 51.12 depends on whether the summing point is additive (i.e., a positive feedback system) or subtractive (i.e., a negative feedback system), respectively. $E(s)$ is the *error transfer function* (*error gain*).

$$\begin{aligned} E(s) &= \mathcal{L}\big(e(t)\big) = V_i(s) \pm V_f(s) \\ &= V_i(s) \pm H(s)V_o(s) \end{aligned} \qquad 51.12$$

Figure 51.4 *Feedback System*

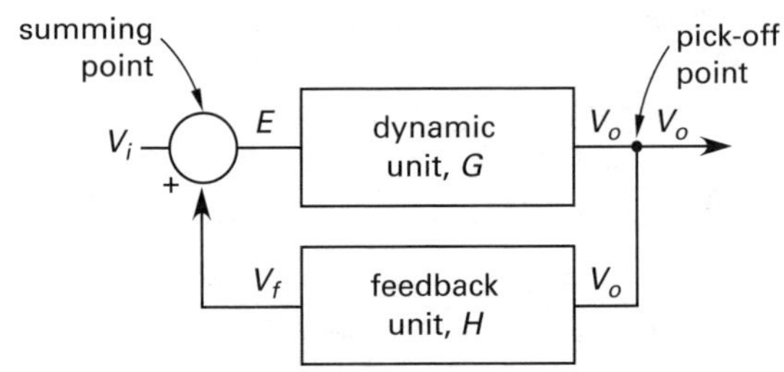

The ratio $E(s)/V_i(s)$ is the *error ratio* (*actuating signal ratio*).

$$\frac{E(s)}{V_i(s)} = \frac{1}{1 + G(s)H(s)} \quad \text{[negative feedback]} \qquad 51.13$$

$$= \frac{1}{1 - G(s)H(s)} \quad \text{[positive feedback]} \qquad 51.14$$

Because the dynamic and feedback units are black boxes, each has an associated transfer function. The transfer function of the dynamic unit is known as the *forward transfer function* (*direct transfer function*), $G(s)$. In most feedback systems—amplifier circuits in particular—the magnitude of the forward transfer function is known as the *forward gain* or *direct gain*. $G(s)$ can be a scalar if the dynamic unit merely scales the error. However, $G(s)$ is normally a complex operator that changes both the magnitude and the phase of the error.

$$V_o(s) = G(s)E(s) \qquad 51.15$$

The pick-off point transmits the output signal, V_o, from the dynamic unit back to the feedback element. The output of the dynamic unit is not reduced by the pickoff point. The transfer function of the feedback unit is the *reverse transfer function* (*feedback transfer function,*

Control Systems

feedback gain), $H(s)$, which can be a simple magnitude-changing scalar or a phase-shifting function.

$$V_f(s) = H(s)\,V_o(s) \qquad 51.16$$

The ratio $V_f(s)/V_i(s)$ is the *feedback ratio* (*primary feedback ratio*).

$$\frac{V_f(s)}{V_i(s)} = \frac{G(s)H(s)}{1+G(s)H(s)} \quad \text{[negative feedback]} \qquad 51.17$$

$$= \frac{G(s)H(s)}{1-G(s)H(s)} \quad \text{[positive feedback]} \qquad 51.18$$

The *loop transfer function* (*loop gain* or *open-loop transfer function*) is the gain after going around the loop one time, $\pm G(s)H(s)$.

The *overall transfer function* (*closed-loop transfer function*, *control ratio*, *system function*, *closed-loop gain*), $G_{\text{loop}}(s)$, is the overall transfer function of the feedback system. The quantity $1+G(s)H(s)=0$ is the *characteristic equation*. The *order of the system* is the largest exponent of s in the characteristic equation. (This corresponds to the highest-order derivative in the system equation.)

$$G_{\text{loop}}(s) = \frac{V_o(s)}{V_i(s)}$$

$$= \frac{G(s)}{1+G(s)H(s)} \quad \text{[negative feedback]} \qquad 51.19$$

$$= \frac{G(s)}{1-G(s)H(s)} \quad \text{[positive feedback]} \qquad 51.20$$

With positive feedback and $G(s)H(s)$ less than 1.0, G_{loop} will be larger than $G(s)$. This increase in gain is a characteristic of positive feedback systems. As $G(s)H(s)$ approaches 1.0, the closed-loop transfer function increases without bound, which is usually an undesirable effect.

In a negative feedback system, the denominator of Eq. 51.19 is greater than 1.0. Although the closed-loop transfer function will be less than $G(s)$, there may be other desirable effects. Generally, a system with negative feedback is less sensitive to variations in temperature, circuit components values, input signal frequency, and signal noise. Other benefits include distortion reduction, increased stability, and impedance matching.[3]

[3]For circuits to be directly connected in series without affecting their performance, all input impedances must be infinite and all output impedances must be zero.

Example 51.3

A high-gain noninverting amplifier has a gain of 10^6. Feedback is provided by a resistor voltage divider. What is the closed-loop gain?

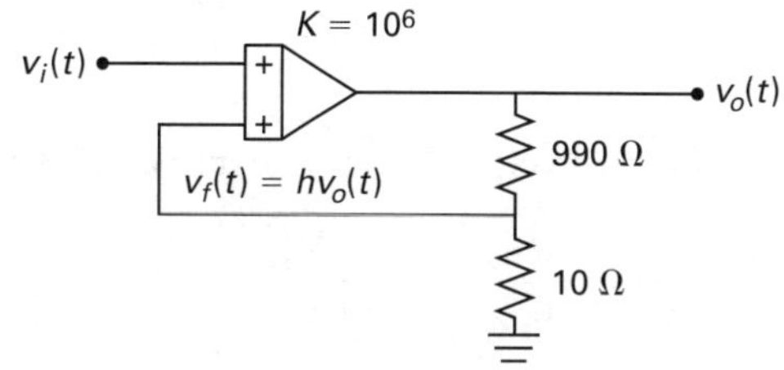

Solution

The fraction of the output signal appearing at the summing point depends on the resistances in the divider circuit.

$$h = \frac{10\ \Omega}{10\ \Omega + 990\ \Omega} = 0.01$$

Because the feedback path only scales the feedback signal, the feedback will be positive. From Eq. 51.20, the closed-loop gain is

$$K_{\text{loop}} = \frac{K}{1-Kh} = \frac{10^6}{1-(10^6)(0.01)} \approx -100$$

5. SENSITIVITY

For large loop gains ($G(s)H(s) \gg 1$) in negative feedback systems, the overall gain is approximately $1/H(s)$. The forward gain is not a factor, and by choosing the correct value of $H(s)$, the output can be made insensitive to variations in $G(s)$.

In general, the *sensitivity*, S, of any variable, A, with respect to changes in another parameter, B, is

$$S_B^A = \frac{d\ln A}{d\ln B} = \frac{\frac{dA}{A}}{\frac{dB}{B}} \approx \left(\frac{\Delta A}{\Delta B}\right)\left(\frac{B}{A}\right) \qquad 51.21$$

The sensitivity of the loop transfer function with respect to the forward transfer function is

$$S_{G(s)}^{G_{\text{loop}}(s)} = \left(\frac{\Delta G_{\text{loop}}(s)}{\Delta G(s)}\right)\left(\frac{G(s)}{G_{\text{loop}}(s)}\right)$$

$$= \frac{1}{1+G(s)H(s)} \quad \begin{bmatrix}\text{negative}\\ \text{feedback}\end{bmatrix} \qquad 51.22$$

$$= \frac{1}{1-G(s)H(s)} \quad \begin{bmatrix}\text{positive}\\ \text{feedback}\end{bmatrix} \qquad 51.23$$

Example 51.4

A closed-loop gain of -100 is required from a circuit, and the output signal must not vary by more than $\pm 1\%$. An amplifier is available, but its output varies by $\pm 20\%$. How can this amplifier be used?

Solution

A closed-loop sensitivity of 0.01 is required. This means

$$\frac{\Delta v_o}{v_o} = \frac{\Delta G_{\text{loop}} v_i}{G_{\text{loop}} v_i} = \frac{\Delta G_{\text{loop}}}{G_{\text{loop}}} = 0.01$$

Similarly, the existing amplifier has a sensitivity of 20%. This means

$$\frac{\Delta G}{G} = 0.20$$

The ratio of these variations corresponds to the definition of sensitivity as in Eq. 51.21. For positive feedback,

$$\frac{\Delta G_{\text{loop}} G}{\Delta G G_{\text{loop}}} = \frac{0.01}{0.20} = \frac{1}{1 - GH}$$

Solving, $GH = -19$.

Solving for G from Eq. 51.20,

$$G_{\text{loop}} = -100 = \frac{G}{1 - GH} = \frac{G}{1 - (-19)}$$

Solving, $G = -2000$. Finally, solve for H.

$$H = \frac{GH}{G} = \frac{-19}{-2000} = 0.0095$$

6. BLOCK DIAGRAM ALGEBRA

The functions represented by several interconnected black boxes (*cascaded blocks*) can be simplified into a single block operation. Some of the most important simplification rules of block diagram algebra are shown in Fig. 51.5. Case 3 represents the standard feedback model.

Example 51.5

A complex block system is constructed from five blocks and two summing points. What is the overall transfer function?

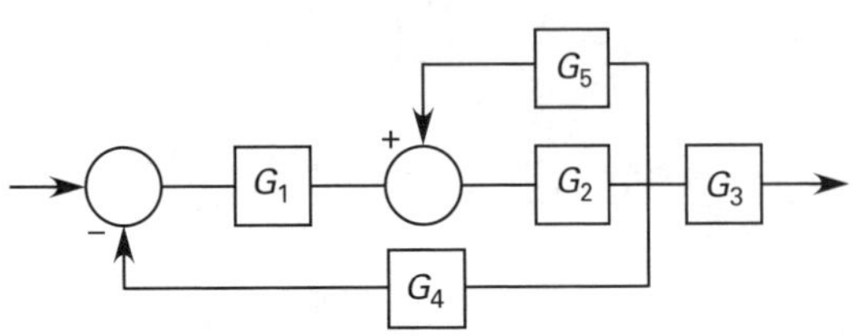

Solution

Use case 5 to move the second summing point back to the first summing point.

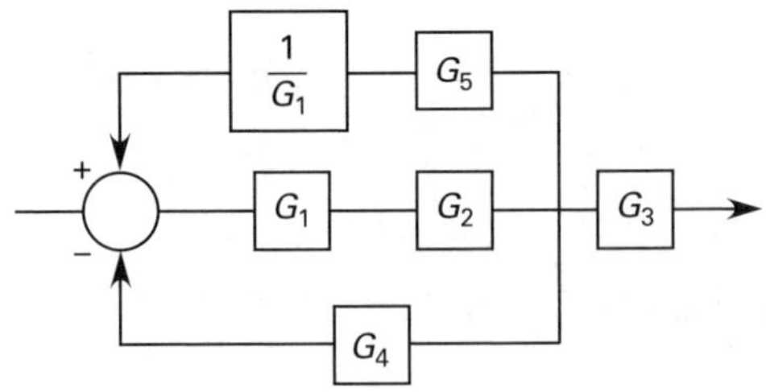

Use case 1 to combine boxes in series.

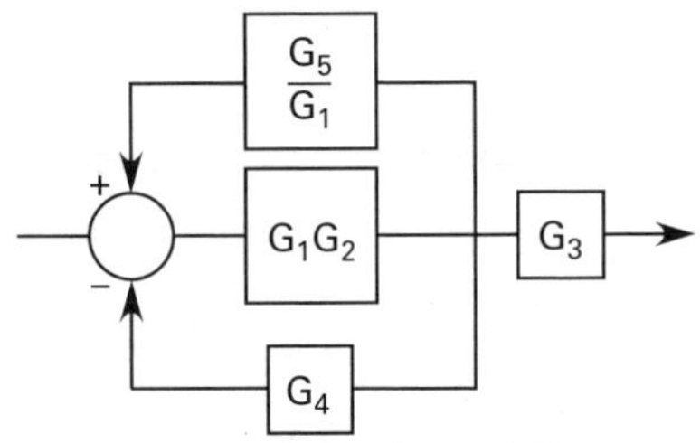

Use case 2 to combine the two feedback loops.

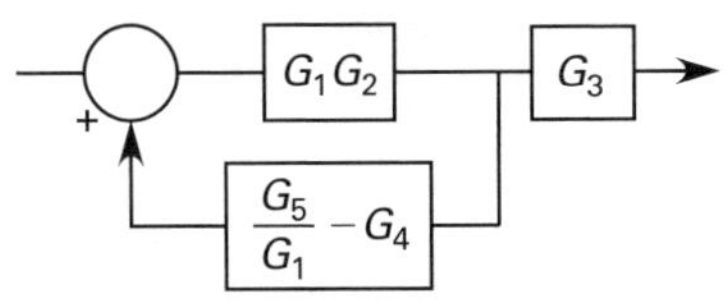

Use case 8 to move the pick-off point outside the G_3 box.

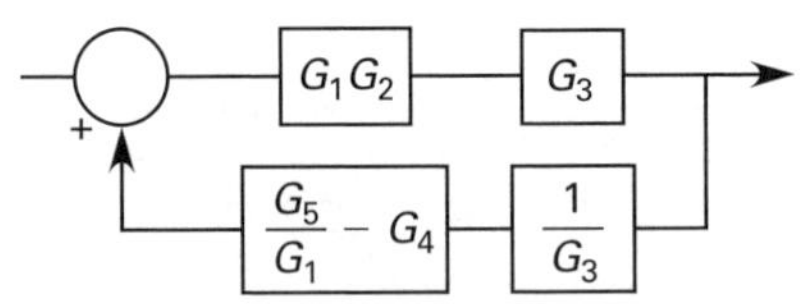

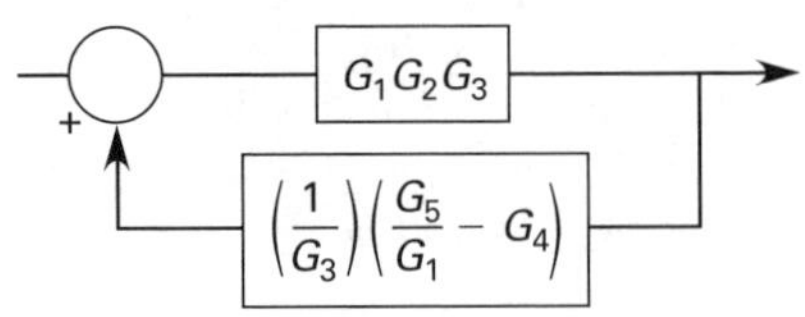

Use case 3 to determine the system gain.

$$G_{\text{loop}} = \frac{G_1 G_2 G_3}{1 - (G_1 G_2 G_3)\left(\frac{1}{G_3}\right)\left(\frac{G_5}{G_1} - G_4\right)}$$

$$= \frac{G_1 G_2 G_3}{1 - G_2 G_5 + G_1 G_2 G_4}$$

Figure 51.5 *Rules of Simplifying Block Diagrams*

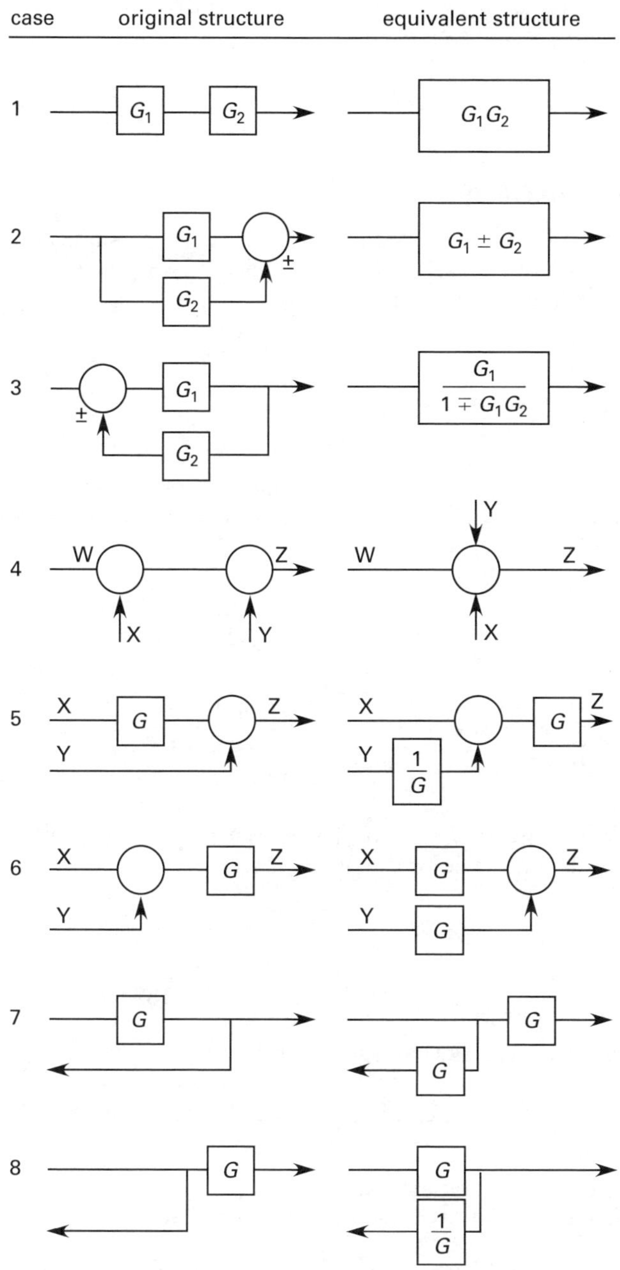

7. PREDICTING SYSTEM TIME RESPONSE

The transfer function is derived without knowledge of the input and is insufficient to predict the time response of the system. The system time response depends on the form of the input function. Because the transfer function is expressed in the *s*-domain, the forcing and response functions must be also. (Laplace transforms of step, pulses, sinusoids, and other functions are evaluated in Chap. 11 and Chap. 15.)

$$R(s) = T(s)F(s) \qquad 51.24$$

The time-based response function, $r(t)$, is found by performing the inverse Laplace transform.

$$r(t) = \mathcal{L}^{-1}\Big(R(s)\Big) \qquad 51.25$$

Example 51.6

A mechanical system is acted upon by a constant force of eight units starting at $t=0$. What is the time-based response function, $r(t)$, if the transfer function is

$$T(s) = \frac{6}{(s+2)(s+4)}$$

Solution

The forcing function is a step of height 8 at $t=0$. The Laplace transform of a unit step is $1/s$. Therefore, $F(s)=8/s$.

From Eq. 51.24,

$$\begin{aligned} R(s) = T(s)F(s) &= \left(\frac{6}{(s+2)(s+4)}\right)\left(\frac{8}{s}\right) \\ &= \frac{48}{s(s+2)(s+4)} \end{aligned}$$

The response function is found from Eq. 51.25 and a table of Laplace transforms (see App. 11.A). A product of linear terms in the denominator of $R(s)$ is equivalent to a sum of terms in $r(t)$, so

$$\begin{aligned} r(t) &= \mathcal{L}^{-1}\left(\frac{48}{s(s+2)(s+4)}\right) \\ &= 6 - 12e^{-2t} + 6e^{-4t} \end{aligned}$$

The last two terms are decaying exponentials, which represent the transient natural response. The first term does not vary with time; it is the steady-state response.

8. PREDICTING TIME RESPONSE FROM A RELATED RESPONSE

In some cases, it may be possible to use a known response to one input to determine the response to another input. For example, the impulse function is the derivative of the step function, so the response to an impulse is the derivative of the response to a step function.

9. INITIAL AND FINAL VALUES

The initial and final (steady-state) values of any function, $P(s)$, can be found from the *initial* and *final value theorems*, respectively, providing limits exist. Equation 51.26 and Eq. 51.27 are particularly valuable in determining the steady-state response (substitute $R(s)$ for $P(s)$) and the steady-state error (substitute $E(s)$ for $P(s)$).

$$\lim_{t\to 0+} p(t) = \lim_{s\to\infty}\left(sP(s)\right) \quad \text{[initial value]} \tag{51.26}$$

$$\lim_{t\to\infty} p(t) = \lim_{s\to 0}\left(sP(s)\right) \quad \text{[final value]} \tag{51.27}$$

Example 51.7

What is the final value of the response function $r(t)$ if

$$R(s) = \frac{1}{s(s+1)}$$

Solution

From Eq. 51.27, $R(s)$ is multiplied by s and the limit taken as s approaches zero.

$$r(\infty) = \lim_{s\to 0}\left(\frac{s}{s(s+1)}\right) = \lim_{s\to 0}\left(\frac{1}{s+1}\right) = 1$$

10. SPECIAL CASES OF STEADY-STATE RESPONSE

In addition to determining the steady-state response from the final value theorem (see Sec. 51.9), the steady-state response to a specific input can be easily derived from the transfer function, $T(s)$, in a few specialized cases. For example, the steady-state response function for a system acted upon by an impulse is simply the transfer function. That is, a pulse has no long-term effect on a system.

$$R(\infty) = T(s) \quad \text{[pulse input]} \tag{51.28}$$

The steady-state response for a *step input* (often referred to as a *DC input*) is obtained by substituting zero for s everywhere in the transfer function. (If the step has magnitude h, the steady-state response is multiplied by h.)

$$R(\infty) = T(0) \quad \text{[unit step input]} \tag{51.29}$$

The steady-state response for a sinusoidal input is obtained by substituting $j\omega_f$ for s everywhere in the transfer function, $T(s)$. The output will have the same frequency as the input. It is particularly convenient to perform sinusoidal calculations using phasor notation (as illustrated in Ex. 51.9).

$$R(\infty) = T(j\omega_f) \tag{51.30}$$

Example 51.8

What is the steady-state response of the system in Ex. 51.6 acted upon by a step of height 8?

Solution

Substitute zero for s in $T(s)$ and multiply by 8.

$$R(\infty) = 8T(0) = (8)\left(\frac{6}{(0+2)(0+4)}\right) = 6$$

Example 51.9

What is the steady-state response when a sinusoidal voltage of $4\sin(2t + 45^\circ)$ is applied to a system with the following transfer function?

$$T(s) = \frac{-1}{7s^2 + 7s + 1}$$

Solution

The angular frequency of the forcing function is $\omega_f = 2$ rad/s. Substitute $j2$ for s in $T(s)$ and simplify the expression by recognizing that $j^2 = -1$.

$$\begin{aligned} T(j2) &= \frac{-1}{7(j2)^2 + 7(j2) + 1} \\ &= \frac{-1}{-28 + 14j + 1} \\ &= \frac{-1}{-27 + 14j} \end{aligned}$$

Next, convert $T(j2)$ to phasor (polar) form. The magnitude and angle of the denominator are

$$\begin{aligned} \text{magnitude} &= \sqrt{(14)^2 + (-27)^2} = 30.4138 \\ \text{angle} &= 180^\circ - \arctan\frac{14}{27} = 152.59^\circ \end{aligned}$$

However, this is the negative reciprocal of $T(j2)$.

$$T(j2) = \frac{-1}{30.4138\angle 152.59^\circ} = -0.03288\angle -152.59^\circ$$

The forcing function expressed in phasor form is $4\angle 45^\circ$. From Eq. 51.11, the steady-state response is

$$\begin{aligned} v(t) &= T(t)f(t) = (-0.03288\angle -152.59^\circ)(4\angle 45^\circ) \\ &= -0.1315\angle -107.6^\circ \end{aligned}$$

11. POLES AND ZEROS

A *pole* is a value of s that makes a function, $P(s)$, infinite. Specifically, a pole makes the denominator of $P(s)$ zero.[4] A zero of the function makes the numerator of $P(s)$ (and hence $P(s)$ itself) zero. Poles and zeros need not be real or unique; they can be imaginary and repeated within a function.

A *pole-zero diagram* is a plot of poles and zeros in the *s-plane*—a rectangular coordinate system with real and

[4]Pole values are the system *eigenvalues*.

imaginary axes. Zero is represented as O; a pole is represented as X. Poles off the real axis always occur in conjugate pairs known as *pole pairs*.

Sometimes it is necessary to derive the function $P(s)$ from its pole-zero diagram. This will be only partially successful because repeating identical poles and zeros are not usually indicated on the diagram. Also, scale factors (scalar constants) are not shown.

Example 51.10

Draw the pole-zero diagram for the transfer function

$$T(s) = \frac{5(s+3)}{(s+2)(s^2+2s+2)}$$

Solution

The numerator is zero when $s = -3$. This is the only zero of the transfer function.

The denominator is zero when $s = -3$ and $s = -1 \pm j$. These three values are the poles of the transfer function.

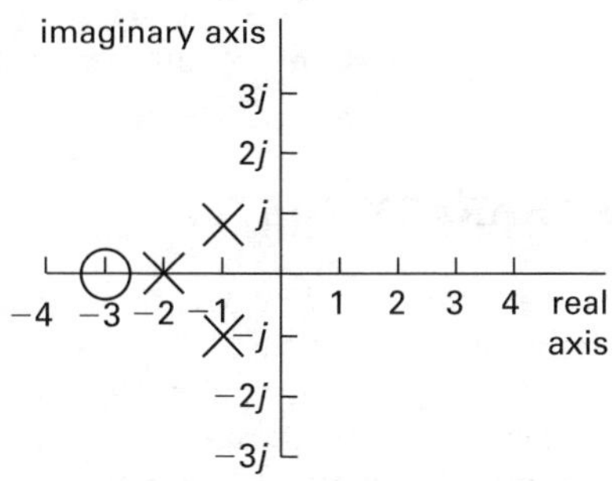

Example 51.11

A pole-zero diagram for a transfer function $T(s)$ has a single pole at $s = -2$ and a single zero at $s = -7$. What is the corresponding function?

Solution

$$T(s) = \frac{K(s+7)}{s+2}$$

The scale factor K must be determined by some other means.

12. PREDICTING SYSTEM TIME RESPONSE FROM RESPONSE POLE-ZERO DIAGRAMS

A response pole-zero diagram based on $R(s)$ can be used to predict how the system responds to a specific input. This pole-zero diagram must be based on the product $T(s)F(s)$ because that is how $R(s)$ is calculated. Plotting the product $T(s)F(s)$ is equivalent to plotting $T(s)$ and $F(s)$ separately on the same diagram. (See Fig. 51.6.)

Figure 51.6 *Types of Response Determined by Pole Location*

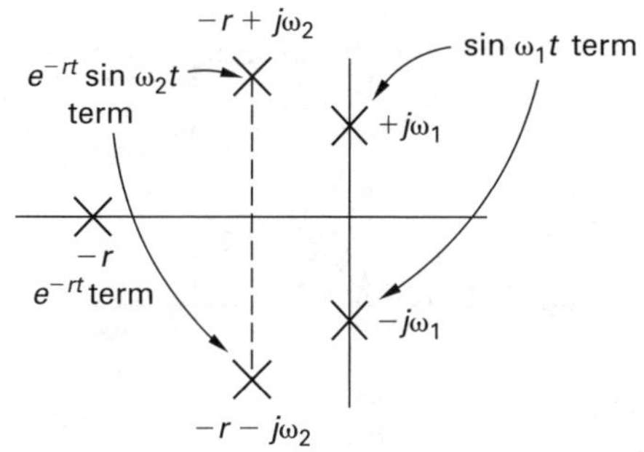

The system will experience an *exponential decay* when a single pole falls on the real axis. A pole with a value of $-r$, corresponding to the linear term $s+r$, will decay at the rate of e^{-rt}. The quantity $1/r$ is the decay *time constant*, the time for the response to achieve approximately 63% of its steady-state value. The farther left the point is located from the vertical imaginary axis, the faster the motion will die out.

Undamped sinusoidal oscillation will occur if a pole pair falls on the imaginary axis. A conjugate pole pair with the value of $\pm j\omega$ indicates oscillation with a natural frequency of ω rad/s.

Pole pairs to the left of the imaginary axis represent *decaying sinusoidal response*. The closer the poles are to the real (horizontal) axis, the slower the oscillations. The closer the poles are to the imaginary (vertical) axis, the slower the decay. The *natural frequency*, ω, of undamped oscillation can be determined from a conjugate pole pair having values of $r \pm \omega_f$.

$$\omega = \sqrt{r^2 + \omega_f^2} \qquad 51.31$$

The magnitude and phase shift can be determined for any input frequency from the pole-zero diagram with the following procedure: Locate the angular frequency, ω_f, on the imaginary axis. Draw a line from each pole (i.e., a pole-line) and zero (i.e., a zero-line) of $T(s)$ to this point. The angle of each of these lines is the angle between it and the horizontal real axis. The overall magnitude is the product of the lengths of the zero-lines divided by the product of the lengths of the pole-lines. (The scale factor must also be included because it is not shown on the pole-zero diagram.) The phase is the sum of the pole-angles less the sum of the zero-angles. (See Fig. 51.7.)

$$|R| = \frac{K\prod\limits_z |L_z|}{\prod\limits_p |L_p|} = \frac{K\prod\limits_z \text{length}}{\prod\limits_p \text{length}} \qquad 51.32$$

$$\angle R = \sum_p \alpha - \sum_z \beta \qquad 51.33$$

Control Systems

Figure 51.7 *Calculating Magnitude and Phase from a Pole-Zero Diagram*

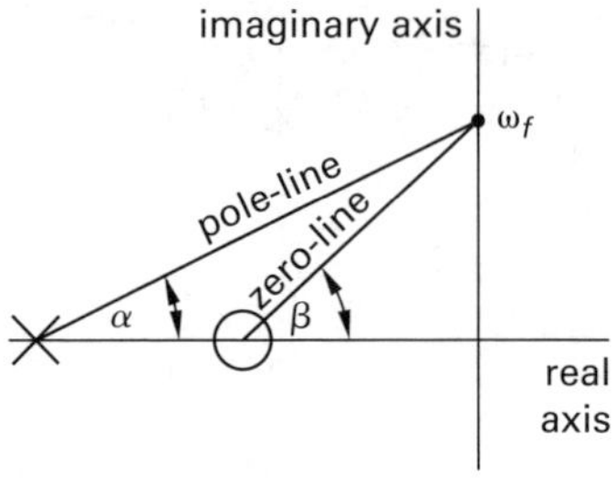

Example 51.12

What is the response of a system with the response pole-zero diagram representing $R(s) = T(s)F(s)$ shown?

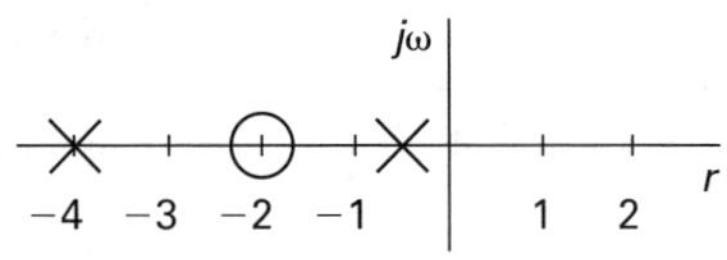

Solution

The poles are at $r = -\frac{1}{2}$ and $r = -4$. The response is

$$r(t) = C_1 e^{(-1/2)t} + C_2 e^{-4t}$$

Constants C_1 and C_2 must be found from other data.

Example 51.13

What is the system response if $T(s) = (s+2)/(s+3)$ and the input is a unit step?

Solution

The transform of a unit step is $1/s$. The response is

$$R(s) = T(s)F(s) = \frac{s+2}{s(s+3)}$$

The response pole-zero diagram is

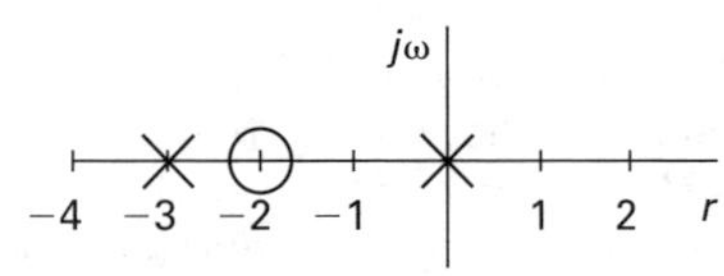

The pole at $r=0$ contributes the exponential $C_1 e^{-0t}$ (or simply C_1) to the total response. The pole at $r=-3$ contributes the term $C_2 e^{-3t}$. The total response is

$$r(t) = C_1 + C_2 e^{-3t}$$

13. FREQUENCY RESPONSE

The gain and phase angle frequency response of a system changes as the forcing frequency is varied. (The dependence of $r(t)$ on ω_f is illustrated in Ex. 51.9.) The *frequency response* is the variation in these parameters, always with a sinusoidal input. *Gain* and *phase characteristics* are plots of the steady-state gain and phase angle responses with a sinusoidal input versus frequency. While a linear frequency scale can be used, frequency response is almost always presented against a logarithmic frequency scale.

The steady-state gain response is expressed in decibels, while the steady-state phase angle response is expressed in degrees. The gain is calculated from Eq. 51.34 where $|T(s)|$ is the absolute value of the steady-state response.

$$\text{gain} = 20 \log|T(j\omega)| \quad \text{[in dB]} \qquad 51.34$$

A doubling of $|T(j\omega)|$ is referred to as an *octave* and corresponds to a 6.02 dB increase. A ten-fold increase in $|T(j\omega)|$ is a *decade* and corresponds to a 20 dB increase.

$$\begin{aligned}\text{number of octaves} &= \frac{\text{gain}_2 - \text{gain}_1}{6.02} \quad \text{[in dB]} \\ &= 3.32 \times \text{number of decades}\end{aligned} \qquad 51.35$$

$$\begin{aligned}\text{number of decades} &= \frac{\text{gain}_2 - \text{gain}_1}{20} \quad \text{[in dB]} \\ &= 0.301 \times \text{number of octaves}\end{aligned} \qquad 51.36$$

14. GAIN CHARACTERISTIC

The *gain characteristic* (*M*-curve for magnitude) is a plot of the gain as ω_f is varied. It is possible to make a rough sketch of the gain characteristic by calculating the gain at a few points (pole frequencies, $\omega=0$, $\omega=\infty$, etc.). The curve will usually be asymptotic to several lines. The frequencies at which these asymptotes intersect are *corner frequencies*. The peak gain, M_p, coincides with the natural (resonant) frequency of the system.[5] Large peak gains indicate reduced stability and large overshoots. The *gain crossover point* is the frequency at which log(gain) = 0.

The *half-power points* (*cut-off frequencies*) are the frequencies for which the gain is 0.707 (i.e., $\sqrt{2}/2$) times the peak value. (This is equivalent to saying the gain is 3 dB less than the peak gain.) The *cut-off rate* is the slope of the gain characteristic in dB/octave at a half-power point. The frequency difference between the half-power points is the *bandwidth*, BW. (See Fig. 51.8.) The *closed-loop bandwidth* is the frequency range over which the closed-loop gain falls 3 dB below its value at $\omega=\omega_n$. (The term bandwidth often means closed-loop bandwidth.) The *quality factor*, Q, is

$$Q = \frac{\omega_n}{\text{BW}} \qquad 51.37$$

[5]The gain characteristic peaks when the forcing frequency equals the natural frequency. It is also said that this peak corresponds to the resonant frequency. Strictly speaking, this is true, although the gain may not actually be resonant (i.e., be infinite).

Figure 51.8 Bandwidth

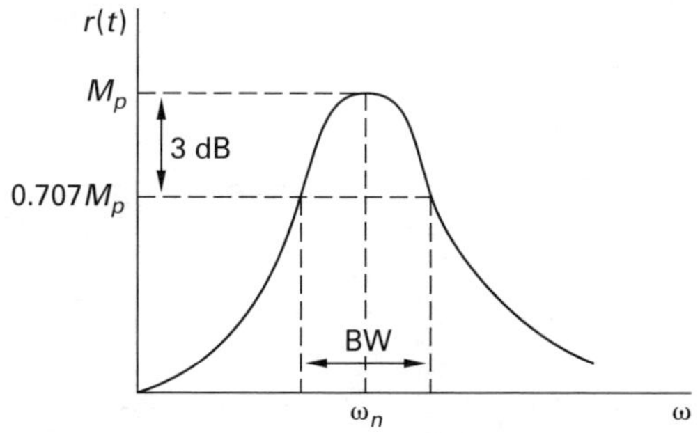

Because a low or negative gain (compared to larger parts of the curve) effectively represents attenuation, the gain characteristic can be used to distinguish between low- and high-pass filters. A low-pass filter has a large gain at low frequencies and a small gain at high frequencies. Conversely, a high-pass filter has a high gain at high frequencies and a low gain at low frequencies.

It may be possible to determine certain parameters (e.g., the natural frequency and BW) from the transfer function directly. For example, when the denominator of $T(s)$ is a single linear term of the form $s+r$, the BW will be equal to r. The BW and time constant are reciprocals.

Another important case is that in which $T(s)$ has the form of Eq. 51.38. (Compare the form of $T(s)$ to the form of Eq. 51.11.) The coefficient of the s^2 term must be unity, and ω_n must be much larger than BW so that the pole is close to the imaginary axis. The zero defined by constants a and b in the numerator is not significant.

$$T(s) = \frac{as+b}{s^2 + (\text{BW})s + \omega_n^2} \qquad \textit{51.38}$$

Example 51.14

What are the maximum gain, BW, upper half-power frequency, and half-power gain of a system whose transfer function is

$$T(s) = \frac{1}{s+5}$$

Solution

The steady-state response to a sinusoidal input is determined by substituting $j\omega$ for s in $T(s)$. When $\omega=0$, $T(s)$ will have a value of $1/5=0.2$, and $T(s)$ decreases thereafter. The maximum gain is therefore 0.2. The BW is 5 rad/s. Because the lower half-power frequency is implicitly 0 rad/s, the upper half-power frequency is 5 rad/s, at which point the system gain will be $(0.707)(0.2)=0.141$.

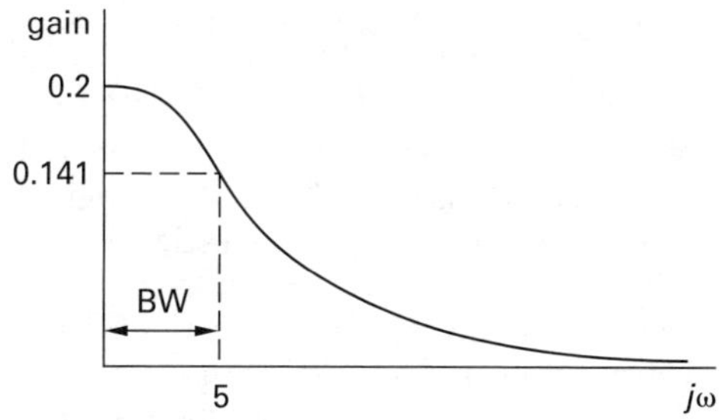

Example 51.15

Predict the natural frequency and BW for the following transfer function.

$$T(s) = \frac{s+19}{s^2+7s+1000}$$

Solution

This equation has the same form as Eq. 51.38. The BW is 7, and the natural frequency is $\sqrt{1000} = 31.62$ rad/s.

15. PHASE CHARACTERISTIC

The phase angle response will also change as the forcing frequency is varied. The *phase characteristic* (*α curve*) is a plot of the phase angle as ω_f is varied.

16. STABILITY

A stable system will remain at rest unless disturbed by external influence and will return to a rest position once the disturbance is removed. A pole with a value of $-r$ on the real axis corresponds to an exponential response of e^{-rt}. Because e^{-rt} is a decaying signal, the system is stable. Similarly, a pole of $+r$ on the real axis corresponds to an exponential response of e^{rt}. Because e^{rt} increases without limit, the system is unstable.

Because any pole to the right of the imaginary axis corresponds to a positive exponential, a *stable system* will have poles only in the left half of the s-plane. If there is an isolated pole on the imaginary axis, the response is stable. However, a conjugate pole pair on the imaginary axis corresponds to a sinusoid that does not decay with time. Such a system is considered to be unstable.

Passive systems (i.e., the homogeneous case) are not acted upon by a forcing function and are always stable. In the absence of an energy source, exponential growth cannot occur. *Active systems* contain one or more energy sources and may be stable or unstable.

There are several *frequency response* (*domain*) *analysis* techniques for determining the stability of a system, including Bode plots, root-locus diagrams, Routh stability criteria, Hurwitz tests, and Nichols charts. The term *frequency response* almost always means the steady-state response to a sinusoidal input.

Control Systems

The value of the denominator of $T(s)$ is the primary factor affecting stability. When the denominator approaches zero, the system increases without bound. In the typical feedback loop, the denominator is $1 \pm GH$, which can be zero only if $|GH| = 1$. It is logical, then, that most of the methods for investigating stability (e.g., Bode plots, root-locus diagrams, Nyquist analyses, and Nichols charts) investigate the value of the open-loop transfer function, GH. Because $\log 1 = 0$, the requirement for stability is that $\log GH$ must not equal 0 dB.

A negative feedback system will also become unstable if it changes to a positive feedback system, which can occur when the feedback signal is changed in phase more than 180°. Therefore, another requirement for stability is that the phase angle change must not exceed 180°.

17. BODE PLOTS

Bode plots are gain and phase characteristics for the open-loop $G(s)H(s)$ transfer function that are used to determine the *relative stability* of a system. The gain characteristic is a plot of $20 \log |G(s)H(s)|$ versus ω for a sinusoidal input. It is important to recognize that Bode plots, though similar in appearance to gain and phase frequency response charts, are used to evaluate stability and do not describe the closed-loop system response.

The *gain margin* is the number of decibels that the open-loop transfer function, $G(s)H(s)$, is below 0 dB at the *phase crossover frequency* (i.e., where the phase angle is −180°). (If the gain happens to be plotted on a linear scale, the gain margin is the reciprocal of the gain at the phase crossover point.) The gain margin must be positive for a stable system, and the larger it is, the more stable the system will be.

The *phase margin* is the number of degrees the phase angle is above −180° at the *gain crossover point* (i.e., where the logarithmic gain is 0 dB or actual gain is 1). (See Fig. 51.9.)

In most cases, large positive gain and phase margins will ensure a stable system. However, the margins could have been measured at other than the crossover frequencies. Therefore, a Nyquist stability plot is needed to verify the absolute stability of a system.

18. ROOT-LOCUS DIAGRAMS

A *root-locus diagram* is a pole-zero diagram showing how the poles of $G(s)H(s)$ move when one of the system parameters (e.g., the gain factor) in the transfer function is varied. The diagram gets its name from the need to find the roots of the denominator (i.e., the poles). The locus of points defined by the various poles is a line or curve that can be used to predict *points of instability* or other critical operating points. A point of instability is reached when the line crosses the imaginary axis into the right-hand side of the pole-zero diagram.

Figure 51.9 *Gain and Phase Margin Bode Plots*

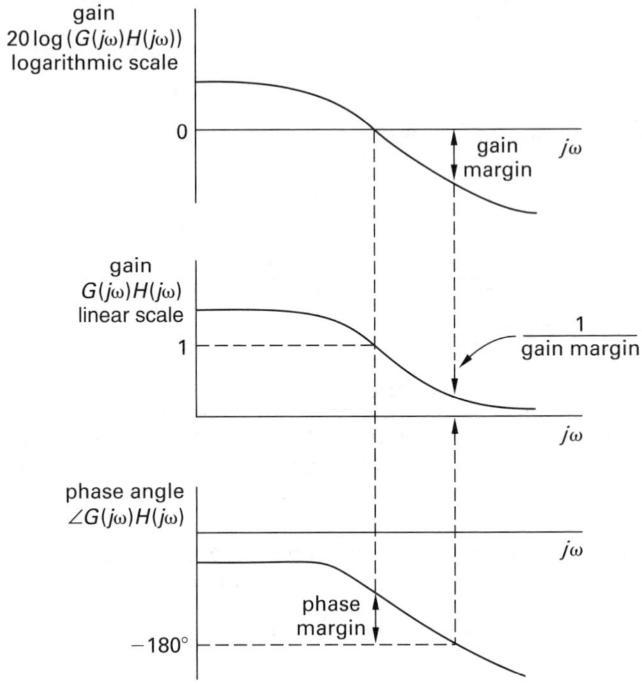

A root-locus curve may not be contiguous, and multiple curves will exist for different sets of roots. Sometimes the curve splits into two branches. In other cases, the curve leaves the real axis at *breakaway points* and continues on with constant or varying slopes approaching asymptotes. One branch of the curve will start at each open-loop pole and end at an open-loop zero.

Example 51.16

Draw the root-locus diagram for a feedback system with open-loop transfer function $G(s)H(s)$. K is a scalar constant that can be varied.

$$G(s)H(s) = \frac{Ks(s+1)(s+2)}{s(s+2) + K(s+1)}$$

Solution

The poles are the zeros of the denominator.

$$s_1, s_2 = -\tfrac{1}{2}(2+K) \pm \sqrt{1 + \tfrac{1}{4}K^2}$$

Because the second term can be either added or subtracted, there are two roots for each value of K. Allowing K to vary from zero to infinity produces a root-locus diagram with two distinct branches. The first branch extends from the pole at the origin to the zero at $s = -1$. The second branch extends from the pole at $s = -2$ to $-\infty$. All poles and zeros are not shown. Because neither branch crosses into the right half, the system is stable for all values of K.

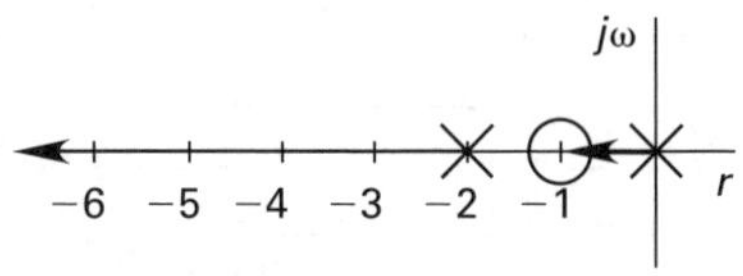

19. HURWITZ TEST

A stable system has poles only in the left half of the s-plane. These poles correspond to roots of the *characteristic equation.* The characteristic equation can be expanded into a polynomial of the form

$$a_0 s^n + a_1 s^{n-1} + \cdots + a_{n-1} s + a_n = 0 \qquad 51.39$$

The *Hurwitz stability criterion* requires that all coefficients be present and be of the same sign (which is equivalent to requiring all coefficients to be positive); if the coefficients differ in sign, the system is unstable. If the coefficients are all alike in sign, the system may or may not be stable. The Routh test should be used in that case.

20. ROUTH CRITERION

The *Routh criterion,* like the Hurwitz test, uses the coefficients of the polynomial characteristic equation. A table (the *Routh table*) of these coefficients is formed. The Routh-Hurwitz criterion states that the number of sign changes in the first column of the table equals the number of positive (unstable) roots. Therefore, the system will be stable if all entries in the first column have the same sign.

The table is organized in the following manner.

a_0	a_2	a_4	a_6	$\cdots$
a_1	a_3	a_5	a_7	$\cdots$
b_1	b_2	b_3	b_4	$\cdots$
c_1	c_2	c_3	c_4	$\cdots$
$\vdots$	$\vdots$	$\vdots$	$\vdots$	

The remaining coefficients are calculated in the following pattern until all values are zero.

$$b_1 = \frac{a_1 a_2 - a_0 a_3}{a_1} \qquad 51.40$$

$$b_2 = \frac{a_1 a_4 - a_0 a_5}{a_1} \qquad 51.41$$

$$b_3 = \frac{a_1 a_6 - a_0 a_7}{a_1} \qquad 51.42$$

$$c_1 = \frac{b_1 a_3 - a_1 b_2}{b_1} \qquad 51.43$$

Special methods are used if there is a zero in the first column but nowhere else in that row. One of the methods is to substitute a small number, represented by ϵ or δ, for the zero and calculate the remaining coefficients as usual.

Example 51.17

Evaluate the stability of a system that has a characteristic equation of

$$s^3 + 5s^2 + 6s + C = 0$$

Solution

All of the polynomial terms are present (Hurwitz criterion), so the Routh table is

s^3	1	6
s^2	5	C
s^1	$\frac{30 - C}{5}$	0
s^0	C	

To be stable, all of the entries in the first column must be positive, which requires $0 < C < 30$.

21. NYQUIST ANALYSIS

Nyquist analysis is another graphical method that is particularly useful when time delays are present in a system or when frequency response data are available. *Nyquist's stability criterion* is $N = P - Z$. P is the number of poles in the right half of the s-plane. Z is the number of zeros in the right half of the s-plane, and N is the number of encirclements (revolutions) of $1 + G(s)H(s)$ around the critical point. N may be positive, negative, or zero.

22. APPLICATION TO CONTROL SYSTEMS

A control system monitors a process and makes adjustments to maintain performance within certain acceptable limits. Feedback (see Fig. 51.10) is implicitly a part of all control systems.[6] The *controller* (*control element*) is the part of the control system that establishes the acceptable limits of performance, usually by setting its own reference inputs. The controller transfer function for a proportional controller is a constant: $G_1(s) = K$.[7] The *plant* (*controlled system*) is the part of the system

[6]Not all controlled systems are feedback systems. The positions of many precision devices (e.g., print heads in dot matrix printers and cutting heads on some numerically controlled machines) are controlled by precision *stepped motors.* However, unless the device has feedback (e.g., a position sensor), it will have no way of knowing if it gets out of control.

[7]Sometimes the notation K_n is used for K where n is the type of the system. In a type 0 system, a constant error signal results in a constant value of the output signal. In a type 1 system, a constant error signal results in a constant rate of change of the output signal. In a type 2 system, a constant error signal produces a constant second derivative of the output variable.

that responds to the controller. Both of these are in the forward loop. The input signal, V_i, in Fig. 51.4, is known in a control system as the *command* or *reference value.*

Figure 51.10 *Typical Feedback Control System*

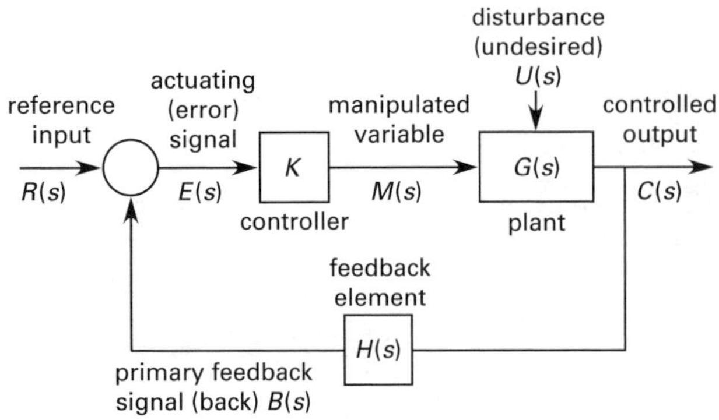

A *servomechanism* is a special type of control system in which the controlled variable is mechanical position, velocity, or acceleration. In many servomechanisms, $H(s) = 1$ (i.e., unity feedback) and it is desired to keep the output equal to the reference input (i.e., maintain a zero error function). If the input, $R(s)$, is constant, the descriptive terms *regulator* and *regulating system* are used.

The output of any given control system may not be in the assigned parameter or metric range. A *compensator*, which is a circuit designed to correct the discrepant parameter (e.g., the phase), may be added in the appropriate location of the control system to provide the necessary correction.

23. CONTROL SYSTEM TYPES/MODES

The method of controlling the desired signal depends on the level and type of desired control. The type of control action used is called the *mode of control.*

The simplest type of controller, and likely the most familiar, is the *on-off* or *bang-bang controller.* As the name implies, the system controlled is turned on when a signal goes above a given setpoint and is turned off when the signal goes below the setpoint. An example is the thermostat controller in most households. Controllers can also be made to respond to the rate that the input parameter or signal is changing. Such controllers are shown in Fig. 51.11.

A *proportional controller* is shown in Fig. 51.11(a). Such a controller adjusts the manipulated variable so that the controlled output matches the input demand. In the case of the household thermostat so designed, the heat input to the house from the furnace would match demand, maintaining a stable temperature.

A *derivative controller* is shown in Fig. 51.11(b). As the name implies, the rate of change is the variable controlled. The advantage is the ability to respond to a

Figure 51.11 *Control System Types*

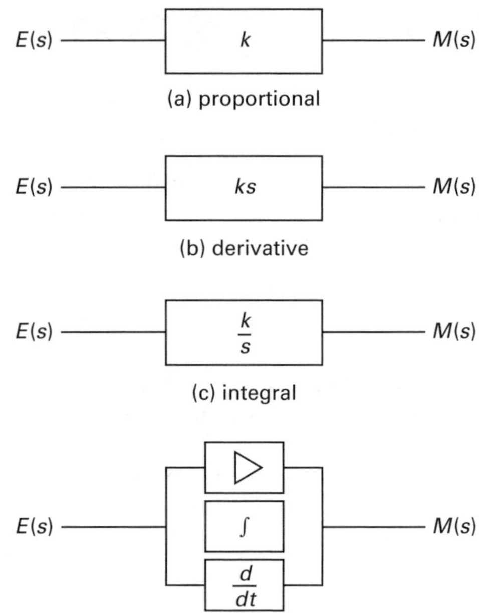

rapidly changing parameter. The *integral controller* is shown in Fig. 51.11(c). The advantage is the ability to respond in a controlled fashion to rapid changes in demand. The disadvantage is a large overshoot if there is a one-time jump in demand.

Figure 51.11(d) shows a *proportional-integral-derivative* (PID) *controller.* The advantage of this controller is its ability to respond to complex demand signals without overshoot of the controlled parameter.

24. STATE MODEL REPRESENTATION

While the classical methods of designing and analyzing control systems are adequate for most situations, state model representations are preferred for more complex cases, particularly those with multiple inputs and outputs or when behavior is nonlinear or varies with time. This method of evaluation is almost always carried out on a digital or analog computer.

The state variables completely define the dynamic state, $x_i(t)$ (i.e., position, voltage, pressure), of the system at time t. In simple problems, the number of state variables corresponds to the number of *degrees of freedom*, n, of the system. The n state variables are written in matrix form as a state vector, **X**.

$$\mathbf{X} = \begin{pmatrix} x_1 \\ x_2 \\ x_3 \\ \vdots \\ x_n \end{pmatrix} \qquad 51.44$$

It is a characteristic of state models that the state vector is acted upon by a first-degree derivative operator, d/dt, to produce a differential term of order 1, $\mathbf{X}'$.

$$\mathbf{X}' = \frac{d\mathbf{X}}{dt} \quad 51.45$$

Equation 51.46 and Eq. 51.47 show the general form of a state model representation: $\mathbf{U}$ is an r-dimensional (i.e., an $r \times 1$ matrix) control vector; $\mathbf{Y}$ is an m-dimensional (i.e., an $m \times 1$ matrix) output vector; $\mathbf{A}$ is an $n \times n$ system matrix; $\mathbf{B}$ is an $n \times r$ control matrix; and $\mathbf{C}$ is an $m \times n$ output matrix. The actual unknowns are the x_i's. The y_i's, which may not be needed in all problems, are only linear combinations of the x_i's. (For example, the x's might represent spring end positions; the y's might represent stresses in the spring. Then, $y = k\Delta x$.) Equation 51.46 is the *state equation*, and Eq. 51.47 is the *response equation*.

$$\mathbf{X}' = \mathbf{AX} + \mathbf{BU} \quad 51.46$$

$$\mathbf{Y} = \mathbf{CX} \quad 51.47$$

A conventional block diagram can be modified to show the multiplicity of signals in a state model, as shown in Fig. 51.12.[8] The actual physical system does not need to be a feedback system. The form of Eq. 51.46 and Eq. 51.47 is the sole reason that a feedback diagram is appropriate.

Figure 51.12 *State Variable Diagram*

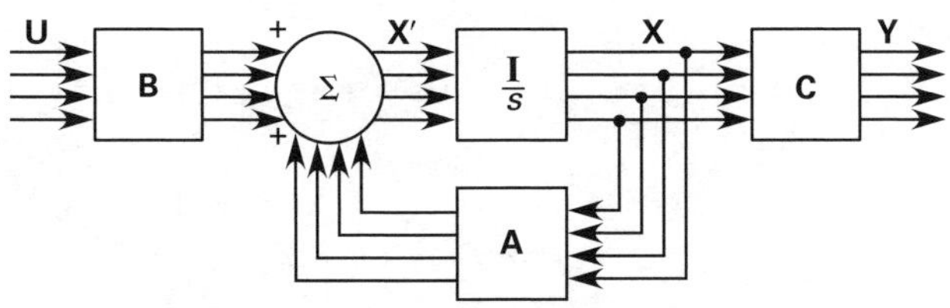

A state variable model permits only first-degree derivatives, so additional x_i state variables are used for higher-order terms (e.g., acceleration).

System controllability exists if all of the system states can be controlled by the inputs, $\mathbf{U}$. In state model language, system controllability means that an arbitrary initial state can be steered to an arbitrary target state in a finite amount of time. *System observability* exists if the initial system states can be predicted by observing the outputs, $\mathbf{Y}$, when the inputs, $\mathbf{U}$, are known.[9]

[8]The block $\mathbf{I}/s$ is a diagonal identity matrix with elements of $1/s$. This effectively is an integration operator.

[9]*Kalman's theorem*, based on matrix rank, is used to determine system controllability and observability.

Example 51.18

Write the state variable formulation of the following mechanical system's transfer function $T(s)$.

$$T(s) = \frac{7s^2 + 3s + 1}{s^3 + 4s^2 + 6s + 2}$$

Solution

Recognize the transfer function as the quotient of two terms, and multiply $T(s)$ by the dimensionless quantity x/x.

$$T(s) = \frac{Y(s)}{U(s)} = \left(\frac{7s^2 + 3s + 1}{s^3 + 4s^2 + 6s + 2}\right)\left(\frac{x}{x}\right)$$

$$Y(s) = 7s^2x + 3sx + x$$

$$U(s) = s^3x + 4s^2x + 6sx + 2x$$

$Y(s)$ and $U(s)$ represent the following differential equations.

$$y(t) = 7x''(t) + 3x'(t) + x(t)$$

$$u(t) = x'''(t) + 4x''(t) + 6x'(t) + 2x(t)$$

Make the following substitutions. ($x_4(t)$ is not needed because one level of differentiation is built into the state model.)

$$x_1(t) = x(t)$$

$$x_2(t) = x'(t) = x_1{}'(t)$$

$$x_3(t) = x''(t) = x_2{}'(t)$$

Write the first derivative variables in terms of the $x_i(t)$ to get the $\mathbf{A}$ matrix entries.

$$\begin{aligned} x_1'(t) &= 0x_1(t) + 1x_2(t) + 0x_3(t) + 0 \\ x_2'(t) &= 0x_1(t) + 0x_2(t) + 1x_3(t) + 0 \\ x_3'(t) &= -2x_1(t) - 6x_2(t) - 4x_3(t) + u(t) \end{aligned}$$

Determine the coefficients of the $\mathbf{C}$ matrix by rewriting $y(t)$ in the same variable order.

$$y(t) = 1x_1(t) + 3x_2(t) + 7x_3(t)$$

From Eq. 51.46 and Eq. 51.47, the state variable representation of $T(s)$ is

$$\begin{bmatrix} x_1' \\ x_2' \\ x_3' \end{bmatrix} = \begin{bmatrix} 0 & 1 & 0 \\ 0 & 0 & 1 \\ -2 & -6 & -4 \end{bmatrix} \begin{bmatrix} x_1 \\ x_2 \\ x_3 \end{bmatrix} + \begin{bmatrix} 0 \\ 0 \\ 1 \end{bmatrix} [u(t)]$$

$$[y(t)] = [1\ 3\ 7] \begin{bmatrix} x_1 \\ x_2 \\ x_3 \end{bmatrix}$$

Topic XIII: Electrical Materials

Chapter

52. Electrical Materials

52 Electrical Materials

1. Overview 52-1
2. Types of Materials 52-1
3. Conducting Materials 52-1
4. Insulating and Dielectric Materials 52-2
5. Magnetic Materials 52-2
6. Radiation-Emitting Materials 52-2

1. OVERVIEW

The electrical properties of materials impact all engineering designs. An understanding of the properties and potential reactions of chosen materials within the component parts and the surrounding environment is necessary both to ensure proper operation over the service lifetime of the components and to minimize long-term effects on the environment.

The fundamental properties of conducting, insulating and dielectric, semiconductor, magnetic, electron emitting, and radiation-emitting materials are covered throughout this book. Materials handbooks are available that provide specific values for these properties for elements and compounds under various conditions. A periodic table with a summary of materials' properties is given in App. 52.A. A table of relative atomic weights is given in App. 52.B.

2. TYPES OF MATERIALS

Materials may be grouped into three main types: metals, polymers (that is, plastics), and ceramics.[1] *Metals* are materials characterized by high thermal and electrical conductivity. They are opaque and can generally be polished to a high luster. Further, they are normally heavy and deformable. Metals are located to the left of the metal-to-nonmetal transition staircase shown in App. 52.A. Elements adjacent to the staircase exhibit characteristics of both metals and nonmetals and are sometimes called *metalloids* or *semiconductors*. Further, in some texts, they are called metals and in others they are called semiconductors. Items to the right of the staircase are called *nonmetals*.

[1]Glass is sometimes listed as a type of material. "Glass," however, is a term applied to noncrystalline materials with specific expansion/contraction characteristics. For example, below a temperature known as the *glass temperature*, further rearrangement of atoms in glass ceases and contraction results from smaller thermal vibrations—rather than the formation of a metallic crystalline structure. Glasses have short-range order with an absence of long-range order. Glass is probably best described as a ceramic.

Polymers, or *plastics*, are materials characterized by low thermal and electrical conductivity. They are generally transparent or translucent and poor reflectors of light. Finally, they are normally flexible and subject to deformation. Polymers are composed of those elements listed in Table 52.1.

Table 52.1 *Elements in Polymers*

element atomic number	element name
1	hydrogen
6	carbon
7	nitrogen
8	oxygen
9	fluorine
14	silicon

Ceramics are compounds that contain metallic and nonmetallic elements. Concrete, rocks, glass, and nuclear fuel (UO_2) are all ceramics, to name but a few. Ceramics have extremely low thermal and electrical conductivity and for this reason are also called *refractory materials*. Indeed, they are highly chemically resistant (inert), can be opaque or translucent, and are very hard and brittle. Ceramics are composed of metallic elements combined with carbon, nitrogen, oxygen, silicon, or phosphorous, or some combination thereof.

3. CONDUCTING MATERIALS

A *conductor* is a wire, cable, or other body or medium that is suitable for carrying an electric current. While no specific dividing lines exist between conductors, semiconductors, and insulators, conductors have a conductivity on the order of 10^6 S/m.

Copper is used extensively in electrical applications because of its high electrical and thermal conductivity and malleability. Aluminum is used when additional conductor strength is required. Silver has the highest conductivity at room temperature. Gold has conductivity similar to aluminum, as well as excellent corrosion and oxidation resistance. Platinum, a precious metal, is used primarily in resistance thermometers and thermocouples. Palladium is similar to platinum and less expensive. Nickel or nickel-coated conductors are used at temperatures up to 300°C with no enhanced corrosion at defective areas. Tungsten has a very high melting point and is erosion resistant, finding use in voltage regulators, repetitive low-current applications, and heating wires or filaments for lamps. Carbon has two crystalline forms: graphite and diamond. Carbon (that is, amorphous carbon) or graphite (that is, crystalline carbon) is used as an electric sliding contact in motor brushes. The diamond structure is used where high heat conduction is desired.

4. INSULATING AND DIELECTRIC MATERIALS

An *insulator* is a device or material having a high electrical resistance (i.e., a low conductivity). A *dielectric* is an electrically insulating material or a material in which an electric field can be maintained with minimum dissipation of power. Insulators generally have a conductivity on the order of 10^{-12} S/m.

Plastics are widely used because of their low cost, ability to be molded into many shapes, and good mechanical properties. Ceramics, because of their very low electrical conductivity, also find many uses in electrical insulating applications. Oils, gases, and papers are inexpensive insulators. Mica, with its stability, arc resistance, and high dielectric strength, is another common insulator.

5. MAGNETIC MATERIALS

A *magnetic material* is generally defined as a material exhibiting ferromagnetism. Magnetic materials are classified into two broad categories: *Nonretentive* or *soft* materials are low-loss materials. *Retentive* or *hard* materials are high-loss, high-retentivity, high-energy product materials.

The soft magnetic materials are used in transformers, motors and generators, cores, and other electromagnetic apparatuses. Iron is the most common soft magnetic material. Iron-silicon, also called *electrical steel*, is a soft material with very low core and eddy current losses and very high permeability. Iron-cobalt alloys are used where high saturation is required. Other alloy combinations are used in special applications.

The hard magnetic materials are used as permanent magnets. Hard materials are composed of a fine structure created during various heat treatments. High-carbon steels with tungsten, chromium, cobalt, aluminum, or vanadium in a martensitic structure are common. Precipitation-hardened alloys also form a fine structure. In this category are *alnico* (aluminum, nickel, and cobalt), *cunife* (copper, nickel, and iron), and *vicalloy* (vanadium, cobalt, and iron), among others.[2]

6. RADIATION-EMITTING MATERIALS

Materials at excited energy levels emit electrons, photons of light, or other radiation. These emissions have numerous electrical applications.

Metal filaments in incandescent lamps emit light when heated. Tungsten is the most common material used in this application. *Discharge lamps*, also called *fluorescent lamps*, use material in vapor form.[3] The noble gases, sodium, and mercury are the materials of choice. *Phosphors* are organic or inorganic luminescent materials.

[2]Ordered alloys and fine-particle magnets are also retentive.

[3]When materials emit light almost instantaneously after excitation, they are called *fluorescent* materials. When the emission is delayed, they are called *phosphorescent* materials.

Topic XIV: Codes and Standards

Chapter

53. National Electrical Code

54. National Electrical Safety Code

53 National Electrical Code

1. History and Overview 53-1
2. Introduction 53-2
3. General 53-2
4. Wiring and Protection 53-3
5. Wiring and Protection: Grounded Conductors 53-3
6. Wiring and Protection: Branch Circuits ... 53-3
7. Wiring and Protection: Feeders 53-4
8. Wiring and Protection: Branch Circuit, Feeder, and Service Calculations 53-5
9. Wiring and Protection: Overcurrent Protection 53-6
10. Wiring and Protection: Grounding 53-6
11. Wiring Methods and Materials 53-7
12. Equipment for General Use 53-7

Nomenclature

A	area	m^2
d	diameter	m
F	factor	–
I	effective or DC current	A
l	length	m
P	power	W
pf	power factor	–
R	resistance	Ω
V	effective or DC voltage	V

Subscripts

l per unit length

1. HISTORY AND OVERVIEW

The first practical incandescent light bulb and lighting system were developed by Thomas Alva Edison in 1879. That same year, the National Association of Fire Engineers met to discuss standards for electrical installations. By 1895 there were five separate codes in the United States. In 1896, a national meeting was held between various concerned organizations and by 1897 the *National Electrical Code* (NEC) (hereafter referred to as the "Code") was adopted. The Code gradually came under the sponsorship and control of the National Fire Protection Association (NFPA). The Code is officially endorsed by the American National Standards Institute (ANSI), and the NFPA *National Electrical Code* committee responsible for the Code is known as ANSI Standards Committee C1. The Code is the most widely adopted code in the United States and the most widely accepted set of electrical safety requirements in the world. Local codes in the United States normally adopt the Code en masse and then add supplemental requirements deemed necessary, or remove and change items to ensure local applicability or uniqueness.

The Code generally applies to the installation of electrical conductors and equipment within or on public and private buildings up to and including the connection to the electrical supply. Its general purpose is to prevent fires and explosions by providing guidance for proper electrical installations.

The Code consists of an introduction followed by nine chapters, which are further subdivided into articles. The outline of the Code follows. Chapter numbers are given in parentheses. Chapter 1 through Chap. 4 generally apply to all situations except as modified by Chap. 5 through Chap. 7.

- Introduction
- (1) General
- (2) Wiring and Protection
- (3) Wiring Methods and Materials
- (4) Equipment for General Use
- (5) Special Occupancies
- (6) Special Equipment
- (7) Special Conditions
- (8) Communications Systems
- (9) Tables
- Annex A: Product Safety Standards
- Annex B: Application Information for Ampacity Calculation
- Annex C: Conduit and Tubing Fill Tables
- Annex D: Examples
- Annex E: Types of Construction
- Annex F: Critical Operators Power Systems
- Annex G: SCADA
- Annex H: Administration and Enforcement

Figure 53.1 NEC Coverage

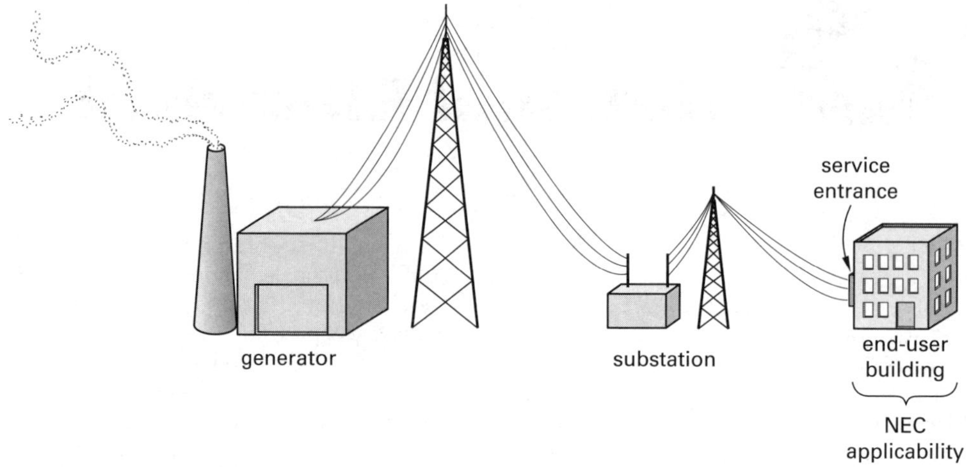

In this chapter, numbers in brackets and article numbers greater than 90 indicate the article or article number in the Code and are meant for ease in referencing the Code directly. Concepts covered and requirements mentioned focus on dwelling units and voltages less than 600 V and will not cover Chap. 5 through Chap. 8 of the Code. A mixture of English engineering units and SI units are used.

2. INTRODUCTION

The Introduction to the Code contains Article 90 and explains that the purpose of the Code is the "practical safeguarding of persons and property from hazards arising from the use of electricity." The intended coverage or applicability of the Code is illustrated in Fig. 53.1.

The Code is meant to be a legal document that can be implemented and interpreted by local governmental bodies that have jurisdiction over electrical installations. Most do just that, with some additions and changes for local needs. A *permissive rule* in the Code is one that is allowed but not required (i.e., an alternative). The term "shall be permitted" is used to indicate these alternatives. Any *fine print note* (FPN) used in the Code is for informational purposes only and does not represent a requirement. Factory-installed internal wiring is not required to be inspected if the equipment has been listed by a qualified electrical testing laboratory that requires installation in accordance with the Code. When metric measurements are used, they conform to the SI System. Most of the values in the Code remain in the English Engineering System, with approximate metric equivalents given. Footnotes are also used to provide the necessary SI conversion factors.

3. GENERAL

The General portion of the Code contains Article 100 (definitions) and Article 110 (requirements for electrical installations).

The definitions are those required for the proper application of the Code and are not all-encompassing. Generally understood terms that can be found in an American language dictionary are not included. Also not included are electrical terms that are not unique to the Code, such as "volt." Some of the terms that are used in a distribution system and found in the definitions article are shown in Fig. 53.2.

Figure 53.2 Typical Distribution System

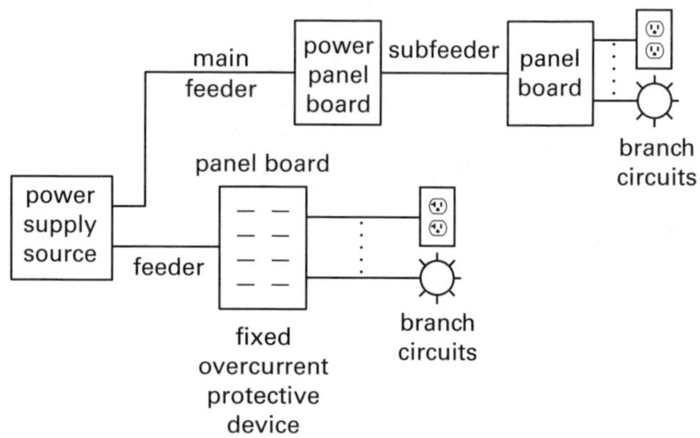

In this section, *overcurrent* is defined as any current in excess of rated current. Overcurrent includes overload, short circuit, and ground fault. An *overload* is defined as the operation of equipment in excess of its normal full-load rating or of a conductor in excess of its rated ampacity, which, after a sufficient time, causes damage or overheating.

Article 110, "Requirements for Electrical Installation," notes that conductor size will be given in *American Wire Gage* (AWG) or in circular mils. A comparison of American Wire Gage diameters to other systems is given in App. 53.A. Conversions between diameters and cross-sectional areas can be accomplished with Eq. 53.1 and Eq. 53.2.

$$A_{\text{cmil}} = \left(\frac{d_{\text{in}}}{0.001}\right)^2 \qquad 53.1$$

$$A_{\text{cmil}} = 1.273 \times 10^6 A_{\text{in}^2} \qquad 53.2$$

Wire gage size up to 4/0 is listed as AWG in the Code. Beyond that, the size is listed in circular mils (cmil) or, more likely, thousands of circular mils (kcmil). The metric standard for marking wire size is square millimeters (mm^2). The conversion is

$$A_{\text{cmil}} = 1973.53 A_{\text{mm}^2} \qquad 53.3$$

Temperature limitations are imposed and ratings coordinated so that conductors and their terminations are rated for the same operating temperature for circuits rated at 100 A or less [Sec. 110.14(C)(1)]. Required working space within and around electrical installations, safety issues, and illumination of such spaces fill out the remaining items in Article 110.

Example 53.1

A wire of unknown gage is measured. The diameter is found to be 0.2057 cm. What is the American Wire Gage (AWG) designation?

Solution

The tables for wire gage comparisons are generally in English Engineering System units, as is App. 53.A. Converting the diameter gives

$$d = (0.2057 \text{ cm})\left(\frac{1 \text{ in}}{2.54 \text{ cm}}\right) = 0.0810 \text{ in}$$

From App. 53.A, this diameter correlates with AWG 12. The generic term used is "twelve gage wire."

4. WIRING AND PROTECTION

The "Wiring and Protection" portion of the Code contains Articles 200 (grounded conductors), 210 (branch circuits), 215 (feeders), 220 (branch circuit, feeder, and service calculations), 225 (outside branch circuits and feeders), 230 (services), 240 (overcurrent protection), 250 (grounding), 280 (surge arresters), and 285 (surge-protective devices).

The term "nominal" is used throughout the Code. Actual power supply voltage specifications, high and low, are set by ANSI standards. Nominal voltage is the desired voltage of the power source and is given in multiples of 120 V. Rated voltage is the desired voltage for operating equipment and is given in multiples of 115 V. Operating equipment is called *utilization equipment* and includes any device that utilizes electric energy for electronic, electromechanical, chemical, heating, lighting, or another similar purpose [Article 100].

Four-wire wye-connected systems normally list the type of connection next to the voltage for phase-to-phase voltage and give the phase-to-neutral voltage after a slash (i.e., 208Y/120 or 480Y/277). Examples of common wye and delta four-wire systems are illustrated in Fig. 53.3.

Figure 53.3 *Four-Wire System Nomenclature*

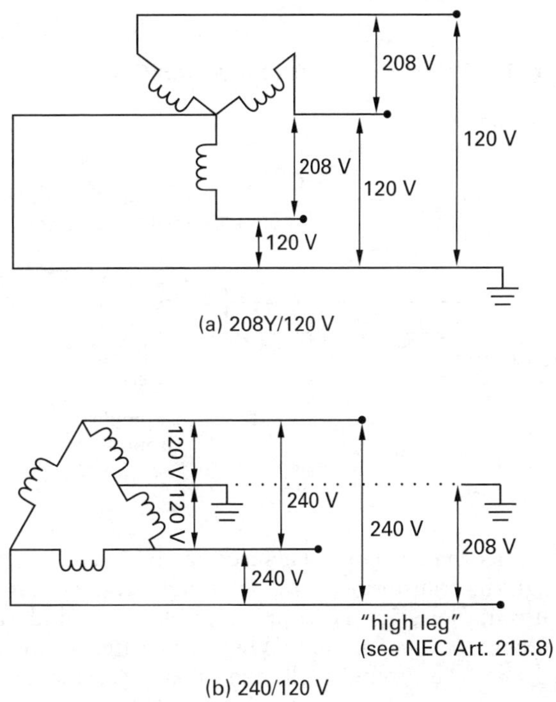

5. WIRING AND PROTECTION: GROUNDED CONDUCTORS

A *grounded conductor*, which is a conductor that is intentionally grounded (for example, the neutral of a system), is identified by a white or natural gray outer finish or by three continuous white stripes on other than green insulation [Sec. 200.6]. This should not be confused with a *grounding conductor*, which is used to connect equipment or the grounded circuit (neutral) of a wiring system to a grounding electrode and which is colored green [Articles 100 and 250.119]. Terminals for connecting grounded conductors must be substantially white, or have the word "white" or the letter "W" adjacent to the terminal.

6. WIRING AND PROTECTION: BRANCH CIRCUITS

A *branch circuit* consists of the circuit conductors between the final overcurrent device protecting the circuit and the outlet. Such circuits are covered in Article 210. However, if the branch circuit supplies only motor loads, it is covered in Article 430. A cross-reference for

Codes and Standards

articles relating to branch circuits for other special conditions is given in Table 210.2.

A *ground-fault circuit interrupter* (GFCI) is a device that de-energizes a circuit within an established period of time when a current to ground exceeds some predetermined value, which is less than that required to operate the overcurrent protective device of the supply circuit [Article 100]. The concept of such a device is illustrated in Fig. 53.4.

Figure 53.4 *Ground-Fault Circuit Interrupter (GFCI)*

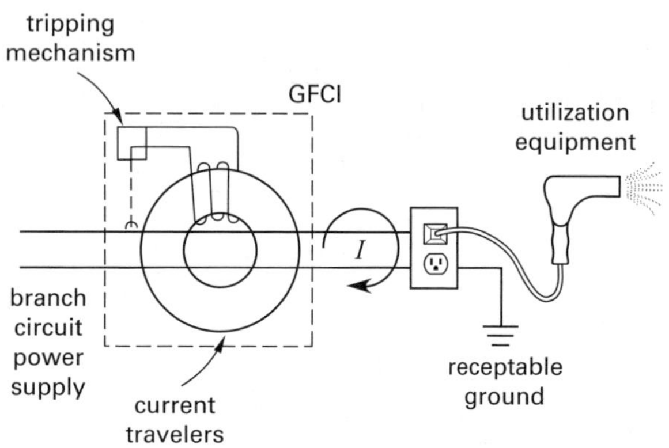

As long as the current is balanced in the lines running through the transformer core, no induced voltage is felt, no current flows in the tripping mechanism, and the GFCI contact remains shut. When the current is unbalanced due to a ground fault, the unbalanced magnetic fields produced induce a net voltage in the transformer, current flows in the tripping mechanism, and the GFCI contact opens, protecting any person holding the offending equipment. GFCIs are required to be installed outdoors and in dwelling unit bathrooms, garages, crawl spaces, unfinished basements, kitchens, wet bar sinks, and boat houses [Sec. 210.8(A)].

The number of branch circuits required in a given building is calculated per Sec. 210.11. A dwelling unit is also required to have two or more 20 A small-appliance branch circuits beyond that calculated, one 20 A branch circuit for laundry, and one 20 A branch circuit for each bathroom [Sec. 210.11(C)]. The conductors that supply branch circuits, before any adjustment is made or derating is allowed, must be able to carry 100% of the noncontinuous load as well as 125% of the continuous load [Sec. 210.19(A)(1)]. The *overcurrent protective device* (OCPD) selected must meet identical requirements [Sec. 210.20(A)]. The allowable conductor ampacities are found in App. 53.B [NEC Table 310.16].[1]

A single receptacle on a branch circuit must have an ampere rating of not less than the branch circuit. If two or more receptacles are used, a table gives their allowed values [Sec. 210.21(B), Table 210.21(B)(3)]. The rating of a branch circuit is based on the overcurrent protective device (OCPD) rating regardless of the actual load or the conductor ampacity. A 15 A or 20 A branch circuit can supply lighting, utilization equipment, or a combination of both. Any single cord-and-plug-connected device cannot exceed 80% of the branch circuit ampere rating. Further, the rating of utilization equipment fastened in place, other than lighting, cannot exceed 50% of the branch circuit ampere rating [Sec. 210.23(d)(2)]. A summarized table of branch circuit requirements is given in the Code in Table 210.24.

Receptacles are spaced in dwellings so that no point along the floor of any wall space is more than 1.8 m (6 ft) from a receptacle. (A distance of 12 ft can exist between receptacles.) Floor receptacles, those higher than 1.68 m (5.5 ft) above the floor, those part of a lighting system or appliance, and those located within cupboards or cabinets, do not count toward the requirement [Sec. 210.52]. On countertops, the receptacle spacing is no greater than 600 mm (2 ft). Receptacles attached and controlled by switches are lighting outlets, the requirements of which are separate from general receptacles [Sec. 210.70].

Example 53.2

A given grocery store branch circuit is meant to carry a continuous lighting load of 8 A. In addition, three intermittent loads designed for display stands are fastened in place and require 2 A each when operating. What is the rating of the overcurrent protective device (OCPD) on the branch circuit?

Solution

By Sec. 210.20(A), the overcurrent device must be rated at 100% of the noncontinuous load and 125% of the continuous load.

$$\begin{aligned}\text{OCPD} &= (1.00)(2\text{ A} + 2\text{ A} + 2\text{ A}) + (1.25)(8\text{ A}) \\ &= 16\text{ A minimum}\end{aligned}$$

The standard ratings for fuses and fixed-trip circuit breakers are given in Sec. 240.6 as 15 A, 20 A, 25 A, 30 A, and so on. Therefore, a 20 A OCPD will suffice.

Utilization equipment "fastened in place" cannot, and does not in this case, exceed 50% of the branch circuit ampere rating [Sec. 210.23(A)(2)]. That is, 6 A is less than 50% of 20 A.

7. WIRING AND PROTECTION: FEEDERS

A *feeder* consists of all circuit conductors between the service equipment, the source of a separately derived system, or another power supply source and the final branch circuit overcurrent device (see Fig. 53.2) [Article 100]. The capacity of the conductors that constitute feeders, before any adjustment or derating allowed,

[1]Table 310.16 is arguably the most used table in the Code. Familiarity with it is essential for proper application of requirements.

must be 100% of the noncontinuous load plus 125% of the continuous load [Sec. 215.2]. The *overcurrent protective device* (OCPD) selected must meet identical requirements [Sec. 215.3]. Several special conditions for sizing the feeders also apply, including the exception that allows dwelling unit feeders carrying the full current of the service conductors to be the same size as those conductors. Though not required, a feeder is recommended to have no more than a 3% voltage drop, with a 5% voltage drop on both feeders and branch circuit conductors to the farthest outlet [Sec. 215.2(A)(3), FPN No. 2]. Useful Code tables for calculating voltage drops based on conductor properties when used in DC or AC circuits are given in App. 53.C, Conductor Properties, and App. 53.D, Conductor AC Properties [NEC Chap. 9, Tables 8 and 9, respectively].

For two-wire circuits, DC or AC, or three-wire single-phase AC circuits, with a power factor equal to one (purely resistive), the voltage drop is given by

$$V_{\text{drop}} = IR_l 2l = \frac{2lIR}{1000\text{ ft}} \qquad 53.4$$

The number 2 in Eq. 53.4 is used because the voltage drop is felt over the entire circuit length. The farthest equation to the right is used only if the resistance per unit length, R_l, is given by the referenced table in "per 1000 ft" quantities.

Example 53.3

A 480Y/277, 80 kW noncontinuous load with a 0.8 pf consists of less than 50% ballast-type lighting.[2] The feeder length is 110 m. Applying the 3% fine print note No. 2 recommendation from Sec. 215.2(A)(3), what is the maximum recommended voltage drop for every 1000 ampere-foot?

Solution

First, it should be recognized that the 1000 ampere-feet unit is asked for since many manufacturers in the United States list conductor information as such. Also, Tables 8 and 9 of Chap. 9 of the Code, which are used for calculating voltage drops, are given in "per 1000 ft" increments (see App. 53.C and App. 53.D).

To determine the line current drawn on a noncontinuous basis, and noting that the voltage given is that of a three-phase four-wire system, use[3]

$$P = \sqrt{3}IV(\text{pf})$$

$$I = \frac{P}{\sqrt{3}V(\text{pf})} = \frac{80\times 10^3\text{ W}}{(\sqrt{3})(480\text{ V})(0.8)} = 120.3\text{ A}$$

The total number of ampere-feet is

$$\begin{aligned}\text{ampere-feet} &= (120.3\text{ A})(110\text{ m})\left(3.281\ \frac{\text{ft}}{\text{m}}\right)\\ &= 43{,}417\text{ A-ft}\quad (43.4\times 1000\text{ A-ft})\end{aligned}$$

The reason for specifying "noncontinuous" is to emphasize that the ampacity is to be calculated at 100%, not 125%. Regardless of the load, continuous or noncontinuous, the voltage drop is calculated using the 100% ampacity. Nevertheless, the conductor must be sized to handle 125% ampacity according to Code standards. The OCPD must be set at 125% ampacity as well.

Since this system has a neutral, the voltage drop of concern is the phase-to-neutral drop.

$$\begin{aligned}V_{\text{drop}} &= (\text{allowed drop})(\text{line-to-neutral voltage})\\ &= (0.03)(277\text{ V})\\ &= 8.31\text{ V}\end{aligned}$$

The voltage drop per 1000 A-ft is then

$$\begin{aligned}V_{\text{drop}} &= \frac{8.31\text{ V}}{(43.4)(1000\text{ A-ft})}\\ &= 0.192\text{ V/1000 A-ft}\end{aligned}$$

This value could then be used with a manufacturer's table of conductors to determine which conductor size meets voltage drop recommendations. Voltage drop specifications on conductor size are in addition to other requirements and must be coordinated to determine the most limiting case. If the conductor size is increased to lower the voltage drop, the overcurrent protection, for the wiring, also increases.

Tables for conductor properties and voltage drops are often given in terms of line-to-neutral values while the recommendation for percentage voltage drop in the Code refers to the line voltage. Calculations should normally be accomplished using line-to-neutral voltages, that is, phase voltages, and then conversions applied to obtain line quantities and the percent voltage drop. By contrast, the line current is used since this is the current causing the voltage drop in the conductor cable, be it a branch circuit or feeder conductor.

8. WIRING AND PROTECTION: BRANCH CIRCUIT, FEEDER, AND SERVICE CALCULATIONS

Article 220 contains the requirements for calculations involving branch circuits, feeders, and service loads. Examples that are representative of the requirements are given in App. D of the Code.

[2]If the ballast-type lighting constitutes greater than 50% of the load, the neutral conductor would be considered a current-carrying conductor and would have to be sized according to the calculations in the example problem [Sec. 310.15(B)(4)(b); cf. Secs. 310.4 Exception No. 2 and 310.10 FPN No. (2)].

[3]Only four-wire systems have a voltage specified after the slash, for example, 208Y/120, 240/120, 480Y/277, and 4160Y/2400.

Codes and Standards

The voltages used in calculations are standardized as 120, 120/240, 208Y/120, 240, 347, 480Y/277, 480, 600Y/347, and 600 V [Sec. 220.2(A)]. Loads are determined in terms of volt-amperes (VA) or kilovolt-amperes (kVA). This is considered equivalent to equipment power ratings, which are given in watts (W) or kilowatts (kW). This concept of equivalence is used since the power factor is unknown and the conductors will carry the volt-ampere load, that is, the apparent power, and so should be rated for such. Fractions of less than 0.5 A are dropped.

Branch circuit lighting loads are determined by the square footage of a building, using the unit loads given in Sec. 220.12. Dwelling units' lighting loads are calculated based on 3 VA per square foot. Additional receptacle loads depend on the equipment attached [Sec. 220.14]. Receptacle outlets are calculated based on 180 VA per outlet minimum, with a minimum of 90 VA per receptacle for outlets with four or more receptacles [Sec. 220.14(I)].

Lighting loads that have ballasts (i.e., fluorescent lamps), transformers, or autotransformers are based on the ampere rating of the units and not the watt ratings [Sec. 220.18(B)]. These types of devices are termed *non-linear loads*and affect the sizing of the neutral conductor [Sec. 310.15(B)(4)(b); cf. Articles 310.4 Exception No. 2 and 310.10 FPN No. (2)].

Requirements for various types of loads, including general lighting, motors, and small appliances, are given in Sec. 220. Demand factors that allow less than the full-load wattage, that is, volt-amperes, to be used in determining the size of conductors and overcurrent devices are also given. Optional calculations for those dwelling units with a total connected load greater than 100 A are described in Sec. 220.82.

Example 53.4

For the circuit shown, what is the total VA load?

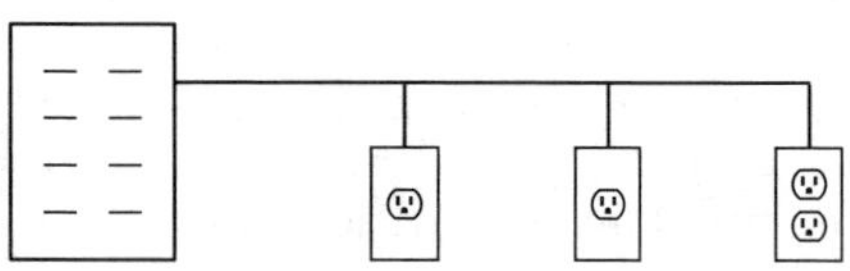

Solution

Regardless of the number of receptacles on an outlet, the minimum load is 180 VA [Sec. 220.14(I)]. The total load is

$$\begin{aligned} VA_{total} &= (\text{number of outlets})\left(180\ \frac{VA}{\text{outlet}}\right) \\ &= (3)\left(180\ \frac{VA}{\text{outlet}}\right) \\ &= 540\ VA \end{aligned}$$

9. WIRING AND PROTECTION: OVERCURRENT PROTECTION

Article 240 of the Code describes the overcurrent protection requirements and the electrical devices that accomplish them. One of the first sections of Part I lists specific types of equipment and the applicable article containing the overcurrent requirements. Table 240.3 is a useful starting point in any determination of overcurrent requirements.

Conductors are protected in accordance with their ampacities [Sec. 240.4]. Standard ampere ratings for fuses and circuit breakers are given in Sec. 240.6 as 15, 20, 25, 30, 35, 40, 45, 50, and so on. Thermal devices not designed to interrupt short circuits cannot be used for overcurrent protection for short circuits or grounds [Sec. 240.9]. The remainder of the article discusses location of overcurrent devices, tap requirements, fuses, and circuit breakers.

10. WIRING AND PROTECTION: GROUNDING

Article 250 of the Code describes general grounding and bonding in Part I, circuit and system grounding in Part II, grounding electrode requirements in Part III, and equipment grounding in Part VI.

Electrical systems are connected to the earth in a manner that will limit voltage from lightning strikes, line surges, or other transients [Sec. 250.4(A)(2)]. Electrical systems are also grounded in order to maintain the voltage at a stable level relative to ground to ensure connected equipment is subjected to a maximum potential difference set by the designer—for example, three-phase four-wire systems (see Sec. 250.26). Equipment is grounded as well, the primary goal being the safety of people in contact with the equipment. The overall requirements of the fault current path and their organization in the Code are given in Sec. 250.4(B)(4), while specific equipment requirements are cross-referenced in Sec. 250.3, which is a good starting point for finding grounding information.

The grounding electrode requirements are described in Sec. 250.50. Electrically continuous water pipes in contact with the earth for 3.0 m (10 ft) or more are used as grounding electrodes [Sec. 250.52(A)(1)]. Grounding electrodes 6.0 m (20 ft) long are allowed to be encased in concrete with a minimum thickness of 50 mm (2 in), if the electrodes are within the concrete and near the foundation, which must be in direct contact with the earth [Sec. 250.52(A)(3)]. Rod- and pipe-type electrodes must be 2.44 m (8 ft) long [Sec. 250.52(A)(5)]. If multiple rods are used, they must be a minimum of 1.8 m (6 ft) apart [Sec. 250.56]. The sizing of grounding electrode conductors for AC systems is given in NEC Table 250.66.

Cord-and-plug-connected equipment is generally grounded unless double-insulated [Sec. 250.114]. Major appliance equipment within dwellings must be grounded [Sec. 250.114(3)]. Insulated equipment grounding conductors are green, or green with one or more yellow stripes [Sec. 250.119]. Grounding terminals use green screws or nuts, the word "green" or "ground", the symbols "G" or "GR", or the grounding symbol [Sec. 250.126]. The sizing of equipment grounding conductors is given in Table 53.1 [NEC Table 250.122].

***Table 53.1** Equipment Grounding (NEC Table 250.122)*

rating or setting of automatic overcurrent device in circuit ahead of equipment, conduit, etc., not exceeding (amperes)	size (AWG or kcmil) copper	size (AWG or kcmil) aluminum or copper-clad aluminum*
15	14	12
20	12	10
30	10	8
40	10	8
60	10	8
100	8	6
200	6	4
300	4	2
400	3	1
500	2	1/0
600	1	2/0
800	1/0	3/0
1000	2/0	4/0
1200	3/0	250
1600	4/0	350
2000	250	400
2500	350	600
3000	400	600
4000	500	800
5000	700	1200
6000	800	1200

(Note: Where necessary to comply with Sec. 250.4(A)(5) or (B)(4), the equipment grounding conductor shall be sized larger than this table.)

*See installation restrictions in Sec. 250.120 of the *National Electrical Code®* (NEC).

11. WIRING METHODS AND MATERIALS

The "Wiring Methods and Materials" portion of the Code contains Articles 300 (wiring methods) and 310 (conductors), several articles on conduit, tubing, and raceways, and many others. The requirements regarding switchboards and panelboards are also in this portion of the Code.

The wiring methods covered in the Code are for all wiring installations, but do not include internal equipment wiring [Sec. 300.1(B)]. All conductors in the same circuit should be contained together [Sec. 300.3(B)]. In general, conductors for systems over 600 V are not mixed with those for systems under 600 V [Sec. 300.3 (C)(2)]. Directly buried conduit must be at least 450 mm (18 in) from the surface [Sec. 300.5(D)(1)].

Conductor application and insulation codes, such as THHN for heat-resistant thermoplastic, are given in NEC Table 310.13(A). The appropriate letter designations are explained, the type of insulation is specified, and the design applications are listed.

Conductor ampacities are determined using App. 53.B (NEC Table 310.16) and numerous other NEC tables given in Sec. 310.15. The loads used to determine the ampacities are calculated in accordance with Article 220 of the Code. *Derating*, or lowering the allowed ampere rating, of a conductor can occur due to temperature limitations of the terminal connections [Sec. 110.14 (C)]; coordination with system overcurrent protection [Sec. 240.4]; carrying more than three current-carrying conductors in a raceway or cable [Sec. 310.15(B)(2)(a) and NEC Table 310.15(B)(2)(a)]; the ambient operating environment [NEC Table Correction Factors at the bottom of each table]; and a variety of other conditions [NEC Annex B]. Under engineering supervision, conductor ampacities may be calculated using the formula given in Sec. 310.15(C) for which sample calculations are given in NEC Annex B.

Example 53.5

A single AWG 10 TW copper conductor is considered for use in an environment expected to have an ambient temperature of 100°F. What is the maximum ampacity?

Solution

A TW cable sized at AWG 10, located in the 60°C column of App. 53.B, has an ampacity of 30 A before derating. The correction factor given at the bottom of the table for 100°F (38°C) is 0.82. The allowable ampacity is

$$\begin{aligned} A_{\text{total}} &= A_{\text{rated}} F_{\text{correction}} \\ &= (30\ \text{A})(0.82) \\ &= 24.6\ \text{A} \end{aligned}$$

12. EQUIPMENT FOR GENERAL USE

The "Equipment for General Use" portion of the Code contains Articles 400 (flexible cords and cables), 402 (fixture wires), 430 (motors, motor circuits, and controllers), and 480 (storage batteries), among others.

The ampacities, insulation information, markings, and other pertinent data regarding flexible cords and cables is contained in NEC Tables 400.4 and 400.5. Flexible cords and cables are limited in use [Sec. 400.7] and can be spliced only under special conditions [Sec. 400.9]. The cords and cables must be protected by bushings or fittings when passing through holes [Sec. 400.14].

The properties of fixture wires are given in NEC Table 402.3 with the allowable ampacity given in Table 402.5.

Specific requirements for lighting fixtures and associated wires are described in Article 410.

Article 430 is the definitive starting point for motor requirements, motor branch circuit and feeder conductor ampacity and overcurrent protection information, motor overload protection, and motor-related topics. Installation requirements are illustrated and cross- referenced in Sec. 430.1, FPN No. 1, Fig. 430.1, which should be the starting point for researching motor requirements.

The size of the conductors supplying motors is determined using the tables of Sec. 310.15(B) unless the conductor is flexible, in which case the requirements are in Sec. 400.5. The current value used to enter the tables in Article 310 is determined from NEC Table 430.247 to Table 430.250, not from the motor nameplate data [Sec. 430.6(A)(1)]. However, the overload protection is determined from the motor nameplate current rating [Sec. 430.6(A)(2)]. The conductors supplying single motors in continuous duty are sized to 125% of the motor's full-load current rating (as determined by the NEC tables listed in Sec. 430.6) [Sec. 430.22(A)]. Motor overload protection is generally set at 115% of the motor nameplate full-load current rating [Sec. 430.32(A) and (B)]. Motors are classed as greater than or less than 1 hp. Motors of less than 1 hp are sometimes called *fractional horsepower motors*.

Storage battery requirements are in Article 480. The voltage used in calculations involving lead-acid batteries is standardized at 2 V per cell. For alkali batteries, the standard voltage is 1.2 V per cell.

54 National Electrical Safety Code

1. Overview 54-1
2. Introduction 54-2
3. Special Terms 54-3
4. References 54-3
5. Grounding 54-3
6. NESC Part 1 54-4
7. NESC Part 2 54-5
8. NESC Part 3 54-5
9. NESC Part 4 54-7

1. OVERVIEW

The *National Electrical Safety Code* (NESC) was initiated in 1913 at the National Bureau of Standards. It is now under the sponsorship and control of the Institute of Electrical and Electronic Engineers (IEEE) and is published as a voluntary IEEE standard. The NESC (hereafter referred to as the "Safety Code") is officially endorsed by the American National Standards Institute (ANSI), and the committee responsible for the Safety Code is known as *ANSI Standards Committee* C2. While voluntary, the Safety Code is often adopted by each individual state, specifically by the state authority with jurisdiction over utilities.

The Safety Code generally applies to "practical safeguarding of persons during the installation, operation, and maintenance of electric supply and communications lines, and associated equipment." Its general purpose is to safeguard utility persons and the general public.

The Safety Code consists of General Sections followed by four parts, all of which are further subdivided into sections and then rules. The outline of the Safety Code follows. Part numbers (single digits) and Section numbers (two digits) are given in parentheses. Rules are three digit numbers.

General Sections

(01) Introduction

(02) Definitions

(03) References

(09) Grounding

Part 1 Rules for the Installation and Maintenance of Electric Supply Stations and Equipment

(10) Purpose and Scope

(11) Protective Arrangements in Electric Supply Stations

(12) Installation and Maintenance of Equipment

(13) Rotating Equipment

(14) Storage Batteries

(15) Transformers and Regulators

(16) Conductors

(17) Circuit Breakers, Reclosers, Switches, and Fuses

(18) Switchgear and Metal-Enclosed Bus

(19) Surge Arrestors

Part 2 Safety Rules for the Installation and Maintenance of Overhead Electric Supply and Communication Lines

(20) Purpose, Scope, and Application of Rules

(21) General Requirements

(22) Relations between Various Classes of Lines and Equipment

(23) Clearance

(24) Grades of Construction

(25) Loading for Grades B and C

(26) Strength Requirements

(27) Line Insulation

Part 3 Safety Rules for the Installation and Maintenance of Underground Electric Supply and Communication Lines

(30) Purpose, Scope, and Application of Rules

(31) General Requirements Applying to Underground Lines

(32) Underground Conduit Systems

(33) Supply Cable

(34) Cable in Underground Structures

(35) Direct Buried Cable

(36) Risers

(37) Supply Cable Terminations

(38) Equipment

(39) Installation in Tunnels

Part 4 Rules for the Operation of Electric Supply and Communication Lines and Equipment

(40) Purpose and Scope

(41) Supply and Communications Systems—Rules for Employers

(42) General Rules for Employees

(43) Additional Rules for Communications Employees

(44) Additional Rules for Supply Employees

Appendices

(A) Uniform System of Clearances

(B) Uniform Clearance Calculations for Conductors Under Ice and Wind Conditions

(C) Example Applications for Rule 250C Tables 250-2 and 250-3

(D) Determining Maximum Anticipated Per-Unit Overvoltage Factor (T) at the Worksite

(E) Bibliography

The Safety Code parts separate into the general areas as shown in Fig. 54.1. In this chapter, numbers in brackets, if used, indicate the rule number in the Safety Code and are meant for ease in referencing the Safety Code directly. Specific limits may be found in the latest code. Though metric is the official system for the Safety Code, a mixture of customary U.S. engineering units and SI units is used.

2. INTRODUCTION

Section 01 of the General Sections is titled "Introduction to the National Electrical Safety Code." Rule 010 specifies the purpose as the safeguarding of persons during installation, operation, or maintenance. The Safety Code is meant to contain the basic guidelines for safety. It covers electric power, which the Safety Code refers to as electric "supply," lines and associated equipment, as well as communication lines and equipment. Communications refers to, but is not limited to, telephone and cable television.

Figure 54.1 *NESC Areas of Concern*

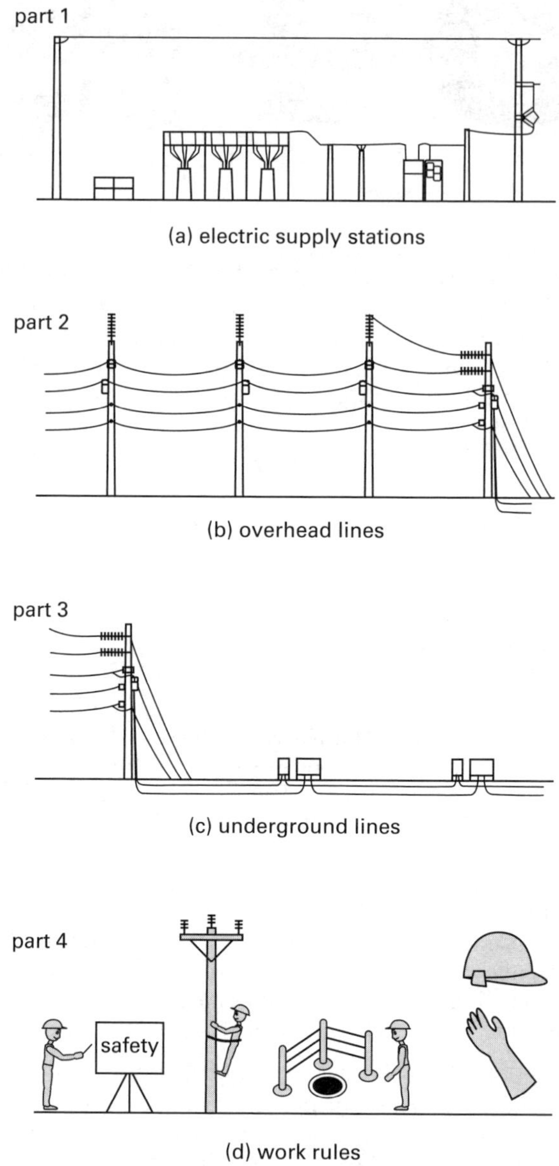

Rule 011 covers the scope of the code. The area of coverage for the Safety Code is shown in Fig. 54.2. The exact demarcation between the NEC and the NESC is shown in Fig. 54.3.

In addition to covering different areas of responsibility, the NESC and NEC differ as indicated in Table 54.1.

Rule 012 requires the use of the Safety Code. If an item or situation is not covered in the Safety Code, then "accepted

Figure 54.2 NESC Coverage

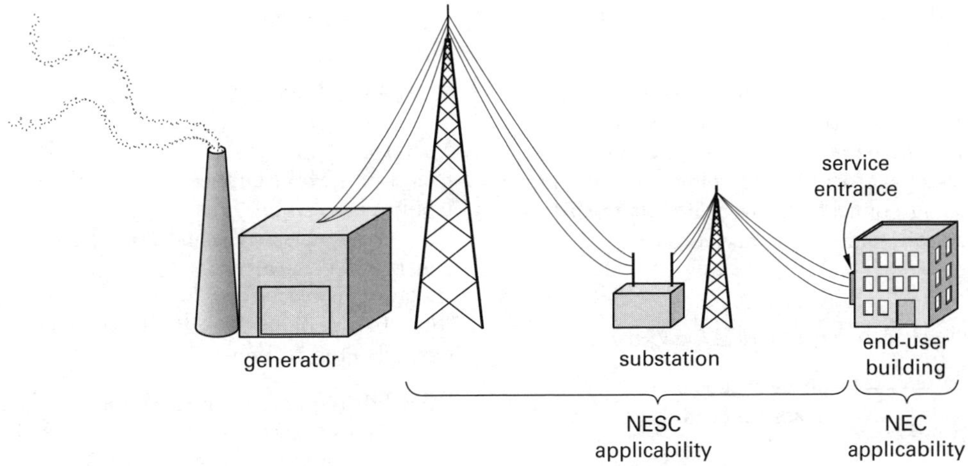

Figure 54.3 NEC/NESC Division of Responsibility

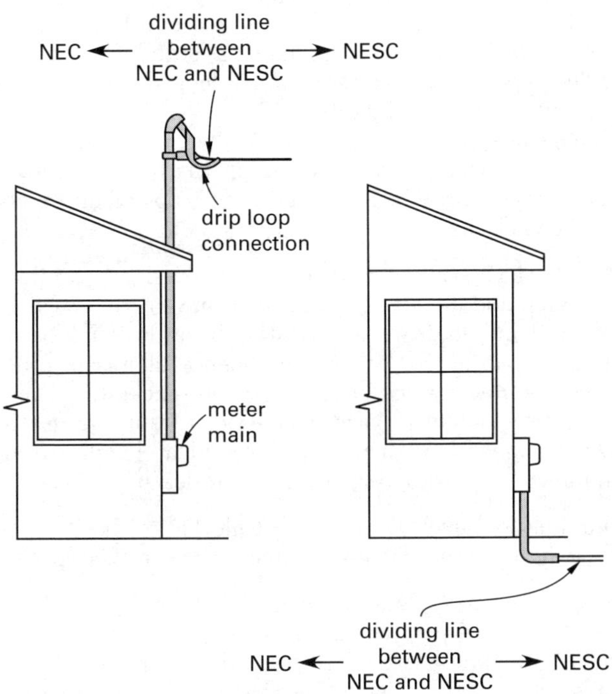

Table 54.1 NEC and NESC Comparison

NEC	NESC
published by NFPA	published by IEEE
code book for building industry	code book for utilities
legal authority by state, city, or local law	legal authority by state utility commissions
revised every 3 yr	revised every 5 yr
used by engineers and electricians	used by engineers and utility linemen

good practice" is allowed. Rule 013 requires all new installations to comply with the Safety Code. Rule 014 provides the conditions for a waiver of requirements.

Rule 015 defines "shall," "should," and "RECOMMENDATION." The term "shall" indicates a rule is mandatory and must be met. The term "should" indicates a rule that is applied normally, and is practical for the given situation. The term "RECOMMENDATION" is used to indicate a rule that is desired but not mandatory. Additionally, Rule 015 specifies that "NOTES" and "EXAMPLES" are not mandatory. However, "Footnotes" carry the force of the table or rule with which they are associated. Interpretation of the scope of "RECOMMENDATION," "NOTE," and "EXCEPTION" is also provided in this rule.

Rule 016 specifies that metric units are the primary units. Customary U.S. units are shown in parenthesis. Rule 017 specifies that values in the Safety Code are nominal and tolerances may exist in other standards.

3. SPECIAL TERMS

Section 02 of the General Sections is titled "Definitions of Special Terms." If a term is not defined in this section, then IEEE Standard 100 is to be used. If the term cannot be found in either, then a standard dictionary is to be used.

4. REFERENCES

Section 03 of the General Sections is titled "References." Any reference in this section is considered part of the Safety Code. If the reference is listed in Appendix B of the code, it is there for information only.

5. GROUNDING

Section 09 of the General Sections is titled "Grounding Methods for Electric Supply and Communications

Facilities." This entire section is devoted to the requirements for properly grounding electrical systems to ensure safety. A distinction is made between systems over 750 V and those under that value.

Rules in the grounding section require lighting loads to use wye and delta circuits that are grounded or else to use a common neutral conductor. The proper ground connections for systems of 750 V and below are shown in Fig. 54.4. Grounding connections are made at both the source and line (load) side of a service entrance to a building.

Figure 54.4 *Allowable Ground Connections (≤ 750 V)*

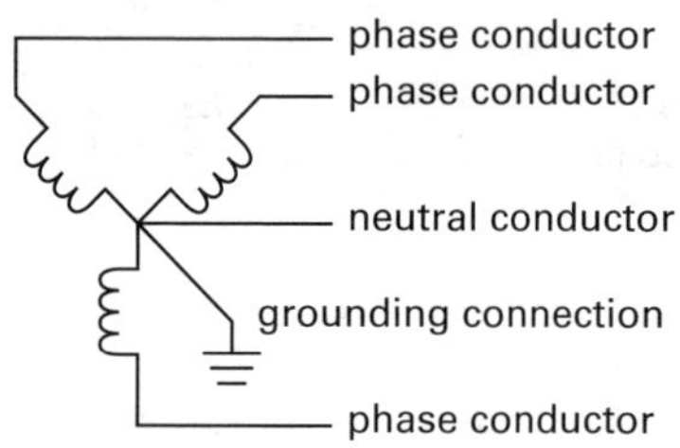

wye connected, 3-phase, 4-wire system
(e.g., 120/208 V, 3ϕ (phase), 4 W, and 277/480 V, 3ϕ (phase), 4 W)

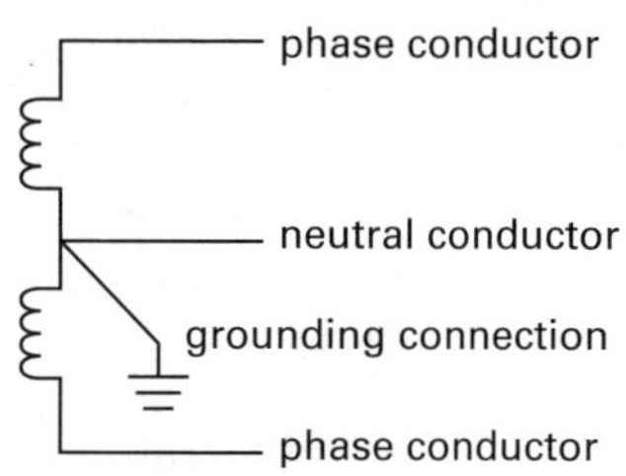

single-phase, 3-wire system
(e.g., 120/240 V, 1ϕ (phase), 3 W)

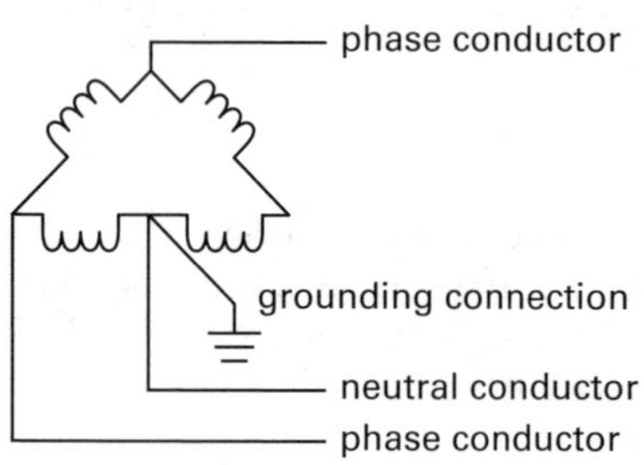

single-phase, 3-wire system on a delta connected 3-phase, 4-wire system (e.g., 120/240 V, 1ϕ (phase), 3 W)

The ampacity of grounding conductors is also specified in this section. The main specification is that fault current of a given time duration cannot melt or damage the grounding conductor. The "given time" is determined by the time-current characteristics of the fault protective device (e.g., the fuse, recluse, relay, or breaker) and the system impedance. The ampacity comes from manufacturers' *short-circuit withstand charts.* If the appropriate value is not determined, the grounding conductor must be sized for the full load continuous current of the supply transformer.

6. NESC PART 1

Part 1 of the Safety Code is titled "Rules for the Installation and Maintenance of Electric Supply Stations and Equipment." This part covers the electric supply stations as shown in Fig. 54.1(a). Electric supply stations consist of generating stations, substations, and switch stations. The rules assume such areas are accessible to "qualified personnel" only. That is, the general public is not expected to be in the area.

Rule 110 requires enclosures for supply stations to consist of fences, screens, partitions, or walls. The electrical components within the enclosure are required to meet clearance requirements to ensure personnel are not likely to come into contact with live electrical components. The enclosure is generally required to be 7 ft high and to limit access. Additionally, the electrical equipment inside the enclosure must have a standoff distance, called *safety clearance zone*, which depends on the highest voltage level in the substation. This rule requires the signage as specified in ANSI Z535. The general approach is that a warning sign is posted on the barrier. And, if the barrier is breached, a danger sign is posted internally to the enclosure.

ANSI Z535 stipulates three types of signs: caution, warning, and danger. *Caution signs* have a yellow background and indicate a potential hazard of minor or moderate injury or other consequence of unsafe practices. *Warning signs* have an orange background and indicate a potential hazard of serious injury or death. *Danger signs* have a red background and indicate an imminent hazard of serious injury or death.

Illumination levels are given in Rule 111, although portable lighting may be used to meet the specified levels.

Rule 112 requires floors to be even and nonslip. Obstructions must be painted, marked, or posted with safety signs. Stairways with three steps do not require handrails; those with four or more steps do require such handrails.

Rule 113 requires exits to be free of obstructions. In addition, double exits are required if an accident can make a single exit inaccessible.

Rule 123 requires protective grounding in accordance with IEEE Standard 80.

Rule 124 requires live parts over 150 V to be guarded. Guarding implies a physical barrier. Additionally, if the electrical potential of a given electrical part is unknown, the vertical clearance must be at least 8 ft 6 in.

Working space, that is, horizontal clearance, is specified in Rule 125. The space required is dependent on the voltage level. The requirements differ for voltages greater than 600 V and voltages of 600 V and less. For

600 V or less, at least 7 ft of head clearance and 30 in of width are the minimum. The working spaces in NESC Table 125-1 are for exposed energized parts, and do not apply if the parts are de-energized during maintenance or servicing.

Rule 131 requires that motors do not restart automatically after a power outage if such starting creates a hazard for personnel. This means that such motors should have *low-voltage protection* (LVP) that requires an operator to physically restart the motor following an outage. *Low-voltage release* (LVR) is a protective scheme that allows automatic restart without interaction with a human operator. Such LVR schemes are allowed if warning signals and time delays are incorporated.

Rule 140 governs the installation and use of batteries. The main thrust of the rule, and associated rules, is that hydrogen accumulation must be kept below "explosive mixtures" though the code does not specify an explosive percentage. (The hydrogen percentage that is explosive is approximately 4%.)

Conductors are covered in Section 16 of Part 1, in Rules 160–164. In general, the conductors must meet four conditions as follows. Conductors must

- be suitable for the intended use
- be suitable for the environmental location in which they are used
- have a voltage rating (i.e., insulation rating) adequate for the intended use
- have an ampacity (i.e., current rating) adequate for the intended use

According to Rule 174 fuses of more than 150 V to ground, or carrying more than 60 A, must be classified as disconnecting, or be arranged to be able to disconnect from the power source, or be removed with insulated handles.

Surge arrestors are common in the utility industry and are used to protect equipment from switching surges and lightning strikes. Arrestors are mentioned in Section 19, Rules 190–193. The basic requirements, however, come from IEEE Standard C62.1 and C62.11.

7. NESC PART 2

Part 2 of the Safety Code is titled "Safety Rules for the Installation and Maintenance of Overhead Electric Supply and Communication Lines." This part covers overhead lines as shown in Fig. 54.1(b). The requirements in Part 2 focus on spacing (center-to-center measurements) between conductors, clearances (surface-to-surface measurements), and the strength of construction.

Accessibility for servicing and maintenance is a concern of this part. The accessibility issue is covered in Rule 213 and throughout Part 2 under the topics "climbing space," "working space," "working facilities," and "clearance between conductors."

A wide variety of clearances are required depending on the situation. However, two important arguments come from the sag and tension chart and the effects of weather conditions, called *loadings*, on the overhead lines. The *sag and tension chart* uses the type and temperature of the cable, the span length, and the weather conditions (including ice accumulation and wind speed) as inputs. The output is the maximum expected sag and tension on the overhead line that is then used to determine the construction grade. The sag and tension chart details are in Rule 230. The clearances required by Rule 230 vary with regions of the country, with Zone 1, 2, and 3 corresponding to weather designations in Rule 250 or heavy, medium, and light, respectively.

The standard weather conditions for U.S. areas are in Rule 250. The classification is light, medium, and heavy. The southern parts of the United States are in the light category, the northwestern and central portions are in the medium category, and the northeastern portions are in the heavy category.

The resultant construction grades required are in Rules 240–243, and are Grade B (the highest grade), Grade C (the medium grade), and Grade N (the lowest grade). The strength requirements for each construction grade are in Rules 260–264.

8. NESC PART 3

Part 3 of the Safety Code is titled "Safety Rules for the Installation and Maintenance of Underground Electric Supply and Communication Lines." This part covers underground installations as shown in Fig. 54.1(c). The requirements of Part 3 focus on underground installations for supply (power) and communications (telephone and cable television, and others). There is overlap at the underground riser and overhead pole between Part 2 and Part 3.

Accessibility is a concern of this part. The accessibility issue is covered in Rule 313 and throughout Part 3 under the topics working space, working facilities, and clearance between conductors.

Definitions are in Rule 320, and are shown in Fig. 54.5.

Surprisingly, the Safety Code does not specify conduit burial depths but instead defers to "accepted good practice," which generally means that the requirements of the NEC are applicable.

Rule 320 requires that conduit running longitudinally under a road should be installed on the shoulder, and if not practical, under one lane only. This rule also requires that supply (power) and communications lines be separated as shown in Table 54.2.

Figure 54.5 *Underground Installation Technology*

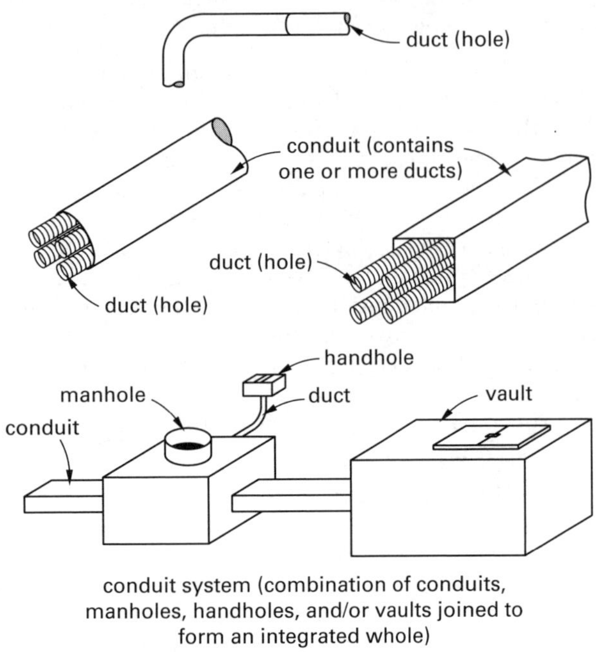

Table 54.2 *Underground Supply (Power) and Communication Conduit Separation Requirements*

burial method	separation requirement (in)
concrete	3
masonry	4
tamped earth	12

Manhole dimensions and requirements are in Rule 323. Cable separation requirements in manholes and vaults are covered in Rule 341. Rule 341 requires communications cables to be a minimum distance from supply (power) cables depending upon their voltage as given in Table 54.3.

Table 54.3 *Supply (Power) and Communication Cable Separation in Joint-Use Manholes and Vaults*

supply cable phase-to-phase voltage (V)	communication cable minimum separation (in)
0–15,000	6
15,001–50,000	9
50,001–120,000	12
120,001 and greater	24

The Safety Code does contain burial (depth) requirements for directly buried (that is, non-conduit encased) supply (power) cables, but not communication cables. These depth requirements are shown in Fig. 54.6(a), whose source is Rule 352. The depth varies with the voltage and is measured surface to surface and not center to center. Nevertheless, communications cables must meet the general burial requirements of Rule 352. Directly buried lighting cable is an exception as is shown in Fig. 54.6(b).

Figure 54.6 *Underground Installation Technology*

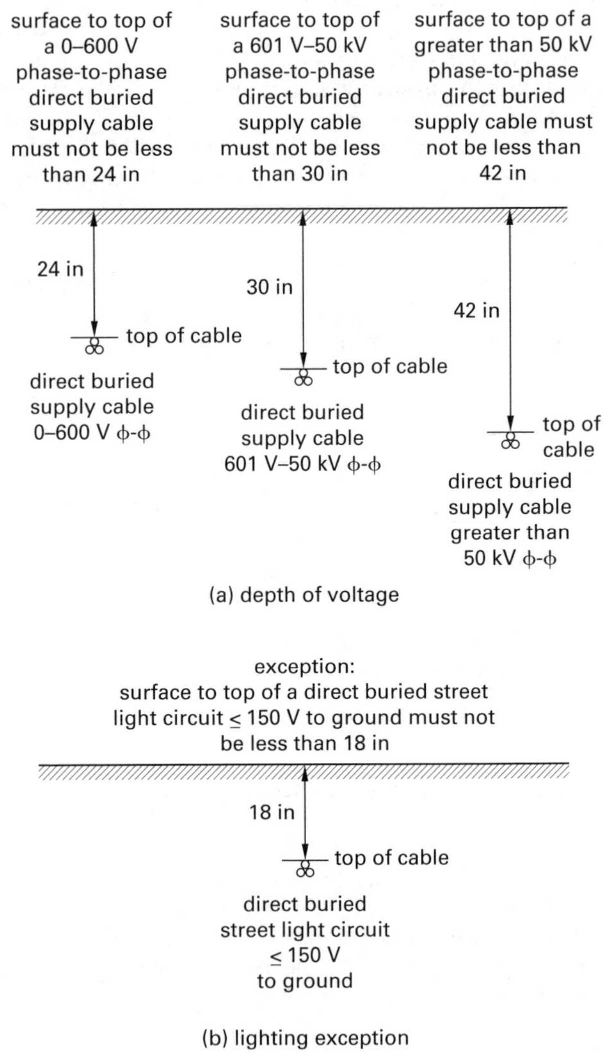

Rules 353 and 354 are similar. Both levy radial separation requirements from underground structures, including sewer, water, natural gas, fuel, and steam lines. These rules also apply to separation from building foundations. Rule 353 covers the specifications for 12 in or greater radial separation. If the 12 in separation cannot be maintained, Rule 354 applies or the cables must be installed in a conduit system, in which case Rule 320 (covered earlier) would apply. The intent of Rule 353 is to allow for maintenance on either system without interference with the other. The intent of Rule 354 is to prevent interference between supply (power) and communication cables.

Risers bring underground cabling to the vault or pole. Riser requirements are in Rules 360–363. The supply cable terminations located in the same area are covered by Rules 370–374. The location of equipment, both on the surface and in vaults is covered in Section 38, Rules 380–385. Many of the separation distances are not specified in Section 38, instead the Safety Code defers to accepted good practice. Here again, the NEC provides a guide to much of that good practice. The Safety Code does require that supply and communications equipment be located horizontally at least 4 ft from fire hydrants, though 3 ft is allowed where conditions do not permit the 4 ft separation.

9. NESC PART 4

Part 4 of the Safety Code is titled "Rules for the Operation of Electric Supply and Communications Liens and Equipment." This part covers work rules, as indicated in Fig. 54.1(d). Part 5 focuses on practical *work rules* designed to safeguard both the employees performing the work and the general public. The work rules of the Safety Code have been coordinated with Occupation Safety and Health Administration (OSHA) regulations related to electric supply and communications systems.

Topic XV: Professional

Chapter

55. Engineering Economic Analysis
56. Engineering Law
57. Engineering Ethics
58. Electrical Engineering Frontiers
59. Engineering Licensing in the United States

55 Engineering Economic Analysis

1. Irrelevant Characteristics 55-2
2. Multiplicity of Solution Methods 55-2
3. Precision and Significant Digits 55-2
4. Nonquantifiable Factors 55-2
5. Year-End and Other Conventions 55-3
6. Cash Flow Diagrams 55-3
7. Types of Cash Flows 55-3
8. Typical Problem Types 55-4
9. Implicit Assumptions 55-4
10. Equivalence 55-5
11. Single-Payment Equivalence 55-5
12. Standard Cash Flow Factors and Symbols 55-6
13. Calculating Uniform Series Equivalence . . . 55-7
14. Finding Past Values 55-8
15. Times to Double and Triple an Investment 55-8
16. Varied and Nonstandard Cash Flows 55-9
17. The Meaning of Present Worth and i 55-11
18. Simple and Compound Interest 55-11
19. Extracting the Interest Rate: Rate of Return 55-12
20. Rate of Return versus Return on Investment 55-14
21. Minimum Attractive Rate of Return 55-14
22. Typical Alternative-Comparison Problem Formats 55-14
23. Durations of Investments 55-14
24. Choice of Alternatives: Comparing One Alternative with Another Alternative . . . 55-14
25. Choice of Alternatives: Comparing an Alternative with a Standard 55-16
26. Ranking Mutually Exclusive Multiple Projects 55-17
27. Alternatives with Different Lives 55-17
28. Opportunity Costs 55-18
29. Replacement Studies 55-18
30. Treatment of Salvage Value in Replacement Studies 55-18
31. Economic Life: Retirement at Minimum Cost 55-19
32. Life-Cycle Cost 55-20
33. Capitalized Assets versus Expenses 55-20
34. Purpose of Depreciation 55-20
35. Depreciation Basis of an Asset 55-20
36. Depreciation Methods 55-21
37. Accelerated Depreciation Methods 55-23
38. Book Value 55-23
39. Amortization 55-24
40. Depletion 55-24
41. Basic Income Tax Considerations 55-24
42. Taxation at the Times of Asset Purchase and Sale 55-25
43. Depreciation Recovery 55-26
44. Other Interest Rates 55-28
45. Rate and Period Changes 55-28
46. Bonds 55-30
47. Probabilistic Problems 55-30
48. Fixed and Variable Costs 55-31
49. Accounting Costs and Expense Terms 55-32
50. Accounting Principles 55-33
51. Cost Accounting 55-36
52. Cost of Goods Sold 55-37
53. Break-Even Analysis 55-37
54. Pay-Back Period 55-38
55. Management Goals 55-38
56. Inflation 55-39
57. Consumer Loans 55-39
58. Forecasting 55-42
59. Learning Curves 55-42
60. Economic Order Quantity 55-43
61. Sensitivity Analysis 55-43
62. Value Engineering 55-44

Nomenclature

A	annual amount	\$
B	present worth of all benefits	\$
BV_j	book value at end of the jth year	\$
C	cost or present worth of all costs	\$
d	declining balance depreciation rate	decimal
D	demand	various
D	depreciation	\$
DR	present worth of after-tax depreciation recovery	\$
e	constant inflation rate	decimal
$\mathcal{E}$	expected value	various
E_0	initial amount of an exponentially growing cash flow	\$
EAA	equivalent annual amount	\$
EUAC	equivalent uniform annual cost	\$
f	federal income tax rate	decimal
F	forecasted quantity	various
F	future worth	\$
g	exponential growth rate	decimal
G	uniform gradient amount	\$
i	effective interest rate	decimal
$i\%$	effective interest rate	%
i'	effective interest rate corrected for inflation	decimal
j	number of years	–
k	number of compounding periods per year	–

Professional

m	an integer	–
n	number of compounding periods or years in life of asset	–
p	probability	decimal
P	present worth	$
r	nominal rate per year (rate per annum)	decimal per unit time
ROI	return on investment	$
ROR	rate of return	decimal per unit time
s	state income tax rate	decimal
S_n	expected salvage value in year n	$
t	composite tax rate	decimal
t	time	years (typical)
T	a quantity equal to $\frac{1}{2}n(n+1)$	–
TC	tax credit	$
z	a quantity equal to $\frac{1+i}{1-d}$	decimal

Symbols

α	smoothing coefficient for forecasts	–
ϕ	effective rate per period	decimal

Subscripts

0	initial
j	at time j
n	at time n
t	at time t

1. IRRELEVANT CHARACTERISTICS

In its simplest form, an *engineering economic analysis* is a study of the desirability of making an investment.[1] The decision-making principles in this chapter can be applied by individuals as well as by companies. The nature of the spending opportunity or industry is not important. Farming equipment, personal investments, and multimillion dollar factory improvements can all be evaluated using the same principles.

Similarly, the applicable principles are insensitive to the monetary units. Although *dollars* are used in this chapter, it is equally convenient to use pounds, yen, or euros.

Finally, this chapter may give the impression that investment alternatives must be evaluated on a year-by-year basis. Actually, the *effective period* can be defined as a day, month, century, or any other convenient period of time.

2. MULTIPLICITY OF SOLUTION METHODS

Most economic conclusions can be reached in more than one manner. There are usually several different analyses that will eventually result in identical answers.[2] Other than the pursuit of elegant solutions in a timely manner, there is no reason to favor one procedural method over another.[3]

3. PRECISION AND SIGNIFICANT DIGITS

The full potential of electronic calculators will never be realized in engineering economic analyses. Considering that calculations are based on estimates of far-future cash flows and that unrealistic assumptions (no inflation, identical cost structures of replacement assets, etc.) are routinely made, it makes little sense to carry cents along in calculations.

The calculations in this chapter have been designed to illustrate and review the principles presented. Because of this, greater precision than is normally necessary in everyday problems may be used. Though used, such precision is not warranted.

Unless there is some compelling reason to strive for greater precision, the following rules are presented for use in reporting final answers to engineering economic analysis problems.

- Omit fractional parts of the dollar (i.e., cents).
- Report and record a number to a maximum of four significant digits unless the first digit of that number is 1, in which case, a maximum of five significant digits should be written. For example,

$49	not	$49.43
$93,450	not	$93,453
$1,289,700	not	$1,289,673

4. NONQUANTIFIABLE FACTORS

An engineering economic analysis is a quantitative analysis. Some factors cannot be introduced as numbers into the calculations. Such factors are known as *non-quantitative factors*, *judgment factors*, and *irreducible factors*. Typical nonquantifiable factors are

- preferences
- political ramifications
- urgency
- goodwill
- prestige
- utility
- corporate strategy

[1]This subject is also known as *engineering economics* and *engineering economy*. There is very little, if any, true economics in this subject.

[2]Because of round-off errors, particularly when factors are taken from tables, these different calculations will produce slightly different numerical results (e.g., $49.49 versus $49.50). However, this type of divergence is well known and accepted in engineering economic analysis.

[3]This does not imply that approximate methods, simplifications, and rules of thumb are acceptable.

- environmental effects
- health and safety rules
- reliability
- political risks

Since these factors are not included in the calculations, the policy is to disregard the issues entirely. Of course, the factors should be discussed in a final report. The factors are particularly useful in breaking ties between competing alternatives that are economically equivalent.

5. YEAR-END AND OTHER CONVENTIONS

Except in short-term transactions, it is simpler to assume that all receipts and disbursements (cash flows) take place at the end of the year in which they occur.[4] This is known as the *year-end convention*. The exceptions to the year-end convention are initial project cost (purchase cost), trade-in allowance, and other cash flows that are associated with the inception of the project at $t = 0$.

On the surface, such a convention appears grossly inappropriate since repair expenses, interest payments, corporate taxes, and so on seldom coincide with the end of a year. However, the convention greatly simplifies engineering economic analysis problems, and it is justifiable on the basis that the increased precision associated with a more rigorous analysis is not warranted (due to the numerous other simplifying assumptions and estimates initially made in the problem).

There are various established procedures, known as *rules* or *conventions*, imposed by the Internal Revenue Service on U.S. taxpayers. An example is the *half-year rule*, which permits only half of the first-year depreciation to be taken in the first year of an asset's life when certain methods of depreciation are used. These rules are subject to constantly changing legislation and are not covered in this book. The implementation of such rules is outside the scope of engineering practice and is best left to accounting professionals.

6. CASH FLOW DIAGRAMS

Although they are not always necessary in simple problems (and they are often unwieldy in very complex problems), *cash flow diagrams* can be drawn to help visualize and simplify problems having diverse receipts and disbursements.

The following conventions are used to standardize cash flow diagrams.

- The horizontal (time) axis is marked off in equal increments, one per period, up to the duration (or *horizon*) of the project.
- Two or more transfers in the same period are placed end to end, and these may be combined.
- Expenses incurred before $t = 0$ are called *sunk costs*. Sunk costs are not relevant to the problem unless they have tax consequences in an after-tax analysis.
- *Receipts* are represented by arrows directed upward. *Disbursements* are represented by arrows directed downward. The arrow length is proportional to the magnitude of the cash flow.

Example 55.1

A mechanical device will cost \$20,000 when purchased. Maintenance will cost \$1000 each year. The device will generate revenues of \$5000 each year for five years, after which the salvage value is expected to be \$7000. Draw and simplify the cash flow diagram.

Solution

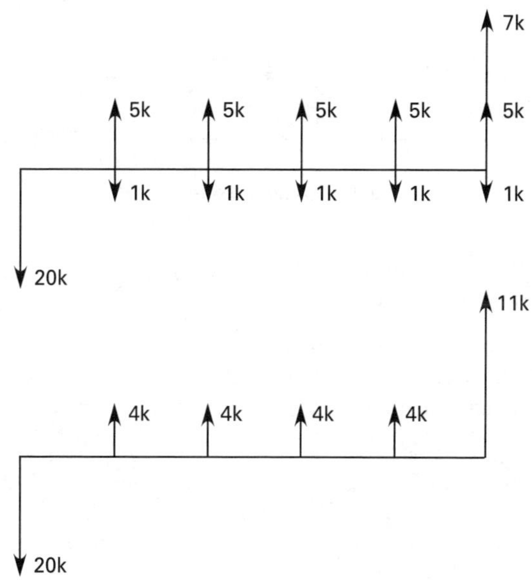

7. TYPES OF CASH FLOWS

To evaluate a real-world project, it is necessary to present the project's cash flows in terms of standard cash flows that can be handled by engineering economic analysis techniques. The standard cash flows are single payment cash flow, uniform series cash flow, gradient series cash flow, and the infrequently encountered exponential gradient series cash flow. (See Fig. 55.1.)

A *single payment cash flow* can occur at the beginning of the time line (designated as $t = 0$), at the end of the time line (designated as $t = n$), or at any time in between.

The *uniform series cash flow*consists of a series of equal transactions starting at $t = 1$ and ending at $t = n$. The

[4]A *short-term transaction* typically has a lifetime of five years or less and has payments or compounding that are more frequent than once per year.

Figure 55.1 *Standard Cash Flows*

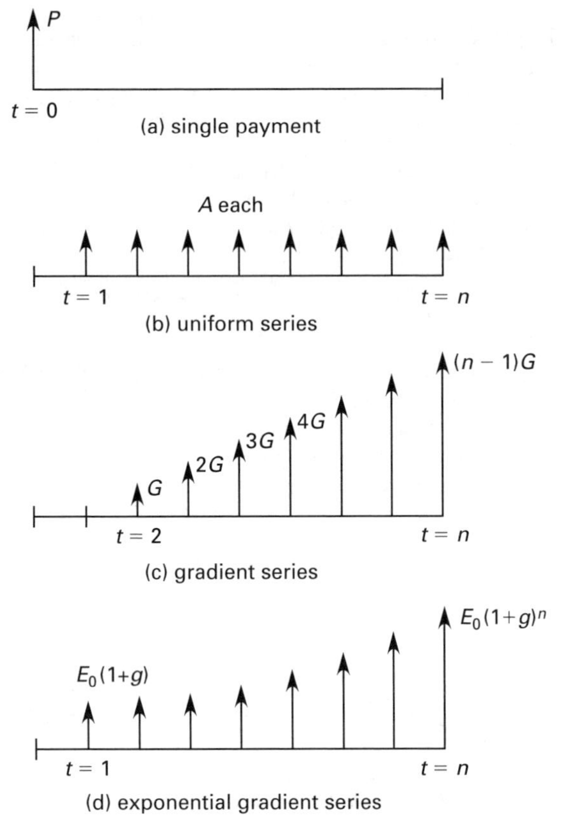

symbol A is typically given to the magnitude of each individual cash flow.[5]

The *gradient series cash flow* starts with a cash flow (typically given the symbol G) at $t = 2$ and increases by G each year until $t = n$, at which time the final cash flow is $(n - 1)G$.

An *exponential gradient series cash flow* is based on a phantom value (typically given the symbol E_0) at $t = 0$ and grows or decays exponentially according to the following relationship.[6]

$$\text{amount at time } t = E_t = E_0(1+g)^t \quad [t = 1, 2, 3, \ldots, n] \qquad 55.1$$

In Eq. 55.1, g is the *exponential growth rate*, which can be either positive or negative. Exponential gradient cash flows are rarely seen in economic justification projects assigned to engineers.[7]

[5]Notice that the cash flows do not begin at $t = 0$. This is an important concept with all of the series cash flows. This convention has been established to accommodate the timing of annual maintenance (and similar) cash flows for which the year-end convention is applicable.

[6]Notice the convention for an exponential cash flow series: The first cash flow, E_0, is at $t = 1$, as in the uniform annual series. However, the first cash flow is $E_0(1 + g)$. The cash flow of E_0 at $t = 0$ is absent (i.e., is a *phantom cash flow*).

[7]For one of the few discussions on exponential cash flow, see *Capital Budgeting*, Robert V. Oakford, The Ronald Press Company, New York, 1970.

8. TYPICAL PROBLEM TYPES

There is a wide variety of problem types that, collectively, are considered to be engineering economic analysis problems.

By far, the majority of engineering economic analysis problems are *alternative comparisons*. In these problems, two or more mutually exclusive investments compete for limited funds. A variation of this is a *replacement/retirement analysis*, which is repeated each year to determine if an existing asset should be replaced. Finding the percentage return on an investment is a *rate of return problem*, one of the alternative comparison solution methods.

Investigating interest and principal amounts in loan payments is a *loan repayment problem*. An *economic life analysis* will determine when an asset should be retired. In addition, there are miscellaneous problems involving economic order quantity, learning curves, break-even points, product costs, and so on.

9. IMPLICIT ASSUMPTIONS

Several assumptions are implicitly made when solving engineering economic analysis problems. Some of these assumptions are made with the knowledge that they are or will be poor approximations of what really will happen. The assumptions are made, regardless, for the benefit of obtaining a solution.

The most common assumptions are the following.

- The year-end convention is applicable.
- There is no inflation now, nor will there be any during the lifetime of the project.
- Unless otherwise specifically called for, a before-tax analysis is needed.
- The effective interest rate used in the problem will be constant during the lifetime of the project.
- Nonquantifiable factors can be disregarded.
- Funds invested in a project are available and are not urgently needed elsewhere.
- Excess funds continue to earn interest at the effective rate used in the analysis.

This last assumption, like most of the assumptions listed, is almost never specifically mentioned in the body of a solution. However, it is a key assumption when comparing two alternatives that have different initial costs.

For example, suppose two investments, one costing \$10,000 and the other costing \$8000, are to be compared at 10%. It is obvious that \$10,000 in funds is available, otherwise the costlier investment would not be under consideration. If the smaller investment is chosen, what is done with the remaining \$2000? The last assumption yields the answer: The \$2000 is "put to work" in some investment earning (in this case, 10%).

10. EQUIVALENCE

Industrial decision makers using engineering economic analysis are concerned with the magnitude and timing of a project's cash flow as well as with the total profitability of that project. In this situation, a method is required to compare projects involving receipts and disbursements occurring at different times.

By way of illustration, consider \$100 placed in a bank account that pays 5% effective annual interest at the end of each year. After the first year, the account will have grown to \$105. After the second year, the account will have grown to \$110.25.

Assume that you will have no need for money during the next two years, and any money received will immediately go into your 5% bank account. Then, which of the following options would be more desirable?

option A: \$100 now

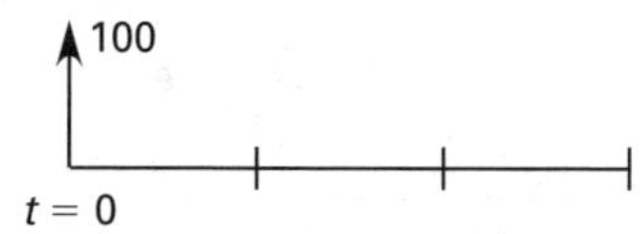

option B: \$105 to be delivered in one year

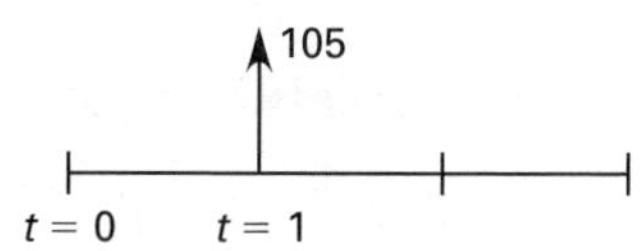

option C: \$110.25 to be delivered in two years

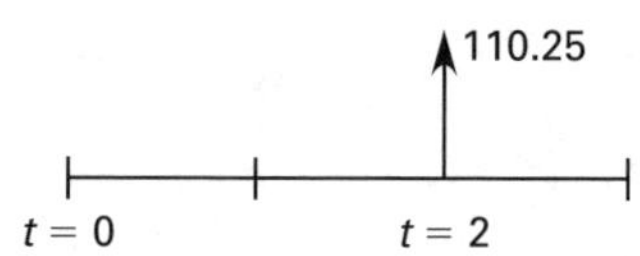

As illustrated, none of the options is superior under the assumptions given. If the first option is chosen, you will immediately place \$100 into a 5% account, and in two years the account will have grown to \$110.25. In fact, the account will contain \$110.25 at the end of two years regardless of the option chosen. Therefore, these alternatives are said to be *equivalent*.

Equivalence may or may not be the case, depending on the interest rate, so an alternative that is acceptable to one decision maker may be unacceptable to another. The interest rate that is used in actual calculations is known as the *effective interest rate*.[8] If compounding is once a year, it is known as the *effective annual interest rate*. However, effective quarterly, monthly, daily, and so on, interest rates are also used.

[8]The adjective *effective* distinguishes this interest rate from other interest rates (e.g., nominal interest rates) that are not meant to be used directly in calculating equivalent amounts.

The fact that \$100 today grows to \$105 in one year (at 5% annual interest) is an example of what is known as the *time value of money* principle. This principle simply articulates what is obvious: Funds placed in a secure investment will increase to an equivalent future amount. The procedure for determining the present investment from the equivalent future amount is known as *discounting*.

11. SINGLE-PAYMENT EQUIVALENCE

The equivalence of any present amount, P, at $t = 0$, to any future amount, F, at $t = n$, is called the *future worth* and can be calculated from Eq. 55.2.

$$F = P(1+i)^n \qquad 55.2$$

The factor $(1 + i)^n$ is known as the single payment *compound amount factor* and has been tabulated in App. 55.B for various combinations of i and n.

Similarly, the equivalence of any future amount to any present amount is called the *present worth* and can be calculated from Eq. 55.3.

$$P = F(1+i)^{-n} = \frac{F}{(1+i)^n} \qquad 55.3$$

The factor $(1 + i)^{-n}$ is known as the *single payment present worth factor*.[9]

The interest rate used in Eq. 55.2 and Eq. 55.3 must be the effective rate per period. Also, the basis of the rate (annually, monthly, etc.) must agree with the type of period used to count n. Therefore, it would be incorrect to use an effective annual interest rate if n was the number of compounding periods in months.

Example 55.2

How much should you put into a 10% (effective annual rate) savings account in order to have \$10,000 in five years?

Solution

This problem could also be stated: What is the equivalent present worth of \$10,000 five years from now if money is worth 10% per year?

$$\begin{aligned} P &= F(1+i)^{-n} = (\$10{,}000)(1+0.10)^{-5} \\ &= \$6209 \end{aligned}$$

The factor 0.6209 would usually be obtained from the tables.

[9]The *present worth* is also called the *present value* and *net present value*. These terms are used interchangeably and no significance should be attached to the terms *value*, *worth*, and *net*.

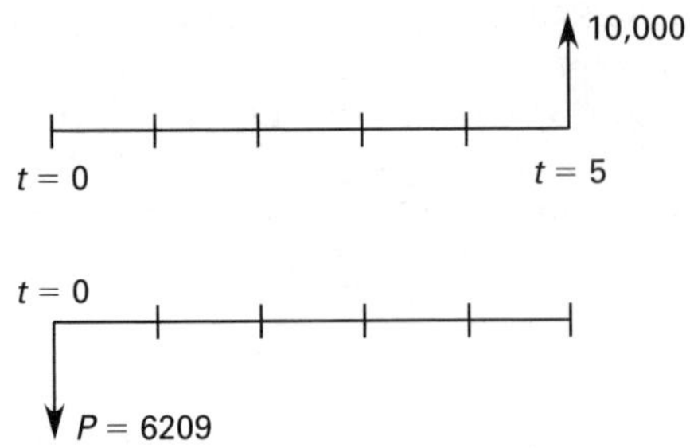

12. STANDARD CASH FLOW FACTORS AND SYMBOLS

Equation 55.2 and Eq. 55.3 may give the impression that solving engineering economic analysis problems involves a lot of calculator use, and, in particular, a lot of exponentiation. Such calculations may be necessary from time to time, but most problems are simplified by the use of tabulated values of the factors.

Rather than actually writing the formula for the compound amount factor (which converts a present amount to a future amount), it is common convention to substitute the standard functional notation of $(F/P, i\%, n)$. Therefore, the future value in n periods of a present amount would be symbolically written as

$$F = P(F/P, i\%, n) \qquad 55.4$$

Similarly, the present worth factor has a functional notation of $(P/F, i\%, n)$. The present worth of a future amount n periods hence would be symbolically written as

$$P = F(P/F, i\%, n) \qquad 55.5$$

Values of these *cash flow (discounting) factors* are tabulated in App. 55.B. There is often initial confusion about whether the (F/P) or (P/F) column should be used in a particular problem. There are several ways of remembering what the functional notations mean.

One method of remembering which factor should be used is to think of the factors as conditional probabilities. The conditional probability of event **A** given that event **B** has occurred is written as $p\{\mathbf{A}|\mathbf{B}\}$, where the given event comes after the vertical bar. In the standard notational form of discounting factors, the given amount is similarly placed after the slash. What you want comes before the slash. (F/P) would be a factor to find F given P.

Another method of remembering the notation is to interpret the factors algebraically. The (F/P) factor could be thought of as the fraction F/P. Algebraically, Eq. 55.4 would be

$$F = P\left(\frac{F}{P}\right) \qquad 55.6$$

This algebraic approach is actually more than an interpretation. The numerical values of the discounting factors are consistent with this algebraic manipulation. The (F/A) factor could be calculated as $(F/P) \times (P/A)$. This consistent relationship can be used to calculate other factors that might be occasionally needed, such as (F/G) or (G/P). For instance, the annual cash flow that would be equivalent to a uniform gradient may be found from

$$A = G(P/G, i\%, n)(A/P, i\%, n) \qquad 55.7$$

Formulas for the compounding and discounting factors are contained in Table 55.1. Normally, it will not be necessary to calculate factors from the formulas. Appendix 55.B is adequate for solving most problems.

Example 55.3

What factor will convert a gradient cash flow ending at $t = 8$ to a future value at $t = 8$? (That is, what is the $(F/G, i\%, 8)$ factor?) The effective annual interest rate is 10%.

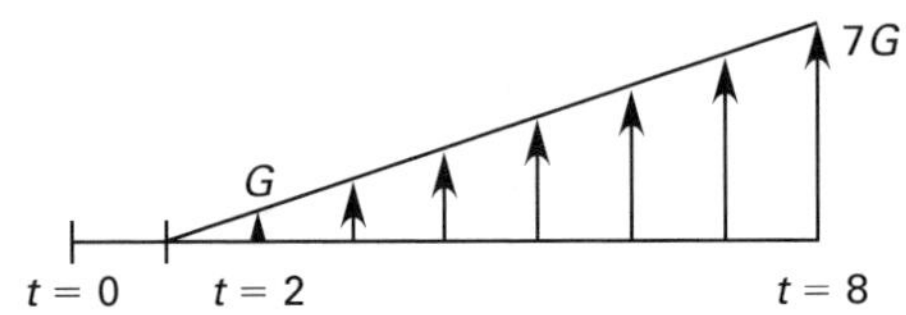

Solution

method 1:

From Table 55.1, the $(F/G, 10\%, 8)$ factor is

$$\begin{aligned}(F/G, 10\%, 8) &= \frac{(1+i)^n - 1}{i^2} \\ &= \frac{(1+0.10)^8 - 1}{(0.10)^2} - \frac{8}{0.10} \\ &= 34.3589\end{aligned}$$

method 2:

The tabulated values of (P/G) and (F/P) in App. 55.B can be used to calculate the factor.

$$\begin{aligned}(F/G, 10\%, 8) &= \Big(P/G, 10\%, 8)(F/P, 10\%, 8) \\ &= \Big(16.0287)(2.1436) \\ &= 34.3591\end{aligned}$$

The (F/G) factor could also have been calculated as the product of the (A/G) and (F/A) factors.

Professional

Table 55.1 *Discount Factors for Discrete Compounding*

factor name	converts	symbol	formula
single payment compound amount	P to F	$(F/P, i\%, n)$	$(1+i)^n$
single payment present worth	F to P	$(P/F, i\%, n)$	$(1+i)^{-n}$
uniform series sinking fund	F to A	$(A/F, i\%, n)$	$\frac{i}{(1+i)^n - 1}$
capital recovery	P to A	$(A/P, i\%, n)$	$\frac{i(1+i)^n}{(1+i)^n - 1}$
uniform series compound amount	A to F	$(F/A, i\%, n)$	$\frac{(1+i)^n - 1}{i}$
uniform series present worth	A to P	$(P/A, i\%, n)$	$\frac{(1+i)^n - 1}{i(1+i)^n}$
uniform gradient present worth	G to P	$(P/G, i\%, n)$	$\frac{(1+i)^n - 1}{i^2(1+i)^n} - \frac{n}{i(1+i)^n}$
uniform gradient future worth	G to F	$(F/G, i\%, n)$	$\frac{(1+i)^n - 1}{i^2} - \frac{n}{i}$
uniform gradient uniform series	G to A	$(A/G, i\%, n)$	$\frac{1}{i} - \frac{n}{(1+i)^n - 1}$

13. CALCULATING UNIFORM SERIES EQUIVALENCE

A cash flow that repeats each year for n years without change in amount is known as an *annual amount* and is given the symbol A. As an example, a piece of equipment may require annual maintenance, and the maintenance cost will be an annual amount. Although the equivalent value for each of the n annual amounts could be calculated and then summed, it is more expedient to use one of the uniform series factors. For example, it is possible to convert from an annual amount to a future amount by use of the (F/A) factor.

$$F = A(F/A, i\%, n) \qquad 55.8$$

A *sinking fund* is a fund or account into which annual deposits of A are made in order to accumulate F at $t = n$ in the future. Since the annual deposit is calculated as $A = F(A/F, i\%, n)$, the (A/F) factor is known as the *sinking fund factor*. An *annuity* is a series of equal payments (A) made over a period of time.[10] Usually, it is necessary to "buy into" an investment (a bond, an insurance policy, etc.) in order to ensure the annuity. In the simplest case of an annuity that starts at the end of the first year and continues for n years, the purchase price (P) is

$$P = A(P/A, i\%, n) \qquad 55.9$$

The present worth of an *infinite (perpetual) series* of annual amounts is known as a *capitalized cost*. There is no $(P/A, i\%, \infty)$ factor in the tables, but the capitalized cost can be calculated simply as

$$P = \frac{A}{i} \quad [i \text{ in decimal form}] \qquad 55.10$$

Alternatives with different lives will generally be compared by way of *equivalent uniform annual cost* (EUAC). An EUAC is the annual amount that is equivalent to all of the cash flows in the alternative. The EUAC differs in sign from all of the other cash flows. Costs and expenses expressed as EUACs, which would normally be considered negative, are actually positive. The term *cost* in the designation EUAC serves to make clear the meaning of a positive number.

Example 55.4

Maintenance costs for a machine are \$250 each year. What is the present worth of these maintenance costs over a 12-year period if the interest rate is 8%?

[10] An annuity may also consist of a lump sum payment made at some future time. However, this interpretation is not considered in this chapter.

Professional

Solution

$$P = A(P/A, 8\%, 12) = (-\$250)(7.5361)$$
$$= -\$1884$$

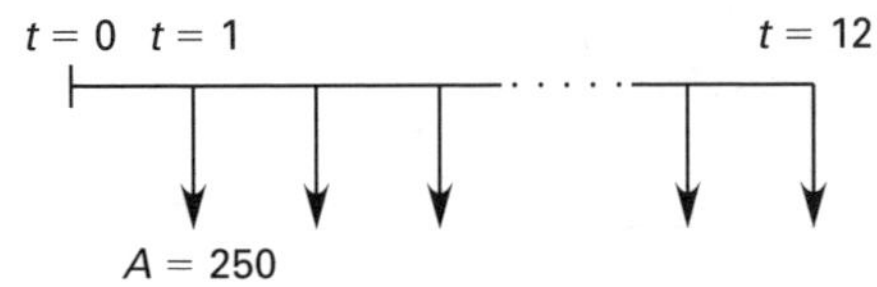

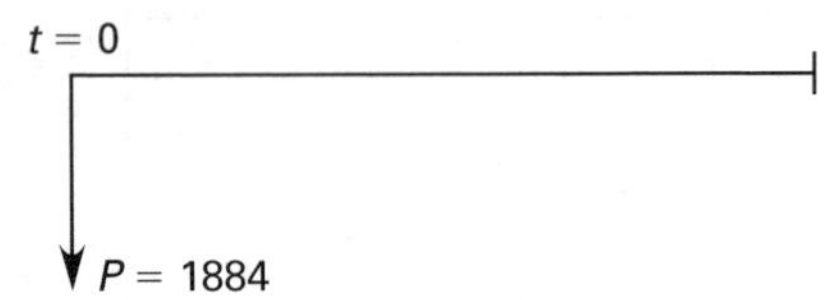

14. FINDING PAST VALUES

From time to time, it will be necessary to determine an amount in the past equivalent to some current (or future) amount. For example, you might have to calculate the original investment made 15 years ago given a current annuity payment.

Such problems are solved by placing the $t = 0$ point at the time of the original investment, and then calculating the past amount as a P value. For example, the original investment, P, can be extracted from the annuity, A, by using the standard cash flow factors.

$$P = A(P/A, i\%, n) \qquad 55.11$$

The choice of $t = 0$ is flexible. As a general rule, the $t = 0$ point should be selected for convenience in solving a problem.

Example 55.5

You currently pay \$250 per month to lease your office phone equipment. You have three years (36 months) left on the five-year (60-month) lease. What would have been an equivalent purchase price two years ago? The effective interest rate per month is 1%.

Solution

The solution of this example is not affected by the fact that investigation is being performed in the middle of the horizon. This is a simple calculation of present worth.

$$P = A(P/A, 1\%, 60)$$
$$= (-\$250)(44.9550)$$
$$= -\$11{,}239$$

15. TIMES TO DOUBLE AND TRIPLE AN INVESTMENT

If an investment doubles in value (in n compounding periods and with $i\%$ effective interest), the ratio of current value to past investment will be 2.

$$F/P = (1+i)^n = 2 \qquad 55.12$$

Similarly, the ratio of current value to past investment will be 3 if an investment triples in value. This can be written as

$$F/P = (1+i)^n = 3 \qquad 55.13$$

It is a simple matter to extract the number of periods, n, from Eq. 55.12 and Eq. 55.13 to determine the *doubling time* and *tripling time*, respectively. (See Table 55.2.) For example, the doubling time is

$$n = \frac{\log 2}{\log(1+i)} \qquad 55.14$$

When a quick estimate of the doubling time is needed, the *rule of 72* can be used. The doubling time is approximately $72/i$.

The tripling time is

$$n = \frac{\log 3}{\log(1+i)} \qquad 55.15$$

***Table 55.2** Doubling and Tripling Times for Various Interest Rates*

interest rate (%)	doubling time (periods)	tripling time (periods)
1	69.7	110.4
2	35.0	55.5
3	23.4	37.2
4	17.7	28.0
5	14.2	22.5
6	11.9	18.9
7	10.2	16.2
8	9.01	14.3
9	8.04	12.7
10	7.27	11.5
11	6.64	10.5
12	6.12	9.69
13	5.67	8.99
14	5.29	8.38
15	4.96	7.86
16	4.67	7.40
17	4.41	7.00
18	4.19	6.64
19	3.98	6.32
20	3.80	6.03

16. VARIED AND NONSTANDARD CASH FLOWS

Gradient Cash Flow

A common situation involves a uniformly increasing cash flow. If the cash flow has the proper form, its present worth can be determined by using the *uniform gradient factor*, $(P/G, i\%, n)$. The uniform gradient factor finds the present worth of a uniformly increasing cash flow that starts in year two (not in year one).

There are three common difficulties associated with the form of the uniform gradient. The first difficulty is that the initial cash flow occurs at $t = 2$. This convention recognizes that annual costs, if they increase uniformly, begin with some value at $t = 1$ (due to the year-end convention) but do not begin to increase until $t = 2$. The tabulated values of (P/G) have been calculated to find the present worth of only the increasing part of the annual expense. The present worth of the base expense incurred at $t = 1$ must be found separately with the (P/A) factor.

The second difficulty is that, even though the factor $(P/G, i\%, n)$ is used, there are only $n - 1$ actual cash flows. It is clear that n must be interpreted as the *period number* in which the last gradient cash flow occurs, not the number of gradient cash flows.

Finally, the sign convention used with gradient cash flows may seem confusing. If an expense increases each year (as in Ex. 55.6), the gradient will be negative, since it is an expense. If a revenue increases each year, the gradient will be positive. (See Fig. 55.2.) In most cases, the sign of the gradient depends on whether the cash flow is an expense or a revenue.[11]

Example 55.6

Maintenance on an old machine is \$100 this year but is expected to increase by \$25 each year thereafter. What is the present worth of five years of the costs of maintenance? Use an interest rate of 10%.

Solution

In this problem, the cash flow must be broken down into parts. (The five-year gradient factor is used even though there are only four nonzero gradient cash flows.)

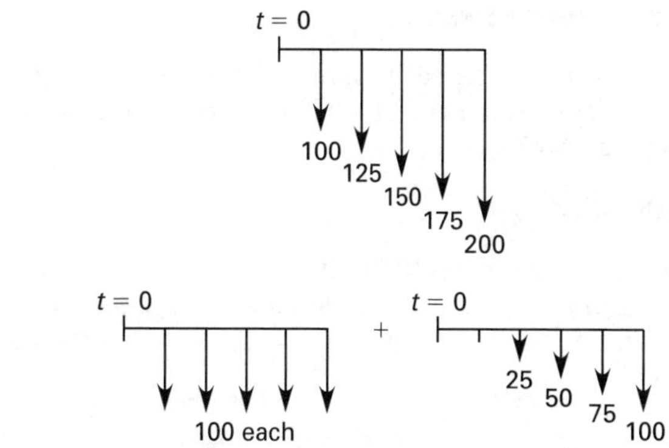

$$
\begin{aligned}
P &= A(P/A, 10\%, 5) + G(P/G, 10\%, 5) \\
&= (-\$100)(3.7908) - (\$25)(6.8618) \\
&= -\$551
\end{aligned}
$$

Figure 55.2 *Positive and Negative Gradient Cash Flows*

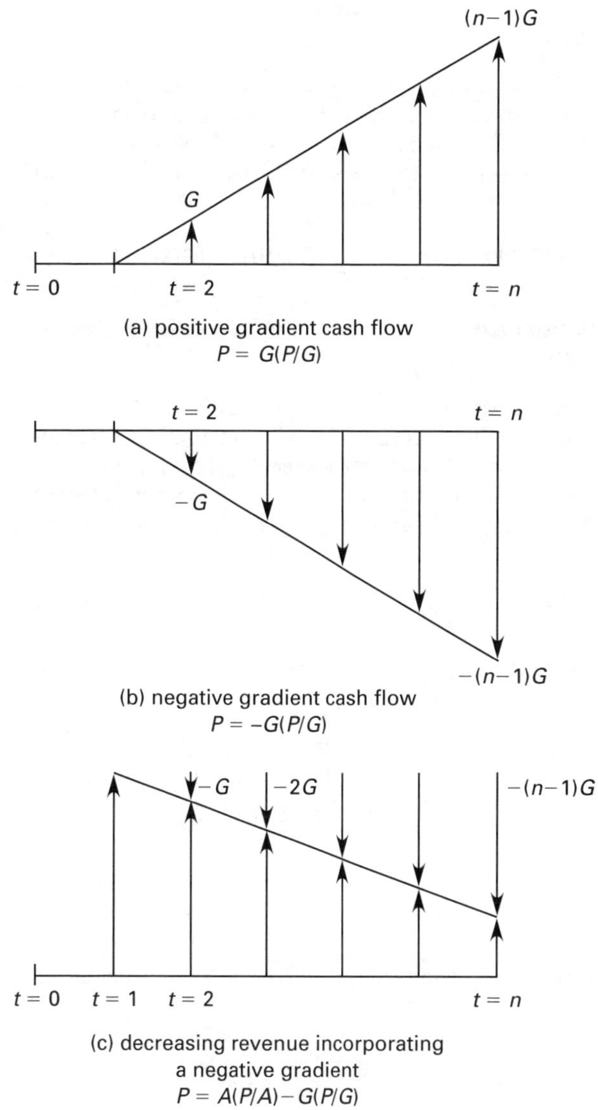

[11]This is not a universal rule. It is possible to have a uniformly decreasing revenue as in Fig. 55.2(c). In this case, the gradient would be negative.

Stepped Cash Flows

Stepped cash flows are easily handled by the technique of *superposition of cash flows.* This technique is illustrated by Ex. 55.7.

Example 55.7

An investment costing \$1000 returns \$100 for the first five years and \$200 for the following five years. How would the present worth of this investment be calculated?

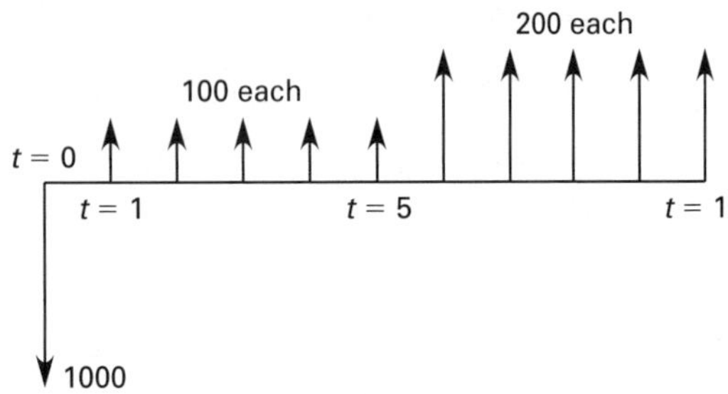

Solution

Using the principle of superposition, the revenue cash flow can be thought of as \$200 each year from $t = 1$ to $t = 10$, with a negative revenue of \$100 from $t = 1$ to $t = 5$. Superimposed, these two cash flows make up the actual performance cash flow.

$$P = -\$1000 + (\$200)(P/A, i\%, 10) - (\$100)(P/A, i\%, 5)$$

Missing and Extra Parts of Standard Cash Flows

A missing or extra part of a standard cash flow can also be handled by superposition. For example, suppose an annual expense is incurred each year for ten years, except in the ninth year. (The cash flow is illustrated in Fig. 55.3.) The present worth could be calculated as a subtractive process.

$$P = A(P/A, i\%, 10) - A(P/F, i\%, 9) \quad 55.16$$

Figure 55.3 *Cash Flow with a Missing Part*

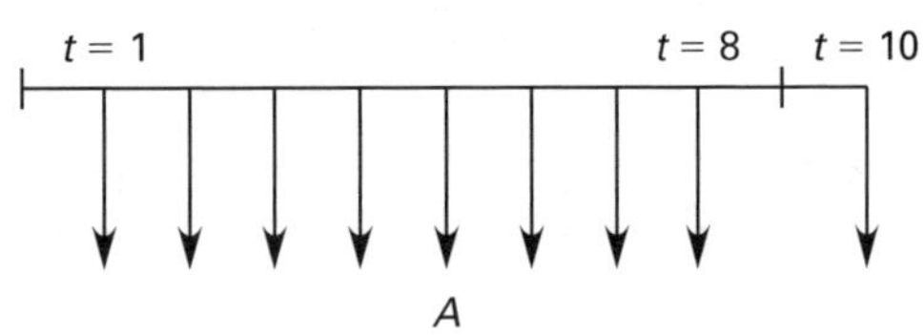

Alternatively, the present worth could be calculated as an additive process.

$$P = A(P/A, i\%, 8) + A(P/F, i\%, 10) \quad 55.17$$

Delayed and Premature Cash Flows

There are cases when a cash flow matches a standard cash flow exactly, except that the cash flow is delayed or starts sooner than it should. Often, such cash flows can be handled with superposition. At other times, it may be more convenient to shift the time axis. This shift is known as the *projection method.* Example 55.8 illustrates the projection method.

Example 55.8

An expense of \$75 is incurred starting at $t = 3$ and continues until $t = 9$. There are no expenses or receipts until $t = 3$. Use the projection method to determine the present worth of this stream of expenses.

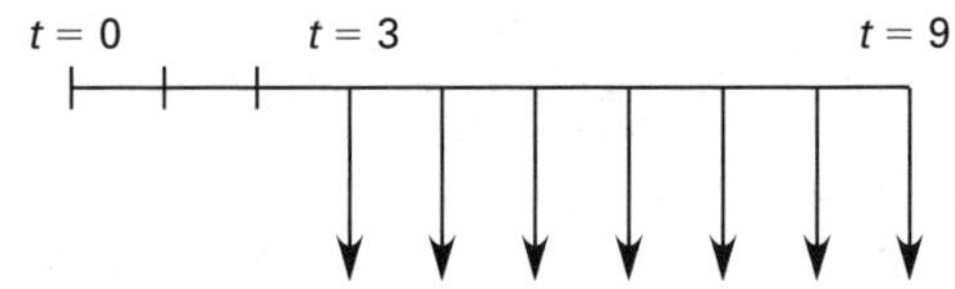

Solution

First, determine a cash flow at $t = 2$ that is equivalent to the entire expense stream. If $t = 0$ was where $t = 2$ actually is, the present worth of the expense stream would be

$$P' = (-\$75)(P/A, i\%, 7)$$

P' is a cash flow at $t = 2$. It is now simple to find the present worth (at $t = 0$) of this future amount.

$$P = P'(P/F, i\%, 2) = (-\$75)(P/A, i\%, 7)(P/F, i\%, 2)$$

Cash Flows at Beginnings of Years: The Christmas Club Problem

This type of problem is characterized by a stream of equal payments (or expenses) starting at $t = 0$ and ending at $t = n - 1$. (See Fig. 55.4.) It differs from the standard annual cash flow in the existence of a cash flow at $t = 0$ and the absence of a cash flow at $t = n$. This problem gets its name from the service provided by some savings institutions whereby money is automatically deposited each week or month (starting immediately, when the savings plan is opened) in order to accumulate money to purchase Christmas presents at the end of the year.

Figure 55.4 *Cash Flows at Beginnings of Years*

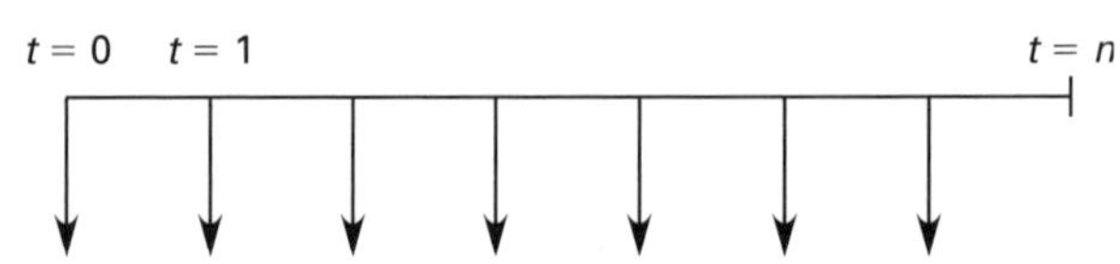

It may seem that the present worth of the savings stream can be determined by directly applying the (P/A) factor. However, this is not the case, since the Christmas Club cash flow and the standard annual cash flow differ. The Christmas Club problem is easily handled by superposition, as illustrated by Ex. 55.9.

Example 55.9

How much can you expect to accumulate by $t = 10$ for a child's college education if you deposit \$300 at the beginning of each year for a total of ten payments?

Solution

Notice that the first payment is made at $t = 0$ and that there is no payment at $t = 10$. The future worth of the first payment is calculated with the (F/P) factor. The absence of the payment at $t = 10$ is handled by superposition. Notice that this "correction" is not multiplied by a factor.

$$\begin{aligned} F &= (\$300)(F/P, i\%, 10) + (\$300)(F/A, i\%, 10) - \$300 \\ &= (\$300)(F/A, i\%, 11) - \$300 \end{aligned}$$

17. THE MEANING OF PRESENT WORTH AND *i*

If \$100 is invested in a 5% bank account (using annual compounding), you can remove \$105 one year from now; if this investment is made, you will receive a *return on investment* (ROI) of \$5. The cash flow diagram (see Figure 55.5) and the present worth of the two transactions are

$$\begin{aligned} P &= -\$100 + (\$105)(P/F, 5\%, 1) \\ &= -\$100 + (\$105)(0.9524) \\ &= 0 \end{aligned}$$

Figure 55.5 *Cash Flow Diagram*

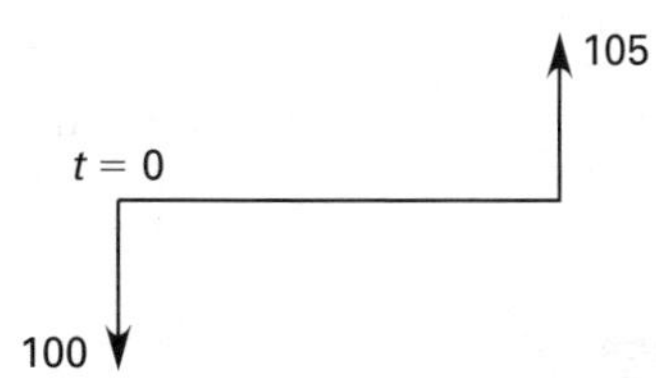

Notice that the present worth is zero even though you will receive a \$5 return on your investment.

However, if you are offered \$120 for the use of \$100 over a one-year period, the cash flow diagram (see Fig. 55.6) and present worth (at 5%) would be

$$\begin{aligned} P &= -\$100 + (\$120)(P/F, 5\%, 1) \\ &= -\$100 + (\$120)(0.9524) \\ &= \$14.29 \end{aligned}$$

Figure 55.6 *Cash Flow Diagram*

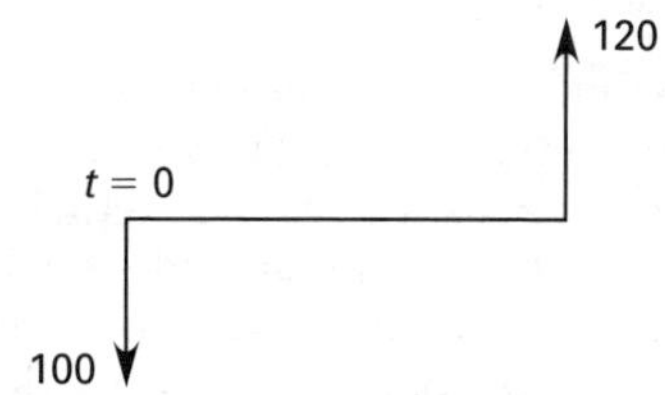

Therefore, the present worth of an alternative is seen to be equal to the equivalent value at $t = 0$ of the increase in return above that which you would be able to earn in an investment offering $i\%$ per period. In the previous case, \$14.29 is the present worth of (\$20 – \$5), the difference in the two ROIs.

The present worth is also the amount that you would have to be given to dissuade you from making an investment, since placing the initial investment amount along with the present worth into a bank account earning $i\%$ will yield the same eventual return on investment. Relating this to the previous paragraphs, you could be dissuaded from investing \$100 in an alternative that would return \$120 in one year by a $t = 0$ payment of \$14.29. Clearly, (\$100 + \$14.29) invested at $t = 0$ will also yield \$120 in one year at 5%.

Income-producing alternatives with negative present worths are undesirable, and alternatives with positive present worths are desirable because they increase the average earning power of invested capital. (In some cases, such as municipal and public works projects, the present worths of all alternatives are negative, in which case, the least negative alternative is best.)

The selection of the interest rate is difficult in engineering economics problems. Usually, it is taken as the average rate of return that an individual or business organization has realized in past investments. Alternatively, the interest rate may be associated with a particular level of risk. Usually, i for individuals is the interest rate that can be earned in relatively *risk-free investments.*

18. SIMPLE AND COMPOUND INTEREST

If \$100 is invested at 5%, it will grow to \$105 in one year. During the second year, 5% interest continues to be accrued, but on \$105, not on \$100. This is the principle of *compound interest*: The interest accrues interest.[12]

If only the original principal accrues interest, the interest is said to be *simple interest.* Simple interest is rarely encountered in engineering economic analyses, but the concept may be incorporated into short-term transactions.

[12]This assumes, of course, that the interest remains in the account. If the interest is removed and spent, only the remaining funds accumulate interest.

19. EXTRACTING THE INTEREST RATE: RATE OF RETURN

An intuitive definition of the *rate of return* (ROR) is the effective annual interest rate at which an investment accrues income. That is, the rate of return of a project is the interest rate that would yield identical profits if all money were invested at that rate. Although this definition is correct, it does not provide a method of determining the rate of return.

It was previously seen that the present worth of a \$100 investment invested at 5% is zero when $i = 5\%$ is used to determine equivalence. Therefore, a working definition of rate of return would be the effective annual interest rate that makes the present worth of the investment zero. Alternatively, rate of return could be defined as the effective annual interest rate that will discount all cash flows to a total present worth equal to the required initial investment.

It is tempting, but impractical, to determine a rate of return analytically. It is simply too difficult to extract the interest rate from the equivalence equation. For example, consider a \$100 investment that pays back \$75 at the end of each of the first two years. The present worth equivalence equation (set equal to zero in order to determine the rate of return) is

$$P = 0 = -\$100 + (\$75)(1+i)^{-1} + (\$75)(1+i)^{-2} \quad \text{55.18}$$

Solving Eq. 55.18 requires finding the roots of a quadratic equation. In general, for an investment or project spanning n years, the roots of an nth-order polynomial would have to be found. It should be obvious that an analytical solution would be essentially impossible for more complex cash flows. (The rate of return in this example is 31.87%.)

If the rate of return is needed, it can be found from a trial-and-error solution. To find the rate of return of an investment, proceed as follows.

step 1: Set up the problem as if to calculate the present worth.

step 2: Arbitrarily select a reasonable value for i. Calculate the present worth.

step 3: Choose another value of i (not too close to the original value), and again solve for the present worth.

step 4: Interpolate or extrapolate the value of i that gives a zero present worth.

step 5: For increased accuracy, repeat steps 2 and 3 with two more values that straddle the value found in step 4.

A common, although incorrect, method of calculating the rate of return involves dividing the annual receipts or returns by the initial investment. (See Sec. 55.74.) However, this technique ignores such items as salvage, depreciation, taxes, and the time value of money. This technique also is inadequate when the annual returns vary.

It is possible that more than one interest rate will satisfy the zero present worth criteria. This confusing situation occurs whenever there is more than one change in sign in the investment's cash flow.[13] Table 55.3 indicates the numbers of possible interest rates as a function of the number of sign reversals in the investment's cash flow.

Table 55.3 *Multiplicity of Rates of Return*

number of sign reversals	number of distinct rates of return
0	0
1	0 or 1
2	0, 1, or 2
3	0, 1, 2, or 3
4	0, 1, 2, 3, or 4
m	$0, 1, 2, 3, \ldots, m-1, m$

Difficulties associated with interpreting the meaning of multiple rates of return can be handled with the concepts of external investment and external rate of return. An *external investment* is an investment that is distinct from the investment being evaluated (which becomes known as the internal investment). The *external rate of return*, which is the rate of return earned by the external investment, does not need to be the same as the rate earned by the internal investment.

Generally, the multiple rates of return indicate that the analysis must proceed as though money will be invested outside of the project. The mechanics of how this is done are not covered here.

Example 55.10

What is the rate of return on invested capital if \$1000 is invested now with \$500 being returned in year 4 and \$1000 being returned in year 8?

[13]There will always be at least one change of sign in the cash flow of a legitimate investment. (This excludes municipal and other tax-supported functions.) At $t = 0$, an investment is made (a negative cash flow). Hopefully, the investment will begin to return money (a positive cash flow) at $t = 1$ or shortly thereafter. Although it is possible to conceive of an investment in which all of the cash flows were negative, such an investment would probably be classified as a *hobby*.

Professional

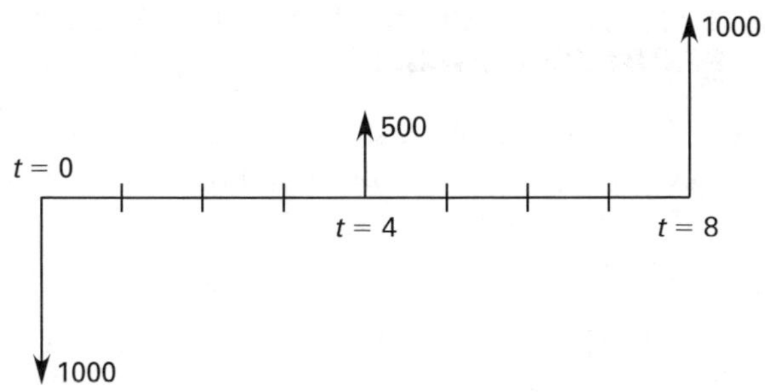

Solution

First, set up the problem as a present worth calculation. Try $i = 5\%$.

$$\begin{aligned} P &= -\$1000 + (\$500)(P/F, 5\%, 4) \\ &\quad + (\$1000)(P/F, 5\%, 8) \\ &= -\$1000 + (\$500)(0.8227) + (\$1000)(0.6768) \\ &= \$88 \end{aligned}$$

Next, try a larger value of i to reduce the present worth. If $i = 10\%$,

$$\begin{aligned} P &= -\$1000 + (\$500)(P/F, 10\%, 4) \\ &\quad + (\$1000)(P/F, 10\%, 8) \\ &= -\$1000 + (\$500)(0.6830) + (\$1000)(0.4665) \\ &= -\$192 \end{aligned}$$

Using simple interpolation, the rate of return is

$$\begin{aligned} \text{ROR} &= 5\% + \left(\frac{\$88}{\$88 + \$192}\right)(10\% - 5\%) \\ &= 6.57\% \end{aligned}$$

A second iteration between 6% and 7% yields 6.39%.

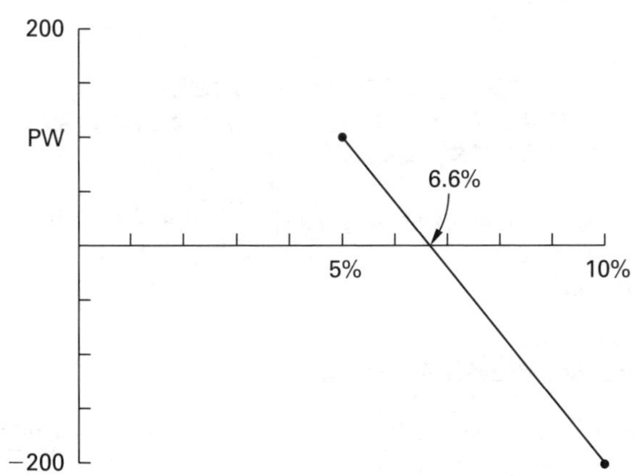

Example 55.11

A biomedical company is developing a new drug. A venture capital firm gives the company \$25 million initially and \$55 million more at the end of the first year. The drug patent will be sold at the end of year 5 to the highest bidder, and the biomedical company will receive \$80 million. (The venture capital firm will receive everything in excess of \$80 million.) The firm invests unused money in short-term commercial paper earning 10% effective interest per year through its bank. In the meantime, the biomedical company incurs development expenses of \$50 million annually for the first three years. The drug is to be evaluated by a government agency and there will be neither expenses nor revenues during the fourth year. What is the biomedical company's rate of return on this investment?

Solution

Normally, the rate of return is determined by setting up a present worth problem and varying the interest rate until the present worth is zero. Writing the cash flows, though, shows that there are two reversals of sign: one at $t = 2$ (positive to negative) and the other at $t = 5$ (negative to positive). Therefore, there could be two interest rates that produce a zero present worth. (In fact, there actually are two interest rates: 10.7% and 41.4%.)

time	cash flow (millions)
0	+25
1	+55 − 15 = +5
2	−50
3	−50
4	0
5	+80

However, this problem can be reduced to one with only one sign reversal in the cash flow series. The initial \$25 million is invested in commercial paper (an *external investment* having nothing to do with the drug development process) during the first year at 10%. The accumulation of interest and principal after one year is

$$(25)(1 + 0.10) = 27.5$$

This 27.5 is combined with the 5 (the money remaining after all expenses are paid at $t = 1$) and invested externally, again at 10%. The accumulation of interest and principal after one year (i.e., at $t = 2$) is

$$(27.5 + 5)(1 + 0.10) = 35.75$$

This 35.75 is combined with the development cost paid at $t = 2$.

The cash flow for the development project (the internal investment) is

time	cash flow (millions)
0	0
1	0
2	35.75 − 50 = − 14.25
3	− 50
4	0
5	+80

Professional

Now, there is only one sign reversal in the cash flow series. The *internal rate of return* on this development project is found by the traditional method to be 10.3%. Notice that this is different from the rate the company can earn from investing externally in commercial paper.

20. RATE OF RETURN VERSUS RETURN ON INVESTMENT

Rate of return (ROR) is an effective annual interest rate, typically stated in percent per year. *Return on investment* (ROI) is a dollar amount. *Rate of return* and *return on investment* are not synonymous.

Return on investment can be calculated in two different ways. The accounting method is to subtract the total of all investment costs from the total of all net profits (i.e., revenues less expenses). The time value of money is not considered.

In engineering economic analysis, the return on investment is calculated from equivalent values. Specifically, the present worth (at $t = 0$) of all investment costs is subtracted from the future worth (at $t = n$) of all net profits.

When there are only two cash flows, a single investment amount and a single payback, the two definitions of return on investment yield the same numerical value. When there are more than two cash flows, the returns on investment will be different depending on which definition is used.

21. MINIMUM ATTRACTIVE RATE OF RETURN

A company may not know what effective interest rate, i, to use in engineering economic analysis. In such a case, the company can establish a minimum level of economic performance that it would like to realize on all investments. This criterion is known as the *minimum attractive rate of return* (MARR). Unlike the effective interest rate, i, the minimum attractive rate of return is not used in numerical calculations.[14] It is used only in comparisons with the rate of return.

Once a rate of return for an investment is known, it can be compared to the minimum attractive rate of return. To be a viable alternative, the rate of return must be greater than the minimum attractive rate of return.

The advantage of using comparisons to the minimum attractive rate of return is that an effective interest rate, i, never needs to be known. The minimum attractive rate of return becomes the correct interest rate for use in present worth and equivalent uniform annual cost calculations.

[14]Not everyone adheres to this rule. Some people use "minimum attractive rate of return" and "effective interest rate" interchangeably.

22. TYPICAL ALTERNATIVE-COMPARISON PROBLEM FORMATS

With the exception of some investment and rate of return problems, the typical problem involving engineering economics will have the following characteristics.

- An interest rate will be given.
- Two or more alternatives will be competing for funding.
- Each alternative will have its own cash flows.
- It is necessary to select the best alternative.

23. DURATIONS OF INVESTMENTS

Because they are handled differently, short-term investments and short-lived assets need to be distinguished from investments and assets that constitute an infinitely lived project. Short-term investments are easily identified: a drill press that is needed for three years or a temporary factory building that is being constructed to last five years.

Investments with perpetual cash flows are also (usually) easily identified: maintenance on a large flood control dam and revenues from a long-span toll bridge. Furthermore, some items with finite lives can expect renewal on a repeated basis.[15] For example, a major freeway with a pavement life of 20 years is unlikely to be abandoned; it will be resurfaced or replaced every 20 years.

Actually, if an investment's finite lifespan is long enough, it can be considered an infinite investment because money 50 or more years from now has little impact on current decisions. The $(P/F, 10\%, 50)$ factor, for example, is 0.0085. Therefore, one dollar at $t = 50$ has an equivalent present worth of less than one penny. Since these far-future cash flows are eclipsed by present cash flows, long-term investments can be considered finite or infinite without significant impact on the calculations.

24. CHOICE OF ALTERNATIVES: COMPARING ONE ALTERNATIVE WITH ANOTHER ALTERNATIVE

Several methods exist for selecting a superior alternative from among a group of proposals. Each method has its own merits and applications.

Present Worth Method

When two or more alternatives are capable of performing the same functions, the superior alternative will have the largest present worth. The *present worth method* is restricted to evaluating alternatives that are mutually exclusive and that have the same lives. This

[15]The term *renewal* can be interpreted to mean replacement or repair.

method is suitable for ranking the desirability of alternatives.

Example 55.12

Investment A costs \$10,000 today and pays back \$11,500 two years from now. Investment B costs \$8000 today and pays back \$4500 each year for two years. If an interest rate of 5% is used, which alternative is superior?

Solution

$$\begin{aligned} P(\text{A}) &= -\$10{,}000 + (\$11{,}500)(P/F, 5\%, 2) \\ &= -\$10{,}000 + (\$11{,}500)(0.9070) \\ &= \$431 \\ P(\text{B}) &= -\$8000 + (\$4500)(P/A, 5\%, 2) \\ &= -\$8000 + (\$4500)(1.8594) \\ &= \$367 \end{aligned}$$

Alternative A is superior and should be chosen.

Capitalized Cost Method

The present worth of a project with an infinite life is known as the *capitalized cost* or *life cycle cost*. Capitalized cost is the amount of money at $t = 0$ needed to perpetually support the project on the earned interest only. Capitalized cost is a positive number when expenses exceed income.

In comparing two alternatives, each of which is infinitely lived, the superior alternative will have the lowest capitalized cost.

Normally, it would be difficult to work with an infinite stream of cash flows since most economics tables do not list factors for periods in excess of 100 years. However, the (A/P) discounting factor approaches the interest rate as n becomes large. Since the (P/A) and (A/P) factors are reciprocals of each other, it is possible to divide an infinite series of equal cash flows by the interest rate in order to calculate the present worth of the infinite series. This is the basis of Eq. 55.19.

$$\text{capitalized cost} = \text{initial cost} + \frac{\text{annual costs}}{i} \qquad 55.19$$

Equation 55.19 can be used when the annual costs are equal in every year. If the operating and maintenance costs occur irregularly instead of annually, or if the costs vary from year to year, it will be necessary to somehow determine a cash flow of equal annual amounts (EAA) that is equivalent to the stream of original costs.

The equal annual amount may be calculated in the usual manner by first finding the present worth of all the actual costs and then multiplying the present worth by the interest rate (the (A/P) factor for an infinite series). However, it is not even necessary to convert the present worth to an equal annual amount since Eq. 55.20 will convert the equal amount back to the present worth.

$$\begin{aligned} \text{capitalized cost} &= \text{initial cost} + \frac{\text{EAA}}{i} \\ &= \text{initial cost} + \begin{array}{l}\text{present worth}\\ \text{of all expenses}\end{array} \end{aligned} \qquad 55.20$$

Example 55.13

What is the capitalized cost of a public works project that will cost \$25,000,000 now and will require \$2,000,000 in maintenance annually? The effective annual interest rate is 12%.

Solution

Worked in millions of dollars, from Eq. 55.19, the capitalized cost is

$$\begin{aligned} \text{capitalized cost} &= 25 + (2)(P/A, 12\%, \infty) \\ &= 25 + \frac{2}{0.12} = 41.67 \end{aligned}$$

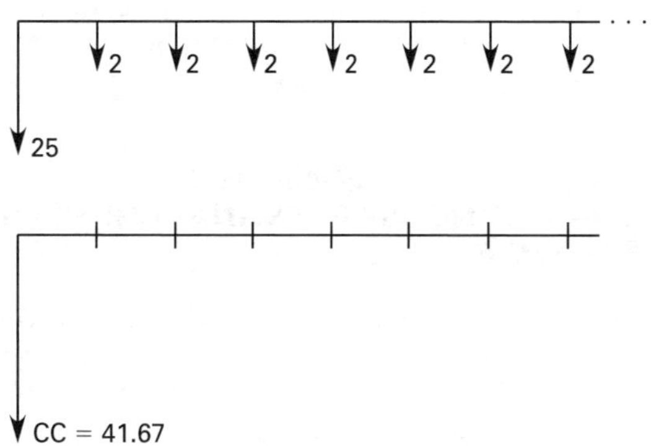

Annual Cost Method

Alternatives that accomplish the same purpose but that have unequal lives must be compared by the *annual cost method*.[16] The annual cost method assumes that each alternative will be replaced by an identical twin at the end of its useful life (infinite renewal). This method, which may also be used to rank alternatives according to their desirability, is also called the *annual return method* or *capital recovery method*.

Restrictions are that the alternatives must be mutually exclusive and repeatedly renewed up to the duration of the longest-lived alternative. The calculated annual cost is known as the *equivalent uniform annual cost* (EUAC) or just *equivalent annual cost*. Cost is a positive number when expenses exceed income.

[16]Of course, the annual cost method can be used to determine the superiority of assets with identical lives as well.

Professional

Example 55.14

Which of the following alternatives is superior over a 30-year period if the interest rate is 7%?

	alternative A	alternative B
type	brick	wood
life	30 years	10 years
initial cost	\$1800	\$450
maintenance	\$5/year	\$20/year

Solution

$$\begin{aligned}\text{EUAC(A)} &= (\$1800)(A/P, 7\%, 30) + \$5\\ &= (\$1800)(0.0806) + \$5\\ &= \$150\\ \text{EUAC(B)} &= (\$450)(A/P, 7\%, 10) + \$20\\ &= (\$450)(0.1424) + \$20\\ &= \$84\end{aligned}$$

Alternative B is superior since its annual cost of operation is the lowest. It is assumed that three wood facilities, each with a life of 10 years and a cost of \$450, will be built to span the 30-year period.

25. CHOICE OF ALTERNATIVES: COMPARING AN ALTERNATIVE WITH A STANDARD

With specific economic performance criteria, it is possible to qualify an investment as acceptable or unacceptable without having to compare it with another investment. Two such performance criteria are the benefit-cost ratio and the minimum attractive rate of return.

Benefit-Cost Ratio Method

The *benefit-cost ratio* method is often used in municipal project evaluations where benefits and costs accrue to different segments of the community. With this method, the present worth of all benefits (irrespective of the beneficiaries) is divided by the present worth of all costs. The project is considered acceptable if the ratio equals or exceeds 1.0, that is, if $B/C \geq 1.0$.

When the benefit-cost ratio method is used, disbursements by the initiators or sponsors are *costs*. Disbursements by the users of the project are known as *disbenefits*. It is often difficult to determine whether a cash flow is a cost or a disbenefit (whether to place it in the numerator or denominator of the benefit-cost ratio calculation).

Regardless of where the cash flow is placed, an acceptable project will always have a benefit-cost ratio greater than or equal to 1.0, although the actual numerical result will depend on the placement. For this reason, the benefit-cost ratio method should not be used to rank competing projects.

The benefit-cost ratio method of comparing alternatives is used extensively in transportation engineering where the ratio is often (but not necessarily) written in terms of annual benefits and annual costs instead of present worths. Another characteristic of highway benefit-cost ratios is that the route (road, highway, etc.) is usually already in place and that various alternative upgrades are being considered. There will be existing benefits and costs associated with the current route. Therefore, the *change* (usually an increase) in benefits and costs is used to calculate the benefit-cost ratio.[17]

$$B/C = \frac{\Delta\begin{array}{c}\text{user}\\\text{benefits}\end{array}}{\Delta\begin{array}{c}\text{investment}\\\text{cost}\end{array} + \Delta\text{maintenance} - \Delta\begin{array}{c}\text{residual}\\\text{value}\end{array}} \qquad 55.21$$

The change in *residual value (terminal value)* appears in the denominator as a negative item. An increase in the residual value would decrease the denominator.

Example 55.15

By building a bridge over a ravine, a state department of transportation can shorten the time it takes to drive through a mountainous area. Estimates of costs and benefits (due to decreased travel time, fewer accidents, reduced gas usage, etc.) have been prepared. Should the bridge be built? Use the benefit-cost ratio method of comparison.

	millions
initial cost	40
capitalized cost of perpetual annual maintenance	12
capitalized value of annual user benefits	49
residual value	0

Solution

If Eq. 55.21 is used, the benefit-cost ratio is

$$B/C = \frac{49}{40 + 12 - 0} = 0.942$$

Since the benefit-cost ratio is less than 1.00, the bridge should not be built.

[17]This discussion of highway benefit-cost ratios is not meant to imply that everyone agrees with Eq. 55.21. In *Economic Analysis for Highways* (International Textbook Company, Scranton, PA, 1969), author Robley Winfrey took a strong stand on one aspect of the benefits versus disbenefits issue: highway maintenance. According to Winfrey, regular highway maintenance costs should be placed in the numerator as a subtraction from the user benefits. Some have called this mandate the *Winfrey method.*

Professional

If the maintenance costs are placed in the numerator (per Ftn. 17), the benefit-cost ratio value will be different, but the conclusion will not change.

$$B/C_{\text{alternate method}} = \frac{49 - 12}{40} = 0.925$$

Rate of Return Method

The minimum attractive rate of return (MARR) has already been introduced as a standard of performance against which an investment's actual *rate of return* (ROR) is compared. If the rate of return is equal to or exceeds the minimum attractive rate of return, the investment is qualified. This is the basis for the *rate of return method* of alternative selection.

Finding the rate of return can be a long, iterative process. Usually, the actual numerical value of rate of return is not needed; it is sufficient to know whether or not the rate of return exceeds the minimum attractive rate of return. This *comparative analysis* can be accomplished without calculating the rate of return simply by finding the present worth of the investment using the minimum attractive rate of return as the effective interest rate (i.e., $i = \text{MARR}$). If the present worth is zero or positive, the investment is qualified. If the present worth is negative, the rate of return is less than the minimum attractive rate of return.

26. RANKING MUTUALLY EXCLUSIVE MULTIPLE PROJECTS

Ranking of multiple investment alternatives is required when there is sufficient funding for more than one investment. Since the best investments should be selected first, it is necessary to place all investments into an ordered list.

Ranking is relatively easy if the present worths, future worths, capitalized costs, or equivalent uniform annual costs have been calculated for all the investments. The highest ranked investment will be the one with the largest present or future worth, or the smallest capitalized or annual cost. Present worth, future worth, capitalized cost, and equivalent uniform annual cost can all be used to rank multiple investment alternatives.

However, neither rates of return nor benefit-cost ratios should be used to rank multiple investment alternatives. Specifically, if two alternatives both have rates of return exceeding the minimum acceptable rate of return, it is not sufficient to select the alternative with the highest rate of return.

An *incremental analysis*, also known as a *rate of return on added investment study*, should be performed if rate of return is used to select between investments. An incremental analysis starts by ranking the alternatives in order of increasing initial investment. Then, the cash flows for the investment with the lower initial cost are subtracted from the cash flows for the higher-priced alternative on a year-by-year basis. This produces, in effect, a third alternative representing the costs and benefits of the added investment. The added expense of the higher-priced investment is not warranted unless the rate of return of this third alternative exceeds the minimum attractive rate of return as well. The choice criterion is to select the alternative with the higher initial investment if the incremental rate of return exceeds the minimum attractive rate of return.

An incremental analysis is also required if ranking is to be done by the benefit-cost ratio method. The incremental analysis is accomplished by calculating the ratio of differences in benefits to differences in costs for each possible pair of alternatives. If the ratio exceeds 1.0, alternative 2 is superior to alternative 1. Otherwise, alternative 1 is superior.[18]

$$\frac{B_2 - B_1}{C_2 - C_1} \geq 1 \quad \text{[alternative 2 superior]} \qquad 55.22$$

27. ALTERNATIVES WITH DIFFERENT LIVES

Comparison of two alternatives is relatively simple when both alternatives have the same life. For example, a problem might be stated: "Which would you rather have: car A with a life of three years, or car B with a life of five years?"

However, care must be taken to understand what is going on when the two alternatives have different lives. If car A has a life of three years and car B has a life of five years, what happens at $t = 3$ if the five-year car is chosen? If a car is needed for five years, what happens at $t = 3$ if the three-year car is chosen?

In this type of situation, it is necessary to distinguish between the length of the need (the *analysis horizon*) and the lives of the alternatives or assets intended to meet that need. The lives do not have to be the same as the horizon.

Finite Horizon with Incomplete Asset Lives

If an asset with a five-year life is chosen for a three-year need, the disposition of the asset at $t = 3$ must be known in order to evaluate the alternative. If the asset is sold at $t = 3$, the salvage value is entered into the analysis (at $t = 3$) and the alternative is evaluated as a three-year investment. The fact that the asset is sold when it has some useful life remaining does not affect the analysis horizon.

Similarly, if a three-year asset is chosen for a five-year need, something about how the need is satisfied during the last two years must be known. Perhaps a rental asset will be used. Or, perhaps the function will be "farmed out" to an outside firm. In any case, the costs of satisfying the need during the last two years enter the analysis, and the alternative is evaluated as a five-year investment.

[18]It goes without saying that the benefit-cost ratios for all investment alternatives by themselves must also be equal to or greater than 1.0.

If both alternatives are "converted" to the same life, any of the alternative selection criteria (present worth method, annual cost method, etc.) can be used to determine which alternative is superior.

Finite Horizon with Integer Multiple Asset Lives

It is common to have a long-term horizon (need) that must be met with short-lived assets. In special instances, the horizon will be an integer number of asset lives. For example, a company may be making a 12-year transportation plan and may be evaluating two cars: one with a three-year life, and another with a four-year life.

In this example, four of the first car or three of the second car are needed to reach the end of the 12-year horizon.

If the horizon is an integer number of asset lives, any of the alternative selection criteria can be used to determine which is superior. If the present worth method is used, all alternatives must be evaluated over the entire horizon. (In this example, the present worth of 12 years of car purchases and use must be determined for both alternatives.)

If the equivalent uniform annual cost method is used, it may be possible to base the calculation of annual cost on one lifespan of each alternative only. It may not be necessary to incorporate all of the cash flows into the analysis. (In the running example, the annual cost over three years would be determined for the first car; the annual cost over four years would be determined for the second car.) This simplification is justified if the subsequent asset replacements (renewals) have the same cost and cash flow structure as the original asset. This assumption is typically made implicitly when the annual cost method of comparison is used.

Infinite Horizon

If the need horizon is infinite, it is not necessary to impose the restriction that asset lives of alternatives be integer multiples of the horizon. The superior alternative will be replaced (renewed) whenever it is necessary to do so, forever.

Infinite horizon problems are almost always solved with either the annual cost or capitalized cost method. It is common to (implicitly) assume that the cost and cash flow structure of the asset replacements (renewals) are the same as the original asset.

28. OPPORTUNITY COSTS

An *opportunity cost* is an imaginary cost representing what will not be received if a particular strategy is rejected. It is what you will lose if you do or do not do something. As an example, consider a growing company with an existing operational computer system. If the company trades in its existing computer as part of an upgrade plan, it will receive a *trade-in allowance.* (In other problems, a *salvage value* may be involved.)

If one of the alternatives being evaluated is not to upgrade the computer system at all, the trade-in allowance (or, salvage value in other problems) will not be realized. The amount of the trade-in allowance is an opportunity cost that must be included in the problem analysis.

Similarly, if one of the alternatives being evaluated is to wait one year before upgrading the computer, the *difference in trade-in allowances* is an opportunity cost that must be included in the problem analysis.

29. REPLACEMENT STUDIES

An investigation into the retirement of an existing process or piece of equipment is known as a *replacement study.* Replacement studies are similar in most respects to other alternative comparison problems: An interest rate is given, two alternatives exist, and one of the previously mentioned methods of comparing alternatives is used to choose the superior alternative. Usually, the annual cost method is used on a year-by-year basis.

In replacement studies, the existing process or piece of equipment is known as the *defender.* The new process or piece of equipment being considered for purchase is known as the *challenger.*

30. TREATMENT OF SALVAGE VALUE IN REPLACEMENT STUDIES

Since most defenders still have some market value when they are retired, the problem of what to do with the salvage arises. It seems logical to use the salvage value of the defender to reduce the initial purchase cost of the challenger. This is consistent with what would actually happen if the defender were to be retired.

By convention, however, the defender's salvage value is subtracted from the defender's present value. This does not seem logical, but it is done to keep all costs and benefits related to the defender with the defender. In this case, the salvage value is treated as an opportunity cost that would be incurred if the defender is not retired.

If the defender and the challenger have the same lives and a present worth study is used to choose the superior alternative, the placement of the salvage value will have no effect on the net difference between present worths for the challenger and defender. Although the values of the two present worths will be different depending on the placement, the difference in present worths will be the same.

If the defender and the challenger have different lives, an annual cost comparison must be made. Since the salvage value would be "spread over" a different number of years depending on its placement, it is important to abide by the conventions listed in this section.

There are a number of ways to handle salvage value in retirement studies. The best way is to calculate the cost of keeping the defender one more year. In addition to the usual operating and maintenance costs, that cost includes an opportunity interest cost incurred by not selling the defender, and also a drop in the salvage value if the defender is kept for one additional year. Specifically,

$$\begin{aligned}\text{EUAC (defender)} &= \text{next year's maintenance costs} \\ &\quad + i(\text{current salvage value}) \\ &\quad + \text{current salvage} \\ &\quad - \text{next year's salvage}\end{aligned} \qquad 55.23$$

It is important in retirement studies not to double count the salvage value. That is, it would be incorrect to add the salvage value to the defender and at the same time subtract it from the challenger.

Equation 55.23 contains the difference in salvage value between two consecutive years. This calculation shows that the defender/challenger decision must be made on a year-by-year basis. One application of Eq. 55.23 will not usually answer the question of whether the defender should remain in service indefinitely. The calculation must be repeatedly made as long as there is a drop in salvage value from one year to the next.

31. ECONOMIC LIFE: RETIREMENT AT MINIMUM COST

As an asset grows older, its operating and maintenance costs typically increase. Eventually, the cost to keep the asset in operation becomes prohibitive, and the asset is retired or replaced. However, it is not always obvious when an asset should be retired or replaced.

As the asset's maintenance cost is increasing each year, the amortized cost of its initial purchase is decreasing. It is the sum of these two costs that should be evaluated to determine the point at which the asset should be retired or replaced. Since an asset's initial purchase price is likely to be high, the amortized cost will be the controlling factor in those years when the maintenance costs are low. Therefore, the EUAC of the asset will decrease in the initial part of its life. (See Fig. 55.7.)

Figure 55.7 *EUAC Versus Age at Retirement*

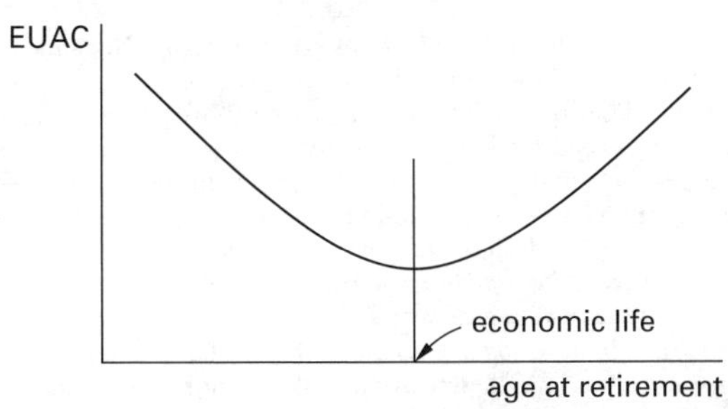

However, as the asset grows older, the change in its amortized cost decreases while maintenance cost increases. Eventually, the sum of the two costs reaches a minimum and then starts to increase. The age of the asset at the minimum cost point is known as the *economic life* of the asset. The economic life generally is less than the length of need and the technological lifetime of the asset.

The determination of an asset's economic life is illustrated by Ex. 55.16.

Example 55.16

Buses in a municipal transit system have the characteristics listed. In order to minimize its annual operating expenses, when should the city replace its buses if money can be borrowed at 8%?

initial cost of bus: $120,000

year	maintenance cost	salvage value
1	35,000	60,000
2	38,000	55,000
3	43,000	45,000
4	50,000	25,000
5	65,000	15,000

Solution

The annual maintenance is different each year. Each maintenance cost must be spread over the life of the bus. This is done by first finding the present worth and then amortizing the maintenance costs. If a bus is kept for one year and then sold, the annual cost will be

$$\begin{aligned}\text{EUAC}(1) &= (\$120{,}000)(A/P, 8\%, 1) \\ &\quad + (\$35{,}000)(A/F, 8\%, 1) \\ &\quad - (\$60{,}000)(A/F, 8\%, 1) \\ &= (\$120{,}000)(1.0800) + (\$35{,}000)(1.000) \\ &\quad - (\$60{,}000)(1.000) \\ &= \$104{,}600\end{aligned}$$

If a bus is kept for two years and then sold, the annual cost will be

$$\begin{aligned}\text{EUAC}(2) &= \big(\$120{,}000 + (\$35{,}000)(P/F, 8\%, 1)\big) \\ &\quad \times (A/P, 8\%, 2) \\ &\quad + (38{,}000 - 55{,}000)(A/F, 8\%, 2) \\ &= \big(120{,}000 + 35{,}000)(0.9259)\big)(0.5608) \\ &\quad + 38{,}000 - 55{,}000)(0.4808) \\ &= 77{,}296\end{aligned}$$

If a bus is kept for three years and then sold, the annual cost will be

$$\begin{aligned}\text{EUAC}(3) &= \big(\$120{,}000 + (35{,}000)(P/F, 8\%, 1) \\ &\quad + (\$38{,}000)(P/F, 8\%, 2)\big)(A/P, 8\%, 3) \\ &\quad + (\$43{,}000 - \$45{,}000)(A/F, 8\%, 3) \\ &= \big(\$120{,}000 + (\$35{,}000)(0.9259) \\ &\quad + (\$38{,}000)(0.8573)\big)(0.3880) \\ &\quad - (\$2000)(0.3080) \\ &= \$71{,}158\end{aligned}$$

This process is continued until the annual cost begins to increase. In this example, EUAC(4) is \$71,700. Therefore, the buses should be retired after three years.

32. LIFE-CYCLE COST

The *life-cycle cost* of an alternative is the equivalent value (at $t = 0$) of the alternative's cash flow over the alternative's lifespan. Since the present worth is evaluated using an effective interest rate of i (which would be the interest rate used for all engineering economic analyses), the life-cycle cost is the same as the alternative's present worth. If the alternative has an infinite horizon, the life-cycle cost and capitalized cost will be identical.

33. CAPITALIZED ASSETS VERSUS EXPENSES

High expenses reduce profit, which in turn reduces income tax. It seems logical to label each and every expenditure, even an asset purchase, as an expense. As an alternative to this *expensing the asset*, it may be decided to capitalize the asset. *Capitalizing the asset* means that the cost of the asset is divided into equal or unequal parts, and only one of these parts is taken as an expense each year. Expensing is clearly the more desirable alternative, since the after-tax profit is increased early in the asset's life.

There are long-standing accounting conventions as to what can be expensed and what must be capitalized.[19] Some companies capitalize everything—regardless of cost—with expected lifetimes greater than one year. Most companies, however, expense items whose purchase costs are below a cut-off value. A cut-off value in the range of \$250–500, depending on the size of the company, is chosen as the maximum purchase cost of an expensed asset. Assets costing more than this are capitalized.

It is not necessary for a large corporation to keep track of every lamp, desk, and chair for which the purchase price is greater than the cut-off value. Such assets, all of which have the same lives and have been purchased in the same year, can be placed into groups or *asset classes*. A group cost, equal to the sum total of the purchase costs of all items in the group, is capitalized as though the group was an identifiable and distinct asset itself.

34. PURPOSE OF DEPRECIATION

Depreciation is an *artificial expense* that spreads the purchase price of an asset or other property over a number of years.[20] Depreciating an asset is an example of capitalization, as previously defined. The inclusion of depreciation in engineering economic analysis problems will increase the after-tax present worth (profitability) of an asset. The larger the depreciation, the greater will be the profitability. Therefore, individuals and companies eligible to utilize depreciation want to maximize and accelerate the depreciation available to them.

Although the entire property purchase price is eventually recognized as an expense, the net recovery from the expense stream never equals the original cost of the asset. That is, depreciation cannot realistically be thought of as a fund (an annuity or sinking fund) that accumulates capital to purchase a replacement at the end of the asset's life. The primary reason for this is that the depreciation expense is reduced significantly by the impact of income taxes, as will be seen in later sections.

35. DEPRECIATION BASIS OF AN ASSET

The *depreciation basis* of an asset is the part of the asset's purchase price that is spread over the *depreciation period*, also known as the *service life*.[21] Usually, the depreciation basis and the purchase price are not the same.

A common depreciation basis is the difference between the purchase price and the expected salvage value at the end of the depreciation period. That is,

$$\text{depreciation basis} = C - S_n \qquad 55.24$$

There are several methods of calculating the year-by-year depreciation of an asset. Equation 55.24 is not universally compatible with all depreciation methods. Some methods do not consider the salvage value. This is known as an *unadjusted basis*. When the depreciation

[19]For example, purchased vehicles must be capitalized; payments for leased vehicles can be expensed. Repainting a building with paint that will last five years is an expense, but the replacement cost of a leaking roof must be capitalized.

[20]In the United States, the tax regulations of Internal Revenue Service (IRS) allow depreciation on almost all forms of *business property* except land. The following types of property are distinguished: *real* (e.g., buildings used for business), *residential* (e.g., buildings used as rental property), and *personal* (e.g., equipment used for business). Personal property does *not* include items for personal use (such as a personal residence), despite its name. *Tangible personal property* is distinguished from *intangible property* (goodwill, copyrights, patents, trademarks, franchises, agreements not to compete, etc.).

[21]The *depreciation period* is selected to be as short as possible within recognized limits. This depreciation will not normally coincide with the *economic life* or *useful life* of an asset. For example, a car may be capitalized over a depreciation period of three years. It may become uneconomical to maintain and use at the end of an economic life of nine years. However, the car may be capable of operation over a useful life of 25 years.

method is known, the depreciation basis can be rigorously defined.[22]

36. DEPRECIATION METHODS

Generally, tax regulations do not allow the cost of an asset to be treated as a deductible expense in the year of purchase. Rather, portions of the depreciation basis must be allocated to each of the n years of the asset's depreciation period. The amount that is allocated each year is called the *depreciation.*

Various methods exist for calculating an asset's depreciation each year.[23] Although the depreciation calculations may be considered independently (for the purpose of determining book value or as an academic exercise), it is important to recognize that depreciation has no effect on engineering economic analyses unless income taxes are also considered.

Straight Line Method

With the *straight line method*, depreciation is the same each year. The depreciation basis $(C - S_n)$ is allocated uniformly to all of the n years in the depreciation period. Each year, the depreciation will be

$$D = \frac{C - S_n}{n} \qquad 55.25$$

Constant Percentage Method

The *constant percentage method*[24] is similar to the straight line method in that the depreciation is the same each year. If the fraction of the basis used as depreciation is $1/n$, there is no difference between the constant percentage and straight line methods. The two methods differ only in what information is available. (With the straight line method, the life is known. With the constant percentage method, the depreciation fraction is known.)

Each year, the depreciation will be

$$\begin{aligned} D &= (\text{depreciation fraction})(\text{depreciation basis}) \\ &= (\text{depreciation fraction})(C - S_n) \end{aligned} \qquad 55.26$$

Sum-of-the-Years' Digits Method

In *sum-of-the-years' digits* (SOYD) depreciation, the digits from 1 to n inclusive are summed. The total, T, can also be calculated from

$$T = \tfrac{1}{2}n(n+1) \qquad 55.27$$

The depreciation in year j can be found from Eq. 55.28. Notice that the depreciation in year j, D_j, decreases by a constant amount each year.

$$D_j = \frac{(C - S_n)(n - j + 1)}{T} \qquad 55.28$$

Double Declining Balance Method[25]

Double declining balance[26] (DDB) depreciation is independent of salvage value. Furthermore, the book value never stops decreasing, although the depreciation decreases in magnitude. Usually, any book value in excess of the salvage value is written off in the last year of the asset's depreciation period. Unlike any of the other depreciation methods, double declining balance depends on accumulated depreciation.

$$D_{\text{first year}} = \frac{2C}{n} \qquad 55.29$$

$$D_j = \frac{2\left(C - \sum_{m=1}^{j-1} D_m\right)}{n} \qquad 55.30$$

Calculating the depreciation in the middle of an asset's life appears particularly difficult with double declining balance, because all previous years' depreciation amounts seem to be required. It appears that the depreciation in the sixth year, for example, cannot be calculated unless the values of depreciation for the first five years are calculated. However, this is not true.

Depreciation in the middle of an asset's life can be found from the following equations. (d is known as the *depreciation rate.*)

$$d = \frac{2}{n} \qquad 55.31$$

$$D_j = dC(1 - d)^{j-1} \qquad 55.32$$

Statutory Depreciation Systems

In the United States, property placed into service in 1981 and thereafter must use the *Accelerated Cost Recovery System* (ACRS), and after 1986, the *Modified*

Professional

[22]For example, with the Accelerated Cost Recovery System (ACRS) the *depreciation basis* is the total purchase cost, regardless of the expected salvage value. With declining balance methods, the depreciation basis is the purchase cost less any previously taken depreciation.

[23]This discussion gives the impression that any form of depreciation may be chosen regardless of the nature and circumstances of the purchase. In reality, the IRS tax regulations place restrictions on the higher-rate (accelerated) methods, such as declining balance and sum-of-the-years' digits methods. Furthermore, the *Economic Recovery Act of 1981* and the *Tax Reform Act of 1986* substantially changed the laws relating to personal and corporate income taxes.

[24]The *constant percentage method* should not be confused with the declining balance method, which used to be known as the *fixed percentage on diminishing balance method.*

[25]In the past, the *declining balance method* has also been known as the *fixed percentage of book value* and *fixed percentage on diminishing balance method.*

[26]Double declining balance depreciation is a particular form of *declining balance depreciation*, as defined by the IRS tax regulations. Declining balance depreciation includes 125% declining balance and 150% declining balance depreciations that can be calculated by substituting 1.25 and 1.50, respectively, for the 2 in Eq. 55.29.

Accelerated Cost Recovery System (MACRS) or other statutory method. Other methods (straight line, declining balance, etc.) cannot be used except in special cases.

Property placed into service in 1980 or before must continue to be depreciated according to the method originally chosen (e.g., straight line, declining balance, or sum-of-the-years' digits). ACRS and MACRS cannot be used.

Under ACRS and MACRS, the cost recovery amount in the jth year of an asset's cost recovery period is calculated by multiplying the initial cost by a factor. (See Table 55.4.)

$$D_j = C \times \text{factor} \qquad 55.33$$

Table 55.4 *Representative ACRS and MACRS Depreciation Factors**

	recovery period, n		
	3 yr	5 yr	10 yr
year j	ACRS/MACRS	ACRS/MACRS	ACRS/MACRS
1	0.25/0.3333	0.15/0.2000	0.08/0.100
2	0.38/0.4445	0.22/0.3200	0.14/0.180
3	0.37/0.1481	0.21/0.1920	0.12/0.1440
4	0/0.0741	0.21/0.1152	0.10/0.1152
5		0.21/0.1152	0.10/0.0922
6		0/0.0576	0.10/0.0737
7			0.09/0.0655
8			0.09/0.0655
9			0.09/0.0656
10			0.09/0.0655
11			0/0.0328

*MACRS values are for the "half-year" convention. This table gives typical values only. Because these factors are subject to continuing revision, they should not be used without consulting an accounting professional.

The initial cost used is not reduced by the asset's salvage value for ACRS and MACRS calculations. The factor used depends on the asset's cost recovery period. Such factors are subject to continuing legislation changes. Current tax publications should be consulted before using this method.

Production or Service Output Method

If an asset has been purchased for a specific task and that task is associated with a specific lifetime amount of output or production, the depreciation may be calculated by the fraction of total production produced during the year. The depreciation is not expected to be the same each year.

$$D_j = (C - S_n)\left(\frac{\text{actual output in year } j}{\text{estimated lifetime output}}\right) \qquad 55.34$$

Sinking Fund Method

The *sinking fund method* is seldom used in industry because the initial depreciation is low. The formula for sinking fund depreciation (which increases each year) is

$$D_j = (C - S_n)(A/F, i\%, n)(F/P, i\%, j-1) \qquad 55.35$$

Disfavored Methods

Three other depreciation methods are mentioned here, not because they are currently accepted or in widespread use, but because they are still occasionally encountered in the literature.[27]

The *sinking fund plus interest on first cost* depreciation method, like the following two methods, is an attempt to include the *opportunity interest cost* on the purchase price with the depreciation. That is, the purchasing company not only incurs an annual expense from the drop in book value, but it also loses the interest on the purchase price. The formula for this method is

$$D = (C - S_n)(A/F, i\%, n) + Ci \qquad 55.36$$

The *straight line plus interest on first cost* method is similar. Its formula is

$$D = \left(\frac{1}{n}\right)(C - S_n) + Ci \qquad 55.37$$

The *straight line plus average interest method* assumes that the opportunity interest cost should be based on the book value only, not on the full purchase price. Because the book value changes each year, an average value is used. The depreciation formula is

$$D = \left(\frac{C - S_n}{n}\right)\left(1 + \frac{i(n+1)}{2}\right) + iS_n \qquad 55.38$$

Example 55.17

An asset is purchased for \$9000. Its estimated economic life is ten years, after which it will be sold for \$200. Find the depreciation in the first three years using the straight line (SL), double declining balance (DDB), and sum-of-the-years' digits (SOYD) depreciation methods.

[27]The three disfavored depreciation methods should not be used in the usual manner (e.g., in conjunction with the income tax rate). These methods are attempts to calculate a more accurate annual cost of an alternative, but sometimes they give misleading answers. Their use is not recommended, and they are included in this chapter only for the sake of completeness.

Professional

Solution

SL:
$$D = \frac{C - S_n}{n} = \frac{\$9000 - \$200}{10} = \$880 \text{ each year}$$

DDB:
$$D_1 = \frac{2C}{n} = \frac{(2)(\$9000)}{10} = \$1800 \text{ in year 1}$$

$$D_j = \frac{2\left(C - \sum_{m=1}^{j-1} D_m\right)}{n}$$

$$D_2 = \frac{(2)(\$9000 - \$1800)}{10} = \$1440 \text{ in year 2}$$

$$D_3 = \frac{(2)(\$9000 - \$3240)}{10} = \$1152 \text{ in year 3}$$

SOYD:
$$T = \tfrac{1}{2}n(n+1) = \left(\tfrac{1}{2}\right)(10)(11) = 55$$

$$D_1 = \left(\tfrac{10}{55}\right)(\$9000 - \$200) = \$1600 \text{ in year 1}$$

$$D_2 = \left(\tfrac{9}{55}\right)(\$8800) = \$1440 \text{ in year 2}$$

$$D_3 = \left(\tfrac{8}{55}\right)(\$8800) = \$1280 \text{ in year 3}$$

37. ACCELERATED DEPRECIATION METHODS

An *accelerated depreciation method* is one that calculates a depreciation amount greater than a straight line amount. Double declining balance and sum-of-the-years' digits methods are accelerated methods. The ACRS and MACRS methods are explicitly accelerated methods. Straight line and sinking fund methods are not accelerated methods.

Use of an accelerated depreciation method may result in unexpected tax consequences when the depreciated asset or property is disposed of. Professional tax advice should be obtained in this area.

38. BOOK VALUE

The difference between original purchase price and accumulated depreciation is known as *book value*.[28] At the end of each year, the book value (which is initially equal to the purchase price) is reduced by the depreciation in that year.

It is important to properly synchronize depreciation calculations. It is difficult to answer the question, "What is the book value in the fifth year?" unless the timing of the book value change is mutually agreed upon. It is better to be specific about an inquiry by identifying when the book value change occurs. For example, the following question is unambiguous: "What is the book value at the end of year 5, after subtracting depreciation in the fifth year?" or "What is the book value after five years?"

Unfortunately, this type of care is seldom taken in book value inquiries, and it is up to the respondent to exercise reasonable care in distinguishing between beginning-of-year book value and end-of-year book value. To be consistent, the book value equations in this chapter have been written in such a way that the year subscript (j) has the same meaning in book value and depreciation calculations. That is, BV_5 means the book value at the end of the fifth year, after five years of depreciation, including D_5, have been subtracted from the original purchase price.

There can be a great difference between the book value of an asset and the *market value* of that asset. There is no legal requirement for the two values to coincide, and no intent for book value to be a reasonable measure of market value.[29] Therefore, it is apparent that book value is merely an accounting convention with little practical use. Even when a depreciated asset is disposed of, the book value is used to determine the consequences of disposal, not the price the asset should bring at sale.

The calculation of book value is relatively easy, even for the case of the declining balance depreciation method.

For the straight line depreciation method, the book value at the end of the jth year, after the jth depreciation deduction has been made, is

$$BV_j = C - \frac{j(C - S_n)}{n} = C - jD \qquad 55.39$$

For the sum-of-the-years' digits method, the book value is

$$BV_j = (C - S_n)\left(1 - \frac{j(2n + 1 - j)}{n(n+1)}\right) + S_n \qquad 55.40$$

For the declining balance method, including double declining balance (see Ftn. 26), the book value is

$$BV_j = C(1 - d)^j \qquad 55.41$$

For the sinking fund method, the book value is calculated directly as

$$BV_j = C - (C - S_n)(A/F, i\%, n)(F/A, i\%, j) \qquad 55.42$$

Of course, the book value at the end of year j can always be calculated for any method by successive subtractions

[28]The balance sheet of a corporation usually has two asset accounts: the *equipment account* and the *accumulated depreciation account*. There is no book value account on this financial statement, other than the implicit value obtained from subtracting the accumulated depreciation account from the equipment account. The book values of various assets, as well as their original purchase cost, date of purchase, salvage value, and so on, and accumulated depreciation appear on detail sheets or other peripheral records for each asset.

[29]Common examples of assets with great divergences of book and market values are buildings (rental houses, apartment complexes, factories, etc.) and company luxury automobiles (Porsches, Mercedes, etc.) during periods of inflation. Book values decrease, but actual values increase.

Professional

(i.e., subtraction of the accumulated depreciation), as Eq. 55.43 illustrates.

$$\text{BV}_j = C - \sum_{m=1}^{j} D_m \qquad 55.43$$

Figure 55.8 illustrates the book value of a hypothetical asset depreciated using several depreciation methods. Notice that the double declining balance method initially produces the fastest write-off, while the sinking fund method produces the slowest write-off. Also, the book value does not automatically equal the salvage value at the end of an asset's depreciation period with the double declining balance method.[30]

Figure 55.8 *Book Value with Different Depreciation Methods*

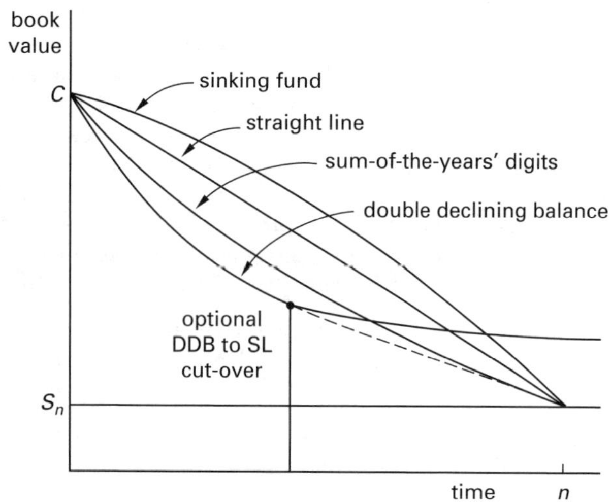

Example 55.18

For the asset described in Ex. 55.17, calculate the book value at the end of the first three years if sum-of-the-years' digits depreciation is used. The book value at the beginning of year 1 is $9000.

Solution

From Eq. 55.43,

$$\text{BV}_1 = \$9000 - \$1600 = \$7400$$
$$\text{BV}_2 = \$7400 - \$1440 = \$5960$$
$$\text{BV}_3 = \$5960 - \$1280 = \$4680$$

[30]This means that the straight line method of depreciation may result in a lower book value at some point in the depreciation period than if double declining balance is used. A *cut-over* from double declining balance to straight line may be permitted in certain cases. Finding the *cut-over point*, however, is usually done by comparing book values determined by both methods. The analytical method is complicated.

39. AMORTIZATION

Amortization and depreciation are similar in that they both divide up the cost basis or value of an asset. In fact, in certain cases, the term "amortization" may be used in place of the term "depreciation." However, depreciation is a specific form of amortization.

Amortization spreads the cost basis or value of an asset over some base. The base can be time, units of production, number of customers, and so on. The asset can be tangible (e.g., a delivery truck or building) or intangible (e.g., goodwill or a patent).

If the asset is tangible, if the base is time, and if the length of time is consistent with accounting standards and taxation guidelines, then the term "depreciation" is appropriate. However, if the asset is intangible, if the base is some other variable, or if some length of time other than the customary period is used, then the term "amortization" is more appropriate.[31]

Example 55.19

A company purchases complete and exclusive patent rights to an invention for $1,200,000. It is estimated that once commercially produced, the invention will have a specific but limited market of 1200 units. For the purpose of allocating the patent right cost to production cost, what is the amortization rate in dollars per unit?

Solution

The patent should be amortized at the rate of

$$\frac{\$1{,}200{,}000}{1200 \text{ units}} = \$1000 \text{ per unit}$$

40. DEPLETION

Depletion is another artificial deductible operating expense, designed to compensate mining organizations for decreasing mineral reserves. Since original and remaining quantities of minerals are seldom known accurately, the *depletion allowance* is calculated as a fixed percentage of the organization's gross income. These percentages are usually in the 10–20% range and apply to such mineral deposits as oil, natural gas, coal, uranium, and most metal ores.

41. BASIC INCOME TAX CONSIDERATIONS

The issue of income taxes is often overlooked in academic engineering economic analysis exercises. Such a position is justifiable when an organization (e.g., a nonprofit school, a church, or the government) pays no income taxes. However, if an individual or organization

[31]From time to time, the U.S. Congress has allowed certain types of facilities (e.g., emergency, grain storage, and pollution control) to be written off more rapidly than would otherwise be permitted in order to encourage investment in such facilities. The term "amortization" has been used with such write-off periods.

is subject to income taxes, the income taxes must be included in an economic analysis of investment alternatives.

Assume that an organization pays a fraction f of its profits to the federal government as income taxes. If the organization also pays a fraction s of its profits as state income taxes and if state taxes paid are recognized by the federal government as tax-deductible expenses, then the composite tax rate is

$$t = s + f - sf \qquad 55.44$$

The basic principles used to incorporate taxation into engineering economic analyses are the following.

- Initial purchase expenditures are unaffected by income taxes.
- Salvage revenues are unaffected by income taxes.
- Deductible expenses, such as operating costs, maintenance costs, and interest payments, are reduced by the fraction t (i.e., multiplied by the quantity $(1 - t)$).
- Revenues are reduced by the fraction t (i.e., multiplied by the quantity $(1 - t)$).
- Since tax regulations allow the depreciation in any year to be handled as if it were an actual operating expense, and since operating expenses are deductible from the income base prior to taxation, the after-tax profits will be increased. If D is the depreciation, the net result to the after-tax cash flow will be the addition of tD. Depreciation is multiplied by t and added to the appropriate year's cash flow, increasing that year's present worth.

For simplicity, most engineering economics practice problems involving income taxes specify a single income tax rate. In practice, however, federal and most state tax rates depend on the income level. Each range of incomes and its associated tax rate are known as *income bracket* and *tax Bracket*, respectively. For example, the state income tax rate might be 4% for incomes up to and including \$30,000, and 5% for incomes above \$30,000. The income tax for a taxpaying entity with an income of \$50,000 would have to be calculated in two parts.

$$\begin{aligned}\text{tax} &= (0.04)(\$30{,}000) + (0.05)(\$50{,}000 - \$30{,}000)\\ &= \$2200\end{aligned}$$

Income taxes and depreciation have no bearing on municipal or governmental projects since municipalities, states, and the U.S. government pay no taxes.

Example 55.20

A corporation that pays 53% of its profit in income taxes invests \$10,000 in an asset that will produce a \$3000 annual revenue for eight years. If the annual expenses are \$700, salvage after eight years is \$500, and 9% interest is used, what is the after-tax present worth? Disregard depreciation.

Solution

$$\begin{aligned}P &= -\$10{,}000 + (\$3000)(P/A, 9\%, 8)(1 - 0.53)\\ &\qquad - (\$700)(P/A, 9\%, 8)(1 - 0.53)\\ &\qquad + (\$500)(P/F, 9\%, 8)\\ &= -\$10{,}000 + (\$3000)(5.5348)(0.47)\\ &\qquad - (\$700)(5.5348)(0.47) + (\$500)(0.5019)\\ &= -\$3766\end{aligned}$$

42. TAXATION AT THE TIMES OF ASSET PURCHASE AND SALE

There are numerous rules and conventions that governmental tax codes and the accounting profession impose on organizations. Engineers are not normally expected to be aware of most of the rules and conventions, but occasionally it may be necessary to incorporate their effects into an engineering economic analysis.

Tax Credit

A *tax credit* (also known as an *investment tax credit* or *investment credit*) is a one-time credit against income taxes.[32] Therefore, it is added to the after-tax present worth as a last step in an engineering economic analysis. Such tax credits may be allowed by the government from time to time for equipment purchases, employment of various classes of workers, rehabilitation of historic landmarks, and so on.

A tax credit is usually calculated as a fraction of the initial purchase price or cost of an asset or activity.

$$\text{TC} = \text{fraction} \times \text{initial cost} \qquad 55.45$$

When the tax credit is applicable, the fraction used is subject to legislation. A professional tax expert or accountant should be consulted prior to applying the tax credit concept to engineering economic analysis problems.

Since the investment tax credit reduces the buyer's tax liability, a tax credit should be included only in after-tax engineering economic analyses. The credit is assumed to be received at the end of the year.

Gain or Loss on the Sale of a Depreciated Asset

If an asset that has been depreciated over a number of prior years is sold for more than its current book value, the difference between the book value and selling price is taxable income in the year of the sale. Alternatively, if the asset is sold for less than its current book value, the difference between the selling price and book value is an expense in the year of the sale.

[32] Strictly, *tax credit* is the more general term, and applies to a credit for doing anything creditable. An *investment tax credit* requires an investment in something (usually real property or equipment).

Example 55.21

One year, a company makes a \$5000 investment in a historic building. The investment is not depreciable, but it does qualify for a one-time 20% tax credit. In that same year, revenue is \$45,000 and expenses (exclusive of the \$5000 investment) are \$25,000. The company pays a total of 53% in income taxes. What is the after-tax present worth of this year's activities if the company's interest rate for investment is 10%?

Solution

The tax credit is

$$\text{TC} = (0.20)(\$5000) = \$1000$$

This tax credit is assumed to be received at the end of the year. The after-tax present worth is

$$\begin{aligned} P &= -\$5000 + (\$45{,}000 - \$25{,}000)(1 - 0.53) \\ &\quad \times (P/F, 10\%, 1) + (\$1000)(P/F, 10\%, 1) \\ &= -\$5000 + (\$20{,}000)(0.47)(0.9091) \\ &\quad + (\$1000)(0.9091) \\ &= \$4455 \end{aligned}$$

43. DEPRECIATION RECOVERY

The economic effect of depreciation is to reduce the income tax in year j by tD_j. The present worth of the asset is also affected: The present worth is increased by $tD_j(P/F, i\%, j)$. The after-tax present worth of all depreciation effects over the depreciation period of the asset is called the *depreciation recovery* (DR).[33]

$$\text{DR} = t\sum_{j=1}^{n} D_j(P/F, i\%, j) \qquad 55.46$$

There are multiple ways depreciation can be calculated, as summarized in Table 55.5. *Straight line depreciation recovery* from an asset is easily calculated, since the depreciation is the same each year. Assuming the asset has a constant depreciation of D and depreciation period of n years, the depreciation recovery is

$$\text{DR} = tD(P/A, i\%, n) \qquad 55.47$$

$$D = \frac{C - S_n}{n} \qquad 55.48$$

Sum-of-the-years' digits depreciation recovery is also relatively easily calculated, since the depreciation decreases uniformly each year.

$$\text{DR} = \left(\frac{t(C - S_n)}{T}\right)\left(n(P/A, i\%, n) - (P/G, i\%, n)\right) \qquad 55.49$$

Finding *declining balance depreciation recovery* is more involved. There are three difficulties. The first (the apparent need to calculate all previous depreciations in order to determine the subsequent depreciation) has already been addressed by Eq. 55.32.

The second difficulty is that there is no way to ensure (that is, to force) the book value to be S_n at $t = n$. Therefore, it is common to write off the remaining book value (down to S_n) at $t = n$ in one lump sum. This assumes $\text{BV}_n \geq S_n$.

The third difficulty is that of finding the present worth of an *exponentially decreasing cash flow*. Although the proof is omitted here, such exponential cash flows can be handled with the *exponential gradient factor*, (P/EG).[34]

$$(P/EG, z - 1, n) = \frac{z^n - 1}{z^n(z - 1)} \qquad 55.50$$

$$z = \frac{1 + i}{1 - d} \qquad 55.51$$

Then, as long as $\text{BV}_n > S_n$, the declining balance depreciation recovery is

$$\text{DR} = tC\left(\frac{d}{1 - d}\right)(P/EG, z - 1, n) \qquad 55.52$$

Example 55.22

For the asset described in Ex. 55.17, calculate the after-tax depreciation recovery with straight line and sum-of-the-years' digits depreciation methods. Use 6% interest with 48% income taxes.

Solution

SL:

$$\begin{aligned} \text{DR} &= (0.48)(\$880)(P/A, 6\%, 10) \\ &= (0.48)(\$880)(7.3601) \\ &= \$3109 \end{aligned}$$

[33]Since the depreciation benefit is reduced by taxation, depreciation cannot be thought of as an annuity to fund a replacement asset.

[34]The (P/A) columns in App. 55.B can be used for (P/EG) as long as the interest rate is assumed to be $z - 1$.

Table 55.5 Depreciation Calculation Summary

method	depreciation basis	depreciation in year j (D_j)	book value after jth depreciation (BV_j)	after-tax depreciation recovery (DR)	supplementary formulas
straight line (SL)	$C - S_n$	$\frac{C - S_n}{n}$ (constant)	$C - jD$	$tD(P/A, i\%, n)$	
constant percentage	$C - S_n$	fraction $\times (C - S_n)$ (constant)	$C - jD$	$tD(P/A, i\%, n)$	
sum-of-the-years' digits (SOYD)	$C - S_n$	$\frac{(C - S_n) \times (n - j + 1)}{T}$	$(C - S_n) \times \left(1 - \frac{j(2n + 1 - j)}{n(n+1)}\right) + S_n$	$\frac{t(C - S_n)}{T} \times \big(n(P/A, i\%, n) - (P/G, i\%, n)\big)$	$T = \frac{1}{2}n(n+1)$
double declining balance (DDB)	C	$dC(1-d)^{j-1}$	$C(1-d)^j$	$tC\left(\frac{d}{1-d}\right) \times (P/EG, z-1, n)$	$d = \frac{2}{n}$; $z = \frac{1+i}{1-d}$; $(P/EG, z-1, n) = \frac{z^n - 1}{z^n(z-1)}$
sinking fund (SF)	$C - S_n$	$(C - S_n) \times (A/F, i\%, n) \times (F/P, i\%, j-1)$	$C - (C - S_n) \times (A/F, i\%, n) \times (F/A, i\%, j)$	$\frac{t(C - S_n)(A/F, i\%, n)}{1 + i}$	
accelerated cost recovery system (ACRS/MACRS)	C	$C \times$ factor	$C - \sum_{m=1}^{j} D_m$	$t\sum_{j=1}^{n} D_j(P/F, i\%, j)$	
units of production or service output	$C - S_n$	$(C - S_n) \times \left(\frac{\text{actual output in year } j}{\text{lifetime output}}\right)$	$C - \sum_{m=1}^{j} D_m$	$t\sum_{j=1}^{n} D_j(P/F, i\%, j)$	

SOYD: The depreciation series can be thought of as a constant \$1600 term with a negative \$160 gradient.

$$\begin{aligned} DR &= (0.48)(\$1600)(P/A, 6\%, 10) \\ &\quad - (0.48)(\$160)(P/G, 6\%, 10) \\ &= (0.48)(\$1600)(7.3601) \\ &\quad - (0.48)(\$160)(29.6023) \\ &= \$3379 \end{aligned}$$

Notice that the ten-year (P/G) factor is used even though there are only nine years in which the gradient reduces the initial \$1600 amount.

Example 55.23

What is the after-tax present worth of the asset described in Ex. 55.20 if straight line, sum-of-the-years' digits, and double declining balance depreciation methods are used?

Solution

Using SL, the depreciation recovery is

$$\begin{aligned} DR &= (0.53)\left(\frac{\$10{,}000 - \$500}{8}\right)(P/A, 9\%, 8) \\ &= (0.53)\left(\frac{\$9500}{8}\right)(5.5348) \\ &= \$3483 \end{aligned}$$

Professional

Using SOYD, the depreciation recovery is calculated as follows.

$$T = \left(\tfrac{1}{2}\right)(8)(9) = 36$$

$$\text{depreciation base} = \$10{,}000 - \$500 = \$9500$$

$$D_1 = \left(\tfrac{8}{36}\right)(\$9500) = \$2111$$

$$G = \left(\tfrac{1}{36}\right)(\$9500) = \$264$$

$$\begin{aligned} \text{DR} &= (0.53)\Big((\$2111)(P/A, 9\%, 8) \\ &\quad - (\$264)(P/G, 9\%, 8)\Big) \\ &= (0.53)\Big((\$2111)(5.5348) \\ &\quad - (\$264)(16.8877)\Big) \\ &= \$3830 \end{aligned}$$

Using DDB, the depreciation recovery is calculated as follows.[35]

$$d = \frac{2}{8} = 0.25$$

$$z = \frac{1 + 0.09}{1 - 0.25} = 1.4533$$

$$(P/EG, z-1, n) = \frac{(1.4533)^8 - 1}{(1.4533)^8(0.4533)} = 2.095$$

From Eq. 55.52,

$$\begin{aligned} \text{DR} &= (0.53)\left(\frac{(0.25)(\$10{,}000)}{0.75}\right)(2.095) \\ &= \$3701 \end{aligned}$$

The after-tax present worth, neglecting depreciation, was previously found to be −$3766.

The after-tax present worths, including depreciation recovery, are

$$\begin{aligned} &\text{SL:} && P = -\$3766 + \$3483 = -\$283 \\ &\text{SOYD:} && P = -\$3766 + \$3830 = \$64 \\ &\text{DDB:} && P = -\$3766 + \$3701 = -\$65 \end{aligned}$$

44. OTHER INTEREST RATES

The *effective interest rate per period*, i (also called *yield* by banks), is the only interest rate that should be used in equivalence equations. The interest rates at the top of the factor tables in App. 55.B are implicitly all effective interest rates. Usually, the period will be one year, hence the name *effective annual interest rate*. However, there are other interest rates in use as well.

The term *nominal interest rate*, r (*rate per annum*), is encountered when compounding is more than once per year. The nominal rate does not include the effect of compounding and is not the same as the effective rate. And, since the effective interest rate can be calculated from the nominal rate only if the number of compounding periods per year is known, nominal rates cannot be compared unless the method of compounding is specified. The only practical use for a nominal rate per year is for calculating the effective rate per period.

45. RATE AND PERIOD CHANGES

If there are k compounding periods during the year (two for semiannual compounding, four for quarterly compounding, twelve for monthly compounding, etc.) and the nominal rate is r, the *effective rate per compounding period* is

$$\phi = \frac{r}{k} \qquad \text{55.53}$$

The effective annual rate, i, can be calculated from the effective rate per period, ϕ, by using Eq. 55.54.

$$\begin{aligned} i &= (1+\phi)^k - 1 \\ &= \left(1 + \frac{r}{k}\right)^k - 1 \end{aligned} \qquad \text{55.54}$$

Sometimes, only the effective rate per period (e.g., per month) is known. However, that will be a simple problem since compounding for n periods at an effective rate per period is not affected by the definition or length of the period.

The following rules may be used to determine which interest rate is given.

- Unless specifically qualified, the interest rate given is an annual rate.
- If the compounding is annual, the rate given is the effective rate. If compounding is other than annual, the rate given is the nominal rate.

The effective annual interest rate determined on a *daily compounding basis* will not be significantly different than if *continuous compounding* is assumed.[36] In the case of continuous (or daily) compounding, the discounting factors can be calculated directly from the nominal interest rate and number of years, without having to find the effective interest rate per period. Table 55.6 can be used to determine the discount factors for continuous compounding.

[35]This method should start by checking that the book value at the end of the depreciation period is greater than the salvage value. In this example, such is the case. However, the step is not shown.

[36]The number of *banking days in a year* (250, 360, etc.) must be specifically known.

Table 55.6 Discount Factors for Continuous Compounding (n is the number of years)

symbol	formula
$(F/P, r\%, n)$	e^{rn}
$(P/F, r\%, n)$	e^{-rn}
$(A/F, r\%, n)$	$\frac{e^r - 1}{e^{rn} - 1}$
$(F/A, r\%, n)$	$\frac{e^{rn} - 1}{e^r - 1}$
$(A/P, r\%, n)$	$\frac{e^r - 1}{1 - e^{-rn}}$
$(P/A, r\%, n)$	$\frac{1 - e^{-rn}}{e^r - 1}$

Example 55.24

A savings and loan offers a nominal rate of 5.25% compounded daily over 365 days in a year. What is the effective annual rate?

Solution

method 1: Use Eq. 55.54.

$$r = 0.0525,\ k = 365$$

$$i = \left(1 + \frac{r}{k}\right)^k - 1 = \left(1 + \frac{0.0525}{365}\right)^{365} - 1 = 0.0539$$

method 2: Assume daily compounding is the same as continuous compounding.

$$\begin{aligned} i &= (F/P, r\%, 1) - 1 \\ &= e^{0.0525} - 1 \\ &= 0.0539 \end{aligned}$$

Example 55.25

A real estate investment trust pays \$7,000,000 for an apartment complex with 100 units. The trust expects to sell the complex in ten years for \$15,000,000. In the meantime, it expects to receive an average rent of \$900 per month from each apartment. Operating expenses are expected to be \$200 per month per occupied apartment. A 95% occupancy rate is predicted. In similar investments, the trust has realized a 15% effective annual return on its investment. Compare to those past investments the expected present worth of this investment when calculated assuming (a) annual compounding (i.e., the year-end convention), and (b) monthly compounding. Disregard taxes, depreciation, and all other factors.

Solution

(a) The net annual income will be

$$\begin{aligned} &(0.95)(100 \text{ units})\left(\frac{\$900}{\text{unit-mo}} - \frac{\$200}{\text{unit-mo}}\right)\left(12\ \frac{\text{mo}}{\text{yr}}\right) \\ &= \$798{,}000/\text{yr} \end{aligned}$$

The present worth of ten years of operation is

$$\begin{aligned} P &= -\$7{,}000{,}000 + (\$798{,}000)(P/A, 15\%, 10) \\ &\quad + (\$15{,}000{,}000)(P/F, 15\%, 10) \\ &= -\$7{,}000{,}000 + (\$798{,}000)(5.0188) \\ &\quad + (\$15{,}000{,}000)(0.2472) \\ &= \$713{,}000 \end{aligned}$$

(b) The net monthly income is

$$\begin{aligned} &(0.95)(100 \text{ units})\left(\frac{\$900}{\text{unit-mo}} - \frac{\$200}{\text{unit-mo}}\right) \\ &= \$66{,}500/\text{mo} \end{aligned}$$

Equation 55.54 is used to calculate the effective monthly rate, ϕ, from the effective annual rate, $i = 15\%$, and the number of compounding periods per year, $k = 12$.

$$\begin{aligned} \phi &= (1 + i)^{1/k} - 1 \\ &= (1 + 0.15)^{1/12} - 1 = 0.011715\ (1.1715\%) \end{aligned}$$

The number of compounding periods in ten years is

$$n = (10 \text{ yr})\left(12\ \frac{\text{mo}}{\text{yr}}\right) = 120 \text{ mo}$$

The present worth of 120 months of operation is

$$\begin{aligned} P &= -\$7{,}000{,}000 + (\$66{,}500)(P/A, 1.1715\%, 120) \\ &\quad + (\$15{,}000{,}000)(P/F, 1.1715\%, 120) \end{aligned}$$

Since table values for 1.1715% discounting factors are not available, the factors are calculated from Table 55.1.

$$\begin{aligned} (P/A, 1.1715\%, 120) &= \frac{(1 + i)^n - 1}{i(1 + i)^n} \\ &= \frac{(1 + 0.011715)^{120} - 1}{(0.011715)(1 + 0.011715)^{120}} \\ &= 64.261 \end{aligned}$$

$$\begin{aligned} (P/F, 1.1715\%, 120) &= (1 + i)^{-n} = (1 + 0.011715)^{-120} \\ &= 0.2472 \end{aligned}$$

The present worth over 120 monthly compounding periods is

$$\begin{aligned} P &= -\$7{,}000{,}000 + (\$66{,}500)(64.261) \\ &\quad + (\$15{,}000{,}000)(0.2472) \\ &= \$981{,}357 \end{aligned}$$

Professional

46. BONDS

A *bond* is a method of long-term financing commonly used by governments, states, municipalities, and very large corporations.[37] The bond represents a contract to pay the bondholder specific amounts of money at specific times. The holder purchases the bond in exchange for specific payments of interest and principal. Typical municipal bonds call for quarterly or semiannual interest payments and a payment of the *face value of the bond* on the *date of maturity* (end of the bond period).[38] Due to the practice of discounting in the bond market, a bond's face value and its purchase price generally will not coincide.

In the past, a bondholder had to submit a coupon or ticket in order to receive an interim interest payment. This has given rise to the term *coupon rate*, which is the nominal annual interest rate on which the interest payments are made. Coupon books are seldom used with modern bonds, but the term survives. The coupon rate determines the magnitude of the semiannual (or otherwise) interest payments during the life of the bond. The bondholder's own effective interest rate should be used for economic decisions about the bond.

Actual *bond yield* is the bondholder's actual rate of return of the bond, considering the purchase price, interest payments, and face value payment (or, value realized if the bond is sold before it matures). By convention, bond yield is calculated as a nominal rate (rate per annum), not an effective rate per year. The bond yield should be determined by finding the effective rate of return per payment period (e.g., per semiannual interest payment) as a conventional rate of return problem. Then, the nominal rate can be found by multiplying the effective rate per period by the number of payments per year, as in Eq. 55.54.

Example 55.26

What is the maximum amount an investor should pay for a 25-year bond with a $20,000 face value and 8% coupon rate (interest only paid semiannually)? The bond will be kept to maturity. The investor's effective annual interest rate for economic decisions is 10%.

Solution

For this problem, take the compounding period to be six months. Then, there are 50 compounding periods. Since 8% is a nominal rate, the effective bond rate per period is calculated from Eq. 55.53 as

$$\phi_{\text{bond}} = \frac{r}{k} = \frac{8\%}{2} = 4\%$$

The bond payment received semiannually is

$$(0.04)(\$20{,}000) = \$800$$

10% is the investor's effective rate per year, so Eq. 55.54 is again used to calculate the effective analysis rate per period.

$$\begin{aligned} 0.10 &= (1+\phi)^2 - 1 \\ \phi &= 0.04881 \quad (4.88\%) \end{aligned}$$

The maximum amount that the investor should be willing to pay is the present worth of the investment.

$$\begin{aligned} P &= (\$800)(P/A, 4.88\%, 50) \\ &\quad + (\$20{,}000)(P/F, 4.88\%, 50) \end{aligned}$$

Table 55.1 can be used to calculate the following factors.

$$\begin{aligned} (P/A, 4.88\%, 50) &= \frac{(1+i)^n - 1}{i(1+i)^n} \\ &= \frac{(1+0.0488)^{50} - 1}{(0.0488)(1.0488)^{50}} \\ &= 18.600 \\ (P/F, 4.88\%, 50) &= \frac{1}{(1+i)^n} \\ &= \frac{1}{(1+0.0488)^{50}} \\ &= 0.09233 \end{aligned}$$

Then, the present worth is

$$\begin{aligned} P &= (\$800)(18.600) + (\$20{,}000)(0.09233) \\ &= \$16{,}727 \end{aligned}$$

47. PROBABILISTIC PROBLEMS

If an alternative's cash flows are specified by an implicit or explicit probability distribution rather than being known exactly, the problem is *probabilistic*.

Probabilistic problems typically possess the following characteristics.

- There is a chance of loss that must be minimized (or, rarely, a chance of gain that must be maximized) by selection of one of the alternatives.
- There are multiple alternatives. Each alternative offers a different degree of protection from the loss. Usually, the alternatives with the greatest protection will be the most expensive.
- The magnitude of loss or gain is independent of the alternative selected.

Probabilistic problems are typically solved using annual costs and expected values. An *expected value* is similar to an *average value* since it is calculated as the mean

[37] In the past, 30-year bonds were typical. Shorter term 10-year, 15-year, 20-year, and 25-year bonds are also commonly issued.

[38] A *fully amortized bond* pays back interest and principal throughout the life of the bond. There is no balloon payment.

Professional

of the given probability distribution. If cost 1 has a probability of occurrence, p_1, cost 2 has a probability of occurrence, p_2, and so on, the expected value is

$$\mathcal{E}\{\text{cost}\} = p_1(\text{cost } 1) + p_2(\text{cost } 2) + \cdots \quad 55.55$$

Example 55.27

Flood damage in any year is given according to the following table. What is the present worth of flood damage for a ten-year period? Use 6% as the effective annual interest rate.

damage	probability
0	0.75
\$10,000	0.20
\$20,000	0.04
\$30,000	0.01

Solution

The expected value of flood damage in any given year is

$$\begin{aligned}\mathcal{E}\{\text{damage}\} &= (0)(0.75) + (\$10{,}000)(0.20) \\ &\quad + (\$20{,}000)(0.04) + (\$30{,}000)(0.01) \\ &= \$3100\end{aligned}$$

The present worth of ten years of expected flood damage is

$$\begin{aligned}\text{present worth} &= (\$3100)(P/A, 6\%, 10) \\ &= (\$3100)(7.3601) \\ &= \$22{,}816\end{aligned}$$

Example 55.28

A dam is being considered on a river that periodically overflows and causes \$600,000 damage. The damage is essentially the same each time the river causes flooding. The project horizon is 40 years. A 10% interest rate is being used.

Three different designs are available, each with different costs and storage capacities.

design alternative	cost	maximum capacity
A	\$500,000	1 unit
B	\$625,000	1.5 units
C	\$900,000	2.0 units

The National Weather Service has provided a statistical analysis of annual rainfall runoff from the watershed draining into the river.

units annual rainfall	probability
0	0.10
0.1–0.5	0.60
0.6–1.0	0.15
1.1–1.5	0.10
1.6–2.0	0.04
2.1 or more	0.01

Which design alternative would you choose assuming the dam is essentially empty at the start of each rainfall season?

Solution

The sum of the construction cost and the expected damage should be minimized. If alternative A is chosen, it will have a capacity of 1 unit. Its capacity will be exceeded (causing \$600,000 damage) when the annual rainfall exceeds 1 unit. Therefore, the expected value of the annual cost of alternative A is

$$\begin{aligned}\mathcal{E}\{\text{EUAC(A)}\} &= (\$500{,}000)(A/P, 10\%, 40) \\ &\quad + (\$600{,}000)(0.10 + 0.40 + 0.01) \\ &= (\$500{,}000)(0.1023) + (\$600{,}000)(0.15) \\ &= \$141{,}150\end{aligned}$$

Similarly,

$$\begin{aligned}\mathcal{E}\{\text{EUAC(B)}\} &= (\$625{,}000)(A/P, 10\%, 40) \\ &\quad + (\$600{,}000)(0.04 + 0.01) \\ &= (\$625{,}000)(0.1023) + (\$600{,}000)(0.05) \\ &= \$93{,}938\end{aligned}$$

$$\begin{aligned}\mathcal{E}\{\text{EUAC(C)}\} &= (\$900{,}000)(A/P, 10\%, 40) \\ &\quad + (\$600{,}000)(0.01) \\ &= (\$900{,}000)(0.1023) + (\$600{,}000)(0.01) \\ &= \$98{,}070\end{aligned}$$

Alternative B should be chosen.

48. FIXED AND VARIABLE COSTS

The distinction between fixed and variable costs depends on how these costs vary when an independent variable changes. For example, factory or machine production is frequently the independent variable. However, it could just as easily be vehicle miles driven, hours of operation, or quantity (mass, volume, etc.).

If a cost is a function of the independent variable, the cost is said to be a *variable cost*. (See Table 55.7.) The change in cost per unit variable change (i.e., what is

***Table 55.7** Summary of Fixed and Variable Costs*

fixed costs
- rent
- property taxes
- interest on loans
- insurance
- janitorial service expense
- tooling expense
- setup, cleanup, and tear-down expenses
- depreciation expense
- marketing and selling costs
- cost of utilities
- general burden and overhead expense

variable costs
- direct material costs
- direct labor costs
- cost of miscellaneous supplies
- payroll benefit costs
- income taxes
- supervision costs

usually called the *slope*) is known as the *incremental cost.* Material and labor costs are examples of variable costs. They increase in proportion to the number of product units manufactured.

If a cost is not a function of the independent variable, the cost is said to be a *fixed cost.* Rent and lease payments are typical fixed costs. These costs will be incurred regardless of production levels.

Some costs have both fixed and variable components, as Fig. 55.9 illustrates. The fixed portion can be determined by calculating the cost at zero production.

***Figure 55.9** Fixed and Variable Costs*

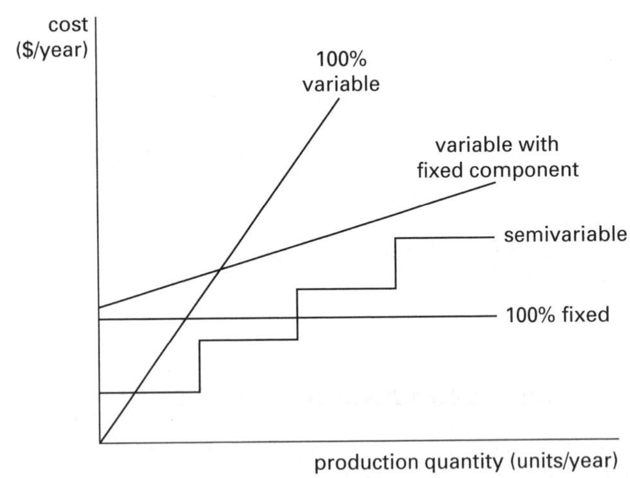

An additional category of cost is the *semivariable cost.* This type of cost increases stepwise. Semivariable cost structures are typical of situations where *excess capacity* exists. For example, supervisory cost is a stepwise function of the number of production shifts. Also, labor cost for truck drivers is a stepwise function of weight (volume) transported. As long as a truck has room left (i.e., excess capacity), no additional driver is needed. As soon as the truck is filled, labor cost will increase.

49. ACCOUNTING COSTS AND EXPENSE TERMS

The accounting profession has developed special terms for certain groups of costs. When annual costs are incurred due to the functioning of a piece of equipment, they are known as *operating and maintenance (O&M) costs.* The annual costs associated with operating a business (other than the costs directly attributable to production) are known as *general, selling, and administrative (GS&A) expenses.*

Direct labor costs are costs incurred in the factory, such as assembly, machining, and painting labor costs. *Direct material costs*are the costs of all materials that go into production.[39] Typically, both direct labor and direct material costs are given on a per-unit or per-item basis. The sum of the direct labor and direct material costs is known as the *prime cost.*

There are certain additional expenses incurred in the factory, such as the costs of factory supervision, stock-picking, quality control, factory utilities, and miscellaneous supplies (cleaning fluids, assembly lubricants, routing tags, etc.) that are not incorporated into the final product. Such costs are known as *indirect manufacturing expenses* (IME) or *indirect material and labor costs.*[40] The sum of the per-unit indirect manufacturing expense and prime cost is known as the *factory cost.*

Research and development(R&D) costs and *administrative expenses* are added to the factory cost to give the *manufacturing cost* of the product.

Additional costs are incurred in marketing the product. Such costs are known as *selling expenses* or *marketing expenses.* The sum of the selling expenses and manufacturing cost is the *total cost* of the product.

Figure 55.10 illustrates these terms.[41]Table 55.8 lists typical classifications of expenses.

The distinctions among the various forms of cost (particularly with overhead costs) are not standardized. Each company must develop a classification system to deal with the various cost factors in a consistent manner. There are also other terms in use (e.g., *raw materials, operating supplies, general plant overhead*), but these terms must be interpreted within the framework of each company's classification system. Table 55.8 is typical of such classification systems.

[39]There may be problems with pricing the material when it is purchased from an outside vendor and the stock on hand derives from several shipments purchased at different prices.

[40]The *indirect material and labor costs* usually exclude costs incurred in the office area.

[41]Notice that *total cost* does not include income taxes.

Figure 55.10 Costs and Expenses Combined

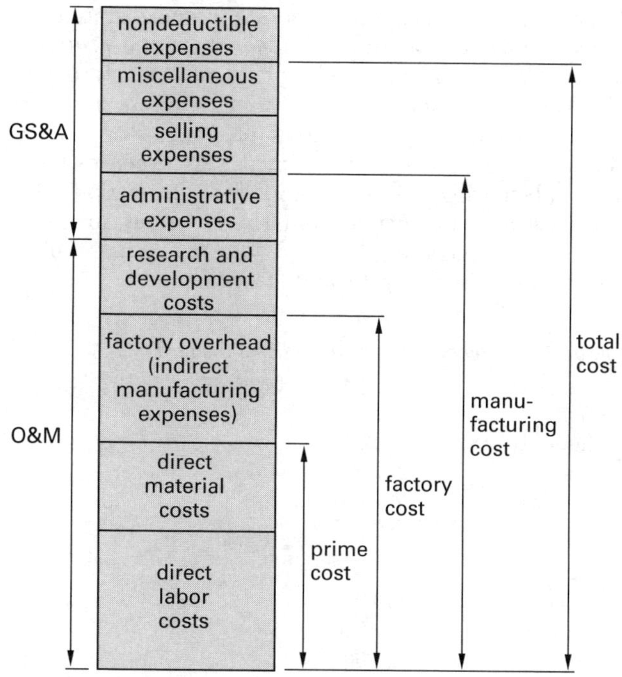

50. ACCOUNTING PRINCIPLES

Basic Bookkeeping

An accounting or *bookkeeping system* is used to record historical financial transactions. The resultant records are used for product costing, satisfaction of statutory requirements, reporting of profit for income tax purposes, and general company management.

Bookkeeping consists of two main steps: recording the transactions, followed by categorization of the transactions.[42] The transactions (receipts and disbursements) are recorded in a *journal (book of original entry)* to complete the first step. Such a journal is organized in a simple chronological and sequential manner. The transactions are then categorized (into interest income, advertising expense, etc.) and posted (i.e., entered or written) into the appropriate *ledger account.*[43]

The ledger accounts together constitute the *general ledger* or *ledger.* All ledger accounts can be classified into one of three types: *asset accounts, liability accounts,* and *owners' equity accounts.* Strictly speaking, income and expense accounts, kept in a separate journal, are included within the classification of owners' equity accounts.

Together, the journal and ledger are known simply as "the books" of the company, regardless of whether bound volumes of pages are actually involved.

[42]These two steps are not to be confused with the *double-entry bookkeeping method.*

[43]The two-step process is more typical of a *manual bookkeeping system* than a computerized *general ledger system.* However, even most computerized systems produce reports in journal entry order, as well as account summaries.

Table 55.8 Typical Classification of Expenses

direct labor expenses
- machining and forming
- assembly
- finishing
- inspection
- testing

direct material expenses
- items purchased from other vendors
- manufactured assemblies

factory overhead expenses (indirect manufacturing expenses)
- supervision
- benefits
  - pension
  - medical insurance
  - vacations
- wages overhead
  - unemployment compensation taxes
  - social security taxes
  - disability taxes
- stock-picking
- quality control and inspection
- expediting
- rework
- maintenance
- miscellaneous supplies
  - routing tags
  - assembly lubricants
  - cleaning fluids
  - wiping cloths
  - janitorial supplies
- packaging (materials and labor)
- factory utilities
- laboratory
- depreciation on factory equipment

research and development expenses
- engineering (labor)
- patents
- testing
- prototypes (material and labor)
- drafting
- O&M of R&D facility

administrative expenses
- corporate officers
- accounting
- secretarial/clerical/reception
- security (protection)
- medical (nurse)
- employment (personnel)
- reproduction
- data processing
- production control
- depreciation on nonfactory equipment
- office supplies
- office utilities
- O&M of offices

selling expenses
- marketing (labor)
- advertising
- transportation (if not paid by customer)
- outside sales force (labor and expenses)
- demonstration units
- commissions
- technical service and support
- order processing
- branch office expenses

miscellaneous expenses
- insurance
- property taxes
- interest on loans

nondeductible expenses
- federal income taxes
- fines and penalties

Balancing the Books

In a business environment, *balancing the books* means more than reconciling the checkbook and bank statements. All accounting entries must be posted in such a way as to maintain the equality of the *basic accounting equation*,

$$\text{assets} = \text{liability} + \text{owner's equity} \qquad 55.56$$

In a *double-entry bookkeeping system*, the equality is maintained within the ledger system by entering each transaction into two balancing ledger accounts. For example, paying a utility bill would decrease the cash account (an asset account) and decrease the utility expense account (a liability account) by the same amount.

Transactions are either *debits* or *credits*, depending on their sign. Increases in asset accounts are debits; decreases are credits. For liability and equity accounts, the opposite is true: Increases are credits, and decreases are debits.[44]

Cash and Accrual Systems[45]

The simplest form of bookkeeping is based on the *cash system*. The only transactions that are entered into the journal are those that represent cash receipts and disbursements. In effect, a checkbook register or bank deposit book could serve as the journal.

During a given period (e.g., month or quarter), expense liabilities may be incurred even though the payments for those expenses have not been made. For example, an invoice (bill) may have been received but not paid. Under the *accrual system*, the obligation is posted into the appropriate expense account before it is paid.[46] Analogous to expenses, under the accrual system, income will be claimed before payment is received. Specifically, a sales transaction can be recorded as income when the customer's order is received, when the outgoing invoice is generated, or when the merchandise is shipped.

[44]There is a difference in sign between asset and liability accounts. An increase in an expense account is actually a decrease. The accounting profession, apparently, is comfortable with the common confusion that exists between debits and credits.

[45]There is also a distinction made between cash flows that are known and those that are expected. It is a *standard accounting principle* to record losses in full, at the time they are recognized, even before their occurrence. In the construction industry, for example, losses are recognized in full and projected to the end of a project as soon as they are foreseeable. Profits, on the other hand, are recognized only as they are realized (typically, as a percentage of project completion). The difference between cash and accrual systems is a matter of *bookkeeping*. The difference between loss and profit recognition is a matter of *accounting convention*. Engineers seldom need to be concerned with the accounting tradition.

[46]The expense for an item or service might be accrued even *before* the invoice is received. It might be recorded when the purchase order for the item or service is generated, or when the item or service is received.

Financial Statements

Each period, two types of corporate financial statements are typically generated: the *balance sheet* and *profit and loss (P&L) statements*.[47] The profit and loss statement, also known as a *statement of income and retained earnings*, is a summary of sources of *income* or *revenue* (interest, sales, fees charged, etc.) and *expenses* (utilities, advertising, repairs, etc.) for the period. The expenses are subtracted from the revenues to give a *net income* (generally, before taxes).[48] Figure 55.11 illustrates a simple profit and loss statement.

Figure 55.11 *Simplified Profit and Loss Statement*

revenue				
interest	2000			
sales	237,000			
returns	(23,000)			
net revenue		216,000		
expenses				
salaries	149,000			
utilities	6000			
advertising	28,000			
insurance	4000			
supplies	1000			
net expenses		188,000		
period net income			28,000	
beginning retained earnings			63,000	
net year-to-date earnings				91,000

The *balance sheet* presents the *basic accounting equation* in tabular form. The balance sheet lists the major categories of assets and outstanding liabilities. The difference between asset values and liabilities is the *equity*, as defined in Eq. 55.56. This equity represents what would be left over after satisfying all debts by liquidating the company.

There are several terms that appear regularly on balance sheets.

- *current assets:* cash and other assets that can be converted quickly into cash, such as accounts receivable, notes receivable, and merchandise (inventory). Also known as *liquid assets*.
- *fixed assets:* relatively permanent assets used in the operation of the business and relatively difficult to convert into cash. Examples are land, buildings, and equipment. Also known as *nonliquid assets*.
- *current liabilities:* liabilities due within a short period of time (e.g., within one year) and typically paid

[47]Other types of financial statements (*statements of changes in financial position, cost of sales statements*, inventory and asset reports, etc.) also will be generated, depending on the needs of the company.

[48]Financial statements also can be prepared with percentages (of total assets and net revenue) instead of dollars, in which case they are known as *common size financial statements*.

out of current assets. Examples are accounts payable, notes payable, and other accrued liabilities.

- *long-term liabilities:* obligations that are not totally payable within a short period of time (e.g., within one year).

Figure 55.12 is a simplified balance sheet.

***Figure 55.12** Simplified Balance Sheet*

ASSETS			
current assets			
cash	14,000		
accounts receivable	36,000		
notes receivable	20,000		
inventory	89,000		
prepaid expenses	3000		
total current assets		162,000	
plant, property, and equipment			
land and buildings	217,000		
motor vehicles	31,000		
equipment	94,000		
accumulated depreciation	(52,000)		
total fixed assets		290,000	
total assets			452,000
LIABILITIES AND OWNERS' EQUITY			
current liabilities			
accounts payable	66,000		
accrued income taxes	17,000		
accrued expenses	8000		
total current liabilities		91,000	
long-term debt			
notes payable	117,000		
mortgage	23,000		
total long-term debt		140,000	
owners' and stockholders' equity			
stock	130,000		
retained earnings	91,000		
total owners' equity		221,000	
total liabilities and owners' equity			452,000

Analysis of Financial Statements

Financial statements are evaluated by management, lenders, stockholders, potential investors, and many other groups for the purpose of determining the *health of the company* The health can be measured in terms of *liquidity* (ability to convert assets to cash quickly), *solvency* (ability to meet debts as they become due), and *relative risk* (of which one measure is *leverage*—the portion of total capital contributed by owners).

The analysis of financial statements involves several common ratios, usually expressed as percentages. The following are some frequently encountered ratios.

- *current ratio:* an index of short-term paying ability.

$$\text{current ratio} = \frac{\text{current assets}}{\text{current liabilities}} \tag{55.57}$$

- *quick* (or *acid-test*) *ratio:* a more stringent measure of short-term debt-paying ability. The *quick assets* are defined to be current assets minus inventories and prepaid expenses.

$$\text{quick ratio} = \frac{\text{quick assets}}{\text{current liabilities}} \tag{55.58}$$

- *receivable turnover:* a measure of the average speed with which accounts receivable are collected.

$$\text{receivable turnover} = \frac{\text{net credit sales}}{\text{average net receivables}} \tag{55.59}$$

- *average age of receivables:* number of days, on the average, in which receivables are collected.

$$\text{average age of receivables} = \frac{365}{\text{receivable turnover}} \tag{55.60}$$

- *inventory turnover:* a measure of the speed with which inventory is sold, on the average.

$$\text{inventory turnover} = \frac{\text{cost of goods sold}}{\text{average cost of inventory on hand}} \tag{55.61}$$

- *days supply of inventory on hand:* number of days, on the average, that the current inventory would last.

$$\text{days supply of inventory on hand} = \frac{365}{\text{inventory turnover}} \tag{55.62}$$

- *book value per share of common stock:* number of dollars represented by the balance sheet owners' equity for each share of common stock outstanding.

$$\text{book value per share of common stock} = \frac{\text{common shareholders' equity}}{\text{number of outstanding shares}} \tag{55.63}$$

- *gross margin:* gross profit as a percentage of sales. (Gross profit is sales less cost of goods sold.)

$$\text{gross margin} = \frac{\text{gross profit}}{\text{net sales}} \tag{55.64}$$

- *profit margin ratio:* percentage of each dollar of sales that is net income.

$$\text{profit margin} = \frac{\text{net income before taxes}}{\text{net sales}} \tag{55.65}$$

- *return on investment ratio:* shows the percent return on owners' investment.

$$\text{return on investment} = \frac{\text{net income}}{\text{owners' equity}} \tag{55.66}$$

- *price-earnings ratio:* indication of relationship between earnings and market price per share of common stock, useful in comparisons between alternative investments.

$$\text{price-earnings} = \frac{\text{market price per share}}{\text{earnings per share}} \qquad 55.67$$

51. COST ACCOUNTING

Cost accounting is the system that determines the cost of manufactured products. Cost accounting is called *job cost accounting* if costs are accumulated by part number or contract. It is called *process cost accounting* if costs are accumulated by departments or manufacturing processes.

Cost accounting is dependent on historical and recorded data. The unit product cost is determined from actual expenses and numbers of units produced. Allowances (i.e., budgets) for future costs are based on these historical figures. Any deviation from historical figures is called a *variance*. Where adequate records are available, variances can be divided into *labor variance* and *material variance*.

When determining a unit product cost, the direct material and direct labor costs are generally clear-cut and easily determined. Furthermore, these costs are 100% variable costs. However, the indirect cost per unit of product is not as easily determined. Indirect costs (*burden*, *overhead*, etc.) can be fixed or semivariable costs. The amount of indirect cost allocated to a unit will depend on the unknown future overhead expense as well as the unknown future production (*vehicle size*).

A typical method of allocating indirect costs to a product is as follows.

step 1: Estimate the total expected indirect (and overhead) costs for the upcoming year.

step 2: Determine the most appropriate vehicle (basis) for allocating the overhead to production. Usually, this vehicle is either the number of units expected to be produced or the number of direct hours expected to be worked in the upcoming year.

step 3: Estimate the quantity or size of the overhead vehicle.

step 4: Divide expected overhead costs by the expected overhead vehicle to obtain the unit overhead.

step 5: Regardless of the true size of the overhead vehicle during the upcoming year, one unit of overhead cost is allocated per unit of overhead vehicle.

Once the prime cost has been determined and the indirect cost calculated based on projections, the two are combined into a *standard factory cost* or *standard cost*, which remains in effect until the next budgeting period (usually a year).

During the subsequent manufacturing year, the standard cost of a product is not generally changed merely because it is found that an error in projected indirect costs or production quantity (vehicle size) has been made. The allocation of indirect costs to a product is assumed to be independent of errors in forecasts. Rather, the difference between the expected and actual expenses, known as the *burden (overhead) variance*, experienced during the year is posted to one or more *variance accounts*.

Burden (overhead) variance is caused by errors in forecasting both the actual indirect expense for the upcoming year and the overhead vehicle size. In the former case, the variance is called *burden budget variance*; in the latter, it is called *burden capacity variance*.

Example 55.29

A company expects to produce 8000 items in the coming year. The current material cost is \$4.54 each. Sixteen minutes of direct labor are required per unit. Workers are paid \$7.50 per hour. 2133 direct labor hours are forecasted for the product. Miscellaneous overhead costs are estimated at \$45,000.

Find the per-unit (a) expected direct material cost, (b) direct labor cost, (c) prime cost, (d) burden as a function of production and direct labor, and (e) total cost.

Solution

(a) The direct material cost was given as \$4.54.

(b) The direct labor cost is

$$\left(\frac{16 \text{ min}}{60 \frac{\text{min}}{\text{hr}}}\right)\left(\frac{\$7.50}{\text{hr}}\right) = \$2.00$$

(c) The prime cost is

$$\$4.54 + \$2.00 = \$6.54$$

(d) If the burden vehicle is production, the burden rate is \$45,000/8000 = \$5.63 per item.

If the burden vehicle is direct labor hours, the burden rate is \$45,000/2133 = \$21.10 per hour.

(e) If the burden vehicle is production, the total cost is

$$\$4.54 + \$2.00 + \$5.63 = \$12.17$$

If the burden vehicle is direct labor hours, the total cost is

$$\$4.54 + \$2.00 + \left(\frac{16 \text{ min}}{60 \frac{\text{min}}{\text{hr}}}\right)\left(\frac{\$21.10}{\text{hr}}\right) = \$12.17$$

Example 55.30

The actual performance of the company in Ex. 55.29 is given by the following figures.

actual production: 7560

actual overhead costs: $47,000

What are the burden budget variance and the burden capacity variance?

Solution

The burden capacity variance is

$$\$45{,}000 - (7560)(\$5.63) = \$2437$$

The burden budget variance is

$$\$47{,}000 - \$45{,}000 = \$2000$$

The overall burden variance is

$$\$47{,}000 - (7560)(\$5.63) = \$4437$$

The sum of the burden capacity and burden budget variances should equal the overall burden variance.

$$\$2437 + \$2000 = \$4437$$

52. COST OF GOODS SOLD

Cost of goods sold (COGS) is an accounting term that represents an inventory account adjustment.[49] Cost of goods sold is the difference between the starting and ending inventory valuations. That is,

$$\text{COGS} = \text{starting inventory valuation} - \text{ending inventory valuation} \quad 55.68$$

Cost of goods sold is subtracted from *gross profit* to determine the *net profit* of a company. Despite the fact that cost of goods sold can be a significant element in the profit equation, the inventory adjustment may not be made each accounting period (e.g., each month) due to the difficulty in obtaining an accurate inventory valuation.

With a *perpetual inventory system*, a company automatically maintains up-to-date inventory records, either through an efficient stocking and stock-releasing system or through a *point of sale(POS) system* integrated with the inventory records. If a company only counts its inventory (i.e., takes a *physical inventory*) at regular intervals (e.g., once a year), it is said to be operating on a *periodic inventory system.*

Inventory accounting is a source of many difficulties. The inventory value is calculated by multiplying the quantity on hand by the standard cost. In the case of completed items actually assembled or manufactured at the company, this standard cost usually is the manufacturing cost, although factory cost also can be used. In the case of purchased items, the standard cost will be the cost per item charged by the supplying vendor. In some cases, delivery and transportation costs will be included in this standard cost.

It is not unusual for the elements in an item's inventory to come from more than one vendor, or from one vendor in more than one order. Inventory valuation is more difficult if the price paid is different for these different purchases. There are four methods of determining the cost of elements in inventory. Any of these methods can be used (if applicable), but the method must be used consistently from year to year. The four methods are as follows.

- *specific identification method:* Each element can be uniquely associated with a cost. Inventory elements with serial numbers fit into this costing scheme. Stock, production, and sales records must include the serial number.
- *average cost method:* The standard cost of an item is the average of (recent or all) purchase costs for that item.
- *first-in, first-out(FIFO) method:* This method keeps track of how many of each item were purchased each time and the number remaining out of each purchase, as well as the price paid at each purchase. The inventory system assumes that the oldest elements are issued first.[50] Inventory value is a weighted average dependent on the number of elements from each purchase remaining. Items issued no longer contribute to the inventory value.
- *last-in, first-out(LIFO) method:* This method keeps track of how many of each item were purchased each time and the number remaining out of each purchase, as well as the price paid at each purchase.[51] The inventory value is a weighted average dependent on the number of elements from each purchase remaining. Items issued no longer contribute to the inventory value.

53. BREAK-EVEN ANALYSIS

Special Nomenclature

a	incremental cost to produce one additional item (also called *marginal cost* or *differential cost*)
C	total cost
f	fixed cost that does not vary with production
p	incremental value (price)
Q	quantity sold
Q^*	break-even quantity sold
R	total revenue

[49]The cost of goods sold inventory adjustment is posted to the COGS *expense account.*

[50]If all elements in an item's inventory are identical, and if all shipments of that item are agglomerated, there will be no way to guarantee that the oldest element in inventory is issued first. But, unless *spoilage* is a problem, it really does not matter.

[51]See previous footnote.

Professional

Break-even analysis is a method of determining when the value of one alternative becomes equal to the value of another. A common application is that of determining when costs exactly equal revenue. If the manufactured quantity is less than the break-even quantity, a loss is incurred. If the manufactured quantity is greater than the break-even quantity, a profit is made. (See Fig. 55.13.)

Figure 55.13 *Break-Even Quantity*

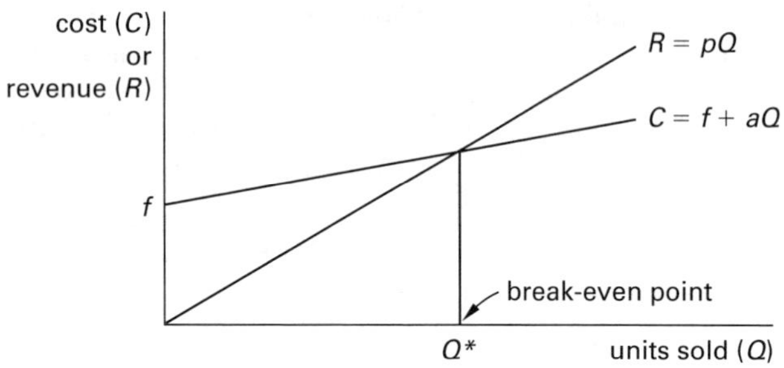

Assuming no change in the inventory, the *break-even point* can be found by setting costs equal to revenue ($C = R$).

$$C = f + aQ \tag{55.69}$$

$$R = pQ \tag{55.70}$$

$$Q^* = \frac{f}{p - a} \tag{55.71}$$

An alternative form of the break-even problem is to find the number of units per period for which two alternatives have the same total costs. Fixed costs are to be spread over a period longer than one year using the equivalent uniform annual cost (EUAC) concept. One of the alternatives will have a lower cost if production is less than the break-even point. The other will have a lower cost for production greater than the break-even point.

Example 55.31

Two plans are available for a company to obtain automobiles for its salesmen. How many miles must the cars be driven each year for the two plans to have the same costs? Use an interest rate of 10%. (Use the year-end convention for all costs.)

plan A: Lease the cars and pay \$0.15 per mile.

plan B: Purchase the cars for \$5000. Each car has an economic life of three years, after which it can be sold for \$1200. Gas and oil cost \$0.04 per mile. Insurance is \$500 per year.

Solution

Let x be the number of miles driven per year. Then, the EUAC for both alternatives is

$$\begin{aligned} \text{EUAC(A)} &= 0.15x \\ \text{EUAC(B)} &= \$0.04x + \$500 + (\$5000)(A/P, 10\%, 3) \\ &\quad - (\$1200)(A/F, 10\%, 3) \\ &= \$0.04x + \$500 + (\$5000)(0.4021) \\ &\quad - (\$1200)(0.3021) \\ &= \$0.04x + \$2148 \end{aligned}$$

Setting EUAC(A) and EUAC(B) equal and solving for x yields 19,527 miles per year as the break-even point.

54. PAY-BACK PERIOD

The *pay-back period* is defined as the length of time, usually in years, for the cumulative net annual profit to equal the initial investment. It is tempting to introduce equivalence into pay-back period calculations, but by convention, this is generally not done.[52]

$$\text{pay-back period} = \frac{\text{initial investment}}{\text{net annual profit}} \tag{55.72}$$

Example 55.32

A ski resort installs two new ski lifts at a total cost of \$1,800,000. The resort expects the annual gross revenue to increase by \$500,000 while it incurs an annual expense of \$50,000 for lift operation and maintenance. What is the pay-back period?

Solution

From Eq. 55.72,

$$\text{pay-back period} = \frac{\$1{,}800{,}000}{\$500{,}000 - \$50{,}000} = 4 \text{ years}$$

55. MANAGEMENT GOALS

Depending on many factors (market position, age of the company, age of the industry, perceived marketing and sales windows, etc.), a company may select one of many production and marketing strategic goals. Three such strategic goals are

- maximization of product demand
- minimization of cost
- maximization of profit

[52]Equivalence (i.e., interest and compounding) generally is not considered when calculating the "pay-back period." However, if it is desirable to include equivalence, then the term *pay-back period* should not be used. Other terms, such as *cost recovery period* or *life of an equivalent investment*, should be used. Unfortunately, this convention is not always followed in practice.

Professional

Such goals require knowledge of how the dependent variable (e.g., demand quantity or quantity sold) varies as a function of the independent variable (e.g., price). Unfortunately, these three goals are not usually satisfied simultaneously. For example, minimization of product cost may require a large production run to realize economies of scale, while the actual demand is too small to take advantage of such economies of scale.

If sufficient data are available to plot the independent and dependent variables, it may be possible to optimize the dependent variable graphically. (See Fig. 55.14.) Of course, if the relationship between independent and dependent variables is known algebraically, the dependent variable can be optimized by taking derivatives or by use of other numerical methods.

Figure 55.14 *Graphs of Management Goal Functions*

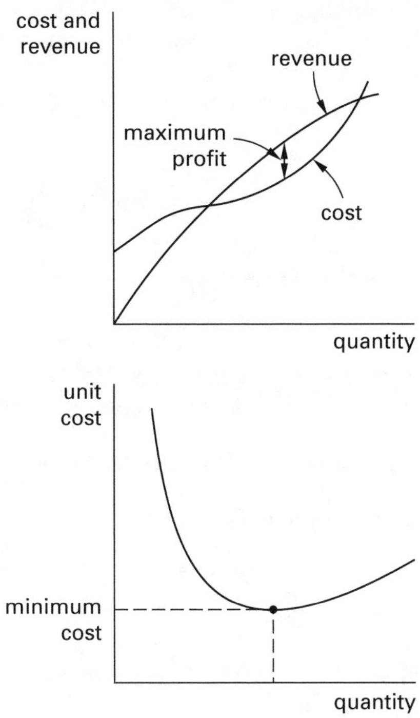

56. INFLATION

It is important to perform economic studies in terms of *constant value dollars*. One method of converting all cash flows to constant value dollars is to divide the flows by some annual *economic indicator* or price index.

If indicators are not available, cash flows can be adjusted by assuming that inflation is constant at a decimal rate (e) per year. Then, all cash flows can be converted to $t = 0$ dollars by dividing by $(1 + e)^n$, where n is the year of the cash flow.

An alternative is to replace the effective annual interest rate (i) with a value corrected for inflation. This corrected value (i') is

$$i' = i + e + ie \qquad 55.73$$

This method has the advantage of simplifying the calculations. However, precalculated factors are not available for the non-integer values of i'. Therefore, Table 55.1 must be used to calculate the factors.

Example 55.33

What is the uninflated present worth of a \$2000 future value in two years if the average inflation rate is 6% and i is 10%?

Solution

$$\begin{aligned} P &= \frac{F}{(1+i)^n(1+e)^n} \\ &= \frac{\$2000}{(1+0.10)^2(1+0.06)^2} \\ &= \$1471 \end{aligned}$$

Example 55.34

Repeat Ex. 55.33 using Eq. 55.73.

Solution

$$\begin{aligned} i' &= i + e + ie \\ &= 0.10 + 0.06 + (0.10)(0.06) = 0.166 \end{aligned}$$

$$P = \frac{F}{(1+i')^2} = \frac{\$2000}{(1+0.166)^2} = \$1471$$

57. CONSUMER LOANS

Special Nomenclature

BAL_j	balance after the jth payment
j	payment or period number
LV	principal total value loaned (cost minus down payment)
N	total number of payments to pay off the loan
PI_j	jth interest payment
PP_j	jth principal payment
PT_j	jth total payment
ϕ	effective rate per period (r/k)

Many different arrangements can be made between a borrower and a lender. With the advent of creative financing concepts, it often seems that there are as many variations of loans as there are loans made. Nevertheless, there are several traditional types of transactions. Real estate or investment texts, or a financial consultant, should be consulted for more complex problems.

Simple Interest

Interest due does not compound with a *simple interest loan*. The interest due is merely proportional to the

Professional

length of time that the principal is outstanding. Because of this, simple interest loans are seldom made for long periods (e.g., more than one year). (For loans less than one year, it is commonly assumed that a year consists of 12 months of 30 days each.)

Example 55.35

A $12,000 simple interest loan is taken out at 16% per annum interest rate. The loan matures in two years with no intermediate payments. How much will be due at the end of the second year?

Solution

The interest each year is

$$\text{PI} = (0.16)(\$12{,}000) = \$1920$$

The total amount due in two years is

$$\text{PT} = \$12{,}000 + (2)(\$1920) = \$15{,}840$$

Example 55.36

$4000 is borrowed for 75 days at 16% per annum simple interest. How much will be due at the end of 75 days?

Solution

$$\begin{aligned}\text{amount due} &= \$4000 + (0.16)\left(\frac{75 \text{ days}}{360 \frac{\text{days}}{\text{bank yr}}}\right)(\$4000)\\ &= \$4133\end{aligned}$$

Loans with Constant Amount Paid Toward Principal

With this loan type, the payment is not the same each period. The amount paid toward the principal is constant, but the interest varies from period to period. (See Fig. 55.15.) The equations that govern this type of loan are

$$\text{BAL}_j = \text{LV} - (j)(\text{PP}) \qquad 55.74$$

$$\text{PI}_j = \phi(\text{BAL})_{j-1} \qquad 55.75$$

$$\text{PT}_j = \text{PP} + \text{PI}_j \qquad 55.76$$

$$\text{PP} = \frac{\text{LV}}{N} \qquad 55.77$$

$$N = \frac{\text{LV}}{\text{PP}} \qquad 55.78$$

$$\begin{aligned}\text{LV} &= (\text{PP} + \text{PI}_1)(P/A, \phi, N)\\ &\quad - \text{PI}_N(P/G, \phi, N)\end{aligned} \qquad 55.79$$

$$\begin{aligned}1 &= \left(\frac{1}{N} + \phi\right)(P/A, \phi, N)\\ &\quad - \left(\frac{\phi}{N}\right)(P/G, \phi, N)\end{aligned} \qquad 55.80$$

Figure 55.15 Loan with Constant Amount Paid Toward Principal

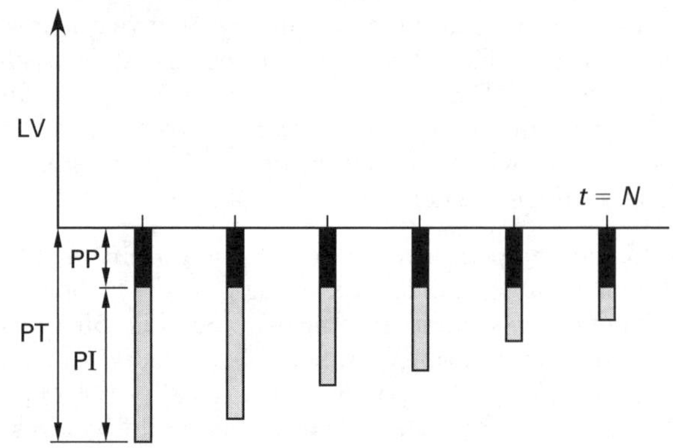

Example 55.37

A $12,000 six-year loan is taken from a bank that charges 15% effective annual interest. Payments toward the principal are uniform, and repayments are made at the end of each year. Tabulate the interest, total payments, and the balance remaining after each payment is made.

Solution

The amount of each principal payment is

$$\text{PP} = \frac{\text{LV}}{N} = \frac{\$12{,}000}{6} = \$2000$$

At the end of the first year (before the first payment is made), the principal balance is $12,000 (i.e., BAL_0 = $12,000). From Eq. 55.75, the interest payment is

$$\text{PI}_1 = \phi(\text{BAL})_0 = (0.15)(\$12{,}000) = \$1800$$

The total first payment is

$$\begin{aligned}\text{PT}_1 &= \text{PP} + \text{PI} = \$2000 + \$1800\\ &= \$3800\end{aligned}$$

The following table is similarly constructed.

	BAL_j	PP_j	PI_j	PT_j
j	(in dollars)			
0	12,000	–	–	–
1	10,000	2000	1800	3800
2	8000	2000	1500	3500
3	6000	2000	1200	3200
4	4000	2000	900	2900
5	2000	2000	600	2600
6	0	2000	300	2300

Direct Reduction Loans

This is the typical "interest paid on unpaid balance" loan. The amount of the periodic payment is constant, but the amounts paid toward the principal and interest both vary. (See Fig. 55.16.)

$$\text{BAL}_{j-1} = \text{PT}\left(\frac{1-(1+\phi)^{j-1-N}}{\phi}\right) \quad 55.81$$

$$\text{PI}_j = \phi(\text{BAL})_{j-1} \quad 55.82$$

$$\text{PP}_j = \text{PT} - \text{PI}_j \quad 55.83$$

$$\text{BAL}_j = \text{BAL}_{j-1} - \text{PP}_j \quad 55.84$$

$$N = \frac{-\ln\left(1-\frac{\phi(\text{LV})}{\text{PT}}\right)}{\ln(1+\phi)} \quad 55.85$$

Figure 55.16 *Direct Reduction Loan*

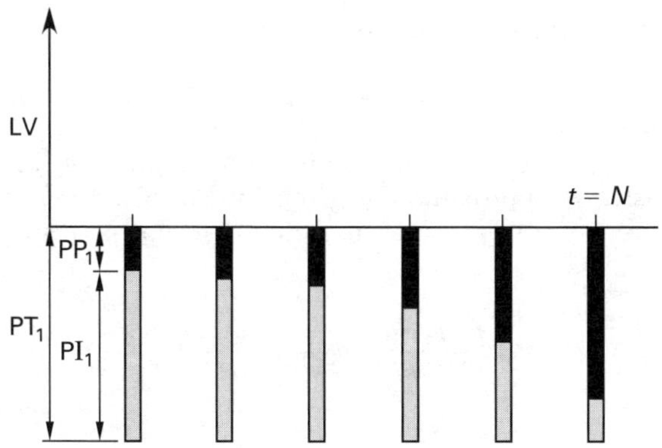

Equation 55.85 calculates the number of payments necessary to pay off a loan. This equation can be solved with effort for the total periodic payment (PT) or the initial value of the loan (LV). It is easier, however, to use the $(A/P,\ I\%,\ n)$ factor to find the payment and loan value.

$$\text{PT} = \text{LV}(A/P, \phi\%, N) \quad 55.86$$

If the loan is repaid in yearly installments, then i is the effective annual rate. If the loan is paid off monthly, then i should be replaced by the effective rate per month (ϕ from Eq. 55.54). For monthly payments, N is the number of months in the loan period.

Example 55.38

A \$45,000 loan is financed at 9.25% per annum. The monthly payment is \$385. What are the amounts paid toward interest and principal in the 14th period? What is the remaining principal balance after the 14th payment has been made?

Solution

The effective rate per month is

$$\begin{aligned}\phi &= \frac{r}{k} = \frac{0.0925}{12} \\ &= 0.0077083\ldots \quad \text{[say 0.007708]}\end{aligned}$$

$$\begin{aligned}N &= \frac{-\ln\left(1-\frac{\phi(\text{LV})}{\text{PT}}\right)}{\ln(1+\phi)} \\ &= \frac{-\ln\left(1-\frac{(0.007708)(45{,}000)}{385}\right)}{\ln(1+0.007708)} \\ &= 301\end{aligned}$$

$$\begin{aligned}\text{BAL}_{14-1} &= \text{PT}\left(\frac{1-(1+\phi)^{14-1-N}}{\phi}\right) \\ &= (\$385)\left(\frac{1-(1+0.007708)^{14-1-301}}{0.007708}\right) \\ &= \$44{,}476.39\end{aligned}$$

$$\begin{aligned}\text{PI}_{14} &= \phi(\text{BAL})_{14-1} \\ &= (0.007708)(\$44{,}476.39) \\ &= \$342.82\end{aligned}$$

$$\text{PP}_{14} = \text{PT} - \text{PI}_{14} = \$385 - \$342.82 = \$42.18$$

$$\begin{aligned}\text{BAL}_{14} &= \text{BAL}_{14-1} - \text{PP}_{14} \\ &= \$44{,}476.39 - \$42.18 \\ &= \$44{,}434.21\end{aligned}$$

Direct Reduction Loans with Balloon Payments

This type of loan has a constant periodic payment, but the duration of the loan is insufficient to completely pay back the principal (i.e, the loan is not fully amortized). Therefore, all remaining unpaid principal must be paid back in a lump sum when the loan matures. This large payment is known as a *balloon payment.*[53] (See Fig. 55.17.)

Equation 55.81 through Eq. 55.85 also can be used with this type of loan. The remaining balance after the last payment is the balloon payment. This balloon

[53]The term *balloon payment* may include the final interest payment as well. Generally, the problem statement will indicate whether the balloon payment is inclusive or exclusive of the regular payment made at the end of the loan period.

Professional

Figure 55.17 *Direct Reduction Loan with Balloon Payment*

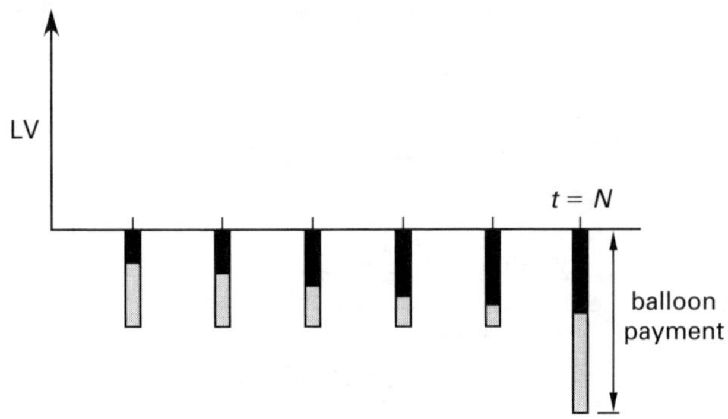

payment must be repaid along with the last regular payment calculated.

58. FORECASTING

There are many types of forecasting models, although most are variations of the basic types.[54] All models produce a *forecast* (F_{t+1}) of some quantity (*demand* is used in this section) in the next period based on actual measurements (D_j) in current and prior periods. All of the models also try to provide *smoothing* (or *damping*) of extreme data points.

Forecasts by Moving Averages

The method of *moving average forecasting* weights all previous demand data points equally and provides some smoothing of extreme data points. The amount of smoothing increases as the number of data points, n, increases.

$$F_{t+1} = \frac{1}{n}\sum_{m=t+1-n}^{t} D_m \qquad 55.87$$

Forecasts by Exponentially Weighted Averages

With *exponentially weighted forecasts*, the more current (most recent) data points receive more weight. This method uses a *weighting factor* (α), also known as a *smoothing coefficient*, which typically varies between 0.01 and 0.30. An initial forecast is needed to start the method. Forecasts immediately following are sensitive to the accuracy of this first forecast. It is common to choose $F_0 = D_1$ to get started.

$$F_{t+1} = \alpha D_t + (1-\alpha)F_t \qquad 55.88$$

[54]For example, forecasting models that take into consideration steady (linear), cyclical, annual, and seasonal trends are typically variations of the exponentially weighted model. A truly different forecasting tool, however, is *Monte Carlo simulation*, which predicts trajectories for variables based on assumed interactions.

59. LEARNING CURVES

Special Nomenclature

b	learning curve constant
n	total number of items produced
R	decimal learning curve rate (2^{-b})
T_1	time or cost for the first item
T_n	time or cost for the nth item

The more products that are made, the more efficient the operation becomes due to experience gained. Therefore, direct labor costs decrease.[55] Usually, a *learning curve* is specified by the decrease in cost each time the cumulative quantity produced doubles. If there is a 20% decrease per doubling, the curve is said to be an 80% learning curve (i.e., the *learning curve rate*, R, is 80%).

Then, the time to produce the nth item is

$$T_n = T_1 n^{-b} \qquad 55.89$$

The total time to produce units from quantity n_1 to n_2 inclusive is approximately given by Eq. 55.90. T_1 is a constant, the time for item 1, and does not correspond to n unless $n_1 = 1$.

$$\int_{n_1}^{n_2} T_n\,dn \approx \left(\frac{T_1}{1-b}\right)\left(\left(n_2+\tfrac{1}{2}\right)^{1-b} - \left(n_1-\tfrac{1}{2}\right)^{1-b}\right) \qquad 55.90$$

The *average time per unit* over the production from n_1 to n_2 is the above total time from Eq. 55.90 divided by the quantity produced, $(n_2 - n_1 + 1)$.

$$T_{\text{ave}} = \frac{\int_{n_1}^{n_2} T_n\,dn}{n_2 - n_1 + 1} \qquad 55.91$$

Table 55.9 lists representative values of the *learning curve constant* (b). For learning curve rates not listed in the table, Eq. 55.92 can be used to find b.

$$b = \frac{-\log_{10} R}{\log_{10}(2)} = \frac{-\log_{10} R}{0.301} \qquad 55.92$$

Table 55.9 *Learning Curve Constants*

learning curve rate (R)	b
0.70 (70%)	0.515
0.75 (75%)	0.415
0.80 (80%)	0.322
0.85 (85%)	0.234
0.90 (90%)	0.152
0.95 (95%)	0.074

[55]Remember that learning curve reductions apply only to direct labor costs. They are not applied to indirect labor or direct material costs.

Example 55.39

A 70% learning curve is used with an item whose first production time is 1.47 hr. (a) How long will it take to produce the 11th item? (b) How long will it take to produce the 11th through 27th items?

Solution

(a) From Eq. 55.89,

$$T_{11} = T_1 11^{-b} = (1.47 \text{ hr})(11)^{-0.515} = 0.428 \text{ hr}$$

(b) The time to produce the 11th item through 27th item is given by Eq. 55.90.

$$T \approx \left(\frac{T_{11}}{1-b}\right)\left(\left(n_{27}+\tfrac{1}{2}\right)^{1-b} - \left(n_{11}-\tfrac{1}{2}\right)^{1-b}\right)$$

$$T \approx \left(\frac{1.47 \text{ hr}}{1-0.515}\right)\left((27.5)^{1-0.515} - (10.5)^{1-0.515}\right)$$
$$= 5.643 \text{ hr}$$

60. ECONOMIC ORDER QUANTITY

Special Nomenclature

a	constant depletion rate (items/unit time)
h	inventory storage cost ($/item-unit time)
H	total inventory storage cost between orders ($)
K	fixed cost of placing an order ($)
Q	order quantity (original quantity on hand)
t^*	time at depletion

The *economic order quantity* (EOQ) is the order quantity that minimizes the inventory costs per unit time. Although there are many different EOQ models, the simplest is based on the following assumptions.

- Reordering is instantaneous. The time between order placement and receipt is zero. (See Fig. 55.18.)
- Shortages are not allowed.
- Demand for the inventory item is deterministic (i.e., is not a random variable).
- Demand is constant with respect to time.
- An order is placed when the inventory is zero.

Figure 55.18 *Inventory with Instantaneous Reorder*

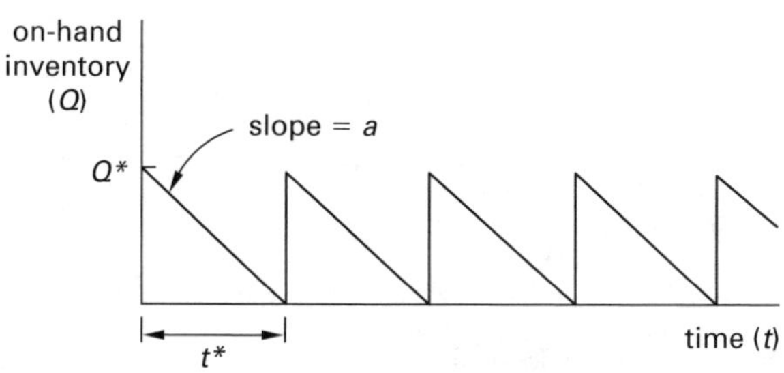

If the original quantity on hand is Q, the stock will be depleted at

$$t^* = \frac{Q}{a} \qquad 55.93$$

The total inventory storage cost between t_0 and t^* is

$$H = \tfrac{1}{2}Qht^* = \frac{Q^2h}{2a} \qquad 55.94$$

The total inventory and ordering cost per unit time is

$$C_t = \frac{aK}{Q} + \frac{hQ}{2} \qquad 55.95$$

C_t can be minimized with respect to Q. The economic order quantity and time between orders are

$$Q^* = \sqrt{\frac{2aK}{h}} \qquad 55.96$$

$$t^* = \frac{Q^*}{a} \qquad 55.97$$

61. SENSITIVITY ANALYSIS

Data analysis and forecasts in economic studies require estimates of costs that will occur in the future. There are always uncertainties about these costs. However, these uncertainties are insufficient reason not to make the best possible estimates of the costs. Nevertheless, a decision between alternatives often can be made more confidently if it is known whether or not the conclusion is sensitive to moderate changes in data forecasts. Sensitivity analysis provides this extra dimension to an economic analysis.

The sensitivity of a decision is determined by inserting a range of estimates for critical cash flows and other parameters. If radical changes can be made to a cash flow without changing the decision, the decision is said to be *insensitive* to uncertainties regarding that cash flow. However, if a small change in the estimate of a cash flow will alter the decision, that decision is said to be very *sensitive* to changes in the estimate. If the decision is sensitive only for a limited range of cash flow values, the term *variable sensitivity* is used. Figure 55.19 illustrates these terms.

An established semantic tradition distinguishes between risk analysis and uncertainty analysis. *Risk analysis* addresses variables that have a known or estimated probability distribution. In this regard, statistics and probability theory can be used to determine the probability of a cash flow varying between given limits. On the other hand, *uncertainty analysis* is concerned with situations in which there is not enough information to determine the probability or frequency distribution for the variables involved.

Figure 55.19 *Types of Sensitivity*

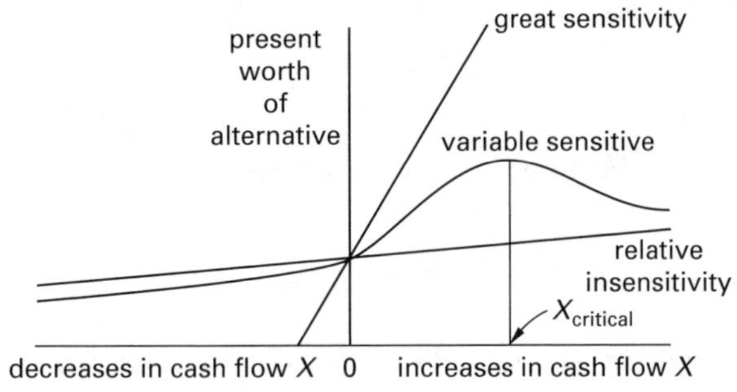

As a first step, sensitivity analysis should be applied one at a time to the dominant factors. Dominant cost factors are those that have the most significant impact on the present value of the alternative.[56] If warranted, additional investigation can be used to determine the sensitivity to several cash flows varying simultaneously. Significant judgment is needed, however, to successfully determine the proper combinations of cash flows to vary. It is common to plot the dependency of the present value on the cash flow being varied in a two-dimensional graph. Simple linear interpolation is used (within reason) to determine the critical value of the cash flow being varied.

62. VALUE ENGINEERING

The *value* of an investment is defined as the ratio of its return (performance or utility) to its cost (effort or investment). The basic object of *value engineering* (VE, also referred to as *value analysis*) is to obtain the maximum per-unit value.[57]

Value engineering concepts often are used to reduce the cost of mass-produced manufactured products. This is done by eliminating unnecessary, redundant, or superfluous features, by redesigning the product for a less expensive manufacturing method, and by including features for easier assembly without sacrificing utility and function.[58] However, the concepts are equally applicable to one-time investments, such as buildings, chemical processing plants, and space vehicles. In particular, value engineering has become an important element in all federally funded work.[59]

Typical examples of large-scale value engineering work are using stock-sized bearings and motors (instead of custom manufactured units), replacing rectangular concrete columns with round columns (which are easier to form), and substituting custom buildings with prefabricated structures.

Value engineering is usually a team effort. And, while the original designers may be on the team, usually outside consultants are utilized. The cost of value engineering is usually returned many times over through reduced construction and life-cycle costs.

[56]In particular, engineering economic analysis problems are sensitive to the choice of effective interest rate (i) and to accuracy in cash flows at or near the beginning of the horizon. The problems will be less sensitive to accuracy in far-future cash flows, such as salvage value and subsequent generation replacement costs.

[57]Value analysis, the methodology that has become today's value engineering, was developed in the early 1950s by Lawrence D. Miles, an analyst at General Electric.

[58]Some people say that value engineering is "the act of going over the plans and taking out everything that is interesting."

[59]U.S. Government Office of Management and Budget Circular A-131 outlines value engineering for federally funded construction projects.

56 Engineering Law

1. Forms of Company Ownership 56-1
2. Sole Proprietorships 56-1
3. Partnerships 56-1
4. Corporations 56-2
5. Limited Liability Entities 56-2
6. Piercing the Corporate Veil 56-3
7. Agency 56-3
8. General Contracts 56-3
9. Standard Boilerplate Clauses 56-4
10. Subcontracts 56-4
11. Parties to a Construction Contract 56-4
12. Standard Contracts for Design Professionals 56-5
13. Consulting Fee Structure 56-5
14. Mechanic's Liens 56-5
15. Discharge of a Contract 56-6
16. Torts 56-6
17. Breach of Contract, Negligence, Misrepresentation, and Fraud 56-6
18. Strict Liability in Tort 56-7
19. Manufacturing and Design Liability 56-7
20. Damages 56-7
21. Insurance 56-8

1. FORMS OF COMPANY OWNERSHIP[1]

There are three basic forms of company ownership in the United States: (a) sole proprietorship, (b) partnership, and (c) corporation.[2] Each of these forms of ownership has advantages and disadvantages.

2. SOLE PROPRIETORSHIPS

A *sole proprietorship (single proprietorship)* is the easiest form of ownership to establish. Other than the necessary licenses, filings, and notices (which apply to all forms of ownership), no legal formalities are required to start business operations. A sole proprietor (the owner) has virtually total control of the business and makes all supervisory and management decisions.

Legally, there is no distinction between the sole proprietor and the sole proprietorship (the business). This is the greatest disadvantage of this form of business. The owner is solely responsible for the operation of the business, even if the owner hires others for assistance. The owner assumes personal, legal, and financial liability for all acts and debts of the company. If the company debts remain unpaid, or in the event there is a legal judgment against the company, the owner's personal assets (home, car, savings, etc.) can be seized or attached.

Another disadvantage of the sole proprietorship is the lack of significant organizational structure. In times of business crisis or trouble, there may be no one to share the responsibility or to help make decisions. When the owner is sick or dies, there may be no way to continue the business.

There is also no distinction between the incomes of the business and the owner. Therefore, the business income is taxed at the owner's income tax rate. Depending on the owner's financial position, the success of the business, and the tax structure, this can be an advantage or a disadvantage.[3]

3. PARTNERSHIPS

A *partnership* (also known as a *general partnership*) is ownership by two or more persons known as *general partners*. Legally, this form is very similar to a sole proprietorship, and the two forms of business have many of the same advantages and disadvantages. For example, with the exception of an optional *partnership agreement*, there are a minimum of formalities to setting up business. The partners make all business and management decisions themselves according to an agreed-upon process. The business income is split among the partners and taxed at the partners' individual tax rates.[4] Continuity of the business is still a problem since most partnerships are automatically dissolved upon the withdrawal or death of one of the partners.[5]

One advantage of a partnership over a sole proprietorship is the increase in available funding. Not only do more partners bring in more start-up capital, but the

[1]This chapter is not intended to be a substitute for professional advice. Law is not always black and white. For every rule there are exceptions. For every legal principle, there are variations. For every type of injury, there are numerous legal precedents. This chapter covers the superficial basics of a small subset of U.S. law affecting engineers.

[2]The discussion of forms of company ownership in Sec. 56.2, Sec. 56.3, and Sec. 56.4 applies equally to service-oriented companies (e.g., consulting engineering firms) and product-oriented companies.

[3]To use a simplistic example, if the corporate tax rates are higher than the individual tax rates, it would be *financially* better to be a sole proprietor because the company income would be taxed at a lower rate.

[4]The percentage split is specified in the partnership agreement.

[5]Some or all of the remaining partners may want to form a new partnership, but this is not always possible.

resource pool may make business credit easier to obtain. Also, the partners bring a diversity of skills and talents.

Unless the partnership agreement states otherwise, each partner can individually obligate (i.e., *bind*) the partnership without the consent of the other partners. Similarly, each partner has personal responsibility and liability for the acts and debts of the partnership company, just as sole proprietors do. In fact, each partner assumes the *sole* responsibility, not just a proportionate share. If one or more partners are unable to pay, the remaining partners shoulder the entire debt. The possibility of one partner having to pay for the actions of another partner must be considered when choosing this form of business ownership.

A *limited partnership* differs from a general partnership in that one or more of the partners are silent. The *limited partners* make a financial contribution to the business and receive a share of the profit but do not participate in the management and cannot bind the partnership. While *general partners* have unlimited personal liabilities, limited partners are generally liable only to the extent of their investment.[6] A written partnership agreement is required, and the agreement must be filed with the proper authorities.

4. CORPORATIONS

A corporation is a legal entity (i.e., a legal person) distinct from the founders and owners. The separation of ownership and management makes the corporation a fundamentally different business form than a sole proprietorship or partnership, with very different advantages and disadvantages.

A corporation becomes legally distinct from its founders upon formation and proper registration. Ownership of the corporation is through shares of stock, distributed to the founders and investors according to some agreed-upon investment and distribution rule. The founders and investors become the stockholders (i.e., owners) of the corporation. A *closely held* (*private*) *corporation* is one in which all stock is owned by a family or small group of co-investors. A *public corporation* is one whose stock is available for the public-at-large to purchase.

There is no mandatory connection between ownership and management functions. The decision-making power is vested in the executive officers and a *board of directors* that governs by majority vote. The stockholders elect the board of directors which, in turn, hires the executive officers, management, and other employees. Employees of the corporation may or may not be stockholders.

Disadvantages (at least for a person or persons who could form a partnership or sole proprietorship) include the higher corporate tax rate, difficulty and complexity of formation (some states require a minimum number of persons on the board of directors), and additional legal and accounting paperwork.

However, since a corporation is distinctly separate from its founders and investors, those individuals are not liable for the acts and debts of the corporation. Debts are paid from the corporate assets. Income to the corporation is not taxable income to the owners. (Only the salaries, if any, paid to the employees by the corporation are taxable to the employees.) Even if the corporation were to go bankrupt, the assets of the owners would not ordinarily be subject to seizure or attachment.

A corporation offers the best guarantee of continuity of operation in the event of the death, incapacity, or retirement of the founders since, as a legal entity, it is distinct from the founders and owners.

5. LIMITED LIABILITY ENTITIES

A variety of other legal entities have been established that blur the lines between the three traditional forms of business (i.e., proprietorship, partnership, and corporation). The *limited liability partnership*, LLP, extends a measure of corporate-like protection to professionals while permitting partnership-like personal participation in management decisions. Since LLPs are formed and operated under state laws, actual details vary from state to state. However, most LLPs allow the members to participate in management decisions, while not being responsible for the misdeeds of other partners. As in a corporation, the debts of the LLP do not become debts of the members.[7] The *double taxation* characteristic of traditional corporations is avoided, as profits to the LLP flow through to the members.

For engineers and architects (as well as doctors, lawyers, and accountants), the *professional corporation*, PC, offers protection from the actions (e.g., malpractice) of other professionals within a shared environment, such as a design firm. While a PC does not shield the individual from responsibility for personal negligence or malpractice, it does permit the professional to be associated with a larger entity, such as a partnership of other PCs, without accepting responsibility for the actions of the other members. In that sense, the protection is similar to that of an LLP. Unlike a traditional corporation, a PC may have a board of directors consisting of only a single individual, the professional.

The *limited liability company*[8], LLC, also combines advantages from partnerships and corporations. In an LLC, the members are shielded from debts of the LLC

[6]That is, if the partnership fails or is liquidated to pay debts, the limited partners lose no more than their initial investments.

[7]Depending on the state, the shield may be complete or limited. It is common that the protection only applies to negligence-related claims, as opposed to intentional tort claims, contract-related obligations, and day-to-day operating expenses such as rent, utilities, and employees.

[8]LLC does not mean *limited liability corporation*. LLCs are not corporations.

while enjoying the pass-through of all profits.[9] Like a partnership or shareholder, a member's obligation is limited to the *membership interest* in (i.e., contribution to) the LLC. LLCs are directed and controlled by one or more managers who may also be members. A variation of the LLC specifically for design, medical, and other professionals is the *professional limited liability company*, PLLC.

The traditional corporation, as described in Sec. 56.4, is referred to as a *Subchapter C corporation* or "*C corp*."[10] A variant is the *S corporation* ("*S corp*") which combines characteristics of the C corporation with pass-through for taxation. S corporations can be limited or treated differently than C corporations by state and federal law.

6. PIERCING THE CORPORATE VEIL

An individual operating as a corporation, LLP, LLC, or PC entity may lose all protection if his or her actions are fraudulent, or if the court decides the business is an "alter ego" of the individual. Basically, this requires the business to be run as a business. Business and personal assets cannot be intermingled, and business decisions must be made and documented in a business-like manner. If operated fraudulently or loosely, a court may assign liability directly to an individual, an action known as *piercing the corporate veil*.

7. AGENCY

In some contracts, decision-making authority and right of action are transferred from one party (the owner, or *principal*) who would normally have that authority to another person (the *agent*). For example, in construction contracts, the engineer may be the agent of the owner for certain transactions. Agents are limited in what they can do by the scope of the agency agreement. Within that scope, however, an agent acts on behalf of the principal, and the principal is liable for the acts of the agent and is bound by contracts made in the principal's name by the agent.

Agents are required to execute their work with care, skill, and diligence. Specifically, agents have *fiduciary responsibility* toward their principal, meaning that agent must be honest and loyal. Agents are liable for damages resulting from a lack of diligence, loyalty, and/or honesty. If the agents misrepresented their skills when obtaining the agency, they can be liable for breach of contract or fraud.

[9]LLCs enjoy *check the box taxation*, which means they can elect to be taxed as sole proprietorships, partnerships, or corporations.

[10]The reference is to subchapter C of the Internal Revenue Code.

8. GENERAL CONTRACTS

A *contract* is a legally binding agreement or promise to exchange goods or services.[11] A written contract is merely a documentation of the agreement. Some agreements must be in writing, but most agreements for engineering services can be verbal, particularly if the parties to the agreement know each other well.[12] Written contract documents do not need to contain intimidating legal language, but all agreements must satisfy three basic requirements to be enforceable (binding).

- There must be a clear, specific, and definite *offer* with no room for ambiguity or misunderstanding.
- There must be some form of conditional future *consideration* (i.e., payment).[13]
- There must be an *acceptance* of the offer.

There are other conditions that the agreement must meet to be enforceable. These conditions are not normally part of the explicit agreement but represent the conditions under which the agreement was made.

- The agreement must be *voluntary* for all parties.
- All parties must have *legal capacity* (i.e., be mentally competent, of legal age, and uninfluenced by drugs).
- The purpose of the agreement must be *legal*.

For small projects, a simple *letter of agreement* on one party's stationery may suffice. For larger, complex projects, a more formal document may be required. Some clients prefer to use a *purchase order*, which can function as a contract if all basic requirements are met.

Regardless of the format of the written document—letter of agreement, purchase order, or standard form—a contract should include the following features.[14]

- introduction, preamble, or preface indicating the purpose of the contract
- name, address, and business forms of both contracting parties
- signature date of the agreement
- effective date of the agreement (if different from the signature date)

[11]Not all agreements are legally binding (i.e., enforceable). Two parties may agree on something, but unless the agreement meets all of the requirements and conditions of a contract, the parties cannot hold each other to the agreement.

[12]All states have a *statute of frauds* that, among other things, specifies what types of contracts must be in writing to be enforceable. These include contracts for the sale of land, contracts requiring more than one year for performance, contracts for the sale of goods over $500 in value, contracts to satisfy the debts of another, and marriage contracts. Contracts to provide engineering services do not fall under the statute of frauds.

[13]Actions taken or payments made prior to the agreement are irrelevant. Also, it does not matter to the courts whether the exchange is based on equal value or not.

[14]*Construction contracts* are unique unto themselves. Items that might also be included as part of the *contract documents* are the agreement form, the general conditions, drawings, specifications, and addenda.

- duties and obligations of both parties
- deadlines and required service dates
- fee amount
- fee schedule and payment terms
- agreement expiration date
- standard boilerplate clauses
- signatures of parties or their agents
- declaration of authority of the signatories to bind the contracting parties
- supporting documents

9. STANDARD BOILERPLATE CLAUSES

It is common for full-length contract documents to include important *boilerplate clauses*. These clauses have specific wordings that should not normally be changed, hence the name "boilerplate." Some of the most common boilerplate clauses are paraphrased here.

- Delays and inadequate performance due to war, strikes, and acts of God and nature are forgiven (*force majeure*).
- The contract document is the complete agreement, superseding all prior verbal and written agreements.
- The contract can be modified or canceled only in writing.
- Parts of the contract that are determined to be void or unenforceable will not affect the enforceability of the remainder of the contract (*severability*). Alternatively, parts of the contract that are determined to be void or unenforceable will be rewritten to accomplish their intended purpose without affecting the remainder of the contract.
- None (or one, or both) of the parties can (or cannot) assign its (or their) rights and responsibilities under the contract (*assignment*).
- All notices provided for in the agreement must be in writing and sent to the address in the agreement.
- Time is of the essence.[15]
- The subject headings of the agreement paragraphs are for convenience only and do not control the meaning of the paragraphs.
- The laws of the state in which the contract is signed must be used to interpret and govern the contract.
- Disagreements shall be arbitrated according to the rules of the American Arbitration Association.
- Any lawsuits related to the contract must be filed in the county and state in which the contract is signed.
- Obligations under the agreement are unique, and in the event of a breach, the defaulting party waives the defense that the loss can be adequately compensated by monetary damages (*specific performance*).
- In the event of a lawsuit, the prevailing party is entitled to an award of reasonable attorneys' and court fees.[16]
- Consequential damages are not recoverable in a lawsuit.

10. SUBCONTRACTS

When a party to a contract engages a third party to perform the work in the original contract, the contract with the third party is known as a *subcontract*. Whether or not responsibilities can be subcontracted under the original contract depends on the content of the *assignment clause* in the original contract.

11. PARTIES TO A CONSTRUCTION CONTRACT

A specific set of terms has developed for referring to parties in consulting and construction contracts. The *owner* of a construction project is the person, partnership, or corporation that actually owns the land, assumes the financial risk, and ends up with the completed project. The *developer* contracts with the architect and/or engineer for the design and with the contractors for the construction of the project. In some cases, the owner and developer are the same, in which case the term *owner-developer* can be used.

The *architect* designs the project according to established codes and guidelines but leaves most stress and capacity calculations to the *engineer*.[17] Depending on the construction contract, the engineer may work for the architect, or vice versa, or both may work for the developer.

Once there are approved plans, the developer hires *contractors* to do the construction. Usually, the entire construction project is awarded to a *general contractor*. Due to the nature of the construction industry, separate *subcontracts* are used for different tasks (electrical, plumbing, mechanical, framing, fire sprinkler installation, finishing, etc.). The general contractor who hires all of these different *subcontractors* is known as the *prime contractor* (or *prime*). (The subcontractors can also work directly for the owner-developer, although this is less common.) The prime contractor is responsible for all acts of the subcontractors and is liable for any damage suffered by the owner-developer due to those acts.

[15]Without this clause in writing, damages for delay cannot be claimed.

[16]Without this clause in writing, attorneys' fees and court costs are rarely recoverable.

[17]On simple small projects, such as wood-framed residential units, the design may be developed by a *building designer*. The legal capacities of building designers vary from state to state.

Construction is managed by an agent of the owner-developer known as the *construction manager*, who may be the engineer, the architect, or someone else.

12. STANDARD CONTRACTS FOR DESIGN PROFESSIONALS

Several professional organizations have produced standard agreement forms and other standard documents for design professionals.[18] Among other standard forms, notices, and agreements, the following standard contracts are available.[19]

- standard contract between engineer and client
- standard contract between engineer and architect
- standard contract between engineer and contractor
- standard contract between owner and construction manager

Besides completeness, the major advantage of a standard contract is that the meanings of the clauses are well established, not only among the design professionals and their clients but also in the courts. The clauses in these contracts have already been litigated many times. Where a clause has been found to be unclear or ambiguous, it has been rewritten to accomplish its intended purpose.

13. CONSULTING FEE STRUCTURE

Compensation for consulting engineering services can incorporate one or more of the following concepts.

- *lump-sum fee:* This is a predetermined fee agreed upon by client and engineer. This payment can be used for small projects where the scope of work is clearly defined.
- *cost plus fixed fee:* All costs (labor, material, travel, etc.) incurred by the engineer are paid by the client. The client also pays a predetermined fee as profit. This method has an advantage when the scope of services cannot be determined accurately in advance. Detailed records must be kept by the engineer in order to allocate costs among different clients.
- *per diem fee:* The engineer is paid a specific sum for each day spent on the job. Usually, certain direct expenses (e.g., travel and reproduction) are billed in addition to the per diem rate.
- *salary plus:* The client pays for the employees on an engineer's payroll (the salary) plus an additional percentage to cover indirect overhead and profit plus certain direct expenses.
- *retainer:* This is a minimum amount paid by the client, usually in total and in advance, for a normal amount of work expected during an agreed-upon period. None of the retainer is returned, regardless of how little work the engineer performs. The engineer can be paid for additional work beyond what is normal, however. Some direct costs, such as travel and reproduction expenses, may be billed directly to the client.
- *percentage of construction cost:* This method, which is widely used in construction design contracts, pays the architect and/or the engineer a percentage of the final total cost of the project. Costs of land, financing, and legal fees are generally not included in the construction cost, and other costs (plan revisions, project management labor, value engineering, etc.) are billed separately.

14. MECHANIC'S LIENS

For various reasons, providers and material, labor, and design services to construction sites may not be promptly paid or even paid at all. Such providers have, of course, the right to file a lawsuit demanding payment, but due to the nature of the construction industry, such relief may be insufficient or untimely. Therefore, such providers have the right to file a *mechanic's lien* (also known as a *construction lien, materialman's lien, supplier's lien,* or *laborer's lien*) against the property. Although there are strict requirements for deadlines, filing, and notices, the procedure for obtaining (and removing) such a lien is simple. The lien establishes the supplier's security interest in the property. Although the details depend on the state, essentially the property owner is prevented from transferring title (i.e., selling) the property until the lien has been removed by the supplier. The act of filing a lawsuit to obtain payment is known as "perfecting the lien." Liens are perfected by forcing a judicial foreclosure sale. The court orders the property sold, and the proceeds are used to pay off any lien-holders.

[18]There are two main sources of standardized construction and design agreements: EJCDC and AIA. Consensus documents, known as *ConsensusDOCS*, for every conceivable situation have been developed by the *Engineers Joint Contracts Documents Committee*, EJCDC. EJCDC includes the American Society of Civil Engineers (ASCE), the American Council of Engineering Companies (ACEC), National Society of Professional Engineers' (NSPE's) Professional Engineers in Private Practice Division, Associated General Contractors of America (AGC), and more than fifteen other participating professional engineering design, construction, owner, legal, and risk management organizations, including the Associated Builders and Contractors; American Subcontractors Association; Construction Users Roundtable; National Roofing Contractors Association; Mechanical Contractors Association of America; and National Plumbing, Heating-Cooling Contractors Association. The American Institute of Architects, AIA, has developed its own standardized agreements in a less-collaborative manner. Though popular with architects, AIA provisions are considered less favorable to engineers, contractors, and subcontractors who believe the AIA documents assign too much authority to architects, too much risk and liability to contractors, and too little flexibility in how construction disputes are addressed and resolved.

[19]The Construction Specifications Institute (CSI) has produced standard specifications for materials. The standards have been organized according to a UNIFORMAT structure consistent with ASTM Standard E1557.

15. DISCHARGE OF A CONTRACT

A contract is normally discharged when all parties have satisfied their obligations. However, a contract can also be terminated for the following reasons:

- mutual agreement of all parties to the contract
- impossibility of performance (e.g., death of a party to the contract)
- illegality of the contract
- material breach by one or more parties to the contract
- fraud on the part of one or more parties
- failure (i.e., loss or destruction) of consideration (e.g., the burning of a building one party expected to own or occupy upon satisfaction of the obligations)

Some contracts may be dissolved by actions of the court (e.g., bankruptcy), passage of new laws and public acts, or a declaration of war.

Extreme difficulty (including economic hardship) in satisfying the contract does not discharge it, even if it becomes more costly or less profitable than originally anticipated.

16. TORTS

A *tort* is a civil wrong committed by one person causing damage to another person or person's property, emotional well-being, or reputation.[20] It is a breach of the rights of an individual to be secure in person or property. In order to correct the wrong, a civil lawsuit (*tort action* or *civil complaint*) is brought by the alleged injured party (the *plaintiff*) against the *defendant*. To be a valid *tort action* (i.e., lawsuit), there must have been injury (i.e., damage). Generally, there will be no contract between the two parties, so the tort action cannot claim a breach of contract.[21]

Tort law is concerned with compensation for the injury, not punishment. Therefore, tort awards usually consist of general, compensatory, and special damages and rarely include punitive and exemplary damages. (See Sec. 56.20 for definitions of these damages.)

[20]The difference between a *civil tort (lawsuit)* and a *criminal lawsuit* is the alleged injured party. A *crime* is a wrong against society. A criminal lawsuit is brought by the state against a defendant.

[21]It is possible for an injury to be both a breach of contract and a tort. Suppose an owner has an agreement with a contractor to construct a building, and the contract requires the contractor to comply with all state and federal safety regulations. If the owner is subsequently injured on a stairway because there was no guardrail, the injury could be recoverable both as a tort and as a breach of contract. If a third party unrelated to the contract was injured, however, that party could recover only through a tort action.

17. BREACH OF CONTRACT, NEGLIGENCE, MISREPRESENTATION, AND FRAUD

A *breach of contract* occurs when one of the parties fails to satisfy all of its obligations under a contract. The breach can be *willful* (as in a contractor walking off a construction job) or *unintentional* (as in providing less than adequate quality work or materials). A *material breach* is defined as nonperformance that results in the injured party receiving something substantially less than or different from what the contract intended.

Normally, the only redress that an *injured party* has through the courts in the event of a breach of contract is to force the breaching party to provide *specific performance*—that is, to satisfy all remaining contract provisions and to pay for any damage caused. Normally, *punitive damages* (to punish the breaching party) are unavailable.

Negligence is an action, willful or unwillful, taken without proper care or consideration for safety, resulting in damages to property or injury to persons. "Proper care" is a subjective term, but in general it is the diligence that would be exercised by a reasonably prudent person.[22] Damages sustained by a negligent act are recoverable in a tort action. (See Sec. 56.16.) If the plaintiff is partially at fault (as in the case of *comparative negligence*), the defendant will be liable only for the portion of the damage caused by the defendant.

Punitive damages are available, however, if the breaching party was fraudulent in obtaining the contract. In addition, the injured party has the right to void (nullify) the contract entirely. A *fraudulent act* is basically a special case of misrepresentation (i.e., an intentionally false statement known to be false at the time it is made). Misrepresentation that does not result in a contract is a tort. When a contract is involved, misrepresentation can be a breach of that contract (i.e., *fraud*).

Unfortunately, it is extremely difficult to prove *compensatory fraud* (i.e., fraud for which damages are available). Proving fraud requires showing *beyond a reasonable doubt* (a) a reckless or intentional misstatement of a material fact, (b) an intention to deceive, (c) it resulted in misleading the innocent party to contract, and (d) it was to the innocent party's detriment.

For example, if an engineer claims to have experience in designing steel buildings but actually has none, the court might consider the misrepresentation a fraudulent action. If, however, the engineer has some experience, but an insufficient amount to do an adequate job, the engineer probably will not be considered to have acted fraudulently.

[22]Negligence of a design professional (e.g., an engineer or architect) is the absence of a *standard of care* (i.e., customary and normal care and attention) that would have been provided by other engineers. It is highly subjective.

18. STRICT LIABILITY IN TORT

Strict liability in tort means that the injured party wins if the injury can be proven. It is not necessary to prove negligence, breach of explicit or implicit warranty, or the existence of a contract (*privity of contract*). Strict liability in tort is most commonly encountered in product liability cases. A defect in a product, regardless of how the defect got there, is sufficient to create strict liability in tort.

Case law surrounding defective products has developed and refined the following requirements for winning a strict liability in tort case. The following points must be proved:

- The product was defective in manufacture, design, labeling, and so on.
- The product was defective when used.
- The defect rendered the product unreasonably dangerous.
- The defect caused the injury.
- The specific use of the product that caused the damage was reasonably foreseeable.

19. MANUFACTURING AND DESIGN LIABILITY

Case law makes a distinction between *design professionals* (architects, structural engineers, building designers, etc.) and manufacturers of consumer products. Design professionals are generally consultants whose primary product is a design service sold to sophisticated clients. Consumer product manufacturers produce specific product lines sold through wholesalers and retailers to the unsophisticated public.

The law treats design professionals favorably. Such professionals are expected to meet a *standard of care* and skill that can be measured by comparison with the conduct of other professionals. However, professionals are not expected to be infallible. In the absence of a contract provision to the contrary, design professionals are not held to be guarantors of their work in the strict sense of legal liability. Damages incurred due to design errors are recoverable through tort actions, but proving a breach of contract requires showing negligence (i.e., not meeting the standard of care).

On the other hand, the law is much stricter with consumer product manufacturers, and perfection is (essentially) expected of them. They are held to the standard of strict liability in tort without regard to negligence. A manufacturer is held liable for all phases of the design and manufacturing of a product being marketed to the public.[23]

Prior to 1916, the court's position toward product defects was exemplified by the expression *caveat emptor* ("let the buyer beware").[24] Subsequent court rulings have clarified that "... a manufacturer is strictly liable in tort when an article [it] places on the market, knowing that it will be used without inspection, proves to have a defect that causes injury to a human being."[25]

Although all defectively designed products can be traced back to a design engineer or team, only the manufacturing company is usually held liable for injury caused by the product. This is more a matter of economics than justice. The company has liability insurance; the product design engineer (who is merely an employee of the company) probably does not. Unless the product design or manufacturing process is intentionally defective, or unless the defect is known in advance and covered up, the product design engineer will rarely be punished by the courts.[26]

20. DAMAGES

An injured party can sue for *damages* as well as for specific performance. Damages are the award made by the court for losses incurred by the injured party.

- *General* or *compensatory damages* are awarded to make up for the injury that was sustained.
- *Special damages* are awarded for the direct financial loss due to the breach of contract.
- *Nominal damages* are awarded when responsibility has been established but the injury is so slight as to be inconsequential.
- *Liquidated damages* are amounts that are specified in the contract document itself for nonperformance.
- *Punitive* or *exemplary damages* are awarded, usually in tort and fraud cases, to punish and make an example of the defendant (i.e., to deter others from doing the same thing).
- *Consequential damages* provide compensation for indirect losses incurred by the injured party but not directly related to the contract.

[23]The reason for this is that the public is not considered to be as sophisticated as a client who contracts with a design professional for building plans.

[24]1916, *McPherson vs. Buick*. McPherson bought a Buick from a car dealer. The car had a defective wooden steering wheel, and there was evidence that reasonable inspection would have uncovered the defect. The steering wheel injured McPherson, who then sued Buick. Buick defended itself under the ancient *prerequisite of privity* (i.e., the requirement of a face-to-face contractual relationship in order for liability to exist), since the dealer, not Buick, had sold the car to McPherson, and no contract between Buick and McPherson existed. The judge disagreed, thus establishing the concept of *third party liability* (i.e., manufacturers are responsible to consumers even though consumers do not buy directly from manufacturers).

[25]1963, *Greenman vs. Yuba Power Products.* Greenman purchased and was injured by an electric power tool.

[26]The engineer can expect to be discharged from the company. However, for strategic reasons, this discharge probably will not occur until after the company loses the case.

Professional

21. INSURANCE

Most design firms and many independent design professionals carry *errors and omissions insurance* to protect them from claims due to their mistakes. Such policies are costly, and for that reason, some professionals choose to "go bare."[27] Policies protect against inadvertent mistakes only, not against willful, knowing, or conscious efforts to defraud or deceive.

[27]Going bare appears foolish at first glance, but there is a perverted logic behind the strategy. One-person consulting firms (and perhaps, firms that are not profitable) are "judgment-proof." Without insurance or other assets, these firms would be unable to pay any large judgments against them. When damage victims (and their lawyers) find this out in advance, they know that judgments will be uncollectable. So, often the lawsuit never makes its way to trial.

57 Engineering Ethics

1. Creeds, Codes, Canons, Statutes, and Rules . . . 57-1
2. Purpose of a Code of Ethics . . . 57-1
3. Ethical Priorities . . . 57-2
4. Dealing with Clients and Employers . . . 57-2
5. Dealing with Suppliers . . . 57-2
6. Dealing with Other Engineers . . . 57-3
7. Dealing with (and Affecting) the Public . . . 57-3
8. Competitive Bidding . . . 57-3

1. CREEDS, CODES, CANONS, STATUTES, AND RULES

It is generally conceded that an individual acting on his or her own cannot be counted on to always act in a proper and moral manner. Creeds, statutes, rules, and codes all attempt to complete the guidance needed for an engineer to do "...the correct thing."

A *creed* is a statement or oath, often religious in nature, taken or assented to by an individual in ceremonies. For example, the *Engineers' Creed* adopted by the National Society of Professional Engineers in 1954 is[1]

> As a professional engineer, I dedicate my professional knowledge and skill to the advancement and betterment of human welfare.
>
> I pledge...
>
> ...to give the utmost of performance;
>
> ...to participate in none but honest enterprise;
>
> ...to live and work according to the laws of man and the highest standards of professional conduct;
>
> ...to place service before profit, the honor and standing of the profession before personal advantage, and the public welfare above all other considerations.
>
> In humility and with need for Divine Guidance, I make this pledge.

A *code* is a system of nonstatutory, nonmandatory canons of personal conduct. A *canon* is a fundamental belief that usually encompasses several rules. For example, the code of ethics of the American Society of Civil Engineers (ASCE) contains seven canons.

[1]The *Faith of an Engineer* adopted by the Accreditation Board for Engineering and Technology (ABET), formerly the Engineer's Council for Professional Development (ECPD), is a similar but more detailed creed.

1. Engineers shall hold paramount the safety, health, and welfare of the public and shall strive to comply with the principles of sustainable development in the performance of their professional duties.
2. Engineers shall perform services only in areas of their competence.
3. Engineers shall issue public statements only in an objective and truthful manner.
4. Engineers shall act in professional matters for each employer or client as faithful agents or trustees and shall avoid conflicts of interest.
5. Engineers shall build their professional reputation on the merit of their service and shall not compete unfairly with others.
6. Engineers shall act in such a manner as to uphold and enhance the honor, integrity, and dignity of the engineering profession, and shall act with zero tolerance for bribery, fraud, and corruption.
7. Engineers shall continue their professional development throughout their careers and shall provide opportunities for the professional development of those engineers under their supervision.

A *rule* is a guide (principle, standard, or norm) for conduct and action in a certain situation. A *statutory rule* is enacted by the legislative branch of a state or federal government and carries the weight of law. Some U.S. engineering registration boards have statutory *rules of professional conduct.*

2. PURPOSE OF A CODE OF ETHICS

Many different sets of *codes of ethics* (*canons of ethics, rules of professional conduct*, etc.) have been produced by various engineering societies, registration boards, and other organizations.[2] The purpose of these ethical guidelines is to guide the conduct and decision making of engineers. Most codes are primarily educational. Nevertheless, from time to time they have been used

[2]All of the major engineering technical and professional societies in the United States (ASCE, IEEE, ASME, AIChE, NSPE, etc.) and throughout the world have adopted codes of ethics. Most U.S. societies have endorsed the *Code of Ethics of Engineers* developed by the Accreditation Board for Engineering and Technology (ABET). The National Council of Examiners for Engineering and Surveying (NCEES) has developed its *Model Rules of Professional Conduct* as a guide for state registration boards in developing guidelines for the professional engineers in those states.

by the societies and regulatory agencies as the basis for disciplinary actions.

Fundamental to ethical codes is the requirement that engineers render faithful, honest, professional service. In providing such service, engineers must represent the interests of their employers or clients and, at the same time, protect public health, safety, and welfare.

There is an important distinction between what is legal and what is ethical. Many legal actions can be violations of codes of ethical or professional behavior.[3] For example, an engineer's contract with a client may give the engineer the right to assign the engineer's responsibilities, but doing so without informing the client would be unethical.

Ethical guidelines can be categorized on the basis of who is affected by the engineer's actions—the client, vendors and suppliers, other engineers, or the public at large.[4]

3. ETHICAL PRIORITIES

There are frequently conflicting demands on engineers. While it is impossible to use a single decision-making process to solve every ethical dilemma, it is clear that ethical considerations will force engineers to subjugate their own self-interests. Specifically, the ethics of engineers dealing with others need to be considered in the following order, from highest to lowest priority.

- society and the public
- the law
- the engineering profession
- the engineer's client
- the engineer's firm
- other involved engineers
- the engineer personally

4. DEALING WITH CLIENTS AND EMPLOYERS

The most common ethical guidelines affecting engineers' interactions with their employer (the *client*) can be summarized as follows.[5]

- Engineers should not accept assignments for which they do not have the skill, knowledge, or time.
- Engineers must recognize their own limitations. They should use associates and other experts when the design requirements exceed their abilities.
- The client's interests must be protected. The extent of this protection exceeds normal business relationships and transcends the legal requirements of the engineer-client contract.
- Engineers must not be bound by what the client wants in instances where such desires would be unsuccessful, dishonest, unethical, unhealthy, or unsafe.
- Confidential client information remains the property of the client and must be kept confidential.
- Engineers must avoid conflicts of interest and should inform the client of any business connections or interests that might influence their judgment. Engineers should also avoid the *appearance* of a conflict of interest when such an appearance would be detrimental to the profession, their client, or themselves.
- The engineers' sole source of income for a particular project should be the fee paid by their client. Engineers should not accept compensation in any form from more than one party for the same services.
- If the client rejects the engineer's recommendations, the engineer should fully explain the consequences to the client.
- Engineers must freely and openly admit to the client any errors made.

All courts of law have required an engineer to perform in a manner consistent with normal professional standards. This is not the same as saying an engineer's work must be error-free. If an engineer completes a design, has the design and calculations checked by another competent engineer, and an error is subsequently shown to have been made, the engineer may be held responsible, but the engineer will probably not be considered negligent.

5. DEALING WITH SUPPLIERS

Engineers routinely deal with manufacturers, contractors, and vendors (*suppliers*). In this regard, engineers have great responsibility and influence. Such a relationship requires that engineers deal justly with both clients and suppliers.

An engineer will often have an interest in maintaining good relationships with suppliers since this often leads to future work. Nevertheless, relationships with suppliers must remain highly ethical. Suppliers should not be encouraged to feel that they have any special favors coming to them because of a long-standing relationship with the engineer.

[3]Whether the guidelines emphasize ethical behavior or professional conduct is a matter of wording. The intention is the same: to provide guidelines that transcend the requirements of the law.

[4]Some authorities also include ethical guidelines for dealing with the employees of an engineer. However, these guidelines are no different for an engineering employer than they are for a supermarket, automobile assembly line, or airline employer. Ethics is not a unique issue when it comes to employees.

[5]These general guidelines contain references to contractors, plans, specifications, and contract documents. This language is common, though not unique, to the situation of an engineer supplying design services to an owner-developer or architect. However, most of the ethical guidelines are general enough to apply to engineers in industry as well.

The ethical responsibilities relating to suppliers are listed as follows.

- The engineer must not accept or solicit gifts or other valuable considerations from a supplier during, prior to, or after any job. An engineer should not accept discounts, allowances, commissions, or any other indirect compensation from suppliers, contractors, or other engineers in connection with any work or recommendations.
- The engineer must enforce the plans and specifications (i.e., the *contract documents*) but must also interpret the contract documents fairly.
- Plans and specifications developed by the engineer on behalf of the client must be complete, definite, and specific.
- Suppliers should not be required to spend time or furnish materials that are not called for in the plans and contract documents.
- The engineer should not unduly delay the performance of suppliers.

6. DEALING WITH OTHER ENGINEERS

Engineers should try to protect the engineering profession as a whole, to strengthen it, and to enhance its public stature. The following ethical guidelines apply.

- An engineer should not attempt to maliciously injure the professional reputation, business practice, or employment position of another engineer. However, if there is proof that another engineer has acted unethically or illegally, the engineer should advise the proper authority.
- An engineer should not review someone else's work while the other engineer is still employed unless the other engineer is made aware of the review.
- An engineer should not try to replace another engineer once the other engineer has received employment.
- An engineer should not use the advantages of a salaried position to compete unfairly (i.e., moonlight) with other engineers who have to charge more for the same consulting services.
- Subject to legal and proprietary restraints, an engineer should freely report, publish, and distribute information that would be useful to other engineers.

7. DEALING WITH (AND AFFECTING) THE PUBLIC

In regard to the social consequences of engineering, the relationship between an engineer and the public is essentially straightforward. Responsibilities to the public demand that the engineer place service to humankind above personal gain. Furthermore, proper ethical behavior requires that an engineer avoid association with projects that are contrary to public health and welfare or that are of questionable legal character.

- Engineers must consider the safety, health, and welfare of the public in all work performed.
- Engineers must uphold the honor and dignity of their profession by refraining from self-laudatory advertising, by explaining (when required) their work to the public, and by expressing opinions only in areas of knowledge.
- When engineers issue a public statement, they must clearly indicate if the statement is being made on anyone's behalf (i.e., if anyone is benefitting from their position).
- Engineers must keep their skills at a state-of-the-art level.
- Engineers should develop public knowledge and appreciation of the engineering profession and its achievements.
- Engineers must notify the proper authorities when decisions adversely affecting public safety and welfare are made.[6]

8. COMPETITIVE BIDDING

The ethical guidelines for dealing with other engineers presented here and in more detailed codes of ethics no longer include a prohibition on *competitive bidding.* Until 1971, most codes of ethics for engineers considered competitive bidding detrimental to public welfare, since cost-cutting normally results in a lower quality design.

However, in a 1971 case against the National Society of Professional Engineers that went all the way to the U.S. Supreme Court, the prohibition against competitive bidding was determined to be a violation of the Sherman Antitrust Act (i.e., it was an unreasonable restraint of trade).

The opinion of the Supreme Court does not *require* competitive bidding—it merely forbids a prohibition against competitive bidding in NSPE's code of ethics. The following points must be considered.

- Engineers and design firms may individually continue to refuse to bid competitively on engineering services.
- Clients are not required to seek competitive bids for design services.
- Federal, state, and local statutes governing the procedures for procuring engineering design services, even those statutes that prohibit competitive bidding, are not affected.

[6]This practice has come to be known as *whistle-blowing.*

- Any prohibitions against competitive bidding in individual state engineering registration laws remain unaffected.
- Engineers and their societies may actively and aggressively lobby for legislation that would prohibit competitive bidding for design services by public agencies.

58 Electrical Engineering Frontiers

1. Artificial Intelligence 58-1
2. CAD/CAM 58-1
3. Communication 58-1
4. Integrated Circuits 58-1
5. Engineering Management 58-1
6. Superconductivity 58-1
7. Quantum Electrodynamics 58-2
8. Education 58-2

1. ARTIFICIAL INTELLIGENCE

Artificial intelligence (AI) is the property of machines capable of reason by which they are capable of learning functions normally associated with human intelligence. Improvements in integrated circuits and computer technology have enabled concurrent advances in the application of artificial intelligence to real-world engineering devices and projects.

2. CAD/CAM

CAD is an acronym for *computer-aided design* or *drafting*.[1] CAM is an acronym for *computer-aided manufacturing*. The "CA" portion of CAD/CAM denotes the principal components, such as the hardware and software (e.g., AutoCAD). The "D" for drafting indicates the graphic language that uses lines, symbols, and words to describe components. The standards for the line conventions used in drafting are governed by the American National Standards Institute (ANSI). Advances in computer capacity and speed, driven by advances in IC technology and software improvements, have greatly improved the efficiency and accuracy of computerized designs.

3. COMMUNICATION

Communication capability is now both global and individualistic in nature and use. Electronic engineering improvements and advances in space technology are leading to worldwide connectivity.

4. INTEGRATED CIRCUITS

The size of integrated circuits continues to shrink as design complexity increases. Wafer and chip sizes have increased while individual feature size has decreased. According to *Moore's Law*, the number of transistors that can be placed inexpensively on an integrated circuit doubles approximately every two years. Such increases have driven advances in manufacturing automation, clean-room engineering, robotics, and so on.

The analysis of such circuits has also become more automated with SPICE and PSPICE. SPICE is the *simulation program with integrated circuit emphasis*, while PSPICE is a similar version for personal computers.

5. ENGINEERING MANAGEMENT

As many aspects of life evolve technologically, engineering management challenges grow and change. When and where to apply technological advances, and whether it is necessary to apply some advances at all, are but a few of the questions managers and individuals face. An understanding of the underlying principles of engineering and the capacities and limitations of the devices, as well as an appreciation for the human component, is vital to the decision-making process.

6. SUPERCONDUCTIVITY

Superconductivity is a property of certain materials such that at low temperatures their electrical resistivity vanishes. Resistance occurs due to thermal scattering of electrons and due to scattering by lattice impurities. These types of resistivity are called *thermal resistivity* and *residual resistivity*, respectively.[2] At absolute zero, the total electrical resistivity is residual and minimal—that is, very near zero. This, however, is not the phenomenon of superconductivity. Below a transition temperature, called the *critical temperature*, certain materials exhibit a *superconducting state* in which electron interactions fundamentally change, and resistance goes to zero.

Advances in materials science have led to compounds that have transition temperatures greater than 100K—a great increase from the first single-digit transition temperatures.

[1]The acronym is sometimes given as CADD.

[2]The sum of the two resistivities gives the total resistivity according to *Matthiessen's rule*.

7. QUANTUM ELECTRODYNAMICS

Quantum electrodynamics (QED), also called the *quantum theory of light* or the *quantum theory of radiation*, successfully incorporates relativity into the explanation of electromagnetic (i.e., radiation) interactions with charged matter. This theory unites the quantum mechanical- and relativity-based view of such interactions. Further, the theory synthesizes the wave and particle views.

QED theory serves as the prototype for the quantum theory used to explain *electroweak* interactions. *Electroweak theory* is the explanation of electromagnetic and weak nuclear interactions as the same phenomena.[3] With the addition of the electroweak theory, QED theory explains all physical phenomena of the world except those involving gravity or the nuclear force.

The force particle in electromagnetic interactions is the *photon* (sometimes called the *virtual photon*). The force particles for the weak interactions are the Z and W *bosons*, also known as *weakons* or *intermediate vector bosons*. The force particle for the strong nuclear force is the *gluon*. The hypothesized force particle for the gravitational force is the *graviton*.

As understanding of electrical phenomena at the quantum level deepens, the capacity, speed, and types of electrical devices increase. Research is currently under way in an attempt to create quantum mechanical computers.

8. EDUCATION

To fully appreciate engineering advances and apply them effectively, people must understand the basics of science and engineering, which begins with a foundation of mathematics and reading comprehension. The challenge of educating people throughout their lives is enormous, and requires technical and managerial skill. The ability to explain complex subjects and processes in logical terms applicable to the targeted audience, in both verbal and written formats, is a vital skill necessary for the advancement of technology and society itself.

[3]With this combination, three general forces remain: the *electroweak force*, the *strong nuclear* (or *color*) *force*, and the *gravitational force*. A theory combining the electroweak force and strong nuclear force is called the *grand unified theory* (GUT). A theory combining all three, that is, the GUT with gravitational theory, is referred to as the *theory of everything*.

59 Engineering Licensing in the United States

1. About Licensing 59-1
2. The U.S. Licensing Procedure 59-1
3. National Council of Examiners for Engineering and Surveying 59-1
4. Uniform Examinations 59-2
5. Reciprocity Among States 59-2
6. Applying for the Examination 59-2
7. Examination Dates 59-2
8. FE Examination Format 59-3
9. PE Examination Format 59-3

1. ABOUT LICENSING

Engineering licensing (also known as *engineering registration*) in the United States is an examination process by which a state's *board of engineering licensing* (typically referred to as the "engineers' board" or "board of registration") determines and certifies that an engineer has achieved a minimum level of competence.[1] This process is intended to protect the public by preventing unqualified individuals from offering engineering services.

Most engineers in the United States do not need to be licensed.[2] In particular, most engineers who work for companies that design and manufacture products are exempt from the licensing requirement. This is known as the *industrial exemption*, something that is built into the laws of most states.[3]

Nevertheless, there are many good reasons to become a licensed engineer. For example, you cannot offer consulting engineering services in any state unless you are licensed in that state. Even within a product-oriented corporation, you may find that employment, advancement, and managerial positions are limited to licensed engineers.

Once you have met the licensing requirements, you will be allowed to use the titles *Professional Engineer* (PE), *Registered Engineer* (RE), and *Consulting Engineer* (CE) as permitted by your state.

Although the licensing process is similar in each of the 50 states, each has its own licensing law. Unless you offer consulting engineering services in more than one state, however, you will not need to be licensed in the other states.

2. THE U.S. LICENSING PROCEDURE

The licensing procedure is similar in all states. You will take two eight-hour written examinations. The full process requires you to complete two applications, one for each of the two examinations. The first examination is the *Fundamentals of Engineering* (FE) *examination*, formerly known (and still commonly referred to) as the *Engineer-In-Training* (E-I-T) *examination*.[4] This examination covers basic subjects from all of the mathematics, physics, chemistry, and engineering courses you took during your university years.

The second examination is the *Professional Engineering* (PE) *examination*, also known as the *Principles and Practices* (P&P) *examination*. This examination covers only the subjects in your engineering discipline (e.g., civil, mechanical, electrical, and others).

The actual details of licensing qualifications, experience requirements, minimum education levels, fees, and examination schedules vary from state to state. Contact your state's licensing board for more information. You will find contact information (websites, telephone numbers, email addresses, etc.) for all U.S. state and territorial boards of registration at **www.ppi2pass.com**. Click on the State Boards link.

3. NATIONAL COUNCIL OF EXAMINERS FOR ENGINEERING AND SURVEYING

The *National Council of Examiners for Engineering and Surveying* (NCEES) in Clemson, South Carolina, writes, prints, distributes, and scores the national FE and PE examinations.[5] The individual states purchase the

[1]Licensing of engineers is not unique to the United States. However, the practice of requiring a degreed engineer to take an examination is not common in other countries. Licensing in many countries requires a degree and may also require experience, references, and demonstrated knowledge of ethics and law, but no technical examination.

[2]Less than one-third of the degreed engineers in the United States are licensed.

[3]Only one or two states have abolished the industrial exemption. There has always been a lot of "talk" among engineers about abolishing it, but there has been little success in actually trying to do so. One of the reasons is that manufacturers' lobbies are very strong.

[4]The terms *engineering intern* (EI) and *intern engineer* (IE) have also been used in the past to designate the status of an engineer who has passed the first exam. These uses are rarer but may still be encountered in some states.

[5]National Council of Examiners for Engineering and Surveying, P.O. Box 1686, Clemson, SC 29633, (800) 250-3196.

examinations from NCEES and administer them in a uniform, controlled environment dictated by NCEES.

4. UNIFORM EXAMINATIONS

Although each state has its own licensing law and is, theoretically, free to administer its own exams, none does so for the major disciplines. All states have chosen to use the NCEES exams. The exams from all the states are sent to NCEES to be graded. Each state adopts the cut-off passing scores recommended by NCEES. These practices have led to the term *uniform examination.*

5. RECIPROCITY AMONG STATES

With minor exceptions, having a license from one state will not permit you to practice engineering in another state. You must have a professional engineering license from each state in which you work. For most engineers, this is not a problem, but for some it is. Luckily, it is not too difficult to get a license from every state you work in once you have a license from one of them.

All states use the NCEES examinations. If you take and pass the FE or PE examination in one state, your certificate or license will be honored by all of the other states. Upon proper application, payment of fees, and proof of your license, you will be issued a license by the new state. Although there may be other special requirements imposed by a state, it will not be necessary to retake the FE or PE examinations.[6] The issuance of an engineering license based on another state's licensing is known as *reciprocity* or *comity.*

6. APPLYING FOR THE EXAMINATION

While the FE exam and the environment in which it is administered to examinees have been standardized by NCEES in all states, the application process may vary from state to state. For many years, each state had its own application form, which you would have had to obtain directly from your state's licensing board or from a designated agent, such as your university if you were still a student. Each state was (and still is) able to charge its own fee and have different age, experience, and educational requirements from other states, and many do just that. Traditionally, when taking the FE exam, all of your contact and communications throughout the application, acceptance, and notification stages would be directly with your state licensing board.

However, at least 39 states, particularly those with the fewest number of annual applicants, have chosen to allow NCEES to administer a partially and, in some cases, unilateral, online application process. Your state will determine your eligibility to take the exam, while NCEES will coordinate your presence. A list of participating states is included on the NCEES website, www.ncees.org.

For some of these participating states, you will complete an online pre-approval application on the NCEES website and pay the appropriate fee to NCEES. Upon notification that your education and other qualifications meet minimum requirements, in order to reserve your seat in the examination room, you will be required to establish an online account with *NCEES Exam Administration Services* (formerly known as ELSES), submit the required information, pay a separate examination fee to NCEES, and wait for NCEES to send you an email or postal Exam Authorization for Admittance notice 3–4 weeks prior to the examination date.

In other states participating in the NCEES Exam Administration Services program, the process for taking the FE exam consists of applying directly to your state in the traditional fashion—either online at your state's website or with a paper form, paying the appropriate application fee directly to your state, and waiting for your state to confirm receipt of a complete application. At that point, you may establish an online account with NCEES Exam Administration Services, as previously described.

Your Exam Authorization for Admittance notice will include information about the date, time, and location of the exam, as well as information about what you may bring to the exam room, how you are expected to act, and what you can expect to experience.

Approximately 10–12 weeks after the exam, NCEES will release the results of your examination to your state, which will notify you directly. Your notification may include information about additional state-specific steps in the process. Some states have a fee for the application.

As with any other important document, it is a good idea to keep a copy of your examination application and send the original application by certified mail, requesting a receipt of delivery. Keep your proof of mailing and delivery receipt with your copy of the application.

All states make special accommodations for persons who are physically challenged or who have religious or other special needs. Be sure to communicate your need to the state board well in advance of the examination day.

7. EXAMINATION DATES

The national FE and PE examinations are administered twice a year (usually in mid-April and late October), on the same weekends in all states. For a current exam schedule, check **www.ppi2pass.com**. Click on the Exam FAQs link.

[6]For example, California requires all civil engineering applicants to pass special examinations in seismic design and surveying in addition to their regular eight-hour PE exams. Licensed engineers from other states only have to pass these two special exams. They do not need to retake the PE exam.

8. FE EXAMINATION FORMAT

The FE examination consists of two four-hour sessions, separated by a one-hour lunch period. There are 120 multiple-choice problems in the morning and 60 multiple-choice problems in the afternoon. All problems have four answer choices, from which you are to choose the best single answer. Afternoon problems are slightly more difficult and have double weight.

Questions from all undergraduate technical courses appear in the examination, including mathematics, chemistry, physics, and engineering.[7]

The FE exam is essentially entirely in SI units. There may be a few non-SI problems and some opportunities to choose between identical SI and non-SI problems.

The FE exam is a limited-reference exam. NCEES provides its own reference handbook for use in the examination. You may bring your own approved calculator, but you are not allowed to bring in any of your own books or notes. Pencils are provided during the exam.

9. PE EXAMINATION FORMAT

There are three separate electrical examinations: Power, Electrical and Electronics, and Computer. You will choose which exam you will take the day that you register for the exam.

Each exam consists of two four-hour sessions separated by a one-hour lunch period. The material covered in the morning session is very similar to the material covered in the afternoon. Both sessions will require an understanding of the basics as well as a detailed study of the appropriate topics contained within your specific discipline. This is an open-book exam. Although some states have a few restrictions, generally you may bring in the books of your choice. Visit **www.ppi2pass.com/eefaqs** to get the most up-to-date exam details.

The exam requires the use of a scientific calculator. NCEES has a very limited list of calculators that they will allow into the exam room. Check the PPI website (**www.ppi2pass.com/calculators**) for a current list of permissible devices.

The questions on the PE exam will be multiple choice. Most questions will be unique; that is, they will be scenario-based with one question per scenario. There may be some multi-part questions, where one problem statement (i.e., scenario) is followed by two or three questions. The multi-part questions will nevertheless be independent; that is, you will not have to answer the first question correctly in order to get the second or third correct (they do not "cascade"). A total of 40 questions are expected in the morning, and another 40 in the afternoon. You must answer them all to get full credit. The subjects likely to be tested on each exam are given in Table 1 of the Introduction to this book.

You may be required to call upon nonacademic knowledge that stems from practical experience. The exam may include terminology used in the field without explanation. You may be asked to display current knowledge (within the last five years) of your chosen field.

To simplify the problems, you may want to use shortcuts known to practicing engineers. Additionally, the exam may require the knowledge of use of codes such as the *National Electrical Code* (NEC), the Code of Federal Regulations (CFR), Occupational Safety and Health Administration (OSHA) rules, and various computer codes and standards.

The electrical engineering field primarily uses SI units (the mks, or metric, system). Accordingly, SI units are used extensively on the exam. There are a few exceptions, however. For example, many of the questions relating to the National Electrical Code and wiring in general will utilize U.S. units. Additionally, magnetic questions may use the cgs system.

Professional

[7]The format of the FE exam, other valuable information, and tips on passing are given in greater detail in the *FE Review Manual* and *Core Engineering Concepts for Students and Professionals*, both published by PPI.

Appendices Table of Contents

1.A Conversion Factors A-1
1.B Common SI Unit Conversion Factors A-3
8.A Mensuration of Two-Dimensional Areas A-7
8.B Mensuration of Three-Dimensional Volumes . . . A-9
10.A Abbreviated Table of Indefinite Integrals A-10
11.A Laplace Transforms A-11
12.A Areas Under the Standard Normal Curve A-12
15.A Gamma Function Values A-13
15.B Bessel Functions J_0 and J_1 . . . . A-14
15.C Properties of Fourier Series for Periodic Signals A-15
15.D Properties of Fourier Transform for Aperiodic Signals A-16
15.E Fourier Transform Pairs A-17
15.F Properties of Laplace Transform A-18
15.G Laplace Transforms A-19
15.H Properties of Discrete-Time Fourier Series for Periodic Signals A-20
15.I Properties of Discrete-Time Fourier Transform for Aperiodic Signals A-21
15.J Properties of z-Transforms A-22
15.K z-Transforms A-23
17.A Room Temperature Properties of Silicon, Germanium, and Gallium Arsenide (at 300K) A-24
18.A Electromagnetic Spectrum A-25
25.A Comparison of Electric and Magnetic Equations A-26
27.A Impedance of Series-Connected Circuit Elements A-28
27.B Impedance of Parallel-Connected Circuit Elements A-29
29.A Two-Port Parameter Conversions A-31
30.A Resonant Circuit Formulas A-32
38.A Electrical Characteristics of Multilayer, Steel-Reinforced, Bare Aluminum Conductors A-33
43.A Periodic Table of the Elements A-34
43.B Semiconductor Symbols and Abbreviations . . A-35
46.A Standard Zener Diodes A-39
49.A Units, Symbols, and Defining Equations for Fundamental Photometric and Radiometric Quantities A-41
49.B Luminance Conversion Factors A-43
49.C Illuminance Conversion Factors A-44
52.A Periodic Table: Materials' Properties Summary A-45
52.B Table of Relative Atomic Weights A-46
54.A Wire (and Sheet-Metal Gage) Diameters (Thickness) in Inches A-47
54.B Conductor Ampacities A-49
54.C Conductor Properties A-50
54.D Conductor AC Properties A-51
56.A Standard Cash Flow Factors A-52
56.B Cash Flow Equivalent Factors A-53

APPENDIX 1.A
Conversion Factors

multiply	by	to obtain	multiply	by	to obtain
acres	0.40468	hectares	feet	30.48	centimeters
	43,560.0	square feet		0.3048	meters
	1.5625×10^{-3}	square miles		1.645×10^{-4}	miles (nautical)
ampere-hours	3600.0	coulombs		1.894×10^{-4}	miles (statute)
angstrom units	3.937×10^{-9}	inches	feet/min	0.5080	centimeters/sec
	1×10^{-4}	microns	feet/sec	0.592	knots
astronomical units	1.496×10^{8}	kilometers		0.6818	miles/hr
atmospheres	76.0	centimeters of mercury	foot-pounds	1.285×10^{-3}	Btu
				5.051×10^{-7}	horsepower-hours
atomic mass unit	9.3149×10^{8}	electron-volts		3.766×10^{-7}	kilowatt-hours
	1.4924×10^{-10}	joules	foot-pound/sec	4.6272	Btu/hr
	1.6605×10^{-27}	kilograms		1.818×10^{-3}	horsepower
BeV (also GeV)	1×10^{9}	electron-volts		1.356×10^{-3}	kilowatts
Btu	3.93×10^{-4}	horsepower-hours	furlongs	660.0	feet
	778.2	foot-pounds		0.125	miles (statute)
	2.931×10^{-4}	kilowatt hours	gallons	0.1337	cubic feet
	1.0×10^{-5}	therms		3.785	liters
Btu/hr	0.2161	foot-pounds/sec	gallons H_2O	8.3453	pounds H_2O
	3.929×10^{-4}	horsepower	gallons/min	8.0208	cubic feet/hr
	0.293	watts		0.002228	cubic feet/sec
bushels	2150.4	cubic inches	GeV (also BeV)	1×10^{9}	electron-volts
calories, gram (mean)	3.9683×10^{-3}	Btu (mean)	grams	1×10^{-3}	kilograms
centares	1.0	square meters		3.527×10^{-2}	ounces (avoirdupois)
centimeters	1×10^{-5}	kilometers		3.215×10^{-2}	ounces (troy)
	1×10^{-2}	meters		2.205×10^{-3}	pounds
	10.0	millimeters	hectares	2.471	acres
	3.281×10^{-2}	feet		1.076×10^{5}	square feet
	0.3937	inches	horsepower	2545.0	Btu/hr
chains	792.0	inches		42.44	Btu/min
coulombs	1.036×10^{-5}	faradays		550	foot-pounds/sec
cubic centimeters	0.06102	cubic inches		0.7457	kilowatts
	2.113×10^{-3}	pints (U.S. liquid)		745.7	watts
cubic feet	0.02832	cubic meters	horsepower-hours	2545.0	Btu
	7.4805	gallons		1.976×10^{-6}	foot-pounds
cubic feet/min	62.43	pounds H_2O/min		0.7457	kilowatt-hours
cubic feet/sec	448.831	gallons/min	hours	4.167×10^{-2}	days
	0.64632	millions of gallons per day		5.952×10^{-3}	weeks
			inches	2.540	centimeters
cubits	18.0	inches		1.578×10^{-5}	miles
days	86,400.0	seconds	inches, H_2O	5.199	pounds force/ft^2
degrees (angle)	1.745×10^{-2}	radians		0.0361	psi
degrees/sec	0.1667	revolutions/min		0.0735	inches, mercury
dynes	1×10^{-5}	newtons	inches, mercury	70.7	pounds force/ft^2
electron-volts	1.074×10^{-9}	atomic mass units		0.491	pounds force/in^2
	1×10^{-9}	BeV (also GeV)		13.60	inches, H_2O
	1.60218×10^{-19}	joules	joules	6.705×10^{9}	atomic mass units
	1.78266×10^{-36}	kilograms		9.478×10^{-4}	Btu
	1×10^{-6}	MeV		1×10^{7}	ergs
faradays/sec	96,485	amperes (absolute)		6.2415×10^{18}	electron-volts
fathoms	6.0	feet		1.1127×10^{-17}	kilograms

(continued)

APPENDIX 1.A *(continued)*
Conversion Factors

multiply	by	to obtain	multiply	by	to obtain
kilograms	6.0221×10^{26}	atomic mass units	pascal-sec	1000	centipoise
	5.6096×10^{35}	electron-volts		10	poise
	8.9875×10^{16}	joules		0.02089	pound force-sec/ft^2
	2.205	pounds		0.6720	pound mass/ft-sec
kilometers	3281.0	feet		0.02089	slug/ft-sec
	1000.0	meters	pints (liquid)	473.2	cubic centimeters
	0.6214	miles		28.87	cubic inches
kilometers/hr	0.5396	knots		0.125	gallons
kilowatts	3412.9	Btu/hr		0.5	quarts (liquid)
	737.6	foot-pounds/sec	poise	0.002089	pound-sec/ft^2
	1.341	horsepower	pounds	0.4536	kilograms
kilowatt-hours	3413.0	Btu		16.0	ounces
knots	6076.0	feet/hr		14.5833	ounces (troy)
	1.0	nautical miles/hr		1.21528	pounds (troy)
	1.151	statute miles/hr	pounds/ft^2	0.006944	pounds/in^2
light years	5.9×10^{12}	miles	pounds/in^2	2.308	feet, H_2O
links (surveyor)	7.92	inches		27.7	inches, H_2O
liters	1000.0	cubic centimeters		2.037	inches, mercury
	61.02	cubic inches		144	pounds/ft^2
	0.2642	gallons (U.S. liquid)	quarts (dry)	67.20	cubic inches
	1000.0	milliliters	quarts (liquid)	57.75	cubic inches
	2.113	pints		0.25	gallons
MeV	1×10^6	electron-volts		0.9463	liters
meters	100.0	centimeters	radians	57.30	degrees
	3.281	feet		3438.0	minutes
	1×10^{-3}	kilometers	revolutions	360.0	degrees
	5.396×10^{-4}	miles (nautical)	revolutions/min	6.0	degrees/sec
	6.214×10^{-4}	miles (statute)	rods	16.5	feet
	1000.0	millimeters		5.029	meters
microns	1×10^{-6}	meters	rods (surveyor)	5.5	yards
miles (nautical)	6076	feet	seconds	1.667×10^{-2}	minutes
	1.853	kilometers	square meters/sec	1×10^6	centistokes
	1.1516	miles (statute)		10.76	square feet/sec
miles (statute)	5280.0	feet		1×10^4	stokes
	1.609	kilometers	slugs	32.174	pounds mass
	0.8684	miles (nautical)	stokes	0.0010764	square feet/sec
miles/hr	88.0	feet/min	tons (long)	1016.0	kilograms
milligrams/liter	1.0	parts/million		2240.0	pounds
milliliters	1×10^{-3}	liters		1.120	tons (short)
millimeters	3.937×10^{-2}	inches	tons (short)	907.1848	kilograms
newtons	1×10^5	dynes		2000.0	pounds
ohms (international)	1.0005	ohms (absolute)		0.89287	tons (long)
ounces	28.349527	grams	volts (absolute)	3.336×10^{-3}	statvolts
	6.25×10^{-2}	pounds	watts	3.4129	Btu/hr
ounces (troy)	1.09714	ounces (avoirdupois)		1.341×10^{-3}	horsepower
parsecs	3.086×10^{13}	kilometers	yards	0.9144	meters
	1.9×10^{13}	miles		4.934×10^{-4}	miles (nautical)
				5.682×10^{-4}	miles (statute)

APPENDIX 1.B
Common SI Unit Conversion Factors

multiply	by	to obtain
AREA		
circular mil	506.7	square micrometer
square foot	0.0929	square meter
square kilometer	0.3861	square mile
square meter	10.764	square foot
	1.196	square yard
square micrometer	0.001974	circular mil
square mile	2.590	square kilometer
square yard	0.8361	square meter
ENERGY		
Btu (international)	1.0551	kilojoule
erg	0.1	microjoule
foot-pound	1.3558	joule
horsepower-hour	2.6485	megajoule
joule	0.7376	foot-pound
	0.10197	meter-kilogram force
kilogram-calorie (international)	4.1868	kilojoule
kilojoule	0.9478	Btu
	0.2388	kilogram-calorie
kilowatt-hour	3.6	megajoule
megajoule	0.3725	horsepower-hour
	0.2778	kilowatt-hour
	0.009478	therm
meter-kilogram force	9.8067	joule
microjoule	10.0	erg
therm	105.506	megajoule
FORCE		
dyne	10.0	micronewton
kilogram force	9.8067	newton
kip	4448.2	newton
micronewton	0.1	dyne
newton	0.10197	kilogram force
	0.0002248	kip
	3.597	ounce force
	0.2248	pound force
ounce force	0.2780	newton
pound force	4.4482	newton
HEAT		
Btu/ft^2-hr	3.1546	watt/m^2
Btu/ft^2-hr-°F	5.6783	watt/m^2·°C
Btu/ft^3	0.0373	megajoule/m^3
Btu/ft^3-°F	0.06707	megajoule/m^3·°C
Btu/hr	0.2931	watt
Btu/lbm	2326	joule/kg
Btu/lbm-°F	4186.8	joule/kg·°C
Btu-inch/ft^2-hr-°F	0.1442	watt/meter·°C
joule/kg	0.000430	Btu/lbm
joule/kg·°C	0.0002388	Btu/lbm-°F
megajoule/m^3	26.839	Btu/ft^3
megajoule/m^3·°C	14.911	Btu/ft^3-°F
watt	3.4121	Btu/hr
watt/m·°C	6.933	Btu-in/ft^2-hr-°F

(continued)

APPENDIX 1.B *(continued)*
Common SI Unit Conversion Factors

multiply	by	to obtain
HEAT *(continued)*		
watt/m^2	0.3170	Btu/ft^2-hr
watt/m^2·°C	0.1761	Btu/hr-ft^2-°F
LENGTH		
angstrom	0.1	nanometer
foot	0.3048	meter
inch	25.4	millimeter
kilometer	0.6214	mile
	0.540	mile (nautical)
meter	3.2808	foot
	1.0936	yard
micrometer	1.0	micron
micron	1.0	micrometer
mil	0.0254	millimeter
mile	1.6093	kilometer
mile (nautical)	1.852	kilometer
millimeter	0.0394	inch
	39.370	mil
nanometer	10.0	angstrom
yard	0.9144	meter
MASS		
grain	64.799	milligram
gram	0.0353	ounce (avoirdupois)
	0.03215	ounce (troy)
kilogram	2.2046	pound mass
	0.068522	slug
	0.0009842	ton (long—2240 lbm)
	0.001102	ton (short—2000 lbm)
milligram	0.0154	grain
ounce (avoirdupois)	28.350	gram
ounce (troy)	31.1035	gram
pound mass	0.4536	kilogram
slug	14.5939	kilogram
ton (long—2240 lbm)	1016.047	kilogram
ton (short—2000 lbm)	907.185	kilogram
PRESSURE		
bar	100.0	kilopascal
inch, H_2O (20°C)	0.2486	kilopascal
inch, Hg (20°C)	3.3741	kilopascal
kilogram force/cm^2	98.067	kilopascal
kilopascal	0.01	bar
	4.0219	inch, H_2O (20°C)
	0.2964	inch, Hg (20°C)
	0.0102	kilogram force/cm^2
	7.528	millimeter Hg (20°C)
	0.1450	pound force/in^2
	0.009869	standard atmosphere (760 torr)
	7.5006	torr
millimeter Hg (20°C)	0.13332	kilopascal
pound force/in^2	6.8948	kilopascal
standard atmosphere (760 torr)	101.325	kilopascal
torr	0.13332	kilopascal

(continued)

APPENDIX 1.B *(continued)*
Common SI Unit Conversion Factors

multiply	by	to obtain
POWER		
Btu (international)/hr	0.2931	watt
foot-pound/sec	1.3558	watt
horsepower	0.7457	kilowatt
kilowatt	1.341	horsepower
	0.2843	ton of refrigeration
meter-kilogram force/second	9.8067	watt
ton of refrigeration	3.517	kilowatt
watt	3.4122	Btu (international)/hr
	0.7376	foot-pound/sec
	0.10197	meter-kilogram force/second
TEMPERATURE		
Celsius	$\frac{9}{5}°C + 32°$	Fahrenheit
Fahrenheit	$\frac{5}{9}(°F - 32°)$	Celsius
Kelvin	$\frac{9}{5}$	Rankine
Rankine	$\frac{5}{9}$	Kelvin
TORQUE		
gram force centimeter	0.098067	millinewton·meter
kilogram force meter	9.8067	newton·meter
millinewton	10.197	gram force centimeter
newton-meter	0.10197	kilogram force meter
	0.7376	foot pound
	8.8495	inch pound
foot-pound	1.3558	newton-meter
inch-pound	0.1130	newton-meter
VELOCITY		
feet/sec	0.3048	meters/second
kilometers/hr	0.6214	miles/hr
meters/second	3.2808	feet/sec
	2.2369	miles/hr
miles/hr	1.60934	kilometers/hr
	0.44704	meters/second
VISCOSITY		
centipoise	0.001	pascal·seconds
centistoke	1×10^{-6}	meter^2/second
pascal·second	1000	centipoise
meter^2/second	1×10^{6}	centistoke
VOLUME (capacity)		
cubic centimeter	0.06102	cubic inch
cubic foot	28.3168	liter
cubic inch	16.3871	cubic centimeter
cubic meter	1.308	cubic yard
cubic yard	0.7646	cubic meter
gallon (U.S.)	3.785	liter
liter	0.2642	gallon (U.S.)
	2.113	pint (U.S. fluid)
	1.0567	quart (U.S. fluid)
	0.03531	cubic foot
milliliter	0.0338	ounce (U.S. fluid)
ounce (U.S. fluid)	29.574	milliliter

(continued)

Appendices

APPENDIX 1.B *(continued)*
Common SI Unit Conversion Factors

multiply	by	to obtain
VOLUME (capacity) *(continued)*		
pint (U.S. fluid)	0.4732	liter
quart (U.S. fluid)	0.9464	liter
VOLUME FLOW (gas-air)		
cubic meter/sec	2119	standard cubic foot/min
liter/sec	2.119	standard cubic foot/min
microliter/sec	0.000127	standard cubic foot/hr
milliliter/sec	0.002119	standard cubic foot/min
	0.127133	standard cubic foot/hr
standard cubic foot/min	0.0004719	cubic meter/sec
	0.4719	liter/sec
	471.947	milliliter/sec
standard cubic foot/hr	7866	microliter/sec
	7.8658	milliliter/sec
VOLUME FLOW (liquid)		
gallon/hr (U.S.)	0.001052	liter/sec
gallon/min (U.S.)	0.06309	liter/sec
liter/sec	951.02	gallon/hr (U.S.)
	15.850	gallon/min (U.S.)

APPENDIX 8.A
Mensuration of Two-Dimensional Areas

Nomenclature

A	total surface area
b	base
c	chord length
d	distance
h	height
L	length
p	perimeter
r	radius
s	side (edge) length, arc length
θ	vertex angle, in radians
ϕ	central angle, in radians

Triangle

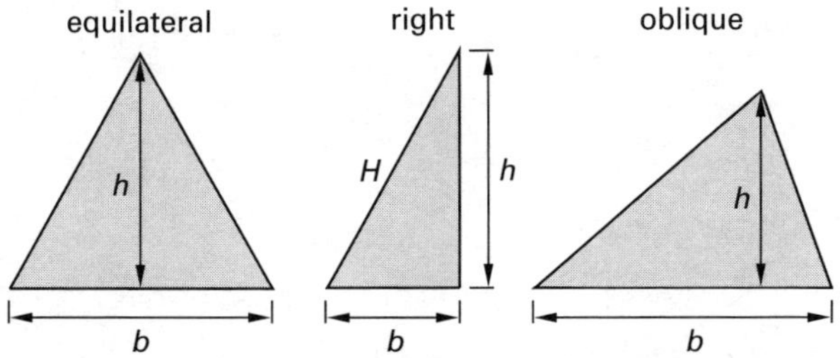

$$A = \tfrac{1}{2}bh = \frac{\sqrt{3}}{4}b^2 \qquad A = \tfrac{1}{2}bh \qquad A = \tfrac{1}{2}bh$$

$$h = \frac{\sqrt{3}}{2}b \qquad H^2 = b^2 + h^2$$

Circle

$$p = 2\pi r$$

$$A = \pi r^2 = \frac{p^2}{4\pi}$$

Circular Segment

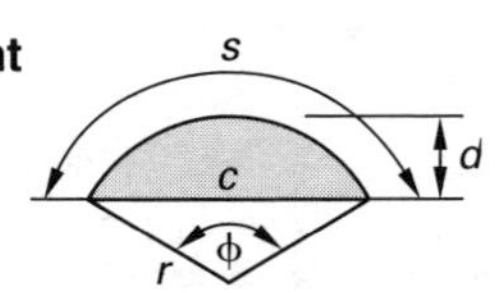

$$A = \tfrac{1}{2}r^2(\phi - \sin\phi)$$

$$\phi = \frac{s}{r} = 2\left(\arccos\frac{r-d}{r}\right)$$

$$c = 2r\sin\left(\frac{\phi}{2}\right)$$

Circular Sector

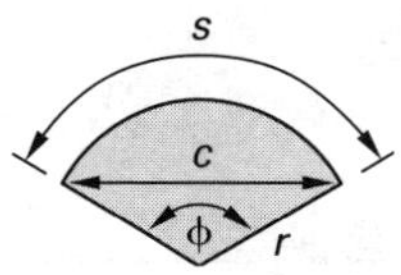

$$A = \tfrac{1}{2}\phi r^2 = \tfrac{1}{2}sr$$

$$\phi = \frac{s}{r}$$

$$s = r\phi$$

$$c = 2r\sin\left(\frac{\phi}{2}\right)$$

Parabola

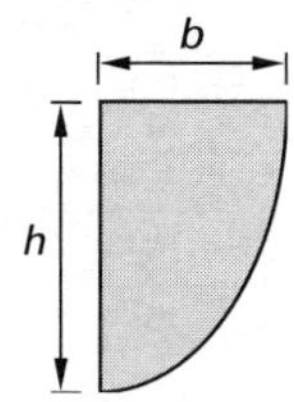

$$A = \tfrac{2}{3}bh$$

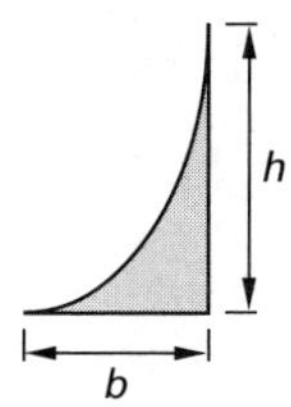

$$A = \tfrac{1}{3}bh$$

Ellipse

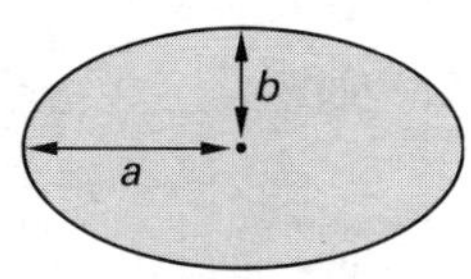

$$A = \pi ab$$

$$p \approx 2\pi\sqrt{\tfrac{1}{2}(a^2+b^2)} \qquad \left[\begin{array}{c}\text{Euler's}\\ \text{upper bound}\end{array}\right]$$

(continued)

Appendices

APPENDIX 8.A *(continued)*
Mensuration of Two-Dimensional Areas

Trapezoid

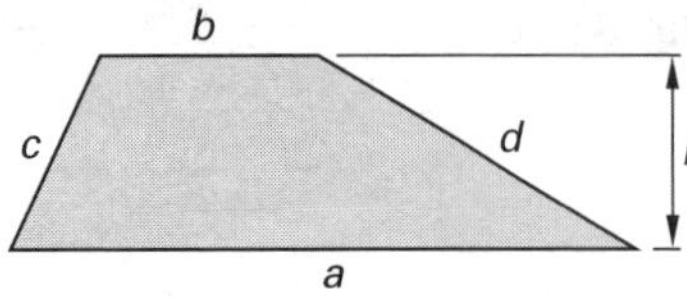

$$p = a + b + c + d$$

$$A = \tfrac{1}{2}h(a + b)$$

The trapezoid is isosceles if $c = d$.

Parallelogram

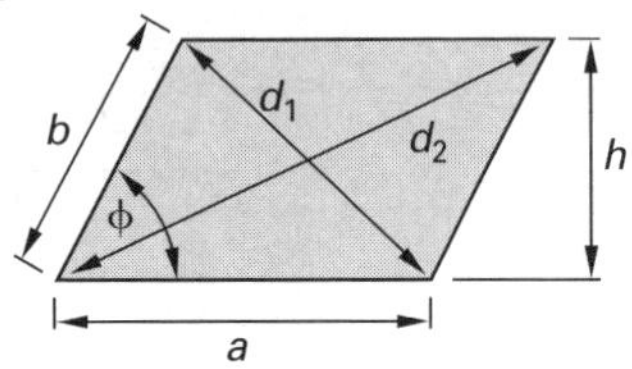

$$p = 2(a + b)$$

$$d_1 = \sqrt{a^2 + b^2 - 2ab\cos\phi}$$

$$d_2 = \sqrt{a^2 + b^2 + 2ab\cos\phi}$$

$$d_1^2 + d_2^2 = 2(a^2 + b^2)$$

$$A = ah = ab\sin\phi$$

If $a = b$, the parallelogram is a rhombus.

Regular Polygon (*n* equal sides)

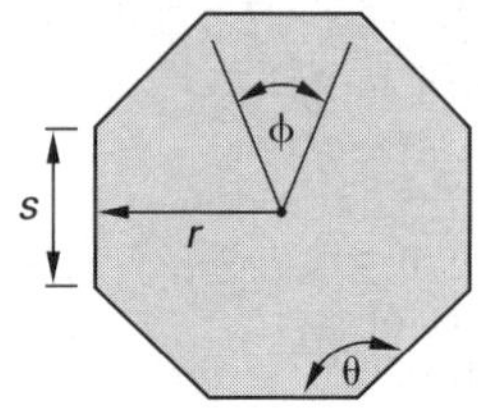

$$\phi = \frac{2\pi}{n}$$

$$\theta = \frac{\pi(n-2)}{n} = \pi - \phi$$

$$p = ns$$

$$s = 2r\tan\left(\frac{\theta}{2}\right)$$

$$A = \tfrac{1}{2}nsr$$

regular polygons		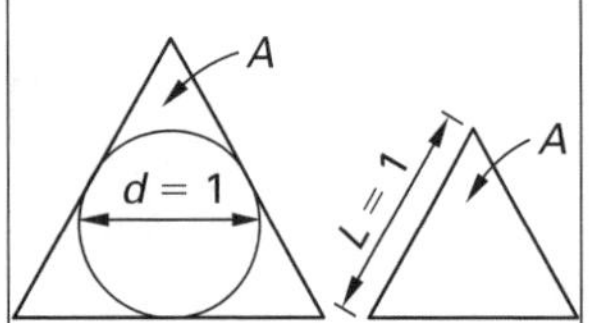		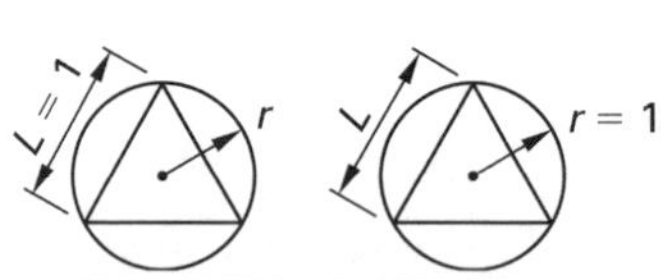		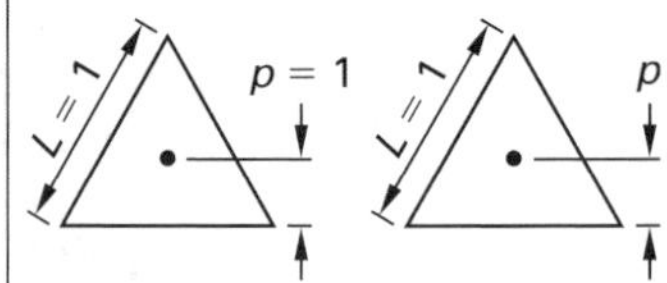	
sides	name	area (A) when diameter of inscribed circle = 1	area (A) when side = 1	radius (r) of circumscribed circle when side = 1	length (L) of side when radius (r) of circumscribed circle = 1	length (L) of side when perpendicular to center = 1	perpendicular (p) to center when side = 1
3	triangle	1.299	0.433	0.577	1.732	3.464	0.289
4	square	1.000	1.000	0.707	1.414	2.000	0.500
5	pentagon	0.908	1.720	0.851	1.176	1.453	0.688
6	hexagon	0.866	2.598	1.000	1.000	1.155	0.866
7	heptagon	0.843	3.634	1.152	0.868	0.963	1.038
8	octagon	0.828	4.828	1.307	0.765	0.828	1.207
9	nonagon	0.819	6.182	1.462	0.684	0.728	1.374
10	decagon	0.812	7.694	1.618	0.618	0.650	1.539
11	undecagon	0.807	9.366	1.775	0.563	0.587	1.703
12	dodecagon	0.804	11.196	1.932	0.518	0.536	1.866

APPENDIX 8.B
Mensuration of Three-Dimensional Volumes

Nomenclature

A	area
b	base
h	height
r	radius
R	radius
s	side (edge) length
V	volume

Sphere

$$V = \frac{4\pi r^3}{3}$$
$$A = 4\pi r^2$$

Right Circular Cone

$$V = \frac{\pi r^2 h}{3}$$
$$A = \pi r\sqrt{r^2 + h^2}$$

(does not include base area)

Right Circular Cylinder

$$V = \pi r^2 h$$
$$A = 2\pi r h$$

(does not include end area)

Spherical Segment (Spherical Cap)

Surface area of a spherical segment of radius r cut out by an angle θ_0 rotated from the center about a radius, r, is

$$A = 2\pi r^2 (1 - \cos\theta_0)$$
$$\omega = \frac{A}{r^2} = 2\pi (1 - \cos\theta_0)$$

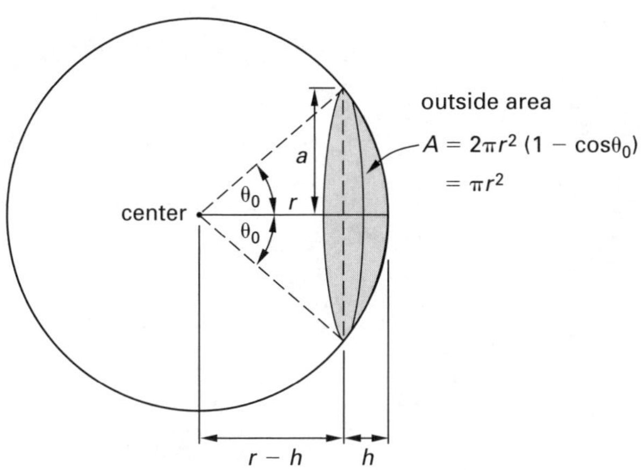

$$V_{cap} = \frac{\pi}{6}h(3a^2 + h^2)$$
$$= \frac{\pi}{3}h^2(3r - h)$$
$$a = \sqrt{h(2r - h)}$$

Paraboloid of Revolution

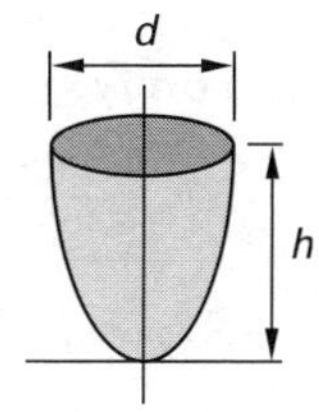

$$V = \frac{\pi d^2 h}{8}$$

Torus

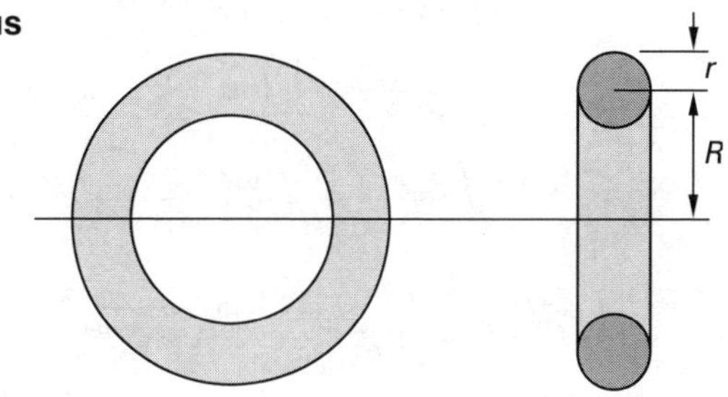

$$A = 4\pi^2 R r$$
$$V = 2\pi^2 R r^2$$

Regular Polyhedra (identical faces)

name	number of faces	form of faces	total surface area	volume
tetrahedron	4	equilateral triangle	$1.7321s^2$	$0.1179s^3$
cube	6	square	$6.0000s^2$	$1.0000s^3$
octahedron	8	equilateral triangle	$3.4641s^2$	$0.4714s^3$
dodecahedron	12	regular pentagon	$20.6457s^2$	$7.6631s^3$
isosahedron	20	equilateral triangle	$8.6603s^2$	$2.1817s^3$

The radius of a sphere inscribed within a regular polyhedron is

$$r = \frac{3V_{\text{polyhedron}}}{A_{\text{polyhedron}}}$$

Appendices

APPENDIX 10.A

Abbreviated Table of Indefinite Integrals
(In each case, add a constant of integration.
All angles are measured in radians.)

General Formulas

1. $\int dx = x$
2. $\int c\,dx = c\int dx$
3. $\int (dx + dy) = \int dx + \int dy$
4. $\int u\,dv = uv - \int v\,du$ (integration by parts)

Algebraic Forms

5. $\displaystyle\int x^n\,dx = \frac{x^{n+1}}{n+1} \quad [n \neq -1]$
6. $\displaystyle\int x^{-1}\,dx = \int \frac{dx}{x} = \ln|x|$
7. $\displaystyle\int (ax+b)^n\,dx = \frac{(ax+b)^{n+1}}{a(n+1)} \quad [n \neq -1]$
8. $\displaystyle\int \frac{dx}{ax+b} = \frac{1}{a}\ln(ax+b)$
9. $\displaystyle\int \frac{x\,dx}{ax+b} = \frac{1}{a^2}\Big(ax+b-b\ln(ax+b)\Big)$
10. $\displaystyle\int \frac{x\,dx}{(ax+b)^2} = \frac{1}{a^2}\left(\frac{b}{ax+b}+\ln(ax+b)\right)$
11. $\displaystyle\int \frac{dx}{x(ax+b)} = \frac{1}{b}\ln\left(\frac{x}{ax+b}\right)$
12. $\displaystyle\int \frac{dx}{x(ax+b)^2} = \frac{1}{b(ax+b)} + \frac{1}{b^2}\ln\left(\frac{x}{ax+b}\right)$
13. $\displaystyle\int \frac{dx}{x^2+a^2} = \frac{1}{a}\tan^{-1}\left(\frac{x}{a}\right)$
14. $\displaystyle\int \frac{dx}{a^2-x^2} = \frac{1}{a}\tanh^{-1}\left(\frac{x}{a}\right)$
15. $\displaystyle\int \frac{x\,dx}{ax^2+b} = \frac{1}{2a}\ln(ax^2+b)$
16. $\displaystyle\int \frac{dx}{x(ax^n+b)} = \frac{1}{bn}\ln\left(\frac{x^n}{ax^n+b}\right)$
17. $\displaystyle\int \frac{dx}{ax^2+bx+c} = \frac{1}{\sqrt{b^2-4ac}}\ln\left(\frac{2ax+b-\sqrt{b^2-4ac}}{2ax+b+\sqrt{b^2-4ac}}\right) \quad [b^2 > 4ac]$
18. $\displaystyle\int \frac{dx}{ax^2+bx+c} = \frac{2}{\sqrt{4ac-b^2}}\tan^{-1}\left(\frac{2ax+b}{\sqrt{4ac-b^2}}\right) \quad [b^2 > 4ac]$
19. $\displaystyle\int \sqrt{a^2-x^2}\,dx = \frac{x}{2}\sqrt{a^2-x^2} + \frac{a^2}{2}\sin^{-1}\left(\frac{x}{a}\right)$
20. $\displaystyle\int x\sqrt{a^2-x^2}\,dx = -\frac{1}{3}(a^2-x^2)^{3/2}$
21. $\displaystyle\int \frac{dx}{\sqrt{a^2-x^2}} = \sin^{-1}\left(\frac{x}{a}\right)$
22. $\displaystyle\int \frac{x\,dx}{\sqrt{a^2-x^2}} = -\sqrt{a^2-x^2}$

Appendices

APPENDIX 11.A
Laplace Transforms
(also see App. 15.G)

$f(t)$	$\mathcal{L}(f(t))$	$f(t)$	$\mathcal{L}(f(t))$
$\delta(t)$ [unit impulse at $t = 0$]	1	e^{at}	$\frac{1}{s-a}$
$\delta(t-c)$ [unit impulse at $t = c$]	e^{-cs}	$e^{at}\sin bt$	$\frac{b}{(s-a)^2+b^2}$
1 or u_0 [unit step at $t = 0$]	$\frac{1}{s}$	$e^{at}\cos bt$	$\frac{s-a}{(s-a)^2+b^2}$
u_c [unit step at $t = c$]	$\frac{e^{-cs}}{s}$	$e^{at}t^n$ (n is a positive integer)	$\frac{n!}{(s-a)^{n+1}}$
t [unit ramp at $t = 0$]	$\frac{1}{s^2}$	$1-e^{-at}$	$\frac{a}{s(s+a)}$
$\frac{t^{n-1}}{(n-1)!}$	$\frac{1}{s^n}$	$e^{-at}+at-1$	$\frac{a^2}{s^2(s+a)}$
$\sin at$	$\frac{a}{s^2+a^2}$	$\frac{e^{-at}-e^{-bt}}{b-a}$	$\frac{1}{(s+a)(s+b)}$
$at-\sin at$	$\frac{a^3}{s^2(s^2+a^2)}$	$\frac{(c-a)e^{-at}-(c-b)e^{-bt}}{b-a}$	$\frac{s+c}{(s+a)(s+b)}$
$\sinh at$	$\frac{a}{s^2-a^2}$	$\frac{1}{ab}+\frac{be^{-at}-ae^{-bt}}{ab(a-b)}$	$\frac{1}{s(s+a)(s+b)}$
$t\sin at$	$\frac{2as}{(s^2+a^2)^2}$	$t\sinh at$	$\frac{2as}{(s^2-a^2)^2}$
$\cos at$	$\frac{s}{s^2+a^2}$	$t\cosh at$	$\frac{s^2+a^2}{(s^2-a^2)^2}$
$1-\cos at$	$\frac{a^2}{s(s^2+a^2)}$	rectangular pulse, magnitude M, duration a	$\left(\frac{M}{s}\right)(1-e^{-as})$
$\cosh at$	$\frac{s}{s^2-a^2}$	triangular pulse, magnitude M, duration $2a$	$\left(\frac{M}{as^2}\right)(1-e^{-as})^2$
$t\cos at$	$\frac{s^2-a^2}{(s^2+a^2)^2}$	sawtooth pulse, magnitude M, duration a	$\left(\frac{M}{as^2}\right)(1-(as+1)e^{-as})$
t^n (n is a positive integer)	$\frac{n!}{s^{n+1}}$	sinusoidal pulse, magnitude M, duration π/a	$\left(\frac{Ma}{s^2+a^2}\right)(1+e^{-\pi s/a})$

Appendices

APPENDIX 12.A
Areas Under the Standard Normal Curve
(0 to z)

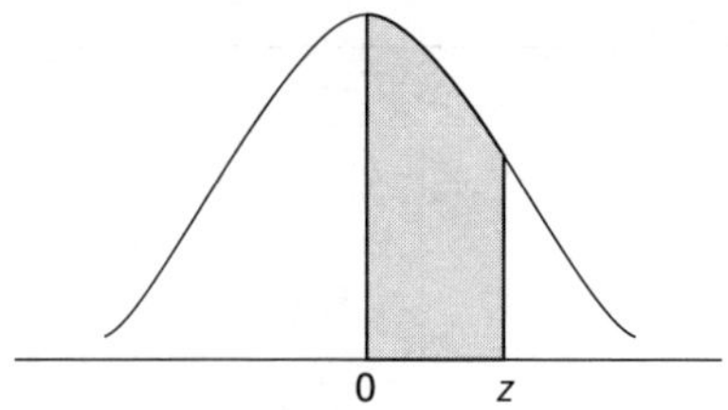

z	0	1	2	3	4	5	6	7	8	9
0.0	0.0000	0.0040	0.0080	0.0120	0.0160	0.0199	0.0239	0.0279	0.0319	0.0359
0.1	0.0398	0.0438	0.0478	0.0517	0.0557	0.0596	0.0636	0.0675	0.0714	0.0754
0.2	0.0793	0.0832	0.0871	0.0910	0.0948	0.0987	0.1026	0.1064	0.1103	0.1141
0.3	0.1179	0.1217	0.1255	0.1293	0.1331	0.1368	0.1406	0.1443	0.1480	0.1517
0.4	0.1554	0.1591	0.1628	0.1664	0.1700	0.1736	0.1772	0.1808	0.1844	0.1879
0.5	0.1915	0.1950	0.1985	0.2019	0.2054	0.2088	0.2123	0.2157	0.2190	0.2224
0.6	0.2258	0.2291	0.2324	0.2357	0.2389	0.2422	0.2454	0.2486	0.2518	0.2549
0.7	0.2580	0.2612	0.2642	0.2673	0.2704	0.2734	0.2764	0.2794	0.2823	0.2852
0.8	0.2881	0.2910	0.2939	0.2967	0.2996	0.3023	0.3051	0.3078	0.3106	0.3133
0.9	0.3159	0.3186	0.3212	0.3238	0.3264	0.3289	0.3315	0.3340	0.3365	0.3389
1.0	0.3413	0.3438	0.3461	0.3485	0.3508	0.3531	0.3554	0.3577	0.3599	0.3621
1.1	0.3643	0.3665	0.3686	0.3708	0.3729	0.3749	0.3770	0.3790	0.3810	0.3830
1.2	0.3849	0.3869	0.3888	0.3907	0.3925	0.3944	0.3962	0.3980	0.3997	0.4015
1.3	0.4032	0.4049	0.4066	0.4082	0.4099	0.4115	0.4131	0.4147	0.4162	0.4177
1.4	0.4192	0.4207	0.4222	0.4236	0.4251	0.4265	0.4279	0.4292	0.4306	0.4319
1.5	0.4332	0.4345	0.4357	0.4370	0.4382	0.4394	0.4406	0.4418	0.4429	0.4441
1.6	0.4452	0.4463	0.4474	0.4484	0.4495	0.4505	0.4515	0.4525	0.4535	0.4545
1.7	0.4554	0.4564	0.4573	0.4582	0.4591	0.4599	0.4608	0.4616	0.4625	0.4633
1.8	0.4641	0.4649	0.4656	0.4664	0.4671	0.4678	0.4686	0.4693	0.4699	0.4706
1.9	0.4713	0.4719	0.4726	0.4732	0.4738	0.4744	0.4750	0.4756	0.4761	0.4767
2.0	0.4772	0.4778	0.4783	0.4788	0.4793	0.4798	0.4803	0.4808	0.4812	0.4817
2.1	0.4821	0.4826	0.4830	0.4834	0.4838	0.4842	0.4846	0.4850	0.4854	0.4857
2.2	0.4861	0.4864	0.4868	0.4871	0.4875	0.4878	0.4881	0.4884	0.4887	0.4890
2.3	0.4893	0.4896	0.4898	0.4901	0.4904	0.4906	0.4909	0.4911	0.4913	0.4916
2.4	0.4918	0.4920	0.4922	0.4925	0.4927	0.4929	0.4931	0.4932	0.4934	0.4936
2.5	0.4938	0.4940	0.4941	0.4943	0.4945	0.4946	0.4948	0.4949	0.4951	0.4952
2.6	0.4953	0.4955	0.4956	0.4957	0.4959	0.4960	0.4961	0.4962	0.4963	0.4964
2.7	0.4965	0.4966	0.4967	0.4968	0.4969	0.4970	0.4971	0.4972	0.4973	0.4974
2.8	0.4974	0.4975	0.4976	0.4977	0.4977	0.4978	0.4979	0.4979	0.4980	0.4981
2.9	0.4981	0.4982	0.4982	0.4983	0.4984	0.4984	0.4985	0.4985	0.4986	0.4986
3.0	0.4987	0.4987	0.4987	0.4988	0.4988	0.4989	0.4989	0.4989	0.4990	0.4990
3.1	0.4990	0.4991	0.4991	0.4991	0.4992	0.4992	0.4992	0.4992	0.4993	0.4993
3.2	0.4993	0.4993	0.4994	0.4994	0.4994	0.4994	0.4994	0.4995	0.4995	0.4995
3.3	0.4995	0.4995	0.4996	0.4996	0.4996	0.4996	0.4996	0.4996	0.4996	0.4997
3.4	0.4997	0.4997	0.4997	0.4997	0.4997	0.4997	0.4997	0.4997	0.4997	0.4998
3.5	0.4998	0.4998	0.4998	0.4998	0.4998	0.4998	0.4998	0.4998	0.4998	0.4998
3.6	0.4998	0.4998	0.4999	0.4999	0.4999	0.4999	0.4999	0.4999	0.4999	0.4999
3.7	0.4999	0.4999	0.4999	0.4999	0.4999	0.4999	0.4999	0.4999	0.4999	0.4999
3.8	0.4999	0.4999	0.4999	0.4999	0.4999	0.4999	0.4999	0.4999	0.4999	0.4999
3.9	0.5000	0.5000	0.5000	0.5000	0.5000	0.5000	0.5000	0.5000	0.5000	0.5000

APPENDIX 15.A
Gamma Function Values*

$$\text{Values of } \Gamma(\alpha) = \int_0^\infty e^{-t}t^{\alpha-1}\,dt;\ \Gamma(\alpha+1) = \alpha\Gamma(\alpha)$$

α	$\Gamma(\alpha)$	α	$\Gamma(\alpha)$	α	$\Gamma(\alpha)$	α	$\Gamma(\alpha)$
1.00	1.00000	1.25	0.90640	1.50	0.88623	1.75	0.91906
1.01	0.99433	1.26	0.90440	1.51	0.88659	1.76	0.92137
1.02	0.98884	1.27	0.90250	1.52	0.88704	1.77	0.92376
1.03	0.98355	1.28	0.90072	1.53	0.88757	1.78	0.92623
1.04	0.97844	1.29	0.89904	1.54	0.88818	1.79	0.92877
1.05	0.97350	1.30	0.89747	1.55	0.88887	1.80	0.93138
1.06	0.96874	1.31	0.89600	1.56	0.88964	1.81	0.93408
1.07	0.96415	1.32	0.89464	1.57	0.89049	1.82	0.93685
1.08	0.95973	1.33	0.89338	1.58	0.89142	1.83	0.93969
1.09	0.95546	1.34	0.89222	1.59	0.89243	1.84	0.94261
1.10	0.95135	1.35	0.89115	1.60	0.89352	1.85	0.94561
1.11	0.94740	1.36	0.89018	1.61	0.89468	1.86	0.94869
1.12	0.94359	1.37	0.88931	1.62	0.89592	1.87	0.95184
1.13	0.93993	1.38	0.88854	1.63	0.89724	1.88	0.95507
1.14	0.93642	1.39	0.88785	1.64	0.89864	1.89	0.95838
1.15	0.93304	1.40	0.88726	1.65	0.90012	1.90	0.96177
1.16	0.92980	1.41	0.88676	1.66	0.90167	1.91	0.96523
1.17	0.92670	1.42	0.88636	1.67	0.90330	1.92	0.96877
1.18	0.92373	1.43	0.88604	1.68	0.90500	1.93	0.97240
1.19	0.92089	1.44	0.88581	1.69	0.90678	1.94	0.97610
1.20	0.91817	1.45	0.88566	1.70	0.90864	1.95	0.97988
1.21	0.91558	1.46	0.88560	1.71	0.91057	1.96	0.98374
1.22	0.91311	1.47	0.88563	1.72	0.91258	1.97	0.98768
1.23	0.91075	1.48	0.88575	1.73	0.91466	1.98	0.999171
1.24	0.90852	1.49	0.88595	1.74	0.91683	1.99	0.99581
						2.00	1.00000

*For large positive values of t, $\Gamma(t)$ approximates Stirling's asymptotic series.

$$\Gamma(t) \approx t^t e^{-t}\sqrt{\frac{2\pi}{t}}\left(1 + \frac{1}{12t} + \frac{1}{288t^2} - \frac{139}{51{,}840t^3} - \frac{571}{2{,}488{,}320t^4} + \cdots\right)$$

APPENDIX 15.B
Bessel Functions J_0 and J_1

x	$J_0(x)$	$J_1(x)$	x	$J_0(x)$	$J_1(x)$	x	$J_0(x)$	$J_1(x)$
0.0	1.0000	0.0000	5.0	−0.1776	−0.3276	10.0	−0.2459	0.0435
0.1	0.9975	0.0499	5.1	−0.1443	−0.3371	10.1	−0.2490	0.0184
0.2	0.9900	0.0995	5.2	−0.1103	−0.3432	10.2	−0.2496	−0.0066
0.3	0.9776	0.1483	5.3	−0.0758	−0.3460	10.3	−0.2477	−0.0313
0.4	0.9604	0.1960	5.4	−0.0412	−0.3453	10.4	−0.2434	−0.0555
0.5	0.9385	0.2423	5.5	−0.0068	−0.3414	10.5	−0.2366	−0.0789
0.6	0.9120	0.2867	5.6	0.0270	−0.3343	10.6	−0.2276	−0.1012
0.7	0.8812	0.3290	5.7	0.0599	−0.3241	10.7	−0.2164	−0.1224
0.8	0.8463	0.3688	5.8	0.0917	−0.3110	10.8	−0.2032	−0.1422
0.9	0.8075	0.4059	5.9	0.1220	−0.2951	10.9	−0.1881	−0.1603
1.0	0.7652	0.4401	6.0	0.1506	−0.2767	11.0	−0.1712	−0.1768
1.1	0.7196	0.4709	6.1	0.1773	−0.2559	11.1	−0.1528	−0.1913
1.2	0.6711	0.4983	6.2	0.2017	−0.2329	11.2	−0.1330	−0.2039
1.3	0.6201	0.5220	6.3	0.2238	−0.2081	11.3	−0.1121	−0.2143
1.4	0.5669	0.5419	6.4	0.2433	−0.1816	11.4	−0.0902	−0.2225
1.5	0.5118	0.5579	6.5	0.2601	−0.1538	11.5	−0.0677	−0.2284
1.6	0.4554	0.5699	6.6	0.2740	−0.1250	11.6	−0.0446	−0.2320
1.7	0.3980	0.5778	6.7	0.2851	−0.0953	11.7	−0.0213	−0.2333
1.8	0.3400	0.5815	6.8	0.2931	−0.0652	11.8	0.0020	−0.2323
1.9	0.2818	0.5812	6.9	0.2981	−0.0349	11.9	0.0250	−0.2290
2.0	0.2239	0.5767	7.0	0.3001	−0.0047	12.0	0.0477	−0.2234
2.1	0.1666	0.5683	7.1	0.2991	0.0252	12.1	0.0697	−0.2157
2.2	0.1104	0.5560	7.2	0.2951	0.0543	12.2	0.0908	−0.2060
2.3	0.0555	0.5399	7.3	0.2882	0.0826	12.3	0.1108	−0.1943
2.4	0.0025	0.5202	7.4	0.2786	0.1096	12.4	0.1296	−0.1807
2.5	−0.0484	0.4971	7.5	0.2663	0.1352	12.5	0.1469	−0.1655
2.6	−0.0968	0.4708	7.6	0.2516	0.1592	12.6	0.1626	−0.1487
2.7	−0.1424	0.4416	7.7	0.2346	0.1813	12.7	0.1766	−0.1307
2.8	−0.1850	0.4097	7.8	0.2154	0.2014	12.8	0.1887	−0.1114
2.9	−0.2243	0.3754	7.9	0.1944	0.2192	12.9	0.1988	−0.0912
3.0	−0.2601	0.3391	8.0	0.1717	0.2346	13.0	0.2069	−0.0703
3.1	−0.2921	0.3009	8.1	0.1475	0.2476	13.1	0.2129	−0.0489
3.2	−0.3202	0.2613	8.2	0.1222	0.2580	13.2	0.2167	−0.0271
3.3	−0.3443	0.2207	8.3	0.0960	0.2657	13.3	0.2183	−0.0052
3.4	−0.3643	0.1792	8.4	0.0692	0.2708	13.4	0.2177	0.0166
3.5	−0.3801	0.1374	8.5	0.0419	0.2731	13.5	0.2150	0.0380
3.6	−0.3918	0.0955	8.6	0.0146	0.2728	13.6	0.2101	0.0590
3.7	−0.3992	0.0538	8.7	−0.0125	0.2697	13.7	0.2032	0.0791
3.8	−0.4026	0.0128	8.8	−0.0392	0.2641	13.8	0.1943	0.0984
3.9	−0.4018	−0.0272	8.9	−0.0653	0.2559	13.9	0.1836	0.1165
4.0	−0.3971	−0.0660	9.0	−0.0903	0.2453	14.0	0.1711	0.1334
4.1	−0.3887	−0.1033	9.1	−0.1142	0.2324	14.1	0.1570	0.1488
4.2	−0.3766	−0.1386	9.2	−0.1367	0.2174	14.2	0.1414	0.1626
4.3	−0.3610	−0.1719	9.3	−0.1577	0.2004	14.3	0.1245	0.1747
4.4	−0.3423	−0.2028	9.4	−0.1768	0.1816	14.4	0.1065	0.1850
4.5	−0.3205	−0.2311	9.5	−0.1939	0.1613	14.5	0.0875	0.1934
4.6	−0.2961	−0.2566	9.6	−0.2090	0.1395	14.6	0.0679	0.1999
4.7	−0.2693	−0.2791	9.7	−0.2218	0.1166	14.7	0.0476	0.2043
4.8	−0.2404	−0.2985	9.8	−0.2323	0.0928	14.8	0.0271	0.2066
4.9	−0.2097	−0.3147	9.9	−0.2403	0.0684	14.9	0.0064	0.2069

APPENDIX 15.C
Properties of Fourier Series for Periodic Signals

periodic signal	Fourier series coefficients
$x(t)$, $y(t)$ } periodic with period T_0	a_k, b_k
$Ax(t) + By(t)$	$Aa_k + Bb_k$
$x(t - t_0)$	$a_k e^{-jk(2\pi/T_0)t_0}$
$e^{jM(2\pi/T_0)t}x(t)$	a_{k-M}
$x^*(t)$	a^*_{-k}
$x(-t)$	a_{-k}
$x(\alpha t)$, $\alpha > 0$ [periodic with period T_0/α]	a_k
$\int_{T_0} x(\tau)y(t-\tau)\,d\tau$	$T_0 a_k b_k$
$x(t)y(t)$	$\sum_{l=-\infty}^{+\infty} a_l b_{k-l}$
$\frac{dx(t)}{dt}$	$jk\frac{2\pi}{T_0}a_k$
$\int_{-\infty}^{t} x(t)\,dt$ [finite-valued and periodic only if $a_0 = 0$]	$\left(\frac{1}{jk(2\pi/T_0)}\right)a_k$
$x(t)$ real	$\begin{cases} a_k = a^*_{-k} \\ \mathcal{R}e\{a_k\} = \mathcal{R}e\{a_{-k}\} \\ \mathcal{I}m\{a_k\} = -\mathcal{I}m\{a_{-k}\} \\ \lvert a_k\rvert = \lvert a_{-k}\rvert \\ \angle a_k = -\angle a_{-k} \end{cases}$
$x_e(t) = \text{even}\{x(t)\}$ [$x(t)$ real]	$\mathcal{R}e\{a_k\}$
$x_o(t) = \text{odd}\{x(t)\}$ [$x(t)$ real]	$j\mathcal{I}m\{a_k\}$

Parseval's relation for periodic signals

$$\frac{1}{T_0}\int_{T_0} \lvert x(t)\rvert^2\,dt = \sum_{k=-\infty}^{+\infty} \lvert a_k\rvert^2$$

Appendices

APPENDIX 15.D
Properties of Fourier Transform for Aperiodic Signals

aperiodic signal	Fourier transform
$x(t)$	$X(\omega)$
$y(t)$	$Y(\omega)$
$ax(t) + by(t)$	$aX(t) + bY(t)$
$x(t - t_0)$	$e^{-j\omega t_0}X(\omega)$
$e^{j\omega_0 t}x(t)$	$X(\omega - \omega_0)$
$x^*(t)$	$X^*(-\omega)$
$x(-t)$	$X(-\omega)$
$x(at)$	$\frac{1}{\lvert a\rvert}X\left(\frac{\omega}{a}\right)$
$x(t) * y(t)$	$X(\omega)Y(\omega)$
$x(t)y(t)$	$\frac{1}{2\pi}X(\omega) * Y(\omega)$
$\frac{d}{dt}x(t)$	$j\omega X(\omega)$
$\int_{-\infty}^{t} x(t)\,dt$	$\frac{1}{j\omega}X(\omega) + \pi X(0)\delta(\omega)$
$tx(t)$	$j\frac{d}{d\omega}X(\omega)$
$x(t)$ real	$\begin{cases} X(\omega) = X^*(-\omega) \\ \mathcal{R}e\{X(\omega)\} = \mathcal{R}e\{X(-\omega)\} \\ \mathcal{I}m\{X(\omega)\} = -\mathcal{I}m\{X(-\omega)\} \\ \lvert X(\omega)\rvert = \lvert X(-\omega)\rvert \\ \angle X(\omega) = -\angle X(-\omega) \end{cases}$
$x_e(t) = \text{even}\{x(t)\}$ [$x(t)$ real]	$\mathcal{R}e\{X(\omega)\}$
$x_o(t) = \text{odd}\{x(t)\}$ [$x(t)$ real]	$j\mathcal{I}m\{X(\omega)\}$

duality

$$f(u) = \int_{-\infty}^{+\infty} g(v)e^{juv}\,dv$$

$$g(t) \overset{\mathcal{F}}{\leftrightarrow} f(\omega)$$

$$f(t) \overset{\mathcal{F}}{\leftrightarrow} 2\pi g(-\omega)$$

Parseval's relation for aperiodic signals

$$\int_{-\infty}^{+\infty} \lvert x(t)\rvert^2\,dt = \frac{1}{2\pi}\int_{-\infty}^{+\infty} \lvert X(\omega)\rvert^2\,d\omega$$

APPENDIX 15.E
Fourier Transform Pairs

signal	Fourier transform
$\sum_{k=-\infty}^{+\infty} a_k e^{jk\omega_0 t}$	$2\pi \sum_{k=-\infty}^{+\infty} a_k \delta(\omega - k\omega_0)$
$e^{j\omega_0 t}$	$2\pi\delta(\omega - \omega_0)$
$\cos\omega_0 t$	$\pi\Big(\delta(\omega-\omega_0)+\delta(\omega+\omega_0)\Big)$
$\sin\omega_0 t$	$\frac{\pi}{j}\Big(\delta(\omega-\omega_0)+\delta(\omega+\omega_0)\Big)$
$x(t) = 1$	$2\pi\delta(\omega)$
Periodic square wave $x(t) = \begin{cases} 1, & \lvert t\rvert < T_1 \\ 0, & T_1 < \lvert t\rvert \le \frac{T_0}{2} \end{cases}$ and $x(t+T_0) = x(t)$	$\sum_{k=-\infty}^{+\infty} \frac{2\sin k\omega_0 T_1}{k}\delta(\omega - k\omega_0)$
$\sum_{k=-\infty}^{+\infty} \delta(t - nT)$	$\frac{2\pi}{T}\sum_{k=-\infty}^{+\infty} \delta\left(\omega - \frac{2\pi k}{T}\right)$
$x(t) = \begin{cases} 1, & \lvert t\rvert < T_1 \\ 0, & \lvert t\rvert > T_1 \end{cases}$	$2T_1 \operatorname{sinc}\frac{\omega T_1}{\pi} = \frac{2\sin\omega T_1}{\omega}$
$\frac{W}{\pi}\operatorname{sinc}\frac{Wt}{\pi} = \frac{\sin Wt}{\pi t}$	$X(\omega) = \begin{cases} 1, & \lvert\omega\rvert < W \\ 0, & \lvert\omega\rvert > W \end{cases}$
$\delta(t)$	1
$u(t)$	$\frac{1}{j\omega} + \pi\delta(\omega)$
$\delta(t - t_0)$	$e^{-j\omega t_0}$
$e^{-at}u(t),\ \mathcal{R}e\{a\} > 0$	$\frac{1}{a + j\omega}$
$te^{-at}u(t),\ \mathcal{R}e\{a\} > 0$	$\frac{1}{(a+j\omega)^2}$
$\frac{t^{n-1}}{(n-1)!}e^{-at}u(t),$ $\mathcal{R}e\{a\} > 0$	$\frac{1}{(a+j\omega)^n}$

Appendices

APPENDIX 15.F
Properties of Laplace Transform

signal	transform	ROC (region of convergence)
$x(t)$	$X(s)$	R
$x_1(t)$	$X_1(s)$	R_1
$x_2(t)$	$X_2(s)$	R_2
$ax_1(t) + bx_2(t)$	$aX_1(s) + bX_2(s)$	at least $R_1 \cap R_2$
$x(t - t_0)$	$e^{-st_0}X(s)$	R
$e^{s_0 t}x(t)$	$X(s - s_0)$	shifted version of R [i.e., s is in the ROC if $(s - s_0)$ is in R]
$x(at)$	$\frac{1}{\lvert a\rvert}X\left(\frac{s}{a}\right)$	"scaled" ROC [i.e., s is in the ROC if s/a is in the ROC of $X(s)$]
$x_1(t) * x_2(t)$	$X_1(s)X_2(s)$	at least $R_1 \cap R_2$
$\frac{d}{dt}x(t)$	$sX(s)$	at least R
$-tx(t)$	$\frac{d}{ds}X(s)$	R
$\int_{-\infty}^{t} x(\tau)\,d\tau$	$\frac{1}{s}X(s)$	at least $R \cap \{\mathcal{R}e\{s\} > 0\}$

APPENDIX 15.G
Laplace Transforms
(also see App. 11.A)

transform pair	signal	transform	ROC
1	$\delta(t)$	1	all s
2	$u(t)$	$\frac{1}{s}$	$\mathcal{R}e\{s\} > 0$
3	$-u(-t)$	$\frac{1}{s}$	$\mathcal{R}e\{s\} < 0$
4	$\frac{t^{n-1}}{(n-1)!}u(t)$	$\frac{1}{s^n}$	$\mathcal{R}e\{s\} > 0$
5	$-\frac{t^{n-1}}{(n-1)!}u(-t)$	$\frac{1}{s^n}$	$\mathcal{R}e\{s\} < 0$
6	$e^{-\alpha t}u(t)$	$\frac{1}{s+\alpha}$	$\mathcal{R}e\{s\} > \alpha$
7	$-e^{-\alpha t}u(-t)$	$\frac{1}{s+\alpha}$	$\mathcal{R}e\{s\} < -\alpha$
8	$\frac{t^{n-1}}{(n-1)!}e^{-\alpha t}u(t)$	$\frac{1}{(s+\alpha)^n}$	$\mathcal{R}e\{s\} > -\alpha$
9	$-\frac{t^{n-1}}{(n-1)!}e^{-\alpha t}u(-t)$	$\frac{1}{(s+\alpha)^n}$	$\mathcal{R}e\{s\} < -\alpha$
10	$\delta(t-T)$	e^{-sT}	all s
11	$(\cos\omega_0 t)u(t)$	$\frac{s}{s^2+\omega_0^2}$	$\mathcal{R}e\{s\} > 0$
12	$(\sin\omega_0 t)u(t)$	$\frac{\omega_0}{s^2+\omega_0^2}$	$\mathcal{R}e\{s\} > 0$
13	$(e^{-\alpha t}\cos\omega_0 t)u(t)$	$\frac{s+\alpha}{(s+\alpha)^2+\omega_0^2}$	$\mathcal{R}e\{s\} > -\alpha$
14	$(e^{-\alpha t}\sin\omega_0 t)u(t)$	$\frac{\omega_0}{(s+\alpha)^2+\omega_0^2}$	$\mathcal{R}e\{s\} > -\alpha$

APPENDIX 15.H
Properties of Discrete-Time Fourier Series for Periodic Signals

periodic signal	Fourier series coefficients
$\left.\begin{matrix} x(n) \\ y(n) \end{matrix}\right\}$ periodic with period N	$\left.\begin{matrix} a_k \\ b_k \end{matrix}\right\}$ periodic with period N
$Ax[n] + By[n]$	$Aa_k + Bb_k$
$x[n - n_0]$	$a_k e^{-jk(2\pi/N)n_0}$
$e^{jM(2\pi/N)n}x[n]$	a_{k-M}
$x^*[n]$	a^*_{-k}
$x^*[-n]$	a_{-k}
$x_{(m)}[n] = \begin{cases} x[n/m] \text{ if } n \text{ is a multiple of } m \\ 0 \text{ if } n \text{ is not a multiple of } m \end{cases}$ (periodic with mN)	$\frac{1}{m}a_k$ $\left[\begin{matrix}\text{viewed as periodic} \\ \text{with period } mN\end{matrix}\right]$
$\sum_{r=\langle N\rangle} x[r]y[n-r]$	$Na_k b_k$
$x[n]y[n]$	$\sum_{l-\langle N\rangle} a_l b_{k-1}$
$x[n] - x[n-1]$	$(1 - e^{-jk(2\pi/N)})a_k$
$\sum_{k=-\infty}^{n} x[k]$ $\left[\begin{matrix}\text{finite-valued and periodic} \\ \text{only if } a_0 = 0\end{matrix}\right]$	$\left(\frac{1}{1 - e^{-jk(2\pi/N)}}\right)a_k$
$x[n]$ real	$\begin{cases} a_k = a^*_{-k} \\ \mathcal{R}e\{a_k\} = \mathcal{R}e\{a_{-k}\} \\ \mathcal{I}m\{a_k\} = -\mathcal{I}m\{a_{-k}\} \\ \lvert a_k\rvert = \lvert a_{-k}\rvert \\ \angle a_k = -\angle a_{-k} \end{cases}$
$x_e[n] = \text{even}\{x[n]\}$ [$x[n]$ real]	$\mathcal{R}e\{a_k\}$
$x_o[n] = \text{odd}\{x[n]\}$ [$x[n]$ real]	$j\mathcal{I}m[a_k]$

Parseval's relation for periodic signals

$$\frac{1}{N}\sum_{n=\langle N\rangle} \lvert x[n]\rvert^2 = \sum_{k=\langle N\rangle} \lvert a_k\rvert^2$$

APPENDIX 15.I

Properties of Discrete-Time Fourier Transform for Aperiodic Signals

aperiodic signal	Fourier transform
$x(n)$ $y(n)$	$\left.\begin{matrix} X(\Omega) \\ Y(\Omega) \end{matrix}\right\}$ periodic with period 2π
$ax[n]+by[n]$	$aX(\Omega)+bY(\Omega)$
$x[n-n_0]$	$e^{-j\Omega n_0}X(\Omega)$
$e^{j\Omega_0 n}x[n]$	$X(\Omega-\Omega_0)$
$x^*[n]$	$X^*(-\Omega)$
$x[-n]$	$X(-\Omega)$
$x_{(k)}[n]=\begin{cases} x[n/k], & \text{if } n \text{ is a multiple of } k \\ 0, & \text{if } n \text{ is not a multiple of } k \end{cases}$	$X(k\Omega)$
$x[n]*y[n]$	$X(\Omega)\,Y(\Omega)$
$x[n]y[n]$	$\dfrac{1}{2\pi}\displaystyle\int_{2\pi} X(\theta)\,Y(\Omega-\theta)\,d\theta$
$x[n]-x[n-1]$	$1-e^{-j\Omega}X(\Omega)$
$\displaystyle\sum_{k=-\infty}^{n} x[k]$	$\dfrac{1}{1-e^{j\Omega}}X(\Omega)+\pi X(0)\displaystyle\sum_{k=-\infty}^{+\infty}\delta(\Omega-2\pi k)$
$nx[n]$	$j\dfrac{dX(\Omega)}{d\Omega}$
$x[n]$ real	$\begin{cases} X(\Omega)=X^*(-\Omega) \\ \mathcal{R}e\{X(\Omega)\}=\mathcal{R}e\{X(-\Omega)\} \\ \mathcal{I}m\{X(\Omega)\}=-\mathcal{I}m\{X(-\Omega)\} \\ \lvert X(\Omega)\rvert=\lvert X(-\Omega)\rvert \\ \angle X(\Omega)=-\angle X(-\Omega) \end{cases}$
$x_e[n]=\text{even}\{x[n]\}$ $[x[n]$ real$]$	$\mathcal{R}e\{X(\Omega)\}$
$x_o[n]=\text{odd}\{x[n]\}$ $[x[n]$ real$]$	$j\mathcal{I}m\{X(\Omega)\}$

Parseval's relation for aperiodic signals

$$\sum_{n=-\infty}^{+\infty}\lvert x[n]\rvert^2=\frac{1}{2\pi}\int_{2\pi}\lvert X(\Omega)\rvert^2\,d\Omega$$

Appendices

APPENDIX 15.J
Properties of z-Transforms

sequence	transform	region of convergence
$x[n]$	$X(z)$	R_x
$x_1[n]$	$X_1(z)$	R_1
$x_2[n]$	$X_2(z)$	R_2
$ax_1[n] + bx_2[n]$	$aX_1(z) + bX_2(z)$	at least the intersection of R_1 and R_2
$x[n - n_0]$	$z^{-n_0}X(z)$	R_x except for the possible addition or deletion of the origin
$e^{j\Omega_0 n}x[n]$	$X(e^{-j\Omega_0}z)$	R_x
$z_0^n x[n]$	$X\left(\frac{z}{z_0}\right)$	z_0R_x
$a^n x[n]$	$X(a^{-1}z)$	scaled version of R_x (i.e., $\lvert a\rvert R_x =$ the set of points $\{\lvert a\rvert z\}$ for z in R_x)
$x[-n]$	$X(z^{-1})$	inverted R_x (i.e., $R_x^{-1} =$ the set of points z^{-1} where z is in R_x)
$w[n] = \begin{cases} x[r], & n = rk \\ 0, & n \neq rk \text{ for some } r \end{cases}$	$X(z^k)$	$R_x^{1/k}$ (i.e., the set of points $z^{1/k}$ where z is in R_x)
$x_1[n]x_2[n]$	$X_1(z)X_2(z)$	at least the intersection of R_1 and R_2
$nx[n]$	$-z\dfrac{dX(z)}{dz}$	R_x except for the possible addition or deletion of the origin
$\sum_{k=-\infty}^{n} x[k]$	$\dfrac{1}{1 - z^{-1}}X(z)$	at least the intersection of R_x and $\lvert z\rvert > 1$

Appendices

APPENDIX 15.K
z-Transforms

$f(t)$	$F(z)$
$\delta(t)$	1
$\delta(t-mT)$	$\dfrac{1}{z^m}$
$H(t)$	$\dfrac{z}{z-m}$
t	$\dfrac{Tz}{(z-1)^2}$
t^2	$\dfrac{T^2 z(z+1)}{(z-1)^3}$
t^3	$\dfrac{T^3 z(z^2+4z+1)}{(z-1)^4}$
t^n	$(-1)^n \lim\limits_{x\to 0} \dfrac{\delta^n}{\delta_{x^n}}\left(\dfrac{z}{z-e^{-xT}}\right)$
$1-a^{\Omega\tau}$	$\dfrac{z(1-a^{\Omega\tau})}{(z-1)(z-a^{\Omega\tau})}$
$a^{\Omega\tau}$	$\dfrac{z}{z-a^{\Omega\tau}}$
$ta^{\Omega\tau}$	$\dfrac{Tza^{\Omega\tau}}{(z-a^{\Omega\tau})^2}$
$t^2 a^{\Omega\tau}$	$\dfrac{T^2 a^{\Omega\tau} z(z+a^{\Omega\tau})}{(z-a^{\Omega\tau})^3}$
$\sin \Omega\tau$	$\dfrac{z \sin \Omega T}{z^2 - 2z \cos \Omega T + 1}$
$\cos \Omega t$	$\dfrac{z(z-\cos \Omega T)}{z^2 - 2z \cos \Omega T + 1}$
$\sinh \Omega t$	$\dfrac{z \sinh \Omega T}{z^2 - 2z \cosh \Omega T + 1}$
$\cosh \Omega t$	$\dfrac{z(z-\cosh \Omega T)}{z^2 - 2z \cosh \Omega T + 1}$
$e^{-at} \sin \Omega t$	$\dfrac{ze^{-a\tau} \sin \Omega T}{z^2 - 2ze^{-\tau} \cos \Omega T + e^{-2a\tau}}$
$e^{-at} \cos \Omega t$	$\dfrac{z(z-e^{-a\tau} \cos \Omega T)}{z^2 - 2ze^{-\tau} \cos \Omega T + e^{-2a\tau}}$
$e^{-at} \sinh \Omega t$	$\dfrac{ze^{-a\tau} \sinh \Omega T}{z^2 - 2ze^{-a\tau} \cosh \Omega T + e^{-2a\tau}}$
$e^{-at} \cosh \Omega t$	$\dfrac{z(z-e^{-a\tau} \cosh \Omega T)}{z^2 - 2ze^{-a\tau} \cosh \Omega T + e^{-2a\tau}}$

Appendices

APPENDIX 17.A
Room Temperature Properties of Silicon, Germanium, and Gallium Arsenide (at 300K)

property	Si	Ge	GaAs
atoms/cm^3	5.0×10^{22}	4.42×10^{22}	2.21×10^{22}
atomic weight	28.08	72.6	144.63
breakdown field	$\approx 3 \times 10^5$ V/cm	$\approx 10^5$ V/cm	$\approx 4 \times 10^5$ V/cm
crystal structure	diamond	diamond	zincblende
density	2.328 g/m^3	5.3267 g/m^3	5.32 g/m^3
dielectric constant	11.8	16	10.9
effective density of states			
in conduction band, N_c	2.8×10^{19} cm^{-3}	1.04×10^{19} cm^{-3}	4.7×10^{17} cm^{-3}
in valence band, N_v	1.02×10^{19} cm^{-3}	6.1×10^{18} cm^{-3}	7.0×10^{18} cm^{-3}
effective mass m^*/m_0 (electrons)			
longitudinal	0.98 kg	1.64 kg	1.98 kg
transverse	0.19 kg	0.082 kg	0.37 kg
effective mass, m^*/m_0 (holes)			
light	0.16 kg	0.044 kg	0.12 kg
heavy	0.49 kg	0.28 kg	0.5 kg
electron affinity, x	4.05 V	4.0 V	4.07 V
energy gap	1.12 eV	0.803 eV	1.43 eV
intrinsic carrier of concentration	1.6×10^{10} cm^{-3}	2.5×10^{12} cm^{-3}	1.1×10^{7} cm^{-3}
lattice constant	0.543086 nm	0.565748 nm	0.56534 nm
linear coefficient of thermal expansion, $\delta = \frac{\Delta L}{L \Delta T}$	2.6×10^{-6} K^{-1}	5.8×10^{-6} K^{-1}	5.9×10^{-6} K^{-1}
melting point	1420°C	937°C	1238°C
minority-carrier lifetime	2.5×10^{-3} s	10^{-3} s	$\approx 10^{-8}$ s
mobility (drift)			
electrons, μ_n	1500 cm^2/V·s	3900 cm^2/V·s	8500 cm^2/V·s
holes, μ_p	600 cm^2/V·s	1900 cm^2/V·s	400 cm^2/V·s
specific heat	0.7 J/g·°C	0.31 J/g·°C	0.35 J/g·°C
thermal conductivity	1.45 W/cm·K	0.64 W/cm·K	0.46 W/cm·K
thermal diffusivity	0.9 cm^2/s	0.36 cm^2/s	0.44 cm^2/s
work function	4.8 V	4.4 V	4.7 V

APPENDIX 18.A
Electromagnetic Spectrum

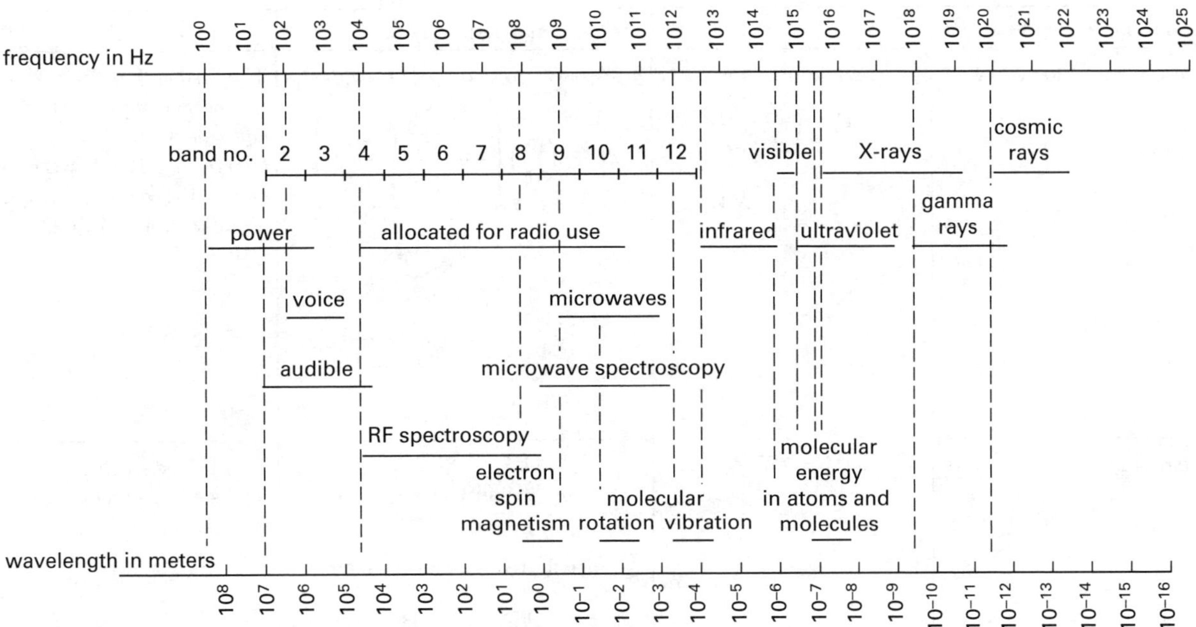

APPENDIX 25.A
Comparison of Electric and Magnetic Equations

equation description	electric version	magnetic version	remarks
experimental force law	Coulomb's law $\mathbf{F} = \left(\frac{Q_1 Q_2}{4\pi r^2}\right)\mathbf{r}$ 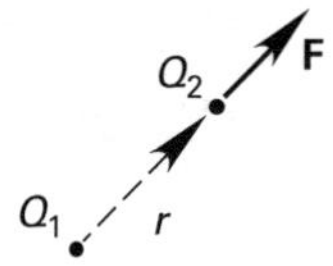	force between two current elements $d\mathbf{F} = \left(\frac{\mu_0}{4\pi}\right)\left(\frac{I_2 d\mathbf{l}_2 \times (I_1 d\mathbf{l}_1 \times r)}{r^2}\right)$ 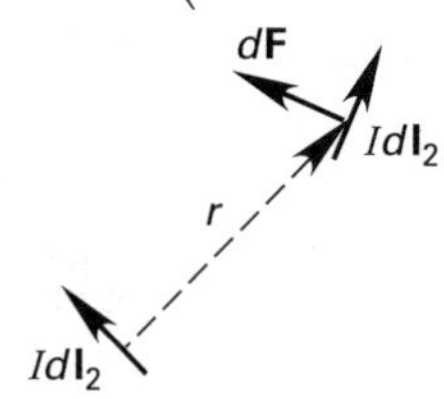	The term $Id\mathbf{l}$ in the magnetic column is the equivalent of a "magnetic charge" q_m. The I or the $d\mathbf{l}$ can be the vector. The $\mathbf{r}$ is a unit vector pointing from 1 to 2.
field definitions from force law	$\mathbf{F} = Q\mathbf{E}$	$d\mathbf{F} = \mathbf{I} \times \mathbf{B}dl$ current element $d\mathbf{F} = \mathbf{J} \times \mathbf{B}dV$ distributed current element $d\mathbf{F} = q\mathbf{v} \times \mathbf{B}$ moving charge	The V used in this row represents volume, not voltage. The $\mathbf{v}$ is the velocity.
general force law	$\mathbf{F} = q(\mathbf{E} + \mathbf{v} \times \mathbf{B})$ $d\mathbf{F} = (\rho\mathbf{E} + \mathbf{J} \times \mathbf{B})dV$ where $dQ = \rho dV$		The V in this row represents the volume, not voltage. The $\mathbf{v}$ is the velocity.
definition of scalar and vector potential	$E = -\nabla V$	$\mathbf{B} = \nabla \times \mathbf{A}$	$\mathbf{A}$ is the magnetic vector potential.
Poisson's equation for the potential function	$\nabla^2 V = -\frac{\rho}{\epsilon}$	$\nabla^2 \mathbf{A} = -\mu_0 \mathbf{J}$	From a knowledge of the charge distribution, the potential can be found and then the $\mathbf{E}$ and $\mathbf{B}$ fields determined.
Gauss's law enclosing charge and Ampere's law enclosing current	$\iint \mathbf{D} \cdot d\mathbf{A} =$ $\iiint \rho dV = Q$ $\nabla \cdot \mathbf{D} = \rho$	$\oint H \cdot d\mathbf{l} = I$ $\nabla \times \mathbf{H} = \mathbf{J}$	The V in this row represents volume.
constitutive relations	$\mathbf{D} = \epsilon\mathbf{E}$ $\mathbf{D} = \epsilon_0\mathbf{E} + \mathbf{P}$	$\mathbf{B} = \mu\mathbf{H}$ $\mathbf{B} = \mu_0\mathbf{H} + \mu_0\mathbf{M}$	The second set of equations is always valid. The first set assumes the medium is linear and isotropic.
definitions of relative permittivity and permeability	$\epsilon_r = \frac{\epsilon}{\epsilon_0}$ $\epsilon_0 =$ 8.854×10^{-12} F/m	$\mu_r = \frac{\mu}{\mu_0}$ $\mu_0 = 4\pi \times 10^{-7}$ H/m	

(continued)

Appendices

APPENDIX 25.A *(continued)*
Comparison of Electric and Magnetic Equations

equation description	electric version	magnetic version	remarks
capacitance and inductance of a field cell	$\epsilon_0 = \frac{C}{l}$	$\mu_0 = \frac{L}{l}$	Field cells are a construct designed to represent free space in terms of a parallel plate capacitor and an inductor. This capacitance and inductance exist regardless of the presence of an electric or magnetic field.
capacitance and inductance	$C = \frac{Q}{V}$	$L = \frac{\Lambda}{I}$	Λ is the flux linkage.
energy density of a field	$U = \frac{1}{2}\epsilon E^2$	$U = \frac{1}{2}\mu H^2$	Both energy and momentum are carried by a field.
energy stored by capacitance and inductance	$W = \frac{1}{2}CV^2$	$W = \frac{1}{2}LI^2$	
electromotive and magnetomotive force with sources present	$\oint \mathcal{E} \cdot d\mathbf{l} = \mathcal{E} = V$	$\oint \mathbf{H} \cdot d\mathbf{l} = NI = F_m = V_m$	The $\mathcal{E}$ is the emf, not the permittivity. Without sources present, both line integrals are equal to zero.
dipole moments	$\mathbf{p} = q\mathbf{d}$ 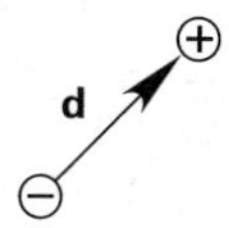	$\mathbf{m} = I\mathbf{A}$ 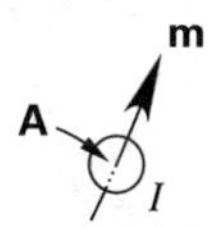	
dipole torque	$\mathbf{T} = \mathbf{p} \times \mathbf{E}$	$\mathbf{T} = \mathbf{m} \times \mathbf{B}$	This torque occurs due to the dipole being immersed in an external **E** or **B** field.
dipole potential energy	$W = -\mathbf{p} \cdot \mathbf{E}$	$W = -\mathbf{m} \cdot \mathbf{B}$	

APPENDIX 27.A
Impedance of Series-Connected Circuit Elements

circuit	impedance $Z = R + jX$ (Ω)	magnitude of impedance $\lvert Z\rvert = \sqrt{R^2 + X^2}$ (Ω)	phase angle $\theta = \tan^{-1}\frac{X}{R}$ (radians)	admittance $Y = \frac{1}{Z}$ (S)
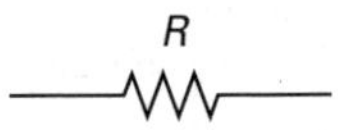	R	R	0	$\frac{1}{R}$
	$j\omega L$	ωL	$+\frac{\pi}{2}$	$-j\left(\frac{1}{\omega L}\right)$
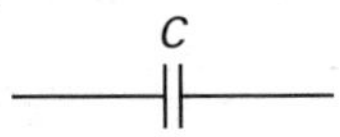	$-j\left(\frac{1}{\omega C}\right)$	$\frac{1}{\omega C}$	$-\frac{\pi}{2}$	$j\omega C$
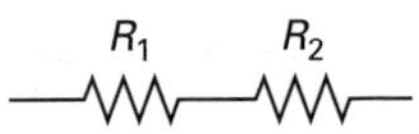	$R_1 + R_2$	$R_1 + R_2$	0	$\frac{1}{R_1 + R_2}$
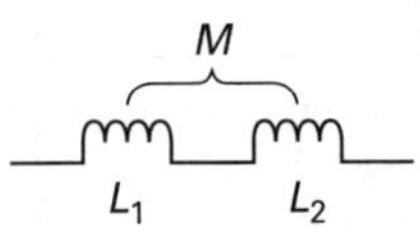	$j\omega(L_1 + L_2 \pm 2M)$	$\omega(L_1 + L_2 \pm 2M)$	$+\frac{\pi}{2}$	$-j\left(\frac{1}{\omega(L_1 + L_2 \pm 2M)}\right)$
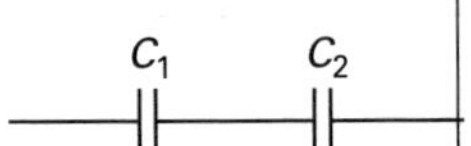	$-j\left(\frac{1}{\omega}\right)\left(\frac{C_1 + C_2}{C_1 C2}\right)$	$\frac{1}{\omega}\left(\frac{C_1 + C_2}{C_1 C2}\right)$	$-\frac{\pi}{2}$	$j\omega\left(\frac{C_1 C2}{C_1 + C_2}\right)$
	$R + j\omega L$	$\sqrt{R^2 + \omega^2 L^2}$	$\tan^{-1}\frac{\omega L}{R}$	$\frac{R - j\omega L}{R^2 + \omega^2 L^2}$
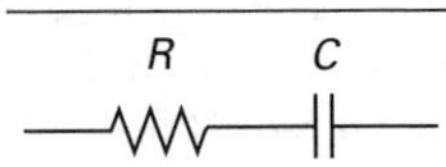	$R - j\left(\frac{1}{\omega C}\right)$	$\sqrt{\frac{\omega^2 C^2 R^2 + 1}{\omega^2 C^2}}$	$\tan^{-1}\frac{1}{\omega Rc}$	$\frac{\omega^2 C^2 R + j\omega C}{\omega^2 C^2 R^2 + 1}$
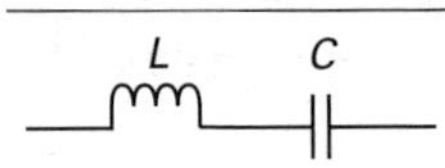	$j\left(\omega L - \frac{1}{\omega C}\right)$	$\omega L - \frac{1}{\omega C}$	$\pm\frac{\pi}{2}$	$-j\left(\frac{\omega C}{\omega^2 LC - 1}\right)$
	$R + j\left(\omega L - \frac{1}{\omega C}\right)$	$\sqrt{R^2 + \left(\omega L - \frac{1}{\omega C}\right)^2}$	$\tan^{-1}\frac{\omega L - \frac{1}{\omega C}}{R}$	$\frac{R - j\left(\omega L - \frac{1}{\omega C}\right)}{R^2 + \left(\omega L - \frac{1}{\omega C}\right)^2}$

Appendices

APPENDIX 27.B
Impedance of Parallel-Connected Circuit Elements

circuit	impedance, $Z = R + jX$ (Ω)	magnitude of impedance, $\lvert Z\rvert = \sqrt{R^2 + X^2}$ (Ω)
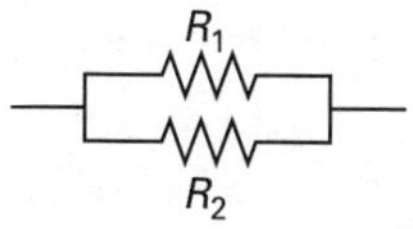	$\dfrac{R_1R_2}{R_1 + R_2}$	$\dfrac{R_1R_2}{R_1 + R_2}$
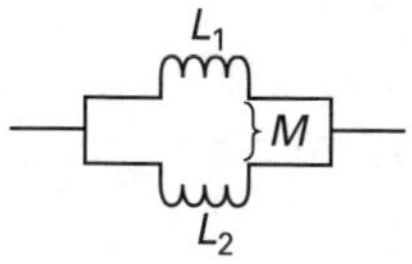	$+j\omega\left(\dfrac{L_1L_2 - M^2}{L_1 + L_2 \mp 2M}\right)$	$\omega\left(\dfrac{L_1L_2 - M^2}{L_1 + L_2 \mp 2M}\right)$
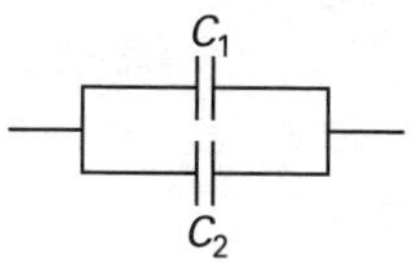	$-j\left(\dfrac{1}{\omega(C_1 + C_2)}\right)$	$\dfrac{1}{\omega(C_1 + C_2)}$
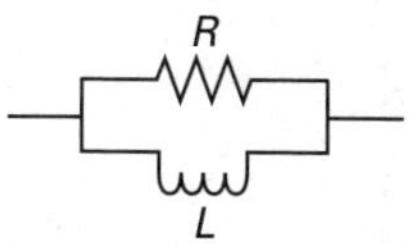	$\dfrac{\omega^2L^2R + j\omega LR^2}{\omega^2L^2 + R^2}$	$\dfrac{\omega LR}{\sqrt{\omega^2L^2 + R^2}}$
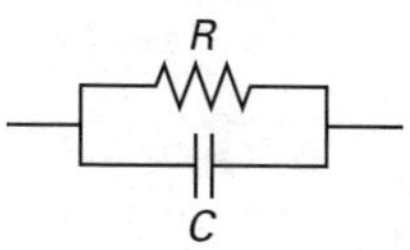	$\dfrac{R - j\omega R^2C}{1 + \omega^2R^2C^2}$	$\dfrac{R}{\sqrt{1 + \omega^2R^2C^2}}$
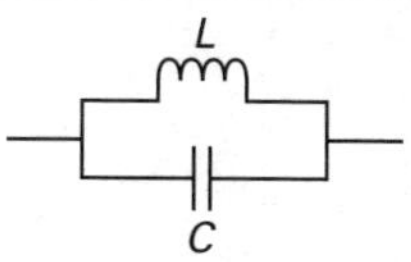	$j\left(\dfrac{\omega L}{1 - \omega^2LC}\right)$	$\dfrac{\omega L}{1 - \omega^2LC}$
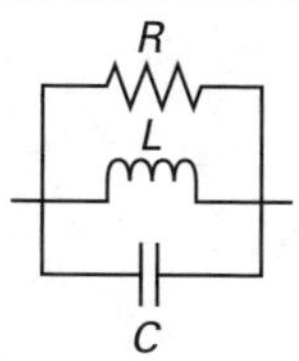	$\dfrac{\dfrac{1}{R} - j\left(\omega C - \dfrac{1}{\omega L}\right)}{\left(\dfrac{1}{R}\right)^2 + \left(\omega C - \dfrac{1}{\omega L}\right)^2}$	$\dfrac{1}{\sqrt{\left(\dfrac{1}{R}\right)^2 + \left(\omega C - \dfrac{1}{\omega L}\right)^2}}$
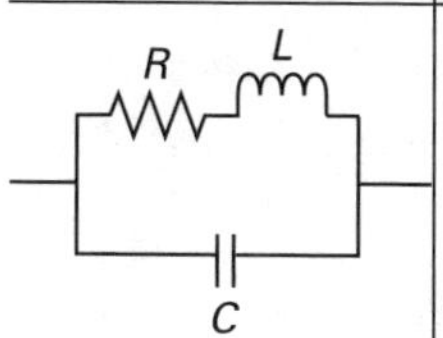	$\dfrac{\dfrac{R}{\omega^2C^2} - j\left(\dfrac{R^2}{\omega C} + \dfrac{L}{C}\left(\omega L - \dfrac{1}{\omega C}\right)\right)}{R^2 + \left(\omega L - \dfrac{1}{\omega C}\right)^2}$	$\dfrac{\sqrt{\left(\dfrac{R}{\omega^2C^2}\right)^2 + \left(\dfrac{R^2}{\omega C} + \left(\dfrac{L}{C}\right)\left(\omega L - \dfrac{1}{\omega C}\right)\right)^2}}{R^2 + \left(\omega L - \dfrac{1}{\omega C}\right)^2}$
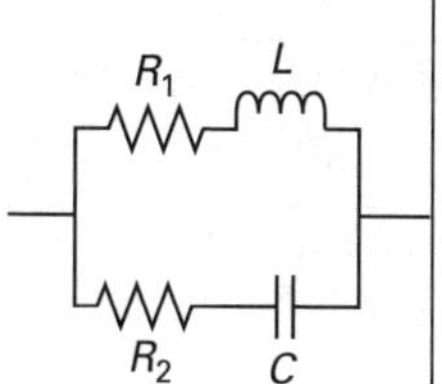	$\dfrac{R_1R_2(R_1 + R_2) + \omega^2L^2R_2 + \dfrac{R_1}{\omega^2C^2}}{(R_1 + R_2)^2 + \left(\omega L - \dfrac{1}{\omega C}\right)^2} + j\left(\dfrac{\omega R_2^2L - \dfrac{R_1^2}{\omega C} - \left(\dfrac{L}{C}\right)\left(\omega L - \dfrac{1}{\omega C}\right)}{(R_1 + R_2)^2 + \left(\omega L - \dfrac{1}{\omega C}\right)^2}\right)$	$\sqrt{\left(\dfrac{R_1R_2(R_1 + R_2) + \omega^2L^2R_2 + \dfrac{R_1}{\omega^2C^2}}{(R_1 + R_2)^2 + \left(\omega L - \dfrac{1}{\omega C}\right)^2}\right)^2 + \left(\dfrac{\omega LR_2^2 - \dfrac{R_1^2}{\omega C} - \dfrac{L}{C}\left(\omega L - \dfrac{1}{\omega C}\right)}{(R_1 + R_2)^2 + \left(\omega L - \dfrac{1}{\omega C}\right)^2}\right)^2}$

(continued)

APPENDIX 27.B *(continued)*
Impedance of Parallel-Connected Circuit Elements

circuit	phase angle, $\theta = \tan^{-1}\frac{X}{R}$ (radians)	admittance, $Y = \frac{1}{Z}$ (S)
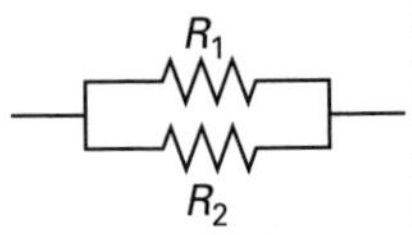	0	$\frac{R_1 + R_2}{R_1 R_2}$
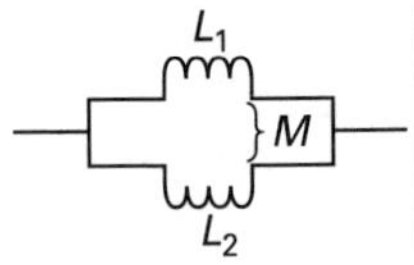	$+\frac{\pi}{2}$	$-j\left(\frac{1}{\omega}\right)\left(\frac{L_1 + L_2 \mp 2M}{L_1 L_2 - M^2}\right)$
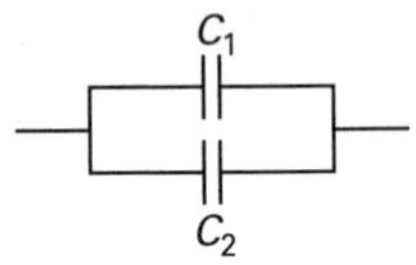	$-\frac{\pi}{2}$	$+j\omega(C_1 + C_2)$
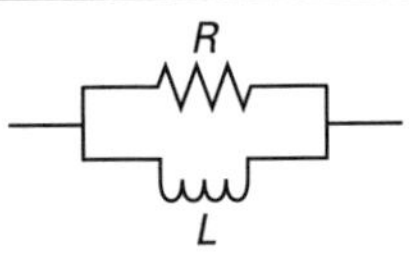	$\tan^{-1}\frac{R}{\omega L}$	$\frac{\omega L - jR}{\omega L R}$
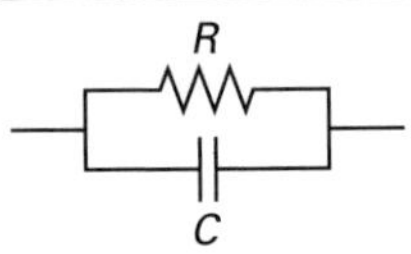	$\tan^{-1}(-\omega R C)$	$\frac{1}{R} + j\omega C$
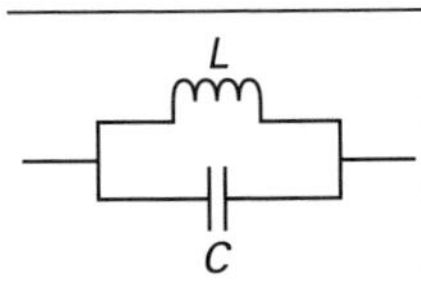	$\pm\frac{\pi}{2}$	$-j\left(\frac{1 - \omega^2 L C}{\omega L}\right)$
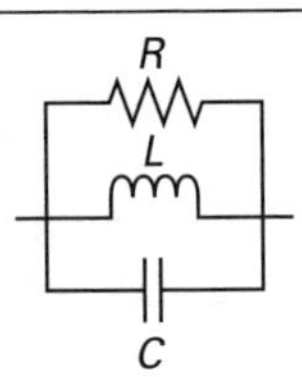	$\tan^{-1} R\left(\frac{1}{\omega L} - \omega C\right)$	$\frac{1}{R} + j\left(\omega C - \frac{1}{\omega L}\right)$
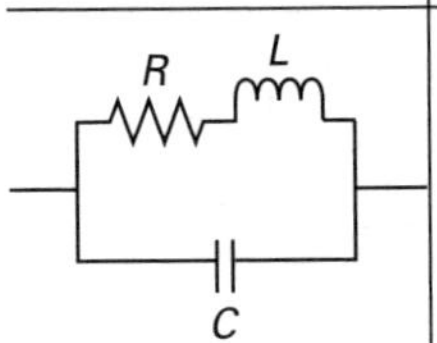	$\tan^{-1}\left(\frac{\frac{R^2}{\omega C} + \left(\frac{L}{C}\right)\left(\omega L - \frac{1}{\omega C}\right)}{\frac{R}{\omega^2 C^2}}\right)$	$\frac{R + j\omega\,(R^2 C - L + \omega^2 L^2 C)}{R^2 + \omega^2 L^2}$
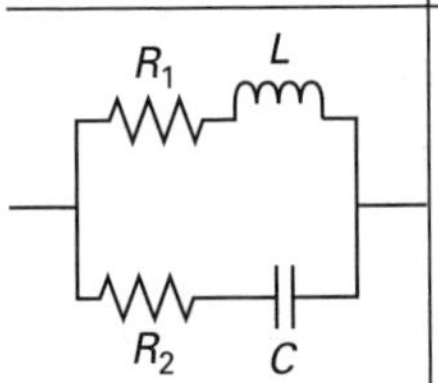	$\tan^{-1}\left(\frac{\omega L R^2 - \frac{R_1^2}{\omega C} - \left(\frac{L}{C}\right)\left(\omega L - \frac{1}{\omega C}\right)}{R_1 R_2 (R_1 + R_2) + \omega^2 L^2 R_2 + \frac{R_1}{\omega^2 C^2}}\right)$	$\frac{R_1 + \omega^2 R_1 R_2 C^2 (R_1 + R_2) + \omega^4 L^2 C^2 R_2}{(R_1^2 + \omega^2 L^2)(\omega^2 R_2^2 C^2 + 1)} + j\left(\frac{\omega\left(R_1^2 C - L + \omega^2 L C (L - R_2^2 C)\right)}{(R_1^2 + \omega^2 L^2)(\omega^2 R_2^2 C^2 + 1)}\right)$

APPENDIX 29.A
Two-Port Parameter Conversions

	$[z]$	$[y]$	$[h]$	$[g]$	$[TL]$	$[TLI]$
$[z]$	$\begin{bmatrix} z_{11} & z_{12} \\ z_{21} & z_{22} \end{bmatrix}$	$\begin{bmatrix} \frac{y_{22}}{\lvert y\rvert} & -\frac{y_{12}}{\lvert y\rvert} \\ -\frac{y_{21}}{\lvert y\rvert} & \frac{y_{11}}{\lvert y\rvert} \end{bmatrix}$	$\begin{bmatrix} \frac{\lvert h\rvert}{h_{22}} & \frac{h_{12}}{h_{22}} \\ -\frac{h_{22}}{h_{22}} & \frac{1}{h_{22}} \end{bmatrix}$	$\begin{bmatrix} \frac{1}{g_{11}} & -\frac{g_{12}}{g_{11}} \\ \frac{g_{21}}{g_{11}} & \frac{\lvert g\rvert}{g_{11}} \end{bmatrix}$	$\begin{bmatrix} \frac{A}{C} & \frac{\lvert TL\rvert}{C} \\ \frac{1}{C} & \frac{D}{C} \end{bmatrix}$	$\begin{bmatrix} \frac{\delta}{\gamma} & \frac{1}{\gamma} \\ \frac{\lvert TLI\rvert}{\gamma} & \frac{\alpha}{\gamma} \end{bmatrix}$
$[y]$	$\begin{bmatrix} \frac{z_{22}}{\lvert z\rvert} & -\frac{z_{12}}{\lvert z\rvert} \\ -\frac{z_{21}}{\lvert z\rvert} & \frac{z_{11}}{\lvert z\rvert} \end{bmatrix}$	$\begin{bmatrix} y_{11} & y_{12} \\ y_{21} & y_{22} \end{bmatrix}$	$\begin{bmatrix} \frac{1}{h_{11}} & -\frac{h_{12}}{h_{11}} \\ \frac{h_{21}}{h_{11}} & \frac{\lvert h\rvert}{h_{11}} \end{bmatrix}$	$\begin{bmatrix} \frac{\lvert g\rvert}{g_{22}} & \frac{g_{12}}{g_{22}} \\ -\frac{g_{21}}{g_{22}} & \frac{1}{g_{22}} \end{bmatrix}$	$\begin{bmatrix} \frac{D}{B} & -\frac{\lvert TL\rvert}{B} \\ -\frac{1}{B} & \frac{A}{B} \end{bmatrix}$	$\begin{bmatrix} \frac{\alpha}{\beta} & -\frac{1}{\beta} \\ -\frac{\lvert TLI\rvert}{\beta} & \frac{\delta}{\beta} \end{bmatrix}$
$[h]$	$\begin{bmatrix} \frac{\lvert z\rvert}{z_{22}} & \frac{z_{21}}{z_{22}} \\ -\frac{z_{21}}{z_{22}} & \frac{1}{z_{22}} \end{bmatrix}$	$\begin{bmatrix} \frac{1}{y_{11}} & -\frac{y_{12}}{y_{11}} \\ \frac{y_{21}}{y_{11}} & \frac{\lvert y\rvert}{y_{11}} \end{bmatrix}$	$\begin{bmatrix} h_{11} & h_{12} \\ h_{21} & h_{22} \end{bmatrix}$	$\begin{bmatrix} \frac{g_{22}}{\lvert g\rvert} & -\frac{g_{12}}{\lvert g\rvert} \\ -\frac{g_{21}}{\lvert g\rvert} & \frac{g_{11}}{\lvert g\rvert} \end{bmatrix}$	$\begin{bmatrix} \frac{B}{D} & \frac{\lvert TL\rvert}{D} \\ -\frac{1}{D} & \frac{C}{D} \end{bmatrix}$	$\begin{bmatrix} \frac{\beta}{\alpha} & \frac{1}{\alpha} \\ -\frac{\lvert TLI\rvert}{\alpha} & \frac{\gamma}{\alpha} \end{bmatrix}$
$[g]$	$\begin{bmatrix} \frac{1}{z_{11}} & -\frac{z_{12}}{z_{11}} \\ \frac{z_{21}}{z_{11}} & \frac{\lvert z\rvert}{z_{11}} \end{bmatrix}$	$\begin{bmatrix} \frac{\lvert y\rvert}{y_{22}} & \frac{y_{12}}{y_{22}} \\ -\frac{y_{21}}{y_{22}} & \frac{1}{y_{22}} \end{bmatrix}$	$\begin{bmatrix} \frac{h_{22}}{\lvert h\rvert} & -\frac{h_{12}}{\lvert h\rvert} \\ -\frac{h_{21}}{\lvert h\rvert} & \frac{h_{11}}{\lvert h\rvert} \end{bmatrix}$	$\begin{bmatrix} g_{11} & g_{12} \\ g_{21} & g_{22} \end{bmatrix}$	$\begin{bmatrix} \frac{C}{A} & \frac{\lvert TL\rvert}{A} \\ \frac{1}{A} & \frac{B}{A} \end{bmatrix}$	$\begin{bmatrix} \frac{\gamma}{\delta} & -\frac{1}{\delta} \\ \frac{\lvert TLI\rvert}{\delta} & \frac{\beta}{\delta} \end{bmatrix}$
$[TL]$	$\begin{bmatrix} \frac{z_{11}}{z_{21}} & \frac{\lvert z\rvert}{z_{21}} \\ \frac{1}{z_{21}} & \frac{z_{22}}{z_{21}} \end{bmatrix}$	$\begin{bmatrix} -\frac{y_{22}}{y_{21}} & -\frac{1}{y_{21}} \\ -\frac{\lvert y\rvert}{y_{21}} & -\frac{y_{11}}{y_{21}} \end{bmatrix}$	$\begin{bmatrix} -\frac{\lvert h\rvert}{h_{21}} & -\frac{h_{11}}{h_{21}} \\ -\frac{h_{22}}{h_{21}} & -\frac{1}{h_{21}} \end{bmatrix}$	$\begin{bmatrix} \frac{1}{g_{21}} & \frac{g_{22}}{g_{21}} \\ \frac{g_{21}}{g_{21}} & \frac{\lvert g\rvert}{g_{21}} \end{bmatrix}$	$\begin{bmatrix} A & B \\ C & D \end{bmatrix}$	$\begin{bmatrix} \frac{\delta}{\lvert TLI\rvert} & \frac{\beta}{\lvert TLI\rvert} \\ \frac{y}{\lvert TLI\rvert} & \frac{\alpha}{\lvert TLI\rvert} \end{bmatrix}$
$[TLI]$	$\begin{bmatrix} \frac{z_{22}}{z_{12}} & \frac{\lvert z\rvert}{z_{12}} \\ \frac{1}{z_{12}} & \frac{z_{11}}{z_{12}} \end{bmatrix}$	$\begin{bmatrix} -\frac{y_{11}}{y_{12}} & -\frac{1}{y_{12}} \\ -\frac{\lvert y\rvert}{y_{12}} & -\frac{y_{22}}{y_{12}} \end{bmatrix}$	$\begin{bmatrix} \frac{1}{h_{12}} & \frac{h_{11}}{h_{12}} \\ \frac{h_{22}}{h_{12}} & \frac{\lvert h\rvert}{h_{12}} \end{bmatrix}$	$\begin{bmatrix} -\frac{\lvert g\rvert}{g_{12}} & -\frac{g_{22}}{g_{12}} \\ -\frac{g_{11}}{g_{12}} & -\frac{1}{g_{12}} \end{bmatrix}$	$\begin{bmatrix} \frac{D}{\lvert TL\rvert} & \frac{B}{\lvert TL\rvert} \\ \frac{C}{\lvert TL\rvert} & \frac{A}{\lvert TL\rvert} \end{bmatrix}$	$\begin{bmatrix} \alpha & \beta \\ \gamma & \delta \end{bmatrix}$

Note: $|x| = \det[x] = \det\begin{bmatrix} x_{11} & x_{12} \\ x_{21} & x_{22} \end{bmatrix} = x_{11}x_{22} - x_{12}x_{21}$ where x is any of the listed parameters, z, y, h, g, TL, or TLI.

Appendices

APPENDIX 30.A
Resonant Circuit Formulas

<table>
<tr><th>unknown quantity</th><th>symbol</th><th>units</th><th>series
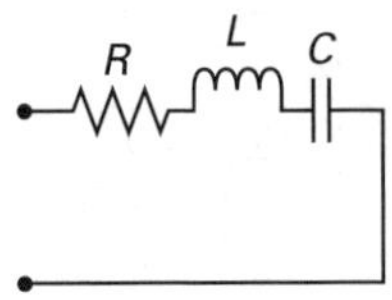</th><th>parallel
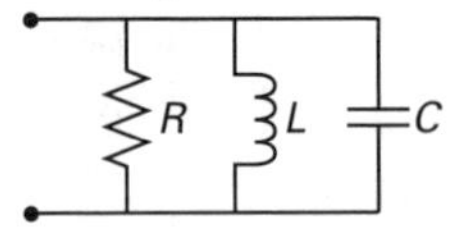</th></tr>
<tr><td rowspan="4">resonant frequency</td><td rowspan="2">f_0</td><td rowspan="2">Hz</td><td colspan="2">$\frac{1}{2\pi\sqrt{LC}}$</td></tr>
<tr><td>$\frac{QR}{2\pi L} = \frac{1}{2\pi QRC}$</td><td>$\frac{R}{2\pi QL} = \frac{Q}{2\pi RC}$</td></tr>
<tr><td rowspan="2">ω_0</td><td rowspan="2">$\frac{\text{rad}}{\text{s}}$</td><td colspan="2">$\frac{1}{\sqrt{LC}}$</td></tr>
<tr><td>$\frac{QR}{L} = \frac{1}{QRC}$</td><td>$\frac{R}{QL} = \frac{Q}{RC}$</td></tr>
<tr><td rowspan="4">bandwidth</td><td rowspan="2">BW</td><td rowspan="2">Hz</td><td colspan="2">$f_2 - f_1 = \frac{f_0}{Q}$</td></tr>
<tr><td>$\frac{R}{2\pi L} = \frac{1}{2\pi Q^2 RC}$</td><td>$\frac{1}{2\pi RC} = \frac{R}{2\pi Q^2 L}$</td></tr>
<tr><td rowspan="2">BW</td><td rowspan="2">$\frac{\text{rad}}{\text{s}}$</td><td colspan="2">$\omega_2 - \omega_1 = \frac{\omega_0}{Q}$</td></tr>
<tr><td>$\frac{R}{L} = \frac{1}{Q^2 RC}$</td><td>$\frac{1}{CR} = \frac{R}{Q^2 L}$</td></tr>
<tr><td rowspan="2">quality factor</td><td rowspan="2">Q</td><td rowspan="2">–</td><td>$\frac{1}{R}\sqrt{\frac{L}{C}}$</td><td>$R\sqrt{\frac{C}{L}}$</td></tr>
<tr><td>$\frac{\omega_0 L}{R} = \frac{1}{\omega_0 RC}$</td><td>$\omega_0 RC = \frac{R}{\omega_0 L}$</td></tr>
<tr><td rowspan="4">lower half-power point</td><td rowspan="2">f_1</td><td rowspan="2">Hz</td><td colspan="2">$f_0 - \frac{\text{BW}}{2} = f_0\left(1 - \frac{1}{2Q}\right)$</td></tr>
<tr><td>$f_0 - \frac{R}{4\pi L}$</td><td>$f_0 - \frac{1}{4\pi CR}$</td></tr>
<tr><td rowspan="2">ω_1</td><td rowspan="2">$\frac{\text{rad}}{\text{s}}$</td><td colspan="2">$\omega_0 - \frac{\text{BW}}{2} = \omega_0\left(1 - \frac{1}{2Q}\right)$</td></tr>
<tr><td>$\omega_0 - \frac{R}{2L}$</td><td>$\omega_0 - \frac{1}{2CR}$</td></tr>
<tr><td rowspan="4">upper half-power point</td><td rowspan="2">f_2</td><td rowspan="2">Hz</td><td colspan="2">$f_0 + \frac{\text{BW}}{2} = f_0\left(1 + \frac{1}{2Q}\right)$</td></tr>
<tr><td>$f_0 + \frac{R}{4\pi L}$</td><td>$f_0 + \frac{1}{4\pi CR}$</td></tr>
<tr><td rowspan="2">ω_2</td><td rowspan="2">$\frac{\text{rad}}{\text{s}}$</td><td colspan="2">$\omega_0 + \frac{\text{BW}}{2} = \omega_0\left(1 + \frac{1}{2Q}\right)$</td></tr>
<tr><td>$\omega_0 + \frac{R}{2L}$</td><td>$\omega_0 + \frac{1}{2CR}$</td></tr>
</table>

APPENDIX 37.A
Periodic Table of the Elements

THE PERIODIC TABLE OF ELEMENTS (Long Form)

The number of electrons in filled shells is shown in the column at the extreme left; the remaining electrons for each element are shown immediately below the symbol for each element. Atomic numbers are enclosed in brackets. Atomic weights (rounded, based on carbon-12) are shown above the symbols. Atomic weight values in parentheses are those of the isotopes of longest half-life for certain radioactive elements whose atomic weights cannot be precisely quoted without knowledge of origin of the element.

METALS — TRANSITION METALS — NON-METALS

periods	1 IA	2 IIA	3 IIIB	4 IVB	5 VB	6 VIB	7 VIIB	8	9 VIII	10	11 IB	12 IIB	13 IIIA	14 IVA	15 VA	16 VIA	17 VIIA	18 VIIIA
1 0	1.0079 H[1] 1																	4.0026 He[2] 2
2 2	6.939 Li[3] 1	9.0122 Be[4] 2											10.81 B[5] 3	12.01115 C[6] 4	14.0067 N[7] 5	15.9994 O[8] 6	18.994 F[9] 7	20.183 Ne[10] 8
3 2,8	22.9898 Na[11] 1	24.312 Mg[12] 2											26.9815 Al[13] 3	28.086 Si[14] 4	30.9738 P[15] 5	32.064 S[16] 6	35.453 Cl[17] 7	39.948 Ar[18] 8
4 2,8	39.098 K[19] 8,1	40.08 Ca[20] 8,2	44.956 Sc[21] 9,2	47.90 Ti[22] 10,2	50.942 V[23] 11,2	51.996 Cr[24] 13,1	54.938 Mn[25] 13,2	55.847 Fe[26] 14,2	58.933 Co[27] 15,2	58.71 Ni[28] 16,2	63.546 Cu[29] 18,1	65.38 Zn[30] 18,2	69.72 Ga[31] 18,3	72.59 Ge[32] 18,4	74.922 As[33] 18,5	78.96 Se[34] 18,6	79.904 Br[35] 18,7	83.80 Kr[36] 18,8
5 2,8,18	85.47 Rb[37] 8,1	87.62 Sr[38] 8,2	88.905 Y[39] 9,2	91.22 Zr[40] 10,2	92.906 Nb[41] 12,1	95.94 Mo[42] 13,1	(98) Tc[43] 14,1	101.07 Ru[44] 15,1	102.905 Rh[45] 16,1	106.4 Pd[46] 18	107.868 Ag[47] 18,1	112.40 Cd[48] 18,2	114.82 In[49] 18,3	118.69 Sn[50] 18,4	121.75 Sb[51] 18,5	127.60 Te[52] 18,6	126.904 I[53] 18,7	131.30 Xe[54] 18,8
6 2,8,18	132.905 Cs[55] 18,8,1	137.34 Ba[56] 18,8,2	* [57-71]	178.49 Hf[72] 32,10,2	180.948 Ta[73] 32,11,2	183.85 W[74] 32,12,2	186.2 Re[75] 32,13,2	190.2 Os[76] 32,14,2	192.2 Ir[77] 32,15,2	195.09 Pt[78] 32,17,1	196.967 Au[79] 32,18,1	200.59 Hg[80] 32,18,2	204.37 Tl[81] 32,18,3	207.19 Pb[82] 32,18,4	208.980 Bi[83] 32,18,5	(210) Po[84] 32,18,6	(210) At[85] 32,18,7	(222) Rn[86] 32,18,8
7 2,8,18,32	(223) Fr[87] 18,8,1	226.025 Ra[88] 18,8,2	† [89-103]	Rf[104] 32,10,2	Ha[105] 32,11,2	[106] 32,12,2	[107]	[108]										

*LANTHANIDE SERIES	138.91 La[57] 18,9,2	140.12 Ce[58] 20,8,2	140.907 Pr[59] 21,8,2	144.24 Nd[60] 22,8,2	(147) Pm[61] 23,8,2	150.35 Sm[62] 24,8,2	151.96 Eu[63] 25,8,2	157.25 Gd[64] 25,9,2	158.924 Tb[65] 27,8,2	162.50 Dy[66] 28,8,2	164.930 Ho[67] 29,8,2	167.26 Er[68] 30,8,2	168.934 Tm[69] 31,8,2	173.04 Yb[70] 32,8,2	174.97 Lu[71] 32,9,2
†ACTINIDE SERIES	(227) Ac[89] 18,9,2	232.038 Th[90] 18,10,2	231.036 Pa[91] 20,9,2	238.03 U[92] 21,9,2	237.048 Np[93] 23,8,2	(242) Pu[94] 24,8,2	(243) Am[95] 25,8,2	(247) Cm[96] 25,9,2	(247) Bk[97] 26,9,2	(249) Cf[98] 28,8,2	(254) Es[99] 29,8,2	(253) Fm[100] 30,8,2	(256) Md[101] 31,8,2	(254) No[102] 32,8,2	(257) Lr[103] 32,9,2

APPENDIX 37.B
Semiconductor Symbols and Abbreviations*

a, A—anode
b, B—base
b_{fs}—common-source small-signal forward transfer susceptance
b_{is}—common-source small-signal input susceptance
b_{os}—common-source small-signal output susceptance
b_{rs}—common-source small-signal reverse transfer susceptance
c, C—collector
C_{cb}—collector-base interterminal capacitance
C_{ce}—collector-emitter inter-terminal capacitance
C_{ds}—drain-source capacitance
C_{du}—drain-substrate capacitance
C_{eb}—emitter-base interterminal capacitance
C_{ibo}—common-base open-circuit input capacitance
C_{ibs}—common-base short-circuit input capacitance
C_{ieo}—common-emitter open-circuit input capacitance
C_{ies}—common-emitter short-circuit input capacitance
C_{iss}—common-source short-circuit input capacitance
C_{obo}—common-base open-circuit output capacitance
C_{obs}—common-base short-circuit output capacitance
C_{oeo}—common-emitter open-circuit output capacitance
C_{oes}—common-emitter short-circuit output capacitance
C_{oss}—common-source short-circuit output capacitance
C_{rbs}—common-base short-circuit reverse transfer capacitance
C_{res}—common-collector short-circuit reverse transfer capacitance
C_{rss}—common-source short-circuit reverse transfer capacitance
C_{tc}—collector depletion-layer capacitance
C_{te}—emitter depletion-layer capacitance
d, D—drain
e, E—emitter
η—intrinsic standoff ratio
f_{hfb}—common-base small-signal short-circuit forward current transfer ratio cutoff frequency
f_{hfc}—common-collector small-signal short-circuit forward current transfer ratio cutoff frequency
f_{hfe}—common-emitter small-signal short-circuit forward current transfer ratio cutoff frequency
f_{max}—maximum frequency of oscillation
f_T—transition frequency (frequency at which common-emitter small-signal forward current transfer ratio extrapolates to unity)
g, G—gate
g_{fs}—common-source small-signal forward transfer conductance
g_{is}—common-source small-signal input conductance
g_{MB}—common-base static transconductance
g_{MC}—common-collector static transconductance
g_{ME}—common-emitter static transconductance
g_{os}—common-source small-signal output conductance
g_{rs}—common-source small-signal reverse transfer conductance
G_{pb}—common-base small-scale insertion power gain
G_{PB}—common-base large-signal insertion power gain
G_{pc}—common-collector small-signal insertion power gain
G_{PC}—common-collector large-signal insertion power gain
G_{pe}—common-emitter small-signal insertion power gain
G_{PE}—common-emitter large-signal insertion power gain
G_{pg}—common-gate small-signal insertion power gain
G_{ps}—common-source small-signal insertion power gain
G_{tb}—common-base small-signal transducer power gain
G_{TB}—common-base large-signal transducer power gain
G_{tc}—common-collector small-signal transducer power gain
G_{TC}—common-collector large-signal transducer power gain
G_{te}—common-emitter small-signal transducer power gain
G_{TE}—common-emitter large-signal transducer power gain
G_{tg}—common-gate small-signal transducer power gain
G_{ts}—common-source small-signal transducer power gain
h_{fb}—common-base small-signal short-circuit forward current transfer ratio
h_{FB}—common-base static forward current transfer ratio
h_{fc}—common-collector small-signal short-circuit forward current transfer ratio
h_{FC}—common-collector static forward current transfer ratio
h_{fe}—common-emitter small-signal short-circuit forward current transfer ratio
h_{FE}—common-emitter static forward current transfer ratio
h_{FEL}—inherent large-signal forward current transfer ratio
h_{ib}—common-base small-signal short-circuit input impedance
h_{IB}—common-base static input resistance
h_{ic}—common-collector small-signal short-circuit input impedance
h_{IC}—common-collector static input resistance
h_{ie}—common-emitter small-signal short-circuit input impedance
h_{IE}—common-emitter static input resistance
$h_{ie(imag)}$—imaginary part of common-emitter small-signal short-circuit input impedance
$h_{ie(real)}$—real part of common-emitter small-signal short-circuit inputimpedance
h_{ob}—common-base small-signal open-circuit output admittance
h_{oc}—common-collector small-signal open-circuit output admittance
h_{oe}—common emmitter small-signal open-circuit output admittance
$h_{oe(imag)}$—imaginary part of common-emitter small-signal open-circuit output admittance
$h_{oe(real)}$—real part of common-emitter small-signal open-circuit output admittance
h_{rb}—common-base small-signal open-circuit reverse voltage transfer ratio
h_{rc}—common-collector small-signal open-circuit reverse voltage transfer ratio
h_{re}—common-emitter small-signal open-circuit reverse voltage transfer ratio

(continued)

APPENDIX 37.B *(continued)*
Semiconductor Symbols and Abbreviations*

i_B—instantaneous total value of base-terminal current
i_C—instantaneous total value of collector-terminal current
i_E—instantaneous total value of emitter-terminal current
i_R—instantaneous total reverse current
I_0—average forward current, 180° conduction angle, 60 Hz half sine wave
I_b—alternating component (rms value) of base-terminal current
I_B—base-terminal DC current
$I_{B2(\text{mod})}$—interbase modulated current
I_{BEV}—base cutoff current, DC
I_c—alternating component (rms value) of collector-terminal current
I_C—collector-terminal DC current
I_{CBO}—collector cutoff current (DC), emitter open
I_{CEO}—collector cutoff current (DC), base open
I_{CER}—collector cutoff current (DC), specified resistance between base and emitter
I_{CES}—collector cutoff current (DC), base shorted to emitter
I_{CEV}—collector cutoff current, DC, specified voltage between base and emitter
I_{CEX}—collector cutoff current, DC, specified circuit between base and emitter
I_D—drain current, DC
$I_{D,\text{off}}$—drain cutoff current
$I_{D,\text{on}}$—on-state drain current
I_{DSS}—zero-gate-voltage drain current
I_e—alternating component (rms value) of emitter-terminal current
I_E—emitter-terminal DC current
I_{EBO}—emitter cutoff current (DC), collector open
I_{ERR}—emitter reverse current
$I_{\text{EC,ofs}}$—emitter-collector offset current
I_{ECS}—emitter cutoff current (DC), base short-circuited to collector
$I_{\text{EIE2,off}}$—emitter cutoff current
I_f—alternating component of forward current (rms value)
I_F—For voltage-regulator and voltage-reference diodes: DC forward current. For signal diodes and rectifier diodes: DC forward current (no alternating component).
I_F—instantaneous total forward current
$I_{F,\text{AV}}$—forward current, DC (with alternating component)
I_{FM}—maximum (peak) total forward current
$I_{F,\text{OV}}$—forward current, overload
I_{FRM}—maximum (peak) forward current, repetitive
I_{FRMS}—total rms forward current
I_{FSM}—maximum (peak) forward current, surge
I_G—gate current, DC
I_{GF}—forward gate current
I_{GR}—reverse gate current
I_{GSS}—reverse gate current, drain short-circuited to source
I_{GSSF}—forward gate current, drain short-circuited to source
I_{GSSR}—reverse gate current, drain short-circuited to source
I_I—inflection-point current
I_P—peak-point current
I_r—alternating component of reverse current (rms value)
I_R—for voltage-regulator and voltage-reference diodes: DC reverse current. For signal diodes and rectifier diodes: DC reverse current (no alternating component).
$I_{R,\text{AV}}$—reverse current, DC (with alternating component)
I_{RM}—maximum (peak) total reverse current
I_{RRM}—maximum (peak) reverse current, repetitive
$I_{R,\text{RMS}}$—total rms reverse current
I_{RSM}—maximum (peak) surge reverse current
I_S—source current, DC
I_{SDS}—zero-gate-voltage source current
$I_{S,\text{off}}$—source cutoff current
I_V—valley-point current
I_Z—regulator current, reference current, DC
I_{ZK}—regulator current, reference current (DC near breakdown knee)
I_{ZM}—regulator current, reference current (DC maximum rated current)
$\text{Im}(h_{\text{oe}})$—imaginary part of common-emitter small-signal short-circuit input impedance
$\text{Im}(h_{\text{oe}})$—imaginary part of common-emitter small-signal open-circuit output admittance
k, K—cathode
L_c—conversion loss
M—figure of merit
NF_o—overall noise figure
NR_o—output noise ratio
p_{BE}—instantaneous total power input to base, common emitter
p_{CB}—instantaneous total power input to collector, common base
p_{CE}—power input (DC) to collector, common emitter
p_{EB}—instantaneous total power input to emitter, common base
p_F—instantaneous total forward power dissipation
p_{ib}—common-base small-signal input power
p_{ic}—common collector small-signal input power
p_{ie}—common-emitter small-signal input power
p_{ob}—common-base small-signal output power
p_{oc}—common-collector small-signal output power
p_{oe}—common-emitter small-signal output power
p_R—instantaneous total reverse power dissipation
p_T—nonreactive power input, instantaneous total, to all terminals
P_{BE}—power input (DC) to base, common emitter
P_{CB}—power input (DC) to collector, common base
P_{CE}—instantaneous total power input to collector, common emitter
P_{EB}—power input (DC) to emitter, common base
P_F—forward power dissipation, DC (no alternating component)
$P_{F,\text{AV}}$—forward power dissipation, DC (with alternating component)
P_{FM}—maximum (peak) total forward power dissipation
P_{IB}—common-base large-signal input power
P_{IC}—common-collector large-signal input power
P_{IE}—common-emitter large-signal input power

(continued)

APPENDIX 37.B *(continued)*
Semiconductor Symbols and Abbreviations*

P_{OB}—common-base large-signal output power
P_{OC}—common-collector large-signal output power
P_{OE}—common-emitter large-signal output power
P_R—reverse power dissipation, DC (no alternating component)
$P_{R(AV)}$—reverse power dissipation, DC (with alternating component)
P_{RM}—maximum (peak) total reverse power dissipation
P_T—total nonreactive power input to all terminals
Q_S—stored charge
r'_b, C_c—collector-base time constant
r_{BB}—interbase resistance
$r_{CE,sat}$—saturation resistance, collector-to-emitter
$r_{ds,on}$—small-signal drain-source on-state resistance
$r_{DS,on}$—static drain-source on-state resistance
$r_{ele2,on}$—small-signal emitter-emitter on- state resistance
r_i—dynamic resistance at inflection point
R_θ—thermal resistance
$R_{\theta CA}$—thermal resistance, case to ambient
$R_{\theta JA}$—thermal resistance, junction to ambient
$R_{\theta JC}$—thermal resistance, junction to case
$\mathrm{Re}(h_{ie})$—real part of common-emitter small-signal short-circuit input impedance
$\mathrm{Re}(h_{oe})$—real part of common-emitter small-signal open-circuit output admittance
s, S—source
t_d—delay time
$t_{d,off}$—turn-off delay time
$t_{d,on}$—turn-on delay time
t_f—fall time
t_{fr}—forward recovery time
t_{off}—turn-off time
t_{on}—turn-on time
t_p—pulse time
t_r—rise time
t_{rr}—reverse recovery time
t_s—storage time
t_w—pulse average time
T_j—junction temperature
T_{stg}—storage temperature
T_A—ambient temperature or free-air temperature
T_C—case temperature
TSS—tangential signal sensitivity
u, U—bulk (substrate)
v_{bc}—instantaneous value of alternating component of base-collector voltage
v_{be}—instantaneous value of alternating component of base-emitter voltage
v_{BR}—breakdown voltage (instantaneous total)
v_{cb}—instantaneous value of alternating component of collector-base voltage
v_{ce}—instantaneous value of alternating component of collector-emitter voltage
v_{eb}—instantaneous value of alternating component of emitter-base voltage
v_{ec}—instantaneous value of alternating component of emitter-collector voltage
v_F—instantaneous total forward voltage
v_R—instantaneous total reverse voltage
V_{B2B1}—interbase voltage
V_{BB}—base supply voltage, DC
V_{BC}—average or DC voltage, base to collector
V_{BE}—average or DC voltage, base to emitter
V_{BR}—breakdown voltage, DC
$V_{BR,CBO}$—collector-base breakdown voltage, emitter open
$V_{BR,CEO}$—collector-emitter breakdown voltage, base open
$V_{BR,CER}$—collector-emitter breakdown voltage, resistance between base and emitter
$V_{BR,CES}$—collector-emitter breakdown voltage, base shorted to emitter
$V_{BR,CEV}$—collector-emitter breakdown voltage, specified voltage between base and emitter
$V_{BR,CEX}$—collector-emitter breakdown voltage, specified circuit between base and emitter
$V_{BR,E1E2}$—emitter-emitter breakdown voltage
$V_{BR,EBO}$—emitter-base breakdown voltage, collector open
$V_{BR,ECO}$—emitter-collector breakdown voltage, base open
$V_{BR,GSS}$—gate-source breakdown voltage
$V_{BR,GSSF}$—forward gate-source breakdown voltage
$V_{BR,GSSR}$—reverse gate-source breakdown voltage
V_{CB}—average or DC voltage, collector to base
V_{CBO}—collector-base voltage, DC, emitter open
$V_{CB,fl}$—collector-base DC open-circuit voltage (floating potential)
V_{CC}—collector supply voltage, DC
V_{CE}—average or DC voltage, collector to emitter
$V_{CE,fl}$—collector-emitter DC open-circuit voltage (floating potential)
V_{CEO}—collector-emitter voltage (DC), base open
$V_{CE,ofs}$—collector-emitter offset voltage
V_{CER}—collector-emitter voltage (DC), resistance between base and emitter
V_{CES}—collector-emitter voltage (DC), base shorted to emitter
$V_{CE,at}$—collector-emitter DC saturation voltage
V_{CEV}—collector-emitter voltage, DC, specified voltage between base and emitter
V_{CEX}—collector-emitter voltage, DC, specified circuit between base and emitter
V_{DD}—drain supply voltage, DC
V_{DG}—drain-gate voltage
V_{DS}—drain-source voltage
$V_{DS,on}$—drain-source on-state voltage
V_{DU}—drain-substrate voltage
V_{EB}—average or DC voltage, emitter to base
$V_{EB,fl}$—emitter-base DC open-circuit voltage (floating potential)
V_{EBO}—emitter-base voltage, DC, collector open
$V_{EB,sat}$—emitter saturation voltage
V_{EC}—average or DC voltage, emitter to collector
$V_{EC,fl}$—emitter-collector DC open-circuit voltage (floating potential)
$V_{EC,ofs}$—emitter-collector offset voltage
V_{EE}—emitter supply voltage, DC
V_f—alternating component of forward voltage (rms value)

(continued)

APPENDIX 37.B *(continued)*

Semiconductor Symbols and Abbreviations*

$V_{F,\mathrm{AV}}$—forward voltage, DC (with alternating component)
V_{FM}—maximum (peak) total forward voltage
$V_{F,\mathrm{RMS}}$—total rms forward voltage
V_{GG}—gate supply voltage, DC
V_{GS}—gate-source voltage
V_{GSF}—forward gate-source voltage
$V_{\mathrm{GS,off}}$—gate-source cutoff voltage
V_{GSR}—reverse gate-source voltage
$V_{\mathrm{GS,th}}$—gate-source threshold voltage
V_{GU}—gate-substrate voltage
V_I—inflection-point voltage
V_{OB1}—base-1 peak voltage
V_P—peak-point voltage
V_{PP}—projected peak-point voltage
V_r—alternating component of reverse voltage (rms value)
V_R—for voltage-regulator and voltage-reference diodes: DC reverse voltage. For signal diodes and rectifier diodes: DC reverse voltage (no alternating component).
$V_{R,\mathrm{AV}}$—reverse voltage, DC (with alternating component)
V_{RM}—maximum (peak) total reverse voltage
V_{RRM}—repetitive peak reverse voltage
$V_{R,\mathrm{RMS}}$—total rms reverse voltage
V_{RSM}—nonrepetitive peak reverse voltage
V_{RT}—reach-through voltage
V_{RWM}—working peak reverse voltage
V_{SS}—source supply voltage, DC
V_{SU}—source-substrate voltage
V_{TO}—threshold voltage
V_V—valley-point voltage
V_Z—regulator voltage, reference voltage, DC
V_{ZM}—regulator voltage, reference voltage (DC at maximum rated current)
VF—for voltage-regulator and voltage-reference diodes: DC forward voltage. For signal diodes and rectifier diodes: DC forward voltage (no alternating component).

y_{fb}—common-base small-signal short-circuit forward transfer admittance
y_{fc}—common-collector small-signal short-circuit forward transfer admittance
y_{fe}—common-emitter small-signal short-circuit forward transfer admittance
y_{fs}—common-source small-signal short-circuit forward transfer admittance
$y_{\mathrm{fs,imag}}$—common-source small-signal forward transfer susceptance
$y_{\mathrm{fs,real}}$—common-source small-signal forward transfer conductance
y_{ib}—common-base small-signal short-circuit input admittance
y_{ic}—common-collector small-signal short-circuit input admittance
y_{ie}—common-emitter small-signal short-circuit input admittance
$y_{\mathrm{ie,imag}}$—imaginary part of small-signal short-circuit input admittance (common emitter)
$y_{\mathrm{ie,real}}$—real part of small-signal short-circuit input admittance (common emitter)
y_{is}—common-source small-signal short-circuit input admittance
$y_{\mathrm{is,imag}}$—common-source small-signal input susceptance
$y_{\mathrm{is,real}}$—common-source small-signal input conductance
y_{ob}—common-base small-signal short-circuit output admittance
y_{oc}—common-collector small-signal short-circuit output admittance
y_{oe}—common-emitter small-signal short-circuit output admittance
$y_{\mathrm{oe,imag}}$—imaginary part of small-signal short-circuit output admittance (common emitter)

$y_{\mathrm{oe,real}}$—real part of small-signal short-circuit output admittance (common emitter)
y_{os}—common-source small-signal short-circuit output admittance
$y_{\mathrm{os,imag}}$—common-source small-signal output susceptance
$y_{\mathrm{os,real}}$—common-source small-signal output conductance
y_{rb}—common-base small-signal short-circuit reverse transfer admittance
y_{rc}—common-collector small-signal short-circuit reverse transfer admittance
y_{re}—common-emitter small-signal short-circuit reverse transfer admittance
y_{rs}—common-source small-signal short-circuit reverse transfer admittance
$y_{\mathrm{rs,imag}}$—common-source small-signal reverse transfer susceptance
$y_{\mathrm{rs,real}}$—common-source small-signal reverse transfer conductance
z_{if}—intermediate-frequency impedance
z_m—modulator-frequency load impedance
z_{rf}—radio-frequency impedance
z_{v}—video impedance
z_{z}—regulator impedance, reference impedance (small signal at $\mathrm{I_z}$)
z_{zk}—regulator impedance, reference impedance (small signal at $\mathrm{I_{ZK}}$)
$Z_{\theta,t}$—transient thermal impedance
$Z_{\theta\mathrm{JA,t}}$—junction-to-case transient thermal impedance
$Z_{\theta\mathrm{JC},t}$—junction-to-case transient thermal impedance
Z_{zm}—regulator impedance, reference impedance (small signal at $\mathrm{I_{ZM}}$)

*Recommended by the Joint Electron Device Engineering Council of the Electronic Industries Association and the National Electrical Manufacturers Associaton for use in semiconductor device data sheets and specifications.

APPENDIX 40.A
Standard Zener Diodes

zener voltage (nominal)	rated power: 500 mW		1 W		5 W		50 W	
	Z_Z[a] (Ω)	I_{ZK}[b] (mA)	Z_Z[a] (Ω)	I_{ZK}[b] (mA)	Z_Z[a] (Ω)	I_{ZK}[b] (mA)	Z_Z[a] (Ω)	I_{ZK}[b] (mA)
2.4	30	1.0						
2.7	30	1.0						
3.0	29	1.0						
3.3	28	1.0	10	1.0	3.0	1.0		
3.6	24	1.0	10	1.0	2.5	1.0		
3.9	23	1.0	9.0	1.0	2.0	1.0	0.16	5.0
4.3	22	1.0	9.0	1.0	2.0	1.0	0.16	5.0
4.7	19	1.0	8.0	1.0	2.0	1.0	0.12	5.0
5.1	17	1.0	7.0	1.0	1.5	1.0	0.12	5.0
5.6	11	1.0	5.0	1.0	1.0	1.0	0.12	5.0
6.2	7.0	1.0	2.0	1.0	1.0	1.0	0.14	5.0
6.8	5.0	1.0	1.5	1.0	1.0	1.0	0.16	5.0
7.5	5.5	0.5	1.5	1.0	1.5	1.0	0.24	5.0
8.2	6.5	0.5	4.5	0.5	1.5	1.0	0.4	5.0
8.7					2.0	1.0		
9.1	7.5	0.5	5.0	0.5	2.0	1.0	0.5	5.0
10	8.5	0.25	7.0	0.25	2.0	1.0	0.6	5.0
11	9.5	0.25	8.0	0.25	2.5	1.0	0.8	5.0
12	11.5	0.25	9.0	0.25	2.5	1.0	1.0	5.0
13	13	0.25	10	0.25	2.5	1.0	1.1	5.0
14					2.5	1.0	1.2	5.0
15	16	0.25	14	0.25	2.5	1.0	1.4	5.0
16	17	0.25	16	0.25	2.5	1.0	1.6	5.0
17					2.5	1.0	1.8	5.0
18	21	0.25	20	0.25	2.5	1.0	2.0	5.0
19					3.0	1.0	2.2	5.0
20	25	0.25	22	0.25	3.0	1.0	2.4	5.0
22	29	0.25	23	0.25	3.5	1.0	2.5	5.0
24	33	0.25	25	0.25	3.5	1.0	2.6	5.0
25					4.0	1.0	2.7	5.0
27	41	0.25	35	0.25	5.0	1.0	2.8	5.0
28					6.0	1.0		
30	49	0.25	40	0.25	8.0	1.0	3.0	5.0
33	58	0.25	45	0.25	10	1.0	3.2	5.0
36	70	0.25	50	0.25	11	1.0	3.5	5.0
39	80	0.25	60	0.25	14	1.0	4.0	5.0
43	93	0.25	70	0.25	20	1.0	4.5	5.0
45							4.5	5.0
47	105	0.25	80	0.25	25	1.0	5.0	5.0
50							5.0	5.0
51	125	0.25	95	0.25	27	1.0	5.2	5.0
56	150	0.25	110	0.25	35	1.0	6.0	5.0
60					40	1.0		
62	185	0.25	125	0.25	42	1.0	7.0	5.0
68	230	0.25	150	0.25	44	1.0	8.0	5.0
75	270	0.25	175	0.25	45	1.0	9.0	5.0
82	330	0.25	200	0.25	65	1.0	11	5.0
87					75	1.0		
91	400	0.25	250	0.25	75	1.0	15	5.0
100			350	0.25	90	1.0	20	5.0
105							25	5.0

(continued)

Appendices

APPENDIX 40.A *(continued)*
Standard Zener Diodes

zener voltage (nominal)	rated power 500 mW		1 W		5 W		50 W	
	Z_Z[a] (Ω)	I_{ZK}[b] (mA)	Z_Z[a] (Ω)	I_{ZK}[b] (mA)	Z_Z[a] (Ω)	I_{ZK}[b] (mA)	Z_Z[a] (Ω)	I_{ZK}[b] (mA)
110					125	1.0	30	5.0
120			550	0.25	170	1.0	40	5.0
130			700	0.25	190	1.0	50	5.0
140					230	1.0	60	5.0
150			1000	0.25	330	1.0	75	5.0
160			1100	0.25	350	1.0	80	5.0
170					380	1.0		
175							85	5.0
180			1200	0.25	430	1.0	90	5.0
190					450	1.0		
200			1500	0.25	480	1.0	100	5.0

[a]Z_Z is the zener impedance.
[b]I_{ZK} is the zener keep-alive or knee current.

APPENDIX 45.A

Smith Chart

Impedance or Admittance Coordinates

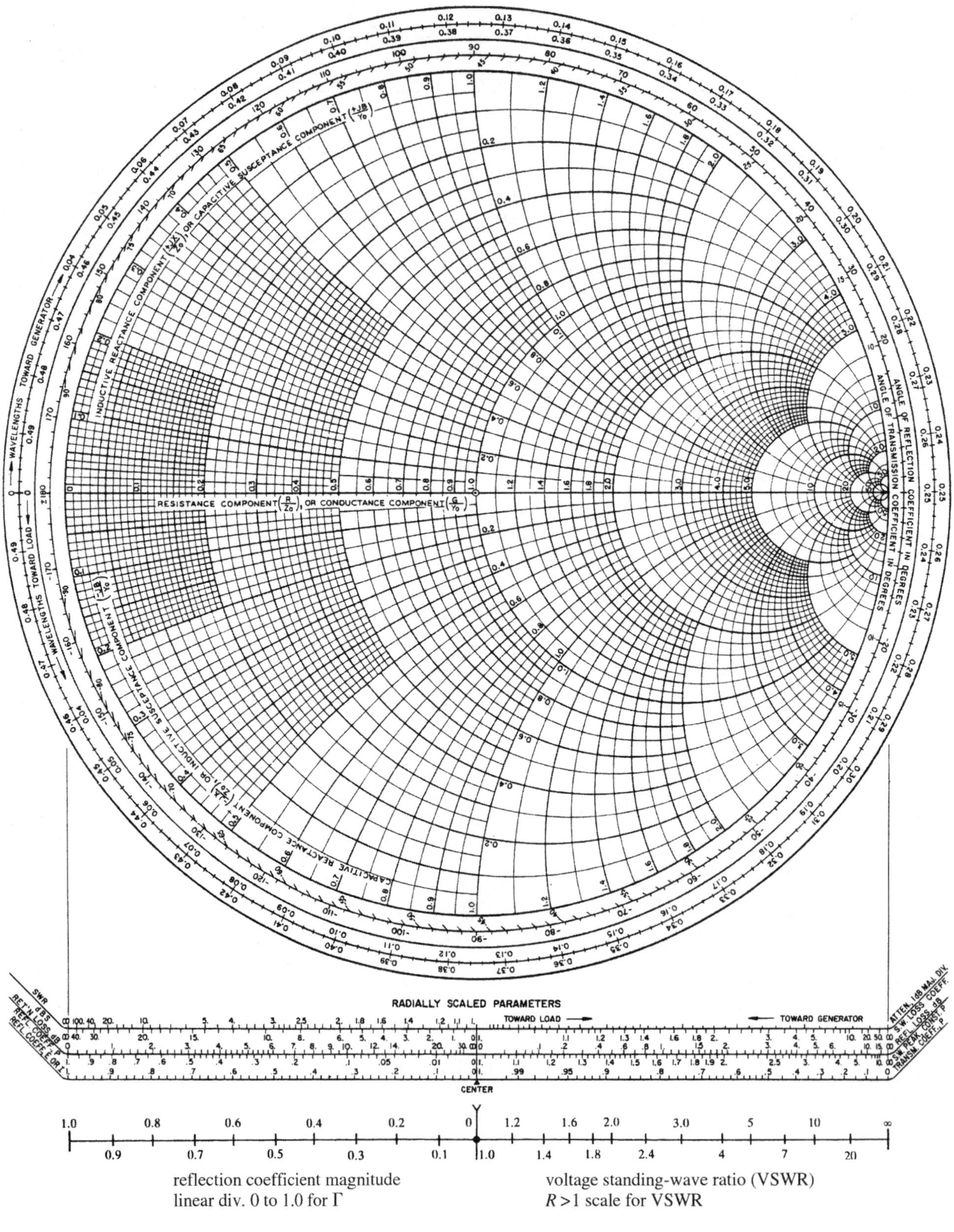

reflection coefficient magnitude
linear div. 0 to 1.0 for Γ

voltage standing-wave ratio (VSWR)
$R>1$ scale for VSWR

APPENDIX 48.A

Error Function Values: $\text{erf}(x) = \frac{2}{\sqrt{\pi}} \int_0^x e^{-t^2}\,dt$

x	0	1	2	3	4	5	6	7	8	9
0.00	0.0000	0011	0023	0034	0045	0056	0068	0079	0090	0102
1	0113	0124	0135	0147	0158	0169	0181	0192	0203	0214
2	0226	0237	0248	0260	0271	0282	0293	0305	0316	0327
3	0338	0350	0361	0372	0384	0395	0406	0417	0429	0440
4	0451	0462	0474	0485	0496	0507	0519	0530	0541	0553
5	0564	0575	0586	0598	0609	0620	0631	0643	0654	0665
6	0676	0688	0699	0710	0721	0732	0744	0755	0766	0777
7	0789	0800	0811	0822	0834	0845	0856	0867	0878	0890
8	0901	0912	0923	0934	0946	0957	0968	0979	0990	1002
9	1013	1024	1035	1046	1058	1069	1080	1091	1102	1113
10	1125	1136	1147	1158	1169	1180	1192	1203	1214	1225
1	1236	1247	1259	1270	1281	1292	1303	1314	1325	1336
2	1348	1359	1370	1381	1392	1403	1414	1425	1436	1448
3	1459	1470	1481	1492	1503	1514	1525	1536	1547	1558
4	1569	1581	1592	1603	1614	1625	1636	1647	1658	1669
5	1680	1691	1702	1713	1724	1735	1746	1757	1768	1779
6	1790	1801	1812	1823	1834	1845	1856	1867	1878	1889
7	1900	1911	1922	1933	1944	1955	1966	1977	1988	1998
8	2009	2020	2031	2042	2053	2064	2075	2086	2097	2108
9	2118	2129	2140	2151	2162	2173	2184	2194	2205	2216
20	2227	2238	2249	2260	2270	2281	2292	2303	2314	2324
1	2335	2346	2357	2368	2378	2389	2400	2411	2421	2432
2	2443	2454	2464	2475	2486	2497	2507	2518	2529	2540
3	2550	2561	2572	2582	2593	2604	2614	2625	2636	2646
4	2657	2668	2678	2689	2700	2710	2721	2731	2742	2753
5	2763	2774	2784	2795	2806	2816	2827	2837	2848	2858
6	2869	2880	2890	2901	2911	2922	2932	2943	2953	2964
7	2974	2985	2995	3006	3016	3027	3037	3047	3058	3068
8	3079	3089	3100	3110	3120	3131	3141	3152	3162	3172
9	3183	3193	3204	3214	3224	3235	3245	3255	3266	3276
30	3286	3297	3307	3317	3327	3338	3348	3358	3369	3379
1	3389	3399	3410	3420	3430	3440	3450	3461	3471	3481
2	3491	3501	3512	3522	3532	3542	3552	3562	3573	3583
3	3593	3603	3613	3623	3633	3643	3653	3663	3674	3684
4	3694	3704	3714	3724	3734	3744	3754	3764	3774	3784
5	3794	3804	3814	3824	3834	3844	3854	3864	3873	3883
6	3893	3903	3913	3923	3933	3943	3953	3963	3972	3982
7	3992	4002	4012	4022	4031	4041	4051	4061	4071	4080
8	4090	4100	4110	4119	4129	4139	4149	4158	4168	4178
9	4187	4197	4207	4216	4226	4236	4245	4255	4265	4274
40	4284	4294	4303	4313	4322	4332	4341	4351	4361	4370
1	4380	4389	4399	4408	4418	4427	4437	4446	4456	4465
2	4475	4484	4494	4503	4512	4522	4531	4541	4550	4559
3	4569	4578	4588	4597	4606	4616	4625	4634	4644	4653
4	4662	4672	4681	4690	4699	4709	4718	4727	4736	4746
5	4755	4764	4773	4782	4792	4801	4810	4819	4828	4837
6	4847	4856	4865	4874	4883	4892	4901	4910	4919	4928
7	4937	4946	4956	4965	4974	4983	4992	5001	5010	5019
8	5027	5036	5045	5054	5063	5072	5081	5090	5099	5108
9	5117	5126	5134	5143	5152	5161	5170	5179	5187	5196
0.50	0.5205	5214	5223	5231	5240	5249	5258	5266	5275	5284

APPENDIX 48.B

Error Function Values: $\text{erf}(x) = \frac{2}{\sqrt{\pi}}\left(x - \frac{x^2}{3 \cdot 1!} + \frac{x^5}{5 \cdot 2!} - \frac{x^7}{7 \cdot 3!} + \cdots\right)$

x	0	1	2	3	4	5	6	7	8	9
0.50	0.5205	5214	5223	5231	5240	5249	5258	5266	5275	5284
1	5292	5301	5310	5318	5327	5336	5344	5353	5362	5370
2	5379	5388	5396	5405	5413	5422	5430	5439	5448	5456
3	5465	5473	5482	5490	5499	5507	5516	5524	5533	5541
4	5549	5558	5566	5575	5583	5591	5600	5608	5617	5625
5	5633	5642	5650	5658	5667	5675	5683	5691	5700	5708
6	5716	5724	5733	5741	5749	5757	5765	5774	5782	5790
7	5798	5806	5814	5823	5831	5839	5847	5855	5863	5871
8	5879	5887	5895	5903	5911	5919	5927	5935	5943	5951
9	5959	5967	5975	5983	5991	5999	6007	6015	6023	6031
60	6039	6046	6054	6062	6070	6078	6086	6093	6101	6109
1	6117	6125	6132	6140	6148	6156	6163	6171	6179	6186
2	6194	6202	6209	6217	6225	6232	6240	6248	6255	6263
3	6270	6278	6286	6293	6301	6308	6316	6323	6331	6338
4	6346	6353	6361	6368	6376	6383	6391	6398	6405	6413
5	6420	6428	6435	6442	6450	6457	6464	6472	6479	6486
6	6494	6501	6508	6516	6523	6530	6537	6545	6552	6559
7	6566	6573	6581	6588	6595	6602	6609	6616	6624	6631
8	6638	6645	6652	6659	6666	6673	6680	6687	6694	6701
9	6708	6715	6722	6729	6737	6743	6750	6757	6764	6771
70	6778	6785	6792	6799	6806	6812	6819	6826	6833	6840
1	6847	6853	6860	6867	6874	6881	6887	6894	6901	6908
2	6914	6921	6928	6934	6941	6948	6954	6961	6968	6974
3	6981	6988	6994	7001	7007	7014	7021	7027	7034	7040
4	7047	7053	7060	7066	7073	7079	7086	7092	7099	7105
5	7112	7118	7124	7131	7137	7144	7150	7156	7163	7169
6	7175	7182	7188	7194	7201	7207	7213	7219	7226	7232
7	7238	7244	7251	7257	7263	7269	7275	7282	7288	7294
8	7300	7306	7312	7318	7325	7331	7337	7343	7349	7355
9	7361	7367	7373	7379	7385	7391	7397	7403	7409	7415
80	7421	7427	7433	7439	7445	7451	7457	7462	7468	7474
1	7480	7486	7492	7498	7503	7509	7515	7521	7527	7532
2	7538	7544	7550	7555	7561	7567	7572	7578	7584	7590
3	7595	7601	7607	7612	7618	7623	7629	7635	7640	7646
4	7651	7657	7663	7668	7674	7679	7685	7690	7696	7701
5	7707	7712	7718	7723	7729	7734	7739	7745	7750	7756
6	7761	7766	7772	7777	7782	7788	7793	7798	7804	7809
7	7814	7820	7825	7830	7835	7841	7846	7851	7856	7862
8	7867	7872	7877	7882	7888	7893	7898	7903	7908	7913
9	7918	7924	7929	7934	7939	7944	7949	7954	7959	7964
90	7969	7974	7979	7984	7989	7994	7999	8004	8009	8014
1	8019	8024	8029	8034	8038	8043	8048	8053	8058	8063
2	8068	8073	8077	8082	8087	8092	8097	8101	8106	8111
3	8116	8120	8125	8130	8135	8139	8144	8149	8153	8158
4	8163	8167	8172	8177	8181	8186	8191	8195	8200	8204
5	8209	8213	8218	8223	8227	8232	8236	8241	8245	8250
6	8254	8259	8263	8268	8272	8277	8281	8285	8290	8294
7	8299	8303	8307	8312	8316	8321	8325	8329	8334	8338
8	8342	8347	8351	8355	8360	8364	8368	8372	8377	8381
9	8385	8389	8394	8398	8402	8406	8410	8415	8419	8423
1.00	0.8427	8431	8435	8439	8444	8448	8452	8456	8460	8464

APPENDIX 52.A

Periodic Table: Materials' Properties Summary

Atomic radius decreases (increases from group VIIA to group 0); boiling and freezing point increases to group VB and then decreases; density increases up to group 8 (elements 28, 46, 78) and then decreases; electronegativity and ionization energy increase; ionic radius decreases (increases across groups VIA and VIIA). →

↓ atomic radius increases; electronegativity decreases; ionization energy decreases; ionic radius increases

↓ boiling and freezing point increases (except for rare earths and groups IB through VIA); density increases

IA	IIA		IIIB	IVB	VB	VIB	VIIB	VIII	VIII	VIII	IB	IIB	IIIA	IVA	VA	VIA	VIIA	0
1																		2
3	4												5	6	7	8	9	10
11	12												13	14	15	16	17	18
19	20		21	22	23	24	25	26	27	28	29	30	31	32	33	34	35	36
37	38		39	40	41	42	43	44	45	46	47	48	49	50	51	52	53	54
55	56	*	71	72	73	74	75	76	77	78	79	80	81	82	83	84	85	86
87	88	**	103	104	105	106	107	108	109	110	111	112	113	114	115	116	117	118

s block: IA–IIA; d block: IIIB–IIB; p block: IIIA–0

f block

lanthanides *	57	58	59	60	61	62	63	64	65	66	67	68	69	70
actinides **	89	90	91	92	93	94	95	96	97	98	99	100	101	102

coding

- alkali metals
- alkaline-earth metals
- transition metals
- inner transition metals
- halogens
- noble gases

(transition metals and inner transition metals: rare earths)

legend

metal	solid
semi-metal	liquid
nonmetal	gas
(staircase line)	metal-to-nonmetal transition staircase

Note: Columns are *groups*; rows are *periods*; elements are listed by atomic number; helium (element 2) is part of the *s* block; grey atomic numbers indicate temporary names and relatively unknown properties for newly discovered elements; The *spdf* blocks refer to the orbitals, filling; the states of the elements are based on STP conditions.

APPENDIX 52.B
Table of Relative Atomic Weights
(based on the atomic mass of $^{12}C = 12$)

name	symbol	atomic number	atomic weight	name	symbol	atomic number	atomic weight
actinium	Ac	89	–	mercury	Hg	80	200.59
aluminum	Al	13	26.9815	molybdenum	Mo	42	95.94
americium	Am	95	–	neodymium	Nd	60	144.24
antimony	Sb	51	121.75	neon	Ne	10	20.183
argon	Ar	18	39.948	neptunium	Np	93	–
arsenic	As	33	74.9216	nickel	Ni	28	58.71
astatine	At	85	–	niobium	Nb	41	92.906
barium	Ba	56	137.34	nitrogen	N	7	14.0067
berkelium	Bk	97	–	nobelium	No	102	–
beryllium	Be	4	9.0122	osmium	Os	76	190.2
bismuth	Bi	83	208.980	oxygen	O	8	15.9994
boron	B	5	10.811	palladium	Pd	46	106.4
bromine	Br	35	79.904	phosphorus	P	15	30.9738
cadmium	Cd	48	112.40	platinum	Pt	78	195.09
calcium	Ca	20	40.08	plutonium	Pu	94	–
californium	Cf	98	–	polonium	Po	84	–
carbon	C	6	12.01115	potassium	K	19	39.102
cerium	Ce	58	140.12	praseodymium	Pr	59	140.907
cesium	Cs	55	132.905	promethium	Pm	61	–
chlorine	Cl	17	35.453	protactinium	Pa	91	–
chromium	Cr	24	51.996	radium	Ra	88	–
cobalt	Co	27	58.9332	radon	Rn	86	–
copper	Cu	29	63.546	rhenium	Re	75	186.2
curium	Cm	96	–	rhodium	Rh	45	102.905
dysprosium	Dy	66	162.50	rubidium	Rb	37	85.47
einsteinium	Es	99	–	ruthenium	Ru	44	101.07
erbium	Er	68	167.26	samarium	Sm	62	150.35
europium	Eu	63	151.96	scandium	Sc	21	44.956
fermium	Fm	100	–	selenium	Se	34	78.96
fluorine	F	9	18.9984	silicon	Si	14	28.086
francium	Fr	87	–	silver	Ag	47	107.868
gadolinium	Gd	64	157.25	sodium	Na	11	22.9898
gallium	Ga	31	69.72	strontium	Sr	38	87.62
germanium	Ge	32	72.59	sulfur	S	16	32.064
gold	Au	79	196.967	tantalum	Ta	73	180.948
hafnium	Hf	72	178.49	technetium	Tc	43	–
helium	He	2	4.0026	tellurium	Te	52	127.60
holmium	Ho	67	164.930	terbium	Tb	65	158.924
hydrogen	H	1	1.00797	thallium	Tl	81	204.37
indium	In	49	114.82	thorium	Th	90	232.038
iodine	I	53	126.9044	thulium	Tm	69	168.934
iridium	Ir	77	192.2	tin	Sn	50	118.69
iron	Fe	26	55.847	titanium	Ti	22	47.90
krypton	Kr	36	83.80	tungsten	W	74	183.85
lanthanum	La	57	138.91	uranium	U	92	238.03
lead	Pb	82	207.19	vanadium	V	23	50.942
lithium	Li	3	6.939	xenon	Xe	54	131.30
lutetium	Lu	71	174.97	ytterbium	Yb	70	173.04
magnesium	Mg	12	24.312	yttrium	Y	39	88.905
manganese	Mn	25	54.9380	zinc	Zn	30	65.37
mendelevium	Md	101	–	zirconium	Zr	40	91.22

APPENDIX 53.A

Wire (and Sheet-Metal Gage) Diameters (Thickness) in Inches

gage no.	(1) **American Wire Gage (AWG)***	(2) steel wire gage	(3) Birmingham iron wire (Stubs)	(4) Stubs steel wire gage	(5) British standard wire gage	(6) steel music wire gage	(7) manufacturer's standard gage	(8) American zinc	(9) Birmingham
15/0	–	–	–	–	–	–	–	–	1.0000
14/0	–	–	–	–	–	–	–	–	0.9583
13/0	–	–	–	–	–	–	–	–	0.9167
12/0	–	–	–	–	–	–	–	–	0.8750
11/0	–	–	–	–	–	–	–	–	0.8333
10/0	–	–	–	–	–	–	–	–	0.7917
9/0	–	–	–	–	–	–	–	–	0.7500
8/0	–	–	–	–	–	–	–	–	0.7083
7/0	–	0.4900	–	–	0.5000	–	0.5000	–	0.6666
6/0	–	0.4615	–	–	0.4640	–	0.4687	–	0.6250
5/0	–	0.4305	–	–	0.4320	–	0.4375	–	0.5883
4/0	**0.4600**	0.3938	0.454	–	0.4000	–	0.4062	–	0.5416
3/0	**0.4100**	0.3625	0.425	–	0.3720	–	0.3750	–	0.5000
2/0	**0.3650**	0.3310	0.380	–	0.3480	0.0087	0.3437	–	0.4452
1/0	**0.3250**	0.3065	0.340	–	0.3240	0.0039	0.3125	–	0.3964
1	**0.2890**	0.2830	0.300	0.227	0.3000	0.0098	0.2812	0.002	0.3532
2	**0.2580**	0.2625	0.284	0.219	0.2760	0.0106	0.2656	0.004	0.3147
3	**0.2290**	0.2437	0.259	0.212	0.2520	0.0114	0.2500	0.006	0.2804
4	**0.2040**	0.2253	0.238	0.207	0.2320	0.0122	0.2344	0.008	0.2500
5	**0.1820**	0.2007	0.220	0.204	0.2120	0.0138	0.2187	0.010	0.2225
6	**0.1620**	0.1920	0.203	0.201	0.1920	0.0157	0.2035	0.012	0.1981
7	**0.1440**	0.1770	0.180	0.199	0.1760	0.0177	0.1875	0.014	0.1764
8	**0.1280**	0.1620	0.165	0.197	0.1600	0.0197	0.1719	0.016	0.1570
9	**0.1140**	0.1483	0.148	0.194	0.1440	0.0216	0.1562	0.018	0.1398
10	**0.1020**	0.1350	0.134	0.191	0.1280	0.0236	0.1406	0.020	0.1250
11	**0.0910**	0.1205	0.120	0.188	0.1160	0.0260	0.1250	0.024	0.1113
12	**0.0810**	0.1055	0.109	0.185	0.1040	0.0283	0.1094	0.028	0.0991
13	**0.0720**	0.0915	0.095	0.182	0.0920	0.0305	0.0937	0.032	0.0882
14	**0.0640**	0.0800	0.083	0.180	0.0800	0.0323	0.0821	0.036	0.0785
15	**0.0570**	0.0720	0.072	0.178	0.0720	0.0342	0.0703	0.040	0.0699
16	**0.0510**	0.0625	0.065	0.175	0.0640	0.0362	0.0625	0.045	0.0625
17	**0.0450**	0.0540	0.058	0.172	0.0560	0.0382	0.0562	0.050	0.0556
18	**0.0400**	0.0475	0.049	0.168	0.0480	0.0400	0.0500	0.055	0.0495
19	**0.0360**	0.0410	0.042	0.164	0.0400	0.0420	0.0437	0.060	0.0440
20	**0.0320**	0.0348	0.035	0.161	0.0360	0.0440	0.0375	0.070	0.0392
21	**0.0285**	0.0317	0.032	0.157	0.0320	0.0460	0.0344	0.080	0.0349
22	**0.0253**	0.0286	0.028	0.155	0.0280	0.0480	0.0312	0.090	0.0312
23	**0.0226**	0.0258	0.025	0.153	0.0240	0.0510	0.0281	0.100	0.0278
24	**0.0210**	0.0230	0.022	0.151	0.0220	0.0550	0.0250	0.125	0.0247
25	**0.0179**	0.0204	0.020	0.148	0.0200	0.0590	0.0219	0.250	0.0220
26	**0.0159**	0.0181	0.018	0.146	0.0180	0.0630	0.0187	0.375	0.0196
27	**0.0142**	0.0173	0.016	0.143	0.0164	0.0670	0.0172	0.500	0.0175
28	**0.0126**	0.0162	0.014	0.139	0.0148	0.0710	0.0156	1.000	0.0156
29	**0.0113**	0.0150	0.013	0.134	0.0136	0.0740	0.0141	–	0.0139
30	**0.0100**	0.0140	0.012	0.127	0.0124	0.0780	0.0125	–	0.0123

(continued)

APPENDIX 53.A *(continued)*
Wire (and Sheet-Metal Gage) Diameters (Thickness) in Inches

gage no.	(1) **American Wire Gage (AWG)***	(2) steel wire gage	(3) Birmingham iron wire (Stubs)	(4) Stubs steel wire gage	(5) British standard wire gage	(6) steel music wire gage	(7) manufacturer's standard gage	(8) American zinc	(9) Birmingham
31	**0.0089**	0.0132	0.010	0.120	0.0116	0.0820	0.0109	–	0.0110
32	**0.0080**	0.0128	0.009	0.115	0.0108	0.0860	0.0101	–	0.0098
33	**0.0071**	0.0118	0.008	0.112	0.0100	–	0.0094	–	0.0087
34	**0.0063**	0.0104	0.007	0.110	0.0092	–	0.0086	–	0.0077
35	**0.0056**	0.0095	0.005	0.108	0.0084	–	0.0078	–	0.0069
36	**0.0050**	0.0090	0.004	0.106	0.0076	–	0.0070	–	0.0061
37	**0.0045**	0.0085	–	0.103	0.0068	–	0.0066	–	0.0054
38	**0.0040**	0.0080	–	0.101	0.0060	–	0.0062	–	0.0048
39	**0.0035**	0.0075	–	0.099	0.0052	–	–	–	0.0043
40	**0.0031**	0.0070	–	0.097	0.0048	–	–	–	0.0039
41	–	0.0066	–	0.095	0.0044	–	–	–	0.0034
42	–	0.0062	–	0.092	0.0040	–	–	–	0.0031
43	–	0.0060	–	0.088	0.0036	–	–	–	0.0027
44	–	0.0058	–	0.085	0.0032	–	–	–	0.0024
45	–	0.0055	–	0.081	0.0028	–	–	–	0.00215
46	–	0.0052	–	0.079	0.0024	–	–	–	0.0019
47	–	0.0050	–	0.077	0.0020	–	–	–	0.0017
48	–	0.0048	–	0.075	0.0016	–	–	–	0.0015
49	–	0.0046	–	0.072	0.0012	–	–	–	0.00135
50	–	0.0044	–	0.069	0.0010	–	–	–	0.0012
51	–	–	–	–	–	–	–	–	0.0011
52	–	–	–	–	–	–	–	–	0.00095

*The American wire gage sizes have been rounded off to the usual limits of commercial accuracy.

Appendices

APPENDIX 53.B
Conductor Ampacities
(NEC Table 310.16)

Allowable ampacities of insulated conductors rated 0 V through 2000 V, 60°C through 90°C (140°F through 194°F), not more than three current-carrying conductors in raceway, cable, or earth (directly buried), based on ambient temperature of 30°C (86°F).

size	temperature rating of conductor (see NEC Table 310-13)						size
	60°C (140°F)	75°C (167°F)	90°C (194°F)	60°C (140°F)	75°C (167°F)	90°C (194°F)	
AWG or kcmil	types TW, UF	types RHW, THHW, THW, THWN, XHHW, USE, ZW	types TBS, SA, SIS, FEP, FEPB, MI, RHH, RHW-2, THHN, THHW, THW-2, THWN-2 USE-2, XHH, XHHW, XHHW-2, ZW-2	types TW,UF	types RHW, THHW, THW, THWN, XHHW, USE	types TBS, SA SIS, THHN, THHW, THW-2 THWN-2, RHH, RHW-2, USE-2, XHH, XHHW, XHHW-2, ZW-2	AWG or kcmil
	copper			aluminum or copper-clad aluminum			
18	–	–	14	–	–	–	–
16	–	–	18	–	–	–	–
14*	20	20	25	–	–	–	–
12*	25	25	30	20	20	25	12*
10*	30	35	40	25	30	35	10*
8	40	50	55	30	40	45	8
6	55	65	75	40	50	60	6
4	70	85	95	55	65	75	4
3	85	100	110	65	75	85	3
2	95	115	130	75	90	100	2
1	110	130	150	85	100	115	1
1/0	125	150	170	100	120	135	1/0
2/0	145	175	195	115	135	150	2/0
3/0	165	200	225	130	155	175	3/0
4/0	195	230	260	150	180	205	4/0
250	215	255	290	170	205	230	250
300	240	285	320	190	230	255	300
350	260	310	350	210	250	280	350
400	280	335	380	225	270	305	400
500	320	380	430	260	310	350	500
600	355	420	475	285	340	385	600
700	385	460	520	310	375	420	700
750	400	475	535	320	385	435	750
800	410	490	555	330	395	450	800
900	435	520	585	355	425	480	900
1000	455	545	615	375	445	500	1000
1250	495	590	665	405	485	545	1250
1500	520	625	705	435	520	585	1500
1750	545	650	735	455	545	615	1750
2000	560	665	750	470	560	630	2000

correction factors

ambient temp. (°C)	For ambient temperatures other than 30°C (86°F), multiply the allowable ampacities shown by the appropriate factor shown below.						ambient temp. (°F)
21-25	1.08	1.05	1.04	1.08	1.05	1.04	70-77
26-30	1.00	1.00	1.00	1.00	1.00	1.00	78-86
31-35	0.91	0.94	0.96	0.91	0.94	0.96	87-95
36-40	0.82	0.88	0.91	0.82	0.88	0.91	96-104
41-45	0.71	0.82	0.87	0.71	0.82	0.87	105-113
46-50	0.58	0.75	0.82	0.58	0.75	0.82	114-122
51-55	0.41	0.67	0.76	0.41	0.67	0.76	123-131
56-60	–	0.58	0.71	–	0.58	0.71	132-140
61-70	–	0.33	0.58	–	0.33	0.58	141-158
71-80	–	–	0.41	–	–	0.41	159–176

*See Section 240.4(D) of the *National Electrical Code* (NEC).

APPENDIX 53.C
Conductor Properties
(NEC Chapter 9, Table 8)

size			conductors							direct current resistance at 75°C (167°F)					
										copper					
(AWG	area,		stranding			overall				uncoated		coated		aluminum	
or	circular			diameter		diameter		area		Ω/	Ω/	Ω/	Ω/	Ω/	Ω/
kcmil)	mm^2	mils	quantity	mm	in	mm	in	mm^2	in^2	km	1000 ft	km	1000 ft	km	1000 ft
18	0.823	1620	1	–	–	1.02	0.040	0.823	0.001	25.5	7.77	26.5	8.08	42.0	12.8
18	0.823	1620	7	0.39	0.015	1.16	0.046	1.06	0.002	26.1	7.95	27.7	8.45	42.8	13.1
16	1.31	2580	1	–	–	1.29	0.051	1.31	0.002	16.0	4.89	16.7	5.08	26.4	8.05
16	1.31	2580	7	0.49	0.019	1.46	0.058	1.68	0.003	16.4	4.99	17.3	5.29	26.9	8.21
14	2.08	4110	1	–	–	1.63	0.064	2.08	0.003	10.1	3.07	10.4	3.19	16.6	5.06
14	2.08	4110	7	0.62	0.024	1.85	0.073	2.68	0.004	10.3	3.14	10.7	3.26	16.9	5.17
12	3.31	6530	1	–	–	2.05	0.081	3.31	0.005	6.34	1.93	6.57	2.01	10.45	3.18
12	3.31	6530	7	0.78	0.030	2.32	0.092	4.25	0.006	6.50	1.98	6.73	2.05	10.69	3.25
10	5.261	10380	1	–	–	2.588	0.102	5.26	0.008	3.984	1.21	4.148	1.26	6.561	2.00
10	5.261	10380	7	0.98	0.038	2.95	0.116	6.76	0.011	4.070	1.24	4.226	1.29	6.679	2.04
8	8.367	16510	1	–	–	3.264	0.128	8.37	0.013	2.506	0.764	2.579	0.786	4.125	1.26
8	8.367	16510	7	1.23	0.049	3.71	0.146	10.76	0.017	2.551	0.778	2.653	0.809	4.204	1.28
6	13.30	26240	7	1.56	0.061	4.67	0.184	17.09	0.027	1.608	0.491	1.671	0.510	2.652	0.808
4	21.15	41740	7	1.96	0.077	5.89	0.232	27.19	0.042	1.010	0.308	1.053	0.321	1.666	0.508
3	26.67	52620	7	2.20	0.087	6.60	0.260	34.28	0.053	0.802	0.245	0.833	0.254	1.320	0.403
2	33.62	66360	7	2.47	0.097	7.42	0.292	43.23	0.067	0.634	0.194	0.661	0.201	1.045	0.319
1	42.41	83690	19	1.69	0.066	8.43	0.332	55.80	0.087	0.505	0.154	0.524	0.160	0.829	0.253
1/0	53.49	105600	19	1.89	0.074	9.45	0.372	70.41	0.109	0.399	0.122	0.415	0.127	0.660	0.201
2/0	67.43	133100	19	2.13	0.084	10.62	0.418	88.74	0.137	0.3170	0.0967	0.329	0.101	0.523	0.159
3/0	85.01	167800	19	2.39	0.094	11.94	0.470	111.9	0.173	0.2512	0.0766	0.2610	0.0797	0.413	0.126
4/0	107.2	211600	19	2.68	0.106	13.41	0.528	141.1	0.219	0.1996	0.0608	0.2050	0.0626	0.328	0.100
250	127	–	37	2.09	0.082	14.61	0.575	168	0.260	0.1687	0.0515	0.1753	0.0535	0.2778	0.0847
300	152	–	37	2.29	0.090	16.00	0.630	201	0.312	0.1409	0.0429	0.1463	0.0446	0.2318	0.0707
350	177	–	37	2.47	0.097	17.30	0.681	235	0.364	0.1205	0.0367	0.1252	0.0382	0.1984	0.0605
400	203	–	37	2.64	0.104	18.49	0.728	268	0.416	0.1053	0.0321	0.1084	0.0331	0.1737	0.0529
500	253	–	37	2.95	0.116	20.65	0.813	336	0.519	0.0845	0.0258	0.0869	0.0265	0.1391	0.0424
600	304	–	61	2.52	0.099	22.68	0.893	404	0.626	0.0704	0.0214	0.0732	0.0223	0.1159	0.0353
700	355	–	61	2.72	0.107	24.49	0.964	471	0.730	0.0603	0.0184	0.0622	0.0189	0.0994	0.0303
750	380	–	61	2.82	0.111	25.35	0.998	505	0.782	0.0563	0.0171	0.0579	0.0176	0.0927	0.0282
800	405	–	61	2.91	0.114	26.16	1.030	538	0.834	0.0528	0.0161	0.0544	0.0166	0.0868	0.0265
900	456	–	61	3.09	0.122	27.79	1.094	606	0.940	0.0470	0.0143	0.0481	0.0147	0.0770	0.0235
1000	507	–	61	3.25	0.128	29.26	1.152	673	1.042	0.0423	0.0129	0.0434	0.0132	0.0695	0.0212
1250	633	–	91	2.98	0.117	32.74	1.289	842	1.305	0.0338	0.0103	0.0347	0.0106	0.0554	0.0169
1500	760	–	91	3.26	0.128	35.86	1.412	1011	1.566	0.02814	0.00858	0.02814	0.00883	0.0464	0.0141
1750	887	–	127	2.98	0.117	38.76	1.526	1180	1.829	0.02410	0.00735	0.02410	0.00756	0.0397	0.0121
2000	1013	–	127	3.19	0.126	41.45	1.632	1349	2.092	0.02109	0.00643	0.02109	0.00662	0.0348	0.0106

Notes:

1. These resistance values are valid *only* for the parameters as given. Using conductors having coated strands, different stranding type, and, especially, other temperatures changes the resistance.
2. Formula for temperature change: $R_2 = R_1(1 + \alpha(T_2 - 75°))$ where $\alpha_{\mathrm{Cu}} = 0.00323$ 1/°C, $\alpha_{\mathrm{Al}} = 0.00330$ 1/°C at 75°C.
3. Conductors with compact and compressed stranding have about 9% and 3%, respectively, smaller bare conductor diameters than those shown. See Table 5A for actual compact cable dimensions.
4. The IACS conductivities used: bare copper = 100%, aluminum = 61%.
5. Class B stranding is listed as well as solid for some sizes. Its overall diameter and area is that of its circumscribing circle.

FPN: The construction information is per NEMA WC8-1992 or ANSI/UL 1581-2001. The resistance is calculated per National Bureau of Standards Handbook 100, dated 1966, and Handbook 109, dated 1972.

APPENDIX 53.D

Conductor AC Properties
(NEC Chapter 9, Table 9)
Alternating-Current Resistance and Reactance for 600-Volt Cables, 3-Phase, 60 Hz, 75°C (167°F)— Three Single Conductors in Conduit

	resistance to neutral per kilometer (Ω/km) resistance to neutral per 1000 ft (Ω/1000 ft)															
	X_L (reactance) for all wires		alternating current resistance for uncoated copper wires			alternating current resistance for aluminum wires			effective Z at 0.85 pf for uncoated copper wires			effective Z at 0.85 pf for aluminum wires			
size (AWG or kcmil)	**PVC, aluminum conduits**	**steel conduit**	**PVC conduit**	**aluminum conduit**	**steel conduit**	**PVC conduit**	**aluminum conduit**	**steel conduit**	**PVC conduit**	**aluminum conduit**	**steel conduit**	**PVC conduit**	**aluminum conduit**	**steel conduit**	**size (AWG or kcmil)**
14	0.190 0.058	0.240 0.073	10.2 3.1	10.2 3.1	10.2 3.1	— —	— —	— —	8.9 2.7	8.9 2.7	8.9 2.7	— —	— —	— —	14
12	0.177 0.054	0.223 0.068	6.6 2.0	6.6 2.0	6.6 2.0	10.5 3.2	10.5 3.2	10.5 3.2	5.6 1.7	5.6 1.7	5.6 1.7	9.2 2.8	9.2 2.8	9.2 2.8	12
10	0.164 0.050	0.207 0.063	3.9 1.2	3.9 1.2	3.9 1.2	6.6 2.0	6.6 2.0	6.6 2.0	3.6 1.1	3.6 1.1	3.6 1.1	5.9 1.8	5.9 1.8	5.9 1.8	10
8	0.171 0.052	0.213 0.065	2.56 0.78	2.56 0.78	2.56 0.78	4.3 1.3	4.3 1.3	4.3 1.3	2.26 0.69	2.26 0.69	2.30 0.70	3.6 1.1	3.6 1.1	3.6 1.1	8
6	0.167 0.051	0.210 0.064	1.61 0.49	1.61 0.49	1.61 0.49	2.66 0.81	2.66 0.81	2.66 0.81	1.44 0.44	1.48 0.45	1.48 0.45	2.33 0.71	2.36 0.72	2.36 0.72	6
4	0.157 0.048	0.197 0.060	1.02 0.31	1.02 0.31	1.02 0.31	1.67 0.51	1.67 0.51	1.67 0.51	0.95 0.29	0.95 0.29	0.98 0.30	1.51 0.46	1.51 0.46	1.51 0.46	4
3	0.154 0.047	0.194 0.059	0.82 0.25	0.82 0.25	0.82 0.25	1.31 0.40	1.35 0.41	1.31 0.40	0.75 0.23	0.79 0.24	0.79 0.24	1.21 0.37	1.21 0.37	1.21 0.37	3
2	0.148 0.045	0.187 0.057	0.62 0.19	0.66 0.20	0.66 0.20	1.05 0.32	1.05 0.32	1.05 0.32	0.62 0.19	0.62 0.19	0.66 0.20	0.98 0.30	0.98 0.30	0.98 0.30	2
1	0.151 0.046	0.187 0.057	0.49 0.15	0.52 0.16	0.52 0.16	0.82 0.25	0.85 0.26	0.82 0.25	0.52 0.16	0.52 0.16	0.52 0.16	0.79 0.24	0.79 0.24	0.82 0.25	1
1/0	0.144 0.044	0.180 0.055	0.39 0.12	0.43 0.13	0.39 0.12	0.66 0.20	0.69 0.21	0.66 0.20	0.43 0.13	0.43 0.13	0.43 0.13	0.62 0.19	0.66 0.20	0.66 0.20	1/0
2/0	0.141 0.043	0.177 0.054	0.33 0.10	0.33 0.10	0.33 0.10	0.52 0.16	0.52 0.16	0.52 0.16	0.36 0.11	0.36 0.11	0.36 0.11	0.52 0.16	0.52 0.16	0.52 0.16	2/0
3/0	0.138 0.042	0.171 0.052	0.253 0.077	0.269 0.082	0.259 0.079	0.43 0.13	0.43 0.13	0.43 0.13	0.289 0.088	0.302 0.092	0.308 0.094	0.43 0.13	0.43 0.13	0.46 0.14	3/0
4/0	0.135 0.041	0.167 0.051	0.203 0.062	0.220 0.067	0.207 0.063	0.33 0.10	0.36 0.11	0.33 0.10	0.243 0.074	0.256 0.078	0.262 0.080	0.36 0.11	0.36 0.11	0.36 0.11	4/0
250	0.135 0.041	0.171 0.052	0.171 0.052	0.187 0.057	0.177 0.054	0.279 0.085	0.295 0.090	0.282 0.086	0.217 0.066	0.230 0.070	0.240 0.073	0.308 0.094	0.322 0.098	0.33 0.10	250
300	0.135 0.041	0.167 0.051	0.144 0.044	0.161 0.049	0.148 0.045	0.233 0.071	0.249 0.076	0.236 0.072	0.194 0.059	0.207 0.063	0.213 0.065	0.269 0.082	0.282 0.086	0.289 0.088	300
350	0.131 0.040	0.164 0.050	0.125 0.038	0.141 0.043	0.128 0.039	0.200 0.061	0.217 0.066	0.207 0.063	0.174 0.053	0.190 0.058	0.197 0.060	0.240 0.073	0.253 0.077	0.262 0.080	350
400	0.131 0.040	0.161 0.049	0.108 0.033	0.125 0.038	0.115 0.035	0.177 0.054	0.194 0.059	0.180 0.055	0.161 0.049	0.174 0.053	0.184 0.056	0.217 0.066	0.233 0.071	0.240 0.073	400
500	0.128 0.039	0.157 0.048	0.089 0.027	0.105 0.032	0.095 0.029	0.141 0.043	0.157 0.048	0.148 0.045	0.141 0.043	0.157 0.048	0.164 0.050	0.187 0.057	0.200 0.061	0.210 0.064	500
600	0.128 0.039	0.157 0.048	0.075 0.023	0.092 0.028	0.082 0.025	0.118 0.036	0.135 0.041	0.125 0.038	0.131 0.040	0.144 0.044	0.154 0.047	0.167 0.051	0.180 0.055	0.190 0.058	600
750	0.125 0.038	0.157 0.048	0.062 0.019	0.079 0.024	0.069 0.021	0.095 0.029	0.112 0.034	0.102 0.031	0.118 0.036	0.131 0.040	0.141 0.043	0.148 0.045	0.161 0.049	0.171 0.052	750
1000	0.121 0.037	0.151 0.046	0.049 0.015	0.062 0.019	0.059 0.018	0.075 0.023	0.089 0.027	0.082 0.025	0.105 0.032	0.118 0.036	0.131 0.040	0.128 0.039	0.138 0.042	0.151 0.046	1000

Notes:

1. These values are based on the following constants: UL-Type RHH wires with Class B stranding, in cradled configuration. Wire conductivities are 100% IACS copper and 61% IACS aluminum, and aluminum conduit is 45% IACS. Capacitive reactance is ignored, since it is negligible at these voltages. These resistance values are valid only at 75°C (167°F) and for the parameters as given, but are representative for 600-volt wire types operating at 60 Hz.

2. *Effective Z* is defined as $R\cos\theta + X\sin\theta$, where θ is the power factor angle of the circuit. Multiplying current by effective impedance gives a good approximation for line-to-neutral voltage drop. Effective impedance values shown in this table are valid only at 0.85 power factor. For another circuit power factor (pf), effective impedance (Z_e) can be calculated from R and X_L values given in this table as follows: $Z_e = R \times \text{pf} + X_L \sin(\arccos \text{pf})$.

APPENDIX 55.A
Standard Cash Flow Factors

multiply	by	to obtain
F	$P = F(1+i)^{-n}$ $(P/F, i\%, n)$	P
P	$F = P(1+i)^{n}$ $(F/P, i\%, n)$	F
A	$P = A\left(\frac{(1+i)^n - 1}{i(1+i)^n}\right)$ $(P/A, i\%, n)$	P
P	$A = P\left(\frac{i(1+i)^n}{(1+i)^n - 1}\right)$ $(A/P, i\%, n)$	A
A	$F = A\left(\frac{(1+i)^n - 1}{i}\right)$ $(F/A, i\%, n)$	F
F	$A = F\left(\frac{i}{(1+i)^n - 1}\right)$ $(A/F, i\%, n)$	A
G	$P = G\left(\frac{(1+i)^n - 1}{i^2(1+i)^n} - \frac{n}{i(1+i)^n}\right)$ $(P/G, i\%, n)$	P
G	$A = G\left(\frac{1}{i} - \frac{n}{(1+i)^n - 1}\right)$ $(A/G, i\%, n)$	A

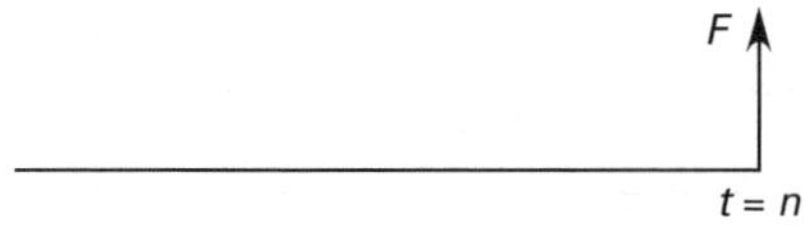

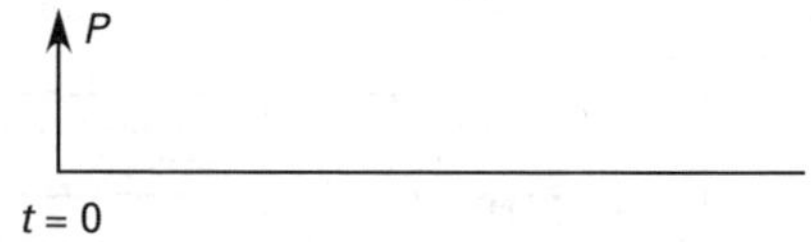

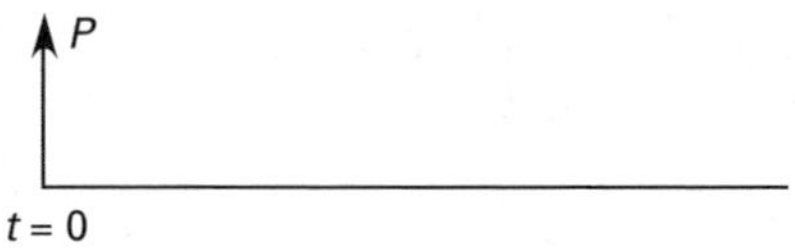

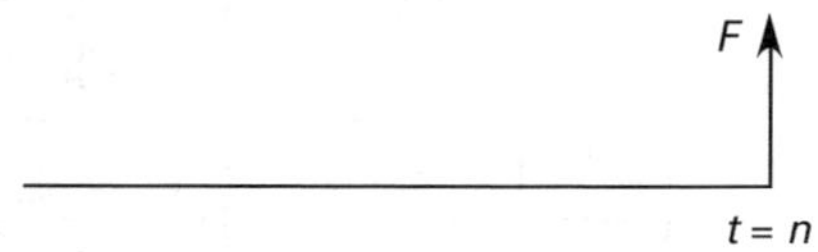

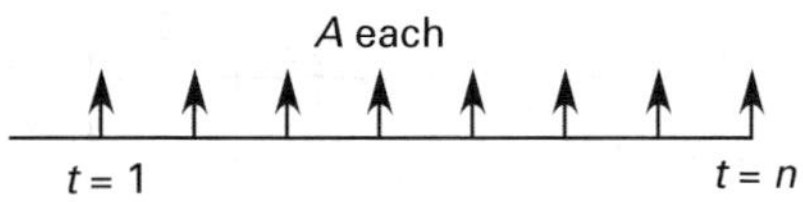

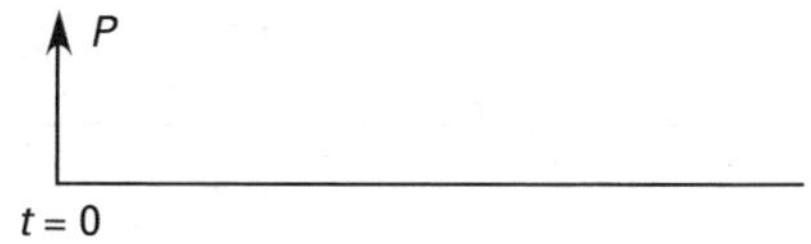

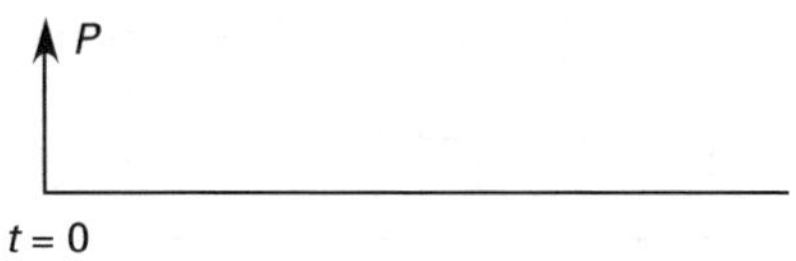

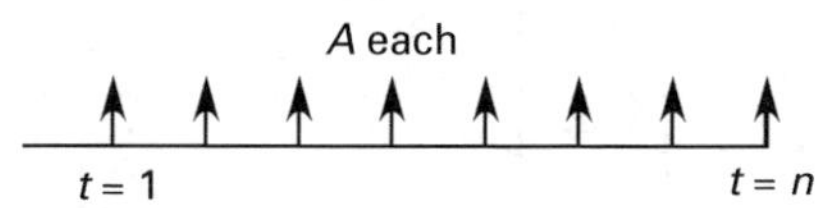

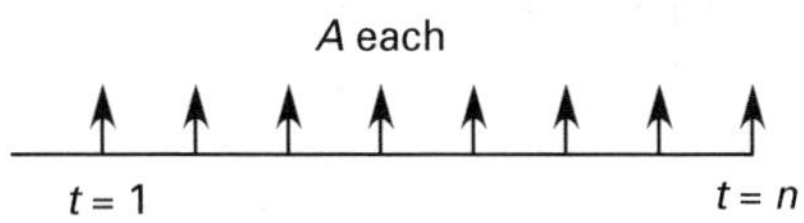

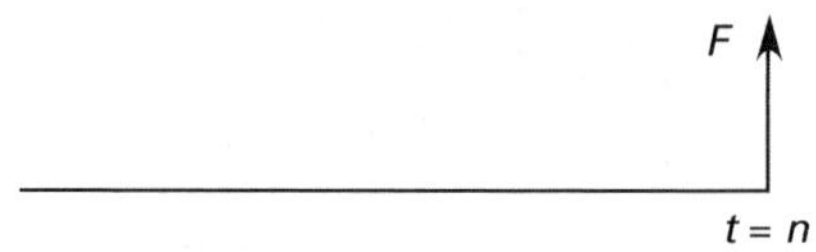

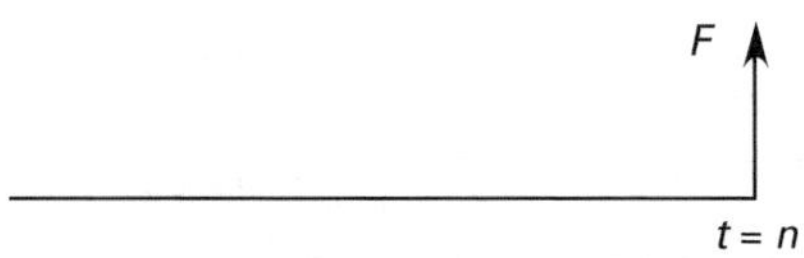

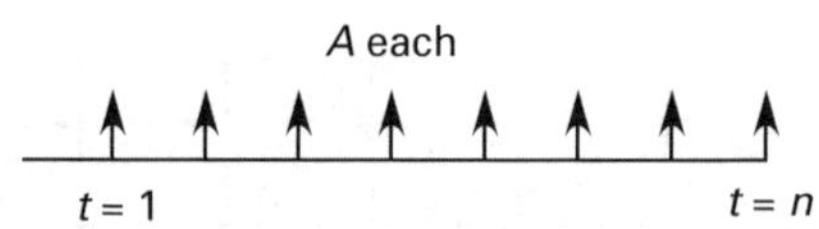

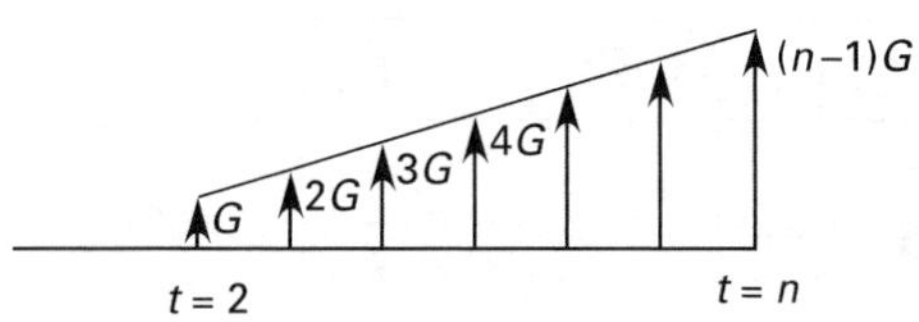

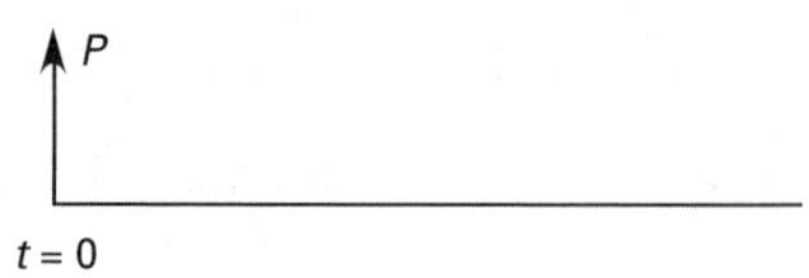

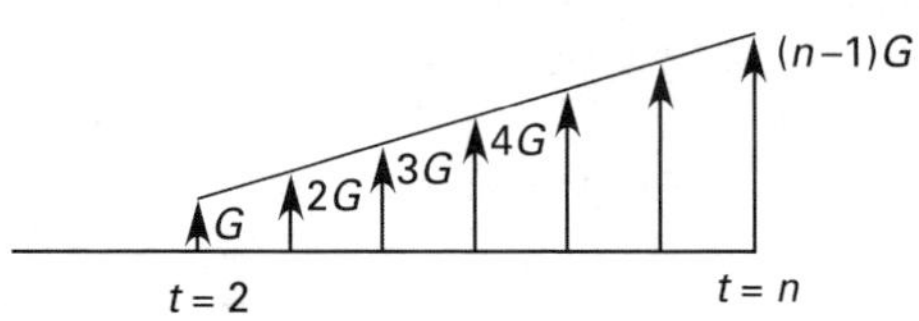

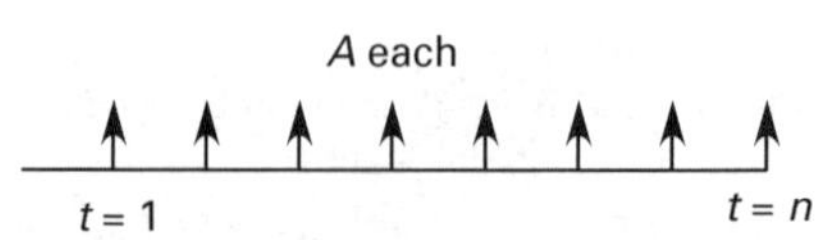

APPENDIX 55.B
Cash Flow Equivalent Factors

$i = 0.50\%$

n	P/F	P/A	P/G	F/P	F/A	A/P	A/F	A/G	n
1	0.9950	0.9950	0.0000	1.0050	1.0000	1.0050	1.0000	0.0000	1
2	0.9901	0.9851	0.9901	1.0100	2.0050	0.5038	0.4988	0.4988	2
3	0.9851	2.9702	2.9604	1.0151	3.0150	0.3367	0.3317	0.9967	3
4	0.9802	3.9505	5.9011	1.0202	4.0301	0.2531	0.2481	1.4938	4
5	0.9754	4.9259	9.8026	1.0253	5.0503	0.2030	0.1980	1.9900	5
6	0.9705	5.8964	14.6552	1.0304	6.0755	0.1696	0.1646	2.4855	6
7	0.9657	6.8621	20.4493	1.0355	7.1059	0.1457	0.1407	2.9801	7
8	0.9609	7.8230	27.1755	1.0407	8.1414	0.1278	0.1228	3.4738	8
9	0.9561	8.7791	34.8244	1.0459	9.1821	0.1139	0.1089	3.9668	9
10	0.9513	9.7304	43.3865	1.0511	10.2280	0.1028	0.0978	4.4589	10
11	0.9466	10.6770	52.8526	1.0564	11.2792	0.0937	0.0887	4.9501	11
12	0.9419	11.6189	63.2136	1.0617	12.3356	0.0861	0.0811	5.4406	12
13	0.9372	12.5562	74.4602	1.0670	13.3972	0.0796	0.0746	5.9302	13
14	0.9326	13.4887	86.5835	1.0723	14.4642	0.0741	0.0691	6.4190	14
15	0.9279	14.4166	99.5743	1.0777	15.5365	0.0694	0.0644	6.9069	15
16	0.9233	15.3399	113.4238	1.0831	16.6142	0.0652	0.0602	7.3940	16
17	0.9187	16.2586	128.1231	1.0885	17.6973	0.0615	0.0565	7.8803	17
18	0.9141	17.1728	143.6634	1.0939	18.7858	0.0582	0.0532	8.3658	18
19	0.9096	18.0824	160.0360	1.0994	19.8797	0.0553	0.0503	8.8504	19
20	0.9051	18.9874	177.2322	1.1049	20.9791	0.0527	0.0477	9.3342	20
21	0.9006	19.8880	195.2434	1.1104	22.0840	0.0503	0.0453	9.8172	21
22	0.8961	20.7841	214.0611	1.1160	23.1944	0.0481	0.0431	10.2993	22
23	0.8916	21.6757	233.6768	1.1216	24.3104	0.0461	0.0411	10.7806	23
24	0.8872	22.5629	254.0820	1.1272	25.4320	0.0443	0.0393	11.2611	24
25	0.8828	23.4456	275.2686	1.1328	26.5591	0.0427	0.0377	11.7407	25
26	0.8784	24.3240	297.2281	1.1385	27.6919	0.0411	0.0361	12.2195	26
27	0.8740	25.1980	319.9523	1.1442	28.8304	0.0397	0.0347	12.6975	27
28	0.8697	26.0677	343.4332	1.1499	29.9745	0.0384	0.0334	13.1747	28
29	0.8653	26.9330	367.6625	1.1556	31.1244	0.0371	0.0321	13.6510	29
30	0.8610	27.7941	392.6324	1.1614	32.2800	0.0360	0.0310	14.1265	30
31	0.8567	28.6508	418.3348	1.1672	33.4414	0.0349	0.0299	14.6012	31
32	0.8525	29.5033	444.7618	1.1730	34.6086	0.0339	0.0289	15.0750	32
33	0.8482	30.3515	471.9055	1.1789	35.7817	0.0329	0.0279	15.5480	33
34	0.8440	31.1955	499.7583	1.1848	36.9606	0.0321	0.0271	16.0202	34
35	0.8398	32.0354	528.3123	1.1907	38.1454	0.0312	0.0262	16.4915	35
36	0.8356	32.8710	557.5598	1.1967	39.3361	0.0304	0.0254	16.9621	36
37	0.8315	33.7025	587.4934	1.2027	40.5328	0.0297	0.0247	17.4317	37
38	0.8274	34.5299	618.1054	1.2087	41.7354	0.0290	0.0240	17.9006	38
39	0.8232	35.3531	649.3883	1.2147	42.9441	0.0283	0.0233	18.3686	39
40	0.8191	36.1722	681.3347	1.2208	44.1588	0.0276	0.0226	18.8359	40
41	0.8151	36.9873	713.9372	1.2269	45.3796	0.0270	0.0220	19.3022	41
42	0.8110	37.7983	747.1886	1.2330	46.6065	0.0265	0.0215	19.7678	42
43	0.8070	38.6053	781.0815	1.2392	47.8396	0.0259	0.0209	20.2325	43
44	0.8030	39.4082	815.6087	1.2454	49.0788	0.0254	0.0204	20.6964	44
45	0.7990	40.2072	850.7631	1.2516	50.3242	0.0249	0.0199	21.1595	45
46	0.7950	41.0022	886.5376	1.2579	51.5758	0.0244	0.0194	21.6217	46
47	0.7910	41.7932	922.9252	1.2642	52.8337	0.0239	0.0189	22.0831	47
48	0.7871	42.5803	959.9188	1.2705	54.0978	0.0235	0.0185	22.5437	48
49	0.7832	43.3635	997.5116	1.2768	55.3683	0.0231	0.0181	23.0035	49
50	0.7793	44.1428	1035.6966	1.2832	56.6452	0.0227	0.0177	23.4624	50
51	0.7754	44.9182	1074.4670	1.2896	57.9284	0.0223	0.0173	23.9205	51
52	0.7716	45.6897	1113.8162	1.2961	59.2180	0.0219	0.0169	24.3778	52
53	0.7677	46.4575	1153.7372	1.3026	60.5141	0.0215	0.0165	24.8343	53
54	0.7639	47.2214	1194.2236	1.3091	61.8167	0.0212	0.0162	25.2899	54
55	0.7601	47.9814	1235.2686	1.3156	63.1258	0.0208	0.0158	25.7447	55
60	0.7414	51.7256	1448.6458	1.3489	69.7700	0.0193	0.0143	28.0064	60
65	0.7231	55.3775	1675.0272	1.3829	76.5821	0.0181	0.0131	30.2475	65
70	0.7053	58.9394	1913.6427	1.4178	83.5661	0.0170	0.0120	32.4680	70
75	0.6879	62.4136	2163.7525	1.4536	90.7265	0.0160	0.0110	34.6679	75
80	0.6710	65.8023	2424.6455	1.4903	98.0677	0.0152	0.0102	36.8474	80
85	0.6545	69.1075	2695.6389	1.5280	105.5943	0.0145	0.0095	39.0065	85
90	0.6383	72.3313	2976.0769	1.5666	113.3109	0.0138	0.0088	41.1451	90
95	0.6226	75.4757	3265.3298	1.6061	121.2224	0.0132	0.0082	43.2633	95
100	0.6073	78.5426	3562.7934	1.6467	129.3337	0.0127	0.0077	45.3613	100

(continued)

APPENDIX 55.B *(continued)*
Cash Flow Equivalent Factors

$i = 0.75\%$

n	P/F	P/A	P/G	F/P	F/A	A/P	A/F	A/G	n
1	0.9926	0.9926	0.0000	1.0075	1.0000	1.0075	1.0000	0.0000	1
2	0.9852	1.9777	0.9852	1.0151	2.0075	0.5056	0.4981	0.4981	2
3	0.9778	2.9556	2.9408	1.0227	3.0226	0.3383	0.3308	0.9950	3
4	0.9706	3.9261	5.8525	1.0303	4.0452	0.2547	0.2472	1.4907	4
5	0.9633	4.8894	9.7058	1.0381	5.0756	0.2045	0.1970	1.9851	5
6	0.9562	5.8456	14.4866	1.0459	6.1136	0.1711	0.1636	2.4782	6
7	0.9490	6.7946	20.1808	1.0537	7.1595	0.1472	0.1397	2.9701	7
8	0.9420	7.7366	26.7747	1.0616	8.2132	0.1293	0.1218	3.4608	8
9	0.9350	8.6716	34.2544	1.0696	9.2748	0.1153	0.1078	3.9502	9
10	0.9280	9.5996	42.6064	1.0776	10.3443	0.1042	0.0967	4.4384	10
11	0.9211	10.5207	51.8174	1.0857	11.4219	0.0951	0.0876	4.9253	11
12	0.9142	11.4349	61.8740	1.0938	12.5076	0.0875	0.0800	5.4110	12
13	0.9074	12.3423	72.7632	1.1020	13.6014	0.0810	0.0735	5.8954	13
14	0.9007	13.2430	84.4720	1.1103	14.7034	0.0755	0.0680	6.3786	14
15	0.8940	14.1370	96.9876	1.1186	15.8137	0.0707	0.0632	6.8606	15
16	0.8873	15.0243	110.2973	1.1270	16.9323	0.0666	0.0591	7.3413	16
17	0.8807	15.9050	124.3887	1.1354	18.0593	0.0629	0.0554	7.8207	17
18	0.8742	16.7792	139.2494	1.1440	19.1947	0.0596	0.0521	8.2989	18
19	0.8676	17.6468	154.8671	1.1525	20.3387	0.0567	0.0492	8.7759	19
20	0.8612	18.5080	171.2297	1.1612	21.4912	0.0540	0.0465	9.2516	20
21	0.8548	19.3628	188.3253	1.1699	22.6524	0.0516	0.0441	9.7261	21
22	0.8484	20.2112	206.1420	1.1787	23.8223	0.0495	0.0420	10.1994	22
23	0.8421	21.0533	224.6682	1.1875	25.0010	0.0475	0.0400	10.6714	23
24	0.8358	21.8891	243.8923	1.1964	26.1885	0.0457	0.0382	11.1422	24
25	0.8296	22.7188	263.8029	1.2054	27.3849	0.0440	0.0365	11.6117	25
26	0.8234	23.5422	284.3888	1.2144	28.5903	0.0425	0.0350	12.0800	26
27	0.8173	24.3595	305.6387	1.2235	29.8047	0.0411	0.0336	12.5470	27
28	0.8112	25.1707	327.5416	1.2327	31.0282	0.0397	0.0322	13.0128	28
29	0.8052	25.9759	350.0867	1.2420	32.2609	0.0385	0.0310	13.4774	29
30	0.7992	26.7751	373.2631	1.2513	33.5029	0.0373	0.0298	13.9407	30
31	0.7932	27.5683	397.0602	1.2607	34.7542	0.0363	0.0288	14.4028	31
32	0.7873	28.3557	421.4675	1.2701	36.0148	0.0353	0.0278	14.8636	32
33	0.7815	29.1371	446.4746	1.2796	37.2849	0.0343	0.0268	15.3232	33
34	0.7757	29.9128	472.0712	1.2892	38.5646	0.0334	0.0259	15.7816	34
35	0.7699	30.6827	498.2471	1.2989	39.8538	0.0326	0.0251	16.2387	35
36	0.7641	31.4468	524.9924	1.3086	41.1527	0.0318	0.0243	16.6946	36
37	0.7585	32.2053	552.2969	1.3185	42.4614	0.0311	0.0236	17.1493	37
38	0.7528	32.9581	580.1511	1.3283	43.7798	0.0303	0.0228	17.6027	38
39	0.7472	33.7053	608.5451	1.3383	45.1082	0.0297	0.0222	18.0549	39
40	0.7416	34.4469	637.4693	1.3483	46.4465	0.0290	0.0215	18.5058	40
41	0.7361	35.1831	666.9144	1.3585	47.7948	0.0284	0.0209	18.9556	41
42	0.7306	35.9137	696.8709	1.3686	49.1533	0.0278	0.0203	19.4040	42
43	0.7252	36.6389	727.3297	1.3789	50.5219	0.0273	0.0198	19.8513	43
44	0.7198	37.3587	758.2815	1.3893	51.9009	0.0268	0.0193	20.2973	44
45	0.7145	38.0732	789.7173	1.3997	53.2901	0.0263	0.0188	20.7421	45
46	0.7091	38.7823	821.6283	1.4102	54.6898	0.0258	0.0183	21.1856	46
47	0.7039	39.4862	854.0056	1.4207	56.1000	0.0253	0.0178	21.6280	47
48	0.6986	40.1848	886.8404	1.4314	57.5207	0.0249	0.0174	22.0691	48
49	0.6934	40.8782	920.1243	1.4421	58.9521	0.0245	0.0170	22.5089	49
50	0.6883	41.5664	953.8486	1.4530	60.3943	0.0241	0.0166	22.9476	50
51	0.6831	42.2496	988.0050	1.4639	61.8472	0.0237	0.0162	23.3850	51
52	0.6780	42.9276	1022.5852	1.4748	63.3111	0.0233	0.0158	23.8211	52
53	0.6730	43.6006	1057.5810	1.4859	64.7859	0.0229	0.0154	24.2561	53
54	0.6680	44.2686	1092.9842	1.4970	66.2718	0.0226	0.0151	24.6898	54
55	0.6630	44.9316	1128.7869	1.5083	67.7688	0.0223	0.0148	25.1223	55
60	0.6387	48.1734	1313.5189	1.5657	75.4241	0.0208	0.0133	27.2665	60
65	0.6153	51.2963	1507.0910	1.6253	83.3709	0.0195	0.0120	29.3801	65
70	0.5927	54.3046	1708.6065	1.6872	91.6201	0.0184	0.0109	31.4634	70
75	0.5710	57.2027	1917.2225	1.7514	100.1833	0.0175	0.0100	33.5163	75
80	0.5500	59.9944	2132.1472	1.8180	109.0725	0.0167	0.0092	35.5391	80
85	0.5299	62.6838	2352.6375	1.8873	118.3001	0.0160	0.0085	37.5318	85
90	0.5104	65.2746	2577.9961	1.9591	127.8790	0.0153	0.0078	39.4946	90
95	0.4917	67.7704	2807.5694	2.0337	137.8225	0.0148	0.0073	41.4277	95
100	0.4737	70.1746	3040.7453	2.1111	148.1445	0.0143	0.0068	43.3311	100

(continued)

APPENDIX 55.B *(continued)*
Cash Flow Equivalent Factors

$i = 1.00\%$

n	P/F	P/A	P/G	F/P	F/A	A/P	A/F	A/G	n
1	0.9901	0.9901	0.0000	1.0100	1.0000	1.0100	1.0000	0.0000	1
2	0.9803	1.9704	0.9803	1.0201	2.0100	0.5075	0.4975	0.4975	2
3	0.9706	2.9410	2.9215	1.0303	3.0301	0.3400	0.3300	0.9934	3
4	0.9610	3.9020	5.8044	1.0406	4.0604	0.2563	0.2463	1.4876	4
5	0.9515	4.8534	9.6103	1.0510	5.1010	0.2060	0.1960	1.9801	5
6	0.9420	5.7955	14.3205	1.0615	6.1520	0.1725	0.1625	2.4710	6
7	0.9327	6.7282	19.9168	1.0721	7.2135	0.1486	0.1386	2.9602	7
8	0.9235	7.6517	26.3812	1.0829	8.2857	0.1307	0.1207	3.4478	8
9	0.9143	8.5660	33.6959	1.0937	9.3685	0.1167	0.1067	3.9337	9
10	0.9053	9.4713	41.8435	1.1046	10.4622	0.1056	0.0956	4.4179	10
11	0.8963	10.3676	50.8067	1.1157	11.5668	0.0965	0.0865	4.9005	11
12	0.8874	11.2551	60.5687	1.1268	12.6825	0.0888	0.0788	5.3815	12
13	0.8787	12.1337	71.1126	1.1381	13.8093	0.0824	0.0724	5.8607	13
14	0.8700	13.0037	82.4221	1.1495	14.9474	0.0769	0.0669	6.3384	14
15	0.8613	13.8651	94.4810	1.1610	16.0969	0.0721	0.0621	6.8143	15
16	0.8528	14.7179	107.2734	1.1726	17.2579	0.0679	0.0579	7.2886	16
17	0.8444	15.5623	120.7834	1.1843	18.4304	0.0643	0.0543	7.7613	17
18	0.8360	16.3983	134.9957	1.1961	19.6147	0.0610	0.0510	8.2323	18
19	0.8277	17.2260	149.8950	1.2081	20.8109	0.0581	0.0481	8.7017	19
20	0.8195	18.0456	165.4664	1.2202	22.0190	0.0554	0.0454	9.1694	20
21	0.8114	18.8570	181.6950	1.2324	23.2392	0.0530	0.0430	9.6354	21
22	0.8034	19.6604	198.5663	1.2447	24.4716	0.0509	0.0409	10.0998	22
23	0.7954	20.4558	216.0660	1.2572	25.7163	0.0489	0.0389	10.5626	23
24	0.7876	21.2434	234.1800	1.2697	26.9735	0.0471	0.0371	11.0237	24
25	0.7798	22.0232	252.8945	1.2824	28.2432	0.0454	0.0354	11.4831	25
26	0.7720	22.7952	272.1957	1.2953	29.5256	0.0439	0.0339	11.9409	26
27	0.7644	23.5596	292.0702	1.3082	30.8209	0.0424	0.0324	12.3971	27
28	0.7568	24.3164	312.5047	1.3213	32.1291	0.0411	0.0311	12.8516	28
29	0.7493	25.0658	333.4863	1.3345	33.4504	0.0399	0.0299	13.3044	29
30	0.7419	25.8077	355.0021	1.3478	34.7849	0.0387	0.0287	13.7557	30
31	0.7346	26.5423	377.0394	1.3613	36.1327	0.0377	0.0277	14.2052	31
32	0.7273	27.2696	399.5858	1.3749	37.4941	0.0367	0.0267	14.6532	32
33	0.7201	27.9897	422.6291	1.3887	38.8690	0.0357	0.0257	15.0995	33
34	0.7130	28.7027	446.1572	1.4026	40.2577	0.0348	0.0248	15.5441	34
35	0.7059	29.4086	470.1583	1.4166	41.6603	0.0340	0.0240	15.9871	35
36	0.6989	30.1075	494.6207	1.4308	43.0769	0.0332	0.0232	16.4285	36
37	0.6920	30.7995	519.5329	1.4451	44.5076	0.0325	0.0225	16.8682	37
38	0.6852	31.4847	544.8835	1.4595	45.9527	0.0318	0.0218	17.3063	38
39	0.6784	32.1630	570.6616	1.4741	47.4123	0.0311	0.0211	17.7428	39
40	0.6717	32.8347	596.8561	1.4889	48.8864	0.0305	0.0205	18.1776	40
41	0.6650	33.4997	623.4562	1.5038	50.3752	0.0299	0.0199	18.6108	41
42	0.6584	34.1581	650.4514	1.5188	51.8790	0.0293	0.0193	19.0424	42
43	0.6519	34.8100	677.8312	1.5340	53.3978	0.0287	0.0187	19.4723	43
44	0.6454	35.4555	705.5853	1.5493	54.9318	0.0282	0.0182	19.9006	44
45	0.6391	36.0945	733.7037	1.5648	56.4811	0.0277	0.0177	20.3273	45
46	0.6327	36.7272	762.1765	1.5805	58.0459	0.0272	0.0172	20.7524	46
47	0.6265	37.3537	790.9938	1.5963	59.6263	0.0268	0.0168	21.1758	47
48	0.6203	37.9740	820.1460	1.6122	61.2226	0.0263	0.0163	21.5976	48
49	0.6141	38.5881	849.6237	1.6283	62.8348	0.0259	0.0159	22.0178	49
50	0.6080	39.1961	879.4176	1.6446	64.4632	0.0255	0.0155	22.4363	50
51	0.6020	39.7981	909.5186	1.6611	66.1078	0.0251	0.0151	22.8533	51
52	0.5961	40.3942	939.9175	1.6777	67.7689	0.0248	0.0148	23.2686	52
53	0.5902	40.9844	970.6057	1.6945	69.4466	0.0244	0.0144	23.6823	53
54	0.5843	41.5687	1001.5743	1.7114	71.1410	0.0241	0.0141	24.0945	54
55	0.5785	42.1472	1032.8148	1.7285	72.8525	0.0237	0.0137	24.5049	55
60	0.5504	44.9550	1192.8061	1.8167	81.6697	0.0222	0.0122	26.5333	60
65	0.5237	47.6266	1358.3903	1.9094	90.9366	0.0210	0.0110	28.5217	65
70	0.4983	50.1685	1528.6474	2.0068	100.6763	0.0199	0.0099	30.4703	70
75	0.4741	52.5871	1702.7340	2.1091	110.9128	0.0190	0.0090	32.3793	75
80	0.4511	54.8882	1879.8771	2.2167	121.6715	0.0182	0.0082	34.2492	80
85	0.4292	57.0777	2059.3701	2.3298	132.9790	0.0175	0.0075	36.0801	85
90	0.4084	59.1609	2240.5675	2.4486	144.8633	0.0169	0.0069	37.8724	90
95	0.3886	61.1430	2422.8811	2.5735	157.3538	0.0164	0.0064	39.6265	95
100	0.3697	63.0289	2605.7758	2.7048	170.4814	0.0159	0.0059	41.3426	100

(continued)

APPENDIX 55.B *(continued)*
Cash Flow Equivalent Factors

$i = 1.50\%$

n	P/F	P/A	P/G	F/P	F/A	A/P	A/F	A/G	n
1	0.9852	0.9852	0.0000	1.0150	1.0000	1.0150	1.0000	0.0000	1
2	0.9707	1.9559	0.9707	1.0302	2.0150	0.5113	0.4963	0.4963	2
3	0.9563	2.9122	2.8833	1.0457	3.0452	0.3434	0.3284	0.9901	3
4	0.9422	3.8544	5.7098	1.0614	4.0909	0.2594	0.2444	1.4814	4
5	0.9283	4.7826	9.4229	1.0773	5.1523	0.2091	0.1941	1.9702	5
6	0.9145	5.6972	13.9956	1.0934	6.2296	0.1755	0.1605	2.4566	6
7	0.9010	6.5982	19.4018	1.1098	7.3230	0.1516	0.1366	2.9405	7
8	0.8877	7.4859	25.6157	1.1265	8.4328	0.1336	0.1186	3.4219	8
9	0.8746	8.3605	32.6125	1.1434	9.5593	0.1196	0.1046	3.9008	9
10	0.8617	9.2222	40.3675	1.1605	10.7027	0.1084	0.0934	4.3772	10
11	0.8489	10.0711	48.8568	1.1779	11.8633	0.0993	0.0843	4.8512	11
12	0.8364	10.9075	58.0571	1.1956	13.0412	0.0917	0.0767	5.3227	12
13	0.8240	11.7315	67.9454	1.2136	14.2368	0.0852	0.0702	5.7917	13
14	0.8118	12.5434	78.4994	1.2318	15.4504	0.0797	0.0647	6.2582	14
15	0.7999	13.3432	89.6974	1.2502	16.6821	0.0749	0.0599	6.7223	15
16	0.7880	14.1313	101.5178	1.2690	17.9324	0.0708	0.0558	7.1839	16
17	0.7764	14.9076	113.9400	1.2880	19.2014	0.0671	0.0521	7.6431	17
18	0.7649	15.6726	126.9435	1.3073	20.4894	0.0638	0.0488	8.0997	18
19	0.7536	16.4262	140.5084	1.3270	21.7967	0.0609	0.0459	8.5539	19
20	0.7425	17.1686	154.6154	1.3469	23.1237	0.0582	0.0432	9.0057	20
21	0.7315	17.9001	169.2453	1.3671	24.4705	0.0559	0.0409	9.4550	21
22	0.7207	18.6208	184.3798	1.3876	25.8376	0.0537	0.0387	9.9018	22
23	0.7100	19.3309	200.0006	1.4084	27.2251	0.0517	0.0367	10.3462	23
24	0.6995	20.0304	216.0901	1.4295	28.6335	0.0499	0.0349	10.7881	24
25	0.6892	20.7196	232.6310	1.4509	30.0630	0.0483	0.0333	11.2276	25
26	0.6790	21.3986	249.6065	1.4727	31.5140	0.0467	0.0317	11.6646	26
27	0.6690	22.0676	267.0002	1.4948	32.9867	0.0453	0.0303	12.0992	27
28	0.6591	22.7267	284.7958	1.5172	34.4815	0.0440	0.0290	12.5313	28
29	0.6494	23.3761	302.9779	1.5400	35.9987	0.0428	0.0278	12.9610	29
30	0.6398	24.0158	321.5310	1.5631	37.5387	0.0416	0.0266	13.3883	30
31	0.6303	24.6461	340.4402	1.5865	39.1018	0.0406	0.0256	13.8131	31
32	0.6210	25.2671	359.6910	1.6103	40.6883	0.0396	0.0246	14.2355	32
33	0.6118	25.8790	379.2691	1.6345	42.2986	0.0386	0.0236	14.6555	33
34	0.6028	26.4817	399.1607	1.6590	43.9331	0.0378	0.0228	15.0731	34
35	0.5939	27.0756	419.3521	1.6839	45.5921	0.0369	0.0219	15.4882	35
36	0.5851	27.6607	439.8303	1.7091	47.2760	0.0362	0.0212	15.9009	36
37	0.5764	28.2371	460.5822	1.7348	48.9851	0.0354	0.0204	16.3112	37
38	0.5679	28.8051	481.5954	1.7608	50.7199	0.0347	0.0197	16.7191	38
39	0.5595	29.3646	502.8576	1.7872	52.4807	0.0341	0.0191	17.1246	39
40	0.5513	29.9158	524.3568	1.8140	54.2679	0.0334	0.0184	17.5277	40
41	0.5431	30.4590	546.0814	1.8412	56.0819	0.0328	0.0178	17.9284	41
42	0.5351	30.9941	568.0201	1.8688	57.9231	0.0323	0.0173	18.3267	42
43	0.5272	31.5212	590.1617	1.8969	59.7920	0.0317	0.0167	18.7227	43
44	0.5194	32.0406	612.4955	1.9253	61.6889	0.0312	0.0162	19.1162	44
45	0.5117	32.5523	635.0110	1.9542	63.6142	0.0307	0.0157	19.5074	45
46	0.5042	33.0565	657.6979	1.9835	65.5684	0.0303	0.0153	19.8962	46
47	0.4967	33.5532	680.5462	2.0133	67.5519	0.0298	0.0148	20.2826	47
48	0.4894	34.0426	703.5462	2.0435	69.5652	0.0294	0.0144	20.6667	48
49	0.4821	34.5247	726.6884	2.0741	71.6087	0.0290	0.0140	21.0484	49
50	0.4750	34.9997	749.9636	2.1052	73.6828	0.0286	0.0136	21.4277	50
51	0.4680	35.4677	773.3629	2.1368	75.7881	0.0282	0.0132	21.8047	51
52	0.4611	35.9287	796.8774	2.1689	77.9249	0.0278	0.0128	22.1794	52
53	0.4543	36.3830	820.4986	2.2014	80.0938	0.0275	0.0125	22.5517	53
54	0.4475	36.8305	844.2184	2.2344	82.2952	0.0272	0.0122	22.9217	54
55	0.4409	37.2715	868.0285	2.2679	84.5296	0.0268	0.0118	23.2894	55
60	0.4093	39.3803	988.1674	2.4432	96.2147	0.0254	0.0104	25.0930	60
65	0.3799	41.3378	1109.4752	2.6320	108.8028	0.0242	0.0092	26.8393	65
70	0.3527	43.1549	1231.1658	2.8355	122.3638	0.0232	0.0082	28.5290	70
75	0.3274	44.8416	1352.5600	3.0546	136.9728	0.0223	0.0073	30.1631	75
80	0.3039	46.4073	1473.0741	3.2907	152.7109	0.0215	0.0065	31.7423	80
85	0.2821	47.8607	1592.2095	3.5450	169.6652	0.0209	0.0059	33.2676	85
90	0.2619	49.2099	1709.5439	3.8189	187.9299	0.0203	0.0053	34.7399	90
95	0.2431	50.4622	1824.7224	4.1141	207.6061	0.0198	0.0048	36.1602	95
100	0.2256	51.6247	1937.4506	4.4320	228.8030	0.0194	0.0044	37.5295	100

(continued)

APPENDIX 55.B *(continued)*
Cash Flow Equivalent Factors

$i = 2.00\%$

n	P/F	P/A	P/G	F/P	F/A	A/P	A/F	A/G	n
1	0.9804	0.9804	0.0000	1.0200	1.0000	1.0200	1.0000	0.0000	1
2	0.9612	1.9416	0.9612	1.0404	2.0200	0.5150	0.4950	0.4950	2
3	0.9423	2.8839	2.8458	1.0612	3.0604	0.3468	0.3268	0.9868	3
4	0.9238	3.8077	5.6173	1.0824	4.1216	0.2626	0.2426	1.4752	4
5	0.9057	4.7135	9.2403	1.1041	5.2040	0.2122	0.1922	1.9604	5
6	0.8880	5.6014	13.6801	1.1262	6.3081	0.1785	0.1585	2.4423	6
7	0.8706	6.4720	18.9035	1.1487	7.4343	0.1545	0.1345	2.9208	7
8	0.8535	7.3255	24.8779	1.1717	8.5830	0.1365	0.1165	3.3961	8
9	0.8368	8.1622	31.5720	1.1951	9.7546	0.1225	0.1025	3.8681	9
10	0.8203	8.9826	38.9551	1.2190	10.9497	0.1113	0.0913	4.3367	10
11	0.8043	9.7868	46.9977	1.2434	12.1687	0.1022	0.0822	4.8021	11
12	0.7885	10.5753	55.6712	1.2682	13.4121	0.0946	0.0746	5.2642	12
13	0.7730	11.3484	64.9475	1.2936	14.6803	0.0881	0.0681	5.7231	13
14	0.7579	12.1062	74.7999	1.3195	15.9739	0.0826	0.0626	6.1786	14
15	0.7430	12.8493	85.2021	1.3459	17.2934	0.0778	0.0578	6.6309	15
16	0.7284	13.5777	96.1288	1.3728	18.6393	0.0737	0.0537	7.0799	16
17	0.7142	14.2919	107.5554	1.4002	20.0121	0.0700	0.0500	7.5256	17
18	0.7002	14.9920	119.4581	1.4282	21.4123	0.0667	0.0467	7.9681	18
19	0.6864	15.6785	131.8139	1.4568	22.8406	0.0638	0.0438	8.4073	19
20	0.6730	16.3514	144.6003	1.4859	24.2974	0.0612	0.0412	8.8433	20
21	0.6598	17.0112	157.7959	1.5157	25.7833	0.0588	0.0388	9.2760	21
22	0.6468	17.6580	171.3795	1.5460	27.2990	0.0566	0.0366	9.7055	22
23	0.6342	18.2922	185.3309	1.5769	28.8450	0.0547	0.0347	10.1317	23
24	0.6217	18.9139	199.6305	1.6084	30.4219	0.0529	0.0329	10.5547	24
25	0.6095	19.5235	214.2592	1.6406	32.0303	0.0512	0.0312	10.9745	25
26	0.5976	20.1210	229.1987	1.6734	33.6709	0.0497	0.0297	11.3910	26
27	0.5859	20.7069	244.4311	1.7069	35.3443	0.0483	0.0283	11.8043	27
28	0.5744	21.2813	259.9392	1.7410	37.0512	0.0470	0.0270	12.2145	28
29	0.5631	21.8444	275.7064	1.7758	38.7922	0.0458	0.0258	12.6214	29
30	0.5521	22.3965	291.7164	1.8114	40.5681	0.0446	0.0246	13.0251	30
31	0.5412	22.9377	307.9538	1.8476	42.3794	0.0436	0.0236	13.4257	31
32	0.5306	23.4683	324.4035	1.8845	44.2270	0.0426	0.0226	13.8230	32
33	0.5202	23.9886	341.0508	1.9222	46.1116	0.0417	0.0217	14.2172	33
34	0.5100	24.4986	357.8817	1.9607	48.0338	0.0408	0.0208	14.6083	34
35	0.5000	24.9986	374.8826	1.9999	49.9945	0.0400	0.0200	14.9961	35
36	0.4902	25.4888	392.0405	2.0399	51.9944	0.0392	0.0192	15.3809	36
37	0.4806	25.9695	409.3424	2.0807	54.0343	0.0385	0.0185	15.7625	37
38	0.4712	26.4406	426.7764	2.1223	56.1149	0.0378	0.0178	16.1409	38
39	0.4619	26.9026	444.3304	2.1647	58.2372	0.0372	0.0172	16.5163	39
40	0.4529	27.3555	461.9931	2.2080	60.4020	0.0366	0.0166	16.8885	40
41	0.4440	27.7995	479.7535	2.2522	62.6100	0.0360	0.0160	17.2576	41
42	0.4353	28.2348	497.6010	2.2972	64.8622	0.0354	0.0154	17.6237	42
43	0.4268	28.6616	515.5253	2.3432	67.1595	0.0349	0.0149	17.9866	43
44	0.4184	29.0800	533.5165	2.3901	69.5027	0.0344	0.0144	18.3465	44
45	0.4102	29.4902	551.5652	2.4379	71.8927	0.0339	0.0139	18.7034	45
46	0.4022	29.8923	569.6621	2.4866	74.3306	0.0335	0.0135	19.0571	46
47	0.3943	30.2866	587.7985	2.5363	76.8172	0.0330	0.0130	19.4079	47
48	0.3865	30.6731	605.9657	2.5871	79.3535	0.0326	0.0126	19.7556	48
49	0.3790	31.0521	624.1557	2.6388	81.9406	0.0322	0.0122	20.1003	49
50	0.3715	31.4236	642.3606	2.6916	84.5794	0.0318	0.0118	20.4420	50
51	0.3642	31.7878	660.5727	2.7454	87.2710	0.0315	0.0115	20.7807	51
52	0.3571	32.1449	678.7849	2.8003	90.0164	0.0311	0.0111	21.1164	52
53	0.3501	32.4950	696.9900	2.8563	92.8167	0.0308	0.0108	21.4491	53
54	0.3432	32.8383	715.1815	2.9135	95.6731	0.0305	0.0105	21.7789	54
55	0.3365	33.1748	733.3527	2.9717	98.5865	0.0301	0.0101	22.1057	55
60	0.3048	34.7609	823.6975	3.2810	114.0515	0.0288	0.0088	23.6961	60
65	0.2761	36.1975	912.7085	3.6225	131.1262	0.0276	0.0076	25.2147	65
70	0.2500	37.4986	999.8343	3.9996	149.9779	0.0267	0.0067	26.6632	70
75	0.2265	38.6771	1084.6393	4.4158	170.7918	0.0259	0.0059	28.0434	75
80	0.2051	39.7445	1166.7868	4.8754	193.7720	0.0252	0.0052	29.3572	80
85	0.1858	40.7113	1246.0241	5.3829	219.1439	0.0246	0.0046	30.6064	85
90	0.1683	41.5869	1322.1701	5.9431	247.1567	0.0240	0.0040	31.7929	90
95	0.1524	42.3800	1395.1033	6.5617	278.0850	0.0236	0.0036	32.9189	95
100	0.1380	43.0984	1464.7527	7.2446	312.2323	0.0232	0.0032	33.9863	100

(continued)

APPENDIX 55.B *(continued)*
Cash Flow Equivalent Factors

$i = 3.00\%$

n	P/F	P/A	P/G	F/P	F/A	A/P	A/F	A/G	n
1	0.9709	0.9709	0.0000	1.0300	1.0000	1.0300	1.0000	0.0000	1
2	0.9426	1.9135	0.9426	1.0609	2.0300	0.5226	0.4926	0.4926	2
3	0.9151	2.8286	2.7729	1.0927	3.0909	0.3535	0.3235	0.9803	3
4	0.8885	3.7171	5.4383	1.1255	4.1836	0.2690	0.2390	1.4631	4
5	0.8626	4.5797	8.8888	1.1593	5.3091	0.2184	0.1884	1.9409	5
6	0.8375	5.4172	13.0762	1.1941	6.4684	0.1846	0.1546	2.4138	6
7	0.8131	6.2303	17.9547	1.2299	7.6625	0.1605	0.1305	2.8819	7
8	0.7894	7.0197	23.4806	1.2668	8.8923	0.1425	0.1125	3.3450	8
9	0.7664	7.7861	29.6119	1.3048	10.1591	0.1284	0.0984	3.8032	9
10	0.7441	8.5302	36.3088	1.3439	11.4639	0.1172	0.0872	4.2565	10
11	0.7224	9.2526	43.5330	1.3842	12.8078	0.1081	0.0781	4.7049	11
12	0.7014	9.9540	51.2482	1.4258	14.1920	0.1005	0.0705	5.1485	12
13	0.6810	10.6350	59.4196	1.4685	15.6178	0.0940	0.0640	5.5872	13
14	0.6611	11.2961	68.0141	1.5126	17.0863	0.0885	0.0585	6.0210	14
15	0.6419	11.9379	77.0002	1.5580	18.5989	0.0838	0.0538	6.4500	15
16	0.6232	12.5611	86.3477	1.6047	20.1569	0.0796	0.0496	6.8742	16
17	0.6050	13.1661	96.0280	1.6528	21.7616	0.0760	0.0460	7.2936	17
18	0.5874	13.7535	106.0137	1.7024	23.4144	0.0727	0.0427	7.7081	18
19	0.5703	14.3238	116.2788	1.7535	25.1169	0.0698	0.0398	8.1179	19
20	0.5537	14.8775	126.7987	1.8061	26.8704	0.0672	0.0372	8.5229	20
21	0.5375	15.4150	137.5496	1.8603	28.6765	0.0649	0.0349	8.9231	21
22	0.5219	15.9369	148.5094	1.9161	30.5368	0.0627	0.0327	9.3186	22
23	0.5067	16.4436	159.6566	1.9736	32.4529	0.0608	0.0308	9.7093	23
24	0.4919	16.9355	170.9711	2.0328	34.4265	0.0590	0.0290	10.0954	24
25	0.4776	17.4131	182.4336	2.0938	36.4593	0.0574	0.0274	10.4768	25
26	0.4637	17.8768	194.0260	2.1566	38.5530	0.0559	0.0259	10.8535	26
27	0.4502	18.3270	205.7309	2.2213	40.7096	0.0546	0.0246	11.2255	27
28	0.4371	18.7641	217.5320	2.2879	42.9309	0.0533	0.0233	11.5930	28
29	0.4243	19.1885	229.4137	2.3566	45.2189	0.0521	0.0221	11.9558	29
30	0.4120	19.6004	241.3613	2.4273	47.5754	0.0510	0.0210	12.3141	30
31	0.4000	20.0004	253.3609	2.5001	50.0027	0.0500	0.0200	12.6678	31
32	0.3883	20.3888	265.3993	2.5751	52.5028	0.0490	0.0190	13.0169	32
33	0.3770	20.7658	277.4642	2.6523	55.0778	0.0482	0.0182	13.3616	33
34	0.3660	21.1318	289.5437	2.7319	57.7302	0.0473	0.0173	13.7018	34
35	0.3554	21.4872	301.6267	2.8139	60.4621	0.0465	0.0165	14.0375	35
36	0.3450	21.8323	313.7028	2.8983	63.2759	0.0458	0.0158	14.3688	36
37	0.3350	22.1672	325.7622	2.9852	66.1742	0.0451	0.0151	14.6957	37
38	0.3252	22.4925	337.7956	3.0748	69.1594	0.0445	0.0145	15.0182	38
39	0.3158	22.8082	349.7942	3.1670	72.2342	0.0438	0.0138	15.3363	39
40	0.3066	23.1148	361.7499	3.2620	75.4013	0.0433	0.0133	15.6502	40
41	0.2976	23.4124	373.6551	3.3599	78.6633	0.0427	0.0127	15.9597	41
42	0.2890	23.7014	385.5024	3.4607	82.0232	0.0422	0.0122	16.2650	42
43	0.2805	23.9819	397.2852	3.5645	85.4839	0.0417	0.0117	16.5660	43
44	0.2724	24.2543	408.9972	3.6715	89.0484	0.0412	0.0112	16.8629	44
45	0.2644	24.5187	420.6325	3.7816	92.7199	0.0408	0.0108	17.1556	45
46	0.2567	24.7754	432.1856	3.8950	96.5015	0.0404	0.0104	17.4441	46
47	0.2493	25.0247	443.6515	4.0119	100.3965	0.0400	0.0100	17.7285	47
48	0.2420	25.2667	455.0255	4.1323	104.4084	0.0396	0.0096	18.0089	48
49	0.2350	25.5017	466.3031	4.2562	108.5406	0.0392	0.0092	18.2852	49
50	0.2281	25.7298	477.4803	4.3839	112.7969	0.0389	0.0089	18.5575	50
51	0.2215	25.9512	488.5535	4.5154	117.1808	0.0385	0.0085	18.8258	51
52	0.2150	26.1662	499.5191	4.6509	121.6962	0.0382	0.0082	19.0902	52
53	0.2088	26.3750	510.3742	4.7904	126.3471	0.0379	0.0079	19.3507	53
54	0.2027	26.5777	521.1157	4.9341	131.1375	0.0376	0.0076	19.6073	54
55	0.1968	26.7744	531.7411	5.0821	136.0716	0.0373	0.0073	19.8600	55
60	0.1697	27.6756	583.0526	5.8916	163.0534	0.0361	0.0061	21.0674	60
65	0.1464	28.4529	631.2010	6.8300	194.3328	0.0351	0.0051	22.1841	65
70	0.1263	29.1234	676.0869	7.9178	230.5941	0.0343	0.0043	23.2145	70
75	0.1089	29.7018	717.6978	9.1789	272.6309	0.0337	0.0037	24.1634	75
80	0.0940	30.2008	756.0865	10.6409	321.3630	0.0331	0.0031	25.0353	80
85	0.0811	30.6312	791.3529	12.3357	377.8570	0.0326	0.0026	25.8349	85
90	0.0699	31.0024	823.6302	14.3005	443.3489	0.0323	0.0023	26.5667	90
95	0.0603	31.3227	853.0742	16.5782	519.2720	0.0319	0.0019	27.2351	95
100	0.0520	31.5989	879.8540	19.2186	607.2877	0.0316	0.0016	27.8444	100

(continued)

APPENDIX 55.B *(continued)*
Cash Flow Equivalent Factors

$i = 4.00\%$

n	P/F	P/A	P/G	F/P	F/A	A/P	A/F	A/G	n
1	0.9615	0.9615	0.0000	1.0400	1.0000	1.0400	1.0000	0.0000	1
2	0.9246	1.8861	0.9246	1.0816	2.0400	0.5302	0.4902	0.4902	2
3	0.8890	2.7751	2.7025	1.1249	3.1216	0.3603	0.3203	0.9739	3
4	0.8548	3.6299	5.2670	1.1699	4.2465	0.2755	0.2355	1.4510	4
5	0.8219	4.4518	8.5547	1.2167	5.4163	0.2246	0.1846	1.9216	5
6	0.7903	5.2421	12.5062	1.2653	6.6330	0.1908	0.1508	2.3857	6
7	0.7599	6.0021	17.0657	1.3159	7.8983	0.1666	0.1266	2.8433	7
8	0.7307	6.7327	22.1806	1.3686	9.2142	0.1485	0.1085	3.2944	8
9	0.7026	7.4353	27.8013	1.4233	10.5828	0.1345	0.0945	3.7391	9
10	0.6756	8.1109	33.8814	1.4802	12.0061	0.1233	0.0833	4.1773	10
11	0.6496	8.7605	40.3772	1.5395	13.4864	0.1141	0.0741	4.6090	11
12	0.6246	9.3851	47.2477	1.6010	15.0258	0.1066	0.0666	5.0343	12
13	0.6006	9.9856	54.4546	1.6651	16.6268	0.1001	0.0601	5.4533	13
14	0.5775	10.5631	61.9618	1.7317	18.2919	0.0947	0.0547	5.8659	14
15	0.5553	11.1184	69.7355	1.8009	20.0236	0.0899	0.0499	6.2721	15
16	0.5339	11.6523	77.7441	1.8730	21.8245	0.0858	0.0458	6.6720	16
17	0.5134	12.1657	85.9581	1.9479	23.6975	0.0822	0.0422	7.0656	17
18	0.4936	12.6593	94.3498	2.0258	25.6454	0.0790	0.0390	7.4530	18
19	0.4746	13.1339	102.8933	2.1068	27.6712	0.0761	0.0361	7.8342	19
20	0.4564	13.5903	111.5647	2.1911	29.7781	0.0736	0.0336	8.2091	20
21	0.4388	14.0292	120.3414	2.2788	31.9692	0.0713	0.0313	8.5779	21
22	0.4220	14.4511	129.2024	2.3699	34.2480	0.0692	0.0292	8.9407	22
23	0.4057	14.8568	138.1284	2.4647	36.6179	0.0673	0.0273	9.2973	23
24	0.3901	15.2470	147.1012	2.5633	39.0826	0.0656	0.0256	9.6479	24
25	0.3751	15.6221	156.1040	2.6658	41.6459	0.0640	0.0240	9.9925	25
26	0.3607	15.9828	165.1212	2.7725	44.3117	0.0626	0.0226	10.3312	26
27	0.3468	16.3296	174.1385	2.8834	47.0842	0.0612	0.0212	10.6640	27
28	0.3335	16.6631	183.1424	2.9987	49.9676	0.0600	0.0200	10.9909	28
29	0.3207	16.9837	192.1206	3.1187	52.9663	0.0589	0.0189	11.3120	29
30	0.3083	17.2920	201.0618	3.2434	56.0849	0.0578	0.0178	11.6274	30
31	0.2965	17.5885	209.9556	3.3731	59.3283	0.0569	0.0169	11.9371	31
32	0.2851	17.8736	218.7924	3.5081	62.7015	0.0559	0.0159	12.2411	32
33	0.2741	18.1476	227.5634	3.6484	66.2095	0.0551	0.0151	12.5396	33
34	0.2636	18.4112	236.2607	3.7943	69.8579	0.0543	0.0143	12.8324	34
35	0.2534	18.6646	244.8768	3.9461	73.6522	0.0536	0.0136	13.1198	35
36	0.2437	18.9083	253.4052	4.1039	77.5983	0.0529	0.0129	13.4018	36
37	0.2343	19.1426	261.8399	4.2681	81.7022	0.0522	0.0122	13.6784	37
38	0.2253	19.3679	270.1754	4.4388	85.9703	0.0516	0.0116	13.9497	38
39	0.2166	19.5845	278.4070	4.6164	90.4091	0.0511	0.0111	14.2157	39
40	0.2083	19.7928	286.5303	4.8010	95.0255	0.0505	0.0105	14.4765	40
41	0.2003	19.9931	294.5414	4.9931	99.8265	0.0500	0.0100	14.7322	41
42	0.1926	20.1856	302.4370	5.1928	104.8196	0.0495	0.0095	14.9828	42
43	0.1852	20.3708	310.2141	5.4005	110.0124	0.0491	0.0091	15.2284	43
44	0.1780	20.5488	317.8700	5.6165	115.4129	0.0487	0.0087	15.4690	44
45	0.1712	20.7200	325.4028	5.8412	121.0294	0.0483	0.0083	15.7047	45
46	0.1646	20.8847	332.8104	6.0748	126.8706	0.0479	0.0079	15.9356	46
47	0.1583	21.0429	340.0914	6.3178	132.9454	0.0475	0.0075	16.1618	47
48	0.1522	21.1951	347.2446	6.5705	139.2632	0.0472	0.0072	16.3832	48
49	0.1463	21.3415	354.2689	6.8333	145.8337	0.0469	0.0069	16.6000	49
50	0.1407	21.4822	361.1638	7.1067	152.6671	0.0466	0.0066	16.8122	50
51	0.1353	21.6175	367.9289	7.3910	159.7738	0.0463	0.0063	17.0200	51
52	0.1301	21.7476	374.5638	7.6866	167.1647	0.0460	0.0060	17.2232	52
53	0.1251	21.8727	381.0686	7.9941	174.8513	0.0457	0.0057	17.4221	53
54	0.1203	21.9930	387.4436	8.3138	182.8454	0.0455	0.0055	17.6167	54
55	0.1157	22.1086	393.6890	8.6464	191.1592	0.0452	0.0052	17.8070	55
60	0.0951	22.6235	422.9966	10.5196	237.9907	0.0442	0.0042	18.6972	60
65	0.0781	23.0467	449.2014	12.7987	294.9684	0.0434	0.0034	19.4909	65
70	0.0642	23.3945	472.4789	15.5716	364.2905	0.0427	0.0027	20.1961	70
75	0.0528	23.6804	493.0408	18.9453	448.6314	0.0422	0.0022	20.8206	75
80	0.0434	23.9154	511.1161	23.0498	551.2450	0.0418	0.0018	21.3718	80
85	0.0357	24.1085	526.9384	28.0436	676.0901	0.0415	0.0015	21.8569	85
90	0.0293	24.2673	540.7369	34.1193	827.9833	0.0412	0.0012	22.2826	90
95	0.0241	24.3978	552.7307	41.5114	1012.7846	0.0410	0.0010	22.6550	95
100	0.0198	24.5050	563.1249	50.5049	1237.6237	0.0408	0.0008	22.9800	100

(continued)

APPENDIX 55.B *(continued)*
Cash Flow Equivalent Factors

$i = 5.00\%$

n	P/F	P/A	P/G	F/P	F/A	A/P	A/F	A/G	n
1	0.9524	0.9524	0.0000	1.0500	1.0000	1.0500	1.0000	0.0000	1
2	0.9070	1.8594	0.9070	1.1025	2.0500	0.5378	0.4878	0.4878	2
3	0.8638	2.7232	2.6347	1.1576	3.1525	0.3672	0.3172	0.9675	3
4	0.8227	3.5460	5.1028	1.2155	4.3101	0.2820	0.2320	1.4391	4
5	0.7835	4.3295	8.2369	1.2763	5.5256	0.2310	0.1810	1.9025	5
6	0.7462	5.0757	11.9680	1.3401	6.8019	0.1970	0.1470	2.3579	6
7	0.7107	5.7864	16.2321	1.4071	8.1420	0.1728	0.1228	2.8052	7
8	0.6768	6.4632	20.9700	1.4775	9.5491	0.1547	0.1047	3.2445	8
9	0.6446	7.1078	26.1268	1.5513	11.0266	0.1407	0.0907	3.6758	9
10	0.6139	7.7217	31.6520	1.6289	12.5779	0.1295	0.0795	4.0991	10
11	0.5847	8.3064	37.4988	1.7103	14.2068	0.1204	0.0704	4.5144	11
12	0.5568	8.8633	43.6241	1.7959	15.9171	0.1128	0.0628	4.9219	12
13	0.5303	9.3936	49.9879	1.8856	17.7130	0.1065	0.0565	5.3215	13
14	0.5051	9.8986	56.5538	1.9799	19.5986	0.1010	0.0510	5.7133	14
15	0.4810	10.3797	63.2880	2.0789	21.5786	0.0963	0.0463	6.0973	15
16	0.4581	10.8378	70.1597	2.1829	23.6575	0.0923	0.0423	6.4736	16
17	0.4363	11.2741	77.1405	2.2920	25.8404	0.0887	0.0387	6.8423	17
18	0.4155	11.6896	84.2043	2.4066	28.1324	0.0855	0.0355	7.2034	18
19	0.3957	12.0853	91.3275	2.5270	30.5390	0.0827	0.0327	7.5569	19
20	0.3769	12.4622	98.4884	2.6533	33.0660	0.0802	0.0302	7.9030	20
21	0.3589	12.8212	105.6673	2.7860	35.7193	0.0780	0.0280	8.2416	21
22	0.3418	13.1630	112.8461	2.9253	38.5052	0.0760	0.0260	8.5730	22
23	0.3256	13.4886	120.0087	3.0715	41.4305	0.0741	0.0241	8.8971	23
24	0.3101	13.7986	127.1402	3.2251	44.5020	0.0725	0.0225	9.2140	24
25	0.2953	14.0939	134.2275	3.3864	47.7271	0.0710	0.0210	9.5238	25
26	0.2812	14.3752	141.2585	3.5557	51.1135	0.0696	0.0196	9.8266	26
27	0.2678	14.6430	148.2226	3.7335	54.6691	0.0683	0.0183	10.1224	27
28	0.2551	14.8981	155.1101	3.9201	58.4026	0.0671	0.0171	10.4114	28
29	0.2429	15.1411	161.9126	4.1161	62.3227	0.0660	0.0160	10.6936	29
30	0.2314	15.3725	168.6226	4.3219	66.4388	0.0651	0.0151	10.9691	30
31	0.2204	15.5928	175.2333	4.5380	70.7608	0.0641	0.0141	11.2381	31
32	0.2099	15.8027	181.7392	4.7649	75.2988	0.0633	0.0133	11.5005	32
33	0.1999	16.0025	188.1351	5.0032	80.0638	0.0625	0.0125	11.7566	33
34	0.1904	16.1929	194.4168	5.2533	85.0670	0.0618	0.0118	12.0063	34
35	0.1813	16.3742	200.5807	5.5160	90.3203	0.0611	0.0111	12.2498	35
36	0.1727	16.5469	206.6237	5.7918	95.8363	0.0604	0.0104	12.4872	36
37	0.1644	16.7113	212.5434	6.0814	101.6281	0.0598	0.0098	12.7186	37
38	0.1566	16.8679	218.3378	6.3855	107.7095	0.0593	0.0093	12.9440	38
39	0.1491	17.0170	224.0054	6.7048	114.0950	0.0588	0.0088	13.1636	39
40	0.1420	17.1591	229.5452	7.0400	120.7998	0.0583	0.0083	13.3775	40
41	0.1353	17.2944	234.9564	7.3920	127.8398	0.0578	0.0078	13.5857	41
42	0.1288	17.4232	240.2389	7.7616	135.2318	0.0574	0.0074	13.7884	42
43	0.1227	17.5459	245.3925	8.1497	142.9933	0.0570	0.0070	13.9857	43
44	0.1169	17.6628	250.4175	8.5572	151.1430	0.0566	0.0066	14.1777	44
45	0.1113	17.7741	255.3145	8.9850	159.7002	0.0563	0.0063	14.3644	45
46	0.1060	17.8801	260.0844	9.4343	168.6852	0.0559	0.0059	14.5461	46
47	0.1009	17.9810	264.7281	9.9060	178.1194	0.0556	0.0056	14.7226	47
48	0.0961	18.0772	269.2467	10.4013	188.0254	0.0553	0.0053	14.8943	48
49	0.0916	18.1687	273.6418	10.9213	198.4267	0.0550	0.0050	15.0611	49
50	0.0872	18.2559	277.9148	11.4674	209.3480	0.0548	0.0048	15.2233	50
51	0.0831	18.3390	282.0673	12.0408	220.8154	0.0545	0.0045	15.3808	51
52	0.0791	18.4181	286.1013	12.6428	232.8562	0.0543	0.0043	15.5337	52
53	0.0753	18.4934	290.0184	13.2749	245.4990	0.0541	0.0041	15.6823	53
54	0.0717	18.5651	293.8208	13.9387	258.7739	0.0539	0.0039	15.8265	54
55	0.0683	18.6335	297.5104	14.6356	272.7126	0.0537	0.0037	15.9664	55
60	0.0535	18.9293	314.3432	18.6792	353.5837	0.0528	0.0028	16.6062	60
65	0.0419	19.1611	328.6910	23.8399	456.7980	0.0522	0.0022	17.1541	65
70	0.0329	19.3427	340.8409	30.4264	588.5285	0.0517	0.0017	17.6212	70
75	0.0258	19.4850	351.0721	38.8327	756.6537	0.0513	0.0013	18.0176	75
80	0.0202	19.5965	359.6460	49.5614	971.2288	0.0510	0.0010	18.3526	80
85	0.0158	19.6838	366.8007	63.2544	1245.0871	0.0508	0.0008	18.6346	85
90	0.0124	19.7523	372.7488	80.7304	1597.6073	0.0506	0.0006	18.8712	90
95	0.0097	19.8059	377.6774	103.0347	2040.6935	0.0505	0.0005	19.0689	95
100	0.0076	19.8479	381.7492	131.5013	2610.0252	0.0504	0.0004	19.2337	100

(continued)

APPENDIX 55.B *(continued)*
Cash Flow Equivalent Factors

$i = 6.00\%$

n	P/F	P/A	P/G	F/P	F/A	A/P	A/F	A/G	n
1	0.9434	0.9434	0.0000	1.0600	1.0000	1.0600	1.0000	0.0000	1
2	0.8900	1.8334	0.8900	1.1236	2.0600	0.5454	0.4854	0.4854	2
3	0.8396	2.6730	2.5692	1.1910	3.1836	0.3741	0.3141	0.9612	3
4	0.7921	3.4651	4.9455	1.2625	4.3746	0.2886	0.2286	1.4272	4
5	0.7473	4.2124	7.9345	1.3382	5.6371	0.2374	0.1774	1.8836	5
6	0.7050	4.9173	11.4594	1.4185	6.9753	0.2034	0.1434	2.3304	6
7	0.6651	5.5824	15.4497	1.5036	8.3938	0.1791	0.1191	2.7676	7
8	0.6274	6.2098	19.8416	1.5938	9.8975	0.1610	0.1010	3.1952	8
9	0.5919	6.8017	24.5768	1.6895	11.4913	0.1470	0.0870	3.6133	9
10	0.5584	7.3601	29.6023	1.7908	13.1808	0.1359	0.0759	4.0220	10
11	0.5268	7.8869	34.8702	1.8983	14.9716	0.1268	0.0668	4.4213	11
12	0.4970	8.3838	40.3369	2.0122	16.8699	0.1193	0.0593	4.8113	12
13	0.4688	8.8527	45.9629	2.1329	18.8821	0.1130	0.0530	5.1920	13
14	0.4423	9.2950	51.7128	2.2609	21.0151	0.1076	0.0476	5.5635	14
15	0.4173	9.7122	57.5546	2.3966	23.2760	0.1030	0.0430	5.9260	15
16	0.3936	10.1059	63.4592	2.5404	25.6725	0.0990	0.0390	6.2794	16
17	0.3714	10.4773	69.4011	2.6928	28.2129	0.0954	0.0354	6.6240	17
18	0.3503	10.8276	75.3569	2.8543	30.9057	0.0924	0.0324	6.9597	18
19	0.3305	11.1581	81.3062	3.0256	33.7600	0.0896	0.0296	7.2867	19
20	0.3118	11.4699	87.2304	3.2071	36.7856	0.0872	0.0272	7.6051	20
21	0.2942	11.7641	93.1136	3.3996	39.9927	0.0850	0.0250	7.9151	21
22	0.2775	12.0416	98.9412	3.6035	43.3923	0.0830	0.0230	8.2166	22
23	0.2618	12.3034	104.7007	3.8197	46.9958	0.0813	0.0213	8.5099	23
24	0.2470	12.5504	110.3812	4.0489	50.8156	0.0797	0.0197	8.7951	24
25	0.2330	12.7834	115.9732	4.2919	54.8645	0.0782	0.0182	9.0722	25
26	0.2198	13.0032	121.4684	4.5494	59.1564	0.0769	0.0169	9.3414	26
27	0.2074	13.2105	126.8600	4.8223	63.7058	0.0757	0.0157	9.6029	27
28	0.1956	13.4062	132.1420	5.1117	68.5281	0.0746	0.0146	9.8568	28
29	0.1846	13.5907	137.3096	5.4184	73.6398	0.0736	0.0136	10.1032	29
30	0.1741	13.7648	142.3588	5.7435	79.0582	0.0726	0.0126	10.3422	30
31	0.1643	13.9291	147.2864	6.0881	84.8017	0.0718	0.0118	10.5740	31
32	0.1550	14.0840	152.0901	6.4534	90.8898	0.0710	0.0110	10.7988	32
33	0.1462	14.2302	156.7681	6.8406	97.3432	0.0703	0.0103	11.0166	33
34	0.1379	14.3681	161.3192	7.2510	104.1838	0.0696	0.0096	11.2276	34
35	0.1301	14.4982	165.7427	7.6861	111.4348	0.0690	0.0090	11.4319	35
36	0.1227	14.6210	170.0387	8.1473	119.1209	0.0684	0.0084	11.6298	36
37	0.1158	14.7368	174.2072	8.6361	127.2681	0.0679	0.0079	11.8213	37
38	0.1092	14.8460	178.2490	9.1543	135.9042	0.0674	0.0074	12.0065	38
39	0.1031	14.9491	182.1652	9.7035	145.0585	0.0669	0.0069	12.1857	39
40	0.0972	15.0463	185.9568	10.2857	154.7620	0.0665	0.0065	12.3590	40
41	0.0917	15.1380	189.6256	10.9029	165.0477	0.0661	0.0061	12.5264	41
42	0.0865	15.2245	193.1732	11.5570	175.9505	0.0657	0.0057	12.6883	42
43	0.0816	15.3062	196.6017	12.2505	187.5076	0.0653	0.0053	12.8446	43
44	0.0770	15.3832	199.9130	12.9855	199.7580	0.0650	0.0050	12.9956	44
45	0.0727	15.4558	203.1096	13.7646	212.7435	0.0647	0.0047	13.1413	45
46	0.0685	15.5244	206.1938	14.5905	226.5081	0.0644	0.0044	13.2819	46
47	0.0647	15.5890	209.1681	15.4659	241.0986	0.0641	0.0041	13.4177	47
48	0.0610	15.6500	212.0351	16.3939	256.5645	0.0639	0.0039	13.5485	48
49	0.0575	15.7076	214.7972	17.3775	272.9584	0.0637	0.0037	13.6748	49
50	0.0543	15.7619	217.4574	18.4202	290.3359	0.0634	0.0034	13.7964	50
51	0.0512	15.8131	220.0181	19.5254	308.7561	0.0632	0.0032	13.9137	51
52	0.0483	15.8614	222.4823	20.6969	328.2814	0.0630	0.0030	14.0267	52
53	0.0456	15.9070	224.8525	21.9387	348.9783	0.0629	0.0029	14.1355	53
54	0.0430	15.9500	227.1316	23.2550	370.9170	0.0627	0.0027	14.2402	54
55	0.0406	15.9905	229.3222	24.6503	394.1720	0.0625	0.0025	14.3411	55
60	0.0303	16.1614	239.0428	32.9877	533.1282	0.0619	0.0019	14.7909	60
65	0.0227	16.2891	246.9450	44.1450	719.0829	0.0614	0.0014	15.1601	65
70	0.0169	16.3845	253.3271	59.0759	967.9322	0.0610	0.0010	15.4613	70
75	0.0126	16.4558	258.4527	79.0569	1300.9487	0.0608	0.0008	15.7058	75
80	0.0095	16.5091	262.5493	105.7960	1746.5999	0.0606	0.0006	15.9033	80
85	0.0071	16.5489	265.8096	141.5789	2342.9817	0.0604	0.0004	16.0620	85
90	0.0053	16.5787	268.3946	189.4645	3141.0752	0.0603	0.0003	16.1891	90
95	0.0039	16.6009	270.4375	253.5463	4209.1042	0.0602	0.0002	16.2905	95
100	0.0029	16.6175	272.0471	339.3021	5638.3681	0.0602	0.0002	16.3711	100

(continued)

APPENDIX 55.B *(continued)*
Cash Flow Equivalent Factors

$i = 7.00\%$

n	P/F	P/A	P/G	F/P	F/A	A/P	A/F	A/G	n
1	0.9346	0.9346	0.0000	1.0700	1.0000	1.0700	1.0000	0.0000	1
2	0.8734	1.8080	0.8734	1.1449	2.0700	0.5531	0.4831	0.4831	2
3	0.8163	2.6243	2.5060	1.2250	3.2149	0.3811	0.3111	0.9549	3
4	0.7629	3.3872	4.7947	1.3108	4.4399	0.2952	0.2252	1.4155	4
5	0.7130	4.1002	7.6467	1.4026	5.7507	0.2439	0.1739	1.8650	5
6	0.6663	4.7665	10.9784	1.5007	7.1533	0.2098	0.1398	2.3032	6
7	0.6227	5.3893	14.7149	1.6058	8.6540	0.1856	0.1156	2.7304	7
8	0.5820	5.9713	18.7889	1.7182	10.2598	0.1675	0.0975	3.1465	8
9	0.5439	6.5152	23.1404	1.8385	11.9780	0.1535	0.0835	3.5517	9
10	0.5083	7.0236	27.7156	1.9672	13.8164	0.1424	0.0724	3.9461	10
11	0.4751	7.4987	32.4665	2.1049	15.7836	0.1334	0.0634	4.3296	11
12	0.4440	7.9427	37.3506	2.2522	17.8885	0.1259	0.0559	4.7025	12
13	0.4150	8.3577	42.3302	2.4098	20.1406	0.1197	0.0497	5.0648	13
14	0.3878	8.7455	47.3718	2.5785	22.5505	0.1143	0.0443	5.4167	14
15	0.3624	9.1079	52.4461	2.7590	25.1290	0.1098	0.0398	5.7583	15
16	0.3387	9.4466	57.5271	2.9522	27.8881	0.1059	0.0359	6.0897	16
17	0.3166	9.7632	62.5923	3.1588	30.8402	0.1024	0.0324	6.4110	17
18	0.2959	10.0591	67.6219	3.3799	33.9990	0.0994	0.0294	6.7225	18
19	0.2765	10.3356	72.5991	3.6165	37.3790	0.0968	0.0268	7.0242	19
20	0.2584	10.5940	77.5091	3.8697	40.9955	0.0944	0.0244	7.3163	20
21	0.2415	10.8355	82.3393	4.1406	44.8652	0.0923	0.0223	7.5990	21
22	0.2257	11.0612	87.0793	4.4304	49.0057	0.0904	0.0204	7.8725	22
23	0.2109	11.2722	91.7201	4.7405	53.4361	0.0887	0.0187	8.1369	23
24	0.1971	11.4693	96.2545	5.0724	58.1767	0.0872	0.0172	8.3923	24
25	0.1842	11.6536	100.6765	5.4274	63.2490	0.0858	0.0158	8.6391	25
26	0.1722	11.8258	104.9814	5.8074	68.6765	0.0846	0.0146	8.8773	26
27	0.1609	11.9867	109.1656	6.2139	74.4838	0.0834	0.0134	9.1072	27
28	0.1504	12.1371	113.2264	6.6488	80.6977	0.0824	0.0124	9.3289	28
29	0.1406	12.2777	117.1622	7.1143	87.3465	0.0814	0.0114	9.5427	29
30	0.1314	12.4090	120.9718	7.6123	94.4608	0.0806	0.0106	9.7487	30
31	0.1228	12.5318	124.6550	8.1451	102.0730	0.0798	0.0098	9.9471	31
32	0.1147	12.6466	128.2120	8.7153	110.2182	0.0791	0.0091	10.1381	32
33	0.1072	12.7538	131.6435	9.3253	118.9334	0.0784	0.0084	10.3219	33
34	0.1002	12.8540	134.9507	9.9781	128.2588	0.0778	0.0078	10.4987	34
35	0.0937	12.9477	138.1353	10.6766	138.2369	0.0772	0.0072	10.6687	35
36	0.0875	13.0352	141.1990	11.4239	148.9135	0.0767	0.0067	10.8321	36
37	0.0818	13.1170	144.1441	12.2236	160.3374	0.0762	0.0062	10.9891	37
38	0.0765	13.1935	146.9730	13.0793	172.5610	0.0758	0.0058	11.1398	38
39	0.0715	13.2649	149.6883	13.9948	185.6403	0.0754	0.0054	11.2845	39
40	0.0668	13.3317	152.2928	14.9745	199.6351	0.0750	0.0050	11.4233	40
41	0.0624	13.3941	154.7892	16.0227	214.6096	0.0747	0.0047	11.5565	41
42	0.0583	13.4524	157.1807	17.1443	230.6322	0.0743	0.0043	11.6842	42
43	0.0545	13.5070	159.4702	18.3444	247.7765	0.0740	0.0040	11.8065	43
44	0.0509	13.5579	161.6609	19.6285	266.1209	0.0738	0.0038	11.9237	44
45	0.0476	13.6055	163.7559	21.0025	285.7493	0.0735	0.0035	12.0360	45
46	0.0445	13.6500	165.7584	22.4726	306.7518	0.0733	0.0033	12.1435	46
47	0.0416	13.6916	167.6714	24.0457	329.2244	0.0730	0.0030	12.2463	47
48	0.0389	13.7305	169.4981	25.7289	353.2701	0.0728	0.0028	12.3447	48
49	0.0363	13.7668	171.2417	27.5299	378.9990	0.0726	0.0026	12.4387	49
50	0.0339	13.8007	172.9051	29.4570	406.5289	0.0725	0.0025	12.5287	50
51	0.0317	13.8325	174.4915	31.5190	435.9860	0.0723	0.0023	12.6146	51
52	0.0297	13.8621	176.0037	33.7253	467.5050	0.0721	0.0021	12.6967	52
53	0.0277	13.8898	177.4447	36.0861	501.2303	0.0720	0.0020	12.7751	53
54	0.0259	13.9157	178.8173	38.6122	537.3164	0.0719	0.0019	12.8500	54
55	0.0242	13.9399	180.1243	41.3150	575.9286	0.0717	0.0017	12.9215	55
60	0.0173	14.0392	185.7677	57.9464	813.5204	0.0712	0.0012	13.2321	60
65	0.0123	14.1099	190.1452	81.2729	1146.7552	0.0709	0.0009	13.4760	65
70	0.0088	14.1604	193.5185	113.9894	1614.1342	0.0706	0.0006	13.6662	70
75	0.0063	14.1964	196.1035	159.8760	2269.6574	0.0704	0.0004	13.8136	75
80	0.0045	14.2220	198.0748	224.2344	3189.0627	0.0703	0.0003	13.9273	80
85	0.0032	14.2403	199.5717	314.5003	4478.5761	0.0702	0.0002	14.0146	85
90	0.0023	14.2533	200.7042	441.1030	6287.1854	0.0702	0.0002	14.0812	90
95	0.0016	14.2626	201.5581	618.6697	8823.8535	0.0701	0.0001	14.1319	95
100	0.0012	14.2693	202.2001	867.7163	12381.6618	0.0701	0.0001	14.1703	100

(continued)

APPENDIX 55.B *(continued)*
Cash Flow Equivalent Factors

$i = 8.00\%$

n	P/F	P/A	P/G	F/P	F/A	A/P	A/F	A/G	n
1	0.9259	0.9259	0.0000	1.0800	1.0000	1.0800	1.0000	0.0000	1
2	0.8573	1.7833	0.8573	1.1664	2.0800	0.5608	0.4808	0.4808	2
3	0.7938	2.5771	2.4450	1.2597	3.2464	0.3880	0.3080	0.9487	3
4	0.7350	3.3121	4.6501	1.3605	4.5061	0.3019	0.2219	1.4040	4
5	0.6806	3.9927	7.3724	1.4693	5.8666	0.2505	0.1705	1.8465	5
6	0.6302	4.6229	10.5233	1.5869	7.3359	0.2163	0.1363	2.2763	6
7	0.5835	5.2064	14.0242	1.7138	8.9228	0.1921	0.1121	2.6937	7
8	0.5403	5.7466	17.8061	1.8509	10.6366	0.1740	0.0940	3.0985	8
9	0.5002	6.2469	21.8081	1.9990	12.4876	0.1601	0.0801	3.4910	9
10	0.4632	6.7101	25.9768	2.1589	14.4866	0.1490	0.0690	3.8713	10
11	0.4289	7.1390	30.2657	2.3316	16.6455	0.1401	0.0601	4.2395	11
12	0.3971	7.5361	34.6339	2.5182	18.9771	0.1327	0.0527	4.5957	12
13	0.3677	7.9038	39.0463	2.7196	21.4953	0.1265	0.0465	4.9402	13
14	0.3405	8.2442	43.4723	2.9372	24.2149	0.1213	0.0413	5.2731	14
15	0.3152	8.5595	47.8857	3.1722	27.1521	0.1168	0.0368	5.5945	15
16	0.2919	8.8514	52.2640	3.4259	30.3243	0.1130	0.0330	5.9046	16
17	0.2703	9.1216	56.5883	3.7000	33.7502	0.1096	0.0296	6.2037	17
18	0.2502	9.3719	60.8426	3.9960	37.4502	0.1067	0.0267	6.4920	18
19	0.2317	9.6036	65.0134	4.3157	41.4463	0.1041	0.0241	6.7697	19
20	0.2145	9.8181	69.0898	4.6610	45.7620	0.1019	0.0219	7.0369	20
21	0.1987	10.0168	73.0629	5.0338	50.4229	0.0998	0.0198	7.2940	21
22	0.1839	10.2007	76.9257	5.4365	55.4568	0.0980	0.0180	7.5412	22
23	0.1703	10.3711	80.6726	5.8715	60.8933	0.0964	0.0164	7.7786	23
24	0.1577	10.5288	84.2997	6.3412	66.7648	0.0950	0.0150	8.0066	24
25	0.1460	10.6748	87.8041	6.8485	73.1059	0.0937	0.0137	8.2254	25
26	0.1352	10.8100	91.1842	7.3964	79.9544	0.0925	0.0125	8.4352	26
27	0.1252	10.9352	94.4390	7.9881	87.3508	0.0914	0.0114	8.6363	27
28	0.1159	11.0511	97.5687	8.6271	95.3388	0.0905	0.0105	8.8289	28
29	0.1073	11.1584	100.5738	9.3173	103.9659	0.0896	0.0096	9.0133	29
30	0.0994	11.2578	103.4558	10.0627	113.2832	0.0888	0.0088	9.1897	30
31	0.0920	11.3498	106.2163	10.8677	123.3459	0.0881	0.0081	9.3584	31
32	0.0852	11.4350	108.8575	11.7371	134.2135	0.0875	0.0075	9.5197	32
33	0.0789	11.5139	111.3819	12.6760	145.9506	0.0869	0.0069	9.6737	33
34	0.0730	11.5869	113.7924	13.6901	158.6267	0.0863	0.0063	9.8208	34
35	0.0676	11.6546	116.0920	14.7853	172.3168	0.0858	0.0058	9.9611	35
36	0.0626	11.7172	118.2839	15.9682	187.1021	0.0853	0.0053	10.0949	36
37	0.0580	11.7752	120.3713	17.2456	203.0703	0.0849	0.0049	10.2225	37
38	0.0537	11.8289	122.3579	18.6253	220.3159	0.0845	0.0045	10.3440	38
39	0.0497	11.8786	124.2470	20.1153	238.9412	0.0842	0.0042	10.4597	39
40	0.0460	11.9246	126.0422	21.7245	259.0565	0.0839	0.0039	10.5699	40
41	0.0426	11.9672	127.7470	23.4625	280.7810	0.0836	0.0036	10.6747	41
42	0.0395	12.0067	129.3651	25.3395	304.2435	0.0833	0.0033	10.7744	42
43	0.0365	12.0432	130.8998	27.3666	329.5830	0.0830	0.0030	10.8692	43
44	0.0338	12.0771	132.3547	29.5560	356.9496	0.0828	0.0028	10.9592	44
45	0.0313	12.1084	133.7331	31.9204	386.5056	0.0826	0.0026	11.0447	45
46	0.0290	12.1374	135.0384	34.4741	418.4261	0.0824	0.0024	11.1258	46
47	0.0269	12.1643	136.2739	37.2320	452.9002	0.0822	0.0022	11.2028	47
48	0.0249	12.1891	137.4428	40.2106	490.1322	0.0820	0.0020	11.2758	48
49	0.0230	12.2122	138.5480	43.4274	530.3427	0.0819	0.0019	11.3451	49
50	0.0213	12.2335	139.5928	46.9016	573.7702	0.0817	0.0017	11.4107	50
51	0.0197	12.2532	140.5799	50.6537	620.6718	0.0816	0.0016	11.4729	51
52	0.0183	12.2715	141.5121	54.7060	671.3255	0.0815	0.0015	11.5318	52
53	0.0169	12.2884	142.3923	59.0825	726.0316	0.0814	0.0014	11.5875	53
54	0.0157	12.3041	143.2229	63.8091	785.1141	0.0813	0.0013	11.6403	54
55	0.0145	12.3186	144.0065	68.9139	848.9232	0.0812	0.0012	11.6902	55
60	0.0099	12.3766	147.3000	101.2571	1253.2133	0.0808	0.0008	11.9015	60
65	0.0067	12.4160	149.7387	148.7798	1847.2481	0.0805	0.0005	12.0602	65
70	0.0046	12.4428	151.5326	218.6064	2720.0801	0.0804	0.0004	12.1783	70
75	0.0031	12.4611	152.8448	321.2045	4002.5566	0.0802	0.0002	12.2658	75
80	0.0021	12.4735	153.8001	471.9548	5886.9354	0.0802	0.0002	12.3301	80
85	0.0014	12.4820	154.4925	693.4565	8655.7061	0.0801	0.0001	12.3772	85
90	0.0010	12.4877	154.9925	1018.9151	12723.9386	0.0801	0.0001	12.4116	90
95	0.0007	12.4917	155.3524	1497.1205	18701.5069	0.0801	0.0001	12.4365	95
100	0.0005	12.4943	155.6107	2199.7613	27484.5157	0.0800	0.0000	12.4545	100

(continued)

APPENDIX 55.B *(continued)*
Cash Flow Equivalent Factors

$i = 9.00\%$

n	P/F	P/A	P/G	F/P	F/A	A/P	A/F	A/G	n
1	0.9174	0.9174	0.0000	1.0900	1.0000	1.0900	1.0000	0.0000	1
2	0.8417	1.7591	0.8417	1.1881	2.0900	0.5685	0.4785	0.4785	2
3	0.7722	2.5313	2.3860	1.2950	3.2781	0.3951	0.3051	0.9426	3
4	0.7084	3.2397	4.5113	1.4116	4.5731	0.3087	0.2187	1.3925	4
5	0.6499	3.8897	7.1110	1.5386	5.9847	0.2571	0.1671	1.8282	5
6	0.5963	4.4859	10.0924	1.6771	7.5233	0.2229	0.1329	2.2498	6
7	0.5470	5.0330	13.3746	1.8280	9.2004	0.1987	0.1087	2.6574	7
8	0.5019	5.5348	16.8877	1.9926	11.0285	0.1807	0.0907	3.0512	8
9	0.4604	5.9952	20.5711	2.1719	13.0210	0.1668	0.0768	3.4312	9
10	0.4224	6.4177	24.3728	2.3674	15.1929	0.1558	0.0658	3.7978	10
11	0.3875	6.8052	28.2481	2.5804	17.5603	0.1469	0.0569	4.1510	11
12	0.3555	7.1607	32.1590	2.8127	20.1407	0.1397	0.0497	4.4910	12
13	0.3262	7.4869	36.0731	3.0658	22.9534	0.1336	0.0436	4.8182	13
14	0.2992	7.7862	39.9633	3.3417	26.0192	0.1284	0.0384	5.1326	14
15	0.2745	8.0607	43.8069	3.6425	29.3609	0.1241	0.0341	5.4346	15
16	0.2519	8.3126	47.5849	3.9703	33.0034	0.1203	0.0303	5.7245	16
17	0.2311	8.5436	51.2821	4.3276	36.9737	0.1170	0.0270	6.0024	17
18	0.2120	8.7556	54.8860	4.7171	41.3013	0.1142	0.0242	6.2687	18
19	0.1945	8.9501	58.3868	5.1417	46.0185	0.1117	0.0217	6.5236	19
20	0.1784	9.1285	61.7770	5.6044	51.1601	0.1095	0.0195	6.7674	20
21	0.1637	9.2922	65.0509	6.1088	56.7645	0.1076	0.0176	7.0006	21
22	0.1502	9.4424	68.2048	6.6586	62.8733	0.1059	0.0159	7.2232	22
23	0.1378	9.5802	71.2359	7.2579	69.5319	0.1044	0.0144	7.4357	23
24	0.1264	9.7066	74.1433	7.9111	76.7898	0.1030	0.0130	7.6384	24
25	0.1160	9.8226	76.9265	8.6231	84.7009	0.1018	0.0118	7.8316	25
26	0.1064	9.9290	79.5863	9.3992	93.3240	0.1007	0.0107	8.0156	26
27	0.0976	10.0266	82.1241	10.2451	102.7231	0.0997	0.0097	8.1906	27
28	0.0895	10.1161	84.5419	11.1671	112.9682	0.0989	0.0089	8.3571	28
29	0.0822	10.1983	86.8422	12.1722	124.1354	0.0981	0.0081	8.5154	29
30	0.0754	10.2737	89.0280	13.2677	136.3076	0.0973	0.0073	8.6657	30
31	0.0691	10.3428	91.1024	14.4618	149.5752	0.0967	0.0067	8.8083	31
32	0.0634	10.4062	93.0690	15.7633	164.0370	0.0961	0.0061	8.9436	32
33	0.0582	10.4644	94.9314	17.1820	179.8003	0.0956	0.0056	9.0718	33
34	0.0534	10.5178	96.6935	18.7284	196.9823	0.0951	0.0051	9.1933	34
35	0.0490	10.5668	98.3590	20.4140	215.7108	0.0946	0.0046	9.3083	35
36	0.0449	10.6118	99.9319	22.2512	236.1247	0.0942	0.0042	9.4171	36
37	0.0412	10.6530	101.4162	24.2538	258.3759	0.0939	0.0039	9.5200	37
38	0.0378	10.6908	102.8158	26.4367	282.6298	0.0935	0.0035	9.6172	38
39	0.0347	10.7255	104.1345	28.8160	309.0665	0.0932	0.0032	9.7090	39
40	0.0318	10.7574	105.3762	31.4094	337.8824	0.0930	0.0030	9.7957	40
41	0.0292	10.7866	106.5445	34.2363	369.2919	0.0927	0.0027	9.8775	41
42	0.0268	10.8134	107.6432	37.3175	403.5281	0.0925	0.0025	9.9546	42
43	0.0246	10.8380	108.6758	40.6761	440.8457	0.0923	0.0023	10.0273	43
44	0.0226	10.8605	109.6456	44.3370	481.5218	0.0921	0.0021	10.0958	44
45	0.0207	10.8812	110.5561	48.3273	525.8587	0.0919	0.0019	10.1603	45
46	0.0190	10.9002	111.4103	52.6767	574.1860	0.0917	0.0017	10.2210	46
47	0.0174	10.9176	112.2115	57.4176	626.8628	0.0916	0.0016	10.2780	47
48	0.0160	10.9336	112.9625	62.5852	684.2804	0.0915	0.0015	10.3317	48
49	0.0147	10.9482	113.6661	68.2179	746.8656	0.0913	0.0013	10.3821	49
50	0.0134	10.9617	114.3251	74.3575	815.0836	0.0912	0.0012	10.4295	50
51	0.0123	10.9740	114.9420	81.0497	889.4411	0.0911	0.0011	10.4740	51
52	0.0113	10.9853	115.5193	88.3442	970.4908	0.0910	0.0010	10.5158	52
53	0.0104	10.9957	116.0593	96.2951	1058.8349	0.0909	0.0009	10.5549	53
54	0.0095	11.0053	116.5642	104.9617	1155.1301	0.0909	0.0009	10.5917	54
55	0.0087	11.0140	117.0362	114.4083	1260.0918	0.0908	0.0008	10.6261	55
60	0.0057	11.0480	118.9683	176.0313	1944.7921	0.0905	0.0005	10.7683	60
65	0.0037	11.0701	120.3344	270.8460	2998.2885	0.0903	0.0003	10.8702	65
70	0.0024	11.0844	121.2942	416.7301	4619.2232	0.0902	0.0002	10.9427	70
75	0.0016	11.0938	121.9646	641.1909	7113.2321	0.0901	0.0001	10.9940	75
80	0.0010	11.0998	122.4306	986.5517	10950.5741	0.0901	0.0001	11.0299	80
85	0.0007	11.1038	122.7533	1517.9320	16854.8003	0.0901	0.0001	11.0551	85
90	0.0004	11.1064	122.9758	2335.5266	25939.1842	0.0900	0.0000	11.0726	90
95	0.0003	11.1080	123.1287	3593.4971	39916.6350	0.0900	0.0000	11.0847	95
100	0.0002	11.1091	123.2335	5529.0408	61422.6755	0.0900	0.0000	11.0930	100

(continued)

Appendices

APPENDIX 55.B *(continued)*
Cash Flow Equivalent Factors

$i = 10.00\%$

n	P/F	P/A	P/G	F/P	F/A	A/P	A/F	A/G	n
1	0.9091	0.9091	0.0000	1.1000	1.0000	1.1000	1.0000	0.0000	1
2	0.8264	1.7355	0.8264	1.2100	2.1000	0.5762	0.4762	0.4762	2
3	0.7513	2.4869	2.3291	1.3310	3.3100	0.4021	0.3021	0.9366	3
4	0.6830	3.1699	4.3781	1.4641	4.6410	0.3155	0.2155	1.3812	4
5	0.6209	3.7908	6.8618	1.6105	6.1051	0.2638	0.1638	1.8101	5
6	0.5645	4.3553	9.6842	1.7716	7.7156	0.2296	0.1296	2.2236	6
7	0.5132	4.8684	12.7631	1.9487	9.4872	0.2054	0.1054	2.6216	7
8	0.4665	5.3349	16.0287	2.1436	11.4359	0.1874	0.0874	3.0045	8
9	0.4241	5.7590	19.4215	2.3579	13.5795	0.1736	0.0736	3.3724	9
10	0.3855	6.1446	22.8913	2.5937	15.9374	0.1627	0.0627	3.7255	10
11	0.3505	6.4951	26.3963	2.8531	18.5312	0.1540	0.0540	4.0641	11
12	0.3186	6.8137	29.9012	3.1384	21.3843	0.1468	0.0468	4.3884	12
13	0.2897	7.1034	33.3772	3.4523	24.5227	0.1408	0.0408	4.6988	13
14	0.2633	7.3667	36.8005	3.7975	27.9750	0.1357	0.0357	4.9955	14
15	0.2394	7.6061	40.1520	4.1772	31.7725	0.1315	0.0315	5.2789	15
16	0.2176	7.8237	43.4164	4.5950	35.9497	0.1278	0.0278	5.5493	16
17	0.1978	8.0216	46.5819	5.0545	40.5447	0.1247	0.0247	5.8071	17
18	0.1799	8.2014	49.6395	5.5599	45.5992	0.1219	0.0219	6.0526	18
19	0.1635	8.3649	52.5827	6.1159	51.1591	0.1195	0.0195	6.2861	19
20	0.1486	8.5136	55.4069	6.7275	57.2750	0.1175	0.0175	6.5081	20
21	0.1351	8.6487	58.1095	7.4002	64.0025	0.1156	0.0156	6.7189	21
22	0.1228	8.7715	60.6893	8.1403	71.4027	0.1140	0.0140	6.9189	22
23	0.1117	8.8832	63.1462	8.9543	79.5430	0.1126	0.0126	7.1085	23
24	0.1015	8.9847	65.4813	9.8497	88.4973	0.1113	0.0113	7.2881	24
25	0.0923	9.0770	67.6964	10.8347	98.3471	0.1102	0.0102	7.4580	25
26	0.0839	9.1609	69.7940	11.9182	109.1818	0.1092	0.0092	7.6186	26
27	0.0763	9.2372	71.7773	13.1100	121.0999	0.1083	0.0083	7.7704	27
28	0.0693	9.3066	73.6495	14.4210	134.2099	0.1075	0.0075	7.9137	28
29	0.0630	9.3696	75.4146	15.8631	148.6309	0.1067	0.0067	8.0489	29
30	0.0573	9.4269	77.0766	17.4494	164.4940	0.1061	0.0061	8.1762	30
31	0.0521	9.4790	78.6395	19.1943	181.9434	0.1055	0.0055	8.2962	31
32	0.0474	9.5264	80.1078	21.1138	201.1378	0.1050	0.0050	8.4091	32
33	0.0431	9.5694	81.4856	23.2252	222.2515	0.1045	0.0045	8.5152	33
34	0.0391	9.6086	82.7773	25.5477	245.4767	0.1041	0.0041	8.6149	34
35	0.0356	9.6442	83.9872	28.1024	271.0244	0.1037	0.0037	8.7086	35
36	0.0323	9.6765	85.1194	30.9127	299.1268	0.1033	0.0033	8.7965	36
37	0.0294	9.7059	86.1781	34.0039	330.0395	0.1030	0.0030	8.8789	37
38	0.0267	9.7327	87.1673	37.4043	364.0434	0.1027	0.0027	8.9562	38
39	0.0243	9.7570	88.0908	41.1448	401.4478	0.0125	0.0025	9.0285	39
40	0.0221	9.7791	88.9525	45.2593	442.5926	0.1023	0.0023	9.0962	40
41	0.0201	9.7991	89.7560	49.7852	487.8518	0.1020	0.0020	9.1596	41
42	0.0183	9.8174	90.5047	54.7637	537.6370	0.1019	0.0019	9.2188	42
43	0.0166	9.8340	91.2019	60.2401	592.4007	0.1017	0.0017	9.2741	43
44	0.0151	9.8491	91.8508	66.2641	652.6408	0.1015	0.0015	9.3258	44
45	0.0137	9.8628	92.4544	72.8905	718.9048	0.1014	0.0014	9.3740	45
46	0.0125	9.8753	93.0157	80.1795	791.7953	0.1013	0.0013	9.4190	46
47	0.0113	9.8866	93.5372	88.1975	871.9749	0.1011	0.0011	9.4610	47
48	0.0103	9.8969	94.0217	97.0172	960.1723	0.1010	0.0010	9.5001	48
49	0.0094	9.9063	94.4715	106.7190	1057.1896	0.1009	0.0009	9.5365	49
50	0.0085	9.9148	94.8889	117.3909	1163.9085	0.1009	0.0009	9.5704	50
51	0.0077	9.9226	95.2761	129.1299	1281.2994	0.1008	0.0008	9.6020	51
52	0.0070	9.9296	95.6351	142.0429	1410.4293	0.1007	0.0007	9.6313	52
53	0.0064	9.9360	95.9679	156.2472	1552.4723	0.1006	0.0006	9.6586	53
54	0.0058	9.9418	96.2763	171.8719	1708.7195	0.1006	0.0006	9.6840	54
55	0.0053	9.9471	96.5619	189.0591	1880.5914	0.1005	0.0005	9.7075	55
60	0.0033	9.9672	97.7010	304.4816	3034.8164	0.1003	0.0003	9.8023	60
65	0.0020	9.9796	98.4705	490.3707	4893.7073	0.1002	0.0002	9.8672	65
70	0.0013	9.9873	98.9870	789.7470	7887.4696	0.1001	0.0001	9.9113	70
75	0.0008	9.9921	99.3317	1271.8954	12708.9537	0.1001	0.0001	9.9410	75
80	0.0005	9.9951	99.5606	2048.4002	20474.0021	0.1000	0.0000	9.9609	80
85	0.0003	9.9970	99.7120	3298.9690	32979.6903	0.1000	0.0000	9.9742	85
90	0.0002	9.9981	99.8118	5313.0226	53120.2261	0.1000	0.0000	9.9831	90
95	0.0001	9.9988	99.8773	8556.6760	85556.7605	0.1000	0.0000	9.9889	95
100	0.0001	9.9993	99.9202	13780.6123	137796.1234	0.1000	0.0000	9.9927	100

(continued)

APPENDIX 55.B *(continued)*
Cash Flow Equivalent Factors

$i = 12.00\%$

n	P/F	P/A	P/G	F/P	F/A	A/P	A/F	A/G	n
1	0.8929	0.8929	0.0000	1.1200	1.0000	1.1200	1.0000	0.0000	1
2	0.7972	1.6901	0.7972	1.2544	2.1200	0.5917	0.4717	0.4717	2
3	0.7118	2.4018	2.2208	1.4049	3.3744	0.4163	0.2963	0.9246	3
4	0.6355	3.0373	4.1273	1.5735	4.7793	0.3292	0.2092	1.3589	4
5	0.5674	3.6048	6.3970	1.7623	6.3528	0.2774	0.1574	1.7746	5
6	0.5066	4.1114	8.9302	1.9738	8.1152	0.2432	0.1232	2.1720	6
7	0.4523	4.5638	11.6443	2.2107	10.0890	0.2191	0.0991	2.5515	7
8	0.4039	4.9676	14.4714	2.4760	12.2997	0.2013	0.0813	2.9131	8
9	0.3606	5.3282	17.3563	2.7731	14.7757	0.1877	0.0677	3.2574	9
10	0.3220	5.6502	20.2541	3.1058	17.5487	0.1770	0.0570	3.5847	10
11	0.2875	5.9377	23.1288	3.4785	20.6546	0.1684	0.0484	3.8953	11
12	0.2567	6.1944	25.9523	3.8960	24.1331	0.1614	0.0414	4.1897	12
13	0.2292	6.4235	28.7024	4.3635	28.0291	0.1557	0.0357	4.4683	13
14	0.2046	6.6282	31.3624	4.8871	32.3926	0.1509	0.0309	4.7317	14
15	0.1827	6.8109	33.9202	5.4736	37.2797	0.1468	0.0268	4.9803	15
16	0.1631	6.9740	36.3670	6.1304	42.7533	0.1434	0.0234	5.2147	16
17	0.1456	7.1196	38.6973	6.8660	48.8837	0.1405	0.0205	5.4353	17
18	0.1300	7.2497	40.9080	7.6900	55.7497	0.1379	0.0179	5.6427	18
19	0.1161	7.3658	42.9979	8.6128	63.4397	0.1358	0.0158	6.8375	19
20	0.1037	7.4694	44.9676	9.6463	72.0524	0.1339	0.0139	6.0202	20
21	0.0926	7.5620	46.8188	10.8038	81.6987	0.1322	0.0122	6.1913	21
22	0.0826	7.6446	48.5543	12.1003	92.5026	0.1308	0.0108	6.3514	22
23	0.0738	7.7184	50.1776	13.5523	104.6029	0.1296	0.0096	6.5010	23
24	0.0659	7.7843	51.6929	15.1786	118.1552	0.1285	0.0085	6.6406	24
25	0.0588	7.8431	53.1046	17.0001	133.3339	0.1275	0.0075	6.7708	25
26	0.0525	7.8957	54.4177	19.0401	150.3339	0.1267	0.0067	6.8921	26
27	0.0469	7.9426	55.6369	21.3249	169.3740	0.1259	0.0059	7.0049	27
28	0.0419	7.9844	56.7674	23.8839	190.6989	0.1252	0.0052	7.1098	28
29	0.0374	8.0218	57.8141	26.7499	214.5828	0.1247	0.0047	7.2071	29
30	0.0334	8.0552	58.7821	29.9599	241.3327	0.1241	0.0041	7.2974	30
31	0.0298	8.0850	59.6761	33.5551	271.2926	0.1237	0.0037	7.3811	31
32	0.0266	8.1116	60.5010	37.5817	304.8477	0.1233	0.0033	7.4586	32
33	0.0238	8.1354	61.2612	42.0915	342.4294	0.1229	0.0029	7.5302	33
34	0.0212	8.1566	61.9612	47.1425	384.5210	0.1226	0.0026	7.5965	34
35	0.0189	8.1755	62.6052	52.7996	431.6635	0.1223	0.0023	7.6577	35
36	0.0169	8.1924	63.1970	59.1356	484.4631	0.1221	0.0021	7.7141	36
37	0.0151	8.2075	63.7406	66.2318	543.5987	0.1218	0.0018	7.7661	37
38	0.0135	8.2210	64.2394	74.1797	609.8305	0.1216	0.0016	7.8141	38
39	0.0120	8.2330	64.6967	83.0812	684.0102	0.1215	0.0015	7.8582	39
40	0.0107	8.2438	65.1159	93.0510	767.0914	0.1213	0.0013	7.8988	40
41	0.0096	8.2534	65.4997	104.2171	860.1424	0.1212	0.0012	7.9361	41
42	0.0086	8.2619	65.8509	116.7231	964.3595	0.1210	0.0010	7.9704	42
43	0.0076	8.2696	66.1722	130.7299	1081.0826	0.1209	0.0009	8.0019	43
44	0.0068	8.2764	66.4659	146.4175	1211.8125	0.1208	0.0008	8.0308	44
45	0.0061	8.2825	66.7342	163.9876	1358.2300	0.1207	0.0007	8.0572	45
46	0.0054	8.2880	66.9792	183.6661	1522.2176	0.1207	0.0007	8.0815	46
47	0.0049	8.2928	67.2028	205.7061	1705.8838	0.1206	0.0006	8.1037	47
48	0.0043	8.2972	67.4068	230.3908	1911.5898	0.1205	0.0005	8.1241	48
49	0.0039	8.3010	67.5929	258.0377	2141.9806	0.1205	0.0005	8.1427	49
50	0.0035	8.3045	67.7624	289.0022	2400.0182	0.1204	0.0004	8.1597	50
51	0.0031	8.3076	67.9169	323.6825	2689.0204	0.1204	0.0004	8.1753	51
52	0.0028	8.3103	68.0576	362.5243	3012.7029	0.1203	0.0003	8.1895	52
53	0.0025	8.3128	68.1856	406.0273	3375.2272	0.1203	0.0003	8.2025	53
54	0.0022	8.3150	68.3022	454.7505	3781.2545	0.1203	0.0003	8.2143	54
55	0.0020	8.3170	68.4082	509.3206	4236.0050	0.1202	0.0002	8.2251	55
60	0.0011	8.3240	68.8100	897.5969	7471.6411	0.1201	0.0001	8.2664	60
65	0.0006	8.3281	69.0581	1581.8725	13173.9374	0.1201	0.0001	8.2922	65
70	0.0004	8.3303	69.2103	2787.7998	23223.3319	0.1200	0.0000	8.3082	70
75	0.0002	8.3316	69.3031	4913.0558	40933.7987	0.1200	0.0000	8.3181	75
80	0.0001	8.3324	69.3594	8658.4831	72145.6925	0.1200	0.0000	8.3241	80
85	0.0001	8.3328	69.3935	15259.2057	127151.7140	0.1200	0.0000	8.3278	85
90	0.0000	8.3330	69.4140	26891.9342	224091.1185	0.1200	0.0000	8.3300	90
95	0.0000	8.3332	69.4263	47392.7766	394931.4719	0.1200	0.0000	8.3313	95
100	0.0000	8.3332	69.4336	83522.2657	696010.5477	0.1200	0.0000	8.3321	100

(continued)

APPENDIX 55.B *(continued)*
Cash Flow Equivalent Factors

$i = 15.00\%$

n	P/F	P/A	P/G	F/P	F/A	A/P	A/F	A/G	n
1	0.8696	0.8696	0.0000	1.1500	1.0000	1.1500	1.0000	0.0000	1
2	0.7561	1.6257	0.7561	1.3225	2.1500	0.6151	0.4651	0.4651	2
3	0.6575	2.2832	2.0712	1.5209	3.4725	0.4380	0.2880	0.9071	3
4	0.5718	2.8550	3.7864	1.7490	4.9934	0.3503	0.2003	1.3263	4
5	0.4972	3.3522	5.7751	2.0114	6.7424	0.2983	0.1483	1.7228	5
6	0.4323	3.7845	7.9368	2.3131	8.7537	0.2642	0.1142	2.0972	6
7	0.3759	4.1604	10.1924	2.6600	11.0668	0.2404	0.0904	2.4498	7
8	0.3269	4.4873	12.4807	3.0590	13.7268	0.2229	0.0729	2.7813	8
9	0.2843	4.7716	14.7548	3.5179	16.7858	0.2096	0.0596	3.0922	9
10	0.2472	5.0188	16.9795	4.0456	20.3037	0.1993	0.0493	3.3832	10
11	0.2149	5.2337	19.1289	4.6524	24.3493	0.1911	0.0411	3.6549	11
12	0.1869	5.4206	21.1849	5.3503	29.0017	0.1845	0.0345	3.9082	12
13	0.1625	5.5831	23.1352	6.1528	34.3519	0.1791	0.0291	4.1438	13
14	0.1413	5.7245	24.9725	7.0757	40.5047	0.1747	0.0247	4.3624	14
15	0.1229	5.8474	26.9630	8.1371	47.5804	0.1710	0.0210	4.5650	15
16	0.1069	5.9542	28.2960	9.3576	55.7175	0.1679	0.0179	4.7522	16
17	0.0929	6.0472	29.7828	10.7613	65.0751	0.1654	0.0154	4.9251	17
18	0.0808	6.1280	31.1565	12.3755	75.8364	0.1632	0.0132	5.0843	18
19	0.0703	6.1982	32.4213	14.2318	88.2118	0.1613	0.0113	5.2307	19
20	0.0611	6.2593	33.5822	16.3665	102.4436	0.1598	0.0098	5.3651	20
21	0.0531	6.3125	34.6448	18.8215	118.8101	0.1584	0.0084	5.4883	21
22	0.0462	6.3587	35.6150	21.6447	137.6316	0.1573	0.0073	5.6010	22
23	0.0402	6.3988	36.4988	24.8915	159.2764	0.1563	0.0063	5.7040	23
24	0.0349	6.4338	37.3023	28.6252	184.1678	0.1554	0.0054	5.7979	24
25	0.0304	6.4641	38.0314	32.9190	212.7930	0.1547	0.0047	5.8834	25
26	0.0264	6.4906	38.6918	37.8568	245.7120	0.1541	0.0041	5.9612	26
27	0.0230	6.5135	39.2890	43.5353	283.5688	0.1535	0.0035	6.0319	27
28	0.0200	6.5335	39.8283	50.0656	327.1041	0.1531	0.0031	6.0960	28
29	0.0174	6.5509	40.3146	57.5755	377.1697	0.1527	0.0027	6.1541	29
30	0.0151	6.5660	40.7526	66.2118	434.7451	0.1523	0.0023	6.2066	30
31	0.0131	6.5791	41.1466	76.1435	500.9569	0.1520	0.0020	6.2541	31
32	0.0114	6.5905	41.5006	87.5651	577.1005	0.1517	0.0017	6.2970	32
33	0.0099	6.6005	41.8184	100.6998	664.6655	0.1515	0.0015	6.3357	33
34	0.0086	6.6091	42.1033	115.8048	765.3654	0.1513	0.0013	6.3705	34
35	0.0075	6.6166	42.3586	133.1755	881.1702	0.1511	0.0011	6.4019	35
36	0.0065	6.6231	42.5872	153.1519	1014.3457	0.1510	0.0010	6.4301	36
37	0.0057	6.6288	42.7916	176.1246	1167.4975	0.1509	0.0009	6.4554	37
38	0.0049	6.6338	42.9743	202.5433	1343.6222	0.1507	0.0007	6.4781	38
39	0.0043	6.6380	43.1374	232.9248	1546.1655	0.1506	0.0006	6.4985	39
40	0.0037	6.6418	43.2830	267.8635	1779.0903	0.1506	0.0006	6.5168	40
41	0.0032	6.6450	43.4128	308.0431	2046.9539	0.1505	0.0005	6.5331	41
42	0.0028	6.6478	43.5286	354.2495	2354.9969	0.1504	0.0004	6.5478	42
43	0.0025	6.6503	43.6317	407.3870	2709.2465	0.1504	0.0004	6.5609	43
44	0.0021	6.6524	43.7235	468.4950	3116.6334	0.1503	0.0003	6.5725	44
45	0.0019	6.6543	43.8051	538.7693	3585.1285	0.1503	0.0003	6.5830	45
46	0.0016	6.6559	43.8778	619.5847	4123.8977	0.1502	0.0002	6.5923	46
47	0.0014	6.6573	43.9423	712.5224	4743.4824	0.1502	0.0002	6.6006	47
48	0.0012	6.6585	43.9997	819.4007	5456.0047	0.1502	0.0002	6.6080	48
49	0.0011	6.6596	44.0506	942.3108	6275.4055	0.1502	0.0002	6.6146	49
50	0.0009	6.6605	44.0958	1083.6574	7217.7163	0.1501	0.0001	6.6205	50
51	0.0008	6.6613	44.1360	1246.2061	8301.3737	0.1501	0.0001	6.6257	51
52	0.0007	6.6620	44.1715	1433.1370	9547.5798	0.1501	0.0001	6.6304	52
53	0.0006	6.6626	44.2031	1648.1075	10980.7167	0.1501	0.0001	6.6345	53
54	0.0005	6.6631	44.2311	1895.3236	12628.8243	0.1501	0.0001	6.6382	54
55	0.0005	6.6636	44.2558	2179.6222	14524.1479	0.1501	0.0001	6.6414	55
60	0.0002	6.6651	44.3431	4383.9987	29219.9916	0.1500	0.0000	6.6530	60
65	0.0001	6.6659	44.3903	8817.7874	58778.5826	0.1500	0.0000	6.6593	65
70	0.0001	6.6663	44.4156	17735.7200	118231.4669	0.1500	0.0000	6.6627	70
75	0.0000	6.6665	44.4292	35672.8680	237812.4532	0.1500	0.0000	6.6646	75
80	0.0000	6.6666	44.4364	71750.8794	478332.5293	0.1500	0.0000	6.6656	80
85	0.0000	6.6666	44.4402	144316.6470	962104.3133	0.1500	0.0000	6.6661	85
90	0.0000	6.6666	44.4422	290272.3252	1935142.1680	0.1500	0.0000	6.6664	90
95	0.0000	6.6667	44.4433	583841.3276	3892268.8509	0.1500	0.0000	6.6665	95
100	0.0000	6.6667	44.4438	1174313.4507	7828749.6713	0.1500	0.0000	6.6666	100

(continued)

APPENDIX 55.B *(continued)*
Cash Flow Equivalent Factors

$i = 20.00\%$

n	P/F	P/A	P/G	F/P	F/A	A/P	A/F	A/G	n
1	0.8333	0.8333	0.0000	1.2000	1.0000	1.2000	1.0000	0.0000	1
2	0.6944	1.5278	0.6944	1.4400	2.2000	0.6545	0.4545	0.4545	2
3	0.5787	2.1065	1.8519	1.7280	3.6400	0.4747	0.2747	0.8791	3
4	0.4823	2.5887	3.2986	2.0736	5.3680	0.3863	0.1863	1.2742	4
5	0.4019	2.9906	4.9061	2.4883	7.4416	0.3344	0.1344	1.6405	5
6	0.3349	3.3255	6.5806	2.9860	9.9299	0.3007	0.1007	1.9788	6
7	0.2791	3.6046	8.2551	3.5832	12.9159	0.2774	0.0774	2.2902	7
8	0.2326	3.8372	9.8831	4.2998	16.4991	0.2606	0.0606	2.5756	8
9	0.1938	4.0310	11.4335	5.1598	20.7989	0.2481	0.0481	2.8364	9
10	0.1615	4.1925	12.8871	6.1917	25.9587	0.2385	0.0385	3.0739	10
11	0.1346	4.3271	14.2330	7.4301	32.1504	0.2311	0.0311	3.2893	11
12	0.1122	4.4392	15.4667	8.9161	39.5805	0.2253	0.0253	3.4841	12
13	0.0935	4.5327	16.5883	10.6993	48.4966	0.2206	0.0206	3.6597	13
14	0.0779	4.6106	17.6008	12.8392	59.1959	0.2169	0.0169	3.8175	14
15	0.0649	4.6755	18.5095	15.4070	72.0351	0.2139	0.0139	3.9588	15
16	0.0541	4.7296	19.3208	18.4884	87.4421	0.2114	0.0114	4.0851	16
17	0.0451	4.7746	20.0419	22.1861	105.9306	0.2094	0.0094	4.1976	17
18	0.0376	4.8122	20.6805	26.6233	128.1167	0.2078	0.0078	4.2975	18
19	0.0313	4.8435	21.2439	31.9480	154.7400	0.2065	0.0065	4.3861	19
20	0.0261	4.8696	21.7395	38.3376	186.6880	0.2054	0.0054	4.4643	20
21	0.0217	4.8913	22.1742	46.0051	225.0256	0.2044	0.0044	4.5334	21
22	0.0181	4.9094	22.5546	55.2061	271.0307	0.2037	0.0037	4.5941	22
23	0.0151	4.9245	22.8867	66.2474	326.2369	0.2031	0.0031	4.6475	23
24	0.0126	4.9371	23.1760	79.4968	392.4842	0.2025	0.0025	4.6943	24
25	0.0105	4.9476	23.4276	95.3962	471.9811	0.2021	0.0021	4.7352	25
26	0.0087	4.9563	23.6460	114.4755	567.3773	0.2018	0.0018	4.7709	26
27	0.0073	4.9636	23.8353	137.3706	681.8528	0.2015	0.0015	4.8020	27
28	0.0061	4.9697	23.9991	164.8447	819.2233	0.2012	0.0012	4.8291	28
29	0.0051	4.9747	24.1406	197.8136	984.0680	0.2010	0.0010	4.8527	29
30	0.0042	4.9789	24.2628	237.3763	1181.8816	0.2008	0.0008	4.8731	30
31	0.0035	4.9824	24.3681	284.8516	1419.2579	0.2007	0.0007	4.8908	31
32	0.0029	4.9854	24.4588	341.8219	1704.1095	0.2006	0.0006	4.9061	32
33	0.0024	4.9878	24.5368	410.1863	2045.9314	0.2005	0.0005	4.9194	33
34	0.0020	4.9898	24.6038	492.2235	2456.1176	0.2004	0.0004	4.9308	34
35	0.0017	4.9915	24.6614	590.6682	2948.3411	0.2003	0.0003	4.9406	35
36	0.0014	4.9929	24.7108	708.8019	3539.0094	0.2003	0.0003	4.9491	36
37	0.0012	4.9941	24.7531	850.5622	4247.8112	0.2002	0.0002	4.9564	37
38	0.0010	4.9951	24.7894	1020.6747	5098.3735	0.2002	0.0002	4.9627	38
39	0.0008	4.9959	24.8204	1224.8096	6119.0482	0.2002	0.0002	4.9681	39
40	0.0007	4.9966	24.8469	1469.7716	7343.8578	0.2001	0.0001	4.9728	40
41	0.0006	4.9972	24.8696	1763.7259	8813.6294	0.2001	0.0001	4.9767	41
42	0.0005	4.9976	24.8890	2116.4711	10577.3553	0.2001	0.0001	4.9801	42
43	0.0004	4.9980	24.9055	2539.7653	12693.8263	0.2001	0.0001	4.9831	43
44	0.0003	4.9984	24.9196	3047.7183	15233.5916	0.2001	0.0001	4.9856	44
45	0.0003	4.9986	24.9316	3657.2620	18281.3099	0.2001	0.0001	4.9877	45
46	0.0002	4.9989	24.9419	4388.7144	21938.5719	0.2000	0.0000	4.9895	46
47	0.0002	4.9991	24.9506	5266.4573	26327.2863	0.2000	0.0000	4.9911	47
48	0.0002	4.9992	24.9581	6319.7487	31593.7436	0.2000	0.0000	4.9924	48
49	0.0001	4.9993	24.9644	7583.6985	37913.4923	0.2000	0.0000	4.9935	49
50	0.0001	4.9995	24.9698	9100.4382	45497.1908	0.2000	0.0000	4.9945	50
51	0.0001	4.9995	24.9744	10920.5258	54597.6289	0.2000	0.0000	4.9953	51
52	0.0001	4.9996	24.9783	13104.6309	65518.1547	0.2000	0.0000	4.9960	52
53	0.0001	4.9997	24.9816	15725.5571	78622.7856	0.2000	0.0000	4.9966	53
54	0.0001	4.9997	24.9844	18870.6685	94348.3427	0.2000	0.0000	4.9971	54
55	0.0000	4.9998	24.9868	22644.8023	113219.0113	0.2000	0.0000	4.9976	55
60	0.0000	4.9999	24.9942	56347.5144	281732.5718	0.2000	0.0000	4.9989	60
65	0.0000	5.0000	24.9975	140210.6469	701048.2346	0.2000	0.0000	4.9995	65
70	0.0000	5.0000	24.9989	348888.9569	1744439.7847	0.2000	0.0000	4.9998	70
75	0.0000	5.0000	24.9995	868147.3693	4340731.8466	0.2000	0.0000	4.9999	75

(continued)

APPENDIX 55.B *(continued)*
Cash Flow Equivalent Factors

$i = 25.00\%$

n	P/F	P/A	P/G	F/P	F/A	A/P	A/F	A/G	n
1	0.8000	0.8000	0.0000	1.2500	1.0000	1.2500	1.0000	0.0000	1
2	0.6400	1.4400	0.6400	1.5625	2.2500	0.6944	0.0444	0.4444	2
3	0.5120	1.9520	1.6640	1.9531	3.8125	0.5123	0.2623	0.8525	3
4	0.4096	2.3616	2.8928	2.4414	5.7656	0.4234	0.1734	1.2249	4
5	0.3277	2.6893	4.2035	3.0518	8.2070	0.3718	0.1218	1.5631	5
6	0.2621	2.9514	5.5142	3.8147	11.2588	0.3383	0.0888	1.8683	6
7	0.2097	3.1611	6.7725	4.7684	15.0735	0.3163	0.0663	2.1424	7
8	0.1678	3.3289	7.9469	5.9605	19.8419	0.3004	0.0504	2.3872	8
9	0.1342	3.4631	9.0207	7.4506	25.8023	0.2888	0.0388	2.6048	9
10	0.1074	3.5705	9.9870	9.3132	33.2529	0.2801	0.0301	2.7971	10
11	0.0859	3.6564	10.8460	11.6415	42.5661	0.2735	0.0235	2.9663	11
12	0.0687	3.7251	11.6020	14.5519	54.2077	0.2684	0.0184	3.1145	12
13	0.0550	3.7801	12.2617	18.1899	68.7596	0.2645	0.0145	3.2437	13
14	0.0440	3.8241	12.8334	22.7374	86.9495	0.2615	0.0115	3.3559	14
15	0.0352	3.8593	13.3260	28.4217	109.6868	0.2591	0.0091	3.4530	15
16	0.0281	3.8874	13.7482	35.5271	138.1085	0.2572	0.0072	3.5366	16
17	0.0225	3.9099	14.1085	44.4089	173.6357	0.2558	0.0058	3.6084	17
18	0.0180	3.9279	14.4147	55.5112	218.0446	0.2546	0.0046	3.6698	18
19	0.0144	3.9424	14.6741	69.3889	273.5558	0.2537	0.0037	3.7222	19
20	0.0115	3.9539	14.8932	86.7362	342.9447	0.2529	0.0029	3.7667	20
21	0.0092	3.9631	15.0777	108.4202	429.6809	0.2523	0.0023	3.8045	21
22	0.0074	3.9705	15.2326	135.5253	538.1011	0.2519	0.0019	3.8365	22
23	0.0059	3.9764	15.3625	169.4066	673.6264	0.2515	0.0015	3.8634	23
24	0.0047	3.9811	15.4711	211.7582	843.0329	0.2512	0.0012	3.8861	24
25	0.0038	3.9849	15.5618	264.6978	1054.7912	0.2509	0.0009	3.9052	25
26	0.0030	3.9879	15.6373	330.8722	1319.4890	0.2508	0.0008	3.9212	26
27	0.0024	3.9903	15.7002	413.5903	1650.3612	0.2506	0.0006	3.9346	27
28	0.0019	3.9923	15.7524	516.9879	2063.9515	0.2505	0.0005	3.9457	28
29	0.0015	3.9938	15.7957	646.2349	2580.9394	0.2504	0.0004	3.9551	29
30	0.0012	3.9950	15.8316	807.7936	3227.1743	0.2503	0.0003	3.9628	30
31	0.0010	3.9960	15.8614	1009.7420	4034.9678	0.2502	0.0002	3.9693	31
32	0.0008	3.9968	15.8859	1262.1774	5044.7098	0.2502	0.0002	3.9746	32
33	0.0006	3.9975	15.9062	1577.7218	6306.8872	0.2502	0.0002	3.9791	33
34	0.0005	3.9980	15.9229	1972.1523	7884.6091	0.2501	0.0001	3.9828	34
35	0.0004	3.9984	15.9367	2465.1903	9856.7613	0.2501	0.0001	3.9858	35
36	0.0003	3.9987	15.9481	3081.4879	12321.9516	0.2501	0.0001	3.9883	36
37	0.0003	3.9990	15.9574	3851.8599	15403.4396	0.2501	0.0001	3.9904	37
38	0.0002	3.9992	15.9651	4814.8249	19255.2994	0.2501	0.0001	3.9921	38
39	0.0002	3.9993	15.9714	6018.5311	24070.1243	0.2500	0.0000	3.9935	39
40	0.0001	3.9995	15.9766	7523.1638	30088.6554	0.2500	0.0000	3.9947	40
41	0.0001	3.9996	15.9809	9403.9548	37611.8192	0.2500	0.0000	3.9956	41
42	0.0001	3.9997	15.9843	11754.9435	47015.7740	0.2500	0.0000	3.9964	42
43	0.0001	3.9997	15.9872	14693.6794	58770.7175	0.2500	0.0000	3.9971	43
44	0.0001	3.9998	15.9895	18367.0992	73464.3969	0.2500	0.0000	3.9976	44
45	0.0000	3.9998	15.9915	22958.8740	91831.4962	0.2500	0.0000	3.9980	45
46	0.0000	3.9999	15.9930	28698.5925	114790.3702	0.2500	0.0000	3.9984	46
47	0.0000	3.9999	15.9943	35873.2407	143488.9627	0.2500	0.0000	3.9987	47
48	0.0000	3.9999	15.9954	44841.5509	179362.2034	0.2500	0.0000	3.9989	48
49	0.0000	3.9999	15.9962	56051.9386	224203.7543	0.2500	0.0000	3.9991	49
50	0.0000	3.9999	15.9969	70064.9232	280255.6929	0.2500	0.0000	3.9993	50
51	0.0000	4.0000	15.9975	87581.1540	350320.6161	0.2500	0.0000	3.9994	51
52	0.0000	4.0000	15.9980	109476.4425	437901.7701	0.2500	0.0000	3.9995	52
53	0.0000	4.0000	15.9983	136845.5532	547378.2126	0.2500	0.0000	3.9996	53
54	0.0000	4.0000	15.9986	171056.9414	684223.7658	0.2500	0.0000	3.9997	54
55	0.0000	4.0000	15.9989	213821.1768	855280.7072	0.2500	0.0000	3.9997	55
60	0.0000	4.0000	15.9996	652530.4468	2610117.7872	0.2500	0.0000	3.9999	60

(continued)

Appendices

APPENDIX 55.B *(continued)*
Cash Flow Equivalent Factors

$i = 30.00\%$

n	*P/F*	*P/A*	*P/G*	*F/P*	*F/A*	*A/P*	*A/F*	*A/G*	*n*
1	0.7692	0.7692	0.0000	1.3000	1.0000	1.3000	1.0000	0.000	1
2	0.5917	1.3609	0.5917	1.6900	2.3000	0.7348	0.4348	0.434	2
3	0.4552	1.8161	1.5020	2.1970	3.9900	0.5506	0.2506	0.827	3
4	0.3501	2.1662	2.5524	2.8561	6.1870	0.4616	0.1616	1.178	4
5	0.2693	2.4356	3.6297	3.7129	9.0431	0.4106	0.1106	1.490	5
6	0.2072	2.6427	4.6656	4.8268	12.7560	0.3784	0.0784	1.765	6
7	0.1594	2.8021	5.6218	6.2749	17.5828	0.3569	0.0569	2.006	7
8	0.1226	2.9247	6.4800	8.1573	23.8577	0.3419	0.0419	2.215	8
9	0.0943	3.0190	7.2343	10.6045	32.0150	0.3312	0.0312	2.396	9
10	0.0725	3.0915	7.8872	13.7858	42.6195	0.3235	0.0235	2.551	10
11	0.0558	3.1473	8.4452	17.9216	56.4053	0.3177	0.0177	2.683	11
12	0.0429	3.1903	8.9173	23.2981	74.3270	0.3135	0.0135	2.795	12
13	0.0330	3.2233	9.3135	30.2875	97.6250	0.3102	0.0102	2.889	13
14	0.0254	3.2487	9.6437	39.3738	127.9125	0.3078	0.0078	2.968	14
15	0.0195	3.2682	9.9172	51.1859	167.2863	0.3060	0.0060	3.034	15
16	0.0150	3.2832	10.1426	66.5417	218.4722	0.3046	0.0046	3.089	16
17	0.0116	3.2948	10.3276	86.5042	285.0139	0.3035	0.0035	3.134	17
18	0.0089	3.3037	10.4788	112.4554	371.5180	0.3027	0.0027	3.171	18
19	0.0068	3.3105	10.6019	146.1920	483.9734	0.3021	0.0021	3.202	19
20	0.0053	3.3158	10.7019	190.0496	630.1655	0.3016	0.0016	3.227	20
21	0.0040	3.3198	10.7828	247.0645	820.2151	0.3012	0.0012	3.248	21
22	0.0031	3.3230	10.8482	321.1839	1067.2796	0.3009	0.0009	3.264	22
23	0.0024	3.3254	10.9009	417.5391	1388.4635	0.3007	0.0007	3.278	23
24	0.0018	3.3272	10.9433	542.8008	1806.0026	0.3006	0.0006	3.289	24
25	0.0014	3.3286	10.9773	705.6410	2348.8033	0.3004	0.0004	3.297	25
26	0.0011	3.3297	11.0045	917.3333	3054.4443	0.3003	0.0003	3.305	26
27	0.0008	3.3305	11.0263	1192.5333	3971.7776	0.3003	0.0003	3.310	27
28	0.0006	3.3312	11.0437	1550.2933	5164.3109	0.3002	0.0002	3.315	28
29	0.0005	3.3317	11.0576	2015.3813	6714.6042	0.3001	0.0001	3.318	29
30	0.0004	3.3321	11.0687	2619.9956	8729.9855	0.3001	0.0001	3.321	30
31	0.0003	3.3324	11.0775	3405.9943	11349.9811	0.3001	0.0001	3.324	31
32	0.0002	3.3326	11.0845	4427.7926	14755.9755	0.3001	0.0001	3.326	32
33	0.0002	3.3328	11.0901	5756.1304	19183.7681	0.3001	0.0001	3.327	33
34	0.0001	3.3329	11.0945	7482.9696	24939.8985	0.3000	0.0000	3.328	34
35	0.0001	3.3330	11.0980	9727.8604	32422.8681	0.3000	0.0000	3.329	35
36	0.0001	3.3331	11.1007	12646.2186	42150.7285	0.3000	0.0000	3.330	36
37	0.0001	3.3331	11.1029	16440.0841	54796.9471	0.3000	0.0000	3.331	37
38	0.0000	3.3332	11.1047	21372.1094	71237.0312	0.3000	0.0000	3.331	38
39	0.0000	3.3332	11.1060	27783.7422	92609.1405	0.3000	0.0000	3.331	39
40	0.0000	3.3332	11.1071	36118.8648	120392.8827	0.3000	0.0000	3.332	40
41	0.0000	3.3333	11.1080	46954.5243	156511.7475	0.3000	0.0000	3.332	41
42	0.0000	3.3333	11.1086	61040.8815	203466.2718	0.3000	0.0000	3.332	42
43	0.0000	3.3333	11.1092	79353.1460	264507.1533	0.3000	0.0000	3.332	43
44	0.0000	3.3333	11.1096	103159.0898	343860.2993	0.3000	0.0000	3.332	44
45	0.0000	3.3333	11.1099	134106.8167	447019.3890	0.3000	0.0000	3.333	45
46	0.0000	3.3333	11.1102	174338.8617	581126.2058	0.3000	0.0000	3.333	46
47	0.0000	3.3333	11.1104	226640.5202	755465.0675	0.3000	0.0000	3.333	47
48	0.0000	3.3333	11.1105	294632.6763	982105.5877	0.3000	0.0000	3.333	48
49	0.0000	3.3333	11.1107	383022.4792	1276738.2640	0.3000	0.0000	3.333	49
50	0.0000	3.3333	11.1108	497929.2230	1659760.7433	0.3000	0.0000	3.333	50

(continued)

APPENDIX 55.B *(continued)*
Cash Flow Equivalent Factors

$i = 40.00\%$

n	*P/F*	*P/A*	*P/G*	*F/P*	*F/A*	*A/P*	*A/F*	*A/G*	*n*
1	0.7143	0.7143	0.0000	1.4000	1.0000	1.4000	1.0000	0.000	1
2	0.5102	1.2245	0.5102	1.9600	2.4000	0.8167	0.4167	0.416	2
3	0.3644	1.5889	1.2391	2.7440	4.3600	0.6294	0.2294	0.779	3
4	0.2603	1.8492	2.0200	3.8416	7.1040	0.5408	0.1408	1.092	4
5	0.1859	2.0352	2.7637	5.3782	10.9456	0.4914	0.0914	1.358	5
6	0.1328	2.1680	3.4278	7.5295	16.3238	0.4613	0.0613	1.581	6
7	0.0949	2.2628	3.9970	10.5414	23.8534	0.4419	0.0419	1.766	7
8	0.0678	2.3306	4.4713	14.7579	34.3947	0.4291	0.0291	1.918	8
9	0.0484	2.3790	4.8585	20.6610	49.1526	0.4203	0.0203	2.042	9
10	0.0346	2.4136	5.1696	28.9255	69.8137	0.4143	0.0143	2.141	10
11	0.0247	2.4383	5.4166	40.4957	98.7391	0.4101	0.0101	2.221	11
12	0.0176	2.4559	5.6106	56.6939	139.2348	0.4072	0.0072	2.284	12
13	0.0126	2.4685	5.7618	79.3715	195.9287	0.4051	0.0051	2.334	13
14	0.0090	2.4775	5.8788	111.1201	275.3002	0.4036	0.0036	2.372	14
15	0.0064	2.4839	5.9688	155.5681	386.4202	0.4026	0.0026	2.403	15
16	0.0046	2.4885	6.0376	217.7953	541.9883	0.4018	0.0018	2.426	16
17	0.0033	2.4918	6.0901	304.9135	759.7837	0.4013	0.0013	2.444	17
18	0.0023	2.4941	6.1299	426.8789	1064.6971	0.4009	0.0009	2.457	18
19	0.0017	2.4958	6.1601	597.6304	1491.5760	0.4007	0.0007	2.468	19
20	0.0012	2.4970	6.1828	836.6826	2089.2064	0.4005	0.0005	2.476	20
21	0.0009	2.4979	6.1998	1171.3556	2925.8889	0.4003	0.0003	2.482	21
22	0.0006	2.4985	6.2127	1639.8978	4097.2445	0.4002	0.0002	2.486	22
23	0.0004	2.4989	6.2222	2295.8569	5737.1423	0.4002	0.0002	2.490	23
24	0.0003	2.4992	6.2294	3214.1997	8032.9993	0.4001	0.0001	2.492	24
25	0.0002	2.4994	6.2347	4499.8796	11247.1990	0.4001	0.0001	2.494	25
26	0.0002	2.4996	6.2387	6299.8314	15747.0785	0.4001	0.0001	2.495	26
27	0.0001	2.4997	6.2416	8819.7640	22046.9099	0.4000	0.0000	2.496	27
28	0.0001	2.4998	6.2438	12347.6696	30866.6739	0.4000	0.0000	2.497	28
29	0.0001	2.4999	6.2454	17286.7374	43214.3435	0.4000	0.0000	2.498	29
30	0.0000	2.4999	6.2466	24201.4324	60501.0809	0.4000	0.0000	2.498	30
31	0.0000	2.4999	6.2475	33882.0053	84702.5132	0.4000	0.0000	2.499	31
32	0.0000	2.4999	6.2482	47434.8074	118584.5185	0.4000	0.0000	2.499	32
33	0.0000	2.5000	6.2487	66408.7304	166019.3260	0.4000	0.0000	2.499	33
34	0.0000	2.5000	6.2490	92972.2225	232428.0563	0.4000	0.0000	2.499	34
35	0.0000	2.5000	6.2493	130161.1116	325400.2789	0.4000	0.0000	2.499	35
36	0.0000	2.5000	6.2495	182225.5562	455561.3904	0.4000	0.0000	2.499	36
37	0.0000	2.5000	6.2496	255115.7786	637786.9466	0.4000	0.0000	2.499	37
38	0.0000	2.5000	6.2497	357162.0901	892902.7252	0.4000	0.0000	2.499	38
39	0.0000	2.5000	6.2498	500026.9261	1250064.8153	0.4000	0.0000	2.499	39
40	0.0000	2.5000	6.2498	700037.6966	1750091.7415	0.4000	0.0000	2.499	40
41	0.0000	2.5000	6.2499	980052.7752	2450129.4381	0.4000	0.0000	2.500	41
42	0.0000	2.5000	6.2499	1372073.8853	3430182.2133	0.4000	0.0000	2.500	42
43	0.0000	2.5000	6.2499	1920903.4394	4802256.0986	0.4000	0.0000	2.500	43
44	0.0000	2.5000	6.2500	2689264.8152	6723159.5381	0.4000	0.0000	2.500	44
45	0.0000	2.5000	6.2500	3764970.7413	9412424.3533	0.4000	0.0000	2.500	45

Index

3 dB, 51-10
bandwidth, 48-8
downpoint, 40-2
frequency, 51-10
half-power points, 30-8, 40-2

A

Abscissa, 8-2
Absolute
convergence, 4-13
English system, 1-4
index of refraction, 46-4
Absorption law, 13-5
Absorptive law, 40-20
AC
circuit fundamentals, 27-1
incremental model, 38-8
load line, 37-6 (ftn)
properties, A-49
switching transient, 30-1
value symbology, 37-2
voltage source symbology, 27-4
waveform characteristics, 27-6
Accelerated
cost recovery system, 55-21, 55-22
depreciation, 55-23
Acceleration
centripetal, 16-20
of gravity, 1-2
Acceptance, 56-3
Acceptor, 17-6, 37-3
Account
accumulated depreciation, 55-23 (ftn)
equipment, 55-23 (ftn)
Accounting, 55-36
convention, 55-34 (ftn)
equation, 55-34
principle, standard, 55-34 (ftn)
Accrual system, 55-34
Accumulation, 9-7
Accuracy, 12-13, 36-1
Acid-test ratio, 55-35
Acknowledge signal, 44-7
Acoustic
/electrical interface, 19-1
energy, 19-1
theory, 19-1
Acoustical/electrical/mechanical elements, 19-4
Action, line of, 6-1
Active, 17-2, 19-2
circuit, 40-9
devicc, 29-3
electron devices, 17-2
logic circuit, 40-19
network, 40-9
power, 27-12
pull-up, 40-25
region, 37-4, 38-2
system, 51-11
waveform shaping, 40-19
Actuating signal ratio, 51-4
Acute angle, 7-1
Adder, 42-11
half-, 42-11
ripple-carry, 42-11
Addition of vectors, 6-3
Additive polarity, 34-5
Adjacency, 42-2
logical, 42-2
Adjacent
angle, 7-1
minterms, 42-3
side, 7-2
-state binary counter, 43-6, 43-7
Adjoint, classical, 5-5
Admittance, 27-3
model, 29-15
model parameters, 29-15
short-circuit parameters, 29-15
transmission line, 45-2
Advanced engineering mathematics, 15-1
Agency, 56-3
Agent, 56-3
Agitation, thermal, 36-1
Agreement, letter of, 56-3
Algebra, 4-1
block diagram, 51-6
block diagram rules, 51-7
Boolean, 13-5, 40-20, 41-3
matrix, 5-4
of feedback, 51-6
switching, 13-5, 41-3
Algebraic equation, 4-2
Algorithm, Steinmetz, 27-7
Alias component, 10-8 (ftn)
Aliasing, 18-2, 48-5
Allowance
depletion, 55-24
trade-in, 55-18
Alnico, 52-2
Alpha curve, 51-11
Alphabet, Greek, 4-1
Alternate-mark inversion, 48-6
Alternating
current, 27-1
sign series, 4-13
waveform, 27-1, 27-6
Alternative
comparison, 55-4
equivalent, 55-5
Aluminum, 16-3, 52-1
American
Society of Civil Engineers, 57-1
Wire Gauge, 16-3, 53-2, A-45
Ammeter
DC, 36-5
DC circuit, 36-5
Amortization, 55-24
Ampacity, A-47
Ampère, 16-13
Ampère's Law, 16-20, 22-8, 25-1
Ampèrian current, 22-6
Amplification factor, 37-4, 37-6
current, 37-4
voltage, 37-4
Amplifier, 37-2, 37-4, 40-2
band-pass, 40-2
bandwidth, 40-2
cascaded, 38-7
cascode, 38-7
class A, 37-5
class AB, 37-5
class B, 37-5
class C, 37-5, 37-6
differential, 40-3, 40-4
gain, 38-7
general, 37-5
high-pass, 40-2
ideal operational, 40-5
input impedance, 37-6
inverting, 37-4 (ftn)
low-pass, 40-2
noise, 40-8
notch, 40-2
optical, 50-4
output impedance, 37-6
power, 38-7, 47-2
region of operation, 40-3
stage, 38-7
symbology, 40-3
Amplitude
distortion, 48-7
factor, 27-6
of voltage, 27-2
shift keying, 48-7
Amplitude modulation, 48-2
band, value, 50-2
bandwidth, 48-4
power, 48-4
Analog
circuit, 36-3
circuit, operation, 38-2
modulation, 48-2
signal, 18-1
-to-digital conversion, 44-2
Analog-to-digital converter
counting, 44-5, 44-6
dual-slope, 44-7
parallel-comparator, 44-6
ratiometric, 44-7
Analogy, magnetic-electric, 16-21
Analysis
circuit, 29-1
circuit, three-phase, 51-1
comparative, 55-17
critically damped, 31-9
dimensional, 1-8
domain, 51-11
economic, engineering, 55-2
economic life, 55-4
equation, 15-14
first-order, 31-2
first-order, switching, 31-3
first-order, systems, 31-16
Fourier, 10-6
frequency, 10-8 (ftn), 31-2
frequency, domain, 31-1
frequency, response, 32-2, 51-11
high-frequency transistor, 38-12
horizon, 55-17
incremental, 55-17
Laplace transform, 31-15
large-signal, 38-2
loop-, 29-12
mesh, 26-12, 29-12
node-, 29-12
of complicated network, 26-7
of transients, 30-1
operational amplifier, 40-5
overdamped, 31-9
pulse transient, 31-6, 31-13
replacement/retirement, 55-4

second-order, 31-7
second-order systems, 31-16
sensitivity, 55-43
signature, 10-8 (ftn)
sinusoidal, 31-1, 31-2
sinusoidal steady-state, 32-1
small-signal, 27-11, 29-7, 37-2, 37-4, 38-3
time-series, 10-8 (ftn)
time domain, 31-1
timing, 43-1
timing design, 43-1
transient, 32-1
transient impedance, 29-14
transistor circuit, 38-9
underdamped, 31-12
Analytic function, 9-1
Analyzer
FFT, 10-8
signal, 10-8
spectrum, 10-8
AND operation, 13-5
Angle
beam, 47-3
between figures, 8-8
between lines, 7-1
critical, 46-4
current, 27-11
direction, 6-2, 8-5
function of, 7-2
impedance, 27-3
modulation, 48-4, 48-5
modulation, bandwidth, 48-5
modulation, power, 48-5
of depression, 7-2
of deviation, 46-4
of elevation, 7-2
of intersection, 8-8
of refraction, 46-4
phase, 27-2
phase difference, 27-3, 27-7
plane, 7-1
power, 27-12
power factor, 27-11
solid, 7-6
solid beam, 47-3
trihedral, 7-5
type, 7-2
Angstrom, A-1
Angular
orientation, 6-1
perspective, 3-3
Anion, 33-1
Anisotropic material, 6-1
Annual
amount, 55-7
cost, 55-7
cost, equivalent uniform, 55-15
cost method, 55-15
return method, 55-15
Annuity, 55-7
Anode, 26-14, 33-1
ANSI, 33-3, 58-1
Standards Committee, 53-1, 54-1
Antenna, 45-1
aperture, 46-2, 47-2
dipole, 46-7, 46-8
directivity, 46-6
efficiencies, values, 47-3
efficiency, 46-2, 46-6
elemental, 46-5
focus, 46-2, 46-6
fundamentals, 46-1
gain, 46-2
gain directivity, 47-3
half-wave, 46-9
horn, 46-2
isotropic, 46-2
microstrip, 46-2
omnidirectional, 46-2
polarization, 46-2
power, receiving, 47-3
quarter-wave, 46-9
receiving, 47-2
reflector, 46-2
short dipole, 46-7
surface wave, 46-2
temperature, 46-2
theory, 46-1
transmitting, 47-1
waveguide, 46-2
wire, 46-2
Antiderivative, 10-1
Antiferromagnetism, 16-4
Antimagnetism, 23-4
Antimony, 37-3
Antiresonant circuit, 40-29
Aperiodic, 15-15
signal, discrete-time Fourier transform, A-21
signal, Fourier transform, A-16
Aperture, 47-2
antenna, 46-2
taper efficiency, 47-3
Apparent power, 27-12
Application
FE exam, 59-2
interface, 44-1
point of, 6-1
Approximation
of gamma function, A-13
series, 9-7
small angle, 7-3
Arc length, by integration, 10-5
Architect, 56-4
Area, A-7
between two curves, 10-4
by integration, 10-4
conversions, SI, A-3
in circular mil, 26-2
irregular, 8-1
mensuration, A-7
negative, 27-4
positive, 27-4
under the standard normal curve, A-12
Argument, 4-8
Arithmetic
hardware, 42-10
logic device, 42-10
logic unit, 42-10
sequence, 4-11
series, 4-12
Array, 5-1, 46-1
circular, 46-2
planar, 46-2
programmable logic, 42-8
simple, 46-1, 46-2
uniform linear, 46-2
Arsenic, 37-3
Artificial
expense, 55-20
intelligence, 58-1
noise, 18-7, 18-8
Aspect ratio, ellipse, 8-11
Assertion
levels, 41-1, 41-2
levels, active high, 41-2
levels, active low, 41-2
Asset
account, 55-33
capitalizing, 55-20
class, 55-20
current, 55-34
expensing, 55-20
fixed, 55-34
Assignment, 56-4
Associative law, 13-5, 40-20
addition, 4-3
convolution, 15-16
matrix, 5-4
multiplication, 4-3
sets, 12-2
Astable
multivibrator, 40-27
oscillator, 40-19
Asymmetric amplitude-shift keying, 48-7
Asymmetrical function, 8-4
Asymptote, 8-2
Asymptotic, 8-2
Asynchronous, 43-5
protocol, 44-7
signal, 43-2
transfer mode, 44-7
Atmosphere, standard, A-3, A-4
Atom, magnetic, 23-3
Atomic spacing, 17-4
Attenuator, programmable, 44-5
Augmented matrix, 5-1
Autocorrelation, 15-16, 15-17
Autonomous network, 43-4
Auxiliary
equation, 11-1
view, 3-2
Avalanche
breakdown, 37-9
region, 37-4, 39-2
Average
AC power, 27-11
area, 27-4
cost, 55-37
of rectified waveform, 27-4
power, 24-3
value, 27-4, 55-30
value, by integration, 10-4
value, in Fourier series, 27-6
Axis
conjugate, 8-11
oblique, 3-3
parabolic, 8-10
Axonometric view, 3-3
Azimuthal quantum number, 16-2

B

B-field, 22-1
sources, 23-1
strength, 16-16
Backward diode, 37-13, 37-14
Balance, 36-3
books, 55-34
mixer, 40-28
sheet, 55-34
Balloon payment, 55-41
Band
conduction, 16-2
energy, 17-4
frequency, 18-6
of communication, 50-3
-pass amplifier, 40-2
-pass filter, 30-8, 30-9
-reject filter, 30-9, 30-10
Bandwidth, 10-8, 18-3, 30-9, 40-2, 48-1, 51-10, 51-11
3 dB, 48-8
AM, 48-4
closed-loop, 51-10
efficiency, 48-8
equivalent noise, 48-8
half-power, 48-8
of frequency analysis, 10-8
Shannon, 48-8
Bang-bang controller, 51-14
Bar magnet, 23-4
Barkhausen effect, 36-2
Barrier diode, 37-12
Base, 13-1, 38-2
-8 system, 13-2
-10 number, 13-1
-16 system, 13-2
-b number, 13-1
clipper circuit, 37-12
spreading resistance, 38-8
unit, 1-5
Baseband, 48-2
digital-data signal, 48-6
digital signals, 48-6
signal, 48-2, 48-5
signal spectrum, typical, 48-2
system, 48-2

INDEX - B

Basic
accounting equation, 55-34
signal processing, 18-3, 18-4
signal theory, 18-2, 18-3
transmission loss, 46-5
transmission loss, free space, 46-5
Basis
daily compounding, 55-28
depreciation, 55-20
unadjusted, 55-20
Bathtub distribution, 12-7
Battery, 26-13, 33-1
advantages, 33-1
bias, 37-11
capacity, 33-3
cell size, 33-3
cell voltages, 33-1
characteristics, 26-13, 33-1
charging, 33-3
cycle, 33-1
cycle life, 33-1
dry cell, 33-3
electrochemical, 26-13
electrolyte, 33-1
energy efficiency, 33-1 (ftn)
equivalent circuit, 33-2
hydrogen production, 33-3
model, 33-1
primary, 26-14, 33-3
reserve, 33-3
secondary, 26-14, 33-3
shelf life, 33-1
theory, 33-1, 33-2
types of, 33-3
Bayes' theorem, 12-4
BBD, 39-2
Beam
angle, 47-3
solid angle, 47-3
width, 47-3
Bell-shaped curve, 12-6
Benefit-cost ratio, 55-16
Bessel equation, 11-5, 15-1, 15-12
Bessel function, 15-5 (ftn), 48-4 (ftn), A-14
first kind, 15-12
hyperbolic, 15-13
second kind, 15-12
third kind, 15-12
Beta function, 15-11 (ftn)
BH curve, 23-2, 28-6
Bias, 12-13, 37-4
battery, 37-11
cutoff, 38-5
fixed, 38-5
forward, 37-4
JFET, 39-3
methods, 38-5
MOSFET, 39-5, 39-6
multiple-battery, 38-5
reverse, 37-4
self-, 38-5
stability, 38-5
switching circuit, 38-5
voltage-divider, 38-5
Biasing transistor, 38-5
biCMOS, 40-19
Bidding, competitive, 57-3
Bilateral
Laplace transform, 15-17
z-transform, 15-18 (ftn)
Billion, 1-7 (tbl)
Binary, 48-8
counter, adjacent-state, 43-6
counter, synchronous, 43-5
digit, 13-1, 18-4
frequency-shift keying, 48-8
modulation, 48-5
number, 13-1
on-off keying, 48-7
operation, 41-2
phase-shift keying, 48-7
scalar, 41-1
system, 13-1
up counter, 43-4
Binomial
coefficient, 12-2
distribution, 12-4
theorem, 4-3
Biot-Savart law, 16-15, 22-3
Biot number, 1-9 (tbl)
Biphase
encoding, 48-7
signal, 48-7
Bipolar, 48-6
Bipolar junction transistor, 38-2
biasing, 38-4, 38-5
characteristics, 38-3
equivalent circuits, 38-9
high-frequency model, 38-12
load line, 38-5, 38-6
model, approximate, 38-8
npn, 38-2
simplified equivalent circuit, 38-10
switch, 40-17
Bisection method, 4-4, 14-1
Bistable
multivibrator, 40-28
oscillator, 40-19
Bit, 13-1, 18-5
error probability, 48-9
error rate, 48-9
interval, 48-6
least significant, 13-1
most significant, 13-1
BJT, 38-2
characteristics, 38-3
equivalent circuits, 38-9
high-frequency model, 38-12
simplified equivalent circuit, 38-10
switch, 40-17, 40-18
Bloch wall, 16-4
Block
cascaded, 51-6
codes, 48-9
protector, 50-2
Block diagram
algebra, 51-6
algebra, rules, 51-7
rules, 51-7
transfer protocol, 44-8
Blowing, whistle, 57-3 (ftn)
Board
foot, 1-8
of directors, 56-2
of engineering licensing, 59-1
Bode diagram, 32-7
magnitude, idealized, 32-9
phase, idealized, 32-11
Bode plot, 32-5, 51-11, 51-12
gain margin, 51-12
magnitude, 32-7
magnitude, idealized, 32-9
method, 32-11
phase, 32-10
phase margin, 51-12
pole at origin, 32-8
pole on negative real axis, 32-9
zero at origin, 32-8
zero on negative real axis, 32-9
Bohr
magneton, 16-14
radius, 22-7
Boilerplate clause, 56-4
Boltzmann's constant, 17-5
Bond, 55-29, 55-30
fully amortized, 55-30 (ftn)
yield, 55-30
Book value, 55-23
of stock, per share, 55-35
Bookkeeping, 55-33
double-entry, 55-34
Boolean algebra, 13-5, 40-20, 41-3
consensus, 13-5
De Morgan's law, 13-5
duality, 13-5
laws, 13-5
postulates, 13-5
simplification, 13-5
theorems, 13-5
Boson intermediate vector, 58-2
Bounce diagram, 45-3 (ftn)
Bound
lower, 10-4
upper, 10-4
vector, 6-1 (ftn)
Boundary
condition, 15-7
irregular, 8-1
value, 11-1
Bracket
income, 55-25
tax, 55-25
Branch circuit, 53-3
calculation, 53-5
receptacles, 53-4
Breach of contract, 56-6
Break
even analysis, 55-38
even point, 55-38
frequency, 32-5, 32-9
Breakaway point, 51-12
Breakdown, 37-9
avalanche, 37-9
region, 37-4, 39-2
region, MOSFET, 39-4
zener, 37-9
Breaker, 34-4
Breakover voltage, 37-15
Bridge
circuit, 36-3
Wheatstone, 36-3
British thermal unit, 2-1, A-1
SI conversion, A-3
Broadcast, 50-1
radio, 50-1, 50-2
television, 50-1, 50-2
Brownian motion, 36-2
Btu, 2-1, A-1
Bucket-brigade device, 39-2
Buckingham pi-theorem, 1-10
Buffer, 44-1
Building designer, 56-4
Bulk modulus, 19-2
Burden, 34-5, 55-36
Burial
depths, conduit, 54-5
method, 54-6
Burst protocol, 44-8
Business property, 55-20 (ftn)

C

C corporation, 56-3
Cabinet projection, 3-3
Cable
coaxial, 16-18, 45-1
communication, 34-7
television, 50-1
Cache coherence protocol, 44-8
Calculus, fundamental theorem of, 10-4
Cancellation, 46-3, 46-4
Canon, 57-1
Canonical
form, row, 5-2
product-of-sum form, 41-5
sum-of-product form, 41-5
Capacitance, 16-11, 29-5
configurations, 16-8
diffusion, 37-10
equation, 16-9
frequency domain, 29-4
in *s* domain, 31-15
time domain, 29-4
transition, 37-10
Capacitive
circuit, 27-7
coupling, 38-7

reactance, 27-9
transduction, 19-3
Capacitor, 16-11, 27-9
characteristics, 27-3
charging, 30-6
discharging, 30-6
energy, 16-12
energy storage, 29-5
equations, 29-5
fringing, 16-12
ideal, 27-9
in parallel, 16-12, 29-5
in series, 16-12, 29-5
initial current source model, 31-16
initial voltage source model, 31-17
parallel plate, 16-11, 16-12
perfect, 27-9
two-source model, 31-16
Capacity
battery, 33-3
channel, 50-1
heat, 2-4
Capital recovery method, 55-15
Capitalized cost, 55-7, 55-15
Capitalizing an asset, 55-20
Carrier, 17-5
concentration, A-24
generation, thermal, 17-5, 37-3
hot, 37-12 (ftn)
inter-exchange, 50-2
majority, 17-5, 17-6, 37-3
minority, 17-6, 37-3
movement, *pn* junction, 37-8
power line, 34-7
signal, 18-4
Carson's rule, 48-5
Cartesian
coordinate system, 8-3
equation form, 4-3 (ftn)
triad, 6-2
unit vector, 6-2
Cascade, 38-7
Cascaded blocks, 51-6
Cascode amplifier, 38-7
Cash flow, 55-3
analysis, 55-8, 55-9
diagram, 55-3
discounting factor, 55-6
exponential gradient series, 55-4
exponentially decreasing, 55-26
factor, 55-6, A-50, A-51
gradient series, 55-4
phantom, 55-4 (ftn)
single payment, 55-3
stepped, 55-10
uniform series, 55-3
Cash system, 55-34
Cathode, 26-14, 33-1
in tube, 17-2
Cation, 33-1
Cauchy
equation, 11-5, 15-3
number, 1-9 (tbl)
-Schwartz theorem, 6-3
Caution signs, 54-4
Cavalier projection, 3-3
Caveat emptor, 56-7
CCD, 39-2
Cell, 33-1, 41-1
communication, 50-3
D/A converter, 44-3
dry, 26-14
fuel, 33-4
primary, 33-1
secondary, 33-1
size, 33-3
state, 41-1
voltages, 33-1
Cellular, 50-1
telephone, 50-3
Center, 8-2
of gravity, 10-6
of series, 15-1
of vision, 3-3
-radius form, 8-9
Central
limit theorem, 12-11
processing unit, 44-1
tendency, 12-10
view, 3-1, 3-2
Centripetal acceleration, 16-20
Centroid, 10-6
Ceramic, 52-1
cgs systems, 1-4
Chain parameter, 29-16
Challenger, 55-18
Channel, 18-1, 18-6
capacity, 50-1
clear, 50-2
communication, 47-1, 50-1
equalization, 48-7
local, 50-2
microwave, 50-1
optical, 50-1
regional, 50-2
Characteristic
corner, 51-10
curve, diode, 37-11
equation, 5-8, 11-1, 31-14, 43-2, 51-13
equation, of feedback, 51-5
gain, 51-10
impedance, 45-2, 46-3
log, 4-5
phase, 51-10, 51-11
polynomial, 5-8
resistance, 45-2 (ftn)
Characteristics
battery, 26-13, 33-1
transducer system, 19-2, 19-3
Charge
carrier, direction, 16-20
carrier, speed, 16-20
-coupled device, 39-2
electric, 16-5
energy, 17-2
force on, 16-12, 17-2
generation, 17-6
in gas, 17-2
in liquid, 17-2
in semiconductors, 17-4
in uniform magnetic field, 22-3
in vacuum, 17-2
line, 20-4
magnetic, 22-5
magnetic test, 22-5
majority carrier, 17-5, 17-6
minority carrier, 17-6
mobility, 16-12
moving, 22-3
plane, 20-4
point, 20-2
positive, 20-3
recombination, 17-6
sheet, 20-4, 20-5
speed, 16-12
storage, 29-5
surface, 20-5
test, 16-7
-transfer device, 39-1
Charging, 33-3
Chart
control, 12-13
sag and tension, 54-5
Smith, 45-3, 45-4
Check code, 18-7
Chip, 37-2
Chirp, 50-4
Chromatic dispersion, 50-4
Circle, 8-9, A-7
formula, A-7
impedance locus, 45-4
segment, A-7
unit, 7-1
Circuit
active, 40-9
analog, 36-3
analysis, 29-1
analysis, method determination, 29-14
antiresonant, 40-29
battery equivalent, 33-2
branch, 53-3
breaker rating, 53-6
clamping, 37-12
clipper, 40-9
clipping, 37-12
digital, 36-3
dimension, 26-6
electromechanical, 36-3
element, equivalent, 19-4
element, lumped, 29-1
equivalent, 29-1
first-order, 31-2
formulas, resonant, A-32
gain, 32-11
GLC, 30-8
higher-order, 31-14
limiter, 40-9
magnetic, 16-21
nonlinear, 27-11, 37-2, 40-3, 40-9
Norton, 26-12
Norton equivalent, 29-9
protection, 53-1
pulse, 40-9
rectifier, 37-12
reduction, 29-8
resonant, 30-8
RLC, 30-8
safety, 53-1
sample-and-hold, 44-5
source-free, 30-2
substitution, 36-3
tank, 40-29
theory, 29-2
Thevenin, 26-10
Thevenin equivalent, 29-9
transducer, 36-3
transformation method, 31-16
variables, 29-1
vs. network, 29-2
Circular
array, 46-2
mil, 16-3, 26-2, 53-3, A-3
permutation, 6-5
polarization, 46-2
transcendental function, 7-2
Circulating current, 22-6
Cis form, 4-8
Civil
complaint, 56-6
tort, 56-6 (ftn)
Clamping, 40-9
circuit, 37-12
Class
A amplifier, 37-5
AB amplifier, 37-5
asset, 55-20
B amplifier, 37-5
C amplifier, 37-5, 37-6
limit, 12-9
Classical
adjoint, 5-5
method, 31-16
Clear
channel, 50-2
signal, 43-2
Clearance safety zone, 54-4
Client, dealing with, 57-2
Clinographic projection, 3-3
Clipper, 40-9
Clipping circuit, 37-12, 40-9
Clock, 43-1
flip-flop, 43-1
skew, 44-8
Closed loop
bandwidth, 51-10
gain, 51-5
transfer function, 51-5
Closed path, 15-19, 15-20
Closely held corporation, 56-2

INDEX - C

Cloud electron, 16-2 (ftn)
CML, 40-26
CMOS, 39-1, 40-18
  inverter, 40-18
  logic, 40-26, 40-27
  switch, 40-18, 40-19
CMRR, 40-3
Coaxial
  cable, 16-18, 45-1
  cable, electric field, 16-8
  line, 45-1
Code, 48-6, 48-7, 53-1, 57-1
  block, 48-9
  check, 18-7
  convolution, 48-9
  -division multiple access, 49-2
  error-correcting, 18-6, 18-7
  even parity, 18-7
  National Electric Safety, 54-1
  odd parity, 18-7
  of Federal Regulations, 50-1, 50-2
  parity check, 18-7
Coding, 18-6, 48-5 (ftn), 48-6
  unit distance, 42-3
Coefficient
  binomial, 12-2
  dielectric, 16-6, 16-9
  diffusion, 17-6
  matrix, 5-6
  of coupling, 28-3
  of series, 15-1
  of transformation, 6-3
  of variation, 12-11
  smoothing, 55-42
Coefficients, method of undetermined, 11-4
Coercive force, 28-6
Cofactor
  matrix, 5-1
  of entry, 5-2
Coherent system, 48-6
Coherent unit system, 1-2 (ftn)
Coil
  cylindrical, 16-18
  toroidal, 16-18
Collector, 38-2
  characteristic, 37-6
  logic, 40-25
Collinear, 8-2
Color force, 58-2
Column
  matrix, 5-1
  rank, 5-5 (ftn)
Combination, 12-2
Combinational logic, 41-2
  design process, 42-1
Comity, 59-2
Command value, 51-14
Common
  base, 38-4, 38-5
  collector, 38-4, 38-5
  emitter, 38-3, 38-5
  emitter, load line, 38-6
  mode, 40-3
  mode, rejection ratio, 40-3
  ratio, 4-11
  terminal, 38-4
Communication, 58-1
  band, 18-6
  cable, 34-7
  cell, 50-3
  channel, 18-1, 47-1, 50-1, 50-1 (ftn)
  cordless, 50-3
  crosstalk, 18-8
  frequency bands, 18-6, 50-3
  link, 47-1, 47-2
  link, transmission, 47-3
  power distribution, 47-1
  theory, 18-1
  uncertainty relationship, 18-3
  wireless, 50-3
  wireline, 45-1
Communication system, 18-1, 47-1, 50-1
  decoding, 18-2
  encoding, 18-2
Communications, optical, 50-3
Commutative law, 13-5, 40-20
  addition, 4-3
  convolution, 15-16
  matrix, 5-4
  multiplication, 4-3
  sets, 12-2
Company health, 55-35
Comparative
  analysis, 55-17
  negligence, 56-6
Comparator, 44-5, 51-4
Comparison
  alternative, 55-4
  of electromagnetic equations, A-26
  test, 4-12
Compatibility, electromagnetic, 50-4
Compensation
  point, 45-6
  reactance, 45-6
  theorem, 26-9
Compensator, 51-14
Compensatory
  damages, 56-7
  fraud, 56-6
Competitive bidding, 57-3
Complaint, 56-6
Complement
  law, sets, 12-1, 12-2
  nines, 13-3
  number, 13-3
  ones, 13-4
  set, 12-1
  tens, 13-3
  twos, 13-4
Complementary
  angle, 7-2
  construction, 40-19
  equation, 11-1
  error function, 15-11
  MOSFET, 39-1, 40-18
  probability, 12-4
  solution, 11-3, 30-2
  transistor logic, 40-19
Complementation law, 13-5
Complete sets, 41-3
Complex
  analytic function, 15-22
  conjugate, 4-8, 15-21, 27-10
  Fourier series, 15-14
  frequency, 31-1, 31-14
  frequency variable, 31-14 (ftn), 31-15
  impedance, 27-3
  matrix, 5-1
  number, 4-1, 4-8
  number operations, 4-8
  plane, 4-8
  power, 27-12
  power, vector, 27-12
  quantities, 27-8
  representation of phasors, 27-8
Complex number, 15-21
  addition, 27-10
  division, 27-10
  imaginary portion, 15-21
  multiplication, 27-10
  properties, 27-9
  real portion, 15-21
Component
  alias, 10-8 (ftn)
  electronic, 37-1
  of a vector, 6-2
Components, passive, 29-3
Compound
  amount factor, 55-5
  interest, 55-11
Compounding
  factors, 55-7
  period, 55-28
Compression wave, 19-1
Computer
  -aided design, 45-3, 58-1
  -aided manufacturing, 58-1
  complement, 13-4
  hardware interface, overview, 44-2
  number system, 41-1
  overhead, 44-7
  processing operands, 42-1
  processing symbology, 42-1
  switching, 44-7
  system correction, 48-9
  system error detection, 48-9
  technical standard, 44-1
  tool, 44-1
  traffic, 44-7
  transmission media, 47-1, 50-1
Concave, 8-2
  down curve, 8-2
  up curve, 8-2
Concavity, 8-2
Concentration of carriers, A-24
Concrete, 52-1
Condenser, 16-11
Condition
  boundary, 15-7
  initial, 15-7
  matched, 45-4
  of steady-state, 27-2
Conditional
  convergence, 4-13
  probability, 12-4
  probability of failure, 12-7
Conductance, 26-3, 27-3
Conducted
  emissions, 50-4
  susceptibility, 50-5
Conduction
  band, 16-2, 17-5
  current, 16-14
  effects, 16-4
  electron, 16-2
  intrinsic, 17-5
Conductivity, 17-2, 26-3, 37-2
  formulas, 17-3
  -modulated field-effect transistor, 37-16
  percent, 26-3
  standard for, 26-3
Conductor, 16-3, 52-1
  AC properties, A-49
  ampacities, 53-7, A-47
  conductivity, value, 52-1
  grounded, 53-3
  grounding, 53-3
  in magnetic field, 16-20
  nominal resistance of, 16-3
  properties, 16-3, A-48
  protection, 53-6
  standards comparison, A-45
Conduit burial depths, 54-5
Confidence
  level, 12-12
  level z-values, 12-12
  limit, 12-12
Configuration, totem-pole, 40-25
Congruency, 8-3
Conic section, 8-8
Conical horn, 46-2
Conjugate
  axis, 8-11
  complex, 4-8, 27-10
Connect/disconnect protocol, 44-8
Consensus theorem, 13-5
ConsensusDOC, 56-5 (ftn)
Consequential damages, 56-7
Conservation of energy, 2-1
Conservative field, 20-7
Consideration, 56-3
Consistent
  system, 4-7
  unit system, 1-2
Constant
  Boltzmann's, 17-5
  coefficient, 11-1, 11-2, 11-3

dielectric, 16-9, 29-5
Euler's, 10-8 (ftn), 15-12
gravitational, 1-2, 16-5
Joule's, 2-1
learning curve, 55-42
logic, 41-2
matrix, 5-6
Newton's universal, 16-5
of integration, 10-1, 11-1
percentage method, 55-21
Planck's, 17-2
propagation, 45-2
specific gas, 19-2
time, 30-2, 51-2, 51-9
universal gas, 19-2
value dollars, 55-39
Constellations of digital modulation, 48-8
Constraint equation, 40-8
Construction
contract, 56-3 (ftn)
lien, 56-5
manager, 56-5
Consulting engineer, 59-1
Contact
noise, 18-7
Schottky, 37-12
Continuity equation, 25-2
Continuous
compounding, 55-28
distribution function, 12-5
-time periodic signal, 15-14
-time signal, 15-14
-time system, 15-13, 15-14, 15-15
wave modulation, 48-3
Contour integral, 15-19
Contract, 56-3
breach, 56-6
construction, 56-3 (ftn)
discharge of, 56-6
documents, 56-3 (ftn)
privity, 56-7
requirements for, 56-3
standard, 56-5
Contractor, 56-4
Contraflexure
point, 8-2
point of, 9-2
Control
chart, 12-13
device, 37-16
element, 51-13
ratio, 51-5
system, 34-1, 51-13
system, analysis, 51-1
system, block diagram rules, 51-7
system, design, 51-1
system, implementation, 51-1
system, modes, 51-14
system, types, 51-14
Controllability, 51-15
Controlled
current source, 29-3
source, 27-3, 27-4
system, 51-13
voltage, oscillator, 40-29
voltage, source, 29-3
Controller, 51-13
bang-bang, 51-14
derivative, 51-14
integral, 51-14
on-off, 51-14
proportional, 51-14
proportional-integral-derivative, 51-14
Convection current, 16-13
Convention
half-year, 55-3
transformer dots, 28-2
year-end, 55-3
Conventional current, 26-2
Convergence
absolute, 4-13
conditional, 4-13
tests for, 4-12
Convergent
sequence, 4-10
series, 4-12
Conversion
analog-to-digital, 44-2
base, 13-3
digital-to-analog, 44-2
factor, SI, A-3
factors, A-1
power, 2-6 (tbl)
two-port, A-31
Converter, 44-5
cells, 44-3
counting A/D, 44-5
D/A, 44-2
dual-slope A/D, 44-7
ladder-type D/A, 44-3
parallel-comparator A/D, 44-6
servo, 44-5
successive-approximation, 44-6
tracking, 44-5
Convex, 8-2
hull, 8-2, 8-3
Convexity, 8-3
Convolution, 15-16
associative law, 15-16
codes, 48-9
commutative law, 15-16
distributive law, 15-16
integral, 11-6, 15-16
properties, 15-16
Cooling, Newton's law of, 11-10
Coordinate system, 8-3
Coplanar, 8-2
Copper, 16-3, 52-1
standard, 26-3
Cordless communication, 50-3
Core
ideal, 28-3
iron losses, 28-3
loss, 28-6
resistance, 28-5
shell, 17-4
transformer, 28-2, 28-3
Corner
characteristic, 51-10
frequency, 32-5, 32-9, 38-12, 51-10
Corporation, 56-2
professional, 56-2
subchapter C, 56-3
Correction, computer system, 48-9
Correlation, 15-16
Corrugated horn, 46-2
Cosine
direction, 6-2, 8-5
function, integral, 10-8
integral, 10-8
Cosines, law of, 7-5, 7-6
Cost, 55-31
accounting, 55-36
capitalized, 55-7
differential, 55-37
life-cycle, 55-20
marginal, 55-37
of goods sold, 55-37
opportunity, 55-18
plus fee, 56-5
sunk, 55-3
total, 55-32
Coulomb gauge, 22-9
Coulomb's Law, 16-6, 20-4, 25-1
magnetic equivalent, 22-8
Counter, 43-4, 43-6
adjacent-state binary, 43-7
analysis procedure, 43-6
binary up, 43-4
modulo, 43-4
ripple, 43-4, 43-5
state diagram, 43-6
timing diagram, 43-6
truth table, 43-8
up-down, 44-5
Counting A/D converter, 44-5, 44-6
Coupling
capacitive, 38-7
coefficient, 28-3
magnetic, 28-1
Coupon rate, 55-30
Covalent bond, 37-2
CPU, 44-1
Cramer's rule, 4-7, 5-7
Credit, 55-34
investment, 55-25
tax, 55-25
Creed, 57-1
Criterion, Routh, 51-13
Critical
angle, 46-4
damping, 31-8
point, 9-2
temperature, superconductivity, 58-1
Critically damped analysis, 31-9
Cross product, vector, 6-4
Crossover
frequency, phase, 51-12
point, gain, 51-10, 51-12
Crosstalk, 18-8
CTD, 39-1
CTL, 40-19
Cumulative frequency, 12-9
Cunife, 52-2
Curie point, 21-3
Curl, 9-7, 20-4, 22-8
electrostatic field, 22-8
Current, 16-13, 26-1, 27-3
ampèrian, 22-6
amplification factor, 37-4
angle, 27-11
asset, 55-34
circulating, 22-6
conduction, 16-14
-controlled voltage source, 29-3
convection, 16-13
conventional, 26-2
density vector, 16-13
diffusion, 17-6, 17-7
displacement, 16-4, 16-13, 21-4
divider, 26-5, 26-6, 29-13
drift, 37-7
eddy, 16-22, 28-6
electron, 26-2
equation, 18-7
equivalent Norton, 26-12
exciting, 28-3, 28-5
flow, 26-2
Foucault, 16-22, 28-6
gain, 38-7
hogging, 40-19
hold, 37-15
in series circuit, 26-7
injection, 37-7
keep-alive, 37-13
law, Kirchhoff's, 26-6, 29-11
liability, 55-34
magnetizing, 28-3, 28-5
-mode logic, 40-26
offset, 40-8
ratio, 38-3, 55-35
recombination, 37-7
reference direction, 26-7
reverse saturation, 37-7
saturation, 37-7, 38-4, 39-2
secondary, 28-2
thermal, 37-7, 38-4
total, 16-13
transformer, 34-4, 36-2
unit sink, 40-21
unit source, 40-21
Current source
controlled, 29-3
dependent, 29-3
ideal, 26-4, 29-2
ideal independent, 29-2
in parallel, 26-5
in series, 26-5
real, 26-4, 29-2

INDEX - C

Curve, 8-3
  alpha, 51-11
  bell-shaped, 12-6
  degree, 8-3
  learning, 55-42
  magnetization, 23-5
  simple, 15-19
  standard normal, A-12
  symmetry, 8-4
Cut-over point, 55-24 (ftn)
Cutoff, 37-4, 38-2
  bias, 38-5
  frequency, 51-10
  mode, 38-3
  rate, 51-10
  region, 38-3
Cycle, 27-2, 33-1
  life, 33-1
Cycloid, 8-4
Cylindrical
  coil, 16-18
  coordinate system, 8-3 (fig), 8-3 (tbl)

D

d'Alembert solution, 15-9
d'Arsonval
  meter movement, 36-4
  movement, 36-4
D flip-flop, 43-3, 43-4
  next-state equation, 43-3
Damage, 56-7
  punitive, 56-6
Damped oscillations, 30-7
Damping, 30-8, 31-14
  critical, 31-8
  over, 31-8
  ratio, 31-14, 51-2
  under-, 31-8
Danger signs, 54-4
Darlington
  pair, 38-12
  transistor, 38-12, 38-13
Data
  flip-flop, 43-2, 43-3
  ready signal, 44-7
  selector, 42-9
Date
  examination, 59-2
  maturity, 55-30
Datum, 26-14
DC
  ammeter, 36-5
  ammeter circuit, 36-5
  biasing methods, 38-6
  circuit fundamentals, 26-1
  input, 51-8
  offset, 27-2 (ftn)
  switching transient, 30-1
  value symbology, 37-2
  voltmeter, 36-4
DCTL, 40-19
de Moivre's theorem, 4-9
de Morgan's
  law, 13-5
  law, sets, 12-2
  theorem, 40-20, 41-3
Debit, 55-34
Decade, 51-10
Decay
  exponential, 11-9
  exponential, pole-zero, 51-9
Decaying sinusoidal response, 51-9
Decibel, 18-6, 26-4, 32-7
  watt, 18-6
Decile, 12-10
Decimal number, 13-1
Declining balance method, 55-26
Decoder, 42-9
Defect
  noise, 18-7
  spherical, 7-6
Defendant, 56-6
Defender, 55-18
Defined quantity, 30-6
Definite integral, 10-1 (ftn), 10-4
Degenerate
  circle, 8-10
  ellipse, 8-11
Degree, 7-1
  of curve, 8-3
  of freedom, 51-14
  of polynomial, 4-3
Del operator, 9-7, 20-5
Delay, 18-3
  distortion, 48-7
  flip-flop, 43-3
  time, 18-3, 40-23, 51-2
Delta-wye
  configuration, 26-8, 29-9
  transformation, 26-8, 29-9
Delta function, 15-10
Demultiplexer, 42-9
Density
  displacement, 16-10
  electric flux, 16-10
  flux, 47-2 (fig)
  function, 12-4
  magnetic flux, 16-17
  mass, 1-3
  power, 46-5, 46-6, 46-7, 46-9, 47-1
  states, effective, 17-5, 37-3
  weight, 1-3
Dependent
  current source, 29-3
  event, 12-3
  source, 26-4, 27-3, 29-3
  variable, 4-2
  voltage source, 29-3
Depletion, 55-24
  allowance, 55-24
  MOSFET, 39-4
  region, 17-7, 37-7
Depreciation, 55-20
  accelerated, 55-23
  basis, 55-20
  disfavored methods, 55-22
  methods, 55-21
  period, 55-20
  production method, 55-22
  rate, 55-21
  recovery, 55-26
  service output method, 55-22
  sinking fund method, 55-22
Depression angle, 7-2
Depth, 3-2
Derating, 53-7
Derivative, 9-1
  controller, 51-14
  directional, 9-6
  Laplace transform of, 11-6
Derived unit, 1-5, 1-6 (tbl)
Deriving equation, 29-14
Descartes' rule of signs, 4-4
Design
  control system, 51-1
  limits, 53-1, 54-2
  professional, 56-7
  timing, 43-1
Destination, 18-1
Detector
  phase frequency, 40-28
  sequence, 43-8
Determinant, 5-3
Developer, 56-4
Deviation
  angle of, 46-4
  standard, 12-10
Device
  active electron, 17-2
  next-state equation, 43-3
  overcurrent, 34-7
  standard modular, 37-2
Diac, 37-15
Diagonal matrix, 5-1
Diagram
  Bode, 32-7
  bounce, 45-3 (ftn)
  cash flow, 55-3
  phasor, 27-8, 27-13
  pole-zero, 51-8
  root-locus, 51-12
  Veitch, 42-5
Dialogue interface, 44-1
Diamagnetic, 16-4
  material, 23-4
Diamagnetism, 16-4, 23-3
Dielectric, 16-2, 16-3, 52-2
  coefficient, 16-6, 16-9
  constant, 16-9, 29-5
  field, 21-3
  lossy, 21-4
  phenomena, 16-4
  polarization, 21-1
  strength, 16-4
  surface charge, 21-3
Difference
  divided, 14-3
  Rth-order, 4-11
  table, divided, 14-3 (ftn)
Differential
  amplifier, 40-3, 40-4
  amplifier, emitter current, 40-4
  encoding, 48-7
  equation, 11-1
  equation, higher order, 11-8
  -mode, 40-3
  term, 10-1
Differentiation
  implicit, 9-4
  partial, 9-4
  unit step, 31-6
Differentiator, 40-6
Diffraction, 46-4
Diffusion, 17-6
  capacitance, 37-10
  coefficient, 17-6
  current, 17-6, 17-7, 37-7
  electron, 37-7
  hole, 37-7
Digit, 13-1
  binary, 13-1
  least significant, 13-1
  most significant, 13-1
  significant, 4-1
Digital, 18-1, 18-2
  baseband signals, 48-6
  circuit, 36-3
  circuit, operation, 38-2
  error probability, 48-8, 48-9
  information representation, 41-1
  instrument, 36-8
  logic, 42-2
  minimization principle, 42-2
  modulation, error probabilities, 48-9
  modulation constellations, 48-8
  multimeter, 36-8
  network, integrated service, 50-2 (ftn)
  signal, 18-1, 18-2
  system, 44-1
  system, architecture, 44-1
  -to-analog conversion, 44-2
Digital-to-analog converter, 44-2
  binary weighted resistors, 44-5
  cell, 44-3
  cells, Thevenin equivalent, 44-4
  inverting-type, 44-4, 44-5
  inverting ladder-type, 44-5
  ladder-type, 44-3
  multiplying-type, 44-5
Digital information
  operations on, 42-1
  scalar, 42-1
  vector, 42-1
Digital modulation, 48-5
  bandwidth, 48-8
  error probability, 48-9
  power, 48-8

INDEX - D

Digitally controlled switch, 44-3
Dimension
  of variables, 1-8, 1-8 (tbl)
  primary, 1-7
Dimensional analysis, 1-8
Dimensionality of signal, 48-8
Dimensionless, 1-8, 1-9 (tbl)
  group, 1-8, 1-9, 1-9 (tbl)
  number, 1-7, 1-9 (tbl)
Dimetric view, 3-3
Diode, 37-13
  applications, 37-12
  backward, 37-13, 37-14
  barrier, 37-12
  characteristic curve, 37-11
  characteristics, 37-9
  circuit, 37-12
  circuit outputs, 37-12
  electroluminescent, 37-14
  equivalent circuit, 37-10
  Esaki, 37-13
  forward bias region, 37-12
  hot-carrier, 37-12
  ideal, 37-9
  light-emitting, 37-14, 50-4
  load line, 37-11
  off region, 37-12
  piecewise linear model, 37-11, 37-12
  precision, 40-10, 40-11
  real, 37-10
  responsivity, 50-4
  reverse breakdown region, 37-12
  Schottky, 37-12, 37-13
  -transistor logic, 40-19, 40-24, 40-25
  tunnel, 37-13, 37-14
  varactor, 37-10
  zener, 37-13, 40-13, A-38, A-39
Dipole, 16-14, 20-3
  alignment, 21-2
  antenna, 46-7, 46-8
  charge, 20-3
  electric, 22-6
  electric field, 16-8, 20-3
  field, 16-15
  half-wave, 46-9
  in magnetic field, 22-5
  magnetic, 16-15, 22-6
  moment, 20-3
  rotation, work of, 21-3
Dipole moment, 16-14, 20-3, 21-1
  comparison, 22-6
  per unit volume, 16-16, 16-17, 22-5
  torque, 20-3
Dirac function, 15-10
Direct
  -coupled transistor logic, 40-19
  gain, 51-4
  labor, 55-32
  manipulation, 44-1
  material, 55-32
  transfer function, 51-4
Direction
  angle, 6-2, 8-5
  cosine, 6-2, 8-5
  number, 8-5
  of magnetic force, 22-4
Directional derivative, 9-6
Directive gain, 47-3
Directivity, 46-2, 46-6, 46-7, 46-9, 47-3
Directrix, 8-10
Disbenefit, 55-16
Disbursement, 55-3
Discharge
  electrostatic, 50-5
  lamp, 52-2
  of contract, 56-6
Discount factor, A-50, A-51
Discounting, 55-5
Discrete
  Fourier series, 15-17
  Fourier transform, 15-17
  inputs, 15-11 (ftn)
Discrete-time
  Fourier series, A-20
  Fourier transform, 15-18, A-21
  periodic signal, 15-17
  signal, 15-17
  system, 15-17
Discriminant, 4-4, 8-8
Dish, 46-2
Disjoint set, 12-1
Dispersion, 12-10
  chromatic, 50-4
  intramodal, 50-4
  material, 50-4
Displacement
  current, 16-4, 16-13, 21-4
  density, 16-10
  flux, 16-10
Distance
  between figures, 8-7
  between points, 8-5, 8-7
Distortion
  amplitude, 48-7
  delay, 48-7
  quantumization, 48-5
Distribution
  bathtub, 12-7
  binomial, 12-4
  continuous, 12-6
  discrete, 12-4
  exponential, 12-6
  frequency, 12-9
  Gaussian, 12-6
  hypergeometric, 12-5
  multiple hypergeometric, 12-5
  normal, 12-6
  Poisson, 12-5
  system, typical, 53-2
Distributive law, 4-3, 12-2, 13-5, 40-20
  convolution, 15-16
  matrix, 5-4
Divergence, 9-7, 20-4
  electric field, 22-8
  of electric flux density, 20-5
  theorem, 20-4, 20-5
Divergent sequence, 4-10
Divided
  difference, 14-3
  difference table, 14-3 (ftn)
Divider
  circuit, 26-6
  current, 26-5, 26-6, 29-13
  voltage, 26-5, 29-13
Division, matrix, 5-4
DMM, 36-8
DMOS, 39-1
Dollars, 55-2
  constant value, 55-39
Domain, 16-4, 23-4, 42-1
  analysis, 51-11
  magnetic, 16-4
Donor, 17-5, 37-3
  n-type, 17-5
  p-type, 17-6
Dopant, 17-5, 37-3
Doping, 17-5
Dot
  convention, transformer, 28-2
  product, 6-3
  rule, 28-3
Double
  -angle formula, 7-4
  declining balance, 55-21, 55-27
  -entry bookkeeping, 55-33
  integral, 10-3
  root, 4-4
  -sideband, AM signal, 18-4
  -stub matching, 45-6
  taxation, 56-2
Double-sideband
  large-carrier, 48-3
  suppressed-carrier, 48-3
  transmitted-carrier, 48-3
Doubling time, 35-1, 55-8
Drain, 38-2, 39-1
Drawing, pictorial, 3-2 (ftn)
Drift
  current, 37-7
  electron, 37-7
  hole, 37-7
  velocity, 16-12, 17-2, 17-3
Driving function, 30-3, 30-5
Drop
  spherical, 11-10
  voltage, 26-3, 26-6
Dry cell, 26-14
  battery, 33-3
DTL, 40-19, 40-24, 40-25
Dual-slope A/D converter, 44-7
Duality, 13-5, A-16
  principle, 41-4
Duration, 18-3
DVOM, 36-8
Dwelling unit, 53-6
Dynamic
  forward resistance, 37-10
  load line, 37-11
  logic, 40-21
  resistance, 38-1, 39-1 (ftn)
  reverse resistance, 37-10
  unit, 51-4
Dyne, 1-5, A-3

E

Ear, human, 26-4 (ftn)
Early effect, 38-8
Earth
  as reference point, 20-5
  functional, 34-2
  ground, 26-14
  orbit, geostationary, 50-3
  orbit, low, 50-3
  orbit, medium, 50-3
  protective, 34-2
Eccentricity, 8-8, 8-11, 8-12
Echelon
  form, row-reduced, 5-2
  matrix, 5-1
Eckert number, 1-9 (tbl)
ECL, 40-26
Economic
  indicator, 55-39
  life, 55-19, 55-20
  order quantity, 55-43
Economic analysis
  break-even analysis, 55-38
  break-even quantity, 55-38
  depreciation methods, 55-21
  engineering, 55-2
  factor tables, A-51
  factors, A-50
  forecasting, 55-42
  life, 55-4
  smoothing, 55-42
Economics
  electric power, 35-2
  engineering, 55-2
Eddy current, 16-22, 28-6
Education, 58-2
Effect
  Early, 38-8
  photoelectric, 17-2
Effective, 27-5, 27-8
  density state, 17-5, 37-3
  electron mass, 16-12 (ftn)
  impedance, 28-4
  interest rate, 55-5, 55-28
  isotropically radiated power, 47-4
  mass, 17-4, A-24
  notation, 27-8
  period, 55-2
  phasor notation, 27-8
  radiated power, 50-2
  value phasor notation, 27-8
  value symbology, 37-2

Effectively grounded, 34-2
Efficiency, 2-7
  antenna, 46-2
  aperture taper, 47-3
  bandwidth, 48-8
  illumination, 47-3
  of antennae, 46-6
Eigenfunction, 15-9
Eigenvalue, 5-8, 15-9, 51-8
Eigenvector, 5-8
Einstein's equation, 17-7
Einstein relationship, 37-10 (ftn)
EJCDC, 56-5 (ftn)
Elastance, 16-11
Elastic modulus, 19-2
Electric
  charges, 16-5
  current, 16-13
  dipole, 20-3
  dipole moment, 22-6
  force, 16-16
  force, line of, 16-8
  force, tube, 16-8
  hysteresis, 21-3
  phenomena, 16-5
  potential, 16-21
  power, 35-2
  susceptibility, 21-2
  susceptibility, as tensor, 21-2
  vs. magnetic circuit, 25-2
  vs. magnetic equations, 25-2, A-26
  vs. magnetic force, 16-16
Electric field, 16-7
  configuration, 16-8
  dielectric, 21-2
  divergence, 22-8
  energy, 20-7
  equation, 16-8
  force, 20-2
  free charge in, 16-12
  intensity, 16-4, 16-7
  interaction, 20-2
  like charges, 20-1
  line charge, 20-4
  positive charge, 20-1
  potential, 20-5
  sign convention, 20-1
  strength, 16-4, 16-7
  uniform, 16-8
  unlike charges, 20-1
  vs. distance, 20-5
  work, 20-6
Electric flux, 16-9
  density, 16-4, 16-10, 20-2, 21-1
  density, divergence, 20-5
  divergence of, 20-5
  line, 20-1
Electrical
  /acoustical/mechanical elements, 19-4
  balance, 36-3
  engineering frontiers, 58-1
  impedance, 27-3
  materials, 52-1
  steel, 52-2
  unit conversions, A-1
Electricity, 16-2
  resinous static, 16-5
  vitreous static, 16-5
Electrochemical
  battery, 26-13
  cell, voltage, 33-1
  reaction, 26-13, 26-14
  reaction, battery, 33-1
Electrode, 33-1
  grounding, 53-6
Electrodynamometer, 36-6
Electrokinetics, 16-12
Electroluminescence, 37-14
Electroluminescent diode, 37-14
Electrolysis, 24-2
  Faraday's law, 24-2
Electrolyte, 33-1
Electromagnetic
  compatibility, 50-4
  effects, 16-3
  induction, 16-22
  induction, Faraday's law, 24-2
  interference, 50-4
  pulse, 50-5
  spectrum, A-25
  theory, 16-1
  theory, force particle, 58-2
Electromagnetic field
  charge in, 17-2
  force on charge, 17-2
  vector, 25-1, 25-2
Electromagnetic wave, 24-3
  polarization, 46-3
  speed, 16-20
Electromagnetism, 16-2
Electromechanical circuit, 36-3
Electromotance, 16-22, 24-1
Electromotive force, 16-22, 24-1, 24-3, 26-1
  flux-cutting method, 24-2
  means of generation, 24-1
  time-varying method, 24-1
  vs. potential, 24-3
  vs. voltage, 24-1
Electron
  charge, 16-2
  cloud, 16-2 (ftn)
  conduction, 16-2
  current, 26-2
  device, active, 17-2
  diffusion, 37-7
  drift, 37-7
  drift velocity, 16-12
  effective mass, 16-12 (ftn), 17-4
  gas theory, 16-12
  mass, 16-2
  mobility, 37-3
  moment, 23-3
  orbital moment, 23-3
  potential energy, 17-4
  properties of, 16-2
  recombination, 37-7
  saturation, 37-7
  shell, 16-2
  spin, 23-3
  spin moment, 23-3
  valence, 37-2
Electronic, 17-1
  component, 37-1
  instrumentation, 36-8
  theory, 17-1
Electronics, power, 36-1
Electrostatic
  discharge, 50-5
  field, 20-1, 21-1
  field curl, 22-8
  potential, 24-3
  system of units, 16-2
Electrostatics, 20-1
  Gauss' law, 16-10
Electroweak theory, 58-2
  force particles, 58-2
Element
  active, 29-3
  lumped, 29-1
  passive, 29-3
  properties, related to periodic table, A-43
  under test, 36-8
Elemental antenna, 46-5
  directivity, 46-6
  efficiency, 46-6
  gain, 46-7
  ohmic resistance, 46-6
  radiated power, 46-6
  radiation resistance, 46-6
Elementary
  function, 15-1 (ftn)
  row operation, 5-2
Elevation
  angle, 7-2
  drawing, 3-2 (ftn)
Elimination, Gauss-Jordan, 5-2
Ellipse, 8-10
  formula, A-7
Elliptical polarization, 46-2
Emanations security, 50-5
EMC issues, 50-4, 50-5
Emissions
  conducted, 50-4
  radiated, 50-4
  security, 50-5
Emitter, 38-2
  -coupled logic, 40-19, 40-26
  follower, 38-4, 40-4
Employer, dealing with, 57-2
EMS, 35-2
Encirclements, 42-5
Encoder, 42-9
Encoding, 18-1
  biphase, 48-7
  differential, 48-7
Encrypting, 18-1 (ftn)
End-around carry, 13-4
Energy, 2-1, 35-1
  acoustic, 19-1
  conservation law, 2-1
  conversions, SI, A-3
  density, electric field, 16-12
  efficiency battery, 33-1 (ftn)
  -efficiency ratio, 2-7 (ftn)
  electric, 16-12
  falling mass, 2-3 (ex)
  flow, 2-4
  gap, 17-4, 37-3, A-24
  gravitational, 2-3
  internal, 2-4
  kinetic, 2-3
  level, Fermi, 17-2
  magnetic potential, 22-8
  management system, 35-2
  mass, 2-1
  of dipole, 20-3
  of electric field, 20-7
  p-V, 2-4
  per unit charge, 26-1
  potential, 2-3
  power unit, 35-1
  pressure, 2-4
  -production efficiency, 2-7
  quanta, 17-4
  specific, 2-1
  spring, 2-3
  static, 2-4
  thermal, 2-1
  total, 2-1
  transfer, maximum, 26-5
  unit of, 2-1
  -use efficiency, 2-7
  -work principle, 2-3, 2-4
Energy band, 17-4
  diagram, 17-4
  n-type, 17-5
  p-type, 17-6
  theory, 17-4
Energy source, 26-4
  ideal, 26-4
  real, 26-4
Engineer-In-Training exam, 59-1
Engineering
  dimensional system, 1-7
  economic analysis, 55-2, 55-21
  economics, 55-2
  economy factors, A-50, A-51
  intern, 59-1 (ftn)
  management, 58-1
  registration, 59-1
  unit, 35-2
Engineers
  dealing with, 57-3
  Joint Contracts Documents Committe, 56-5 (ftn)
Engineers' Creed, 57-1

English
  engineering system, 1-2
  gravitational system, 1-3
Enhancement MOSFET, 39-4
Entropy, 18-5
  information, 18-4
Envelope, 30-7
Environmental noise, 18-8
Eötvös number, 1-9 (tbl)
Epicycloid, 8-4
Epidemic, 11-10
Equal vectors, 6-1
Equality of matrices, 5-4
Equalization of channel, 48-7
Equalizer, 48-7
Equation, 4-2
  Ampère's law, 22-8
  amplitude modulation, 48-3
  antenna directivity, 47-3
  antenna gain, 46-7
  antenna ohmic resistance, 46-6
  auxiliary, 11-1
  average, 27-4
  Bessel, 11-5, 15-1
  Biot-Savart, 16-15
  capacitance, 16-11, 29-5
  Cauchy, 11-5, 15-3
  characteristic, 11-1, 31-14, 43-2, 51-13
  characteristic impedance, 46-3
  charge force, 16-12
  circular mil, 53-3
  complementary, 11-1
  complementary error function, 15-11
  conduction current, 16-14
  conductivity, 17-2
  constraint, 40-8
  continuity, 25-2
  convolution, 15-16
  current, 18-7
  decibel, 26-4
  deriving, parameters, 29-14, 29-15
  differential, 11-1
  digital modulation error, 48-9
  dipole charge, 20-3
  directivity, 46-6
  Einstein's, 17-7
  EIRP, 47-4
  electric field, 20-2
  electric field potential, 20-6
  electric field work, 20-6
  electric flux density, 21-1
  electron drift velocity, 16-12
  error function, 15-11
  Euler's, 4-8, 11-5, 15-3, 27-2
  Euler's constant, 15-12
  even periodic extension, 15-6
  Fourier analysis, 15-14
  Fourier synthesis, 15-15
  Friss, 47-3, 47-4
  Gauss' hypergeometric, 11-5
  impedance, 27-3
  indicial, 15-3
  inductance, 16-19, 29-6
  Laplace's, 20-6
  Laplacian operator, 20-6
  Legendre's, 11-5, 15-1
  line charge, 20-4
  Lorentz force, 16-20, 25-2
  magnetic charge, 22-6
  magnetic field strength, 16-16
  magnetic flux, 16-18, 22-2, 28-1
  magnetic force, 16-16
  magnetic pole, 22-6
  magnetization, 23-1
  mass action law, 37-2
  Maxwell's, 16-2, 25-1, 25-2
  negative feedback, 51-5
  next-state, 43-2
  noise power, 40-8
  noise voltage, 36-2
  non-parametric, 4-3
  odd periodic extension, 15-6
  of feedback characteristics, 51-5
  one-dimensional heat, 15-7
  one-dimensional wave, 15-7
  parametric, 4-2
  plane charge, 20-4
  point change, 20-2
  Poisson's, 20-6
  polarization, 21-1
  positive feedback, 51-5
  power triangle, 27-12
  Poynting's vector, 24-3
  quadratic, 4-3
  radiated power, 46-6
  receiving antenna power, 47-3
  rectified sinusoid average, 27-4
  reduced, 11-1
  resistance, 26-2, 29-4
  response, 51-15
  root-mean-square, 27-5
  Shannon bandwidth, 48-8
  Shannon limit, 50-1
  Shockley, 39-3
  simultaneous linear, 4-7
  state, 51-15
  temperature coefficient, 29-4
  three-dimensional Laplace, 15-7
  transformer coupling coefficient, 28-3
  transition, 43-2
  two-dimensional Laplace, 15-7
  two-dimensional Poisson, 15-7
  voltage drop, 53-5
  wave length, 46-3
  wave velocity, 46-3
Equilateral hyperbola, 8-12
Equipment grounding requirements, 53-6
Equity, 55-34
Equivalence, economic, 55-5
Equivalent
  alternative, 55-5
  current, Norton, 26-12
  factor, A-50, A-51
  isotropically radiated power, 46-2 (ftn)
  Joule, 2-1
  logically, 41-3
  magnetic-electric circuit, 16-21
  noise bandwidth, 48-8
  real transformer, 28-5
  rectangular bandwidth, 18-3
  series voltage, 26-7
  source, 26-5
  uniform annual cost, 55-7, 55-15
  vector, 6-1 (ftn)
  voltage, 26-4
  voltage, Thevenin, 26-10
  weight, 24-2
Equivalent circuit, 29-1
  battery, 33-2
  BJT, 38-9
  diode, 37-10
  elements, 19-4
  Norton, 26-12, 29-9, 29-10
  operational amplifier, 40-4
  parameter, 38-8
  Thevenin, 26-10, 29-9
Equivalent resistance, 26-4
  Norton, 26-12
  parallel, 26-7
  series, 26-7
  Thevenin, 26-10
Erg, A-3
Ergonomic standard, 44-1
Error, 12-13
  correcting, forward, 48-9
  -correcting code, 18-6, 18-7
  detection, computer systems, 48-9
  function, 10-8, 12-6, 15-11, 48-9
  gain, 51-4
  phase, 40-28, 40-29
  phase angle, 34-5
  probabilities, of digital modulation, 48-9
  quantization, 44-2
  random, 36-1
  rate, bit, 48-9
  ratio, 51-4
  signal, 51-4
  systematic, 36-1
  transfer function, 51-4
  transformation, 34-5
Errors and omissions insurance, 56-8
Esaki diode, 37-13
Estimator, unbiased, 12-10
Ethical priority, 57-2
Ethics, 57-1
EU, 35-2
Euler's
  constant, 10-8 (ftn), 15-12
  equation, 4-8, 11-5, 15-3, 27-2
  number, 1-9 (tbl)
  relation, 15-13, 27-2
EUT, 36-8
Evaporation, 11-10
Even
  parity, 18-7
  periodic extension, 15-6
  symmetry, 8-4, 10-7 (tbl)
Event, 12-3
Exact, first-order, 11-3
Exam
  date, 59-2
  licensing, 59-1, 59-2
Excess
  noise, 18-7
  spherical, 7-6
Exchange force, 23-4 (ftn)
Excitation loss, 28-6
Exciting current, 28-3, 28-5
Exemplary damages, 56-7
Expansion
  by cofactor, 5-3
  method, 13-1
  series, 9-7
Expected value, 55-30
Expense, 55-32
  administrative, 55-32
  marketing, 55-32
  selling, 55-32
Expensing an asset, 55-20
Exponent, 4-5
  isentropic, 19-2
  rules, 4-5
Exponential
  behavior, 30-2
  decay, 11-9
  decay, pole-zero, 51-9
  decreasing cash flow, 55-26
  distribution, 12-6, 12-7
  form, 4-8
  form of sinusoid, 27-2
  Fourier series, 15-14
  function, integral, 10-8
  gradient cash flow, 55-4
  gradient factor, 55-26
  growth, 11-9, 55-4
  reliability, 12-7
  weighted forecast, 55-42
Extended
  Ohm's law, 27-3
  power series method, 15-3
Extension, periodic, 15-6
External
  investment, 55-13
  rate of return, 55-12
  work, 2-2
Extraneous root, 4-4
Extranuclear, 16-2
Extreme point, 9-2
Extrinsic semiconductor, 17-5, 37-2, 37-3
Eye pattern mask, 50-4

F

Face value, 55-30
Factor
  cash flow, 55-6, A-50, A-51
  cash flow, discounting, 55-6
  compounding, 55-7
  conversion, A-1

economic analysis, A-50, A-51
integrating, 11-2
irreducible, 55-2
judgment, 55-2
nonquantifiable, 55-2
nonquantitative, 55-2
normalizing, 32-11
of modulation, 48-3
quality, 30-9, 51-10
reactive, 27-12
sinking fund, 55-7
standard cash flow, A-50
table, A-50, A-51
weighting, 55-42
Factorial, 4-2 (tbl, ftn)
function, 15-11
Factory cost, 55-32, 55-36
Failure, 12-3
mean time to, 12-7
Faith of an Engineer, 57-1
Fall time, 40-23
Falling
edge trigger, 43-1
mass energy, 2-3 (ex)
Fan-in, 40-24
Fan-out, 40-20
circuits, 40-21
load, 40-20
sink, 40-23
source, 40-23
unit loads, 40-21
Far field, 46-6 (ftn)
Farad, 16-11
Faraday, 16-2, A-1
Faraday's Law, 16-22, 24-2, 25-1
of electrolysis, 24-2
of electromagnetic induction, 24-2, 26-6
Fast Fourier transform, 10-8, 48-4 (ftn)
FBI rule, 22-4
Federal Communications Commission, 50-1, 50-2
rules, 50-2
Fee structure, 56-5
Feedback
algebra, 51-6
characteristic equation, 51-5
control system, typical, 51-13
factor, 40-16, 40-17
gain, 51-4, 51-5
negative, 51-4
positive, 51-4
ratio, 51-5
sensitivity, 51-5
system, 51-4
theory, 51-4
transfer function, 51-4
unit, 51-4
Feeder, 53-4
calculation, 53-4
Fermi-Dirac
distribution, 37-12 (ftn)
probability function, 37-10
Fermi level, 17-2, 37-13
energy, 37-12 (ftn)
Ferrimagnetism, 16-4
Ferroelectrics, 21-3
Ferromagnetism, 16-4, 23-4
FET equivalent circuit, 39-4
FFT, 10-8
analyzer, 10-8
Fiber, 34-7
optical, 46-4
Fiduciary responsibility, 56-3
Field, 20-1, 22-1
conservative, 20-7
electrostatic, 20-1
force on charge, 17-2
in dielectric, 21-3
irrotational, 22-8
like charges, 20-1
magnetic strength, 16-17, 16-18
polarized, 21-2
radiation, 16-21
solenoidal, 22-8
strength, magnetic, 23-5
strength, unit, 22-6
uniform electric, 16-8
unlike charges, 20-1
Field, electric, 16-7, 20-1
force, 20-2
intensity, 16-7
strength, 16-7
Field, magnetic, 16-16, 22-1
dipole, 16-15
strength of, 16-17
Field-effect transistor, 38-2, 39-1
equivalent circuit, 39-4
high-frequency model, 39-4
model, 39-4
Field vector, 25-1
electromagnetic, 25-2
force, 16-7
Figure of merit, 40-2
Filter
band-pass, 30-8, 30-9
band-reject, 30-9, 30-10
Filters, 40-2
Final value, 51-7
theorem, 15-5
Fine print note, 53-2
Finish, surface, 3-4
Finite series, 4-12
First
derivative, 9-1
-in, first-out, 55-37
moment of a function, 10-6
-order, exact, 11-3
-order, linear, 11-2
-order, separable, 11-2, 11-3
First-order analysis, 31-2
pulse transient, 31-6
switching transient, 31-3
First-order circuit, 31-2
impulse response, 31-8
response, 31-8
step response, 31-8
Fixed
asset, 55-34
bias, 38-5
cost, 55-31
percentage of book value, 55-21
-point iteration, 14-2
vector, 6-1 (ftn)
Flag, 44-1
Flexible
cable, 53-7
cord, 53-7
Flicker noise, 18-7, 40-8
Flip-flop, 43-1
clocks, 43-1
D, 43-3, 43-4
data, 43-3
delay, 43-3
J-K, 43-2
S-R, 43-2, 43-3
T, 43-3
toggle, 43-3
transition table, 43-4
transparent, 43-2
trigger, 43-1
Float, 26-14
Flow
energy, 2-4
through, profit, 56-2
work, 2-4
Fluorescent lamp, 52-2
Flux, 16-10
changing method, 16-22
cutting method, 16-22
displacement, 16-10
divergence of, 20-5
electric, 16-9, 20-1
electric density, 16-4
Gauss' law, 16-10
intensity, unit, 22-6
leakage, 28-1, 28-5
like charges, 20-1
line, 16-9, 16-10
line, convention, 16-9
linkage, 16-19, 24-1
linking, 16-19
magnetic, 16-15, 16-17, 22-2, 28-1
mutual, 28-1
to charge relationship, 16-9
unlike charges, 20-1
Flux density, 16-10, 47-2 (fig)
electric, 16-10, 20-2, 21-1
magnetic, 16-16
residual, 28-6
saturation, 28-6
unit, 22-6
vector, 16-10
Focus, 8-10, 46-2, 46-6
Foldover, 18-2 (ftn)
Foot board, 1-8
Footprint, 46-3 (ftn), 50-3
Forbidden band, 17-4
Force, 58-2 (ftn)
coercive, 28-6
conversions, SI, A-3
electric field, 20-2
electric vs. magnetic, 16-16
electromotive, 16-22, 24-1, 24-3
equation, Lorentz, 25-2
exchange, 23-4 (ftn)
field, vector, 16-7
Lorentz, 16-20
magnetic pole, 22-8
magnetomotive, 16-17, 16-21, 24-3
majeure, 56-4
on a charge, 16-12, 17-2
on current elements, 22-4
per unit volume, 22-5
pound, A-3
response, 32-1
strong nuclear, 58-2
work performed by, 10-4
Forced response, 30-2, 30-4, 51-2
RC circuit, 30-3
RL circuit, 30-5
Forcing
function, 11-1, 11-3, 30-3, 30-5
function, operations on, 31-6
Form
canonical product-of-sum, 41-5
canonical sum-of-product, 41-5
cis, 4-8
exponential, 4-8
factor, 27-6
maxterm, 41-5
minterm, 41-5
Newton, 14-3 (ftn)
of waves, 27-6
operational, log, 4-5
phasor, 4-8
polar, 4-8
rectangular, 4-8
trigonometric, 4-8
Format
of PE exam, 59-3
signal, 48-7
Formula
double-angle, 7-4
half-angle, 7-4
linear spring, 2-2 (ftn)
two-angle, 7-4
work, 2-2
Forward
bias, 37-4
bias region, 37-12
error correcting, 48-9
error correcting, code types, 48-9
gain, 51-4
transfer function, 51-4
Foucault current, 16-22, 28-6

Four-wire
symbology, 53-5
system nomenclature, 53-5
Fourier
analysis, 10-6, 15-14
integral, 15-15
inversion, 10-6
number, 1-9 (tbl)
transformations, 15-15
waveforms, 10-7
Fourier's theorem, 10-6
Fourier series, 10-6, 15-13
complex, 15-14
discrete, 15-17
discrete-time, A-20
exponential, 15-14
of periodic signal, 15-14, 15-15, A-15
Fourier transform, 15-13, 15-15
discrete, 15-17
discrete-time, A-21
fast, 10-8
inverse, 15-15
of aperiodic signal, 15-15, 15-18, A-16
pair, 15-15, 15-18, A-17
relationship to Laplace transform, 15-17
Fourth state of matter, 17-3
Fraction, partial, 4-6, 10-3
Fractional horsepower motor, 53-8
Fraud, 56-6
Frauds, Statute of, 56-3 (ftn)
Fraudulent act, 56-6
Free-space, 46-3
basic transmission loss, 46-5
characteristic impedance, 46-3
characteristics, 46-3
magnetization, 23-5
permeability, 16-9, 16-17, 29-6
permittivity, 29-5
transmission loss, 46-5
Freedom, degrees of, 51-14
Frequency
analysis, 10-8 (ftn), 31-2
band, 18-6, 50-3
band, communication, 50-3
break, 32-5, 32-9
complex, 31-1, 31-14
corner, 32-5, 32-9, 38-12, 51-10
cut-off, 51-10
detector, phase, 40-28
deviation, 48-4, 48-5
distribution, 12-9
-division multiplexing, 49-1, 50-3
domain, 15-16 (ftn)
domain analysis, 31-1
fundamental, 15-14
hopping, 49-2
inputs, 32-1
intermediate, 50-2
modulator constant, 48-4
multiplexing, 49-1
napier, 32-4
natural, 10-6, 31-14, 51-2, 51-9
neper, 31-1, 31-14, 32-4
of oscillation, 31-14
phase crossover, 51-12
polygon, 12-9
resonance, 30-8
response, 15-13 (ftn), 31-2, 32-1, 51-2, 51-10, 51-11, 51-20
response analysis, 32-1, 51-11
shift keying, 48-7
spectrum, 18-4
spectrum, spreading of, 18-4
threshold, 17-2
unity gain, 40-8
variable, complex, 31-15
vs. wavelength, A-25
Frequency modulation, 48-2, 48-4
channel, value, 50-2
index, 48-4
Fresnel integral, 10-8
Friction, 2-2
Fringing, 16-12
Friss transmission formula, 47-3, 47-4
Frobenius' method, 15-3
Frontiers of electrical engineering, 58-1
Froude number, 1-9 (tbl)
Fuel cell, 33-1, 33-4
Full
-wave rectifier, 37-12
-wave symmetry, 10-7 (tbl)
Fully amortized bond, 55-30 (ftn)
Function
autocorrelation, 15-16, 15-17
average value, 27-4
Bessel, A-14
circular transcendental, 7-2
complementary error, 15-11, 48-9
correlation, 15-16
delta, 15-10
Dirac, 15-10
domain, 15-18
driving, 30-3, 30-5
elementary, 15-1 (ftn)
error, 10-8, 12-6, 15-11, 48-9
error transfer, 51-4
factorial, 15-11
first moment of, 10-6
forcing, 11-1, 30-3, 30-5
gamma, 15-11, A-13
Hankel, 15-13
hazard, 12-7
Heaviside, 15-10
holomorphic, 9-1
hyperbolic, 7-4
hyperbolic, Bessel, 15-13
identity, 41-2
impulse, 15-10
integral, 10-8
kernal, 15-18
Neumann's, 15-12
nonelementary, 15-1
of an angle, 7-2
of related angle, 7-3
orthogonal, 15-5
performance, 12-8
probability, 48-9
rational, 51-3
regular, 9-1
second moment of, 10-6
space, 15-18
step, 15-9
symmetrical, 8-4
system, 51-5
transcendental, 15-1 (ftn)
transfer, 32-1, 51-3
trigonometric, 7-2
unit finite impulse, 15-10
unit impulse, 11-6, 15-10
unit step, 11-6, 15-9, 15-10
vector, 6-5
Weber's, 15-12
work, 17-2
Functional earth, 34-2
Functions
integral of combination, 10-2
of related angles, 7-3
Fundamental
frequency, 15-14
mode, 15-9
period, 15-14
theorem of calculus, 10-4
Fundamentals of engineering exam, 59-1
Fuse, 34-7
rating, 53-6
Future
growth, 35-1
worth, 55-5

G

Gage, AWG properties, A-45
Gain, 26-5, 29-13, 32-7, 32-11, 37-4, 37-6
amplifier, 38-7
antenna, 46-2
-bandwidth product, 40-2
characteristic, 51-10
closed-loop, 51-5
crossover point, 51-10, 51-12
current, 38-7
direct, 51-4
elemental antenna, 46-7
error, 51-4
feedback, 51-4, 51-5
forward, 51-4
loop, 51-5
margin, 51-12
margin Bode plot, 51-12
power, 38-7
resistance, 38-7
transistor, 38-7
voltage, 38-7
Gallium, 37-3
Gallium arsenide
energy gap, A-24
properties, A-24
Galvanometer, 36-4
Gamma function, 15-11
series approximation, A-13
values of, A-13
Gap, 17-4
energy, A-24
Gas constant, 19-2
Gate, 39-1, 40-19, 40-20
logic, 40-19
NOT, 40-21
run-off thyristor, 37-16
turn-off thyristor, 37-15
Gauge
condition, 22-9
Coulomb, 22-9
Gauss, 22-6
-Jordan elimination, 5-2
Gauss'
divergence theorem, 15-18, 20-4, 20-5
hypergeometric equation, 11-5, 15-3
Gauss' Law, 25-1, 26-6
for electrostatics, 16-10
for magnetic flux, 16-18
Gaussian
distribution, 12-6
noise, 18-7, 48-8
surface, 16-10
surface, special, 16-11
General
amplifier, 37-5
contractor, 56-4
damages, 56-7
expense, 55-32
ledger, 55-33
partner, 56-1
Purpose Interface Bus, 36-8
term, 4-10
triangle, 7-5
Generation, 17-6
Geometric
mean, 12-10
sequence, 4-11
series, 4-12
Geostationary
Earth orbit, 50-3
satellite, 50-3
Geosynchronous orbit, 50-3
Germanium, 37-2
energy gap, 37-3, A-24
properties, A-24
Giacoletto model, 38-8
Glass, 52-1 (ftn)
temperature, 52-1
GLC circuit, 30-8
Gluon, 58-2
Gold, 52-1
GPIB, 36-8
Graded system, 34-7
Gradient
cash flow, 55-9
series cash flow, 55-4
vector, 9-5
Grading gap, 6-5
Graetz number, 1-9 (tbl)

INDEX - G

Grain, A-3, A-4
  boundary, 28-6
Gram
  -equivalent weight, 26-14
  molecular weight, 19-1
Grand unified theory, 58-2 (ftn)
Graphic user interface, 35-2
Grashof number, 1-9 (tbl)
Gravitational
  constant, 1-2, 16-5
  energy, 2-3
  force, particle, 58-2
Graviton, 58-2
Gravity
  altitude, 1-2
  earth, 1-2
  work of, 2-2
Greek alphabet, 4-1
Green's theorem, 15-19
Grid, 17-2
Gross margin, 55-35
Ground
  connections, 54-4
  -fault circuit interrupter, 53-4
  fault neutralizer, 34-2 (ftn)
  resistance testing, 36-2
  symbology, 26-14
  virtual, 40-4
Grounded, 26-14
  conductor, 53-3
  effectively, 34-2
Grounding, 53-6, 54-3
  conductor, 53-3
  electrodes, 53-6
  NEC requirements, 53-6
  scheme, 34-1
  terminal, 53-6, 53-7
Group, dimensionless, 1-8, 1-9, 1-9 (tbl)
Groupings of minterms, 42-4
Growth
  exponential, 11-9, 55-4
  future, 35-1
Guard slot, 49-1
GUI, 35-2
Guide, optical, 45-1

H

Half
  -adder, 42-11
  -angle formula, 7-4
  -cell, 26-14
  -power bandwidth, 48-8
  -power point, 30-8, 51-10
  -range expansion, 15-6
  -wave antenna, 46-9
  -wave dipole, 46-9
  -wave rectifier, 37-12
  -wave symmetry, 8-4, 10-7 (tbl)
  -year convention, 55-3
Hall voltage, 37-2
Handshaking, 44-1
Hang-up state, 43-6
Hankel function, 15-13
Hard
  ferromagnetic material, 23-5
  magnetic material, 23-2, 52-2
  material, 52-2
Hardware
  arithmetic, 42-10
  interface, overview, 44-2
  programmable, 42-8, 42-9
Harmonic
  hum, 18-8
  sequence, 4-11
  series, 4-12
  terms, 10-6
Hartley, 18-5
Hazard function, 12-7
Health, company, 55-35
Heat
  capacity, 2-4
  conversions, SI, A-3
  equation, 15-7
  specific, 2-4
Heaviside step function, 15-10
Heavy weather, 54-5
Height, 3-2
  above average terrain, 50-2
  waviness, 3-4
Helix, 8-12
  pitch, 8-12
Henrys, 16-19
Heterodyning, 50-2
Hex number, 13-2 (ftn)
Hexadecimal system, 13-2
Hierarchy, synchronous digital, 50-4
High
  -band region, 40-3
  -band VHF, 50-2
  -definition television, 50-3
  -gain amplifier, 37-6
  -impedance region, 37-15
  -pass amplifier, 40-2
  -power bipolar junction transistor, 37-16
  -threshold logic, 40-19, 40-24
High-frequency
  analysis, transistor, 38-12
  BJT transistor model, 38-12
  model, FET, 39-4
  transmission, 45-2
  transmission line, 45-2
Higher-order circuit, 31-14
Histogram, 12-9
HMOS, 39-1
Hobby, 55-12 (ftn)
Hogging current, 40-19
Hold
  current, 37-15
  time, 43-2
  voltage, 37-15
Hole, 17-5, 37-2
  diffusion, 37-7
  drift, 37-7
  effective, mass, 17-4
  mobility, 37-3
  recombination, 37-7
  saturation, 37-7
Holomorphic function, 9-1
Homogeneous, 15-7
  differential equation, 11-1, 11-2
  response, 51-2
  solution, 30-2
  unit system, 1-2 (ftn)
Homogeneous linear equation, 4-7
  differential, extended power series method, 15-3
  differential, power series method, 15-1
Hopping frequency, 49-2
Horizon, 55-3
  analysis, 55-17
Horn antenna, types of, 46-2
Horsepower, 2-6, A-1
  h-parameter, 38-8
  typical values, 38-8
Hot carrier, 37-12 (ftn)
  diode, 37-12
Hour, kilowatt, 35-1
House-area network, 38-3
  h-parameter, 38-4, 38-8
  symbology, 38-9
  value, 38-8
HTL, 40-19, 40-24
Hum, 18-8
Human ear, 26-4
Hurwitz
  stability criterion, 51-13
  test, 51-13
Huygens' source, 46-1
Hybrid
  model, 29-15
  parameter, 29-15
  parameter, inverse, 29-15
  parameter model, 29-15
  -pi model, 38-8, 38-12
Hydrogen, 33-3
  -oxygen fuel cell, 33-4
Hyperbola, 8-11
  equilateral, 8-12
  rectangular, 8-12
Hyperbolic
  Bessel function, 15-13
  function, 7-4
  function, derivative, 9-2
  function, integral of, 10-2
  identity, 7-5
  radian, 7-5
Hypergeometric
  distribution, 12-5
  equation, Gauss', 11-5
Hypocycloid, 8-4
Hypothesis, test, 12-12
Hysteresis, 16-22, 23-2, 28-6
  electric, 21-3
  loop, 16-22, 23-2, 28-6

I

I^2R losses, 26-4
Ideal
  capacitor, 27-9
  core, 28-3
  current source, 26-4, 29-2
  diode, 37-9
  diode limiter circuit, 40-9
  energy source, 26-4
  independent current source, 29-2
  independent voltage source, 29-2
  inductor, 27-9
  machine, 2-7
  operational amplifier, 40-5
  operational amplifier, assumptions, 40-5
  resistor, 27-9, 29-2
  sampled signal, 18-2
  transformer, 28-3
  transformer model, 28-4
  voltage source, 26-4, 29-2
  zener voltage regulator circuit, 40-13
Idealized, 32-9
  Bode magnitude, 32-9
  Bode magnitude plot, 32-9
  Bode phase diagram, 32-11
  Bode phase plot, 32-11
Idempotence laws, 13-5
Idempotent law, sets, 12-1
Identity
  function, 41-2
  hyperbolic, 7-5
  law, sets, 12-1
  logarithm, 4-5
  matrix, 5-1
  trigonometric, 7-2, 7-3, 7-4, 7-5
IEEE-488, 36-8
Illumination efficiency, 47-3
Imaginary number, 4-1
Impedance, 26-9 (ftn), 27-3
  addition, 27-10
  angle, 27-3
  characteristic, 45-2, 46-3
  complex, 27-3
  equation, 27-3
  input, 37-6, 45-2
  intrinsic, 47-1 (ftn)
  locus circle, 45-4
  matching, 28-4, 45-6
  model, 29-14
  model parameters, 29-14
  of elements in a series, A-29
  of elements in parallel, A-29
  open-circuit parameters, 29-14
  output, 37-6
  phasor form, 27-3
  polar form, 27-3
  primary effective, 28-4
  radiation, 46-2 (ftn)
  reciprocal, 27-3
  rectangular form, 27-3
  reflected, 28-4

INDEX - I

s domain, 31-15
transmission line, 45-2
triangle, 27-3
Implementation, control system, 51-1
Implicit differentiation, 9-4
Impulse
first-order circuit, 31-8
function, 11-6, 15-10
noise, 18-8
response, 30-1
Income, 55-34
bracket, 55-25
net, 55-34
tax, 55-24
Incomplete gamma function, 15-11 (ftn)
Incompletely specified, 42-3
Inconsistent system, 4-7
Incremental
analysis, 55-17
cost, 55-31, 55-32
Indefinite integral, 10-1, 10-1 (ftn), 10-2, 10-4, A-10
Independent
current source, ideal, 29-2
event, 12-3
source, 26-4, 27-3, 29-7, 38-10
variable, 4-2
voltage source, 29-2
voltage source, ideal, 29-2
Index, 48-3
of modulation, 48-3, 48-4
of refraction, 46-4
roughness height, 3-4
Indicial equation, 15-3
Indirect
manufacturing expense, 55-32
material and labor cost, 55-32
Indium, 37-3
Induced
dipole moment, 21-1
electromotance, 16-22
Induced voltage
flux-changing method, 16-22
flux-cutting method, 16-22
Inductance, 16-18, 29-6
form, 29-6
frequency domain, 29-4
leakage, 28-5, 28-6
magnetic, 16-19
magnetic equations, 16-16
mutual, 16-19, 28-2, 29-7
reciprocal, 16-18
s domain, 31-16, 31-17
self-, 16-19, 24-2, 28-2
time domain, 29-4
Induction, electromagnetic, 16-22
Inductive
circuit, 27-7
reactance, 27-9
transduction, 19-3
Inductor, 16-19, 27-9
energy storage, 29-6
ideal, 27-9
in parallel, 29-6
in series, 29-6
initial current source model, 31-17
initial voltage source model, 31-17
perfect, 27-9
two-source model, 31-17
Inductor equations, 29-6
characteristics, 27-3
charging, 30-6
discharging, 30-6
initial current source model, 31-17
initial voltage source model, 31-17
two-source model, 31-17
Industrial exemption, 59-1
Infant mortality, 12-7
Infinite series, 4-12, 4-12 (ftn), 55-7
Infinitely long conductor
in magnetic field, 22-4
Inflation, 55-39
Inflection point, 8-2, 9-2
Information, 18-1
entropy, 18-4
management subsystem, 35-2
processing unit, 44-1
set, 18-1
signal, 18-4, 48-5
source, 18-1
Infrared Data Association, 36-8
Initial
condition, 15-7
condition response, 51-2
current source model, of inductor, 31-17
magnetization curve, 28-6
permeability, 28-6
value, 10-4, 11-1, 51-7
value theorem, 15-5
Initial voltage source model
of capacitor, 31-17
of inductor, 31-17
Injection, current, 37-7
Injured party, 56-6
Input
DC, 51-8
impedance, 37-6, 45-2
port, 29-14
signal, 48-5, 51-3
step, 51-8
Insensitivity, 12-13
Instability point, 51-12
Installations, underground, 54-5
Instantaneous
AC power, 27-11
resistance, 38-1, 39-1 (ftn)
voltage, 27-2
Instrument, 36-1
accuracy, 36-1
digital, 36-8
electrodynamometer, 36-6
moving-iron, 36-7
power, 36-7
precision, 36-1
resolution, 36-1
transformer, 36-2
Instrumentation, 36-1
electronic, 36-8
Insulated gate bipolar transistor, 37-16, 39-1
Insulation testing, 36-8
Insulator, 16-2, 16-3, 37-2, 52-2
common, 16-3
conductivity, value, 52-2
Insurance, 56-7, 56-8
Intangible property, 55-20 (ftn)
Integral, 10-2
closed path, 15-19, 15-20
contour, 15-19
controller, 51-14
convolution, 11-6, 15-16
cosine function, 10-8
definite, 10-1 (ftn), 10-4
double, 10-3
exponential function, 10-8
Fourier, 15-15
Fresnel, 10-8
functions, 10-8
indefinite, 10-1, 10-4, A-10
line, 15-19
of combination of functions, 10-2
of hyperbolic function, 10-2
of transcendental function, 10-1
probability, 15-11
sine function, 10-8
surface, 15-19
transform, 15-18
triple, 10-3
Integrand, 10-1
Integrated
circuit, 37-2, 58-1
services digital networks, 50-2 (ftn)
Integrating factor, 11-2
Integration, 10-1
by parts, 10-2
by separation of terms, 10-3
constant of, 10-1, 11-1
large-scale, 37-2
medium-scale, 37-2
small-scale, 37-2
ultra large-scale, 37-2
unit impulse, 31-6
unit step, 31-6
Integrator, 40-7
Intelligence, artificial, 58-1
Intensity
electric field, 16-4, 16-7
of magnetization, 16-16, 22-5
of sound, 19-3
Inter-exchange carriers, 50-2
Interbase resistance, 38-8
Intercept form, 8-5
Interest
compound, 55-11
rate, effective, 55-5, 55-28
rate, nominal, 55-28
simple, 55-11
Interface
application, 44-1
dialogue, 44-1
problem, 44-1
Interference, 18-7, 46-3, 46-4 (ftn)
electromagnetic, 50-4
intersymbol, 48-7
noise, 18-8, 40-8
Intermediate
frequency, 50-2
vector boson, 58-2
Intern engineer, 59-1 (ftn)
Internal
energy, 2-4
rate of return, 55-12
International
Annealed Copper Standard, 26-3
Radio Regulations, 46-2 (ftn)
Interpolating polynomial, Newton's, 14-3
Interpolation, nonlinear, 14-2, 14-3
Intersecting lines, 3-1
Intersection
angle, 8-6, 8-8
line, 8-6
set, 12-1
Intersymbol interference, 48-7
Interval, keying, 48-7
Intervals, bit, 48-6
Intramodal dispersion, 50-4
Intrinsic, 37-2
conduction, 17-5
impedance, 47-1 (ftn)
semiconductor, 17-5, 37-2
Inventory, 55-37
supply, 55-35
turnover, 55-35
value, 55-37
Inverse
Fourier transform, 15-15
function, 7-4
hybrid parameters, 29-15
Laplace transform, 11-6
matrix, 5-6
-square law, 20-2
transmission line parameter, 29-16
transmission parameters, 29-16
Inversion, 48-6
Fourier, 10-6
Inverting
amplifier, 37-4 (ftn), 40-6
ladder-type DA converter, 44-5
terminal, 40-3
-type D/A converter, 44-4, 44-5
Investment
credit, 55-25
external, 55-13
return on, 55-11, 55-14
risk-free, 55-11
IR drop, 26-3
IrDA, 36-8
Iron losses, 28-3, 28-6
Irrational number, 4-1
Irreducible factor, 55-2

INDEX - I

Irregular area, 8-1
Irregular boundary, 8-1
Irrotational field, 22-8
IRS rules, 55-21
Isentropic exponent, 19-2
ISI, 48-7
Isometric view, 3-3
Isometry, 8-3
Isotropic, 21-2, 23-1, 46-2
  power, 46-9
Iteration, fixed-point, 14-2

J

J-K flip-flop, 43-2, 43-3
  next-state equation, 43-3
Job cost accounting, 55-36
Johnson noise, 18-7, 36-1, 36-2, 40-8
Joint probability, 12-3
Joule, 1-5, A-3
Joule's
  constant, 2-1
  equivalent, 2-1
  law, 2-5
  law of hearing effect, 26-4
Journal, 55-33
Judgment factor, 55-2
Junction
  characteristics, *pn*, 37-4
  linearly graded, 37-7 (ftn)
  pn, 17-7
  step-graded, 37-7
Junction field-effect transistor, 39-1, 39-2
  avalanche region, 39-2
  biasing, 39-3
  breakdown region, 39-2
  ohmic region, 39-2
  pinchoff, 39-2
  pinchoff region, 39-2
  saturation current, 39-2
  switch, 40-18
  triode region, 39-2

K

k-space, 31-1
Kalman's theorem, 51-15 (ftn)
Karnaugh
  map, 42-3
  map, groupings, 42-3
  map, minimization, 42-3
  truth table, 42-2
  -Veitch map, 42-5
  -Veitch minterms, 42-5
Keep-alive current, 37-13, 40-13 (ftn)
Kernal, 15-18
Keying, 48-5 (ftn)
  interval, 48-7
Kilocalorie, 2-1
Kilogram molecular weight, 19-1 (ftn)
Kilowatt hour, 35-1
Kinetic energy, 2-3
Kips, A-3
Kirchhoff's
  current law, 26-6, 29-11
  voltage law, 26-6, 29-11
Knudsen number, 1-9 (tbl)

L

Labor
  direct, 55-32
  variance, 55-36
Laborer's lien, 56-5
Ladder-type D/A converter, 44-3
  cells, 44-3
Lag, 27-7
Lagging, 27-11, 27-12
  circuit, 27-7
Lagrangian interpolating polynomial, 14-2
Lamination, 28-6
Laplace's equation, 15-7, 20-6
Laplace analysis
  circuit transformation method, 31-16
  classical method, 31-16
  first-order systems, 31-16
  second-order systems, 31-16
Laplace transform, 11-5, 15-17, A-11, A-18
  analysis, 31-15
  bilateral, 15-17
  final value, 15-5
  initial value, 15-5
  of a derivative, 11-6
  properties, A-18
  region of convergence, 15-17
  relationship to Fourier transform, 15-17
  table, A-11
  unilateral, 15-17
Laplacian operator, 20-6
Large-scale integration, 37-2
Large-signal
  analysis, 38-2
  model, 38-6, 38-10
Laser, 50-4
Last-in, first out, 55-37
Latch, 43-1
Lattice constant, 17-5
Latus rectum, 8-10
Law
  AC Ohm's, 27-11
  Ampère's, 16-20, 22-8, 25-1
  associative, addition, 4-3
  associative, multiplication, 4-3
  Biot-Savart, 16-15, 22-3
  Boolean algebra, 13-5
  commutative, addition, 4-3
  commutative, multiplication, 4-3
  Coulomb's, 16-6, 20-4, 25-1
  distributive, 4-3
  extended Ohm's, 27-3
  Faraday's, 16-22, 24-2, 25-1
  Gauss', 16-10, 16-18, 25-1
  inverse-square, 20-2
  Joule's, 2-5, 26-4
  Kirchhoff's, 26-6, 26-7
  Kirchhoff's current, 29-11
  Kirchhoff's voltage, 29-11
  Lenz's, 16-22, 23-4 (ftn), 24-2
  Lorentz force, 22-5
  mass action, 37-2
  Moore's, 58-1
  of conservation of energy, 2-1
  of cooling, Newton's, 11-10
  of cosines, 7-5, 7-6
  of mass action, 17-6
  of sines, 7-5, 7-6
  of tangents, 7-5
  Ohm's, 26-3
  sets, 12-1
  Snell's, 46-4
Lead, 27-7
Leading, 27-11, 27-12
  circuit, 27-7
  phase angle difference, 27-7
Leakage
  flux, 28-1, 28-5
  inductance, 28-5, 28-6
  method, 36-8
Learning curve, 55-42
Least
  significant bit, 13-1, 13-3
  significant digit, 13-1
Ledger account, 55-33
Left-hand circularly polarized, 46-2
Legendre equation, 11-5, 15-1
Legendre polynomial, 15-5 (ftn)
Length, 3-2
  arc, by integration, 10-5
  conversions, SI, A-3, A-4
Lenz's law, 16-22, 23-4 (ftn), 24-2
Letter of agreement, 56-3
Level
  assertion, 41-1, 41-2
  confidence, 12-12
  of sound, 26-4 (ftn)
  Z, 40-25
Leverage, 55-35
Lewis number, 1-9 (tbl)
Liability
  account, 55-34
  current, 55-34
  in tort, 56-6, 56-7
Licensing, 59-1
Lien
  mechanic's, 56-5
  perfecting, 56-5
Life
  analysis, economical, 55-4
  -cycle cost, 55-15, 55-20
  economic, 55-19, 55-20
  service, 55-20
  useful, 55-20
Light
  diffraction, 46-4
  -emitting diode, 37-14, 50-4
  pipe, 46-4
  polarization, 46-3
  quantum theory, 58-2
  reflection, 46-4
  refraction, 46-4
  weather, 54-5
Lightning strike, 53-6
Like charges, 20-1
Limit, 4-9
  confidence, 12-12
  design, 53-1, 54-2
  safety, 53-1, 54-2
  Shannon, 50-1 (ftn)
  simplifying, 4-10
Limited
  liability company, 56-2
  liability partnership, 56-2
  partner, 56-2
  partnership, 56-2
Limiter, 40-9
  circuit, 40-9
Limiting value, 4-9
Line, 7-1 (ftn), 8-3
  charge, 20-4
  coaxial, 45-1
  high-frequency, 45-2
  integral, 15-19
  intersection, 8-6
  normal vector, 9-6
  -of-site, 46-3
  of action, 6-1
  of electric force, 16-8
  of force, maxwell, 22-6
  perpendicular, 3-1
  slotted, 45-7
  straight, 8-4
  trunk, 50-2
Lineal measurement, 1-8
Linear, 15-7, 21-2, 23-1
  analysis, transistor, 38-9
  differential equation, 11-1
  element, 26-3
  element, passive, 40-9
  equation, 4-7
  equations, simultaneous, 4-7
  first-order, 11-2
  network, 29-3
  polarization, 46-2
  region, 38-2
  second-order, 11-3
  source, 26-3
  spring, 2-2
  spring, formula, 2-2 (ftn)
  time invariant, 15-11 (ftn)
Linear circuit, 26-3
  analysis, 29-1
  element parameters, 29-4
  elements, 29-3
  parameters, frequency domain, 29-3
  parameters, time domain, 29-3
Linearity, 11-6
Linearly graded junction, 37-7 (ftn)

INDEX - L

Line carrier power, 34-7
Lines
  angle between, 7-1
  intersecting, 3-1
  of flux, 16-9
  of magnetic flux, 16-15
  of magnetic force, 22-2
Link
  communication, 47-1
  magnetic, 16-19
Linkage, 28-1
  of flux, 16-19, 24-1
Liquidated damages, 56-7
Liquidity, 55-35
Liter, 1-6, A-1, A-2
Load
  nonlinear, 53-6
  resistor, 37-4
Load line, 37-6
  diode, 37-11
  dynamic, 37-11
  static, 37-11
  transistor, 38-6
Loading, 54-5
Loan, 55-39
  repayment, 55-4
Local
  access transport area, 50-2
  channel, 50-2
Lockout state, 43-6
Locus
  circle, 45-4
  of points, 8-2
Logarithm, 4-5
  identity, 4-5
  Naperian, 31-1
Logic
  array, programmable, 42-8
  CMOS, 40-26, 40-27
  collector, 40-25
  combination, 41-2
  constants, 41-2
  current-mode, 40-26
  delay time, 40-23
  diode-transistor, 40-24
  dynamic, 40-21
  element, sequential, 43-1
  emitter-coupled, 40-26
  fall time, 40-23
  families, 40-19
  flip-flop control equation, 43-1
  gate, 40-19, 40-20
  minimization principle, 42-2
  MOS, 40-26, 40-27
  negative, 40-21, 41-1
  network design, 42-1
  next-state equation, 43-1
  operation, 42-1
  operators, 41-2
  output equation, 43-1
  positive, 40-21, 41-1
  propagation delay, 40-23
  pulse, 40-21
  resistor-transistor, 40-24
  rise time, 40-23
  sequential, 41-2
  state diagram, 43-1
  state table, 43-1
  storage time, 40-23
  timing, 43-1
  transistor-transistor, 40-25
  transition diagram, 43-1
  transition table, 43-1
  tri-state, 40-25
  TTL, 40-25
  types, 40-19
  types, comparison of, 40-24
  variable, 41-2
  wired, 40-25
Logic circuit, 40-19
  delay, 40-23
  fan-out, 40-20
Logic family
  data, 40-24
  unit load, 40-21
  unit sink current, 40-21
  unit source current, 40-21
Logical adjacency, 42-2
Logically equivalent, 41-3
Long-term liability, 55-35
Longitudinal wave, 19-1
Loop
  analysis, 29-12
  gain, 51-5
  hysteresis, 16-22, 23-2, 28-6
  of wire, 22-6
  phase-locked, 40-28
  transfer function, 51-5
  vs. node, 26-13
Loop current
  analysis, 29-12
  method, 26-12
  method, Maxwell, 29-12
Lorentz
  force, equation, 16-20, 25-2
  force, law, 22-5
  transformation, 16-2 (ftn)
Loss
  excitation, 28-6
  iron, 28-6
  of charge method, 36-8
  path, 46-5
Losses
  antenna, 47-2
  core, 28-6
  ohmic, 47-2
  scatter, 47-2
  space, 47-2
  transformer core, 28-3
Lossy dielectric, 21-4
Low
  -band region, 40-3
  -band VHF, 50-2
  Earth orbit, 50-3
  -frequency noise, 40-8
  -pass amplifier, 40-2
  -pass filter, 40-7
  -voltage protection, 54-5
  -voltage release, 54-5
Lower bound, 10-4
Lower sideband, 18-3
L'Hôpital's rule, 4-10
LSB, 13-1
Lump-sum fee, 56-5
Lumped elements, 29-1

M

M-ary, 48-8
Mach number, 1-9 (tbl)
Machine
  ideal, 2-7
  Mealy, 43-8
  Moore, 43-8
  real, 2-7
  state, 43-8
Maclaurin series, 9-8
Magnet
  bar, 23-4
  permanent, 23-5
  theoretical views, 23-3
Magnetic
  atom, 23-3
  cgs units, 22-5
  charge, 22-5
  circuit, 16-21
  Coulomb's law, 22-8
  coupling, 28-1
  Curie temperature, 23-4
  dipole, 22-6
  domain, 16-4, 23-4
  -electric circuit analogy, 16-21
  exchange force, 23-4 (ftn)
  field strength, 16-17, 16-18
  flux, density, 16-17
  force, 16-15
  force, direction, 22-4
  grain boundary, 28-6
  hysteresis, 16-22, 28-6
  hysteresis, loop, 28-6
  inductance, 16-18, 16-19, 28-1
  lines of force, 16-17
  monopole, 16-14 (ftn)
  permeability, 16-16
  phenomena, 16-4
  pole, 16-14
  pole, strength, 22-6
  SI units, 22-5, 22-6
  sink, 16-15
  source, 16-15
  spectrum, 16-17
  spin arrangement, 16-5
  superposition, 16-16
  susceptibility, 16-17
  test charge, 22-5
  unit-pole, 22-6
  units, 22-6
  vector potential, 22-9
  vs. electric circuit, 25-2
  vs. electric equations, 25-2, A-26
  vs. electric force, 16-16
  work, 22-8
Magnetic dipole, 16-15
  field, 16-15
  moment, 22-6
Magnetic field, 16-15, 16-16, 22-1, 23-4
  conductor in, 16-20
  configurations, 16-18
  direction, 16-16
  energy in, 16-19
  equations, 16-16
  -induced voltage, 16-22
  of positive current, 22-2
  strength, 16-17, 22-8, 23-5
Magnetic flux, 16-15, 16-17, 22-2, 28-1, 28-2
  density, 16-16, 22-2
  Gauss' law, 16-18
  linked, 22-1
  links, 16-19
  method, time-varying, 24-1
  solenoid, 16-19
Magnetic material, 16-3, 16-4, 52-2
  hard, 23-2
  properties, 23-1
  soft, 23-2
  temperature dependence, 23-4
Magnetic moment, 22-6, 22-7
  comparison, 22-6
  per unit volume, 23-1
  torque, 22-6
Magnetic pole, 22-6
Magnetic potential, 22-8, 22-9, 24-3
  energy, 22-8
  gradient, 22-8
Magnetically coupled circuit, 28-2
Magnetism, 16-2, 24-3
  antiferro-, 16-4
  dia-, 16-4
  ferri-, 16-4
  ferro-, 16-4
  para-, 16-4
  traditional, 22-5
Magnetization, 16-16 (ftn), 23-1
  curve, 16-4, 23-5, 28-6
  intensity of, 16-16, 22-5
  of free space, 23-5
  unit, 22-6
Magnetizing current, 28-3, 28-5
Magnetokinetics, 16-14, 16-15, 16-20
Magnetomotance, 22-9, 24-3
Magnetomotive force, 16-17, 16-21, 24-3
Magneton, 16-14
Magnetostatic field, 23-1
Magnetostatics, 16-14
Magnitude
  plot, 32-5
  sinusoid, 27-7

INDEX - M

Major axis, 8-10
Majority carrier, 17-5, 17-6, 37-3
Management system energy, 35-2
Manchester, 48-7
  signal, 48-7
Mantissa, 4-5
Manufacturing cost, 55-32
Mapping, 48-5 (ftn)
  function, 8-3
Margin
  gain, 51-12
  phase, 51-12
Mark, 48-7
Market value, 55-23
Mask, 50-4
Mass, 1-1
  action law, 17-6, 37-2
  conversion, SI, A-3, A-4
  effective, 17-4, A-24
  effective electron, 16-12 (ftn)
  energy, 2-1
Matched condition, 45-4
Matching
  double-stub, 45-6
  impedance, 28-4, 45-6
  single-stub, 45-6
Material
  anisotropic, 6-1
  breach, 56-6
  conductivity, 26-2
  diamagnetic, 23-4
  direct, 55-32
  dispersion, 50-4
  magnetic, 16-3, 52-2
  magnetization of, 23-5
  paramagnetic, 23-4
  properties, isotropic, 23-1
  properties, linear, 23-1
  properties of elements, A-43
  properties related to periodic table, A-43
  semiconductor, 37-2
  thermal coefficient, 26-2
  variance, 55-36
Materialman's lien, 56-5
Mathematical analysis of circuits, 29-1
Mathematics, advanced engineering, 15-1
Matrix, 5-1
  algebra, 5-4
  inverse, 5-6
  multiplication, 5-4
  switching, 42-8
  transformation, 6-3
  type, 5-1
Matter, fourth state, 17-3
Matthiessen's rule, 58-1 (ftn)
Maturity date, 55-30
Maximum
  energy transfer, 26-5
  permeability, 28-6
  point, 9-2
  power transfer, 26-5, 27-13, 29-10
  voltage, 27-2
Maxterm, 41-4, 41-5
  form, 41-5
Maxwell, 22-6
  loop current method, 26-12, 29-12
Maxwell's equations, 16-2, 25-1, 25-2, 26-6
Mealy machine, 43-8
Mean, 12-10
  arithmetic, 12-10
  geometric, 12-10
  time to failure, 12-7
Measurand, 19-1, 19-2
Measurement, 36-1
Mechanic's lien, 56-5
Mechanical/acoustical/electrical elements, 19-4
Mechanics, tunneling, 37-13
Media transmission, 47-1, 50-1
Median, 12-10
Medium
  Earth orbit, 50-3
  -scale integration, 37-2
  weather, 54-5
Megger, 36-8
Membership interest, 56-3
Memory, read-only, 42-8 (ftn)
Mensuration, 8-1, A-7
  of areas, A-3, A-5
  of volumes, A-9
  three-dimensional, A-9
  two-dimensional, A-7
Mesh analysis, 26-12, 29-12
Message signal, 18-2
Metal, 37-1, 52-1
Metal-nitride-oxide-silicon, 39-1
Metal-oxide semiconductor, 40-19
Metal-oxide semiconductor field-effect transistor, 37-4, 39-1, 39-4
  biasing, 39-5, 39-6
  complimentary, 39-1
  depletion device, 39-4
  enhancement device, 39-4
  operating regions, 39-4
  symbology, 39-4, 39-5
Metalloid, 52-1
Metastability, 43-2
Meter
  movement, 36-4
  volt-ohm-milliamp, 36-8
  watt-hour, 36-7
Metering transformer, 34-4
Method
  bisection, 4-4, 14-1
  burial, 54-6
  double-dabble, 13-1 (ftn)
  expansion, 13-1
  extended power series, 15-3
  Frobenius', 15-3 (ftn)
  leakage, 36-8
  loss of charge, 36-8
  Newton's, 14-2
  numerical, 4-4, 14-1
  of circuit reduction, 29-8
  of loop-current, 26-12
  of node-voltage, 26-13
  of undetermined coefficients, 4-6, 11-4
  parallelogram, 6-3
  polygon, 6-3
  power series, 15-1
  product, 15-7
  projection, 55-10
  Quine-McClusky, 42-3
  regula falsi, 14-2 (ftn)
  remainder, 13-1
  tabular, 42-3
  two-point, 29-7
  Winfrey, 55-16 (ftn)
Metric system, 1-4
MH curve, 23-2 (ftn)
mho, 26-3
Micrometer, 50-3
Microphone, 29-7
Microstrip antenna, 46-2
Microwave, 34-7, 50-1 (ftn)
  channel, 50-1
  system, 50-1
Mid-band region, 40-3
MIDI, 44-1
Mil, A-3, A-4
Mile, statute, A-1, A-2
Miller's
  theorem, 29-10
  transformation, 29-10
Millman's theorem, 26-5
Minimization principle, 42-2
Minimum
  attractive rate of return, 55-14, 55-16
  point, 9-2
Minor
  axis, 8-10
  of entry, 5-2
Minority carrier, 17-6, 37-3
Minterm, 41-4, 41-5
  adjacent, 42-3
  form, 41-5
  grouping, 42-3
Mixed
  -mode operation, 42-2
  triple product, 6-4
Mixer balance, 40-28
Mixing problem, 11-8
Mks system, 1-4
MNOS, 39-1
Mobile radio, 50-3
Mobility, 17-2
  electron, 37-3
  hole, 37-3
Modbus, 35-2
Mode, 12-10
  asynchronous transfer, 44-7
  fundamental, 15-9
Model
  AC incremental, 38-8
  battery, 33-1
  diode, 37-11
  diode piecewise linear, 37-11
  high-frequency FET, 39-4
  hybrid-pi, 38-8
  large-signal, 38-6, 38-10
  piecewise linear, 37-4
  small-signal, 38-8, 38-10, 39-3, 40-3
Model transformer, 28-2
  real, 28-5
Modem, 48-8 (ftn)
Mode of control, 51-14
Modified Accelerated Cost Recovery System, 55-21, 55-22
Modular devices standard, 37-2
Modulation, 18-8, 48-2
  amplitude, 48-2
  analog, 48-2
  angle, 48-4
  binary, 48-5
  digital, 48-5
  factor, 48-3, 48-4
  frequency, 48-2, 48-4
  index, 48-3
  multilevel, 48-5
  phase, 48-2, 48-4
  pulse amplitude, 48-5, 48-7
  pulse code, 48-7
  pulse code keyed, 48-5
  quadrature amplitude, 48-8
  spread-spectrum, 49-2
  types, 48-2
  wave continuous, 48-3
Modulator
  -demodulator, 48-8
  optical, 50-3
Modulo counter, 43-4
Modulus, 4-8
  bulk, 19-2
  elastic, 19-2
Molar specific heat, 2-4
Mole, Faraday unit, 16-2
Molecular current, 16-15
Molecule
  nonpolar, 21-3
  polar, 21-3
Moment
  dipole, 16-14, 20-3, 21-1
  electron, 23-3
  magnetic, 22-6
  of a function, first, 10-6
  of a function, second, 10-6
Monomial, 4-3
Monopole, 16-14 (ftn), 22-5
Monostable
  multivibrator, 40-28
  oscillator, 40-19
Moore's law, 58-1
Moore machine, 43-8
Mortality, infant, 12-7

INDEX - M

MOS, 40-19
-controlled thyristor, 39-1
logic, 40-26, 40-27
Most
significant bit, 13-1, 13-4
significant digit, 13-1
Motor
branch circuit requirements, 53-8
fractional horsepower, 53-8
overload protection, 53-8
requirements, 53-8
stepped, 51-13
Moving
averages, 55-42
charge, force on, 22-3
charge, magnetic field, 16-14
-iron instrument, 36-6, 36-7
-iron mechanism, 36-6
MSB, 13-1
MTTF, 12-7
Multilevel modulation, 48-5
Multimeter, 36-8
digital, 36-8
Multiple
access, code-division, 49-2
access, time-division, 49-2
-battery bias, 38-5
hypergeometric distribution, 12-5
Multiplexer, 42-10, 42-11
Multiplexing, 18-8, 48-2, 49-1
frequency, 49-1
guard slot, 49-1
optical frequency division , 50-3
polarization, 49-2
quadrature, 48-4
space, 49-2
spread-spectrum, 49-2
time, 49-1
wavelength-division, 50-3
Multiplication
matrix, 5-4
vector, 6-3
Multiplying-type D/A converter, 44-5
Multivibrator, 40-27
Musical instrument digital interface, 44-1
Mutual
flux, 28-1
inductance, 16-19, 28-2, 29-7
reactance, 28-2, 28-5

N

n
-channel MOSFET, 39-1
-tuple, 41-1
-type, 17-5, 37-3
Named equations, 15-9
Naperian logarithm, 31-1
Napier, 31-1
frequency, 32-4
Napierian logarithm, 4-5
Narrowband, 48-5
Nat, 18-5
National Bureau of Standards, 36-1
National Council of Examiners for Engineering and Surveying, 59-1
National Electrical Code, 53-1, 53-3
ambient environment correction, 53-7
branch circuit, 53-3
branch circuit calculation, 53-5
conductor ampacities, 53-7
conductor protection, 53-6
coverage, 53-2
definitions, 53-2
derating, 53-7
dwelling unit requirements, 53-6
electrical installations, 53-2
equipment grounding, 53-6
feeder, 53-4
feeder calculation, 53-4
feeder requirements, 53-8
fine print note, 53-2
flexible cable, 53-7
flexible cord, 53-7
general use equipment, 53-7
grounding, 53-6
grounding requirements, 53-6
illumination, 53-3
introduction, 53-2
lighting calculation, 53-6
motor branch circuit, 53-8
motor overload, 53-8
motor requirements, 53-6, 53-8
noncontinuous duty, 53-4
overcurrent protection, 53-6
overcurrent protective device, 53-4
permissive rule, 53-2
receptacle, 53-4
service calculation, 53-5
small appliance requirement, 53-6
standard circuit breaker rating, 53-6
standard fuse rating, 53-6
table 310-16, A-47
table 8, A-48
table 9, A-49
temperature correction, 53-7
typical distribution system, 53-2
voltage drop, 53-5
wiring and protection, 53-3
wiring method, 53-7
wye-connection terminology, 53-3
National Electrical Safety Code, 54-1, 54-2
areas of concern, 54-2
National Fire Protection Association, 53-1
National Institute of Standards and Technology, 36-1
National Society of Professional Engineers, 57-1, 57-3
Natural
frequency, 10-6, 31-14, 51-2, 51-9, 51-10
logarithm, 4-5
noise, 18-7
Natural response, 30-2, 30-4, 32-1, 51-2
RC circuit, 30-2
RL circuit, 30-4
NCEES exam administration services, 59-2
NEC/NESC comparison, 54-2, 54-3
Negative
area, 27-4
energy, 15-10 (ftn)
exponential distribution, 12-7
feedback, 51-4
feedback, benefits, 51-5
logic, 40-21, 41-1
number, computer representation, 13-4
resistance region, 37-15
Negligence, 56-6
Neper, 31-1
frequency, 31-1, 31-14, 32-4
Network
active, 40-9
analysis, 26-7
autonomous, 43-4
function, 32-2
general synchronous sequential, 43-1
integrated service digital, 50-2 (ftn)
linear, 29-3
personal communication, 50-3
sequential, 43-1
synchronous optical, 50-4
system, 34-2
theory, 29-2
two-port, 29-14
vs. circuit, 29-2
Neumann's function, 15-12
Newton's
interpolating polynomial, 14-3
law of cooling, 11-10
method, 14-2
universal constant, 16-5
Newton form, 14-3 (ftn)
Next-state equation, 43-2
Nickel, 52-1
Nines complement, 13-3
NMOS, 39-1, 40-19
No-signal point, 37-4
Node, 8-2, 26-6
analysis, 29-12
principal, 26-6
simple, 26-6
voltage analysis, 29-12
voltage method, 26-13
vs. loop method, 26-13
Noise, 18-7, 36-1
$1/f$, 18-7
amplifier, 40-8
artificial, 18-7, 18-8
contact, 18-7
current, 18-7
defect, 18-7
environmental, 18-8
excess, 18-7
figure, 18-7, 36-2
flicker, 18-7, 40-8
Gaussian, 18-7, 48-8
impulse, 18-8
interference, 18-8, 40-8
Johnson, 18-7, 36-1, 36-2, 40-8
low-frequency, 40-8
natural, 18-7
power, 40-8, 46-2
quantization, 48-5
quantumization, 48-5
resistance, 40-8
shot, 18-7, 40-8
temperature, 18-7, 36-1, 40-8
thermal, 18-7, 40-8
voltage, 36-2
white, 18-7, 48-8
Nominal, 53-3
damages, 56-7
interest rate, 55-28
voltage, 53-3
Non-SI units, 1-7
Noncontinuous duty, 53-4
Nonelectrochemical reaction, 26-14
Nonelementary function, 15-1
Nonhomogeneous
differential equation, 11-1, 11-3
differential equation, general method, 15-3
linear equation, 4-7
solution, 30-3
Noninverting terminal, 40-3
Nonlinear
circuit, 27-11, 37-2, 40-3, 40-9
device, 27-11
element, passive, 40-9
interpolation, 14-2, 14-3
load, 53-6
operation, 37-4
Nonmetal, 52-1
Nonparametric equation, 4-3
Nonpolar molecule, 21-3
Nonquantifiable factor, 55-2
Nonquantitative factor, 55-2
Nonretentive material, 52-2
Nonreturn-to-zero, 48-6
Nonsingular matrix, 5-5
Non-unit system, 34-7
Norm, 15-5
Normal
curve, A-12
curve, standard, A-12
distribution, 12-6
form, 8-5
line vector, 9-6
magnetization curve, 28-6
toward the, 46-4
vector, 8-6
view, 3-1
Normalized
quantity, 45-2 (ftn)
reactance, 45-4 (ftn)
resistance, 45-5
Normalizing factor, 32-8 (ftn), 32-11
North pole, 16-14
Norton's theorem, 29-9

Norton equivalent
circuit, 26-12, 29-9, 29-10
resistance, 26-5, 26-12
NOT
gate, 40-21
operation, 13-5
Notation
effective, 27-8
effective value phasor, 27-8
of sinks, 29-2
of sources, 29-2
phasor, 27-8
Notch amplifier, 40-2
npn transistor, 38-2, 38-5, 38-8
Nuclear
force, 58-2
fuel, 52-1
Null
event, 12-3
matrix, 5-1
region, 8-3
set, 12-1
Number
base-10, 13-1
base-b, 13-1
binary, 13-1
complement, 13-3
complex, 4-1, 4-8, 27-9
decimal, 13-1
dimensionless, 1-9 (tbl)
direction, 8-5
hex, 13-2 (ftn)
imaginary, 4-1
irrational, 4-1
negative, computer, 13-4
period, 55-9
rational, 4-1
real, 4-1
system, positional, 13-1
type, 4-1
Numbering system, 4-1
Numerical
event, 12-4
methods, 4-4, 14-1
Nusselt number, 1-9 (tbl)
Nyquist
analysis, 51-13
criterion, 48-7
interval, 18-2
rate, 18-2
stability plot, 51-13
Nyquist's stability criterion, 51-13

O

Oblique
axis, 3-3
perspective, 3-3
triangle, 7-5
view, 3-2
Observability, 51-15
Obtuse angle, 7-1
Octagon, relationships, A-7, A-8
Octal
number system, 13-2
system, 13-2
Octave, 51-10
Odd
periodic extension, 15-6
symmetry, 8-4, 10-7 (tbl)
Oersted, 22-6
Off
condition, 37-4
region, 37-15
region, diode, 37-12
state, 37-4
Offer, 56-3
Offset, 27-2
current, 40-8
voltage, 40-8
Ohm, international, A-1, A-2
Ohm's law, 16-13, 26-3, 27-11
extended, 27-3
for AC, 27-11
for magnetic circuits, 16-21
point form, 16-14, 17-3
Ohmic
losses, 47-2
power, 46-7, 46-9
region, 39-2
resistance, 46-2, 46-8, 46-9
resistance, elemental antenna, 46-6
Omnidirectional antenna, 46-2
On-off controller, 51-14
On condition, 37-4
On state, 37-4
One-tail limit, 12-12
Ones complement, 13-4
Op amp, 40-3
Open-loop transfer function, 51-5
Open circuit, 26-2
impedance parameters, 29-14
Open collector TTL gate, 40-25
Operand representation, 42-1
Operating
expense, 55-32
point, 37-6, 37-11
Operation
binary, 41-2
logic, 42-1
mixed-mode, 42-2
nonlinear, 37-4
push-pull, 37-5
relational, 42-1
unary, 41-2
Operational amplifier, 40-3
analysis, 40-5
circuit types, 40-6
circuits, 40-6
configurations, 40-6
constraint equation, 40-8
equivalent circuit, 40-4
frequency response, 40-8
ideal, 40-5
internal configuration, 40-3
limits, 40-8
offset, 40-8
saturation, 40-8
slew rate, 40-8
symbology, 40-3
virtual ground, 40-5
virtual short circuit, 40-5
voltage regulator, 40-16
voltage shunt feedback, 40-5
Operational form, log, 4-5
Operations
on complex numbers, 4-8
universal, 41-3
Operator
del, 20-5
Laplacian, 20-6
logic, 41-2
Opportunity cost, 55-18
Opposite side, 7-2
Optical
amplifier, 50-4
channel, 50-1
chirp, 50-4
communications, 50-3
dispersion, 50-4
eye pattern mask, 50-4
fiber, 46-4
frequency-division multiplexing, 50-3
frequency range, 50-3
guide, 45-1
jitter, 50-4
mask, 50-4
modulator, 50-3
regenerator, 50-4
transmission system, 50-3, 50-4
OR operation, 13-5
Orbit
geostationary Earth, 50-3
geosynchronous, 50-3
low Earth, 50-3
medium Earth, 50-3
Orbital moment, 23-3
Order
differential equation, 11-1
matrix, 5-1
of precedence, 41-2
of system, 51-5
Ordinary differential equation, 15-1 (ftn)
Ordinate, 8-2
Orientation, angular, 6-1
Orthogonal, 8-2, 15-5, 49-1
vectors, 6-3
view, 3-1
Orthographic
oblique view, 3-3
view, 3-1
Oscillations, 30-7
sustained, 30-8
Oscillator
circuit, 40-19
relaxation, 38-12
voltage controlled, 40-28
Ounces, A-1, A-2
Output
characteristic curve, 37-6
impedance, 37-6
port, 29-14
signal, 51-3
Overall transfer function, 51-5
Overcurrent, 53-2
device, 34-7
protection, 53-6
protective device, 53-4
Overdamped analysis, 31-9
Overdamping, 31-8
Overflow bit, 13-4
Overhead, 44-7, 55-36
Overload, 53-2
protection, motor, 53-8
Overshoot, 51-2
Owner, 56-4
Owners' equity, 55-33
Ownership, 56-1
Oxidation
potential, 26-14
reduction, 26-13, 26-14

P

p
-channel MOSFET, 39-1
sequence, 4-11
series, 4-12
-type, 17-6, 37-3
-V work, 2-4
Packet-switched protocol, 44-8
Pair
Darlington, 38-12
pole, 51-9
Palladium, 52-1
Pappus' theorems, 10-5
Parabola, 8-10
formula, A-7
Parabolic axis, 8-10
Paraboloid of revolutions, A-9
Parallel
circuit, 26-7
combination rules, 29-8
comparator, 44-6
-comparator A/D converter, 44-6
current, 26-7
elements, impedance of, A-29
line, 8-8
perspective, 3-3
reliability, 12-8
resistance, 26-7
resonance, 30-9, 30-10
Parallelogram
formula, A-7, A-8
method, 6-3
Paramagnetic, 16-4
material, 23-4
Paramagnetism, 16-4, 23-4
Parameter type, 29-14

INDEX - P

Parameters
  hybrid, 29-15
  inverse hybrid, 29-15
  variation of, 11-4, 15-3
Parametric
  equation, 4-2
  equation, derivative, 9-3
  equation, plane, 8-6
Parity, 18-7
  check code, 18-7
Parseval's relation, A-15, A-16, A-21
Partial
  differential equation, 15-1 (ftn), 15-7
  differentiation, 9-4
  fraction, 4-6, 10-3
Particular solution, 11-3, 11-4 (tbl), 30-2
Partner
  general, 56-1
  limited, 56-2
Partnership, 56-1
  limited, 56-2
  limited liability, 56-2
Parts, integration by, 10-2
Pascal's triangle, 4-3
Passive, 19-2
  components, 29-3
  device, 29-3
  element, 40-9
  linear element, 29-3
  logic circuit, 40-19
  system, 51-11
Path
  closed, 15-19, 15-20
  loss, 46-5
  of integration, 15-19
Payback period, 55-38
Peak
  clipper circuit, 37-12
  factor, 27-6
  inverse voltage, 37-9
Peak value sinusoid, 27-7
Peclet number, 1-9 (tbl)
Peltier effect, 16-5, 26-4
Pentagon relationships, A-7, A-8
Per diem fee, 56-5
Percent conductivity, 26-3
Percentile, 12-10
  rank, 12-10
Perfect
  capacitor, 27-9
  inductor, 27-9
Perfecting a lien, 56-5
Performance function, 12-8
Period, 27-2, 55-2
  depreciation, 55-20
  fundamental, 15-14
  number, 55-9
  of waveform, 10-6
Periodic
  extension, 15-6
  inventory, 55-37
  signal, continuous-time, 15-14
  signal, discrete-time, 15-17
  signal, discrete-time Fourier series, A-20
  signal, Fourier series, A-15
  table, material properties, A-43
  waveform, 10-6
Permanent magnet, 23-5
  theoretical views, 23-3
Permeability, 16-17, 29-6
  cgs vs. mks, 22-6 (ftn)
  initial, 28-6
  magnetic, 16-16
  maximum, 28-6
  of free space, 16-17, 29-6
  ratio, 29-6
  relative, 16-17, 23-2, 29-6
  spectrum, 16-17
  units of, 22-6
Permissive rule, 53-2
Permittivity, 16-6, 16-9, 29-5
  as a tensor, 16-10
  of free space, 16-9, 29-5
  relative, 16-6, 16-9, 29-5
  units of, 16-9
Permutation, 12-2
  circular, 6-5
Perpendicular
  line, 8-8
  line principle, 3-1
  lines, 3-1
Perpetual inventory system, 55-37
Personal
  communication network, 50-3
  communication system, 50-3
  property, 55-20 (ftn)
Perspective
  angular, 3-3
  oblique, 3-3
  parallel, 3-3
  view, 3-1, 3-2, 3-3
Petersen coil, 34-2 (ftn)
Phase, 27-2 (ftn)
  angle, 27-2
  angle, difference, 27-3, 27-7
  angle error, 34-5
  characteristic, 51-10, 51-11
  crossover frequency, 51-12
  deviation, 48-5
  difference, 27-2 (ftn)
  error, 40-28, 40-29
  factor, 27-12
  frequency detector, 40-28
  -locked loop, 40-28
  margin, 51-12
  margin Bode plot, 51-12
  modulation, 48-2, 48-4
  plot, 32-5
  rotation, 27-8
  -shift keying, 48-7
  velocity, 45-2
  vs. phase difference, 27-2 (ftn)
Phasor, 27-7, 27-8
  addition, 27-8
  complex representation, 27-8
  diagram, 27-8, 27-13
  form, 4-8, 27-3
  form of sinusoid, 27-2
  notation, 27-8
  notation, effective value, 27-8
  reference, 27-7
  rotation, 27-8
  steady-state, 32-3
Phonon, 16-12 (ftn), 37-14 (ftn)
Phosphors, 52-2
Phosphorus, 37-3
Photoconductive transduction, 19-3, 19-4
Photocurrent, 37-14
Photodiode, 29-7, 37-14
Photoelectric effect, 17-2
Photon
  as a force particle, 58-2
  virtual, 58-2
Photovoltaic, 19-3
Physical inventory, 55-37
Pi
  -group, 1-8, 1-10
  -T configuration, 26-8, 29-9
Pickup setting, 34-5
Pictorial drawing, 3-2 (ftn)
Piecewise linear model, 37-4
  diode, 37-11, 37-12
Piercing the corporate veil, 56-3
Piezoelectric, 19-3, 24-1
Pilot protection, 34-7
Pinchoff
  region, 37-4, 39-2
  region, MOSFET, 39-4
  theory, 39-3
  voltage, 39-2
Pipe light, 46-4
Pipelined protocol, 44-8
Pitch helix, 8-12
Plain old telephone service, 50-2
Plaintiff, 56-6
Planar
  array, 46-2
  view, 3-2
Planck constant, 17-2
Plane, 8-6
  angle, 7-1
  charge, 20-4
  complex, 4-8
  polarization, 46-2
  tangent, 9-5
Plant, 51-13
Plasma, 17-3
Plastic, 52-1
Plate, 17-2
Platinum, 52-1
Plot
  Bode, 32-7
  time-distance, 45-3 (ftn)
PMOS, 39-1, 40-19
pn junction, 17-7, 37-7
  carrier movement, 37-8
  characteristics, 37-4
  current, 37-8
  space-charge region, 37-8
pnp transistor, 38-5
pnpn device, 37-14
Point
  3 dB, 30-8
  breakaway, 51-12
  charge, 20-2
  compensation, 45-6
  contraflexure, 9-2
  critical, 9-2
  cut-over, 55-24 (ftn)
  extreme, 9-2
  form of Ohm's law, 16-14, 17-3
  half-power, 30-8
  inflection, 9-2
  no-signal, 37-4
  of application, 6-1
  of contraflexure, 8-2
  of inflection, 8-2
  of instability, 51-12
  of presence, 50-2
  of sale, 55-37
  operating, 37-11
  quiescent, 37-4, 37-6, 37-11
  singular, 9-1
  -slope form, 8-5
  terminal, 6-1
Poise, A-1, A-2
Poisson's
  distribution, 12-5
  equation, 15-7, 20-6
Polar
  coordinate system, 8-3 (fig), 8-3 (tbl)
  diagram, 45-3
  form, 4-8, 8-5, 27-3
  molecule, 21-3
  nonreturn-to-zero, 48-6
  substance, 20-3
Polarity
  additive, 34-5
  subtractive, 34-5
Polarization, 21-1, 33-3, 46-3
  antenna, 46-2
  circular, 46-2
  elliptical, 46-2
  in conductors, 21-4
  in dielectrics, 21-4
  linear, 46-2
  multiplexing, 49-2
  plane, 46-2
Polarized field, 21-2
Pole, 22-6, 32-2, 51-8
  at origin, 32-5
  location response, 51-9
  magnetic, 16-14, 22-6
  north, 16-14
  on negative real axis, 32-6
  pair, 51-9
  south, 16-14
  strength, 16-14, 22-6

INDEX - P

strength, unit, 22-6
transient response effect, 32-4
Pole-zero diagram, 51-8
calculations from, 51-10
exponential decay, 51-9
procedure for, 51-9
response from, 51-9
sinusoidal oscillation, 51-9
Poles, 31-14
Polygon
formula, A-7, A-8
mensuration, A-7
method, 6-3
Polyhedra properties, A-9
Polymer, 52-1
elements in, 52-1
Polynomial, 4-3
characteristic, 5-8
degree, 4-3
Lagrangian interpolating, 14-2
Newton's interpolating, 14-3
standard form, 4-3
Population, 12-3
Port, 29-14
input, 29-14
output, 29-14
Portable radio, 50-3
Positional numbering systems, 13-1
Positive
area, 27-4
current, magnetic field of, 22-2
feedback, 51-4
logic, 40-21, 41-1
Positive charge, 20-3
dipole alignment, 21-2
electric field of, 20-1
in electric field, 20-6
test, 20-1
unit, 20-3
Postulate, Boolean algebra, 13-5
Potential, 20-2, 20-5, 24-3
difference, 26-1
electric, 16-21
electrostatic, 24-3
energy, 2-3
energy, electron, 17-4
energy, magnetic, 22-8
magnetic, 22-8, 22-9, 24-3
theory, 15-7, 15-22
transformer, 36-2
vs. distance, 20-5
vs. electromotive force, 24-3
Potentiometer, 26-2
Potentiometric transduction, 19-3
Pound force, A-3
Poundal, 1-4
Power, 2-6, 26-3, 27-11, 35-1
active, 27-12
AM, 48-4
amplifier, 38-7, 47-2
angle, 27-11
apparent, 27-12
average, 24-3
complex, 27-12
conversion, SI, A-3, A-5
conversion factors, A-1
conversions, 2-6, 2-7
density, 46-5, 46-6, 46-7, 46-9, 47-1
effective radiated, 50-2
electric, 35-2
electronics, 36-1, 37-15
factor, 27-12
factor measurement, 36-7
gain, 38-7
instrument, 36-7
isotropic, 46-9
line carrier, 34-7
metal-oxide semiconductor field-effect transistor, 37-16
noise, 40-8
ohmic, 46-7, 46-9
reactive, 27-12
real, 27-12
receiver, 47-2
receiving antenna, 47-3
series, 9-8
series method, 15-1
transfer, maximum, 26-5, 27-13, 29-10
transistor, 38-7
triangle, 27-12, 27-13
true, 27-12
unit, 35-1
unit, energy, 35-1
wattless, 27-12
Power, AC
average, 27-11
instantaneous, 27-11
real, 27-11
Power semiconductor, 37-15
control device, 37-16
trigger device, 37-16
Power supply, 26-4, 33-4
current source, 26-4
voltage source, 26-4
Poynting's
theorem, 24-3
vector, 24-3, 46-3
Prandtl number, 1-9 (tbl)
Precedence, order of, 41-2
Precision, 12-13, 36-1
diode, 40-10
diode circuit, 40-10
Prefix, SI, 1-7
Prerequisite of privity, 56-7 (ftn)
Presence, point of, 50-2
Present worth, 55-5, 55-11
method, 55-14
Preset signal, 43-2
Pressure
conversions, SI, A-3, A-4
level, of sound, 26-4 (ftn)
sound, 19-3
Price-earnings ratio, 55-35
Primary, 28-2
battery, 26-14, 33-3
cell, 33-1
dimension, 1-7
feedback ratio, 51-5
impedance, effective, 28-4
unit, 1-7
winding, 28-2
Prime
contractor, 56-4
cost, 55-32
Principal
node, 26-6
view, 3-2
Principle
of duality, 41-4
of transformers, 28-1
quantum number, 16-2
work-energy, 2-3, 2-4
Priority, ethical, 57-2
Privity of contract, 56-7
probability, 12-2
Probabilistic problem, 55-30
Probability
bit error, 48-9
density function, 12-4
function, 48-9
integral, 15-11
of occurrence, 18-4
Problem
interface, 44-1
mixing, 11-8
Process, cost accounting, 55-36
Processing
of signals, 27-7
symbology, 42-2
task representation, 42-2
Product
method, 15-7
mixed triple, 6-4
term, 41-4
triple scalar, 6-4
vector, cross, 6-4
vector, dot, 6-3
vector, triple, 6-5
Professional
corporation, 56-2
engineer, 59-1
engineering exam, 59-1
engineering exam format, 59-3
limited liability company, 56-3
Profile, auxiliary view, 3-2
Profit, 55-37
and loss statement, 55-34
flow through, 56-2
margin, 55-35
Programmable
attenuator, 44-5
hardware, 42-8, 42-9
logic array, 42-8, 42-9
Projection
cabinet, 3-3
cavalier, 3-3
method, 55-10
Projector, 3-1
Propagation
constant, 45-2
delay, 40-23
Proper
polynomial fraction, 4-6 (ftn)
subset, 12-1
Properties
of AC conductors, A-49
of complex numbers, 27-9, 27-10
of conductors, A-49
of gallium arsenide, A-24
of germanium, A-24
of semiconductors, A-24
of silicon, A-24
related to periodic table, A-43
Property
business, 55-20 (ftn)
personal, 55-20
real, 55-20
residential, 55-20
Proportional
controller, 51-14
-integral-derivative controller, 51-14
Proprietorship, sole, 56-1
Protection
low-voltage, 54-5
motor overload, 53-8
non-pilot, 34-7
pilot, 34-7
system, 34-1
Protective
device, overcurrent, 53-4
earth, 34-2
relay, 34-1
Protector block, 50-2
Protocol, 44-7
asynchronous, 44-7
block-transfer, 44-8
burst, 44-8
cache coherence, 44-8
connect/disconnect, 44-8
packet-switched, 44-8
pipelined, 44-8
split-transaction, 44-8
synchronous, 44-8
Proton
charge, 16-2
mass, 16-2
properties of, 16-2
PSPICE, 58-1
Public
corporation, 56-2
dealing with, 57-3
switched telephone network, 50-2
Pulse, 31-6
amplitude modulation, 48-5, 48-7
circuit, 40-9
code modulation, 48-5, 48-7
electromagnetic, 50-5
generator, 40-9
logic, 40-21

INDEX - P

step, 31-6
transfer, 44-1
transient, 30-1, 31-6
transient analysis, 31-13
unit, 31-6
Punitive damages, 56-6, 56-7
Purchase order, 56-3
Pure resistor, 27-9
Push-pull operation, 37-5
Pyramid horn, 46-2
Pythagorean theorem, 7-2

Q

Q
-point, 37-6, 37-11, 38-2, 38-6
value, 30-7, 30-8, 30-9, 31-14, 51-10
Quad, 2-1, 35-1 (tbl)
Quadrants, 7-3
signs of functions, 7-3
Quadratic equation, 4-3
Quadrature, 27-12
amplitude modulation, 48-8
multiplexing, 48-4
Quality factor, 30-7, 30-8, 30-9, 31-14, 51-10
Quanta, 17-4
Quantity
defined, 30-6
economic order, 55-43
Quantization, 44-2, 48-5
error, 44-2
noise, 48-5
resolution, 44-2
Quantum electrodynamics, 58-2
Quantum number, 16-2
azimuthal, 16-2
principle, 16-2
spatial, 16-2
spin, 16-2
Quantum theory
of light, 58-2
of radiation, 58-2
Quantumization
distortion, 48-5
noise, 48-5
Quarter
-wave antenna, 46-9
-wave symmetry, 10-7 (tbl)
Quartile, 12-10
Quasistatic, 24-2
Quick ratio, 55-35
Quiescent
current, 37-6
point, 37-4, 37-6, 37-11, 38-6
Quine-McClusky method, 42-3

R

Rad, 7-1 (ftn), A-1, A-2
Radial
system, 34-2
-vane mechanism, 36-6
Radian, 7-1, A-1, A-2
hyperbolic, 7-5
Radiated
emissions, 50-4
susceptibility, 50-5
Radiation
-emitting material, 52-2
field, 16-21
impedance, 46-2 (ftn)
polarization, 46-3
quantum theory, 58-2
reactance, 46-2 (ftn)
resistance, 46-2, 46-6, 46-9
Radio
communication bands, 50-3
mobile, 50-3
portable, 50-3
system, 50-3
system, satellite, 50-3
Radius, Bohr, 22-7
Radix, 13-1
Rail voltage, 40-3
Random error, 36-1
Range, 12-10
Rank matrix, 5-5
Ranking alternatives, 55-17
Rarefaction, 19-1
Rate
cut-off, 51-10
of change, 9-1
of return, 55-4, 55-12, 55-14
of return method, 55-17
per annum, 55-28
Ratio
actuating signal, 51-4
benefit-cost, 55-16
common, 4-11
control, 51-5
current, 55-35
damping, 31-14, 51-2
energy-efficiency, 2-7 (ftn)
error, 51-4
feedback, 51-5
of transformation, 28-4
permeability, 29-6
price-earnings, 55-35
signal-to-noise, 48-8
test, 4-12
turns, 28-4
Ratiometric A/D converter, 44-7
Rational
function, 51-3
number, 4-1
Rationalization, 4-8
Rationalized, 16-6
Ray, 7-1 (ftn)
RC
and *RL* circuit solution, 30-5
circuit, forced response, 30-3
circuit, natural response, 30-2
differentiator, 40-9
RCTL, 40-19
Reactance
capacitive, 27-9
compensation, 45-6
inductive, 27-9
mutual, 28-2, 28-5
radiation, 46-2 (ftn)
Reaction
electrochemical, 26-13, 26-14
nonelectrochemical, 26-14 (ftn)
Reactive
factor, 27-12
power, 27-12, 34-3
power, vector, 27-12
Read only (R/O) memory, 39-1, 42-8 (ftn)
Real
AC power, 27-11
current source, 26-4, 29-2
diode, 37-10
energy source, 26-4
machine, 2-7
number, 4-1
power, 27-12
resistor, 29-2
transformer, 28-5
transformer model, 28-5
voltage source, 26-4, 29-2
Realization, 41-6
Receipt, 55-3
Receivable
average age, 55-35
turnover, 55-35
Receiver, 18-1
power, 47-2
superheterodyne, 50-2
Receiving
antenna, 47-2
antenna power, 47-3
Receptacles, 53-4
Reciprocal, 26-3
inductance, 16-19
Reciprocity, 26-9, 59-2
theorem, 26-9
Recloser, 34-7
Recombination, 17-6
current, 37-7
electron, 37-7
hole, 37-7
Recommended Standard 232, 36-8
Recovery, depreciation, 55-26
Rectangular
coordinate system, 8-3
form, 4-8, 27-3
form of sinusoid, 27-2
hyperbola, 8-12
Rectified waveform, 27-4
Rectifier, 37-9
circuit, 37-12
full-wave, 37-12
half-wave, 37-12
Rectum, latus, 8-10
Redbook, 50-2
Reduced equation, 11-1
Redundancy reliability, 12-8
Redundant view, 3-2
Reference, 26-14
direction, current, 26-7
direction, voltage, 26-6
phasor, 27-7
sinusoid, 27-2
standard, 36-1
value, 51-14
Reflected impedance, 28-4
Reflection, 46-4
coefficient, 45-3
coefficient, VSWR equation, 45-3
from surface, 46-4
Reflector antenna, 46-2
Reflex angle, 7-1
Refraction, 46-4
angle of, 46-4
index, 46-4
Refractory material, 52-1
Regenerator, 50-4
optical, 50-4
Region
active, 37-4, 38-2
avalanche, 37-4, 39-2
breakdown, 37-9, 39-2
cutoff, 38-3
depletion, 17-7, 37-7
diode forward bias, 37-12
diode off, 37-12
diode reverse breakdown, 37-12
high-band, 40-3
high-impedance, 37-15
linear, 38-2
low-band, 40-3
mid-band, 40-3
negative resistance, 37-15
of convergence, 15-17
ohmic, 39-2
pinchoff, 37-4, 39-2
saturation, 38-3
space-charge, 17-7, 37-7
transition, 17-7, 37-7
triode, 39-2
Regional channel, 50-2
Register, 41-1, 43-1
Registered engineer, 59-1
Registration, 59-1
Regula falsi method, 14-2 (ftn)
Regular function, 9-1
Regulating system, 51-14
Regulation, 26-4
voltage, 29-2, 34-5
Regulator, 51-14
Reinforcement, 46-3, 46-4
Related angle, 7-1
functions of, 7-3
Relational operation, 42-1
Relative
permeability, 16-16, 23-2, 29-6
permittivity, 16-6, 16-9, 29-5

INDEX - R

risk, 55-35
stability, 51-12
Relatively selective system, 34-7
Relativistic effect, 22-3
Relativity, 25-1 (ftn)
Relaxation oscillator, 38-12
Relay, 34-3, 34-4
computer, 34-6
dependable, 34-3
differential comparison, 34-6
distance, 34-3 (tbl), 34-6
equation, universal, 34-6
frequency, 34-6
harmonic, 34-6
high-speed, 34-4
impedance, 34-6
instantaneous, 34-4
level detection, 34-5
magnitude, 34-5
mho, 34-6
overcurrent, 34-5, 34-7
phase angle, 34-6
pilot, 34-6
protective, 34-1
ratio, 34-6
reliability, 34-3
secure, 34-3
solid-state, 34-6, 34-7
time-delay, 34-4
ultra high-speed, 34-4
Release, low-voltage, 54-5
Reliability, 12-7, 12-8, 12-13
parallel, 12-8
relay, 34-3
serial, 12-8
Reluctance, 16-21
Reluctive transduction, 19-3
Remainder method, 13-1
Remote, 34-3
Renewal, 55-14 (ftn)
Repayment, loan, 55-4
Replacement
/retirement analysis, 55-4
study, 55-18
Research and development cost, 55-32
Reserve battery, 33-3
Residual
flux, 16-22
flux density, 28-6
resistivity, 58-1
value, 55-16
Resinous static electricity, 16-5
Resistance, 26-2, 26-9 (ftn), 29-3
characteristic, 45-2
core, 28-5
dynamic, 38-1, 39-1 (ftn)
equivalence, 26-5
equivalent, 26-12
equivalent Thevenin, 26-10
forward, 37-9
gain, 38-7
instantaneous, 38-1, 39-1 (ftn)
noise, 40-8
ohmic, 46-2, 46-8, 46-9
radiation, 46-2, 46-6, 46-9
reciprocal of, 26-3
reverse, 37-10
temperature coefficient, 26-2
temperature dependence, 29-3
testing, ground, 36-2
transfer, 26-9
Resistive
region, MOSFET, 39-2
transduction, 19-4
Resistivity, 16-13, 26-2
reciprocal of, 26-3
residual, 58-1
temperature coefficient, 29-4
thermal, 58-1
Resistor, 27-9, 29-2
-capacitor circuit, 30-2, 30-3
-capacitor transistor logic, 40-19
characteristic, 27-3
equations, 29-4
ideal, 27-9, 29-2
in series, 29-4
-inductor circuit, 30-4
load, 37-4
parallel, 29-4, 29-5
pure, 27-9
real, 29-2
swamping, 36-4
-transistor logic, 40-19, 40-24
types of, 26-2
Resolution, 4-6, 36-1
of frequency analysis, 10-8
quantization, 44-2
Resonance, 27-13, 30-8
parallel, 30-9, 30-10
series, 30-8, 30-9
Resonant
circuit, 30-8
circuit formulas, A-32
frequency, 30-8, 51-10
Response
equation, 51-15
forced, 30-2, 30-4, 51-2
frequency, 31-2, 51-10, 51-11
homogeneous, 51-2
impulse, 30-1
initial condition, 51-2
natural, 30-2, 30-4, 51-2
second-order, 51-2
steady-state, 51-2
step, 30-1, 30-6, 31-6
time, 18-3, 51-2
to pole location, 51-9
total, 51-2
to unit step, 51-2
transient, 30-5, 30-6, 51-2
unforced, 51-2
Responsivity, 50-4
Resultant vector, 6-2, 6-3
Retainer fee, 56-5
Retentive material, 52-2
Return
on investment, 55-11, 55-14, 55-35
rate of, 55-4, 55-12, 55-14
-to-zero, 48-6
Revenue, 55-34
Reverse
bias, 37-4
breakdown region, 37-12
saturation current, 37-7
transfer function, 51-4
voltage, peak, 37-9
Revolution
surface of, 10-5
volume of, 10-5
Reynolds number, 1-9 (tbl)
Rheostat, 26-2
Right
angle, 7-1
circular cone, formula, A-9
circular cylinder, A-9
-hand bit, 13-3
-hand circularly polarized, 46-2
-hand rule, 6-4, 22-3
triangle, 7-2
Ring permutation, 12-2
Ringing, 30-7
circuit, 30-7
Ripple
-carry adder, 42-11
counter, 43-4, 43-5
Rise
in temperature, 2-4
time, 18-3, 30-6, 40-23, 51-2
voltage, 26-6
Rising edge trigger, 43-1
Risk
analysis, 55-43
-free investment, 55-11
RL circuit
forced response, 30-4, 30-5
natural response, 30-4
RLC circuit, 30-8
Room temperature, 37-10
Root, 4-3, 14-1
double, 4-4
extraneous, 4-4
finding, 14-1
-locus diagram, 51-12
-mean-square value, 12-10, 27-5
Rotation
of dipole, 21-3
of phasor, 27-8
Rotational symmetry, 8-4
Roughness
height index, 3-4
weight, 3-4
width, 3-4
Routh
criterion, 51-13
-Hurwitz criterion, 51-13
table, 51-13
Row
canonical form, 5-2
equivalent matrix, 5-2
matrix, 5-1
operation, 5-2
rank, 5-5 (ftn)
-reduced echelon form, 5-2
-reduced echelon matrix, 5-1
RS-232, 36-8
*R*th-order difference, 4-11
RTL, 40-19, 40-24
Rule, 57-1
Carson's, 48-5
Cramer's, 4-7, 5-7
FBI, 22-4
L'Hôpital's, 4-10
of professional conduct, 57-1
of signs, Descartes', 4-4
of transformer dots, 28-3
right-hand, 6-4, 22-3
Simpson's, 8-2
trapezoidal, 8-1
Rules
SI system, 1-5, 1-6
work, 54-7

S

s
domain impedance, 31-15
domain signal representation, 31-15
-plane, 51-8
-space, 31-1
S-R flip-flop, 43-2
next-state equation, 43-2
Saddle point, 8-2
Safety
clearance zone, 54-4
Code, 54-1
grounding, 53-3, 54-4
issues, 53-3
limits, 53-1, 54-2
Sag and tension chart, 54-5
Salary plus fee, 56-5
Salvage value, 55-18
Sample, 12-3
-and-hold circuit, 44-5
space, 12-3
standard deviation, 12-10
variance, 12-11
Sampling, 18-2
theorem, 18-2
with replacement, 12-3
Satellite, 50-1
communication bands, 50-3
footprint, 46-3 (ftn), 50-3
geostationary, 50-3
radio system, 50-3
system, 50-3
Saturation, 37-4, 37-7, 38-2
current, 37-7, 38-4, 39-2
electron, 37-7
flux density, 28-6
region, 38-3

Sawtooth
  wave, 27-6
  waveform, 27-6
SCADA, 35-2
Scalar, 6-1, 41-1
  binary, 41-1
  matrix, 5-1
  of digital information, 42-1
  operation, example, 42-2
  product, triple, 6-4
  source, 22-8
Scatter losses, 47-2
Scattering, 46-4
Scheduling, 35-2
Scheme, grounding, 34-1
Schmidt number, 1-9 (tbl)
Schottky
  contact, 37-12
  diode, 37-12, 37-13
  noise, 18-7
Second
  derivative, 9-1
  moment of a function, 10-6
Second-order
  analysis, 31-7
  analysis, critically damped, 31-11
  analysis, pulse transients, 31-13
  analysis, underdamped, 31-12
  circuit, 31-7
  circuit, step change response, 31-9
  linear, 11-3
  response, 51-2
  time response, 51-3
  transient, 31-7
Secondary, 28-2
  battery, 26-14, 33-3
  cell, 33-1
  current, 28-2
  voltage, 28-1, 28-2
  winding, 28-2
Section
  conic, 8-8
  drawing, 3-3
Sectionalizer, 34-7
Sector, A-7
Sectoral horn, 46-2
Security
  emanations, 50-5
  emissions, 50-5
Seebeck effect, 16-5, 26-4
Selector data, 42-9
Self
  -bias, 38-5
  -biasing, 37-4
  -generating transducer, 19-3
  -inductance, 16-19, 24-2, 28-2
  -information, 18-4
Semi-
  major distance, 8-11
  minor distance, 8-11
  variable cost, 55-32
Semiconductor, 16-3, 52-1
  charge, 17-4
  common, 16-3
  conductivity, 37-1
  extrinsic, 17-5, 37-2, 37-3
  intrinsic, 17-5, 37-2
  material, 37-2
  n-type, 37-3
  p-type, 37-3
  performance, typical, 37-4
  power, 37-16
  principles, summary, 17-7
  properties, A-24
  symbology, A-34
Sending voltage, 45-2
Sense, 6-1
Sensitivity, 12-13, 36-4, 51-5
  analysis, 55-43
Separable, first-order, 11-2, 11-3
Separation
  of terms, 10-3
  of variables, 15-7
Sequence, 4-10
  arithmetic, 4-11
  convergent, 4-10
  detection, 43-8
  divergent, 4-10
  geometric, 4-11
  harmonic, 4-11
  p-, 4-11
  standard, 4-11
Sequence detector, 43-8, 43-10
  adjacency weighting, 43-9
  control equations, 43-9
  state diagram, 43-8
  state table, 43-9
Sequential
  logic, 41-2
  logic, element, 43-1
  network, 43-1
  network, general synchronous, 43-1
Serial reliability, 12-8
Series, 4-11, 4-12
  arithmetic, 4-12
  circuit, 26-7
  combination rules, 29-8
  elements, impedance of, A-28
  expansion, 9-7
  finite, 4-12
  Fourier, 10-6, 15-13
  geometric, 4-12
  harmonic, 4-12
  infinite, 4-12, 55-7
  Maclaurin, 9-8
  of alternating sign, 4-13
  power, 9-8
  resistance, 26-7
  resonance, 30-8, 30-9
  Taylor's, 9-7, 9-8
  voltage, 26-7
Service
  calculation, 53-5
  life, 55-20
Servo converter, 44-5
Servomechanism, 51-14
Set, 12-1, 48-7
  laws, 12-1
  null, 12-1
  universal, 12-1
Sets, complete, 41-3
Setting, pickup, 34-5
Settling time, 51-2
Setup time, 43-2
Severability, 56-4
Shannon
  bandwidth, 48-8
  coding theorem, 18-6, 50-1 (ftn)
  limit, 50-1
Shape, mensuration, A-7
Sheet
  charge, 20-4, 20-5
  metal gage, A-45
Shelf life, 33-1
Shell, 16-2
  core, 17-4
  transformer, 28-3
  valence, 17-4
Sherwood number, 1-9 (tbl)
Shifting, 27-2
  source, 29-12
Shockley's equation, 39-3
Short
  dipole antenna, 46-7
  -term transaction, 55-3
Short circuit, 26-2
  admittance parameters, 29-15
  virtual, 40-5
Shortwave, 50-2
Shot noise, 18-7, 40-8
SI
  base units, 1-5 (tbl)
  conversion factors, A-3
  derived units, 1-6
  prefix, 1-7
  system, 1-5
  system rules, 1-5, 1-6
Sideband, 18-3
  lower, 18-3
  upper, 18-3
Sifting, 15-11
Sign
  alternating, series of, 4-13
  caution, 54-4
  danger, 54-4
  of the function, 7-3
  warning, 54-4
Signal, 18-1, 18-2, 32-1
  acknowledge, 44-7
  AMI, 48-6
  analog, 18-1
  analysis, small, 27-11
  analyzer, 10-8
  asynchronous, 43-2
  bandwidth, 18-3
  baseband, 48-2, 48-5
  biphase, 48-7
  carrier, 18-4
  clear, 43-2
  common-mode, 40-3
  continuous-time, 15-14
  data-ready, 44-7
  delay, 18-3
  delay time, 18-3
  differential-mode, 40-3
  digital, 48-6
  digital-data, 48-6
  dimensionality, 48-8
  discrete-time, 15-17
  duration, 18-3
  error, 51-4
  format, 48-1, 48-7
  Fourier series of, 36-2
  ideal sampled, 18-2
  in s domain, 31-15
  information, 18-4, 48-5
  input, 48-5, 51-3
  Manchester, 48-7
  message, 18-2
  modulation, types of, 48-2
  multiplexing, 49-1
  NRZ, 48-6
  output, 51-3
  preset, 43-2
  processing, 27-7
  representation, 36-2
  response, in s domain, 31-15
  response time, 18-3
  rise time, 18-3
  RZ, 48-6
  spectrum, baseband, 48-1
  steering, 46-2
  theory, 18-2, 18-3
  -to-noise ratio, 40-9, 48-8
Signature analysis, 10-8 (ftn)
Significant
  bit, least, 13-1
  bit, most, 13-1
  digit, 4-1, 13-1
  digit, least, 13-1
  digit, most, 13-1
Silicon, 37-2
  crystal, 17-5
  energy gap, 37-3, A-24
  properties, A-24
Silicon-controlled rectifier, 37-14, 37-15
  negative resistance region, 37-15
  off region, 37-15
Silver, 52-1
Similar triangle, 7-2
Simple
  array, 46-1, 46-2
  curve, 15-19
  interest, 55-11, 55-39
  node, 26-6
Simplification theorem, 13-5
Simpson's rule, 8-2

Simulation program with integrated circuit emphasis, 58-1
Simultaneous
  equations, matrix, 5-6
  equations, solving, 4-7
  linear equations, 4-7
Sine
  function, integral, 10-8
  integral, 10-8
  wave, 27-2
Sines, law of, 7-5, 7-6
Single
  payment cash flow, 55-3
  payment present worth factor, 55-5
  sideband, 48-3
  -stub matching, 45-6
Singular
  matrix, 5-1, 5-5
  point, 9-1, 9-3 (ftn)
Singularity, 5-5, 15-3
Sink
  fan-out, 40-23
  magnetic, 16-15
  notation, 29-2
Sinking fund, 55-7, 55-22
  factor, 55-7
Sinusoid, 27-7
  amplitude, 27-2
  cycle, 27-2
  exponential form, 27-2
  magnitude, 27-7
  peak value, 27-7
  period, 27-2
  phasor form, 27-2
  rectangular form, 27-2
  rectified average, 27-4
  reference, 27-2
  representation, equivalent forms, 27-2
  shifting, 27-2
  trigonometric form, 27-2
  waveform, 27-1
Sinusoidal
  analysis, 31-1, 31-2
  oscillation, from pole-zero diagram, 51-9
  response, decaying, 51-9
  response, undamped, 51-9
  steady-state analysis, 32-1
  waveform, 27-1
Skew
  clock, 44-8
  symmetric matrix, 5-2
Skewness, 12-11
Slew rate, 40-8
Sliding vector, 6-1 (ftn)
Slope, 9-1
  form, 8-5
Slotted line, 45-7
Slug, 1-3, A-1, A-2, A-3, A-4
Small
  angle approximation, 7-3
  -scale integration, 37-2
Small-signal
  analysis, 27-11, 29-7, 37-2, 37-4, 38-3
  base bulk resistance, 38-8
  base input resistance, 38-8
  feedback resistance, 38-8
  model, 38-8, 38-10, 39-3, 40-3
  output resistance, 38-8
  terminology, 38-4
Smith chart, 45-3
  alternate positioning, 45-5
  compensation reactance, 45-6
  components, 45-4
  electrical conditions, 45-5
  fundamentals, 45-3
  impedance determination, 45-7
  procedure, 45-5
  unknown impedance, 45-7
Smoothing
  coefficient, 55-42
  economic analysis, 55-42
Snell's law, 46-4
SNR, 40-9, 48-8
Soft material, 23-2, 52-2
  ferromagnetic, 23-5
  magnetic, 23-5, 52-2
Software surface, 44-1
Solar cell, 37-14
Sole proprietorship, 56-1
Solenoid, 16-19
  flux lines, 16-19
Solenoidal field, 22-8
Solid angle, 7-6
  beam, 47-3
Solution
  complementary, 11-3, 30-2
  homogeneous, 30-2
  nonhomogeneous, 30-3
  particular, 11-3, 30-2
  trivial, 4-7 (ftn)
Solvency, 55-35
Sonic velocity, 19-2
Sound
  intensity, 19-3
  pressure, 19-3
  pressure level, 19-3, 26-4 (ftn)
  speed of, 19-2
Source, 38-2, 39-1
  AC voltage, 27-3
  controlled, 27-3, 27-4, 29-3
  current, 40-21
  dependent, 26-4, 27-3
  dependent voltage, 29-3
  equivalent, 26-5
  fan-out, 40-23
  -free circuit, 30-2
  Huygens', 46-1
  ideal, 26-3
  ideal current, 26-4, 29-2
  ideal voltage, 26-4, 29-2
  independent, 26-4, 27-3, 29-7, 38-10
  information, 18-1
  magnetic, 16-15
  notation, 29-2
  of energy, 26-4
  real current, 26-4, 29-2
  real voltage, 26-4, 29-2
  scalar, 22-8
  shifting, 29-12
  transformation, 26-5, 29-8
  transformation model, 29-8
South pole, 16-14
Space
  -charge region, 17-7, 37-7, 37-8
  -charge region, pn junction, 37-8
  -division multiplexing, 49-2
  losses, 47-2
  multiplexing, 49-2
  working, 54-4
Spatial quantum number, 16-2
Special
  damages, 56-7
  Gaussian surface, 16-11
  relativity, 16-2 (ftn)
  Theory of Relativity, 16-2
  triangle, 7-2
Specific
  energy, 2-1
  gas constant, 19-2
  heat, 2-4
  heat, molar, 2-4
  identification, 55-37
  performance, 56-4, 56-6
Specified incompletely, 42-3
Spectral
  coefficient, 15-14 (ftn)
  inputs, 32-1
Spectrum
  analyzer, 10-8
  electromagnetic, A-25
  magnetic, 16-17
  permeability, 16-17
Speed, 16-2 (ftn)
  of electromagnetic wave, 16-20
  of sound, 19-2
Sphere, 8-12
  formula, A-9
  unit, 7-6
Spherical
  coordinate system, 8-3 (fig), 8-3 (tbl)
  defect, 7-6
  drop, 11-10
  excess, 7-6
  segment, A-9
  triangle, 7-5
  trigonometry, 7-5
SPICE, 58-1
Spin, 23-3
  moment, 23-3
  quantum number, 16-2
Split-transaction protocol, 44-8
Spread
  -spectrum modulation, 49-2
  -spectrum multiplexing, 49-2
Spreading, 18-3, 18-4
Spring
  energy, 2-3
  linear, 2-2
  work, 2-2
Square
  matrix, 5-2
  wave, 27-6
SSB, 48-3
Stability, 12-13, 51-11
  bias, 38-5
  criterion, Hurwitz, 51-13
  criterion, Nyquist's, 51-13
  criterion, Roth, 51-13
  relative, 51-12
Stable system, 51-11
Stage amplifier, 38-7
Standard
  atmosphere, SI, A-3, A-4
  cash flow factor, A-50
  contract, 56-5
  cost, 55-36
  deviation, 12-10
  ergonomic, 44-1
  error of the mean, 12-11
  fuse rating, 53-6
  gravity, 1-2
  modular devices, 37-2
  normal curve, area of, A-12
  normal value, 12-6, 12-12
  of care, 56-6 (ftn), 56-7
  orthographic view, 3-2
  polynomial form, 4-3
  sequence, 4-11
  technical, 44-1
  user interface, 44-1
  weather, 54-5
  zener diode, properties, A-38, A-39
Stanton number, 1-9 (tbl)
Start-up state, 43-8
State
  cell, 41-1
  diagram, 43-1
  equation, 51-15
  hang-up, 43-6
  level Z, 40-25
  lockout, 43-6
  machine, 43-8
  model, 51-14
  of super-conduction, 58-1
  of the register, 41-1
  start-up, 43-8
  steady-, 27-1, 29-13
  table, 43-1, 43-5
  variable, 31-6, 31-6 (ftn), 41-1
  variable diagram, 51-15
Static, 48-6
  characteristic, 37-6
  electricity, resinous, 16-5
  electricity, vitreous, 16-5
  energy, 2-4
  load line, 37-11
Statistical process control, 12-13

Statute, 57-1
  mile, A-1, A-2
  of Frauds, 56-3 (ftn)
Steady state, 27-1, 29-13
  response, 51-2
  response, 32-3, 51-7
  response, phasor, 32-3
Steel, 16-3
  electrical, 52-2
Steering signal, 46-2
Steinmetz algorithm, 27-7, 27-8
Step
  -down transformer, 28-4
  function, 11-6, 15-9
  -graded junction, 37-7
  input, 51-8
  pulse, 31-6
  response, 30-1, 30-6, 31-6
  response, first-order circuit, 31-8
  unit, 15-9
  -up transformer, 28-4
Stepped
  cash flow, 55-10
  motor, 51-13 (ftn)
Steradian, 7-6
Stirling's asymptotic series of gamma function, A-13
Stokes'
  number, 1-9 (tbl)
  theorem, 15-19
Storage time, 40-23
Straddle, 14-1
Straight
  angle, 7-1
  line, forms, 8-4
  line method, 55-21, 55-26
Strain-gauge transduction, 19-4
Strength
  electric field, 16-4, 16-7
  magnetic, 23-5
  magnetic field, 16-17
  of dielectric, 16-4
  pole, 16-14
Strict liability in tort, 56-6, 56-7
Strong nuclear force, 58-2
  force particle, 58-2
Strouhal number, 1-9 (tbl)
Subchapter C corporation, 56-3
Subcontract, 56-4
Submatrix, 5-1
Subset, 12-1
  proper, 12-1
Substitution
  circuit, 36-3
  theorem, 26-9
Subtractive polarity, 34-5
Success, 12-3
Successive approximation converter, 44-6
Sum
  -of-the-years' digits, 55-21
  -of-the-years' digits method, 55-26
  term, 41-4
Summer, 51-4
Sunk
  cost, 55-3
  current, 40-21
Superconducting state, 58-1
Superconductivity, 58-1
  critical temperature, 58-1
Superheterodyne receiver, 50-2
Superposition, 26-9, 29-10, 46-4
  magnetic, 16-16
  of cash flow, 55-10
  theorem, 11-6, 26-9, 27-1
Supervisory control and data acquisition, 35-2
Supplementary
  angle, 7-1
  unit, 1-6 (tbl)
Supplier's lien, 56-5
Supplier (dealing with), 57-2
Supply power, 33-4
Suppressed-carrier amplitude modulation system, 18-3
Surface
  charge, 20-4, 20-5
  charge, dielectric, 21-3
  finish, 3-4
  Gaussian, 16-10
  integral, 15-19
  of revolution, 10-5
  reflection, 46-4
  temperature, 11-10
  wave antenna, 46-2
Susceptance, 27-3
Susceptibility, 16-17
  conducted, 50-5
  radiated, 50-5
Sustained oscillations, 30-8
Swamping resistor, 36-4
Switch
  BJT, 40-17
  CMOS, 40-18
  digitally controlled, 44-3
  JFET, 40-18
  transistor, 40-17
Switching
  algebra, 13-5, 41-3
  circuit bias, 38-5
  computer, 44-7
  matrix, 42-8
  telecommunication, 44-7
  time, 37-11
  transient, 30-1, 31-3
Symbol error probability, 48-9
Symbology
  AC value, 37-2
  AC voltage source, 27-4
  amplifier, 40-3
  DC value, 37-2
  effective value, 37-2
  h-parameter, 38-9
  MOSFET, 39-4
  of semiconductor, A-34
Symbols, 4-1
Symmetrical
  even, 10-7 (tbl)
  full-wave, 10-7 (tbl)
  function, 8-4
  half-wave, 10-7 (tbl)
  matrix, 5-2
  odd, 10-7 (tbl)
  quarter-wave, 10-7 (tbl)
  waveform, 10-7 (tbl)
Symmetry curve, types, 8-4
Synchronization, 48-7
Synchronous, 43-5
  binary counter, 43-5
  digital hierarchy, 50-4
  optical network, 50-4
  protocol, 44-8
  sequential network, 43-1
Synthesis equation, 15-15
System
  active, 51-11
  baseband, 48-2
  binary, 13-1
  coherent, 48-6
  communication, 18-1, 47-1, 50-1
  consistent, 4-7
  control, 34-1
  controllability, 51-15
  function, 51-5
  graded, 34-7
  hexadecimal, 13-2
  inconsistent, 4-7
  microwave, 50-1
  network, 34-2
  non-unit, 34-7
  numbering, 4-1
  observability, 51-15
  octal, 13-2
  of units, 1-2
  order, 51-5
  passive, 51-11
  personal communication, 50-3
  protection, 34-1
  radial, 34-2
  relatively selective, 34-7
  satellite, 50-3
  SI, 1-5
  stable, 51-11
  time response, 51-7
  TN, 34-2 (ftn)
Systematic error, 36-1

T

T^2L, 40-19, 40-24, 40-25
Table
  divided difference, 14-3
  factor, A-50, A-51
  of Indefinite Integrals, A-10
  of integrals, 10-1 (ftn)
  Routh, 51-13
  state, 43-5
  truth, 41-2
Tabular method, 42-3
Tail, curve, 12-6
Tangent
  plane, 9-5
  point, 8-2
Tangents, law of, 7-5
Tangible property, 55-20 (ftn)
Tank circuit, 40-29
Taper efficiency, 47-3
Tax
  bracket, 55-25
  credit, 55-25
Taxation, double, 56-2
Taylor's
  formula, 9-7
  series, 9-7, 9-8
Technical standard, 44-1
Technology, 57-1 (ftn)
Telecommunication, 47-1, 50-1
  switching, 44-7
  traffic, 44-7
Telegraph, 50-1
  terminology, 48-7
Telephone, 50-2
  cellular, 50-3
  network, 50-2
  service, 50-2
Telephony, 50-2
Television
  broadcast, 50-2
  high-definition, 50-3
Temperature
  antenna, 46-2
  conversions, SI, A-3, A-5
  Curie, 23-4
  figure, 18-7, 36-2
  noise, 18-7, 36-1, 40-8
  rise, 2-4
  room, 37-10
  surface, 11-10
  volt-equivalent, 17-7
  voltage equivalent, 37-10
Temperature coefficient, 37-13
  of resistance, 26-3
  of resistivity, 29-4
TEMPEST, 50-5
Tendency, central, 12-10
Tens complement, 13-3
Tension and sag chart, 54-5
Tensor, 6-1
Term
  general, 4-10
  in a sequence, 4-10
  product, 41-4
  sum, 41-4
Terminal
  common, 38-4
  grounding, 53-6, 53-7
  inverting, 40-3
  noninverting, 40-3

INDEX - T

point, 6-1
value, 55-16
Terms
harmonic, 10-6
separation of, 10-3
Terrestrial radio system, 50-3
Test
charge, 16-7
comparison, 4-12
element under, 36-8
financial, 55-34, 55-35
for convergence, 4-12
ratio, 4-12
Testing
ground resistance, 36-2
insulation, 36-8
Theorem, 11-6
binomial, 4-3
Boolean algebra, 13-5
Buckingham, pi-, 1-10
Cauchy-Schwartz, 6-3
central limit, 12-11
compensation, 26-9
consensus, 13-5
de Moivre's, 4-9
De Morgan's, 40-20, 41-3
divergence, 20-4, 20-5
final value, 15-5
Gauss' divergence, 15-18, 20-4, 20-5
Green's, 15-19
initial value, 15-5
linearity, 11-6
Miller's, 29-10, 29-11
Millman's, 26-5
Norton's, 26-12, 29-9
of calculus, fundamental, 10-4
of Fourier, 10-6
of Pappus, 10-5
or logical adjacency, 42-2
Poynting's, 24-3
Pythagorean, 7-2
reciprocity, 26-9
sampling, 18-2
Shannon coding, 18-6, 50-1 (ftn)
simplification, 13-5
Stokes', 15-19
substitution, 26-9
superposition, 11-6, 26-9, 27-1
Thevenin's, 26-10, 29-9
time-shifting, 11-6
Theory
acoustic, 19-1
antenna, 46-1
basic signal, 18-2, 18-3
communication, 18-1
electromagnetic, 16-1
electron-gas, 16-12
electronic, 17-1
energy band, 17-4
of everything, 58-2
of feedback, 51-4
of light, quantum, 58-2
of networks, 29-2
of radiation, quantum, 58-2
relativity, 16-2
transducer, 19-1
Therm, 2-1
Thermal
agitation, 36-1
carrier generation, 17-5, 37-3
coefficient of resistivity, 26-2, 26-3
current, 37-7, 38-4
energy, 2-1, 2-4 (ftn)
noise, 18-7, 40-8
resistivity, 58-1
runaway, 38-5
voltage, 37-10
Thermocouple, 29-7
Thermoelectric
effect, 26-4
phenomena, 16-5
Thermoelement instrument, 36-4
Thevenin's theorem, 26-10, 29-9
Thevenin equivalent, 26-5
circuit, 26-10, 29-9
resistance, 26-10
voltage, 26-10
Thomson effect, 16-5, 26-4
Three-dimensional mensuration, A-9
Threshold frequency, 17-2
Thyratron, 37-14 (ftn)
Thyristor, 37-14, 37-14 (ftn)
gate turn-off, 37-15
MCT, 39-1
Time
delay, 30-7, 51-2
-distance plot, 45-3 (ftn)
-division multiple access, 49-2
-division multiplexing, 49-1
domain analysis, 31-1
doubling, 35-1, 55-8
hold, 43-2
multiplexing, 49-1
response, 51-2
response, of system, 51-7
-series analysis, 10-8 (ftn)
setup, 43-2
shifting, 27-2
-shifting theorem, 11-6
to double, 55-8
to triple, 55-8
tripling, 55-8
value of money, 55-5
Time constant, 30-2, 51-2, 51-9
RC circuit, 30-2
RL circuit, 30-2
Timing, 43-1
design, 43-1
Toggle flip-flop, 43-3
Tolerance, 3-4
Ton
long, A-3, A-4
of refrigeration, A-3, A-5
short, A-3, A-4
Tonne, 1-7
Tool, 44-1
Toroid, 29-6
Toroidal coil, 16-18
Torque
conversions, SI, A-3, A-5
on magnetic charge, 22-5
Torr, A-3, A-4
Tort, 56-6
Torus, 16-18
formula, A-9
Total
cost, 55-32
current, 16-13
energy, 2-1
response, 51-2
Totem-pole configuration, 40-25
Toward the normal, 46-4
Tracking converter, 44-5
Trade-in allowance, 55-18
Traditional magnetism, 22-5
Traffic
computer, 44-7
telecommunication, 44-7
Trailing edge, 43-1
Transaction, short-term, 55-3
Transcendental function, 15-1 (ftn)
circular, 7-2
derivative, 9-2
integral of, 10-1
Transconductance, 38-8, 39-3
Transducer, 19-1, 19-3, 29-7, 34-4
active element, 19-2
circuit, 36-3
passive element, 19-2
self-generating, 19-2, 19-3
system, characteristics, 19-2, 19-3
theory, 19-1
Transduction, 19-3
capacitive, 19-3
electromagnetic, 19-3
inductive, 19-3
photoconductive, 19-3
photovoltaic, 19-3
piezoelectric, 19-3
potentiometric, 19-3
reluctive, 19-3
resisitive, 19-4
strain-gauge, 19-4
Transfer, 26-9
of maximum energy, 26-5
of maximum power, 26-5
pulse, 44-1
ratio, 32-2
resistance, 26-9
Transfer function, 31-14, 32-1, 32-2, 51-3
closed-loop, 51-5
direct, 51-4
feedback, 51-4
forward, 51-4
loop, 51-5
open-loop, 51-5
overall, 51-5
reverse, 51-4
transform of, 51-3
voltage ratio, 26-5
Transform, 15-18
fast Fourier, 10-8
Fourier, 15-13, 15-15
Laplace, 11-5, A-11
of the transfer function, 51-3
pairs, Fourier, A-17
Transformation
coefficient of, 6-3
error, 34-5
matrix, 6-3
of integrals, 15-18
of sources, 29-8
Transformer, 28-1, 28-3
core, 28-2, 28-3
coupling coefficient, 28-3
current, 34-4, 36-2
dot convention, 28-2
dot rule, 28-3
exciting current, 28-3
ideal, 28-3
ideal model, 28-4
impedance matching, 28-4
instrument, 36-2
losses, 28-6
magnetizing current, 28-3
metering, 34-4
model, 28-2
model, real, 28-5
potential, 36-2
properties of, 28-3
real, 28-5
step-down, 28-4
step-up, 28-4
turns ratio, 28-4
two-port network, 29-14
voltage, 34-5
Transient, 29-13, 30-1
analysis, 30-1, 32-1
impedance analysis, 29-14
pulse, 30-1, 31-6
response, 30-5, 30-6, 32-4, 51-2
switching, 30-1, 31-3
Transistor, 38-2
active region, 38-2, 38-3
analog operation, 38-2
analysis, 38-9
bias, 38-5
biasing methods, 38-5
bipolar junction, 38-2
BJT, 38-2
capacitive coupling, 38-7
characteristics, 38-3
circuit linear analysis, 38-9
closed, 38-3
common base, 38-4
common collector, 38-4
common emitter, 38-4
common terminal, 38-4

configuration, 38-4
configuration properties, 38-5
current ratio, 38-3
cutoff, 38-2, 38-3
cutoff region, 38-3
Darlington, 38-12, 38-13
digital operation, 38-2
emitter follower, 38-4
equivalent circuit, 38-8
equivalent circuit parameters, 38-8
field-effect, 38-2, 39-1
gain, 38-7
high-frequency analysis, 38-12
load line, 38-6
logic, types of, 40-19
logic family data, 40-24
model, approximate, 38-8
npn, 37-15
off, 38-3
on, 38-3
open, 38-3
output characteristic curve, 38-6
power, 38-7
saturation, 38-2
saturation current, 38-4
saturation region, 38-2
small-signal terminology, 38-4
switch, 40-17
thermal current, 38-4
-transistor logic, 40-19, 40-24, 40-25
two-port network, 29-14, 38-7
unijunction, 38-12
wired logic, 40-25
Transition
capacitance, 37-10
diagram, 43-1
equation, 43-2
region, 17-7, 37-7
table, 43-1
Transmissible vector, 6-1 (ftn)
Transmission
formula, Friss, 47-3, 47-4
fundamentals, 45-1
high-frequency, 45-2
loss, basic, 46-5
loss, free space, 46-5
media, 47-1, 50-1
parameter, 29-16
Transmission line
admittance, 45-2
impedance, 45-2
input impedance, 45-2
parameter, inverse, 29-16
propagation constant, 45-2
Transmitting antenna, 47-1
Transparent flip-flop, 43-2
Transpose, 5-4, 5-5
Transverse axis, 8-11
Trapezoid formula, A-7, A-8
Trapezoidal rule, 8-1
Tri-state logic, 40-25
Triac, 37-15, 38-2
Triad, Cartesian, 6-2
Trial, 12-2
Triangle, 7-2
formula, A-7
general, 7-5
oblique, 7-5
Pascal's, 4-3
power, 27-12, 27-13
right, 7-2
similar, 7-2
special, 7-2
spherical, 7-5
Triangular
matrix, 5-2
wave, 27-6
Trigger, 43-1
device, 37-16
falling edge, 43-1
rising edge, 43-1
Trigonometric
form, 4-8
form of sinusoid, 27-2
function, 7-2
identity, 7-2, 7-3, 7-4, 7-5
Trigonometry, spherical, 7-5
Trihedral angle, 7-5
Trimetric view, 3-3
Triode, 17-2
region, 39-2
Triple
integral, 10-3
product, mixed, 6-4
scalar product, 6-4
Tripling time, 55-8
Trivial solution, 4-7 (ftn)
Troy ounces, A-1, A-2
True power, 27-12
Trunk line, 50-2
Truth table, 40-19, 41-2
Karnaugh map, 42-2
TTL, 40-19, 40-24
gate, open collector, 40-25
logic, 40-25
Tube
of force, 16-8
vacuum, 17-2
Tuned circuit, 40-3 (ftn)
Tungsten, 52-1
Tunnel diode, 37-13, 37-14
Tunneling, 37-13
Turns ratio, 28-4
TV, 50-2
Two
-angle formula, 7-4
-dimensional mensuration, A-7
lines, angle, 8-8
-point form, 8-5
-point method, 29-7
points, distance, 8-7
-port network, 29-14
-port network parameters, 29-16
-port network transistor, 29-14, 38-7
-port parameter conversions, A-31
-source model of capacitor, 31-16
-source model of inductor, 31-17
-tail limit, 12-12
-wattmeter method, 36-7
Twos complement, 13-4
Type
number, 4-1
parameter, 29-14
Types of views, 3-1

U

UHF channels value, 50-2
UJT, 38-12, 38-13
ULSI, 37-2
Ultra large scale integration, 37-2
Unadjusted basis, 55-20
Unary operation, 41-2
Unbiased estimator, 12-10
Uncertainty
analysis, 55-43
relationship, 18-3
Undamped sinusoidal response, 51-9
Underdamped analysis, 31-12
Underdamping, 31-8
Underground installations, 54-5
Undetermined coefficients, method of, 4-6, 11-4
Unforced response, 51-2
Ungrounded, 34-2
Uniform
cost, annual, 55-7
electric field, 16-8
exam, 59-2
gradient factor, 55-9
linear array, 46-2
series cash flow, 55-3
Unijunction transistor, 38-12, 38-13
Unilateral
Laplace transform, 15-17
z-transform, 15-18 (ftn)
Uninterruptable power supply, 34-4
Union, set, 12-1
Unipolar, 48-6
Unit
Angstrom, A-1
base, 1-5
bit, 18-5
circle, 7-1
distance coding, 42-3
dynamic, 51-4
engineering, 35-2
feedback, 51-4
finite impulse function, 15-10
impulse, 31-6
impulse function, 11-6, 15-10
load, 40-21
magnetic, 22-6
matrix, 5-2
non-SI, 1-7
-pole, 22-6
positive charge, 20-3
power, 35-1
primary, 1-7
pulse, 31-6
quad, 2-1
SI conversions, A-3
sink current, 40-21
source current, 40-21
sphere, 7-6
step, 31-6
step function, 11-6, 15-9
step response, 51-2
therm, 2-1
vector, 6-2
vector, Cartesian, 6-2
volt, 26-1
Units
consistent system of, 1-2
SI derived, 1-5, 1-6
Unity gain frequency, 40-8
Universal
gas constant, 19-2
operations, 41-3
Serial Bus, 36-8
set, 12-1
Universe, 12-3
Unlike charges, 20-1
Up-down counter, 44-5
Upper
bound, 10-4
sideband, 18-3
USB, 36-8
Useful
life, 55-20
User interface
graphics, 35-2
standard, 44-1
Utilization equipment, 53-3

V

Vacuum tube, 17-2
cathode, 17-2
grid, 17-2
plate, 17-2
Valence
band, 17-4, 17-5
electron, 37-2
shell, 17-4
Value
analysis, 55-44
average, 55-30
average (by integration), 10-4
book, 55-23
boundary, 11-1
command, 51-14
engineering, 55-44
expected, 55-30
face, 55-30
initial, 10-4, 11-1

limiting, 4-9
market, 55-23
phasor notation, 27-8
reference, 51-14
residual, 55-16
salvage, 55-18
standard normal, 12-6
Varactor, 37-10
Variable, 4-2
circuit, 29-1
complex frequency, 31-15
cost, 55-31
dimension of, 1-8 (tbl)
frequency inputs, 32-1
logic, 41-2
matrix, 5-6
separation of, 15-7
state, 31-6, 31-6 (ftn), 41-1, 51-14
-threshold logic, 40-19
Variance, 12-11
account, 55-36
burden, 55-36
Variation
coefficient of, 12-11
of parameters, 11-4, 15-3
VCO, 40-29
Vector, 6-1
addition, 6-3
addition for phasors, 27-8
bound, 6-1 (ftn)
Cartesian unit, 6-2
cross product, 6-4
curl, 15-19
current density, 16-13
digital information, 42-1
dot product, 6-3
electromagnetic field, 25-1, 25-2
equal, 6-1
equivalent, 6-1 (ftn)
field, curl, 9-7
field, divergence, 9-7
fixed, 6-1 (ftn)
flux density, 16-10
force field, 16-7
function, 6-5
gradient, 9-5
magnetic potential, 22-9
multiplication, 6-3
normal line, 9-6
normal to plane, 8-6
operation, example, 42-2
orthogonal, 6-3
Poynting's, 24-3, 46-3
resultant, 6-3
sliding, 6-1 (ftn)
transmissible, 6-1 (ftn)
triple product, 6-5
unit, 6-2
Veil, corporate, piercing, 56-3
Veitch diagram, 42-5, 42-6
Velocity, 16-2 (ftn), 22-1
conversions, SI, A-3, A-5
drift, 17-2, 17-3
electron drift, 16-12
electrons in plasma, 17-3
phase, 45-2
sonic, 19-2
Vertex, parabola, 8-10
Vertical angle, 7-2
Very large-scale integration, 37-2
Vestigial sideband, 48-3
Vicalloy, 52-2
View
auxiliary, 3-2
axonometric, 3-3
central, 3-1, 3-2
dimetric, 3-3
isometric, 3-3
normal, 3-1
oblique, 3-2
orthogonal, 3-1
orthographic, 3-1
orthographic oblique, 3-3
perspective, 3-1, 3-2, 3-3
planar, 3-2
principal, 3-2
profile auxiliary, 3-2
redundant, 3-2
trimetric, 3-3
types, 3-1
Virtual
ground, 40-4
photon, 58-2
short circuit, 40-5
Viscosity conversions, SI, A-3, A-5
Vision, center of, 3-3
Vitreous static electricity, 16-5
VLSI, 37-2
VMOS, 39-1
Void region, 8-3
Volt, 16-21, 26-1, A-1, A-2
-equivalent of temperature, 17-7
-ohm-milliamp meter, 36-8
Voltage, 16-21, 24-3, 26-1
amplification factor, 37-4
amplitude, 27-2
breakover, 37-15
cell, 33-1
-controlled oscillator, 40-28, 40-29
-controlled voltage source, 29-3
divider, 26-5, 29-13
-divider bias, 38-5
drop, 26-3, 26-6, 53-5
equivalence, 26-5
equivalent of temperature, 37-10
equivalent Thevenin, 26-10
floating, 26-14
gain, 38-7
Hall, 37-2
hold, 37-15
in parallel circuit, 26-7
induced, 16-22
instantaneous, 27-2
law, Kirchhoff's, 26-6, 29-11
maximum, 27-2
noise, 36-2
nominal, 53-3
offset, 40-8
peak inverse, 37-9
pinchoff, 39-2
rail, 40-3
-ratio transfer function, 26-5, 29-13
reference direction, 26-6
regulation, 29-2, 34-5
regulator, operational amplifier, 40-16
regulator, zener, 40-13
reverse bias, 37-9
rise, 26-6
secondary, 28-1, 28-2
sending, 45-2
shunt feedback, 40-5
standing wave ratio, 45-3
thermal, 37-10
transformer, 34-5
vs. electromotive force, 24-1
zener, 37-13
Voltage source, 26-4, 29-3
AC types of, 27-3
controlled, 29-3
dependent, 29-3
ideal, 26-4, 29-2
ideal independent, 29-2
in parallel, 26-5
in series, 26-4
model, of capacitor, 31-17
real, 26-4, 29-2
symbology, 27-4
Voltmeter, DC, 36-4
Volume
conversions, SI, A-3, A-5
flow conversions, SI, A-3, A-6
mensuration, A-9
of revolution, 10-5
three-dimensional, mensuration of, A-9
VOM, 36-4, 36-8
Vorticity, 9-7
VSB, 48-3
VSWR, 45-3
VTL, 40-19

W

Wafer, 58-1
Wall, Bloch, 16-4
Warning signs, 54-4
Water, 20-3
vapor, 46-3
Watt, 1-5, A-1, A-2
-hour meter, 36-7
Wattless power, 27-12
Wave
cancellation, 46-3, 46-4
compression, 19-1
electrical phenomena of, 24-4
electromagnetic, 24-3
energy front, 24-3
equation, 15-7
longitudinal, 19-1
propagation fundamentals, 46-2
reflection, 46-4
reinforcement, 46-3, 46-4
superposition, 46-4
types, 27-6
Waveform
damping, 30-7
envelope, 30-7
Fourier, 10-7
frequency, 10-6
period, 10-6
periodic, 10-6
processing, 27-7
rectified, 27-4
shaping, 40-19
sinusoidal, 27-1
symmetrical, 10-7 (tbl)
Waveguide, 18-6, 45-1
antenna, 46-2
Wavelength, 19-1
-division multiplexing, 50-3
Waviness height, 3-4
WDM, 50-3
Weakons, 58-2
Weather standard, 54-5
Weber's function, 15-12
Weber number, 1-9 (tbl)
Weight, 1-1, 1-3
density, 1-3
equivalent, 24-2
roughness, 3-4
Weighting factor, 55-42
Weld slot, 7-2
Wheatstone bridge, 36-3
Whistle blowing, 57-3 (ftn)
White noise, 18-7, 48-8
Wideband, 48-5
Width, 3-2
beam, 47-3
Winding
primary, 28-2
secondary, 28-2
Winfrey method, 55-16 (ftn)
Wire
antenna, 46-2
loop, 22-6
standard comparison, A-45
Wired logic, 40-25
Wireless, 50-1
communication, 50-3
Wireline communication, 45-1
Wiring method, 53-7
Word length, 41-1
Work, 2-2
constant force, 2-2
-energy principle, 2-3, 2-4
external, 2-2
field, 20-2
flow, 2-4
friction, 2-2
function, 17-2

gravity, 2-2
in electric field, 20-6
internal, 2-2
magnetic, 22-8
negative, 2-2
p-V, 2-4
performed by a force, 10-4
positive, 2-2
rotating dipole, 21-3
rules, 54-7
spring, 2-2
Working space, 54-4
Worth
future, 55-5
method, present, 55-14
present, 55-5
Wronskian, 15-4
Wye
connection terminology, 53-3
-delta, 26-8, 29-9

Y

y-parameters, 29-15
Year-end convention, 55-3
Yield, 55-28
bond, 55-30

Z

Z
-boson, 58-2
-level, 40-25
-parameters, 29-14
-transform, 15-18, A-23
-transform, bilateral, 15-18 (ftn)
-transform, properties, A-23
-transform, unilateral, 15-18 (ftn)
-values for confidence levels, 12-12
Zener, 37-9 (ftn), 37-13
breakdown, 37-9
diode, 37-13, 40-13
diode, properties of, A-38, A-39
regulator circuit, 40-13
voltage, 37-13
voltage regulator circuit, 40-13
voltage regulator circuit, design, 40-14
Zero, 4-3, 14-1, 32-2, 51-8
at origin, 32-5
intersymbol interference, 48-7
matrix, 5-2
on negative real axis, 32-5
Zone
absolutely selective, 34-4
closed, 34-4
differential, 34-4
non-unit, 34-4
of protection, 34-4
open, 34-4
relatively selective, 34-4
unrestricted, 34-4

CONVERSION FACTORS

(Atmospheres are standard; calories are gram-calories; gallons are U.S. liquid; miles are statute; pounds-mass are avoirdupois.)

multiply	by	to obtain
acre	43,560	ft^2
angstrom	1×10^{-10}	m
atm	1.01325	bar
atm	76.0	cm Hg
atm	33.90	ft water
atm	29.92	in Hg
atm	14.696	lbf/in^2
atm	101.3	kPa
atm	1.013×10^5	Pa
bar	0.9869	atm
bar	10^5	Pa
Btu	778.17	ft-lbf
Btu	1055	J
Btu	2.928×10^{-4}	kW-hr
Btu	10^{-5}	therm
Btu/hr	0.216	ft-lbf/sec
Btu/hr	3.929×10^{-4}	hp
Btu/hr	0.2931	W
Btu/lbm	2.326	kJ/kg
Btu/lbm-°R	4.1868	kJ/kg·K
cal	3.968×10^{-3}	Btu
cal	4.1868	J
cm	0.03281	ft
cm	0.3937	in
eV	1.602×10^{-19}	J
ft	0.3048	m
ft^2	2.2957×10^{-5}	acre
ft^3	7.481	gal
ft-lbf	1.285×10^{-3}	Btu
ft-lbf	1.35582	J
ft-lbf	3.766×10^{-7}	kW-hr
ft-lbf	1.3558	N·m
gal	0.13368	ft^3
gal	3.785	L
gal	3.7854×10^{-3}	m^3
gal/min	0.002228	ft^3/sec
g/cm^3	1000	kg/m^3
g/cm^3	62.428	lbm/ft^3
hp	2545	Btu/hr
hp	33,000	ft-lbf/min
hp	550	ft-lbf/sec
hp	0.7457	kW
hp-hr	2545	Btu
in	2.54	cm
J	9.478×10^{-4}	Btu
J	6.2415×10^{18}	eV
J	0.73756	ft-lbf
J	1.0	N·m
J/s	1.0	W

multiply	by	to obtain
kg	2.20462	lbm
kip	1000	lbf
kip	4448	N
kJ	0.9478	Btu
kJ	737.56	ft-lbf
kJ/kg	0.42992	Btu/lbm
kJ/kg·K	0.23885	Btu/lbm-°R
km	3280.8	ft
km	0.6214	mi
km/hr	0.6214	mi/hr
kPa	9.8693×10^{-3}	atm
kPa	0.01	bar
kPa	0.14504	lbf/in^2
ksi	6894.8	kPa
kW	3413	Btu/hr
kW	0.9478	Btu/sec
kW	737.6	ft-lbf/sec
kW	1.341	hp
kW-hr	3413	Btu
kW-hr	3.6×10^6	J
L	0.03531	ft^3
L	61.02	in^3
L	0.2642	gal
L	0.001	m^3
L/s	2.119	ft^3/min
L/s	15.85	gal/min
lbf	4.4482	N
lbf/in^2	0.06805	atm
lbf/in^2	2.307	ft water
lbf/in^2	2.036	in Hg
lbf/in^2	6894.8	Pa
lbm	0.4536	kg
lbm/ft^3	0.016018	g/cm^3
lbm/ft^3	16.018	kg/m^3
m	3.28083	ft
m/sec	196.8	ft/min
mi	5280	ft
mi	1.6093	km
micron	1×10^{-6}	m
N	0.22481	lbf
N·m	0.7376	ft-lbf
N·m	1.0	J
Pa	1.4504×10^{-4}	lbf/in^2
therm	10^5	Btu
W	3.4121	Btu/hr
W	0.7376	ft-lbf/sec
W	1.341×10^{-3}	hp
W	1.0	J/s